Handbook of Research on Advanced Intelligent Control Engineering and Automation

Ahmad Taher Azar
Benha University, Egypt

Sundarapandian Vaidyanathan
Vel Tech University, India

A volume in the Advances in Computational Intelligence and Robotics (ACIR) Book Series

ENGINEERING SCIENCE REFERENCE
An Imprint of IGI Global

Managing Director:	Lindsay Johnston
Managing Editor:	Austin DeMarco
Director of Intellectual Property & Contracts:	Jan Travers
Acquisitions Editor:	Kayla Wolfe
Production Editor:	Christina Henning
Typesetter:	Michael Brehm
Cover Design:	Jason Mull

Published in the United States of America by
Engineering Science Reference (an imprint of IGI Global)
701 E. Chocolate Avenue
Hershey PA, USA 17033
Tel: 717-533-8845
Fax: 717-533-8661
E-mail: cust@igi-global.com
Web site: http://www.igi-global.com

Library of Congress Cataloging-in-Publication Data

Handbook of research on advanced intelligent control engineering and automation / Ahmad Taher Azar and Sunddarapandian Vaidyanathan, editors.
pages cm
ISBN 978-1-4666-7248-2 (hardcover) -- ISBN 978-1-4666-7249-9 (ebook) -- ISBN 978-1-4666-7251-2 (print & perpetual access) 1. Automatic control. I. Azar, Ahmad Taher. II. Vaidyanathan, Sunddarapandian, 1967-
TJ213.H3145 2014
629.8'9563--dc23
2014037073

This book is published in the IGI Global book series Advances in Computational Intelligence and Robotics (ACIR) (ISSN: 2327-0411; eISSN: 2327-042X)

British Cataloguing in Publication Data
A Cataloguing in Publication record for this book is available from the British Library.

For electronic access to this publication, please contact: eresources@igi-global.com.

Advances in Computational Intelligence and Robotics (ACIR) Book Series

ISSN: 2327-0411
EISSN: 2327-042X

Mission

While intelligence is traditionally a term applied to humans and human cognition, technology has progressed in such a way to allow for the development of intelligent systems able to simulate many human traits. With this new era of simulated and artificial intelligence, much research is needed in order to continue to advance the field and also to evaluate the ethical and societal concerns of the existence of artificial life and machine learning.

The **Advances in Computational Intelligence and Robotics (ACIR) Book Series** encourages scholarly discourse on all topics pertaining to evolutionary computing, artificial life, computational intelligence, machine learning, and robotics. ACIR presents the latest research being conducted on diverse topics in intelligence technologies with the goal of advancing knowledge and applications in this rapidly evolving field.

Coverage

- Intelligent Control
- Cognitive Informatics
- Computational Logic
- Heuristics
- Robotics
- Computer Vision
- Automated Reasoning
- Adaptive and Complex Systems
- Evolutionary Computing
- Natural Language Processing

IGI Global is currently accepting manuscripts for publication within this series. To submit a proposal for a volume in this series, please contact our Acquisition Editors at Acquisitions@igi-global.com or visit: http://www.igi-global.com/publish/.

The Advances in Computational Intelligence and Robotics (ACIR) Book Series (ISSN 2327-0411) is published by IGI Global, 701 E. Chocolate Avenue, Hershey, PA 17033-1240, USA, www.igi-global.com. This series is composed of titles available for purchase individually; each title is edited to be contextually exclusive from any other title within the series. For pricing and ordering information please visit http://www.igi-global.com/book-series/advances-computational-intelligence-robotics/73674. Postmaster: Send all address changes to above address.

Titles in this Series

For a list of additional titles in this series, please visit: www.igi-global.com

Handbook of Research on Advancements in Robotics and Mechatronics
Maki K. Habib (The American University in Cairo, Egypt)
Engineering Science Reference • copyright 2015 • 953pp • H/C (ISBN: 9781466673878) • US $515.00 (our price)

Handbook of Research on Advanced Intelligent Control Engineering and Automation
Ahmad Taher Azar (Benha University, Egypt) and Sundarapandian Vaidyanathan (Vel Tech University, India)
Engineering Science Reference • copyright 2015 • 567pp • H/C (ISBN: 9781466672482) • US $335.00 (our price)

Handbook of Research on Artificial Intelligence Techniques and Algorithms
Pandian Vasant (Universiti Teknologi Petronas, Malaysia)
Information Science Reference • copyright 2015 • 694pp • H/C (ISBN: 9781466672581) • US $345.00 (our price)

Handbook of Research on Synthesizing Human Emotion in Intelligent Systems and Robotics
Jordi Vallverdú (Universitat Autònoma de Barcelona, Spain)
Information Science Reference • copyright 2015 • 426pp • H/C (ISBN: 9781466672789) • US $245.00 (our price)

Recent Advances in Ambient Intelligence and Context-Aware Computing
Kevin Curran (University of Ulster, UK)
Information Science Reference • copyright 2015 • 376pp • H/C (ISBN: 9781466672840) • US $225.00 (our price)

Recent Advances in Intelligent Technologies and Information Systems
Vijayan Sugumaran (Oakland University, USA & Sogang University, Seoul, Korea)
Information Science Reference • copyright 2015 • 300pp • H/C (ISBN: 9781466666399) • US $200.00 (our price)

Emerging Research on Swarm Intelligence and Algorithm Optimization
Yuhui Shi (Xi'an Jiaotong-Liverpool University, China)
Information Science Reference • copyright 2015 • 341pp • H/C (ISBN: 9781466663282) • US $225.00 (our price)

Face Recognition in Adverse Conditions
Maria De Marsico (Sapienza University of Rome, Italy) Michele Nappi (University of Salerno, Italy) and Massimo Tistarelli (University of Sassari, Italy)
Information Science Reference • copyright 2014 • 480pp • H/C (ISBN: 9781466659667) • US $235.00 (our price)

Computer Vision and Image Processing in Intelligent Systems and Multimedia Technologies
Muhammad Sarfraz (Kuwait University, Kuwait)
Information Science Reference • copyright 2014 • 312pp • H/C (ISBN: 9781466660304) • US $215.00 (our price)

IGI GLOBAL
DISSEMINATOR OF KNOWLEDGE
www.igi-global.com

701 E. Chocolate Ave., Hershey, PA 17033
Order online at www.igi-global.com or call 717-533-8845 x100
To place a standing order for titles released in this series, contact: cust@igi-global.com
Mon-Fri 8:00 am - 5:00 pm (est) or fax 24 hours a day 717-533-8661

Editorial Advisory Board

List of Contributors

Table of Contents

Detailed Table of Contents

Atomic Force Microscopes (AFMs) are used in many nanopositioning applications in order to measure the topography of various specimens at an atomic level through surface imaging. The imaging of the samples in AFMs is carried out by using a three degree-of-freedom positioning unit called Piezoelectric Tube Scanner (PTS). The performance of the AFM mostly depends on the performance of the PTS. However, the PTS of the AFM suffers from the problem of vibration. This chapter presents a design of a damping controller to compensate the induced vibration of the scanner. Experimental results are presented to show the effectiveness of the proposed controller. The experimental results show that the proposed controller is able to compensate 90% of the vibration of the PTS.

This chapter deals with the description, the parametric estimation, the state estimation, and the parametric and state estimation conjointly of nonlinear systems. The focus is on the class of nonlinear systems, which are described by Wiener state-space discrete-time mathematical models. Thus, the authors develop a new recursive parametric estimation algorithm, which is based on least squares techniques. The stability conditions of the developed parametric estimation scheme are analyzed using the Lyapunov method. The state estimation problem of the considered nonlinear systems is formulated. Thus, the authors propose a recursive state estimation algorithm, which is based on Kalman Filter. A new recursive algorithm is proposed, which permits one to estimate conjointly the parameters and the state variables of nonlinear systems described by Wiener mathematical models, with unknown parameters and state variables. The efficiency and performance of the proposed recursive estimation algorithms are tested on numerical simulation examples.

This chapter studies the control problem for linear singular systems. Firstly a proportional plus derivative feedback controller for continuous and discrete cases for descriptor systems is given, satisfying the closed loop systems normal and stable states. A necessary and sufficient condition for the solvability of this problem should be obtained in terms of linear matrix inequality. In second part, the designed controller is required to yield a cost function with guaranteed upper bound for the continuous and discrete case and also an extension for robust control is developed. Numerical examples are included to show the effectiveness of the proposed result.

attractors for some suitable chosen system parameters. In order to confirm the correction and feasibility of the theoretical model, the system is also implemented with analog electronic circuit. Finally, a practical application of such circuit is discussed at the end of this chapter.

This chapter deals with the control of chaos exhibited in the behavior of a one-degree-of-freedom impact mechanical oscillator with a single rigid constraint. The mathematical model of such impact oscillator is represented by an impulsive hybrid linear non-autonomous dynamics. The proposed control approach is based chiefly on the OGY method. First, an analytical expression of a constrained controlled Poincaré map is derived. Secondly, the linearized controlled Poincaré map around the fixed point of the constrained map is determined. Relying on the linearized map, a state feedback controller is designed. It is shown that the proposed control strategy is efficient for the control of chaos for the desired fixed point and for the fixed parameter. It is shown also that, by changing the system parameter, the behavior of the impact mechanical oscillator changes radically. Thus, the drawback of the developed OGY control method is revealed and some remedies are given.

This chapter presents further investigations into the period-three route to chaos exhibited in the passive dynamic walking of the compass-gait biped robot as it goes down an inclined surface. This discovered kind of route in the passive bipedal locomotion was found to coexist with the conventional period-one passive hybrid limit cycle. The further analysis on the period-three route chaos is realized by means of the Lyapunov exponents and the fractal Lyapunov dimension. Numerical computation method of these two tools is presented. The first return Poincaré map of the chaotic attractor and its basin of attraction are presented. Furthermore, the further study of the period-three passive gait is realized. The analysis of the period-three hybrid limit cycle is given. The balance between the potential energy and the kinetic energy of the biped robot is illustrated. In addition, the basin of attraction of the period-three passive gait is also presented.

This chapter deals with the analysis of the dynamic behavior of a controlled single-phase H-bridge inverter. The authors show that in addition to border collision bifurcation, when it is controlled with a time-delayed controller or with a dynamic controller that increases the system dimension, the H-bridge inverter can exhibit several other types of behaviors such as Neimark-Sacker bifurcation, quasi-periodicity, and coexistence of different periodic behaviors, as well as coexistence between periodic and chaotic behaviors. The proposed controllers are of different types. In addition to the Fixed-Point Induced Controller (FPIC), the authors also present the Time-Delayed Feedback Controller (TDFC) and the dynamic linear controller, such as the proportional-integral controller. The main issue of this chapter is to perform analysis within and beyond the stability region. Analytic calculation and numerical simulations are presented to confirm the obtained results.

Section 3
Intelligent Control Applications

This chapter presents two fuzzy adaptive variable structure controllers for a class of uncertain multi-input multi-output nonlinear systems with actuator nonlinearities (i.e. with sector nonlinearities and dead-zones). The design of the first controller concerns systems with symmetric and positive definite control-gain matrix, while the design of the second one is extended to the case of non-symmetric control-gain matrix thanks to an appropriate matrix decomposition, namely the product of a symmetric positive-definite matrix, a diagonal matrix with diagonal entries +1 or -1, and a unity upper triangular matrix. An appropriate adaptive fuzzy-logic system is used to reasonably approximate the uncertain functions. A Lyapunov approach is adopted to derive the parameter adaptation laws and prove the stability of the closed-loop control system. Finally, some simulation results are carried out to show the effectiveness of the proposed controllers.

This work deals with the fault detection and localization in the case of uncertain nonlinear systems. The presented method uses the diagnosis based on mathematical models. To model nonlinear systems, the multiple model approach is used. This method uses the Takagi-Sugeno fuzzy systems principle to obtain a nonlinear system named multiple models. This modeling principle has the advantage of obtaining a general model that can describe any class of nonlinear systems. This modeling principle also allows one to obtain the generalization of many results that are already obtained for linear systems to the nonlinear systems. To model the system uncertainties, the interval approach is used because the faults or disturbances are generally unknown, but it is possible to know their upper and lower bounds. The proposed technique is insensitive to measurement uncertainties and highly reliable in case of a fault affecting the outputs system.

Many works are related to the analysis and control of either continuous or else discrete time-delay systems. In general, discrete-time controls have become more and more preferable in engineering because of their easy implementation and simple computation. However, the available discretization approaches for the systems having time delays increase the system dimensions and have a high computational cost. The case studies in this chapter support the efficiency of the two methods. However, the discretization of continuous time-delay systems has not been sufficiently/extensively studied in many works. In this work, the authors present two methods of the effective discretization approach for the continuous-time systems with an input and output delays. Sampled-data time-delay systems with internal and external point delays are described by approximate discrete time-delay systems in the discrete domain. These approximate discrete systems allow the hybrid control of time-delay systems.

This chapter aims to study the problem of stability analysis, and robust exponential stabilization for a class of switched linear systems with polytopic uncertainties is reviewed. A sufficient condition based on the average dwell time that guarantees the exponential stability of uncertain switched linear systems is given. First, the iterative learning control is presented to build a formulation ensuring the exponential stability of the given system. The integrated design of this ILC scheme is transformed into a robust control problem of an uncertain 2D Roesser system. The results are obtained through original connection with the notion of stability along the pass for 2D repetitive systems. An overview of the stabilization methods of switched discrete systems found in the literature is outlined. All the given formulations are presented in terms of LMIs. A numerical simulation example is established, showing the effectiveness of the proposed method.

In this chapter, two computational algorithms are proposed and applied on an estimation algorithm, in order to improve the global performance of the estimation phase. The proposed system is studied based on the Model Reference Adaptive System (MRAS). The importance of the estimation phase in a large applications number is basically observed on the applications applied on electrical motors, where a lot number of parameters are measured with real measurement equipments, as Tesla Meter, speed shaft, and others. The idea is based generally on the software applications, where it is possible to guarantee the desired estimation phase using a software algorithm. In this chapter the MRAS technique is proposed as the software algorithm, for replacing the measurement materials for online estimate the overall characteristic PMSM parameters. Our approach aims to ameliorate the MRAS technique with intelligent optimization methods called BFO and PSO.

Attention has been paid by many researchers to address the various challenges of grid connection of DFIG-based Wind Energy Conversion Systems (WECS). This chapter focuses on the design of a robust $H\infty$ controller for the power flow between the stator of the Doubly-Fed Induction Generator (DFIG) and the grid. The robust $H\infty$ controller design is formulated as a mixed-sensitivity problem. A mathematical model of the DFIG written in an appropriate d-q reference frame is established to carry out simulations. The proposed power control scheme is elaborated and compared with a conventional Proportional-Integral (PI) controller based on vector control technique. The results show interesting performance of the controlled system in terms of the power reference tracking (the active and reactive power) and robustness against parameter variations compared with the conventional PI controller.

The basic idea of this chapter is to implement a fixed low order controller on an electronic system using the STM32 microcontroller. The principal aim of this controller is to guarantee some time response specifications as the settling time and the overshoot. The controller parameters are obtained by minimizing a non-convex optimization problem while taking into account the desired closed-loop performances. Accordingly, the resolution of this sort of problem with classical optimization method may lead to a local solution and the achieved control law is not optimal. Hence, with an aim of obtaining an optimal solution that will be able to satisfy the desired specifications, the use a global optimization method is recommended. In this chapter, the Generalized Geometric Programming (GGP) method is exploited. The practical implementation, on a fast electronic system, of the designed control law and a Proportional Integral (PI) controller illustrates the effectiveness of the proposed algorithm.

Due to different concerns, renewable energies stand out as an opportunity for a sustainable future. Wind/Photovoltaic Hybrid Systems are one important type of renewable energy power systems. In order to increase renewable energies integration rate, this chapter proposes a utility interactive grid-connected wind-PV hybrid system with storage batteries. Facing modern electricity industries, the hybrid system considers the dispatch-ability and the quality of its power injection into the grid. The adopted modeling approach has been based on energy exchange considerations taking into account the converters structures. Particular attention has been given to cases where the hybrid system is connected to an unbalanced grid. This chapter develops a new control strategy, which aims to isolate the hybrid system of the adverse impact of the grid fault. The results point out thoroughly the applicability of the proposed control scheme under unbalanced grid conditions.

Multilevel inverters have seen an increasing popularity in the last few years for medium- and high-voltage applications. The most popular has been the three-level neutral clamped inverter. Multilevel inverters synthesize output voltage from more than two voltage levels. Thus, the output signal spectrum is significantly improved in comparison with the classical two-level converters. This chapter discusses modelling and control of a Neutral Point Clamped (NPC) inverter which operates with the PWM switching pattern using a DSP. The mathematical model of the NPC inverter is carried out using conversion and connection functions for an easier understanding of the system operation. Simulation results using MATLAB program are reported, and it is shown that the performances obtained for driving an asynchronous motor using this inverter are very promising. Finally, analysis of the theoretical and the experimental results is carried out in order to validate the effectiveness of the proposed control solution.

Preface

Control engineering is concerned with the understanding and control of machines, processes, and industrial automation systems to provide useful economic products for society. Control engineering is based on foundations of feedback theory and linear systems analysis. The subject is multidisciplinary in nature since it deals with dynamic systems drawn from the disciplines of electrical, electronics, chemical, mechanical, aerospace, and instrumentation engineering.

Control and automation in its broadest sense plays a fundamental role in process industries. Automatic control systems involve mathematics more than it is usual in other engineering disciplines. Even though the subject has strong mathematical foundation, emphasis throughout the text is not on mathematical rigour or formal derivation (unless they contribute to understanding the concept), but instead, on the methods of application associated with the analysis and design of feedback system. This book presents a comprehensive analysis and design of various types of control systems such as intelligence control. The prerequisite for the reader is some elementary knowledge of differential equations, vector-matrix analysis, and mechanics.

This book is intended to be a major reference book for scientists and engineers interested in applying new computational and mathematical tools for solving the complicated problems of mathematical modeling, simulation, and control. This book presents a comprehensive analysis and design of various types of control systems such as adaptive control, nonlinear control, and intelligent control, which includes fuzzy logic, neural networks, and evolutionary algorithms.

The main advantage of this book that it is multi-disciplinary and it will attract a lot of researchers working in control, automation, simulations, modeling, and related fields. With a broad coverage of the contents essential for the analysis and design of various control problems in control engineering practice, this book can also be used as a handy desktop reference during control applications.

ORGANIZATION OF THE BOOK

This well-structured book consists of 25 full chapters. They are organized into 6 sections:

Section 1: Nonlinear Control Systems
Section 2: Chaos and Control Applications
Section 3: Intelligent Control Applications
Section 4: Robotics and Control Applications
Section 5: Time-Delay and Switched Control Systems
Section 6: Power Systems and Engineering Applications

BOOK FEATURES

- The book chapters deal with the recent research problems in the areas of computer science, information technology, automation, control engineering, intelligent control, and applications.
- The book chapters are well written with detailed theory and block diagrams.
- The book chapters are lucidly illustrated with numerical examples and simulations.
- The book chapters give a good literature survey with a long list of references.
- The book chapters discuss details of engineering applications and future research areas.

SUMMARY OF THE BOOK CHAPTERS

Section 1 contains Chapters 1-6 focusing on the area of Nonlinear Control Systems.

Chapter 1 presents new results for the stabilization and control of mechanical systems with backlash. Backlash is one of the physical nonlinearities found in different types of actuators such as mechanical and hydraulic actuators. In Chapter 1, three different control approaches are derived for the stabilization and control of mechanical systems with backlash. As the first control approach, an independent joint control is designed from a SISO point of view, explaining how to deal with backlash in the actuator due to the interactions of gears. As the second control approach, a multivariable approach is considered. As the third control approach, the optimal control of mechanical systems with backlash is used, obtaining the optimal control law and frequency oscillations, when backlash is found in mechanical systems using a describing function.

Chapter 2 presents a family of kernel affine projection algorithms for nonlinear adaptive filtering. Linear adaptive filters have been widely used in a variety of applications such as mobile communications, video processing, etc. In system identification problems, the main goal is to identify an unknown system using an adaptive filer. As an extension of the linear counterparts, kernel adaptive filters have been proposed in the literature to identify nonlinear control systems. Chapter 2 explains the similarities between the linear algorithms and their corresponding kernel versions. Chapter 2 features a detailed discussion of several simulation results.

Chapter 3 presents new results on the advanced vibration control of atomic force microscope scanner. Atomic Force Microscopes (AFMs) are used in many nanopositioning applications in order to measure the topography of various specimens at an atomic level through surface imaging. The imaging of the samples in AFMs is carried out by using a three degree-of-freedom positioning unit called Piezoelectric Tube Scanner (PTS). The performance of the AFM mostly depends on the performance of the PTS. However, the PTS of the AFM is subject to the problem of vibration. Chapter 3 presents a design of a damping controller to compensate the vibration of the PTS. Chapter 3 gives details of the experimental results in order to show the effectiveness of the proposed damping controller.

Chapter 4 deals with the description, the parametric estimation, the state estimation, and the parametric and state estimation conjointly of nonlinear systems. Chapter 4 focuses on the class of nonlinear systems operating in a stochastic environment, which are described by Wiener state-space discrete-time mathematical models. Chapter 4 describes a new recursive parametric estimation algorithm, which is based on least squares techniques. The stability conditions of the developed parametric estimation scheme are analyzed using the Lyapunov method. The state estimation problem of nonlinear systems described by Wiener mathematical model is formulated. Chapter 4 details a new recursive state estimation algorithm, which is based on Kalman Filter, in order to estimate the unknown state variables. The new recursive algorithm permits to estimate conjointly the parameters and the state variables of nonlinear systems described by Wiener mathematical models, with unknown parameters and state variables.

Chapter 5 describes new results on the Proportional-Plus-Derivative (PD) state feedback H_2 control for descriptor systems. Chapter 5 gives a detailed study of the stabilization problem for singular system by proportional plus derivative feedback controller, which avoids developing the issue of admissibility (stable, regular, and impulse free) of singular systems and also to develop the H_2 control similarly to state space systems. Chapter 5 gives simulation results and illustrates the feasibility of the approaches followed in this chapter when an explicit expression of the desired proportional plus derivative state feedback controller is given.

Chapter 6 describes new results for the approximate input-output feedback linearization of non-minimum phase control systems using vanishing perturbation theory. Chapter 6 presents a new tracking control method for the nonlinear non-minimum phase system. Indeed, the main idea here is to dismiss a part of the system dynamics in order to make the approximate system input-state feedback linearizable. The neglected part is then considered as a perturbation part that vanishes at the origin. Finally, a linear controller is designed to control the approximate system. Stability is analyzed using the vanishing perturbation theory.

Section 2 contains Chapters 7-11 focusing on the area of Chaos and Control Applications.

Chapter 7 presents a novel Chaotic Random Bit Generator (CRBG), which is based on the Poincaré map of a non-autonomous chaotic dynamical system, viz. the Duffing-van der Pol system. As a result of the close relationship between cryptography and chaos theory in the last decade, several chaos-based cryptosystems, especially of using autonomous chaotic dynamical systems, have been put forward. Chapter 7 presents a novel Chaotic Random Bit Generator (CRBG), which is based on the Poincaré map of the Duffing-van der Pol system. The proposed CRBG also uses the X-OR function for improving the "randomness" of the produced bit streams, which are confirmed by using the four statistical tests, Monobit test, Poker test, Runs test, and Long run test, which are part of the FIPS-140-2 suite.

Chapter 8 proposes a novel nonlinear time-delayed system having a saturation function, which can exhibit chaotic behavior. Motivated by attract features and potential applications of time-delay models, Chapter 8 proposes a new chaotic system with a single scalar time delay and a nonlinearity described by a saturation function. Chapter 8 exhibits that the novel time-delayed chaotic system shows double-scroll chaotic attractors for some suitable chosen system parameters. The novel time-delayed chaotic system is also implemented with analog electronic circuit using off-the-shelf circuit components.

Chapter 9 deals with the control of chaos exhibited in the behavior of a one-degree-of-freedom impact mechanical oscillator with a single rigid constraint. First, an analytical expression of a constrained controlled Poincaré map is derived. Secondly, the linearized controlled Poincaré map around the fixed point is determined. Relying on this linearized map, a state feedback controller to stabilize the desired identified fixed point is designed. Chapter 9 shows that the proposed control strategy is efficient for the control of chaos for the desired fixed point and for the fixed parameter. Chapter 9 also shows that by changing the system parameter the oscillatory behavior of the mechanical system changes radically and becomes chaotic. Thus, the drawback of the developed OGY control method is revealed and some remedies are provided.

Chapter 10 presents some new investigations into the period-three route to chaos exhibited in the passive dynamic walking of the compass-gait biped robot as it goes down an inclined surface. This discovered kind of route in the passive bipedal locomotion was found to coexist with the conventional period-1 passive hybrid limit cycle. The further analysis on the period-three route chaos is realized by means of the Lyapunov exponents and the fractal dimension. Calculation method of the spectrum of Lyapunov exponents and the associated fractal dimension for the impulsive hybrid nonlinear dynamics is presented. Different attractive structures of the chaotic attractor in the phase plane are illustrated. Its first return Poincaré map and its basin of attraction are also presented. Furthermore, the further study of the period-three passive gait is also realized. The analysis of the period-three hybrid limit cycle is given. In addition, the basin of attraction of the period-three passive gait is also presented.

Chapter 11 investigates the bifurcation, quasi-periodicity, chaos, and co-existence of different behaviors in the controlled H-bridge inverter. Chapter 11 deals with the analysis of the dynamic behavior of a controlled single phase H-bridge inverter. Most of the control methods used to create the AC signal out of the DC source are based on the Pulse Width Modulation (PWM) with a switching period T.

Section 3 contains Chapters 12-14 focusing on the area of Intelligent Control Applications.

Chapter 12 presents two fuzzy adaptive variable structure controllers for a class of uncertain Multi-Input Multi-Output (MIMO) nonlinear systems with actuator nonlinearities (i.e. with sector nonlinearities and dead-zones). The design of the first controller concerns systems with symmetric and positive definite control-gain matrix, while the design of the second controller is extended to the case of non-symmetric control-gain matrix with the help of an appropriate matrix decomposition, namely the product of a symmetric positive-definite matrix, a diagonal matrix with diagonal entries +1 or -1 and a unity upper triangular matrix. An appropriate adaptive fuzzy-logic system is used to reasonably approximate the uncertain functions.

Chapter 13 investigates the fault detection and localization in the case of uncertain nonlinear systems. The multiple model approach is used to model nonlinear systems. This method uses the Takagi-Sugeno fuzzy systems principle to obtain a nonlinear system named multiple models. To model the system uncertainties, the interval approach is used in Chapter 13 because the faults or disturbances are generally unknown, but it is possible to know their upper and lower bounds. The main contribution in this work is to use the approach of the interval analysis to model the measurement uncertainties and propose a method to detect and locate faults in the case of non-linear uncertain systems described by Takagi-Sugeno multiple models. The proposed technique is insensitive to measurement uncertainties and highly reliable in case of a fault affecting the outputs system.

Chapter 14 presents a novel system for controlling intelligent building systems using natural user interfaces. In recent years, intelligent building systems have become standard in residential buildings as well as commercial buildings. A control system for an intelligent building system consists of a set of depth cameras installed in buildings rooms and a central processing computer, which analyses that depth data. Due to advanced logic, it is capable of identifying people occupying defined areas and recognizing their gestures. The system cooperates with a Domatiq System – Intelligent Building Management System, to provide an overall control of the building. Together with the Domatiq, users can associate specific gestures with Domatiq actions and events to realize specialized scenarios. This allows the user to control particular building components in a more natural way. Embedded feedback mechanism enhances the overall performance by informing users about current state of their commands and actions.

Section 4 contains Chapters 15-17 focusing on the area of Robotics and Control Applications.

Chapter 15 provides a practical and intuitive way of cooperative task planning and execution for complex robotic systems using multiple robots in automated manufacturing applications. Based on hierarchical representation of discrete event robotic activities, independent and cooperative robotic actions are represented by extended Petri nets. Net models representing inter-robot cooperation with synchronized interaction are presented to achieve distributed autonomous coordinated activities in an example multi-robot cell. An implementation of control software on hierarchical and distributed architecture for two-robot task planning, and execution is also presented using Web service on an Internet environment, where the higher level controller executes an activity-based global net model of task plan representing cooperative behaviors performed by the robots, and the transitions of parallel activities of the robots are synchronously coordinated through the transmission of requests and the reception of status through gate arcs connecting between the net based controllers.

Chapter 16 investigates the problem of fault estimation using a fast adaptive fault diagnosis observer. A new form of the estimator bloc considered for this purpose is an Unknown Input Observer (UIO), which is subsequently used for a robust fault detection scheme and also as an adaptive detection design for an additive actuator fault. This observer is designed for an unknown input and fault free system, which is obtained by coordinate transformations of original systems with unknown inputs (disturbance) and faults. The observer information is devoted to the fault estimation for fault detection and isolation. The fault estimates can be used to form an additional control input to accommodate the fault. Stability of the adaptive estimation is provided by a Lyapunov function ending with solving the Linear Matrix Inequalities (LMI).

Chapter 17 proposes a fuzzy controller for a two-wheeled mobile robot that moves following a Genetic Algorithmic-based plan in its configuration space. The fuzzy controller aims at giving the accurate acceleration so that the controlled robot can move efficiently from one position to another in a limited time window frame. First, a fuzzy motion control structure starts with two-input-and-one-output in which the fuzzy systems are proposed. Then a Genetic Algorithmic-based plan is examined where a path solution is obtained with the combination of the Genetic Algorithms and the Constraint Satisfaction Problem techniques.

Section 5 contains Chapters 18-20 focusing on the area of Time-Delay and Switched Control Systems.

Chapter 18 focuses on the stability analysis problem for a class of continuous-time switched time-delay systems modelled by delay differential equations under arbitrary switching. By using a constructed Lyapunov function, the aggregation techniques, the Kotelyanski lemma associated with the M-matrix proprieties, and the vector norms notion, new delay-dependent sufficient conditions are derived for the considered system to be asymptotically stable. The obtained results corresponding to the Lyapunov vector function provide a solution to one of the basic problems in continuous-time switched time-delay systems. The results presented in Chapter 18 are explicit, simple to use, and allow us to avoid the problem of searching a common Lyapunov function. On the other hand, this proposed approach could be further used as a constructive solution to the problems of state and static output feedback stabilization. Chapter 18 provides two examples with numerical simulations to demonstrate the effectiveness of the proposed method.

Chapter 19 provides new results on the discrete-time approximation of multi-variable continuous-time delay systems. The major tools used in Chapter 19 are state-transition method and a method based on the trapezoidal rule for integration. Chapter 19 presents two methods of the effective discretization approach for the continuous-time systems with input and output delays. Sampled-data time-delay systems with internal and external point delays are described by approximate discrete time-delay systems in the discrete domain. These approximate discrete systems allow the hybrid control of time-delay systems. Chapter 19 provides two numerical examples in order to illustrate the correctness of the discretization process.

Chapter 20 provides a brief survey of the recent research results on robust iterative learning control for linear discrete-time switched systems. First, the problem of stability analysis and robust exponential stabilization for a class of switched linear systems with polytopic uncertainties is reviewed. A sufficient condition based on the average dwell time that guarantees the exponential stability of uncertain switched linear systems is given. Next, the Iterative Learning Control (ILC) is presented to build a formulation ensuring the exponential stability of the given system. The integrated design of this ILC scheme is transformed into a robust control problem of an uncertain 2D Roesser system. The results are obtained through original connection with the notion of stability along the pass for 2D repetitive systems. After a brief review of the stability analysis under restricted switching and the multiple Lyapunov function theory, the switching stabilization problem is studied, and a variety of switching stabilization methods found in the literature are outlined. Then the switching stabilizability problem is investigated.

Section 6 contains Chapters 21-25 focusing on the area of Power Systems and Engineering Applications.

In Chapter 21, two computational algorithms are proposed and applied on an estimation algorithm in order to improve the global performance of the estimation phase. The proposed system is studied based on the Model Reference Adaptive System (MRAS). In this chapter, the MRAS technique is proposed as the software algorithm, for replacing the Tesla meter and the other measurement components for online estimate the overall characteristic PMSM parameters. In the conventional MRAS, the specific PI parameters are manually tuned, due to the complicated mathematical expressions. Here, the approach aims to ameliorate the MRAS technique with intelligent optimization methods. Thereby, a comparative analysis between two intelligent optimization methods is achieved to define the best one. These two methods esteemed the natural bacterial and insect movement and called, respectively, BFO and PSO algorithms. This novel MRAS technique is applied in the PMSM motor type to approximate the three main parameters deviation, in terms of stator resistance, inductance, and the permanent magnet flux.

Chapter 22 focuses on the design of a robust H∞ controller for the power flow between the stator of the Doubly-Fed Induction Generator (DFIG) and the grid. The robust H∞ controller design is formulated as a mixed-sensitivity problem. A mathematical model of the DFIG written in an appropriate d-q reference frame is established to carry out simulations. The proposed power control scheme is elaborated and compared with a conventional Proportional- Integral (PI) controller based on vector control technique.

Chapter 23 describes the design of a controller with time response specifications on STM32 microcontroller. The basic idea of this work is to implement a fixed low order controller on a real electronic system by using the STM32 microcontroller. The principal aim of this controller is to guarantee some time response specifications such as the settling time and the overshoot. In this work, the Generalized Geometric Programming (GGP) method is exploited as a global optimization method. The key concept of this method is to transform an optimization problem originally non-convex to a convex one by mean of some mathematical transformations. The practical implementation, on a fast electronic system characterized by a little constant time, of the proposed control law and a Proportional Integral (PI) controller illustrates the effectiveness of the proposed algorithm.

Chapter 24 develops a new control strategy, for the grid side converter, based on symmetrical components theory. This strategy named Symmetrical Components Control Strategy (SCCS) aims to isolate the hybrid system of the adverse impact of the grid fault while fulfilling all necessary interconnection requirements. Dynamic modeling and simulations of the studied power system have been accomplished using MATLAB/Simulink. The results point out thoroughly the applicability of the proposed control scheme under unbalanced grid conditions.

Chapter 25 discusses modelling and control of a Neutral Point Clamped (NPC) inverter that operates with the PWM switching pattern using a DSP. The mathematical model of the NPC inverter is carried out using conversion and connection functions for an easier understanding of the system operation. Simulation results using MATLAB program are reported in this chapter. It is shown that the performances obtained for driving an asynchronous motor using this inverter are very promising. Finally, analysis of the theoretical and the experimental results is carried out in order to validate the effectiveness of the proposed control solution.

AUDIENCE

This book will serve as a reference handbook for scientists and engineers who are interested in applying new computational and mathematical tools for solving recent research problems in Computer Science, Information Technology, Engineering, Automation, and Chaos and Control Engineering. This book can also be used at the graduate or advanced undergraduate level as a textbook or major reference for courses such as Mathematical Modeling, Computational Science, Numerical Simulation, Nonlinear Dynamical Systems and Chaos, Control Systems, Applied Artificial Intelligence, and many others.

Ahmad Taher Azar
Benha University, Egypt

Sundarapandian Vaidyanathan
Vel Tech University, India

Acknowledgment

As the editors, we hope that the chapters in this book will stimulate further research into control systems and be utilized in real-world applications. We hope that this book, covering so many different aspects, will be of value for all readers.

We would like to thank also the reviewers for their diligence in reviewing the chapters.

Special thanks go to IGI Global, especially the book editorial team.

Ahmad Taher Azar
Benha University, Egypt

Sunddarapandian Vaidyanathan
Vel Tech University, India

Section 1
Nonlinear Control Systems

Chapter 1
Stabilization and Control of Mechanical Systems with Backlash

Ahmad Taher Azar
Benha University, Egypt

Fernando E. Serrano
Florida International University, USA

ABSTRACT

Backlash is one of several discontinuities found in different kinds of systems; it can be found in actuators of different types, such as mechanical and hydraulic, giving way to unwanted effects in the system behavior. In this chapter, three different control approaches are derived to stabilize mechanical systems in which this phenomenon is present in the actuators of the system. First, an independent joint control approach when backlash is found in the actuators is derived; then a PI loop shaping control design implementing a describing function to find the limit cycle oscillations and the appropriate control gain is developed. Finally, an optimal controller for mechanical systems with backlash is derived, obtaining the optimal control law and oscillations frequency when this nonlinearity is found implementing a describing function to model the backlash effects.

1. INTRODUCTION

Backlash is a phenomena found in different kinds of actuators such as mechanical and hydraulic, generally it occurs when the contact of two mating gears do not match and this give way to many unwanted effects provoking problems to the whole mechanical system. Thus some authors have proposed several solutions respect to the control and stability issues of these systems with input nonlinearities; taking in count that is not a trivial task due to the complexity of the model, traditional control strategies fail in most of the cases, and therefore it is necessary to design nonlinear control strategies or implement modified traditional ones. The objective of this book chapter is to explain diverse control strategies for

DOI: 10.4018/978-1-4666-7248-2.ch001

systems with this kind of nonlinearity. The intention is to show some nonlinear control techniques that have been developed and propose different approaches to solve this problem.

Robust and output feedback controllers (Barreiro & Baños, 2006; Azar & Serrano 2014; Azar & Serrano 2015 a,b; Azar and Zhu 2015; Zhu and Azar 2015; Mekki et al. 2015; Vaidyanathan et al. 2015; Vaidyanathan and Azar 2015a, b, c, d) are some approaches that has been used for decades taking in count that most of the mechanical systems are nonlinear and multivariable, thus these techniques are explained as a preamble for some novel control procedures developed for the stabilization of mechanical systems with this kind of nonlinearity. Adaptive backlash control is another control strategy developed by some authors; this approach has the advantage of cancelling the backlash effects using a backlash inverse model (Tao & Kokotovic, 1993; Tao & Kokotovic, 1993; Tao & Kokotovik, 1995, Zhou, Er, & Wen, 2005; Jing & Wen, 2007; Guo, Yao, Chen, & Wu, 2009). The implementation of optimal nonlinear controller with backlash is another different approach to solve this problem, in which an optimal control law is found to keep the system about the equilibrium point by the use of the Riccati equations (Tao & Ma, 1999; Tao, Ma, & Ling,2001). Sometimes, this problem is solved by using a deadzone or hysteresis model of the backlash, even when this approach is not so accurate it yields acceptable control performances (Ma & Yang, 2008; Ahmad & Khorrami, 1999). Other authors propose an input-output stable controllers using an H_∞ and L_∞ approach to deal with the backlash nonlinearity (Barreiro & Baos, 2006; Bentsman, Orlov, & Aguilar, 2013). Intelligent control (Azar and Vaidyanathan 2015a,b) is a good choice for the design of efficient control strategies, some authors have proposed different control approaches with this characteristic as explained in (Jang, Lee, Chung, & Jeon, 2003) where a fuzzy controller (Azar, 2010) is designed by modeling the backlash nonlinearity and suppressing the unwanted effects yielded by this phenomenon. This control strategy is suitable when this discontinuity is found in mechanical and other kinds of systems due to the estimation properties of the intelligent controller allowing the designer to conceive an efficient control system. There are several mechanical and mechatronics systems that possess backlash nonlinearities, in this chapter some control strategies for n degrees of freedom robotics and multibody systems are developed and analyzed starting with the derivations of the dynamic equations obtained by the Euler Lagrange formulation, including the backlash nonlinearity, then some control strategies are obtained for the control and stabilization of these models under the effects of input backlash. The effects of backlash is common in the above mentioned systems, considering the contact of different actuators and sensors specially gears; this is a non desirable characteristic that yields unwanted effects, so it is necessary to devise controllers that deal with this kind of feature in the design process. The second section of this chapter is devoted to depict the different kinds of backlash models found in literature; these models are explained and analyzed because it is necessary to obtain an adequate discontinuity model before the controller design process. The kinds of mathematical models that can be implemented could be approximate and accurate models (Nordin & Gutman, 2002), so the designer could select an appropriate mathematical model for the controller design. In this section, the inverse backlash model is explained in detail, although it is not implemented in the proposed control strategies of this chapter, it is important to cite this model because is very useful in the design of adaptive and intelligent controllers (Jang, Lee, Chung, & Jeon, 2003).The intention of this chapter is to elucidate three kinds of control approaches when backlash is present in the actuators, to show how to deal with this nonlinearity by independent or multivariable control of the mechanical systems. These approaches are done by frequency domain and state space control techniques with their respective analysis as explained in each of the following sections. An independent control approach of mechanical systems with backlash is developed in Sect. 3. This control technique is still in use due to its simplicity and the diversity in tuning methods when backlash is present in the outputs of the mechanical systems, specially, when there is contact of two gears. In this section, the design of PD independent joint controller with output backlash is shown taking into account the actuators effects. An example of the independent control of a two links robotic manipulator with backlash is shown. In Sect. 4, a multivariable PI H_{∞} loop shaping design for mechanical systems with backlash is shown. This technique is implemented due to their frequency domain characteristics; therefore it is suitable for the implementation of the backlash describing function model to analyze the limit cycles due to the backlash effects. The tuning procedure is explained in this section and an example of the stabilization of a cart-pendulum system with output backlash is shown. The optimal control of mechanical systems with backlash is shown in Sect. 5, where the implementation of a describing function to obtain the optimal control law by the Riccati equations is done. Similar as the controller explained in Sect. 4, the idea is to analyze the limit cycles effects yielded by the backlash discontinuity. In Sect. 6, a discussion is done about the proposed control strategies to elucidate the advantages and disadvantages of each model and, finally, the conclusions of this chapter are explained in Sect. 7.

2. BACKLASH MODELS

In this section, some mathematical models of the backlash discontinuity are shown; these models are commonly used before the derivations of suitable controllers when this nonlinearity appears. Backlash occurs when adjacent mechanical parts are not directly connected, for example, when the actuator, that in this case could be a DC motor, and the load do not have a perfect contact. For this reason, it is important to obtain a mathematical model for this non differentiable nonlinearity. There are some approximated and exact models that could be used in the design of an appropriate controller in mechanical systems where this discontinuity is found.

Apart from the backlash models presented in this section, a backlash inverse model is shown due to this mathematical model is used by some control strategies (Tao & Kokotovic, 1993; Tao & Kokotovic, 1993; Tao & Kokotovik, 1995), and it is important to mention this model due to its properties.

2.1 Approximate Model

Sometimes it is necessary to implement approximate backlash models because of its complex characteristics that make it difficult in the design of appropriate controllers.

Therefore a dead zone model is proposed by some authors (Nordin & Gutman, 2002; Bentsman, Orlov, & Aguilar, 2013) because this model is non linear and static.

The mathematical model is given in (1)

$$D_a(x) = \begin{cases} x - \alpha & x > \alpha \\ 0 & |x| < \alpha \\ x + \alpha & x < -\alpha \end{cases} \tag{1}$$

where $\alpha > 0$

Figure 1. Backlash model

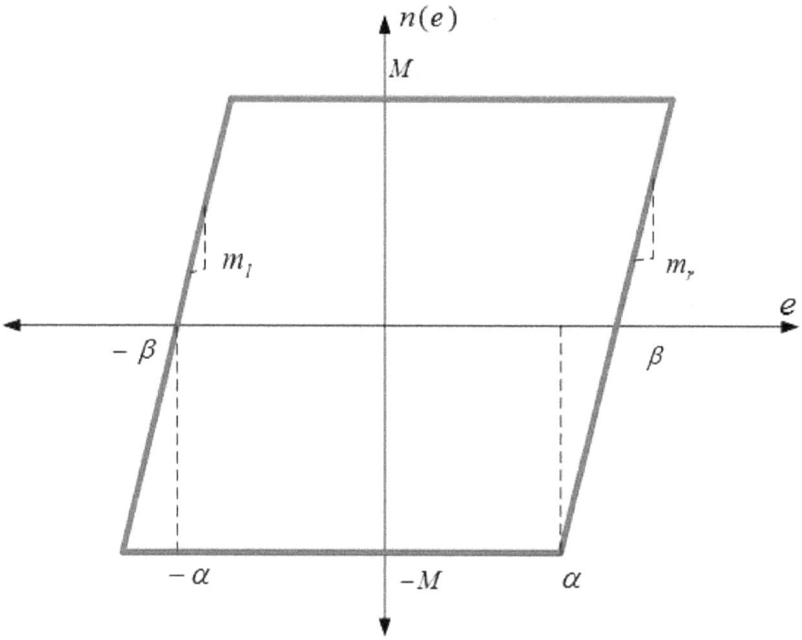

This model is useful and easy to implement in the design of controller with backlash, although it depends on the actuator characteristics of the model.

2.2 Exact Model

Sometimes it is necessary to use an exact backlash model due to the accuracy of the controller design. Therefore a model as represented in Figure 1 is used, where e is the input to the backlash and $n(e)$ is the output of the model (Tao & Kokotovik, 1995).

The mathematical description of the backlash model is shown in (2)

$$n(t) = \begin{cases} m_l\left(e(t) - \beta\right) & e(t) \leq e_l(t) \\ m_r\left(e(t) - \beta\right) & e(t) \geq e_r(t) \\ n(t-1) & e_l(t) < e(t) < e_r(t) \end{cases} \tag{2}$$

where $\beta > 0$, $m_l, m_r > 0$ are the backlash slope, $e_l(t) = \frac{n(t)}{m_l} + \beta$ and $e_r(t) = \frac{n(t)}{m_r} + \beta$.

The backlash input is e and the backlash output is n(e). This backlash model is often used in the design of anti backlash controllers due to their accuracy in comparison with the dead zone model, but it is quite more difficult than other simplified mathematical models.

In the controller design of Sect. 4 and Sect. 5, this mathematical model is implemented, with the derivation of a describing function (Chyung, 1992; Tierno, Kim, Lacy, & Bernstein, 2000) that is an efficient backlash representation for this complex discontinuity.

2.3 Inverse Backlash Model

Even when this model is not treated in this chapter, it is important to understand the inverse backlash model. This model simplifies the control systems design when this nonlinearity is found; several authors have proposed different control techniques from output feedback (Chyung, 1992) to adaptive control (Tao & Kokotovik, 1995; Ma & Yang, 2008) to deal with this difficult problem.

This model is represented in Figure 2; a description of the backlash inverse model can be found in (Tao & Kokotovik, 1995) as follows:

$$\chi(x) = \begin{cases} 1 & if\ x\ is\ true \\ 0 & otherwise \end{cases} \tag{3}$$

Then the following equations are necessary to derive the backlash inverse

$$\chi_{rm}(t) = \chi\left[n(t) > n(t-1) or\, n(t) = n(t-1), \chi_{rm}(t) = 1\right] \tag{4}$$

Figure 2. Inverse backlash model

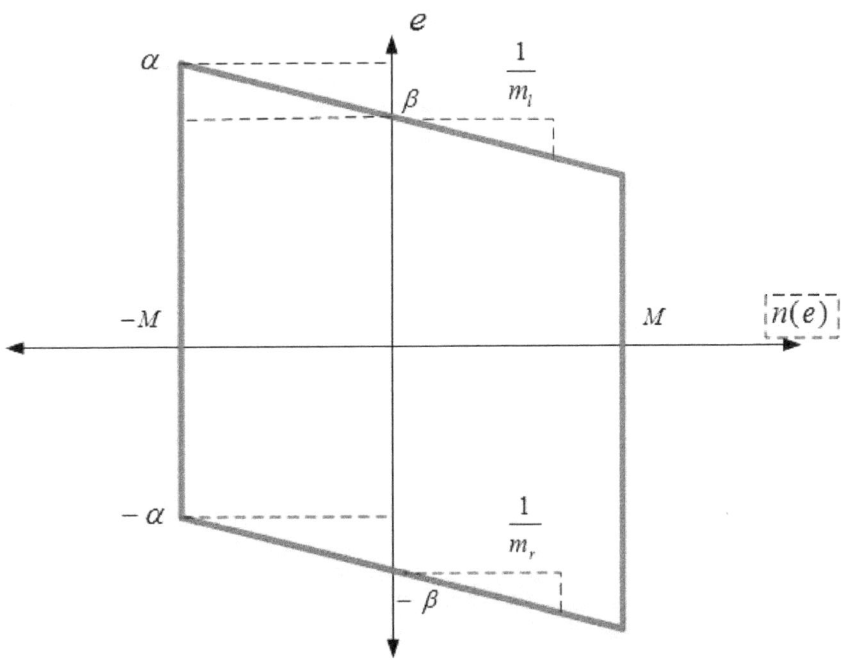

$$\chi_{lm}(t) = \chi\left[n(t) < n(t-1) or n(t) = n(t-1), \chi_{lm}(t) = 1\right] \tag{5}$$

Then the inverse backlash model is represented by:

$$e_m(t) = \frac{n_m(t)}{m} + \chi_{rm}(t)\beta + + \chi_{lm}(t)\beta \tag{6}$$

where $m = m_l = m_r$.

This function is useful to find a suitable control law cancelling the effects of backlash when this discontinuity is found in the input of the system.

3. INDEPENDENT JOINT CONTROLS OF MECHANICAL SYSTEMS WITH BACKLASH

As explained in the first section of this chapter, independent joint control of mechanical systems is still very useful, even when in the latest years multivariable control has become a novel alternative for the control of mechanical systems.

Independent joint control consist in the design of independent controllers for each joint, which sometimes they could be PI (Nordin & Gutman, 2002; Azar & Serrano, 2014, 2015c) or PD (Indri & Tornambe, 1997; Spong, Hutchinson, & M., 2006) controllers, that can be tuned by conventional tuning procedures. The advantage of this control strategy is that there is not interaction between each joint, so the controller design is simple, taking into account that these kind of mechanical systems are coupled and nonlinear. Even when this control technique is not so novel at all and modern multivariable control methods are implemented nowadays, they are still useful and applied in the control of mechanical systems.

In this section the effects of backlash when there is interaction of the actuator shaft, that in this case the actuator is a DC motor, and the link gear is investigated. The derivations of PD control laws are done for mechanical systems with or without backlash, in order to make a comparison of the system performance for both cases. The actuator dynamics is derived first to obtain the respective transfer function that is related to the mechanical systems, and then the effects of backlash are analyzed when this effect is present in the systems interconnections. The effect of backlash and other nonlinearities cause that the selection of a suitable control law would be determined by the kind of nonlinearity found; for this reason, the proposed control law is analyzed to test the stability of mechanical system, or the joint dynamics, and the performance of the system when this nonlinearity appears in the joint interconnection.

This section is a preamble of the control strategies explained in the following sections, where in this circumstance the problem is treated as a SISO problem, and then later this problem is solved by using multivariable control techniques analyzing the system behavior when this discontinuity is found.

3.1 PD Control of Mechanical Systems with Backlash

In this subsection the derivation of a PD control law for each joint of an n degrees of freedom mechanical system is derived. There are two perspectives, as mentioned in this subsection, the first case is when backlash is not present in each joint of the mechanical system; and the second perspective is when backlash is present in all the joints of the system.

Case 1: PD Control of Mechanical Systems with no Backlash

Before analyzing the mechanical systems with no backlash, a PD controller when no backlash is present is derived. The reason to choose this control law is because of its typical implementation in the independent joint control of mechanical systems, due to their performance improvement and the stability assurance (Spong, Hutchinson, & M., 2006; Kashiwagi, Harada, & Toyozawa, 2004), then as it is proved later in this section, this control law can stabilize mechanical systems with backlash. Sometimes a PI controller for systems with backlash is suitable for this kind of problems (Nordin & Gutman, 2002) but in this chapter we chose a PD control alternative.

First, the dynamical model of an n degrees of freedom mechanical system is shown before the PD controller is derived, along with the actuator model. An n degrees of freedom mechanical model is defined in (Seraji, 1989) (7)

$$D(q_s)\ddot{q}_s + C\left(q_s,\dot{q}_s\right)\dot{q}_s + g\left(q_s\right) = \tau \tag{7}$$

where $q_s \in \mathbb{R}^n$ is the joint angle vector, $\tau \in \mathbb{R}^n$ is the input vector torque, $D(q_s) \in \mathbb{R}^{n\times n}$ is the inertia matrix, $C\left(q_s,\dot{q}_s\right)\dot{q}_s \in \mathbb{R}^n$ is the coriolis vector and $g\left(q_s\right) \in \mathbb{R}^n$ is the gravity vector. This mathematical model can be subdivided in n parts for each independent joint as follow. For $\mathrm{i} = 1 \ldots \mathrm{n}$ the system can be described as:

$$\begin{pmatrix} D_{11}\left(q_s\right) \\ \vdots \\ D_{nn}\left(q_s\right) \end{pmatrix}^T \begin{pmatrix} \ddot{q}_{s1} \\ \vdots \\ \ddot{q}_{sn} \end{pmatrix} + \sum_{\substack{j=1 \\ j\neq i}}^{n} D_{ij}\left(q_s\right)\ddot{q}_{sj} + H_i\left(q_s,\dot{q}_s\right) + g_i\left(q_s\right) = \tau_i \tag{8}$$

where $H_i\left(q_s,\dot{q}_s\right) = C\left(q_s,\dot{q}_s\right)\dot{q}_{s1}$

Then the proposed PD control law is required for the stabilization of this problem, as explained in (9)

$$\tau_i = -k_{pi}q_s - k_{Di}\dot{q}_s \tag{9}$$

where $k_{pi} > 0$ is the proportional gain and $k_{Di} > 0$ is the derivative gain. Then, the independent joint dynamics for control purpose is defined as follows:

$$D_{ii}\left(q_s\right)\ddot{q}_s + d_i\left(q_s,\dot{q}_s,\ddot{q}_s\right) = \tau_i\left(t\right) \tag{10}$$

where

$$d_i\left(q_s, \dot{q}_s, \ddot{q}_s\right) = \sum_{\substack{j=1 \\ j \neq i}}^{n} D_{ij}\left(q_s\right)\ddot{q}_{sj} + H_i\left(q_s, \dot{q}_s\right) + g_i\left(q_s\right) \tag{11}$$

After the system is subdivided into n independents parts, the proposed control law can be applied, but the following condition is necessary before implementing the proposed control technique.

Definition 1: The function $d_i\left(q_s, \dot{q}_s, \ddot{q}_s\right)$ is assumed as a torque perturbation of the system represented by τ_{li} usually divided by the gear ratio r as defined later.

The defined vectors q_s, $\dot{q}_s$ and $\ddot{q}_s$ are the respective position, angular velocity and angular acceleration of the joint adjacent to the motor shaft that are connected by gears. The next step is to define the position, angular velocity and angular acceleration of the motor shaft related to the respective joint variables by the following relation:

$$q_m = \frac{r_s}{r_m} q_s \tag{12}$$

where q_m is the motor shaft angular velocity, r_s is the joint gear constant and r_m is the motor shaft gear constant.

Usually, the relation $\frac{r_s}{r_m}$ is called the gear ratio r; typical values of this ratio are between 20 and 200 or even more. In Figure 3 is depicted the relation of the motor shaft and the joint, coupled by two gears in which backlash can occur, but this case will be analyzed in Case 2.

Keeping in mind the relation of the actuator shaft and the joint coupled by a gear train, the electrical (actuator/DC motor) and mechanical (joint coupling) systems can be related by a set of equations as described in (Nordin & Gutman, 2002; Spong, Hutchinson, & M., 2006) and depicted in Figure 3.

Then the equation of motion of the system is denoted by

$$J_{mi}\ddot{q}_{mi} + b_{mi}\dot{q}_{mi} = \tau_{mi} - \frac{\tau_{li}}{r}$$

$$= k_{mi} i_a - \frac{\tau_{li}}{r} \tag{13}$$

where J_{mi} is the motor inertia, b_{mi} is the coefficient of the motor friction and $\tau_{li}\left(t\right)$ is the joint torque, usually considered as a disturbance. The mechanical model of the motor described in (13) is necessary before deriving the equations of the electrical model, implemented in the design of a PD controller for the first case when backlash is not found in the system.

The electrical model of the actuator is given by (14)

Figure 3. Model of a single link with actuator/gear train

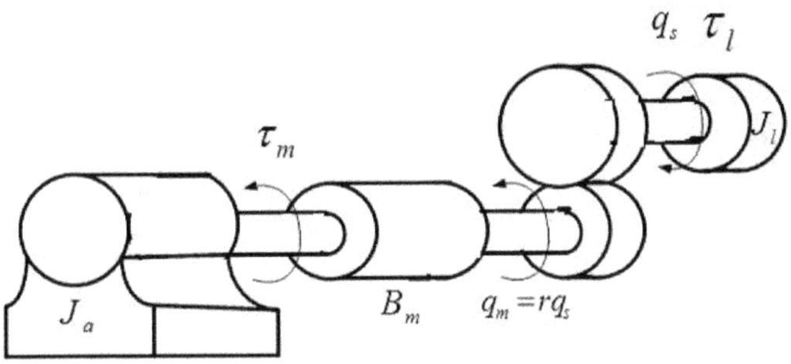

$$L_j \frac{di_{aj}}{dt} + R_j i_{aj} = V_j - V_{bj} \tag{14}$$

where L_j is the armature inductance, R_j is the armature resistance, i_{aj} is the armature current, V_j is the armature voltage and V_{bj} is the back emf for the j joint. Given that the flux ϕ is constant, the developed torque by the motor is

$$\tau_{mj} = K_1 \phi \omega_{mj} = K_{mj} i_a \tag{15}$$

where K_{mj} is the torque constant for the actuator j. Then

$$V_{bj} = K_2 \phi \omega_{mj} = K_{bj} \frac{dq_{mj}}{dt} \tag{16}$$

where K_{bj} is the back emf constant for the j joint. Then the transfer functions are obtained as shown in (17) and (18) setting $\tau_j = 0$ and $V_j = 0$ respectively (Nordin and Gutman, 2002; Spong, et al., 2006).

$$\frac{q_{mi}(s)}{V_i(s)} = \frac{K_{mi}}{s\left[(L_i s + R_i)(J_{mi} s + B_{mi}) + K_{bi} K_{mi}\right]} \tag{17}$$

$$\frac{q_{mi}(s)}{\tau_i(s)} = \frac{(L_i s + R_i)/r_i}{s\left[(L_i s + R_i)(J_{mi} s + B_{mi}) + K_{bi} K_{mi}\right]} \tag{18}$$

The resulting open loop system for the tuning procedure is shown in Figure 4. Based on the model of the actuator it is necessary to test the stability of the system when a PD control law is implemented.

Figure 4. Block diagram of a DC motor

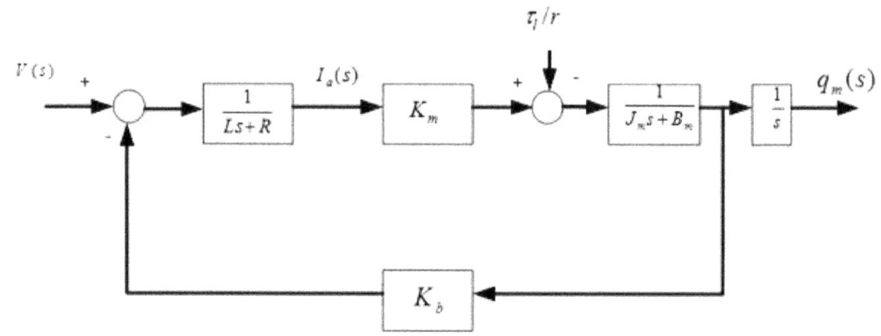

Now, it is necessary to test the stability of the proposed control law, as described in (9), by the construction of a Lyapunov function.

For this objective, (10) is transformed in (19); for simplicity the variable q_s is denoted by q_i for the joint i

$$\ddot{q}_i(t) = \left(\tau_{mi} - \frac{\tau_{li}}{r}\right) D_{ii}^{-1}(q_i) \tag{19}$$

and assuming the disturbance torque $\tau_{li} = 0$, then (19) becomes in

$$\ddot{q}_i(t) = \tau_{mi} D_{ii}^{-1}(q_i) \tag{20}$$

Converting (20) to state space yield:

$$\dot{x}_{1i} = x_{2i}$$

$$\dot{x}_{2i} = \tau_{mi} D_{ii}^{-1}(q_i) \tag{21}$$

where

$$x_{1i} = q_{1i}$$

$$x_{2i} = \dot{q}_{1i} \tag{22}$$

Considering that $\frac{r_s}{r_m} = r$ (13) the following result is obtained:

$$J_{mi}\ddot{q}_{mi} + b_{mi}\dot{q}_{mi} = \tau_{mi}$$

$$rJ_{mi}\dot{x}_{2i} + b_{mi}\dot{q}_{mi} = \tau_{mi}$$

$$rJ_{mi}\left(\tau_{mi}D_{ii}^{-1}\left(q_i\right)\right) + b_{mi}\dot{q}_{mi} = \tau_{mi} \tag{23}$$

Therefore from (22) $\dot{q}_{mi}$ is converted into

$$\dot{q}_{mi} = \frac{\tau_{mi}}{b_{mi}}\left(1 - rJ_{mi}D_{ii}^{-1}\left(q_i\right)\right) \tag{24}$$

where r is the gear ratio.

Theorem 1: The PD control law (9) is asymptotically stable, when the mechanical system does not have backlash according to the following stability test.

Proof. Consider the following Lyapunov function candidate:

$$V\left(q_{mi}\right) = \frac{1}{2}q_{mi}Pq_{mi} \tag{25}$$

where $P>0$ is a positive constant. Now, considering (9) and the fact that $\frac{r_s}{r_m} = r$ the following equation is obtained:

$$\dot{q}_{mi} = \frac{-q_{mi}\left(K_{pi} - rK_{pi}J_{mi}D_{ii}^{-1}\left(q_i\right)\right)}{rb_{mi} + K_{Di} - rK_{Di}J_{mi}D_{ii}^{-1}\left(q_i\right)} \tag{26}$$

Then deriving (25) along the trajectory of (24)

$$\dot{V}\left(q_{mi}\right) = \dot{q}_{mi}Pq_{mi} \tag{27}$$

Then the following equation is obtained

$$\dot{V}\left(q_{mi}\right) = -\alpha q_{mi}Pq_{mi}$$

$$\dot{V}\left(q_{mi}\right) \le -\left|\alpha\right|\left\|q_{mi}Pq_{mi}\right\| \tag{28}$$

where

$$\alpha = \frac{\left(K_{pi} - rK_{pi}J_{mi}D_{ii}^{-1}\left(q_i\right)\right)}{rb_{mi} + K_{Di} - rK_{Di}J_{mi}D_{ii}^{-1}\left(q_i\right)} \tag{29}$$

Therefore, asymptotical stability is assured and this completes the proof. With this proof, it is clear that the n degrees of freedom mechanical system can be stabilized by the PD control law implementing an independent controller with each joint. In this case backlash is not found in any joint so the problem solution is straightforward and the PD controllers for each joint can be tuned by any of the existent tuning methods for this kind of controller.

In Case 2, the effects of backlash on the system performance are analyzed proving that the system can be stabilized by a PD control law as shown in Case 1.

Case 2: PD Control of Mechanical Systems with Backlash

In the Case 1 an independent joint controller for n degrees of freedom mechanical system was derived.

Even when this control structure is plain, the stabilization of this kind of systems is simple and very effective. In Case 2 a similar control strategy is followed proving that in the presence of backlash, the independent joint controller can be stabilized implementing a PD controller. The control strategy is derived taking into account that there is backlash between two coupled gears, that in this case, is the interaction of the actuator shaft gear and the joint (Indri & Tornambe, 1997; Nordin & Gutman, 2002).

When there is no contact between the two coupled gears q_{si} and q_{mi}, a backlash of $\pm\Delta$ occurs and then a function δ is defined

$$\delta_i\left(t\right) = \begin{cases} r_{mi}q_{mi} - r_{si}q_{si} - \Delta & if\, r_{mi}q_{mi} - r_{si}q_{si} \geq \Delta \\ r_{mi}q_{mi} - r_{si}q_{si} + \Delta & if\, r_{mi}q_{mi} - r_{si}q_{si} \leq -\Delta \\ 0 & Otherwise \end{cases} \tag{30}$$

Then the torque equation is defined as (Boukhnifer & Ferreira, 2005)

$$\tau_i = k_i\delta_i + b_i\dot{\delta}_i \tag{31}$$

where $k_i, b_i > 0$ are constants that meet these conditions. Equation (31) is denoted by:

$$\tau_i = k_i\left(r_{mi}q_{mi} - r_{si}q_{si} - S_i\Delta_i\right) + b_i\left(r_{mi}\dot{q}_{mi} - r_{si}\dot{q}_{si}\right) \tag{32}$$

where $S_i = sign\left(r_{mi}q_{mi} - r_{si}q_{si} - \Delta_i\right) = sign\left(\varphi\right)$ is the sign function (Tao, Ma, & Ling, 2001).

Implementing the control law (9), as τ_{mi}, and (31) in (13) yields

$$J_{mi}\ddot{q}_{mi} + b_{mi}\dot{q}_{mi} + K_p q_{si} + K_d \dot{q}_{si} = -\frac{\delta}{r} \tag{33}$$

In order to simplify the stability analysis the following assumption is needed.

Definition 2: The sign function S is substituted by a continuous and differentiable $\tan sig(\varphi)$ function, denoted as

$$tansig(\varphi) = \frac{1 - e^{-2\varphi}}{1 + e^{-2\varphi}} \tag{34}$$

where φ is defined previously.

Therefore (10) is converted to:

$$D_{ii}(q_s)\ddot{q}_{si} + K_p q_{si} + K_D \dot{q}_{si} + \frac{\delta_i}{r} = 0 \tag{35}$$

substituting δ_i in (35)

$$D_{ii}(q_s)\ddot{q}_{si} + K_p q_{si} + K_D \dot{q}_{si} + \frac{r_{mi}q_{mi}}{r} - \frac{r_{si}q_{si}}{r} - \frac{\Delta_i \tan sig(\varphi)}{r} = 0 \tag{36}$$

Equation (36) is transformed to state space by:

$$x_1 = q_{si}$$

$$x_2 = \dot{q}_{si}$$

$$\dot{x}_1 = x_2$$

$$\dot{x}_2 = \ddot{q}_{si} \tag{37}$$

Then the state space form of (36) is given in:

$$\begin{pmatrix} \dot{x}_1 \\ \dot{x}_2 \end{pmatrix} = \begin{pmatrix} 0 & 1 \\ -\left(K_p + r_{mi} - \frac{r_{si}}{r}\right) D_{ii}^{-1}(q_{si}) & -K_D D_{ii}^{-1}(q_{si}) \end{pmatrix} \begin{pmatrix} x_1 \\ x_2 \end{pmatrix} + \begin{pmatrix} 0 \\ \frac{D_{ii}^{-1}(q_{si}) \tan sig(\varphi)}{r} \end{pmatrix} \Delta_i \tag{38}$$

Then to prove the stability of the system the following Lyapunov function is proposed:

$$V(x) = \frac{1}{2}x_2^2 \tag{39}$$

Deriving (39) the following result is obtained:

$$\dot{V}(x) = x_2 . \dot{x}_2 \tag{40}$$

and substituting from the state space model (38) the following result is obtained:

$$\dot{V}(x) = -x_2\left(K_p + r_{mi} - \frac{r_{si}}{r}\right)D_{ii}^{-1}(q_{si})x_1 - x_2 K_D D_{ii}^{-1}(q_{si})x_2 + x_2\frac{D_{ii}^{-1}(q_{si})\tan sig(\varphi)\Delta_i}{r} \tag{41}$$

Applying the norm to (41) yields:

$$\dot{V}(x) \leq -\|x_2\|\left|\left(K_p + r_{mi} - \frac{r_{si}}{r}\right)D_{ii}^{-1}(q_{si})\right|\|x_1\| - \|x_2\|\left|K_D D_{ii}^{-1}(q_{si})\right|\|x_2\| + \|x_2\|\left|\frac{D_{ii}^{-1}(q_{si})\tan sig(\varphi)\Delta_i}{r}\right| \tag{42}$$

The independent joint PD control is proved to stabilized any mechanical system when backlash is present in its coupled gears, assuming that Δ is sufficiently small and K_p and K_D are selected to make (42) a negative semi definite function. Even when this control strategy is applied in mechanisms in which the desired trajectory or path does not change rapidly, it is still very useful in many mechanical applications.

In the next subsections the stability and response of a mechanical system with and without backlash are analyzed. The performance of the two examples are tested and compared to observe the effects of backlash in the closed loop system and the behavior of the measured variables.

3.2 Example 1: PD Control of a Two Links Robotic Manipulator with No Backlash

In this subsection the design of a PD controller for a two links robotic manipulator, as shown in Figure 5, with no backlash is explained. An independent joint approach is implemented when there is no backlash in the system. Even when this could be a trivial case, the objective of this example is to be used as a comparative benchmark for the problem in which backlash is found. The dynamic equations for this mechanism is denoted by (7) with n = 2, the components of the matrices $D(q_s)$, $c(q_s, \dot{q}_s)$ are derived using the Euler Lagrange formulation and can be found in the literature (Spong, Hutchinson, & M., 2006).

Figure 5. Two links robotic manipulator

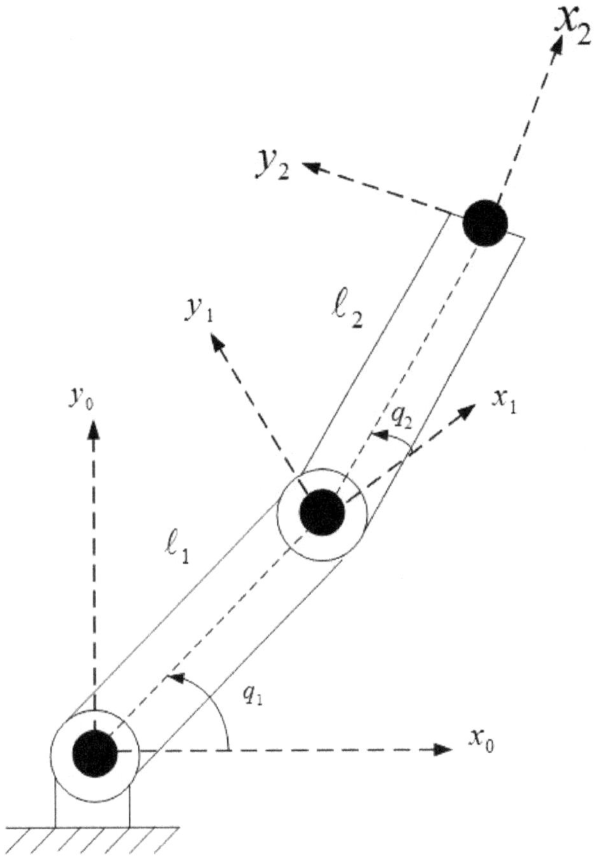

Table 1. Parameters of the two links robotic manipulator

Joint	Mass (Kgs)	Length (m)	Mass Center (m)	Inertia (Kg•m²)
1	4.0	0.5	0.25	1.0
2	2.0	0.25	0.15	0.8

Obviously, this problem have two joints because is a two degrees of freedom mechanical system, see Figure 5. Therefore, $i=1,2$ for each controlled variable.

A PD controller is implemented in an independent fashion. The parameters of the mechanical systems are shown in Table 1. The Parameters of the DC motor actuators are shown in Table 2 and they are the same for each joint actuator.

The relations of the coupled gears for each joint is $r=1$, which means that $r_m = r_s$. This is valid for Example 1 and Example 2; where in the first case no backlash occurs.

In Figure 6 a block diagram of the mechanical system is shown, so it can be seen the independent controllers for each joint, and the controller signals v1 and v2 applied on the actuators or DC motors as explained in Subsection 1.

Table 2. Parameters of the DC motors actuators

Parameter	Value
R_a	2.581 Ω
L_a	0.028 H
J_m	0.02215 $Kg \cdot m^2$
B_m	0.002953 $N \cdot m \cdot s$

The objective is the regulation of the controlled variables q_1 and q_2 which are the angular velocities of joint 1 and joint 2 respectively.

The parameters K_{Di} and K_{pi} are found by the Ziegler-Nichols procedure on the transfer function (17), a different tuning procedure could be implemented as long as $K_{Di} > 0$ and $K_{pi} > 0$, so the stability condition (28) is met for each joint. In Table 3, the parameters of the PD controllers are shown.

The simulations and discussions are shown in Subsection 3 for comparison purposes with the mechanical system with backlash.

Figure 6. Block diagram of the controlled system

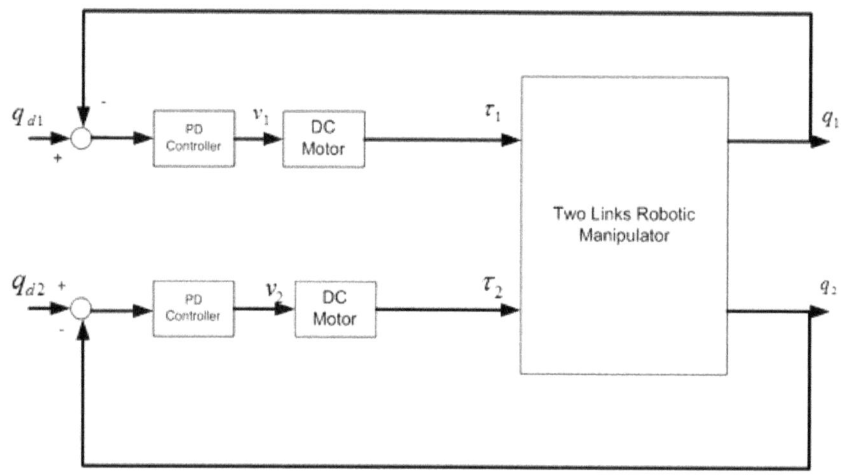

Table 3. Controller parameters for each joint

Joint	K_p	K_D
1	19.4	1.94
2	19.4	1.94

3.3 Example 2: PD Control of a Two Links Robotic Manipulator with Backlash

In this section an independent PD controller of a two links robotic manipulator with backlash is developed to deal with this nonlinearity that causes unwanted effects in the system. In this case, the backlash parameter is $\Delta = 0.003$, the motor and the manipulator parameters are the same as in Example 1.

In Figure 7 the block diagram of the independent joint controller is shown. The actuator dynamics is considered for the development of the PD controller. In this case, there is backlash between the motor shaft and the joint coupling, yielding unwanted effects that can be reduced and eliminated by selecting an appropriate controller gains.

The controlled variables are q_1 and q_2, which denote the angle positions of each joint. The controller signals are v_1 and v_2 which are the applied voltages for each joint actuator.

The controllers constant are found by the Ziegler-Nichols tuning procedure. These constants stabilize the systems as long as condition (42) is met. In Table 4 the required gain constants are shown. These constants are obtained considering the relations given by (17) as long as they meet condition (42) to assure stability.

The simulations were done using MATLAB®, SimMechanics® and SimPowerSystems®. These simulations were done applying a step function reference to be followed by the outputs of the system. The comparisons of the model with and without backlash are done, analyzing the performance and responses of both examples.

Figure 7. Block diagram of the controlled system with backlash

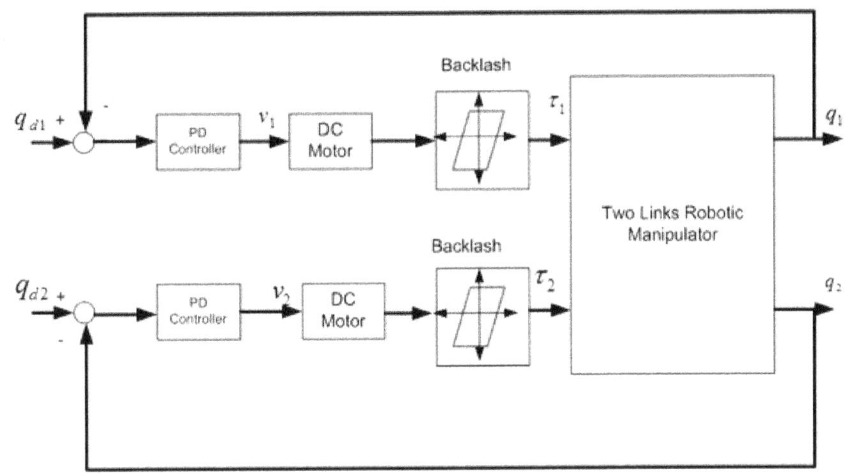

Table 4. Controller parameters for each joint with input backlash

Joint	K_p	K_D
1	15.52	8
2	15.52	8

In Figure 8 and Figure 9 the angular positions for q_1 and q_2 are shown, when a step changes in their inputs are applied respectively. As can be noticed the angular positions reach the desired values in a considerable low settling time, in both cases, with backlash and no backlash respectively.

The PD controllers stabilize the controlled variables with a small overshoot when no backlash is present. When backlash is present, the angular position reaches the desired value with a small increasing in settling time and no overshoot is yielded. This outcome is provoked by the effects of backlash in the coupling gears, but this effect is cancelled by selecting appropriate controller gains to meet the stability requirements.

As expected by the definitions of the PD controller design for both cases, the controlled variables asymptotically tends to the required variables. This condition depends on a proper selection of the PD controller constants K_p and K_D assuring that the systems reach the desired values in a finite time. With these conditions, as it seen in both examples, the mechanical system outputs are regulated with the design specifications with a simple but effective control law.

The angular velocities for link 1 and link 2 are shown in Figure 10 and Figure 11, as it is expected the angular velocities for each link reach asymptotically the desired values in finite time, this means that the angular velocities tend to zero as the angular positions reach the desired values. It can be noticed the effects of backlash in the angular velocities, but as it is expected, with an appropriate gain selection the velocity when backlash is found reaches the final value of zero with some differences in comparison when there is no backlash in the coupling gears.

Figure 8. Angular position of link 1

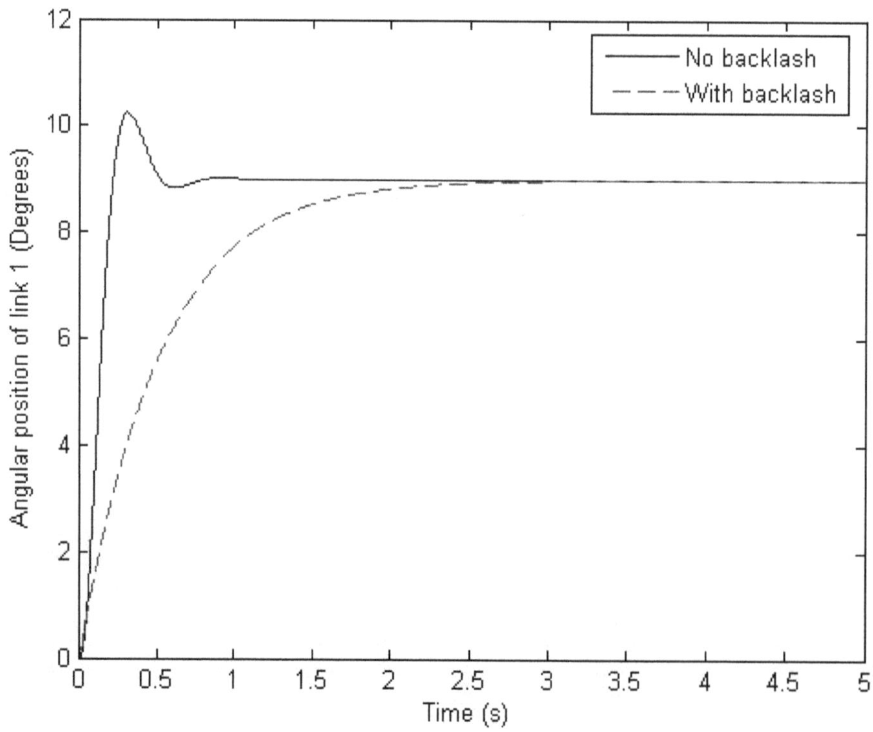

Figure 9. Angular position of link 2

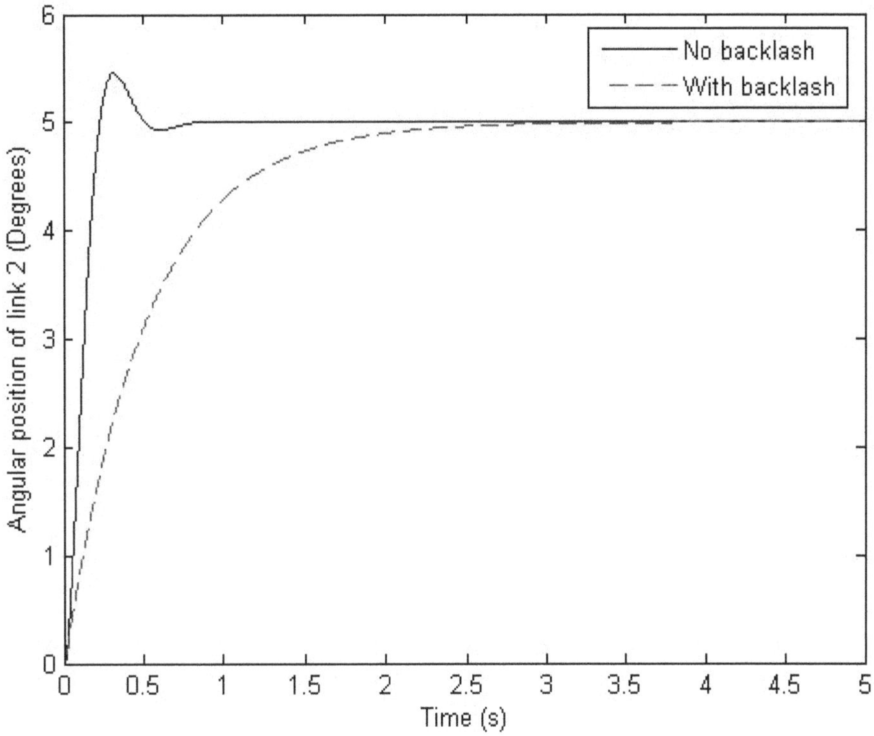

As it is proved, the backlash effects in the angular velocities is minimized or cancelled by an appropriate selection of the controller parameters, assuring asymptotically stability of these variables reaching the zero value in finite time.

In Figure 12 and Figure 13 the input voltages of the actuators for joint 1 and joint 2 are shown. As it is expected the applied voltages excite the motor until the required torque is yield to move the actuators until the required positions are reached.

Even when there is backlash actuator/gear train the control effort (or voltage) is less than the control effort in the case when no backlash is present. This means that is not necessary an actuator limiter in both cases, therefore the actuators are not saturated and are not forced to work in a dangerous operating point.

In Figure 14 and Figure 15 the applied control torque for each joint is shown. As can be noticed the torque applied on each joint when backlash is present is less than the torque applied when there is no backlash due to a small control effort generated by the control law. As a conclusion of this section, a PD controller is able to stabilize an n degrees of freedom mechanical system by an independent control approach when backlash is present in the actuator /gear train. Even when this technique is simple, it has some advantages and properties that make it ideal for this kind of problem. This strategy can be implemented in mechanical systems when their output response is slow, and an independent joint control is necessary to eliminate or reduced the backlash effects. It is proven that a standard stabilizing control law can be implemented independently meeting some conditions and reducing the unwanted effects produced by backlash. In this section an independent joint control was implemented by a SISO controller approach. In the following sections, some multivariable control approaches are explained to deal with the control of mechanical systems with backlash.

Figure 10. Angular velocity of link 1

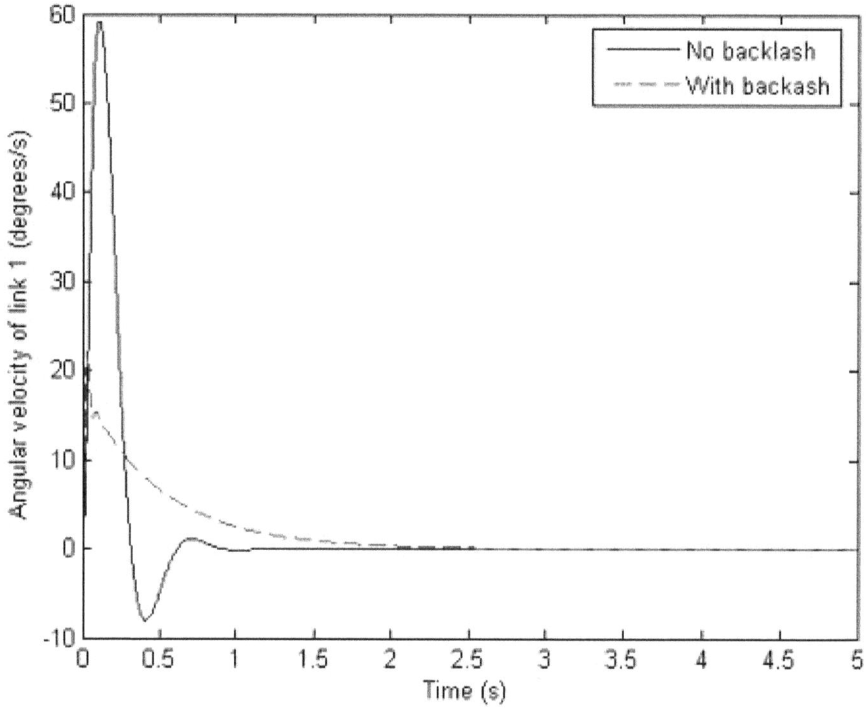

Figure 11. Angular velocity of link 2

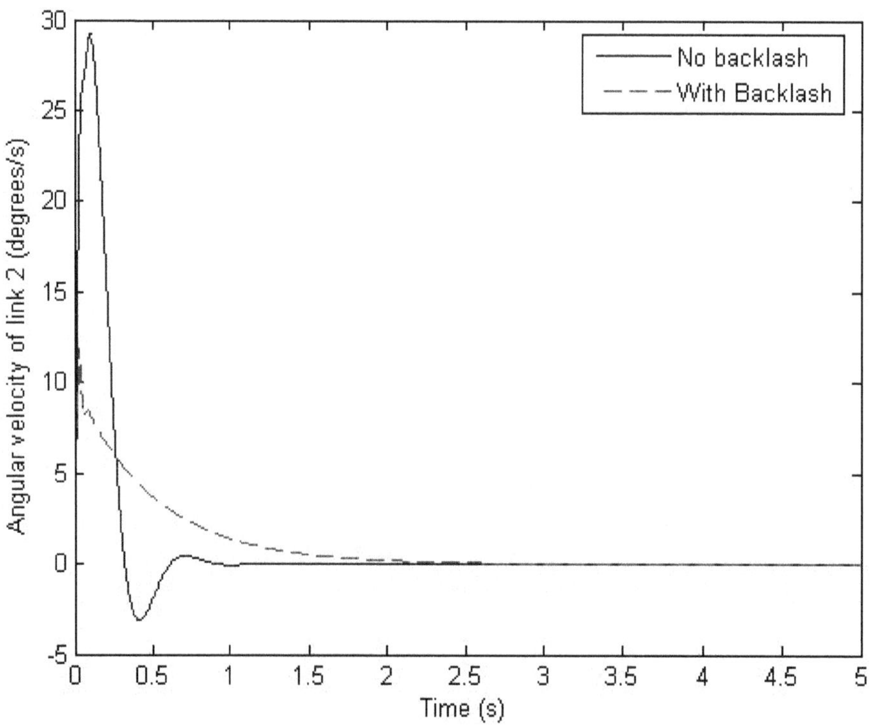

Figure 12. Input voltage of the actuator of joint 1

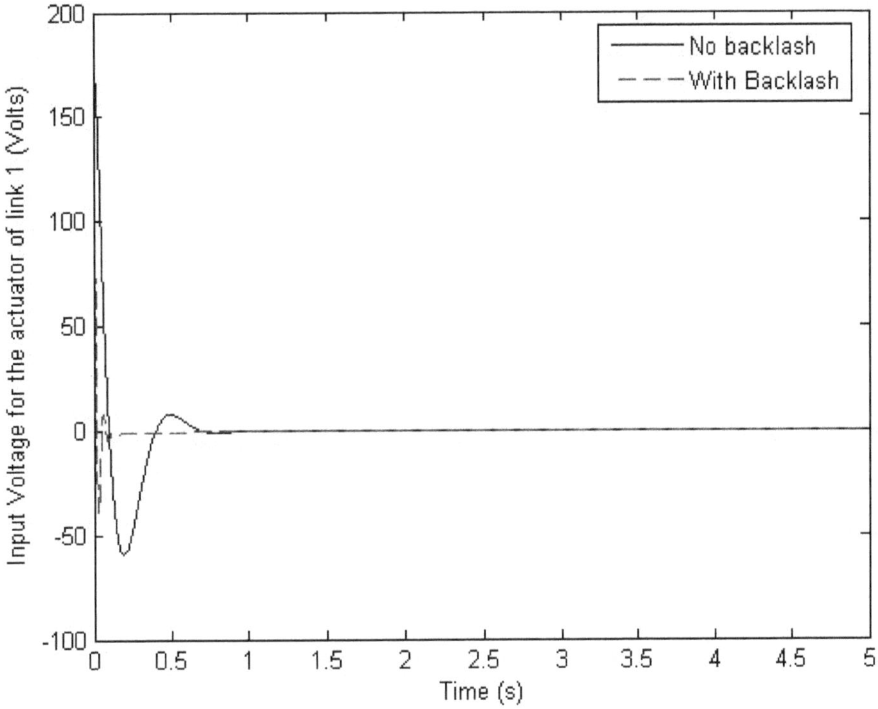

Figure 13. Input voltage of the actuator of joint 2

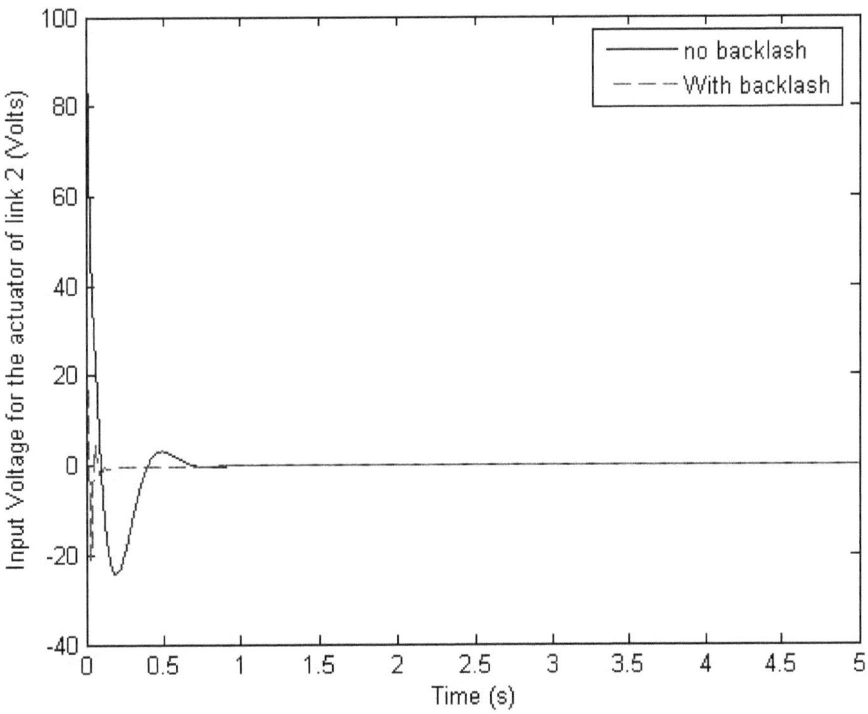

Figure 14. Input torque for joint 1

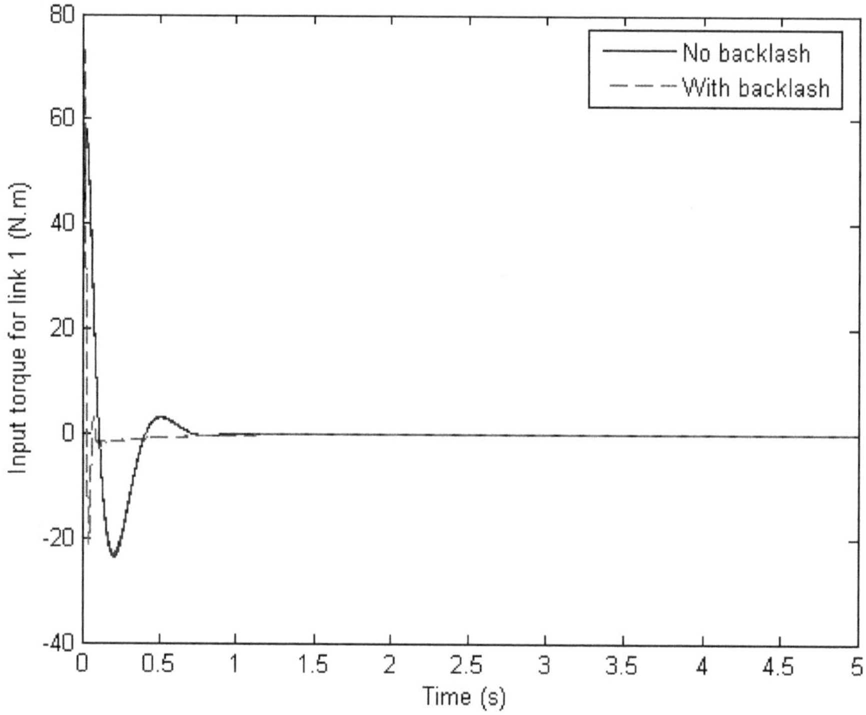

Figure 15. Input torque for joint 2

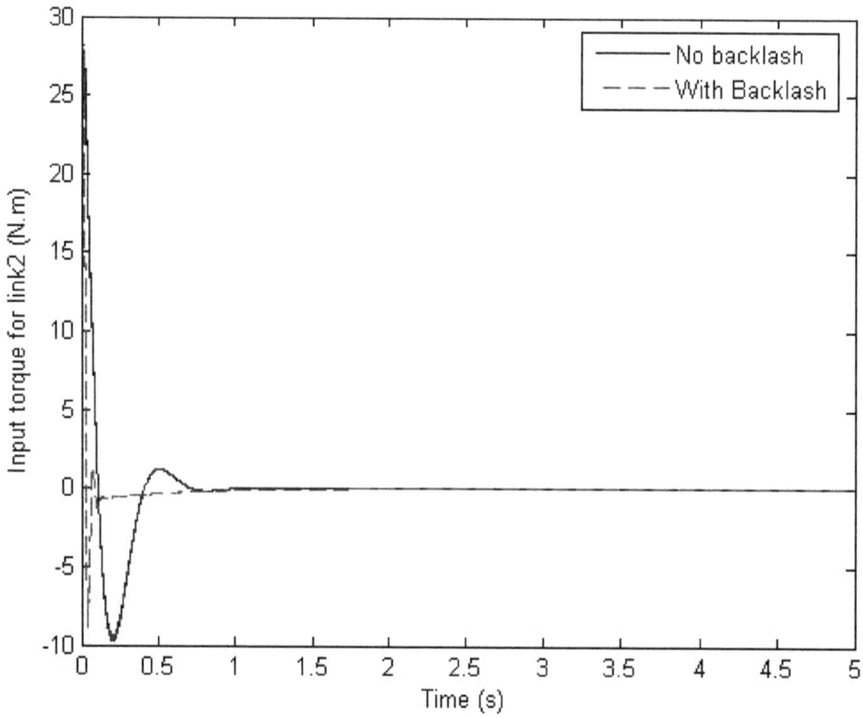

4. LOOP SHAPING CONTROL OF MECHANICAL SYSTEMS WITH BACKLASH

Loop shaping is a well known approach used in the control of multivariable systems. It consists in designing the closed loop specifications of the system in terms of the open loop characteristics (McFarlane & Glover, 1992). This is done by selecting an appropriate control gain with pre specified closed loop frequency response obtaining the desired performance. H_{∞} loop shaping control design consist on minimizing the infinity norm of the system, therefore some aspects such as sensitivity and disturbance rejection can be included in the controller design (Adams & Baron,1998).

Loop shaping consists on selecting two weight functions W_1 and W_2 that is used as a pre compensator and post compensator respectively to meet the required bandwidth characteristics. Then a K_{∞} gain is calculated based on the minimization of the infinity norm of the system. To implement this control strategy it is necessary to obtain the coprime factorization of the system (Hippe, 1994; Gao & So, 2003) for multivariable systems.

Then each factor is used in the design of the K_{∞} gain which stabilizes the system solving the specified Riccati equations (Adams & Bar-on, 1998; Boukhnifer & Ferreira, 2005). In this section a H_{∞} loop shaping control approach for mechanical systems with backlash is proposed. Considering the characteristics of mechanical systems this strategy is suitable for these models due to its frequency domain properties taking in count the backlash effects in the system response. For this purpose a describing function is implemented to model the backlash nonlinearity with a sinusoidal input, assuming that the system outputs oscillate in the presence of this discontinuity. These oscillations yield a limit cycle that is a non trivial periodic orbit when there exist a neighborhood of the periodic orbit that does not contain any other periodic solution (Haddad & Chellaboina, s.f.). This effect is analyzed in this section and then included in the loop shaping design procedure obtaining a controller which deal with backlash with minimal effects on the output of the system. In this section, a describing function model is explained first to be used in the loop shaping control design for mechanical systems as shown in Subsection 4.1. Then, the loop shaping design for an n degrees of freedom mechanical system, when backlash is present, is explained to develop an efficient controller with the desired frequency domain specifications as explained in Subsection 4.2. Later, a frequency domain analysis is done for the loop shaping controller, in which an analysis with the positivity and Nyquist theorem is done to validate the stability of the system from a frequency domain point of view, is shown in subsection 4.3.

The characteristics of the limit cycle oscillations, such as angular frequency and amplitude of the periodic orbit is found by this frequency domain analysis.

Finally, in subsection 4.4, an example is shown by the stabilization by loop shaping controller design of a cart pendulum system with backlash in its actuator/joint coupling. An analysis of the system performance is done to interpret the system output and analyze the limit cycle oscillations in the mechanical system produced by the nonlinearities of the system.

4.1 Backlash Model by Describing Function

Describing functions are a representation of different kinds of nonlinearities and discontinuities (Brogan; Tierno, Kim, Lacy, & Bernstein, 2000; Nordin & Gutman, 2002). It consists in representing a discontinuity by a harmonic decomposition using the Fourier series of this nonlinearity when the input $e(t)$ is a sinusoidal function of time.

The describing function representation consist in deriving a mathematical representation by the Fourier series coefficients of the backlash with a sinusoidal input, as seen in Figure 16, finding the times t_1, t_2, t_3 and t_4. The input signal has a magnitude of E and for this nonlinearity representation the slopes m_r and m_l are assumed to be equal $m_r = m_l$ and then the output of this discontinuity n(e) is divided according to their respective time instants.

Figure 16. Backlash representation with a sinusoidal input

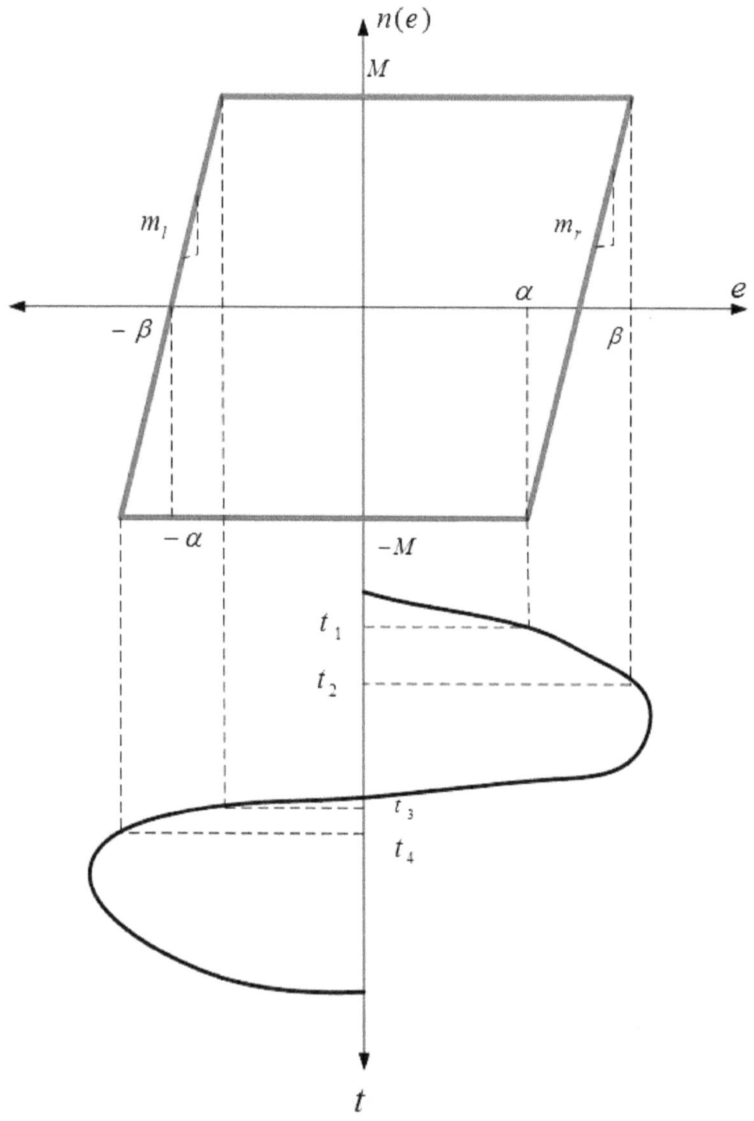

The backlash input is:

$$e(t) = E\sin(t) \tag{43}$$

and the backlash output function is defined as:

$$n(t) = \pm\beta + m(e(t)) \tag{44}$$

where the sign of β depends if it is the right or the left intersection with the e axis.

The Fourier coefficients of $n(t)$ are defined by:

$$a_k = \frac{1}{T}\int_{-T}^{T} \sin(k\omega t)\, n(t)\, dt$$

$$b_k = \frac{1}{T}\int_{-T}^{T} \cos(k\omega t)\, n(t)\, dt$$

$$b_0 = \frac{1}{2T}\int_{-T}^{T} n(t)\, dt \tag{45}$$

Usually b_0 is ignored in the describing function, and only a_1 and b_1 are considered for the describing function model. If ω is set as $\omega=1$, because the signal is 2π periodic, the output backlash function is divided according to the integration time as follows:

$$\begin{array}{ll} 0 < t < t_1 & -M \\ t_1 < t < t_2 & \beta + m\sin(t) \\ t_2 < t < t_3 & M \\ t_3 < t < t_4 & -(\beta + m\sin(t)) \\ t_4 < t < 2\pi & -M \end{array} \tag{46}$$

where $m=m_r = m_l$ with the values for t_1, t_2, t_3 and t_4

$$t_1 = sin^{-1}\left(\frac{\alpha}{E}\right)$$

$$t_2 = sin^{-1}\left(\frac{M-\beta}{mE}\right)$$

$$t_3 = t_1 + \pi$$

$$t_4 = t_2 + \pi \tag{47}$$

The coefficient a_1 is found using the subdivisions explained in (46) for integration purposes:

$$a_1 = \frac{M}{T}\left(\int_{t_2}^{t_3} \sin(t)\,dt - \int_0^{t_1} \sin(t)\,dt - \int_{t_4}^{2\pi} \sin(t)\,dt\right) \tag{48}$$

Then the coefficient a_1 is:

$$a_1 = \frac{M}{T}\left(2\cos(t_1) + 2\cos(t_2)\right) \tag{49}$$

To find the coefficient b_1 the subdivisions shown in (46) are applied for integration purposes

$$b_1 = \frac{M}{T}\left(\int_{t_2}^{t_3} \cos(t)\,dt - \int_0^{t_1} \cos(t)\,dt - \int_{t_4}^{2\pi} \cos(t)\,dt\right) \tag{50}$$

Then the following coefficient b_1 is obtained

$$b_1 = \quad \frac{M}{T}\left(\sin(t_1) + \sin(t_2)\right) + 2\beta\sin(t_2) - 2\beta\sin(t_1) + 2mcos^2(t_2) - \quad 2mcos^2(t_1) \tag{51}$$

Then the ratio of the Laplace transform of the input $e(t)$ and the output $n(t)$ is given by $N(s)$ as (Brogan; Nordin & Gutman, 2002):

$$N(s) = \frac{a_1\omega + b_1 s}{E\omega} \tag{52}$$

The function (52) is of significant importance for the design of loop shaping controller explained in the next section, it gives a frequency domain based description of the backlash nonlinearity, and therefore it is easier to design a suitable controller with a less complex representation.

4.2 Loop Shaping PI Control of Mechanical Systems with Backlash

In this subsection a loop shaping PI control for n degrees of freedom mechanical systems is developed. As explained before, loop shaping is a control technique which consists in obtaining a controller minimizing the H_∞ norm of the system (McFarlane & Glover, 1992; Zhou, Han, Wang, & Liu, 2004; Boukhnifer & Ferreira, 2005) solving a set of Riccati equations (Zhou, Han, Wang, & Liu, 2004; Bentsman,

Orlov, & Aguilar, 2013). The design of loop shaping controllers for this kind of systems consist in implementing a pre compensator and a post compensator to the plant (the mechanical system) to design a controller that meets the required frequency domain specifications. In order to design the specified controller, a linearized model of the plant (mechanical) is needed, so the first part of this subsection is devoted to obtain a linearized model by the Jacobian matrix representation of the model. The second part of this subsection consists in the design of the loop shaping PI controller when there is backlash in the system.

Consider the following n degrees of freedom mechanical system:

$$D(q)\ddot{q} + C(q,\dot{q})\dot{q} + g(q) = \tau \tag{53}$$

where $q \in \mathbb{R}^n$ is the joint angle vector, $\tau \in \mathbb{R}^n$ is the input vector torque, $D(q) \in \mathbb{R}^{n\times n}$ is the inertia matrix, $C(q,\dot{q})\dot{q} \in \mathbb{R}^n$ is the coriolis vector and $g(q) \in \mathbb{R}^n$ is the gravity vector.

Transforming (53) into state space with $x_1 = q$ and $x_2 = \dot{q}$, the following model is obtained:

$$\dot{x}_1 = x_2$$

$$\dot{x}_2 = D^{-1}(x_1)\tau - D^{-1}(x_1)C(x_1,x_2)x_2 - D^{-1}(x_1)g(x_1) \tag{54}$$

So, this model is in the following form

$$\dot{X} = f(X,\tau,t) \tag{55}$$

Linearizing this system about the equilibrium point yields (Brogan; Silva & Erraz, 2006)

$$\delta\dot{X} = A\delta X + B\delta\tau$$

$$\delta Y = C\delta X + D\delta\tau \tag{56}$$

Then the transfer function is defined as

$$A = \frac{\partial f}{\partial x}$$

$$B = \frac{\partial f}{\partial \tau}$$

$$C = I_n$$

$$D = 0 \tag{57}$$

where

$$G(s) \sim \begin{pmatrix} A\,B \\ C\,D \end{pmatrix} \tag{58}$$

The loop shaping control strategy when there is backlash consist of implementing the describing function (52) and finding a control gain K_∞ which stabilizes the system with a required closed loop performance. In Figure 17 a block diagram of the proposed control strategy is shown, where the equivalent open loop system is defined as $G_n(s)$, that is, the combination of the describing input function $N(s)$ and the mechanical system transfer matrix $G(s)$.

As explained before, the describing function of the backlash nonlinearity is very helpful for the loop shaping controller design, because, the resulting system can be modeled in the frequency domain. The performance of the system can be obtained by traditional or nonlinear frequency analysis, investigating the limit cycle produced by the backlash discontinuity effects on the model. The describing function implementation avoids the use of a backlash inverse model as explained in (Tierno, Kim, Lacy, & Bernstein, 2000) and allows a frequency domain analysis of the model.

The resulting transfer matrix G_n is:

$$G_n(s) \sim \begin{pmatrix} A & b_{1n}ACB + a_{1n}CB \\ I & b_{1n}CB \end{pmatrix} \tag{59}$$

where I is the identity matrix, $b_{1n} = \frac{b_1}{E\omega}$ and $a_{1n} = \frac{a_1}{E}$.

Figure 17. Equivalent closed loop block diagram

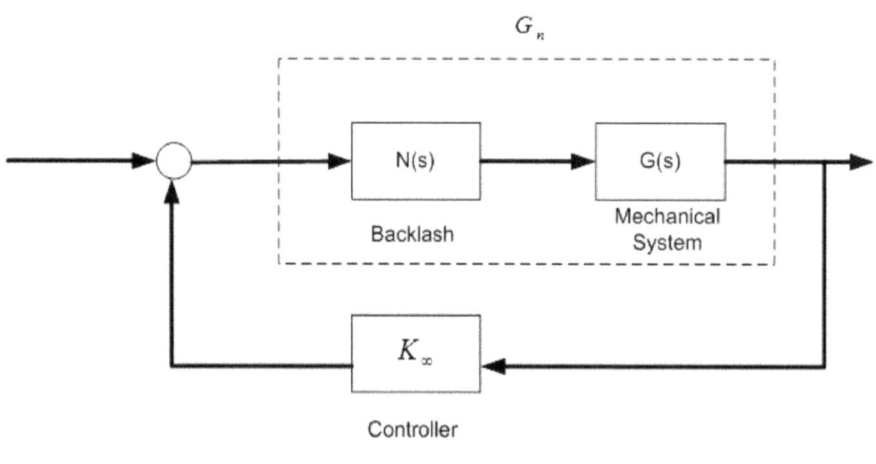

Figure 18. Perturbed model of the plant

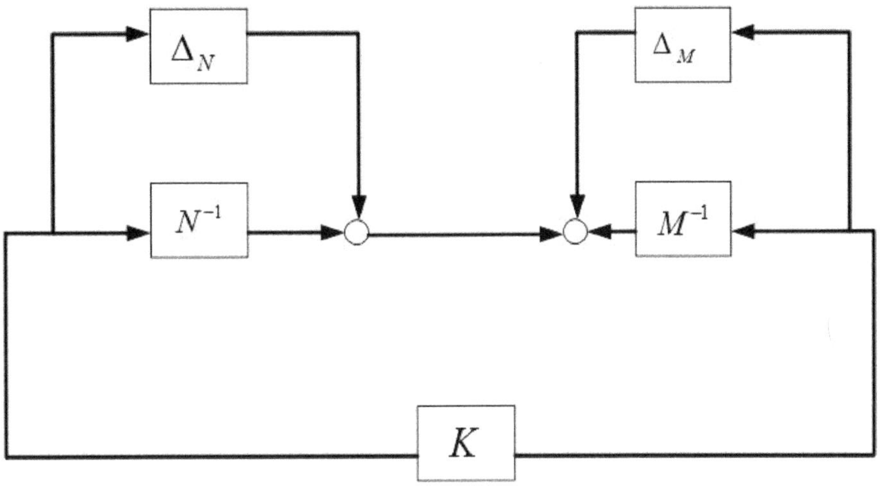

The idea behind the loop shaping control is to find a factorized model of the plant and then using a coprime factorization method therefore the resulting model $G_{n\Delta}$ is found as:

$$G_{n\Delta} = M^{-1}N \tag{60}$$

where the condition $MM' + NN' = I$ then N and M are the normalized coprime factorization of G_n. In Figure 18 a perturbed model of the plant is shown where the coprime factors are M and N (Hippe, 1994; Gao & So, 2003).

Therefore the perturbed model of the plant is given in (61) as:

$$G_{n\Delta} = \left(M + \Delta M\right)^{-1}\left(N + \Delta N\right) \tag{61}$$

The loop shaping design procedure is divided in two steps. In step 1 the loop shaping design procedure is described along with the PI controller definition. In Step 2 the robust stabilization of the model is explained with the definition of the system norm.

Step 1: Loop Shaping PI Controller Design

The first step in the loop shaping design is to establish the pre compensator and post compensator functions, to be used in the controller design. The objective is to specify the closed loop characteristics in terms of the requirements of the weighted transfer matrix. Usually the weighting matrices are described as W_1 and W_2 for the pre compensator and post compensator respectively. In Figure 19 the loop shaping design procedure is shown.

The weighted transfer matrix, denoted as $\mathrm{G_s = W_1 G_n(s) W_2}$, is used to specify the open loop performance of the system, then it is implemented later in the design of the K_∞ controller gain. The loop

Figure 19. Loop shaping design procedure: weighted transfer matrix

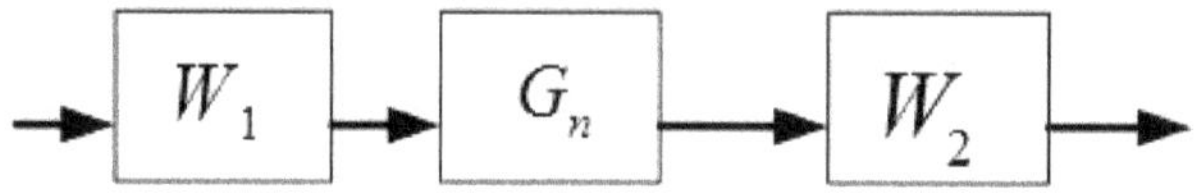

shaping procedure has the property that the design specifications are valid for all frequencies, while the open loop shaping are restricted to frequencies of low and high gain (McFarlane & Glover, 1992). The loop shaping procedure consists in finding the optimal K_∞ closed loop gain by weighting the transfer matrix $G_n(s)$ as shown in Figure 20.

After finding the optimal closed loop gain K_∞, this gain is weighted with the functions W_1 and W_2 to obtain the final controller gain K as shown in Figure 21, where $K = W_1 K_\infty W_2$ and it is the definitive controller gain.

The selected weighting functions for this loop shaping procedure are for the sake of simplicity $W_1 = I$, where I is the identity matrix, and W_2 is the PI controller given by the following state space representation

Figure 20. Loop shaping design procedure: weighted transfer matrix with K_∞ gain

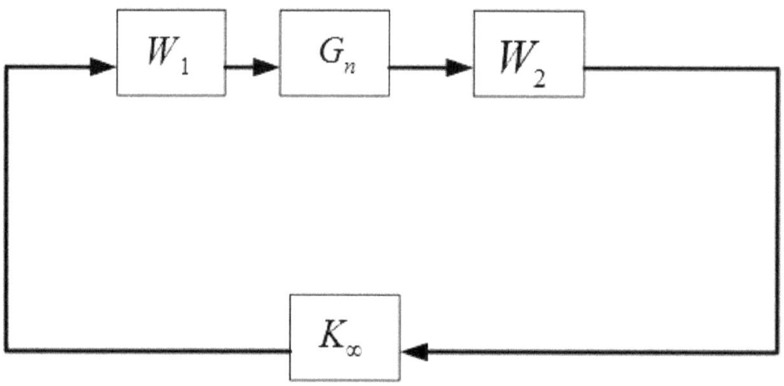

Figure 21. Loop shaping design procedure: weighted K_∞ gain

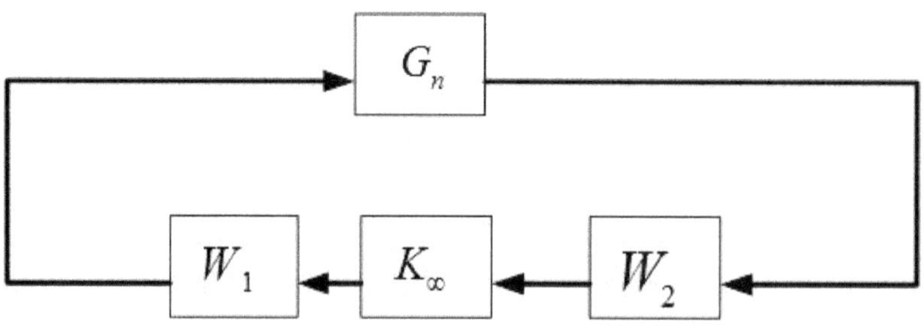

$$W_2 = \begin{cases} \dot{x}_c = K_I u \\ y_c = x_c + K_p u \end{cases} \quad (62)$$

where K_I is the integral $\mathbb{R}^{n\times n}$ gain matrix and K_p is the proportional $\mathbb{R}^{n\times n}$ gain matrix. The corresponding transfer matrix for W_2 is shown in:

$$W_2(s) = \begin{pmatrix} 0_n \, K_I \\ I_n \, K_p \end{pmatrix} \quad (63)$$

Step 2: Robust Stabilization

With the definition of the specified weighting functions, W_1 and W_2, the robust stabilization of the system is done by a H_∞ loop shaping design. Then the robust problem of stability is to find the greatest value of $\varepsilon = \varepsilon_{\max}$, where ε is defined later, such as the systems model is stabilized by a gain K (McFarlane & Glover, 1992; Zhou, Han, Wang, & Liu, 2004; Boukhnifer & Ferreira, 2005). The robust stabilization problem is described in the following procedure, the first step of this procedure is to calculate $\varepsilon_{\max}$ by (64)

$$\varepsilon_{\max}^{-1} \triangleq \underset{Stabilizing\ K}{Inf} \left\| \begin{pmatrix} I \\ K \end{pmatrix} \left(I - G_s K\right)^{-1} \right\|_\infty \quad (64)$$

where

$$\gamma_{\min} = \varepsilon_{\max}^{-1} = \sqrt{1 - \left\|N - M\right\|_H^2} < 1 \quad (65)$$

or (64) must meet the condition (66)

$$\left\| \begin{pmatrix} I \\ K_\infty \end{pmatrix} \left(I - G_{n\Delta} K_\infty\right)^{-1} M^{-1} \right\|_\infty \leq \varepsilon^{-1} \quad (66)$$

If $\varepsilon_{\max} \ll 1$, then go back (64) and adjust W_2. $\gamma_{\min}$ and $\varepsilon_{\max}$ can be found by:

$$\gamma_{\min} = \varepsilon_{\max}^{-1} = \sqrt{1 - \lambda_{\max}\left(XZ\right)} \quad (67)$$

where $\left\|\left[\Delta_N \Delta_M\right]\right\|_\infty < \gamma^{-1}$, X and Z are the solutions of the following Riccati equations (Zhou, Han, Wang, & Liu, 2004; Boukhnifer & Ferreira, 2005; Bentsman, Orlov, & Aguilar, 2013)

$$\left(A_n - B_n S^{-1} D_n^T C_n\right)^T X + X\left(A_n - B_n S^{-1} D_n^T C_n\right) - X B_n S^{-1} B_n{}^T X + C_n{}^T R^{-1} C_n = 0$$

$$\left(A_n - B_n S^{-1} D_n^T C_n\right) Z + Z\left(A_n - B_n S^{-1} D_n^T C_n\right)^T - Z C_n{}^T R^{-1} C_n Z + B_n S^{-1} B_n{}^T = 0 \tag{68}$$

where $R = I + DD^T$, $S = I + D^T D$, $A_n = A$, $B_n = b_{1n} ACB + a_{1n} CB$, $C_n = I$ and $D_n = b_{1n} CB$ from the transfer matrix realization of $G_n(s) = C_n\left(SI - A_n\right)^{-1} B_n + D_n$ Then after selecting the appropriate values of γ, ε, Z and X; a robust control gain K_∞ can be found for the stabilization of the mechanical system as shown in (69) (Zhou, Han, Wang, & Liu, 2004; Boukhnifer & Ferreira, 2005)

$$K_\infty = \left(\begin{array}{c|c} A_n + B_n F + \gamma^2\left(L^T\right)^{-1} Z C_n\left(C_n + D_n F\right) & \gamma^2\left(L^T\right)^{-1} Z C_n \\ \hline B_n^T X & -D_n \end{array}\right) \tag{69}$$

where $F = -S^{-1}\left(D_n^T C_n + B_n^T X\right)$ and $L = \left(1 - \gamma^2\right) I + XZ$.

The robust control gain K_∞ shown in (69) can be expressed in transfer matrix form as shown in (70)

$$K_\infty = \left(\begin{array}{c|c} A_k & B_k \\ \hline C_k & D_k \end{array}\right) \tag{70}$$

Then with the gain K_∞, the controller gain K is obtained as shown in Figure 21 with $K = W_1 K_\infty W_2$.

After obtaining the appropriate gain by the loop shaping procedure of an n degrees of freedom mechanical system, it is necessary to attain a frequency domain analysis to study the limit cycles produced by the backlash nonlinearity modeled by a describing function.

In the next subsection, this analysis is done by the implementation of the positivity theorem and the Nyquist theorem, finding the limit cycle frequency for a mechanical implementing a describing function to model the backlash discontinuity with defined oscillation amplitude.

4.3 Frequency Analysis of Mechanical Systems with Backlash

In this subsection the analysis of an n degrees of freedom mechanical system with backlash is done. There are several approaches to analyze the limit cycles produced by the backlash nonlinearities, such as the Nyquist theorem, Popov theorem and positivity theorem.

The idea behind this analysis is to find analytically the oscillatory behavior of the systems. This objective is reached implementing two important theorems to analyze nonlinear systems in the frequency domain. These approaches used to analyze the absolute stability of nonlinear systems are the positivity theorem and the Nyquist theorem.

To analyze the absolute stability of the system under control and the limit cycle analysis is necessary to use the positivity theorem along with the Nyquist theorem. These theorems are derived assuming the following statements. Taking into account the following state matrix representation of the mechanical system (58):

$$G(s) = \left(\begin{array}{c|c} A & B \\ \hline C & D \end{array}\right) \textit{is minimum} \tag{71}$$

where $I + N(s)G(s)$ is strictly positive real. $N(s)$ (52) is the backlash describing function which could be considered as gain for the mechanical system transfer matrix representation $G(s)$. Then the closed loop system $I+N(s)K(s)G(s)$ is strictly positive real and the following theorem applies (Haddad & Chellaboina).

Theorem 2: Consider the mechanical system with the following state space realization:

$$\dot{x}(t) = Ax(t) + B\tau(t)$$

$$y(t) = Cx(t) + D\tau(t) \tag{72}$$

and

$$G(s) = \left(\begin{array}{c|c} A & B \\ \hline C & D \end{array}\right) \textit{is minimum} \tag{73}$$

If $I + N(s)K(s)G(s)$ is strictly positive real. Then the zero solution $x(t)=0$ of the negative feedback interconnection, as shown in Figure 21, is asymptotically stable.

Then by the Nyquist and positivity theorems the limit cycle oscillation frequency can be found analyzing the critical point:

$$-\frac{1}{N(s)K(s)} \tag{74}$$

Then the number of the closed loop unstable poles Z_r is

$$Z_r = P_r - N_c \tag{75}$$

where P_r is the number of RHP open loop poles and N_c is the number of counterclockwise rounding around the critical point $-\frac{1}{N(s)K(s)}$ as established by the Nyquist and positivity theorem. Then the frequency of the limit cycle oscillations yielded by the backlash nonlinearity in the input of an n degrees of freedom mechanical system is obtained by solving the following equation as established by the Nyquist and positivity theorem to obtain a suitable controller which assures the asymptotical stability of the system.

$$\left| I + N(s)K(s)G(s) \right| = 0 \tag{76}$$

4.4 Example 3: Stabilization of a Cart Pendulum System with Backlash

In this subsection the design of controller by the loop shaping procedure for a cart pendulum (Iqbal, Bhatti, Akhtar, & Ullah, 2007; Glück, Eder, & Kugi, 2013) system with input backlash is shown. The simulations and calculations were done using MATLAB ®, and SimMechanics ® obtaining the following results. The dynamic equations of the cart pendulum system are:

$$(M + m)\ddot{x} + m\ell\left(\ddot{\theta} + \dot{\theta}^2 \sin(\theta)\right) = u + d\cos(\theta)$$

$$m\left(\ddot{x}\cos(\theta) + \ell\ddot{\theta} - g\sin(\theta)\right) = d \tag{77}$$

The parameters of the mechanical system and the respective values used in this example as shown in Figure 22 is described in Table 5.

Figure 22. Cart-pendulum system

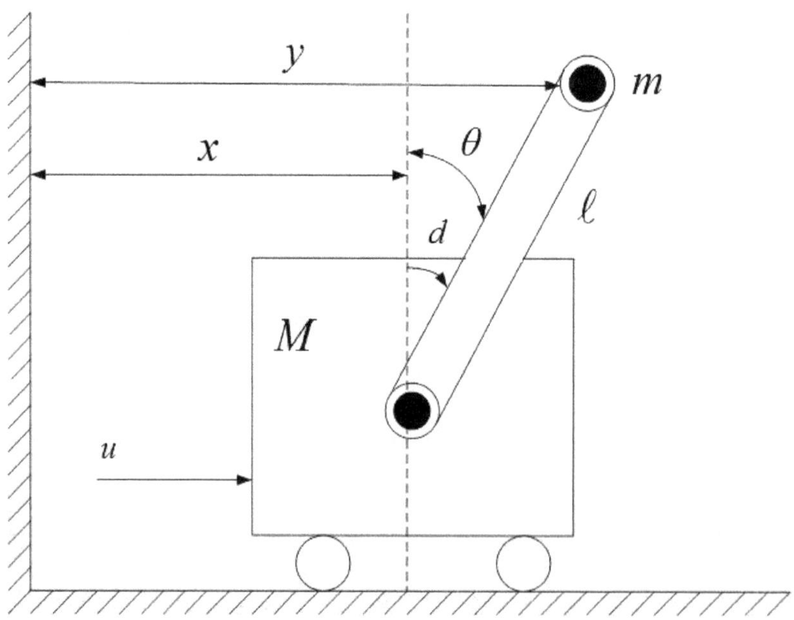

Table 5. Parameters of the cart pendulum system

Parameter	Description	Value
M	Mass of the cart	1 Kg
m	Mass of the Pendulum	1 Kg
ℓ	Length of the pendulum	0.85 m
θ	Pendulum rotating angle	degrees
x	Cart displacement	(m)
u	Cart force	N
d	Pendulum torque	N.m
g	Gravity constant	9.81 m/s^2

The cart is moved by a rotating actuator so the measurement is converted from translational force (*N*) to rotational torque (*N.m*).

The linearized model about the equilibrium point $(0,0)$ on (x,θ) is shown in (78):

$$(M+m)\ddot{x}+m\ell\ddot{\theta}=u+d$$

$$\ddot{x}+\ell\ddot{\theta}-g\theta=\frac{1}{m}d \tag{78}$$

Then using the following phase variables, the system is converted to state space form:

$$\begin{aligned} q_1 &= x \\ q_2 &= \dot{x} \\ q_3 &= \theta \\ q_4 &= \dot{\theta} \end{aligned} \tag{79}$$

Then the following state space matrix is obtained:

$$\begin{pmatrix} 0 & (M+m) & 0 & m\ell \\ 1 & 0 & 0 & 0 \\ 0 & 1 & 0 & \ell \\ 0 & 0 & 1 & 0 \end{pmatrix}\begin{pmatrix} \dot{q}_1 \\ \dot{q}_2 \\ \dot{q}_3 \\ \dot{q}_4 \end{pmatrix}=\begin{pmatrix} 0 & 0 & 0 & 0 \\ 0 & 1 & 0 & 0 \\ 0 & 0 & g & 0 \\ 0 & 0 & 0 & 1 \end{pmatrix}\begin{pmatrix} q_1 \\ q_2 \\ q_3 \\ q_4 \end{pmatrix}+\begin{pmatrix} 1 & 1 \\ 0 & 0 \\ 0 & 1/m \\ 0 & 0 \end{pmatrix}\begin{pmatrix} u \\ d \end{pmatrix} \tag{80}$$

Substituting the parameters values, the state space matrices, as shown in (58), are:

$$A = \begin{pmatrix} 0 & 0 & 0 & 0 \\ 0 & 0 & -2.45 & 0 \\ 0 & 0 & 0 & 1 \\ 0 & 0 & 14.43 & 0 \end{pmatrix}$$

$$B = \begin{pmatrix} 0 & 0 \\ 0.25 & 0 \\ 0 & 1/m \\ -0.29 & 1.17 \end{pmatrix}$$

$$C = \begin{pmatrix} 1 & 0 & 0 & 0 \\ 0 & 1 & 0 & 0 \\ 0 & 0 & 1 & 0 \\ 0 & 0 & 0 & 1 \end{pmatrix}$$

$$D = 0 \tag{81}$$

Then the equivalent system with input backlash $G_n(s)$ is given by (59). Therefore, applying (76) and solving for ω and E the following oscillation limit cycle frequency is $\omega = 1$, $E = 1$; and the time values for the description function as shown in (47) are $t_1 = 0.0945$, $t_2 = 0.3869$, $t_3 = 3.2361$ and $t_4 = 3.5285$. Then the constant values are $a_1 = 7.3400$ and $b_1 = -0.0358$ as shown in (49) and (51) and applied in the model (59).

The PI controller function W_2 has the following gains:

$$K_I = \begin{pmatrix} 20 & 0 \\ 0 & 30 \end{pmatrix}$$

$$K_p = \begin{pmatrix} 1000 & 0 \\ 0 & 1000 \end{pmatrix} \tag{82}$$

Then the matrices for the K_∞ gain function shown in (70) are:

$$A_k = \begin{pmatrix} -2.0440 & 0.6075 & 0.0118 & -0.3488 \\ -0.3925 & -21.2334 & -4.3115 & -19.0145 \\ 0.0118 & -1.8590 & -3.6897 & -0.8334 \\ -0.3488 & -19.0145 & 12.5930 & -17.9578 \end{pmatrix}$$

$$B_k = \begin{pmatrix} -2.0440 & -0.3925 & 0.0118 & -0.3488 \\ -0.3925 & -21.2334 & -1.8590 & -19.0145 \\ 0.0118 & -1.8590 & -3.6897 & -1.8334 \\ -0.3488 & -19.0145 & -1.8334 & -17.9578 \end{pmatrix}$$

$$C_k = \begin{pmatrix} 0.9911 & 2.8877 & -7.9548 & -1.6685 \\ -0.1334 & -0.5685 & 23.5952 & 6.1909 \end{pmatrix}$$

$$D_k = 0 \tag{83}$$

The respective matrices A_k, B_k, C_k and D_k were obtained by solving the Riccati equations shown in (68), obtaining the expected results for X and Z.

The value of $\gamma_{\min}$ is $\gamma_{\min} = 14.67$, then the selected value of γ for the design of the loop shaping controller is $\gamma = 19.7$, therefore the selection of the PI controller W_2 meets the requirements and it is not necessary a redesign of K_∞ because the obtained γ and ϵ are greater than the expected values.

In Figure 23 and Figure 24 the respective linear displacement and linear velocity of the cart are shown. As it is expected, the backlash effect yields limit cycle oscillations that are approximated equal to the oscillation frequency ω obtained analytically and the amplitude E by solving (76).

Figure 23. Linear displacement of the cart

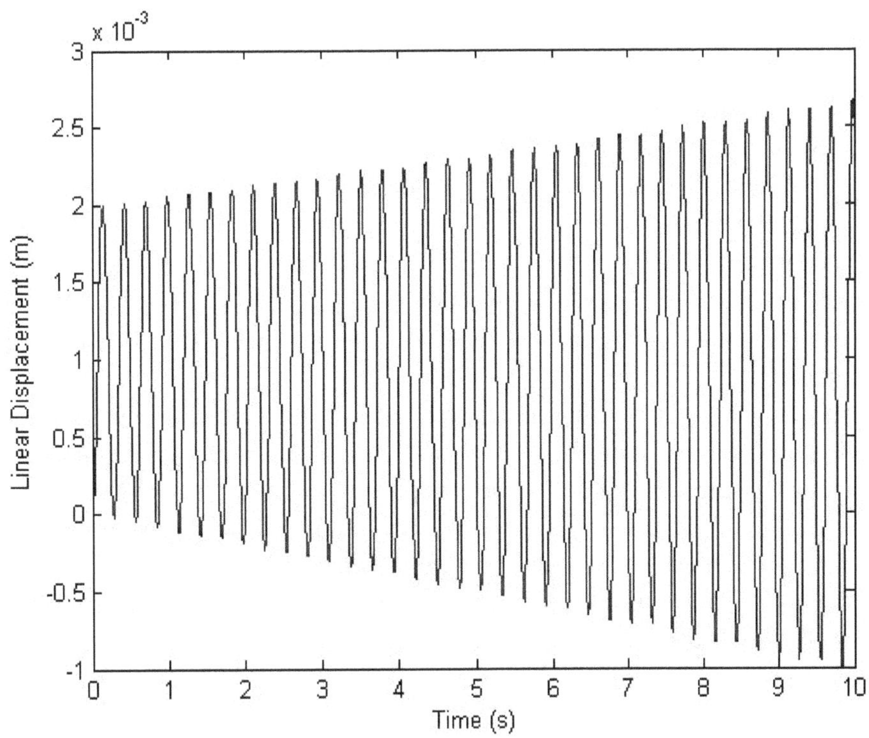

Figure 24. Cart linear velocity

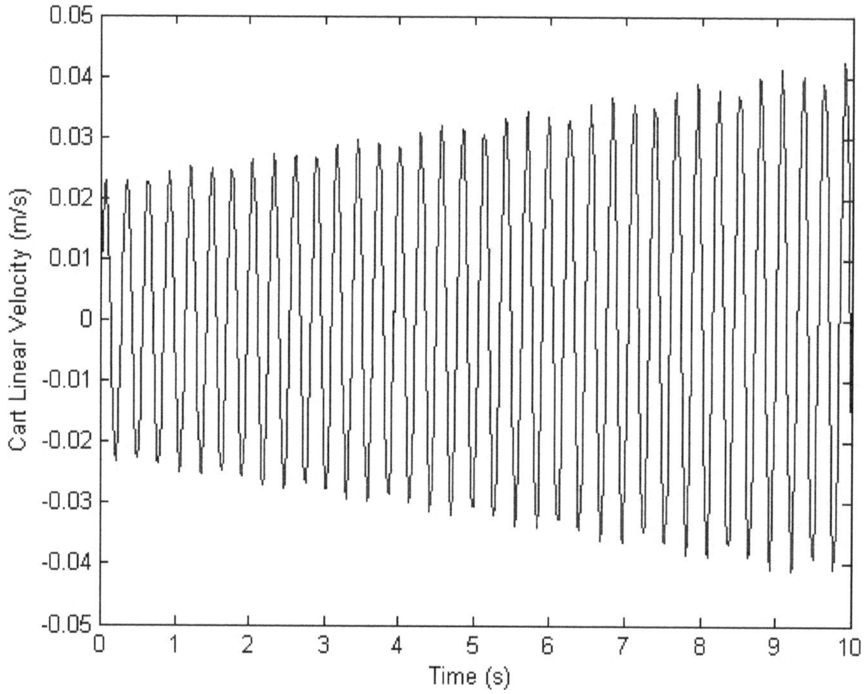

The respective rotation angle and angular velocity of the pendulum are shown in Figure 25 and Figure 26, and similar to the linear displacement and linear velocity, this variable presents an oscillatory behavior generating a limit cycle and, as it is expected, the amplitude of the oscillation are small.

As it is shown in Figure 27 and Figure 28 the actuator torque for the cart and the pendulum is oscillatory due to the effects of backlash in the input of the system. As can be noticed, there is a small control effort generated for the two actuators, so the actuator is not saturated with high torque values. As it is explained later, the backlash effects yield an oscillatory result on the system, for this reason, the control torque behaves in the same manner, but as it is explained later, these torque actuators values keep the system variables asymptotically stable.

In Figure 29 and Figure 30 the phase portraits for (q_1, q_2), for the linear variables and (q_3, q_4) for the angular variables are shown respectively. It can be noticed in both figures, that the limit cycles are asymptotically stable about the equilibrium points (0,0), which means that the system variables works on the operating point.

In this case, the loop shaping PI control strategy avoids that the system presents a chaotic behavior, then the describing function modeling allow the control designer to understand and analyze a possible chaotic behavior on the system. In Figure 31, the cart pendulum system is depicted when the system is working during simulation. The cart is moved along the X - axis and then the pendulum rotates about the Y - axis, mounted on the center of mass of the cart. During simulation the pendulum angle is kept on the desired rotating range, and maintained near to the equilibrium point (zero) the resulting Nyquist functions are shown in Figure 32, where asymptotic stability is assured by the loop shaping PI control strategy. In Figure 16 is proved that the mechanical system of this example is stable and meet condition (16) as specified by the positivity and nyquist theorems.

Figure 25. Pendulum rotation angle

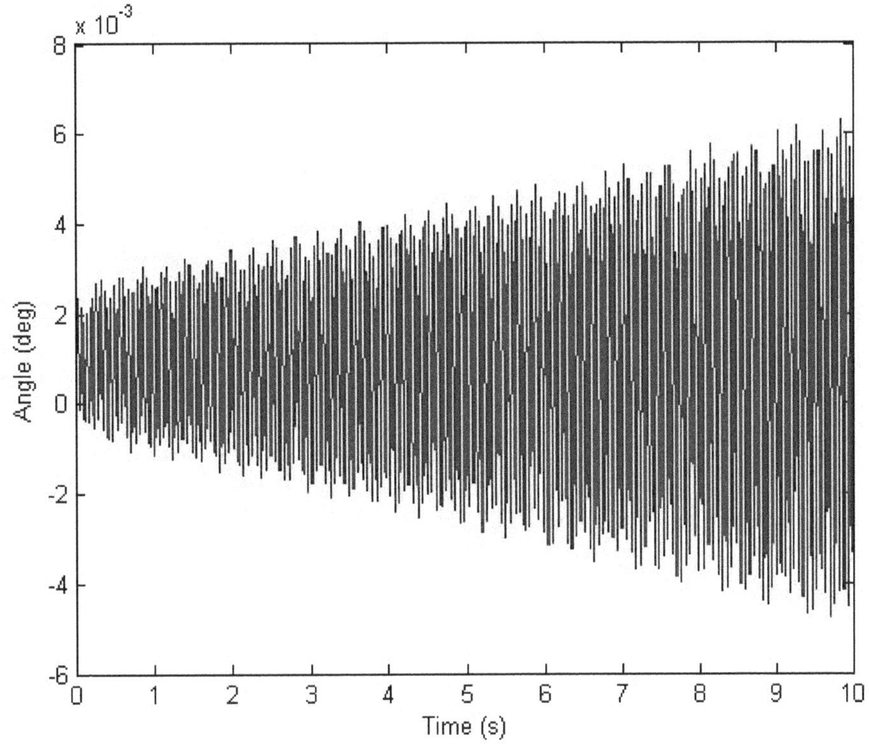

Figure 26. Pendulum angular velocity

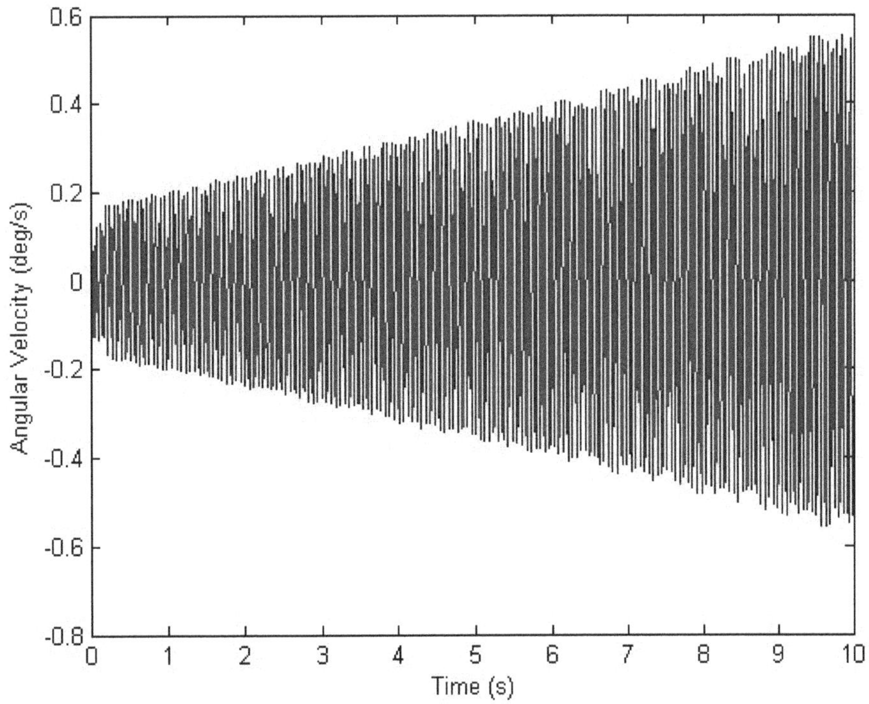

Figure 27. Cart actuator torque

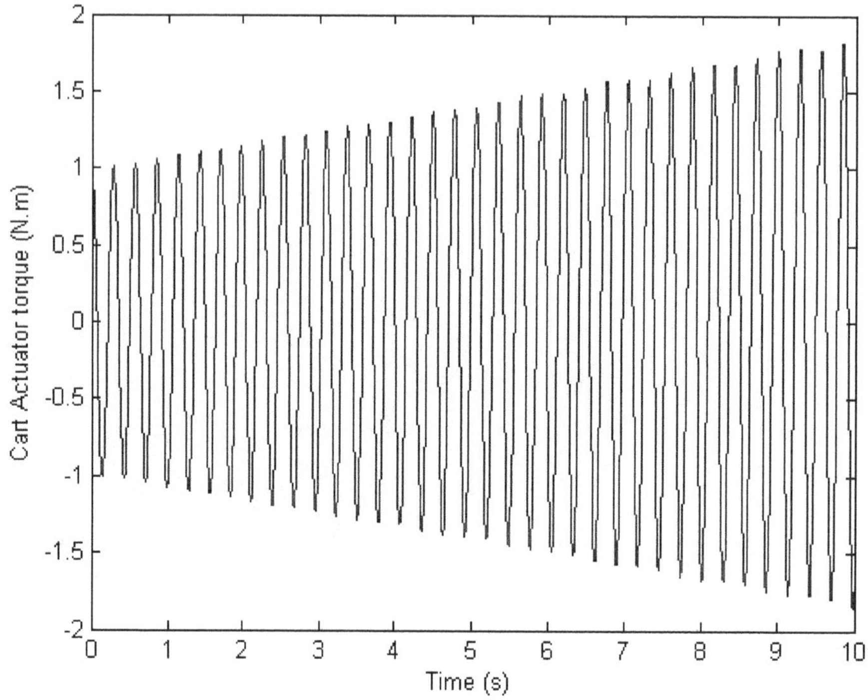

Figure 28. Pendulum actuator torque

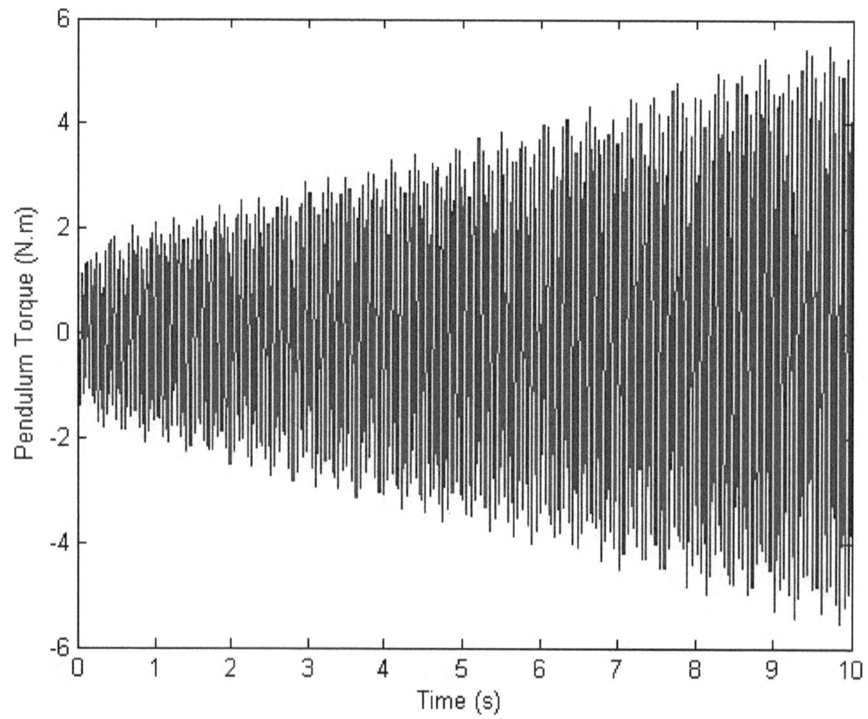

Figure 29. Phase portrait of x

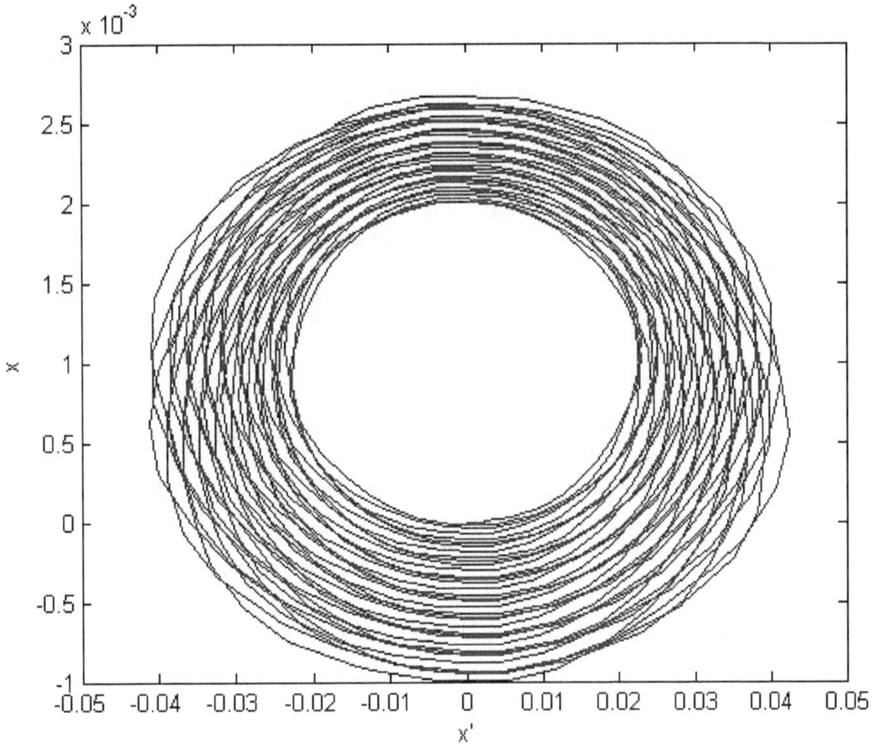

Figure 30. Phase portrait of θ

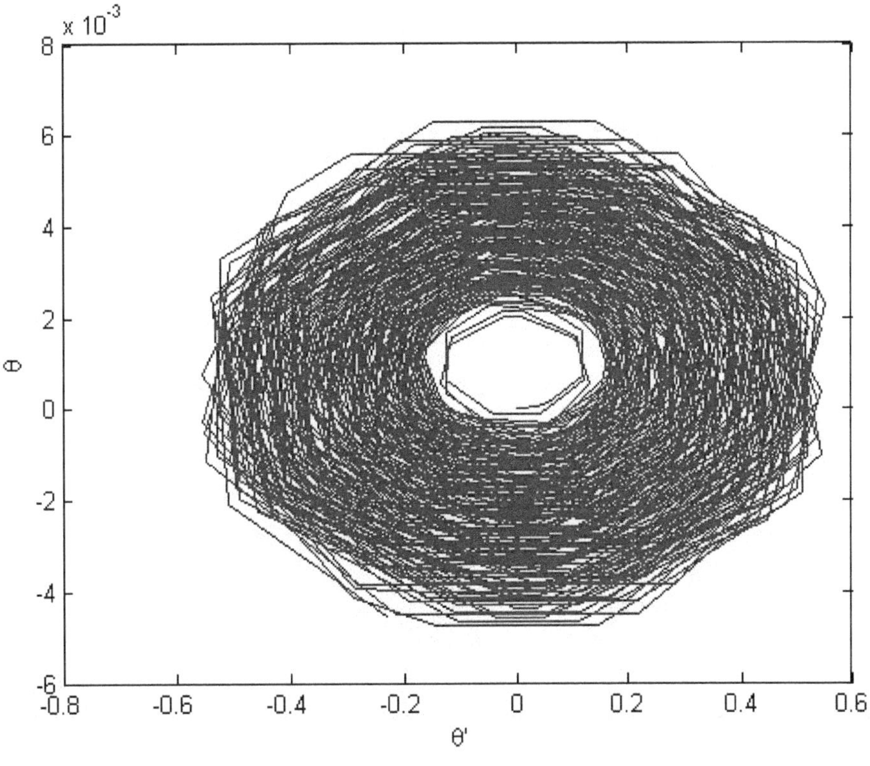

Figure 31. Cart pendulum system visualization

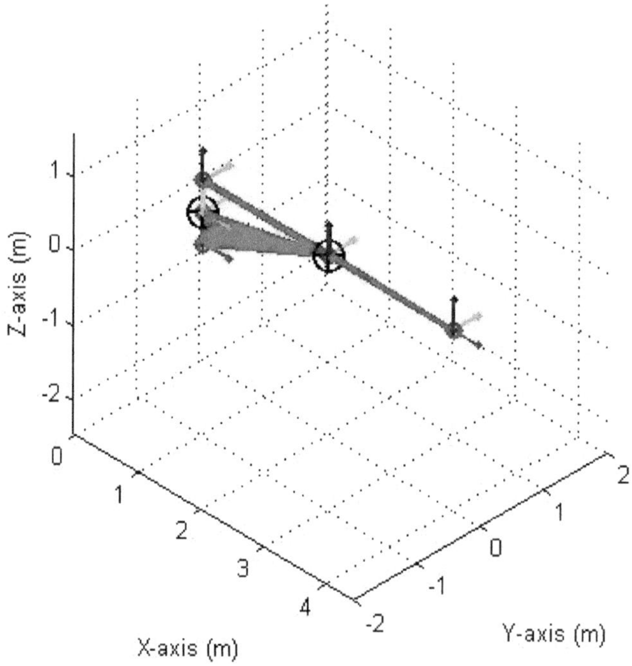

Figure 32. Nyquist Diagram of the cart pendulum controlled system

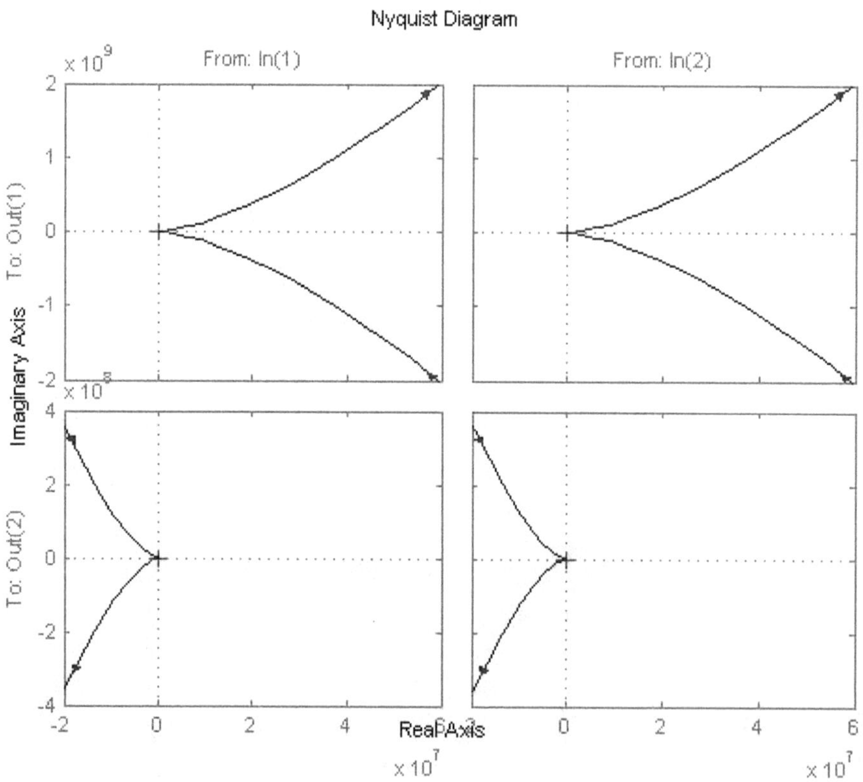

As a conclusion of this section, a loop shaping PI controller is proposed as an efficient technique for the control of mechanical systems with backlash. It is shown that a describing function model of the backlash nonlinearity is very useful for the design and analysis of different kinds of control approaches, allowing the study of limit cycles produced by this discontinuity in mechanical systems. The describing function analysis allows the designer to choose an appropriate control strategy to make the system asymptotically stable about the equilibrium point. An important result obtained by this control strategy, is that the oscillation frequency can be found analytically as long as by simulation, before deriving the K_∞ matrix using the Riccati equations that is implemented in the path of the closed loop system. The advantage of loop shaping control design in this kind of problem is that is a control technique which improves the open loop characteristics of the system in the frequency domain, so the analysis of the control system design with a describing function model is very straightforward.

The frequency domain analysis of mechanical system with input backlash is done by the positivity and Nyquist theorem, which are; appropriate for this kind of linearized system with a backlash discontinuity that adds nonlinearities to the system and producing unwanted effects. The analysis with the positivity and Nyquist theorem, are an appropriate option to analyze the absolute stability of this kind of model, obtaining the frequency of the oscillations analytically or graphically.

5. OPTIMAL CONTROL OF MECHANICAL SYSTEMS WITH BACKLASH

In this section an optimal control approach (Lin & Brandt, 1998; Riedinger & Iung, 1999) is implemented on systems with input backlash. Optimal control consists in solving the linear quadratic regulator problem. This is a multivariable approach that consists in finding an optimal control law which minimizes a performance index based on the input and states of the system.

Optimal control approach has been extensively used in diverse kinds of problems due to their efficient control properties providing a minimum control effort strategy. In this section this approach is implemented in the control of mechanical systems with backlash. This technique has been demonstrated to be effective in the presence of different kinds of nonlinearities and discontinuities. In (Goebel, 2004) an optimal control approach for systems with saturation is proposed, where the stabilization of linear systems with saturated actuator is shown, in which the optimal control law is solved avoiding the effects of saturation on the inputs. In this approach, the saturation is modeled by a mathematical representation in which an approximated model of saturation is obtained containing information about the model.

Another interesting approach of optimal control with an input nonlinearity is shown in (Riedinger & Iung, 1999), where the optimization of a system with hysteresis is proposed where the Pontryagin maximum principle is implemented to solve this complex problem in which a discontinuity is found in its inputs. Some authors propose a H_∞ with quadratic constraints to solve an optimal control problem (Petersen, 2008; Pannocchia, Rawlings, Mayne, & Marquardt, 2010). The integral quadratics constraints are used in the optimization problem to obtain an optimal control law by a H_∞ approach.

The optimal control law is found by solving the Ricatti equation and then the controller gain is obtained with the required stability and optimality characteristics (Cimen, 2010; Pannocchia, Rawlings, Mayne, & Marquardt, 2010) There are a vast amount of methods to solve this equation, it could be done numerically or analytically as found in literature.

In this section, the development of an optimal control for mechanical systems with input backlash is shown. As found in different articles such as (Tao, Ma, & Ling, 2001; Tao & Ma, 1999), this method can be applied in systems where this discontinuity is found. The proposed approach differs from other optimal control techniques with discontinuities in which a describing function, as applied in Sect. 4, is implemented in the derivation of an optimal control law when backlash is found in the system inputs.

The describing function is implemented to model this system's discontinuity and as it is shown in Sect. 4, with this model a limit cycle analysis is done to study the system stability and output behavior of mechanical systems with input backlash.

A frequency domain analysis is performed to test the absolute stability and find the limit cycle characteristics, appropriate for this kind of model when there is a discontinuity in its input.

In this section the definition of an optimal control strategy and their frequency domain analysis is done, with an example of the optimal control of a mechanical system is explained to elucidate the performance of this approach in a real mechanical system example.

5.1 Optimal Control Design

The optimal control of mechanical systems with backlash is explained in this subsection. Before deriving the formulae for the optimal control of this kind of systems, the model of an n degrees of freedom must be considered as shown in (53), with a state space model shown in (54). Then a linearized model is needed for the limit cycle analysis where the Laplace transform of the describing function $N(s)$ is shown in (52). In Figure 33, the optimal control system is depicted where $N(s)$ is the describing function of the backlash nonlinearity, $G(s)$ is the linearized mechanical system, ϕ is the optimal control law and $G_n(s)$ is the combined transfer matrix of the backlash describing function and the linearized mechanical system model.

Figure 33. Block diagram of the optimal controller

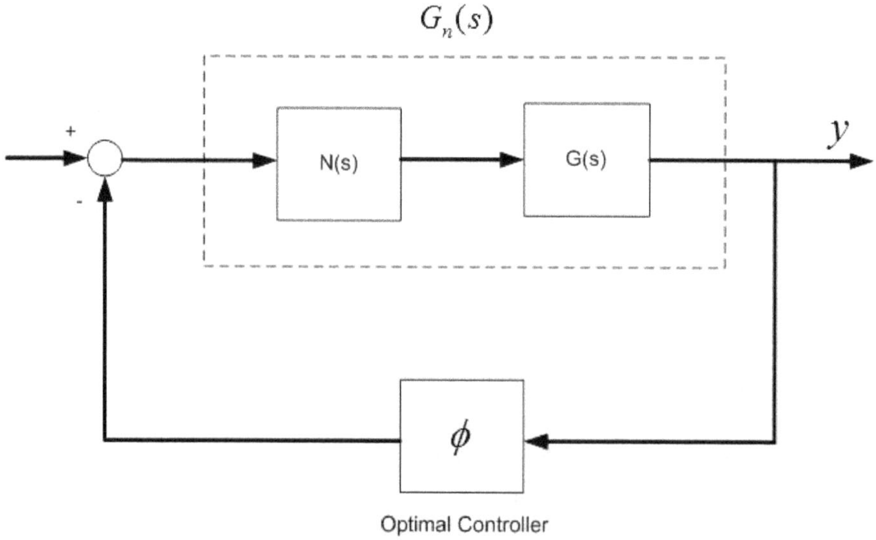

The optimal control approach shown in this section, is based on the study of the limit cycles oscillation to find an optimal controller which minimizes a performance index. This approach is similar as the analysis shown in Sect. 4, so some assumptions are similar to the expressed ones shown in the prior section.

The equivalent transfer matrix of the mechanical systems with input backlash is:

$$G_n(s) \sim \begin{pmatrix} A'B' \\ C'D' \end{pmatrix} \tag{84}$$

where $A' = A$, $B' = b_{1n}ACB + a_{1n}CB$, $C' = I$, $D' = b_{1n}CB$, $b_{1n} = \frac{b_1}{E\omega}$ and $a_{1n} = \frac{a_1}{E}$.

To find the optimal control law for this problem, it is necessary to define a Lyapunov function V

$$V(x) = x^T Px \tag{85}$$

where $P>0$ is a positive definite matrix.

Then a performance index J is denoted as:

$$J(x_0, u(\cdot)) = \int_0^\infty \left[x^T R_1 x + u^T R_2 u\right] dt \tag{86}$$

where $R_1, R_2>0$ are positive definite matrices selected for the design of the optimal control law. The respective Hamiltonian function is given by (87) using (85):

$$H = x^T R_1 x + u^T R_2 u + x^T PA'x + x^T PB'u + x^T A'P^T x + PxB'u \tag{87}$$

Then the optimal control law is obtained by solving $\frac{\partial H}{\partial u} = 0$. The resulting optimal control law is:

$$u^* = \phi = -(R_2^{-1}) * B'^T * P * x \tag{88}$$

which is the optimal control law that minimizes (86), and then the Ricatti equations are found using the Hamilton - Jacobi - Bellman equations as shown in:

$$0 = \frac{\partial V(x(t),t)}{\partial t} + \min_{u(t)\in u} H\left(x, u, \frac{\partial V(x(t),t)}{\partial x}, t\right) \tag{89}$$

Usually, $\frac{\partial \mathrm{V}(\mathrm{x}(\mathrm{t}),\mathrm{t})}{\partial \mathrm{t}} = 0$ then the Riccati equation is:

$$0 = R_1 + PA' + A'P^T - PB'R_2^{-1}PB' \tag{90}$$

This equation could be solved numerically by any of the methods shown in (Cimen, 2010; Pannocchia, Rawlings, Mayne, & Marquardt, 2010)

5.2 Frequency Domain Analysis

The frequency domain analysis and the frequency parameters of mechanical systems with backlash is similar to the frequency domain analysis shown in Sect. 4 for the loop shaping controller design. The frequency domain analysis to study the absolute stability of the system can be done by the positivity and Nyquist theorem as explained in the previous section. To perform this analysis it is necessary to find the transfer matrix which relates the system states and the control law. This transfer matrix is given by:

$$\frac{X(s)}{u(s)} = (SI - A')^{-1} B' \tag{91}$$

Then the transfer matrix in terms of the control law input is:

$$X(s) = (SI - A')^{-1} B'u(s) \tag{92}$$

and assuming a sinusoidal input for the combined system, the input $u(s)$ of the system with a frequency ω is given by:

$$u(s) = \frac{\omega}{s^2 + \omega^2} \tag{93}$$

Then the control law $\phi(\mathrm{s})$ is given by:

$$\phi(s) = -(R_2^{-1}) * B'^T * P * x(s) \tag{94}$$

The following condition is required to make the solution periodic:

$$|I + \phi(s)G_n(s)| = 0 \tag{95}$$

where $s = j\omega$ Therefore the limit cycle oscillating parameters can be found by solving (95) and the absolute stability can be analyzed by analyzing the system about the critical point $\frac{-1}{\phi(s)}$

5.3 Example 4: Optimal Control of a Four Bar Linkage with Input Backlash

In this subsection the optimal control of a four bar linkage with input backlash is shown. This mechanism is depicted in Figure 34 and it consist of four bars connected with two actuators for the link 1 and Link 2 respectively. The simulations were done in MATLAB ® and SimMechanics ®.

The parameters of the four bar linkage is shown in Table 6, as can be noticed $\ell_1 = \ell_3$ and $\ell_2 = \ell_4$. The backlash parameters are $\beta = 0.01$, $\alpha = 0.0075$ and $M = 0.08$.

The dynamic equations of the four bar linkage are shown in Figure 34 (Spong, Hutchinson, & M., 2006).

$$d_{11}\ddot{q}_1 + g_1(q_1) = \tau_1$$

$$d_{22}\ddot{q}_2 + g_2(q_2) = \tau_2 \quad (96)$$

Figure 34. Diagram of the four bar linkage mechanism

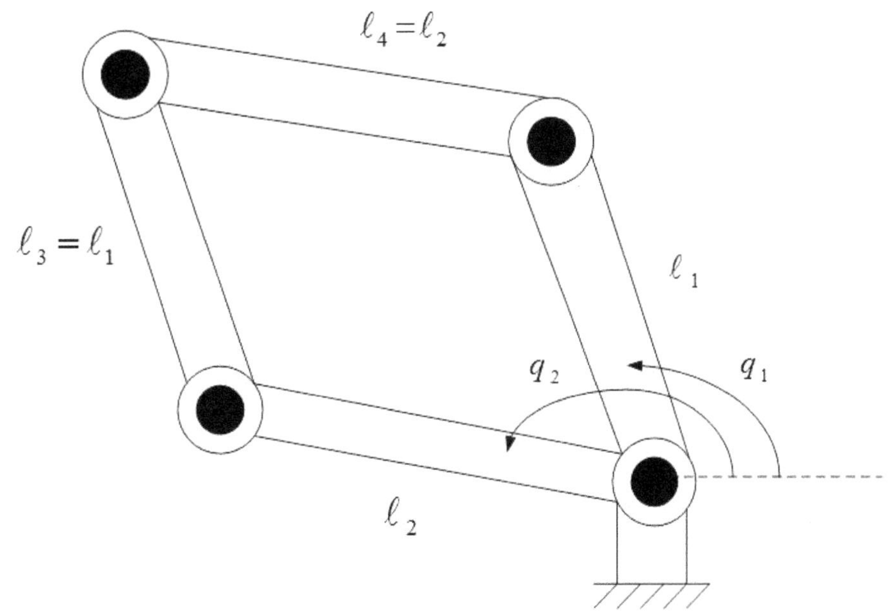

Table 6. Parameters of the four bar linkage

Joint	Mass (Kgs)	Length (m)	Mass Center (m)	Inertia (Kgs•m2)
1	2	0.7	0.35	0.23
2	3	0.9	0.45	0.37
3	2	0.7	0.35	0.23
4	3	0.9	0.45	0.37

where

$$g_1(q_1) = g\eta\cos(q_1)$$

$$g_2(q_2) = g\rho\cos(q_2)$$

$$d_{11} = m_1\ell_{c1}^2 + m_3\ell_{c3}^2 + m_4\ell_1^2 + I_1 + I_3$$

$$d_{22} = m_2\ell_{c2}^2 + m_3\ell_2^2 + m_4\ell_{c4}^2 + I_2 + I_4 \tag{97}$$

with $\eta = m_1\ell_{c1} + m_3\ell_{c3} + m_4\ell_1$ and $\rho = m_2\ell_{c2} + m_3\ell_2 - m_4\ell_{c4}$. Converting this system to state space, substituting the mechanism parameters and linearizing about the equilibrium point ($\pi / 2$, π) as shown in (56) and (57) the following state space matrix are obtained:

$$A = \begin{pmatrix} 0 & 0 \\ 0 & 1.2766 \end{pmatrix}$$

$$B = \begin{pmatrix} 0.4423 & 0 \\ 0 & 0.3965 \end{pmatrix}$$

$$C = \begin{pmatrix} 1 & 0 \\ 0 & 1 \end{pmatrix}$$

$$D = 0 \tag{98}$$

For the optimal controller design with a limit cycle oscillation frequency of $\omega = 1$ found with the following matrices employed for the solution of the Riccati equation

$$R_1 = \begin{pmatrix} 0.6 & 0 \\ 0 & 0.2 \end{pmatrix}$$

$$R_2 = \begin{pmatrix} 0.2 & 0 \\ 0 & 0.9 \end{pmatrix} \tag{99}$$

Then the solution of the Riccati equation for the matrix P and K are

$$P = \begin{pmatrix} 0.6066 & 0 \\ 0 & 14.6315 \end{pmatrix}$$

$$K = \begin{pmatrix} 1.3416 & 0 \\ 0 & 6.4462 \end{pmatrix} \tag{100}$$

The describing function parameters for the backlash are $t_1 = 0.0028$, $t_2 = 0.3365$, $t_3 = 3.1444$, $t_4 = 3.4781$, $a_1 = 0.0990$ and $b_1 = -0.2200$.

In Figure 35 and Figure 36 the angular position and velocity of q_1 are shown. It can be noticed that this system output q_1 oscillates about the equilibrium point ($\pi / 2$)

Similar to the angular velocity that oscillates about zero. This effect is caused by the backlash nonlinearity, but as it is explained later, the system is asymptotically stable. In Figure 37 and Figure 38 the angular position and velocity of q_2 is shown. It can be noticed that this system output q_2 oscillates about the equilibrium point $\left(\pi\right)$ similar to the angular velocity that oscillates about zero. This effect is caused by the backlash nonlinearity, but as it is explained later, the system is asymptotically stable. In Figure 39 and Figure 40 the input torques generated by the actuator one and two respectively are shown. It can be noticed that a small control effort is necessary to assure the stability of the system due to the optimal control law implemented to control this system. The effects of backlash are diminished by an appropriate selection of R_1, R_2 obtaining the desired P matrix by the Riccati equation. Then the optimal control law which minimizes the performance index (86) is obtained for an optimal control performance.

In Figure 41 and Figure 42 the phase portraits for q_1 and q_2 are shown respectively. It can be noticed that the limit cycle oscillations are asymptotically stable about the equilibrium point ($\pi / 2, \pi$), keeping the system stable in the presence of backlash. The describing function model is very helpful in the design of an optimal control law for mechanical systems in the presence of input backlash. As in the previous section the limit cycle oscillation parameters can be obtained analytically and by testing the system performance. This analysis allows the designer to create an efficient control strategy avoiding the unwanted effects caused by this nonlinearity. In Figure 44 the Nyquist plot of the system is shown, where as it is corroborated the system variables are asymptotically stables according to the Nyquist theorem.

As it is explained before, this analysis can be done by the circle criterion or the Popov theorem, but an analysis with the Nyquist criterion offers enough information to analyze the absolute stability of mechanical systems in the presence of input backlash as explained in (95).

In Figure 43 the simulation diagram of the four bar linkages is shown. As can be noticed the final position is kept in the required values by the optimal control law strategy, proving that it is an efficient control approach to solve this kind of problems.

As a conclusion of this section, an optimal control approach is proposed to stabilize an n degrees of freedom mechanical system. Again, the describing function mathematical model of the backlash nonlinearity was used to design an optimal control law for this kind of systems assuring the asymptotical stability of the system.

The proposed control strategy was implemented in a real benchmark system, a four bar linkages, in which the optimal control approach of mechanical systems with backlash is implemented to find the

Figure 35. Angular position of q_1

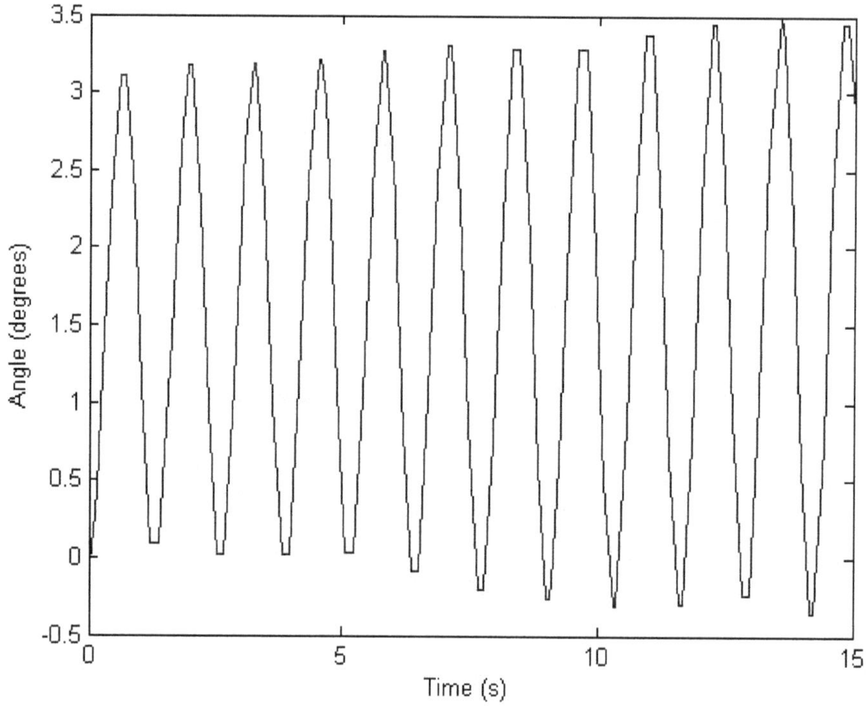

Figure 36. Angular velocity $\dot{q}_1$

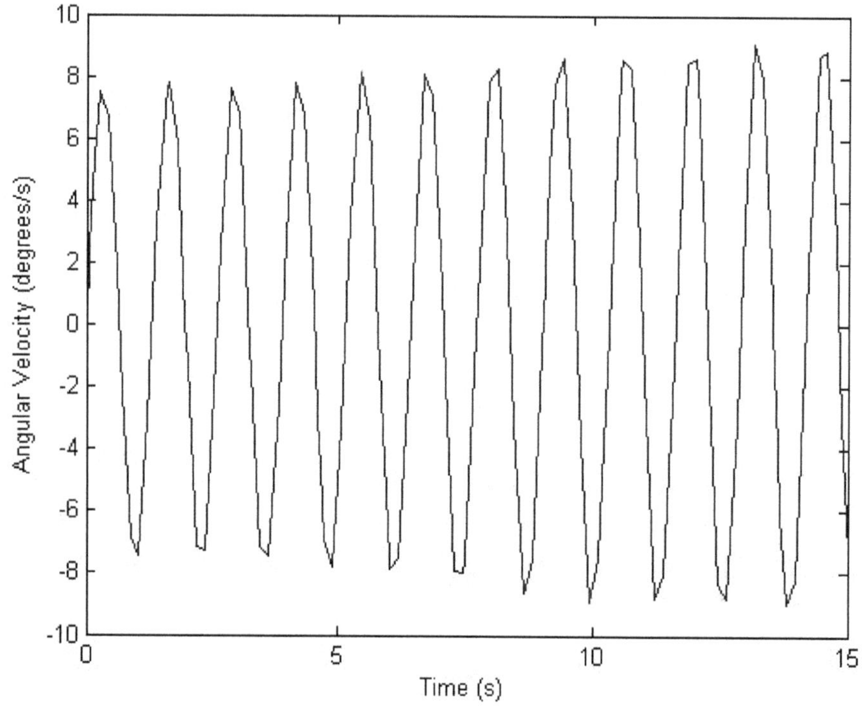

Figure 37. Angular position of q_2

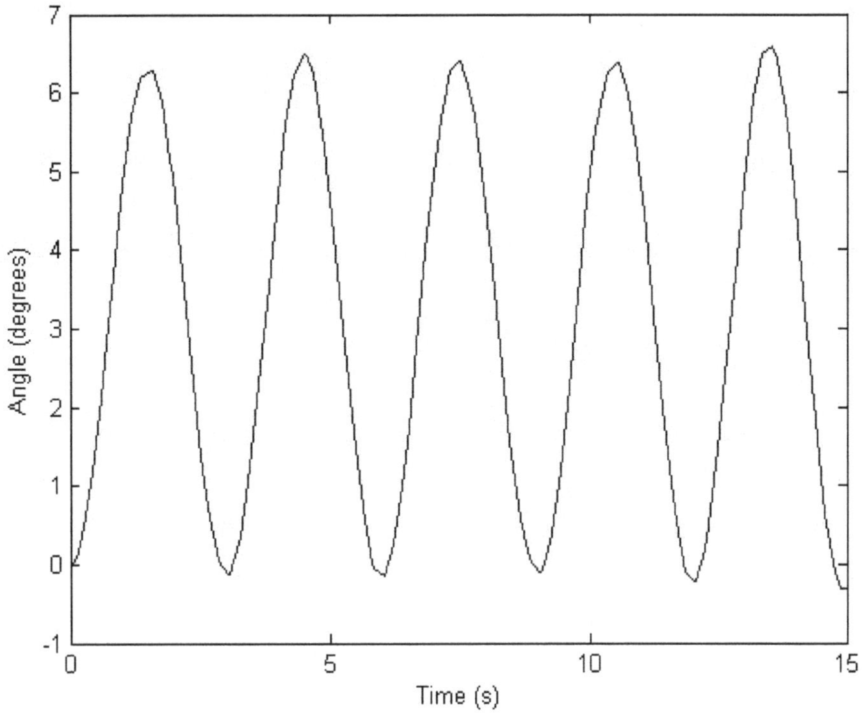

Figure 38. Angular velocity $\dot{q}_2$

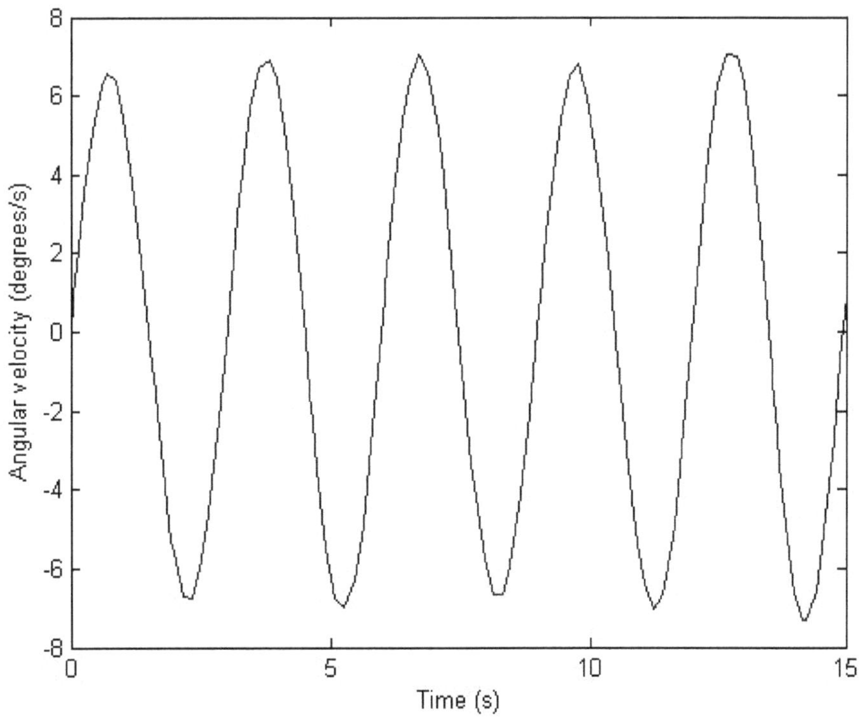

Figure 39. Actuator torque for q_1

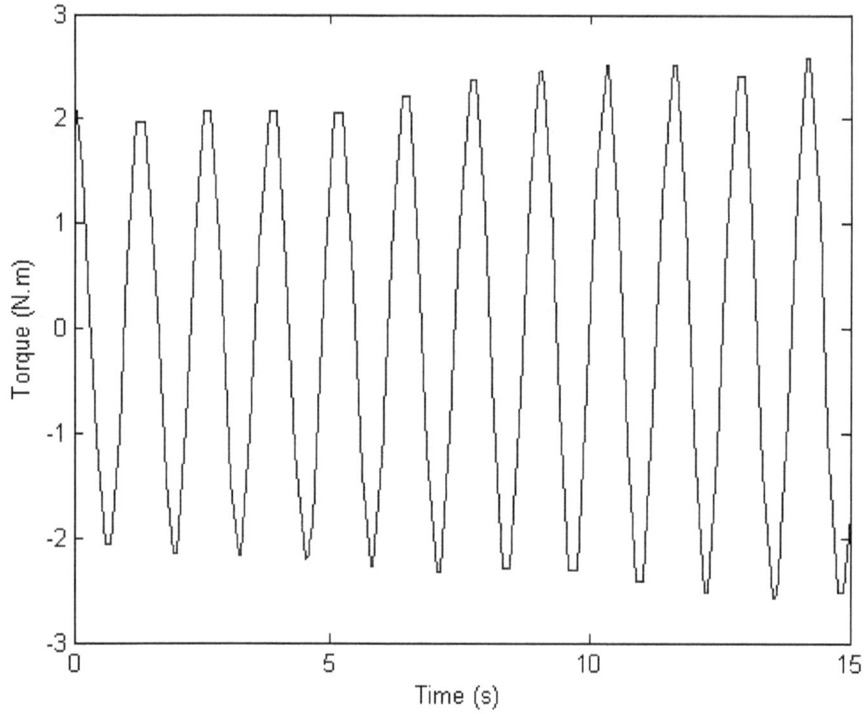

Figure 40. Actuator torque for q_2

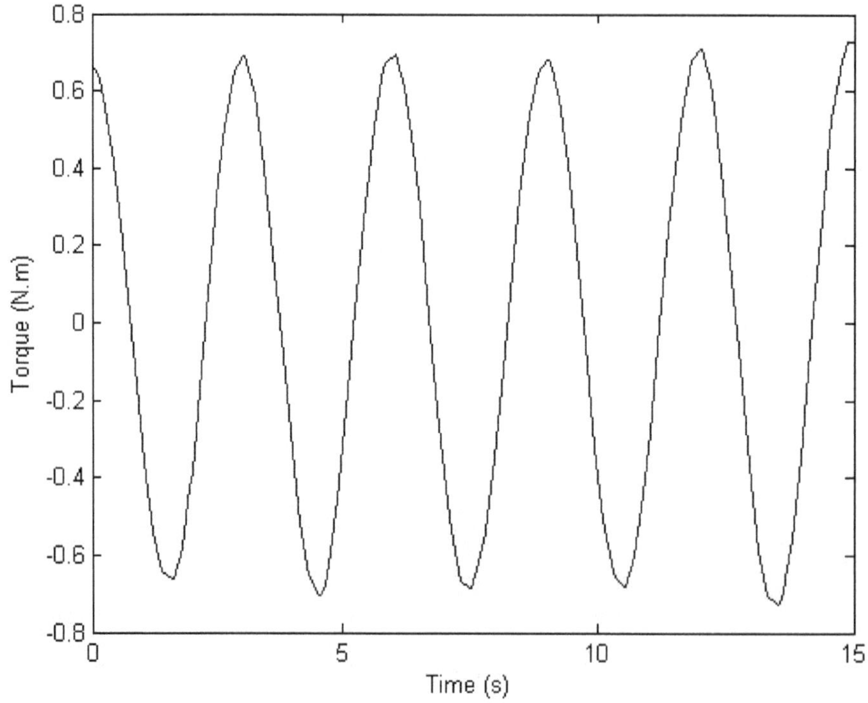

Figure 41. Phase portrait for q_1

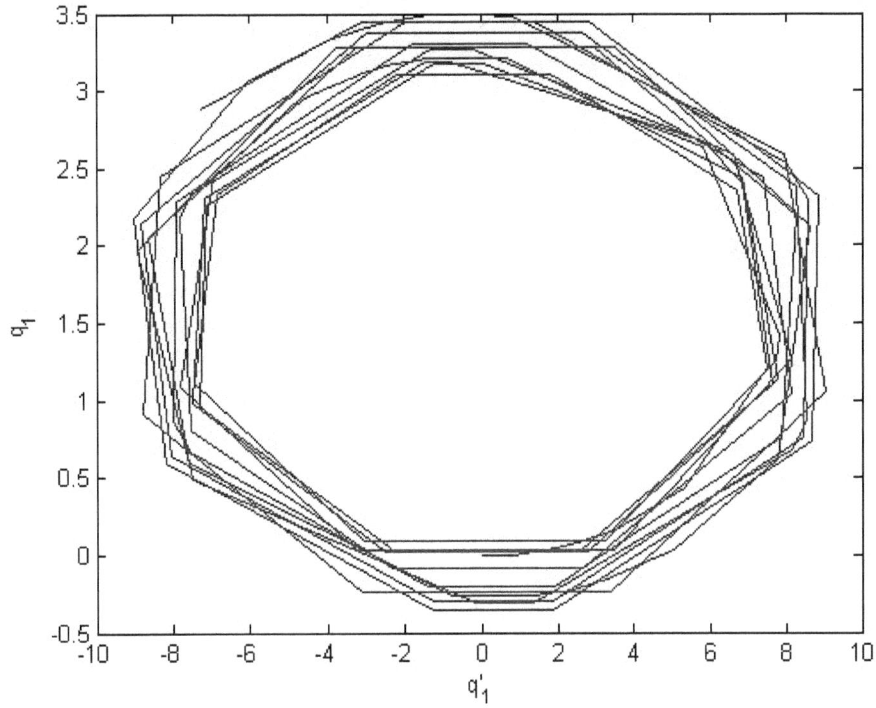

Figure 42. Phase portrait for q_2

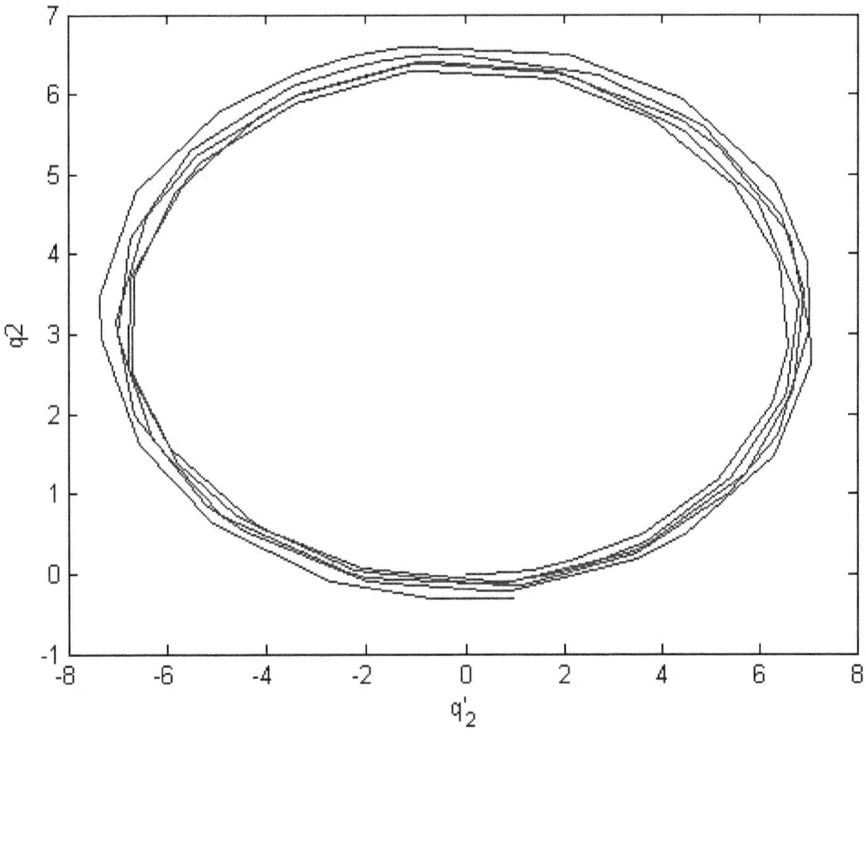

Figure 44. Nyquist diagram of the system

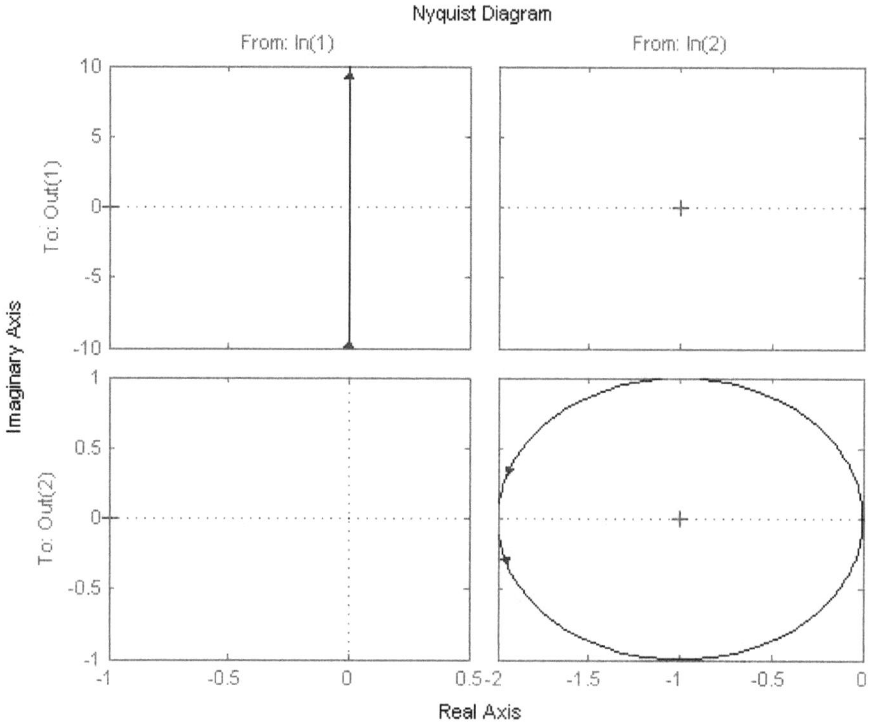

Figure 43. Simulation diagram of the four bar linkage

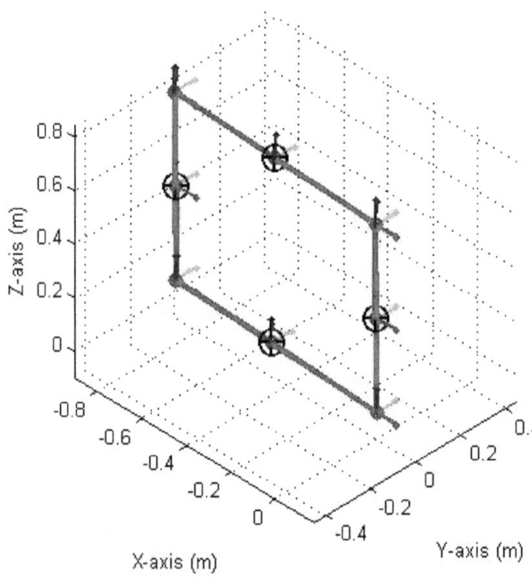

limit cycle oscillation frequency by the describing function modeling of backlash. This parameters can be found analytically or by simulation, to corroborate if the presence of backlash in the system does not produce chaotic behavior or instability in the systems measured variables.

6. DISCUSSION

In this chapter several control approaches for the stabilization of mechanical systems with backlash is presented. The different control strategies explained in this chapter provide diverse point of view for the design of an efficient approach to solve this kind of problem. The first control strategy explained in this chapter solve this problem in an independent control point of view, even when this technique is implemented when it is not necessary a faster response of the mechanical system output in the presence of backlash, it is still useful to be implemented when this nonlinearity is found in the actuators of the mechanical producing unwanted effects in the systems response. A straightforward control approach is implemented first when backlash is not present in the actuators, even when this is a trivial case it is important to develop an efficient control strategy, that in this case, a PD control law control approach is implemented to stabilize mechanical systems with backlash in the actuator/joint gear train. The second case, when there is backlash in the actuator/joint gear train is shown, where a similar PD control strategy is proposed when backlash is present; this control strategy is proved to be effective for this kind of problem, and it was proven that the mechanical system is stabilized as long the controller meet some parameters requirements. The advantages of this control strategy in comparison with multivariable control strategies; is that is very simple to apply, some classical tuning methods could be applied as long as they meet the stability requirement and the motor/joint could be treated as an independent problem.

After defining an independent joint control approach for mechanical systems with input backlash two multivariable control approaches are explained. The first multivariable control strategy shown in this chapter is a loop shaping controller design which provide an efficient control strategy to design an appropriate controller improving the closed loop characteristics. This approach is done by implementing a describing function model of the backlash nonlinearity, which is very important to be implemented due to the analysis of the limit cycle oscillations and the stability analysis of the mechanical system loop shaping controller design. The advantage of this method over independent joint controller is that a decentralized controller for mechanical systems with backlash is proposed, where the input backlash is present in all the joints. A frequency analysis of the system is proposed by the positivity theorem and the Nyquist criterion to test the stability of the system and analyze the limit cycles oscillations allowing the design of an appropriate controller that avoids a chaotic behavior of the controlled variables.

Another multivariable controller approach is proposed by the implementation of an optimal controller (Linear Quadratic Regulator) to find an optimal control law for mechanical systems with input backlash. Again this technique is a modern control theory approach that allows the designer to implement a decentralized control law for this kind of mechanical systems. The describing function is very important in the design of an optimal controller to analyze the limit cycle oscillations produced by the backlash effects on the system and obtain an optimal control law that avoid a chaotic behavior of the controlled variables. Even when similar control approaches are found in literature, the main difference is that the frequency domain characteristics are considered for the controller design, and then a frequency analysis allows the analysis of the absolute stability of the system and the limit cycle oscillation analysis.

The three control strategies proposed in this chapter, offer different points of view of the controller design for mechanical systems with backlash. The independent control approach is suitable due to its simplicity and can be applied in systems in which the output response is not important at all, but an acceptable control design is obtained when there is backlash in the actuator/joint gear train as proved analytically and corroborated by the illustrative example shown in this chapter. The describing function modeling of the backlash nonlinearity simplifies the multivariable controller design and it is not necessary to implement a backlash inverse model as defined in literature. The multivariable loop shaping design for mechanical systems with backlash is an efficient control strategy that can be used along with describing functions due to their frequency domain characteristics. The limit cycles are studied analytically and by simulation proving the efficiency of the proposed control strategy. The optimal control strategy is commonly implemented for this kind of problem, but in this chapter a frequency domain analysis is proposed to study the absolute stability of the system and the limit cycle oscillations of the mechanical model in the presence of backlash. Then an appropriate and stabilizing control law is obtained ensuring the asymptotical stability of the model and avoiding a chaotic behavior of the measured variables.

7. CONCLUSION

In this chapter several control strategies for the control of mechanical systems in the presence of backlash is proposed. These approaches offer different point of views when this nonlinearity is present in different kinds of n degrees of freedom mechanical systems. A simple and straightforward independent control approach is proved to be an effective control strategy to deal with this discontinuity avoiding unwanted effects and the decline of the system output. A multivariable control loop shaping approach is implemented with the use of a describing function model; this control strategy has the advantage that allows the designer to select the desired frequency domain characteristics of the closed loop system making it ideal to be implemented along with describing functions. An absolute stability test is performed to check if the system is asymptotically stable and find the limit cycle oscillations parameters.

Finally, an optimal controller approach for this kind of model is implemented to find the optimal controller that stabilizes mechanical systems in the presence of backlash.

Again, the limit cycle oscillations produced by this nonlinearity are studied analytically and by simulation in an illustrative example obtaining asymptotically stable systems.

REFERENCES

Adams, R. J., & Bar-on, J. R. (1998). Loop Shaping to Improve the Multivariable Phase Margin. In *Proceedings of the 37th IEEE Conference on Decision & Control. Decision and Control.* Tampa, FL: IEEE. doi:10.1109/CDC.1998.757872

Ahmad, N. J., & Khorrami, F. (1999). Adaptive Control of Systems with Backlash Hysteresis at the Input. In *Proceedings of the American Control Conference. American Control Conference.* San Diego, CA: IEEE. doi:10.1109/ACC.1999.782315

Azar, A. T., & Serrano, F. E. (2014). Robust IMC–PID tuning for cascade control systems with gain and phase margin specifications. *Neural Computing & Applications*, *25*(5), 983–995. doi:10.1007/s00521-014-1560-x

Azar, A. T., & Serrano, F. E. (2015a). Adaptive Sliding mode control of the Furuta pendulum. In A. T. Azar & Q. Zhu (Eds.), *Advances and Applications in Sliding Mode Control systems, Studies in Computational Intelligence* (Vol. 576, pp. 1–42). Berlin: Springer-Verlag.

Azar, A. T., & Serrano, F. E. (2015b). Deadbeat Control for Multivariable Systems with Time Varying Delays. In A. T. Azar & S. Vaidyanathan (Eds.), *Chaos Modeling and Control Systems Design, Studies in Computational Intelligence* (Vol. 581, pp. 97–132). Berlin: Springer-Verlag.

Azar, A. T., & Serrano, F. E. (2015c). Design and Modeling of Anti Wind Up PID Controllers. In Q. Zhu & A. T. Azar (Eds.), *Complex system modelling and control through intelligent soft computations, Studies in Fuzziness and Soft Computing* (Vol. 319, pp. 1–44). Berlin: Springer-Verlag.

Azar, A. T., & Zhu, Q. (2015). *Advances and Applications in Sliding Mode Control systems. Studies in Computational Intelligence* (Vol. 576). Berlin: Springer-Verlag.

Azar, A. T., & Vaidyanathan, S. (2015a). *Computational Intelligence applications in Modeling and Control. Studies in Computational Intelligence* (Vol. 575). Berlin: Springer-Verlag.

Azar, A. T., & Vaidyanathan, S. (2015b). *Chaos Modeling and Control Systems Design. Studies in Computational Intelligence* (Vol. 581). Berlin: Springer-Verlag.

Azar, A. T. (2010). *Fuzzy Systems*. Vienna, Austria: IN-TECH.

Barreiro, A., & Baños, A. (2006). Input output stability of systems with backlash. *Automatica, 42*(6), 1017–1024. doi:10.1016/j.automatica.2006.02.017

Bentsman, J., Orlov, Y., & Aguilar, L. T. (2013). Nonsmooth H-infinity Output Regulation with Application to a Coal-Fired Boiler/Turbine Unit with Actuator Deadzone. In *Proceedings of 2013 American Control Conference (ACC)*. Washington, DC: IEEE. doi:10.1109/ACC.2013.6580434

Boukhnifer, M., & Ferreira, A. (2005). H Infinity Loop Shaping for Stabilization and Robustness of a Tele-Micromanipulation System. In *Proceedings of IEEE/RSJ International Conference on Intelligent Robots and Systems* (IROS 2005). IEEE.

Brogan, W. L. (n.d.). *Modern Control Theory*. Prentice Hall.

Chyung, D. H. (1992). Output Feedback Controller for Systems Containing a Backlash. In *Proceedings of the 31st Conference on Decision and Control*. Tucson, AZ: IEEE. doi:10.1109/CDC.1992.371218

Cimen, T. (2010). Systematic and effective design of nonlinear feedback controllers via the state-dependent Riccati equation (SDRE) method. *Annual Reviews in Control, 34*(1), 32–51. doi:10.1016/j.arcontrol.2010.03.001

Gao, Z. W., & So, A. T. (2003). A general doubly coprime factorization for descriptor systems. *Systems & Control Letters, 49*(3), 213–224. doi:10.1016/S0167-6911(02)00325-0

Glück, T., Eder, A., & Kugi, A. (2013). Swing-up control of a triple pendulum on a cart with experimental validation. *Automatica, 49*(3), 801–808. doi:10.1016/j.automatica.2012.12.006

Goebel, R. (2004). Stabilizing linear systems with saturation through optimal control. In *Proceedings of IEEE Conference on Decision and Control. Decision and Control*. IEEE.

Guo, J., Yao, B., Chen, Q., & Wu, X. (2009). High Performance Adaptive Robust Control for Nonlinear System with Unknown Input Backlash. In *Proceedings of Joint 48th IEEE Conference on Decision and Control and 28th Chinese Control Conference.* Shanghai, China: IEEE.

Haddad, W. M., & Chellaboina, V. (n.d.). *Nonlinear Dynamical Systems and Control - A Lyapunov Based Approach.* Princeton University Press.

Hippe, P. (1994). Strictly Doubly Coprime Factorizations and all Stabilizing Compensators Related to Reduced-order Observer. *Automatica, 30*(12), 1955–1959. doi:10.1016/0005-1098(94)90056-6

Indri, M., & Tornambe, A. (1997). Application of a PD controller on two mating gears with elasticity and backlash. In *Proceedings of the 36th Conference on Dectston & Control. Decision and Control.* San Diego, CA: IEEE. doi:10.1109/CDC.1997.657892

Iqbal, S., Bhatti, A. I., Akhtar, M., & Ullah, S. (2007). Design and Robustness Evaluation of an H Infinity Loop Shaping Controller for a 2DOF Stabilized Platform. In *Proceedings of the European Control Conference.* Kos, Greece: Academic Press.

Jang, J. O., Lee, P. G., Chung, H. T., & Jeon, G. J. (2003). Output Backlash Compensation of Systems Using Fuzzy Logic. In *Proceedings of American Control Conference.* Academic Press. doi:10.1109/ACC.2003.1243449

Jing, Z., & Wen, C. (2007). Adaptive Inverse Control of a Magnetic Suspension System with Input Backlash. In *Proceedings of 16th IEEE International Conference on Control Applications. Control Applications.* IEEE.

Kashiwagi, H., Harada, H., & Toyozawa, Y. (2004). Identification of nonlinear mechatronic servo system having backlash. *SICE Annual Conference in Sapporo.* In *Proceedings of SICE 2004 Annual Conference.* IEEE.

Lin, F., & Brandt, R. (1998). An Optimal Control Approach to Robust Control of Robot Manipulators. *IEEE Transactions on Robotics and Automation, 14*(1), 69–77. doi:10.1109/70.660845

Ma, H.-J., & Yang, G.-H. (2008). Adaptive regulation of uncertain nonlinear systems with dead-zone. In *Proceedings of the 47th IEEE Conference on Decision and Control.* Cancun, Mexico: IEEE.

McFarlane, D., & Glover, K. (1992). A Loop Shaping Design Procedure Using H Infinity Synthesis. *IEEE Transactions on Automatic Control, 37*(6), 759–769. doi:10.1109/9.256330

Mekki, H., Boukhetala, D., & Azar, A. T. (2015). Sliding Modes for Fault Tolerant Control. In A. T. Azar & Q. Zhu (Eds.), *Advances and Applications in Sliding Mode Control systems, Studies in Computational Intelligence book Series* (Vol. 576, pp. 407–433). Berlin: Springer-Verlag.

Nordin, M., & Gutman, P.-O. (2002). Controlling mechanical systems with backlash—a survey. *Automatica, 38*(10), 1633–1649. doi:10.1016/S0005-1098(02)00047-X

Pannocchia, G., Rawlings, J. B., Mayne, D. Q., & Marquardt, W. (2010). On Computing Solutions to the Continuous Time Constrained Linear Quadratic Regulator. *IEEE Transactions on Automatic Control, 55*(9), 2192–2198. doi:10.1109/TAC.2010.2053478

Petersen, I. R. (2008). Robust adaptive H infinity control using integral quadratic constraints. *Automatica*, *44*(10), 2661–2668. doi:10.1016/j.automatica.2008.03.007

Riedinger, P., & Iung, C. (1999). Optimal Control for Hybrid Systems: An Hysteresis Example. ystems, Man, and Cybernetics, 1999. In *Proceedings of IEEE SMC '99 Conference*. IEEE.

Seraji, H. (1989). Decentralized Adaptive Control of Manipulators: Theory, Simulation, and Experimentation. *IEEE Transactions on Robotics and Automation*, *5*(2), 183–201. doi:10.1109/70.88039

Silva, E. I., & Erraz, D. A. (2006). An LQR Based MIMO PID Controller Synthesis Method for Unconstrained Lagrangian Mechanical Systems. In *Proceedings of the 45th IEEE Conference on Decision & Control. Decision and Control*. San Diego, CA: IEEE. doi:10.1109/CDC.2006.377348

Spong, M. W., & Hutchinson, S. (2006). *Robot Modeling and Control*. John Wiley and Sons.

Tao, G., & Kokotovic, P. V. (1993). Adaptive Control of Systems with Backlash. *Automatica*, *29*(2), 323–335. doi:10.1016/0005-1098(93)90126-E

Tao, G., & Kokotovic, P. V. (1993). Continuous-time Adaptive Control of Systems with Unknown Backlash. In *Proceedings of the American Control Conference. American Control Conference*. San Francisco, CA: IEEE.

Tao, G., & Kokotovik, P. (1995). Adaptive Control of Systems with Unknown Output Backlash. *IEEE Transactions on Automatic Control*, *40*(2), 326–330. doi:10.1109/9.341803

Tao, G., & Ma, X. (1999). Backlash Compensation for Multivariable Nonlinear Systems with Actuator Dynamics. In *Proceedings of the 38th Conference on Decision & Control. Decision and Control*. Phoenix, AZ: IEEE.

Tao, G., Ma, X., & Ling, Y. (2001). Optimal and nonlinear decoupling control of systems with sandwiched backlash. *Automatica*, *37*(2), 165–176. doi:10.1016/S0005-1098(00)00153-9

Tierno, J. E., Kim, K. Y., Lacy, S. L., & Bernstein, D. S. (2000). Describing Function Analysis of an Anti-Backlash Controller. In *Proceedings of the American Control Conference*. Academic Press. doi:10.1109/ACC.2000.877005

Vaidyanathan, S., & Azar, A. T. (2015a). Anti-Synchronization of Identical Chaotic Systems using Sliding Mode Control and an Application to Vaidyanathan-Madhavan Chaotic Systems. In A. T. Azar & Q. Zhu (Eds.), *Advances and Applications in Sliding Mode Control systems, Studies in Computational Intelligence book Series* (Vol. 576, pp. 527–547). Berlin: Springer-Verlag.

Vaidyanathan, S., & Azar, A. T. (2015b). Hybrid Synchronization of Identical Chaotic Systems using Sliding Mode Control and an Application to Vaidyanathan Chaotic Systems. In A. T. Azar & Q. Zhu (Eds.), *Advances and Applications in Sliding Mode Control systems, Studies in Computational Intelligence book Series* (Vol. 576, pp. 549–569). Berlin: Springer-Verlag.

Vaidyanathan, S., & Azar, A. T. (2015c). Analysis, Control and Synchronization of a Nine-Term 3-D Novel Chaotic System. In A. T. Azar & S. Vaidyanathan (Eds.), *Chaos Modeling and Control Systems Design, Studies in Computational Intelligence* (Vol. 581, pp. 3–17). Berlin: Springer-Verlag.

Vaidyanathan, S., & Azar, A. T. (2015d). Analysis and Control of a 4-D Novel Hyperchaotic System. In A. T. Azar & S. Vaidyanathan (Eds.), *Chaos Modeling and Control Systems Design, Studies in Computational Intelligence* (Vol. 581, pp. 19–38). Berlin: Springer-Verlag.

Vaidyanathan, S., Idowu, B. A., & Azar, A. T. (2015). Backstepping Controller Design for the Global Chaos Synchronization of Sprott's Jerk Systems. In A. T. Azar & S. Vaidyanathan (Eds.), *Chaos Modeling and Control Systems Design, Studies in Computational Intelligence* (Vol. 581, pp. 39–58). Berlin: Springer-Verlag.

Zhou, J., Er, M. J., & Wen, C. (2005). Adaptive Control of Nonlinear Systems with Uncertain Dead-zone Nonlinearity. In *Proceedings of the 44th IEEE Conference on Decision and Control, and the European Control Conference. IEEE.* doi:10.1109/CDC.2005.1582254

Zhou, S.-L., Han, P., Wang, D.-F., & Liu, Y.-Y. (2004). A Kind of Multivariable PID Design Method for Chaos System using H infinity loop shaping design procedure. In *Proceedings of the Third International Conference on Machine Learning and Cybernetics.* IEEE.

Zhu, Q., & Azar, A. T. (2015). *Complex system modelling and control through intelligent soft computations. Studies in Fuzziness and Soft Computing* (Vol. 319). Berlin: Springer-Verlag.

KEY TERMS AND DEFINITIONS

Backlash: A nonlinearity found in many mechanical, hydraulic and other kinds of systems; due to the interconnections of elements such as gears and other elements in which there is an imperfect contact between two mating elements.

H_∞**: Control:** A control technique which consists in obtaining a controller minimizing the H_∞ norm of the system.

Independent Joint Control: A control approach implemented for the stabilization and control of robotics and other mechanical systems in which the control of each joint is designed independently.

Limit Cycles: A self-excited self-sustaining periodic oscillation.

Mechanical Systems: A kind of system which consists or can be represented by a series of mechanical parts.

Multivariable Control: A control strategy for multivariable systems generally represented by a state space model.

Nonlinearities: Different phenomena found in many kinds of systems which introduce a nonlinear behavior to the model.

Optimal Control: A control technique which consist in find the optimal controller input among all admissible control functions while minimizing a performance criteria.

Robotics: A science field that deals with informatics, mechanical engineering, electrical engineering and industrial engineering to develop and design mechanisms to substitute operators or persons in many industrial and other dangerous operations.

Chapter 2
Nonlinear Adaptive Filtering with a Family of Kernel Affine Projection Algorithms

Felix Albu
Valahia University of Targoviste, Romania

Kiyoshi Nishikawa
Tokyo Metropolitan University, Japan

ABSTRACT

In this chapter, the family of kernel affine projection algorithms with coherence criterion is presented. The proportionality principle is translated to the kernel-based version. A new algorithm called Kernel Proportionate Affine Projection Algorithm (KPAPA) is proposed. It is proved that the additional computational increase burden is independent of the order of the algorithm, being dependent only on the order of the kernel expansion. The Dichotomous Coordinate Descent (DCD) method and an example of an efficient implementation of KAPA using DCD are presented. This chapter also discusses the influence of the coherence value, the step size value, and the dictionary size on the performance of KAPA and KPAPA algorithms. The effectiveness of the proposed algorithms and the effect of different parameters are confirmed by computer simulations for nonlinear system identification application.

INTRODUCTION

Linear adaptive filters have been widely used in a variety of applications, e.g., mobile communications, video processing etc. In system identification applications, the main goal is to identify an unknown system using an adaptive filter. The well known normalized least-mean-square (NLMS) algorithm has been widely used, but converges very slowly (Sayed, 2003). The affine projection algorithm (APA) provides a much improved convergence speed (Sayed, 2003) and in some situations close to that of the more complex recursive least-squares (RLS) algorithm. Another related family of linear algorithms is the proportionate APA (PAPA) family. The update of each coefficient of the filter is made independently of the others, by adjusting the adaptation step-size in proportion to the magnitude of the estimated filter coefficient

DOI: 10.4018/978-1-4666-7248-2.ch002

(Deng, 2006). The proportionate normalized least-mean-square (PNLMS) algorithm (Duttweiler, 2000) was one of the first proportionate-type algorithms. All of the APA/PAPA algorithms require the computation of the inverse of a matrix, which is a complex operation and can lead to algorithm instability if finite precision is used. One of the most efficient methods to avoid the matrix inversion is the use of the Dichotomous Coordinate Descent (DCD) methods (Zakharov & Tozer, 2004), (Zakharov, White & Liu, 2008) and (Albu, Bouchard & Zakharov, 2007).

As an extension of the linear counterparts, kernel adaptive filters have been proposed to identify non-linear systems (Liu, Principe & Haykin, 2010). Kernel adaptive filters are derived by applying the kernel methods to linear adaptive filters, and many such algorithms were proposed, (e.g., the kernel recursive least squares (KRLS) (Engel, Mannor & Meir, 2004), the kernel least mean square (KLMS) (Liu, Pokharel & Principe, 2008), the kernel normalized LMS (KNLMS) (Richard, Bermudez, & Honeine, 2009), kernel affine projection (KAPA) (Richard et al., 2009) and its efficient implementation (Albu, Coltuc, Rotaru & Nishikawa, 2013), etc. The application of a reproducing kernel Hilbert space (RKHS) methodology blended with many signal processing techniques is a powerful tool in solving numerous practical artificial intelligence or control systems problems (Müller, Adali, Fukumizu, Principe, & Theodoridis, 2013).

This chapter explains the similarities between the linear algorithms and their corresponding kernel versions. We start with a short description of the related work and of the linear NLMS, APA, and PAPA algorithms. Next, the kernel AP algorithm with a coherence criterion is presented. Following the resemblance of APA with KAPA, the proportionality criterion is used in order to derive a new algorithm called, kernel PAPA (KPAPA). The KNLMS and a new proportionate version of it are mentioned as particular cases of KAPA and KPAPA respectively. The recent research directions in the development of the reduced complexity algorithms based on DCD iterations are also discussed. Next, the chapter investigates the reduction of the numerical complexity of the kernel algorithms, the fine tuning of the parameters of the algorithms, etc. Their importance and significance for improving the overall performance is discussed. Several simulation results are shown in the Simulation section. Finally, the conclusions and future research direction conclude this work.

RELATED WORK

The algorithms described in this chapter have been derived from the kernel affine projection algorithm with coherence criterion, described in (Richard et al., 2009). The similarities between linear and nonlinear algorithms have been pointed out in (Liu et al., 2010). The novel proposed algorithms uses the proportionality principle proposed in (Benesty & Gay, 2002) and (Hoshuyama et al., 2004). The first time this principle was used for a kernel based algorithm was in (Albu & Nishikawa, 2013). The proposed KPNLMS algorithm showed improved convergence characteristics over KNLMS for nonlinear system identification and forward prediction applications (Albu & Nishikawa, 2013). The divisionless and multiplier-less method, called dichotomous coordinate descent (DCD) was proposed in 2004 (Zakharov & Tozer, 2004). The DCD method was used many linear adaptive algorithms for numerous applications in order to reduce the numerical complexity (Zakharov et al., 2008), (Albu et al., 2007). Also, a lot of work has been dedicated to develop linear adaptive algorithms with variable parameters (e.g., (Arablouei, Dogançay, & Perreau, 2012, Liu, Fukumoto, Saiki & Zhang, 2009, etc.) To the best of our knowledge, the only kernel based algorithm using a variable step size has been recently proposed in (Wang, Feng and Tse, 2013). In (Albu & Nishikawa, 2013) the effect of the variable step size on the KPNLMS algorithm was firstly investigated.

LINEAR ADAPTIVE ALGORITHMS

A short introduction to the NLMS, APA and PAPA is made in the following lines (Sayed, 2003). In the context of system identification the desired signal $d(n)$ is obtained as

$$d(n) = \mathbf{h}^T \mathbf{x}(n) + w(n) \tag{1}$$

where the impulse response of the unknown system is the L-length vector

$$\mathbf{h}(n) = \left[h_0(n), h_1(n), \ldots, h_{L-1}(n)\right]^T,\ x(n)$$

is the input signal, $\mathbf{x}(n) = \left[x(n), x(n-1), \ldots, x(n-L+1)\right]^T$ is the input signal vector, $w(n)$ is a zero-mean white noise and the superscript T shows the transpose of a vector or a matrix.

The impulse response of the system is identified with the adaptive filter $\hat{\mathbf{h}}(n) = \left[\hat{h}_0(n), \hat{h}_1(n), \ldots, \hat{h}_{L-1}(n)\right]^T$ whose real coefficients are updated with an adaptive algorithm. Figure 1 shows a typical structure of a linear adaptive filter used for system identification.

The APA is one of the most popular algorithms and is described by the following equations (Sayed, 2003):

Figure 1. System identification with a linear filter

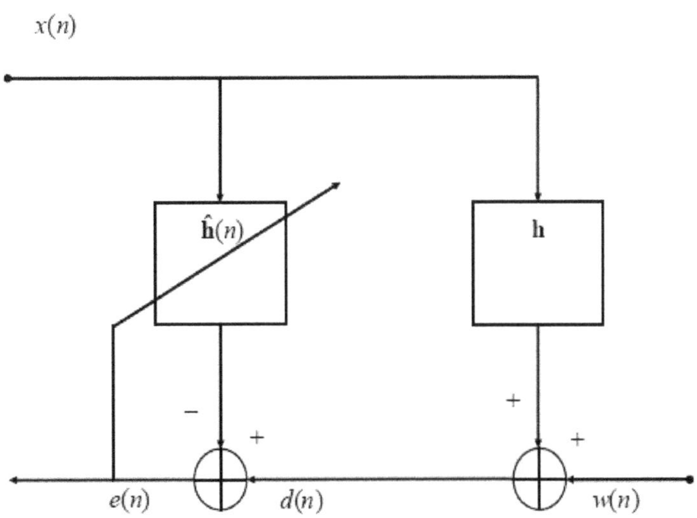

$$e(n) = d(n) - \hat{\mathbf{h}}^T(n-1)\mathbf{X}(n), \tag{2}$$

$$\mathbf{R}(n) = \delta \mathbf{I}_p + \mathbf{X}^T(n)\mathbf{X}(n), \tag{3}$$

$$\hat{\mathbf{h}}(n) = \hat{\mathbf{h}}(n-1) + \mu \mathbf{X}(n)\mathbf{R}^{-1}(n)\mathbf{e}(n) \tag{4}$$

where $\mathbf{e}(n) = [e(n), e(n-1), \ldots, e(n-P+1)]^T$ is the error vector of length *P*, *P* is the projection order, $\mathbf{d}(n) = [d(n), d(n-1), \ldots, x(n-P+1)]^T$, $\mathbf{X}(n) = [\mathbf{x}(n), \mathbf{x}(n-1), \ldots, \mathbf{x}(n-P+1)]$ is the data matrix, $\mathbf{R}(n)$ is the correlation matrix, δ is the regularization parameter, $\mathbf{I}_P$ is the $P \times P$ unity matrix and μ is the step size. If Equation (2) and Equation (3) are combined, the following equation update is obtained:

$$\hat{\mathbf{h}}(n) = \hat{\mathbf{h}}(n-1) + \mu \mathbf{X}(n)\left(\delta + \mathbf{X}^T(n)\mathbf{X}(n)\right)^{-1}\mathbf{e}(n) \tag{5}$$

The NLMS is a particular case of APA for $P = 1$.

It can be noticed from Equation (5) that the algorithm has three variable parameters: μ, δ and *P*. Many of the developed algorithms with variable parameters are based on APA's MSE estimation formula proposed in (Shin et al., 2004). For example, the variable projection order APA proposed in (Kim et al., 2009) and the variable regularization APA proposed in (Yin & Mehr, 2010).

Variable step size (VSS) APA versions can be obtained by varying μ. Examples of VSS APA versions are presented in (Shin, Sayed, & Song, 2004) and (Albu, Paleologu & Benesty, 2011). Computational simpler versions can be based on the output error. Such possibilities are (Haweel, 2004):

$$\mu(n) = \rho\mu(n-1) + \tau sign\left(e(n-1)\right) \tag{6}$$

(Kwong &Johnston, 1992)

$$\mu(n) = \rho\mu(n-1) + \tau e^2(n-1) \tag{7}$$

(Aboulnasr & Mayyas, 1997)

$$\mu(n) = \rho\mu(n-1) + \tau q^2(n-1) \tag{8}$$

with $q(n)$ given by (Aboulnasr & Mayyas, 1997)

$$q(n-1) = \nu q(n-2) + (1-\nu)e(n-1)e(n-2) \tag{9}$$

where ρ and ν are constants between 0 and 1, and τ is a positive constant.

Also, if δ is varied, variable regularization versions are obtained (Yin & Mehr, 2010). If an APA has a variable P value in time, an evolutionary APA version is obtained (Kim, Kong, & Song, 2009). It is known that if the projection order increases, the convergence speed is faster, but the steady-state error also increases (Sayed, 2003). The projection order is modified depending on the relationship between the output error and a threshold based on the MSE estimation formula from (Shin et al., 2004).

Proportionate adaptive algorithms have been proposed as an approach to accelerate convergence in case of sparse impulse response. A proportionate-type APA algorithm updates its coefficients according to:

$$\hat{\mathbf{h}}(n) = \hat{\mathbf{h}}(n-1) + \mu\mathbf{G}(n-1)\mathbf{X}(n)\left(\delta + \mathbf{X}^T(n)\mathbf{G}(n-1)\mathbf{X}(n)\right)^{-1}\mathbf{e}(n) \tag{10}$$

where $\mathbf{G}(n-1)$ is an $L \times L$ diagonal matrix which assigns an individual step-size to each filter coefficient. The IPNLMS (improved proportionate NLMS) algorithm (Benesty & Gay, 2002) is obtained when $P = 1$ in equation (10) and the diagonal elements of $\mathbf{G}(n-1)$, denoted in the following by $g_l(n-1)$, with $0 \le l \le L-1$, are evaluated as:

$$g_l(n-1) = \frac{1-\beta}{2L} + \frac{\left|\hat{h}_l(n-1)\right|(1+\beta)}{2\sum_{i=0}^{L-1}\left|\hat{h}_i(n-1)\right| + \xi}, \tag{11}$$

where $-1 \le \beta \le 1$ and the small positive constant ξ avoids division by zero. Therefore, the improved PAPA (IPAPA) has the equations (2) and (10) if (11) is used for computing the proportionate coefficients. Other versions of PAPA are possible in case of using a different approach for computing the $g_l(n-1)$ coefficients.

It was shown in (Benesty & Gay, 2002) and (Hoshuyama, Goubran, & Sugiyama, 2004) that proportionate algorithms obtain a better convergence speed in the case of a sparse impulse response. Such typical cases are the acoustic echo cancellation or network echo cancellation. PAPA improved the slow convergence speed of PNLMS for colored excitation signals. Many approaches have been proposed to improve PAPA's steady-state performance. One of these approaches was to use a variable step size by cancelling the a posteriori estimation error (Liu et al., 2009). It was shown in (Liu et al., 2009) that a high convergence speed was obtained using a large step size and then a considerably decreased steady-state misalignment was achieved by using a small step size after the adaptive filter has converged.

It was shown in (Paleologu, Benesty & Albu, 2012) that the regularization parameter is highly influenced by the level of the system noise (the near-end background noise), i.e., the more the noise, the larger the value of the regularization parameter. Therefore the choice of this parameter for IPAPA can attenuate the effects of the noise in the adaptive filter estimate.

Another possibility to improve the convergence characteristics of PAPA is to use a variable projection order and take into account the proportionate history (Arablouei, Dogançay, & Perreau, 2012). Also

Figure 2. Diagram of a kernel adaptive filter

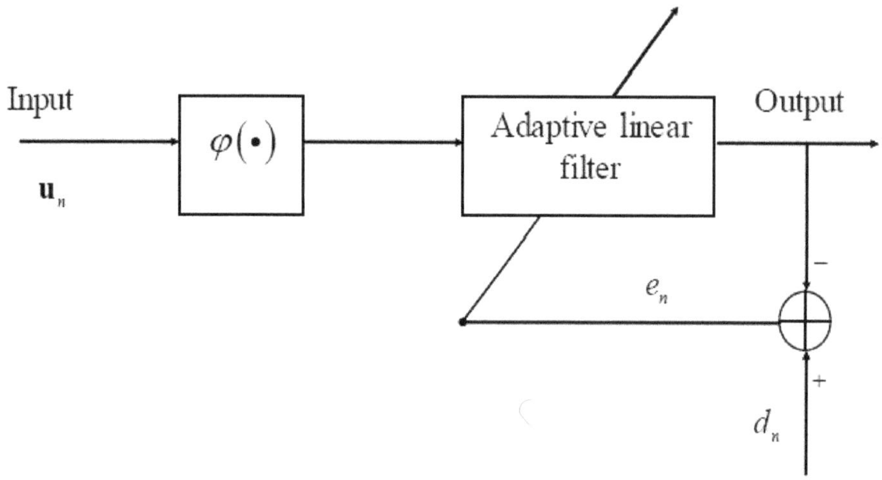

a logarithmic proportionate updating scheme can be used together with approximations for the output errors (Albu, 2012). Other methods for reducing algorithm's numerical complexity are the use of fast filtering techniques and dichotomous coordinate descent iterations (Albu, 2011). Various techniques for improving the convergence characteristics of PAPA were initially used for the APA family of algorithms. These techniques can be adapted for the PAPA assuming that the MSE estimation error formula for APA still holds.

Kernel Affine Projection Algorithm

The kernel methods are powerful nonlinear techniques based on a nonlinear transformation $\varphi(\bullet)$ of the input data $\mathbf{u}_i$ into a high-dimensional feature space in which we then apply linear adaptive algorithms to the transformed input signal $\varphi(\mathbf{u}_i)$ (Richard et al., 2009). Figure 2 shows a typical structure of the kernel adaptive filter.

Kernel-based methods lead to manageable dimensions for some mappings. The kernel has to satisfy Mercer's conditions (Liu et al., 2010). The following "kernel trick" (Vapnik, 1995) allows us to obtain nonlinear versions of the linear adaptive algorithms using inner products by replacing them with kernel evaluation (Rzepka, 2012).

$$k(\mathbf{u}_i, \mathbf{u}_j) = \langle \varphi(\mathbf{u}_i), \varphi(\mathbf{u}_j) \rangle \tag{12}$$

There is no need to have an explicit knowledge of $\varphi(\bullet)$. The most popular kernels are the Gaussian kernel $k(\mathbf{u}_i, \mathbf{u}_j) = \exp(-\|\mathbf{u}_i - \mathbf{u}_j\| / 2\beta_0^2)$
and the Laplacian kernel $k(\mathbf{u}_i, \mathbf{u}_j) = \exp(-\|\mathbf{u}_i - \mathbf{u}_j\| / \beta_0)$ with β_0 the kernel bandwidth.

In the kernel adaptive signal processing, the unknown system in the feature space is modeled as the following form

$$\psi_n(\bullet) = \sum_{j=1}^{m} \alpha_j k(\bullet, \mathbf{u}_{\omega_j}) \tag{13}$$

at the time *n*, where ω_j 's form an *m*-element subset of I_n of $\{1,...,n\}$, $\{k(\bullet,\mathbf{u}_{\omega_j})\}_{j=1}^{m}$ is called the dictionary and *m* is the order of the kernel expansion (Richard et al., 2009). The objective of the kernel adaptive filter is to estimate α_j in the equation (13) using an adaptive algorithm.

Let us briefly explain the general concepts of the kernel adaptive algorithms. The kernel adaptive algorithms are derived by replacing the inner products in the linear algorithms by the kernel function as shown in Equation (12). Besides, the kernel algorithms have some additional processes which do not exist in the linear ones. The first process is that of inserting or removing kernel functions from the dictionary. The second process is the parameter update. The insertion of $k(\bullet,\mathbf{u}_n)$ into the dictionary is made if

$$\max_{\omega_j \in I_{n-1}} | k(\mathbf{u}_n, \mathbf{u}_{\omega_j}) | \leq \mu_0 \tag{14}$$

where μ_0 is a predefined parameter in the [0,1] range that influences both the sparsity level and the coherence of the dictionary (Richard et al., 2009). Also, the order of the kernel filters will increase in time (Richard et al., 2009) and this is the special feature of the kernel adaptive algorithms whose counterpart does not exist in the linear ones.

The kernel AP algorithm with coherence criterion was proposed in (Richard et al., 2009).The kernel output error vector is

$$\mathbf{e}_n = \mathbf{d}_n - \mathbf{H}_n \hat{\boldsymbol{\alpha}}_{n-1} \tag{15}$$

where $\mathbf{d}_n = \left[d_n,...,d_{n-p+1}\right]^T$ is the observations vector, $\mathbf{e}_n = \left[e_n,...,e_{n-p+1}\right]^T$ is the output error, *p* is the order of the algorithm and $\mathbf{H}_n$ denotes the *p*-by-*m* matrix whose $(i,j)^{th}$ entry is $\mathbf{H}_n = \left[k(\mathbf{u}_{n-i+1}, \mathbf{u}_{\omega_j})\right]_{\substack{i=1,...,p \\ j=1,...,m}}$.

The update equations for $\hat{\alpha}_n$ are obtained from the affine projection problem at time step *n* (Richard et al., 2009)

$$\min_{\alpha} || \boldsymbol{\alpha} - \hat{\boldsymbol{\alpha}}_{n-1} ||^2 \qquad \text{subject to } \mathbf{d}_n = \mathbf{H}_n \boldsymbol{\alpha} \tag{16}$$

If $\max_{j=1,...,m} | k(\mathbf{u}_n, \mathbf{u}_{\omega_j}) | > \mu_0$, $k(\bullet,\mathbf{u}_n)$ can be well represented by the kernel functions of the dictionary and the solution to Equation (16) can be determined by minimizing the Lagrangian function (Richard, 2009)

$$J(\boldsymbol{\alpha},\boldsymbol{\lambda}) = || \boldsymbol{\alpha} - \hat{\boldsymbol{\alpha}}_{n-1} ||^2 + \boldsymbol{\lambda}^t(\mathbf{d}_n - \mathbf{H}_n \boldsymbol{\alpha}) \tag{17}$$

where $\boldsymbol{\lambda}$ is the vector of Langrange multipliers.

The recursive update equation for $\hat{\alpha}_n$ is (Richard, 2009)

$$\hat{\alpha}_n = \hat{\alpha}_{n-1} + \eta \mathbf{H}_n^t \left(\varepsilon \mathbf{I} + \mathbf{H}_n \mathbf{H}_n^t\right)^{-1} \mathbf{e}_n \tag{18}$$

It is easy to notice now the similarity between Equation (5) and Equation (18). However, there are major differences between regarding the size and time evolution of the elements of the mentioned equations. Note that, in this case, the order of the filter and the dictionary remains the same.

On the other hand, if $\max_{j=1,\ldots,m} | k(\mathbf{u}_n, \mathbf{u}_{\omega_j}) | \leq \mu_0$ $k(\bullet, \mathbf{u}_n)$ is inserted into the dictionary where it is denoted by $k(\bullet, \mathbf{u}_{\omega_{m+1}})$ (Richard et al., 2009). This new element imposes to modify (16) as

$$\min_{\alpha} || \boldsymbol{\alpha}_{1,\ldots,m} - \hat{\boldsymbol{\alpha}}_{n-1} ||^2 + \alpha_{m+1}^2 \qquad \text{subject to } \mathbf{d}_n = \mathbf{H}_n \boldsymbol{\alpha} \tag{19}$$

One entry is added to the vector $\hat{\alpha}_n$ and the number of columns of $\mathbf{H}_n$ is increased by one by appending $[k(\mathbf{u}_n, \mathbf{u}_{\omega_{m+1}} \ldots k(\mathbf{u}_{n-p+1}, \mathbf{u}_{\omega_{m+1}})]^t$ (Richard et al., 2009).

If we define $\mathbf{e}_n = \mathbf{d}_n - \mathbf{H}_n \begin{bmatrix} \hat{\alpha}_{n-1} \\ 0 \end{bmatrix}$ then the update equation is the following

$$\hat{\alpha}_n = \begin{bmatrix} \hat{\alpha}_{n-1} \\ 0 \end{bmatrix} + \eta \mathbf{H}_n^t \left(\varepsilon \mathbf{I} + \mathbf{H}_n \mathbf{H}_n^t\right)^{-1} \mathbf{e}_n \tag{20}$$

Therefore the KAPA equations are shown in Table 1 (Richard et al., 2009).

Table 1. The KAPA with coherence criterion

Equations
Initialization: • Fix the memory length *p*, the step-size η, the regularization parameter ε, $m = 1$, $\hat{\alpha}_p = 0$ • Insert $k(\bullet, \mathbf{u}_p)$ into the dictionary, denote it by $k(\bullet, \mathbf{u}_{\omega_1})$. $\mathbf{H}_p = \left[k(\mathbf{u}_p, \mathbf{u}_{\omega_1})..k(\mathbf{u}_1, \mathbf{u}_{\omega_1})\right]^t$
for $n > p$: • Get $\left(\mathbf{u}_n, d_n\right)$ • If $\max_{j=1,\ldots,m} | k(u_n, u_{\omega_j}) | > \mu_0$, compute $\mathbf{H}_n$ and $\hat{\alpha}_n$ using equation (18) • If $\max_{j=1,\ldots,m} | k(u_n, u_{\omega_j}) | \leq \mu_0$, $m = m + 1$, insert $k\left(\bullet, \mathbf{u}_n\right)$ into the dictionary, denote it by $k\left(\bullet, \mathbf{u}_{\omega_m}\right)$, calculate $\hat{\alpha}_n$ using equation (20)

If *p* is set to one, the KNLMS algorithm is obtained (Richard et al., 2009).

KERNEL PROPORTIONATE AFFINE PROJECTION ALGORITHM

We propose a novel algorithm called kernel proportionate AP algorithm (KPAPA). It uses the proportionate coefficients based on the $\hat{\alpha}_n$ coefficients in a similar way to Equation (11):

$$c_{n-1}^{l} = \frac{1-\beta}{2m} + \frac{\left|\hat{\alpha}_{n-1}^{l}\right|(1+\beta)}{2\sum_{i=0}^{m-1}\left|\hat{\alpha}_{n-1}^{i}\right| + \xi}, \tag{21}$$

where $\hat{\alpha}_{n-1}^{l}$ are the coefficients of $\hat{\alpha}_n$. The matrix $\mathbf{C}_{n-1}$ is formed as an $m \times m$ diagonal matrix with null coefficients except the diagonal terms given by c_{n-1}^{l}, i.e. $\mathbf{C}_{n-1} = diag\left\{c_{n-1}^{0},\ldots,c_{n-1}^{m-1}\right\}$. The updating of the equation is

$$\hat{\alpha}_n = \hat{\alpha}_{n-1} + \eta\mathbf{C}_{n-1}\mathbf{H}_n^{t}\left(\varepsilon\mathbf{I} + \mathbf{H}_n\mathbf{C}_{n-1}\mathbf{H}_n^{t}\right)^{-1}\mathbf{e}_n \tag{22}$$

where η is the normalized step-size parameter in the range $0 < \eta < 2$. The value of the threshold μ_0 (Richard et al., 2009) is determined according to the sparseness of the signal, and ranges between 0 and 1. For the order increase case the updating is given by

$$\hat{\alpha}_n = \begin{bmatrix}\hat{\alpha}_{n-1}\\ 0\end{bmatrix} + \eta\mathbf{C}_{n-1}\mathbf{H}_n^{t}\left(\varepsilon\mathbf{I} + \mathbf{H}_n\mathbf{C}_{n-1}\mathbf{H}_n^{t}\right)^{-1}\mathbf{e}_n \tag{23}$$

If p is set to one, a new proportionate KNLMS algorithm is obtained. A similar variant using the proportionate coefficients $\mathbf{h}_{n-1} = [k(\mathbf{u}_{n-1},\mathbf{u}_{\omega_j})]_{j=1,\ldots,m}$ was also proposed in (Albu & Nishikawa, 2013).

The proportionate coefficients are computed as follows:

Table 2. The KPAPA with coherence criterion

Equations
Initialization: • Fix the memory length p, the step-size η, the regularization parameter ε, $m = 1$, $\hat{\alpha}_p = 0$ • Insert $k(\bullet,\mathbf{u}_p)$ into the dictionary, denote it by $k(\bullet,\mathbf{u}_{\omega_1})$. $\mathbf{H}_p = \left[k(\mathbf{u}_p,\mathbf{u}_{\omega_1})..k(\mathbf{u}_1,\mathbf{u}_{\omega_1})\right]^t$
for $n > p$: • Get $(\mathbf{u}_n, d_n)$ • Compute c_{n-1}^{l} using equation (21) and form the matrix $\mathbf{C}_{n-1} = diag\left\{c_{n-1}^{0},\ldots,c_{n-1}^{m-1}\right\}$ • If $\max_{j=1,\ldots,m} \mid k(u_n,u_{\omega_j}) \mid > \mu_0$, compute $\mathbf{H}_n$ and $\hat{\alpha}_n$ using equation (22) • If $\max_{j=1,\ldots,m} \mid k(u_n,u_{\omega_j}) \mid \le \mu_0$, $m = m + 1$, insert $k(\bullet,\mathbf{u}_n)$ into the dictionary, denote it by $k(\bullet,\mathbf{u}_{\omega_m})$, calculate $\hat{\alpha}_n$ using equation (23)

$$c_{n-1}^{l} = \frac{1-\beta}{2m} + \frac{\left|h_{n-1}^{l}\right|\left(1+\beta\right)}{2\sum_{i=0}^{m-1}\left|h_{n-1}^{i}\right| + \xi}, \tag{24}$$

Therefore the KPAPA equations are shown in Table 2.

The KPAPA requires additional $2m$ multiplications, $2m$ additions and one division if compared with the KAPA algorithm. This complexity increase is small if compared with $O\left(mp^2\right)$ of KAPA (Richard et al., 2009). Also, it can be easily seen that this additional complexity is dependent only on the order of kernel expansion. It is the same for the identical kernel expansion order regardless on the order of the algorithm.

Other KAPA versions were obtained in (Liu et al., 2010) by following a different approach starting from the linear filter theory. The proportionality principle can be applied as well to the algorithms employing a matrix inversion using the same principle as above.

DCD BASED VERSIONS FOR FAST IMPLEMENTATION

It can be seen in Table 1 that the KAPA involves a matrix inversion operation. This is a numerically expensive operation, especially for a high algorithm order. If we note that

$$\mathbf{s}_n = \left(\varepsilon \mathbf{I} + \mathbf{H}_n \mathbf{H}_n^t\right)^{-1} \mathbf{e}_n \tag{25}$$

then the filter coefficients of KAPA are updated as follows with no order increase

$$\hat{\alpha}_n = \hat{\alpha}_{n-1} + \eta \mathbf{H}_n^t \mathbf{s}_n \tag{26}$$

or with order increase

$$\hat{\alpha}_n = \begin{bmatrix} \hat{\alpha}_{n-1} \\ 0 \end{bmatrix} + \eta \mathbf{H}_n^t \mathbf{s}_n \tag{27}$$

In the case of a direct method for solving $\hat{\mathbf{s}}_n = \mathbf{Q}_n^{-1}\mathbf{e}_n$, where $\mathbf{Q}_n = \varepsilon \mathbf{I} + \mathbf{H}_n \mathbf{H}_n^t$ the numerical complexity is proportional with p^3 (Golub & Loan, 1996).

For implementing KAPA in applications, we should reduce its computational complexity. One possibility is to replace the matrix inversion operation with an efficient implementation of a linear system of equation solving method.

A division-less and multiplier-less solution, called dichotomous coordinate descent (DCD) was published in 2004 (Zakharov & Tozer, 2004). The solution of a linear system of equations is updated in directions of Euclidian coordinates in the cyclic order and with a step size that takes one of (number of bits) predefined values corresponding to a binary representation bounded by an interval $[-S, S]$.

Table 3. The cyclic DCD algorithm and its numerical complexity in terms of additions

Step	Equations	+
	Initialization: $\hat{\mathbf{s}}_n = \mathbf{0}, \mathbf{r}_n = \mathbf{e}_n, q = 0, \sigma = S$	
	$for\ i = 1,2,...M_b$	
1	$\sigma = \sigma / 2$	
2	Flag=0	
	$for\ j = 1,2,...p$	
3	$if\ \lvert r_j \rvert > (\sigma / 2)Q_{j,j}$	1
4	$\hat{s}_j = \hat{s}_j + sign(r_j)\sigma$	1
5	$\mathbf{r} = \mathbf{r} - sign(r_j)\sigma\mathbf{Q}^{(j)}$	*p*
6	*q*=*q*+1, Flag=1	
7	*if* $q > N_u$, the algorithm *stops*	
8	*if* Flag = 1, *then repeat* step 2	
Total:	$\leq p(2N_u + M_b - 1) + N_u$ additions	

The algorithm starts the iterative search from the most significant bits of the solution and continues until the least significant bits were updated. The algorithm complexity is limited by *Nu*, the maximum number of "successful" iterations.

This version (shown in Table 3) is called the cyclic DCD (Zakharov et al., 2008).

The DCD algorithm introduces the following parameters: σ, M_b and N_u. The selection of σ is made as follow: For "unsuccessful" iterations it is 0 and for "successful" iterations it is a power of two. M_b can be considered the number of bits used for fixed point representation of the solution.

The complexity is given by the number of successful updates and a limit to the number of updates N_u is imposed.

All the multiplications and divisions are replaced by bit shifts and the upper complexity of the cyclic DCD is $p\ (2N_u + M_b - 1) + N_u$ additions in the worst case scenario (Zakharov et al., 2008).

It can be seen from Table 3 that if $N_u \gg M_b$ the maximum complexity is about $2pN_u$ additions. However, if $N_u < M_b / 2$, the dominant complexity term is pM_b. The leading element DCD algorithm (Zakharov et al., 2008) can eliminate this term. The Leading Element Dichotomous Coordinate Descent (LEDCD) algorithm equations are shown in Table 4.

This version finds a 'leading' (t^{th}) element of the solution to be updated. It is preferable to choose *S* higher than $S_{\max}$, the absolute value of the maximum value of the solution.

If $S > S_{\max}$ is chosen, the performances are not affected, only M_b values would increase with $\log_2(S / S_{\max})$ (Zakharov et al., 2008). However, this corresponds to an insignificant complexity increase of the algorithm.

If $S < S_{\max}$ is chosen, the performance of an adaptive algorithm using DCD could be reduced a bit in the initial convergence stage (Zakharov et al., 2008).

Table 4. The leading element DCD (LEDCD) and its numerical complexity in terms of additions

Step	Equation	+
	Initialization: $\hat{\mathbf{s}}_n = \mathbf{0}, \mathbf{r}_n = \mathbf{e}_n, q = 1, \sigma = S/2$	
	$for\ j = 1,2,...p$	
1	$t = \arg\max_{i=1:p}\{\lvert r_i \rvert\}$	$p-1$
2	$while\ \lvert r_t \rvert \le \frac{\sigma}{2}[\mathbf{Q}_j]_{(t,t)}\ and\ i \le M_b\ do$ $q = q+1, \sigma = \sigma/2$	1
3	$if\ q > M_b$ then STOP	
4	$\hat{s}_t = \hat{s}_t + sign(r_t)\sigma$	1
5	$\mathbf{r} = \mathbf{r} - sign(r_t)\sigma[\mathbf{Q}]_{(t,:)}$	p
Total:	$\le 2N_u p + M_b$ additions	

If S is chosen as a power of two, all multiplications with σ can be implemented with bit shifts, therefore simplifying a lot the hardware implementation. Each update requires p bit shifts, p additions and p comparisons. If the comparions are counted as additions the maximum LEDCD complexity is $2pN_u + M_b$ additions (Zakharov et al., 2008). It is obvious that there is an important complexity reduction if compared with the direct method reduction especially if the number of updates is small.

The computation of the matrix $\mathbf{Q}_n$ can be made in an efficient way as in (Albu & Nishikawa, 2013) and (Albu, Paleologu, & Benesty, 2011). There are two possible situations. In the case without order increase, the matrix $\mathbf{Q}_n$ is updated by replacing only the first row and column, while the bottom $(p-1)\times p$ sub-matrix is replaced with the top $(p-1)\times p$ sub-matrix of $\mathbf{Q}_{n-1}$. For the first p iterations the full $\mathbf{Q}_{n-1}$ is used. In the second situation, when there is an order increase, the matrix $\mathbf{Q}_n$ is updated by replacing the first row and column with a modified vector. The last $p-1$ elements of the last column of $\mathbf{H}_n$ are taken into account in order to update $\mathbf{Q}_n$ (Albu et al., 2013).

It was shown that the implementation of KAPA with cyclic DCD iterations led to important computational savings depending on the N_u value. The number of addition-operations for KAPA without order increase is $(p^2+2p)m + p^3 + p^2$ while it is $(p^2+2p)m + p^3 + p^2 + p - 1$ in case of an order increase. The number of addition-operations of the efficiently DCD based DCD-KAPA (Albu et al. 2013) is $3pm + p(2N_u + M_b + 1)$ without order increase and $3pm + p^2 + 1 + p(2N_u + M_b)$ with an order increase. The number of multiplication-operations for KAPA without order increase is $(p^2+2p)m + p^3 + p$ while it is $(p^2+2p)m + p^3 + 2p^2 + p$ in case of an order increase.

The number of multiplication-operations of the efficiently DCD based DCD-KAPA (Albu et al. 2013) is $3pm + p$ without order increase and $3pm + p^2 + 2p$ with an order increase. The savings were substantial in many cases (e.g. around 50% in the cases investigated in (Albu et al., 2013). As expected, a lower N_u value led to higher complexity savings.

The leading element DCD can be used instead of the cyclic DCD. A smaller overall complexity could be expected, but a decrease in performance is possible as well. The DCD algorithms can be used in all the kernel affine projection or kernel proportionate affine projection algorithms where there is a matrix inversion operation. The matrix inversion operation is replaced by a linear system solved with one of the DCD variants.

For example, in case of KPAPA, if we denote by

$$\mathbf{s}_n = \left(\varepsilon\mathbf{I} + \mathbf{H}_n\mathbf{C}_{n-1}\mathbf{H}_n^t\right)^{-1}\mathbf{e}_n \tag{28}$$

then the filter coefficients of KPAPA are updated as follows with no order increase

$$\hat{\alpha}_n = \hat{\alpha}_{n-1} + \eta\mathbf{C}_{n-1}\mathbf{H}_n^t\mathbf{s}_n \tag{29}$$

or with order increase

$$\hat{\alpha}_n = \begin{bmatrix}\hat{\alpha}_{n-1}\\ 0\end{bmatrix} + \eta\mathbf{C}_{n-1}\mathbf{H}_n^t\mathbf{s}_n \tag{30}$$

It is expected that more DCD iterations would lead to a closer performance to that of an ideal system solution. However, a compromise of the performance versus overall complexity has to be made. The number of bits is less important than the number of updates. Typically, M_b can have values 16 or 24 for most applications (Albu et al., 2013), (Zakharov & Albu, 2005).

ALGORITHMS PARAMETER INFLUENCE

An important parameter of KAPA and KPAPA is the coherence value μ_0. It is useful to find the most suitable value for each specific application. The steady-state performance of the algorithm is measured by the mean-square prediction error on the steady-state region and averaged on many independent runs. The value of μ_0 is varied from 0 to 1 and the best coherence value corresponds to its minimum. It was shown in (Richard et al., 2009) and (Albu et al., 2013) that this value depends on the application and the filter length.

Another important parameter that influences the convergence behavior of these algorithms is the step size. It was shown in (Albu & Nishikawa, 2013) that a smaller step size leads to a slower convergence speed of the KPNLMS algorithm. These findings were similar to those obtained for the linear adaptive algorithms (Sayed, 2003). In the Simulation section the influence of the above mentioned parameters is exemplified for KAPA and KPAPA with coherence criterion.

There are many alternatives to reducing the dimension of the dictionary or keeping it at a fixed size. A simple criterion for pruning the dictionary was proposed in (Suykens et al., 2002). The input vector with the minimum weight of $\hat{\alpha}_n$ is removed. This pruning method does not minimize error (Kruif &Vries, 2003). To maintain the minimal pruning error, a better is the following (Kruif &Vries, 2003)

$$\min_n \frac{|\alpha_n|}{[\mathbf{K}^{-1}]_{nn}} \tag{31}$$

where $[\mathbf{K}^{-1}]_{nn}$ is n-th diagonal coefficient of kernel matrix inverse (Van Vaerenbergh, Santamaria, Liu & Principe, 2010) whose elements are the inner products of dictionary vectors (Repzka, 2012) .

Similar techniques were also proposed in (Nishikawa, & Makizaki, 2011) and (Nishikawa, Ogawa, & Albu, 2013). Other techniques for curbing the growth of the kernel matrix are the ALD criterion (Engel et al., 2004), sliding window techniques (Van Vaerenbergh et al., 2010), surprise information measure (Liu et al., 2010) etc.

SIMULATIONS

In this section, we compare the convergence characteristics of KAPA and KPAPA, examine the influence of the parameters of the algorithms and prove the advantages of the DCD based versions.

In the following simulations, we consider the nonlinear system described by

$$d_n = \left(0.8 - 0.4\exp\left(-d_{n-1}^2\right)\right)d_{n-1} - \left(0.3 + 0.8\exp\left(-d_{n-1}^2\right)\right)d_{n-2} + 0.05\sin\left(d_{n-1}\pi\right) \tag{32}$$

where d_n is the desired signal. This condition is similar with that of (Richard et al., 2009). The data was generated by iterating the above equation from the initial condition (0.1, 0.1) and d_n were corrupted by a measurement noise sampled from a zero-mean Gaussian distribution with standard deviation equal to 0.1 (Richard et al., 2009). The Gaussian kernel $k\left(\mathbf{u}_i, \mathbf{u}_j\right) = \exp\left(-3.73 \cdot \|\mathbf{u}_i - \mathbf{u}_j\|^2\right)$, $\xi = 10^{-8}$ and the regularization parameter $\varepsilon = 0.07$ were used.

Figure 3a shows the MSE difference in dB between KAPA and KPAPA for $p = 30$, $\mu_0 = 0.6$ and $\beta = -0.9$. It can be noticed that most of the values are positive. Therefore, the MSE performance of KPAPA is superior by about 1 dB average to that of KAPA for most iterations.

The influence of the coherence value μ_0 is examined in Figure 3b. Steady-state performance of KPAPA was measured by the mean-square prediction error over the last 500 samples and averaged over 100 independent runs:

$$\text{Mean MSE} = \frac{1}{100}\sum_{i=1}^{100} \frac{\sum_{n=2501}^{300} (d_n - \psi_{n-1}(\mathbf{u}_n))^2}{\sum_{n=2501}^{300} d_n^2} \tag{33}$$

Figure 3. a) MSE difference of learning curves of KAPA and KPAPA; b) mean MSE of KPAPA for the last 500 iterations for different μ_0 *values*

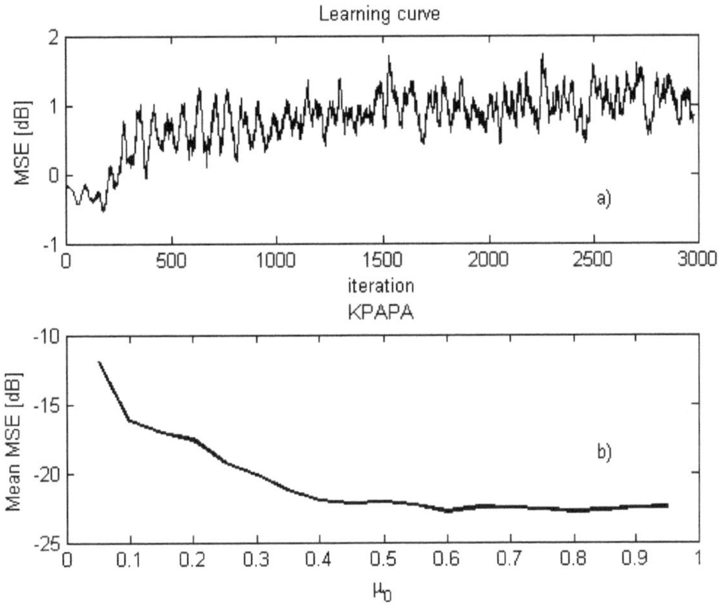

Figure 4. a) Learning curves of KAPA with different μ_0 *values ; b) learning curves of KAPA with different* η *values*

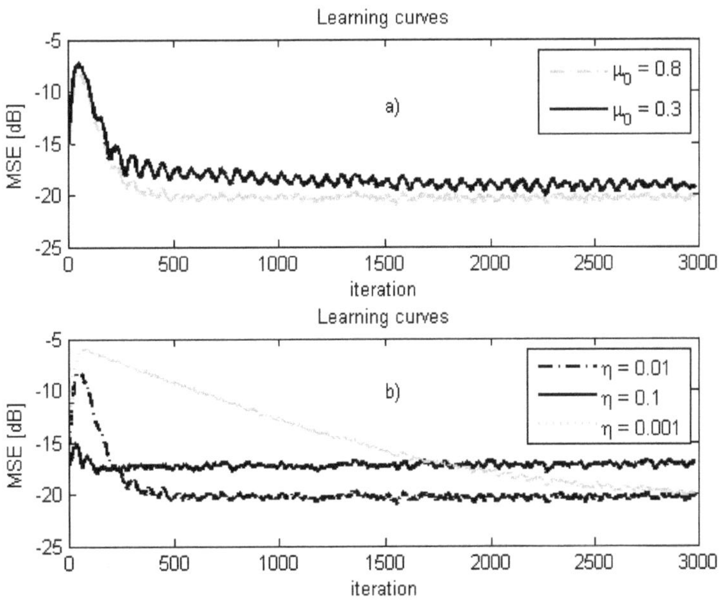

Figure 5. a) MSE difference of learning curves of KAPA and DCD-KAPA with different number of updates; b) complexity ratios of DCD-KAPA and KAPA in terms of multiplications and additions and different number of updates - without order increase (solid line), with order increase (dotted line)

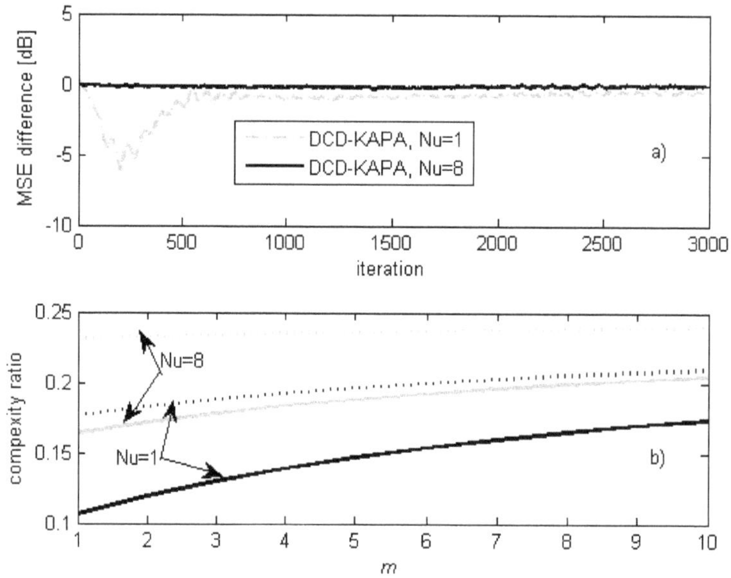

The dictionary was initialized with $k(\bullet, u_1)$. The value of μ_0 was varied from 0.05 to 0.95 in increments of 0.05. It can be seen from Figure 3b that the best coherence value that provides the minimum mean MSE is 0.6.

Figure 4 shows the effect on the learning curves of KAPA for different μ_0 and η value and $p = 10$. Figure 4a shows that KAPA with a coherence value of 0.8 has improved steady-state characteristics relative to KAPA with a coherence value of 0.3.

Figure 4b compares the convergence characteristics of KAPA for three η values. It can be seen that KAPA with a very small step size, $\eta = 0.001$ has the lowest convergence speed. The highest MSE steady-state value is obtained for $\eta = 0.1$, while the best compromise between initial convergence speed and low MSE steady-state value is obtained for $\eta = 0.01$. Therefore, suitable parameters have to be found for the investigated algorithm in order to obtain the best performance for a particular application.

The computational complexity of KAPA using DCD iterations and its effect on convergence characteristics is shown in Figure 5. It can be seen from Figure 5a that DCD-KAPA with 8 DCD iterations obtains a very close performance to KAPA because the MSE difference from KAPA is very close to zero. As expected increasing the number of updates from 1 to 8 leads to an improved approximation. Figure 5b shows the ratios of numerical complexities of DCD-KAPA and KAPA for different N_u values, $S = 4$, $p = 10$ and $M_b = 16$. It can be seen that the savings depend on the N_u value and if there is or is not an order increase. The savings are substantial (e.g. more than 76% for $N_u = 8$). Also, it can be noticed that a lower N_u value leads to higher complexity savings.

DISCUSSION ABOUT THE SIMULATION RESULTS

The simulations above prove that the performance of KPAPA can be superior to that of KAPA with convergence criterion. Like for the linear case (Hoshuyama et. al., 2004) the size of the improvement is dependent on the sparsity of the solution. In case of very dispersive solution, no MSE improvement is expected. This behavior is similar to that of the proportionate affine projection filter (Hoshuyama et. al., 2004).

The use of variable step size can increase a lot the convergence speed of the kernel affine projection or kernel proportionate affine projection algorithms. Also, a lower MSE misalignment can be obtained. The approaches used for linear algorithms can be cautiously applied for the nonlinear versions. The implicit nonlinearity increases the challenge of developing a theoretical analysis of the MSE that could indicate how to adjust the step size values.

The DCD iterations can bring important computational complexity reduction if used in other implementations of the KAPA/KPAPA family of algorithms. However, this complexity reduction depends on the order the algorithm. If p has a value of 2 or 3, it might be not worth replacing the inverse operation with DCD iterations. In these cases it is easier to find the inverse of the matrix. For higher p values the DCD can reduce the overall complexity a lot as shown in the example above. However, a compromise has to be made between the overall complexity reduction and the performance losses. The DCD method is particularly useful if a hardware solution is envisaged.

To conclude, in recent years, there are numerous publications on kernel adaptive filtering algorithms. Thus, we envisage their use in other domains, together or not with the classical adaptive filtering algorithms. The techniques described above can represent the base for developing new kernel algorithms suited to particular applications.

CONCLUSION AND FUTURE RESEARCH DIRECTIONS

In this chapter, several nonlinear kernel based versions of the adaptive algorithms using a coherence criterion were presented. The kernel affine projection algorithm (KAPA) with coherence criterion was presented. The novel kernel proportionate affine projection algorithm (KPAPA) was introduced using the same principle that was used to derive PAPA from APA. In the same way, the proportionate versions of KNLMS and KPNLMS were obtained as a particularization of KAPA and KPAPA. The influence of several parameters of these algorithms has been discussed. Their importance and significance for improving the overall performance was emphasized. Several possibilities of reducing the numerical complexity of these algorithms have been presented. One of the most promising techniques is the use of the DCD method in order to avoid the matrix inversion operation in the investigated algorithms. Simulations have proved the effectiveness of the proposed approaches for a nonlinear system identification application.

Our future work will be focused on further extending the techniques used for the linear adaptive algorithms into the nonlinear domain and design new kernel based algorithms. Developing variable step, variable regularization kernel based versions of APA and PAPA is also envisaged. Another direction of research is to use sparsification methods that can make kernel methods better suited for online learning. It is expected that the use of DCD iterations would lead to a reduction of the numerical complexity of numerous kernel based algorithms which use matrix inversion operations.

REFERENCES

Aboulnasr, T., & Mayyas, K. (1997). A robust variable step-size LMS-type algorithm analysis and simulations. *IEEE Transactions on Signal Processing*, *45*(3), 631–639. doi:10.1109/78.558478

Albu, F. (2011, August). A Proportionate Affine Projection Algorithm using Fast Recursive Filtering and Dichotomous Coordinate Descent Iterations. *Proceedings of SPAMEC, 2011*, 93–96.

Albu, F. (2012, April). Simplified proportionate affine projection algorithms. In *Proceedings of IWSSIP*, (pp. 382-385). IWSSIP.

Albu, F., Bouchard, M., & Zakharov, Y. (2007). Pseudo Affine Projection Algorithms for Multichannel Active Noise Control. *Audio. IEEE Transaction on Speech and Language Processing*, *15*(3), 1044–1052. doi:10.1109/TASL.2006.881677

Albu, F., & Nishikawa, K. (September 2013). The Kernel Proportionate NLMS Algorithm. In *Proceedings of EUSIPCO 2013* (no. TU-P6.1). EUSIPCO.

Albu, F., Nishikawa, K., Rotaru, M., & Coltuc, D. (September 2013). An efficient implementation of the kernel affine projection algorithm. In *Proceedings of IEEE ISPA* (pp. 342-346). IEEE. doi:10.1109/ISPA.2013.6703765

Albu, F., Paleologu, C., & Benesty, J. (2011). A Variable Step Size Evolutionary Affine Projection algorithm. In *Proceedings of ICASSP 2011* (pp. 429-432). ICASSP.

Albu, F., Paleologu, C., Benesty, J., & Ciochina, S. (2010, August). A low complexity proportionate affine projection algorithm for echo cancellation. In *Proceedings of EUSIPCO*, (pp. 6–10). EUSIPCO.

Arablouei, R., Dogançay, K., & Perreau, S. (October 2010). Proportionate affine projection algorithm with selective projections for sparse system identification. In *Proceedings of APSIPA ASC 2010* (pp. 362). APSIPA.

Benesty, J., & Gay, S. L. (2002, May). An improved PNLMS algorithm. *Proceedings of ICASSP*, *2*, 1881–1884.

Deng, H., & Doroslovački, M. (2006). Proportionate adaptive algorithms for network echo cancellation. *IEEE Transactions on Signal Processing*, *54*(5), 1794–1803. doi:10.1109/TSP.2006.872533

Duttweiler, D. L. (2000). Proportionate normalized least-mean-squares adaptation in echo cancellers. *IEEE Transactions on Speech and Audio Processing*, *8*(5), 508–518. doi:10.1109/89.861368

Engel, Y., Mannor, S., & Meir, R. (2004). The Kernel Recursive Least-Squares Algorithm. *IEEE Transactions on Signal Processing*, *52*(8), 2275–2285. doi:10.1109/TSP.2004.830985

Golub, G. H., & Van Loan, C. F. (1996). *Matrix Computation*. Baltimore, MD: Johns Hopkins University Press.

Haweel T.I., (2004). A simple variable step size LMS adaptive algorithm. *International Journal of Circuit Theory and Applications*, (32), 523-536.

Hoshuyama, O., Goubran, R. A., & Sugiyama, A. (2004, May). A generalized proportionate variable step-size algorithm for fast changing acoustic environments. In *Proceedings of ICASSP*, (pp. 161–164). ICASSP.

Kim, S.-E., Kong, S.-J., & Song, W.-J. (2009). An affine projection algorithm with evolving order. *IEEE Signal Processing Letters*, *16*(11), 937–940. doi:10.1109/LSP.2009.2027638

Kruif, B., & Vries, T. (2003). Pruning error minimization in least squares vector support machines. *IEEE Transactions on Neural Networks*, *14*(3), 696–702.

Kwong, R. H., & Johnston, E. W. (1992). A variable step size LMS algorithm. *IEEE Transactions on Signal Processing*, *40*(7), 1633–1642. doi:10.1109/78.143435

Liu, L., Fukumoto, M., Saiki, S., & Zhang, S. (2009). A variable step-size proportionate Affine Projection Algorithm for identification of sparse impulse response. *EURASIP Journal on Advances in Signal Processing*, Article ID 150914.

Liu, W., Pokharel, P. P., & Principe, J. C. (2008). The Kernel Least-Mean-Square Algorithm. *IEEE Transactions on Signal Processing*, *56*(2), 543–554. doi:10.1109/TSP.2007.907881

Liu, W., Principe, J. C., & Haykin, S. (2010). *Kernel Adaptive Filtering*. Hoboken, NJ: Wiley. doi:10.1002/9780470608593

Müller, K.-R., Adali, T., Fukumizu, K., Principe, J. C., & Theodoridis, S. (2013, July). Special Issue on Advances in Kernel-Based Learning for Signal Processing. *IEEE Signal Processing Magazine*, *30*(4), 14–15. doi:10.1109/MSP.2013.2253031

Nishikawa, K., & Makizaki, K. (October 2011). Fixed Order Implementation Method of Kernel Adaptive Filters with Lower Computational Complexity. In *Proceedings of APSIPA ASC 2011*. APSIPA.

Nishikawa, K., Ogawa, Y., & Albu, F. (2013, October). Fixed Order Implementation of Kernel RLS-DCD Adaptive Filters. In *Proceedings of APSIPA ASC*, (pp. 1–6). APSIPA.

Ozeki, K., & Umeda, T. (1984). An adaptive filtering algorithm using an orthogonal projection to an affine subspace and its properties. *Electronic Communications Japan*, *67-A*(5), 19–27.

Paleologu, C., Benesty, J., & Albu, F. (2012). Regularization of the improved proportionate affine projection algorithms. In *Proceedings of ICASSP*, (pp. 169–172). ICASSP.

Richard, C., Bermudez, J. C. M., & Honeine, P. (2009). Online Prediction of Time Series Data with Kernels. *IEEE Transactions on Signal Processing*, *57*(3), 1058–1067. doi:10.1109/TSP.2008.2009895

Rzepka, D. (September 2012). Fixed-budget kernel least mean squares. In *Proceedings of IEEE 17th Conference on Emerging Technologies & Factory Automation* (pp. 1-4). IEEE.

Sayed, A. H. (2003). *Fundamentals of Adaptive Filtering*. John Wiley & Sons.

Shin, H. C., Sayed, A. H., & Song, W. J. (2004). Variable Step-Size NLMS and Affine Projection Algorithms. *IEEE Signal Processing Letters*, *11*(2), 132–135. doi:10.1109/LSP.2003.821722

Suykens, J., Brabanter, J. D., Lukas, L., & Vandewalle, J. (2002). Weighted least squares support vector machines: Robustness and sparse approximation. *Neurocomputing*, *48*(14), 85–105. doi:10.1016/S0925-2312(01)00644-0

Van Vaerenbergh, S., Santamaria, I., Liu, W., & Principe, J. (2010, March). Fixed-Budget Kernel Recursive Least-Squares. In *Proceedings of ICASSP*, (pp. 1882–1885). ICASSP.

Vapnik, V. N. (1995). *The Nature of Statistical Learning Theory*. New York: Springer-Verlag. doi:10.1007/978-1-4757-2440-0

Wang, S., Feng, J., & Tse, C. K. (2013). Kernal affine projection sign algorithms for combating impulse interference. *IEEE Transactions on Circuits and Systems II, Express Briefs*, *60*(11), 811–815. doi:10.1109/TCSII.2013.2281712

Yang, J., & Sobelman, G. E. (2010). Efficient μ -law improved proportionate affine projection algorithm for echo cancellation. *Electronics Letters*, *47*(2), 73–74. doi:10.1049/el.2010.7937 PMID:21079852

Yin, W., & Mehr, A. S. (2010). A variable regularization method for affine projection algorithm. *IEEE Transactions on Circuits and Systems II*, *57*(6), 476–480. doi:10.1109/TCSII.2010.2048355

Zakharov, Y., & Albu, F. (2005). Coordinate descent iterations in fast affine projection algorithm. *IEEE Signal Processing Letters*, *12*(5), 353–356. doi:10.1109/LSP.2005.843765

Zakharov, Y., & Tozer, T. (2004). Multiplication-free iterative algorithm for LS problem. *Electronics Letters*, *40*(9), 567–569. doi:10.1049/el:20040353

Zakharov, Y., White, G., & Liu, J. (2008). Low complexity RLS algorithms using dichotomous coordinate descent iterations. *IEEE Transactions on Signal Processing*, *56*(7), 3150–3161. doi:10.1109/TSP.2008.917874

ADDITIONAL READING

An, S., Liu, W., & Venkatesh, S. (2007). Fast cross - validation algorithms for least squares support vector machine and kernel ridge regression. *Pattern Recognition*, *40*(8), 2154–2162. doi:10.1016/j.patcog.2006.12.015

Bordes, A., Ertekin, S., Weston, J., & Bottou, L. (2005). Fast kernel classifiers with online and active learning. *Journal of Machine Learning Research*, *6*, 1579–1619.

Chen, B., Zhao, S., Zhu, P., & Principe, J. C. (2012). Quantized Kernel Least Mean Square Algorithm. *Neural Networks and Learning Systems IEEE Transactions on*, *23*(1), 22–32. doi:10.1109/TNNLS.2011.2178446 PMID:24808453

Cristianini, N., & Taylor, J. S. (2000). *An Introduction to Support Vector Machines and Other Kernel - based Learning Methods*. Cambridge, MA: Cambridge University Press. doi:10.1017/CBO9780511801389

Cruz, P. F., Van Vaerenbergh, S., Murillo-Fuentes, J. J., Lazaro-Gredilla, M., & Santamaria, I. (2013). Gaussian processes for nonlinear signal processing: An overview of recent advances. *IEEE Signal Processing Magazine*, *30*(4), 40–50. doi:10.1109/MSP.2013.2250352

Diniz, P. S. R. (2008). *Adaptive Filtering: Algorithms and Practical Implementation* (3rd ed.). Springer Publishers. doi:10.1007/978-0-387-68606-6

Fan, H., & Song, Q. (2013). A sparse kernel algorithm for online time series data prediction. *Expert Systems with Applications*, *40*(6), 2174–2181. doi:10.1016/j.eswa.2012.10.046

Gao, W., Chen, J., Richard, C., Huang, J., & Flamary, R. (2013, May). Kernel LMS algorithm with forward-backward splitting for dictionary learning. *Proceedings of ICASSP*, *2013*, 5735–5739.

Gonzales, A., Albu, F., Ferrer, M., & Diego, M. (2013). Evolutionary and variable step size affine projection algorithms for active noise control. *IET Signal Processing*, *7*(6), 471–476. doi:10.1049/iet-spr.2012.0213

Gonzalez, A., Ferrer, M., Albu, F., & de Diego, M. (2012, August). Affine projection algorithms: Evolution to smart and fast multichannel algorithms and applications. *Proceedings of Eusipco*, *2012*, 1965–1969.

Haykin, S. (1996). *Adaptive Filter Theory* (3rd ed.). Upper Saddle River, NJ: Prentice Hall.

Haykin, S. (1998). Neural Networks: A Comprehensive Foundation, 2nd edition. Upper Saddle, River, NJ, Prentice Hall.

Haykin, S. (2009). *Neural Networks and Learning Machines* (3rd ed.). Upper Saddle River, NJ: Prentice Hall.

Herbrich, R. (2002). *Learning Kernel Classifi ers: Theory and Algorithms*. Cambridge, MA: MIT Press.

Hoegaerts, L., Suykens, J. A. K., Vandewalle, J., & De Moor, B. (2004). A comparison of pruning algorithms for sparse least squares support vector machines. *Lecture Notes in Computer Science*, *3316*, 1247–1253. doi:10.1007/978-3-540-30499-9_194

Kivinen, J., Smola, A., & Williamson, C. (2004). Online Learning with Kernels. *Signal Processing IEEE Transactions on*, *52*(8), 2165–2176. doi:10.1109/TSP.2004.830991

Leung, C., & Tsoi, A. (2006). Combined learning and pruning for recurrent radial basis function networks based on recursive least square algorithms. *Neural Computing & Applications*, *15*(1), 62–78. doi:10.1007/s00521-005-0009-7

Liu, W., Park, I., Wang, Y., & Principe, J. C. (2009). Extended kernel recursive least squares algorithm. *Signal Processing IEEE Transactions on*, *57*(10), 3801–3814. doi:10.1109/TSP.2009.2022007

Liu, W., & Principe, J. C. (2008). Kernel Affine Projection Algorithms. *Advances in Signal Processing EURASIP Journal on*, Article ID 784292, 12 pages.

Malipatil, V., Huang, Y.-F., Andra, S., & Bennett, K. (2005, March). Kernelized set-membership approach to nonlinear adaptive filtering. *Proceedings of ICASSP*, *2005*, 149–152.

Micchelli, C., & Pontil, M. (2005). Learning the kernel function via regularization. *Journal of Machine Learning Research, 6*, 1099–1125.

Nishikawa, K., & Nakazato, H. (December 2012). Mixture structure of kernel adaptive filters for improving the convergence characteristics. *Proceedings of APSIPA ASC* 2012.

Ogawa, Y., & Nishikawa, K. (June 2011). A Kernel Adaptive Filter based on ERLS-DCD Algorithm. *Proceedings of International Technology Conference on. Circuits, Systems, Computer, Communications 2011*, no.P4-13, (pp.1228–1231).

Ong, C. S., Smola, A. J., & Williamson, R. C. (2005). Learning the kernel with hyperkernels. *Journal of Machine Learning Research, 6*, 1043–1071.

Schölkopf, B., Herbrich, R., & Smola, A. (2001). Lecture notes in computer science: Vol. 2111. *A generalized representer theorem. Computational learning theory* (pp. 416–426). Springer. doi:10.1007/3-540-44581-1_27

Schölkopf, B., & Smola, A. (2001). *Learning with Kernels*. Cambridge, MA: MIT Press.

Singh, A., Ahuja, N., & Moulin, P. (September 2012) Online learning with kernels: Overcoming the growing sum problem. *Proceedings of International Workshop on Machine Learning for Signal Processing*. doi:10.1109/MLSP.2012.6349811

Slavakis, K., Theodoridis, S., & Yamada, I. (2008). Online kernel-based classification using adaptive projection algorithms. *Signal Processing IEEE Transactions on, 56*(7), 2781–2796. doi:10.1109/TSP.2008.917376

Takizawa, M., & Yukawa, M. (2013, May). An Efficient Data-Reusing Kernel Adaptive Filtering Algorithm Based on Parallel Hyperslab Projection Along Affine Subspace. *Proceedings of ICASSP, 2013*, 3557–3561.

Taylor, J. S., & Cristianini, N. (2004). *Kernel Methods for Pattern Analysis*. Cambridge, UK: Cambridge University Press. doi:10.1017/CBO9780511809682

Van Vaerenbergh, S., Lazaro-Gredilla, M., & Santamaria, I. (2012). Kernel Recursive Least-Squares Tracker for Time-Varying Regression. *Neural Networks and Learning Systems IEEE Transactions on, 23*(8), 1313–1326. doi:10.1109/TNNLS.2012.2200500 PMID:24807527

Van Vaerenbergh, S., Santamaria, I., & Lazaro-Gredilla, M. (September 2012). Estimation of the forgetting factor in kernel recursive least squares. *Proceedings of International Workshop on Machine Learning for Signal Processing*. doi:10.1109/MLSP.2012.6349749

Van Vaerenbergh, S., Via, J., & Santamaria, I. (2006, May). A sliding-window kernel RLS algorithm and its application to nonlinear channel identification. *Proceedings of ICASSP, 5*, 789–792.

Yukawa, M. (2012). Multikernel Adaptive Filtering. *Signal Processing IEEE Transactions on, 60*(9), 4672–4682. doi:10.1109/TSP.2012.2200889

KEY TERMS AND DEFINITIONS

Adaptive Filters: Systems using linear filters whose coefficients adapts according to an optimization algorithm.

Affine Projection Algorithm: It is a generalization of the NLMS algorithms where the projections are made in multiple dimensions.

Dichotomous Coordinate Descend: A multiplier-less and division-less coordinate descent method useful for finding the optimal solution of the normal equations.

Kernel Adaptive Filters: A nonlinear adaptive filter that uses kernel methods.

Kernel Function: A kernel function defines the distribution of similarities of points around a given point.

Least Mean Square Algorithms: A stochastic gradient descent family of adaptive algorithms that tries to minimize the mean squares of the difference between the desired and the actual signal.

Normalized Least Mean Square Algorithm: A variant of the LMS algorithm whose learning rate is normalized with the power of the input signal.

Proportionate Adaptive Algorithms: Variants of the adaptive algorithms whose coefficients are updated proportionally to the sparsity of the system.

Chapter 3
Advanced Vibration Control of Atomic Force Microscope Scanner

Sajal K. Das
University of New South Wales, Australia

Hemanshu R. Pota
University of New South Wales, Australia

Ian R. Petersen
University of New South Wales, Australia

ABSTRACT

Atomic Force Microscopes (AFMs) are used in many nanopositioning applications in order to measure the topography of various specimens at an atomic level through surface imaging. The imaging of the samples in AFMs is carried out by using a three degree-of-freedom positioning unit called Piezoelectric Tube Scanner (PTS). The performance of the AFM mostly depends on the performance of the PTS. However, the PTS of the AFM suffers from the problem of vibration. This chapter presents a design of a damping controller to compensate the induced vibration of the scanner. Experimental results are presented to show the effectiveness of the proposed controller. The experimental results show that the proposed controller is able to compensate 90% of the vibration of the PTS.

1. INTRODUCTION

Scanning probe microscopy (SPM) (Binning et al., 1986; Meyer et al., 2004) refers to a group of imaging techniques that collect images of sample surfaces through scanning. Many types of SPM systems such as scanning tunneling microscope (STM) (Binning et al., 1987), atomic force microscope (AFM) (Bhikkaji et al., 2008; Das et al., 2012; Ratnam et al., 2005), and magnetic force microscope (MFM) (Rugar et al., 1990) are used now a day to view sample surfaces. The SPM was first found in the form of STM. The STM uses only conducting samples to scan which limits the use of STM for non-conducting samples. This leads to the invention of the AFM. The AFM is a type of SPM system which is used to measure the

DOI: 10.4018/978-1-4666-7248-2.ch003

Figure 1. Basic component of the AFM system

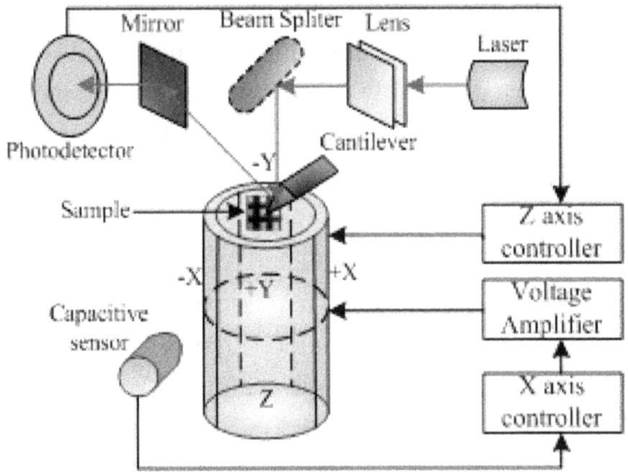

topography of various specimens down to atomic level. The AFM generates three dimensional images of both conducting and non-conducting samples in vacuum and non-vacuum environments.

G. Binning and his colleagues invented the AFM in the year of 1986 based on their design of STM. The AFM is also known as scanning force microscopy (SFM). Since from the invention the AFM has emerged as an important tool in nanotechnology research. The AFM can be used in the vacuum and non-vacuum environments. It can also operate at different temperatures.

In the imaging process of the AFM a sharp tip is used to scan the sample. The tip is brought into the closed proximity of the sample. The AFM works based on the interactive forces between the tip and sample surface. The interactive forces include various kinds of forces such as capillary, van-der-Waals, electrostatic and magnetic. These forces are composed of long and short range components.

The basic component of the AFM system is illustrated in Figure 1. The AFM consists of a cantilever with a sharp tip, a laser source, a laser photo detector, and a scanner called piezoelectric tube scanner. The sample is placed on the scanner during the imaging process. A capacitive sensor is placed with the scanner to measure the displacement of the scanner. When the cantilever is brought close to the sample surface at a distance in the order of few nanometers or less, the cantilever tip deflects. The deflection of the cantilever is measured by the photo detector. Depending on the amount of the deflection of the cantilever an image of the sample is generated.

There are basically two operating modes of the AFM and the modes are static and dynamic modes. The static mode measures the interactive forces by using static bending of the micro cantilever. In the dynamic mode the cantilever is excited at its resonance frequency. Static mode is also known as contact mode or repulsive mode. In contact mode the sample and cantilever tip are placed very close to each other nearly few Angstroms. The interactive force works in contact mode is the repulsive force.

The resolution of the scanned images in static mode is limited by the tip-sample contact area which is typically in the range of 1-10 nm. In order to obtain good image of the sample it is required that the tip is placed within tenth of a nanometer to the sample surface. However, it is difficult to achieve such a tip-sample distance in static mode. In this mode the tip exerts a large lateral force from the sample surface which can break the tip of the cantilever. This mode is also not suitable to scan soft biological sample.

Dynamic mode of the AFM is of interest because of its non-contact nature between the sample and tip of the cantilever. In this technique the tip does not remain contact with the sample. The interactive force acts in the dynamic mode of the AFM is short-range attractive force. The dynamic mode operation of the AFM is performed by vibrating the cantilever at or near to it resonance frequency. This dynamic mode is operated in two forms, namely amplitude modulation mode and frequency modulation mode. In amplitude modulation mode an external signal with constant amplitude and phase is applied to the piezo actuator of the cantilever to excite and vibrate the cantilever. In this mode the amplitude of the cantilever is affected by the repulsive force acting on the tip during the operation. In frequency modulation mode the cantilever is always excited to vibrate at its resonance frequency.

In the imaging process of the AFM the sample is placed on a positioning unit termed as nanopositioner (Schitter et al., 2007). Nanopositioners are designed to achieve fine mechanical displacements at atomic scale. Different types nanopositioners such as piezoelectric tube scanners (PTSs) (Bhikkaji et al., 2008; Das et al., 2012; Ratnam et al., 2005), serial-kinematic scanners (Wadikhaye et al., 2012), flexure based scanners (Schitter et al., 2007; Li et al., 2009) are used in SPM systems depending on the applications. The nanopositioners used in most of the commercial SPM systems are of two types and the types are:

1. Scan-by-sample scanner (Das et al., 2012), and
2. Scan-by-head scanner (Chuang et al., 2011).

Both types of nanopositioners are equally used in various nanopositioning applications. In the case of scan-by-sample scanner the scanner moves during the operation whereas scan-by-head scanner remains fixed during the operation.

In most of the cases the nanopositioner such as the PTS is made using thin coatings of copper or silver on its inner and outer walls which act as electrodes (Schitter et al., 2001). There are four electrodes in the outer axes of the scanner. Two of the outer axes electrodes are referred as X−X electrodes and the other two are referred as Y−Y electrodes. The electrodes are connected to external voltage signals to move the piezo in the X and Y directions. There is one electrode in the inner axis of the PTS. The inner axis electrode is grounded. One electrode each of the X−X and Y−Y pairs, referred to as X+ and Y+ respectively, are chosen as the input ends of the piezoelectric tube, and the corresponding opposite ends, referred to as X− and Y− respectively, are chosen as the output ends of the piezoelectric tube.

Nanopositioners are designed to achieve large scanning area which is accomplished by having large length-to-diameter ratio of the nanopositioner (Bhikkaji et al., 2008; Das et al., 2012, 2012b, 2013a, 2013b,2013c,2013d; Ratnam et al., 2005). A major disadvantage of having large length-to-diameter ratio is the low mechanical resonance frequency of the scanner. In practice, most of the cases it is seen that, the nanopositioners are actuated in a raster pattern which uses triangular signal in the fast axis, i.e., the X-axis of the scanner (Fleming et al., 2010) as illustrated in Figure 2.

One of the limitations of using triangular signal in the X-axis of the scanner is that the triangular signal contains all odd harmonics of its fundamental frequency. Therefore, when PTSs are actuated by triangular signal the odd harmonics of the fundamental frequency of the triangular signal excites the mechanical resonance of the nanopositioner. This excitation of the resonant modes of the nanopositioners results in loss of precision positioning of the scanner at high scanning rate.

An illustration of the effect of vibration of the PTS is presented in Figure 3. It can be seen that the scanned images obtained at high scanning rates are more distorted as compared to the scanned images obtained at low scanning rate. The scanned images presented in Figure 3 are obtained for a reference grating sample by using open-loop. The reference grating sample is composed of number of squares. Each square is 1.5 micro meter.

Figure 2. Illustration of (a) staircase and (b) triangular wave signals applied to the y and x axis during rater scanning, (c) raster scanning method

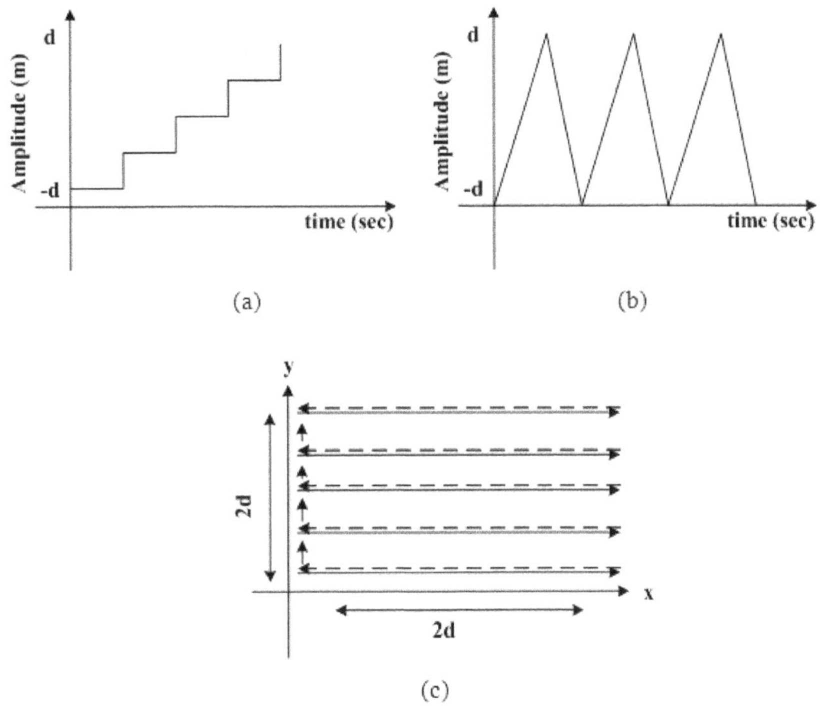

Figure 3. Illustration of the effect of vibration in the PTS: (a) 10.42 Hz, (b) 31.25 Hz, (c) 62.5 Hz, (d) 125 Hz scanned images

The rest of the chapter is organized as follows. In the first section of this Chapter, a review of existing feedback and feedforward control techniques, and the motivation for this work are presented. Section 3 discusses the mathematical tools needed to examine the main stability results. The main stability result is discussed in Section 4. After that the dynamics identification of the system is presented in the Section 5. The design of the controller is presented in Section 6. Section 7 presents the performance of the proposed controller. The future direction of the proposed work and brief remarks with conclusion is presented in the Section 8 and Section 9.

2. BACKGROUND

There has been a consistent effort in recent years to improve accuracy and speed of piezoelectric tube scanners using feedback control techniques (Fleming et al., 2010; Kodera et al., 2005; Leang et al., 2007). The approaches to the design of damping controller to control the resonant mode of the nanopositioners can be classified into two categories and the categories are (a) open-loop control (Croft et al., 1999; Croft et al., 2000) and (b) closed-loop control) (Bhikkaji et al., 2008; Das et al., 2012; Ratnam et al., 2005; Salapaka et al., 2002; Sebastian et al., 2003; Schitter et al., 2001; Schitter et al., 2003).

The design of open-loop control (Croft et al., 1999; Croft et al., 2000; Leang et al., 2007) damping controller is popular because of their noise efficiency as they do not require any additional sensor during the implementation. However, the identification of the dynamics of the nanopositioners in open-loop control is done by using additional sensor and hardware. One of the major drawbacks of the use of open-loop control techniques is their low performance against the variation of plant transfer functions. The transfer functions of the nanopositioners can be varied during the service for many reasons such as due to the changing of loads on the nanopositioners if the nanopositioners are scan-by-sample scanners. The transfer functions of the scan-by-head scanners can also be changed due to the changes in environmental condition such as temperature and air.

Feedback controllers design to damp the resonant modes of scanners is of more interest as they provide robustness against the changes in resonance frequencies of scanners. High bandwidth robust controllers such as linear quadratic Gaussian (LQG) controller (Petersen et al., 2003) and H_∞ controller (Salapaka et al., 2002; Sebastian et al., 2003; Schitter et al., 2001; Schitter et al., 2003) are designed to achieve a bandwidth near to the resonance frequency of systems. However, the order of controllers using LQG and H_∞ methods depend on the order of systems which increase complexity in the design process for high order systems. The implementation of high order controllers requires advance DSP system. Also, the design of LQG and H_∞ controllers are done for finite number of resonant modes of systems which can disturb the unmodeled dynamics of the scanner during the implementation (Fleming et al., 2010; Sebastian et al., 2005).

Most of the commercial SPM systems use integral or proportional integral (PI) controllers in the axes of the nanopositoners because of their simplicity and ease of implementation (Sebastian et al., 2005). However, the closed-loop systems using PI controllers are sensitive against the changes in transfer functions of the plants and can make the closed-loop systems unstable for a very small-change of variation of the plant transfer functions. Another major drawback of the integral controllers is the achievable closed-loop bandwidths. Due to the low gain of integral controllers at high frequencies, the resultant closed-loop systems between the nanopositioners and integral controllers result in low closed-loop bandwidth. This limits the use of integral controllers for high speed nanopositioning of the PTSs.

Piezoelectric shunt damping (Eielsen et al., 2010) is another type of damping technique where a shunt circuit which consists of an inductor is added in series with the nanopositioners to damp their resonant modes. The amount of damping in shunt control depends on the inductor. This high value of inductor sometimes reaches hundreds of Henries in low-frequency domain. This shunt damping method is also sensitive to environmental factors such as outside temperature, which may cause drift in the structure's resonance frequencies.

Positive position feedback (PPF) controllers (Mahmood et al., 2009) and integral resonant controllers (IRCs) (Bhikkaji et al., 2008) are damping controllers used to damp the resonant modes of PTSs. PPF controllers and IRCs are low pass controllers with sharp roll-off at high frequencies. In spite of providing good damping, PPF controllers and IRCs may suffer from the problem of low gain and phase margins. One of the objectives of the proposed controller is to suppress the first resonant mode of the scanner with little effect at other frequencies in order to achieve large gain and phase margins. The resonant controller (Das et al., 2012; Pota et al., 1999; Pota et al., 2002) provides excellent damping with large gain and phase margins. The resonant controller is a high pass controller for which the resultant closed-loop system may add additional sensor noise at high frequency.

Two important properties of PPF controllers, integral resonant controllers, and resonant controllers are that the phases of the transfer functions of these damping controllers lie between -180^0 and 0^0 and the imaginary parts of these controllers transfer functions are negative at each frequency. A transfer function which has negative-imaginary part at each frequency is to referred as negative-imaginary (NI) (Patra et al., 2011; Petersen et al., 2010; Petersen et al., 2011). A multi-input multi-output transfer function matrix G(s) is said to be NI if $j[G(j\omega) - G^*(j\omega)] \geq 0$ for all $\omega \in (0, \infty)$. Here, $G^*(j\omega)$ is the complex conjugate transpose of the matrix $G(j\omega)$. The transfer function matrix G(s) is said to have the negative-imaginary property at frequency ω if $j[G(j\omega) - G^*(j\omega)] \geq 0$ at frequency ω.

PPF controllers, integral resonant controllers, and resonant controllers can be used to damp the resonant modes of structures with collocated position sensors and force actuators. Structures with collocated position sensors and force actuators are also negative-imaginary systems. The transfer functions of lightly damped structures with collocated position sensors and force actuators can be modeled by a sum of second order transfer functions of the form $N(s) = \sum_{i=1}^{M} \frac{\phi\phi^T}{s^2 + 2\xi_i\omega_i s + \omega_i^2}$, where $\xi_i > 0$ is the damping coefficient of ith mode, $\omega_i > 0$ is the resonance frequency of ith mode, M is the number of modes of the structure and ϕ is a column vector (Pota et al., 1999; Pota et al., 2002). The feedback interconnection between a structure with collocated position sensors and force actuators and PPF controllers, integral resonant controllers, and resonant controllers is guaranteed to be finite-gain stable if the DC loop-gain is strictly less than unity as the plant and the controller are both NI with one of them being strictly NI.

Passive damping controllers such as velocity feedback controllers (Balas et al., 1979) are another types of damping controllers used to damp the resonant modes of collocated smart structures between force actuators and velocity sensors. The transfer functions of lightly damped smart structures with collocated velocity sensors and force actuators are modeled by a sum of second order transfer functions $G(s) = \sum_{i=1}^{M} \frac{\phi\phi^T s}{s^2 + 2\xi_i\omega_i s + \omega_i^2}$, where $\xi_i > 0$ is the damping coefficient associated for ith mode, $\omega_i > 0$

is the resonance frequency of the ith mode, M is the total number of modes and ϕ is a column vector obtained from the boundary conditions on the corresponding partial differential equations.

Passive damping controllers to damp the resonant modes of structures are of interest because of their band pass nature which not only results in large gain and phase margins but also reduce the high frequency sensor noise. By definition, the passive systems are stable systems with equal number of inputs and outputs. A linear stable time invariant system $G(j\omega)$ is said to have passivity properties at ω if the real part of $G(j\omega)$ is positive at ω (Lanzon et el., 2007). The phase of a SISO passive linear system lies between $[-\pi / 2, \pi / 2]$ and the Nyquist plot lies to the right of the imaginary axis. A multi-input multi-input transfer function matrix $G(j\omega)$ is said to be passive if the eigenvalues of the matrix $[G(j\omega)+G^{*}(j\omega)]$ are greater than or equal to zero and strictly passive if the eigenvalues of the matrix $[G(j\omega)+G^{*}(j\omega)]$ are greater than zero for all $\omega \in (0,\infty)$. The negative feedback interconnection of two passive systems is finite-gain stable if one of the system is strictly passive.

However, in practice, the transfer functions between the force actuators and position sensors of piezoelectric tube scanners are neither NI nor passive (Lanzon et al., 2008). The PTSs are systems with mixed negative-imaginary, passivity and small-gain properties. Typically, the PTSs have NI property (i.e., the phases of the PTSs lie between 0^0 and -180^0) at low frequencies, passivity property (i.e., the phases of the PTSs lie between -90^0 and 90^0) at intermediate frequencies and small-gain property (i.e., the gain of the PTSs is less than unity) at high frequencies. Possible reasons for the PTS system not being NI or passive is delays in the sensors or actuators or the collocation of the sensors and actuators may not be perfect. The electronic systems to which the PTSs are connected can also add additional phase lag to the system.

Since the piezoelectric tube scanners are not passive in practice, therefore, the design of passive damping controller based on passivity theorem cannot guarantee the closed-loop stability of the interconnection between the PTSs and the passive damping controllers. A small-gain approach (Desoer et al., 1975; Zhou et al., 1988) is a possible solution to establish the finite-gain stability between the PTS and the passive damping controllers. The small-gain theorem states that, the feedback interconnection between two stable linear systems is stable if the loop-gain between the two systems is strictly less than one. Typically, piezoelectric tube scanners have large gain near the resonance frequency for which the gains of the controllers have to select low in order to make the loop-gain less than one to satisfy the stability condition of the small-gain theorem. The low gains of the controllers result in low level damping of the resonant modes of the scanners.

The motivation for designing a passive damping controller for a nanopositioner is its bandpass nature which not only results in large gain and phase margins but also ended with low sensor noise. Two other motivations of passive damping controllers are high bandwidth and robustness against the changes in resonance frequency due to the load change on the scanner. In this Chapter, an analytical framework is presented to establish the finite-gain stability between the interconnected systems with mixed negative-imaginary, passivity and small-gain properties. The proposed analytical framework is exploited to design the controller. The motivation to propose the analytical framework as compared to small-gain approach is to achieve large gain of the controller which allows large level of damping of the resonant mode of the scanner.

3. MATHEMATICAL DEFINITIONS OF THE SYSTEMS

As discussed in the previous section that, the piezoelectric tube scanners are neither NI nor passive systems. They are systems with mixed NI, passivity and small-gain properties. In the next section in order to find the finite gain stability between the PTS and passive damping controller, an analytical framework is proposed to examine the finite gain stability between two stable, linear time-invariant systems where one system has mixed passivity, negative-imaginary and small-gain properties and the other system has mixed negative-passivity, negative-imaginary, and small-gain properties. The proposed analytical framework is exploited to design the controller. Following section shows the notation and mathematical definitions required to establish the framework:

Let $\mathcal{R}$ be the set of all real rational transfer function matrices and $\mathcal{RH}_{\infty}^{m\times n}$ be the set of all real rational stable transfer functions matrices with m rows and n columns. Let, $\mathcal{RL}_{\infty}$ be the set of all proper, real-rational transfer function matrices with no pole in the origin and $\mathcal{L}_{\infty}$ is a Banach space of matrix. For a transfer function matrix $G(s) \in \mathcal{R}$, $G^{*}(s)$ is the complex conjugate transpose of $G(s)$ which is defined as $G(-s)^{T}$. The largest and smallest singular values of a matrix A will be denoted by $\bar{\sigma}(A)$ and $\underline{\sigma}(A)$.

Definition 1: A linear causal time invariant system with transfer function matrix $M \in \mathcal{RH}_{\infty}^{m\times m}$ is said to be passive if there exists $\gamma, \delta \geq 0$ such that

$$-\gamma M^{*}(j\omega)M(j\omega) + M^{*}(j\omega) + M(j\omega) - \delta I \geq 0$$

for all $\omega \in (-\infty, \infty)$.

Definition 2: A linear causal time invariant system with transfer function matrix $M \in \mathcal{RH}_{\infty}^{m\times m}$ is said to be negative passive if there exists $\gamma, \delta \geq 0$ such that

$$-\gamma M^{*}(j\omega)M(j\omega) - M^{*}(j\omega) - M(j\omega) - \delta I \geq 0$$

for all $\omega \in (-\infty, \infty)$.

Definition 3: A linear causal time invariant system with transfer function matrix $M \in \mathcal{RH}_{\infty}^{m\times m}$ is said to be negative-imaginary if there exists $\alpha, \beta \geq 0$ such that

$$-\alpha\omega^{2} M^{*}(j\omega)M(j\omega) + j\omega M(j\omega) - j\omega M^{*}(j\omega) - \beta I \geq 0$$

for all $\omega \in (-\infty, \infty)$.

Definition 4: A linear causal time invariant system with transfer function matrix $M \in \mathcal{RH}_{\infty}^{m\times m}$ is said to be finite-gain system bounded by a gain $k \geq 0$ if

$$k := inf\{\bar{k} \in \mathbb{R}_{+} : -M^{*}(j\omega)M(j\omega) + \bar{k}^{2} I \geq 0\}$$

for all $\omega \in (-\infty, \infty)$.

Definition 5: A linear causal time invariant system with transfer function matrix $M \in \mathcal{RH}_\infty^{m\times m}$ is said to be a system with mixed negative-imaginary, passivity and finite-gain properties bounded by a gain $k \geq 0$ if, for each frequency $\omega \in \mathbb{R} \cup \{\pm\infty\}$, either:

1. $-\alpha\omega^2 M^*(j\omega)M(j\omega) + j\omega M(j\omega) - j\omega M^*(j\omega) - \beta I \geq 0$; or
2. $-\gamma M^*(j\omega)M(j\omega) + M^*(j\omega) + M(j\omega) - \delta I \geq 0$; and/or
3. $-M^*(j\omega)M(j\omega) + k^2 I \geq 0$, where $\alpha, \beta, \gamma, \delta, k \geq 0$.

Definition 6: A linear causal time invariant system with transfer function matrix $M \in \mathcal{RH}_\infty^{m\times m}$ is said to be a system with mixed negative-imaginary, negative-passivity and finite-gain properties bounded by a gain $k \geq 0$ if, for each frequency $\omega \in \mathbb{R} \cup \{\pm\infty\}$: either (i) $-\alpha\omega^2 M^*(j\omega)M(j\omega) + j\omega M(j\omega) - j\omega M^*(j\omega) - \beta I \geq 0$ or (ii) $-\gamma M^*(j\omega)M(j\omega) - M^*(j\omega) - M(j\omega) - \delta I \geq 0$; and/or (iii) $-M^*(j\omega)M(j\omega) + k^2 I \geq 0$, where $\alpha, \beta, \gamma, \delta, k \geq 0$.

Now, in order to describe the mixed passivity, negative-imaginary and finite-gain properties and the mixed negative-imaginary, negative-passivity and finite-gain properties of the systems being considered, the frequency sets are defined in Figure 4.

1. Z_{NI}: Given a linear system with transfer function matrix $M(s)$, let $\mathcal{N}_k$, $k = 1, \ldots, l$ be the frequency interval for which the system has negative-imaginary property and let $\mathcal{N} = (\cup_{k=1}^{l} \mathcal{N}_k) \cup (\cup_{k=-l}^{-1} \mathcal{N}_k)$ where $\mathcal{N}_{-k}$ is the negative reflected regions of $\mathcal{N}_k$.
2. Z_P: Let $\mathcal{P}_k$, $k = 1, \ldots, m$ be the frequency interval for which the system has passivity property and let $\mathcal{P} = (\cup_{k=1}^{m} \mathcal{P}_k) \cup \{0\} \cup (\cup_{k=-m}^{-1} \mathcal{P}_k)$ where $\mathcal{P}_{-k}$ is the negative reflected regions of $\mathcal{P}_k$.
3. Z_q: Let $\mathcal{Q}_k$, $k = 1, \ldots, m$ be the frequency interval for which the system has passivity property and let $\mathcal{Q} = (\cup_{k=1}^{m} \mathcal{Q}_k) \cup \{0\} \cup (\cup_{k=-m}^{-1} \mathcal{Q}_k)$ where $\mathcal{Q}_{-k}$ is the negative reflected regions of $\mathcal{Q}_k$.
4. Z_{FG}: Let $\mathcal{S}_k$, $k = 1, \ldots, q$ be the frequency interval where the system is bounded by a gain $k \geq 0$ where $k > \bar{\sigma}(M(0))$ and $\mathcal{S} = (\cup_{k=1}^{q} \mathcal{S}_k) \cup \{0\} \cup (\cup_{k=-q}^{-1} \mathcal{S}_k)$, where $\mathcal{S}_{-k}$ is the negative reflected region of $\mathcal{S}_k$.

Figure 4. Block diagram of the closed-loop system

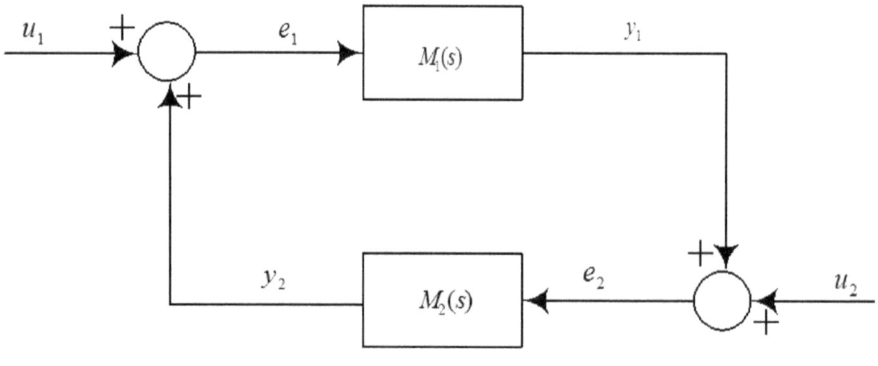

In the above definitions it should be noted that zero frequency and its neighborhoods are included in the passivity set $\mathcal{Z}_P$ or negative-passivity set $\mathcal{Z}_q$ or finite-gain set $\mathcal{Z}_{FG}$ and zero frequency is always excluded from the negative-imaginary set $\mathcal{Z}_{NI}$.

In the following we will show some preliminary results in order to establish the main results:

Lemma 1: Suppose $M_1(j\omega) \in \mathcal{RL}_\infty$ and $M_2(j\omega) \in \mathcal{RL}_\infty$ are two transfer function matrices where at some $\omega \in \mathbb{R} \cup \{\pm\infty\}$, $M_1^*(j\omega) + M_1(j\omega) > 0$ and $-M_2^*(j\omega) - M_2(j\omega) \geq 0$, then $det[I - M_1(j\omega)M_2(j\omega)] \neq 0$ and $det[I - \frac{1}{g}M_1(j\omega)M_2(j\omega)] \neq 0$ for any $g \geq 1$.

Proof: See (Lanzon et al., 2008).

Lemma 2: Suppose $M_1(j\omega) \in \mathcal{RL}_\infty$ and $M_2(j\omega) \in \mathcal{RL}_\infty$ are two transfer function matrices where at some $\omega \in \mathbb{R} \cup \{\pm\infty\}$, $j(M_1(j\omega) - M_1^*(j\omega)) > 0$ and $j(M_2(j\omega) - M_2^*(j\omega)) \geq 0$, then $det[I - M_1(j\omega)M_2(j\omega)] \neq 0$ and $det[I - \frac{1}{g}M_1(j\omega)M_2(j\omega)] \neq 0$ for any $g \geq 1$.

Proof: See (Lanzon et al., 2008).

Theorem 3: (Nobale et al., 1988) Consider the positive feedback interconnection as shown in Figure 1 where $M_1(s) \in \mathcal{RH}_\infty^{m\times m}$, $M_2(s) \in \mathcal{RH}_\infty^{m\times m}$ and that interconnection is well-posed. The feedback interconnection of Figure 1 is said to be stable if and only if the Nyquist plot of $det[I - M_1(j\omega)M_2(j\omega)]$ does not make any encirclements of the origin for $-\infty \leq \omega \leq \infty$.

The Nyquist plot of $det[I - M_1(j\omega)M_2(j\omega)]$ belongs to a family of Nyquist plot of $det[I - \frac{1}{g}M_1(j\omega)M_2(j\omega)]$, where $g \in [1, \infty)$ and the sufficient condition so that $det[I - M_1(j\omega)M_2(j\omega)]$ does not make any encirclement with the origin for all $\omega \in \mathbb{R} \cup \{\pm\infty\}$ and $g \in [1, \infty)$ is $det[I - \frac{1}{g}M_1(j\omega)M_2(j\omega)] \neq 0$. In the following theorem, we will explore the conditions for which $det[I - \frac{1}{g}M_1(j\omega)M_2(j\omega)] \neq 0$ for all $\omega \in \mathbb{R} \cup \{\pm\infty\}$ and $g \in [1, \infty)$ to show the finite-gain stability of the interconnected systems as shown in Figure 1 where $M_1(s)$ is a system with mixed passivity, negative-imaginary and finite-gain properties bounded by gain k_1 and $M_2(s)$ is a system with mixed negative-imaginary, negative-passivity and finite-gain properties bounded by gain k_2.

4. INTERCONNECTION OF SYSTEMS WITH POSITIVE FEEDBACK INTERCONNECTION

Consider the positive feedback interconnection between system $M_1(s)$ and system $M_2(s)$ as shown in Figure 4 where $M_1(s)$ is a system with mixed passivity, negative-imaginary, and finite-gain properties bounded by gain k_1 and $M_2(s)$ is a system with mixed negative-imaginary, negative-passivity and finite-gain properties bounded by gain k_1 on different frequency intervals.

Theorem 4: Let $M_1(s) \in \mathcal{RH}_\infty^{m\times m}$ and $M_2(s) \in \mathcal{RH}_\infty^{m\times m}$ be two strictly proper, causal and linear time invariant systems which are connected as shown in Figure 4 where $M_1(s)$ is mixed passive, negative-imaginary, and finite-gain system bounded by a gain $k_1 > \bar{\sigma}(M_1(0))$ and $M_2(s)$ is "mixed" negative-imaginary, negative-passive, and finite-gain system bounded by a gain $k_2 > \bar{\sigma}(M_2(0))$. The interconnection of Figure 1 is finite-gain stable if

1. $k_1 k_2 < 1$,
2. There exists three sets frequency bands:
   a. A set denoted by $\Phi(\omega)_{NI}$ that consists of frequency interval over which both $M_1(s)$ and $M_2(s)$ are associated with the frequency set $\mathcal{Z}_{NI}$,
   b. A set denoted by $\Psi(\omega)_P$ that consists of frequency interval over which $M_1(s)$ is associated with the frequency set $\mathcal{Z}_P$ and $M_2(s)$ is associated with the frequency set $\mathcal{Z}_q$, and
   c. A set denoted by $\Omega(\omega)_{FG}$ that consists of frequency interval over which both $M_1(s)$ and $M_2(s)$ are associated with the frequency set $\mathcal{Z}_{FG}$, and
3. $\Phi(\omega)_{NI} \cup \Psi(\omega)_P \cup \Omega(\omega)_{FG} = \mathbb{R} \cup \{\pm\infty\}$.

Proof: The proof of Theorem 4 is given by showing $det[I - \frac{1}{g} M_1(j\omega)M_2(j\omega)] \neq 0$ for all $g \in [1,\infty)$ and all $\omega \in (-\infty,\infty)$. In order to do so, the overall proof is shown by splitting into three parts:

1. First, it is shown that $det[I - \frac{1}{g} M_1(j\omega)M_2(j\omega)] \neq 0$ for all $g \in [1,\infty)$ when both $M_1(s)$ and $M_2(s)$ belongs to the frequency set $\Phi(\omega)_{NI}$;
2. Then, it is shown that $det[I - \frac{1}{g} M_1(j\omega)M_2(j\omega)] \neq 0$ for all $g \in [1,\infty)$ when both $M_1(s)$ and $M_2(s)$ belongs to the frequency set $\Psi(\omega)_P$; and
3. Finally, it is shown that $det[I - \frac{1}{g} M_1(j\omega)M_2(j\omega)] \neq 0$ for all $g \in [1,\infty)$ if $k_1 > \bar{\sigma}(M_1(0))$, $k_2 > \bar{\sigma}(M_2(0))$ and $k_1 k_2 < 1$ when both $M_1(s)$ and $M_2(s)$ belongs to the frequency set $\Omega(\omega)_{FG}$.

Proof of Part 1

Since both $M_1(s)$ and $M_2(s)$ belongs to the frequency set $\Phi(\omega)_{NI}$, then there exists $\alpha_i, \beta_i > 0$ for $i = 1,2$ such that $-\alpha_i\omega^2 M_i^*(j\omega)M_i(j\omega) + j\omega M_i(j\omega) - j\omega M_i^*(j\omega) - \beta_i I \geq 0$, which also implies that, $j\omega M_i(j\omega) - j\omega M_i^*(j\omega) > 0$. Since, $j\omega M_i(j\omega) - j\omega M_i^*(j\omega) > 0$, hence, $\frac{1}{\sqrt{g}} j\omega M_i(j\omega) - \frac{1}{\sqrt{g}} j\omega M_i^*(j\omega) > 0$, for $g > 0$. Hence, from Lemma 2, $det[I - \frac{1}{g} M_1(j\omega)M_2(j\omega)] \neq 0$ for all $g \in [1, \infty)$.

Proof of Part 2

When the systems belong to the frequency set $\Psi(\omega)_P$, i.e., in the frequency interval when system $M_1(s)$ has passivity property and system $M_2(s)$ has negative-passivity property then there exists $\gamma_1, \gamma_2, \delta_1, \delta_2 > 0$ such that $-\gamma_1 M_1^*(j\omega)M_1(j\omega) + M_1^*(j\omega) + M_1(j\omega) - \delta_1 I \geq 0$ and $-\gamma_2 M_2^*(j\omega)M_2(j\omega) - M_2^*(j\omega) - M_2(j\omega) - \delta_2 I \geq 0$. This implies that, $M_1^*(j\omega) + M_1(j\omega) > 0$ and $-M_2^*(j\omega) - M_2(j\omega) > 0$. Now, since $M_1^*(j\omega) + M_1(j\omega) > 0$ and $-M_2^*(j\omega) - M_2(j\omega) > 0$, therefore, $\frac{1}{\sqrt{g}} M_1^*(j\omega) + \frac{1}{\sqrt{g}} M_1(j\omega) > 0$ and $-\frac{1}{\sqrt{g}} M_2^*(j\omega) - \frac{1}{\sqrt{g}} M_2(j\omega) > 0$ for $g > 0$. Hence from Lemma 1, $det[I - \frac{1}{g} M_1(j\omega)M_2(j\omega)] \neq 0$ for all $g \in [1, \infty)$.

Proof of Part 3

Since both $M_1(s)$ and $M_2(s)$ belongs to the frequency set $\Omega(\omega)_{FG}$, then there exists $k_i > 0$ for i=1,2, such that $-M_i^*(j\omega)M_i(j\omega) + \overline{k}_i^{2} I \geq 0$. Since $M_1(s)$ and $M_2(s)$ are bounded by gain k_1 and k_2, i.e., $k_1 > \bar{\sigma}(M_1(0))$ and $k_2 > \bar{\sigma}(M_2(0))$ hence, $det[I - \frac{1}{g} M_1(j\omega)M_2(j\omega)] \neq 0$ if and only if $k_1 k_2 < 1$ for all $g \in [1, \infty)$. This concludes the proof of Theorem 4.

5. EXPERIMENTAL SETUP AND SYSTEM IDENTIFICATION

The nanopositioner used in this paper in order to implement the proposed controller is a piezoelectric tube scanner. The considered piezoelectric tube scanner is a scan-by-sample scanner with thin coating of copper or silver in its inner and outer wall which acts as electrodes. The controller implementation is done to damp the first resonant mode of the scanner in the lateral axis. A dynamic signal analyzer (DSA) is used to measure the frequency response of the scanner. Signal to the lateral axis is applied thorough a high voltage amplifier (HVA) of gain 15. A signal access module is placed between the HVA and PTS to allow direct access to the electrodes of the PTS. A rapid prototype system dSPACE is used for digital implementation of the controller.

A linear model of the piezo scanner in the X-axis is obtained by applying swept sinusoid voltages of 100 mV rms between 10 Hz to 2kHz from the one channel of an HP35760A dual channel DSA to the X-axis electrodes of the scanner and recording the X sensor output from the other channel of the DSA. A system identification process using prediction-error minimization method (Ljung et al., 2000; Ljung et al., 2002) is performed to obtain transfer function of the dynamics in the X-axis of the piezo scanner. The aim of the present work is to damp the first resonant mode of the scanner. The system identification is done to capture the first resonant mode of the scanner using low order transfer function. The following transfer function is obtained in the identification process to capture the first resonant mode of the scanner using low order transfer function:

$$M_{xx}(s) = \frac{M_{dx}(s)}{M_{vx}(s)} = \frac{-186.6s^2 + 1.348\times 10^6 s - 2.412\times 10^{10}}{s^3 + 1755s^2 + 3.452\times 10^7 s + 4.459\times 10^{10}}$$

where $M_{dx}(s)$ is the Laplace transform of the output voltage from the X sensor and $M_{vx}(s)$ is the Laplace transform of the input voltage to the HVA driving the X-axis of the piezo. A matching between the measured and identified model frequency responses given in Figure 5 shows that the system $M_{xx}(s)$ has negative-imaginary property between 918.3 Hz and 1.91×10^3 Hz as the phase of the identified model is between 0 and -180° in that range of frequencies, passivity property between 582.5 Hz and 1.07×10^3 Hz as the phase of the identified model is between -90° and $+90^{\circ}$ in that range of frequencies.

Figure 5. Measured and identified model frequency response of $M_{xx}(s)$

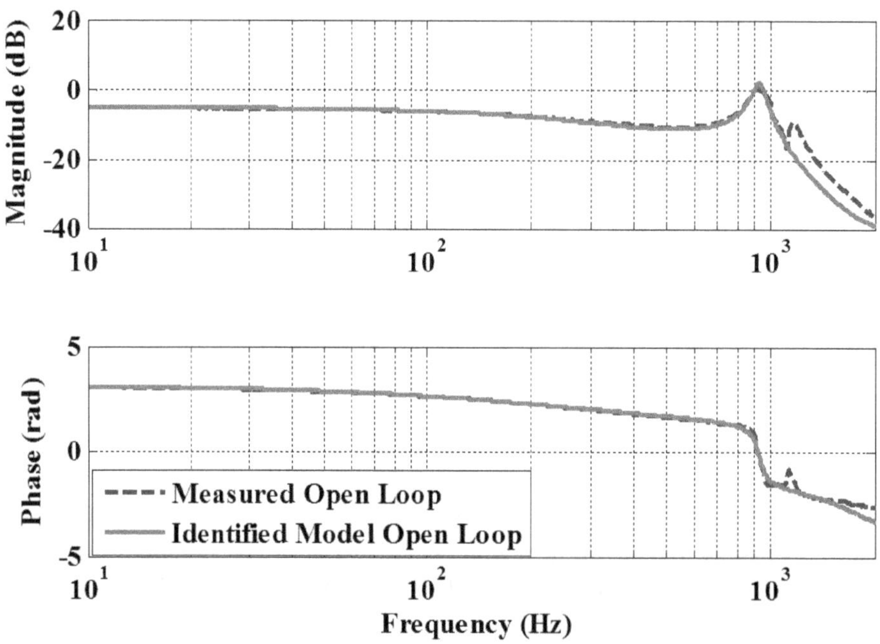

6. CONTROLLER DESIGN, SELECTION, AND STABILITY OF THE CLOSED-LOOP SYSTEM

Consider the block diagram of the closed-loop system shown in Figure 4 where $M_1(s)$ is the plant transfer function, i.e., the transfer function of the positioning system of the PTS and $M_2(s)$ is the transfer function of a passive damping controller. A passive damping controller can be realized by the following transfer function $M_2(s)$:

$$M_2(s) = -k_p \frac{s}{s^2 + 2\xi_p \omega_p s + \omega_p^2}$$

Here, $k_p > 0$ is the gain of the passive damping controller, $\xi_p > 0$ is the damping constant of the controller and $\omega_p > 0$ is the frequency at which resonant mode needs to be damped. The selection of the controller parameters for the PTS is based on the desired aim. The frequency at which the controller provides large gain is frequency ω_p. It is desired that in order to damp the resonant mode of the scanner the value of ω_p is chosen equal or nearly equal to the first resonant mode of the scanner. The damping constant is chosen in a way to achieve large level of damping. A low value of damping constant ξ_p introduces undesirable phase shift in the closed-loop and a high value of the damping constant hardly has any effect on damping. The controller gain k_p plays a major role to damp the resonant mode of a piezo scanner. The controller gain k_p is selected as large as possible to achieve large level of damping of the first resonant mode of the scanner. It is also important that, the controller gain k_p is chosen such that closed-loop system is stable and achieves large level of damping of the resonant mode. In order to guarantee the closed-loop stability and large level of damping of the first resonant mode of the piezo scanner, the controller design is presented by exploiting Theorem 4 of this paper.

The overall design process of the controller using Theorem 4 includes following steps:

1. Find the frequencies at which the system $M_1(s)$ has the negative-imaginary, passivity, and negative-passivity properties.
2. Select a gain k_1 such that at the frequencies where $M_1(s)$ does not have the negative-imaginary, passivity, and negative-passivity properties bounded by a gain k_1.
3. Make the gain k_1 as low as possible such that, $M_1(s)$ has either the negative-imaginary or passivity or negative-passivity properties or has a finite-gain bounded by the gain k_1.
4. Find the frequencies at which $M_1(s)$ has only the negative-imaginary, passivity, negative-passivity, and finite-gain properties bounded by the gain k_1 and both.
5. Select the controller $M_2(s)$ parameters and find the frequencies at which $M_2(s)$ has the negative-imaginary, passivity, and negative-passivity properties.
6. Increase the gain of the controller $M_2(s)$ to be as large as possible and select the gain k_2 for $M_2(s)$ such that, (a) $k_1k_2 < 1$, (b) $M_2(s)$ has the NI property at the frequencies where $M_1(s)$ has the NI

property, (c) $M_2(s)$ has the negative passivity property at the frequencies where $M_1(s)$ has the passivity property.

Using the above selection criteria the controller is chosen as follows:

$$M_2(s) = \frac{-12000s}{s^2 + 6474s + 3.341\times 10^7}$$

The stability analysis between the interconnected system is given below:

First the values of the gains k_1 and k_2 are selected as 0.55 $(> \bar{\sigma}(M_1(0) = 0.54))$ and 1.8 $(> \bar{\sigma}(M_2(0) = 0)$ which satisfy the condition $k_1k_2 < 1$. System 1 has negative-passivity property in the frequency intervals [0, 3660] and [6770, ∞), passivity property in the frequency intervals [3660,6770], negative-imaginary property in the frequency interval [5780, 1.19×10^4] and finite-gain property bounded by gain 0.55 in the frequency intervals [0,5263] and [6140, ∞) $rads^{-1}$. System 2 has negative-imaginary property in the frequency interval [0, 5670], negative-passivity property in the frequency interval [0, ∞) and finite-gain property bounded by gain 1.8 in the frequency intervals [0, 4770] and (6770, ∞) $rads^{-1}$.

The mixed properties are only shown for positive frequencies. However, the mixed properties between the systems for positive feedback interconnection can also be obtained for negative frequencies in a similar way. It can be seen that, at each frequency when system 1 has NI property, system 2 also has NI property; when system 1 has passivity property, system 2 has negative-passivity property or vice versa and when system 1 has finite-gain property bounded by gain k_1, system 2 has finite-gain property bounded by gain k_2 and $k_1k_2 < 1$. Hence, the closed-loop system corresponding to interconnection between $M_1(s)$ and $M_2(s)$ is stable. This is verified by calculating the corresponding closed-loop poles for the positive feedback interconnection between the transfer functions $M_1(s)$ and $M_2(s)$ and the closed-loop poles calculated as -841.1+6210.7i, -841.1-6210.7i, -2314.8+3543.7i, -2314.8-3543.7i and -2116.8 which shows that the closed-loop system is stable.

7. PERFORMANCE OF THE CONTROLLER

The performance of the proposed controller is examined by implementing the controller on an experimental piezoelectric tube scanner. A comparison of the measured open- and closed-loop magnitude frequency responses of $M_{xx}(s)$ is given in Figure 6. The comparison shows that, the proposed controller is able to provide 7 damping of the first resonant mode of the scanner. From the frequency response of $M_{xx}(s)$ in open-loop it can be observed that, there is a dip in the range of frequencies from 100 Hz to 800 Hz, which means that there is a variation of gain in open-loop. The magnitude of the output voltage in that range will vary even though the magnitude of the input voltage remains same. This will certainly reduce the performance of the PTS for high speed nano-positioning. The closed-loop frequency response of $M_{xx}(s)$ shows less amount of dip in the frequency response as compared to the open-loop frequency response. The controller is also able to flatten the frequency response of the PTS. This signifies large bandwidth of the closed-loop system which in turns ensures faster scanning.

Figure 6. Open-and closed-loop magnitude frequency responses of $M_{xx}(s)$

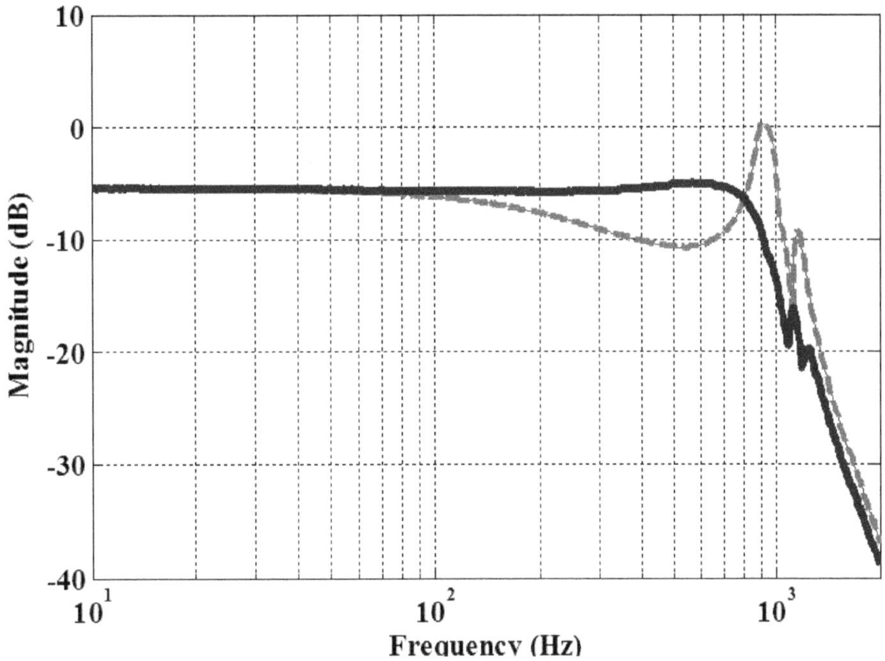

A magnitude Bode diagram of the controller is presented in Figure 7. The magnitude Bode diagram of the controller shows that, the controller only applies high gain near to the first resonance frequency of the scanner which in turn results in large gain and phase margins. The Nyquist plot of the measured loop-gain between the plant and the damping controller shown in Figure 8 provides a good information about the stability, gain margin, and phase margin of the closed-loop system. The damping controller is connected in positive feedback and the measured open loop-gain does not encircle the point 1+j0 which means the closed-loop system between the PTS and the controller is stable.

To illustrate controller performance in time domain, triangular voltage signals of different frequencies (50 Hz, 100 Hz, 150 Hz, 200 Hz) are applied to the X-axis of the scanner in the open- and closed-loop cases and the corresponding X sensor outputs are recorded. Figure 9 shows the outputs from the X sensor in open- and closed-loop cases for different frequencies obtained by using the controller in the X-axis of the scanner. A small time delay was purposely added to the open- and closed-loop sensor outputs in order to clearly display the open- and closed-loop time responses. The closed-loop X sensor output shows less vibration as compared the open-loop sensor output.

The time domain performance of the proposed controller is further examined by applying pulse signals of 1 Hz, 10 Hz, 50 Hz, and 100 Hz to the X-axis of the scanner. The X sensor outputs in open- and closed-loop cases for different frequencies of the pulse signals are given in Figure 10 which shows that the proposed controller is able to reduce the settling time to 20 percent of each pulse period at low frequencies with a significant reduction of vibration.

The advantage of the proposed analytical framework as compared to the small-gain only approach is shown by designing the proposed controller using small-gain only approach. The small-gain approach limits the measured loop-gain to be strictly less than one as shown in Figure 11, whereas, the proposed

Figure 7. Magnitude Bode diagram of the controller

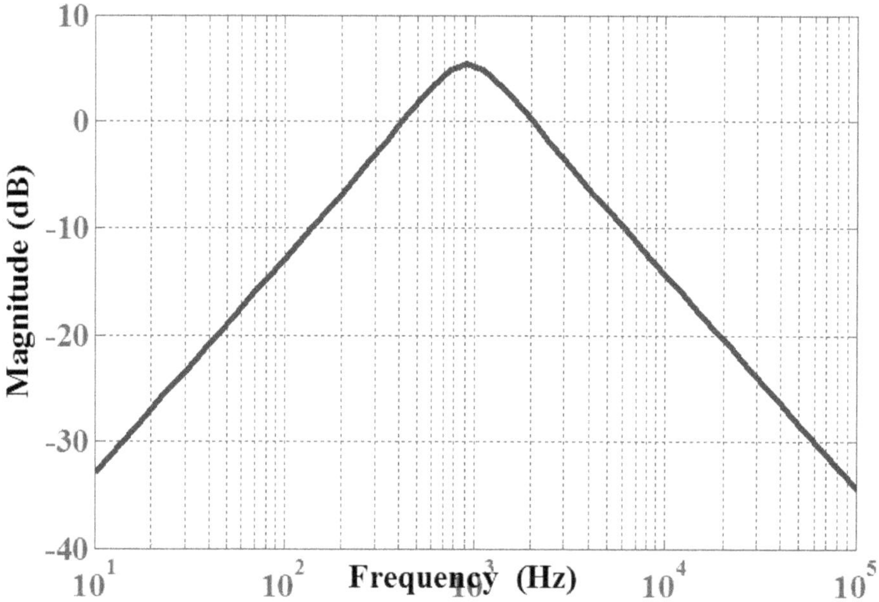

Figure 8. Measured loop gain $M_1(j\omega)M_2(j\omega)$

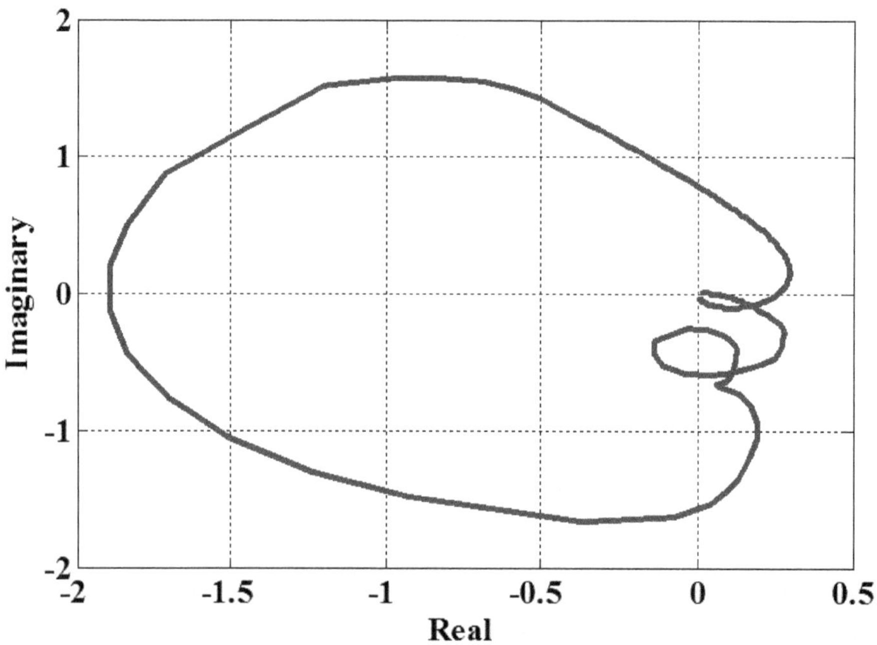

Figure 9. Measured open-and closed-loop X sensor outputs; the solid line (-) represents the open-loop X sensor output and the dashed line represents the closed-loop X sensor output.

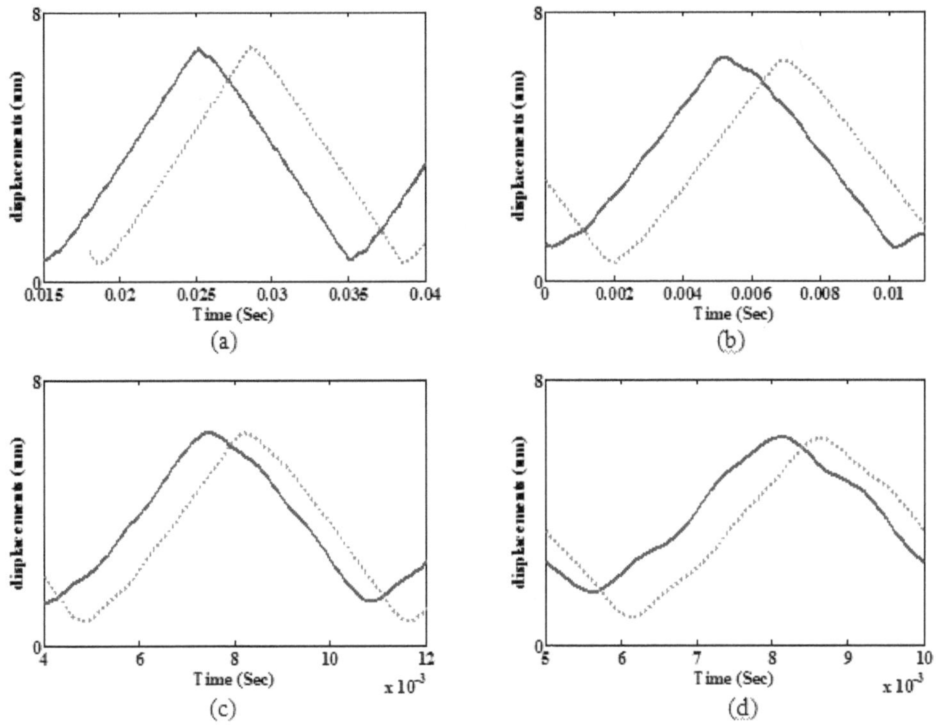

Figure 10. Measured open-and closed-loop X sensor outputs for pulse signals; the solid blue line represents the open-loop X sensor output and the solid red line represents the closed-loop X sensor output.

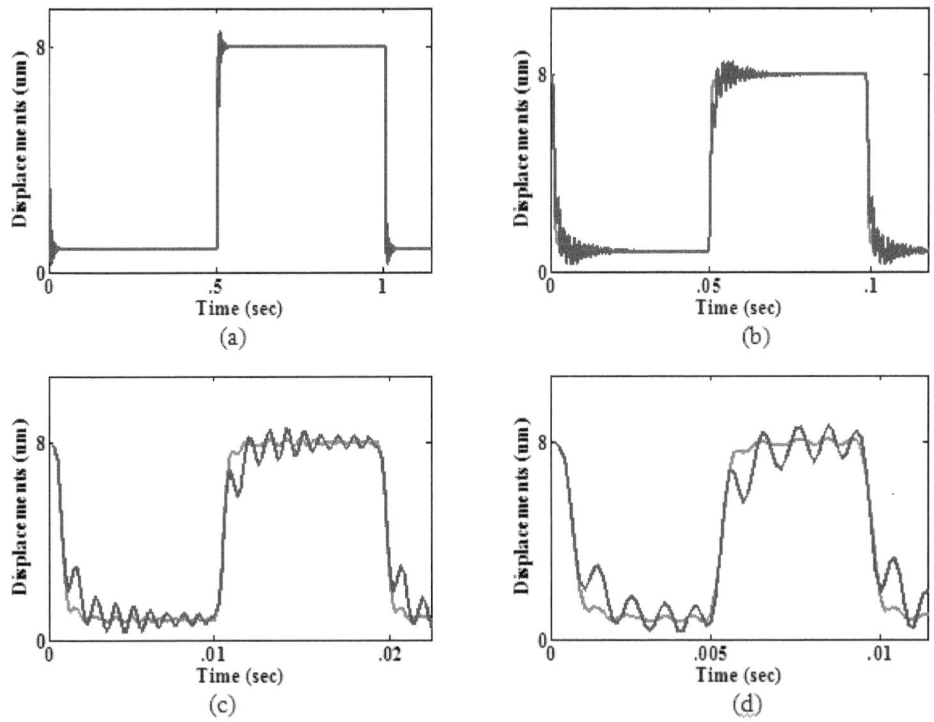

Figure 11. Loop-gains $M_1(j\omega)M_2(j\omega)$ *when the controller is designed using the small-gain only approach (.. the dotted line) and the proposed analytical framework (– the solid line)*

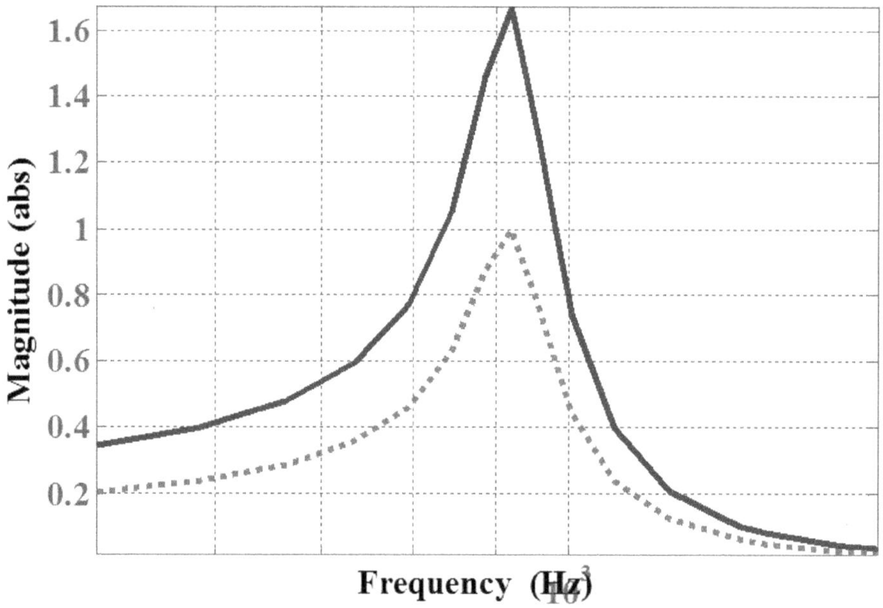

Figure 12. Measured open- (– the solid line) and closed-loop (- - the dashed line) frequency responses of $M_{xx}(s)$ *using small-gain only approach*

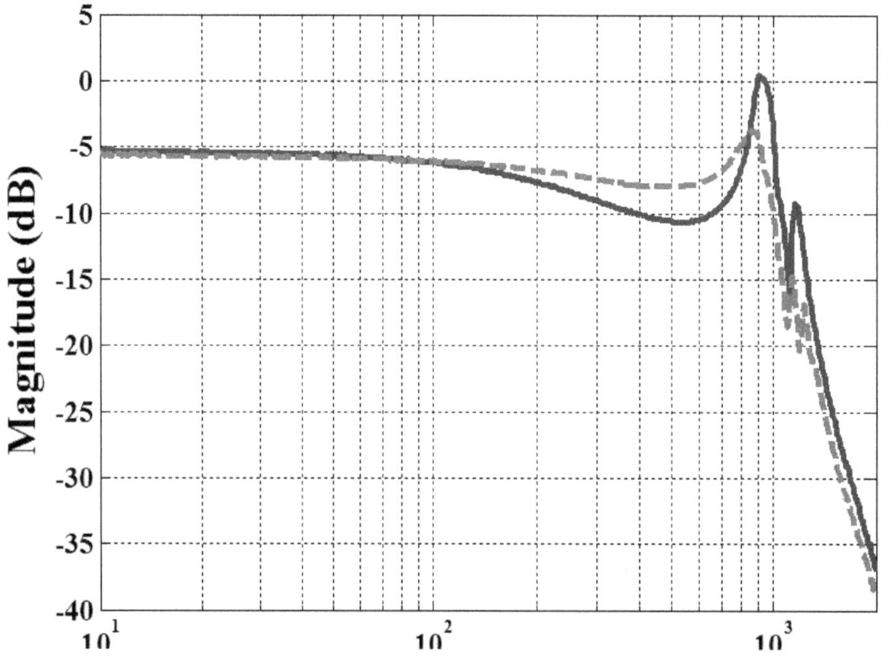

analytical framework allows loop-gain greater than one. The gain of the controller obtained in the small-gain only approach is 7200, whereas the gain of the controller obtained in the proposed analytical framework is 12000. The controller with gain 7500 achieved 3.5 dB damping of the first resonant mode of the scanner and the controller with gain 12000 achieved 7 dB damping of the first resonant mode of the scanner as shown in Figure 12.

8. FUTURE RESEARCH DIRECTIONS

This Chapter presents the design of the passive damping controller to damp the first resonant mode of the PTS in the lateral axis only. Our future aim is to design of the passive damping controller for both the lateral and longitudinal axes of the PTS using multiple-input multiple-output approach.

9. CONCLUSION

Nanopositioners such as piezoelectric tube scanners are used in most of commercial atomic force microscopes to achieve fine mechanical displacement. The performance of nanopositioners is hindered due to their low mechanical resonance frequency. In this Chapter a passive damping controller design is presented to damp the first resonant mode of a piezoelectric tube scanner using a mixed negative-imaginary, passivity and small-gain approach. The experimental results presented in the paper show that the proposed controller is able to flatten the frequency response by achieving a 7 dB damping of the first resonant mode of the scanner.

REFERENCES

Balas, M. J. (1979). Direct velocity feedback control of large space structures. *Journal of Guidance and Control*, *2*(3), 252–253. doi:10.2514/3.55869

Bhikkaji, B., & Moheimani, S. O. (2008). Integral resonant control of a piezoelectric tube actuator for fast nanoscale positioning. *IEEE/ASME Transactions on Mechatronics*, *13*(5), 530–537. doi:10.1109/TMECH.2008.2001186

Binnig, G., & Rohrer, H. (1987). Scanning tunneling microscopy from birth to adolescence. *Reviews of Modern Physics*, *59*(3), 615–625. doi:10.1103/RevModPhys.59.615

Binnig, G., & Smith, D. P. E. (1986). Single-tube three-dimensional scanner for scanning tunneling microscopy. *The Review of Scientific Instruments*, *57*(8), 1688–1689. doi:10.1063/1.1139196

Chuang, N., Petersen, I. R., & Pota, H. R. (2011). Robust H-infinity control in fast atomic force microscopy. In *Proc. of the American Control Conference* (pp. 2258–2265). Academic Press.

Croft, D., & Devasia, S. (1999). Vibration compensation for high speed scanning tunneling microscopy. *The Review of Scientific Instruments*, *70*(12), 4600–4605. doi:10.1063/1.1150119

Croft, D., Shedd, G., & Devasia, S. (2000). Creep, hysteresis, and vibration compensation for piezoactuators: atomic force microscopy application. In *Proc. of the American Control Conference* (pp. 2123–2128). Academic Press. doi:10.1109/ACC.2000.879576

Das, S. K., Pota, H. R., & Petersen, I. R. (2012). Resonant controller for fast atomic force microscopy. In *Proc. of the Conference on Decision and Control* (pp. 2471–2476). Academic Press. doi:10.1109/CDC.2012.6426563

Das, S. K., Pota, H. R., & Petersen, I. R. (2012b). Multi-variable Resonant Controller for Fast Atomic Force Microscopy. In *Proceedings of Australian Control Conference*, (pp. 448-453). Academic Press.

Das, S. K., Pota, H. R., & Petersen, I. R. (2013a). Resonant Control of Atomic Force Microscope Scanner: A "Mixed" Negative-Imaginary and Small-Gain Approach. In *Proceedings of American Control Conference*, (pp. 5476-5481). Academic Press.

Das, S. K., Pota, H. R., & Petersen, I. R. (2013b). Multi-variable Double Resonant Controller for Fast Image Scanning of Atomic Force Microscope. In *Proceedings of Asian Control Conference*, (pp. 1-6). Academic Press. doi:10.1109/ASCC.2013.6606336

Das, S. K., Pota, H. R., & Petersen, I. R. (2013c). Double Resonant Controller for Fast Atomic Force Microscopy. In *Proceedings of Asian Control Conference*, (pp. 1-6). Academic Press. doi:10.1109/ASCC.2013.6606209

Das, S. K., Pota, H. R., & Petersen, I. R. (2013d). (in press). Resonant Controller Design for a Piezoelectric Tube Scanner: A Mixed Negative-Imaginary and Small-Gain Approach. *IEEE Transactions on Control Systems Technology*.

Desoer, C. A., & Vidyasagar, M. (1975). *Feedback Systems: Input-Output Properties*. New York: Academic Press.

Eielsen, A. A., & Fleming, A. J. (2010), Passive shunt damping of a piezoelectric stack nanopositioner. In *Proceedings of American Control Conference*. Academic Press. doi:10.1109/ACC.2010.5530938

Fleming, A. J. (2010). Nanopositioning system with force feedback for highperformance tracking and vibration control. *IEEE/ASME Transactions on Mechatronics*, *15*(3), 433–447. doi:10.1109/TMECH.2009.2028422

Fleming, A. J., Aphale, S. S., & Moheimani, S. O. R. (2010). A new method for robust damping and tracking control of scanning probe microscope positioning stages. *IEEE Transactions on Nanotechnology*, *9*(4), 438–448. doi:10.1109/TNANO.2009.2032418

Kodera, N., Yamashita, H., & Ando, T. (2005). Active damping of the scanner for high-speed atomic force microscopy. *Rev. of Sci. Instrum., 76* (5).

Lanzon, A., & Petersen, I. R. (2007). A modified positive-real type stability condition. In *Proceedings of European Control Conference*. Kos, Greece: Academic Press.

Lanzon, A., & Petersen, I. R. (2008). Stability robustness of a feedback interconnection of systems with negative imaginary frequency response. *IEEE Transactions on Automatic Control*, *53*(4), 1042–1046. doi:10.1109/TAC.2008.919567

Leang, K. K., & Devasia, S. (2007). Feedback-linearized inverse feedforward for creep, hysteresis, and vibration compensation in afm piezoactuators. *IEEE Transactions on Control Systems Technology*, *15*(5), 927–935. doi:10.1109/TCST.2007.902956

Li, Y., & Xu, Q. (2009). Design and analysis of a totally decoupled flexurebased xy parallel micromanipulator. *IEEE Transactions on Robotics*, *25*(3), 645–657. doi:10.1109/TRO.2009.2014130

Ljung, L. (2000). *System Identification Toolbox for use with MATLAB*. Natick, MA: The Math Works, Inc.

Ljung, L. (2002). Prediction error estimation methods. *Circuits. Systems and Signal Processing*, *21*(3), 11–21. doi:10.1007/BF01211648

Mahmood, I. A., & Moheimani, S. O. R. (2009). Making a commercial atomic force microscope more accurate and faster using positive position feedback control. *Rev. of Sci. Instrum., 80*(6).

Meyer, H. E., & Bennewitz, R. (2004). *Scanning Probe Microscopy*. Heidelberg, Germany: Springer. doi:10.1007/978-3-662-09801-1

Nobale, B., & Daniel, J. W. (1988). *Applied Linear Algebra.* (N. P. Hall, Ed.). Englewood Cliffs, NJ: Prentice Hall.

Patra, S., & Lanzon, A. (2011). Stability analysis of interconnected systems with "mixed" negative-imaginary and small-gain properties. *IEEE Transactions on Automatic Control*, *56*(6), 1395–1400. doi:10.1109/TAC.2011.2122670

Petersen, I., & Lanzon, A. (2010). Feedback control of negative-imaginary systems. *IEEE Control Systems Magazine*, *30*(5), 54–72. doi:10.1109/MCS.2010.937676

Petersen, I. R. (2011). Negative imaginary systems theory in the robust control of highly resonant flexible structures. In *Proceedings of Australian Control Conference*. Melbourne, Australia: Academic Press.

Petersen, I. R., & Pota, H. R. (2003). Minimax LQG optimal control of a flexible beam. *Control Engineering Practice*, *11*(11), 1273–1287. doi:10.1016/S0967-0661(02)00240-X

Pota, H., Moheimani, S. R., & Smith, M. (1999). Resonant controllers for flexible structures. In *Proc. Conf. of Decision and Control*, (vol. 1, pp. 631–636). Academic Press.

Pota, H., Moheimani, S. R., & Smith, M. (2002). Resonant controller for smart structures. *Smart Materials and Structures*, *11*(1), 1–8. doi:10.1088/0964-1726/11/1/301

Ratnam, M., Bhikkaji, B., Fleming, A., & Moheimani, S. (2005). PPF control of a piezoelectric tube scanner. In *Proc. of the 44th IEEE Conference on Decision and Control and European Control Conference* (pp. 1168 – 1173). IEEE. doi:10.1109/CDC.2005.1582316

Rugar, D., Mamin, H. J., Guethner, P., Lambert, S. E., Stern, J. E., McFadyen, I., & Yogi, T. (1990). Magnetic force microscopy: General principles and application to longitudinal recording media. *Journal of Applied Physics*, *68*(3), 1169–1183. doi:10.1063/1.346713

Salapaka, S., Sebastian, A., Cleveland, J. P., & Salapaka, M. V. (2002). High bandwidth nano-positioner: A robust control approach. *The Review of Scientific Instruments*, *73*(9), 3232–3241. doi:10.1063/1.1499533

Schitter, G. (2007). Advanced mechanical design and control methods for atomic force microscopy in real-time. In *Proc. of the American Control Conference* (pp. 3503–3508). Academic Press. doi:10.1109/ACC.2007.4282253

Schitter, G., Menold, P., Knapp, H. F., Allgwer, F., & Stemmer, A. (2001). High performance feedback for fast scanning atomic force microscopes. *The Review of Scientific Instruments*, *8*, 72.

Schitter, G., Stemmer, A., & Allgower, F. (2003). Robust 2 DoF-control of a piezoelectric tube scanner for high speed atomic force microscopy. In *Proc. of the American Control Conference* (*vol. 5*, pp. 3720–3725). Academic Press. doi:10.1109/ACC.2003.1240413

Sebastian, A., Salapaka, M. V., & Cleveland, J. P. (2003). Robust control approach to atomic force microscopy. In *Proc. of the 42nd IEEE Conf. Decision and Control* (vol. 4, pp. 3443–3444). IEEE. doi:10.1109/CDC.2003.1271677

Sebastian, A., & Salapaka, S. M. (2005). Design methodologies for robust nano-positioning. *IEEE Transactions on Control Systems Technology*, *13*(6), 868–876. doi:10.1109/TCST.2005.854336

Wadikhaye, S., Yong, Y., & Moheimani, S. (2012). Design of a compact serial-kinematic scanner for high-speed atomic force microscopy: An analytical approach. *Micro Nano Letters, IET*, *7*(4), 309–313. doi:10.1049/mnl.2011.0477

Zhou, K., Doyle, J. C., & Glover, K. (1996). *Robust and Optimal Control.* (N. P. Hall, Ed.). Upper Saddle River, NJ: Prentice Hall.

KEY TERMS AND DEFINITIONS

Atomic Force Microscope: A microscope use to map the topography of specimens.

Cantilever Beam: A cantilever beam is used to measure surface property.

Capacitive Sensor: A capacitive sensor is placed with the scanner to measure the displacement of the scanner.

Negative Imaginary System: A multi-input multi-input transfer function matrix $G(j\omega)$ is said to be passive if the eigenvalues of the matrix $j[G(j\omega) - G^*(j\omega)] \geq 0$ are greater than or equal to zero for all $\omega \in (0, \infty)$.

Passive System: A multi-input multi-input transfer function matrix $G(j\omega)$ is said to be passive if the eigenvalues of the matrix $[G(j\omega) + G^*(j\omega)]$ are greater than or equal to zero for all $\omega \in (0, \infty)$.

Photo Detector: A photo detector is used to measure the cantilever deflection.

Piezoelectric Tube Scanner: A positioning unit in atomic force microscope.

Small Gain Theorem: The small-gain theorem states that, the feedback interconnection between two stable linear systems is stable if the loop-gain between the two systems is strictly less than one.

Chapter 4
State and Parametric Estimation of Nonlinear Systems Described by Wiener Sate-Space Mathematical Models

Houda Salhi
University of Sfax, Tunisia

Samira Kamoun
University of Sfax, Tunisia

ABSTRACT

This chapter deals with the description, the parametric estimation, the state estimation, and the parametric and state estimation conjointly of nonlinear systems. The focus is on the class of nonlinear systems, which are described by Wiener state-space discrete-time mathematical models. Thus, the authors develop a new recursive parametric estimation algorithm, which is based on least squares techniques. The stability conditions of the developed parametric estimation scheme are analyzed using the Lyapunov method. The state estimation problem of the considered nonlinear systems is formulated. Thus, the authors propose a recursive state estimation algorithm, which is based on Kalman Filter. A new recursive algorithm is proposed, which permits one to estimate conjointly the parameters and the state variables of nonlinear systems described by Wiener mathematical models, with unknown parameters and state variables. The efficiency and performance of the proposed recursive estimation algorithms are tested on numerical simulation examples.

1. INTRODUCTION

The development of a mathematical model (an equation or set of equations) to describe correctly a dynamical system is a crucial step before embarking the formulation of a control scheme (Kamoun, 2003). This step can present some difficulties when developing a good mathematical model, especially in the case of nonlinear systems. Indeed, the synthesis of a control law is generally based on the quality of the

DOI: 10.4018/978-1-4666-7248-2.ch004

used mathematical model. More works on the description and the parametric estimation of nonlinear systems have been published in the literature (Haber & Keviczky, 1999; Kamoun, 2003; Billings, 2013).

During the two past decades, some approaches to describing the nonlinear systems have been developed, which are based mainly on the concepts of neural networks (Chen & Billings, 1992; Janczak, 2005) and fuzzy logic (Lee & Teng, 2000; Hong & Harris, 2001). In fact, the description of the nonlinear systems on the basis of these new approaches can be carried by using learning algorithms (backpropagation algorithm, etc.) and from linguistic rules, which can be given by the experts from specific fields (Nelles, 2001).

Problems relating to the description and the estimation of nonlinear systems have preoccupied many researchers over the past three decades. Thus, extensive works have been published in the literature. In this sense, we can mention the works of Haber and Keviczky (1976) and Billings (1980), which are the oldest works dealing with the description and the identification of nonlinear systems.

Identification and control of nonlinear dynamic systems have been widely studied and different model structures and identification and control methods have been reported in the literature. Many models structure are used to describe the nonlinear systems. Several studies related to the estimation of nonlinear systems have been developed by considering different types of mathematical models on the basis of various analytical methods (Haber & Keviczky, 1999; Belforte & Gay, 2001; Vörös, 2002; Zhao & Kanellakopoulos, 2002a, 2002b; Kamoun, 2003). In parallel, other studies dealing with the identification of nonlinear dynamic systems are developed, and that, especially using the concepts of neural networks (Yingwei et al., 1997; Li et al., 2000).

The important developments in several fields of science and technology (information processing, microelectronics, computer engineering, etc.) and the availability of numerical components (DSP cards, calculation software, etc.) with moderate prices have permit the study of discrete-time mathematical models. Computer systems may be connected, without any difficulty, to dynamic systems which are described by discrete-time mathematical models for the practical implementation (in real time) of estimation and control schemes. In this context, we can use interfacing cards, data acquisition and computer software (MATLAB, etc.), which are related to the estimation and control systems. For example, we can estimate the system parameters by implementation of a recursive parametric estimation algorithm on the basis of the computer software MATLAB. In this chapter, we will focus only on the description, the parametric estimation and the state estimation of nonlinear systems, which can be described by discrete-time mathematical models.

Four types of mathematical models can essentially describe nonlinear systems, namely: the functions series mathematical models, the state-space mathematical models, the input-output mathematical models and the block-oriented mathematical models. Note that the block-oriented mathematical models are widely used in industrial applications. The block-oriented models are often used to describe adequately nonlinear systems, over the entire range of operating conditions. This class of mathematical models is composed of linear dynamic parts and nonlinear static parts. The well-known models of the block-oriented models are the Hammerstein models and Wiener models, which correspond to processes with linear dynamic, but a nonlinear gain (Glaria Lopez & Sbarbaro, 2011).

The structure of Hammerstein mathematical models is consists of a nonlinear static part followed by a linear dynamic part. A Wiener mathematical model is defined as a linear dynamic part precedes a nonlinear static part. These two mathematical models are useful in representing the nonlinearity of a dynamic system without introducing the complications associated with general nonlinear operator (Norquay & Palazoglu, 1998). Hammerstein and Wiener mathematical models reveal the capability of describing a wide class of different systems and apart from industrial examples, such as distillation columns, pH neutralization process (Biagiola & Figueroa, 2009; Tang et al., 2010), hydraulic plants, pneumatic valve and pimp-valve systems.

The parametric or state estimation problem of a dynamic system described by a state-space mathematical model consists of the development of an estimation procedure, which permits to estimate the parameters or the state variables of this mathematical model. In this case, the formulation of this parametric or state estimation problem will be carried on the basis of several sequences of experimental measurements (input, output, state variables) from the considered system, while using an adequate estimation method.

The rest of this chapter is organized as follows: Section 2 treats the description of nonlinear systems by mathematical models. Thus, four types of nonlinear mathematical models are presented: the mathematical models in series of functions, the state-space mathematical models, the input-output mathematical models and the block-oriented mathematical models. Section 3 presents the considered Wiener state-space discrete-time mathematical model, which can be described nonlinear systems. The parametric estimation problem of Wiener mathematical model is formulated in Section 4. A recursive parametric estimation algorithm is proposed. The convergence conditions of this algorithm are established, on the basis of the Lyapunov method. We develop, in Section 5, a recursive state estimation algorithm for the considered Wiener mathematical model. The convergence conditions of this algorithm are given, on the basis of the Lyapunov method. The section 6 is reserved to the study of the parametric and state estimation conjointly problem for the considered Wiener mathematical model. Thus, a recursive algorithm is developed, which can be estimate conjointly the parameters and the state variables of nonlinear system described by Wiener mathematical model. The convergence conditions of this algorithm are given. Section 7 treats a numerical simulation example to show the effectiveness of the proposed estimation algorithms. Finally, Section 7 gives some conclusions of the developed results and presents our future work.

2. NONLINEAR SYSTEMS DESCRIPTION

The study of a control law, which can be applied to a dynamic system to meet desired performance indices (stability, precision, etc.), is based on the description of the considered system by an adequate mathematical model. Let us note that a mathematical model is characterized by three elements, namely: class (nonlinear, deterministic, etc.), structure (order, delay) and the parameters. The parameters intervening in a mathematical model are related by a mathematical relationship, such as a difference equation in the discrete-time case. In addition, this equation may be linear or nonlinear relative or parameters or to observations (input, output, state variables) of the system.

This paragraph is devoted to the analysis of mathematical models for the description of nonlinear systems. In this way, different types of mathematical models are given. The problem of description of nonlinear systems by mathematical models has attracted the attention of many researchers (Haber & Keviczky, 1999; Kamoun, 2003; Giri & Bai, 2010; Billings, 2013).

Different types of representations of nonlinear systems are reported in the literature (Haber & Keviczky, 1999). The traditional representation is based on the approximation of a nonlinear system by a functions series of Volterra or Wiener (Schetzen, 1980). However, this type of representation has a large number of parameters, which requires powerful estimation and control methods (Ogunfunmi, 2007). Other types of representations are used in industrial applications, such as: state-space models, input-output models and block-oriented models.

2.1. Functions Series Mathematical Models

The representation of a nonlinear system by a functions series consists of the description of the system output, at a given time, by an infinite sum of dependent polynomial functions of the system input at the same time and earlier times. In this sense, we can use the representation of Volterra series, which is very popular in the analysis and description of nonlinear systems (Haber & Keviczky, 1999: Billings, 2013).

Using the Taylor series, nonlinear system can be described by polynomial functions. This description approach was suggested by Volterra for nonlinear systems described by continuous mathematical models (Volterra, 1959). The Volterra series expansion for nonlinear systems, which are described by discrete-time mathematical models, is widely used in the literature (Kamoun, 2003). Thus, we can describe a nonlinear system by the following discrete-time mathematical model:

$$y(k) = \mu + \sum_{i=0}^{\infty} \upsilon_i u(k-i) + \sum_{i=0}^{\infty}\sum_{j=0}^{\infty} \upsilon_{ij} u(k-i)u(k-j) + \cdots + \sum_{i=0}^{\infty} \cdots \sum_{h=0}^{\infty} \upsilon_{i\ldots h} u(k-h) \tag{1}$$

where $u(k)$ and $y(k)$ are respectively the system input and output respectively at the discrete-time k, μ is a constant, and υ_i and υ_{ij} are positive parameters. These parameters are called Volterra kernels.

A quadratic approximation of the mathematical model (1) can be obtained using the second order of Volterra kernels, such that:

$$y(k) = \mu + \sum_{i=1}^{q} \upsilon_i u(k-i) + \sum_{i=1}^{q}\sum_{j=1}^{q} \upsilon_{ij} u(k-i)u(k-j) \tag{2}$$

where q is a finite parameter.

The determination of the parameter q intervening in the mathematical model (2) can be made using an appropriate structural estimation method (Kortmann & Unbehauen, 1988), on the basis of a certain performance criterion.

It should be noted that the functions series representation of nonlinear systems contain multiple values of nonlinearity (e.g., hysteresis) is not possible.

2.2. State-Space Mathematical Models

Since several decades, state-space mathematical models were used in the description of nonlinear systems, to formulate estimation or control schemes. General state-space mathematical model, which can be described nonlinear systems, is given by:

$$\begin{aligned} x(k+1) &= f_1[x(k),\ u(k)] \\ y(k) &= f_2[x(k),\ u(k)] \end{aligned} \tag{3}$$

where $x(k)$, $u(k)$ and $y(k)$ are respectively the state vector, the system input and output at the discrete-time k, and $f_1[.]$ and $f_2[.]$ represent nonlinear functions with finite elements.

We can define different types of nonlinear state-space mathematical models from the general mathematical model (3); include in particular the following mathematical models:

State-space mathematical model linearized around a certain operating point, such as:

$$\begin{aligned} \Delta x(k+1) &= A\Delta x(k) + B\Delta u(k) \\ \Delta y(k) &= C^T \Delta x(k) + D\Delta u(k) \end{aligned} \tag{4}$$

Mathematical model described by a nonlinear state-space equation and a linear output equation, such that:

$$\begin{aligned} x(k+1) &= f_1[x(k),\ u(k)] \\ y(k) &= C^T x(k) + Du(k) \end{aligned} \tag{5}$$

Mathematical model described by a linear state-space equation and a nonlinear output equation, such that:

$$\begin{aligned} x(k+1) &= Ax(k) + Bu(k) \\ y(k) &= f_2[x(k),\ u(k)] \end{aligned} \tag{6}$$

Mathematical model described by separable state-space and output nonlinear equations, such that:

$$\begin{aligned} x(k+1) &= f_3[x(k)] + f_4[u(k)] \\ y(k) &= f_5[x(k)] + f_6[u(k)] \end{aligned} \tag{7}$$

where $f_3[.]$, $f_4[.]$, $f_5[\cdot]$ and $f_6[\cdot]$ represent nonlinear functions with finite elements.

2.3. Input-Output Mathematical Models

The description of nonlinear systems by input-output mathematical models is a common approach in industrial applications for the formulation of parametric estimation or control schemes (Haber & Unbehauen, 1990; Haber & Keviczky, 1999; Kamoun, 2008; Billings, 2013). Leontaritis and Billings (1985a, 1985b) studied the problems related to the description of a nonlinear system with an input-output mathematical model, and this, from the representation of this system by a state-space mathematical model of the type (3). Conditions for passing from the state-space representation to the input-output representation are given. The input-output mathematical model can be classified into deterministic models and stochastic models.

Input-output mathematical models of type NDARMA (Deterministic Nonlinear Autoregressive Moving Average) can describe nonlinear systems operating in a deterministic environment, such that:

$$y(k) = f_d[y(k-1),\cdots,y(k-n_A),u(k-1),\cdots,\ u(k-n_B)] \tag{8}$$

where $f_d[.]$ represents a deterministic nonlinear function, which depends on the input and output sequence of the system.

In addition, input-output mathematical models of type NARMAX (Nonlinear Autoregressive Moving Average with eXogenous) can describe nonlinear systems operating in a stochastic environment, such that:

$$y(k) = f_s[y(k-1),\cdots,y(k-n_A),\ u(k-1),\cdots,u(k-n_B),\ e(k-1),\cdots,e(k-n_C)] + e(k) \tag{9}$$

where $e(k)$ denotes the noise which acts on the system and $f_s[.]$ represents a nonlinear stochastic function, which depends on the input, the output and the noise of the system.

Description of nonlinear systems operating in a stochastic environment by mathematical models NARMAX is a widespread approach in various industrial applications (Kamoun, 2003; Billings, 2013).

Note that the input-output mathematical models of type NARMAX or NDARMA, which have only a linearity with respect to the parameters, are widely used in the literature (Kamoun, 2003; Giri & Bai, 2010; Billings, 2013).

The output $y(k)$ of a nonlinear system described by an input-output mathematical model, which is nonlinear with respect to observations but linear in relation to the parameters, can be described as follows:

$$y(k) = \theta^T \psi(k) + e(k) \tag{10}$$

where θ is the parameter vector (linear parameters) and $\psi(k)$ is the observation vector. In this case, some or all components of the system input and the system output intervening in the observation vector are nonlinear functions.

2.4. Block-Oriented Mathematical Models

Some classes of nonlinear systems can be represented by mathematical models of type (2) which correspond to a series of linear functions with respect to parameters. This type of representation leads to connected systems, which are constituted by linear dynamic parts and nonlinear static parts; hence the name of block-oriented mathematical models. It should be noted that the block-oriented mathematical models are widespread in the literature (Wigren, 1993; Vörös, 2002; Giri & Bai, 2010). In fact, the study of a nonlinear dynamical system in terms of connected systems can further simplify the formulation of estimation or control system schemes. Add that, in this case, we can apply the developed results for linear systems, with some techniques for practical implementation.

During the last three decades, the study of the identification or control problem of nonlinear systems, which can be described by block-oriented mathematical models, has been an active research area in several laboratories. We can distinguish two main types of block-oriented mathematical models, i.e., which consist of separated linear dynamic parts and static nonlinear parts. This corresponds to the Hammerstein mathematical model and the Wiener mathematical model, which are widespread in many industrial applications. Let us indicate that in block-oriented mathematical models, each block (linear dynamic part and static nonlinearity part) can be connected in many different ways (series, parallel, feedback) and this flexibility provides the block-oriented modeling approach with an ability to capture

Figure 1. Structure of the considered Hammerstein mathematical model

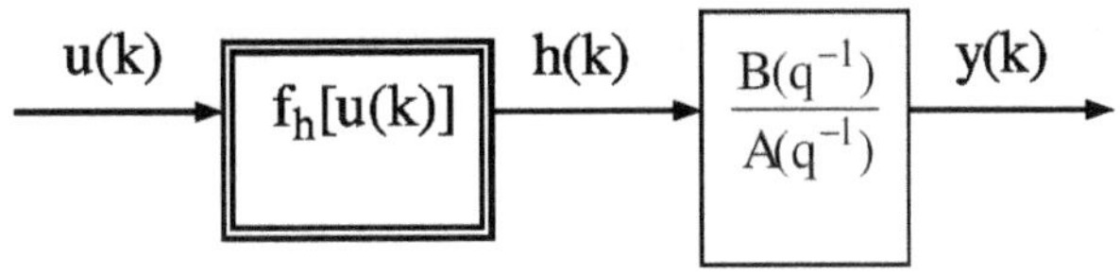

a large class of nonlinear systems. Thus, we can use other types of block-oriented mathematical models, which correspond to the Hammerstein and Wiener mathematical models (e.g., Wiener-Hammerstein mathematical models, Hammesrtein-Wiener-Hammerstein mathematical models), for the description of nonlinear systems (Haber & Keviczky 1999; Giri & Bai, 2010; Billings, 2013).

It is important to indicate that intermediate signals in such block-oriented mathematical models are not measurable and the inaccessibility of such measurements is the main difficulty in the formulation of parametric or state estimation schemes for nonlinear systems. To overcome this difficulty, some approaches and assumptions can be used (Kamoun, 2003).

In what follows, we present discrete-time Hammerstein and Wiener mathematical models.

Hammerstein Mathematical Models

The description of a nonlinear dynamic system by a Hammerstein mathematical model corresponds to a shape of the mathematical model, which is constituted by a static nonlinear part followed by a dynamic linear part.

The structure of the considered discrete-time deterministic Hammerstein mathematical model is represented in Figure 1.

Let us note that the form of the Hammerstein mathematical model, as illustrated in Figure 1, is widely used in many industrial applications. For example, this mathematical model form can be used to describe an electric motor driving a mechanical load and driven by a relay. In this case, the electric motor is described by a dynamic linear input-output mathematical model and the relay is defined by a static nonlinear function.

The dynamic linear part of the considered Hammerstein mathematical model is described by:

$$A(q^{-1})y(k) = B(q^{-1})h(k) \tag{11}$$

where $h(k)$ and $y(k)$ are respectively the input and the output of the dynamic linear part at the discrete-time k, and $A(q^{-1})$ and $B(q^{-1})$ are polynomials, which are defined by:

$$A(q^{-1}) = 1 + a_1 q^{-1} + \cdots + a_{n_A} q^{-n_A} \tag{12}$$

and

$$B(q^{-1}) = b_1 q^{-1} + \cdots + b_{n_B} q^{-n_B} \tag{13}$$

where n_A and n_B are the polynomials orders of $A(q^{-1})$ and $B(q^{-1})$, respectively.

The static nonlinear part of the Hammerstein mathematical model is given by the following equation:

$$h(k) = f_h[u(k)] \tag{14}$$

where $f_h[.]$ represents the nonlinear function.

We can use a polynomial approximation of equation (14), such that:

$$h(k) = \sum_{r=1}^{p} \beta_r u^r(k) + \Delta h[u(k)] \tag{15}$$

where $\Delta h[u(k)]$ is an approximation error of the nonlinear function, which can be likened to a noise acting on the system output. The variance of this approximation error depends on the selected value of the nonlinearity degree p. Thus, the value of this variance decreases as the value of the nonlinearity degree p increases

From equations (11) and (15), we can define the system output $y(k)$ by the following expression:

$$y(k) = -\sum_{i=1}^{n_A} a_i y(k-i) + \sum_{j=1}^{n_B} \sum_{r=1}^{p} b_j \beta_r u^r(k-j) + \sum_{j=1}^{n_B} \Delta h[u(k-j)] \tag{16}$$

Several types of Hammerstein mathematical models are used in the description and the parametric estimation of nonlinear systems based on different configurations of the static nonlinear part (Zi-Qiang, 1994; Al-Duwaish Nazmul Karim, 1997; Li, 1999; Belforte & Gay, 2001; Giri & Bai, 2010; Billings, 2013).

Wiener Mathematical Models

The description of a nonlinear system by a Wiener mathematical model leads to a form of the mathematical model, which consists of a dynamic linear part followed by a static nonlinear part.

The structure of the considered discrete-time deterministic Wiener mathematical model can be seen in Figure 2.

Note that the form of the Wiener mathematical model, as illustrated in Figure 2, is widespread in many industrial applications. For example, this type of mathematical model may be used in the description of an electric oven provided with a temperature sensor. In this case, a dynamic linear mathematical model describes the electric oven and the temperature sensor is defined by a static nonlinear function.

The dynamic linear part of the considered Wiener mathematical model, which is represented in Figure 2, is described by the following equation:

$$A(q^{-1})w(k) = B(q^{-1})u(k) \tag{17}$$

Figure 2. Structure of the considered Wiener mathematical model

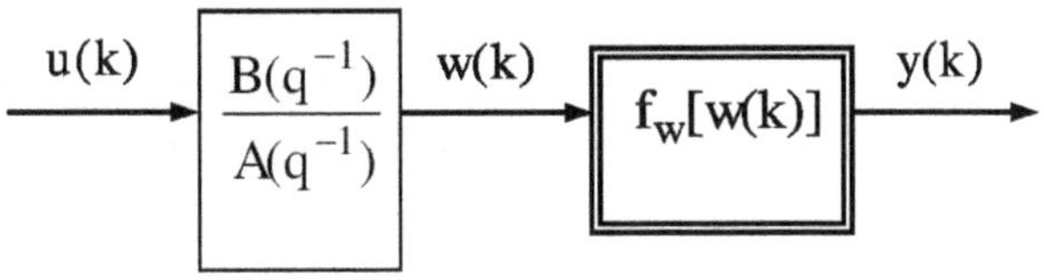

where $u(k)$ and $w(k)$ are respectively the input and the output of the dynamic linear part at the discrete-time k, and $A(q^{-1})$ and $B(q^{-1})$ are polynomials, which are defined respectively by (12) and (13).

The static nonlinear part of the considered Wiener mathematical model is defined by the following equation:

$$y(k) = f_w[w(k)] \tag{18}$$

where $f_w[.]$ represents the nonlinear function.

Then, equation (18) can be defined by a polynomial approximation, such that:

$$y(k) = \sum_{r=1}^{p} \beta_r w^r(k) + \Delta y[w(k)] \tag{19}$$

where p denotes the nonlinearity degree, which can be selected suitably, and β_r, $r = 1,\cdots,p$, are unknown parameters and $\Delta y[w(k)]$ represents the approximation error of the nonlinear function. This approximation error, which depends on the selected value of the nonlinearity degree p, can be regarded as additive noise in the system output. Note that for a suitable choice of the value of the nonlinearity degree p, we can assume that the approximation error can be neglected. This assumption is often accepted in several industrial applications.

Given the dynamic linear part of the considered Wiener mathematical model, as described by equation (17), the system output $y(k)$ can be defined in the following form:

$$y(k) = \sum_{r=1}^{p} \beta_r [-\sum_{i=1}^{n_A} a_i w(k-i) + \sum_{j=1}^{n_B} b_j u(k-j)]^r + \Delta y[w(k)] \tag{20}$$

Different types of Wiener mathematical models, based on various forms of static nonlinear part, are used for the formulation of parametric estimation schemes of nonlinear systems (Wigren, 1993; Sbarbaro & Johansen, 1997; Kamoun, 2003; Giri & Bai, 2010; Billings, 2013).

3. WIENER STATE-SPACE DISCRETE-TIME MATHEMATICAL MODEL

Since several years, description of dynamical systems by state-space mathematical models has been widely used by several authors and applied in various industrial fields. In this sense, different types of state-space mathematical models, with various canonical forms, have been developed and published in the literature (Kamoun, 2003). In addition, the description of dynamic systems by state-space mathematical models is a common approach in many areas of science and engineering technology (e.g., automatic control area). In this context, the first important works are developed by Kalman (1960), where state estimation method and optimal control systems are proposed.

Since the extraordinary evolution of computer technology during the past two decades and the remarkable development of digital computing and data processing (large memory, computational speed, easy implementation and flexible of estimation or control algorithms, etc.), state-space continuous mathematical models are less used, leaving room for state-space discrete-time mathematical models. Thus, we consider here only dynamic systems, which are described by state-space discrete-time mathematical models. The analysis and synthesis problems of parametric and state estimation schemes can be applied to dynamic systems, which are described by linear state-space discrete-time mathematical models were discussed by many researchers, and interest various studies are published in the literature (Kamoun, 2007).

We can distinguish two types of state-space mathematical models. The first type concerns the state-space mathematical models, which is represented in a canonical form. The second type concerns the state-space mathematical models, which is represented in a general form. In this section, we consider state-space discrete-time mathematical models with unknown parameters that are represented in a general form. We deal in particular nonlinear systems operating in a stochastic environment, which can be described by the class of mathematical models

Let us consider a nonlinear system described by a Wiener discrete-time state-space mathematical model, which is composed by a linear dynamic part followed by a static nonlinear part.

The linear dynamic part and the nonlinear static part of the considered Wiener mathematical model are described by the two following mathematical models (21) and (22), respectively:

$$\begin{aligned} x(k+1) &= Ax(k) + Bu(k) + v(k) \\ z(k) &= C^T x(k) + du(k) + w(k) \end{aligned} \tag{21}$$

and

$$y(k) = f[z(k)] + e(k) \tag{22}$$

where $u(k) \in \Re$ and $y(k) \in \Re$ are the system input and output, $x(k) \in \Re^n$ and $z(k) \in \Re$ are the internal variables, $v(k)$, $w(k)$ and $e(k)$ are stochastic white noises with zero means, $f[.]$ is a nonlinear function, which can be described by a matrix form, such as:

$$f[z(k)] = \Theta^T \Psi(k) \tag{23}$$

where Θ is the parameters vector and $\Psi(k)$ is the nonlinear observation vector, the matrix A and the vectors B and C are definite respectively by:

$$A = \begin{bmatrix} a_{11} & \cdots & a_{1n} \\ \vdots & \ddots & \vdots \\ a_{n1} & \cdots & a_{nn} \end{bmatrix} \tag{24}$$

$$B^T = [b_1 \dots b_n] \tag{25}$$

and

$$C^T = [c_1 \dots c_n] \tag{26}$$

The parameters a_{ij}, $i, j = 1, \cdots, n$, in the matrix A, and the parameters b_j and c_i in the vectors B and C, respectively, are supposed to be unknown. Also, the parameters in the vector Θ are supposed also unknown. Nothing that the internal variables $x(k)$ and $z(k)$ are inaccessible for measurement. Let us retain the following assumptions:

1. The orders n and the nonlinearity degree p are known,
2. The characteristic of the nonlinear element can be continuous as can be discontinuous,
3. The linear dynamic part is asymptotically stable.

The purpose of this chapter is to present a recursive algorithm to estimate the state and the parameters of the considered Wiener mathematical model (21) and (22), from given input-output data $\{u(k), y(k)\}$. First, we will formulate the parametric estimation problem assuming that the internal variables $x(k)$ and $z(k)$ are measurable. Second, we treat the inverse case, i.e., we suppose that the parameters of the considered model are known but the internal variables are inaccessible for measurement. Finally, we combined these two problems and we suppose that the state vector and the parameters are unknown.

4. DEVELOPMENT OF PARAMETRIC ESTIMATION METHOD

Several recursive algorithms have been proposed to solve the parametric estimation problem, which are based on the prediction error method and such as least squares technique (Kamoun, 2003). The least squares technique is the simplest technique, which can be extended for an online implementation (Wang et al., 2010).

4.1. Recursive Parametric Estimation Algorithm

In this section, we will propose a recursive parametric estimation algorithm for parameters estimation, based on the adjustable model and the Least Squares technique.

Referring to the Wiener mathematical model (21) and (22), we propose the following adjustable mathematical model:

$$\begin{aligned} x_p(k+1) &= \hat{A}(k+1)x(k) + \hat{B}(k+1)u(k) \\ z_p(k) &= \hat{C}^T(k)x(k) + \hat{d}(k)u(k) \\ y_p(k) &= \hat{\Theta}^T(k)\Psi(k) \end{aligned} \quad (27)$$

where $\hat{A}(k+1)$ and $\hat{B}(k+1)$ represent the estimated parameters of the matrix A and the vector B, respectively, at the discrete-time $k+1$, $\hat{C}(k)$, $\hat{d}(k)$ and $\hat{\Theta}(k)$ is the estimated vectors of C, d and Θ, respectively.

The parametric estimation problem can be formulated as follows: Given the available system input-output measurements and the internal variables measurements, estimate the parameters of the Wiener mathematical model (21) and (22) minimizing the following prediction errors

$$\delta_x(k) = x(k) - x_p(k) \quad (28)$$

$$\delta_z(k) = z(k) - z_p(k) \quad (29)$$

and

$$\delta_y(k) = y(k) - y_p(k) \quad (30)$$

These equations can be written as the following forms

$$\delta_x(k) = x(k) - \hat{A}(k)x(k-1) - \hat{B}(k)u(k-1) \quad (31)$$

$$\delta_z(k) = z(k) - \hat{C}^T(k)x(k) - \hat{d}(k)u(k) \quad (32)$$

and

$$\delta_y(k) = y(k) - \hat{\Theta}^T(k)\Psi(k) \quad (33)$$

The recursive parametric estimation algorithm allowing estimating the various parameters intervening in the matrix A, in the vectors B, C, d, and Θ, is given by:

$$\begin{aligned}
&\hat{A}(k+1) = \hat{A}(k) + \zeta_x(k) G_x \delta_x(k) x^T(k-1) \\
&\hat{B}(k+1) = \hat{B}(k) + \zeta_x(k) G_x \delta_x(k) u(k-1) \\
&\hat{C}(k+1) = \hat{C}(k) + \zeta_z(k) \alpha I_{nxn} \delta_z(k) x(k) \\
&\hat{d}(k+1) = \hat{d}(k) + \zeta_z(k) \alpha \delta_z(k) u(k) \\
&\hat{\Theta}(k+1) = \hat{\Theta}(k) + \zeta_y(k) G_y \delta_y(k) \Psi(k) \\
&\delta_x(k) = x(k) - \hat{A}(k) x(k-1) - \hat{B}(k) u(k-1) \\
&\delta_z(k) = z(k) - \hat{C}^T(k) x(k) - \hat{d}(k) u(k) \\
&\delta_y(k) = y(k) - \hat{\Theta}^T(k) \Psi(k) \\
&\zeta_x(k) = \frac{\ell_x}{\lambda_{G_x}[u^2(k-1) + x^T(k-1)x(k-1)]} \\
&\zeta_z(k) = \frac{\ell_z}{\alpha[u^2(k) + x^T(k)x(k)]} \\
&\zeta_y(k) = \frac{\ell_y}{\lambda_{G_y}[\Psi^T(k)\Psi(k)]}
\end{aligned} \tag{34}$$

where G_x and G_y are definite positive symmetrical matrices, λ_{G_x} and λ_{G_y} are, respectively the maximum eigenvalue of the matrix G_x and the maximum eigenvalue of the matrix G_y, α is a positive constant and I_{nxn} is a unit matrix with dimension nxn, ℓ_x, ℓ_z and ℓ_y are three parametric gains, that must be chosen in a proper way to ensure the stability of the parametric estimation scheme.

4.2. Convergence Analysis of the Developed Parametric Estimation Algorithm

This section deals with the analysis of the convergence conditions of the developed parametric estimation algorithm (34) of the considered Wiener mathematical model. The convergence study was achieved using the Lyapunov method.

The prediction errors related to the state variables $x_i(k)$, the intermediate output $z(k)$ and the system output $y(k)$ can be written as follows:

$$\delta_x(k) = \tilde{A}(k) x(k-1) + \tilde{B}(k) u(k-1) \tag{35}$$

$$\delta_z(k) = \tilde{C}^T(k) x(k) + \tilde{d}(k) u(k) \tag{36}$$

and

$$\delta_y(k) = \tilde{\Theta}^T(k)\Psi(k) \tag{37}$$

where $\tilde{A}(k) = A - \hat{A}(k)$, $\tilde{B}(k) = B - \hat{B}(k)$, $\tilde{C}(k) = C - \hat{C}(k)$, $\tilde{d}(k) = d - \hat{d}(k)$ and $\tilde{\Theta}(k) = \Theta - \hat{\Theta}(k)$ are the estimated parametric errors. These errors must tend asymptotically to minimal values close to zero, in order to ensure a good estimation quality.

To analyze the convergence conditions of the developed parametric estimation algorithm (34), we will use the following Lyapunov functions:

$$L_x(k) = \mathrm{tr}[\tilde{A}^T(k)\tilde{A}(k)] + \tilde{B}^T(k)\tilde{B}(k) \tag{38}$$

$$L_z(k) = \tilde{C}^T(k)\tilde{C}(k) + \tilde{d}^2(k) \tag{39}$$

and

$$L_y(k) = \tilde{\Theta}^T(k)\tilde{\Theta}(k) \tag{40}$$

where the symbol $\mathrm{tr}[A]$ indicates the trace of the matrix A.

Define the following variations of Lyapunov functions

$$\Delta L_x(k+1) = \tilde{B}^T(k+1)\tilde{B}(k+1) + \mathrm{tr}[\tilde{A}^T(k+1)\tilde{A}(k+1)] - \tilde{B}^T(k)\tilde{B}(k) - \mathrm{tr}[\tilde{A}^T(k)\tilde{A}(k)] \tag{41}$$

$$\Delta L_z(k+1) = \tilde{C}^T(k+1)\tilde{C}(k+1) + \tilde{d}^2(k+1) - \tilde{C}^T(k)\tilde{C}(k) - \tilde{d}^2(k) \tag{42}$$

and

$$\Delta L_y(k+1) = \tilde{\Theta}^T(k+1)\tilde{\Theta}(k+1) - \tilde{\Theta}^T(k)\tilde{\Theta}(k) \tag{43}$$

Referring to the parametric estimation algorithm of state-space linear systems in Glaria Lopez and Sbarbaro (2011), we obtain the following recursive relations of the parametric estimation errors $\tilde{A}(k+1)$, $\tilde{B}(k+1)$, $\tilde{C}(k+1)$, $\tilde{d}(k+1)$ and $\tilde{\Theta}(k+1)$.

$$\tilde{A}(k+1) = \tilde{A}(k) - \varsigma_x(k)G_x\delta_x(k)x^T(k-1) \tag{44}$$

$$\tilde{B}(k+1) = \tilde{B}(k) - \varsigma_x(k)G_x\delta_x(k)u(k-1) \tag{45}$$

$$\tilde{C}(k+1) = \tilde{C}(k) - \varsigma_z(k)G_{1z}\delta_z(k)x(k) \tag{46}$$

$$\tilde{d}(k+1) = \tilde{d}(k) - \zeta_z(k) G_{2z} \delta_z(k) u(k) \tag{47}$$

and

$$\tilde{\Theta}(k+1) = \tilde{\Theta}(k) - \zeta_y(k) G_y \delta_y(k) \Psi(k) \tag{48}$$

Using the two following equations

$$\text{tr}\left[\tilde{A}(k) G_x \delta_x(k) x^T(k-1)\right] = x^T(k-1) \tilde{A}^T(k) G_x \delta_x(k) \tag{49}$$

and

$$\text{tr}\left[x(k-1) \delta_x{}^T(k) G_x{}^T G_x \delta_x x^T(k-1)\right] = x^T(k-1) \delta_x{}^T(k) G_x{}^T G_x \delta_x(k) x(k-1) \tag{50}$$

We can rewrite equations (41), (42) and (43) in the following forms:

$$\begin{aligned} \Delta L_x(k+1) = & -2\xi_x(k)\left[u(k-1)\tilde{B}^T(k) + x^T(k-1)\tilde{A}^T(k)\right] G_x \delta_x(k) \\ & + \xi_x{}^2(k)\left[u^2(k-1)\delta_x{}^T(k) G_x{}^T G_x + x^T(k-1)\delta_x{}^T(k) G_x{}^T G_x x(k-1)\right]\delta_x(k) \end{aligned} \tag{51}$$

$$\begin{aligned} \Delta L_z(k+1) = & -2\xi_z(k)\left[\tilde{C}^T(k) G_{1z} x(k) + \tilde{d}(k) G_{2z} u(k)\right]\delta_z(k) \\ & + \xi_z{}^2(k)\delta_z(k)\left[x^T(k) G_{1z}{}^T(k) G_{1z} x(k) + u(k) G_{2z}{}^T G_{2z} u(k)\right]\delta_z(k) x(k) \end{aligned} \tag{52}$$

and

$$\Delta L_y(k+1) = -2\xi_y(k)\delta_y{}^T(k) G_y \delta_y(k) + \xi_y{}^2(k)\psi^T(k) G_y{}^T G_y \psi(k)\delta_y{}^2(k) \tag{53}$$

In order to simplify the calculation of equation (52), we propose to choose G_{1z} and G_{2z} as follows:

$$G_{1z} = \alpha I_{nxn} \tag{54}$$

and

$$G_{2z} = \alpha \tag{55}$$

where α is a positive constant and I_{nxn} is a unit matrix of dimension nxn.

Thus, we can rewrite equation (52) in the following form:

$$\begin{aligned} \Delta L_z(k+1) = & -2\xi_z(k)\alpha\left[\tilde{C}^T(k) x(k) + \tilde{d}(k) u(k)\right]\delta_z(k) \\ & + \xi_z{}^2(k)\delta_z(k)\alpha^2\left[x^T(k) x(k) + u^2(k)\right]\delta_z(k) x(k) \end{aligned} \tag{56}$$

Then, we can write equations (51), (56) and (53) as follows:

$$\Delta L_x(k+1) = -2\delta_x^{\ T}(k)\Phi_x(k)\delta_x(k) \tag{57}$$

$$\Delta L_z(k+1) = -\delta_z^{\ T}(k)\Phi_z(k)\delta_z(k) \tag{58}$$

and

$$\Delta L_y(k+1) = -\delta_y^{\ T}(k)\Phi_y(k)\delta_y(k) \tag{59}$$

where $\Phi_x(k)$, $\Phi_z(k)$ and $\Phi_y(k)$ are given by, respectively:

$$\Phi_x(k) = -2\xi_x(k)G_x - \xi_x^{\ 2}(k)u^2(k-1)G_x^{\ T}G_x - \xi_x^{\ 2}(k)x^T(k-1)x(k-1)G_x^{\ T}G_x \tag{60}$$

$$\Phi_z(k) = -2\xi_z(k)\alpha - \xi_z^{\ 2}(k)\alpha^2\left[x^T(k)x(k) + u^2(k)\right] \tag{61}$$

and

$$\Phi_y(k) = -2\xi_y(k)G_y - \xi_z^{\ 2}(k)\psi^T(k)\psi(k)G_y^{\ T}G_y \tag{62}$$

The convergence conditions, based on the Lyapunov method, of the developed recursive parametric estimation scheme are given by:

$$\delta_x^{\ T}(k)\Phi_x(k)\delta_x(k) > 0 \tag{63}$$

$$\delta_z^{\ T}(k)\Phi_z(k)\delta_z(k) > 0 \tag{64}$$

and

$$\delta_y^{\ T}(k)\Phi_y(k)\delta_y(k) > 0 \tag{65}$$

From the conditions (63), (64) and (65), we can deduce that the gains $\xi_x(k)$, $\xi_z(k)$ and $\xi_y(k)$ satisfy the following inequalities:

$$\zeta_x(k) = \frac{\ell_x}{\lambda_{G_x}\left[x^T(k-1)x(k-1) + u^2(k-1)\right]} \tag{66}$$

$$\zeta_z(k) = \frac{\ell_z}{\alpha\left[x^T(k)x(k) + u^2(k)\right]} \tag{67}$$

and

$$\zeta_y(k) = \frac{\ell_y}{\lambda_{G_y}\psi^T(k)\psi(k)} \tag{68}$$

where the parameters ℓ_x, ℓ_z and ℓ_y are defined in the following Theorem 1.

Theorem 1: Consider a nonlinear dynamic system described by Wiener mathematical model, which is composed of a dynamic linear part, modulated by a state-space equation, and a static nonlinear part. The estimation of the parameters in the matrix A and in the vectors B, C and Θ of this mathematical model can be made using the recursive estimation algorithm (46). The choice of parameters ℓ_x, ℓ_z and ℓ_y must satisfy the following conditions:

$$0 < \ell_x < 2 \tag{69}$$

$$0 < \ell_z < 2 \tag{70}$$

and

$$0 < \ell_y < 2 \tag{71}$$

to ensure the stability of the proposed scheme.

Nothing that the parameters of the gains ℓ_x, ℓ_z and ℓ_y can be chosen time-varying, such that: $0 < \ell_x(k) < 2$, $0 < \ell_z(k) < 2$, $0 < \ell_y(k) < 2$.

5. DEVELOPMENT OF STATE ESTIMATION METHOD

This section is reserved for state estimation of the considered Wiener mathematical models, as described by (21) and (22).

The development of a control law (optimal, adaptive, etc.), which can applied to a dynamic system described by a state-space mathematical model, is based on the knowledge of the state variables values of this mathematical model. However, in many practical situations, some or all of these state variables cannot be determined by direct measurement. Note that the dynamic system can be subjected to various types of random disturbances.

Thus, in order to develop a state feedback control law, we must proceed, in a first step, to estimate the state variables system. This state estimation step will be conducted from the knowledge of several measurements carried out on the controlled system and *a priori* information that we have on the statistical characteristics of noise that affects the measured output value and the noise acting the state variables. In order to formulate the state estimation problem, we assume that the parameters in the state-space mathematical model are known. However, in many industrial applications, these parameters are partially or totally unknown. Consequently, a step of parametric estimation is necessary.

It should be noted that the term state estimation used here corresponds to the term filtering, which is used by the founders of the filtering theory of stochastic systems described by state-space equations (Kalman, 1960; Kalman & Bucy, 1961). Let us note that the first significant work concerning the filtering of stationary random signals is developed by Norbert Wiener around 1940, see his book published in 1949 (Wiener, 1949). He proposed a stationary optimal linear filter, which currently bears his name Wiener filter. This filter has some restrictions (difficulty to develop a recursive version, difficulty to generalize to non-stationary signals, etc.).

During the past three decades, several studies related to filtering have been developed and published in the literature (Nørgaard et al., 2000; Rao et al., 2003; Alessandria, et al., 2008).

The purpose of this section is the analysis of a recursive method for estimating state variables of nonlinear systems operating in a stochastic environment, which are described by state-space mathematical models, as given by (21). In this case, we will develop a recursive algorithm, which permits to estimate the state variables of this system.

5.1. Recursive State Estimation Algorithm

The well-known state estimation filter is the Kalman Filter (KF), which has been widely used in the literature due to its simplicity, optimality and robustness. In the case of general nonlinear system, the application of the linear KF can be difficult because of the inherent nonlinear dynamic.

Several research activities have been developed in the literature to solve the state estimation problem of nonlinear dynamical systems. The most popular estimation approach is the Extended Kalman Filter (EKF), which utilizes a linearized model along the estimated state trajectory. However, this linearization may produce highly unstable filters and the derivation of the Jacobian matrices required in this filter, lead to significant implementation difficulties (Xiong et al., 2012).

In this chapter, a new state estimation algorithm is developed. The linearization steps and Jacobian calculation, used in the EKF, are not required in this developed filter. Indeed, this state estimation filter is similar to the linear KF, generalizes to nonlinear systems, and more particularly, which can be described by blocks oriented models. Recall that the studies of a dynamic nonlinear system in term of connected systems simplify the formulation of the identification diagram and control system. Since, one can apply the developed results for the linear systems, with some techniques of practical implementation (Lopez and Yuz, 2012).

In this section, we will present a state estimation algorithm based on the linear Kalman filter, in order to estimate the state vector of the considered system, which is described by a wiener mathematical model (21).

First, we define $\hat{x}(k)$ the *a priori* estimate of the state $x(k)$, which can be formulated at discrete-time k, given the data at discrete-time $k-1$. We define also $\hat{x}_{\circ}(k)$ the *a posteriori* estimate of the state $x(k)$, developed at discrete-time k, based on the data at the same time k.

We can define the *a priori* estimation error $\xi_x(k)$ and the *a posteriori* estimation error $\xi_{x\circ}(k)$, by the following expressions, respectively:

$$\xi_x(k) = x(k) - \hat{x}(k) \tag{72}$$

and

$$\xi_{x\circ}(k) = x(k) - \hat{x}_{\circ}(k) \tag{73}$$

The variance-covariance matrix $P_x(k)$ of the *a priori* estimation error $\xi_x(k)$ can be defined by:

$$P_x(k) = \mathrm{E}[\xi_x(k)\xi_x^{\,T}(k)] \tag{74}$$

where the operator $\mathrm{E}[.]$ denotes the expectation.

The variance-covariance matrix $P_{x\circ}(k)$ of the *a posteriori* estimation error $\xi_{x\circ}(k)$ is defined as follows:

$$P_{x\circ}(k) = \mathrm{E}[\xi_{x\circ}(k)\xi_{x\circ}^{\,T}(k)] \tag{75}$$

Based on the equation of the Kalman filter, we can express the *a posteriori* estimate $\hat{x}_{\circ}(k)$ of the considered model (21) as follows:

$$\hat{x}_{\circ}(k) = \hat{x}(k) + K(k)[z(k) - C^T\hat{x}(k) - du(k)] \tag{76}$$

with $K_x(k)$ an adaptation gain vector, which can be calculated by the following expression:

$$K_x(k) = \frac{P_x(k)C^T}{\sigma_w^{\,2} + CP_x(k)C^T} \tag{77}$$

where $\sigma_w^{\,2}$ is the variance of the noise $w(k)$ affecting the output of the linear part. The problem is that the output of the linear part $z(k)$ is inaccessible to measurement. For that, it must be estimated using measured data of the considered system.

The problem is that the output of the linear part $z\left(k\right)$ is inaccessible to measurement. For that, it must be estimated using measured data of the considered system.

Define the *a priori* estimation error

$$\xi_z(k) = z(k) - \hat{z}(k) \tag{78}$$

which can be rewritten as:

$$\xi_z(k) = C^T \xi_x(k) \tag{79}$$

This *a priori* estimation error, called also innovation or residual, corresponds to the best estimate of the noise $w(k)$ affecting the intermediate output $z(k)$. Thus, if the estimation procedure is done on a sufficiently large measurement horizon, allowing the estimate state $\hat{x}(k)$ to get close to the state $x(k)$, then the variance $\sigma_z^{\ 2}(k)$ of the *a priori* prediction error $\xi_z(k)$ approximate the noise variance $\sigma_w^{\ 2}(k)$.

The problem consist on the estimation of the intermediate variable $z(k)$, which is supposed inaccessible for the measurement. In order to resolve this problem, we begin by rewriting the nonlinear function (23) in the following form:

$$y(k) = \gamma z(k) + \theta^T \psi(k) \tag{80}$$

Consider the following equations:

$$\begin{aligned} z(k+1) &= C^T x(k+1) + du(k+1) + w(k+1) \\ y(k) &= \gamma z(k) + \theta^T \psi(k) + e(k) \end{aligned} \tag{81}$$

This can be written in the following form:

$$\begin{aligned} z(k+1) &= C^T A x(k) + C^T B u(k) + w(k+1) \\ y(k) &= \gamma C^T x(k) + \gamma d u(k) + \theta^T \psi(k) + e(k) \end{aligned} \tag{82}$$

Define the *a priori* estimation error $P_z(k)$ of the *a priori* estimation error by the following equation:

$$P_z(k) = \mathrm{E}[\xi_z(k) \xi_z^{\ T}(k)] \tag{83}$$

We can express the estimate $\hat{z}(k+1)$ of the considered model (21) as follows:

$$\hat{z}(k+1) = C^T A \hat{x}_{\circ}(k) + C^T B u(k) + du(k+1) + K_z(k)[y(k) - \gamma C^T \hat{x}_{\circ}(k) - \gamma du(k) - \theta^T \widehat{\psi}(k)] \tag{84}$$

where $K_z(k)$ is an adaptation gain vector, which is defined by the following expression:

$$K_z(k) = \frac{C^T A P_x(k) \gamma C}{\gamma C^T P_x(k) \gamma C + \sigma_e^{\ 2} + \theta^T G(k) \theta} \tag{85}$$

where $\sigma_e^{\ 2}$ is the variance of the noise $e(k)$ affecting the output system and $G(k)$ is the variance-covariance matrix, defined by:

$$G(k) = \mathrm{E}\left[[\psi(k) - \widehat{\psi}(k)][\psi(k) - \widehat{\psi}(k)^{T}\right] \tag{86}$$

where $\widetilde{\psi}(k)$ is the approximated vector of $\psi(k)$.

The resulting recursive state estimation algorithm of the wiener mathematical model, in the sense of Kalman filter, consists of the following steps:

1. Step 1 (Estimation at Discrete-Time k):
   a. Determining the *a posteriori* estimate $\hat{x}_{\circ}(k)$ of the state $x(k)$:

$$\begin{aligned} \hat{x}_{\circ}(k) &= \hat{x}(k) + K_{x}(k)[\hat{z}(k) - C^{T}\hat{x}(k) - du(k)] \\ K(k) &= \frac{P_{x}(k)C}{\sigma_{w}^{2} + C^{T}P_{x}(k)C} \end{aligned} \tag{87}$$

   b. Determining the variance-covariance matrix $P_{x\circ}(k)$ of the *a posteriori* estimation error $\xi_{x\circ}(k)$:

$$P_{x\circ}(k) = P_{x}(k) - K_{x}(k)C^{T}P_{x}(k) \tag{88}$$

2. Step 2 (Prediction):
   a. Determining the *a priori* estimate $\hat{x}(k+1)$ of the state vector $x(k+1)$:

$$\hat{x}(k+1) = A\hat{x}_{\circ}(k) + Bu(k) \tag{89}$$

   b. Determining the variance-covariance matrix $P_{x}(k+1)$ of the *a priori* estimation error $\xi_{x}(k)$:

$$P_{x}(k+1) = AP_{x\circ}(k)A^{T} + Q \tag{90}$$

   c. Determining the estimate $\hat{z}(k+1)$ of the output linear part:

$$\begin{aligned} \hat{z}(k+1) &= C^{T}A\hat{x}_{\circ}(k) + C^{T}Bu(k) + du(k+1) + K_{z}(k)[y(k) - \gamma C^{T}\hat{x}_{\circ}(k) - \gamma du(k) - \theta^{T}\widehat{\psi}(k)] \\ K_{z}(k) &= \frac{C^{T}AP_{x}(k)\gamma C}{\gamma C^{T}P_{x}(k)\gamma C + \sigma_{e}^{2} + \theta^{T}G(k)\theta} \end{aligned} \tag{91}$$

Determining the approximate vector $\widehat{\psi}(k+1)$ of the vector $\psi(k+1)$ at the discrete-time $k+1$:

$$\widehat{\psi}(k+1) = h(\hat{z}(k+1)) \tag{92}$$

The practical implantation of the first step of the proposed state estimation algorithm requires the knowledge of two types of variables. The first type uses the relative variables of the system, especially, the system input $u(k)$ and the noise variance $\sigma_e^{\,2}(k)$. The second type includes the calculated variables such as the *a priori* estimate $\hat{x}(k)$, the variance-covariance matrix $P_x(k)$ and the estimate $\hat{z}(k)$, which is calculated in the second step of the proposed state estimation algorithm, and this, at discrete-time $k-1$.

Theorem 2: Consider a nonlinear dynamic system described by Wiener mathematical model, which is composed of a dynamic linear part, modulated by a state-space equation, and a static nonlinear part. The estimation of the state variables and the intermediate output of this mathematical model can be made using the recursive state estimation algorithm (87)-(92). The stability condition of the proposed algorithm provided by satisfying the following conditions:

$$\begin{bmatrix} \Lambda_1 & 2K_x(k)[P_{x\circ}(k)]^{-1} \\ 0 & K_x(k)[P_{x\circ}(k)]^{-1}K_x(k) \end{bmatrix} < 0 \tag{93}$$

$$\begin{bmatrix} \Lambda_2 & \Lambda_3 & -K_z(k)\theta^T \end{bmatrix} \begin{bmatrix} \xi_x(k) \\ \xi_z(k) \\ \xi_\psi(k) \end{bmatrix} < 0 \tag{94}$$

where

$$\begin{aligned} \Lambda_1 = {} & A^T[P_x(k+1)]^{-1}A - A^T C[\sigma_w^{\,2}(k+1)]^{-1}C^T A + A^T C[\sigma_w^{\,2}(k+1)]^{-2}C^T P_x(k+1)CC^T A \\ & - A^T C[\sigma_w^{\,2}(k+1)]^{-2}C^T K_x(k+1)C^T P_x(k+1)CC^T A - [P_{x\circ}(k)]^{-1} \end{aligned} \tag{95}$$

$$\Lambda_2 = -K_z(k)\gamma C^T + C^T + C^T A \tag{96}$$

and

$$\Lambda_3 = K_z(k)\gamma C^T K_x(k) - C^T A K_x(k) \tag{97}$$

5.2. Convergence Analysis of the Proposed Recursive State Estimation Algorithm

The convergence analysis of the proposed state estimation algorithm will be made based on the Lyapunov method. Thus, we propose to use the two following functions:

$$V(k+1) = \xi_x^{\,T}(k+1)[P_x(k+1)]^{-1}\xi_x(k+1) \tag{98}$$

and

$$V_{\circ}(k+1) = \xi_{x\circ}^{\ T}(k+1)[P_{x\circ}(k+1)]^{-1}\xi_{x\circ}(k+1) \tag{99}$$

The state prediction error $\xi_x(k+1)$ and the output prediction error $\xi_z(k+1)$ can be written in the following forms:

$$\xi_x(k+1) = A\xi_x(k) \tag{100}$$

and

$$\xi_z(k+1) = C^T\xi_x(k+1) \tag{101}$$

Using equation (87) and adding $x(k)$, we can express the *a posteriori* estimate error as follows:

$$\xi_{x\circ}(k) = \xi_x(k) - \frac{P_x(k)C}{C^T P_x(k)C + \sigma_w^{\ 2}(k)}\xi_z(k) \tag{102}$$

which can be rewritten at discrete-time $k+1$ as follows:

$$\begin{aligned} \xi_{x\circ}(k+1) &= \xi_x(k+1) - K_x(k+1)\xi_z(k) \\ K_x(k+1) &= \frac{P_x(k+1)C}{C^T P_x(k+1)C + \sigma_w^{\ 2}(k+1)} \end{aligned} \tag{103}$$

Using equation (103), we can deduce the following form:

$$K_x(k+1)C^T P_x(k+1)C + K_x(k+1)\sigma_w^{\ 2}(k+1) = P_x(k+1)C\,. \tag{104}$$

Then, the following relation is deduced

$$K_x(k+1)\sigma_w^{\ 2}(k+1) = -[K_x(k+1)C^T - I]P_x(k+1)C \tag{105}$$

Using equation (88), we can write:

$$P_{x\circ}(k+1) = [I - K_x(k+1)C^T]P_x(k+1) \tag{106}$$

By multiplying (106) by C, we can write:

$$P_{x\circ}(k+1)C = -[K_x(k+1)C^T - I]P_x(k+1)C \tag{107}$$

We can express $K_x(k+1)$ by the following equation:

$$K_x(k+1) = P_{x\circ}(k+1)C[\sigma_w^{\,2}(k+1)]^{-1} \tag{108}$$

Using (106) and multiplying by $[P_{x\circ}(k+1)]^{-1}$ and $[P_x(k+1)]^{-1}$, we deduce:

$$[P_{x\circ}(k+1)]^{-1} = [P_x(k+1)]^{-1} + C[\sigma_w^{\,2}(k+1)]^{-1}C^T \tag{109}$$

Using these relations, the *a posteriori* Lyapunov function $V_\circ(k+1)$ can be written as follows:

$$\begin{aligned} V_\circ(k+1) = \xi_x^{\,T}(k)\big[&A^T[P_x(k+1)]^{-1}A - A^TC[\sigma_w^{\,2}(k+1)]^{-1}C^TA \\ &+ A^TC[\sigma_w^{\,2}(k+1)]^{-1}C^TP_x(k+1)CC^TA \\ &- A^TC[\sigma_w^{\,2}(k+1)]^{-1}C^TK_x(k+1)C^TP_x(k+1)CC^TA\big]\xi_x(k) \end{aligned} \tag{110}$$

The time variation $\Delta V_\circ(k) = V_\circ(k+1) - V_\circ(k)$ of the Lyapunov function $V_\circ(k+1)$ is given by:

$$\begin{aligned} \Delta V_\circ(k) = \xi_x^{\,T}(k)\big[&A^T[P_x(k+1)]^{-1}A - A^TC[\sigma_w^{\,2}(k+1)]^{-1}C^TA \\ &+ A^TC[\sigma_w^{\,2}(k+1)]^{-1}C^TP_x(k+1)CC^TA \\ &- A^TC[\sigma_w^{\,2}(k+1)]^{-1}C^TK_x(k+1)C^TP_x(k+1)CC^TA\big]\xi_x(k) \\ &- \xi_x^{\,T}(k)[P_{x\circ}(k)]^{-1}\xi_x(k) + 2\xi_x^{\,T}(k)K_x(k)[P_{x\circ}(k)]^{-1}\xi_z(k) \\ &- \xi_z^{\,T}(k)K_x(k)[P_{x\circ}(k)]^{-1}K_x(k)\xi_z(k) \end{aligned} \tag{111}$$

Developing equation (72), we can deduce the following matrix form:

$$\Delta V_\circ(k) = \begin{bmatrix}\xi_x^{\,T}(k) & \xi_z^{\,T}(k)\end{bmatrix}\begin{bmatrix}\Lambda_1 & 2K_x(k)[P_{x\circ}(k)]^{-1} \\ 0 & K_x(k)[P_{x\circ}(k)]^{-1}K_x(k)\end{bmatrix}\begin{bmatrix}\xi_x(k) \\ \xi_z(k)\end{bmatrix} \tag{112}$$

where Λ_1 is given by (95).

The first convergence condition of the proposed state estimation algorithm (87)-(92) corresponds to the following inequality: $\Delta V_\circ(k) < 0$.

Using (112), we can conclude that the convergence condition is given by:

$$\begin{bmatrix}\Lambda_1 & 2K_x(k)[P_{x\circ}(k)]^{-1} \\ 0 & K_x(k)[P_{x\circ}(k)]^{-1}K_x(k)\end{bmatrix} < 0 \tag{113}$$

Let us consider now a second Lyapunov function $V_z(k)$, which is defined by:

$$V_z(k) = \xi_z^{\,2}(k) \tag{114}$$

The time variation $\Delta V_z(k) = V_z(k+1) - V_z(k)$ of this Lyapunov function $V_z(k)$ is given by the following equation:

$$\Delta V_z(k) = 2\xi_z(k)\Delta\xi_z(k) + \Delta\xi_z^{\,2}(k) \tag{115}$$

where the time variation $\Delta\xi_z(k) = \xi_z(k+1) - \xi_z(k)$ of the gain $\xi_z(k)$ is given by:

$$\Delta\xi_z(k) = C^T A\xi_{x\circ}(k) - K_z(k)\left[\gamma C^T\xi_{x\circ}(k) + \theta^T[\psi(k) - \tilde{\psi}(k)]\right] - C^T\xi_x(k) \tag{116}$$

Multiplying (116) by $\xi_z(k)$, we obtain:

$$\xi_z(k)\Delta\xi_z(k) = C^T\xi_x(k)\left[C^T A\xi_{x\circ}(k) - K_z(k)x\left[\gamma C^T\xi_{x\circ}(k) + \theta^T[\psi(k) - \tilde{\psi}(k)]\right] - C^T\xi_x(k)\right] \tag{117}$$

Using equations (94) and (117), we deduce the following relation:

$$\frac{1}{2}C^T\xi_x(k) < -\frac{1}{2}\left[C^T A\xi_{x\circ}(k) - K_z(k)\left[\gamma C^T\xi_{x\circ}(k) + \theta^T[\psi(k) - \tilde{\psi}(k)]\right]\right] \tag{118}$$

which can be rewritten as

$$C^T\xi_x(k) < -\frac{1}{2}\left[C^T A\xi_{x\circ}(k) - K_z(k)\left[\gamma C^T\xi_{x\circ}(k) + \theta^T[\psi(k) - \tilde{\psi}(k)]\right] - C^T\xi_x(k)\right] \tag{119}$$

Using the following inequality:

$$\begin{aligned} C^T\xi_x(k)\left[C^T A\xi_{x\circ}(k) - K_z(k)\left[\gamma C^T\xi_{x\circ}(k) + \theta^T\xi_{\psi}(k) - C^T\xi_x(k)\right]\right] \\ < -\frac{1}{2}\left[C^T A\xi_{x\circ}(k) - K_z(k)\left[\gamma C^T\xi x_{z\circ}(k) + \theta^T\xi_{\psi}(k)\right] - C^T\xi_x(k)\right]^2 \end{aligned} \tag{120}$$

which can finally be written in the following form:

$$\xi_z(k)\Delta\xi_z(k) < -\frac{1}{2}\Delta\xi_z^{\,2}(k) \tag{121}$$

Thus, we can conclude that the time variation $\Delta\xi_z(k)$ of the Lyapunov function $V_z(k)$, as given by (115), is defined negative.

Finally, we conclude that (93) and (94) give the convergence conditions of the proposed state estimation algorithm (87)-(92).

6. DEVELOPMENT OF PARAMETRIC AND STATE ESTIMATION METHOD

In this section, we present a recursive algorithm that combines the state estimation method with the recursive parametric estimation algorithm, in order to estimate the state and the parameters of the Wiener mathematical model (21) and (22).

6.1. Recursive Parametric and State Estimation Algorithm

The steps of the proposed algorithm are as follows:

Initialization: An initial estimate is given for various parameters of the considered model, i.e., $\hat{A}(0)$, $\hat{A}(0)$, $\hat{B}(0)$, $\hat{C}(0)$, $\hat{\Theta}(0)$, $\hat{x}(0)$, $\hat{x}(1)$, $\hat{x}(2)$, $\hat{x}_{\circ}(0)$, $\hat{z}(0)$, $\hat{z}(1)$, $\tilde{\psi}(0)$, $P_x(0)$, $G(0)$ and $P_{x\circ}(0)$. The good choice of initial values can make the recursive algorithm more stable and can give convergence to the global minimum.

Parametric Estimation: given $\hat{x}(k-1)$, $\hat{x}(k-2)$, $\tilde{\psi}(k-1)$, we can use the following algorithm:

$$\begin{aligned}
&\hat{A}(k) = \hat{A}(k-1) + \zeta_x(k)G_x\delta_x(k-1)\hat{x}^T(k-2)\\
&\hat{B}(k) = \hat{B}(k) + \zeta_x(k)G_x\delta_x(k-1)u(k-2)\\
&\hat{C}(k) = \hat{C}(k) + \zeta_z(k)\alpha I_{nxn}\delta_z(k-1)\hat{x}(k-1)\\
&\hat{d}(k) = \hat{d}(k) + \zeta_z(k)\alpha\delta_z(k)u(k-1)\\
&\hat{\Theta}(k) = \hat{\Theta}(k) + \zeta_y(k)G_y\delta_y(k-1)\tilde{\psi}(k-1)\\
&\delta_x(k-1) = \hat{x}(k-1) - \hat{A}(k-1)\hat{x}(k-2) - \hat{B}(k-1)u(k-2)\ .\\
&\delta_z(k-1) = \hat{z}(k-1) - \hat{C}^T(k-1)\hat{x}(k-1) - \hat{d}(k-1)u(k-1)\\
&\delta_y(k-1) = y(k-1) - \hat{\Theta}^T(k-1)\tilde{\psi}(k-1)\\
&\zeta_x(k) = \frac{\ell_x}{\lambda_{G_x}[u^2(k-2) + \hat{x}^T(k-2)\hat{x}(k-2)]}\\
&\zeta_z(k) = \frac{\ell_z}{\alpha[\hat{x}^T(k-1)\hat{x}(k-1) + u^2(k-1)]}\\
&\zeta_y(k) = \frac{\ell_y}{\lambda_{G_y}[\tilde{\psi}^T(k-1)\tilde{\psi}(k-1)]}
\end{aligned} \tag{122}$$

1. **State Estimation:** This step consists of:
   Calculate the Gain Filter:

$$K_x(k) = \frac{P_x(k)\hat{C}^T(k)}{\sigma^2 + \hat{C}(k)P_x(k)\hat{C}^T(k)} \tag{123}$$

Determine the *A Posteriori* Estimate $\hat{x}_\circ(k)$ of the State $x(k)$:

$$\hat{x}_\circ(k) = \hat{x}(k) + K(k)[\hat{z}(k) - \hat{C}^T(k)\hat{x}(k) - \hat{d}(k)u(k)] \tag{124}$$

Determine the Variance-Covariance Matrix $P_{x\circ}(k)$:

$$P_{x\circ}(k) = P_x(k) - K_x(k)\hat{C}^T(k)P_x(k) \tag{125}$$

Determine the *A Priori* Estimate $\hat{x}(k)$:

$$\hat{x}(k+1) = \hat{A}(k)\hat{x}_\circ(k) + \hat{B}(k)u(k) \tag{126}$$

Determine the Variance-Covariance Matrix $P_x(k+1)$:

$$P_x(k+1) = \hat{A}(k)P_{x\circ}(k)\hat{A}^T(k) + Q \tag{127}$$

Calculate the Gain $K_z(k)$:

$$K_z(k) = \frac{\hat{C}^T(k)\hat{A}(k)P_x(k)\hat{\gamma}(k)\hat{C}(k)}{\hat{\gamma}(k)\hat{C}^T(k)P_x(k)\hat{\gamma}(k)\hat{C}(k) + \sigma_e^{\ 2} + \hat{\theta}^T(k)G(k)\hat{\theta}(k)} \tag{128}$$

Determine the Estimate $\hat{z}(k+1)$:

$$\hat{z}(k+1) = \hat{C}^T(k)\hat{x}(k+1) + K_z(k)[y(k) - \hat{\gamma}(k)\hat{C}^T(k)\hat{x}_\circ(k) - \hat{\theta}^T(k)\hat{\psi}(k)] \tag{129}$$

Determine the Estimate Vector $\hat{\psi}(k+1)$:

$$\hat{\psi}(k+1) = h(\hat{z}(k+1)). \tag{130}$$

In this section, we propose to use the parameters gains ℓ_x, ℓ_z and ℓ_y as the variables in discrete-time k, in order to give more robustness of the state and parametric estimation algorithm. The variation range of each gain should be chosen adequately, in order to verify the convergences conditions (69), (70) and (71) of the parametric estimation algorithm.

Calculation of the three parametric gains can be made as follows (Kamoun, 2003):

$$\ell_i(k) = \ell_{oi}(1 - j_i(k))$$
$$i = x, z, y \tag{131}$$

where ℓ_{oi}, a positive parameter, must be selected as: $0 < \ell_{oi} < 2$ and the parameter $j_i(k)$ can be calculated using the following recursive equation:

$$j_i(k) = j_{oi}j_i(k-1) + j_{oi}(1 - j_{oi})$$
$$i = x, z, y \tag{132}$$

We can easily show that:

$$\lim_{k \to \infty} j_i(k) = j_{oi} \tag{133}$$

and

$$\lim_{k \to \infty} \ell_i(k) = \ell_{oi}(1 - j_{oi}) \tag{134}$$

We represent in Figure 3 an example of evolution curve of the parameter gain $\ell_i(k)$, $i = x, z, y$, which is described by (131).

It can be noted that the proposed recursive algorithm of state and parametric estimation can be divided into three blocks: a parametric estimation block, a state parametric block and a prediction block.

There is a coupling between the three blocks, in the sense that the practical implementation of one block requires a transfer of information and data with the other blocks.

7. ILLUSTRATIVE NUMERICAL EXAMPLE

In this section, we treat a numerical example to illustrate the performance of the proposed recursive parametric and state estimation algorithm and demonstrate its benefits over a recursive EKF.

Let us consider a nonlinear system, which can be described by the following Wiener state-space discrete-time mathematical model:

$$x(k+1) = \begin{bmatrix} a_{11} & a_{12} & a_{13} \\ a_{21} & a_{22} & a_{23} \\ a_{31} & a_{32} & a_{33} \end{bmatrix} x(k) + \begin{bmatrix} b_1 \\ b_2 \\ b_3 \end{bmatrix} u(k) + v(k)$$
$$z(k) = [c_1 \ c_2 \ c_3]x(k) + w(k) \tag{135}$$
$$y(k) = f[z(k)] + e(k)$$

Figure 3. Evolution curve of the parametric gain $\ell_i(k)$, $i = x, z, y$

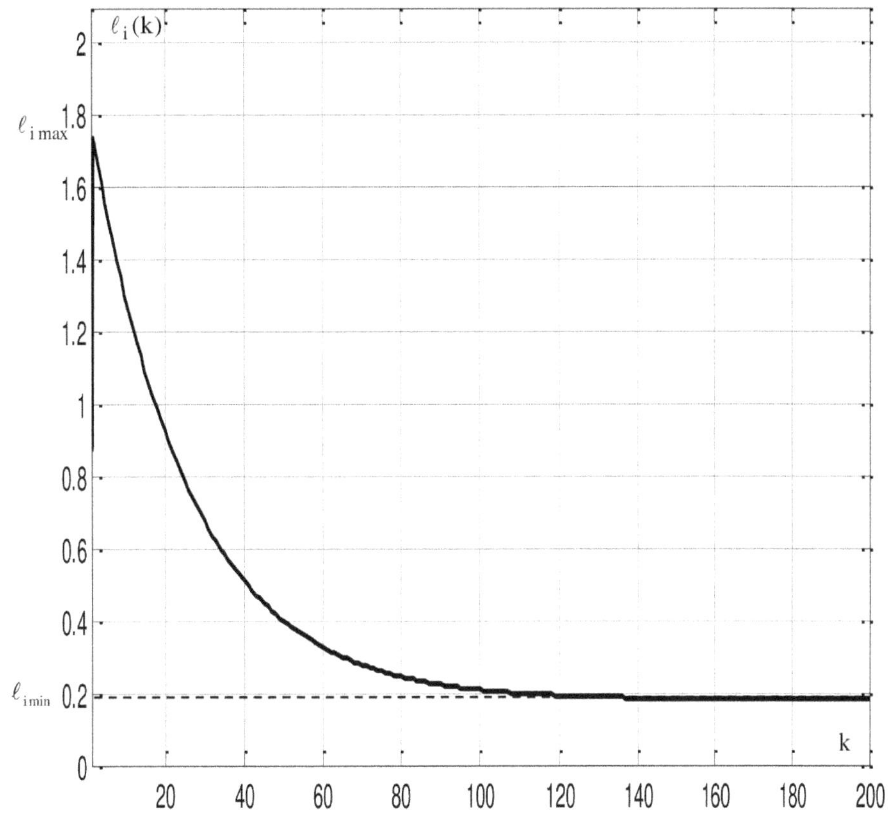

where $x^T(k) = [x_1(k)\ x_2(k)\ x_3(k)]$, $u(k)$, $z(k)$ and $y(k)$ represent, respectively, the state vector, the system input, the output of the linear part and the system output at the discrete-time k, $v^T(k) = [v_1(k)\ v_2(k)\ v_3(k)]$ is the noise vector that acts the state system $x(k)$, $w(k)$ and $e(k)$ denote the noises affecting, respectively, the output of the linear part $z(k)$ and the measured value of the output system $y(k)$ and $f : R \to R$ is a nonlinear function (see Figure 4), with hysteresis-relay nonlinearity such that:

$$f[z(k)] = \begin{cases} +\beta, & z(k) > -\tau \\ -\beta, & z(k) < \tau \end{cases} \tag{136}$$

Note that the noises $v(k)$ $w(k)$ and $e(k)$ intervening in the Wiener mathematical model (135) are often caused by the measuring sensors and the elements of the acquisition of the experimental measurements, which are performed on the signals $x(k)$, $z(k)$ and $y(k)$ of the considered nonlinear system. In fact, these noises represent a set of random disturbances, corresponding to measurement errors. Furthermore, the values of these measurement errors can be given by the manufacturers of the various elements used in the measurement of experimental tests.

The function $f[z(k)]$, as given by (136), can be expressed as follows:

$$f[z(k)] = \frac{\beta}{2\tau}[(z(k)+\tau)\operatorname{sgn}(z(k)+\tau)-(z(k)-\tau)\operatorname{sgn}(z(k)-\tau)] \tag{137}$$

where

$$\operatorname{sgn}[\eta] = \begin{cases} +1, & \eta \geq 0 \\ -1, & \eta < 0 \end{cases} \tag{138}$$

Then, (141) can be written as:

$$f[z(k)] = \Theta^T \Psi(k) \tag{139}$$

with

$$\Theta^T = \begin{bmatrix} 1 & -1 & \frac{\beta}{2\tau} & \frac{\beta}{2} & -\frac{\beta}{2\tau} & \frac{\beta}{2} \end{bmatrix} \tag{140}$$

and

$$\Psi^T(k) = [z(k) \;\; -z(k) \;\; z(k)\operatorname{sgn}(z(k)+\tau) \;\; \operatorname{sgn}(z(k)+\tau) \;\; z(k)\operatorname{sgn}(z(k)-\tau) \;\; \operatorname{sgn}(z(k)-\tau)] \tag{141}$$

The parameters a_{ij}, b_i and c_i of the linear part are supposed unknown, for $i, j = 1, 2, 3$. The parameters intervening in the vector Θ are supposed also unknown. The state variables $x_1(k)$, $x_2(k)$ and $x_3(k)$ are assumed immeasurable for all values of the discrete-time k. The input signal $u(k)$ and the output signal $y(k)$ are measurable. These measured values are assumed independent of the noise components $v_1(k)$, $v_2(k)$ and $v_3(k)$.

In order to estimate the state variables and the parameters of the mathematical model (139), we will use the proposed state and parametric estimation algorithm given by (122) to (130).

The relative data to this example of numerical simulation, for the practical implementation of the proposed algorithm, are given as hereafter:

1. The input $u(k)$ is taken as a random sequence with zero mean and constant variance.
2. The noise components $v_1(k)$, $v_2(k)$ and $v_3(k)$ of the noise vector $v(k)$ consist of independent random variables with zero mean. The variance-covariance matrix of the noise vector is given by $Q = 0.0058 I_{3x3}$.
3. The noise $w(k)$ and $e(k)$ are consists of an independent random sequences with zero mean and variances $\sigma_w^2 = 0.0020$ and $\sigma_e^2 = 0.0815$.
4. The number of measurements M being chosen, such as: $M = 1, \cdots, 400$.

Figure 4. Hysteresis-relay function

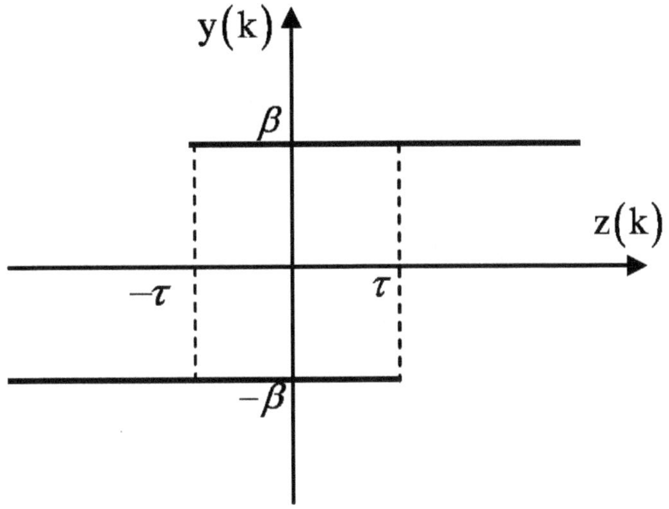

5. The adaptation gain and the initial conditions of the various quantities involved in the proposed state and parameters estimation algorithm are selected with an adequate way.

The quality estimation of the parameters intervening in the Wiener model (135) can be made by considering the global parametric distance $d\mathrm{G}(k)$ given by:

$$d\mathrm{G}(k)=\left[\sum_{i=1}^{3}\sum_{j=1}^{3}\left[\frac{a_{ij}-\hat{a}_{ij}(k)}{a_{ij}}\right]^{2}+\sum_{i=1}^{3}\left[\frac{b_{i}-\hat{b}_{i}(k)}{b_{i}}\right]^{2}+\sum_{i=1}^{3}\left[\frac{c_{i}-\hat{c}_{i}(k)}{c_{i}}\right]^{2}+\sum_{p=1}^{6}\left[\frac{\Theta_{p}-\hat{\Theta}_{p}(k)}{\Theta_{p}}\right]^{2}\right]^{0.5} \tag{142}$$

The parameters of the nonlinear part can be deduced from the following expression:

$$\hat{\beta}(k)=\hat{\Theta}(4,k)+\hat{\Theta}(6,k) \tag{143}$$

and

$$\hat{\tau}(k)=\frac{1}{2}\left[\frac{\hat{\beta}(k)}{2\hat{\Theta}(3,k)}-\frac{\hat{\beta}(k)}{2\hat{\Theta}(5,k)}\right] \tag{144}$$

The evolution curves of the three state variables $x_1(k)$, $x_2(k)$ and $x_3(k)$ with the estimated state variables $\hat{x}_1(k)$, $\hat{x}_2(k)$ and $\hat{x}_3(k)$, and the system output $y(k)$ are given in Figure 5. The evolutions curves of the three state estimation errors $\delta_{x_1}(k)$, $\delta_{x_2}(k)$ and $\delta_{x_3}(k)$ and the evolution curve of the output system $\delta_{y}(k)$ are illustrated in Figure 6. Figure 7 shows the evolutions curves of the variance of the

Figure 5. Evolution curves of the state variables, and with the estimated state variables $\hat{x}_1(k)$, $\hat{x}_2(k)$ and $\hat{x}_3(k)$, and the output system $y(k)$

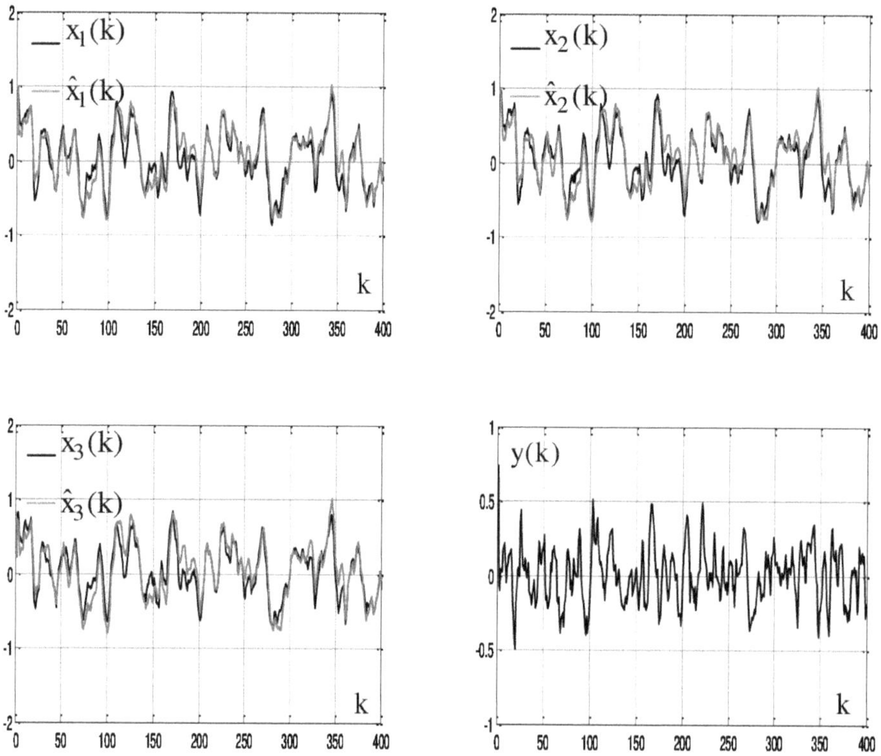

state estimation errors $\sigma_{x_1}^{\ 2}(k)$, $\sigma_{x_2}^{\ 2}(k)$ and $\sigma_{x_3}^{\ 2}(k)$ and the evolution curve of the global parametric distance $d\mathrm{G}(k)$.

Tables 1 and 2 illustrate the real parameters values, the estimated parameters values and the average of its estimated (for constant values of the parametric gains ℓ_x, ℓ_z and ℓ_y). The real parameters values, the estimated parameters values and the average of its estimated are tabulated in Tables 3 and 4 (for variable values of the parametric gains $\ell_x(k)$, $\ell_z(k)$ and $\ell_y(k)$).

Figure 5 shows the state estimation performance using the proposed state estimation method based on linear Kalman filter. The averages of the state estimation error of the three state variables $x_1(k)$, $x_2(k)$ and $x_3(k)$ are, respectively, $\bar{m}_{\delta_1(k)} = 0.1241$, $\bar{m}_{\delta_2(k)} = 0.1125$ and $\bar{m}_{\delta_3(k)} = 0.1139$.

Moreover, the variances $\sigma_{x_1}^{\ 2}(k)$, $\sigma_{x_2}^{\ 2}(k)$ and $\sigma_{x_3}^{\ 2}(k)$ of the state estimation errors and the parametric distance in Figure 7, decreases asymptotically towards a low minimum.

In order to compare the performance of the developed state estimation method, a recursive EKF is implemented with the same initial conditions and the same noises variances. For the EKF, a linearization of the nonlinear function $f[z(k)]$, given by (137), is necessary. Let us consider the error $\delta_{e_x}(k)$ depends on the norm of the a priori prediction errors, such as:

Figure 6. Evolution curves of the three states estimation errors $\delta_{x_1}(k)$, $\delta_{x_2}(k)$ and $\delta_{x_3}(k)$, and the output system $\delta_y(k)$

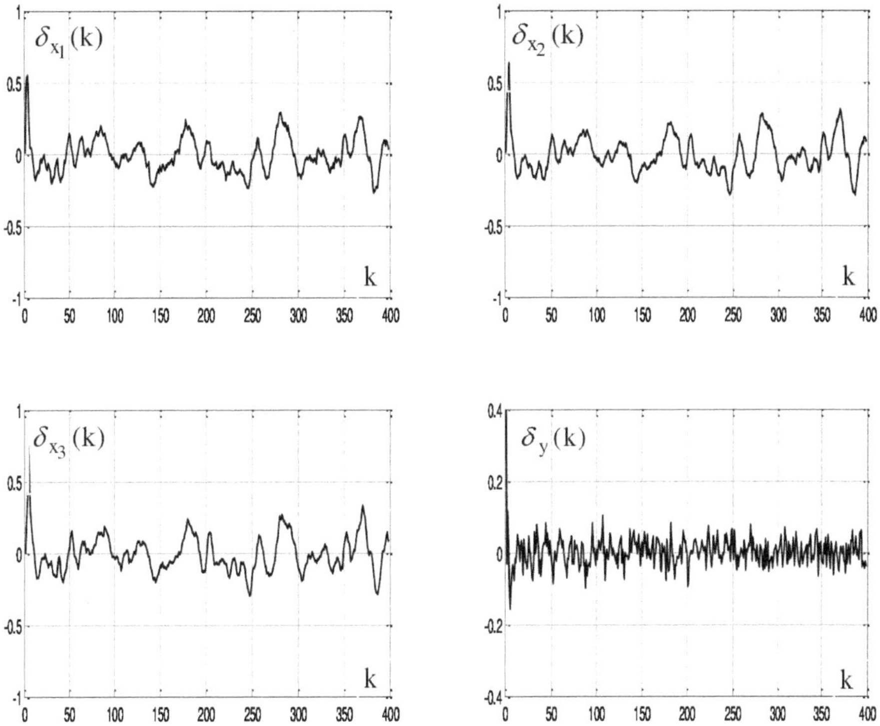

$$\delta_{e_x}(k) = \left\|\delta_x(k)\right\| = \sqrt{\delta_{x_1}^{\ 2}(k) + \delta_{x_2}^{\ 2}(k) + \delta_{x_3}^{\ 2}(k)} \tag{145}$$

Figure 8 shows the evolution curves of the estimated state variables calculated by a recursive EKF and by the developed recursive algorithm and the error $\delta_{e_x}(k)$.

In all the graphs, the black line is the real state variable, the blue line is the estimated state variable calculates by the EKF and the red line is the estimated state variable calculated by the developed algorithm.

From these simulation results, we can draw the following conclusions:

1. The estimates parameters given by the proposed parameters and state estimation algorithm converge to their true values as the parameters gains ℓ_x, ℓ_z and ℓ_y are variables in discrete-time k.
2. The state estimation errors depend on the quality of the parametric estimation, i.e., it become smaller when the estimated parameters converge rapidly to their real parameters.
3. The performance of the developed algorithm is superior to that of the EKF. It can be seen that the estimated state variables, calculated by EKF, are biased, and this, is the direct result of linearization errors.

Finally, we can conclude that the proposed algorithm of the parametric and state estimation is effective.

Figure 7. Evolution curves of the variances of the state estimation errors $\sigma_{x_1}^{\ 2}(k)$ $\sigma_{x_2}^{\ 2}(k)$ *and* $\sigma_{x_3}^{\ 2}(k)$*, and the global parametric distance* $d\mathrm{G}(k)$

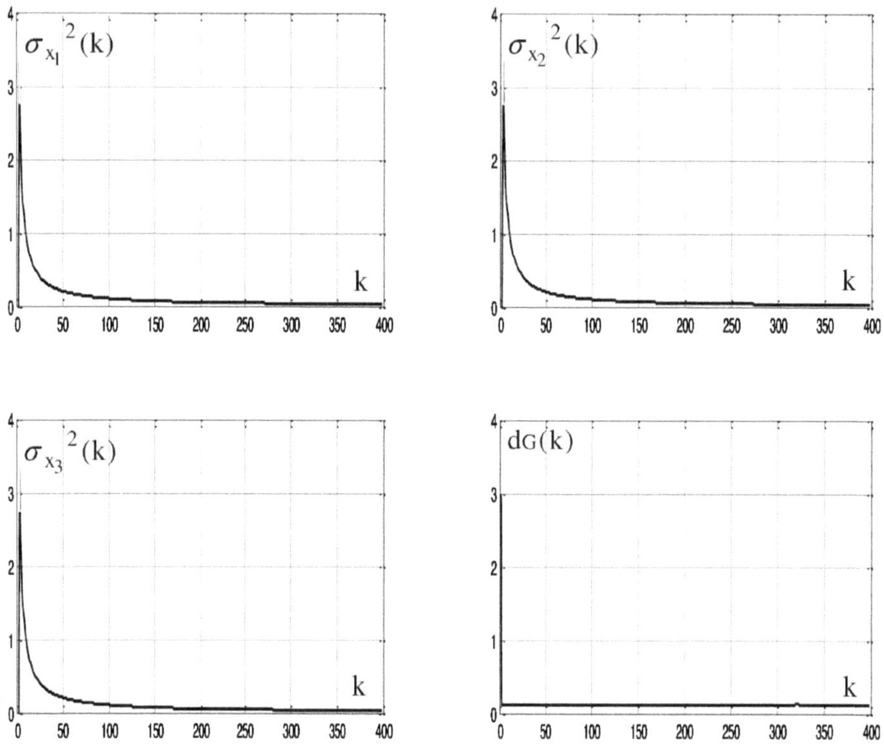

Table 1. Real parameters values and estimated parameters values (for constant values of the parametric gains ℓ_x*,* ℓ_z *and* ℓ_y*)*

	a_{11}	a_{12}	a_{13}	a_{21}	a_{22}	a_{13}	a_{31}	a_{32}	a_{33}	b_1	b_2	b_3
Real Parameters	1.800	-1.040	0.192	1.000	0.000	0.000	0.000	1.000	0.000	1.000	0.000	0.000
Estimated Parameters	1.506	-1.006	0.176	0.853	-0.009	-0.009	-0.001	0.848	-0.001	0.897	-0.001	-0.000

Table 2. Real parameters values and estimated parameters values and the average of the states estimation errors (for constant values of the parametric gains ℓ_x*,* ℓ_z *and* ℓ_y*)*

	Θ_1	Θ_2	Θ_3	Θ_4	Θ_5	Θ_6	c_1	c_2	c_3	$\bar{\delta}_{x_1}$	$\bar{\delta}_{x_2}$	$\bar{\delta}_{x_3}$
Real Parameters	1.000	-1.000	1.000	0.500	-1.000	0.500	1.000	-0.224	-0.647			
Estimated Parameters	0.879	-0.823	0.978	0.238	-0.855	0.278	0.857	-0.026	-0.555	0.234	0.212	0.194

Table 3. Real parameters values and estimated parameters values (for variable values of the parametric gains $\ell_x(k)$, $\ell_z(k)$ and $\ell_y(k)$)

	a_{11}	a_{12}	a_{13}	a_{21}	a_{22}	a_{13}	a_{31}	a_{32}	a_{33}	b_1	b_2	b_3
Real Parameters	1.800	-1.040	0.192	1.000	0.000	0.000	0.000	1.000	0.000	1.000	0.000	0.000
Estimated Parameters	1.799	-1.051	0.186	0.998	-0.005	-0.001	0.002	0.995	0.00	0.998	0.001	0.000

Table 4. Real parameters values and estimated parameters values and the average of the states estimation errors (for variable values of the parametric gains $\ell_x(k)$, $\ell_z(k)$ and $\ell_y(k)$)

	Θ_1	Θ_2	Θ_3	Θ_4	Θ_5	Θ_6	c_1	c_2	c_3	$\overline{\delta}_{x_1}$	$\overline{\delta}_{x_2}$	$\overline{\delta}_{x_3}$
Real Parameters	1.000	-1.000	1.000	0.500	-1.000	0.500	1.000	-0.224	-0.647			
Estimated Parameters	0.914	-0.971	0.973	0.381	-0.986	0.460	1.003	-0.223	-0.649	-0.057	-0.049	-0.044

Figure 8. Evolution curves of the estimated state variables calculated by a recursive EKF and by the developed recursive algorithm the error $\delta_{e_x}(k)$

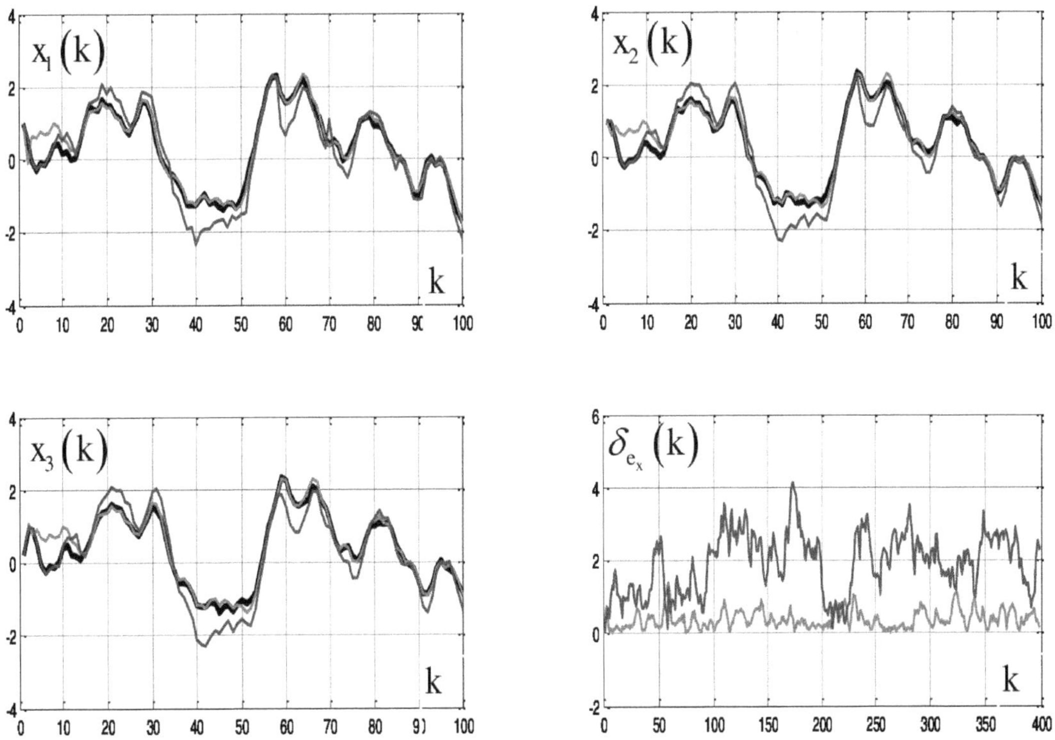

8. CONCLUSION AND FUTURE WORK

This chapter has treated the parametric and state estimation problem of nonlinear systems, which can be described by Wiener discrete-time mathematical models. At the beginning of this chapter, we have presented different types of mathematical models for the description of nonlinear systems. Thus, four types of nonlinear mathematical models have been defined: the functions series mathematical models, the state-space mathematical models, the input-output mathematical models and the block-oriented mathematical models. We have defined the considered Wiener state-space discrete-time mathematical model, which can described nonlinear systems.

The parametric estimation problem of the considered Wiener mathematical model was formulated by using least squares method. In this case, we have assumed that the state variables of this mathematical model are measurable. Thus, a recursive parametric estimation algorithm was proposed. The convergence conditions of this algorithm were established, on the basis of the Lyapunov method. We have developed a recursive state estimation algorithm for the considered Wiener mathematical model, where the parameters in this mathematical model are supposed known. The convergence conditions of this algorithm are given, on the basis of Lyapunov method. The study of the parametric and state estimation conjointly problem for the Wiener mathematical model was treated. In this case, the parameters and the state variables intervening in this mathematical model are unknown. A recursive algorithm was developed, which can be estimate conjointly the parameters and the state variables of nonlinear systems described by Wiener mathematical models.

We have treated a numerical simulation example to show the effectiveness of the proposed estimation algorithms. The obtained numerical results are very satisfactory.

We consider the extension of the developed results in this chapter to nonlinear systems with unknown time-varying parameters. Our future work involves parametric and state estimation of nonlinear systems, which can be described by Hammerstein mathematical models. We will address issues related to the parametric and state estimation of nonlinear large-scale systems that can be decomposed into several interconnected systems. Problems related to adaptive control of nonlinear systems to be studied.

REFERENCES

Al-Duwaish, H., & Nazmul Karim, M. (1997). A new method for the identification of Hammerstein model. *Automatica*, *33*(10), 1871–1875.

Alessandria, A., Baglietto b, M., & Battistellic, G. (2008). Moving-horizon state estimation for nonlinear discrete-time systems: New stability results and approximation schemes. *Automatica*, *44*(7), 1753–1765.

Belforte, G., & Gay, P. (2001). Discrete time Hammerstein model identification with unknown but bounded errors. *IEE Proceedings. Control Theory and Applications*, *148*(6), 523–529.

Biagiola, S. I., & Figueroa, J. L. (2009). Wiener and Hammerstein uncertain models identification. *Mathematics and Computers in Simulation*, *79*(11), 3296–3313.

Billings, S. A. (1980). Identification of nonlinear systems. *IEE Proceedings Pt. D*, *127*(6), 272-285.

Billings, S. A. (2013). *Nonlinear System Identification NARMAX Methods in the Time, Frequency, and Spatio –Temporal Domains*. Wiley.

Chen, S., & Billings, S. A. (1992). Neural networks for nonlinear dynamic system modelling and identification. *International Journal of Control*, *56*(2), 319–346.

Giri, F., & Bai, E.-W. (Eds.). (2010). *Block-oriented Nonlinear System Identification*. Lecture Notes in Control and Information Sciences.

Glaria Lopez, T. A., & Sbarbaro, D. (2011). Observer design for nonlinear processes with Wiener structure. In *Proceedings of the IEEE Conference on Decision and Control and European Control Conference* (pp. 12-15). Orlando, FL: IEEE.

Haber, R., & Keviczky, L. (1976). Identification of nonlinear dynamics systems – A survey paper. In *Proceedings of the* 4th *International Federation of Automatic Control (IFAC) Symposium on Identification and System Parameter Estimations* (pp. 62-112). Tbilisi.

Haber, R., & Keviczky, L. (1999). *Nonlinear System Identification: Input-output Modelling Approach*. Boston: Kluwer.

Haber, R., & Unbehauen, H. (1990). Structure identification of nonlinear dynamic system – a survey on input/output approaches. *Automatica*, *26*(4), 651–677.

Hong, X., & Harris, C. J. (2001). Nerofuzzy design and model construction of nonlinear dynamical processes from data. *IEE Proceedings Control Theory and Applications*, *148*(6), 530–538.

Janczak, A. (2005). *Identification of Nonlinear Systems Using Neural Networks and Polynomial Models: A Block-Oriented Approach*. Lecture Notes in Control and Information Sciences.

Kalman, R. E. (1960). A new approach to linear filtering and prediction problems. *Journal of Basic Engineering*, *82*(1), 35–45.

Kalman, R. E., & Bucy, R. S. (1961). New results in linear filtering and prediction theory. *Journal of Basic Engineering*, *83*(1), 95–108.

Kamoun, S. (2003). *Contribution to identification and adaptive control of complex systems* (in French). (Ph.D. dissertation). National School of Engineering of Sfax, University of Sfax, Sfax, Tunisia.

Kamoun, S. (2007). Design of parametric and state estimation algorithms for stochastic systems. *International Journal of Sciences and Techniques of Automatic Control & Computer Engineering*, *1*(2), 120–135.

Kamoun, S. (2008). Parametric estimation of non-linear stochastic systems described by input-output mathematical models. *International Review of Automatic Control*, *1*(4), 422–434.

Kortmann, M., & Unbehauen, H. (1988). Structure detection in the identification of nonlinear systems. *RAIRO–APII*, *22*(1), 5–25.

Lee, C.-H., & Teng, C.-C. (2000). Identification and control of dynamic systems using recurrent fuzzy neural networks. *IEEE Transactions on Fuzzy Systems*, *8*(4), 349–366.

Leontaritis, I. J., & Billings, S. A. (1985a). Input-output parametric models for non-linear systems. Part 1: Deterministic non-linear systems. *International Journal of Control*, *41*(2), 303–328.

Leontaritis, I. J., & Billings, S. A. (1985b). Input-output parametric models for non-linear systems. Part 2: Stochastic non-linear systems. *International Journal of Control*, *41*(2), 329–344.

Li, H.-X. (1999). Identification of Hammerstein models using genetic algorithms. *IEE Proceedings Control Theory and Applications*, *146*(6), 499–504.

Li, Y., Sundararajan, N., & Saratchandran, P. (2000). Analysis of minimal radial basis function network algorithm for real-time identification of nonlinear dynamic systems. *IEE Proceedings Control Theory and Applications*, *147*(4), 476–484.

Lopez, R., & Yuz, J. (2012). Recursive parameter and state estimation for a mining industry process. In *Proceedings of the 20th Mediterranean Conference on Control & Automation* (pp. 30-35). Barcelona, Spain: Academic Press.

Nelles, O. (2001). *Nonlinear System Identification From Classical Approaches to Neural Networks and Fuzzy Models*. Springer.

Nørgaard, M., Poulsen, N. K., & Ravn, O. (2000). New developments in state estimation for nonlinear systems. *Automatica*, *36*(11), 1627–1638.

Norquay, S., & Palazoglu, A. (1998). Model predictive control based on Wiener models. *Chemical Engineering Science*, *53*(1), 75–84.

Ogunfunmi, T. (2007). *Adaptive Nonlinear System Identification: The Volterra and Wiener Model Appraches*. Springer.

Rao, C. V., Rawlings, J. B., & Mayne, D. Q. (2003). Constrained state estimation for nonlinear discrete-time systems: Stability and moving horizon approximations. *IEEE Transactions on Automatic Control*, *48*(2), 246–258.

Sbarbaro, D., & Johansen, T. A. (1997). Multiple local Laguerre models for modelling nonliner dynamic systems of the Wiener class. *IEE Proceedings. Control Theory and Applications*, *144*(5), 375–380.

Schetzen, M. (1980). *The Volterra and Wiener Theories of Nonlinear Systems*. Wiley.

Tang, Y., Giao, L., & Guan, X. (2010). Identification of Wiener model using Step signals and particle swarm optimization. *Expert Systems with Applications*, *37*(4), 3398–3404.

Volterra, V. (1959). Theory of Functional of Integral and Integro-Differential Equations. New York: Dover.

Vörös, J. (2002). Modeling and parameter identification of systems with multisegment piecewise-linear characteristics. *IEEE Transactions on Automatic Control*, *47*(1), 184–188.

Wang, D., Chu, Y., & Ding, F. (2010). Auxiliary model based recursive generalized least squares parameter estimation for Hammerstein OEAR systems. *Mathematical and Computer Modelling*, *52*(1-2), 309–317.

Wiener, N. (1949). *Extrapolation, Interpolation and Smoothing of Stationary Time Series, with Engineering Applications*. Wiley.

Wigren, T. (1993). Recursive prediction error identification using the nonlinear Wiener model. *Automatica, 29*(4), 1011–1025.

Xiong, K., Wei, C. L., & Liu, L. (2012). Robust Kalman filtering for discrete-time nonlinear systems with parameter uncertainties. *Aerospace Science and Technology, 18*(1), 15–24.

Yingwei, L., Sundararajan, N., & Saratchandran, P. (1997). Identification of time-varying nonlinear systems using minimal radial basis function neural networks. *IEE Proceedings Control Theory and Applications, 144*(2), 202–208.

Zhao, J., & Kanellakopoulos, I. (2002a). Active identification for discrete-time nonlinear control – Part 1: Output-feedback systems. *IEEE Transactions on Automatic Control, 47*(2), 210–224.

Zhao, J., & Kanellakopoulos, I. (2002b). Active identification for discrete-time nonlinear control – Part 2: Strict-feedback systems. *IEEE Transactions on Automatic Control, 47*(2), 225–240.

Zi-Qiang, L. (1994). On identification of the controlled plants described by the Hammerstein system. *IEEE Transactions on Automatic Control, 39*(3), 569–573.

KEY TERMS AND DEFINITIONS

Block-Oriented Mathematical Models: These models lead to connected systems, which are constituted by linear dynamic parts and nonlinear static parts.

Kalman Filter: The Kalman filter, also known as Linear Quadratic Estimation (LQE), is an algorithm for estimating the state variables of stochastic systems that are described by state-space mathematical models, based on experimental measurements.

Mathematical Model: A mathematical model consists of an equation or a set of equations belonging to a certain class of mathematical models to describe the dynamic behavior of the corresponding system. The parameters involved in this mathematical model are related to a certain mathematical structure. This mathematical model is characterized by its class, its structure and its parameters.

Nonlinear System: A dynamic system is called linear if the relation between its input and output is linear. As against, if this relationship contains one or more non-linear terms, while the said system is nonlinear.

Parametric Estimation: The parametric estimation of a dynamic system is the determination, from the knowledge of several experimental measurements, which are derived from this system, the parameters involved in the equation of a mathematical model that can describe this system, where its class and its structure are assumed well known.

Recursive Parametric Estimation Algorithm: A recursive parametric estimation algorithm used to solve the parametric estimation problem of dynamical systems. It is based on the adjustable model and the least squares techniques from the knowledge of several experimental measurements.

Recursive State Estimation Algorithm: A recursive state estimation algorithm is an algorithm that matches the Kalman filter, allows estimating the state variables of a dynamic system described by a state-space mathematical model based on several experimental measurements.

Wiener Mathematical Model: Nonlinear model defined as a linear dynamic part precedes a nonlinear static part.

Chapter 5
On Proportional Plus Derivative State Feedback H_2 Control for Descriptor Systems

Rim Zaghdoud
National School of Engineering of Tunis, Tunisia

Salah Salhi
National School of Engineering of Tunis, Tunisia

Moufida Ksouri
National School of Engineering of Tunis, Tunisia

ABSTRACT

This chapter studies the H_2 control problem for linear singular systems. Firstly a proportional plus derivative feedback controller for continuous and discrete cases for descriptor systems is given, satisfying the closed loop systems normal and stable states. A necessary and sufficient condition for the solvability of this problem should be obtained in terms of linear matrix inequality. In second part, the designed controller is required to yield a H_2 cost function with guaranteed upper bound for the continuous and discrete case and also an extension for robust H_2 control is developed. Numerical examples are included to show the effectiveness of the proposed result.

1. INTRODUCTION

The singular systems have attracted the attention of a lot of researchers due to their capacity to preserve the structure of physical systems and also encompass impulsive behaviors and non-dynamic constraints. There has been a growing interest in the control of descriptor systems since such systems have extensive applications in economic systems, large scale systems and other areas (Dai, 1989; Lewis, 1986). Many fundamental concepts and results based on the theory of ordinary systems have been successfully developed to descriptor systems. Let's us quote the notion of observability, controllability, stability and

DOI: 10.4018/978-1-4666-7248-2.ch005

stabilization technique (Dai, 1989; Xu & Lam, 2006) as well as results including robustness aspects (Xu, & Lam, 2006; Chaabane et al., 2011; Sari et al., 2008; Ren & Zhang, 2010; Faria et al., 2009; Bara, 2011).

Since the stability of dynamic system is well regarded as one of the most important properties of systems, much attention has been focused on the problems of robust stability and robust stabilization of descriptor systems. It is known that the robust stability problem for singular systems is much more complicated than that for ordinary systems because it requires to consider not only stability but also absence of impulses and regularity simultaneously (Fang & Chang, 1993; Fang & Chang, 1994; Lin et al., 1997; Xu & Lam, 2003; Xu, 2002) while the latter two do not arise in the conventional state space context. This justifies the two approaches used in relevant literature to resolve the stability problem: one is to develop the admissibility concept which requires verifying the stability, regularity and non-impulsiveness of the descriptor system. The other is limited to ensure the stability and the regularity of the system by the design of controller which make the system be normal or in other words the non-singularity of matrix E. In Xu et al. (2002) an algebraic approach was proposed to solve the robust stabilization problem via a linear matrix inequality (LMI), as well as to singular systems with time delays was developed. For the generalized quadratic stabilization, necessary and sufficient conditions were given in terms of Riccati inequality in Lin et al. (2000). The robust stabilization problem for uncertain discrete time singular systems was investigated in Xu et al. (2001), where a sufficient condition for generalized quadratic stabilization was developed. Under some assumptions about the system matrices, this sufficient condition was modified to be necessary. In Xu and Lam (2004), the robust stabilization problem is solved through state feedback controller where the parameters uncertainties appearing in both the state and input matrices and the concepts of generalized quadratic stability and stabilization are introduced. In the same paper, a necessary and sufficient condition is given in terms of a strict LMI such the system will be admissible. Again, the robust stabilization problem through state feedback for uncertain singular systems is given in the work of Xu et al. (2002). In Xu and Yang (1999), to reduce the conservatism of the stabilization quadratic, a PDL approach is used to salve the robust static output feedback admissibility problem for the descriptor systems case. However, in practical control system design, the robust performance is also required as well as robust stability against the uncertainty of the plant model (Boukas, 2005; Boukas et al., 2003; Dragan, 2005; Dragan, et al., 1999; Mahmoud and Shi, 2003; Shi, 1999; Xia & Jia, 2003a; Xia & Jia, 2003b). Both H_2 and H_∞ of the system have been widely used to bring clarity to the system performance in past studies and the problems of robust H_2 and H_∞ control for systems with uncertain parameters have been extensively studied in the past decade (Ikeda et al., 2000; Masubuchi, 1997; Takaba, 1998). In More recent times, much work has been done on multiobjective optimization problems and specially, the mixed H_2 / H_∞ control problem has received considerable attention (Bernstein & Haddad, 1981; Doyle et al., 1994; Geromel et al., 1992; Khargonekar & Rotea, 1991; Rotea & Khargonekar, 1991; Zhou et al., 1994). In work Xu and Lam (2004), necessary conditions were given to minimize an upper bound on the H_2 performance under a H_∞ constraint.

Lead by its critical role played in a wide range of topics in modern control theory (Feng et al., 2010; Xia et al., 2005; Yan et al., 2012), the famous H_2 drew much attention during last decades. Moreover this H_2 norm system is seen as one of the main control performance indices measured in the control system design and the best associated with performance index. As a result the robust H_2 control was a subject of ultimate interest for the last years. In Renquan et al. (2008), a robust control for uncertain continuous descriptor system is being studied based on new bounded real lemma, the result is the elaboration of a static state feed-back controller that guarantees the robust performance.

Another work dealing with the state feed-back control in Feng et al. (2011) treated the problem of regulation with guaranteed H_2 performance, where parameterizations of all optimal controllers solving the problem is given. The same idea is developed in Yagoubi et al. (2007) where the problem of state feedback H_2 control with presence of unstable exo-systems has been considered. An extended H_2 output feedback control problem for continuous descriptor systems has investigated where the non-proper and unstable weights are considered (Feng et al., 2012). The existence of an optimal solution of this problem is based on two GAREs and two generalized Sylvester-type equation. In Ishihara et al. (2003), an explicit formulation of all optimal controllers is given for the full information and the state feedback gain. The existence of a solution of this problem is characterized in terms of GAREs, where sufficient conditions were given. In Takaba and Katayama (1998) an explicit way out to the output feedback control problem for continuous descriptor systems using to the model matching technique was proposed where an optimal controller is set as solution of generalized algebraic Riccati equation. Takaba and Katayama (1997) used the generalized Lyapunov equation in order to design a robust state feed-back control gain for uncertainty descriptor system in terms of LMIs, while Ikeda et al. (2000) proposed strict LMI approach for robust control. To the authors' best knowledge, the problem of H_2 proportional plus derivative state feedback (PD) for singular systems has not been investigated yet. This motivates us for this study.

So this paper therefore aims to develop a new approach which should hopefully guarantee the stability of the closed loop singular systems, in its first part, and furthermore to guarantee the H_2 performance in second part, by a proportional and derivative state feedback controller where an explicit expression of the desired control law is also given.

1.1 Examples of Descriptor Linear Systems (Duan, 2010)

Descriptor systems appear in many fields, such as power systems, electrical networks, aerospace engineering, chemical processes, social economic systems, network analysis, biological systems, time-series analysis, and so on. In this section, some examples of descriptor systems are presented, from which readers can indeed see the existence of descriptor linear systems in our real world.

1.1.1 Electrical Circuit Systems

Many electrical circuit systems can be described by descriptor linear systems. Considering two such simple circuit networks (Dai, 1989).

Consider a simple circuit network as shown in Figure 1,where *R*; *L*; and C_0 stand for the resistor, inductor, and capacity, respectively, and their voltages are denoted by $V_R(t)$; $V_L(t)$; $V_C(t)$; respectively. $V_e(t)$ is the voltage source which is taken as the control input. Following basic circuit theory and the Kirchoff's law, we have the following equations, which describe the system:

$$L\dot{I}(t) = V_L(t) \tag{1}$$

Figure 1. A single-loop circuit network

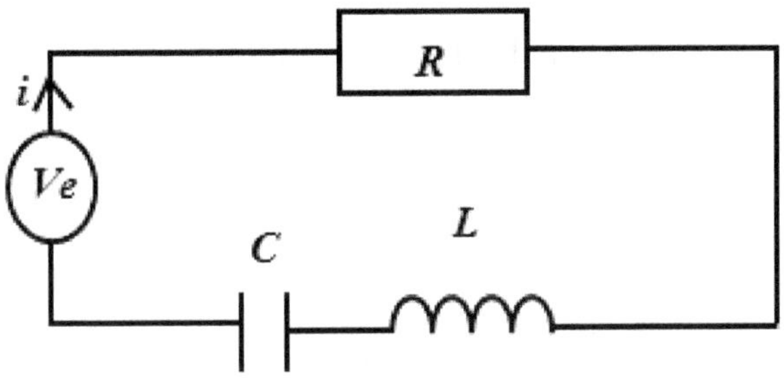

$$\dot{V}_C(t) = \frac{1}{C_0} I(t) \tag{2}$$

$$RI(t) = V_R(t) \tag{3}$$

$$V_L(t) + V_C(t) + V_R(t) = V_e(t) \tag{4}$$

By choosing the state vector x, the input vector u and the output vector y as follows:

$$x = \begin{bmatrix} I(t) \\ V_L(t) \\ V_C(t) \\ V_R(t) \end{bmatrix}, \quad u = V_e(t),\ y = V_C(t)$$

The above (1)-(4) can be written in the following descriptor linear system form:

$$\begin{cases} E\dot{x}(t) = Ax + Bu \\ y = Cx \end{cases} \tag{5}$$

$$E = \begin{bmatrix} L & 0 & 0 & 0 \\ 0 & 0 & 1 & 0 \\ 0 & 0 & 0 & 0 \\ 0 & 0 & 0 & 0 \end{bmatrix}, \quad B = \begin{bmatrix} 0 \\ 0 \\ 0 \\ -1 \end{bmatrix}, \quad A = \begin{bmatrix} 0 & 1 & 0 & 0 \\ \frac{1}{C_0} & 0 & 0 & 0 \\ -R & 0 & 0 & 1 \\ 0 & 1 & 1 & 1 \end{bmatrix}, \quad C = \begin{bmatrix} 0 & 0 & 1 & 0 \end{bmatrix}$$

1.1.2 Constrained Mechanical Systems

Constrained linear mechanical systems can be described as follows:

$$M\ddot{z} + D\dot{z} = Lf + J\mu \tag{6}$$

$$G\dot{z} + Hz = 0 \tag{7}$$

where $z \in \mathbb{R}^n$ is the displacement vector, $f \in \mathbb{R}^n$ is the vector of known input forces, $\mu \in \mathbb{R}^q$ is the vector of Lagrangian multipliers, *M* is the inertial matrix, which is usually symmetric and positive definite, *D* is the damping and gyroscopic matrix, *K* is the stiffness and circulator matrix, L is the force distribution matrix, *J* is the Jacobian of the constraint equation, G and H are the coefficient matrices of the constraint equation. All matrices in (6)-(7) are known and constant ones of appropriate dimensions. Equation (6) is the dynamical equation, while (7) is the constraint equation.

Assume that a linear combination of displacements and velocities is measurable, then, the output equation is of the form:

$$y = C_p z + C_v \dot{z} \tag{8}$$

where $C_p, C_v \in \mathbb{R}^{m \times n}$ By further choosing the state vector and the input vector as:

$$x = \begin{bmatrix} z \\ \dot{z} \\ \mu \end{bmatrix} \quad and \;\; u = f$$

respectively, then the above (6)-(8)can be written in the following descriptor linear system form:

$$\begin{cases} E\dot{x}(t) = Ax + Bu \\ y = Cx \end{cases} \tag{9}$$

with:

$$E = \begin{bmatrix} I & 0 & 0 \\ 0 & M & 0 \\ 0 & 0 & 0 \end{bmatrix}, B = \begin{bmatrix} 0 \\ L \\ 0 \end{bmatrix}, A = \begin{bmatrix} 0 & I & 0 \\ -K & -D & J \\ H & G & 0 \end{bmatrix}, C = \begin{bmatrix} C_p & C_v & 0 \end{bmatrix}$$

Consider a mechanical system shown in Figure 2, this system consists of two one-mass oscillators connected by a dashpot element.

Figure 2. Two connected one-mass oscillators

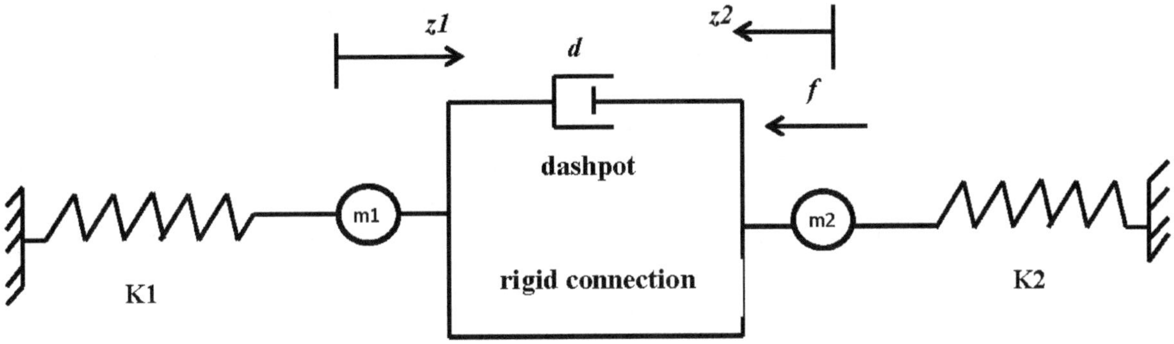

Let $m_1 = m_2 = 11kg,\ d = 1Ns\ /\ m,\ K_1 = 2N\ /\ m,\ K_2 = 1N\ /\ m$:

Then the three equations (1.17)–(1.19) for this system can be obtained as follows:

$$\begin{bmatrix} 1 & 0 \\ 0 & 1 \end{bmatrix}\begin{bmatrix} \ddot{z}_1 \\ \ddot{z}_2 \end{bmatrix} + \begin{bmatrix} 1 & 1 \\ 1 & 1 \end{bmatrix}\begin{bmatrix} \dot{z}_1 \\ \dot{z}_2 \end{bmatrix} + \begin{bmatrix} 2 & 0 \\ 0 & 1 \end{bmatrix}\begin{bmatrix} z_1 \\ z_2 \end{bmatrix} = \begin{bmatrix} -1 \\ 1 \end{bmatrix} f + \begin{bmatrix} 1 \\ 1 \end{bmatrix} \mu$$

$$\begin{bmatrix} 1 & 1 \end{bmatrix}\begin{bmatrix} z_1 \\ z_2 \end{bmatrix} = 0,\ y = \begin{bmatrix} 1 & 0 \end{bmatrix}\begin{bmatrix} z_1 \\ z_2 \end{bmatrix},$$

Therefore, we have:

$$M = \begin{bmatrix} 1 & 0 \\ 0 & 1 \end{bmatrix},\ D = \begin{bmatrix} 1 & 1 \\ 1 & 1 \end{bmatrix},\ K = \begin{bmatrix} 2 & 0 \\ 0 & 1 \end{bmatrix},\ L = \begin{bmatrix} -1 \\ 1 \end{bmatrix}$$

$$J = \begin{bmatrix} 1 \\ 1 \end{bmatrix}, G = \begin{bmatrix} 0 & 0 \end{bmatrix},\ H = \begin{bmatrix} 1 & 1 \end{bmatrix}$$

$$C_p = \begin{bmatrix} 1 & 0 \end{bmatrix},\ C_v = \begin{bmatrix} 0 & 0 \end{bmatrix}$$

Based on these matrices, the descriptor linear system form (9) for the system can be readily written out.

1.2 Problem Statement

Control system designs and their meanings are often seen to be very broad. Within the context of linear systems, control systems design usually tends to mean; to design a controller of a particular kind such that the closed-loop system, and oft times the actual controller itself, meet a specific design criteria. Concerning analysis of descriptor linear systems, there can often be many problems such as that, it is important to consider not merely the stability alone, but also to investigate regularity and non-impulsiveness (continuous time case) or causality (discrete time case) simultaneously as the two are delicately and intricately related. To help us to avoid the problem of the regularity, which presents a very important property, or in other words to investigate the said condition of existence and uniqueness of solution to a singular systems in the concept of regularity et and also to avoid the problem of l impulsiveness which can be seen on the answer of singular systems and because the gravity terms which cause a saturation of control and may even, in some cases, destroy the system, a proportional plus derivative state feedback controller will be constructed. The conception of such a controller makes the singular systems normal in such way that the following is so: rank (E) $=n$ (order system). In such a case as this the conception and the opening analysis of singular systems will invariably be similar to those that can be seen in ordinary systems.

Firstly it should be mentioned that the goal of this paper is to study a proportional plus derivative feedback controller for both continuous and discrete cases for descriptor systems, therefore or hopefully, satisfying the closed loop systems normal and stable states. As a direct result of this, a necessary and sufficient condition for the solvability of this problem should be obtained in terms of linear matrix inequality thus solving a common quagmire. Following on from this into our second part is the fact that the designed controller is required to yield a H_2 cost function with a guaranteed upper bound for the continuous and discrete case and also an extension for robust H_2 control is developed.

The remainder of the paper will then proceed as stated in the following: The second section shall aim to treat the stabilization problem via proportional plus derivative feedback controller, wherein a necessary and sufficient condition under an LMI formalism is given for the continuous and discrete case and also an extension of robust stabilization for polytopic systems will be looked at and developed with some amount of detail. In section three, the infinite horizon of the H_2 control problem is described for both continuous and discrete singular systems and also an extension for uncertain systems is derived, where sufficient conditions for the construction of a constant gain (PD) state feedback controller that shall minimize an upper bound of the H_2 norm is hence obtained in terms of the previously talked about linear matrix inequalities (LMIs). Following these three sections an illustrative example is then worked in section four. The last section shall also contain but not be limited to some concluding remarks and following this future works are presented.

Notation: Throughout this paper, the following notations will be used. For two matrices A and B, $A > B$ means that $A - B$ is positive definite. Denotes the transpose of A and the transpose of the inverse of A. Identity and null matrices will be denoted respectively by I and 0. Furthermore, in the case of portioned symmetric matrices, the symbol denotes generally each of its symmetric blocks and denotes.

2. STABILIZATION PROBLEM FOR LINEAR DESCRIPTOR SYSTEMS

The aim and desired consequence of this section is to study in some details a proportional plus derivative (PD) state feedback controller for descriptor systems, which can be seen to satisfy the closed loop systems stability. The present study exploits the stabilization concept for the standard usual systems. Following on from this and consequential to it, a proportional plus derivative feedback (PD) controller will be designed to translate the study from descriptor systems to a standard one. So we shall therefore avoid studying the some of the concepts of singular systems such as its regularity, impulsive behavior and its admissibility. In order to present this as stated, the normalization concept will be in turn presented, which consists of designing a proportional plus derivative feedback controller (PD) such that the closed loop system is normal: rank(E)=n (order system).

Let's consider the following continuous-time linear descriptor system described by:

$$E\dot{x} = Ax + Bu \tag{10}$$

where $x(t) \in \mathbb{R}^n$ and $u(t) \in \mathbb{R}^m$ are the state and control input vector respectively. Matrices A and B are constant matrices of appropriate dimensions. The matrix E may be singular. It is then assumed that $rank(E) = q \leq n$.

Considering the state proportional plus derivative control expressed as:

$$u(t) = K_p x(t) - K_d \dot{x}(t) \tag{11}$$

where gains $K_p, K_d \in \mathbb{R}^{m\times n}$ are constant matrixes of appropriate dimensions. Applying the control law (11) to the descriptor system (10), the closed loop system is expressed as follows:

$$(E + BK_d)\dot{x}(t) = (A + BK_p)x(t) \tag{12}$$

- **Remark 1:** The role that the proportional plus derivative feedback control law (11) plays is to supply a constant gain K_p which acts on the stability of the system (10) and a constant gain K_d which make the expression (12) well defined. In other words find a gain K_p which makes the system (10) be normalizable or again write the equation (12) as follows:

$$\dot{x}(t) = (E + BK_d)^{-1}(A + BK_p)x(t) \tag{13}$$

Before wording the problem that will be studied in this paper, we shall give the following concept for the description systems which are playing a key role in solving the aforementioned problem.

- **Definition 1:** The descriptor system $E\dot{x}(t) = Ax(t)$ is said to be normal and stable (NS) if E is non-singular $(\det E \neq 0)$ and there exists a matrix P >0 such that:

$$E^{-1}AP + PA^T E^{-T} < 0 \tag{14}$$

- **Lemma 1 (Projection Lemma):** Given a symmetric matrix $\psi \in \mathbb{R}^{m\times m}$ and two matrices P, Q of column dimension m there exists X such that the following LMI holds:

$$\psi + P^T X^T Q + Q^T XP < 0$$

If and only if the projection inequalities with respect to X are satisfied:

$$N_P^T \psi N < 0, \qquad N_Q \psi N_Q^T < 0,$$

where N_P and N_Q denote arbitrary bases of the null spaces of P and Q, respectively.

- **Lemma 2** (Scherer and Weiland, 2005)**:** Let $T_{wz}(s) = C(sI - A)^{-1} B$ and A is stable, then we have the following equivalence:
  - $\left\| T_{wz}(s) \right\|_2 < \gamma$.
  - There exists $P > 0$ such that: $PA + A^T P + C^T C < 0, \quad trace(B^T PB) < \gamma$.
  - There exists $X > 0$ and T with appropriate dimension such that:

$$\begin{bmatrix} XA^T + AX & XC^T \\ CX & -1 \end{bmatrix} < 0, \quad \begin{bmatrix} X & B \\ B^T & T \end{bmatrix} < 0, \quad trace(T) < \gamma.$$

2.1. Stabilization Condition for Continuous Time Descriptor Systems

In this section, a necessary condition for the solvability of the PD state feedback controller is given in terms of linear matrix inequalities. The theorem below follows directly from the previous definition.

- **Theorem 2.1:** The following statements are equivalent:
  - System (10) is normal and stable.
  - There is seen to exist a positive definite matrix Y and matrices F, R_1 and R_2 of appropriate dimensions such that:

$$\begin{bmatrix} F^T A + AF + BR_1 + R_1^T B & Y + F^T A^T + R_1^T B^T - EF - BR_2 \\ * & -F^T E^T - EF - BR_2 - R_2^T B^T \end{bmatrix} \quad (15)$$

The gain given by: $K_p = R_1 F^{-1}$and $K_d = R_2 F^{-1}$ solves problem.

Proof: Since the system (12) is normal and stable, from Definition 1, then E_c is non-singular and $E_c^{-1} A_c$ is stable. In this case, the stability stressed under Lyapunov's sense is written in the following way $(Q = P^{-1})$, Q>0.

$$(A + BK_p)^{-T} Q(E + BK_d)^{-T} + (E + BK_d)^{-T} Q(A + BK_p)^{-T} < 0$$

which can be written as it is in the following:

$$\begin{bmatrix} A_c^{-T} & E_c^{-T} \end{bmatrix} \begin{bmatrix} 0 & Q \\ Q & 0 \end{bmatrix} \begin{bmatrix} A_c^{-1} \\ E_c^{-1} \end{bmatrix} < 0.$$

Applying Lemma 1 (projection lemma), this is equivalent to the existence of matrices G such that:

By taking $Q = P^{-1}$,Q>0, and applying the projection lemma, there exists a matrix G of appropriate dimension such that:

$$\begin{bmatrix} 0 & Q \\ Q & 0 \end{bmatrix} + \underbrace{\begin{bmatrix} A^T \\ -E^T \end{bmatrix} G^T \begin{bmatrix} I & I \end{bmatrix}}_{*} + sym(*) < 0 =$$

$$\begin{bmatrix} \left(A + BK_p\right)^T G^T + G\left(A + BK_p\right) & \left(A + BK_p\right)^T G^T + Q - G\left(E + BK_d\right) \\ * & -\left(E + BK_d\right)^T G^T - G\left(E + BK_d\right) \end{bmatrix} < 0$$

Taking now $F = G^{-T}$, and applying the congruence transformation $diag(G^{-1}, G^{-1})$ and denoting $Y = G^{-1}QG^{-T}$, $R_1 = K_pF$, $R_2 = K_dF$, one gets the inequality (15). This completes the proof of theorem.

2. 2. Stabilization Condition for Discrete Time Descriptor Systems

In this subsection, the methodology can be extended for the discrete-time.

Let a given discrete-time descriptor system be seen as follows:

$$Ex(k+1) = Ax(k) + Bu(k) \tag{16}$$

where $x(k) \in \mathbb{R}^n$, $u(k) \in \mathbb{R}^m$ and $y(k) \in \mathbb{R}^r$ are its state, control input vector, and measurement output respectively. $E, A \in \mathbb{R}^{nxn}$ $B \in \mathbb{R}^{nxm}$, and $C \in \mathbb{R}^{rxn}$ are constant matrices of appropriate dimensions. The matrix E may be singular. It is assumed that $rank(E) = q \leq n$.

Similar to the results obtained in the previous section, the (PD) state feedback law is as such:

$$u(k) = K_p x(k) + K_d u(k+1)$$

Applied to the discrete system (16), succeeded to the closed loop system expressed as follows:

$$(E + BK_d)x(\mathrm{k}+1) = (A + BK_p)x(\mathrm{k}) \tag{17}$$

Hence, before giving the result of this subsection, we would in cohesion with this outcome seek to expose the preliminary definition.

- **Definition 2:** The descriptor system $Ex(k+1) = Ax(k)$ is said to be normal and stable (NS) if E is non-singular $(\det E \neq 0)$ and there exists a matrix P >0 such that:

$$A^T E^{-T} P E^{-1} A - P < 0$$

The solvability of the stabilization problem of the system (16) is given by the following theorem.

- **Theorem 2.2:** The following statements are equivalent:
  - System (16) is normalizable and stabilizable.
  - There exist positive definite matrixes *P*,matrices *G*, R_1 and R_2 of appropriate dimensions and scalar $\alpha > 0$, sufficiently large, such that:

$$\begin{bmatrix} P + \alpha G^T E^T + \alpha B R_2 + \alpha R_2^T B & -\alpha G^T A^T - \alpha R_1^T B^T \\ * & -P \end{bmatrix} \quad (18)$$

- **Proof:** Since the system (17) is normal stable, from Definition 1, then E_c is non-singular and $E_c^{-1} A_c$ is stable. In this case, the stability that is stressed under Lyapunov's sense is written in the following way:

$$E_c^{-T} P E_c^{-1} - A_c^{-T} P A_c^{-1} < 0$$

which can furthermore be written as:

$$\begin{bmatrix} E_c^{-T} & A_c^{-T} \end{bmatrix} \begin{bmatrix} Q & 0 \\ 0 & -Q \end{bmatrix} \begin{bmatrix} E_c^{-1} \\ A_c^{-1} \end{bmatrix} < 0.$$

Applying Lemma 1 (projection lemma), this is equivalent to the existence of matrices *G* such that:

$$\begin{bmatrix} P + \alpha G^T E^T + \alpha B K_d G + \alpha G^T K_d^T B & -\alpha G^T A^T - \alpha G^T K_p^T B^T \\ * & -P \end{bmatrix} < 0.$$

Taking now $R_1 = K_p F$ and $R_2 = K_d F$, the inequality (18) follows.

2.3 Extension to Robust Stabilization

The outcome or result of the previous section can be extended to robust stabilization for uncertain descriptor systems with polytopic coefficient matrices.

Let us consider the following linear uncertain descriptor system:

$$E(\theta)\dot{x}(t) = A(\beta)x(t) + B(\zeta)u(t) \quad (19)$$

where $x \in \mathbb{R}^n$ is the state vector, we assume that *E*, *A* and *B* are constant and respectively belong to the classes:

$$
\begin{aligned}
E \in \mathcal{E} &= \left\{ E(\theta) : E(\theta) = \sum_{i=1}^{N_E} \theta_i E_i,\ \sum_{i=1}^{N_E} \theta_i = 1,\ \theta_i \geq 0 \right\} \\
A \in \mathcal{A} &= \left\{ A(\beta) : A(\beta) = \sum_{j=1}^{N_A} \beta_j A_j,\ \sum_{j=1}^{N_A} \beta_j = 1,\ \beta_j \geq 0 \right\}. \\
B \in \mathcal{B} &= \left\{ B(\zeta) : B(\zeta) = \sum_{k=1}^{N_B} \zeta_k B_k,\ \sum_{k=1}^{N_B} \zeta_k = 1,\ \zeta_k \geq 0 \right\}
\end{aligned}
$$

To extend the result of Theorem 2.1 for this polytopic system, we give the following theorem:

- **Theorem 2.3:** The following statements are equivalent:
  - System (19) is normalizable and stabilizable.
  - There exist a positive definite matrix Y and matrices F, R_1 and R_2 of appropriate dimensions such that:

$$
\begin{bmatrix} F^T A_i^T + A_i F + B_j R_1 + R_1^T B_j^T & Y_{i,j,k} + F^T A_i^T + R_1^T B_j^T - E_k F - B_j R_2 \\ * & -F^T E_k^T - E_k F - B_j R_2 - R_2^T B_j^T \end{bmatrix} < 0 \tag{20}
$$

$$
i = 1 \ldots N_E,\quad j = 1 \ldots N_A,\quad k = 1 \ldots N_B
$$

The gain given by: $K_p = R_1 F^{-1}$ and $K_d = R_2 F^{-1}$ solves problem.

- **Proof:** The proof follows by way of simple convexity arguments. Note that $\sum_{i=1}^{N_A} \sum_{j=1}^{N_B} \sum_{k=1}^{N_E} \beta_i \theta_j \zeta_k Y_{i,j,k}$ is a Lyapunov function for the closed-loop system.
- **Remark 2:** Compared with the previous result obtained in Tian and Hou (2009) for the robust stabilization, the necessary and sufficient condition proposed for the determination of the (PD) controller structure, is just limited for the proportional gain (K_2). However, the derivative gain must be chosen from the beginning for an adequate resolution to the problem. On the other hand, the present contribution is different for the reason that both the constants K_p and K_d will be induced by LMI solvers, which in turn can be seen to highlight the reason for the interest in Theorem 3.

After achieving with the normalization stabilization problem for singular systems, a proportional plus derivative feedback H_2 control problem will be developed in the next sectionμ. A sufficient condition to the existence of a H_2 (PD) feedback controller is given in terms of LMIs such that the closed loop system is normal stable and guarantees the H_2 performance.

3. H_2 CONTROL PROBLEM FOR LINEAR DESCRIPTOR SYSTEMS

3.1. H_2 Control Condition for Continuous Time Descriptor Systems

Let's consider the following continuous-time linear descriptor system described by:

$$\begin{cases} E\dot{x}(t) = Ax(t) + Bu(t) + B_w w(t) \\ z(t) = Cx(t) \end{cases}, \qquad x(0) = 0 \tag{21}$$

where $x(t) \in \mathbb{R}^n$ is the state vector and $z(t) \in \mathbb{R}^p$ is the controlled output vector. Matrices A, B and C are constant matrices of appropriate dimensions. The matrix E may be singular: $rank(E) = q \leq n$.

In this section, we aim to find a proportional plus derivative state feed-back controller (PD) given by the equation (11) in order to ensure the stability of the system (21) and minimizes the upper bound of performance measures.

So, applying the control law (11) to the system (21), the closed loop system is expressed as follows:

$$(E + BK_d)\dot{x}(t) = (A + BK_p)x(t) + B_w w(t) \tag{22}$$

Then denote the transfer function of the closed loop system (21) from w to z as:

$$T_{wz}(s) = C(sE_c - A_c)^{-1} B_w \tag{23}$$

where $E_c = E + BK_d$, $A_c = A + BK_p$.

Hence, the main results are obtained as the following theorem.

- **Theorem 3.1:** For a given positive number g and a scalar $\alpha > 0$ sufficiently large, there is a PD controller such that the closed-loop system (22) is stable and satisfies $\left\| C(sI - E_c^{-1}A_c)^{-1} E_c^{-1} B_w \right\|_2 < \gamma$ if there exist positive definite matrices Q and matrices T, F, R_1, R_2 of appropriate dimensions solution of the optimization problem given in (24).

$$\min_{Q,T,F,R_1,R_2} \gamma \tag{24}$$

$$trace\, T < \gamma$$

$$\begin{bmatrix} \alpha F^T A^T + \alpha AF + \alpha BR_1 + \alpha R_1^T B^T & Q + F^T A^T + R_1^T B^T - \alpha EF & F^T C^T \\ * & -F^T E^T - R_2^T B^T - EF - BR_2 & 0 \\ * & * & -I \end{bmatrix} < 0$$

$$\begin{bmatrix} -T & B_w^T \\ * & Q - EF - BR_2 - F^T E^T - R_2^T B^T \end{bmatrix} < 0$$

Hence, the control law: $u = R_1 F^{-1} x - R_2 F^{-1} \dot{x}$ stabilizes asymptotically the system and minimizes the H_2 cost.

- **Proof:** Since the system (22) is normal stable, from Definition 1, E_c is non-singular and $E_c^{-1} A_c$ is stable.

According to Lemma 1,

$$C(sI - (E + BK_d)^{-1}(A + BK_p))^{-1}(E + BK_d)^{-1} B_w < \gamma$$

is equivalent to that there exists a matrix $P > 0$ such that:

$$(A + BK_p)^T (E + BK_d)^{-T} P + P(E + BK_d)^{-1}(A + BK_p) + C^T C < 0 \tag{25}$$

$$trace(B^T (E + BK_d)^{-T} P(E + BK_d)^{-1} B) < 0 \tag{26}$$

We note that (25) can be written as:

$$\begin{bmatrix} (A + BK_p)^{-T} & (E + BK_d)^{-T} \end{bmatrix} \begin{bmatrix} C^T C & P \\ P & 0 \end{bmatrix} \begin{bmatrix} (A + BK_p)^{-1} \\ (E + BK_d)^{-1} \end{bmatrix} < 0$$

Then by applying the projection lemma, there exist a matrix G of appropriate dimensions such that:

$$\begin{bmatrix} C^T C & P \\ P & 0 \end{bmatrix} + \underbrace{\begin{bmatrix} (A + BK_p)^T \\ -(E + BK_d)^T \end{bmatrix} G \begin{bmatrix} \alpha & I \end{bmatrix}}_{*} + sym(*) < 0$$

Equal to:

$$\begin{bmatrix} C^T C + \alpha(A + BK_p)^T G + \alpha G^T (A + BK_p) & P + (A + BK_p)^T G - \alpha G^T (E + BK_d) \\ * & -(E + BK_d)^T G - G^T (E + BK_d) \end{bmatrix}$$

where $\alpha > 0$ is a sufficiently large number.

Denoting $F = G^{-1}$, $Q = F^T PF$, $R_1 = K_p F, R_2 = K_d F$ and multiplying this inequality with $diag(F^T, F^T)$ to the right and its transpose to the left yields to:

$$\begin{bmatrix} F^T C^T CF + \alpha F^T A^T + \alpha R_1^T B^T + \alpha BR_1 + \alpha AF & Q + F^T A^T + R_1^T B^T - \alpha EF - \alpha BR_2 \\ * & -F^T E^T - R_2^T B^T - EF - BR_2 \end{bmatrix} < 0$$

by using Schur complement argument, we leads to second inequality of Theorem 1.

On the other hand, (26) also can be written as:

$$B_w^T (E + BK_d)^{-T} P(E + BK_d)^{-1} B_w < T \Leftrightarrow \begin{bmatrix} -T & B_w^T (E + BK_d)^{-T} \\ * & -P^{-1} \end{bmatrix}$$

where T is a symmetric matrix, multiplying this inequality with $diag(I, E + BK_d)$ to the left and in the right par its transpose, we obtain:

$$\begin{bmatrix} -T & B_w^T \\ * & -(E + BK_d)P^{-1}(E + BK_d)^T \end{bmatrix} < 0$$

furthermore, the (2,2) bloc implies:

$$(E + BK_d)P^{-1}(E + BK_d)^T \leq F^T PF - (E + BK_d)F - F^T (E + BK_d)^T$$

And by taking $Q = F^T PF$, the last inequality can be written as:

$$\begin{bmatrix} -T & B_w^T \\ * & Q - (E + BK_d)F - F^T (E + BK_d)^T \end{bmatrix} < 0$$

This completes the proof and implies a sufficient condition for Theorem 1.

3.2. H_2 Control Condition for Discrete Time Descriptor Systems

Let us consider the following linear discrete singular system as such:

$$\begin{cases} Ex(k+1) = Ax(k) + Bu(k) + B_w w(k) \\ z(k) = Cx(k) \end{cases} \tag{27}$$

where $x(k) \in \mathbb{R}^n$ is the state vector and $z(k) \in \mathbb{R}^p$ is the controlled output vector. Matrices A, B and C are constant matrices of appropriate dimensions. The matrix E may be singular $rank(E) = q \leq n$.

In the same way as we saw expressed in the continuous case, in this subsection we want to find a proportional plus derivative state feedback controller such that the closed loop system is H_2 performance.

Let the PD state feedback control expressed as follows:

$$u(k) = K_p x(k) - K_d x(k+1) \tag{28}$$

where gain $K_p, K_d \in \mathbb{R}^{m \times n}$ are constant matrix of appropriate dimensions.

The resulting closed loop system is as follows:

$$(E + BK_d)x(k+1) = (A + BK_p)x(k) + B_w w(k) \tag{29}$$

and the resulting closed loop transfer function matrix is as follows:

$$T_{wz}(z) = C(z(E + BK_d) - (A + BK_p))^{-1} B_w$$

We present a strict LMI condition for the H_2 norm to be smaller than a given positive number, for this purpose the following definition will be needed:

- **Definition 3:** The descriptor system (27), when the disturbance input *w*(*k*) = 0, is said to be normal and stable (NS) if E is non-singular $(\det E \neq 0)$ and there exists a matrix *P* >0 such that:

$$(E^{-1}A)^T P(E^{-1}A) - P < 0 \tag{30}$$

- **Theorem 3.2:** For a given positive number $\gamma > 0$ and a scalar ε sufficiently small, there is a PD controller such that the closed loop system (29) is normal stable and satisfies $\left\| C(zI - (E + BK_d)^{-1}(A + BK_p))^{-1}(E + BK_d)^{-1} B_w \right\|_2 < \gamma$, if there exist positive definite matrices Q and matrices T, F, R_1, R_2 of appropriate dimensions solution of the following optimization problem:

$$\min_{Q,T,F,R_1,R_2} \gamma$$

$$trace\, T < \gamma$$

$$\begin{bmatrix} \varepsilon Q + F^T E^T + EF + BR_2 + R_2^T B^T & AF - BR_1 & 0 \\ * & -Q & F^T C^T \\ * & * & -I \end{bmatrix} < 0 \tag{31}$$

$$\begin{bmatrix} -T & B_w^T \\ * & Q - EF - BR_2 - F^T E^T - R_2^T B^T \end{bmatrix} < 0$$

Then the control law: $u(k) = R_1 F^{-1} x(k) - R_2 F^{-1} x(k+1)$ stabilizes asymptotically the system and minimizes the H_2 cost.

- **Proof:** Since the system (29) is NS, from Definition 1, E_c is non-singular and $E_c^{-1} A_c$ is stable.

According to Lemma 1, but in discrete version,

$$C(sI - (E + BK_d)^{-1}(A + BK_p))^{-1}(E + BK_d)^{-1} B_w < \gamma,$$

is equivalent to that there exists a matrix $P > 0$ such that:

$$((E + BK_d)^{-1}(A + BK_p))^T P (E + BK_d)^{-1}(A + BK_p) - P + C^T C < 0 \tag{32}$$

$$trace\, B^T P B < \gamma \tag{33}$$

For a given scalar $\varepsilon > 0$ sufficiently small, the (32) can be rewritten as follows:

$$\varepsilon((E + BK_d)^{-1}(A + BK_p))^T P (E + BK_d)^{-1}(A + BK_p) - P + \varepsilon C^T C < 0 \tag{34}$$

Note that this modification causes no loss of generality, since the stability of the closed loop system did not affect. Hence, (34) can be written:

$$\begin{bmatrix} (E + BK_d)^{-T} & (A + BK_p)^{-T} \end{bmatrix} \begin{bmatrix} \varepsilon P & 0 \\ 0 & -P + \varepsilon C^T C \end{bmatrix} \begin{bmatrix} (A + BK_p)^{-1} \\ (E + BK_d)^{-1} \end{bmatrix} < 0$$

Then by applying the projection lemma, there exist a matrix G of appropriate dimensions such that:

$$\begin{bmatrix} \varepsilon P & 0 \\ 0 & -P + \varepsilon C^T C \end{bmatrix} + \underbrace{\begin{bmatrix} (A + BK_p)^T \\ -(E + BK_d)^T \end{bmatrix} G \begin{bmatrix} I & 0 \end{bmatrix}}_{*} + sym(*) < 0$$

$$= \begin{bmatrix} \varepsilon P + E^T G + G^T E + K_d^T B^T G + G^T B K_d & -G^T A - G^T B K_p \\ * & -P + \varepsilon C^T C \end{bmatrix} < 0$$

Denoting $F = G^{-1}, Q = F^T P F$, $R_1 = K_p F, R_2 = K_d F$ and multiplying this inequality with $diag(F^T, F^T)$ to the right and its transpose to the left yields to:

$$\begin{bmatrix} \varepsilon Q + F^T E^T + R_2^T B^T + EF + BR_2 & -AF - BR_1 \\ * & -Q + \varepsilon F^T C^T C F \end{bmatrix} < 0$$

by using Schur we can complement the argument, which in turn leads us on to the second inequality of Theorem 2. For the last equality, the proof is similar to the continuous case.

3.3. Extension to Robust H$_2$ Control

The result of the previous section can be extended to robust H_2 control for uncertain descriptor systems with polytopic coefficient matrices.

Considering the following linear uncertain descriptor system:

$$\begin{cases} E(\theta)x(t) = A(\beta)x(t) + B(\zeta)u(t) + B_w w(t) \\ z(t) = Cx(t) \end{cases} \tag{35}$$

where $x(k) \in \mathbb{R}^n$ is the state vector, $w(t) \in \mathbb{R}^q$ is the disturbance vector and $z(t) \in \mathbb{R}^p$ is the controlled output vector is the state vector. We assume that *E*, *A* and *B* are constant and respectively belong to the outlined classes:

$$E \in \mathcal{E} = \left\{ E(\theta) : E(\theta) = \sum_{i=1}^{N_E} \theta_i E_i, \sum_{i=1}^{N_E} \theta_i = 1, \theta_i \geq 0 \right\}$$
$$A \in \mathcal{A} = \left\{ A(\beta) : A(\beta) = \sum_{j=1}^{N_A} \beta_j A_j, \sum_{j=1}^{N_A} \beta_j = 1, \beta_j \geq 0 \right\}$$
$$B \in \mathcal{B} = \left\{ B(\zeta) : B(\zeta) = \sum_{k=1}^{N_B} \zeta_k B_k, \sum_{k=1}^{N_B} \zeta_k = 1, \zeta_k \geq 0 \right\}$$

To extend the result of Theorem 3.1 for this polytopic system, we give the following theorem:

- **Theorem 3.3:** Consider the uncertain singular system (35). There exists a PD state feedback controller (10) such that the resulting closed loop system is NS and the H_2 performance bound constraint $\left\| T_{zw}(s) \right\|_2 < \gamma$ is satisfied, if there exist a scalar $\alpha > 0$, matrices *Q* >0, *T* >0, *F*, *Y*, R_1 and R_2 of appropriate dimensions such that the LMIs condition (36) is feasible and can be carried out.

$$\min_{Q,T,F,R_1,R_2} \gamma$$

$$trace\, T < \gamma \tag{36}$$

$$\begin{bmatrix} \alpha F^T A_j^T + \alpha A_j F + \alpha B_k R_1 + \alpha R_1^T B_k^T & Q_{i,j,k} + F^T A_j^T + R_1^T B_k^T - \alpha E_i F & F^T C^T \\ * & -F^T E_i^T - R_2^T B_k^T - E_i F - B_k R_2 & 0 \\ * & * & -I \end{bmatrix} < 0$$

$$\begin{bmatrix} -T & B_w^T \\ * & Q_{i,j,k} - E_i F - B_k R_2 - F^T E_i^T - R_2^T B_k^T \end{bmatrix} < 0$$

Then the control law: $u(k) = R_1 F^{-1} x(k) - R_2 F^{-1} x(k+1)$ stabilizes asymptotically the uncertain system and minimizes the H_2 cost.

- **Proof:** The proof follows along a line of simple convexity arguments. Note that $\sum_{i=1}^{N_E} \sum_{j=1}^{N_A} \sum_{k=1}^{N_B} \theta_i \beta_j \zeta_k Q_{ijk}$ is a Lyapunov function for the closed loop system.

3. NUMERICAL ILLUSTRATION

To help us to demonstrate the effectiveness and applicability of the proposed method, we shall provide three distinct examples. In the first simulation the (PD) controller is applied to a DC motor modeled by singular systems, in the second one, we use the proposed method for discrete singular system and in the last example, we will complete our examples by using the proposed method for an uncertain polytopic system.

- **Example 1:** The Figure 3 shows a simple DC motor. Since typically L is small, if we neglect the armature inductance, we can write the electrical and mechanical equations as what is outlined here:

$$\begin{cases} v(t) = Ri(t) + k_e w(t) \\ J \dfrac{dw}{dt} = k_c i(t) - bw(t) \end{cases}$$

where *i*(*t*), *v*(*t*) and *R* are respectively, and each in their own right, the armature current, voltage and resistance. J is the moment of inertia. k_e, k_c, are respectively, the torque and the back e.m.f developed with constant excitation flux. The constant b is a damping coefficient and *w*(*t*) is the angular speed. The model of the motor consists of a differential equation and an additionally algebraic equation which is part of the results of Kirchhoff's voltage law in armature circuit. It is possible to model the DC motor using just one of our differential equations in ordinary state space, but instead, by using the singular model of the motor we can reach a more precise model which can also describe the behavior of armature current.

Figure 3. Electrical circuit of DC motor

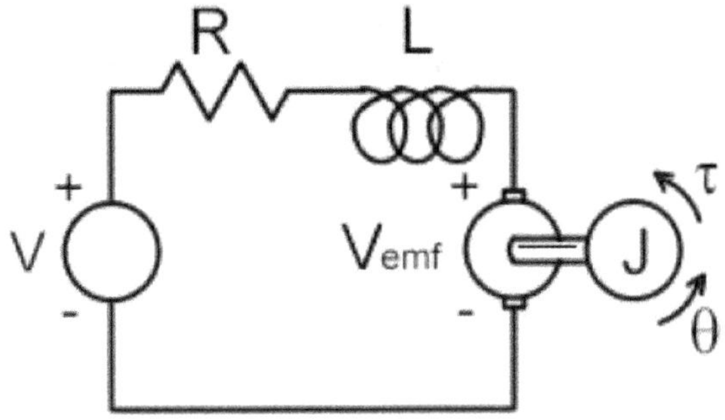

By choosing: $x_1(t) = i(t)$, $x_2(t) = w(t)$ and $u(t) = v(t)$, we can rewrite the descriptor system as (Azarfar et al., 2013):

$$\begin{bmatrix} 0 & 0 \\ 0 & J \end{bmatrix} \begin{bmatrix} \dot{x}_1 \\ \dot{x}_2 \end{bmatrix} = \begin{bmatrix} R & k_e \\ k_c & -b \end{bmatrix} \begin{bmatrix} x_1 \\ x_2 \end{bmatrix} + \begin{bmatrix} -1 \\ 0 \end{bmatrix} u + \begin{bmatrix} 1 \\ 0 \end{bmatrix} w(t)$$

This is, as you will see, in the form of (1). Now we will apply the proposed control method to the singular model of DC motor.

The system parameters are defined in Table 1.

The purpose is to design and furthermore outline in full a (PD) state feedback such that the resulting closed loop system is robust H_2 performance. As a result, by Theorem 1, a (PD) feedback controller can be obtained as:

$$u(t) = 10^5 \begin{bmatrix} -2.087 & -0.0173 \end{bmatrix} x(t) - 10^4 \begin{bmatrix} 3.7 & 0.0306 \end{bmatrix} \dot{x}(t)$$

Because of this, by using this controller and if the state's initial values are set to $x_1(0) = 3A$, $x_2(0) = 500\ rpm$ and by simulating this closed- loop system, we can get the state trajectory shown as Figure 4 which in turn proves the stability of the system and. Following on from this the upper bounded then attained is as follows: $\gamma = 10^3\ 2.7886$.

- **Example 2:** It is now our task to consider the following discrete descriptor system:

Table 1. Electrical circuit of DC motor

B	*J*	*Ke*	*Kc*	*R*
1.510^{-5} NM / rpm	810^{-3} NMS	810^{-3} r/rpm	810^{-3} r/rpm	810^{-3} r/rpm

Figure 4. The evolution of the DC motor state

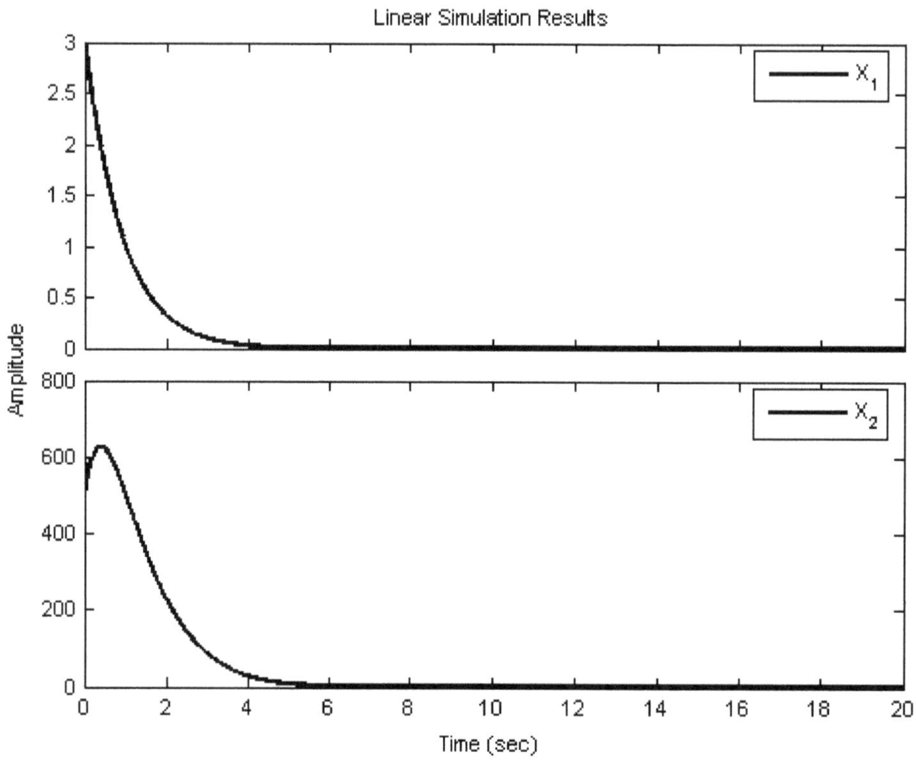

$$E=\begin{bmatrix}1&0&0\\0&1&0\\0&0&0\end{bmatrix},\ A=\begin{bmatrix}1&2&1\\1&1&1\\0&0&0\end{bmatrix},\ B=\begin{bmatrix}2\\2\\2\end{bmatrix},$$

$$C=\begin{bmatrix}0&0&1\end{bmatrix},\ B_1=\begin{bmatrix}1\\0.1\\1\end{bmatrix}$$

Our objective is to design a controller such that the resultant closed loop system is normal stable and an upper bound of H_2 norm is achieved. Using the Theorem 2 and choosing the parameter $\alpha = 0.6$, a feasible solution can be obtained and the corresponding state feedback control law is given as this:

$$u(k) = 10^3\begin{bmatrix}-3.3442 & 4.729 & 2.4377\end{bmatrix}x(k) \\ -10^3\begin{bmatrix}-3.9442 & -5.776 & -3.8751\end{bmatrix}x(k+1)$$

By using this controller and when the state's initial values are set to $x(0)=\begin{bmatrix}1 & 0 & 1\end{bmatrix}^T$ and by simulating this closed- loop system, we can then get the state trajectory shown by Figure 5 which prove the stability of the system and the upper bound which is attained is outlined in the following: $\gamma = 6.0937$

Figure 5. The evolution of the system state

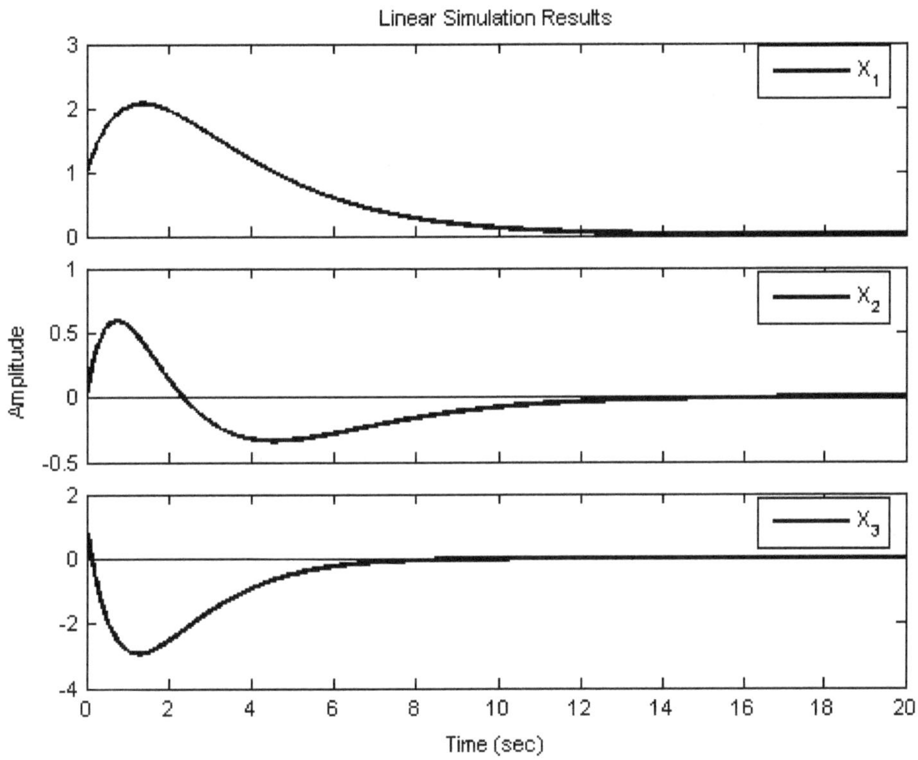

- **Example 3:** A second example is presented for polytopic system with parameters as follows:

$$E_1 = \begin{bmatrix} 10 & 0 \\ 0 & 0 \end{bmatrix},\ E_2 = \begin{bmatrix} 0.8 & 0 \\ 0 & 0 \end{bmatrix},\ A_1 = \begin{bmatrix} 3 & 1 \\ 1 & 0 \end{bmatrix},\ A_2 = \begin{bmatrix} 6 & 1 \\ 1 & 0 \end{bmatrix}$$

$$B_1 = \begin{bmatrix} 0.5 \\ 1 \end{bmatrix}, B_2 = \begin{bmatrix} 1 \\ 1 \end{bmatrix},\ B_w = \begin{bmatrix} 0.5 \\ 0.1 \end{bmatrix},\ C = \begin{bmatrix} 1 & 0 \end{bmatrix}$$

Using the Theorem 3, and choosing the $\alpha = 30$ parameter, the corresponding derivative plus proportional feedback controller such that the closed loop system is NS while satisfying a prescribed H_2 norm bound g is given by:

$$K_d = 10^3 \begin{bmatrix} 3.8549 & 0.00275 \end{bmatrix},\ K_p = 10^4 \begin{bmatrix} -9.7591 & -0.0695 \end{bmatrix}$$

and the upper bounded is attained as: $\gamma = 0.1517$.

4. DISCUSSION

The problem of stabilization of descriptor systems is of both practical and theoretical importance and has attracted the attention of many authors in the past years. In the literature, there are two kinds of stabilization problems of descriptor continuous time systems. One is to design state feedback controllers such that the closed loop systems is regular, impulse free and stable (admissible), the other to construct state feedback controllers to make the closed loop systems stable and regular. On other hand, many of existing techniques for the control of descriptor systems are based on the assumption that the system is regular. However, in practical applications, descriptor systems are note always regular, even if a descriptor systems is regular, then the regularity may be destroyed by state feedback inputs. For this reason, to develop a direct design approach to stabilize descriptor systems without the regularity assumption is important.

This paper, considers the proportional plus derivative state feedback (PD) stabilization problem of singular systems as a solution to the regularity and also to the non-impulsiveness problems. The main purpose is to design proportional plus derivative state feedback controllers to make the closed loop system normal and stable. As normal linear system its response is impulse free and does not involve infinite frequencies. Then, it is desirable in singular linear systems design, to make the constructed feedback control system be normal. This leads to normalization problem developed in the first part of this paper: find the conditions of existence of a feedback controllers for the singular systems such that the closed loop system is normal.

The idea of this paper is to design a proportional plus derivative controller which the derivative gain K_d acts on the normalization of the system and the proportional gain K_p acts on the stability.

After resolved the normalization stabilization problem, a H_2 performance control is further integrated with the same controller used below. A matrix inequalities approach is given to derive the conditions under which the desired (PD) exist. Hence, the expected control law can easily obtained in terms of the solution of this matrix inequality where the design procedure of H_2 (PD) controller is direct and simple. This work has not been investigated yet. All the approaches studied in the literature used the output feedback controller or the state feedback controller to resolve the H_2 control problem for singular systems

4. CONCLUSION AND FUTURE WORK

In The present paper, the problem of H_2 proportional plus derivative feedback control for a class of descriptor systems, in both continuous and discrete case has been studied. Based on LMI conditions with regard to normal stability and H_2 properties are derived for both continuous and discrete time setting and also an extension to uncertain system is considered. A sufficient condition to the existence of a H_2 proportional plus derivative state feedback controller is given and the desired controller is also achieved. To show the effectiveness of the proposed approaches, numerical examples are included.

The main ides used in this paper is to find a structure of controller which makes the normalization of the system and it is the rule of the derivative feedback. However, it is well known from classical control theory that derivative feedback is used to improve the system performance. For singular systems, the derivative feedback can offers many features of the system, such as the dynamical order of the system, the regularity, and the finite and infinite frequencies. Due to this reason, derivative feedback will be used in future work to discuss the admissibility of the system without any assumption on the system matrix for both continuous and discrete case and also an extension to uncertain system will be given. A sufficient condition in terms of LMIs matrix will be developed and an explicit expression of the desired derivative feedback controller will be achieved.

REFERENCES

Azarfar, A., Shandiz, T. H., & Shafiee, M. (2013). Adaptive feedback control for linear singular systems. *Turkish Journal of Electrical Engineering & Computer Sciences, 22*(2014), 132-142.

Bara, L. G. (2011). Robust analysis and control of parameter-dependent uncertain descriptor systems. *Elsevier Systems & Control Letters, 60*(6), 356–364. doi:10.1016/j.sysconle.2011.03.001

Bernstein, S. D., & Haddad, M. W. (1989). LQG control with H_∞. *IEEE Transactions on Automatic Control, 34*(3), 293–305. doi:10.1109/9.16419

Boukas, K. E. (2005). Stabilization of stochastic nonlinear hybrid systems. *International Journal of Innovative Computing, Information, & Control, 1*(1), 131–141.

Boukas, K. E., Shi, P., & Nguang, S. (2003). Robust Hinfini control for linear Markovian jump systems with unknown non linearrities. *Journal of Mathematical Analysis and Applications, 282*(1), 241–255. doi:10.1016/S0022-247X(03)00144-6

Chaabane, M., Tadeo, F., Mehdi, D., & Souissi, M. (2011). Robust Admissibilization of Descriptor Systems by Static Output-Feedback: An LMI Approach. Hindawi Publishing Corporation, 2012(2012).

Dai, L. (1989). *Singular Control Systems. Lecture Notes in Control and Information Sciences.* Germany: Springer.

Doyle, J., Glover, K., Khargonekar, P. P., & Francis, B. (1989). State space solutions to standards H_2 H_∞. *IEEE Transactions on Automatic Control, 34*(8), 831–847. doi:10.1109/9.29425

Doyle, J., Zhou, K., Glover, K., & Bodenheimer, B. (1994). Mixed H_2 / H_∞. *IEEE Transactions on Automatic Control, 39*(8), 1575–1587. doi:10.1109/9.310031

Dragan, V. (2005). The linear quadratic optimization problem for a class of singularly perturbed stochastic systems. *International Journal of Innovative Computing, Information, & Control, 1*(1), 53–63.

Dragan, V. Sh. P., & Boukas, K. E. (1999). Control of singularly perturbed systems with Markovian jump parameters: An H infini approach. *Automatica, 35*(8), 1369–1378. doi:10.1016/S0005-1098(99)00047-3

Duan, R. G. (2010). Analysis and Design of linear descriptor systems. New York, NY: Springer. doi:10.1007/978-1-4419-6397-0

Fang, H. C., & Chang, R. F. (1993). Analysis of stability robustness for generalized state space systems with structured perturbations. *Systems & Control Letters, 21*(2), 109–114. doi:10.1016/0167-6911(93)90113-K

Fang, H. C., Lee, L., & Chang, F.-R. (1994). Robust control analysis and design for discrete time singular systems. *Automatica*, *30*(11), 1741–1750. doi:10.1016/0005-1098(94)90076-0

Faria, A. F., Assunçao, E., Teixeira, M., Cardim, R., & da Silva, N. A. P. (2009). Robust state-derivative pole placement LMI-based designs for linear systems. *International Journal of Control*, *82*(1), 1–12. doi:10.1080/00207170801942188

Feng, Y., Yagoubi, M., & Chevrel, P. (2010). Extended output Control for Continuous Descriptor Systems. In *Proceedings of 49th IEEE Conference on Decision and Control* (pp. 6016-6021). Atlanta, GA: IEEE. doi:10.1109/CDC.2010.5717866

Feng, Y., Yagoubi, M., & Chevrel, P. (2011). State Feedback H_2 optimal Controllers under regulation constraints for Descriptor Systems. *International Journal of Innovative Computing, Information, & Control*, *7*(10), 5761–5770.

Feng, Y., Yagoubi, M., & Chevrel, P. (2012). Extended H_2 Controller Synthesis for Continuous Descriptor Systems. Automatic Control. *IEEE Transactions*, *57*(6), 1559–1564.

Geromel, C. J., Peres, D. L. P., & Souza, S. R. (1992). Mixed control for continuous time linear systems. In *Proceedings of 31th IEEE Conference Decision and control* (pp. 3717-3722). IEEE.

Ikeda, M., Lee, T., & Uezato, E. (2000). A Strict LMI Condition for Control of Descriptor Systems. In *Proceedings of 38th IEEE Conference on Decision and Control* (pp. 4092-4097). Sydney: IEEE.

Ishihara, Y. J., Terra, M. H., & Sales, R. M. (2003). The full information and state feedback H_2. *Automatica*, *39*(3), 391–402. doi:10.1016/S0005-1098(02)00243-1

Khargonekar, P., & Rotea, A. M. (1991). Mixed H_2 / H_∞. *IEEE Transactions on Automatic Control*, *36*(7), 824–837. doi:10.1109/9.85062

Lewis, L. F. (1986). Asurvey of linear singular systems. *Circuits System Signal Processing*, *5*(1), 3–36. doi:10.1007/BF01600184

Lin, C., Wang, D. J., & Soh, B. C. (1997). Robustness of uncertain descriptor systems. *Systems & Control Letters*, *31*(3), 129–138. doi:10.1016/S0167-6911(97)00038-8

Lin, C., Wang, L. J., Yang, H. G., & Lam, J. (2000). Robust stabilization via state feedback for descriptor systems with uncertainties in the derivative matrix. *International Journal of Control*, *73*(5), 407–415. doi:10.1080/002071700219588

Luenberger, D. (1977). Dynamic equations in descriptor form. *IEEE Transactions on Automatic Control*, *22*(3), 312–321. doi:10.1109/TAC.1977.1101502

Mahmoud, M., & Shi, P. (2003). Robust stability, stabilization and H_∞. *International Journal Robust Nonlinear Control*, *13*(8), 755–784. doi:10.1002/rnc.744

Masubuchi, I., Kamitane, Y., Ohara, A., & Suda, N. (1997). H_∞ control for descriptor systems: A matrix inequalities approach. *Automatica*, *33*(4), 669–673. doi:10.1016/S0005-1098(96)00193-8

Ren, J., & Zhang, Q. (2010). Robust H_∞. *International Journal of Control, 83*(1), 89–96. doi:10.1080/00207170903100222

Renquan, L., Wei, D., & Lin, X. (2008). Robust Control for Descriptor System Based on New Bounded Real Lemma. In *Proceedings of 27th Chinese Control Conference* (pp. 636-646). Yunna, China: Academic Press.

Rotea, A. M., & Khargonekar, P. P. (1991). H2-optimal Control with an H∞ constraint The State Feedback Case. *Automatica, 27*(2), 307–316. doi:10.1016/0005-1098(91)90079-H

Sari, B., Bachelier, O., & Mehdi, D. (2006). Robust state feedback admissibilization of discrete linear polytopic descriptor systems: a strict linear matrix inequality approach. *IET Control Theory and Applications, 6*(8), 1097-1108.

Scherer, C., & Weiland, S. (2005). Linear Matrix Inequalities in Control, *Lecture notes course. CiteUlike, 3853526*, 1–232.

Shi, P., Boukas, K. E., & Agarwal, K. R. (1999). Control of Markovian jump discrete time systems with norm bounded uncertainty and unknown delay. *IEEE Transactions on Automatic Control, 44*(11), 2139–2155. doi:10.1109/9.802932

Takaba, K. (1998). Robust H_2. *International Journal of Control, 71*(4), 559–579. doi:10.1080/002071798221678

Takaba, K., & Katayama, T. (1997). Robust performance of uncertain descriptor systems. In *Proceeding of European Control Conference.* Brussels, Belgium: Academic Press.

Takaba, K., & Katayama, T. (1998). H_2 Output Feedback Control for Descriptor Systems. *Automatica, 34*(7), 841–850. doi:10.1016/S0005-1098(98)00025-9

Tian, H., & Hou, Y. (2009). Robust Stabilization for a class of Uncertain Descriptor Systems via both Proportional and Derivative State Feedback. In *Proceedings of the Second International Conference on Intelligent Computation Technology and Automation* (pp. 874-877). Academic Press. doi:10.1109/ICICTA.2009.218

Xia, Y., & Jia, Y. (2003a). Robust sliding mode for uncertain time delay systems: An lmi approach. *IEEE Transactions on Automatic Control, 84*(6), 1086–1092.

Xia, Y., & Jia, Y. (2003b). Robust control of state delayed systems with polytopic type uncertainties via parameter dependent Lyapunov functionals. *Systems & Control Letters, 50*(3), 183–193. doi:10.1016/S0167-6911(03)00153-1

Xia, Y., Shi, P., Liu, G., & Rees, D. (2005). Robust Mixed H_2 / H_∞ Sate feedback Control for Continuous-Time Descriptor Systems with Parameter Uncertainties. *Circuit Systems Signal, 24*(4), 431–443.

Xu, S., Dooren, V. P., Stefan, R., & Lam, J. (2002). Robust stability and stabilization for singular systems with state delay and parameter uncertainty. *IEEE Transactions on Automatic Control, 47*(7), 1112–1128.

Xu, S., & Lam, J. (2003). Robust stability for uncertain discrete singular systems with state delay. *Asian Journal of Control, 5*(3), 399–405. doi:10.1111/j.1934-6093.2003.tb00132.x

Xu, S., & Lam, J. (2004). Robust stability and stabilization for discrete singular systems: An equivalent characterization. *IEEE Transactions on Automatic Control, 49*(4), 568–574. doi:10.1109/TAC.2003.822854

Xu, S., & Lam, J. (2006). *Robust Control and Filtering of Singular Systems*. Berlin: Springer.

Xu, S., Lam, J., & Yang, C. (2002). Robust H_∞ control for discrete singular systems with state delay and parameter uncertainty. *Dynamic Continuous Discrete Impulsive Systems B, 9*, 539–554.

Xu, S., & Yang, C. (1999). Stabilization of discrete time singular systems: A matrix inequalities approach. *Automatica, 35*(9), 1613–1617. doi:10.1016/S0005-1098(99)00061-8

Xu, S., Yang, C., Niu, Y., & Lam, J. (2001). Robust stabilization for uncertain discrete singular systems. *Automatica, 37*(5), 769–774. doi:10.1016/S0005-1098(01)00013-9

Yagoubi, M., Chevrel, P., & Claveau, F. (2007). An Extended state-feedback for Descriptor Systems. In *Proceedings of the 3rd IFAC Symposium on System, Structure and Control* (pp.386-391). Iguassu Falls, Brazil: IFAC.

Yan, Z., Zhang, G., & Wang, J. (2012). Infinite horizon H_2 / H_∞. *Control Theory Application, 10*(2), 159–165. doi:10.1007/s11768-012-0038-6

Zhou, K., Glover, K., Bodenheimer, B., & Doyle, J. (1994). Mixed H_2 / H_∞ performance objective I: Robust performance analysis. *IEEE Transactions on Automatic Control, 39*(8), 1564–1574.

KEY TERMS AND DEFINITIONS

H_2 Control Problem: Consists of finding a controller which stabilizes the system and which minimizes the cost function $\|T_{wz}(s)\|_2^2$, where $\|T_{wz}(s)\|_2$ is the H_2 norm.

LMIs: A linear matrix inequality is a convex constraint. They are a useful tool for solving a wide variety of optimization and control problems.

Normalization: Is concerned with designing a controller for a singular system such that the closed loop system is normal.

PD State Feedback Controller: Is designed for the closed loop system to be normal and stable.

Robust Control: Considers the design of decision or control rules that fare well across a range of alternative models. Thus robust control is inherently about model uncertainty, particularly focusing on the implications of model uncertainty for decisions.

Singular Systems: A dynamic systems that the matrix on the derivative of the state, that is E in (5), is generally singular.

Stabilization: Is concerned with the designing a controller of a prescribed type such that the closed loop system is stable.

Chapter 6
Approximate Input-Output Feedback Linearization of Non-Minimum Phase System using Vanishing Perturbation Theory

Monia Charfeddine
Polytechnic School of Tunisia (EPT), Tunisia

Khalil Jouili
Polytechnic School of Tunisia (EPT), Tunisia

Naceur Benhadj Braiek
Polytechnic School of Tunisia (EPT), Tunisia

ABSTRACT

The inverse of a non-minimum-phase system being unstable, standard input-output feedback linearization is not effective to control such systems. In this chapter is a presentation of a new tracking control method for the nonlinear non-minimum-phase system. Indeed, the main idea here is to dismiss a part of system dynamics in order to make the approximate system input-state feedback linearizable. The neglected part is then considered as a perturbation part that vanishes at the origin. Finally, a linear controller is designed to control the approximate system. Stability is analyzed using the vanishing perturbation theory. The efficacy and usage of the proposed approach is evaluated in an illustrative inverted cart-pendulum example.

1. INTRODUCTION

The problem of tracking the nonlinear systems has been an active area of research in recent years. Several control methodologies have been proposed, which certain feedback structures exploit to achieve the desired tacking performance. In particular, the feedback linearization technique has received important

DOI: 10.4018/978-1-4666-7248-2.ch006

attention. Moreover, the control of nonlinear systems using feedback linearization techniques is divided in two structures: the input-state linearization and the input-output linearization.

Since 1980, the exact conditions under which a nonlinear system can be input-state linearized by static state feedback and coordinate transformation, have stated (Krener, 1973; Isidori, 1990; Hunt et al., 1983 ; Su, 1982). This approach is called input-state feed linearization, where the state equation of the non linear system becomes completely linearized after the usage of a coordinate transformation and a static state feedback. Besides, the conditions for input-state linearization of a nonlinear system are only satisfied for a small number of classes of systems.

For this reason, the idea of linearizing only the dynamics between the inputs and the outputs of the nonlinear system are given by Isidori in 1981 (Isidori, 1990). Such method, called to as input-output feedback linearization. Thus, the state equation may be partially linearized because only the input-output dynamic is linearized. Residual nonlinear dynamics, called internal dynamics, take place. These dynamics do not depend explicitly on the system input, and so are not controlled. This method provides a natural framework for the design of tracking controllers; in fact, it has been successfully implemented in several practical applications, such as flight control (Hauser et al., 1989), synchronous generators (Mielczarski & Zajaczkowsk, 1994) and chemical reactor control (Jouili et al., 2010).

The main limitation of the input-output linearization is that the internal dynamics, which can be unstable at the equilibrium point, are not controlled. These mentioned systems are called non minimum phase systems.

Therefore, the tracking control of nonlinear non minimum systems is a challenging problem in the control theory (Hauser et al., 1992; Liu et al., 2008). The standard input-output linearization (Isidori, 1990; Jouili et al.; 2010) leads then to an unstable closed loop system due to the presence of unstable zero dynamics. Hence, various ideas related to the possibility of using input-output linearization have been explored in the literature dealing with the nonlinear non minimum phase system. In Bortoff (1997), a nonlinear state feedback and a coordinate transformation are used to make the system as close to the linear one as possible. In Allgöwer (1997), the system input output feedback is linearized at the very beginning. Then, the zero dynamics are factorized into stable and unstable parts. The unstable part is approximately linear and independent of the coordinates of the stable part. Deutscher (2005) Proposed a numerical approach applying multivariable legendre polynomials to achieve an exact algebraic expression for the exact linearizing feedback. Moreover, an original technique of control based on an approximation of the method of exact input-output linearization was proposed in the works of Hauser and al (Hauser et al., 1992). In Charfeddine et al. (2010, 2011), the approximation presented in (Hauser et al., 1992) is used to improve the desired control performance. A cascade control scheme has been considered (Charfeddine et al., 2013; Charfeddin et al. 2014; Yakoub et al., 2013) that combines the input-output feedback linearization and the backstepping approach.

As a result, it is clear that the overall idea of tracking control of nonlinear non minimum phase system in the literature is based on a method of approximation. Inspired by the recent papers (Andrieu & Praly, 2008; Marino & Tomei, 2005) on the controller design for nonlinear non minimum phase systems and the work of Charfeddine et al. (2011), in this paper, we address the problem of tracking control of a single input/single output of non minimum phase nonlinear systems. The idea here is to approximate the given system into input-state feedback linearizable system. The system is feedback linearized by neglecting a part of the system dynamics, with the neglected part being considered as a perturbation that is vanished to zero in the origin. Stability analysis is also provided based on the vanishing perturbation theory (khalil, 1996).

The present chapter is organized as follows: In Section 2 some mathematical preliminaries and problem formulation are presented. The development of the proposed approach and the stability analysis are given in Section 3. Section 4 presents the inverted cart-pendulum to illustrate the effectiveness of the proposed approach. Section 5 discusses the simulation results of the proposed approach. Finally, some concluding remarks and future work are provided in Section 6.

2. RELATED WORK

The input-state feedback linearization is aimed to find a state feedback control and change of variables that transform the nonlinear system into an equivalent linear one. The main difficulties with input-state feedback linearization of nonlinear systems are its limited applicability, as well as the complexity, sensitivity and design difficulties of the exact linearizing compensators (Guardabassi & Savaresi, 2001). When the system is not input-state linearizable, passed on to another technique is called the input-output linearization. But, this technique is not effective for non minimum phase systems.

In literature, there has been an increasing amount of work searching for approximate solutions to the problem of linearizing nonlinear systems by state or output feedback. The main approximate linearization methods proposed in the literature can be cited by:

- Partial linearization,
- Linearization-oriented modeling, and
- Linear linearization.

2.1. Partial Linearization

Giving a nonlinear system, these methods consist of local approximations of the system. First, a Taylor series expansion of the system dynamics is computed. Then, the nonlinear system is locally transformed to an approximately linear one around the origin, or within a set of operating points by state feedback and local coordinate transformation. The main disadvantage of partial linearization techniques is that the stability results are only local.

In literature, Research on partial linearization problems are developed along three main streams, all initiated around 1984.they are known as higher-order approximate linearization, pseudo linearization and extended linearization.

Higher-Order Approximate Linearization

Local state feedback and local coordinate transformation are computed such that a number of higher terms of the Taylor series expansion around the operating point become zero. Some important results of higher-order approximate linearization reported in the literature given by:

- Necessary and sufficient conditions for the existence of static state feedback and coordinate transformation are given in (Krener, 1984; Krener et al.,1987; Krener et al.,1988).

- Extensions, symbolic or numeric algorithms, and alternative approaches are discussed in (Kang, 1994; Barbot et al., 1997; Barbot et al., 1999; Lee & Marcus, 1986; Nam et al., 1993; Nam et al., 1994).

The main disadvantage of this control design is that the characteristics of the linearized closed-loop system will change with changes in the operating point.

Pseudo Linearization

The pseudo-linearization problem was first stated in (Reboulet & Champetier, 1984). In this method, state feedback and coordinate change are computed such as the Taylor series linearization of the transformed system at a constant operating point has input-output behavior that is independent from the considered operating point, within a local set that contains it. Some important results of pseudo linearization reported in the literature can be given by:

- Necessary and sufficient conditions for the existence of static state feedback and coordinate transformation are given in (Reboulet & Champetier, 1984; Champetier et al.,1984; Champetier et al., 1985).
- Numerical methods for designing the linearizing feedback are cited in (Bortof & Spong, 1992; Bortof, 1997).
- Generalization, extensions and specific methods to construct state feedback laws are discussed in(Wang & Rugh, 1989; Lawrence, 1995; Lawrence, 1999).

Extended Linearization

The objective of extended linearization is basically the same pursued by pseudo-linearization. Though, extended linearization deals with the input-output feedback linearization problem while pseudo linearization does not require the definition of an output since it focuses on the input-to-state behavior of the control system. It is interesting to notice that pseudo-linearization and extended linearization, first presented in 1984 (Rugh, 1984; Reboulet & Champetier, 1984), were originally developed following different and independent output. Some important results of extended linearization reported in the literature given by:

- First results for the design of an output-feedback linearizing controller are given in (Rugh,1984; Rugh, 1986).
- The conditions under which extended linearization can be achieved are presented and discussed in (Baumann & Rugh,1986; Wang & Rugh, 1987; Wang & Rugh, 1987).

2.2. Linear Model Following

In the previous section, the three most popular methods for the partial linearization of nonlinear plants have been outlined. This last techniques lead to local results, since the system is linearized at an operating point or a local set of operating points. In linear model following techniques, the idea is to use a feedback to match a prescribed linear behavior. In other words, the problem is to find a feedback compensator in a way that the behavior of the closed-loop system gets as close as possible, in a sense to be specified, to a linear model referred to as reference model. The design of such feedback compensator is based on optimization methods. Some important results of linear model following reported in the literature can be summarized as follow:

- **Model-Based Techniques:** If the available description of the plant is a mathematical model, are given in (Bovivento,1998; Hung et al.1993).
- **Data-Based Techniques:** If the description of the plant is only a set of input-output data, are mentioned in (Levin & Narendra, 1996; Savaresi et al.,2000 ; Guardabassi & Savaresi, 2000).

The main difficulty of the feedback compensator design in linear model following techniques, since it is based on optimization methods, is the possibility of nontrivial optimization tasks to be solved and very demanding computations.

2.3. Linearization-Oriented Modeling

In contrast to the partial linearization techniques, the system is approximated everywhere in the linearization-oriented modeling. The approximation can be achieved using:

- Numerical methods (Hauser,1991; Banasuk & Hauser, 1993; Banasuk & Hauser, 1996).
- Rigorous theoretical methods (Chen & Paden, 1996; Niemiec & Kavaris, 2003; Kavaris et al., 1994; Kanter et al., 2001).

Nevertheless, numerical methods of linearization-oriented modeling techniques can be very demanding computations. Thus, the linearization-oriented modeling using theoretical methods is more interesting to approximate globally a nonlinear system to a linear one.

2.4. Linearization-Oriented Modeling of Nonlinear Non Minimum Phase System

Linearization-oriented modeling techniques applied to the case of nonlinear non minimum phase systems have been widely addressed in the literature. Several results on the approximate linearization of non minimum-phase systems are developed by Kravaris and Soroush given in (Kravaris & Daoutidis, 1992 ;Kavaris et al.,1994; Sorouh & Kavaris, 1996; Kater et al.,2001).

In particular, in (Sorouh & Kavaris, 1996; Kanter et al., 2001), the system output is differentiated as much as the order of the system, instead of stopping at the relative degree yet would be done with standard input-output feedback linearization. As an approximation, the input derivatives that appear in the control law are set to zero, in the computation of the state feedback input. In Kanter (2002), the above results are extended to the multiple-inputs multiple-outputs case. The main disadvantage of Kravaris and Soroush results is that they require the open-loop system to be stable.

An extension of the controller is developed for unstable systems in (Panjapornpon et al., 2003). However, the feedback is dynamic, and only asymptotic stability has been provided therein. In Doyle et al. (1995), new outputs are introduced, so as to approximately factor the nonlinear system as a minimum phase factor and an all-pass factor. The system is then linearized using a dynamic state feedback. The main disadvantage of this method is its restrictive applicability domain as the approximate factorization can only be derived for maximum phase systems, i.e. systems with only anti-stable zero dynamics.

In Allgower (1997), the system is first input-output feedback linearized. Then, the zero dynamics are factorized into stable and anti-stable parts. The anti-stable part is approximately linear and independent of the coordinates of the stable part. The main difficulty with this method is that every conservative condition have been stated unless the zero dynamics can be separated into stable and anti-stable parts, so that only a few practically relevant non minimum phase systems can be transformed into this form.

3. BACKGROUND AND PROBLEM FORMULATION

In this paper, we consider a nonlinear Single-Input Single-Output (SISO) system of the form:

$$\begin{cases} \dot{x} = f(x) + g(x)u, \quad x(0) = x_0 \\ y = h(x) \end{cases} \tag{1}$$

where $x \in \Re^n$ is the n-dimensional state variables, $u \in \Re$ is a scalar manipulate input and $y \in \Re$ is a scalar output. $f(.)$, $g(.)$ and $h(.)$ are smooth functions describing the system dynamics.

3.1. Input-Output Linearization

Considering the nonlinear system (1), the problem of the exact input-output linearization is based on the search for a nonlinear coordinate transformation, which transforms the nonlinear system (1) into a linear one using a state feedback. The property of linearity, in this case, should be established between a new control v and the output y. For the case of single-input single-output systems, the approach consists in defining a linear input-output relation by successively deriving the output until the input appears which can be expressed by:

$$\begin{cases} \dot{y} = L_f h(x) \\ \ddot{y} = L_f^2 h(x) \\ \quad\vdots \\ y^{(r)} = L_f^r h(x) + L_g L_f^{r-1} h(x) u \end{cases} \tag{2}$$

where r is the system relative degree.

- **Definition 1** (Isidori, 1990):

Consider the output $y = h(x)$ for system (1). The nonlinear system (1) has relative degree r at the point x_0 if:

$$\begin{cases} L_f^k h(x) = 0 \quad \forall x \neq x_0 \text{ and } \forall k \leq r-1 \\ L_g L_f^{(k-1)} h(x) \neq 0 \end{cases} \tag{3}$$

So, the relative degree r is the number of times we have to differentiate the output y with respect to time before the input appears.

If $r < n$, then system (1) can be feedback linearized into Byrnes-Isidori normal form (Isidori & Byrnes, 1990):

$$\xi = \begin{bmatrix} h(x) \\ \vdots \\ L_f^r h(x) \\ \eta_1 \\ \vdots \\ \eta_{n-r} \end{bmatrix} \tag{4}$$

The resulting system with the transformed variables (1) can be written as:

$$\begin{cases} \dot{\xi}_i = \xi_{i+1} \quad i = 1,\dots,r-1 \\ \dot{\xi}_r = v = L_f^r h(x) + L_g L_f^{r-1} h(x) u \\ \dot{\eta} = q(\xi, \eta) \\ y = \xi_1 \end{cases} \tag{5}$$

where v is the new control law given by:

$$v = y_{ref}^{(r)} + \sum_{i=0}^{r-1} k_{i+1} \left(y_{ref}^{(i)} - y^{(i)} \right) \tag{6}$$

with y_{ref} is the reference trajectory.

Thus, the control law can be written as:

$$u(x) = \frac{v - L_f^r h(x)}{L_g L_f^{r-1} h(x)} \tag{7}$$

3.2. Vanishing Perturbation Theory

In this section, we consider the nonlinear system (1) which is written in the autonomous form for the following perturbed system:

$$\dot{x} = f(x) + \Theta(x), \quad x(0) = x_0 \tag{8}$$

where $f(x)$ represents the nominal dynamics, with $f(0)=0$, and $\Theta(x)$ represents the perturbed dynamics. f and Θ are Lipchitz in x. The vanishing perturbation theory is based on the assumption that the perturbation tends towards zero in the origin $\Theta(0)=0$.

Thus, if the nominal system is exponentially stable and the parameter λ is smaller than a predetermined limit, then the perturbed system is also exponentially stable.

- **Theorem 1** (Khalil, 1996)**:** Let $(x=0)$ is the equilibrium point of the nominal system $\dot{x}=f(x)$ which is exponentially stable and $V(x)$ is a Lyapunov function of the nominal system that satisfies the following conditions:

$$\frac{\partial V(x)}{\partial x} f(x) \leq -c_1 \|x\|^2 \tag{9}$$

$$\left\| \frac{\partial V(x)}{\partial x} \right\| \leq c_2 \|x\| \tag{10}$$

with c_1 and c_2 are two real positive constants.

Let $\exists \lambda > 0 : \|\Theta(x)\| \leq \lambda \|x\|$, so, if $\lambda < \frac{c_1}{c_2}$, the origin $(x=0)$ is an exponentially stable equilibrium point of the perturbed system (8).

4. MAIN RESULT

In this paper, an approach to the tracking control problem of the nonlinear non minimum phase system is proposed based on the vanishing perturbation theory. First, the nonlinear non minimum phase system is transformed into its observability normal form. Next, it is in turn approximated as a chain of integrators by neglecting a part of the dynamics. Finally, it is controlled via a linearizing feedback.

4.1. Control Law Design

For dealing with the SISO system in (1), the following assumptions are first made.

- **Assumption 1**:

$$\forall x \in \Re^n \quad span\left\{ \left[\frac{\partial h}{\partial x} \quad \frac{\partial L_f h}{\partial x} \quad \frac{\partial L_f^2 h}{\partial x} \quad \cdots \quad \frac{\partial L_f^{n-1} h}{\partial x} \right] \right\} = n \tag{11}$$

where $\frac{\partial h}{\partial x}, \frac{\partial L_f h}{\partial x}, \frac{\partial L_f^2 h}{\partial x}, \cdots$ and $\frac{\partial L_f^{n-1} h}{\partial x}$ are linearly independent[10].

This assumption implies that the linearized system is observable around all operating points (Nijmeijer & Schaft,1990).

- **Assumption 2:**

Let the following state transformation:

$$\xi = \Phi(x) = \left[h(x) \quad L_f h(x) \quad L_f^2 h(x) \quad \ldots \quad L_f^{n-1} h(x)\right]^T \tag{12}$$

$\Phi(x)$ is a diffeomorphism, i.e., the inverse transformation $x = \Phi^{-1}(\xi)$ exists and is unique for all $x \in \Re^n$.

We start by defining the following error states:

$$e_i = \xi_i - y_{ref}^{(i-1)} \quad \forall i \in \{1, 2, \ldots, n\} \tag{13}$$

where y_{ref} is a reference trajectory using the state transformation (12) and the linearized feedback control (7), the system (1) is written as follows:

$$\begin{cases} \dot{e}_1 &= e_2 \\ \dot{e}_2 &= e_3 \\ &\vdots \\ \dot{e}_r &= e_{r+1} + L_g L_f^{r-1} h\left(\Phi^{-1}(\xi)\right) \\ \dot{e}_{r+1} &= e_{r+2} + L_g L_f^{r} h\left(\Phi^{-1}(\xi)\right) \\ &\vdots \\ \dot{e}_n &= L_f^n h\left(\Phi^{-1}(\xi)\right) + L_g L_f^{n-1} h\left(\Phi^{-1}(\xi)\right) u - y_{ref}^n, \quad e(0) = e_0 \end{cases} \tag{14}$$

with $L_g L_f^i h\left(\Phi^{-1}(\xi)\right) = 0$, $\forall\, 0 < i < r-1$ and $L_g L_f^i h\left(\Phi^{-1}(\xi)\right) \neq 0, \forall\, r-1 < i < n-1$.

The main idea of the proposed approach is to assume that the terms $L_g L_f^i h\left(\Phi^{-1}(\xi)\right) \neq 0$, $\forall\, r-1 < i < n-1$ are towards zero and are, then, neglected.

Thus, the use of this assumption in (14) yielding the following approximate system:

$$\begin{cases} \dot{e}_1 &= e_2 \\ \dot{e}_2 &= e_3 \\ & \vdots \\ \dot{e}_r &= e_{r+1} \\ \dot{e}_{r+1} &= e_{r+2} \\ & \vdots \\ \dot{e}_n &= L_f^n h\left(\Phi^{-1}\left(\xi\right)\right) + L_g L_f^{n-1} h\left(\Phi^{-1}\left(\xi\right)\right) u - y_d^{(n)}, \quad e\left(0\right) = e_0 \end{cases} \tag{15}$$

The resulting system (15) consists of a chain of n integrators, and then the following linearizing control can be applied:

$$u = \frac{v - L_f^n h\left(\Phi^{-1}\left(\xi\right)\right)}{L_g L_f^{n-1} h\left(\Phi^{-1}\left(\xi\right)\right)} \tag{16}$$

Then, system (15) can be rewritten as follows:

$$\begin{cases} \dot{e}_1 = e_2 \\ \dot{e}_2 = e_3 \\ \quad\vdots \\ \quad\vdots \\ \dot{e}_n = v - y_{ref}^{(n)}, \; e\left(0\right) = e_0 \end{cases} \tag{17}$$

So, it may be written as:

$$\dot{e} = Ae + Bv, \; e\left(0\right) = e_0 \tag{18}$$

with

$$A = \begin{bmatrix} 0 & 1 & 0 & . & 0 \\ 0 & 0 & 1 & . & 0 \\ \vdots & \vdots & \vdots & \vdots & \vdots \\ 0 & 0 & 0 & . & 1 \\ 0 & 0 & . & . & 0 \end{bmatrix}$$

and

$$B = \begin{bmatrix} 0 \\ 0 \\ \vdots \\ 0 \\ 1 \end{bmatrix}.$$

The resulting system (18) is composed of a linear one and corresponds to an approximate input-state input of system (1). Consequently, the following feedback:

$$v = y_{ref}^{(r)} + \sum_{i=0}^{r-1} k_{i+1}\left(y_{ref}^{i} - y^{i}\right) \tag{19}$$

can be applied to the approximate system (18). We obtain, finally, the following linear system in closed loop:

$$\dot{e} = A_{CL}\, e, \qquad e\left(0\right) = 0 \tag{20}$$

with:

$$A_{CL} = \begin{bmatrix} 0 & 1 & 0 & . & 0 \\ 0 & 0 & 1 & . & 0 \\ \vdots & \vdots & \vdots & \vdots & \vdots \\ 0 & 0 & 0 & . & 1 \\ -k_1 & -k_2 & . & . & -k_n \end{bmatrix}$$

and $k_i > 0,\ \forall i \in \left\{1, 2, \ldots, n\right\}$ are the coefficients of a Hurwitz polynomial (Khalil, 1996).

4.2. Stability Analysis

In this section, it's recommended to use the theory of vanishing perturbation to analyze the stability of the closed loop system. The application of the linear feedback (19) to the transformed system (16) leads to:

$$\dot{e} = A_{CL} e + \Theta\left(\Phi^{-1}\left(\xi\right)\right), \qquad e\left(0\right) = e_0 \tag{21}$$

with

$$\Theta\left(\Phi^{-1}\left(\xi\right)\right) = -\begin{bmatrix} 0 \\ \vdots \\ L_g L_f^{r-1} h\left(\Phi^{-1}\left(\xi\right)\right) \\ \vdots \\ L_g L_f^{n-2} h\left(\Phi^{-1}\left(\xi\right)\right) \\ 0 \end{bmatrix} \frac{\left(Ke\left(t\right) + L_f^n h\left(\Phi^{-1}\left(\xi\right)\right)\right)}{L_g L_f^{n-1} h\left(\Phi^{-1}\left(\xi\right)\right)} \tag{22}$$

Then, replacing $\left(\xi, u\right)$ by their values at the equilibrium $\left(\xi, u\right) = \left(0, 0\right)$ in (14) gives $L_f^{n-1} h\left(\Phi^{-1}\left(\xi = 0\right)\right) = 0$. Replacing it in (22) yields $\Theta\left(0\right) = 0$. Therefore, the perturbation is indeed vanishing, and the theory of vanishing perturbations can be used. The following theorem gives sufficient conditions for the exponential stability of system (21).

- **Theorem 2** (Guemghar et al., 2003)**:** Consider system (1) and assume that the following conditions are verified:
  - $\forall x \in \Re^n,\ span\left\{\left[\frac{\partial h\left(x\right)}{\partial x} \quad \frac{\partial L_f h\left(x\right)}{\partial x} \quad \frac{\partial L_f^2 h\left(x\right)}{\partial x} \quad \cdots \quad \frac{\partial L_f^{n-1} h\left(x\right)}{\partial x}\right]\right\} = n,$
  - $\exists \delta_1 > 0,\ \left|L_f^n h\left(\Phi^{-1}\left(\xi\right)\right)\right| \leq \delta_1 \left\|\xi\right\|$,
  - $\exists \delta_2 > 0,\ \left|L_g L_f^{n-1} h\left(\Phi^{-1}\left(\xi\right)\right)\right| \geq \delta_2$
  - The gains k_i, $\left(i = 0, \ldots, n\right)$ are the solutions of Hurwitz polynomial.
  - Let us consider the function $\Theta^*\left(\xi\right)$ defined by:

$$\Theta^*\left(\xi\right) = \left[0 \ \cdots \ 0 \ \ L_g L_f^{r-1} h\left(\Phi^{-1}\left(\xi\right)\right) \ \cdots \ L_g L_f^{n-2} h\left(\Phi^{-1}\left(\xi\right)\right) \ \ 0\right]^T \tag{23}$$

if

$$\left\|\Theta^*\left(\xi\right)\right\| < \frac{\lambda_2}{2\alpha_{\max}\left(P\right)\left(\lambda_1 + \left\|K\right\|\right)} \tag{24}$$

where P is the solution of Lyapunov equation $P A_{CL} + A_{CL}^T P = -I$

Thus, the control law (16)-(19) stabilizes system (14) exponentially.

- **Proof:** Consider Lyapunov function $V = e^T P\ e$ for system (20), where P is a positive symmetric matrix that satisfies Lyapunov equation $P A + A^T P = -I$. Then,

$$\frac{\partial V}{\partial e} A_{CL} e = e^T P A e + e^T A^T P e = -\left\|e\right\|^2 \tag{25}$$

and

$$\begin{aligned} \left\|\frac{\partial V}{\partial e}\right\| &= \left\|e^T P + P e\right\| \leq 2 \left\|e\right\| \left\|P\right\| \\ &\leq 2\alpha_{\max} \left\|e\right\| \end{aligned} \tag{26}$$

Consider the same Lyapunov function for the perturbed system (21), given by:

$$\dot{V} = \frac{\partial V}{\partial e} A_{CL} e + \frac{\partial V}{\partial e} \Theta(\xi) \leq -\left\|e\right\|^2 + 2\alpha_{\max}(P)\left\|e\right\| \left\|\Theta(\xi)\right\| \tag{27}$$

Noting that $\left|L_f^n h\left(\Phi^{-1}(\xi)\right)\right| \leq \lambda_1 \left\|\xi\right\|$ and $\left|L_g L_f^{n-1} h\left(\Phi^{-1}(\xi)\right)\right| \geq \lambda_2 \geq 0$ so by using the expression $\Theta(\xi)$ (22), yields:

$$\left\|\Theta(\xi)\right\| \leq \frac{\lambda_1 + \left\|K\right\|}{\lambda_2} \left\|\Theta^*(\xi)\right\| \left\|\xi\right\| \tag{28}$$

Replacing (28) in expression (27) gives:

$$\dot{V} \leq -\left\|e\right\|^2 \left(1 - 2\alpha_{\max}(P) \frac{\lambda_1 + \left\|K\right\|}{\lambda_2} \left\|\Theta^*(\xi)\right\| \right) \tag{29}$$

Thus, if (24) is verified, the term in the parenthesis is positive and the feedback law (16)-(19) stabilizes (14) exponentially. The following corollaries are related to Theorem 2.

- **Corollary 1:** For a given vector gain K, the closed loop system exponentially stabilizes (21) for all:

$$\left\|\Theta^*(\xi)\right\| < \frac{\lambda_2}{2\alpha_{\max}(P)\left(\lambda_1 + \left\|K\right\|\right)}$$

The corollary above declares that, for any given vector gain K there is a range different from the non-null values of $\Theta^*(\xi)$ for which system (14) can be stabilized. However, the converse is not true, i.e., that is to say, given a perturbation $\Theta^*(\xi)$, it is not always possible to find a gain vector K that stabilizes the system.

- **Corollary 2:** Let λ_1 and λ_2 and $\lambda^* = \max_K \frac{\lambda_2}{2\lambda_{\max}(P)(\lambda_1 + \|K\|)}$, $eig(A_{CL}) < 0$. For $\Theta^*(\xi) < \lambda^*$, there exists a vector gain K that stabilizes system (21) exponentially.

5. ILLUSTRATIVE EXAMPLE

5.1. Description of the Inverted Cart-Pendulum System

Consider the familiar inverted cart-pendulum system (Al-hiddabi, 2005), depicted in Figure. 1. The cart must be moved using the force u so that the pendulum remains in the upright position as the cart tracks any desired trajectory from a class of admissible trajectories.

Let the mass of the cart be M, the mass of the pendulum be m, the length of the stick be L, and the acceleration of the gravity be g. The mass of the stick is small compared with the mass m and will be neglected by choosing the pendulum angle θ and the cart position y_c as the generalized position coordinates for the system. The effect of friction is also neglected.

The inverted cart-pendulum equations are:

$$\begin{cases} \ddot{y}_c = \dfrac{u + m\left(L\dot{\theta}^2 - g\cos\theta\right)\sin\theta}{M + m\left(\sin\theta\right)^2} \\ \ddot{\theta} = \dfrac{1}{L}\left(g\sin\theta - \ddot{y}\cos\theta\right) \end{cases} \tag{30}$$

Figure 1. A Schematic representation of the inverted cart-pendulum system

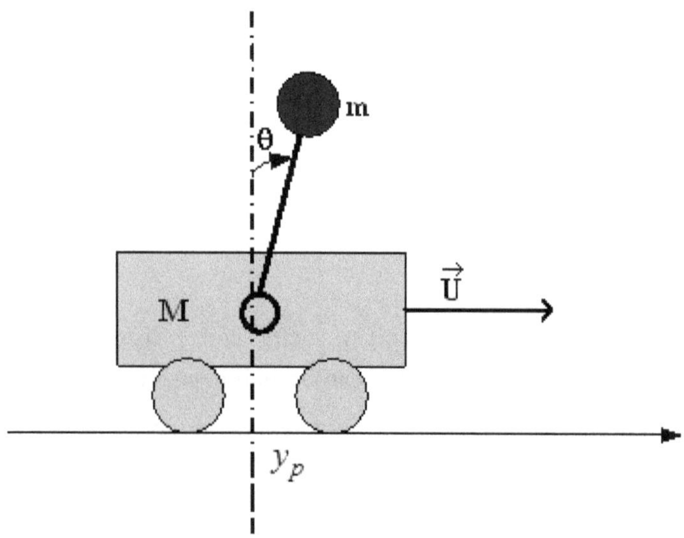

Let $x = \begin{pmatrix} x_1 & x_2 & x_3 & x_4 \end{pmatrix}^T = \begin{pmatrix} \theta & \dot{\theta} & y_c & \dot{y}_c \end{pmatrix}^T$ and $y = x_3$. Then, we obtain the following state space equation.

$$\begin{cases} \dot{x}_1 = x_2 \\ \dot{x}_2 = \dfrac{1}{L}\left(g\sin\left(x_1\right) - \dot{x}_4 \cos\left(x_1\right)\right) \\ \dot{x}_3 = x_4 \\ \dot{x}_4 = \dfrac{u + m\left(Lx_2^2 + g\cos\left(x_1\right)\right)\sin\left(x_1\right)}{M + m\sin\left(x_1\right)^2} \\ y = x_3 \end{cases} \tag{31}$$

The initial conditions are $y_c\left(0\right) = \dot{y}_c\left(0\right) = \dot{\theta}\left(0\right) = 0$ and $\theta\left(0\right) = -\pi$, the downward position for the pendulum. The system parameters are given in Table 1.

5.2. Control Law Design

The objective here is to control the cart displacement y_c along a reference trajectory and to stabilize the pendulum angle θ to the upright position. Thus, the proposed approach of the system (2) is applied. Again, the Lie derivatives are computed up to the order of the system $h = x_3, L_f h = x_4, L_f^2 h = T_1$, $L_f^3 h = T_2$, $L_f^4 h = T_3$, $L_g h = 0$, $L_g L_f h = 1 / (\text{M} + m\sin\left(x\right)^2)$, $L_g L_f^2 h = T_4$ and $L_f^3 h = T_5$.

The length terms T_i, $i = 1, \ldots, 5$ are given by:

$$T_1 = m(lx_2^2 - g\cos(x_1))\sin(x_1) / (l(M + m\sin(x_1)^2))$$

$$\begin{aligned} T_2 = & -(-3x_2 gM + 4x_2 g\cos(x_1)^2 M - 3x_2 gm + 3x_2 gm\cos(x_1)^2 + 3lm\cos(x_1)x_2^3 - \cos(x_1)lx_2^3 M \\ & + \sin(x1)lx_2^2 M - \sin(x_1)g\cos(x_1)M)m / (l(M + m\sin(x_1)^2)^2) \end{aligned}$$

$$\begin{aligned} T_3 = & -(-6x_2^3 gM - 12x_2^2 Mg\sin(x_1)^2 M - 6x_2 gm + 6x_2 gm\cos(x_1)^2 + 2x_2 gm^2\cos(x_1)^2 \\ & - \cos(x_1)l^2 x_2^3 M + 6lm\cos(x_1)x_2^3 - \cos(x_1)lx_2^3 M + \sin(x1)lx_2^3 M + gMsin(x_1)^2\cos(x_1) \\ & - 2\sin(x_1)g\cos(x_1)M)m / (l(M + m\sin(x_1)^2)^4) \end{aligned}$$

$$T_4 = l(M + 2mx_2\sin(x_1)) - 3m\cos(x_1)^3 lx_2^3 - 2x_2\sin(x_1)\cos(x_1) / (l(M + m\sin(x_1)^2)^2)$$

Table 1. Numerical parameters of the inverted cart-pendulum system

Parameters	Signification	Numerical Value
M	Mass of the cart	$0.445kg$
M	Mass of the rod	$0.21kg$
L	Length of the rod	$0.355m$
G	Gravitational acceleration	$9.81m/s^2$

$$\begin{aligned} T_5 = \ & l(2x_2 \sin(x_1) g \cos(x_1) M) - m \sin(x_1) \cos(x_1)^3 l x_2^3 - 3m \cos(x_1)^3 l x_2^4 \\ & - 2x_2^2 \, \mathrm{gM} \sin(x_1) \cos(x_1) / (l(M + m \sin(x_1)^2)^4) \end{aligned}$$

Since $h(x)$, $L_f h(x)$, $L_f^2 h(x)$ and $L_f^3 h(x)$ are independent, the system can be written in the observability normal form (16) given by:

$$\begin{cases} \dot{e}_1 = e_2 \\ \dot{e}_2 = e_3 + L_g L_f h(x) u \\ \dot{e}_3 = e_4 + L_g L_f^2 h(x) u \\ \dot{e}_4 = L_f^4 h(x) + L_g L_f^3 h(x) u - y_{cref}^4, \quad e(0) = e_0 \end{cases} \tag{32}$$

and the approximate system, after neglecting the internal dynamics, is given by:

$$\begin{cases} \dot{e}_1 = e_2 \\ \dot{e}_2 = e_3 \\ \dot{e}_3 = e_4 \\ \dot{e}_4 = v - y_{cref}^{(4)}, \quad e(0) = e_0 \end{cases} \tag{33}$$

Finally, the approximate system is linearized using the following control law:

$$u = \frac{v - L_f^4 h(x)}{L_g L_f^3 h(x)} \tag{34}$$

where v is a linear feedback control:

$$v = y_{cref}^{(4)} - k_1 \left(y_{cref} - y_c \right) - k_2 \left(\dot{y}_{cref} - \dot{y}_c \right) - k_3 \left(\ddot{y}_{cref} - \ddot{y}_c \right) - k_4 \left(y_{cref}^3 - y_c^3 \right) \tag{35}$$

with $y_{cref}(t)$ is the desired trajectory to follow and the gains k_1, k_2, k_3 and k_4 are the solutions of Hurwitz polynomial.

5.3. Simulations Results

In simulation, the reference trajectory used in the proposed approach is $y_{cref} = 1.5\left(\sin(t) + \sin(0.5\,t)\right)$ and the gains values are: $k_1 = -2.71$, $k_2 = -2.72, k_3 = -2.73$ and $k_4 = -2.74$.

The simulation result of the tracking control is shown in Figure 2. This figure presents a perfect agreement between the two trajectories. Figure 3 shows the evolution of the pendulum angle; indeed, it is a small variation around zero. The evolution of the stabilizing control law is shown in Figure 4. The dynamics of this control signal is quite satisfactory.

6. COMPARISON OF THE PROPOSED APPROACH WITH THE TECHNIQUE OF DIFFERENTIAL FLATNESS

The aim of this section is to conduct a comparative study of the proposed approach and the traditional technique of the differential flatness.

6.1. Explicit Differential Flatness Control

Differential flatness is a concept introduced in 1992 by the researchers Fliess, Lévine, Martin and Rouchon (Fliess et al., 1992; Fliess et al., 1995). It is a new approach for the analysis and design of control of nonlinear systems. Differential flatness is a property of a nonlinear system. Differential flatness is an extension of the term 'controllability' for nonlinear systems. For linear systems, it is equivalent to controllability, but for nonlinear systems, it is a self-standing property. A given dynamical system is either flat or not, depending on whether it fits the following definition of flatness or not.

Figure 2. Evolution of the cart displacement y_c and the reference trajectory y_{cref}

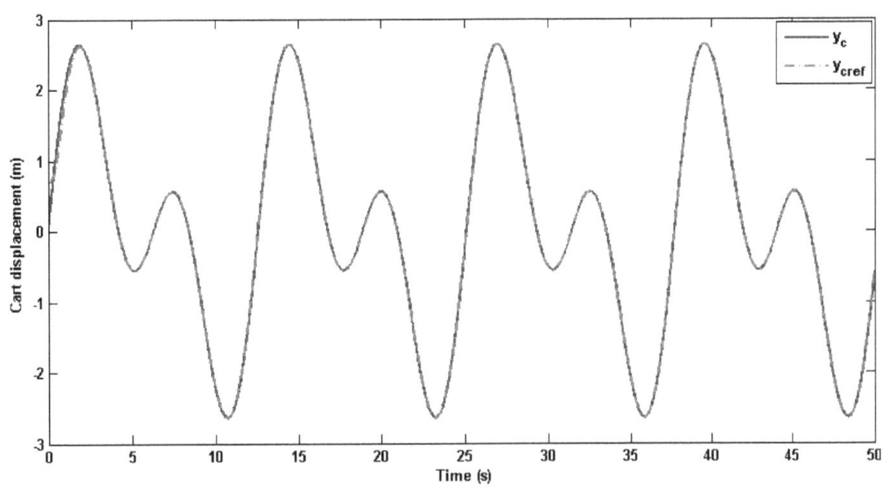

Figure 3. Evolution of the pendulum angle θ

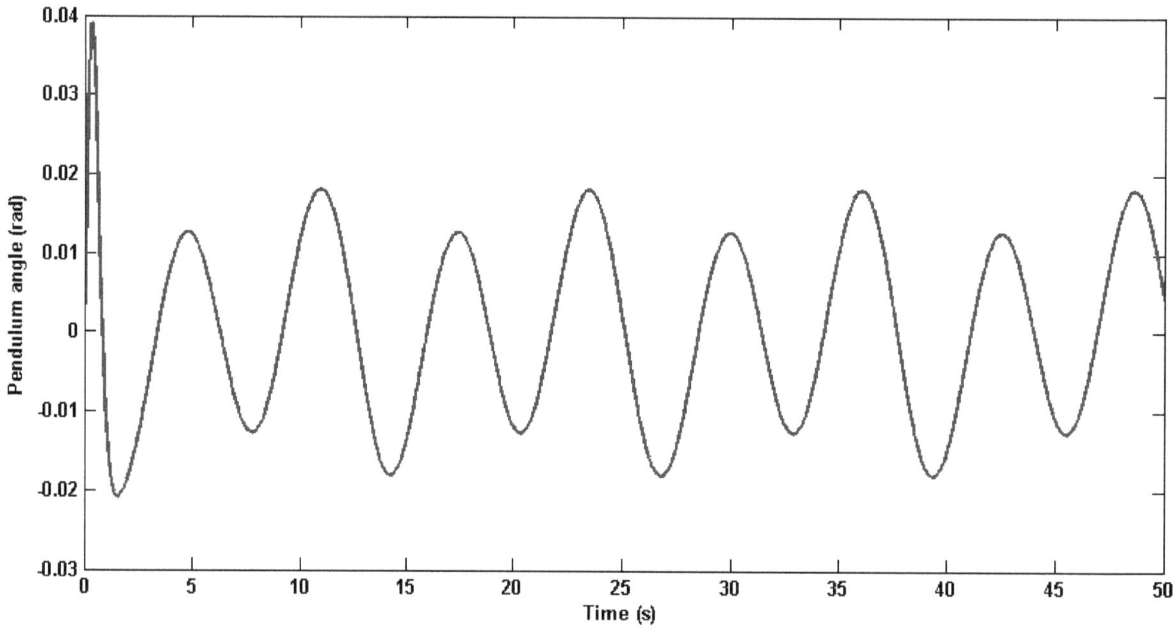

Figure 4. Evolution of the control signal

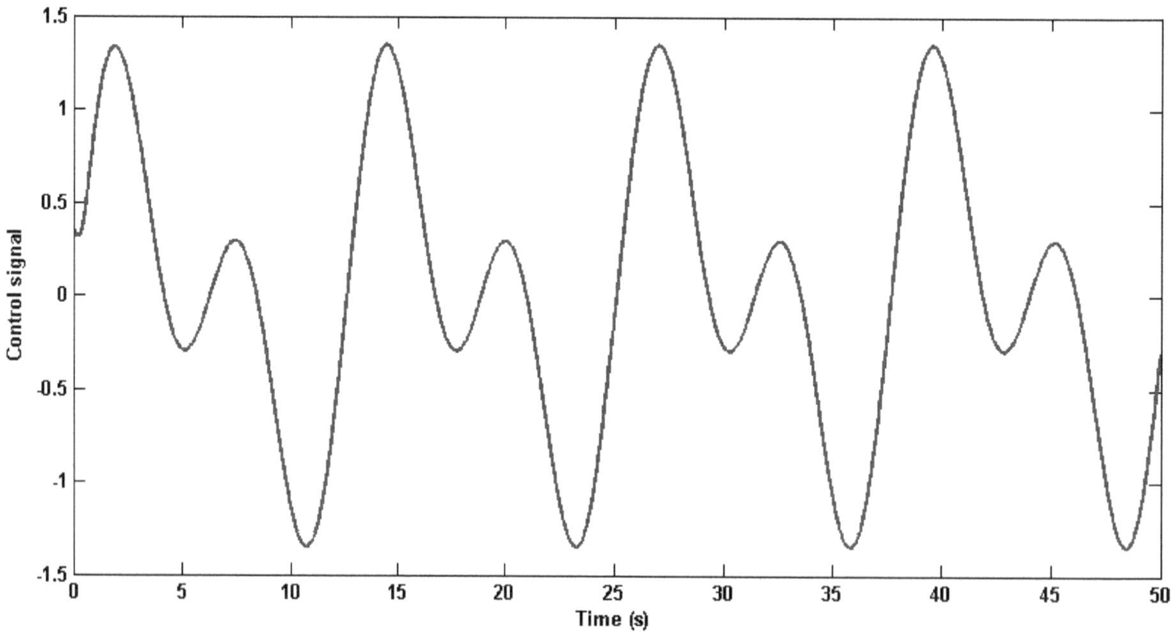

- **Definition 2** (Rotella & Zambettakis, 2008)**:** The system (1.1) is called differentially flat if there exists a vector z called flat output dimension equal to that of the control vector u depending on x, u and s a number of its derivatives:

$$z = \zeta\left(x, u, u^{(1)}, \ldots, u^{(s)}\right) \tag{34}$$

such as the state and the input of the system expressed analytically according to thereof in the following form:

$$x = \phi\left(\xi, \dot{\xi}, \ldots, \xi^{(w)}\right) \tag{35}$$

$$u = \varphi\left(z, \dot{z}, \ldots, z^{(l)}\right) \tag{36}$$

$$y = \psi\left(z, \dot{z}, \ldots, z^{(t)}\right) \tag{37}$$

where $s, w, l,$ and t are four finite parameters.

The relation (36) led to propose the following control law:

$$u = \varphi\left(z, \dot{z}, \ldots, z^{(l-1)}, v\right) \tag{38}$$

where v is a new control Brunovsky given by:

$$v = z^{l} = z^{l}_{ref} + \sum_{i=1}^{l-1} k_i \left(z^{i}_{ref} - z^{i}\right) \tag{39}$$

with $k_i > 0, \ \forall i \in \left\{1, \ldots, l-1\right\}$.

Then, the control law (38) can be written in the following form:

$$u = \varphi\left(z, \dot{z}, \ldots, z^{(l-1)}, z^{l}_{ref} + \sum_{i=1}^{l-1} k_i \left(z^{i}_{ref} - z^{i}\right)\right) \tag{40}$$

that ensures an asymptotic tracking of the reference trajectory with:

$$\lim_{t\to\infty}\left(z_{ref}\left(t\right) - z\left(t\right)\right) = 0 \tag{41}$$

6.2. Explicit Differential Flatness Control of the Inverted Cart-Pendulum System

Under the assumption that $\theta \to 0$, we can assume:

$$\begin{cases} \sin\left(\theta\right) \cong \theta \\ \cos\left(\theta\right) \cong 0 \end{cases} \tag{42}$$

the system (29) becomes:

$$\begin{cases} \ddot{y}_p = \dfrac{u + m\left(L\dot{\theta}^2 - g\right)\theta}{M + m\theta^2} \\ \ddot{\theta} = \dfrac{1}{L}\left(g\theta - \ddot{y}_p\right) \end{cases} . \tag{43}$$

Considering the flat output z defined by:

$$z = y_p + L\theta . \tag{44}$$

By deriving equation (44), we obtain:

$$\ddot{z} = \ddot{y}_c + L\ddot{\theta} \tag{45}$$

where also:

$$\ddot{z} = g\theta \tag{46}$$

Then, the state of the system (43) can be written as follows:

$$\begin{cases} x_1 = \dfrac{\ddot{z}}{g} \\ x_2 = \dfrac{\dddot{z}}{g} \\ x_3 = z - \dfrac{L}{g}\ddot{z} \\ x_4 = \dot{z} - \dfrac{L}{g}\dddot{z} \end{cases} \tag{47}$$

and the control law:

$$u = M\dot{z} - ml\ddot{z} + \frac{m\ddot{z}\left(g\dot{z}\ddot{z} - l\dddot{z}^{2}\right)}{g^{3}} - \frac{lv\left(Mg^{2} + m\ddot{z}^{2}\right)}{g^{3}} \tag{48}$$

with

$$v = z^{4} = z_{ref}^{4} + \sum_{i=1}^{3} k_{i}\left(z_{ref}^{i} - z^{i}\right). \tag{49}$$

6.3. Simulations Results

In order to evaluate the performance of proposed approach of the model of the inverted pendulum, a comparative study is given in this section between the proposed approach and the differential flatness. The results of this study are represented in the Figures (5), (6) and (7).

The parameters used in the both cases (proposed approach and differential flatness) are $k_1 = -2.71$, $k_2 = -2.72$, $k_3 = -2.73$ and $k_4 = -2.74$ andthereferencetrajectoryis $y_{cref} = 1.5\left(\sin(\mathrm{t}) + \sin(0.5\,\mathrm{t})\right)$. The initial conditions are $y_c(0) = 1.5$ and $\theta(0) = \dot{\theta}(0) = \dot{y}_c(0) = 0$.

Figure 5 presents a perfect tracking of the reference trajectory. However, it shows that the proposed control is more efficient than the traditional flatness control in terms of speed of convergence to the reference trajectory. The control signal obtained by the proposed approach is less than the one obtained

Figure 5. Evolution of the cart displacement y_c for the proposed approach (...) and the differential flatness control (---) and the reference trajectory y_{cref}

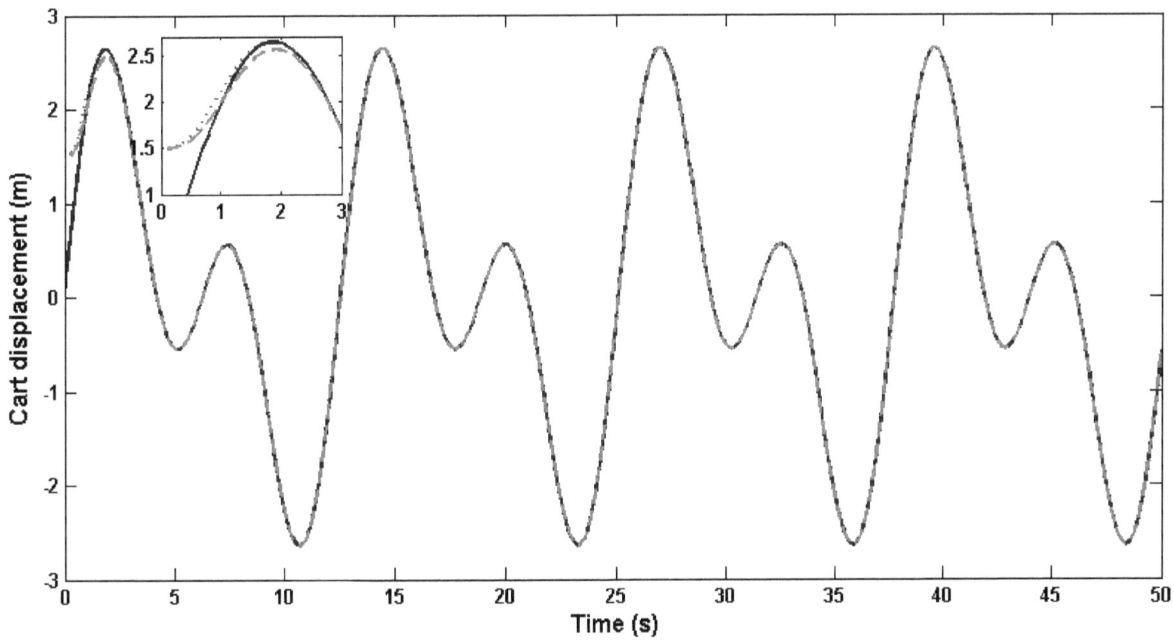

Figure 6. Evolution of the pendulum angle θ for the proposed approach (---)and the differential flatness (__)

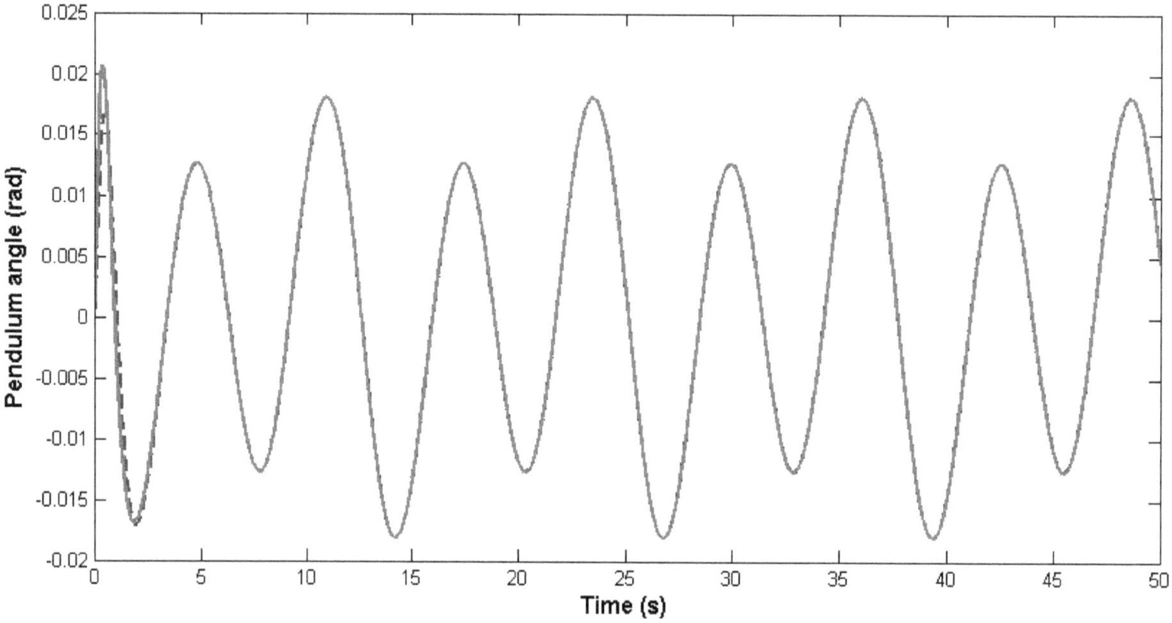

Figure 7. Evolution of the control signal for the proposed approach (---) and the differential flatness(__)

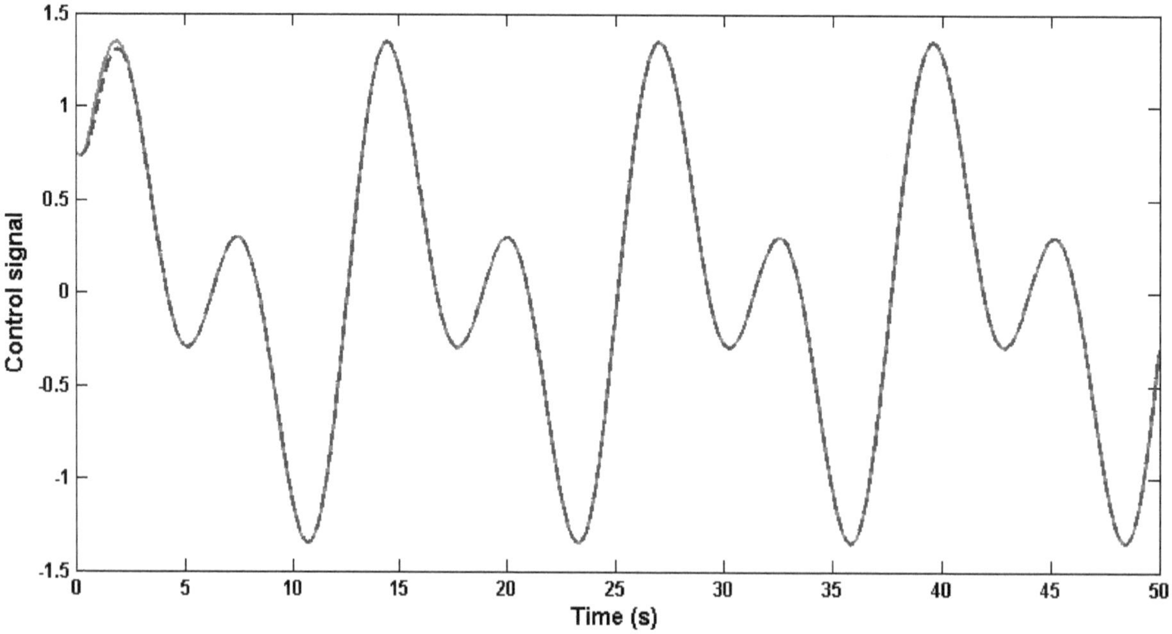

by the differential flatness Figure 7. This highlights a more stable evolution of the dynamic variable θ in Figure 6. These results yet again confirm the performance investigated for both controls while stressing on the advantage of the control based on the input-output linearization and vanishing perturbation theory.

7. DISCUSSION

In the literature, numerous methods based on an approximation of the input-output feedback linearization of a nonlinear system are exposed. In particular, in (Hauser et al.1992), Hauser proposes an approximate input-output linearization method of nonlinear non minimum phase systems which fails to unveil a well defined relative degree. For such systems, a method for constructing approximate systems that are input-output linearizable is provided. Moreover, in (Kanter et al., 2001; Soroush & Kravaris, 1996; Kanter et al., 2002), the outputs are differentiated as much as the order of the system instead of stopping at the relative degree and would be done with standard input-output feedback linearization. For the approximation, the input derivatives that appear in the control law are set to zero in the computation of the state feedback input. Nevertheless global stability can be established, the main disadvantage of this method is that it does require the open-loop system to be stable. In Panjapornpon et al. (2003), the controller design has been extended to a case where the nonlinear system is unstable. Though, the feedback is dynamic, and only asymptotic stability has been provided therein.

The idea of the proposed approach is related to (Guemghar, 2003), though the tracking control problem of non minimum phase system will be addressed. The system is transformed into its observability normal form, neglecting part of the dynamics so as to make the approximate system feedback linearizable, which will be considered as perturbation. The main advantages of the approximate input-output feedback linearization technique developed in this work are notably:

- The nonlinearities of the nonlinear system have not been compensated but simply neglected, there is no necessity that the open-loop system be stable for the stability of the closed loop system.
- The perturbation is vanishing at the origin, thus allowing the use of the vanishing perturbation theory to prove the global exponential stability of the closed loop system.
- The closed loop system provides a good tracking of the reference trajectory.

The main contribution of the proposed method is the solving of the tracking control problem for the nonlinear non minimum phase system. In Guemghar (2003), the stability analysis is local around an operating point. The objective of proposed approach is to ensure the stability of the closed-loop system behavior. Thus, vanishing perturbation theory for the stability analysis is used instead of the small gain theorem, in order to deal with unstable systems.

However, the main limitation of the proposed approach is essentially due to the use of vanishing perturbation theory for the stability analysis, as such the stability results presented here can be satisfied only for a few numbers of class of systems.

The approximate input-output feedback linearization using the input output linearization and vanishing theory has been applied to the inverted cart-pendulum with satisfactory results. In addition, the comparative study carried out in the previous paragraph implemented the importance of the proposed approach such as the flatness control is regarded as a solution for the control of the nonlinear non minimum phase system

8. CONCLUSION AND FUTURE WORK

To conclude, this paper mainly based upon an approximate input-state feedback linearization technique, presented in a new control scheme to solve the problem of tracking control for the nonlinear non minimum phase system. The nonlinear system is first transformed into its observability normal form. The latter is in turn approximated as a chain of integrators, neglecting part of the dynamics, and is finally controlled via a linearizing feedback.

The neglected part is considered as a perturbation, which is vanishing at the origin. Thus, the stability analysis provided is based on the vanishing perturbation theory. The efficiency and the validity of the proposed approach are illustrated through an example of inverted cart-pendulum.

Although the approximate input-output feedback linearization was developed in this work for the control of single-input single-output systems, the obtained results can be easily extended to the Multi Input Multi Output case. On the other hand, the robustness of the proposed approach isn't studied in this paper. This could be investigated in the future work.

REFERENCES

Al-Hiddabi, S. (2009). Quadrotor control using feedback linearization with dynamic extension. In *Proceedings of the Conference on Mechatronics and its Applications*. ISMA. DOI: doi:10.1109/ISMA.2009.5164788

Allgöwer, F. (1997). Approximate input-output linearization of non-minimum phase nonlinear systems. In *Proceedings of the 1997 European Control Conference (ECC)*, (pp. 604–610). ECC.

Andrieu, V., & Praly, L. (2008). Global asymptotic stabilization for nonminimum phase non linear systems admitting a strict normal form. *IEEE Transactions on Automatic Control*, *53*(5), 1120–1132. doi:10.1109/TAC.2008.923657

Banaszuk, A., & Hauser, J. (1993). Least squares approximate feedback linearization: A variational approach. In *Proceedings of the 1993 IEEE Control and Decision Conference (CDC)*, (pp. 2760–2765). IEEE. DOI: doi:10.1109/CDC.1993.325698

Banaszuk, A., & Hauser, J. (1996). Approximate feedback linearization: A homotopy operator approach. *SIAM Journal on Control and Optimization*, *34*(5), 1533–1554. doi:10.1137/S0363012994261306

Barbot, J. P., Monaco, S., & Normand-Cyrot, D. (1997). Quadratic forms and approximate feedback linearization in discrete time. *International Journal of Control*, *67*(4), 567–586. doi:10.1080/002071797224081

Barbot, J. P., Monaco, S., & Normand-Cyrot, D. (1999). Discrete time approximated linearization of SISO systems under output feedback. *IEEE Transactions on Automatic Control*, *44*(9), 1729–1733. doi:10.1109/9.788541

Baumann, W. T., & Rugh, W. J. (1986). Feedback control of nonlinear systems by extended linearization. *IEEE Transactions on Automatic Control*, *31*(1), 40–46. doi:10.1109/TAC.1986.1104100

Bonivento, C., Zanasi, R., & Sandri, M. (1998). Discrete variable structure integral controllers. *Automatica*, *34*(3), 355–36. doi:10.1016/S0005-1098(97)00197-0

Bortoff, S. A. (1997). Approximate state-feedback linearization using spline functions. *Automatica*, *33*(8), 1449–1458. doi:10.1016/S0005-1098(97)00070-8

Bortoff, S. A., & Spong, M. W. (1992). Pseudolinearization of the acrobot using spline functions. In *Proceedings of IEEE Control and Decision Conference (CDC)*, (pp. 593–598). IEEE. DOI: doi:10.1109/CDC.1992.371658

Champetier, C., Mouyon, P., & Magni, J. F. (1985). Pseudolinearization of nonlinear systems by dynamic precompensation. In *Proceedings of the 1985 IEEE Control and Decision Conference (CDC)*, (pp. 1371–1372). IEEE. DOI: doi:10.1109/CDC.1985.268735

Champetier, C., Mouyon, P., & Reboulet, C. (1984). Pseudolinearization of multi-input nonlinear systems. In *Proceedings of the 1984 IEEE Control and Decision Conference (CDC)*, (pp. 96–97). IEEE. DOI: doi:10.1109/CDC.1984.272279

Charfeddine, M., Jouili, K., & Benhadj Braiek, N. (2013) The Output Tracking Control of a Nonlinear Non-Minimum Phase System. In *Proceedings of the 2013 International Conference on Control, Engineering & Information Technology (CEIT)*, (pp. 83-88). CEIT.

Charfeddine, M., Jouili, K., & Benhadj Braiek, N. (2013). Output tracking control of inverted cart-pendulum using input-output linearization and predictive control. In *Proceedings of the 2013 IEEE International conference on Sciences and Techniques of Automatic Control & Computer Engineering*, (pp. 84–88). IEEE. DOI: doi:10.1109/STA.2013.6783110

Charfeddine, M., Jouili, K., & Benhadj Braiek, N. (2014). Cascade control for nonlinear non-minimum phase system based on backstepping approach and input output linearization. *International Journal of Control. Energy and Electrical Engineering*, *1*, 43–49.

Charfeddine, M., Jouili, K., Jerbi, H., & Benhadj Braiek, N. (2010). Linearizing Control with a Robust Relative Degree Based on Lyapunov Function: Case of the Ball and Beam system. *International Review on Modelling and Simulations*, *3*(2), 219–226.

Charfeddine, M., Jouili, K., Jerbi, H., & Benhadj Braiek, N. (2011). Output Tracking Control Design for Non-Minimum Phase Systems: Application to the Ball and Beam Model. *International Review of Automatic Control*, *4*(1), 47–55.

Chen, D., & Paden, B. (1996). Stable inversion of nonlinear nonminimum-phase systems. *International Journal of Control*, *64*(1), 81–97. doi:10.1080/00207179608921618

Davison, D. E., & Bortoff, S. A. (1994). Enlarge your region of attraction using high-gain feedback. In *Proceedings of the 1994 IEEE Control and Decision Conference*, (pp. 634–639). IEEE. DOI: doi:10.1109/CDC.1994.410974

Deutscher, J. (2005). Input-output linearization of nonlinear systems using multivariable Legendre polynomials. *Automatica*, *41*(2), 299–304. doi:10.1016/j.automatica.2004.11.001

Doyle, F. J. III, Allgower, F., & Morari, M. (1995). A normal form approach to approximate input-output linearization for maximum phase nonlinear systems. *IEEE Transactions on Automatic Control*, *41*(2), 305–309. doi:10.1109/9.481553

Fliess, M., Lévine, J., Martin, P., & Rouchon, P. (1992). On differentially flat nonlinear systems. In Proceedings of the 1992 IFAC symposium on Nonlinear Control Systems Design, (pp. 408–412). IFAC.

Fliess, M., Lévine, J., Martin, P., & Rouchon, P. (1995). Flatness and defect of non-linear systems: Introductory theory and examples. *International Journal of Control*, *61*(6), 1327–1361. doi:10.1080/00207179508921959

Guardabassi, G. O., & Savaresi, S. M. (2000). Virtual reference direct design method: An off-line approach to data-based control system design. *IEEE Transactions on Automatic Control*, *45*(5), 954–959. doi:10.1109/9.855559

Guardabassi, G. O., & Savaresi, S. M. (2001). Approximate linearization via feedback - An overview. *Automatica*, *37*(1), 1–15. doi:10.1016/S0005-1098(00)00117-5

Guemghar K., Srinivasan, B. & Bonvin,D. (2005). Analysis of cascade structure with predictive control and feedback linearisation. In *Proceedings of the 2005 Control Theory and Applications* (pp. 317-324). IEEE. DOI:.10.1049/ip-cta:20041309

Hauser, J. (1991). Nonlinear control via uniform system approximation. *Systems & Control Letters*, *17*(2), 145–154. doi:10.1016/0167-6911(91)90041-C

Hauser, J., Sastry, S., & Kokotovic, P. (1992). Nonlinear control via approximates input-output linearization: The ball and beam example. *IEEE Transactions on Automatic Control*, *37*(3), 392–398. doi:10.1109/9.119645

Hauser, J., Sastry, S., & Meyer, G. (1989). Nonlinear controller design for flight control systems. In Proceedings of the 1989 IFAC symposium on Nonlinear Control Systems Design, (pp. 136–141). IFAC.

Hung, J. Y., Gao, W., & Hung, J. C. (1993). Variable structure control: A survey. *IEEE Transactions on Industrial Electronics*, *40*(1), 2–21. doi:10.1109/41.184817

Hunt, L. R., Su, R., & Mayer, G. (1983). Global transformations of nonlinear systems. *IEEE Transactions on Automatic Control*, *28*(1), 24–31. doi:10.1109/TAC.1983.1103137

Isidori, A. (1990). Nonlinear control systems. London: Spring-Verlag.

Isidori, A., & Byrnes, C. I. (1990). Output regulation of nonlinear systems. *IEEE Transactions on Automatic Control*, *35*(2), 131–140. doi:10.1109/9.45168

Jouili, K., Jerbi, H., & Benhadj Braiek, N. (2010). An advanced fuzzy logic gain scheduling trajectory control for nonlinear systems. *Journal of Process Control*, *20*(4), 426–440. doi:10.1016/j.jprocont.2010.01.001

Kang, W. (1994). Approximate linearization of nonlinear control systems. *Systems & Control Letters*, *23*(5), 43–52. doi:10.1016/0167-6911(94)90080-9

Kanter, J. R., Soroush, M., & Seider, W. D. (2001). Continuous time, nonlinear feedback control of stable processes. *Industrial & Engineering Chemistry Research*, *40*(9), 2069–2078. doi:10.1021/ie000406k

Kanter, J. R., Soroush, M., & Seider, W. D. (2002). Nonlinear feedback control of multivariable nonminimum-phase processes. *Journal of Process Control*, *12*(6), 667–686. doi:10.1016/S0959-1524(01)00059-2

Khalil, H. K. (1996). *Nonlinear Systems*. Prentice Hall.

Kravaris, C., & Daoutidis, P. (1992). Output feedback controller realizations for open-loop stable nonlinear processes. In *Proceedings of the 1992 IEEE American control conference*, (pp. 2596–2601). IEEE.

Kravaris, C., Daoutidis, P., & Wright, R. A. (1994). Output feedback control of nonminimum-phase nonlinear processes. *Chemical Engineering Science*, *49*(13), 2107–2122. doi:10.1016/0009-2509(94)E0009-F

Krener, A. J. (1973). On the equivalence of control systems and the linearization of nonlinear systems. *SIAM Journal on Control*, *11*(4), 670–676. doi:10.1137/0311051

Krener, A. J. (1984). Approximate linearization by state feedback and coordinate change. *Systems & Control Letters*, *5*(3), 181–185. doi:10.1016/S0167-6911(84)80100-0

Krener, A. J., Karahan, S., & Hubbard, M. (1988). Approximate normal forms of nonlinear systems. In *Proceedings of the 1988 IEEE Control and Decision Conference*, (pp. 1223–1229). IEEE. DOI: doi:10.1109/CDC.1988.194516

Krener, A. J., Karahan, S., Hubbard, M., & Frezza, F. (1987).Higher order linear approximations to nonlinear control systems. In *Proceedings of the 1987 IEEE Control and Decision Conference*, (pp. 519–523). IEEE. DOI: doi:10.1109/CDC.1987.272895

Lawrence, D. A. (1995). A general approach to input-output pseudo-linearization for nonlinear systems. In *Proceedings of the 1995 IEEE Control and Decision Conference*, (pp. 613–618). IEEE. DOI: doi:10.1109/CDC.1995.478980

Lawrence, D. A. (1999). Approximate model matching for nonlinear systems. In *Proceedings of the 1999 IEEE Control and Decision Conference*, (pp. 720–725). IEEE. DOI: doi:10.1109/CDC.1999.832873

Lee, H. G., & Marcus, S. J. (1986). Approximate and local linearizability of nonlinear discrete-time systems. *International Journal of Control*, *44*(4), 1103–1124. doi:10.1080/00207178608933653

Levin, A. U., & Narendra, K. S. (1996). Control of nonlinear dynamical systems using neural networks. Part ii: Observability, identification, and control. *IEEE Transactions on Neural Networks*, *7*(1), 30–42. doi:10.1109/72.478390 PMID:18255556

Liu, S. J., Jiang, Z. P., & Zhang, J. F. (2008). Globally output-feedback stabilization for a class of stochastic non-minmum-phase nonlinear systems. *Automatica*, *44*(8), 1944–1957. doi:10.1016/j.automatica.2007.11.011

Marino, R., & Tomei, P. (2005). A class of globally output feedback stabilizable nonlinear non minimum phase systems. *IEEE Transactions on Automatic Control*, *50*(12), 451–460. doi:10.1109/TAC.2005.858652

Mielczarski, W., & Zajaczkowsk, A. M. (1994). Nonlinear Field V oltage Control of a Synchronous Generator using Feedback Linearization. *Automatica*, *30*(10), 1625–1630. doi:10.1016/0005-1098(94)90102-3

Nam, K., Arapostathis, A., & Lee, S. (1993). Some numerical aspects of approximate linearization of single-input nonlinear systems. *International Journal of Control*, *57*(2), 463–472. doi:10.1080/00207179308934399

Nam, K., Lee, S., & Won, S. (1994). A local stabilizing control scheme using an approximate feedback linearization. *IEEE Transactions on Automatic Control*, *39*(11), 2311–2314. doi:10.1109/9.333783

Niemiec, M., & Kravaris, C. (2003). Nonlinear model-state feedback control for nonminimum-phase processes. *Automatica*, *39*(7), 1294–1302. doi:10.1016/S0005-1098(03)00103-1

Nijmeijer, H. & Van Der Schaft, A. J. (1990). *Nonlinear Dynamical Control Systems*. Springer-Verlag.

Panjapornpon, C., Soroush, M., & Seider, W. D. (2003). Model based control of unstable nonminimum-phase nonlinear processes. In *Proceedings of the 2003 IEEE Conference on Decision and Control (CDC)*, (pp. 6151–6156). IEEE. DOI: doi:10.1109/CDC.2003.1272256

Reboulet, C., & Champetier, C. (1984). A new method for linearizing nonlinear systems: The pseudolinearization. *International Journal of Control*, *40*(4), 631–638. doi:10.1080/00207178408933297

Rotella, F., & Zambettakis, I. (2008). *Commande des systèmes par platitude*. Available at: http://pierre.rouchon.chez-alice.fr/MVA/FlatSystems.pdf

Rugh, W. J. (1984). Design of nonlinear compensators for nonlinear systems by an extended linearization technique. In *Proceedings of the 1984 IEEE Control and Decision Conference (CDC)*, (pp. 69–73). IEEE. DOI: doi:10.1109/CDC.1984.272275

Rugh, W. J. (1986). An extended linearization approach to nonlinear system inversion. *IEEE Transactions on Automatic Control*, *31*(8), 725–733. doi:10.1109/TAC.1986.1104385

Savaresi, S. M., Nijmeijer, H., & Guardabassi, G. O. (2000). On the design of approximate nonlinear parametric controllers. *International Journal of Robust and Nonlinear Control*, *10*(3), 137–155. doi:10.1002/(SICI)1099-1239(200003)10:3<137::AID-RNC466>3.0.CO;2-I

Soroush, M., & Kravaris, C. (1996). A continuous-time formulation of nonlinear model predictive control. *International Journal of Control*, *63*(1), 121–146. doi:10.1080/00207179608921835

Su, R. (1982). On the linear equivalents of nonlinear systems. *Systems & Control Letters*, *2*(1), 48–52. doi:10.1016/S0167-6911(82)80042-X

Wang, J., & Rugh, W. J. (1987). Feedback linearization families for nonlinear systems. *IEEE Transactions on Automatic Control*, *32*(10), 935–940. doi:10.1109/TAC.1987.1104470

Wang, J., & Rugh, W. J. (1987). Parameterized linear systems and linearization families for nonlinear systems. *IEEE Transactions on Circuits and Systems*, *34*(6), 650–657. doi:10.1109/TCS.1987.1086191

Wang, J., & Rugh, W. J. (1989). On the pseudo-linearization problem for nonlinear systems. *Systems & Control Letters*, *12*(2), 161–167. doi:10.1016/0167-6911(89)90009-1

Yakoub, Z., Charfeddine, M., Jouili, K., & Benhadj Braiek, N. (2013). A combination of backstepping and the feedback linearization for the controller of inverted pendulum. In *Proceedings of the 2013 IEEE International Multi-Conference on Systems, Signals and Devices (SSD)*. IEEE. DOI: 10.1109/SSD.2013.6564123

KEY TERMS AND DEFINITIONS

Input-Output Linearization: A control technique where the output y of the dynamic system is differentiated until the physical input u appears in the rth derivative of y.

Input-State Linearization: A control technique where some new output $y_{new} = \lambda(x)$ is chosen so that with respect to y_{new}, the relative degree of the system is n. Then the design procedure using this new output y_{new} is the same as for input-output linearization.

Non Minimum Phase System: Nonlinear system which the zero dynamics are unstable.

Tracking Control: Control a system to track a reference trajectory.

Vanishing Perturbation Theory: The vanishing perturbation theory is based on the assumption that the perturbation tends towards zero.

Section 2
Chaos and Control Applications

Chapter 7
Random Bit Generator Based on Non-Autonomous Chaotic Systems

Christos Volos
Aristotle University of Thessaloniki, Greece

Ioannis Kyprianidis
Aristotle University of Thessaloniki, Greece

Ioannis Stouboulos
Aristotle University of Thessaloniki, Greece

Sundarapandian Vaidyanathan
Vel Tech University, India

ABSTRACT

In the last decade, a very interesting relationship between cryptography and chaos theory was developed. As a result of this close relationship, several chaos-based cryptosystems, especially using autonomous chaotic dynamical systems, have been put forward. However, this chapter presents a novel Chaotic Random Bit Generator (CRBG), which is based on the Poincaré map of a non-autonomous dynamical system. For this reason, the very-well known Duffing-van der Pol system has been used. The proposed CRBG also uses the X-OR function for improving the "randomness" of the produced bit streams, which are subjected to the most stringent statistical tests, the FIPS-140-2 suite tests, to detect the specific characteristics that are expected from random bit sequences.

1. INTRODUCTION

The rapid development of communication technology such as in Internet, in mobile networks and especially in military networks require enhanced security in the transmission of information. This means, that for any possible "intruder" in these networks the possibility of obtaining the information should be extremely low. For this reason, the "randomness" of the generated quantities in the sense that prob-

DOI: 10.4018/978-1-4666-7248-2.ch007

ability of any particular value being selected must be sufficiently small to preclude an adversary from gaining advantage through optimizing a search strategy based on such probability. In recent years, various techniques including the keystream in the one-time pad, the secret key in the DES encryption algorithm, the primes p, q in the RSA encryption and digital signature schemes, the private key a in the DSA, and the challenges used in challenge-response identification systems, have been developed and are commercially available.

In academic community and industry many research teams work on the design of devices or algorithms which are called Random Number Generators (RNGs) that satisfy the basic demand of generating unpredictable quantities. From these devices or algorithms, sequences of random bits can be obtained, which are very useful in many applications. In literature various implementations of Random Bit Generators (RBGs) have been presented. However, all these generators can be classified into three major types, the True Random Bit Generators (TRBGs), the Pseudo-Random Bit Generators (PRBGs) and the Hybrid Random Bit Generators (HRBGs), based on the source of the randomness (Shu, 1995).

The True Random Bit Generators require a naturally occurring source of randomness, which comes from an unpredictable natural process in a physical or hardware device. Until now, a great number of TRBGs based on various physical or hardware processes have been presented, such as the frequency instability of an oscillator (Fairfield, 1987), the integrating dark current from a metal insulator semiconductor capacitor (Agnew, 1986), the elapsed time during radioactive decay (Guide, 1985), the thermal from a resistor and the shot noise, which is a quantum mechanical noise source in electronic circuits (Holman, 1997), the nuclear decay radiation source detected by a Geiger counter attached to a PC and the mutually exclusive events of traveling photons through a semi-transparent mirror (Trcek, 2006), the environmental noise (Bardis, 2009), the amplification of the signal produced on the base of a reverse-biased transistor and the Zener breakdown noise from a reverse-biased Zener diode (Dube, 2008), the mouse movement (Hu, 2009) and the variations in disk drive response times (Davis, 1994).

However, designing a hardware device to exploit this randomness and produce a bit sequence that is free of biases and correlations is a difficult task. Additionally, for most cryptographic applications, the generator must not be subject to observation or manipulation by an adversary. Due to the fact that the TRBGs are based on natural sources of randomness, they are subject to influence by external factors which cause malfunctions.

All the aforementioned difficulties of obtaining uniform random sequences from TRNGs lead many researchers to the design of Pseudo-Rrandom Bit Generators. A PRBG which is also known as a Deterministic Random Bit Generator (DRBG) is a deterministic algorithm which, outputs a binary sequence of length $l >> k$ that “appears” to be random, if a binary sequence of length k is given. The input to the PRBG is called the “seed”, while the output of the PRBG is called a pseudorandom bit sequence. This bit sequence is not truly random in that it is completely determined by a relatively small set of initial values. Also, PRBGs are very important in practice for their speed in number generation, their portability and their reproducibility, and they are thus central in applications such as in cryptography, procedural generation, in decision making, in simulating natural phenomena and in sampling data (Knuth, 1981; Park, 1988). Common classes of pseudo-random number generators algorithms include linear congruential generators, which are algorithms that yield sequences of randomized numbers calculated with linear equations. Also, lagged Fibonacci generators are improvements on the ‘standard’ linear congruential generators and linear feedback shift registers, that input bits are linear functions of their previous states (Menezes, 1996). More recent instances of pseudo-random number generators with strong randomness guarantees, are based on computational hardness assumptions and include the Blum Blum Shub, proposed in 1986 by Lenore Blum, Manuel Blum and Michael Shub (Blum, 1986), the Fortuna, which is devised by Bruce Schneier and Niels Ferguson and published in 2003 (Ferguson, 2003) and Mersenne Twister algorithm, which name derives from the fact that its period length is chosen to be a Mersenne prime (Matsumoto, 1997). So, a good PRBG for the previous mentioned applications should possess three very important characteristics: long period, high speed and randomness. However, in practice, the output from many common PRBGs exhibit artifacts which cause them to fail statistical pattern-detection tests. These include shorter than expected periods for some seed states (such seed states are called "weak"), lack of uniformity of distribution for large amounts of generated numbers, correlation of successive values, poor dimensional distribution of the output sequence, and the distances between where certain values occur are distributed differently from those in a random sequence distribution.

Also, it is obvious that in PRBGs due to the fact that the output is a function of the seed state, the actual entropy of the output can never exceed the entropy of the seed. Hence, the randomness level of the pseudo-random numbers depends on the level of randomness of the seed. Thus, HRBGs have been proposed to use a random generator as a seed generator and expand it. A seed generator is a hardware-based RNG with or without user's interaction, such as mouse movements, random keystrokes, or hard drive seek times (Koc, 2009).

Also, as it is known, in the last decades, nonlinear systems and especially chaotic systems have aroused tremendous interest because of their applications in several disciplines including meteorology, physics, engineering, economics, biology, and philosophy (Grebogi, 1997). Chaos theory studies the behavior of dynamical systems that are highly sensitive to initial conditions, an effect which is popularly referred to as the "Butterfly Effect". Small differences in initial conditions (such as those due to rounding errors in numerical computation) yield widely diverging outcomes for such dynamical systems, rendering long-term prediction impossible in general. This happens even though these systems are deterministic, meaning that their future behavior is fully determined by their initial conditions, with no random elements involved. In other words, the deterministic nature of these systems does not make them predictable. This was summarized by Edward Lorenz as follows (Lorenz, 1963):

Chaos: When the present determines the future, but the approximate present does not approximately determine the future.

From a mathematical viewpoint a nonlinear dynamical system, in order to be considered as chaotic, must fulfill the following three conditions (Hasselblatt, 2003). It must be topologically mixing, its periodic orbits must be dense and it must be very sensitive on initial conditions.

Firstly, the term topologically mixing means that the chaotic dynamical system, especially the chaotic designated area of the trajectory will eventually cover part of any particular region. The second feature of chaotic systems is that its periodic orbits have to be dense. This means that, the trajectory of a dynamical system is dense, if it comes arbitrarily close to any point in the domain. Finally, the most important feature of chaotic systems, as it is mentioned, is the sensitivity on initial conditions. This means that a small variation on a system's initial conditions will produce a totally different chaotic trajectory.

Nowadays it is known, that cryptography and chaos have a structural relationship (Alvarez, 2206) due to their many similar properties (Table 1). As a result of this close relationship several chaotic cryptosystems have been presented (Kocarev, 2001; Yang, 2004). One of the most interesting way through which chaotic cryptosystems can be realized is via the implementation of Chaotic Random Bit Generators (CRBGs). Until now, the great majority of such generators are based on autonomous nonlinear dynamical

Table 1. Properties of cryptography and its analogous properties of chaos

Cryptography	Chaos
Confusion	Ergodicity
Diffusion with small changes in plaintext / secret keys	Sensitivity to initial conditions / system parameters
Diffusion with a small change within one block of the plaintext	Mixing property
Deterministic pseudo randomness	Deterministic dynamics
Algorithmic complexity	Structural complexity

systems, in order to use the independence of these systems to external sources. However, in the present work a novel CRBG, which is based on a non-autonomous dynamical system, is examined. For ridding from the system the influence of the external source and increasing the security of the proposed generator, the Poincaré section for sampling the signal has been used.

As a first step in this approach, one of the most well-known non-autonomous chaotic systems, the Duffing – van der Pol, has been chosen. The produced, by the proposed CRBG, bit sequence is a result of the X-OR function in the outputs of two threshold circuits that use two variables x which are produced by the two Duffing – van der Pol's Poincaré maps. Finally, this bit sequence is subjected to the de-skewing technique to extract unbiased bits with no correlation and so to increase their complexity, as it is confirmed by the statistical test suite, FIPS-140-2.

This paper is organized as follows. In Section 2, related works with the subject of CRBGs are presented. In Section 3 the non-autonomous dynamical system of Duffing – van der Pol, which is the base of this CRBG, is presented. Section 4 describes the proposed CRBG and especially of using the Poincaré section in the production of the random bit sequences. In Section 5 the results of the statistical tests of FIPS-140-2 which assess the statistical properties of the CRBG, are presented. Finally, Section 6 includes the discussion of the proposed method, while Section 7 presents the conclusion remarks of this work and some thoughts for future work.

2. RELATED WORK

As it was pointed out, one of the methods to obtain aperiodic sequences is to use chaos, which is defined as "random" phenomenon generated by simple deterministic systems. Until now there have been many works on random number generation based on chaos by using either continuous or discrete chaotic systems.

One of the first scientists who proposed such a chaotic number generator was Oishi (1982). He designed a pseudo-random number generator by making use of chaotic first order nonlinear difference equations. In that paper, the specific design problem was formulated mathematically as an inverse problem of the Perron-Frobenius equation. In 1990, Bernstein and Lieberman (1990) proposed a generator implemented by using a chaotic Digital Phase-Locked Loop (DPLL). At the end of 90's Tsuneda, Eguchi and Inoue presented the design of chaotic binary sequences with good statistical properties produced by one-dimensional piecewise linear maps. Such chaotic binary sequences were shown to be balanced, independent and identically distributed. They also realized such sequences by an analog circuit using Switched-Current (SI) techniques.

Next, at the beginning of this century Stojanovski and Kocarev (2001a; 2001b) published two papers which are devoted to the analysis of the application of a chaotic piecewise-linear one-dimensional map as Random Number Generator (RNG). Piecewise linearity of the map enabled them to mathematically find parameter values for which a generating partition was Markov, while the RNG behaved as a Markov information source. In the second paper they presented practical aspects of their chaos-based RNGs. Bernardini and Cortelazzo (2001) explored the possibility for generation of random numbers and gave a simple nonlinear circuit arrangement, together with the necessary tools to assess the random characteristics of its output. At the same year Gerosa, Bernardini and Pietri (2001), presented an efficient architecture, based on a pipeline analog-to-digital converter, in order to realize a simple circuit in 0.8-μm CMOS arrangement technology of a number generator. According to simulation results, the circuit provided sequences of 8-bit numbers that are uniformly distributed and statistically independent, at a rate of 20 Ms/s. In 2003, Kocarev and Jakimoski (2003) proposed different possibilities of using chaotic maps as pseudorandom number generators. In that work, a class of chaos-based pseudorandom bit generators was constructed by them.

Fu, Chen and Zhou (2004) by following the analysis of generating random numbers using piecewise chaotic map, proposed a structure of chaotic random number generator which was based on a digital-analog hybrid circuit. Also in 2004, Huaping, Wang and Gang (2004) showed that with suitable cooperative applications of both chaotic and conventional approaches, the output of a one-way coupled chaotic map lattice could easily meet the practical requirements of random numbers, by presenting excellent random statistical properties, long periodicity of computer realizations, and fast speed of random number generations. In the same year, Yalcin, Suykens and Vandewalle (2004) proposed a novel TRBG based on a double-scroll attractor. The double-scroll attractor was obtained from a simple model which was qualitatively similar to Chua's circuit. This system successfully passed all the used statistical tests and could be implemented in integrated circuits.

Based on this approach, in 2010 Volos, Kyprianidis and Stouboulos (2010) implemented a TRBG for a chaotic image encryption scheme by using a different non-linear electronic circuit, which also produced double-scroll chaotic attractors. Furthermore, the same authors in 2013 (Volos, 2013a) presented a novel chaotic TRBG which was based on the coexistence of two different synchronization phenomena. The first one was the well-known complete chaotic synchronization, while the second one was the recently new proposed synchronization phenomenon, the inverse π-lag synchronization. This coexistence was observed in the case of two mutually coupled identical nonlinear circuits. At the same year, Volos (2013b) also proposed another chaotic random bit generator, which was realized by the Arduino, an open-source physical computing platform based on a simple microcontroller board. The proposed CPRBG, used the X-OR function, in the bit sequences, that were produced by two Logistic maps.

It is a fact, that apart from the above researchers, the subject of designing chaotic random number generators has attracted the interest of several other research groups (Kohda, 1997; Kolesov, 2001; Li, 2001; Gentle, 2003; Tsuneda, 1999; Kennedy, 2000, Pareschi, 2006; Tang, 2006) due to the large number of applications that they have.

3. THE DUFFING – VAN DER POL SYSTEM

In this work the system of Duffing – van der Pol system, which is described by the following set of differential equation (1), is used (Ueda, 1981).

Figure 1. The schematic of the Duffing – van der Pol circuit

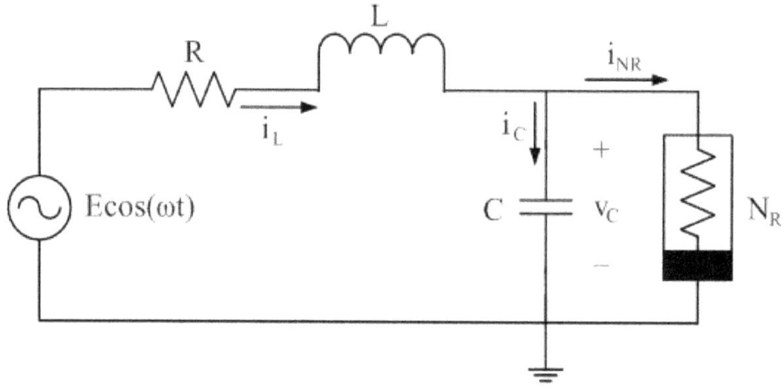

$$\begin{cases} \dfrac{dx}{d\tau} = y \\ \\ \dfrac{dy}{d\tau} = \mu(1 - x^2)y - x^3 + B\cos(\omega_N z) \end{cases} \tag{1}$$

The Duffing – van der Pol is a second order nonlinear, non-autonomous system. It contains in the second equation the term, $\mu(1 - x^2)y$, which is a characteristic feature of the van der Pol oscillator and the cubic term x^3 of Duffing's equation. This is the reason why the system (1) is called Duffing – van der Pol.

The system's equation (1) could be realized by using the circuit topology of Figure 1. Circuital realization of the specific system plays an important role in the case that the proposed generator implemented in hardware as a TRBG. However, in this work, the generator is studied only in simulation environment, in order to examine the effectiveness of the proposed method.

The nonlinear element N_R in Figure 1 realizes the nonlinear $v - i$ characteristic curve of Figure 2, which follows the equation (2),

$$i_N = f(v_C) = -Sv_C\left(1 - \frac{v_C^2}{V_S^2}\right) \tag{2}$$

where V_S is constant and S = 1/R. This nonlinear element can be realized by using the IC multipliers (AD633JN), as it occurred in Refs. (Zhong, 1994; Volos, 2009) (i.e. see Figure 3 in Ref.(Volos, 2009)).

In the aforementioned system's equation (1), $x = v_C$, while the parameters μ, B and ω_N are defined as follows:

$$\mu = \sqrt{\frac{3S\left(LS - RC\right)}{C}} \tag{3}$$

Figure 2. The v – i characteristic curve of the nonlinear element N_R

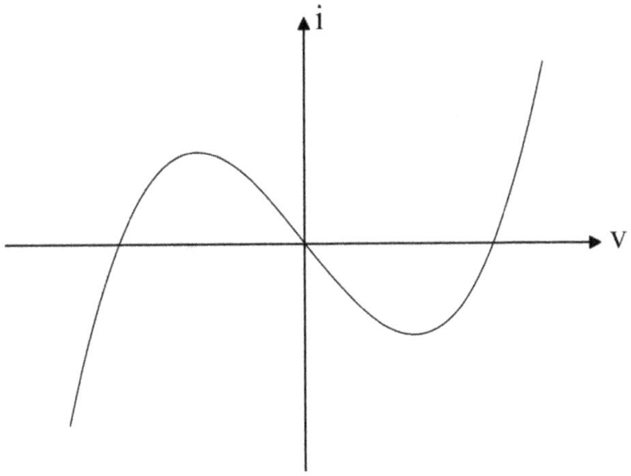

$$B = \frac{\alpha^3 E}{V_S} \tag{4}$$

$$\omega_N = \alpha\omega\sqrt{LC} \tag{5}$$

where,

$$\alpha = \sqrt{\frac{3LS}{LS - RC}} \tag{6}$$

Also, the normalized time (τ) is:

$$\tau = \frac{t}{\alpha\sqrt{LC}} \tag{7}$$

In this work, the dynamic behavior of the Duffing – van der Pol system is investigated numerically by employing the forth order Runge-Kutta algorithm. Figure 3 shows the bifurcation diagrams of x versus the parameter B, for various values of ω_N, while $\mu = 0.2$, which reveals the system's rich dynamical behavior. Periodic and chaotic regions alternate as the parameter B increases while interesting dynamical phenomena, such as routes to chaos through the mechanism of period doubling and crisis phenomena (boundary, internal), is also displayed. This richness of dynamic behavior makes the proposed non-autonomous system a suitable candidate for use in this work.

As it is mentioned, the base of the proposed work is the well-known Poincaré map named by Henri Poincaré. It is the intersection of an orbit in the state space of a continuous dynamical system with a cer-

Figure 3. The bifurcation diagram of x vs. B, for $\mu = 0.2$, while, (a) $\omega_N = 0.6$, (b) $\omega_N = 4.0$, (c) $\omega_N = 0.7$ and (d) $\omega_N = 0.94$

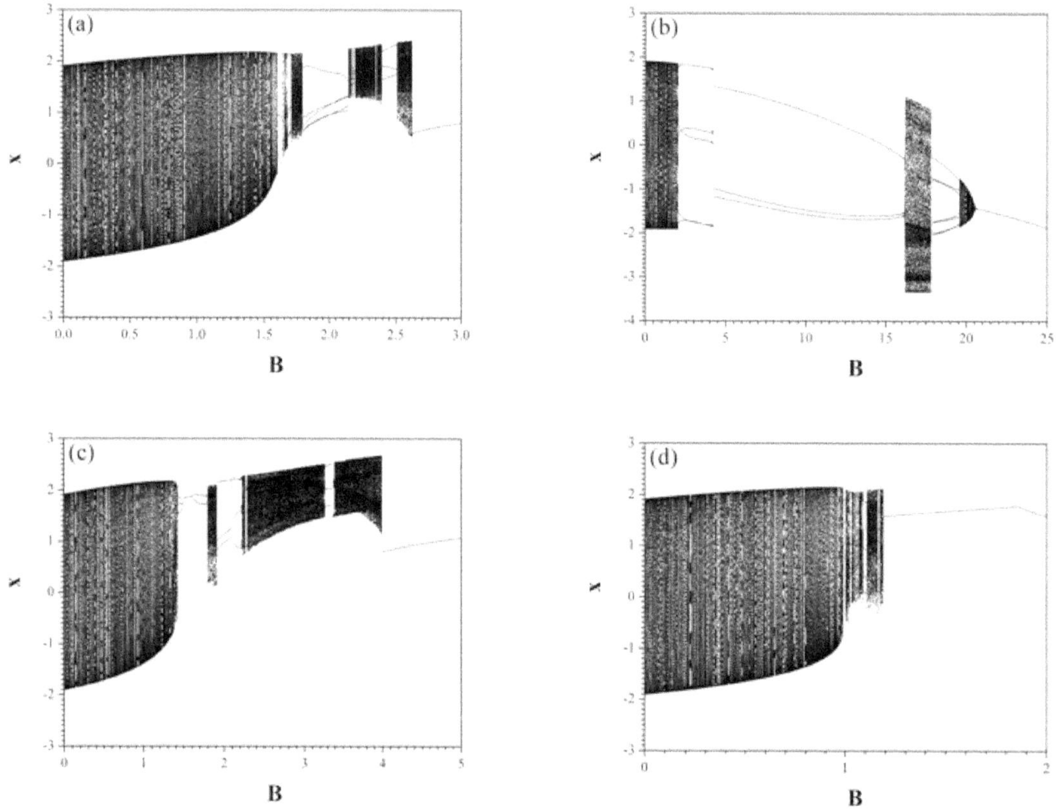

tain lower dimensional subspace, called the Poincaré cross-section, transversal to the flow of the system. So, in this case, the Poincaré map is produced by using the Poincaré cross-section which is defined by

$$\sum = \left\{ \left(x, y, \theta = \omega_n \tau_n \right) \in \mathbb{R}^2 \times S^1 \right\} \tag{8}$$

which $\tau_n = nT + \tau_0$ is the sampling time, τ_0 the initial time determining the location of the Poincaré cross-section on which the coordinates (x, y) of the attractors are expressed and $T = 2\pi/\omega_n$ is the period of the voltage source.

In Figure 4 a number of Poincaré maps for various values of the parameter B of the bifurcation diagram of Figure 3(b), in the case of $\mu = 0.2$ and $\omega_N = 4.0$, are displayed. From this figure the great utility of the Poincaré map can be seen, since for different system's dynamic behavior the Poincaré map has a totally different form. Table 2 shows the various forms of the Poincaré map depending on the dynamic behavior of the system.

Figure 4. Poincaré maps of y vs. x, for $\mu = 0.2$, $\omega_N = 4.0$, while (a) $B = 1$ (Quasi-periodic behavior), (b) $B = 8$ (period-3 state), (c) $B = 17$ (chaotic strange attractor), (d) $B = 19$ (period-7 state), (e) $B = 20$ (Quasi-periodic behavior) and (f) $B = 21$(period-1 state)

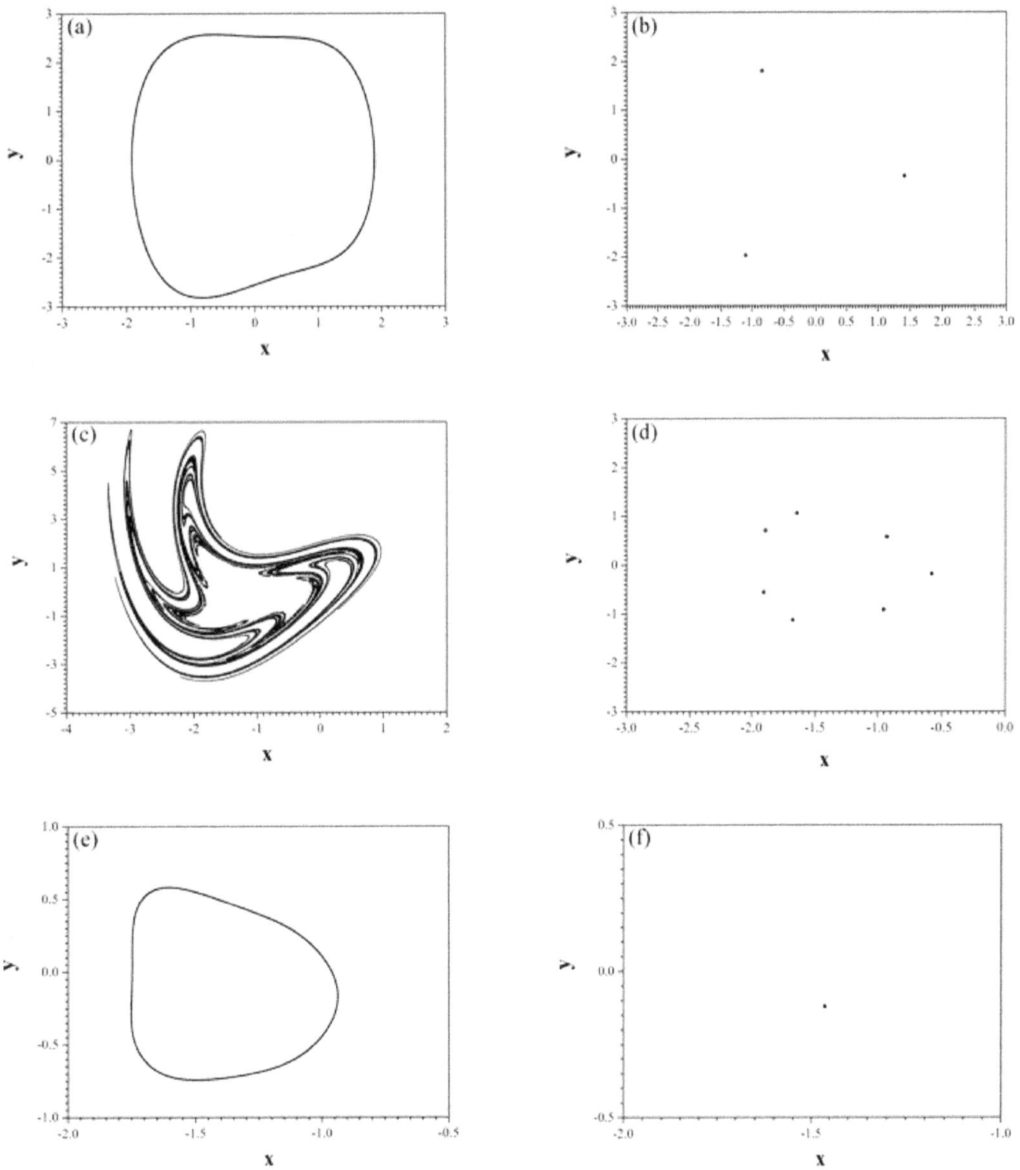

Table 2. Forms of the Poincaré map depending on system's dynamic behavior.

System's Behavior	Poincaré Map
Periodic	Discrete number of points
Quasi-periodic	Closed curve
Chaotic	Strange attractor

4. THE CHAOTIC RANDOM BIT GENERATOR

As it was mentioned, the proposed random bit generator is based on a continuous non-autonomous chaotic system. So, for this reason, the values of system's parameters were selected so that the system is in chaotic regions. In Figure 5 phase portraits and the respective Poincaré maps for various sets of system's parameters are shown.

In Figure 6, the proposed CRBG is presented. This generator consists of five blocks. The first block (S_1) includes two chaotic non-autonomous systems (Duffing – van der Pol) running side by side with different set of parameters (μ_1, B_1, ω_{N1}) and (μ_2, B_2, ω_{N2}) respectively. In the second block (S_2) the Poincaré map for the two systems is used. The state variables x_i (i = 1, 2) from the two systems are partitioned into two subspaces each other by using the two threshold functions in the third block (S_3), which are described by

$$S_3: \quad \begin{array}{l} \sigma^{x1} = \begin{cases} 0, & if \ x_{i1} < x_{n1} \\ 1, & if \ x_{i1} \geq x_{n1} \end{cases} \\ \\ \sigma^{x2} = \begin{cases} 0, & if \ x_{i2} < x_{n2} \\ 1, & if \ x_{i2} \geq x_{n2} \end{cases} \end{array} \tag{9}$$

where x_{n1} and x_{n2} are the threshold values for the variables x_{i1} and x_{i2} respectively.

The fourth block (S_4) produces the bit sequence σ_i by using the X-OR function in σ^{x1} and σ^{x1}, which are finally subjected into the well-known De-skewing technique for eliminating the correlation in the output of the sources of random bits. Von Neumann has been the first author to state this problem (Von Neumann, 1951). He proposed a digital post-processing that balances the distribution of bits. This technique consists of converting the bit pair "01" into the output "0", "10" into the output "1" and of discarding bit pairs "00" and "11".

As it was mentioned in the previous section, for testing reasons in this work the proposed CRBG is implemented by software. However, all the blocks of the generator can be easily realized in hardware.

5. STATISTICAL TESTS

Random number generation has been of great interest, especially from the beginnings of computing. From a practical point of view, "randomness" occurs to the extent that something cannot be predicted. In the literature there are several informal definitions of "randomness", usually based on either a lack of discernible patterns in a sequence, or the unpredictability of the sequence or various aspects of it by, generally, the most puissant possible adversary. Poincaré pointed out that the classic random outcome of a die throwing or a flipping coin comes from sensitive dependence to the initial condition. A small perturbation causes a large difference in the final outcome, thereby making prediction difficult. This sensitive dependence, as it was pointed out, is a hallmark of Chaos.

Also, in the research subject of generating random numbers nobody can be sure, if the produced numbers are really random. That is exactly why a background theory is needed for it, and a set of standardized statistical tests must certify the numbers as random. So, in order to gain the confidence that a

Figure 5. The phase portraits and the Poincaré maps of y vs. x, for $\mu = 0.2$, while, (a), (b) $B = 1.175$, $\omega_N = 0.92$, (c), (d) $B = 17$, $\omega_N = 4.0$, (e), (f) $B = 2.4$, $\omega_N = 0.7$ and (g), (h) $B = 1$, $\omega_N = 0.94$

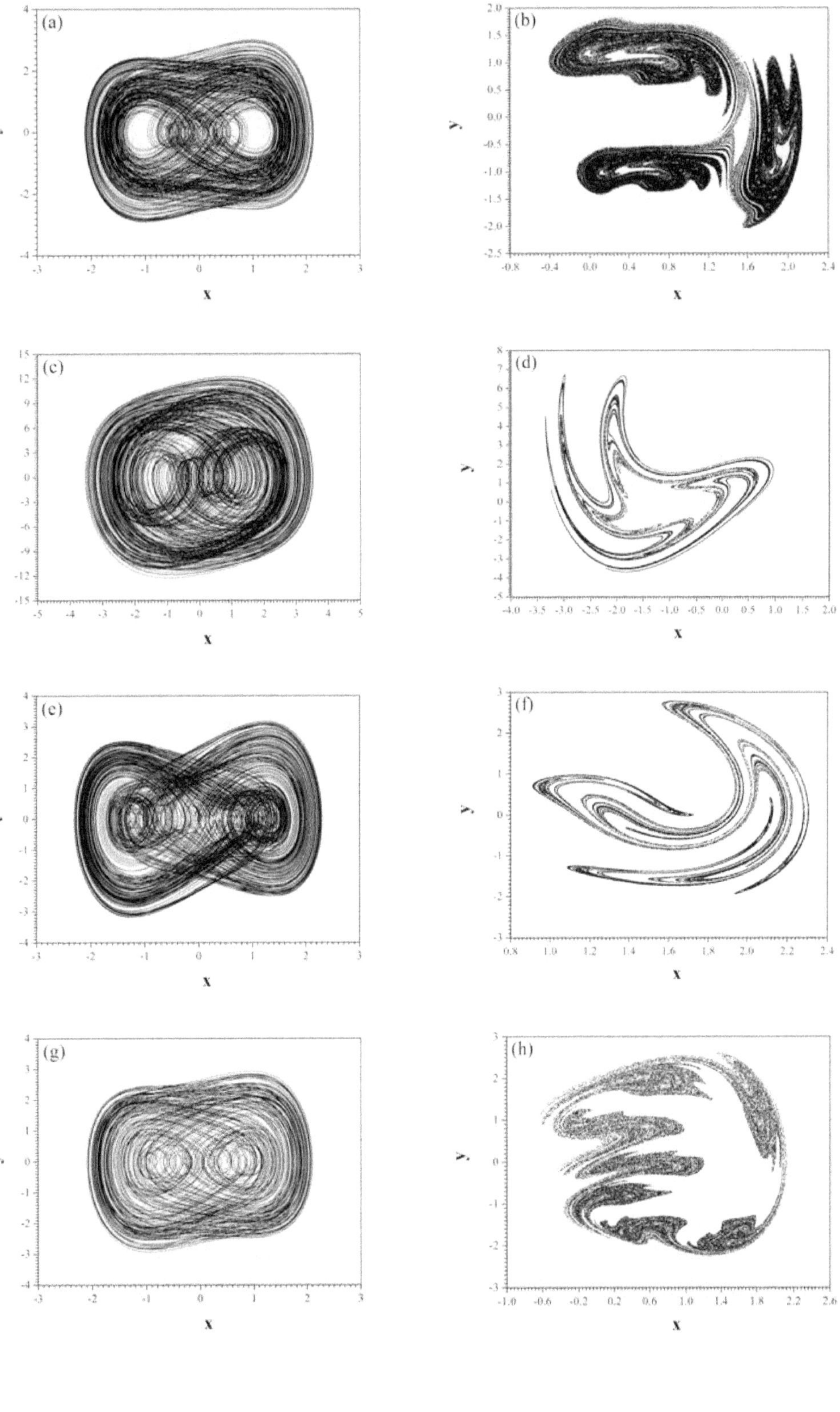

Figure 6. The block diagram of the proposed CRBG

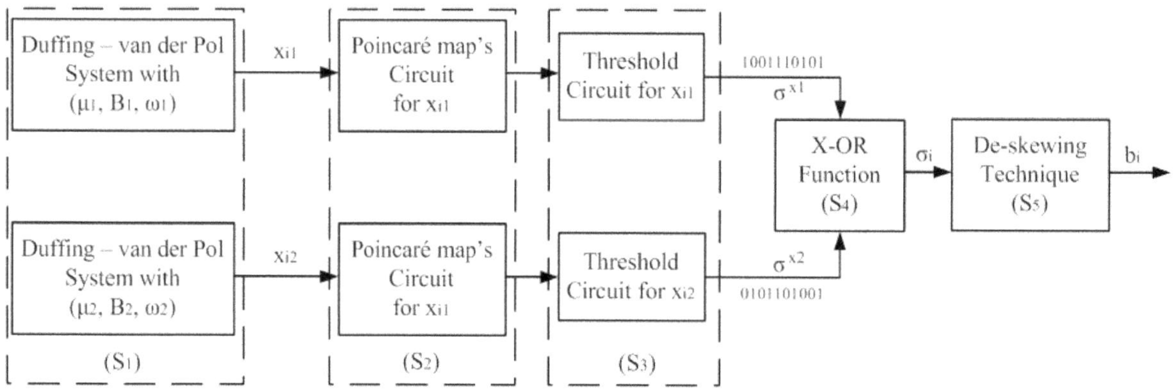

newly developed random bit generator is cryptographically secure, due to its high level of randomness, it should be subjected to a variety of statistical tests designed to detect the specific characteristics expected of truly random sequences. There are several options available in the bibliography for analyzing the randomness of the newly developed random bit generators. The four most popular options are:

1. the FIPS-140-2 (Federal Information Processing Standards) suite of statistical tests of the National Institute of Standards and Technology (NIST) (NIST, 2001),
2. the DIEHARD suite of statistical tests, which was created by the statistician George Marsaglia (Marsaglia, 1995),
3. the Crypt-XS suite of statistical tests, which was developed by researchers at the Information Security Research Centre at Queensland University of Technology in Australia (Gustafson, 1994) and
4. the Donald Knuth's statistical tests set, which includes several empirical statistical tests (Knuth, 1998).

In this work the "randomness" of the produced bit sequences, by the proposed CRBG, is analyzed by using the most stringent tests of randomness: the FIPS-140-2 suite of statistical tests. The results of the use of the four statistical tests, Monobit test, Poker test, Runs test, and Long run test, which are part of the FIPS-140-2, are presented in details. According to FIPS-140-2, the examined CRBG will produce a bitstream, $b_i = b_0, b_1, b_2, \ldots, b_{n-1}$, of length n (at least 20,000 bits), which must satisfy the following standards.

- Monobit Test: The number n_1 of 1's in the bitstream must be $9{,}725 < n_1 < 10{,}275$.
- Poker Test: This test determines whether the sequences of length n (n = 4) show approximately the same number of times in the bitstream. The bounds of this statistic are then $2.16 < X_3 < 46.17$.
- Runs Test: This test determines whether the number of 0's (Gap) and 1's (Block) of various lengths in the bitstream are as expected for a random sequence (Table 3).
- Long Run Test: This test is passed if there are no runs longer than 26 bits.

It is known from the information theory, that the noise has maximum entropy. For this reason, the system's parameters and initial condition are chosen so as measure-theoretic entropy (Fraser, 1989) of the CRBG, which is given by the following equation, is maximum.

$$H_n = \lim_{n \to \infty} \left(-\sum_{B^n} P(B^n) \ln P(B^n) \Big/ n \right) \qquad (10)$$

$P(B^n)$ is the probability of occurrence of a binary subsequence B of length n.

So, from the needs of this work, two Duffing – van der Pol systems with almost identical parameters $(\mu_1, B_1, \omega_1) = (0.2, 2.0, 0.6)$, $(\mu_2, B_2, \omega_2) = (0.21, 2.0, 0.6)$, initial conditions $(x_{01}, y_{01}) = (0.5, 1.0)$, $(x_{02}, y_{02}) = (1.5, -0.5)$ and threshold values $(x_{n1}, x_{n2}) = (1.238, 1.313)$, are running side by side as it was mentioned. For the chosen sets of parameters the two systems present the expected chaotic behavior as it confirmed by the phase portraits and the respective Poincaré maps of Figure 7.

Using the procedure described in the previous section, by using the aforementioned sets of systems' parameters a bitstream of 200,000 bits is obtained which is divided in 10 bit sequences of length 20,000 bits. In Table 4 the measure-theoretic entropy for n = 3 and n = 4 and the detailed results of the 10 bit sequences which were subjected to the four tests of FIPS-140-2 test suite, are presented. As a conclusion, all the bit sequences produced by the CRBG have numerically verified the specific characteristics expected of random bit sequences. Finally, Table 5 presents the analytical results for the four tests of FIPS-140-2, in the case of the first of the 10 bit sequences.

6. DISCUSSION

In this chapter one of the few published works (Ergün, 2007; Ergün 2010) for using the continuous non-autonomous nonlinear systems in the implementation of chaotic random generators has been presented. The problem of the existence of external sources in this nonlinear system's family makes their use in such applications impossible. This occurred, because the influence of the external source on system's dynamical behavior is very easy to be detected by using various techniques, such as the well-known frequency analysis. Thus, by using this kind of systems, the chaotic generator provides no security in the case of attack. This is the reason for which the great majority of works on this subject includes autonomous chaotic systems, as it was presented in details in Section 2.

Table 3. Required intervals for length of runs test

Length of Run	Required Interval
1	2,315 – 2,685
2	1,114 – 1,386
3	527 – 723
4	240 – 384
5 6	103 – 209 103 – 209

Table 4. Results of the measure-theoretic entropy and FIPS-140-2 tests suite, for 10 bit sequences produced by the proposed CRBG

Tests	1	2	3	4	5	6	7	8	9	10
H_3	0.692432	0.692381	0.691921	0.692405	0.692339	0.692510	0.692249	0.692583	0.692279	0.692411
H_4	0.692108	0.691206	0.691510	0.691459	0.691413	0.690966	0.690322	0.691496	0.691421	0.690892
Monobit Test	√	√	√	√	√	√	√	√	√	√
Poker Test	√	√	√	√	√	√	√	√	√	√
Runs Test	√	√	√	√	√	√	√	√	√	√
Long Run Test	√	√	√	√	√	√	√	√	√	√
Result	Passed	Passed	Passed	Passed	Passed	Passed	Passed	Passed	Passed	Passed

Table 5. Analytical results of FIPS-140-2 tests, for the first bit sequence of the proposed CRBG

Monobit Test	Poker Test	Runs Test	Long Run Test
		$B_1 = 2{,}585$	
		$B_2 = 1{,}212$	
$n_1 = 10{,}044$	28.51425	$B_3 = 509$	No
(50.22%)		$B_4 = 289$	
		$B_5 = 146$	
		$B_6 = 201$	
Passed	Passed	Passed	Passed

However, there is a way to delete this influence of the external source. It is a simple but effective method of using the well-known Poincaré map. This map is very useful in the study of nonlinear systems, because it provides information for systems' dynamical behavior. More specific, it allows distinguishing periodic from non-periodic orbits, as it shown in Table 2. The dynamic behavior of a system, which describes an oscillator like the Duffing – van der Poll, presented in a 3-dimensional phase space, where x, is the state variable, $y = dx/dt$ and $\theta = \omega t$ the phase of the external source. Then, the Poincaré map is produced from the sampling the points $(x(t_n), y(t_n), \omega t_n = 2\pi n)$, in the case of external source of period T, where the sampling times are $t_n = nT + \tau_0$, $(n = 1, 2, \ldots)$. So, the study of system's dynamic behavior in a 3-dimensional phase space is converted to the study in a 2-dimensional phase space. Also, the external source is not taken into account but only for sampling the points.

By using the Poincaré map, the application of continuous non-autonomous chaotic systems in implementing CRBGs is possible. This opens new perspectives in the design of such systems, which have a great number of applications, not only in cryptography, but also in gambling, statistical sampling, computer simulation, completely randomized design, and other areas where producing an unpredictable result is desirable.

Figure 7. The phase portraits and the Poincaré maps of y vs. x, for $B = 2$, $\omega_N = 0.6$, while, (a), (b) $\mu = 0.2$, (c), (d) $\mu = 0.21$

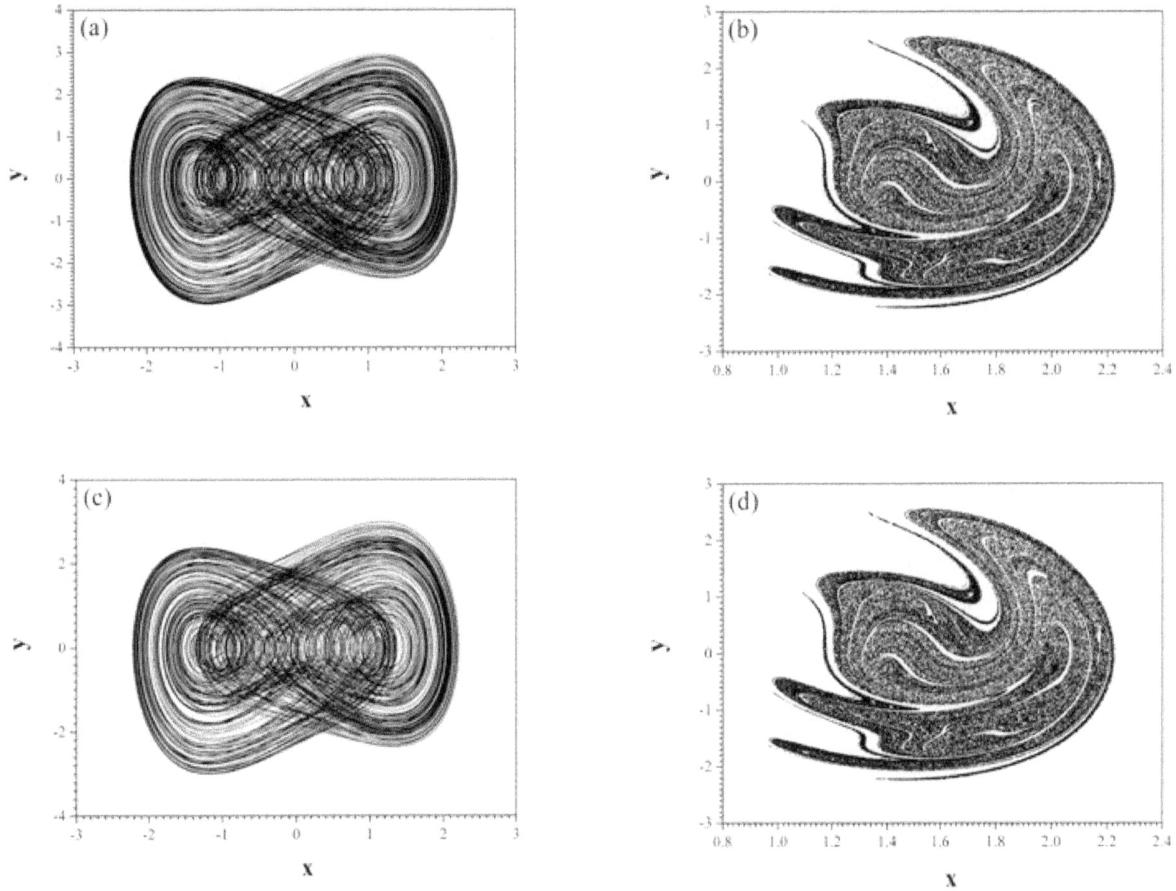

The chosen system in this chapter, the Duffing – van der Pol, is one of the most studied continuous non-autonomous dynamical systems, which shows a variety of phenomena concerning chaos theory. That is the reason for choosing the specific system in this work. However, other known continuous non-autonomous nonlinear systems can be used, such as the van der Pol or Duffing system, which will be studied in the near future.

The use of two Duffing – van der Pol systems running side by side, in the proposed CRBG, with different set of parameters ($\mu_{1,2}$, $B_{1,2}$, $\omega_{N1,2}$), initial conditions ($x_{1,2}$, $y_{1,2}$) and threshold values ($x_{n1,}$ x_{n2}), satisfies the criterion of big key space, which is crucial, especially in the case of using this CRBG in encryption schemes. If we assign 12 bits to each system's parameter, initial condition and threshold values, we get $12 \bullet 12 = 144$ bits. The remaining bits to reach the key length of 256, which is considered as adequate today, can be achieved by using in the proposed CRBG, higher-order dynamical systems. Also, in this work, from the set of system's parameters, only one of them (μ) is different in two systems of the CRNG, while the rest of them remain the same. This occurred, for confirming the significant effect of even a small change of the parameters on the dynamical behavior of the system and hence in the derived sequence of each system, which after the application of X-OR, the final binary sequence is produced.

Furthermore, in this chapter the "randomness" of the specific chaotic random bit generator is confirmed by using the four statistical tests, Monobit test, Poker test, Runs test, and Long run test, which are part of the FIPS-140-2 suite. The very good statistical results of this generator, especially in regard to other previous similar works, confirm not only the applicability of this proposal but also the very good behavior of the derived binary sequence from randomness side. Moreover, a very important development in the study of this CRBG will be the application in the derived binary sequence, and others known from the literature suite of statistical tests, such as the full NIST suite, the DIEHARD suite, the CRYPT-XS suite or the Knuth's statistical tests.

7. CONCLUSION

The great majority of the works, which have been presented till now, on the subject of chaotic random bit generators, was based on autonomous continuous or discrete dynamical systems. This class of nonlinear systems is chosen in order to use the independence of these systems to external sources, which is crucial especially in the case of "attacks" from intruders, when the generators are used in secure communication systems or in image encryption schemes.

For this reason, in this chapter, a chaotic random bit generator, which is based on the Poincaré map of a non-autonomous chaotic system, was presented. The Poincaré map was used for ridding from the system the influence of the external source and increasing the security of the proposed generator. Generator's keys consisted of the system's parameters, initial conditions and the threshold values of the Poincaré sections. The final bit sequence was the result of the application of the X-OR in the outputs of two threshold circuits that use two variables x, which are produced by the two Duffing – van der Pol's Poincaré maps. The use of the X-OR function joint with different values of threshold values of each system increased the complexity, as it was confirmed by the use of a well-known statistical test suite FIPS-140-2 and as a result the chaotic random bit generator of this work proved to be very robust against interference from an intruder.

So, as a future work, the experimental realization of such a random bit generator based on the proposed technique will be studied. Also, the realization of the specific generator by using other non-autonomous dynamical systems could be considered. Furthermore, the use of this generator in various applications, such as in secure communication systems, in image encryption schemes and in robot's motion controllers, will be carried out.

REFERENCES

Agnew, G. B. (1986). *Random sources from cryptographic systems. In Proceedings of Advances in Cryptology-CRYPTO'85* (pp. 77–81). New York: Springer-Verlag.

Alvarez, G., & Li, S. (2006). Some basic cryptographic requirements for chaos based cryptosystems. *International Journal of Bifurcation and Chaos in Applied Sciences and Engineering*, *16*(08), 2129–2151. doi:10.1142/S0218127406015970

Bardis, N. G., Markovskyi, A. P., Doukas, N., & Karadimas, N. V. (2009). True random number generation based on environmental noise measurements for military applications. In *Proc. of 8th WSEAS Int. Conf. on Signal Processing, Robotics and Automation*, (pp. 68-73). Cambridge, UK: WSEAS.

Bernardini, R., & Cortelazzo, G. (2001). Tools for designing chaotic systems for secure random number generation. *IEEE Transactions on Circuits and Systems*, *48*(5), 552–564. doi:10.1109/81.922458

Bernstein, G. M., & Lieberman, M. A. (1990). Secure random number generation using chaotic circuits. *IEEE Transactions on Circuits and Systems*, *37*(9), 1157–1164. doi:10.1109/31.57604

Blum, L., Blum, M., & Shub, M. (1986). A simple unpredictable pseudo-random number generator. *SIAM Journal on Computing*, *15*(2), 364–383. doi:10.1137/0215025

Davis, D., Ihaka, R., & Fenstermacher, P. (1994). *Cryptographic randomness from air turbulence in disk drives. In Advances in Cryptology* (pp. 114–120). Berlin: Springer-Verlag.

Dube, R. R. (2008). *Hardware-based computer security techniques to defeat hackers: From biometrics to quantum cryptography*. Wiley.

Ergün, S. (2007). Truly random number generators based on a non-autonomous chaotic oscillator. *AEÜ. International Journal of Electronics and Communications*, *61*(4), 235–242. doi:10.1016/j.aeue.2006.05.006

Ergün, S., & Özouz, S. (2010). Truly random number generators based on non-autonomous continuous-time chaos. *International Journal of Circuit Theory and Applications*, *38*(1), 1–24. doi:10.1002/cta.520

Fairfield, R. C., Mortenson, R. L., & Coulthart, K. B. (1987). *An LSI random number generator (RNG). In Advances in Cryptology* (pp. 203–230). Berlin: Springer-Verlag.

Ferguson, N., & Schneier, B. (2003). *Practical cryptography*. Wiley.

Fraser, A. M. (1989). Information and entropy in strange attractors. *IEEE Transactions on Information Theory*, *35*(2), 245–262. doi:10.1109/18.32121

Fu, S. M., Chen, Z. Y., & Zhou, Y. A. (2004). Chaos based random number generators. *Computer Research and Development*, *41*, 749–754.

Gentle, J. E. (2003). *Random Number generation and Monte Carlo method*. New York: Springer.

Gerosa, A., Bernardini, R., & Pietri, S. (2001). A fully integrated chaotic system for the generation of truly random numbers. *IEEE Transactions on Circuits and Systems I*, *49*(7), 993–1000. doi:10.1109/TCSI.2002.800833

Grebogi, C., & Yorke, J. (1997). *The impact of chaos on science and society*. United Nations University Press.

Guide, M. (1985). Concept for a high-performance random number generator based on physical random phenomena. *Frequenz*, *39*, 187–190.

Gustafson, H., Dawson, H. E., Nielsen, L., & Caelli, W. (1994). A computer package for measuring the strength of encryption algorithms. *Journal of Computer Security*, *13*(8), 687–697. doi:10.1016/0167-4048(94)90051-5

Hasselblatt, B., & Katok, A. (2003). *A first course in dynamics: With a panorama of recent developments*. Cambridge, UK: Cambridge University Press. doi:10.1017/CBO9780511998188

Holman, W. T., Connelly, J. S., & Downlatabadi, A. B. (1997). An integrated analog-digital random noise source. *IEEE Transactions on Circuits and Systems I*, *44*(6), 521–528. doi:10.1109/81.586025

Hu, Y., Liao, X., Wong, K., & Zhou, Q. (2009). A true random number generator based on mouse movement and chaotic cryptography. *Chaos, Solitons, and Fractals*, *40*(5), 2286–2293. doi:10.1016/j.chaos.2007.10.022

Huaping, L., Wang, S., & Gang, H. (2004). Pseudo-random number generator based on coupled map lattices. *International Journal of Modern Physics B*, *18*(17n19), 2409–2414. doi:10.1142/S0217979204025440

Kennedy, M. P., Rovatti, R., & Setti, G. (2000). *Chaotic electronics in Telecommunications*. CRC Press.

Knuth, D. (1981). *The art of computer programming*. Addison-Wesley.

Knuth, D. (1998). *The art of computer programming: Semiemperical algorithms*. Addison Wesley.

Koc, C. K. (2009). *Cryptographic Engineering*. New York: Springer. doi:10.1007/978-0-387-71817-0

Kocarev, L. (2001). Chaos-based cryptography: A brief overview. *IEEE Circuits and Systems Magazine*, *1*(3), 6–21. doi:10.1109/7384.963463

Kocarev, L., & Jakimoski, G. (2003). Pseudorandom bits generated by chaotic maps. *IEEE Transactions on Circuits and Systems. I, Fundamental Theory and Applications*, *50*(1), 123–126. doi:10.1109/TCSI.2002.804550

Kohda, T., & Tsuneda, A. (1997). Statistics of chaotic binary sequences. *IEEE Transactions on Information Theory*, *43*(1), 104–112. doi:10.1109/18.567654

Kolesov, V. V., Belyaev, R. V., & Voronov, G. M. (2001). Digital random-number generator based on the chaotic signal algorithm. *Journal of Communications Technology and Electronics*, *46*, 1258–1263.

Li, K., Soh, Y. C., & Li, Z. G. (2003). Chaotic cryptosystem with high sensitivity to parameter mismatch. *IEEE Transactions on Circuits and Systems. I, Fundamental Theory and Applications*, *50*(4), 579–583. doi:10.1109/TCSI.2003.809808

Li, S., Mou, X., & Cai, Y. (2001). Pseudo-random bit generator based on coupled chaotic systems and its application in stream-ciphers cryptography. *In Progress in Cryptology – INDOCRYPT 2001. Lecture Notes in Computer Science*, *2247*, 316–329.

Lorenz, E. N. (1963). Deterministic non-periodic flow. *Journal of the Atmospheric Sciences*, *20*(2), 130–141. doi:10.1175/1520-0469(1963)020<0130:DNF>2.0.CO;2

Marsaglia, G. (1995). *DIEHARD Statistical Tests*. Retrieved from http://stst.fsu.edu/pub/diehard

Matsumoto, M., & Nishimura, T. (1998). Mersenne twister: A 623-dimensionally equidistributed uniform pseudo-random number generator. *ACM Transactions on Modeling and Computer Simulation*, *8*(1), 3–30. doi:10.1145/272991.272995

Menezes, A. J., Van Oorschot, P. C., & Vanstone, S. A. (1996). *Handbook of applied cryptography*. CRC Press. doi:10.1201/9781439821916

NIST. (2001). *Security Requirements for Cryptographic Modules* (FIPS PUB 140-2). Retrieved from http://csrc.nist.gov/publications/fips/fips140-2/ fips1402.pdf

Oishi, S., & Inoue, H. (1982). Pseudo-random number generators and chaos. *Transactions of the Institute of Electronics and Communication Engineers of Japan E*, *65*, 534–541.

Pareschi, F., Rovatti, R., & Setti, G. (2006). Simple and effective post-processing stage for random stream generated by a chaos-based RNG. In *Proceedings of 2006 International Symposium in Nonlinear Theory and its Applications*, (pp. 383-386). Academic Press.

Park, S., & Miller, K. (1988). Random numbers generators: Good ones are hard to find. *Communications of the ACM*, *31*(10), 1192–1201. doi:10.1145/63039.63042

Shu, T. (1995). *Uniform random numbers: Theory and practice*. Kluwer Academic Publishers.

Stojanovski, T., & Kocarev, L. (2001a). Chaos-based random number generators - Part I: Analysis. *IEEE Transactions on Circuits and Systems. I, Fundamental Theory and Applications*, *48*(3), 281–288. doi:10.1109/81.915385

Stojanovski, T., Pihl, J., & Kocarev, L. (2001b). Chaos-based random number generators - Part II: Practical realizations. *IEEE Transactions on Circuits and Systems. I, Fundamental Theory and Applications*, *48*(3), 382–385. doi:10.1109/81.915396

Tang, K. W., & Tang, W. (2006). A low cost chaos-based random number generator realized in 8-bit precision environment. In *Proceedings of 2006 International Symposium in Nonlinear Theory and its Applications*, (pp. 395-398). Academic Press.

Trcek, D. (2006). *Managing information systems security and privacy*. Berlin: Springer-Verlag.

Tsuneda, A., Eguchi, K., & Inoue, T. (1999). Design of chaotic binary sequences with good statistical properties based on piecewise linear into maps. In *Proceedings of the 7th International Conference on Microelectronics for Neural, Fuzzy and Bio-Inspired Systems*, (pp. 261-266). Academic Press. doi:10.1109/MN.1999.758873

Ueda, Y., & Akamatsu, N. (1981). Chaotically transitional phenomena in the forced negative-resistance oscillator. *IEEE Transactions on Circuits and Systems*, *CAS-28*(3), 217–224. doi:10.1109/TCS.1981.1084975

Volos, Ch. K. (2013b). Chaotic random bit generator realized with a microcontroller. *Journal of Computations & Modelling*, *3*(4), 115–136.

Volos, Ch. K., Kyprianidis, I. M., & Stouboulos, I. N. (2006). Experimental demonstration of a chaotic cryptographic scheme. *WSEAS Transactions on Circuits and Systems*, *5*, 1654–1661.

Volos, Ch. K., Kyprianidis, I. M., & Stouboulos, I. N. (2010). Fingerprint images encryption process based on a chaotic true random bits generator. *International Journal of Multimedia Intelligence and Security*, *1*(4), 320–335. doi:10.1504/IJMIS.2010.039234

Volos, Ch. K., Kyprianidis, I. M., & Stouboulos, I. N. (2013a). Image encryption process based on chaotic synchronization phenomena. *Signal Processing*, *93*(5), 1328–1340. doi:10.1016/j.sigpro.2012.11.008

Volos, Ch. K., Laftsis, I., & Gkogka, G. D. (2009). Synchronization of two chaotic Chua – type circuits with the inverse system approach. In *Proc. 1st Panhellenic Conference on Electronics and Telecommunications*. Patras, Greece: Academic Press.

Von Neumann, J. (1951). Various techniques used in connection with random digits. Applied Mathematics Series, 12, 36-38.

Wei, J., Liao, X., Wong, K., & Xiang, T. (2006). A new chaotic cryptosystem. *Chaos, Solitons, and Fractals*, *30*(5), 1143–1152. doi:10.1016/j.chaos.2005.09.005

Yalcin, M. E., Suykens, J. A. K., & Vandewalle, J. (2004). True random bit generation from a double-scroll attractor. *IEEE Transactions on Circuits and Systems I*, *51*(7), 1395–1404. doi:10.1109/TCSI.2004.830683

Yang, T. (2004). A survey of chaotic secure communication systems. *International Journal of Computational Cognition*, *2*, 81–130.

Zhong, G. Q. (1994). Implementation of Chua's circuit with a cubic nonlinearity. *IEEE Transactions on Circuits and Systems*, *41*(12), 934–941. doi:10.1109/81.340866

ADDITIONAL READING

Acho, L. (2008). On nonlinear oscillator design. *Journal of Circuits, Systems, and Computers*, *17*(4), 659–662. doi:10.1142/S0218126608004460

Bagini, V., & Bucci, M. (1999). *A design of reliable true random number generator for cryptographic applications*. Cryptographic Hardware and Embedded Systems. *Lecture Notes in Computer Science*, *1717*, 204–218. doi:10.1007/3-540-48059-5_18

Banerjee, S. (2011). *Chaos synchronization and cryptography for secure communications: Applications for encryption*. Hershey: IGI Global.

Banerjee, S., Mitra, M., & Rondoni, L. (2011). *Applications of chaos and nonlinear dynamics in engineering – Part 1*. Berlin, Heidelberg: Springer-Verlag. doi:10.1007/978-3-642-21922-1

Banerjee, T., Biswas, D., & Sarkar, B. C. (2012). Design of chaotic and hyperchaotic time-delayed electronic circuit. *Bonfring International Journal of Power Systems and Integrated Circuits*, *2*(3), 13–17. doi:10.9756/BIJPSIC.3138

Benitez, S., Acho, L., & Guerra, J. R. (2006). Chaotification of the Van der Pol system using Jerk architecture. *IEICE Trans. Fundamentals*, *E89-A*(4), 1088–1091. doi:10.1093/ietfec/e89-a.4.1088

Carroll, T. L., & Pecora, L. M. (1995). *Nonlinear dynamics in circuits*. Singapore: World Scientific Publishing.

Chen, G., & Ueta, T. (2002). *Chaos in circuits and systems*. Singapore: World Scientific Publishing.

Davies, B. (2004). *Exploring Chaos: Theory and experiment*. Oxford, New York: Westview Press – Perseus Books Group.

Eichenauer, J., Grothe, H., & Lehn, J. (1988). Marsaglia's lattice test and non-linear congruential pseudo random number generators. *Metrika*, *35*(1), 241–250. doi:10.1007/BF02613312

Eichenauer, J., & Lehn, J. (1986). A non-linear congruential pseudo random number generator. *Statistical Papers*, *27*(1), 315–326. doi:10.1007/BF02932576

Enayatifar, R., Abdullah, A. H., & Isnin, I. F. (2014). Chaos-based image encryption using a hybrid genetic algorithm and a DNA sequence. *Optics and Lasers in Engineering*, *56*, 83–93. doi:10.1016/j.optlaseng.2013.12.003

Fayos, J. (2001). Emergence of order from chaos by cooperative non-biased learning between random bit strings. *Chaos, Solitons, and Fractals*, *12*(5), 865–873. doi:10.1016/S0960-0779(00)00053-9

Feki, M., Robert, B., Gelle, G., & Colas, M. (2003). Secure digital communication using discrete-time chaos synchronization. *Chaos, Solitons, and Fractals*, *18*(4), 881–890. doi:10.1016/S0960-0779(03)00065-1

Fischer, V., & Drutarovský, M. (2003). *True random number generator embedded in reconfigurable hardware*. Cryptographic Hardware and Embedded Systems-CHES 2002. *Lecture Notes in Computer Science*, *2523*, 415–430. doi:10.1007/3-540-36400-5_30

Gammoudi, I., & Feki, M. (2011). A new look at Chua's circuit. *International Journal of Bifurcation and Chaos in Applied Sciences and Engineering*, *21*(06), 1653–1666. doi:10.1142/S0218127411029343

Hanias, M. P., Avgerinos, Z., & Tombras, G. S. (2009). Period Doubling, Feigenbaum constant and time series prediction in an experimental chaotic RLD circuit. *Chaos, Solitons, and Fractals*, *40*(3), 1050–1059. doi:10.1016/j.chaos.2007.08.061

Hanias, M. P., Giannis, I. L., & Tombras, G. S. (2010). Chaotic operation by a single transistor circuit in the reverse active region. *Chaos (Woodbury, N.Y.)*, *20*(1), 013105. doi:10.1063/1.3293133 PMID:20370260

Hauser, J. R. (2009). Numerical methods for nonlinear engineering models. New York: Springer Science+Business Media. doi:10.1007/978-1-4020-9920-5

He, Q., Xu, W., Rong, H., & Fang, T. Stochastic bifurcation in Duffing–Van der Pol oscillators. *Physica A: Statistical Mechanics and its Applications, 338*, 319-334.

Hilborn, R. C. (2000). *Chaos and nonlinear dynamics: An introduction for scientists and engineers*. Oxford, New York: Oxford University Press. doi:10.1093/acprof:oso/9780198507239.001.0001

Holland, J. H. (1998). *Emergence: From chaos to order*. Oxford, New York: Oxford University Press.

Hu, Y., Liao, X., Wong, K. W., & Zhou, Q. (2009). A true random number generator based on mouse movement and chaotic cryptography. *Chaos, Solitons, and Fractals*, *40*(5), 2286–2293. doi:10.1016/j.chaos.2007.10.022

In, V., Longhini, P., & Palacios, A. (2009). *Applications of nonlinear dynamics: Model and design of complex systems*. Berlin, Heidelberg: Springer-Verlag. doi:10.1007/978-3-540-85632-0

Ivancevic, V. G., & Ivancevic, T. T. (2008). *Complex nonlinearity: Chaos, phase transitions, topology change, and path integrals*. Berlin, Heidelberg: Springer-Verlag.

Jennewein, T., Achleitner, U., Weihs, G., Weinfurter, H., & Zeilinger, A. (2000). A fast and compact quantum random number generator. *The Review of Scientific Instruments*, *71*(4), 1675–1680. doi:10.1063/1.1150518

Ji, J. C., & Hansen, C. H. (2006). Stability and dynamics of a controlled van der Pol–Duffing oscillator. *Chaos, Solitons, and Fractals*, *28*(2), 555–570. doi:10.1016/j.chaos.2005.08.021

Jing, Z., Yang, Z., & Jiang, T. (2006). Complex dynamics in Duffing–Van der Pol equation. *Chaos, Solitons, and Fractals*, *27*(3), 722–747. doi:10.1016/j.chaos.2005.04.044

Jung, C. (1986). Poincaré map for scattering states. *Journal of Physics. A, Mathematical and General*, *19*(8), 1345–1353. doi:10.1088/0305-4470/19/8/016

Kakmeni, F. M. M., Bowong, S., Tchawoua, C., & Kaptouom, E. (2004). Strange attractors and chaos control in a Duffing–Van der Pol oscillator with two external periodic forces. *Journal of Sound and Vibration*, *277*(4-5), 783–799. doi:10.1016/j.jsv.2003.09.051

Kapitaniak, T. (1992). *Chaotic oscillators: Theory and applications*. Singapore: World Scientific Publishing.

Karimi, H., Hosseini, S. M., & Zahan, M. V. (2013). On the combination of self-organized systems to generate pseudo-random numbers. *Information Sciences*, *221*, 371–388. doi:10.1016/j.ins.2012.09.029

Kilic, R. (2010). *A practical guide for studying Chua's circuits*. Singapore: World Scientific Publishing.

Kimiaeifar, A., Saidi, A. R., Bagheri, G. H., Rahimpour, M., & Domairry, D. G. (2009). Analytical solution for Van der Pol–Duffing oscillators. *Chaos, Solitons, and Fractals*, *42*(5), 2660–2666. doi:10.1016/j.chaos.2009.03.145

Kohlbrenner, P., & Gaj, K. (2004). *An embedded true random number generator for FPGAs*. In Proceedings of the 12th international symposium on Field programmable gate arrays (FPGA '04), 71-78, New York, USA. doi:10.1145/968280.968292

Koper, M. T. M. (1995). Bifurcations of mixed-mode oscillations in a three-variable autonomous Van der Pol-Duffing model with a cross-shaped phase diagram. *Physica D. Nonlinear Phenomena*, *80*(1-2), 72–94. doi:10.1016/0167-2789(95)90061-6

Korsch, H. J., Jodl, H.-J., & Hartmann, T. (2008). *Chaos: A program collection for the PC*. Berlin, Heidelberg: Springer-Verlag.

Kyprianidis, I. M., Papachristou, V., Stouboulos, I. N., & Volos, Ch. K. (2012). Dynamics of coupled chaotic Bonhoeffer – van der Pol oscillators. *WSEAS Transactions on Systems*, *11*(9), 516–526.

Kyprianidis, I. M., Volos, Ch. K., Stouboulos, I. N., & Hadjidemetriou, J. (2006). Dynamics of two resistively coupled Duffing – type electrical oscillators. *International Journal of Bifurcation and Chaos in Applied Sciences and Engineering*, *16*(06), 1765–1775. doi:10.1142/S0218127406015660

Lakshmanan, M. (1996). *Chaos in nonlinear oscillators: Controlling and synchronization*. Singapore: World Scientific Publishing.

Liu, G., Li, J., & Liu, H. (2014). Chaos-based color pathological image encryption scheme using one-time keys. *Computers in Biology and Medicine*, *45*, 111–117. doi:10.1016/j.compbiomed.2013.11.010 PMID:24480170

Liu, H., & Wang, X. (2013). Triple-image encryption scheme based on one-time key stream generated by chaos and plain images. *Journal of Systems and Software*, *86*(3), 826–834. doi:10.1016/j.jss.2012.11.026

Luo, A. C. J., Machado, J. A. T., & Baleanu, D. (2011) Dynamical Systems and Methods. New York: Springer Science+Business Media.

Machicao, J., Marco, A. G., & Bruno, O. M. (2012). Chaotic encryption method based on life-like cellular automata. *Expert Systems with Applications*, *39*(16), 12626–12635. doi:10.1016/j.eswa.2012.05.020

Madhavan, K., & Vaidyanathan, S. (2013). A new procedure for the identification of slow and fast states of discrete-time linear systems. *Far East Journal of Mathematical Sciences*, *72*(2), 225–236.

Micali, S., & Schnorr, C. P. (1991). Efficient, perfect polynomial random number generators. *Journal of Cryptology*, *3*(3), 157–172. doi:10.1007/BF00196909

Ogorzałek, M. J. (1997). *Chaos and complexity in nonlinear electronic circuits*. Singapore: World Scientific Publishing.

Orsucci, F. F., & Sala, N. (2006). *New research on chaos and complexity*. New York: Nova Science Publishers.

Pacheco, J. M. M., & Cuautle, E. T. (2010). *Electronic design automation of multi-scroll chaos generators*. Oak Park, Illinois: Bentham Science Publishers.

Pham, V.-T., Fortuna, L., & Frasca, M. (2013). Implementation of chaotic circuits with a digital time-delay block. *Nonlinear Dynamics*, *67*(1), 345–355. doi:10.1007/s11071-011-9982-9

Saied, A. B., Salem, S. B., Feki, M., & Masmoudi, D. S. (2007). A new high frequency second generation current conveyor based chaos generator. *In Proc. of 14th IEEE International Conference on Electronics, Circuits and Systems, 2007 (ICECS 2007)*, Marrakech, Morocco.

Sarasu, P., & Sundarapandian, V. (2011). Active controller design for generalized projective synchronization of four-scroll chaotic systems. *International Journal of Systems Signal Control and Engineering Application*, *4*(2), 26–33.

Sarasu, P., & Sundarapandian, V. (2011). The generalized projective synchronization of hyperchaotic Lorenz and hyperchaotic Qi systems via active control. *International Journal of Soft Computing*, *6*(5), 216–223. doi:10.3923/ijscomp.2011.216.223

Sarasu, P., & Sundarapandian, V. (2012). Generalized projective synchronization of three-scroll chaotic systems via adaptive control. *European Journal of Scientific Research*, *72*(4), 504–522.

Sarasu, P., & Sundarapandian, V. (2012). Generalized projective synchronization of two-scroll systems via adaptive control. *International Journal of Soft Computing*, *7*(4), 146–156. doi:10.3923/ijscomp.2012.146.156

Schenk-Hoppé, K. R. (1996). Bifurcation scenarios of the noisy Duffing-van der Pol oscillator. *Nonlinear Dynamics*, *11*(3), 255–274. doi:10.1007/BF00120720

Scott, A. (2005). *Encyclopedia of nonlinear science*. New York: Taylor & Francis Group.

Som, S., & Sen, S. (2013). A non-adaptive partial encryption of grayscale images based on chaos. *Procedia Technology*, *10*, 663–671. doi:10.1016/j.protcy.2013.12.408

Sundarapandian, V. (2004). Local and global asymptotic stability of nonlinear cascade interconnected systems. *Mathematical and Computer Modelling*, *40*(1-2), 227–232. doi:10.1016/j.mcm.2003.02.013

Sundarapandian, V. (2005). Nonlinear observer design for a general class of nonlinear systems with real parametric uncertainty. *Computers & Mathematics with Applications (Oxford, England)*, *49*(7-8), 1195–1211. doi:10.1016/j.camwa.2004.06.034

Sundarapandian, V. (2006). New results on the parametric stability of nonlinear systems. *Mathematical and Computer Modelling*, *43*(1), 9–15. doi:10.1016/j.mcm.2005.04.010

Sundarapandian, V. (2006). A necessary condition for local asymptotic stability of nonlinear systems with exogenous disturbance. *Computers & Mathematics with Applications (Oxford, England)*, *51*(3-4), 667–672. doi:10.1016/j.camwa.2005.01.032

Sundarapandian, V. (2010). Output regulation of the Lorenz attractor. *Far East Journal of Mathematical Sciences*, *42*(2), 289–299.

Sundarapandian, V. (2013). Analysis and anti-synchronization of a novel chaotic system via active and adaptive controllers. *Journal of Engineering Science and Technology Review*, *6*(4), 45–52.

Sundarapandian, V., & Karthikeyan, R. (2011). Anti-synchronization of Lü and Pan chaotic systems by adaptive nonlinear control. *European Journal of Scientific Research*, *64*(1), 94–106.

Sundarapandian, V., & Karthikeyan, R. (2011). Anti-synchronization of hyperchaotic Lorenz and hyperchaotic Chen systems by adaptive control. *International Journal of Systems Signal Control and Engineering Application*, *4*(2), 18–25.

Sundarapandian, V., & Karthikeyan, R. (2012). Adaptive anti-synchronization of uncertain Tigan and Li systems. *Journal of Engineering and Applied Sciences (Asian Research Publishing Network)*, *7*(1), 45–52.

Sundarapandian, V., & Karthikeyan, R. (2012). Hybrid synchronization of hyperchaotic Lorenz and hyperchaotic Chen systems via active control. *Journal of Engineering and Applied Sciences (Asian Research Publishing Network)*, *7*(3), 254–264. doi:10.3923/jeasci.2012.254.264

Sundarapandian, V., & Pehlivan, I. (2012). Analysis, control, synchronization, and circuit design of a novel chaotic system. *Mathematical and Computer Modelling*, *55*(7-8), 1904–1915. doi:10.1016/j.mcm.2011.11.048

Sundarapandian, V., & Sivaperumal, S. (2011). Sliding controller design of hybrid synchronization of four-wing chaotic systems. *International Journal of Soft Computing*, *6*(5), 224–231. doi:10.3923/ijscomp.2011.224.231

Tchitnga, R., Fotsin, H. B., Nana, B., Fotso, P. H. L., & Woafo, P. (2012). Hartley's oscillator: The simplest chaotic two-component circuit. *Chaos, Solitons, and Fractals*, *45*(3), 306–313. doi:10.1016/j.chaos.2011.12.017

Thamrin, N. M., Witjaksono, G., Nuruddin, A., & Abdullah, M. S. (2009). *An enhanced hardware-based hybrid random number generator for cryptosystem*. In Proc. of International Conference on Information Management and Engineering (ICIME '09), 152-156, Kuala Lumpur. doi:10.1109/ICIME.2009.115

Trejo-Guerra, R., Tlelo-Cuautle, E., Carbajal-Gomez, V. H., & Rodriguez-Gomez, G. A. (2013). Survey on the integrated design of chaotic oscillators. *Applied Mathematics and Computation*, *219*(10), 5113–5122. doi:10.1016/j.amc.2012.11.021

Vaidyanathan, S. (2012). Sliding mode control based global chaos control of Liu-Liu-Liu-Su chaotic system. *International Journal of Control Theory and Applications*, *5*(1), 15–20.

Vaidyanathan, S. (2012). Anti-synchronization of Sprott-L and Sprott-M chaotic systems via adaptive control. *International Journal of Control Theory and Applications*, *5*(1), 41–59.

Vaidyanathan, S. (2013). A new six-term 3-D chaotic system with an exponential nonlinearity. *Far East Journal of Mathematical Sciences*, *79*(1), 135–143.

Vaidyanathan, S. (2013). Analysis and adaptive synchronization of two novel chaotic systems with hyperbolic sinusoidal and cosinusoidal nonlinearity and unknown parameters. *Journal of Engineering Science and Technology Review*, *6*(4), 53–65.

Vaidyanathan, S., & Rajagopal, K. (2012). Global chaos synchronization of hyperchaotic Pang and hyperchaotic Wang systems via adaptive control. *International Journal of Soft Computing*, *7*(1), 28–37. doi:10.3923/ijscomp.2012.28.37

Vaidyanathan, S., & Sampath, S. (2012). Anti-synchronization of four-wing chaotic systems via sliding mode control. *International Journal of Automation and Computing*, *9*(3), 274–279. doi:10.1007/s11633-012-0644-2

Vazirani, U. V., & Vazirani, V. V. (1985). *Efficient and secure pseudo-random number generation*. Advances in Cryptology. *Lecture Notes in Computer Science*, *196*, 193–202. doi:10.1007/3-540-39568-7_17

Venkatesan, A., & Lakshmanan, M. (1997). Bifurcation and chaos in the double-well Duffing–van der Pol oscillator: Numerical and analytical studies. *Physical Review E: Statistical Physics, Plasmas, Fluids, and Related Interdisciplinary Topics*, *56*(6), 6321–6330. doi:10.1103/PhysRevE.56.6321

Volos, Ch. K. (2013). Image encryption using the coexistence of chaotic synchronization phenomena. *Journal of Applied Mathematics and Bioinformatics*, *3*(1), 123–149.

Volos, Ch. K., Kyprianidis, I. M., & Stouboulos, I. N. (2009). Image encryption process based on a chaotic true random bit generator. *In Proc. 16th IEEE International Conference on Digital Signal Processing (DSP 2009)*, *1 & 2*, 1091-1094, Santorini, Greece. doi:10.1109/ICDSP.2009.5201107

Volos, Ch. K., Kyprianidis, I. M., & Stouboulos, I. N. (2011). Various synchronization phenomena in bidirectionally coupled double scroll circuits. *Communications in Nonlinear Science and Numerical Simulation*, *16*(8), 3356–3366. doi:10.1016/j.cnsns.2010.11.015

Volos, Ch. K., Kyprianidis, I. M., & Stouboulos, I. N. (2012). A chaotic path planning generator for autonomous mobile robots. *Robotics and Autonomous Systems*, *60*(4), 651–656. doi:10.1016/j.robot.2012.01.001

Volos, Ch. K., Kyprianidis, I. M., & Stouboulos, I. N. (2012). Synchronization phenomena in coupled nonlinear systems applied in economic cycles. *WSEAS Trans. Systems*, *11*(12), 681–690.

Wang, X., & Luan, D. (2013). A novel image encryption algorithm using chaos and reversible cellular automata. *Communications in Nonlinear Science and Numerical Simulation*, *18*(11), 3075–3085. doi:10.1016/j.cnsns.2013.04.008

Wang, X., Zhang, W., Guo, W., & Zhang, J. (2013). Secure chaotic system with application to chaotic ciphers. *Information Sciences*, *221*, 555–570. doi:10.1016/j.ins.2012.09.037

Wu, C. W. (2002). *Synchronization in coupled chaotic circuits and systems*. Singapore: World Scientific Publishing.

Wu, X., Wang, J., Lu, J., & Iu, H. H. C. (2007). Hyperchaotic behavior in a non-autonomous unified chaotic system with continuous periodic switch. *Chaos, Solitons, and Fractals*, *32*(4), 1485–1490. doi:10.1016/j.chaos.2005.11.100

Yang, D., Liu, Z., & Zhou, J. (2014). Chaos optimization algorithms based on chaotic maps with different probability distribution and search speed for global optimization. *Communications in Nonlinear Science and Numerical Simulation*, *19*(4), 1229–1246. doi:10.1016/j.cnsns.2013.08.017

Zapateiro, M., Vidal, Y., & Acho, L. (2013). A chaotic secure communication scheme based on Duffing oscillators and frequency estimation. *In Proc. of 9th IFAC Symposium on Nonlinear Control Systems*, Toulouse, France.

Zeraoulia, E., & Sprott, J. C. (2011). *Frontiers in the study of chaotic dynamical systems with open problems*. Singapore: World Scientific Publishing.

Zhang, X., & Wang, X. (2013). Chaos-based partial encryption of SPHIT color coded images. *Signal Processing*, *93*(9), 2422–2431. doi:10.1016/j.sigpro.2013.03.017

Zheng, F., Tian, X. J., Song, J. Y., & Li, X. Y. (2008). Pseudo-random sequence generator based on the generalized Henon map. *Journal of China Universities of Posts and Telecommunications*, *15*(3), 64–68. doi:10.1016/S1005-8885(08)60109-0

Zhu, H., Zhao, C., & Zhang, X. (2013). A novel image-compression scheme using hyper-chaos and Chinese remainder theorem. *Signal Processing Image Communication*, *28*(6), 670–680. doi:10.1016/j.image.2013.02.004

Zhu, H., Zhao, C., Zhang, X., & Yang, L. (2013). A novel iris and chaos-based random number generator. *Computers & Security*, *36*, 40–48. doi:10.1016/j.cose.2013.02.003

Zhusubaliyev, Z. T. (2003). *Bifurcations and chaos in piecewise-smooth dynamical systems*. Singapore: World Scientific Publishing.

KEY TERMS AND DEFINITIONS

Bifurcation Diagram: A diagram in nonlinear dynamics, which shows the possible long-term values (equilibria/fixed points or periodic orbits) of a system as a function of a parameter.

Chaos: The type of behavior of a complex system, where tiny changes in a system's initial conditions can lead to very large changes over time.

Chaos Theory: The field of study in Mathematics that deals with nonlinear dynamics, in which seemingly "random" events are actually predictable from simple deterministic equations.

Chaotic Attractor: The attractor for which the approach to its final state in phase space is chaotic.

FIPS Statistical Tests: A suite of statistical tests (Federal Information Processing Standards - FIPS) of the National Institute of Standards and Technology (NIST) for testing the "randomness" of random bit generators.

Non-Autonomous Systems: Non-relativistic dynamical systems subject to time-dependent transformations.

Nonlinear Element: An electrical element which does not have a linear relationship between current and voltage.

Poincaré Map: The intersection of an orbit in the state space of a continuous dynamical system with a certain lower dimensional subspace, called the Poincaré cross-section, transversal to the flow of the system.

Random Bit Generator: A computational or physical device designed to generate a sequence of bits that lack any pattern, i.e. appear "random".

Chapter 8
Chaotic Attractor in a Novel Time-Delayed System with a Saturation Function

Viet-Thanh Pham
Hanoi University of Science and Technology, Vietnam

Christos Volos
Aristotle University of Thessaloniki, Greece

Sundarapandian Vaidyanathan
Vel Tech University, India

ABSTRACT

From the viewpoint of engineering applications, time delay is useful for constructing a chaotic signal generator, which is the major part of diverse potential applications. Although different mathematical models of time-delay systems have been known, few models can exhibit chaotic behaviors. Motivated by attractive features and potential applications of time-delay models, a new chaotic system with a single scalar time delay and a nonlinearity described by a saturation function is proposed in this chapter. Nonlinear behavior of the system is discovered through bifurcation diagrams and the maximum Lyapunov exponent with the variance of system parameters. Interestingly, the system shows double-scroll chaotic attractors for some suitable chosen system parameters. In order to confirm the correction and feasibility of the theoretical model, the system is also implemented with analog electronic circuit. Finally, a practical application of such circuit is discussed at the end of this chapter.

1. INTRODUCTION

After the discovery of the first classical chaotic attractor in 1963 (Lorenz, 1963), chaos theory has received a great deal of attentions (Anishchenko, 1993; Grebogi, 1997; Davies, 2004; Banerjee, 2011). Although the fact that there is not an universal definition of chaos, three remarkable characteristics of a chaotic system are: dynamical instability, topological mixing and dense periodic orbits (Hasselblatt,

DOI: 10.4018/978-1-4666-7248-2.ch008

2003). Dynamical instability is known as the "butterfly effect", which means that a small change in initial conditions of system can create significant differences. In other word, this vital characteristic makes the system highly sensitive to initial conditions (Lorenz, 1963; Strogatz, 1994). Topologically mixing is refers as stretching and folding of the phase space, which means that the chaotic trajectory at the phase space will evolve in time so that each given area of this trajectory will eventually cover part of any particular region. Dense periodic orbits means that the trajectory can come arbitrarily close every possible asymptotic state. Hence, the future behaviors of a chaotic system seem to be arbitrary and unpredictable. In fact, chaos appears naturally in weather and climate, biology, sociology, and physics etc. (Strogatz, 1994; Holland, 1998; Hilborn, 2000).

From the view point of engineering, chaotic systems with their complex characteristics have also been successfully utilized in diverse applications, ranging from secure chaotic communications (Cuomo, 1993), information encryption (Volos, 2013), economics (Medio, 2009), robotics (Volos, 2012), to liquid shakers (Zhang, 2007) and so on. One of the most interesting applications is chaotic cryptography (Kocarev, 2001), where chaotic systems provide alternative techniques capable of enhancing cryptographic features (Koc, 2009). The structural relationship between cryptography and chaos seem quite natural (Alvarez, 2006). It has been known that main properties of cryptography like confusion, diffusion, deterministic pseudo randomness and algorithm complexity connect directly with their analogous properties of chaos like ergodicity, sensitivity to initial conditions or system parameters, deterministic dynamics and structural complexity, respectively (Shannon, 1949; Alvarez, 2006; Kocarev, 2011). By now, a great majority of chaos-based cryptographic systems has been developed (Zhang, 2005; Tong, 2009; Wang, 2012; Seyedzadeh, 2012). Recent obtained results of such cryptographic systems can be found in (Kocarev, 2011).

Time delays have become the subject of active research due to the fact that they almost present in any real dynamical systems, few examples are optical bistable resonator (Ikeda, 1980; Ikeda, 1987), neural networks (Murray, 1990), control systems (Pyragas, 2006), or biological systems (Mackey, 1997; Sun, 2006). Delay differential equations differ from ordinary differential equations in that the evolution of dependent variables at a certain time depend on their values at previous times (Driver, 1977). Therefore, delay differential equation has used widely to study properties of such systems with delays (Kuang, 1993; Balachandran, 2009; Lakshmanan, 2011).

As it has already been established that delay can be exploited for chaos control (Pyragas, 1992). Time-delayed feedback control has reported as an alternative method besides the classical OGY method (Ott, 1990). This effective control method does not require the prior knowledge about the model and can stabilize the chaotic behavior to one of unstable fixed points or unstable periodic orbits embedded within chaotic attractor (Pyragas, 2006). Because the feedback signal is proportional to the difference of output signal and its delayed version, there are two control parameters: the feedback gain and the feedback time delay. It has been known that the feedback time delay is often different from the intrinsic delay of the considered time-delay system which needs to be controlled. In contrast, the presence of delay in dynamical systems may induce instability and complex phenomena (Xia, 2009). Oscillatory instabilities are frequently observed in systems described by delay differential equations. These unexpected oscillations lead to limitations to the performance of practical systems, especially mechanical machines (Hu, 2002). On the opposite, time delays may also be useful to generate chaos in the so-called time-delayed chaotic system which constitutes an importance class of chaotic systems (Lu, 1996). Time-delayed chaotic systems have been discovered in both theory (Lakshmanan, 2011) and realistic applications, i.e. secure communications (Prokhorov, 2008; Kwon, 2011), economic process (Salarieh, 2009), road traffic (Safonov, 2002), or food web (Wang, 2010). That is why, it is significant to investigate time-delayed chaotic models with simple nonlinear functions which can be easily realized (Buscarino, 2011).

In this chapter a novel nonlinear time-delayed system having a saturation function which can exhibit chaotic behavior is investigated. In particular, the saturation function is implemented easily by using operational amplifiers and resistors. This chapter is organized in the following manner. After a brief review of related works in Section 2, the mathematical model of the new system is proposed in Section 3. Dynamics of this new model are introduced through bifurcation diagrams and Lyapunov exponents in Section 4 while Section 5 presents the circuital implementation of the proposed system. Next, obtained results are also reported and discussed in Section 5 and Section 6, respectively. And finally, concluding remarks are given in the last Section.

2. RELATED WORK

According to the Poincare-Bendixson theorem (Bendixson, 1901), it requires at least three state variables for an autonomous continuous-time system to be able to create chaotic attractors. As a result, three-dimensional autonomous chaotic systems have become the conventional ones (Sprott, 2010), i.e. Lorentz system, Rölsser system, Chua system, Chen system, Lü system etc. However, when modeling the production of white blood cells, Mackey and Glass (Mackey, 1977) investigated that a system described by just a first-order delay differential equation can be chaotic because it can be considered as an infinitive-dimensional system. For this reason, some time-delayed chaotic systems (Ikeda, 1987; Lu, 1996; Namajunas, 1995; Tamasevicius, 2006; Duan, 2009) have studied due to their relatively simple configurations. From another view of point, the presence of time delay in practical systems is unavoidable because of memory effects, the limit calculation speed, or finite transmission velocity etc. (Farmer, 1982; Pei, 2011; Yongzhen, 2011; Wang, 2010; Prokhorov, 2008; Tang, 2010; Zhan, 2003). In particular, the presence of time delay may create various complex phenomena which do not exist in original systems. Therefore, exploring the behavior of these time-delayed systems is important from the engineering perspective (Xia, 2009; Kwon, 2011).

In the literature, time-delayed systems have investigated with the general structure as shown in Figure 1 (Buscarino, 2011). It consists of three blocks: a nonlinearity, a integrator and a time-delay block. Since the first publications on time-delayed systems, much effort has been deployed towards the development of nonlinearities and delay bock that can support and sustain efficient and cost-effective time-delayed chaotic systems. Obviously, the nonlinear block has played a very important role in making diverse

Figure 1. General structure of the time-delayed system

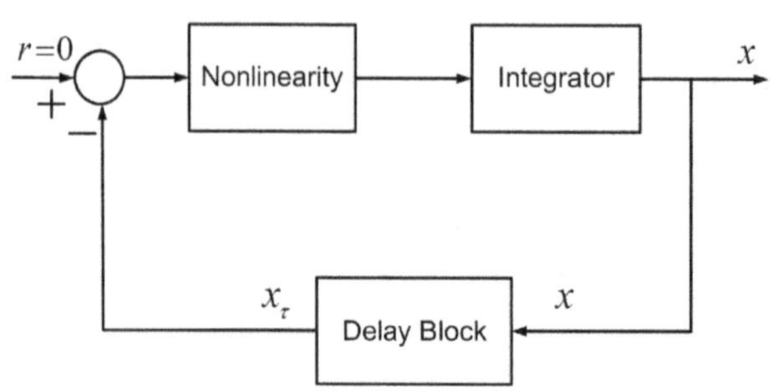

features of designed systems (Namajunas, 1995; Tamasevicius, 2006; Duan, 2009; Busacarino, 2011). Typical reported nonlinear functions, which were used as nonlinearity blocks in literature, and their implementations are summarized here.

Mackey-Glass model (Mackey, 1977) for the production of white blood cells was given by the following function:

$$f(x) = \frac{Ax}{1 + x^{10}}, \tag{1}$$

where $A = 10$. The function (1) was produced easily by coupling in a special way two complementary junction field-effect transistors (Namajunas, 1995). Ikeka proposed a model of optical resonator with the sine function:

$$f(x) = A\sin(x), \tag{2}$$

with $A = -20$ (Ikeda, 1987). A special integrated circuit AD639, universal trigonometric function converter, could be used to implement this function. More details can be found in (Tang, 2001). Voss model (Voss, 2002) used a single-humped smooth nonlinearity:

$$f(x) = A_1x^3 + A_2x^2 + A_3x + A_4, \tag{3}$$

with $A_1 = -10.44$, $A_2 = -13.95$, $A_3 = -3.63$, and $A_4 = 0.85$. The nonlinearity (3) was built in essence from a transistor and four resistors (Horbelt, 2002). In Ucar's works (Ucar, 2002; 2003), the author developed a model containing cubic nonlinearity:

$$f(x) = Ax + Bx^3, \tag{4}$$

with $A = 1$ and $B = 1$. Obviously, the equation (4) is simpler than the equation (3). Two different methods (Mishra, 2006) were tested in order to design the cubic nonlinearity, using log-antilog operational amplifiers and analog multipliers. Banerjee defined a nonlinearity with the function given by:

$$f(x) = 0.5A(|x| + x) + B\tanh(Cx), \tag{5}$$

where $A = 1.15$, $B = 0.97$, and $C = 2.19$ (Banerjee, 2012). This function was realized easily by a half-wave rectifier and an operational amplifier. In the neural model of Duan (Duan, 2009), a non-monotonously increasing transfer function was employed as activation function:

$$f(x) = \sum_{i=1}^{2} A_i[\tanh(x + B_i) - \tanh(x - B_i)], \tag{6}$$

where $A_1 = A_2 = 1.5$, $B_1 = 1$, and $B_2 = 4/3$. The active function was implemented as a linear combinations of four translated hyperbolic tangent functions obtained by dual-transistor pairs. Lu and He introduced an odd piecewise linear function as:

$$f(x) = \begin{cases} 0 & x \leq -\frac{4}{3} \\ Ax - 2 & -\frac{4}{3} < x < -0.8 \\ Bx & -0.8 < x \leq 0.8 \\ Ax + 2 & 0.8 < x \leq \frac{4}{3} \\ 0 & x > \frac{4}{3}, \end{cases} \tag{7}$$

with A = -1.5 and B = 1 (Lu, 1996). Similar to Mackey-Glass nonlinear function, it saturates to zero at large $|x|$. Another three-segment nonlinear function, which was obtained from two diodes, and an operational amplifier, has been studied (Tamasevicius, 2006). It was described as:

$$f(x) = \begin{cases} B(x+1) + A & x < -1 \\ Ax & -1 \leq x \leq 1 \\ B(x-1) + A & x > 1, \end{cases} \tag{8}$$

with A = 2.15 and B = -4.3. In contrast to the previous function (7), it does not saturate to zero at large $|x|$. Furthermore, a piecewise linear approximation of the Ikeda model was also experimental studied (Buscarino, 2011), described by the following equation:

$$f(x) = \begin{cases} \frac{2}{\pi}x & |x| < \frac{\pi}{2} \\ -\frac{2}{\pi}x - 2 & x < -\frac{\pi}{2} \\ -\frac{2}{\pi}x + 2 & x > \frac{\pi}{2} \end{cases} \tag{9}$$

This nonlinearity was constituted by two operational amplifiers and six resistors (Buscarino, 2011).

In some electronic circuits delay blocks are intentionally introduced to implement specific functions. For instance, in measurement applications, the highly integrated time-multiplexing device (Bocci, 2005), the particle detector (Seferiadis, 2006), and the time analog-to-digital converter (Parayandeh, 2008) make use of time-delay blocks. In radio frequency applications, in high-speed microprocessors and in memories, delay circuits play a vital role in delay-locked loop (Chang, 2005) where they allow to improve the performance of the integrated circuits. In the reported low-frequency time-delay chaotic circuits, the typical values of the delay were relatively large (in the order of magnitude of milliseconds). Therefore, time-delay blocks were implemented by delay lines or analog circuits.

A bucket brigade delay-line device (MN3011 integrated circuit with 3328 stages) was employed [Voss, 2002; Horbelt, 2002; Le, 2012] as illustrated in Figure 2a. The delay time may be varied by changing the clock frequency which is provided by a function generator or a special integrated circuit MN3101 that is ideally suited for driving MN3011. In fact, due to the fact that the voltages of the chaotic circuits are not within the input range of MN3011, it required additional subparts to adapt voltages. Furthermore, as the delay-line has a high frequency switch operation, a low-pass filter is required to remove the high frequency noise from the delayed voltage. Similarly, a lossless transmission line was used as a delay line in the another design [Hoang, 2009]. The schematic for lossless transmission line is displayed in Figure 2b.

A delay block consisting of T-type LCL filters (see Figure 2c) was often used. This approach was introduced firstly by Namajunas with matching resister at the input and output [Namajunas, 1995]. The delay can be change by increasing or decreasing the number of LCL filters. The total delay can be calculated approximately as:

$$T_{delay} = n\sqrt{2LC} \tag{10}$$

if n filters are utilized. Because the delay unit is frequency dependent, the cutoff frequency is only 3 kHz. The tuning delay block can be implemented by first order all pass filter [Banerjee, 2012] as displayed in Figure 2d. The transfer function of an all pass filter has the following form:

$$H(s) = -a_1 \frac{s - \omega_0}{s + \omega_0} \tag{11}$$

where the flat gain $a_1 = 1$ is determined by R_1, R_2 and the frequency at which the phase shift is $\frac{\pi}{2}$ is calculated as $\omega_0 = \frac{1}{RC}$. In this case, the output power level does not depend upon input signal frequency; however, the phase depends on the input signal frequency. Because each all pass filter contributes with a delay of value equal to RC, the total delay of the delay block is:

$$T_{delay} = nRC \tag{12}$$

where n is the number of all pass filters. It is clear to see that by changing the value of resistor R, the amount of delay can vary, as the results the resolution of the delay block can be controlled easily.

It is possible to implement the time-delay by using a cascade of n low-pass second-order Bessel filters [Buscarino, 2011]. Each filter is characterized by the Sallen-Key topology [Sedra, 2003] as shown in Figure 2e and by the following transfer function:

$$H(s) = \frac{1}{1 + C_1(R_1 + R_2)s + C_1C_2R_1R_2s^2} \tag{13}$$

Figure 2. Realization of the delay block by using: (a) bucket brigade delay-line device, (b) lossless transmission line, (c) T-type LCL filters, (d) first order all pass filter, (e) low-pass second-order Bessel filter

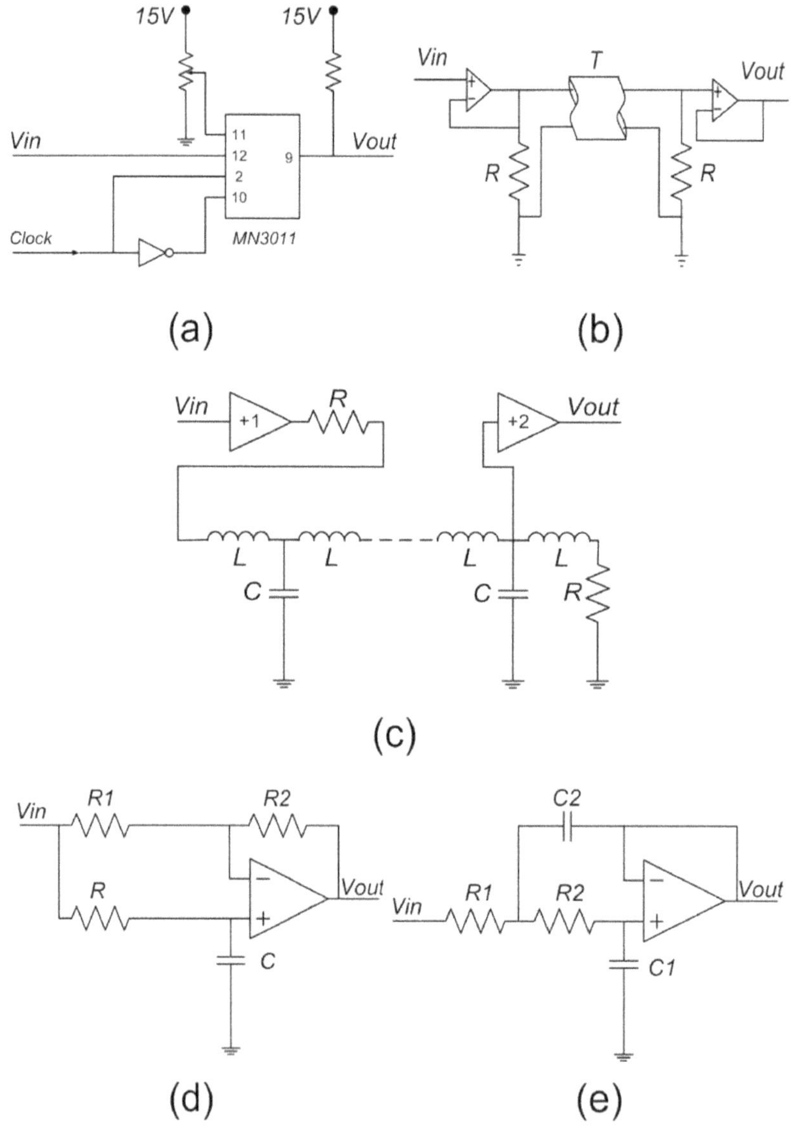

The values of the filter components have been chosen in order to realize a Bessel filter with 3 dB frequency equal to $f_C \simeq 1$ kHz and taking into account off-the-shelf component values. The time-delay introduced by this filter in the band up to f_C can be calculated as $\tau_i = -\dfrac{d\Phi(\omega)}{d\omega}$. In the 3 dB band, $\tau_i \simeq C_1(R_1 + R_2)$. For the values of the components $R_1 = R_2 = 10k\Omega$, $C_1 = 10nF$, and $C_2 = 22nF$ we have $\tau \simeq 0.2ms$. Larger delays are realized by taking into account a cascade of *n* filters.

The approximation of an ideal delay through a cascade of multiple second-order filters or all pass filters allowed to design an efficient and simple circuitry which avoids the drawback of usual delay devices, like delay lines and LCL T-type filters. Furthermore, the filter realization with operational amplifiers, resistors and capacitors is more compact compared to LCL filters as inductors are bulky for large time-delays, although power dissipation is larger. Another approach to this end is to realize digital time delay block (Pham, 2012). In the perspective of this digital approach there are an additional analog-to-digital converter and an additional digital-to-analog converter which interface the time-delay digital block with the remaining parts. The analog signal is transformed into digital one by the analog-to-digital converter. Then, the digital signal from the analog-to-digital converter is delayed in the time-delay digital block before being fed through the digital-to-analog converter to the remaining part of the circuit. Finally, a delayed analog signal is achieved at the output of the digital-to-analog converter. The total delay can be approximate as the delay of the digital time-delay block. However, the conversion process affects the accuracy of the implementation. Hence the parameters of the conversion processes (sampling rate and the number of bits) have to be considered carefully.

3. SYSTEM DESCRIPTION

In this work, a new time-delayed system is considered, which is described by the following equation:

$$\frac{dx}{dt} = ax_{\tau} + bf(x_{\tau}), \tag{14}$$

where a, b are parameters, $x \in \mathbb{R}$ is the state variables, $\tau \in \mathbb{R}^{+}$ is the time-delay, and $x_{\tau} = x\left(t - \tau\right)$. Here the saturation function, which has extensively studied in [Volos, 2011; 2012; 2013], is used. The mathematical form of the saturation function is defined by:

$$f(x_{\tau}) = \begin{cases} 1 & x_{\tau} > \frac{1}{k} \\ kx_{\tau} & -\frac{1}{k} \leq x_{\tau} \leq \frac{1}{k} \\ -1 & x_{\tau} < -\frac{1}{k}. \end{cases} \tag{15}$$

Hence, the saturation plateaus are ± 1 and the slope of the intermediate linear region is k. The nature of the nonlinearity of $f(x_{\tau})$ for different values of k is illustrated in Figure 3.

4. DYNAMICS OF THE TIME-DELAYED SYSTEM WITH A SATURATION FUNCTION

The new time-delayed system (14) is considered for the following values of parameters a = -1, b = 1, $\tau = 2$, and $k \in [2,6]$.

The bifurcation diagram of the system (14) versus parameter k is presented on Figure 4. Clearly, Figure 4 indicates the well-known phenomenon of period doubling route to chaos. When increasing the

Figure 3. The saturation function (15) when changing the value of k

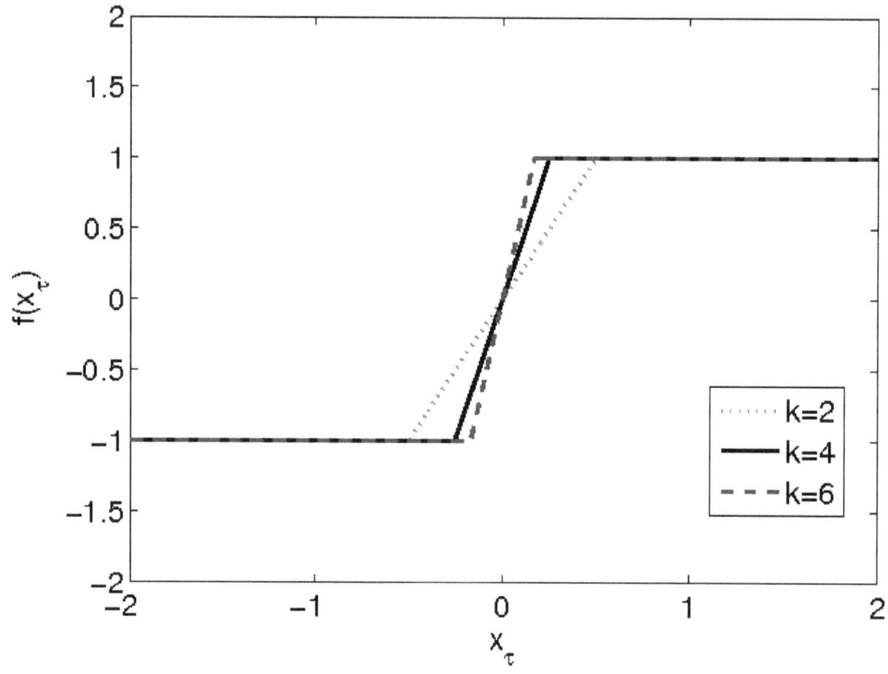

values of the parameter k, the system (14) exhibits the period-1, period-2, period-4 limit cycles (Figures 5a, b, c, respectively), and double-scroll chaotic attractor (Figure 5d).

Lyapunov exponents measure the exponential rates of divergence and convergence of nearby trajectories in state space (Strogatz, 1994; Hilborn, 2000). Time-delayed systems are infinitive-dimensional systems, hence the critical test for chaos in such systems is calculation of the largest Lyapunov exponent. As it is well known, the positive largest Lyapunov exponent indicates chaos . In order to calculate the maximum Lyapunov exponent of the system, the mentioned procedure in (Sprott, 2007) has been applied. The system (14) is approximated in terms of (N+1) ordinary differential equations as follows:

$$\begin{aligned} \dot{x}_0 &= ax_N + bf(x_N) \\ \dot{x}_1 &= \frac{N(x_0 - x_1)}{\tau} \\ &\vdots \\ \dot{x}_N &= \frac{N(x_{N-1} - x_N)}{\tau} \end{aligned} \tag{16}$$

where $x_0, x_1, \dots, x_N \in \mathbb{R}^n$ and $N \to \infty$. In this work, the value of N is fixed ($N = 100$). In other word, the proposed time-delay system (14) is described by (N+1)-dimensional ordinary differential equation:

$$\dot{\mathbf{X}} = \mathbf{F}(\mathbf{X}) \tag{17}$$

Figure 4. Numerical bifurcation diagram of x with k as the control parameter. Other parameters are a = -1, b = 1 and τ = 2

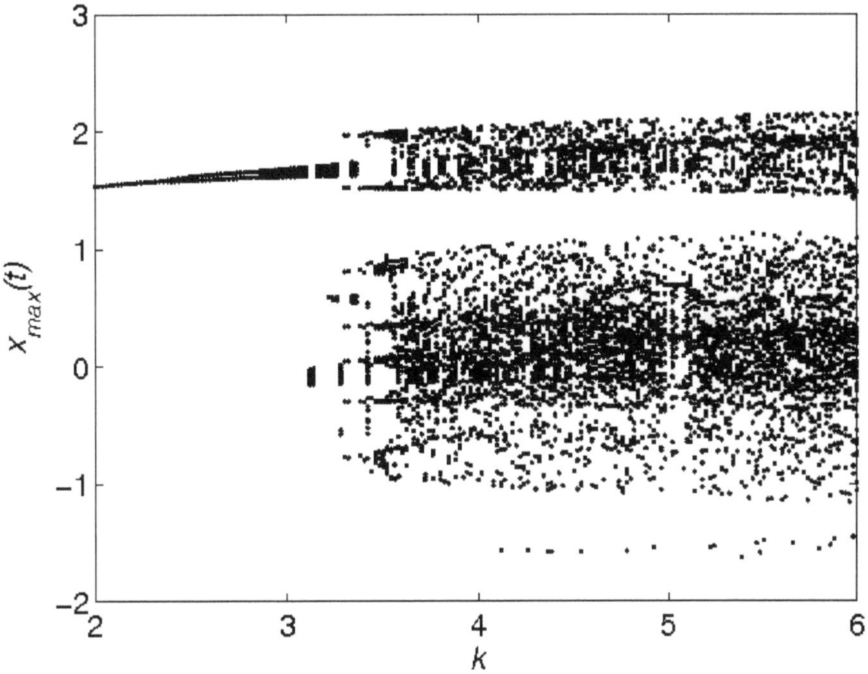

where $\mathbf{X} = (x_0, x_1, \ldots, x_N)$. As a result, the Wolf's algorithm [Wolf, 1985] can be applied to compute the Lyapunov exponents of the flow of a dynamical system (17). The obtained maximum Lyapunov exponent $\lambda_{\max} \approx 0.1187$, confirms further that the system (14) behaves chaotically when $k = 4$.

Here the main steps of the Wolf's algorithm (Woft, 1985; Lakshmanan, 2011) are summarized briefly in order to support the readers, who would like to repeat the calculations by themselves, as follows. Firstly, two trajectories in the (*N*+1) dimensional phase space which start from two nearly initial condition $\mathbf{X}_{ic}$ and $\mathbf{X}'_{ic}$ are considered. It is noting that $\mathbf{X}'_{ic}$ equals $\mathbf{X}_{ic} + \delta\mathbf{X}_{ic}$. Secondary, the distance between these two trajectories is measured as:

$$L(\mathbf{X}_{ic}, t) = \| \delta\mathbf{X}(\mathbf{X}_{ic}, t) \| = \sqrt{\delta x_0^2 + \delta x_1^2 + \ldots + \delta x_N^2} \tag{18}$$

Thirdly, the Jacobian matrix of **F** is found by using the following equation:

$$J = \frac{\partial \mathbf{F}}{\partial \mathbf{X}}\bigg|_{\mathbf{X}=\mathbf{X}_{ic}} \tag{19}$$

As the result the linearized version of the system (17) is given as:

$$\delta\dot{\mathbf{X}} = J(\mathbf{X})\delta\mathbf{X} \tag{20}$$

Figure 5. Behavior of the system (14) in the phase plane $x(t) - x(t - \tau)$ *when a = -1, b = 1, τ = 2 and (a) k = 2, (b) k = 2.8, (c) k = 2.9, (d) k = 4*

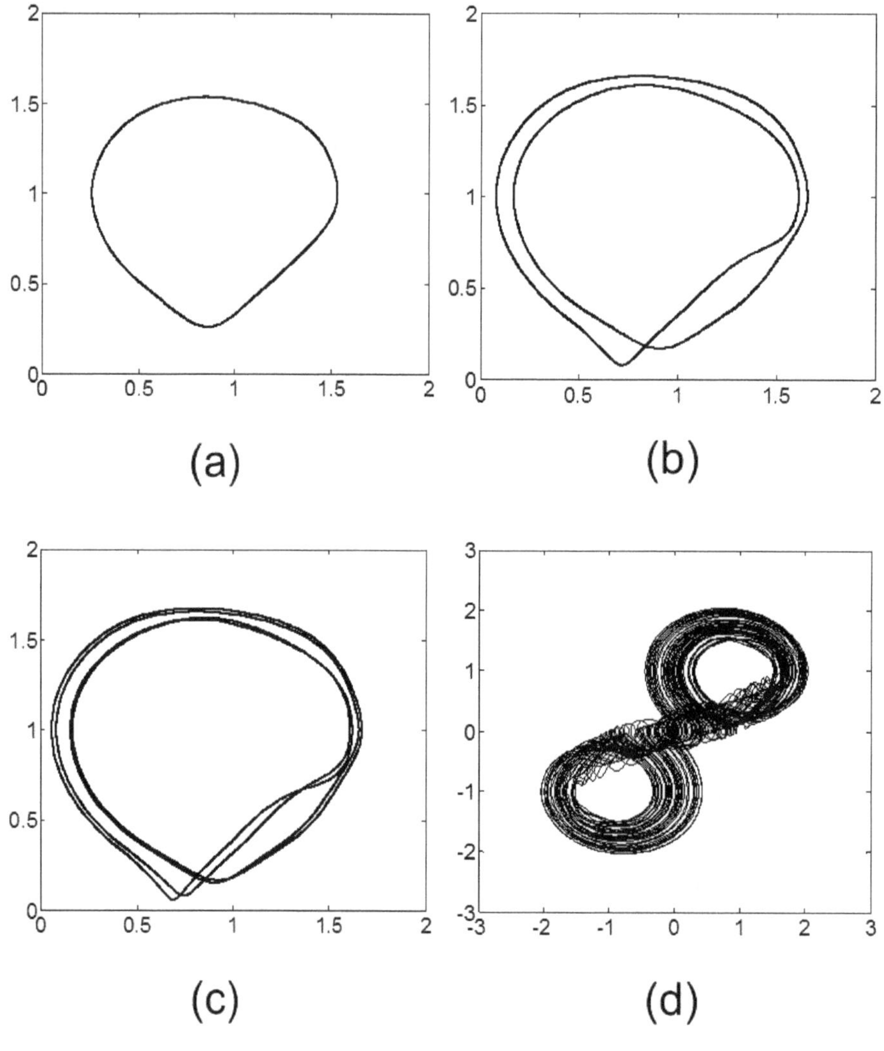

Fourthly, the average rate of divergence of two close trajectories $\mathbf{X}$ and $\mathbf{X}'$ can be written as:

$$\lambda(\mathbf{X}_{ic}, \delta\mathbf{X}) = \lim_{t\to\infty} \frac{1}{t} \log\left(\frac{L(\mathbf{X}_{ic}, t)}{L(\mathbf{X}_{ic}, 0)}\right) \tag{21}$$

Then orthonormal vectors of $\delta\mathbf{X}$ are found as:

$$\delta\dot{\mathbf{e}}_i = J(\mathbf{X}_{ic})\mathbf{e}_i \tag{22}$$

where $J = \text{diag}(\lambda_0, \lambda_1, \ldots, \lambda_N)$.

Finally, Lyapunov exponents are estimated as:

$$\lambda_i(\mathbf{X}_{ic}) = \lambda_i(\mathbf{X}_{ic}, \mathbf{e}_i) \tag{23}$$

with $\lambda_0 \geq \lambda_1 \geq \ldots \geq \lambda_N$. Hence, $\lambda_{\max} = \lambda_1$ is the maximum Lyapunov exponent.

5. CIRCUITRY IMPLEMENTATION OF THE NEW PROPOSED CHAOTIC SYSTEM

The mathematical model (14) can be electronically designed in a circuit, as shown in Figure 6. The circuit is implemented using the general approach based on operational amplifiers (Fortuna, 2009), in which the state variable x of the model is the voltages across the capacitor *C*. Selected components are summarized in Table 1.

TL084 operational amplifiers and a dual voltage supply ($\pm 15\text{V}$) have been used. The operational amplifier U_1 acts as an integrator. The saturation function $f(x_T)$, which is depicted within dashed lines, is built around two operational amplifiers (U_2, U_3) and four resistors. The nonlinear voltage expression of this function is given by:

Figure 6. Circuit schematic of the new chaotic system (14) using the following passive component values: $R = 10k\Omega$, $R_1 = 1k\Omega$, $R_2 = 14.3k\Omega$, $C = 10nF$, *and the value of* R_3 *is made adjustable*

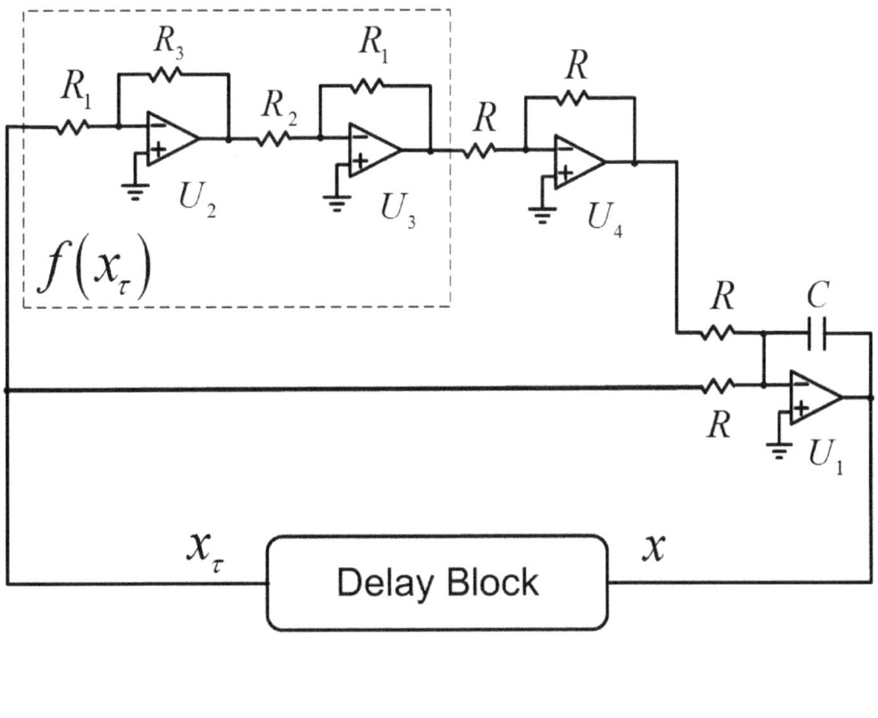

Table 1. Values of selected circuital components

Circuit's Components	Selected Values
R	$10k\Omega$
R_1	$1k\Omega$
R_2	$14.3k\Omega$
R_3	$100k\Omega$
C	$10nF$

$$f\left(V_C\left(t-T_{delay}\right)\right)=\begin{cases}1 & V_C\left(t-T_{delay}\right)>\frac{R_2}{R_3}\cdot 1V\\ \frac{R_3}{R_2}V_C\left(t-T_{delay}\right) & -\frac{R_2}{R_3}\cdot 1V\leq V_C\left(t-T_{delay}\right)\leq\frac{R_2}{R_3}\cdot 1V\\ -1 & V_C\left(t-T_{delay}\right)<-\frac{R_2}{R_3}\cdot 1V.\end{cases} \tag{24}$$

So the slope of the intermediate linear region $k = R_3/R_2$ can be changed when adjusting the value of the $100k\Omega$ trimmer R_3. The circuit equation will be described as:

$$\frac{dV_C\left(t\right)}{dt}=-\frac{1}{RC}V_C\left(t-T_{delay}\right)+\frac{1}{RC}f\left(V_C\left(t-T_{delay}\right)\right), \tag{25}$$

where $T_{delay} = 0.2ms$ is the time-delay of the delay block. In the design and simulation, a lossless transmission line is utilized as a delay block (Hoang, 2009). In practice the delay block can be implemented by either a network of T-type LCL filters (Namajunas, 1995) or a series of Bessel filters in cascade (Buscarino, 2011). The schematic for the delay block was presented in Figure 2b. As a result, the dimensionless delay τ is calculated as:

$$\tau=\frac{T_{delay}}{RC}=2. \tag{26}$$

It can be seen that the circuit equation (25) is equivalent to the model equation (14).

The observations of the designed circuit are obtained by Multisim as shown in Figure 7. Apparently, a good agreement between the numerical simulations (Figure 5) and the Multisim results (Figure 7) is observed. The designed circuit shows a period doubling to chaos and double-scroll chaotic behavior with the change of the value of R_3.

Figure 7. Phase-plane plots in the $V_C(t) - V_C(t - T_{delay})$ *space obtained by using the oscilloscope of Multisim (a) period-1 at* $R_3 = 28.6k\Omega$, *(b) period-2 at* $R_3 = 36.3k\Omega$, *(c) period-4 at* $R_3 = 37.1k\Omega$, *and (d) double-scroll chaos at* $R_3 = 57.2k\Omega$.

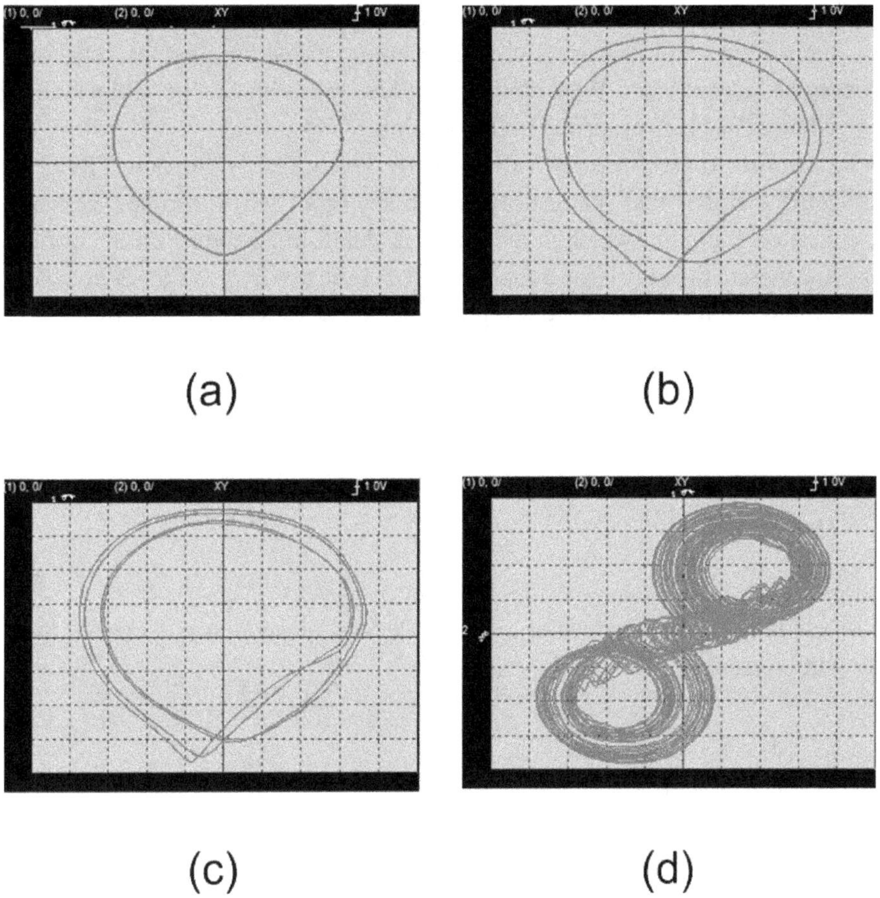

6. DISCUSSION

Random bit generator plays an important role in various areas including computer simulations, games, measurements, evolutionary algorithms and especially cryptography (Shu, 1995; Stojanovski, 2001; Ferguson, 2003). In general, there are three main categories of random bit generator: the true random bit generator, pseudo random bit generator and hybrid random generator. Among them, true random bit generator has received significant attention because it bases on natural source of randomness. It is interesting that a true random bit generator (TRBG) can be constructed from a double-scroll attractor (Yalcin, 2004). The author introduced a straightforward method for creating true random bits and took care of randomness in the sense of unpredictability. Thus the designed TRBG successfully passed all statistical tests of the FIPS-140-1 and Diehard test suite. By applying this effective approach, another TRBG for fingerprint images encryption was represented (Volos, 2010). The high security of the encryption scheme was demonstrated through the security analysis of the encrypted images. In addition, recently Volos has been implemented a TRBG for path planning purpose by using a double-scroll chaotic circuit, in which the nonlinear function is a saturation function (Volos, 2012). The established chaotic path planning generator is useful for autonomous mobile robots. A very essential feature raised from the obtained results is that the scanning rate of the terrain is not only the same but also a bit faster than ones in other similar works (Martins-Filho, 2007). As a result, this TRBG may have diverse potential applications in many social and military fields, like automatic floor-cleaning machines, or military autonomous robots for surveillance. Motivated by previous researches, a possible TRBG based on the novel time-delayed model (14) is proposed in this section.

Similar to the mentioned architecture (Yalcin, 2004; Volos, 2012), the new TRBG is presented in Figure 8 . The TRBG includes four major blocks. The key block (S_1) is the new chaotic time-delayed circuit with a saturation function, which is discussed in detail in Section 5. It is noting that the time-delayed circuit can generate the necessary double-scroll chaotic attractors. The second block (S_2) divides the state variable x into three subspaces by using two threshold functions as follows

$$S_2: \quad \sigma^1(x) = \begin{cases} 0 & x < th_1 \\ 1 & x \geq th_1 \end{cases}$$

$$\sigma^0(x) = \begin{cases} 0 & x > th_2 \\ 1 & x \leq th_2 \end{cases}$$

where th_1 and th_2 are the threshold values of the variable x. A bit sequence is created in the third block (S_3) by the following formula

$$S_3: \quad \sigma_i(\sigma^0, \sigma^1) = \begin{cases} 0 & \sigma^0 = 0, \sigma^1 :_0\uparrow^1 \\ 1 & \sigma^1 = 0, \sigma^0 :_0\uparrow^1 \end{cases}$$

Figure 8. Block diagram of the chaotic true random bit generator based on the new time-delay circuit with a saturation function

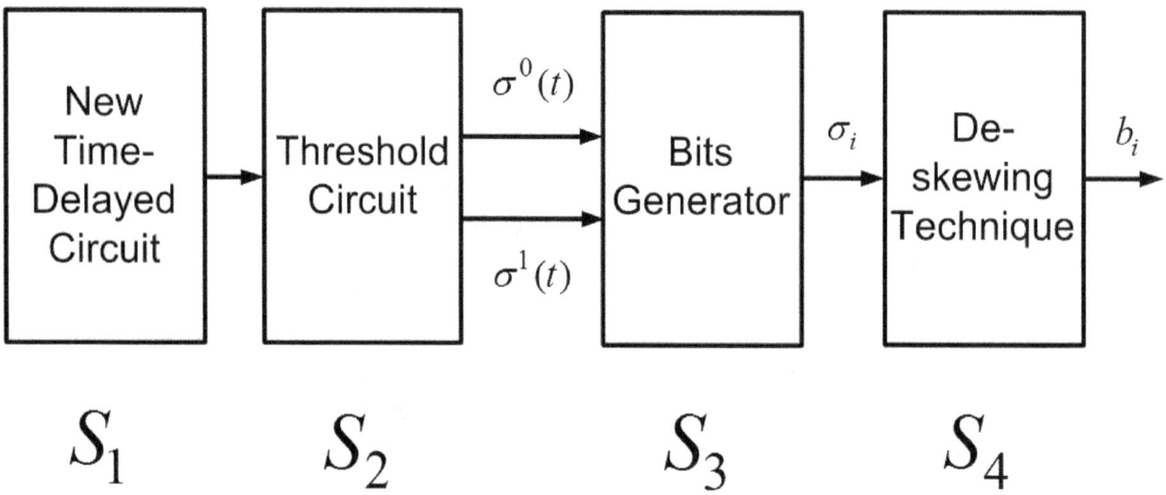

where $\sigma^0 : {}_0\uparrow^1$ and $\sigma^1 : {}_0\uparrow^1$ show the transition from logic "0" to logic "1" of σ^0 and σ^1, respectively. The Von Neumann's technique is implemented in the last block (S_4). As it has already been known that this Von Neumann's technique can produce a balance between bits "0" and bits "1" (Neumann, 1951; Yalcin, 2004). In other word, this digital post-processing converts the bit pairs "01" and "10" into the corresponding output bits "0" and "1", while discards the bit pairs "00" and "11". Thus the distribution of bits is balanced.

7. CONCLUSION

In this work, a novel time-delayed chaotic system based on a saturation function is presented. Numerical investigation demonstrates that the system exhibits a period doubling route to chaos and double scroll chaotic attractor when varying the slope of the intermediate linear region, the parameter k. In addition, discussion and details on calculating the maximum Lyapunov exponent of the novel system with delay are provided to support the readers. In order to confirm the proposed system, its analog electronic circuit is also implemented. The circuit employs simple electronic elements such as resistors, a capacitor and operational amplifiers. It is thus clear that this time-delayed chaotic system can be implemented in integrated circuits.

Although the fact that a possible TRBG based on the novel time-delayed model is proposed, it must be tested by popular suites of statistical tests such as FIPS-140-2 (NIST, 2001), Diehard (Marsaglia, 1995), Donald Knuth (Knuth, 1998) or Crypt-XS (Gustafson, 1994). Furthermore the experimental realization of such TRBG will be considered as a future work.

Synchronization of chaos refers to a process in which chaotic systems adjust a given property of their motion to a common behavior (Boccaletti, 2002; Sundarapandian, 2012). Chaotic synchronization plays an important role in chaos-based communication where the transmitted information signal can be retrieved if the synchronization between chaotic transmitter and chaotic receiver occurs (Pecora, 1990). Thus the synchronization of the new time-delay system with a saturation function will be taken as next works in order to apply it to practical applications such as secure communications.

ACKNOWLEDGMENT

This research is funded by Vietnam National Foundation for Science and Technology Development (NAFOSTED) under grant number 102.99-2013.06.

REFERENCES

Alvarez, G., & Li, S. (2006). Some basic cryptographic requirements for chaos-based cryptosystems. *International Journal of Bifurcation and Chaos in Applied Sciences and Engineering*, *16*(08), 2129–2151. doi:10.1142/S0218127406015970

Anishchenko, V. S., Neiman, A. B., & Chua, L. O. (1993). *A paradigm for chaos*. Singapore: World Scientific.

Balachandran, B., Kalmar-Nagy, T., & Gilsinn, D. E. (2009). *Delay Differential Equations: Recent Advanced and New Directions*. New York: Springer.

Banerjee, S., Mitra, M., & Rondoni, L. (2011). *Applications of chaos and nonlinear dynamics in engineering – Part 1*. Berlin: Springer-Verlag. doi:10.1007/978-3-642-21922-1

Banerjee, T., Biswas, D., & Sarkar, B. C. (2012). Design and analysis of a first order time-delayed chaotic system. *Nonlinear Dynamics*, *70*(1), 721–734. doi:10.1007/s11071-012-0490-3

Bendixson, I. (1901). Sur les courbes definies par des equations differentielles. *Acta Mathematica*, *24*(1), 1–88. doi:10.1007/BF02403068

Boccaletti, S., Kurths, J., Osipov, G., Valladares, D. L., & Zhou, C. S. (2002). The synchronization of chaotic systems. *Physical Review A*, *64*, 821–824. PMID:12443117

Bocci, V., Chiodi, G., Iacoangeli, F., Nobrege, R., Pinci, D., & Rinaldi, W. (2005). Time-multiplexing of signal using highly integrated digital delay: an FPGA implementation. In *Proc. IEEE Nuclear Science Symposium Conference Record*, (pp. 398-402). IEEE. doi:10.1109/NSSMIC.2005.1596279

Buscarino, A., Fortuna, L., Frasca, M., & Sciuto, G. (2011). Design of time-delay chaotic electronic circuits. *IEEE Transactions on Circuits and Systems I, Regular Papers*, *58*(8), 1888–1896. doi:10.1109/TCSI.2011.2107190

Buscarino, A., Fortuna, L., Frasca, M., Sciuto, G., & Xibilia, M. G. (2011) Harmonic balance method for time-delay chaotic systems design. In *Proc. World Congress of the International Federation of Automatic Control (IFAC 2011)*, (pp. 5112-5117). Milan, Italy: IFAC.

Chang, H.-H., & Liu, S.-I. (2005). A wide-range and fast-locking all digital cycle-controlled delay-locked loop. *IEEE Journal of Solid-State Circuits*, *40*(3), 661–670. doi:10.1109/JSSC.2005.843596

Cuomo, K. M., & Oppenheim, A. V. (1993). Circuit implementation of synchronized chaos with applications to communications. *Physical Review Letters*, *71*(1), 65–68. doi:10.1103/PhysRevLett.71.65 PMID:10054374

Davies, B. (2004). *Exploring Chaos: Theory and experiment*. Oxford, UK: Westview Press – Perseus Books Group.

Driver, R. D. (1977). *Ordinary and Delay Differential Equations*. New York: Springer Verlag. doi:10.1007/978-1-4684-9467-9

Duan, S., & Wang, L. (2009). A novel delayed chaotic neural model and its circuitry implementation. *Computers and Mathematics with Application*, *57*(11-12), 1736–1742. doi:10.1016/j.camwa.2008.10.017

Farmer, J. D. (1982). Chaotic attractor of an infinite-dimensional dynamical system. *Physica D Nonlinear Phenomena*, *4*(3), 366–393. doi:10.1016/0167-2789(82)90042-2

Ferguson, N., & Schneier, B. (2003). *Practical cryptography*. Wiley.

Fortuna, L., Frasca, M., & Xibilia, M. G. (2009). *Chua's Circuit Implementations: Yesterday, Today and Tomorrow*. Singapore: World Scientific.

Grebogi, C., & Yorke, J. (1997). *The impact of chaos on science and society*. United Nations University Press.

Gustafson, H., Dawson, H. E., Nielsen, L., & Caelli, W. (1994). A computer package for measuring the strength of encryption algorithms. *Journal of Computer Security*, *13*(8), 687–697. doi:10.1016/0167-4048(94)90051-5

Hasselblatt, B., & Katok, A. (2003). *A first course in dynamics: With a panorama of recent developments*. Cambridge, UK: Cambridge University Press. doi:10.1017/CBO9780511998188

Hilborn, R. C. (2000). *Chaos and nonlinear dynamics: An introduction for scientists and engineers*. Oxford, UK: Oxford University Press. doi:10.1093/acprof:oso/9780198507239.001.0001

Hoang, T.-M., Nguyen, T. D., Nguyen, V. D., Chedjou, J. C., & Kyamakya, K. (2009) Design and simulation of circuit for synchronization of multidelay feeback systems. In *Proc. 25th International Symposium on Theoretical Engineering (ISTET 2009)*, (pp. 1-4). Lubeck, Germany: ISTET.

Holland, J. H. (1998). *Emergence: From chaos to order*. Oxford, UK: Oxford University Press.

Horbelt, W., Timmer, J., & Voss, H. U. (2002). Parameter estimation in nolinear delayed feedback systems from noisy data. *Physics Letters [Part A]*, *299*(5-6), 513–521. doi:10.1016/S0375-9601(02)00748-X

Hu, H. Y., & Wang, Z. H. (2002). *Dynamics of Controlled Mechanical Systems with Delayed Feedback*. Berlin: Springer-Verlag. doi:10.1007/978-3-662-05030-9

Ikeda, K., Daido, H., & Akimoto, O. (1980). Optical turbulence: Chaotic behavior of transmitted light from a ring cavity. *Physical Review Letters*, *45*(9), 709–712. doi:10.1103/PhysRevLett.45.709

Ikeda, K., & Matsumoto, K. (1987). High-dimensional chaotic behavior in systems with time-delayed feedback. *Physica D. Nonlinear Phenomena*, *29*(1-2), 223–235. doi:10.1016/0167-2789(87)90058-3

Knuth, D. (1998). *The art of computer programming: Semiemperical algorithms.* Addison Wesley.

Koc, C. K. (2009). *Cryptographic Engineering*. New York: Springer. doi:10.1007/978-0-387-71817-0

Kocarev, L. (2001). Chaos-based cryptography: A brief overview. *IEEE Circuits and Systems Magazine*, *1*(3), 6–21. doi:10.1109/7384.963463

Kocarev, L., & Lian, S. (2011). *Chaos-Based Cryptography. Theory, Algorithms and Applications.* Berlin: Springer-Verlag.

Kuang, Y. (1993). *Delay Differential Equations: With Applications in Population Dynamics*. New York: Academic Press.

Kwon, O. M., Park, J. H., & Lee, S. M. (2011). Secure communication based on chaotic synchronization via interval time varying delay feedback control. *Nonlinear Dynamics*, *63*(1-2), 239–252. doi:10.1007/s11071-010-9800-9

Lakshmanan, M., & Senthikumar, D. V. (2011). *Dynamics of Nonlinear Time-Delay Systems*. Berlin: Springer. doi:10.1007/978-3-642-14938-2

Le, L. B., Konishi, K., & Hara, N. (2012). Design and experimental verification of multiple delay feedback control for time-delay nonlinear oscillators. *Nonlinear Dynamics*, *67*(2), 1407–1418. doi:10.1007/s11071-011-0077-4

Lorenz, E. N. (1963). Deterministic nonperiodic flow. *Journal of the Atmospheric Sciences*, *29*(2), 130–141. doi:10.1175/1520-0469(1963)020<0130:DNF>2.0.CO;2

Lu, H., & He, Z. (1996). Chaotic behavior in first–order autonomous continuous-time systems with delay. *IEEE Transactions on Circuits and Systems. I, Fundamental Theory and Applications*, *43*(8), 700–702. doi:10.1109/81.526689

Mackey, M. C., & Glass, L. (1977). Oscillation and chaos in physiological control system. *Science*, *197*(4300), 287–289. doi:10.1126/science.267326 PMID:267326

Marsaglia, G. (1995). *DIEHARD Statistical Tests*. Retrieved from http://stst.fsu.edu/pub/diehard

Martins-Filho, L. S., & Macau, E. E. N. (2007). Patrol mobile robots and chaotic trajectories. Mathematical Problems in Engineering, Article ID 61541.

Medio, A., Pireddu, M., & Zanolin, F. (2009). Chaotic dynamics for maps in one and two dimensions: A geometrical method and applications to economics. *International Journal of Bifurcation and Chaos in Applied Sciences and Engineering*, *19*(10), 3283–3309. doi:10.1142/S0218127409024761

Mishra, D., & Ucar, A. (2006). Pspice realization of the Ucar prototype delay model. *International Journal of Nonlinear Sciences and Numerical Simulation*, *7*, 209–214.

Murray, J. D. (1990). *Mathematical Biology*. New York: Springer.

Namajunas, A., Pyragas, K., & Tamasevicius, A. (1995). An electronic analog of the Mackey-Glass system. *Physics Letters [Part A]*, *201*(1), 42–46. doi:10.1016/0375-9601(95)00208-K

Neumann, J. V. (1951). Various techniques used in connection with random digits. In G. E. Forsythe (Ed.), *Applied Mathematical Series-Notes* (Vol. 12, pp. 36–38). National Bureau of Standards.

NIST. (2001). *Security Requirements for Cryptographic Modules* (FIPS PUB 140-2). Retrieved from http://csrc.nist.gov/publications/fips/fips140-2/ fips1402.pdf

Ott, E., Grebogi, C., & Yorke, J. A. (1990). Controlling chaos. *Physical Review Letters*, *170*, 421–428. PMID:10041332

Parayandeh, A., & Prodic, A. (2008). Programmable analog-to-digital converter for low-power DC-DC SMPS. *IEEE Transactions on Power Electronics*, *23*(1), 500–505. doi:10.1109/TPEL.2007.913932

Pecora, L. M., & Caroll, T. L. (1990). Synchronization in chaotic signals. *Physical Review Letters*, *64*(8), 821–824. doi:10.1103/PhysRevLett.64.821 PMID:10042089

Pei, L., Wang, Q., & Shi, H. (2011). Bifurcation dynamics of the modified physiological model of artificial pancreas with insulin secretion delay. *Nonlinear Dynamics*, *63*(3), 417–427. doi:10.1007/s11071-010-9812-5

Pham, V.-T., Fortuna, L., & Frasca, M. (2012). Implementation of chaotic circuits with a digital time-delay block. *Nonlinear Dynamics*, *68*(1), 345–355. doi:10.1007/s11071-011-9982-9

Prokhorov, M. D., & Ponomarenko, V. I. (2008). Encryption and decryption of information in chaotic communication systems governed by delay-differential equations. *Chaos, Solitons, and Fractals*, *35*(5), 871–877. doi:10.1016/j.chaos.2006.05.081

Pyragas, K. (1992). Continuous control of chaos by self-controlling feedback. *Physics Letters. [Part A]*, *170*(6), 421–428. doi:10.1016/0375-9601(92)90745-8

Pyragas, K. (2006). Delayed feedback control of chaos. *Philosophical Transactions of the Royal Society A*, *364*(1846), 2309–2334. doi:10.1098/rsta.2006.1827 PMID:16893790

Safonov, L. A., Tomer, E., Strygin, V. V., Ashkenazy, Y., & Havlin, S. (2002). Multifractal chaotic attractors in a system of delay-differential equations modeling road traffic. *Chaos (Woodbury, N.Y.)*, *12*(4), 1006–1014. doi:10.1063/1.1507903 PMID:12779624

Salarieh, H., & Alasty, A. (2009). Chaos control in an economic model via minimum entropy strategy. *Chaos, Solitons, and Fractals*, *40*(2), 839–847. doi:10.1016/j.chaos.2007.08.045

Sedra, A. S., & Smith, K. C. (2003). *Microelectronic Circuits*. London: Oxford University Press.

Seferiadis, G., Pouchet, M., & Gough, M. P. (2006). FPGA implementation of a delay-line readout system for a particle detector. *Measurement*, *40*(1), 90–99. doi:10.1016/j.measurement.2005.07.003

Seydezadeh, S. M., & Mirzakuchaki, S. (2012). A fast color image encryption algorithm based on coupled two-dimensional piecewise chaotic map. *Signal Processing*, *92*(5), 1202–1215. doi:10.1016/j.sigpro.2011.11.004

Shannon, C. (1949). Communication theory of secrecy systems. *The Bell System Technical Journal*, *28*(4), 656–725. doi:10.1002/j.1538-7305.1949.tb00928.x

Shu, T. (1995). *Uniform random numbers: Theory and practice*. Kluwer Academic Publishers.

Sprott, J. C. (2007). A simple chaotic delay differential equation. *Physics Letters [Part A]*, *366*(4-5), 397–402. doi:10.1016/j.physleta.2007.01.083

Sprott, J. C. (2010). *Elegant Chaos: Algebraically Simple Chaotic Flows*. Singapore: World Scientific.

Stojanovski, T., Pihl, J., & Kocarev, L. (2001). Chaos-based random number generators - Part II: Practical realizations. *IEEE Transactions on Circuits and Systems I, Fundamental Theory and Applications*, *48*(3), 382–385. doi:10.1109/81.915396

Strogatz, S. H. (1994). *Nonlinear Dynamics and Chaos: with Applications to Physics, Biology, Chemistry, and Engineering*, Massachusetts: Perseus Books, Sun, C., Lin, Y., & Han, M. (2006). Stability and Hopf bifurcation for an epidemic disease model with delay. *Chaos, Solitons, and Fractals*, *30*, 204–216.

Sundarapandian, V., & Karthikeyan, R. (2012). Hybrid synchronization of hyperchaotic Lorenz and hyperchaotic Chen systems via active control. *Journal of Engineering and Applied Sciences (Asian Research Publishing Network)*, *7*(3), 254–264. doi:10.3923/jeasci.2012.254.264

Tamasevicius, A., Mykolaitis, G., & Bumeliene, S. (2006). Delayed feedback chaotic oscillator with improved spectral characteristic. *Electronics Letters*, *42*(13), 736–737. doi:10.1049/el:20061245

Tang, K. S., Zhong, G. Q., Chen, G., & Man, K. F. (2001). Generation of n-scroll attractors via sine function. *IEEE Transactions on Circuits and Systems I*, *48*(11), 1369–1372. doi:10.1109/81.964432

Tang, Y., Wang, Z. D., & Fang, J. A. (2010). Image encryption using chaotic coupled map lattices with time-varying delays. *Communications in Nonlinear Science and Numerical Simulation*, *15*(9), 2456–2468. doi:10.1016/j.cnsns.2009.09.023

Tong, X., & Cui, M. (2009). Image encryption scheme based on 3D Baker with dynamical compound chaotic sequence cipher generator. *Signal Processing*, *89*(4), 480–491. doi:10.1016/j.sigpro.2008.09.011

Ucar, A. (2002). A prototype model for chaos studies. *International Journal of Engineering Science*, *40*(3), 251–258. doi:10.1016/S0020-7225(01)00060-X

Ucar, A. (2003). On the chaotic behaviour of a prototype delayed dynamical system. *Chaos, Solitons, and Fractals*, *16*(2), 187–194. doi:10.1016/S0960-0779(02)00160-1

Volos, Ch. K., Kyprianidis, I. M., & Stouboulos, I. N. (2010). Fingerprint images encryption process based on a chaotic true random bits generator. *International Journal of Multimedia Intelligence and Security*, *1*(4), 320–335. doi:10.1504/IJMIS.2010.039234

Volos, C. K., Kyprianidis, I. M., & Stouboulos, I. N. (2011). Various synchronization phenomena in bidirectionally coupled double scroll circuits. *Communications in Nonlinear Science and Numerical Simulation*, *16*(8), 3356–3366. doi:10.1016/j.cnsns.2010.11.015

Volos, C. K., Kyprianidis, I. M., & Stouboulos, I. N. (2012). A chaotic path planning generator for autonomous mobile robots. *Robotics and Autonomous Systems*, *60*(4), 651–656. doi:10.1016/j.robot.2012.01.001

Volos, C. K., Kyprianidis, I. M., & Stouboulos, I. N. (2013). Image encryption process based on chaotic synchronization phenomena. *Signal Processing*, *93*(5), 1328–1340. doi:10.1016/j.sigpro.2012.11.008

Voss, H. U. (2002). Real-time anticipation of chaotic states of an electronic circuit. *International Journal of Bifurcation and Chaos in Applied Sciences and Engineering*, *12*(07), 1619–1625. doi:10.1142/S0218127402005340

Wang, X., Teng, L., & Qin, X. (2012). A novel colour image encryption algorithm based on chaos. *Signal Processing*, *92*(4), 1101–1108. doi:10.1016/j.sigpro.2011.10.023

Wang, X. M., Yu, H. G., Zhong, S. M., & Agarwal, R. P. (2010). Analysis of mathematics and dynamics in a food web system with impulsive perturbations and distributed time delay. *Applied Mathematical Modelling*, *34*(12), 3850–3863. doi:10.1016/j.apm.2010.03.024

Woft, A., Swift, J., Swinney, H., & Vastano, J. (1985). Determining Lyapunov exponents from a time series. *Physica D. Nonlinear Phenomena*, *16*(3), 285–317. doi:10.1016/0167-2789(85)90011-9

Xia, Y., Fu, M., & Shi, P. (2009). Analysis and synthesis of dynamical systems with time-delays. In *Lecture Notes in Control and Information Sciences*. New York: Springer.

Yalcin, M. E., Suykens, J. A. K., & Vandewalle, J. (2004). True random bit generation from a double-scroll attractor. *IEEE Transactions on Circuits and Systems I*, *51*(7), 1395–1404. doi:10.1109/TCSI.2004.830683

Yongzhen, P., Shuping, L., & Changguo, L. (2011). Effect of delay on a predator-prey model with parasitic infection. *Nonlinear Dynamics*, *63*(3), 311–321. doi:10.1007/s11071-010-9805-4

Zhan, M., Wang, X., Gong, X., Wei, G. W., & Lai, C. H. (2003). Complete synchronization and generalized synchronization of one-way coupled time-delay systems. *Physical Review E: Statistical, Nonlinear, and Soft Matter Physics*, *68*(3), 036208. doi:10.1103/PhysRevE.68.036208 PMID:14524869

Zhang, L., Liao, X., & Wang, X. (2005). An image encryption approach based on chaotic maps. *Chaos, Solitons, and Fractals*, *24*(3), 759–765. doi:10.1016/j.chaos.2004.09.035

Zhang, Z., & Chen, G. (2007). Chaotic liquid shaker: Design, implementation and application. *International Journal of Bifurcation and Chaos in Applied Sciences and Engineering*, *17*(12), 4443–4451. doi:10.1142/S0218127407020105

ADDITIONAL READING

Acho, L. (2008). On nonlinear oscillator design. *Journal of Circuits, Systems, and Computers*, *17*(4), 659–662. doi:10.1142/S0218126608004460

Ando, B., & Graziani, S. (2000). *Stochastic Resonance: Theory and Applications*. Norwell: Kluwer. doi:10.1007/978-1-4615-4391-6

Bagini, V., & Bucci, M. (1999). *A design of reliable true random number generator for cryptographic applications*. Cryptographic Hardware and Embedded Systems. *Lecture Notes in Computer Science*, *1717*, 204–218. doi:10.1007/3-540-48059-5_18

Banerjee, S. (2011). *Chaos synchronization and cryptography for secure communications: Applications for encryption*. Hershey: IGI Global.

Banerjee, T., Biswas, D., & Sarkar, B. C. (2012). Design of chaotic and hyperchaotic time-delayed electronic circuit. *Bonfring International Journal of Power Systems and Integrated Circuits*, *2*(3), 13–17. doi:10.9756/BIJPSIC.3138

Carroll, T. L., & Pecora, L. M. (1995). *Nonlinear dynamics in circuits*. Singapore: World Scientific Publishing.

Chen, G., & Ueta, T. (2002). *Chaos in circuits and systems*. Singapore: World Scientific Publishing.

Eichenauer, J., Grothe, H., & Lehn, J. (1988). Marsaglia's lattice test and non-linear congruential pseudo random number generators. *Metrika*, *35*(1), 241–250. doi:10.1007/BF02613312

Eichenauer, J., & Lehn, J. (1986). A non-linear congruential pseudo random number generator. *Statistical Papers*, *27*(1), 315–326. doi:10.1007/BF02932576

Enayatifar, R., Abdullah, A. H., & Isnin, I. F. (2014). Chaos-based image encryption using a hybrid genetic algorithm and a DNA sequence. *Optics and Lasers in Engineering*, *56*, 83–93. doi:10.1016/j.optlaseng.2013.12.003

Fayos, J. (2001). Emergence of order from chaos by cooperative non-biased learning between random bit strings. *Chaos, Solitons, and Fractals*, *12*(5), 865–873. doi:10.1016/S0960-0779(00)00053-9

Feki, M., Robert, B., Gelle, G., & Colas, M. (2003). Secure digital communication using discrete-time chaos synchronization. *Chaos, Solitons, and Fractals*, *18*(4), 881–890. doi:10.1016/S0960-0779(03)00065-1

Fischer, V., & Drutarovský, M. (2003). *True random number generator embedded in reconfigurable hardware*. Cryptographic Hardware and Embedded Systems-CHES 2002. *Lecture Notes in Computer Science*, *2523*, 415–430. doi:10.1007/3-540-36400-5_30

Gammoudi, I., & Feki, M. (2011). A new look at Chua's circuit. *International Journal of Bifurcation and Chaos in Applied Sciences and Engineering*, *21*(06), 1653–1666. doi:10.1142/S0218127411029343

Hassard, B., & Kazarinoff, N. D. (1981). *Theory and Applications of Hopf Bifurcation*. London: Cambridge University Press.

Hauser, J. R. (2009). Numerical methods for nonlinear engineering models. New York: Springer Science+Business Media. doi:10.1007/978-1-4020-9920-5

He, Q., Xu, W., Rong, H., & Fang, T. Stochastic bifurcation in Duffing-Van der Pol oscillators. *Physica A: Statistical Mechanics and its Applications, 338*, 319-334.

Hoang, T. M. (2010). A new secure communication model based on synchronization of coupled multi-delay feedback systems. *International Journal of Electrical Electronics Engineering*, *4*, 240–246.

Hoang, T.M., & Nakagawa, M. (2006). Enhancing security for chaos-based communication system with change in synchronization manifolds' delay and in encoder's parameters. *Journal of Physical Society of Japan, 75*, 064801.1-12.

Hoang, T.M., & Nakagawa, M. (2006). New encoding model for chaos-based secure communication. *Journal of Physical Society of Japan*, **75**, 034801.1-10.

Hoang, T. M., & Nakagawa, M. (2008). A secure communication system using projective-lag and/or projective-anticipating synchronizations of coupled multi-delay feedback systems. *Chaos, Solitons, and Fractals*, *38*(5), 1423–1438. doi:10.1016/j.chaos.2008.02.008

Hu, Y., Liao, X., Wong, K. W., & Zhou, Q. (2009). A true random number generator based on mouse movement and chaotic cryptography. *Chaos, Solitons, and Fractals*, *40*(5), 2286–2293. doi:10.1016/j.chaos.2007.10.022

In, V., Longhini, P., & Palacios, A. (2009). *Applications of nonlinear dynamics: Model and design of complex systems*. Berlin, Heidelberg: Springer-Verlag. doi:10.1007/978-3-540-85632-0

Ivancevic, V. G., & Ivancevic, T. T. (2008). *Complex nonlinearity: Chaos, phase transitions, topology change, and path integrals*. Berlin, Heidelberg: Springer-Verlag.

Jennewein, T., Achleitner, U., Weihs, G., Weinfurter, H., & Zeilinger, A. (2000). A fast and compact quantum random number generator. *The Review of Scientific Instruments*, *71*(4), 1675–1680. doi:10.1063/1.1150518

Jung, C. (1986). Poincaré map for scattering states. *Journal of Physics. A, Mathematical and General*, *19*(8), 1345–1353. doi:10.1088/0305-4470/19/8/016

Kapitaniak, T. (1992). *Chaotic oscillators: Theory and applications*. Singapore: World Scientific Publishing.

Karimi, H., Hosseini, S. M., & Zahan, M. V. (2013). On the combination of self-organized systems to generate pseudo-random numbers. *Information Sciences*, *221*, 371–388. doi:10.1016/j.ins.2012.09.029

Kilic, R. (2010). *A practical guide for studying Chua's circuits*. Singapore: World Scientific Publishing.

Koper, M. T. M. (1995). Bifurcations of mixed-mode oscillations in a three-variable autonomous Van der Pol-Duffing model with a cross-shaped phase diagram. *Physica D. Nonlinear Phenomena*, *80*(1-2), 72–94. doi:10.1016/0167-2789(95)90061-6

Korsch, H. J., Jodl, H.-J., & Hartmann, T. (2008). *Chaos: A program collection for the PC*. Berlin, Heidelberg: Springer-Verlag.

Lakshmanan, M. (1996). *Chaos in nonlinear oscillators: Controlling and synchronization*. Singapore: World Scientific Publishing.

Leonov, G. A., & Kuznetsov, N. V. (2007). Time-varying linearization and the Perron effects. *International Journal of Bifurcation and Chaos in Applied Sciences and Engineering*, *17*(04), 1079–1107. doi:10.1142/S0218127407017732

Liao, X., & Chen, G. (2005). Hopf bifurcation and chaos analysis of Chen's system with distributed delays. *Chaos, Solitons, and Fractals*, *25*(1), 197–220. doi:10.1016/j.chaos.2004.11.007

Liu, G., Li, J., & Liu, H. (2014). Chaos-based color pathological image encryption scheme using one-time keys. *Computers in Biology and Medicine*, *45*, 111–117. doi:10.1016/j.compbiomed.2013.11.010 PMID:24480170

Liu, H., & Wang, X. (2013). Triple-image encryption scheme based on one-time key stream generated by chaos and plain images. *Journal of Systems and Software*, *86*(3), 826–834. doi:10.1016/j.jss.2012.11.026

Luo, A. C. J., Machado, J. A. T., & Baleanu, D. (2011) Dynamical Systems and Methods. New York: Springer Science+Business Media.

Machicao, J., Marco, A. G., & Bruno, O. M. (2012). Chaotic encryption method based on life-like cellular automata. *Expert Systems with Applications*, *39*(16), 12626–12635. doi:10.1016/j.eswa.2012.05.020

Madhavan, K., & Vaidyanathan, S. (2013). A new procedure for the identification of slow and fast states of discrete-time linear systems. *Far East Journal of Mathematical Sciences*, *72*(2), 225–236.

Micali, S., & Schnorr, C. P. (1991). Efficient, perfect polynomial random number generators. *Journal of Cryptology*, *3*(3), 157–172. doi:10.1007/BF00196909

Ogorzałek, M. J. (1997). *Chaos and complexity in nonlinear electronic circuits*. Singapore: World Scientific Publishing.

Orsucci, F. F., & Sala, N. (2006). *New research on chaos and complexity*. New York: Nova Science Publishers.

Pacheco, J. M. M., & Cuautle, E. T. (2010). *Electronic design automation of multi-scroll chaos generators*. Oak Park, Illinois: Bentham Science Publishers.

Peil, M., Jacquot, M., Chembo, Y. K., Larger, L., & Erneux, T. (2009). Routes to chaos and multiple time scale dynamics in broadband bandpass nonlinear delay electro-optic oscillators. *Physical Review E: Statistical, Nonlinear, and Soft Matter Physics*, *79*(2), 026208. doi:10.1103/PhysRevE.79.026208 PMID:19391821

Ryu, J. W., Kye, W. H., Lee, S. Y., Kim, M. W., Choi, M., & Rim, S. et al. (2004). Effects of time-delayed feedback on chaotic oscillators. *Physical Review E: Statistical, Nonlinear, and Soft Matter Physics*, *70*(3), 036220. doi:10.1103/PhysRevE.70.036220 PMID:15524625

Saied, A. B., Salem, S. B., Feki, M., & Masmoudi, D. S. (2007). A new high frequency second generation current conveyor based chaos generator. *In Proc. of 14th IEEE International Conference on Electronics, Circuits and Systems, 2007 (ICECS 2007)*, Marrakech, Morocco.

Sarasu, P., & Sundarapandian, V. (2011). Active controller design for generalized projective synchronization of four-scroll chaotic systems. *International Journal of Systems Signal Control and Engineering Application*, *4*(2), 26–33.

Sarasu, P., & Sundarapandian, V. (2011). The generalized projective synchronization of hyperchaotic Lorenz and hyperchaotic Qi systems via active control. *International Journal of Soft Computing*, *6*(5), 216–223. doi:10.3923/ijscomp.2011.216.223

Sarasu, P., & Sundarapandian, V. (2012). Generalized projective synchronization of three-scroll chaotic systems via adaptive control. *European Journal of Scientific Research*, *72*(4), 504–522.

Sarasu, P., & Sundarapandian, V. (2012). Generalized projective synchronization of two-scroll systems via adaptive control. *International Journal of Soft Computing*, *7*(4), 146–156. doi:10.3923/ijscomp.2012.146.156

Scott, A. (2005). *Encyclopedia of nonlinear science*. New York: Taylor & Francis Group.

Som, S., & Sen, S. (2013). A non-adaptive partial encryption of grayscale images based on chaos. *Procedia Technology*, *10*, 663–671. doi:10.1016/j.protcy.2013.12.408

Sundarapandian, V. (2004). Local and global asymptotic stability of nonlinear cascade interconnected systems. *Mathematical and Computer Modelling*, *40*(1-2), 227–232. doi:10.1016/j.mcm.2003.02.013

Sundarapandian, V. (2005). Nonlinear observer design for a general class of nonlinear systems with real parametric uncertainty. *Computers & Mathematics with Applications (Oxford, England)*, *49*(7-8), 1195–1211. doi:10.1016/j.camwa.2004.06.034

Sundarapandian, V. (2006). New results on the parametric stability of nonlinear systems. *Mathematical and Computer Modelling*, *43*(1), 9–15. doi:10.1016/j.mcm.2005.04.010

Sundarapandian, V. (2006). A necessary condition for local asymptotic stability of nonlinear systems with exogenous disturbance. *Computers & Mathematics with Applications (Oxford, England)*, *51*(3-4), 667–672. doi:10.1016/j.camwa.2005.01.032

Sundarapandian, V. (2010). Output regulation of the Lorenz attractor. *Far East Journal of Mathematical Sciences*, *42*(2), 289–299.

Sundarapandian, V. (2013). Analysis and anti-synchronization of a novel chaotic system via active and adaptive controllers. *Journal of Engineering Science and Technology Review*, *6*(4), 45–52.

Sundarapandian, V., & Karthikeyan, R. (2011). Anti-synchronization of Lü and Pan chaotic systems by adaptive nonlinear control. *European Journal of Scientific Research*, *64*(1), 94–106.

Sundarapandian, V., & Karthikeyan, R. (2011). Anti-synchronization of hyperchaotic Lorenz and hyperchaotic Chen systems by adaptive control. *International Journal of Systems Signal Control and Engineering Application*, *4*(2), 18–25.

Sundarapandian, V., & Karthikeyan, R. (2012). Adaptive anti-synchronization of uncertain Tigan and Li systems. *Journal of Engineering and Applied Sciences (Asian Research Publishing Network)*, *7*(1), 45–52.

Sundarapandian, V., & Pehlivan, I. (2012). Analysis, control, synchronization, and circuit design of a novel chaotic system. *Mathematical and Computer Modelling*, *55*(7-8), 1904–1915. doi:10.1016/j.mcm.2011.11.048

Sundarapandian, V., & Sivaperumal, S. (2011). Sliding controller design of hybrid synchronization of four-wing chaotic systems. *International Journal of Soft Computing*, *6*(5), 224–231. doi:10.3923/ijscomp.2011.224.231

Trejo-Guerra, R., Tlelo-Cuautle, E., Carbajal-Gomez, V. H., & Rodriguez-Gomez, G. A. (2013). Survey on the integrated design of chaotic oscillators. *Applied Mathematics and Computation*, *219*(10), 5113–5122. doi:10.1016/j.amc.2012.11.021

Vaidyanathan, S. (2012). Sliding mode control based global chaos control of Liu-Liu-Liu-Su chaotic system. *International Journal of Control Theory and Applications*, *5*(1), 15–20.

Vaidyanathan, S. (2012). Anti-synchronization of Sprott-L and Sprott-M chaotic systems via adaptive control. *International Journal of Control Theory and Applications*, *5*(1), 41–59.

Vaidyanathan, S. (2013). A new six-term 3-D chaotic system with an exponential nonlinearity. *Far East Journal of Mathematical Sciences*, *79*(1), 135–143.

Vaidyanathan, S. (2013). Analysis and adaptive synchronization of two novel chaotic systems with hyperbolic sinusoidal and cosinusoidal nonlinearity and unknown parameters. *Journal of Engineering Science and Technology Review*, *6*(4), 53–65.

Vaidyanathan, S., & Rajagopal, K. (2012). Global chaos synchronization of hyperchaotic Pang and hyperchaotic Wang systems via adaptive control. *International Journal of Soft Computing*, *7*(1), 28–37. doi:10.3923/ijscomp.2012.28.37

Vaidyanathan, S., & Sampath, S. (2012). Anti-synchronization of four-wing chaotic systems via sliding mode control. *International Journal of Automation and Computing*, *9*(3), 274–279. doi:10.1007/s11633-012-0644-2

Vazirani, U. V., & Vazirani, V. V. (1985). *Efficient and secure pseudo-random number generation*. Advances in Cryptology. *Lecture Notes in Computer Science*, *196*, 193–202. doi:10.1007/3-540-39568-7_17

Venkatesan, A., & Lakshmanan, M. (1997). Bifurcation and chaos in the double-well Duffing–van der Pol oscillator: Numerical and analytical studies. *Physical Review E: Statistical Physics, Plasmas, Fluids, and Related Interdisciplinary Topics*, *56*(6), 6321–6330. doi:10.1103/PhysRevE.56.6321

Volos, Ch. K. (2013). Image encryption using the coexistence of chaotic synchronization phenomena. *Journal of Applied Mathematics and Bioinformatics*, *3*(1), 123–149.

Volos, Ch. K., Kyprianidis, I. M., & Stouboulos, I. N. (2009). Image encryption process based on a chaotic true random bit generator. *In Proc. 16th IEEE International Conference on Digital Signal Processing (DSP 2009)*, *1 & 2*, 1091-1094, Santorini, Greece. doi:10.1109/ICDSP.2009.5201107

Volos, Ch. K., Kyprianidis, I. M., & Stouboulos, I. N. (2012). Synchronization phenomena in coupled nonlinear systems applied in economic cycles. *WSEAS Trans. Systems*, *11*(12), 681–690.

Wang, L., & Duan, S. (2012). A chaotic attractor in delayed memristive system. *Abstract and Applied Analysis*, *2012*, 1–8.

Wang, X., & Luan, D. (2013). A novel image encryption algorithm using chaos and reversible cellular automata. *Communications in Nonlinear Science and Numerical Simulation*, *18*(11), 3075–3085. doi:10.1016/j.cnsns.2013.04.008

Wang, X., Zhang, W., Guo, W., & Zhang, J. (2013). Secure chaotic system with application to chaotic ciphers. *Information Sciences*, *221*, 555–570. doi:10.1016/j.ins.2012.09.037

Wen, S., & Zeng, Z. (2012). Dynamics analysis of a class of memristor-based recurrent networks with time-varying delays in the presence of strong external stimuli. *Neural Processing Letters*, *35*(1), 47–59. doi:10.1007/s11063-011-9203-z

Wu, A., & Zeng, Z. (2012). Dynamic behaviors of memristor-based recurrent neural networks with time-varying delays. *Neural Networks*, *36*, 1–10. doi:10.1016/j.neunet.2012.08.009 PMID:23037770

Wu, A., Zeng, Z., Zhang, J., & Zhu, X. (2011). Exponential synchronization of memristor-based recurrent neural networks with time delays. *Neurocomputing*, *74*(17), 3043–3050. doi:10.1016/j.neucom.2011.04.016

Wu, C. W. (2002). *Synchronization in coupled chaotic circuits and systems*. Singapore: World Scientific Publishing.

Yang, D., Liu, Z., & Zhou, J. (2014). Chaos optimization algorithms based on chaotic maps with different probability distribution and search speed for global optimization. *Communications in Nonlinear Science and Numerical Simulation*, *19*(4), 1229–1246. doi:10.1016/j.cnsns.2013.08.017

Zeraoulia, E., & Sprott, J. C. (2011). *Frontiers in the study of chaotic dynamical systems with open problems*. Singapore: World Scientific Publishing.

Zhang, G., Shen, Y., & Sun, J. (2012). Global exponential stability of a class of memristor-based recurrent neural networks with time-varying delays. *Neurocomputing*, *97*, 233–240. doi:10.1016/j.neucom.2012.05.002

Zhang, G., Shen, Y., Yin, Q., & Sun, J. (2013). Global exponential periodicity and stability of a class of memristor-based recurrent neural networks with multiple delays. *Information Sciences*, *232*, 386–396. doi:10.1016/j.ins.2012.11.023

Zhang, X., & Wang, X. (2013). Chaos-based partial encryption of SPHIT color coded images. *Signal Processing*, *93*(9), 2422–2431. doi:10.1016/j.sigpro.2013.03.017

Zheng, F., Tian, X. J., Song, J. Y., & Li, X. Y. (2008). Pseudo-random sequence generator based on the generalized Henon map. *Journal of China Universities of Posts and Telecommunications*, *15*(3), 64–68. doi:10.1016/S1005-8885(08)60109-0

Zhu, H., Zhao, C., & Zhang, X. (2013). A novel image-compression scheme using hyper-chaos and Chinese remainder theorem. *Signal Processing Image Communication*, *28*(6), 670–680. doi:10.1016/j.image.2013.02.004

Zhu, H., Zhao, C., Zhang, X., & Yang, L. (2013). A novel iris and chaos-based random number generator. *Computers & Security*, *36*, 40–48. doi:10.1016/j.cose.2013.02.003

Zhusubaliyev, Z. T. (2003). *Bifurcations and chaos in piecewise-smooth dynamical systems*. Singapore: World Scientific Publishing.

KEY TERMS AND DEFINITIONS

Attractor: In dynamical systems, it is a region of phase space, towards which neighboring trajectories asymptotically approach.

Bifurcation Diagram: A diagram which displays the possible long-term changes of system behavior when the value of the control parameter is varied.

Bifurcation: A sudden change that accompanies the onset of chaos at a critical value of a varied control parameter.

Chaos: A type of behavior of a deterministic nonlinear system, where tiny changes in initial conditions make huge changes over time.

Delay Differential Equation (DDE): A differential equation in which the evolution of dependent variables at a certain time depend on their values at previous times.

Limit Cycle: A closed trajectory in the phase space associated with a nonlinear system.

Lyapunov Exponent: A quantity that characterizes the rate of separation of infinitesimally close trajectories. Chaotic motion is indicated by at least one positive Lyapunov exponent.

Multisim: An electronic schematic capture and simulation program of National Instruments (NI). Multisim provides an advanced, industry-standard SPICE simulation environment and is used widely in the world.

Phase Portrait: A geometric presentation of the orbits of a dynamical nonlinear system in the phase plane.

Piecewise Linear Function: A function which is composed of piecewise linear segments.

Chapter 9
Chaos Control of an Impact Mechanical Oscillator Based on the OGY Method

Hassène Gritli
Institut Supérieur des Etudes Technologiques de Kélibia, Tunisia

Safya Belghith
Ecole Nationale d'Ingénieurs de Tunis, Tunisia

Nahla Khraief
Ecole Supérieure de Technologie et d'Informatique, Tunisia

ABSTRACT

This chapter deals with the control of chaos exhibited in the behavior of a one-degree-of-freedom impact mechanical oscillator with a single rigid constraint. The mathematical model of such impact oscillator is represented by an impulsive hybrid linear non-autonomous dynamics. The proposed control approach is based chiefly on the OGY method. First, an analytical expression of a constrained controlled Poincaré map is derived. Secondly, the linearized controlled Poincaré map around the fixed point of the constrained map is determined. Relying on the linearized map, a state feedback controller is designed. It is shown that the proposed control strategy is efficient for the control of chaos for the desired fixed point and for the fixed parameter. It is shown also that, by changing the system parameter, the behavior of the impact mechanical oscillator changes radically. Thus, the drawback of the developed OGY control method is revealed and some remedies are given.

1. INTRODUCTION

Chaos is defined as a strange phenomenon with erratic appearance and which arises in deterministic continuous or discrete nonlinear systems. The chaos paradigm provides a better understanding of the inherent properties of natural systems. A chaotic system is a nonlinear dynamic system and its response has certain characteristics:

DOI: 10.4018/978-1-4666-7248-2.ch009

- It has an excessive sensitivity to initial conditions: the very small difference between the same two initial conditions will grow and the two paths from these two initial conditions separate exponentially over time.
- A chaotic system has a wide spectrum of the Fourier transform, and has fractal properties of the motion in the phase space.
- In addition, the chaotic motion of a dynamical system is locally unstable and globally bounded in the phase space. This means that solutions with very similar initial conditions will diverge at a certain distance after a finite time. Thus, chaos denotes a state of disorder and irregularity which reflects that the forecast is only possible in the short term.
- Moreover, a nonlinear chaotic dynamic system is said to be chaotic if it has at least one strange attractor. An attractor is called strange or chaotic if it is bounded and all trajectories starting from this attractor are actually chaotic (Andrievskii & Fradkov, 2003; Fradkov & Evans, 2005).

The study of nonlinear science has evolved into a new era. Fundamental theories have been established and important results have been obtained in the areas of chaotic nonlinear dynamics such as chaos control, namely, the investigation of bringing order to chaos (Wiggins, 2003; Schöll & Schuster, 2007). It is now recognized that chaos occurs commonly in many fields of science and engineering and so in a large variety of nonlinear dynamical systems such as nonlinear oscillators (Lakshmanan & Murali, 1996; Fradkov & Pogromsky, 1996), impact mechanical systems (Wiercigroch & de Kraker, 2000; Awrejcewicz & Lamarque, 2003; Thota & Dankowicz, 2006a, 2006b; Thotaa *et al.*, 2006; Jin *et al.*, 2006), legged walking robots (Kaygisiz *et al.*, 2006; Gritli *et al.*, 2012a, 2012b, 2012c, 2014b; Iqbal *et al.*, 2014), etc. Nonlinear dynamics of impact oscillators and their nonlinear behaviors have received considerable theoretical and experimental attentions in the past. Many mathematical-numerical studies of these impacting systems have been carried out, and particular interest has been focused on the analysis and the control of chaos and bifurcations (Lenci & Rega, 2004). More general studies of multiple-degrees-of-freedom impact mechanical oscillators have also been carried out (Pfeiffer & Glocker, 1996; Pun *et al.*, 1998; Wagg, 2005). However, almost all these studies focus primarily on a single impact constraint. In this chapter, we will consider a one-degree-of-freedom impact mechanical oscillator periodically excited by an external sinusoidal input. This oscillator consists of one mass, coupled with one spring and one dashpot, and the motion of the mass is restricted by only one rigid constraint (Budd, 1996; de Souza & Caldas, 2004a, 2004b; de Souza *et al.*, 2007). This proposed impact system is classified as an impulsive hybrid linear non-autonomous dynamics that is simple enough to be amendable to analysis, yet complex enough to exhibit a wealth of attractive phenomena.

The design of an effective control system of chaos is a central focus in the field of nonlinear science. In literature, many methods and strategies of control of chaos in nonlinear dynamical systems have been proposed (Andrievskii & Fradkov, 2003, 2004; Fradkov & Evans, 2005; Fradkov *et al.*, 2006). Chaos control is chiefly based on the richness of chaotic behavior. It has been known so far that there are an infinite number of unstable periodic orbits (UPOs) embedded into a chaotic attractor. Moreover, the system trajectory often goes in the vicinity of each of them. Generally, chaos control is used in order to stabilize a desired UPO for some set of parameters of the dynamical system.

Chaos control methods may be classified as discrete or continuous techniques. On the first hand, continuous strategies are illustrated by the so-called delayed feedback control which was proposed by Pyragas (1992) and which states that chaotic systems can be stabilized by a feedback perturbation proportional to the difference between the present and the delayed state of the dynamical system. On the other hand, and in the huge interest to the control of chaos, Ott *et al.* (1990) formulated the following two key ideas (Andrievskii & Fradkov, 2003; Fradkov *et al.*, 2006; Gritli *et al.*, 2013, 2014a):

1. Designing controller by the discrete system model based on linearization of the Poincaré map, and
2. Using the property of recurrence of the chaotic trajectories and applying the control action only at the instants when the trajectory returns to some neighborhood of the desired state or given orbit.

This control method is mainly based on the linearization of the Poincaré map and it is nowadays known as the OGY method as a tribute of their authors (Ott–Grebogi–Yorke). The OGY method is a discrete technique that can be applied to either discrete systems or continuous dynamical systems (Boccaletti *et al.*, 2000; Witvoet, 2005). It considers small perturbations applied in the vicinity of the desired orbit when the system trajectory passes through a specific surface called the Poincare section (Grebogi & Lai, 1997).

In this chapter, we use the OGY method in order to control chaos exhibited in the impulsive hybrid non-autonomous linear dynamics of the impact mechanical oscillator. We will derive from this dynamics an explicit expression of a constrained controlled Poincaré map. Relying on this map, we will determine an UPO (or systematically a fixed point of an UPO) where its stability can be analyzed by means of the linearized Poincaré map. Basing on this linearized map, we design a state feedback controller. Relying on the second key of the OGY method, and as the dynamics of the impact oscillator is non-autonomous with some defined period; this controller will be applied at the beginning of and during each period.

This chapter is organized into ten sections. In Section 2, a literature review of the OGY method and the control of the impact mechanical oscillator is presented. Section 3 gives the impulsive hybrid linear dynamics of the impact mechanical oscillator. Periodic and chaotic behaviors of the impact oscillator are also discussed briefly in this section. Section 4 is dedicated to the development of an analytical expression of the constrained controlled Poincaré map. The procedure for determining the fixed point based on the constrained Poincaré map is discussed in Section 5. Section 6 describes the linearized controlled Poincaré map. The problem of stabilizing an unstable fixed point and thus the control of chaos is discussed in Section 7. We present in Section 8 numerical simulations showing the control of chaos in the impulsive hybrid dynamics of the impact mechanical oscillator. In Section 9, we discuss the results of our work. Finally, in Section 10, future directions of research are outlined.

2. BACKGROUND

The Ott-Grebogi-Yorke (OGY) method is a very efficient control approach to stabilize fixed points of maps or limit cycles of nonlinear continuous dynamics. Generally, the OGY control method has been used to control chaos in dynamical systems (Boccaletti *et al.*, 2000; Witvoet, 2005; Schöll & Schuster, 2007). The OGY ideas found experimental applications in several different fields, such as mechanical oscillations (magnetoelastic ribbon), electronic circuits (diode resonator), chemical systems (Belouzov-Zhabotinski reaction), cardiac tissues, and nonlinear optics (multimode laser) (Boccaletti *et al.*, 2000). In fact, two essential features of a chaotic dynamic system make it a candidate for the OGY control: first, the presence within the chaotic attractor at least of one saddle-node-type unstable fixed point, and second an available control parameter through which the location of the unstable fixed point in the phase plane changes.

For the case of discontinuous dynamic systems, the OGY control method was applied also with success. A literature review of different control methods (among them we found the OGY method) of mechanical systems is presented in (Fradkov *et al.*, 2006). The impact mechanical oscillators were also illustrated.

Compared with the subject of chaos control in the impact mechanical oscillators using the OGY control method, the nonlinear phenomena of impact oscillators have been fully investigated. Kalagnanam (1994) applied the OGY method to impact oscillators and stabilized its chaotic attractor on period-1 and period-2 orbits using small time-dependent perturbations of the driving frequency. He also demonstrated the ability to switch the chaotic system between the period-1 and period-2 orbits at will by controlled time-dependent perturbations. Similarly, the OGY method was used in (Begley & Virgin, 2001) to control an impact-friction oscillator using friction force as the control parameter. Besides, de Souza & Caldas (2004b) employed the OGY-based method in order to stabilize desired unstable periodic orbits embedded in the chaotic invariant sets of two vibro-impact systems: an impact oscillator and an impact-pair system. Authors introduced a transcendental map (impact map) for the dynamical variables computed just after the impacts. Furthermore, Gutiérrez & Arrowsmith (2003) applied some control schemes based on the OGY approach to both numerical and experimental models of a double impacting mechanical oscillator by realizing simple -single-degree-of-freedom- numerical and experimental models. Besides, de Weger (2005) presented the impact mechanical oscillator driven by a time-periodic force and he studied it based on the Nordmark map. He studied chaos control by means of the OGY method in a parametrically driven pendulum. In addition, de Souza *et al.* (2007) used a damping control law for a chaotic single-degree-of-freedom impact mechanical oscillator. They applied a feedback control technique to suppress chaotic behavior in the mechanical systems by using a small-amplitude damping signal.

On the other hand, the major drawback of the OGY control method is due to the long waiting time until the system trajectory enters with the Poincaré section a close neighborhood of the target periodic orbit. In literature, several papers have showed the disadvantages of the OGY method in some dynamic systems (Witvoet, 2005; Schöll & Schuster, 2007). It has been shown that a behavior of the system can change and undergoes different behaviors and generally a period-doubling route to chaos is generated. There are many improvements of the OGY method that aim at overcoming some of its original limitations, as for example: control of high periodic and high unstable UPO (Hübinger *et al.*, 1994; Ritz *et al.*, 1997), control using time delay coordinates (Dressler & Nitsche, 1992; de Korte *et al.*, 1995), control using multiparameter approach based on pole placement formalism (Barreto & Grebogi, 1995), and a multiparameter chaos control method based on OGY approach (de Paula & Savi, 2009).

3. IMPULSIVE HYBRID LINEAR DYNAMICS OF THE IMPACT OSCILLATOR

3.1. Impact Mechanical Oscillator

We consider in this work a non-smooth dynamical system: an oscillating mechanical system with impact known as the impact oscillator. Figure 1 provides a schematic representation of the impact mechanical oscillator (Budd, 1996). This impact oscillator is composed of a mass m. This mass is interconnected to the wall by both a spring having a stiffness k and a dashpot with a coefficient c. A second wall introduced as a rigid constraint is strayed from the mass of a distance d. The mass is periodically excited via a sinusoidal input $u(t) = U_m cos(wt)$ where the parameters w and U_m are excitation fre-

Figure 1. The one degree of freedom impact mechanical oscillator with a single rigid constraint

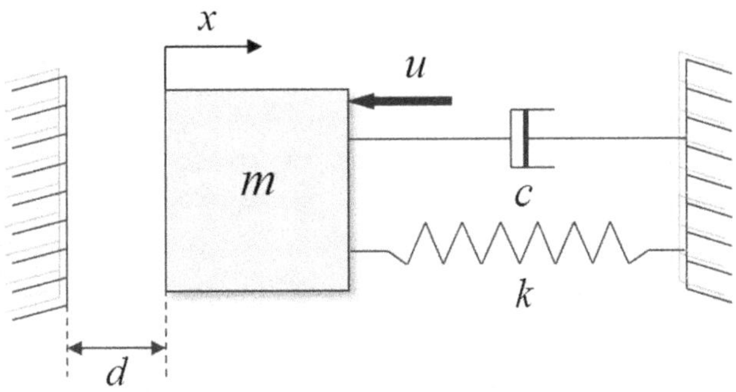

quency and amplitude, respectively. Under this oscillatory input, and from an initial condition at time $t_0 = 0\,s$, the mechanical system will oscillate along the horizontal axis x and hence, the mass will produce impacts with the wall on the left for $x = d$. At each impact, the velocity of the mass will undergo a restitution with a coefficient r. However, the position of the mass is not changed and no sliding occurs at the impact. Let $T = \frac{2\pi}{w}$ be the period of the oscillation and τ is the impact time at which the mass hits the wall.

3.2. Impulsive Hybrid Non-Autonomous Linear Dynamics

As the impact mechanical oscillator moves horizontally, its dynamics is composed of two phases: an oscillation phase and an instantaneous impact phase. Let $z = \begin{bmatrix} x & \dot{x} \end{bmatrix}^T$ be the state vector where x and $\dot{x}$ are the displacement and the velocity components, respectively. The impulsive hybrid linear dynamics of the impact mechanical oscillator is given by:

$$\dot{z} = Az + Bu\left(t\right), \text{if } z \notin \Gamma \tag{1}$$

$$z^+ = Rz^-, \text{if } z \in \Gamma \tag{2}$$

where subscribes $^+$ and $^-$ denotes just after and just before the impact phase, respectively.

In (1) and (2), the different matrices are defined like so: $A = \begin{bmatrix} 0 & 1 \\ -\frac{k}{m} & -\frac{c}{m} \end{bmatrix}$, $B = \begin{bmatrix} 0 \\ -\frac{1}{m} \end{bmatrix}$ and $R = \begin{bmatrix} 1 & 0 \\ 0 & -r \end{bmatrix}$.

The linear differential equation (1) describes the non-autonomous linear dynamics during the oscillation phase. However, the algebraic expression (2) provides the impulsive-effects during the impact phase. All these equations translate the impulsive hybrid non-autonomous linear dynamics of the impact mechanical oscillator. In (1) and (2), Γ defines the impact condition (surface). It is defined by:

$$\Gamma = \left\{ z \in \Re^{2\times 1},\, h\left(z\right) = Cz - d = 0 \right\} \tag{3}$$

where $C = \begin{bmatrix} 1 & 0 \end{bmatrix}$.

3.3. Periodic and Chaotic Behaviors

The impact mechanical oscillator can exhibit periodic and chaotic behaviors with respect to the bifurcation parameters r, d and w (Budd, 1996; de Souza & Caldas, 2004a, 2004b; Jin *et al.*, 2006; de Souza *et al.*, 2007). In this chapter, we choose the following system parameters: $m = 1\ Kg$, $k = 1\ N / m$, $c = 0\ N / \left(m / s\right)$, $r = 0.8$, and $U_m = -1\ N$. In order to study nonlinear dynamics of the impact mechanical oscillator through bifurcation diagrams, it is indispensable to choose a Poincaré section Λ that gathers all intersection points in direct way. Such Poincaré section Λ should be transversal to the orbit (trajectory) of the impulsive hybrid dynamics (1)-(3). These points will be then plotted in bifurcation diagrams as a system parameter changes. As the dynamics is non-autonomous, then the appropriate choice of a Poincaré section Λ is:

$$\Lambda = \left\{ t \geq 0,\, s\left(t\right) = t - \left(n-1\right)T = 0,\, n = 1, 2, \ldots \right\} \tag{4}$$

Figure 2 gives the bifurcation diagrams with respect to the excitation frequency w. They are plotted using the brute force technique (Parker & Chua, 1989) by varying the bifurcation parameter w from 1.5 to 3.5. The left plot shows the velocity of the mass at the Poincaré section (4). In contrast, the right bifurcation diagram reveals the impact time τ of the mass with the rigid constraint. Here, the impact time is: $\tau \equiv \text{mod}\left(\tau,\, T\right)$. In these two bifurcation diagrams, a scenario of period-doubling bifurcations as

Figure 2. Bifurcation diagrams: velocity of the mass (a) and the impact instant of the mass (b) as a function of the excitation frequency w

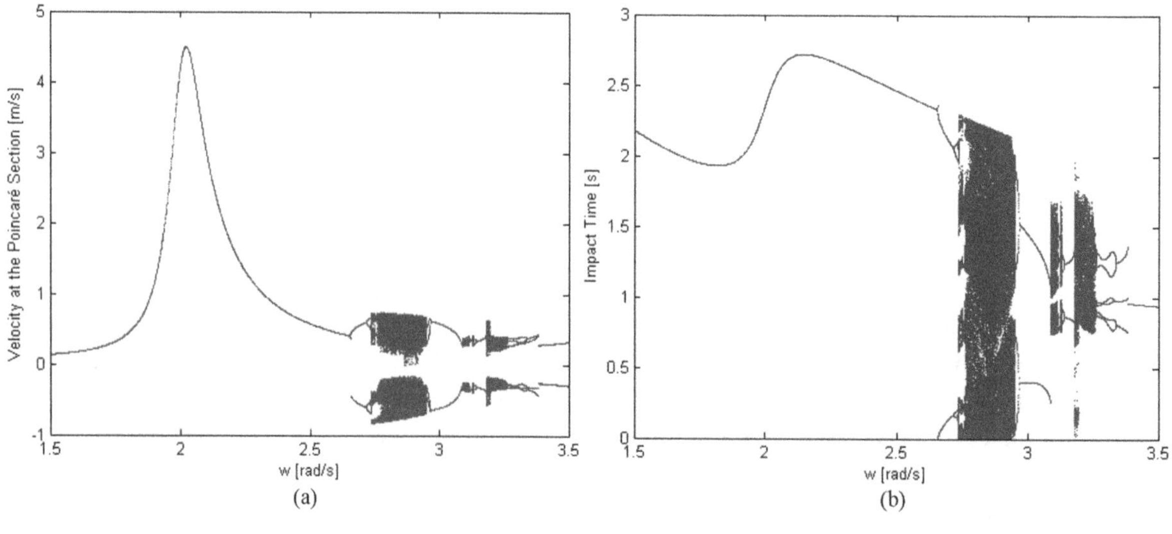

route to chaos is displayed. We will denote by a p-period limit cycle for a periodic system if and only if the state z of the mass m returns to the same state after each period pT. Moreover, the impact oscillator can undergo several impacts q for the p-period limit cycle. Accordingly, a periodic impacting oscillation of the mechanical oscillator can be described by the period number p and the impact number q. Thus, we will denote (p, q)-limit cycle as a p-periodic q-impacting limit cycle.

Form bifurcation diagrams of Figure 2, the scenario of period-doubling route to chaos begins with a $(1, 1)$-limit cycle. However, for a bifurcation parameter w near to 3.5, the oscillation is described by a $(2, 1)$-limit cycle. In addition, with the decrease of w, we reveal a $(4, 3)$-limit cycle that creates an inverse period-doubling route to chaos. In fact, it is well known so far that such simple one-degree-of-freedom impact mechanical oscillator exhibits several complex phenomena such as grazing bifurcations, intermittency and crisis (Budd, 1996; Awrejcewicz & Lamarque, 2003). In this chapter, we will not revisit these nonlinear phenomena.

It is clear that the simplest behavior of the impact mechanical oscillator is that with the $(1, 1)$-limit cycle. Figure 3a shows a limit cycle revealing a periodic behavior with period 1. This limit cycle shows that the mechanical oscillator undergoes only one impact in only one period. In contrast, Figure 3b shows a chaotic attractor depicted for the excitation frequency $w = 2.8\ rad / s$.

In this work, we will not consider the case of multiple impacts. We will interest only in the case where there is only one impact during a period T, i.e. we will deal with the $(1, 1)$-limit cycle of the impact mechanical oscillator in order to control chaos.

Therefore, our main goal is to control chaos exhibited in the impulsive hybrid linear dynamics of the impact mechanical oscillator. Then, our objective is to transform the chaotic behavior of the oscillator

Figure 3. (a) A 1-period nominal hybrid limit cycle of the impact mechanical oscillator for $w = 2.2\ rad / s$. (b) A chaotic attractor of the impact mechanical oscillator for $w = 2.8\ rad / s$

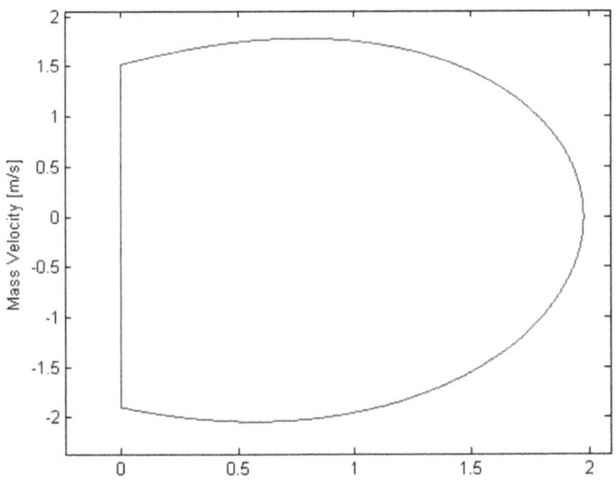

(a)

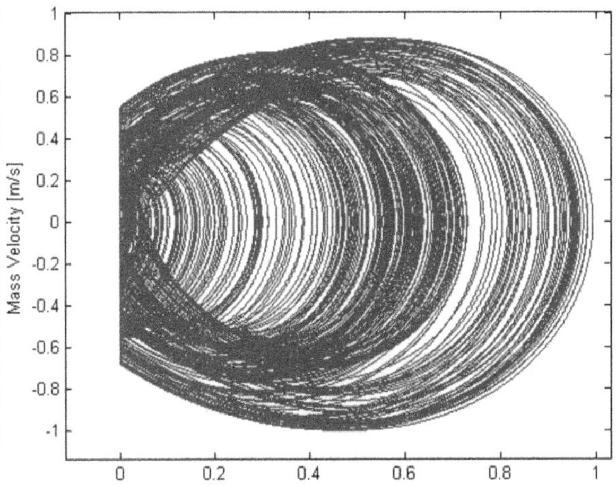

(b)

to a 1-period behavior by stabilizing it. Thus, our strategy of control of chaos will be treated according to the OGY method in two steps:

- The first step deals with the numerical identification of an unstable limit cycle (or an unstable periodic orbit) embedded into a chaotic attractor for some bifurcation parameter w. This will be done by the design of an explicit expression of a controlled Poincaré map.
- The second step lies in the stabilization of this unstable limit cycle by designing a specific controller based on the linearization of the controlled Poincaré map.

4. CONTROLLED POINCARE MAP

A controlled Poincaré map is the one with a controller v which will be designed. Then, to control chaos displayed in the impulsive hybrid linear dynamics (1)-(3), we will add the controller v to the excitation input u. Thus, following our control approach which is based on the OGY method, the proposed controller v must be constant during the nth period T, i.e. $v(t) = v_n$ for all $(n-1)T \leq T \leq nT$, with $n = 1, 2, \ldots$. Thus, the impulsive hybrid dynamics (1)-(2) becomes:

$$\dot{z} = Az + Bu(t) + Bv_n, \text{if } z \notin \Gamma \tag{5}$$

$$z^+ = Rz^-, \text{if } z \in \Gamma \tag{6}$$

In order to determine the analytical expression of the controlled Poincaré map, we will consider the Poincaré section Λ defined by (4). From an initial condition $z_1 = z(t_0 = 0)$, the mass oscillates before the contact with the impact surface Γ defined by (3). Here, we assume that the initial condition z_1 does not belong to Γ. We note τ_1 the impact time and z_1^- is the state vector just before the impact event. At this event, the state z_1^- will be transited to the state just after the impact z_1^+ following expression (6). Thus, from this state, the mass will oscillate again until the state $z_2 = z(T)$ defined at the instant T. Accordingly, we obtain a first cycle that begins with z_1 at 0 seconds and that is terminated at T with z_2. During this cycle, the controller v will be remained constant. Then we will have a first controller v_1 for this first cycle. The state z_2 will be used as the initial condition for the next (second) cycle, and the mass oscillates, and so on. Next in this chapter, we will use the following notations for all $n = 1, 2, \ldots$:

- z_n is the initial condition for the nth cycle,
- τ_n is the impact time for the nth cycle, here τ_n is chosen to be such that: $0 < \tau_n < T$,
- z_n^- is the state just before impact for the nth cycle,
- z_n^+ is the state just after impact for the nth cycle,
- v_n is the controller applied during the nth cycle, it is constant between $(n-1)T$ and nT,

- z_{n+1} is the initial condition for the $(n+1)^{th}$ cycle.

Then, starting from the state z_n and arriving at the next state z_{n+1}, the trajectory passes through three successive phases: 1st phase of oscillation $\Rightarrow$ 2nd phase of impact $\Rightarrow$ 3rd phase of oscillation. Because the dynamics during the phase of oscillation is governed by the linear system (5) with a sinusoidal input u and a constant controller v, it will be easy to determine an explicit expression of a controlled Poincaré map. This map will relate the states z_n at the beginning of each cycle. Then, in order to determine the expression of this map, we must determine the expressions lying the states at the beginning of each phase to those at the end of this phase. We should emphasize that the solution of the linear differential equations in (5) is defined by:

$$z(t) = e^{A(t-t_0)}z(t_0) + e^{At}\int_{t_0}^{t} e^{-A\xi}B\left(u(\xi)+v\right)d\xi \tag{7}$$

For the periodic excitation input $u(t) = U_m \cos(wt)$, and for all parameters of the mechanical oscillator satisfying $\det(A) \neq 0$ and $\det\left(A + w^2A^{-1}\right) \neq 0$, the state vector $z(t)$ is defined by the following expression:

$$z(t) = e^{A(t-t_0)}z(t_0) + U_m\left(A + w^2A^{-1}\right)^{-1}\left(g(t) - e^{A(t-t_0)}g(t_0)\right) + \left(e^{A(t-t_0)} - I_2\right)A^{-1}Bv \tag{8}$$

with $g(t) = \left(wA^{-1}\sin(wt) - I_2\cos(wt)\right)B$, and I_2 is the identity matrix of dimension 2.

1st phase: oscillation phase: In this first phase, we have $z_n = z\left((n-1)T\right)$. At the instant $(n-1)T + \tau_n$, the trajectory undergoes an impact with the state $z_n^- = z\left((n-1)T + \tau_n\right)$. Hence, according to (8), we obtain the following expression:

$$z_n^- = G_1(\tau_n)z_n + H_1(\tau_n) + J_1(\tau_n)v_n \tag{9}$$

with $G_1(\tau_n) = e^{A\tau_n}$, $H_1(\tau_n) = U_m\left(A + w^2A^{-1}\right)^{-1}\left(wA^{-1}\sin(w\tau_n) - I_2\cos(w\tau_n) + e^{A\tau_n}\right)B$, and $J_1(\tau_n) = \left(G_1(\tau_n) - I_2\right)A^{-1}B$.

Relying on the impact condition (3), we emphasize that $h(z_n^-) = Cz_n^- - d = 0$. Thus, in order that an impact occurs, the initial state z_n, the impact time τ_n and the controller v_n must satisfy the following condition:

$$\Psi(z_n, \tau_n, v_n) = G_0(\tau_n)z_n + H_0(\tau_n) + J_0(\tau_n)v_n = 0 \tag{10}$$

with $G_0(\tau_n) = CG_1(\tau_n)$, $H_0(\tau_n) = CH_1(\tau_n) - d$ and $J_0(\tau_n) = CJ_1(\tau_n)$.

2nd phase: impact phase: In this instantaneous phase, and relying on the algebraic equations (6), the state vector just before impact z_n^- is linked to the state vector just after impact z_n^+ according to the following expression:

$$z_n^+ = Rz_n^- \tag{11}$$

3rd phase: oscillation phase: From the state z_n^+, the trajectory evolves according to expression (8). This third phase is terminated at the instant $t = nT$. Hence, we obtain the state vector z_{n+1} as follows:

$$z_{n+1} = G_2(\tau_n)z_n^+ + H_2(\tau_n) + J_2(\tau_n)v_n \tag{12}$$

with $G_2(\tau_n) = e^{A(T-\tau_n)}$, $H_2(\tau_n) = U_m(e^{AT} - I_2)(A + w^2A^{-1})^{-1} B - G_2(\tau_n)H_1(\tau_n)$, and $J_2(\tau_n) = (G_2(\tau_n) - I_2)A^{-1}B$.

Using expressions (9)-(12), the constrained controlled Poincaré map is expressed like so:

$$\begin{cases} z_{n+1} = \mathrm{P}(z_n, \tau_n, v_n) \\ 0 = \Psi(z_n, \tau_n, v_n) \end{cases} \tag{13}$$

with $\mathrm{P}(z_n, \tau_n, v_n) = G(\tau_n)z_n + H(\tau_n) + J(\tau_n)v_n$, where $G(\tau_n) = G_2(\tau_n)RG_1(\tau_n)$, $H(\tau_n) = G_2(\tau_n)RH_1(\tau_n) + H_2(\tau_n)$, $J(\tau_n) = G_2(\tau_n)RJ_1(\tau_n) + J_2(\tau_n)$, and the matrix $\Psi(z_n, \tau_n, v_n)$ is defined in (10).

5. IDENTIFICATION OF AN UNSTABLE PERIODIC ORBIT

The identification of an unstable periodic orbit (UPO) embedded into a chaotic attractor for some defined bifurcation parameter lies in the numerical identification of a fixed point of the constrained uncontrolled Poincaré map (13), this means that for $v_n = 0$, for all $n = 1, 2, \ldots$. Let z_* be the fixed point of the constrained uncontrolled Poincaré map. This can be obtained by enforcing the periodicity $z_{n+1} = z_n = z_*$. For this fixed point z_*, it corresponds an impact instant τ_*. Thus, for $v_* = 0$, and based on (13), the fixed point z_* must verify the following expressions:

$$\mathrm{P}(z_*, \tau_*, v_*) - z_* = 0 \tag{14}$$

and

$$\Psi\left(z_*, \tau_*, v_*\right) = 0 \tag{15}$$

Relying on expressions of the two quantities $P\left(z_*, \tau_*, v_*\right)$ and $\Psi\left(z_*, \tau_*, v_*\right)$, we can show that the nominal impact time τ_* must verify the following function:

$$G_0\left(\tau_*\right)\left(I_2 - G\left(\tau_*\right)\right)^{-1} H\left(\tau_*\right) + H_0\left(\tau_*\right) = 0 \tag{16}$$

and the fixed point z_* is expressed by:

$$z_* = \left(I_2 - G\left(\tau_*\right)\right)^{-1} H\left(\tau_*\right) \tag{17}$$

In fact, τ_* must be calculated numerically from (16) using the well-known iterative scheme of Newton-Raphson (Parker & Chua, 1989). Once the nominal impact time τ_* is identified, the fixed point z_* will be calculated according to the relation (17).

Now, we will analyze the stability of the identified fixed point z_*. Then, in order to accomplish this task, we should determine the linearized controlled Poincaré map around this fixed point.

6. LINEARIZED CONTROLLED POINCARE MAP

Let noting: $\Delta z_{n+1} = z_{n+1} - z_*$, $\Delta z_n = z_n - z_*$, $\Delta v_n = v_n - v_*$, with $v_* = 0$. Using the constrained controlled Poincaré map (13) and relying on expression (14), the linearized controlled Poincaré map is defined by:

$$\Delta z_{n+1} = DP_{z_n}\left(z_*, \tau_*, v_*\right)\Delta z_n + DP_{v_n}\left(z_*, \tau_*, v_*\right)\Delta v_n \tag{18}$$

with DP_{z_n} is the Jacobean matrix of the Poincaré map with respect to z_n, and DP_{v_n} is the derivative of the Poincaré map with respect to the controller v_n. These two matrices are evaluated at the fixed point z_* and at the nominal impact time τ_* and for $v_* = 0$. They are defined as following:

$$\begin{cases} DP_{z_n}\left(z_n, \tau_n, v_n\right) = \dfrac{\partial P\left(z_n, \tau_n, v_n\right)}{\partial z_n} + \dfrac{\partial P\left(z_n, \tau_n, v_n\right)}{\partial \tau_n}\dfrac{\partial \tau_n}{\partial z_n} \\ DP_{v_n}\left(z_n, \tau_n, v_n\right) = \dfrac{\partial P\left(z_n, \tau_n, v_n\right)}{\partial v_n} + \dfrac{\partial P\left(z_n, \tau_n, v_n\right)}{\partial \tau_n}\dfrac{\partial \tau_n}{\partial v_n} \end{cases} \tag{19}$$

By linearizing the second expression in (13) and relying on expression (15), we obtain:

$$\begin{cases} \dfrac{\partial \Psi\left(z_n, \tau_n, v_n\right)}{\partial z_n} + \dfrac{\partial \Psi\left(z_n, \tau_n, v_n\right)}{\partial \tau_n} \dfrac{\partial \tau_n}{\partial z_n} = 0 \\ \dfrac{\partial \Psi\left(z_n, \tau_n, v_n\right)}{\partial v_n} + \dfrac{\partial \Psi\left(z_n, \tau_n, v_n\right)}{\partial \tau_n} \dfrac{\partial \tau_n}{\partial v_n} = 0 \end{cases} \tag{20}$$

We stress that the quantity $\dfrac{\partial \mathrm{P}\left(z_n, \tau_n, v_n\right)}{\partial \tau_n}$ is a scalar number. Assuming that this quantity is non-zero, then using (20) it states that

$$\begin{cases} \dfrac{\partial \tau_n}{\partial z_n} = -\left(\dfrac{\partial \Psi\left(z_n, \tau_n, v_n\right)}{\partial \tau_n}\right)^{-1} \dfrac{\partial \Psi\left(z_n, \tau_n, v_n\right)}{\partial z_n} \\ \dfrac{\partial \tau_n}{\partial v_n} = -\left(\dfrac{\partial \Psi\left(z_n, \tau_n, v_n\right)}{\partial \tau_n}\right)^{-1} \dfrac{\partial \Psi\left(z_n, \tau_n, v_n\right)}{\partial v_n} \end{cases} \tag{21}$$

Substituting these two expressions (21) into (19), we obtain then:

$$\begin{cases} D\mathrm{P}_{z_n}\left(z_n, \tau_n, v_n\right) = \dfrac{\partial \mathrm{P}\left(z_n, \tau_n, v_n\right)}{\partial z_n} - \dfrac{\partial \mathrm{P}\left(z_n, \tau_n, v_n\right)}{\partial \tau_n}\left(\dfrac{\partial \Psi\left(z_n, \tau_n, v_n\right)}{\partial v_n}\right)^{-1} \dfrac{\partial \Psi\left(z_n, \tau_n, v_n\right)}{\partial z_n} \\ D\mathrm{P}_{v_n}\left(z_n, \tau_n, v_n\right) = \dfrac{\partial \mathrm{P}\left(z_n, \tau_n, v_n\right)}{\partial v_n} - \dfrac{\partial \mathrm{P}\left(z_n, \tau_n, v_n\right)}{\partial \tau_n}\left(\dfrac{\partial \Psi\left(z_n, \tau_n, v_n\right)}{\partial v_n}\right)^{-1} \dfrac{\partial \Psi\left(z_n, \tau_n, v_n\right)}{\partial v_n} \end{cases} \tag{22}$$

The calculation of all terms in (22) is very easy using expressions in (13). The stability of the fixed point z_* will be made by analyzing the eigenvalues of the Jacobean matrix $D\mathrm{P}_{z_n}\left(z_*, \tau_*, v_*\right)$. In fact, this Jacobean matrix admits two eigenvalues. Depending on the position of these two eigenvalues from the unit circle, we can investigate the stability of the fixed point z_*. A sufficient condition for stability of z_* is that these two eigenvalues are inside the unit circle.

It is worth noting that for a chaotic attractor or a chaotic behavior of the impact mechanical oscillator, its impulsive hybrid dynamics generates 1-period unstable limit cycles. Then, after the detection of an unstable fixed point of this unstable limit cycle within the chaotic attractor; our goal is to control chaos through the stabilization of this identified unstable fixed point. This task is the subject of the next section.

7. STABILIZATION OF THE FIXED POINT

Stabilizing a fixed point z_* which was previously determined allows to have a 1-period limit cycle for the impact mechanical oscillator. Thus, we will look to determine the appropriate controller $\Delta v_n = v_n - v_*$, with $v_* = 0$, in order to stabilize the linearized controlled Poincaré map defined by (18). Then, we will adopt a state feedback controller as follows:

$$\Delta v_n = K\Delta z_n \tag{23}$$

where K is the gain of the controller Δv_n (or v_n).

By defining a standard candidate Lyapunov function: $V\left(\Delta z_n\right) = \Delta z_n{}^T S \Delta z_n$, with $S = S^T > 0$, then the linearized controlled Poincaré map (18) is asymptotically stable in the closed loop if the following Linear Matrix Inequality (LMI) is feasible:

$$\begin{bmatrix} S & DP_{z_n}\left(z_*, \tau_*, v_*\right) + DP_{v_n}\left(z_*, \tau_*, v_*\right)Q \\ \left(DP_{z_n}\left(z_*, \tau_*, v_*\right) + DP_{v_n}\left(z_*, \tau_*, v_*\right)Q\right)^T & S \end{bmatrix} > 0 \tag{24}$$

with $Q = KS$.

The gain K of the state feedback controller (23) is then defined by $K = QS^{-1}$. This gain K will ensure the asymptotic stability of the fixed point z_* of the constrained Poincaré map (13). Therefore, by applying the control law:

$$v_n = \begin{cases} K\left(z_n - z_*\right) & \text{if} \quad \left\|\Delta z_n\right\| \le \varepsilon \\ 0 & \text{otherwise} \end{cases}, \text{ for some } \varepsilon << 1 \tag{25}$$

in the impulsive hybrid linear dynamics (5)-(6), the identified fixed point z_* and hence the unstable limit cycle is stabilized. Accordingly, the chaotic behavior of the impact oscillator will be controlled.

8. CONTROL OF CHAOS: NUMERICAL SIMULATIONS

In this section, we propose to apply the developed controller v_n (25) to verify its effectiveness in the control of chaos exhibited in the impulsive hybrid dynamics of the impact mechanical oscillator. Then, based on Section 5, we have identified the following fixed point $z_* = \left[0.0083 \quad 0.2982\right]^T$ of some unstable limit cycle embedded into the chaotic attractor of Figure 3b. The associated nominal impact time is $\tau_* = 2.2146\ s$. This fixed point z_* (and its corresponding nominal impact time τ_*) is that of the constrained uncontrolled Poincaré map (13). The eigenvalues of the Jacobian matrix of the constrained

Poincaré map are -0.2043 and -3.7295. Because one of these two eigenvalues lies outside the unit circle, the identified fixed point (the UPO) is well unstable. Its stabilization will occur by means of the state feedback controller (25). Using the LMI in (24), the gain K of the controller v_n is computed to be: $K = \begin{bmatrix} 0.4121 & -0.4291 \end{bmatrix}$.

Using the impulsive hybrid non-autonomous linear dynamics (5)-(6) with the controller (25), Figure 4 shows the controlled period-1 hybrid limit cycle. In this plot, the solid circle is the identified fixed point z_* of the unstable limit cycle. Figure 5a shows the temporal evolution of the state vector of the mass. This plot reveals that the behavior of the impact mechanical oscillator becomes 1-periodic just after 12 seconds. This plot is depicted for an initial condition different to the fixed point. In Figure 5b, the variation in time of the controller v_n is shown. It is obvious that when the chaotic behavior of the impact oscillator is controlled, the value of the controller becomes very small and remains constant around $8 \times 10^{-4} N$. In addition, the maximal value of the designed controller v_n is about $0.5N$. This shows that our controller is feasible and can be applied to the impact mechanical oscillator. In addition, it reveals that chaos is well controlled by adding a small perturbation to the excitation input $u(t)$ of the impact mechanical oscillator.

In contrast, we have applied the same designed controller with the same gain K for a different parameter w in order to see the robustness of our OGY based control method. Then, we have chosen $w = 2.87\ rad / s$. Figure 6 shows the variation of the mass states (the position and the velocity) at the beginning of each period and the applied control law. In the two plots, we have simulated the impact mechanical oscillator for 80 periods (which correspond to a final time $80 \times T$). It is obvious the impact oscillation of the mechanical system is not periodic and seems to be completely chaotic. We emphasize that for the bifurcation parameters w smaller than the desired one $w = 2.8\ rad / s$, the oscillation is well periodic and it is with period 1. However, for higher values of the bifurcation parameter w, some

Figure 4. The controlled hybrid limit cycle of the impact mechanical oscillator for the bifurcation parameter $w = 2.8\ rad / s$

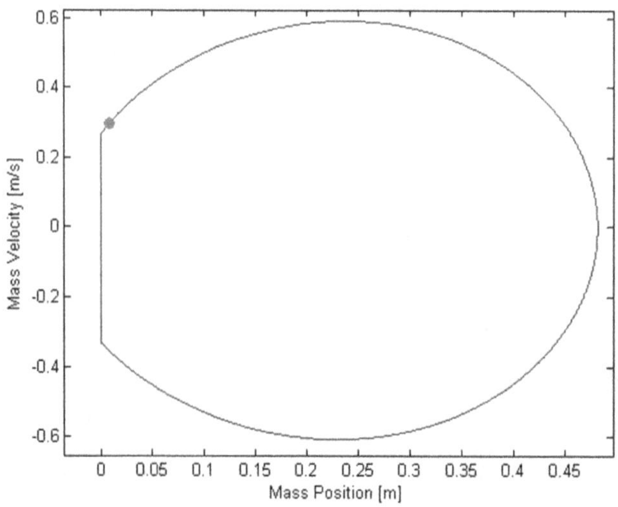

Figure 5. Temporal evolution of (a) the state (the position x and the velocity $\dot{x}$) of the mass, and (b) the designed controller v for the bifurcation parameter $w = 2.8\ rad\ /\ s$

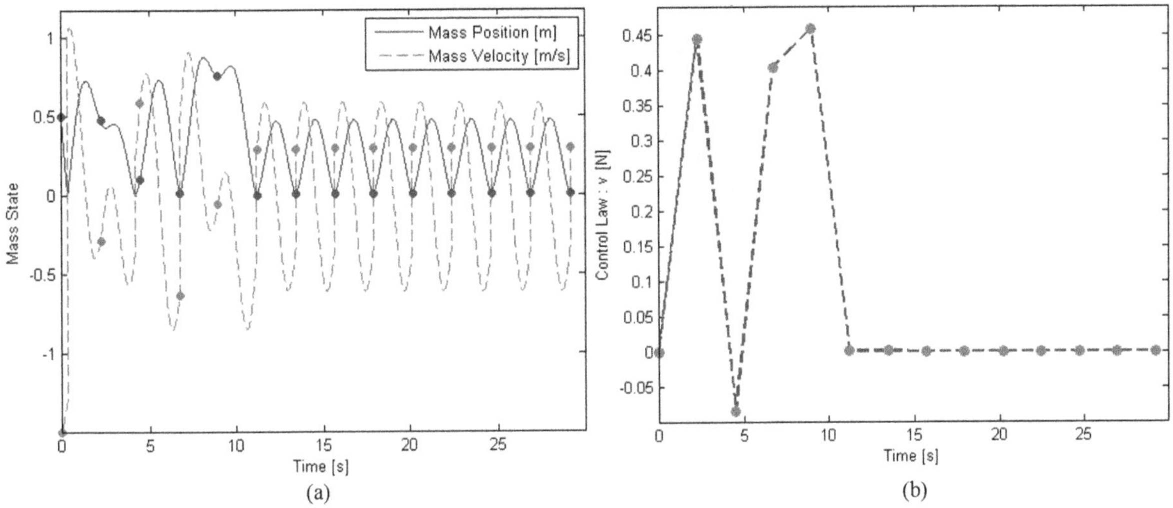

Figure 6. The mass states (a) and the designed control law v (b) with respect to the period number for $w = 2.87\ rad\ /\ s$

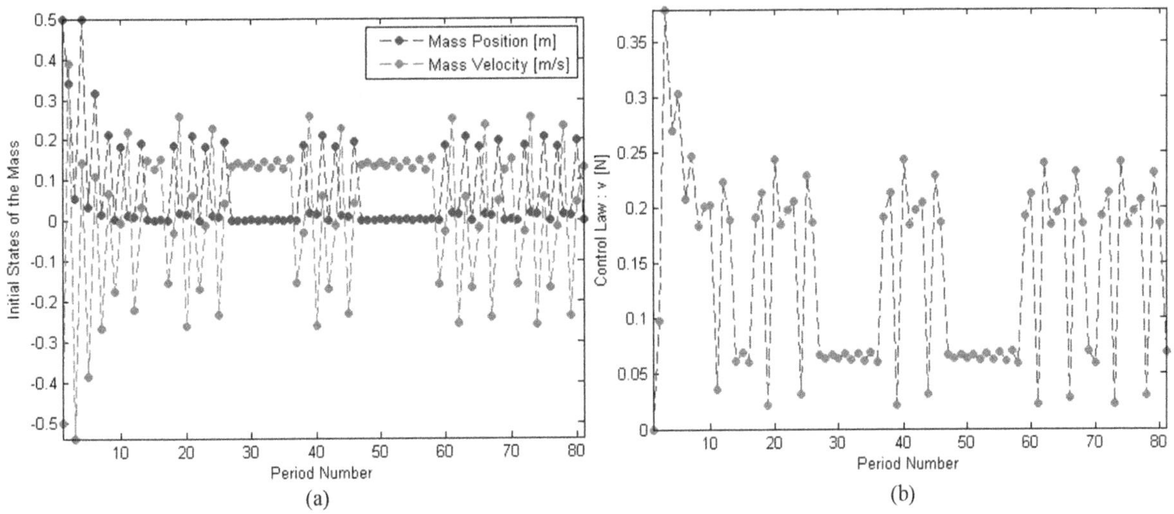

unexpected behaviors can be exhibited such as chaos and bifurcations. These behaviors are not studied in this chapter and they will be of our interest in future works.

Then, a small increase (about 0.07 rad/s) of the bifurcation parameter w from the desired one undergoes a radical change in the behavior of the impact mechanical oscillator and the oscillation is found to be chaotic. Hence, the impulsive hybrid dynamics of the impact oscillator is not robust under the developed OGY control method. Such missing robustness is generally considered as a serious drawback for practical applications and must be solved.

9. DISCUSSION

In this work, we have proposed a control approach of chaos exhibited in the impulsive hybrid linear non-autonomous dynamics of a simple impact mechanical oscillator with one degree of freedom. This mechanical system is constrained with one rigid contact and it is found that such system displays nonlinear complex phenomena such as period-doubling bifurcations, grazing bifurcations, crisis, etc.

Our control strategy is mainly based on the OGY method which in its turn is based on the linearization of the controlled Poincaré map. The designed control law is applied only at the beginning of the oscillation and it is maintained constant during each oscillation period T. Thus, using the impulsive hybrid linear non-autonomous dynamics and based on the Poincaré section, we have built an explicit expression of a constrained controlled Poincaré map. Relying on this constrained map, we have identified its fixed point. We have emphasized that such fixed point of the Poincaré map represents the state that belongs to the Poincaré section and it is the initial condition of some period-one hybrid limit cycle. Then, in order to transform a chaotic attractor into a desired period-1 limit cycle, we must stabilize the fixed point of the constrained Poincaré map. Therefore, and according to the OGY method, we have determined the linearized controlled Poincaré map around the desired unstable fixed point. Thus, we have obtained a classical linear discrete system. Using the linearized Poincaré map, we have designed a simple state feedback controller in order to stabilize the linear discrete system and hence in order to stabilize the desired fixed point of the constrained Poincaré map. We have shown that the design of the state feedback controller lies on the resolution of a linear matrix inequality. Finally, we have applied the designed control law to the impulsive hybrid linear dynamics of the impact mechanical oscillator by choosing a chaotic behavior. We showed that the chaotic hybrid oscillation is controlled and it is then transformed to a period-1 hybrid limit cycle. We showed also that by choosing a completely different departure point (i.e. different to the desired fixed point), the chaotic oscillation is rapidly controlled in only 10 seconds. In addition, we showed that the designed controller converges quickly to a very small value almost zero when it was applied for the desired bifurcation parameter w.

However, we showed that by modifying the pulsation w, the oscillation behavior of the impact mechanical oscillator becomes completely different and it is chaotic. Despite the success brought by the OGY method in experimental applications, the displayed chaotic behavior of the controlled impacting oscillator reveals the drawback of the OGY control method.

10. CONCLUSION

In this chapter, we have focused on the subject of chaos control of an impact mechanical oscillator. We have used the OGY methodology in order to design the controller. In this work and in the control strategy of chaos, we have interested only in the 1-period limit cycle with only one impact. However, for some bifurcation parameter, the impact mechanical oscillator can undergo several impacts in only one period. Then, it is important to control such behavior with multiple impacts that leads also to a 1-period limit cycle. This task will be our main future work. We propose also to apply our strategy of chaos control developed in this chapter to some other mechanical systems with multiple rigid constraints and also to impact electrical oscillators.

It is known nowadays that the OGY method found its success in many experimental applications in several different fields. However, there are some major drawbacks of the OGY control. For example, the control is restricted to one of the embedded unstable periodic orbits within the chaotic attractor. Then, when some system parameter changes, we must look for a new controller (a new gain of the designed controller). Moreover, applicability of the designed controlled for a desired fixed point is efficient in the vicinity of the desired parameter. In addition, the second major drawback of the OGY idea is the usually long reaching time until the system trajectory visits a close neighborhood of the target unstable periodic orbit and the control can be applied. In other worlds, the time to achieve control is unpredictable and could be very large. Thus, this OGY based control method do not deal with unforeseen behaviors (such as noises and perturbations) that can occur and then influence on the control process and hence the generated errors will be hardly remediable. Then, it will be very interesting to remedy these problems. These two points will be considered as our future direction of research. In fact, in literature, there are many improvements of the OGY control method that aim at overcoming some of its original limitations, as for example: control of high periodic and high unstable periodic orbit, control using time delay coordinates and control using multiparameter approach based on pole placement formalism. Thus, we will study these methods in order to improve the OGY control method for the impact mechanical oscillator.

Furthermore, we will interest in the future in the robust control of the constrained controlled Poincaré map that was developed in this chapter. Besides, we will focus on the analysis of the chaotic phenomena of the controlled impact mechanical oscillator under the OGY controller.

In addition, we will investigate in our future works the stable and unstable manifolds of such impact mechanical systems using the constrained Poincaré map in order to demonstrate existence of chaos for some bifurcation parameter. This will open a new topic on analysis and demonstration of chaos in impulsive hybrid nonlinear systems where their study is very difficult and requires some techniques such as the center manifold theory.

REFERENCES

Andrievskii, B. R., & Fradkov, A. L. (2003). Control of Chaos: Methods and Applications I: Methods. *Automation and Remote Control*, *64*(5), 673–713. doi:10.1023/A:1023684619933

Andrievskii, B. R., & Fradkov, A. L. (2004). Control of Chaos: Methods and Applications. II. Applications. *Automation and Remote Control*, *65*(4), 505–533. doi:10.1023/B:AURC.0000023528.59389.09

Awrejcewicz, J., & Lamarque, C. H. (2003). *Bifurcation and Chaos in Nonsmooth Mechanical Systems*. Singapore: World Scientific.

Barreto, E., & Grebogi, C. (1995). Multiparameter control of chaos. *Physical Review E: Statistical Physics, Plasmas, Fluids, and Related Interdisciplinary Topics*, *52*(4), 3553–3557. doi:10.1103/PhysRevE.52.3553 PMID:9963831

Begley, C. J., & Virgin, L. N. (2001). On the OGY Control of an Impact-Friction Oscillator. *Journal of Vibration and Control*, *7*(6), 923–931. doi:10.1177/107754630100700609

Boccaletti, S., Grebogi, C., Lai, Y.-C., Mancini, H., & Maza, D. (2000). The control of chaos: Theory and applications. *Physics Reports*, *329*(3), 103–197. doi:10.1016/S0370-1573(99)00096-4

Budd, C. J. (1996). Non-smooth dynamical systems and the grazing bifurcation. (C. J. Aston, Éd.) Cambridge, UK: Cambridge University Press.

de Korte, R. J., Schouten, J. C., & van den Bleek, C. M. (1995). Experimental control of a chaotic pendulum with unknown dynamics using delay coordinates. *Physical Review E: Statistical Physics, Plasmas, Fluids, and Related Interdisciplinary Topics*, *52*(4), 3358–3365. doi:10.1103/PhysRevE.52.3358 PMID:9963811

de Paula, A. S., & Savi, M. A. (2009). A multiparameter chaos control method based on OGY approach. *Chaos, Solitons, and Fractals*, *40*(3), 1376–1390. doi:10.1016/j.chaos.2007.09.056

de Souza, S. L., & Caldas, I. L. (2004a). Calculation of Lyapunov exponents in systems with impacts. *Chaos, Solitons, and Fractals*, *19*(3), 569–579. doi:10.1016/S0960-0779(03)00130-9

de Souza, S. L., & Caldas, I. L. (2004b). Controlling chaotic orbits in mechanical systems with impacts. *Chaos, Solitons, and Fractals*, *19*(1), 171–178. doi:10.1016/S0960-0779(03)00129-2

de Souza, S. L., Caldas, I. L., & Viana, R. L. (2007). Damping control law for a chaotic impact oscillator. *Chaos, Solitons, and Fractals*, *32*(2), 745–750. doi:10.1016/j.chaos.2005.11.046

de Weger, J. G. (2005). *The Grazing Bifurcation and Chaos Control.* Eindhoven, The Netherlands: Tehnische Universiteit Eindhoven.

Dressler, U., & Nitsche, G. (1992). Controlling chaos using time delay coordinates. *Physical Review Letters*, *68*(1), 1–4. doi:10.1103/PhysRevLett.68.1 PMID:10045097

Fradkov, A. L., & Evans, R. J. (2005). Control of chaos: Methods and applications in engineering. *Annual Reviews in Control*, *29*(1), 33–56. doi:10.1016/j.arcontrol.2005.01.001

Fradkov, A. L., Evans, R. J., & Andrievsky, B. R. (2006). Control of chaos: Methods and applications in mechanics. *Philosophical Transactions of the Royal Society A, 364*(1846), 2279-2307.

Fradkov, A. L., & Pogromsky, A. Y. (1996). *Introduction to Control of Oscillations and Chaos*. Singapore: World Scientific.

Grebogi, C., & Lai, Y. C. (1997). Controlling chaotic dynamical systems. *Systems & Control Letters*, *31*(5), 307–312. doi:10.1016/S0167-6911(97)00046-7

Gritli, H., Belghith, S., & Khraeif, N. (2012a). Intermittency and interior crisis as route to chaos in dynamic walking of two biped robots. International Journal of Bifurcation and Chaos, 22(3).

Gritli, H., Belghith, S., & Khraeif, N. (2012b). Cyclic-fold bifurcation and boundary crisis in dynamic walking of biped robots. International Journal of Bifurcation and Chaos, 22(10).

Gritli, H., Belghith, S., & Khraief, N. (2014a). Control of Chaos in an Impact Mechanical Oscillator. *Proceedings-Copyright IPCO*, 116-122.

Gritli, H., Khraeif, N., & Belghith, S. (2012c). Period-three route to chaos induced by a cyclic-fold bifurcation in passive dynamic walking of a compass-gait biped robot. *Communications in Nonlinear Science and Numerical Simulation*, *17*(11), 4356–4372. doi:10.1016/j.cnsns.2012.02.034

Gritli, H., Khraeif, N., & Belghith, S. (2013). Chaos control in passive walking dynamics of a compass-gait model. *Communications in Nonlinear Science and Numerical Simulation*, *18*(8), 2048–2065. doi:10.1016/j.cnsns.2012.12.009

Gritli, H., Khraief, N., & Belghith, S. (2014b). Further Analysis of the Period-Three Route to Chaos in Passive Dynamic Walking of a Compass-Gait Biped Robot. *Proceedings-Copyright IPCO*, 123-130.

Gutiérrez, E., & Arrowsmith, D. K. (2003). Control of a double impacting mechanical oscillator using displacement feedback. *International Journal of Bifurcation and Chaos in Applied Sciences and Engineering*, *14*(9), 3095–3113. doi:10.1142/S021812740401120X

Hübinger, B., Doerner, R., Martienssen, W., Herdering, M., Pitka, R., & Dressler, U. (1994). Controlling chaos experimentally in systems exhibiting large effective Lyapunov exponents. *Physical Review E: Statistical Physics, Plasmas, Fluids, and Related Interdisciplinary Topics*, *50*(2), 932–948. doi:10.1103/PhysRevE.50.932 PMID:9962053

Iqbal, S., Zang, X. Z., Zhu, Y. H., & Zhao, J. (2014). Bifurcations and chaos in passive dynamic walking: A review. *Robotics and Autonomous Systems*, *62*(6), 889–909. doi:10.1016/j.robot.2014.01.006

Jin, L., Lu, Q. S., & Twizell, E. H. (2006). A method for calculating the spectrum of Lyapunov exponents by local maps in non-smooth impact-vibrating systems. *Journal of Sound and Vibration*, *298*(4-5), 1019–1033. doi:10.1016/j.jsv.2006.06.035

Kalagnanam, J. R. (1994). Controlling Chaos: The Example of an Impact Oscillator. *Journal of Dynamic Systems, Measurement, and Control*, *116*(3), 557–564. doi:10.1115/1.2899253

Kaygisiz, B. H., Erkmen, I., & Erkmen, A. M. (2006). Intelligent analysis of chaos roughness in regularity of walk for a two legged robot. *Chaos, Solitons, and Fractals*, *29*(1), 148–161. doi:10.1016/j.chaos.2005.08.047

Lakshmanan, M., & Murali, K. (1996). *Chaos in Nonlinear Oscillators: Controlling and Synchronization* (Vol. 13). Singapore: World Scientific.

Lenci, S., & Rega, G. (2004). Regular nonlinear dynamics and bifurcations of an impacting system under general periodic excitation. *Nonlinear Dynamics*, *34*(3/4), 249–268. doi:10.1023/B:NODY.0000013507.01603.a3

Ott, E., Grebogi, C., & Yorke, J. A. (1990). Controlling chaos. *Physical Review Letters*, *64*(11), 1196–1199. doi:10.1103/PhysRevLett.64.1196 PMID:10041332

Parker, T. S., & Chua, L. O. (1989). *Practical numerical algorithms for chaotic systems*. New York: Springer-Verlag. doi:10.1007/978-1-4612-3486-9

Pfeiffer, F., & Glocker, C. (1996). *Multibody Dynamics with Unilateral Contacts*. New York: Wiley-VCH Verlag. doi:10.1002/9783527618385

Pun, D., Lua, S. L., Law, S. S., & Cao, D. Q. (1998). Forced vibration of a multidegree impact oscillator. *Journal of Sound and Vibration*, *213*(3), 447–466. doi:10.1006/jsvi.1997.1494

Pyragas, K. (1992). Continuous control of chaos by self-controlling feedback. *Physics Letters. [Part A]*, *170*(6), 421–428. doi:10.1016/0375-9601(92)90745-8

Ritz, T., Chweinsberg, A. S., & Dressler, U. (1997). Chaos control with adjustable control times. *Chaos, Solitons, and Fractals*, *8*(9), 1559–1576. doi:10.1016/S0960-0779(96)00155-5

Schöll, E., & Schuster, H. G. (2007). Handbook of Chaos Control (2nd ed.). Weinheim, Germany: Wiley-VCH.

Thota, P., & Dankowicz, H. (2006a). Continuous and discontinuous grazing bifurcations. *Physica D. Nonlinear Phenomena*, *214*(2), 187–197. doi:10.1016/j.physd.2006.01.006

Thota, P., & Dankowicz, H. (2006b). Analysis of grazing bifurcations of quasiperiodic system attractors. *Physica D. Nonlinear Phenomena*, *220*(2), 163–174. doi:10.1016/j.physd.2006.07.005

Thotaa, P., Zhaob, X., & Dankowicz, H. (2006). Co-dimension-Two Grazing Bifurcations in Single-Degree-of-Freedom Impact Oscillators. *Journal of Computational and Nonlinear Dynamics*, *1*(4), 328–335. doi:10.1115/1.2338658

Wagg, D. J. (2005). Periodic sticking motion in a two-degree-of-freedom impact oscillator. *International Journal of Non-linear Mechanics*, *40*(8), 1076–1087. doi:10.1016/j.ijnonlinmec.2005.03.002

Wiercigroch, M., & de Kraker, B. (2000). *Applied nonlinear dynamics and chaos of mechanical systems with discontinuities*. Singapore: World Scientific.

Wiggins, S. (2003). *Introduction to Applied Nonlinear Dynamical Systems and Chaos (Texts in Applied Mathematics)* (2nd ed.). Heidelberg, Germany: Springer.

Witvoet, G. (2005). *Control of chaotic dynamical systems using OGY*. Eindhoven, The Netherlands: Technische Universiteit Eindhoven.

KEY TERMS AND DEFINITIONS

Chaos: A nonlinear erratic phenomenon that is found to be exhibited in several physical systems.

Hybrid Limit Cycle: A periodic behavior of a dynamical system that has a transition in the state vector.

Hybrid Poincaré Map: A discrete application that is constrained with algebraic conditions.

Impact Mechanical Oscillator: A mechanical physical system that undergoes impacts with a rigid wall.

Linearized Poincaré Map: A Poincaré map linearized around a fixed point.

LMI: Linear Matrix Inequality.

OGY Control: A control method that was invented by Ott, Grebogi andYorke in 1990 to control chaos in nonlinear dynamics.

Chapter 10
Further Investigation of the Period-Three Route to Chaos in the Passive Compass-Gait Biped Model

Hassène Gritli
Institut Supérieur des Etudes Technologiques de Kélibia, Tunisia

Nahla Khraief
Ecole Supérieure de Technologie et d'Informatique, Tunisia

Safya Belghith
Ecole Nationale d'Ingénieurs de Tunis, Tunisia

ABSTRACT

This chapter presents further investigations into the period-three route to chaos exhibited in the passive dynamic walking of the compass-gait biped robot as it goes down an inclined surface. This discovered kind of route in the passive bipedal locomotion was found to coexist with the conventional period-one passive hybrid limit cycle. The further analysis on the period-three route chaos is realized by means of the Lyapunov exponents and the fractal Lyapunov dimension. Numerical computation method of these two tools is presented. The first return Poincaré map of the chaotic attractor and its basin of attraction are presented. Furthermore, the further study of the period-three passive gait is realized. The analysis of the period-three hybrid limit cycle is given. The balance between the potential energy and the kinetic energy of the biped robot is illustrated. In addition, the basin of attraction of the period-three passive gait is also presented.

1. INTRODUCTION

Robotics has evolved in recent years in various fields such as industrial robotics, medical robotics, domestic robotics, military robotics, etc. One of the most interesting applications in robotics is the analysis

DOI: 10.4018/978-1-4666-7248-2.ch010

of human walking through various prototypes of biped robots. The studies have been found to solve some problems related to the stability of human walking and also related to the design of active and passive prostheses of lower limbs of the human being. However, despite its simplicity, human walking is considered quite complex from a dynamic system point-of-view and it is not very well understood. The dynamic walking of biped robots is modeled by an impulsive hybrid nonlinear dynamics. To obtain a synergy between human walking gaits study and biped robots, a simple two-link bipedal mechanism will be a good basic experimental as well as theoretical model.

In the concept of bipedal robotics, passive dynamic walking has attracted the attention of many researchers and has been considered as the starting point for the control of biped robots. Passive dynamic models of biped walking have proven useful in understanding generalized principles that govern walking motions. The term passive dynamics arises from the ability of these models to walk without active control. Thus, passive dynamic walking is a mode of bipedal locomotion for which the biped robot requires no exogenous source of energy but it only uses gravity to walk on an inclined plane (McGeer, 1990; Goswami *et al.*, 1996, 1998; Wisse *et al.*, 2004; Gritli *et al.*, 2012c). This method of walk solves the problem of energy consumption of bipedal robots and get a maximum energetic efficiency. In addition, the use of passive dynamic walking is expected also to obtain additional insights into the design principles of legged locomotion in nature. The best known biped robot using passive dynamic walking is the compass-gait biped robot. This biped is a two-link bipedal mechanism that was originally studied in 1996 by Goswami *et al.* (1996, 1998). These researchers have shown that this type of bipedal walking can generate chaos and period-doubling bifurcations (Wiggins, 2003) based on certain intrinsic geometric and physical parameters of the biped robot. Until nowadays, many researchers are working on the passive dynamic walking of the compass-gait biped robot and other simple passive biped robots in order to find other properties that can help in the understanding of human walking and also for the control of walking gaits of biped robots. The list of publications is very long. We cite for example (Kaygisiz *et al.*, 2006; Safa *et al.*, 2007; Norris *et al.*, 2008; Kai & Shintani, 2011; Li & Yang, 2012; Li *et al.*, 2013).

One of the most important factors for successful and efficient walking is stability. In fact, the physical biped robots that have been designed around passive dynamic walkers are very sensitive to small perturbations. One of fundamental tools used to investigate stability of dynamic systems is the Lyapunov exponents. Since the fundamental paper of Oseledec (1968), the Lyapunov characteristic numbers or exponents are an important tool for the characterization of dynamical systems attractors of finite-dimensional nonlinear dynamic systems and their initial sensitivity to nearby initial conditions. The spectrum of Lyapunov exponents measures in fact the average divergence or convergence of nearby orbits along certain directions in state space. Sekhavat *et al.* (2004) employed the concept of Lyapunov exponents in order to analyze the stability of nonlinear dynamical systems and showed that the method is constructive and powerful. The sign of the largest Lyapunov exponent can infer the stability of systems and can rigorously prove the stability of the nonlinear system if numerical artefacts are under control (Kuo, 1995; Sekhavat *et al.*, 2004).

We showed recently that a cyclic-fold bifurcation is generated in the passive dynamic walking of the compass-gait biped giving rise to a cascade of period-doubling bifurcations and then a period-three route to chaos (Gritli *et al.*, 2012c). We have analyzed this period-doubling route to chaos by means of the spectrum of Lyapunov exponents and the fractal dimension (Gritli *et al.*, 2012a). In this chapter, we will revisit the passive dynamic walking of the compass-gait biped robot. We will give further results on the period-three passive limit cycle and the corresponding route to chaos. Our analysis will be based first on the computation of the Lyapunov exponents and the fractal dimension in order to quantify order and chaos (Gritli *et al.*, 2014). We will present different attractive structures of the chaotic attractor. The Poincaré first return map of the chaotic attractor and its basin of attraction are also presented. Furthermore, we will study the period-three passive limit cycle using the state space and the balance of the kinetic and potential energy. We will study also the basin of attraction of the period-three passive gait compared with that of the conventional period-one passive gait.

This chapter is organized as follows. In Section 2, an iterative review of chaos in passive dynamic walking of biped robots is presented. In Section 3, the compass-gait biped robot and its impulsive hybrid non-linear dynamics are presented. The passive walking patterns of the biped are also illustrated in this Section. Section 4 describes the analysis of the period-three route to chaos by means of the Lyapunov exponents and the fractal dimension. This section provide the calculation procedure of the spectrum of Lyapunov exponents and the Lyapunov dimension. Investigation of the chaotic attractor and the period-three passive gait are presented in Sections 5 and 6, respectively. Section 7 contains the concluding remarks of this work. Finally, in Section 8, some thoughts for the future are given.

2. BACKGROUND

More than a decade ago, there has been much interest stimulated in dynamic walking of passive biped robots. Tad McGeer (1990) showed that a completely unactuated and therefore uncontrolled legged-machine can perform a stable walk on a range of shallow slopes. He is the first who introduced the idea of ''passive dynamic walking'' in robotic vocabulary. After the first passive dynamic walker made by McGeer, many follow-on researchers have resorted to theoretic as well as experimental studies of passive dynamic walking model. They have developed some prototypes of passive biped walkers (Goswami *et al.*, 1996, 1998; Garcia *et al.*, 1998, 2000; Osuka & Kirihara, 2000; Wisse *et al.*, 2004; Collins *et al.*, 2005; Zhang *et al.*, 2008; Asano & Luo, 2009). The famous passive walkers are the point-foot walker, introduced by Garcia *et al.* (1998), and the compass-gait biped model which is studied by Goswami *et al.* (1996, 1998). The last prototype of biped robot is more realistic and more sophisticated than the point-foot walker.

The passive gait model of the compass-like biped robot is qualified as an impulsive hybrid dynamic system defined by algebro-differential equations. Nonlinear phenomena in passive dynamic walking are very complicated, and their investigation has proceeded with difficulty. The most interesting phenomenon that passive-dynamic walkers exhibit is the period-doubling bifurcation route to chaos (Goswami *et al.*, 1996, 1998). Goswami *et al.* were the first to report the occurrence of period-doubling bifurcation for the compass-like biped robot on a gentle slope leading progressively to chaotic motion. They have numerically showed that this biped robot exhibits period-doubling bifurcations by changing the slope angle and even parameters of the compass robot such as leg-mass location and legs' mass. After that, Garcia *et al.* (1998, 2000) showed that the simplest walking model, the "point-foot walker", and the knee model exhibit also period-doubling bifurcations and they showed the existence of both stable and unstable period gaits. In addition, Howell & Baillieul (1998) discovered a stable period-three gait and subsequent period-doublings leading to chaos for the "point-foot walker" without and with torso via homotopy method. Recently, chaos in the passive walking model of a point-foot walker has been investigated in (Li & Yang, 2012; Li *et al.*, 2013). Iqbal *et al.* (2014) provide an overview of previous literature on the chaotic behavior of passive dynamic biped robots and its control.

We showed recently that a cyclic-fold bifurcation is generated in the passive dynamic walking of the compass-gait biped robot giving rise to a cascade of period-doubling bifurcations and then a route to chaos (Gritli *et al.*, 2012c). This local-type bifurcation creates a pair of a period-3 stable gait and a period-3 unstable gait. Furthermore, we showed that the passive gait of a biped robot with unequal leg length exhibits also the cyclic-fold bifurcation with a hysteresis phenomenon (Gritli *et al.*, 2011b). In addition, we showed in (Gritli *et al.*, 2012a) that this walking locomotion exhibits two additional routes to chaos namely the interior crisis and the intermittency. We showed also in (Gritli *et al.*, 2011b, 2012c) that the compass-gait biped robot falls down because of the birth of a global bifurcation known as the boundary crisis. This type of global bifurcation is generated mainly by means of the period-3 unstable limit cycle. In (Gritli *et al.*, 2011a), we used an energy-tracking controller in order to stabilize and to track the period-three passive gait of the compass biped robot. In (Gritli *et al.*, 2013), we controlled chaos exhibited in the passive dynamic walking of the compass-gait biped in order to obtain a period-one stable walking cycle.

For a smooth dynamical system, several methods for calculating Lyapunov exponents have been well established (Parker & Chua, 1989; Ramasubramanian & Sriram, 2000; Lu *et al.*, 2005; Chen *et al.*, 2006). Many numerical algorithms allow an easy estimation of the spectrum of Lyapunov exponents for many smooth continuous systems described by differential equation of motion and for discrete maps described by difference equations. However, there is a numerical problem according to an overflow trouble due to the exponentially diverging solutions of a chaotic system. In order to keep the calculations away from this trouble, the Gram-Schmidth reorthonormalisation procedure is applied to the fundamental solution matrix (Parker & Chua, 1989). Therefore, after some integration steps, the normalization factors of the fundamental solution matrix are summed up over a long integration interval to calculate the Lyapunov exponents.

Study of chaos in robotics by means of Lyapunov exponents is not much investigated. A few of work have been found in literature. For practical robotic systems, it is in general almost impossible to determine Lyapunov exponents analytically and they very often have to be calculated numerically. In (Yang & Wu, 2006; Sun & Wua, 2012), authors investigated the balance control and analyzed the stability of a biped during disturbed standing while satisfying the constraints between the feet and the ground. Recently, we used the spectrum of Lyapunov exponents in order to investigate order and chaos in the passive dynamic walking of the compass-gait biped robot and in the semi-passive dynamic walking of the torso-driven biped robot as they descend some inclined surfaces (Gritli *et al.*, 2012a).

3. HYBRID DYNAMICS AND PASSIVE WALKING PATTERNS OF THE COMPASS-GAIT BIPED ROBOT

3.1. The Compass-Gait Biped Robot

Figure 1 provides a diagrammatic representation of the compass-gait biped robot where the significant parameters in the dynamics description are given in Table 1 (Goswami *et al.*, 1998; Gritli *et al.*, 2012c). Such biped robot is a subclass of rigid mechanical systems subject to natural unilateral constraints. The compass-gait biped model is composed of two completely identical legs: a stance leg and a swing leg.

Figure 1. The model of the compass-gait biped robot down a slope ϕ.

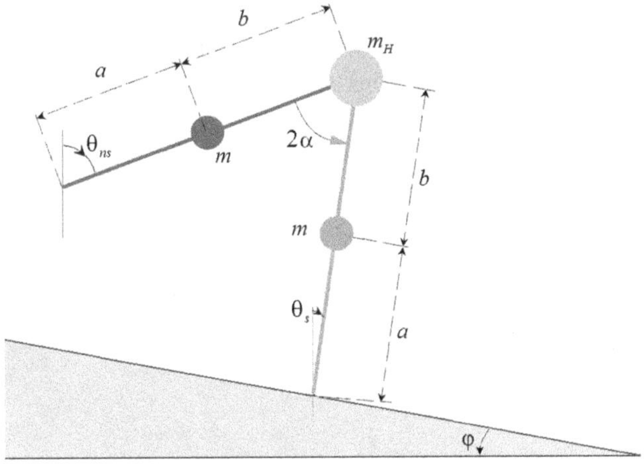

Table 1. Specifications of the compass biped simulation model

Symbol	Description	Value
a	Lower leg segment	0.5 m
b	Upper leg segment	5.0 kg
m	Mass of leg	5 kg
m_H	Mass of hip	10 kg
g	Gravitational constant	9.8m/s^2

In this biped model, the two legs are modeled as rigid bars without knees and feet, and with frictionless hip. Each leg has a mass m concentrated at a distance b from the hip of mass m_H.

For an adequate initial condition and a corresponding slope angle ϕ, the compass robot carries out a passive walk without any external action. This typical biped robot is powered only by gravity. The passive walk of such typical robot is constrained in the sagittal plane and is made up primarily of two phases: a swing phase and a very instantaneous impact phase. In the former case, the compass-like biped can be modeled as a double pendulum of two legs. The latter case occurs when the swing leg touches the ground and the previous support leg leaves the ground. The impact of the swing leg with the ground is assumed to be slipless plastic. The compass walk configuration is determined by the support angle θ_s and the nonsupport angle θ_{ns}. In Figure 1, α is the half-interleg angle.

3.2. Impulsive Hybrid Nonlinear Dynamics

It is well-known so far that the hybrid model of the passive dynamic walking of the compass-gait biped robot consists of nonlinear differential equations for the swing stage and algebraic equations for the impact stage (Gritli *et al.*, 2012c, 2014).

3.2.1. Continuous Dynamics of the Swing Phase

Let $\theta = \begin{bmatrix} \theta_{ns} & \theta_{s} \end{bmatrix}^T$ be the vector of generalized coordinates of the compass-gait biped robot. The motion of the compass-gait biped robot is described by the following Lagrangian system:

$$J(\theta)\ddot{\theta} + H(\theta,\dot{\theta}) + G(\theta) = 0 \tag{1}$$

This continuous dynamics is in fact subject to unilateral constraints:

$$\Omega = \left\{\theta \in \Re^2 : h_1(\theta) = l\left(\cos(\theta_s + \phi) - \cos(\theta_{ns} + \phi)\right) > 0\right\} \tag{2}$$

with $l = a + b$.

Matrices in (1) are defined by:

$$J(\theta) = \begin{bmatrix} mb^2 & -mlb\cos(\theta_s - \theta_{ns}) \\ -mlb\cos(\theta_s - \theta_{ns}) & m_H l^2 + m(l^2 + a^2) \end{bmatrix},$$

$$H(\theta,\dot{\theta}) = \begin{bmatrix} 0 & mlb\sin(\theta_s - \theta_{ns}) \\ -mlb\sin(\theta_s - \theta_{ns}) & 0 \end{bmatrix} \begin{bmatrix} \dot{\theta}_{ns}^2 \\ \dot{\theta}_s^2 \end{bmatrix}$$

and

$$G(\theta) = g\begin{bmatrix} mb\sin(\theta_{ns}) \\ -(m_H l + m(a + l))\sin(\theta_s) \end{bmatrix}.$$

3.2.2. Algebraic Equations of the Impact Phase

Algebraic equations of the impact phase are given by:

$$\begin{cases} \theta^+ = R_e\theta^- \\ \dot{\theta}^+ = S_e\dot{\theta}^- \end{cases} \tag{3}$$

where subscribes $^+$ and $^-$ denotes just after and just before the impact phase, respectively.

Matrices in (Eq3) are defined by:

$$R_e = \begin{bmatrix} 0 & 1 \\ 1 & 0 \end{bmatrix} \text{ and } S_e = Q_p^{-1}(\theta) Q_m(\theta),$$

$$\text{with } Q_p(\theta) = \begin{bmatrix} mb\left(b - l\cos\left(\theta_s - \theta_{ns}\right)\right) & ml\left(l - b\cos\left(\theta_s - \theta_{ns}\right)\right) + ma^2 + m_H l^2 \\ mb^2 & -mbl\cos\left(\theta_s - \theta_{ns}\right) \end{bmatrix}$$

$$\text{and } Q_m(\theta) = \begin{bmatrix} -mab & -mab + \left(m_H l^2 + 2mal\right)\cos\left(\theta_s - \theta_{ns}\right) \\ 0 & -mab \end{bmatrix}.$$

The impact phase occurs when θ and $\dot{\theta}$ belong to the following impact surface:

$$\Gamma = \left\{\theta, \dot{\theta}, \ / \ h_1(\theta) = l\left(\cos\left(\theta_s + \phi\right) - \cos\left(\theta_{ns} + \phi\right)\right) = 0, \quad h_2(\theta) = \frac{d\Phi_1(\theta)}{dt} = \frac{\partial \Phi_1(\theta)}{\partial \theta}\dot{\theta} < 0\right\} \tag{4}$$

3.2.3. Passive Walking Patterns of the Compass Biped

The bifurcation diagram of Figure 2a provides all passive gaits of the compass-gait biped robot while walking down a slope (Gritli *et al.*, 2012b, 2012c, 2013). Here, the bifurcation parameter is the slope angle ϕ. In this bifurcation diagram, the blue attractor A_1 is the conventional behavior exhibited by the passive compass-gait biped robot. A cascade of period-doubling bifurcations leads to chaos. However, the pink attractor A_2 is the one recently found by us (Gritli *et al.*, 2012c). This attractor reveals a scenario of a period-3 route to chaos. Figure 2b shows this route. The attractor A_2 is born at the cyclic-fold bifurcation (marked as CFB). This bifurcation occurs at $\phi = 3.8734°$. At this bifurcation, a period-three stable periodic orbit (p3-SPO) enters in collision with a period-three unstable periodic orbit (p3-UPO). Such p-3 UPO is responsible on the generation of a double boundary crisis (marked as BC) which is found to be the main cause of the falling down of the compass-gait biped robot (Gritli *et al.*, 2011b, 2012c). For the attractor A_1, the period-doubling to chaos is born from a period-1 stable gait (p1-SPO). Chaos is found to be terminated by the boundary crisis at $\phi = 5.201°$. However, in Figure 2b, the period-three stable gait undergoes a period-doubling leading to chaos. Chaos is dead for $\phi = 4.0036°$. Some periodic windows (PW) are appeared. The window which is marked by p9-PW in Figure 2b reveals the existence of a scenario of period-doubling to chaos originated from a period-9 passive gait (Gritli *et al.*, 2012c).

In the next of this chapter, we will interest in the analysis of the scenario of the period-three route to chaos presented in Figure 2b. This analysis is realized first by means of the spectrum of Lyapunov exponents and the fractal Lyapunov dimension. Secondly, we will analyze the chaotic attractor and the period-three passive gait.

Figure 2. Bifurcation diagrams: step period as a function of slope angle ϕ. (b) is an enlargement of (a)

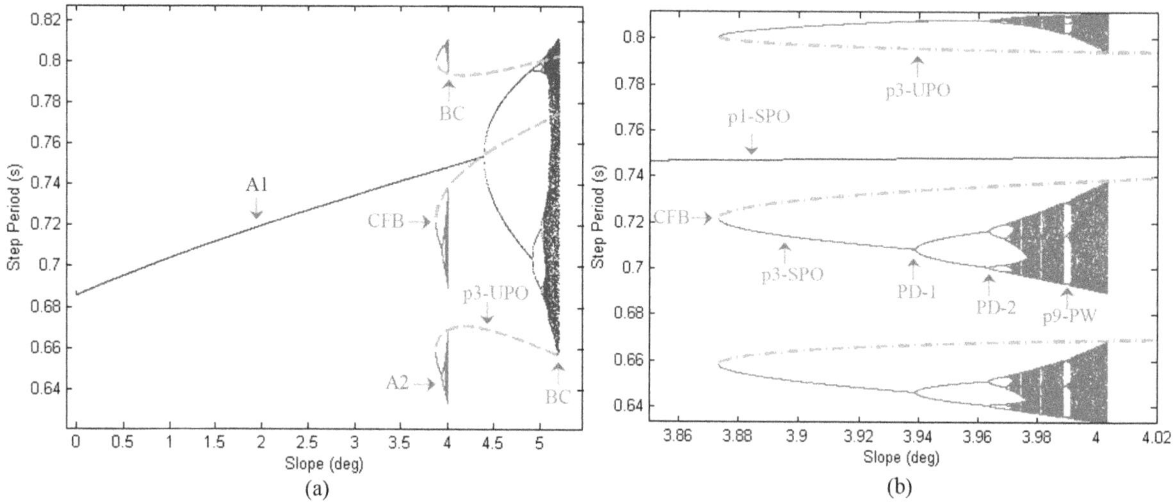

4. ANALYSIS OF ORDER/CHAOS VIA LYAPUNOV EXPONENTS AND FRACTAL DIMENSION

In the present section, we will analyze the variation of Lyapunov exponents and the fractal (Lyapunov) dimension in order to quantify order and chaos of the passive dynamic walking of the compass-gait biped robot.

The spectrum of Lyapunov exponents is one of the most remarkable characterizations of dynamical systems attractors of a finite-dimensional nonlinear dynamic system of dimension n and their initial sensitivity for initial conditions. The determination of Lyapunov exponents provides a qualitative and quantitative characterization of the dynamic behavior through the exponential convergence or divergence of neighboring trajectories evolving in the phase space. A system with ordinary autonomous differential equations of dimension n has in fact n Lyapunov exponents (Parker & Chua, 1989). Positive Lyapunov exponents measure the average exponential spreading of nearby trajectories, and negative exponents measure the exponential convergence of trajectories onto the attractor. The sum of the Lyapunov exponents is the rate of the average volume contraction. In order that the solution of the differential system is an attractor, it is necessary that the sum of Lyapunov exponents is certainly negative. If the attractor is an equilibrium point, then all Lyapunov exponents are negative. If the attractor is a periodic oscillation, then there is at least one zero Lyapunov exponent. If it is periodic, the largest Lyapunov exponent is zero and the other $\left(n-1\right)$ Lyapunov exponents are negative. In the case of a torus, the attractor has another zero Lyapunov exponent and the remaining Lyapunov exponents are negative. If the attractor is chaotic, then there are one or more positive Lyapunov exponents. If there are several (higher than 2) positive Lyapunov exponents, then the attractor is hyperchaotic (Parker & Chua, 1989).

Besides the Lyapunov exponents, the most fundamental property of a chaotic attractor that was considered is its fractal dimension. In fact, the identification of chaos includes finding a strange attractor in the dynamics of the state space, which is characterized by its fractal structure. The dimension of an attractor is a measure of the number of active variables and complexity of necessary equations to model the dynamics of the system dynamics (Sprott, 2003). Typical features of the fractal structure are:

- They have a fine structure and no characteristic length scale,
- They are too irregular to be described by ordinary geometry, both locally and globally, and
- They have a certain degree of self-similarity, which means that small pieces of the object resemble the whole in some respects.

Fractal structures can be either deterministic where they are precisely self-similar, or random they are only statistically self-similar. The self-similarity means that the structure is scale invariant.

Geometric objects with fractional dimensions are called fractals, and in dynamical systems, it has been found that chaotic attractors are fractal objects. The determination of the fractal dimension is therefore a method for characterizing a chaotic attractor (Banerjee & Verghese, 2001). The dimension is a kind of quantifier describing the attractor from the geometrical aspects. The dimension of an attractor gives us an estimate of the number of active degrees of freedom in the system. If the dimension of the attractor is not an integer, hence the attractor is a strange attractor.

Many methods have been proposed to characterize the fractal dimension of strange attractors produced by chaotic flows: the correlation dimension (Grassberger & Procaccia, 1983), the Hausdorff-Besicovitch dimension (Bergé *et al.*, 1984), and the Lyapunov dimension (also called the Kaplan-Yorke dimension) (Kaplan & Yorke, 1978).

In (Gritli *et al.*, 2012a), we calculated the spectrum of Lyapunov exponents and the fractal dimension for only the attractor A_1. In the present section, we will compute them for the attractor A_2 that reveals the period-three route to chaos. In addition, since the passive dynamics of the compass-gait biped robot is four-dimensional, then we have four Lyapunov exponents.

4.1. Calculation of the Lyapunov Exponents

Let $x = \left[\theta^T \quad \dot{\theta}^T\right]^T$ is the state variable. The passive gait model of the compass-gait biped robot defined by (1) and (3) can be rewritten in the following state representation:

$$\dot{x} = f\left(x\right) \text{if } x \in \Omega \tag{5}$$

$$x^+ = g\left(x^-\right) \text{if } x \in \Gamma \tag{6}$$

with Ω and Γ are defined by (2) and (4), respectively.

The solution x of the impulsive hybrid nonlinear model (5)-(6) can be expressed in terms of flow as:

$$x\left(t\right) = \varphi\left(t, x_0\right) \tag{7}$$

with initial condition $x\left(t_0\right) = \varphi\left(t_0, x_0\right)$.

For a smooth dynamical system which is represented by only the dynamics (5), methods for calculating Lyapunov exponents based essentially on the integration of a variational equation have been well established (Parker & Chua, 1989; Ramasubramanian & Sriram, 2000; Lu *et al.*, 2005; Chen *et al.*, 2006). The variational equation of (5) is given by:

$$\delta\dot{x} = J_f(x)\delta x \tag{8}$$

where J_f is the Jacobian matrix of the function f.

The variational equation (8) can be expressed in terms of the fundamental solution matrix Φ as follows (Parker & Chua, 1989):

$$\dot{\Phi}(t, x_0) = J_f(x)\Phi(t, x_0) \tag{9}$$

where $\Phi(t_0, x_0) = I_4$.

The fundamental solution matrix Φ is called very often the sensibility matrix in the initial state x_0 (Hiskens & Pai, 2000) and is given by:

$$\Phi(t, x_0) = \frac{\partial \varphi(t, x_0)}{\partial x_0} \tag{10}$$

For continuous systems, the fundamental solution matrix Φ is obtained by integrating the following system:

$$\begin{cases} \dot{x} = f(x) \\ \dot{\Phi}(t, x_0) = J_f(x)\Phi(t, x_0) \end{cases} \text{for} \begin{cases} x_0 = \varphi(t_0, x_0) \\ \Phi(t_0, x_0) = I_4 \end{cases} \tag{11}$$

However, impulsive hybrid systems defined by (5)-(6) exhibit discontinuities in the Jacobian matrix J_f at the impact instant with the impact surface Γ. The jumps in the fundamental solution matrix will be computed analytically according to (Muller, 1995; Hiskens & Pai, 2000) as follows:

$$\Phi^+(\tau, x_0) = \Pi(x^-, x^+)\Phi^-(\tau, x_0) \tag{12}$$

where:

$$\Pi(x^-, x^+) = G(x^-) + \frac{\left(f(x^+) - G(x^-)f(x^-)\right)H(x^-)^T}{H(x^-)^T f(x^-)} \tag{13}$$

with $G(x) = \dfrac{\partial g(x)}{\partial x}$ and $H(x) = \dfrac{\partial h_1(x)}{\partial x}$. In (12), τ is the impact instant.

Therefore, calculation of the fundamental solution matrix Φ requires mainly a precise knowledge of the impact time τ and of course of the jump condition (12). Accordingly, in order to calculate the fundamental solution matrix Φ for the impulsive hybrid nonlinear dynamic (5)-(6), we must solve the dynamic system (11) and the following jump system:

$$\begin{cases} x^{+} = g\left(x^{-}\right) \\ \Phi^{+}\left(\tau, x_{0}\right) = \Pi\left(x^{-}, x^{+}\right)\Phi^{-}\left(\tau, x_{0}\right) \end{cases} \text{if } x \in \Gamma \tag{14}$$

Contrary to periodic systems, there is a numerical problem for chaotic systems due to the exponential divergence of the fundamental solution matrix Φ. In order to remedy to this divergence problem, the Gram-Schmidth Reorthonormalisation (GSR) procedure is applied to the matrix Φ at some given time. For the case of discontinuous systems, the GSR procedure will be applied at impact instants τ. Then, the GSR procedure will be applied on the matrix Φ^{+}. Therefore, the Lyapunov exponents λ_i will be calculated as follows:

$$\lambda_{i} = \lim_{\tau \to +\infty} \frac{1}{\tau} \sum \left(\Phi^{+}\left(\tau, x_{0}\right) v_{i}\right) \tag{15}$$

with $v_i \in \Re^{4\times 1}$ and the ith element of v_i is 1 and the remaining elements are equal to 0.

4.2. Calculation of the Fractal Lyapunov Dimension

The Lyapunov dimension is named also the Kaplan-Yorke dimension (Kaplan & Yorke, 1978). The fractal Lyapunov dimension is calculated using the spectrum of Lyapunov exponents as follows (Gritli *et al.*, 2012a).

Let λ_1, λ_2, λ_3 and λ_4 be the 4 Lyapunov exponents where λ_1 is the largest one and λ_4 is the smallest one. Let j be an integer such that:

$$S_{j} = \sum_{i=1}^{j} \lambda_{i} \geq 0 \text{ and } S_{j+1} = \sum_{i=1}^{j+1} \lambda_{i} < 0 \tag{16}$$

Accordingly, the fractal Lyapunov dimension d_L is given by:

$$d_{L} = j - \frac{S_{j}}{\lambda_{j+1}} \tag{17}$$

4.3. Spectrum of Lyapunov Exponents and Lyapunov Dimension for the Compass-Gait Biped Robot

Variation of the spectrum of the Lyapunov exponents and the fractal Lyapunov dimension with respect to the slope angle ϕ are given by Figure 3a and Figure 3b, respectively. These diagrams are depicted for slopes between 3.85° and 4.02°. Recall that the cyclic-fold bifurcation is born at $\phi = 3.8734°$, and chaos of the attractor A_2 is terminated at $\phi = 4.0036°$. The phenomenon of the period-three route to chaos occurs between 3.8734° and 4.0035°. Outside of this range, the passive walking pattern is of period-1 (marked as p1-SPO). Obviously, in this range, the first (largest) Lyapunov exponent λ_1 is zero while the three remaining Lyapunov exponents are negative for slopes between 3.8734° and 3.9708°. In this interval of slopes, the passive gait of the compass biped robot is periodic and it exhibits a cascade of period-doubling bifurcations when the slope angle increases. The period-doubling (PD) phenomenon is expressed in each diagram with a cascade of parabolic curves. Furthermore, the attractor dimension remains almost constant at an integer value 1 when the gait is periodic except at the period-doubling bifurcations where the Lyapunov dimension is equal to 2. This translates the fact that the periodic attractor is dimensionally a line and all trajectories expand in state-space in only one direction.

However, for slopes higher than 3.9708, the largest Lyapunov exponent, λ_1 (λ_2), oscillates between positive (zero) and zero (negative) values corresponding to chaotic and periodic gaits. The two other Lyapunov exponents, λ_3 and λ_4, remain always negative. The periodic gaits bring in fact about the existence of periodicity windows in the chaotic regime. In addition, we emphasize that λ_1 reaches its highest value which is equal to 0.3 at $\phi = 4.0035°$. Moreover, when the passive walking pattern is chaotic, the Lyapunov dimension is found to be higher than 2 and increases to reach its maximum fractal value 2.15 at $\phi = 4.0035°$. For a system with a four-dimensional state space, this indicates a strong phase-space volume contraction. It indicates also that the attractor is dimensionally close to a Euclidean plane.

Figure 3. Variation of the spectrum of Lyapunov exponents (a) and the fractal Lyapunov dimension (b) as the slope angle ϕ increases for the period-three route to chaos

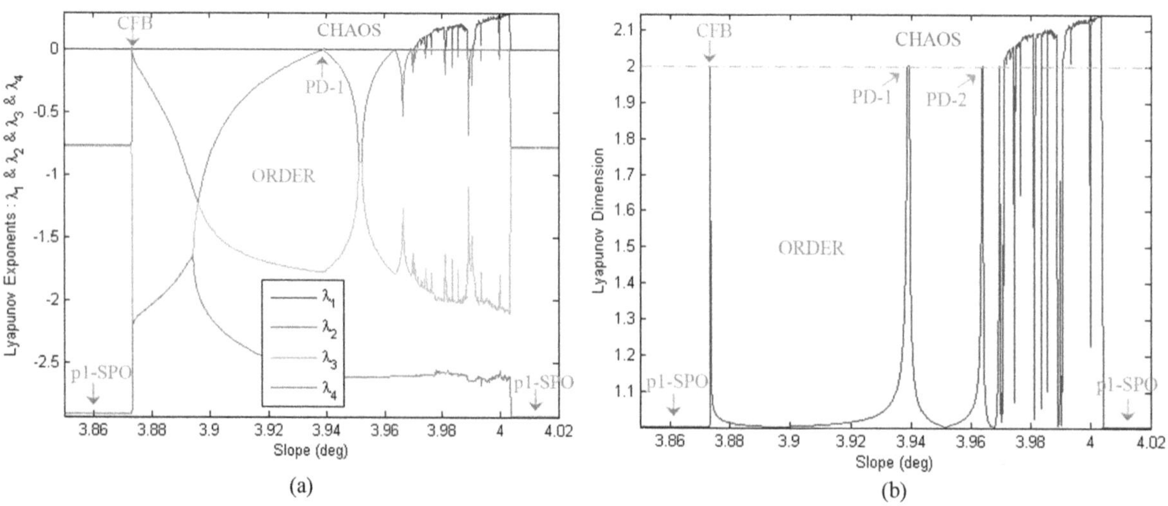

5. ANALYSIS OF THE CHAOTIC ATTRACTOR

In this section, we will analyze the chaotic attractor for the slope angle $\phi = 4.0035°$.

5.1. The Chaotic Attractor and its First Return Map

Figure 4 shows different forms of this chaotic attractor in different phase-spaces. The passive dynamic walking is described by a complete disappearance of order and is an extreme case of an asymmetric gait and affirming thus that the attractor is chaotic.

Obviously, in Figure 4a and Figure 4b, the form of the chaotic attractor is quite attractive. Figure 4a shows the chaotic attractor in 2D plotted with respect to the angular velocities of the two legs of the compass-gait biped robot. The chaotic attractor seems like a butterfly. In addition, Figure 4b manifests another structure of the chaotic attractor plotted in the three-dimensional state-space. The chaotic attractor has the shape of a heart. The largest Lyapunov exponent and the fractal dimension were computed to be about 0.3 and 2.15, respectively.

Figure 5 shows the Poincaré first return map in 2D. Figure 5b is an enlargement of Figure 5a. The first return map is composed of an infinite number of points irregularly distributed into three arcs of curves. By making an enlargement of the right part (Figure 5b), we can note that the first return map consists of several closed lines separated by empty spaces. This confirms that the attractor is well chaotic and it has a fractal dimension.

5.2. Basin of Attraction of the Chaotic Attractor

Localization of the basin of attraction of some limit set is very important. It shows the set of admissible initial conditions allowing the gait of the biped robot to converge to some domain of attraction like the period-3 passive stable gait or the chaotic attractor. Furthermore, coexistence of two different attractors

Figure 4. Chaotic attractor for the slope angle ϕ =4.0035° in different state-spaces

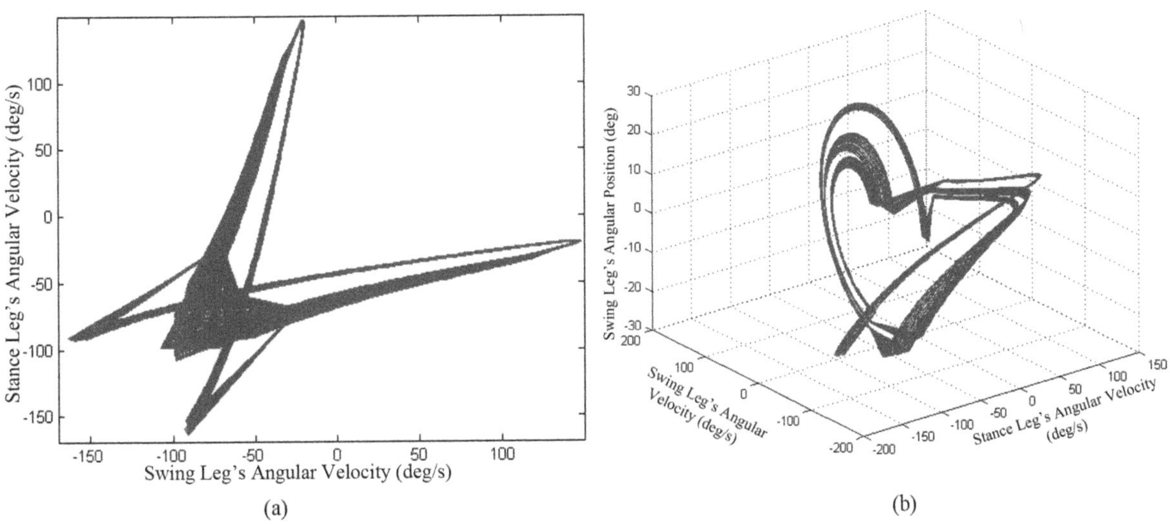

Figure 5. First return map for the slope angle ϕ =4.0035°. (b) is an enlargement of (a)

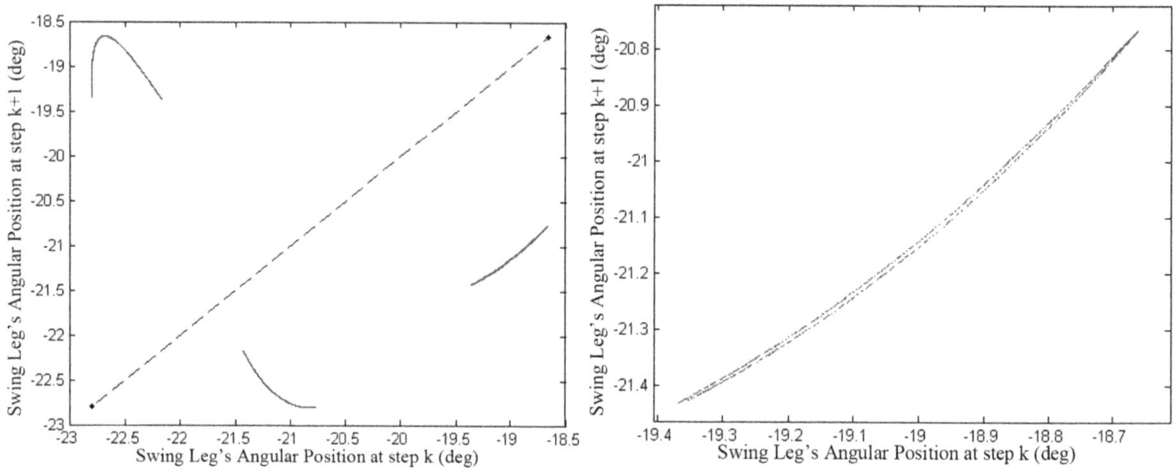

(a) (b)

for the same set of parameters suggests a good knowledge of the basin of attraction of the two attractors and also a good knowledge of the boundary of each basin of attraction.

Since the dynamics of the passive compass-gait biped robot is four-dimensional, graphical visualization of the basin of attraction is very difficult. Then, we should fix two state variables and vary the two others to obtain a 2D picture.

Figure 6 shows the basin of attraction of the chaotic gait and that of the period-one gait. The pink set is the basin of attraction of the chaotic attractor. However, the blue set reveals the basin of attraction of the period-one gait. Here, the basin of attraction is plotted for a fixed initial position of the two legs. We have chosen $\theta_{ns} = -21°$ and $\theta_s = -\theta_{ns} - 2\phi$. It is obvious that the basin of attraction of the chaotic

Figure 6. Basin of attraction of the chaotic attractor compared with that of the period-one limit cycle for the same slope ϕ =4.0035°

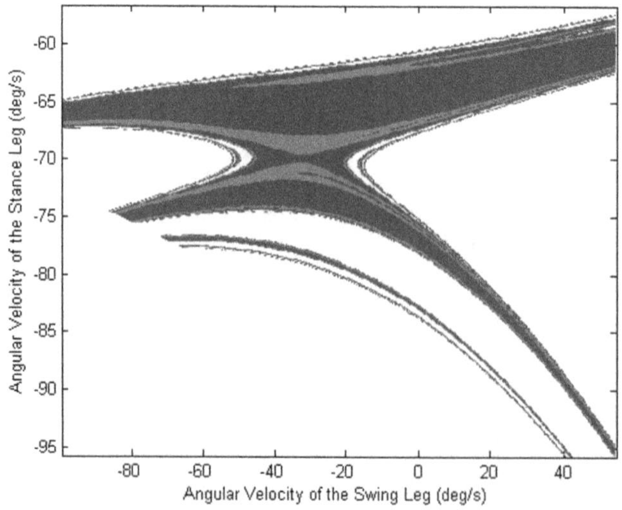

attractor is smaller than that of the period-one passive gait. Thus, a small perturbation on the initial conditions of the compass biped robot can bring the behavior of the period-one passive walk to another completely different form of locomotion, i.e. the chaotic gait.

6. ANALYZE OF THE PERIOD-THREE PASSIVE GAIT

In this subsection, we focus on the analysis of the period-three passive gait of the compass biped robot. Then, we choose the slope angle $\phi = 3.92°$.

6.1. Period-Three Passive Gait in the Phase-Plane

Figure 7 shows the angular velocity of each leg as a function of its angular position. In this state-space, small dark circles represent impact points of each leg with the ground. Whereas, small dark squares reveal states just after impact. Arrows are signs of behavior transition of each leg. The cycle of the passive gait consists of three steps in order that the compass robot returns to its initial state that is marked by solid squares. From an initial condition lying on the first cycle (marked by 1), the compass biped follows the cycles from 1 to 3 before it returns again to the first cycle.

Table 2 summarizes some characteristics of the period-three gait. It is clear that, from the first step (cycle) to the third step, the three quantities (namely the step length, the average velocity and the mechanical energy) of the biped robot increases. We showed in (Gritli *et al.*, 2011a) that the increase in the average velocity and in the step length could have a fundamental rule for the control of the bipedal walking of our compass biped robot to the period-three passive gait.

Tracking of such period-three passive gait has in fact an important relevance. It can be employed as a reference trajectory that must be tracked in order to jump for example some wall, some obstacle, or some undesired territory. Furthermore, this period-three passive gait can be used also in order to speed up or to slow down the biped robot. Accordingly, for the control principle design of passive dynamic walker,

Figure 7. A typical period-three passive limit cycle of the compass-gait biped for ϕ =3.92°

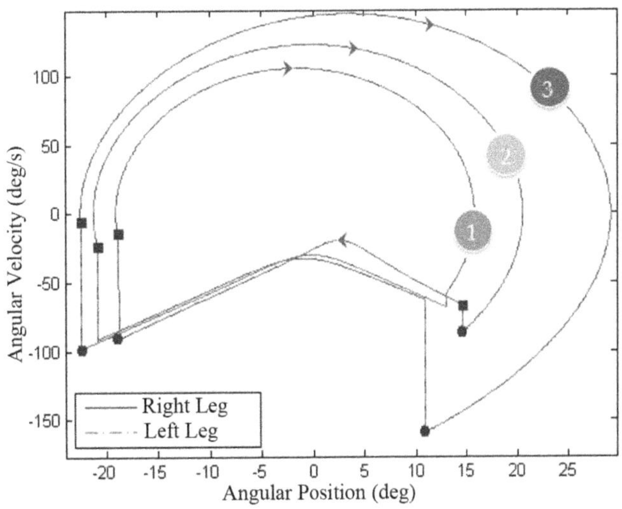

Table 2. Characteristics of the period-three stable passive gait

Cycle Number	Step Length (m)	Average Velocity (m/s)	Mechanical Energy (J)
1	0.5125	0.6355	153.9906
2	0.5824	0.8982	154.3163
3	0.6350	0.8937	154.8982

the aim is to make the walking pattern of a biped robot converges to a stable limit cycle or changes the original limit cycle to a new one by control input in order to take the biped robot to another walking plane. In (Gritli *et al.*, 2011a), we realized the tracking control of the period-three passive stable gait by means of an energy tracking control law.

6.2. Energy Balance

To investigate the period-3 gait, we refer to the energy balance in the passive gait of the compass biped robot. Figure 8 reveals the kinetic energy of the biped versus its potential energy during one cycle (which consists of three steps). As reported by Goswami *et al.* (1996, 1998), this energy balance plot consists of constant potential energies (dashed lines) translating the ground impact, and parallel inclined lines representing the swing phase. Moreover, each energy trajectory (line BC, or EF, or HI) is a straight line making a 135° angle with the kinetic energy axis. This affirms that the mechanical energy, which is the sum of the two energies: potential and kinetic, is constant during each step. The trajectory of the robot starts from the point A and follows the path ABCDEFGHI. The points C, F and I are the touchdown points of the swing leg with the ground. Horizontal lines CD, FG and IA represent the instantaneous loss of kinetic energy due to impact event with the ground.

Figure 8. Potential energy as a function of kinetic energy of the compass-gait biped robot for the period-three passive gait

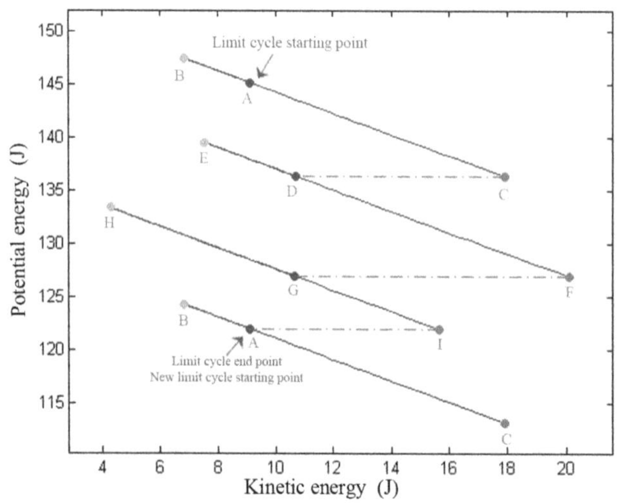

Furthermore, it is shown in (Goswami *et al.*, 1996, 1998) that there exists a limit cycle when the loss of kinetic energy at impact equals the change in potential energy during only one step. However, for the period-3 passive gait, this occurrence is not true. Indeed, the distance between an initial point (A or D or G) of a gait step and its end-point (D or G or A, respectively) is not equal to the distance CD or FG or IA, respectively. Nevertheless, the sum of lost kinetic energy is equal to the lost potential energy during a complete gait cycle that consists of three successive steps. For the studied period-three gait, the loss of kinetic energy during a gait cycle is calculated to be about 23.18 J. The mechanical energy of the compass robot for the three steps is indicated in Table 2. Obviously, the mechanical energies increases from the first step to the third one.

6.3. Basin of Attraction of the Period-Three Passive Gait

The basin of attraction of the period-three gait and that of the period-one gait are depicted in Figure 9. The blue set is the basin of attraction of the period-one gait, whereas the pink set reveals the basin of attraction of the period-three gait. The remaining white set refers to the initial conditions from which the compass robot falls down. The basin of attraction of the period-one gait is depicted for the same slope angle $\phi = 3.92°$. Each basin of attraction is plotted for a two state variables of the compass-gait model. It is depicted for a fixed initial position of the two legs of the compass biped robot. We have chosen $\theta_{ns} = -17°$ and $\theta_s = -\theta_{ns} - 2\phi$, and we have varied the angular velocities of the two legs.

Obviously, in Figure 9, the basin of attraction of the period-three passive gait is very small compared to that of the period-one gait. Furthermore, a small change either on initial push or on the initial position of the biped robot can provoke a different behavior: the robot can fall down or can follow either the period-one gait or the period-three one.

Figure 9. Basin of attraction of the period-one passive gait (blue) and the period-three passive gait (pink)

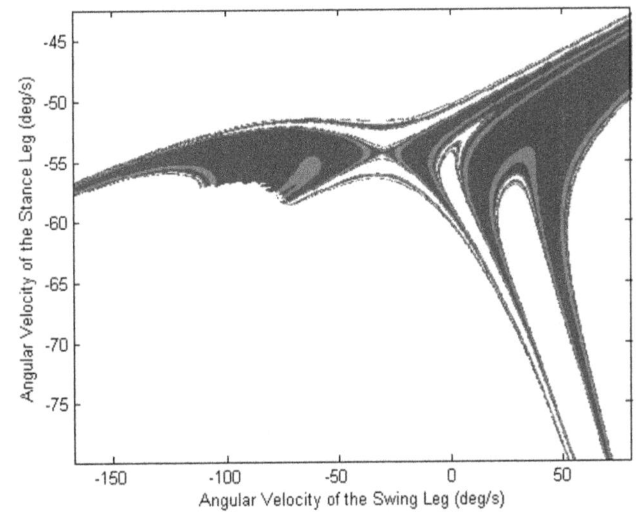

7. DISCUSSION

The focus of the work is a relative further study of chaos in robotics. We investigated the passive dynamic walking of the compass-gait biped robot on inclined slopes. The study was mainly articulated around the period-three route to chaos generated in the passive bipedal locomotion. Our further analysis on chaos and order was made by means of the spectrum of Lyapunov exponents and the fractal Lyapunov dimension. Computing method of these two tools was presented in details. We have demonstrated that the chaotic attractor can reach a fractal value about 2.15 and for which the largest Lyapunov exponent is calculate to be about 0.3. At these highest values, the bipedal chaos was found to be abruptly destructed by means of a boundary crisis. We showed that the fractal dimension reveals a strong volume contraction in the state space and the chaotic attractor is dimensionally close to a Euclidean plane.

In addition, we have analyzed the first return Poincaré map of the chaotic attractor and its basin of attraction. Two different attractive beautiful forms of the chaotic attractor in the phase plane are illustrated.

Moreover, we have analyzed some attractive characteristics of the period-three passive gait. This investigation was made with the state space, the balance of kinetic and potential energies, and the basin of attraction. We showed that the period-three passive gait can be employed as a target that should be tracked in order to increase the speed of the biped robot in only three steps and then in order to jump some undesirable obstacles. However, it is shown that the basin of attraction of this kind of passive gait is very small and any perturbation on the states of the biped robot can switch its actual bipedal locomotion to the conventional passive period-1 gait where its basin of attraction is too large. Then, stabilization and robustness of such period-three passive gait will be needed.

8. CONCLUSION

In this chapter, our analysis on chaos was realized by means of the Lyapunov exponents and the fractal dimension. It is well known that these two tools are used to identify chaos and also to study stability of dynamic system.

In our future direction of research, we will employ the spectrum of Lyaounov exponents in order to study stability of hybrid limit cycle of the passive gait of the compass biped robot and also other models of biped robots like the torso-driven biped robot.

In addition, in our future works, we will interest in the computation of the stable and unstable manifolds of the chaotic attractor in order to demonstrate the existence of chaos in the passive dynamic walking of the compass-gait biped robot for some specified bifurcation parameter such as the slope angle. Furthermore, our intention is to use to center manifold theorem in order to transform the impulsive hybrid nonlinear dynamics of the compass-gait biped robot into a normal form. This will help us to study bifurcations and also chaos.

REFERENCES

Asano, F., & Luo, Z. W. (2009). On efficiency and optimality of asymmetric dynamic bipedal gait. In *Proceedings of IEEE International Conference on Robotics and Automation*, (pp. 1972-1977). Kobe, Japan: IEEE. doi:10.1109/ROBOT.2009.5152322

Banerjee, S., & Verghese, G. C. (2001). *Nonlinear Phenomena in Power Electronics*. New York: IEEE Press. doi:10.1109/9780470545393

Bergé, P., Pomeau, Y., & Vidal, C. (1984). *Order within Chaos*. New York: Wiley.

Chen, Z.-M., Djidjeli, K., & Price, W. G. (2006). Computing Lyapunov exponents based on the solution expression of the variational system. *Applied Mathematics and Computation*, *174*(2), 982–996. doi:10.1016/j.amc.2005.05.016

Collins, S., Ruina, A., Tedrake, R., & Wisse, M. (2005). Efficient bipedal robots based on passive-dynamic walkers. *Science Magazine*, *307*, 1082–1085. PMID:15718465

Garcia, M., Chatterjee, A., & Ruina, A. (2000). Efficiency, speed, and scaling of two-dimensional passive-dynamic walking. *Dynamics and Stability of Systems*, *15*(2), 75–99. doi:10.1080/713603737

Garcia, M., Chatterjee, A., Ruina, A., & Coleman, M. (1998). The simplest walking model: Stability, complexity, and scaling. *Journal of Biomechanical Engineering*, *120*(2), 281–288. doi:10.1115/1.2798313 PMID:10412391

Goswami, A., Thuilot, B., & Espiau, B. (1996). *Compass-like biped robot. Part I: Stability and bifurcation of passive gaits*. France: INRIA.

Goswami, A., Thuilot, B., & Espiau, B. (1998). A study of the passive gait of a compass-like biped robot: Symmetry and chaos. *The International Journal of Robotics Research*, *17*(12), 1282–1301. doi:10.1177/027836499801701202

Grassberger, P., & Procaccia, I. (1983). Characterization of strange attractors. *Physical Review Letters*, *50*(5), 346–349. doi:10.1103/PhysRevLett.50.346

Gritli, H., Belghith, S., & Khraeif, N. (2012a). Intermittency and interior crisis as route to chaos in dynamic walking of two biped robots. International Journal of Bifurcation and Chaos, 22(3).

Gritli, H., Belghith, S., & Khraeif, N. (2012b). Cyclic-fold bifurcation and boundary crisis in dynamic walking of biped robots. International Journal of Bifurcation and Chaos, 22(10).

Gritli, H., Khraeif, N., & Belghith, S. (2011a). A Period-Three Passive Gait Tracking Control for Bipedal Walking of a Compass-Gait Biped Robot. In *Proceedings of the IEEE-2011 International Conference on Communications, Computing and Control Applications*, (pp. 01-06). Hammamet, Tunisia: IEEE.

Gritli, H., Khraeif, N., & Belghith, S. (2011b). Cyclic-fold bifurcation in passive bipedal walking of a compass-gait biped robot with leg length discrepancy. In *Proceedings of IEEE International Conference on Mechatronics*, (pp. 851-856). Istanbul, Turkey: IEEE. doi:10.1109/ICMECH.2011.5971233

Gritli, H., Khraeif, N., & Belghith, S. (2011c). Falling of a Passive Compass-Gait Biped Robot Caused by a Boundary Crisis. In *Proceedings of the 4th Chaotic Modeling and Simulation International Conference* (pp. 155-162). Crete, Greece: Academic Press.

Gritli, H., Khraeif, N., & Belghith, S. (2012c). Period-three route to chaos induced by a cyclic-fold bifurcation in passive dynamic walking of a compass-gait biped robot. *Communications in Nonlinear Science and Numerical Simulation*, *17*(11), 4356–4372. doi:10.1016/j.cnsns.2012.02.034

Gritli, H., Khraeif, N., & Belghith, S. (2013). Chaos control in passive walking dynamics of a compass-gait model. *Communications in Nonlinear Science and Numerical Simulation*, *18*(8), 2048–2065. doi:10.1016/j.cnsns.2012.12.009

Gritli, H., Khraief, N., & Belghith, S. (2014). Further Analysis of the Period-Three Route to Chaos in Passive Dynamic Walking of a Compass-Gait Biped Robot. *Proceedings-Copyright IPCO*, 123-130.

Hiskens, I. A., & Pai, M. A. (2000). Trajectory sensitivity analysis of hybrid systems. *IEEE Transactions on Circuits and Systems-Part I. Fundamental Theory and Applications*, *47*(2), 204–220.

Howell, G. W., & Baillieul, J. (1998). Simple controllable walking mechanisms which exhibit bifurcations. *IEEE Conference on Decision and Control*, (pp. 3027–3032). Tampa, FL: IEEE. doi:10.1109/CDC.1998.757956

Iqbal, S., Zang, X. Z., Zhu, Y. H., & Zhao, J. (2014). Bifurcations and chaos in passive dynamic walking: A review. *Robotics and Autonomous Systems*, *62*(6), 889–909. doi:10.1016/j.robot.2014.01.006

Kai, T., & Shintani, T. (2011). A Gait Generation Method for the Compass-type Biped Robot on Slopes via Discrete Mechanics. In *Proceedings of 50th IEEE Conference on Decision and Control and European Control Conference (CDC-ECC)*, (pp. 675-681). Orlando, FL: IEEE. doi:10.1109/CDC.2011.6161060

Kaplan, J. L., & Yorke, J. A. (1978). Chaotic behavior of multidimensional difference equations, in functional differential equations and approximation of fixed points. *Lecture Notes in Mathematics*, *730*, 228–237. doi:10.1007/BFb0064320

Kaygisiz, B. H., Erkmen, I., & Erkmen, A. M. (2006). Intelligent analysis of chaos roughness in regularity of walk for a two legged robot. *Chaos, Solitons, and Fractals*, *29*(1), 148–161. doi:10.1016/j.chaos.2005.08.047

Kuo, A. D. (1995). An optimal control model for analyzing human postural balance. *IEEE Transactions on Bio-Medical Engineering*, *42*(1), 87–101. doi:10.1109/10.362914 PMID:7851935

Li, Q., Tang, S., & Yang, X. S. (2013). New bifurcations in the simplest passive walking model. *Chaos: An Interdisciplinary Journal of Nonlinear Science*, *23*(4), 043110. doi:10.1063/1.4824975 PMID:24387549

Li, Q., & Yang, X. S. (2012). New walking dynamics in the simplest passive bipedal walking model. *Applied Mathematical Modelling*, *36*(11), 5262–5271. doi:10.1016/j.apm.2011.12.049

Lu, J., Yang, G., Oh, H., & Luo, A. C. (2005). Computing Lyapunov exponents of continuous dynamical systems: Method of Lyapunov vectors. *Chaos, Solitons, and Fractals*, *23*(5), 1879–1892. doi:10.1016/j.chaos.2004.07.034

McGeer, T. (1990). Passive dynamic walking. *The International Journal of Robotics Research*, *9*(2), 62–68. doi:10.1177/027836499000900206

Muller, P. C. (1995). Calculation of Lyapunov exponents for dynamics systems with discontinuities. *Chaos, Solitons, and Fractals*, *5*(9), 1671–1681. doi:10.1016/0960-0779(94)00170-U

Norris, J. A., Marsh, A. P., Granata, K. P., & Ross, S. D. (2008). Revisiting the stability of 2D passive biped walking: Local behavior. *Physica D Nonlinear Phenomena*, *237*(23), 3038–3045. doi:10.1016/j.physd.2008.07.008

Oseledec, V. I. (1968). A multiplicative ergodic theorem: Lyapunov characteristic numbers for dynamical systems. *Transactions of the Moscow Mathematical Society*, *19*, 197–231.

Osuka, K., & Kirihara, K. (2000). Motion Analysis and Experiments of Passive Walking Robot Quartet II. In *Proceedings of IEEE International Conference on Robotics and Automation*, (pp. 3052-3056). San Francisco, CA: IEEE. doi:10.1109/ROBOT.2000.846491

Parker, T. S., & Chua, L. O. (1989). *Practical numerical algorithms for chaotic systems*. New York: Springer-Verlag. doi:10.1007/978-1-4612-3486-9

Ramasubramanian, K., & Sriram, M. S. (2000). A Comparative study of computation of Lyapunov spectral with different algorithms. *Physica D Nonlinear Phenomena*, *139*(1-2), 72–86. doi:10.1016/S0167-2789(99)00234-1

Safa, A. T., Saadat, M. G., & Naraghi, M. (2007). Passive dynamic of the simplest walking model: Replacing ramps with stairs. *Mechanism and Machine Theory*, *42*(10), 1314–1325. doi:10.1016/j.mechmachtheory.2006.11.001

Sekhavat, P., Sepehri, N., & Wu, Q. (2004). Calculation of Lyapunov exponents using nonstandard finite difference discretization scheme: A case study. *Journal of Difference Equations and Applications*, *10*(4), 369–378. doi:10.1080/10236190310001625244

Sprott, J. (2003). *Chaos and Time-Series Analysis*. Oxford, UK: Oxford University Press.

Sun, Y., & Wua, C. Q. (2012). Stability analysis via the concept of Lyapunov exponents: A case study in optimal controlled biped standing. *International Journal of Control*, *85*(12), 1952–1966. doi:10.1080/00207179.2012.713985

Wiggins, S. (2003). *Introduction to Applied Nonlinear Dynamical Systems and Chaos (Texts in Applied Mathematics)* (2nd ed.). Heidelberg, Germany: Springer.

Wisse, M., Schwab, A. L., & Van Der Helm, F. C. (2004). Passive dynamic walking model with upper body. *Robotica*, *22*(6), 681–688. doi:10.1017/S0263574704000475

Yang, C., & Wu, Q. (2006). On stabilization of bipedal robots during disturbed standing using the concept of Lyapunov exponents. *Robotica*, *24*(5), 621–624. doi:10.1017/S0263574706002700

Zhang, P., Tian, Y., Liu, Z., Yang, S., & Tian, R. (2008). Further research and comparison of gaits for compass-like biped and kneed passive dynamic walker. In *Proceedings of International Conference on Intelligent Robotics and Applications*, (pp. 1216-1226). Wuhan, China: Academic Press. doi:10.1007/978-3-540-88513-9_129

KEY TERMS AND DEFINITIONS

Basin of Attraction: It contains the set of admissible initial conditions allowing the system trajectory to converge to some set of attraction.

Chaos: A nonlinear erratic phenomenon that is found to be exhibited in several physical systems.

Compass-Gait Biped Robot: A bipedal mechanical system composed of only two links and that can descend an inclined plane without any source of energy.

Fractal Dimension: Is the most basic property of a chaotic (strange) attractor. The dimension of a strange attractor is a measure of its geometric scaling properties or its "complexity".

Hybrid Limit Cycle: A periodic behavior of a dynamical system that has a transition in the state vector.

Lyapunov Exponents: the Lyapunov exponents are an important tool for the characterization of an attractor of a finite-dimensional nonlinear dynamic system and their excessive sensitivity to initial conditions.

Period-Three Route Chaos: A behavior in bifurcation diagram that shows a transition from a period-three solution to chaos.

Chapter 11
Bifurcation, Quasi-Periodicity, Chaos, and Co-Existence of Different Behaviors in the Controlled H-Bridge Inverter

Yosra Miladi
University of Sfax, Tunisia

Moez Feki
University of Sfax, Tunisia

ABSTRACT

This chapter deals with the analysis of the dynamic behavior of a controlled single-phase H-bridge inverter. The authors show that in addition to border collision bifurcation, when it is controlled with a time-delayed controller or with a dynamic controller that increases the system dimension, the H-bridge inverter can exhibit several other types of behaviors such as Neimark-Sacker bifurcation, quasi-periodicity, and coexistence of different periodic behaviors, as well as coexistence between periodic and chaotic behaviors. The proposed controllers are of different types. In addition to the Fixed-Point Induced Controller (FPIC), the authors also present the Time-Delayed Feedback Controller (TDFC) and the dynamic linear controller, such as the proportional-integral controller. The main issue of this chapter is to perform analysis within and beyond the stability region. Analytic calculation and numerical simulations are presented to confirm the obtained results.

1. INTRODUCTION

Power electronics is a discipline that has emerged from the need to convert electrical energy. Its field of application is wide and concerns industrial, commercial, residential and also aerospace environments. Nowadays, there is a great interest to use green energy such as the solar and wind energy, there is also a need to use mobile electrical devices supplied by batteries or fuel cells. Knowing that such power sources supply a low DC voltage, then the design of electronic devices that convert the low DC signal

DOI: 10.4018/978-1-4666-7248-2.ch011

into a high AC signal similar to power that would be available at an electrical wall outlet becomes a challenging endeavor. This prescribed aim has been achieved using two stages. The first one is a boosting DC to DC converter and the second is the DC to AC inverter (Luo and Ye, 2004; Rashid, 2001). Today, there are a myriad of such power converters ranging from the most basic to the most complicated and yielding to several optimal behaviors in different senses such as wave forms, power efficiency and circuit simplicity and size.

Power converters are the main devices that helped boosting the use of electronic devices. However, converters have been used long before their behavior is completely understood. Indeed, it's only on late eighties that Hamill (Hamill and Jefferies, 1988) reported the existence of several nonlinear phenomena in power converters. These include subharmonic oscillations, quasi-periodic operations, bifurcations and chaos.

Chaos in power electronics have intrigued many researchers (Di Bernardo and Tse, 2002; Deane and Hamill, 1990; Tse and Di Bernardo, 2002; Robert and Robert, 2002). In fact, engineers have frequently encountered chaos in power electronics systems, but more often than not this phenomenon was considered as strange and undesirable, hence engineers usually attempted to avoid chaos. During the last two decades, tools of analyzing bifurcations and chaos have been well developed. Therefore, the investigation of the very peculiar aspect of this phenomenon has become an attractive endeavor.

Power converters are basic switching circuits that are modeled by a number of linear differential equations corresponding to different topologies. The switching procedure imposed by different control schemes causes toggling among a set of linear circuits. Therefore the overall dynamics can be easily described by a piecewise linear model. In the case of simple linear models, a discrete map can be obtained by solving and stacking up solutions (Di Bernardo and Vasca, 2000; Robert and El Aroudi, 2006). The discrete-time modeling of the converters has provided an easy way to conceive controllers by using the discrete-time control theory and that can be applied further using the digital pulse width modulator (DPWM) (Feki et al., 2004; Robert et al., 2003; Hamza et al., 2011; Kaoubaâ et al., 2012).

Controlling the converters has many objectives such as ensuring converter closed loop stability and enhancing efficiency and dynamic performances. The switching nature of the continuous time mathematical model or the discrete time model with saturating controller (duty cycle) makes the control of the electric power converters a challenging topic in itself. In addition, the converter performances are tightly related to the adopted control method. Therefore, the control of electric power converters have attracted the attention of many researchers (Pinard, 2007; Rodriguez et al., 2000; Alvarez-Ramirez and Espinosa-Pérez, 2002).

In addition to linear control methods (Robert and Robert, 2002; Di Bernardo and Tse, 2002), several nonlinear techniques have been used to control the DC/DC and DC/AC converters as well. For instance, passivity based controllers (Sira-Ramirez et al., 1997), sliding mode controllers (Bensaada and Stambouli, 2013 ; Fernandez-Vargas and Ledwich, 2010; Tan et al., 2005, 2007 ; Wai and Shih, 2011), adaptive controllers (Albea et al., 2011), robust controllers (Olalla et al., 2011) and optimal controllers (Kawashima et al., 2012). The non conventional controllers such as neural controllers (Chen et al., 2012; Gnanasaravanan and Rajaram, 2012a), genetic algorithm (Gnanasaravanan and Rajaram, 2012b; Miladi et al., 2014) and fuzzy controllers (Baek and Park, 2012; Mazouz and Midoun, 2011; Nachidi et al., 2013; Guesmi et al., 2008) have also been used to control the power converters.

In this chapter we present an analysis of the dynamic behavior of a current mode controlled single phase H-bridge inverter driving a resistive and inductive load under the action of simple controller methods. We show that the switching nature of the converter provides it with a variety of complex behaviors. If in addition, the controller, though simple, increases the complexity of the synamic system, then additional complex behaviors may appear.

Straightforwardly, we show that if a proportional controller is used, then a steady state error occurs and it decreases when the proportional gain is increased. When it exceeds a stability limit a border collision bifurcation occurs leading to chaos. To improve the inverter performance, we suggest to append the fixed point induced controller (FPIC) to extend the possibility of increasing the proportional controller. The time delayed feedback controller (TDFC) is another method that can be appended to the proportional controller with the aim of stabilizing the periodic behavior for larger values of the proportional gain, however, the TDFC controller increases the system dimension by one and hence showing up several additional complex behaviors such as Neimark-Sacker bifurcation, quasi-periodicity and coexistence of chaotic behavior and periodic behavior. The dynamic linear controller (PI controller) is an other efficient method that may also help to eliminate the steady state error but also increases the system dimension by one and hence leads to the appearance of Neimark-Sacker bifurcation, quasi-periodicity and coexistence of two different periodic behaviors.

The rest of this chapter is outlined as follows. In the second section, we present the discrete model of the H-bridge inverter. In the third section, we investigate the effect of different controllers and we present the stability analysis. In section four, the additional complex behavior occurring outside the stability region is analyzed. Eventually, in section five, a discussion of the presented behaviors and comparison between different controllers' efficiency is presented. Finally, section six includes the concluding remarks and perspectives.

2. H-BRIDGE INVERTER DISCRETE MODEL

A well-known structure, used in most variable speed drives, is the H-bridge. Figure 1 illustrates the circuit of this voltage inverter *i.e.* a DC/AC converter. The inverter is fed by a voltage source E and it supplies a resistive and inductive load (L, R). The output current is controlled by a current loop. The four equivalent switches, named S_1, S_2, S_3, and S_4, are shared among two pairs (S_1, S_2) and (S_3, S_4). Pairs are controlled by a DPWM in a complementary way. States of the switches define two distinct topologies of the inverter, T_1 (S_1, S_2 on and S_3, S_4 off) and T_2 (S_1, S_2 off and S_3, S_4 on), and yield to two opposite voltages across the load. For example, topology T_1 implies that the voltage across the load $v = E$.

Figure 1. H-bridge inverter

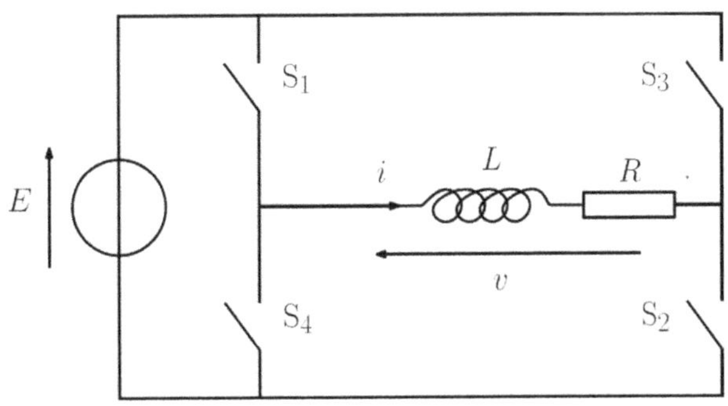

Over each topology, the unique state variable is the current in the load, denoted by *i*, and it can be described by a first order linear differential equation.

$$\frac{di}{dt} = -\frac{R}{L}i + \frac{E}{L} \quad \text{for } T_1 \tag{1}$$

$$\frac{di}{dt} = -\frac{R}{L}i - \frac{E}{L} \quad \text{for } T_2 \tag{2}$$

The normal operation involves a duty cycle $0 < d(n) < 1$ and a sequence of three states: $T_2 - T_1 - T_2$. However, when the controller saturates the modulator, *i.e.* $d(n) = 0$ or $d(n) = 1$, then we have only one state sequence T_2 or T_1. The normal sequence is shown in Figure 2.

Since the differential equations (1) and (2) are linear and simple then we can solve each of them in its corresponding time domain and then by stacking up the solutions, we can obtain a discrete time map expressing the current at the beginning of the nth period in terms of the current at the beginning of the $(n + 1)^{th}$ period. The obtained relation is given by (Robert and Robert, 2002):

$$x(n+1) = \alpha x(n) + \beta(2\sinh(\delta d(n)) - \sinh(\delta)) \tag{3}$$

where $x(n)$ is the normalized current state $\frac{i(n)R}{E}$, $\delta = \frac{RT}{2L}$, $\alpha = e^{-2\delta}$ and $\beta = 2e^{-\delta}$. To guarantee an aspect of the output current close to a sinusoidal shape, the frequency of the normalized reference current I_{ref} should be much smaller than the switching frequency. Considering a reference frequency ranging from zero to 100Hz then we may choose a switching period $T = 0.2$ms. Furthermore, if the switching

Figure 2. Centered DPWM voltage and output current

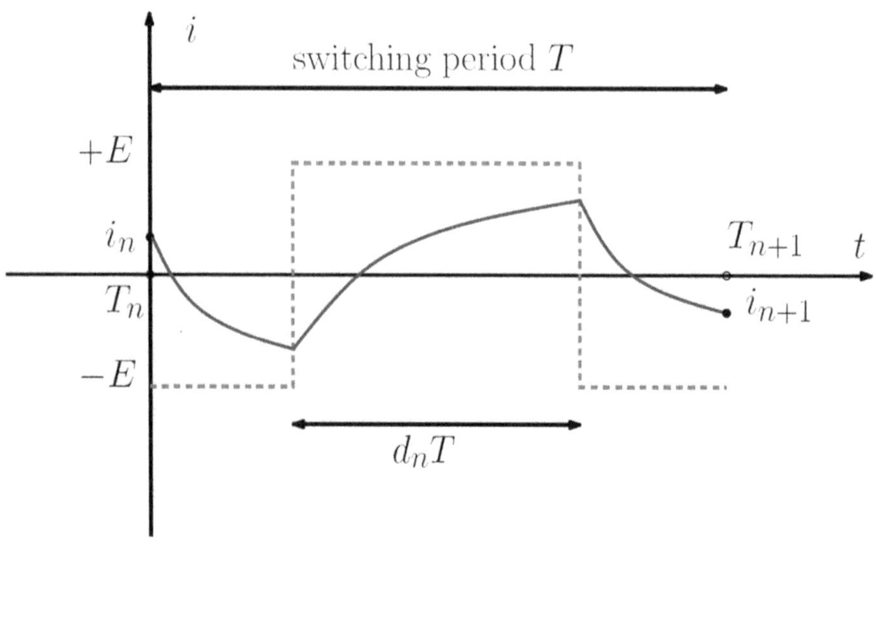

period is much smaller than the main time constant of the load L/R, the ripple current within a period will be low. By choosing $\delta = 0.2$, the ripple current is reduced to an acceptable value for this type of application. In the sequel, the system will be analyzed and simulated using the following values: $R = 40\Omega$, $L = 20\text{mH}$ and $E = 400\text{V}$, thus leading to the following constants: $\delta = 0.2$, $\alpha = 670.32 \times 10^{-3}$ and $\beta = 1.6375$. For simplicity of the analysis, the current reference is considered constant $I_{ref} = 0.5$.

3 CONTROLLING THE H-BRIDGE INVERTER

Figure 3 depicts an H-bridge current-programmed single phase inverter. In this section, several control strategies are applied to the H-bridge inverter and its behavior under those controllers is analyzed with respect to the control parameters variation.

3.1 Proportional Controller

It is always convenient in practice to start with a proportional controller which is the simplest to realize. In this case the controller expression is given by:

$$d(n) = d_0 + K_p\left(I_{ref} - x(n)\right) \tag{4}$$

where K_p is the proportional gain and $d_0 = \frac{1}{2}$ is a bias to the proportional controller that helps to diminish the controller variations. Thus, if we consider that $\sinh(\delta d^*) \simeq \delta d^*$ then the fixed point of the system is given by:

$$x^* = \frac{2\beta\delta K_p}{2\beta\delta K_p - \alpha + 1} I_{ref} \tag{5}$$

which implies that the reference current is only reached if K_p is increased to infinity. However, the closed loop Floquet multiplier is given by:

Figure 3. Controlled H-bridge inverter

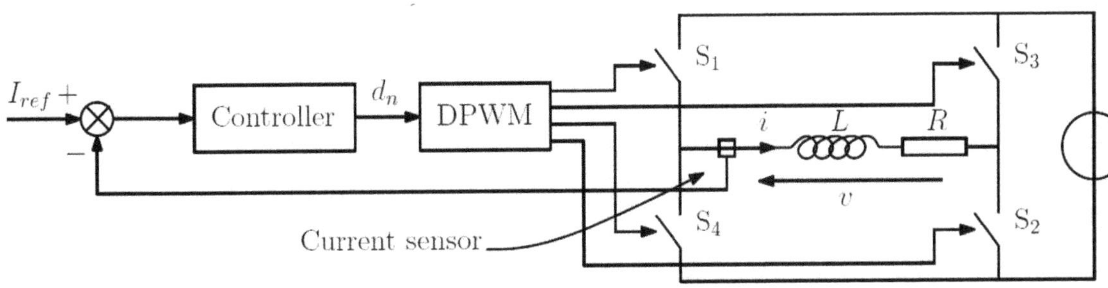

$$z = \alpha - 2\beta\delta K_p \cosh(\delta d^*) \tag{6}$$

where $d^* = d_0 + K_p(I_{ref} - x^*)$ is the steady state duty cycle. Therefore, if we take into account that $\delta d^* << 1$, then we may approximate $\cosh(\delta d^*) \simeq 1$ and thus the system is stable if the proportional gain is chosen to satisfy:

$$\frac{\alpha - 1}{2\beta\delta} < K_p < \frac{\alpha + 1}{2\beta\delta} \tag{7}$$

Figure 4, depicts the H-bridge response for different gain values, namely the gain leading to the least steady state error which is the highest gain it can be applied (we notice here that we did not use the maximum gain in order to achieve convergence) and the gain leading to the dead beat response which is the gain leading to zero Floquet multiplier.

Figure 4. Current behavior of the controlled H-bridge inverter. (a) K_p = 2.5 for least steady state error. (b) $K_p = \dfrac{\alpha}{2\beta\delta}$ for dead beat response

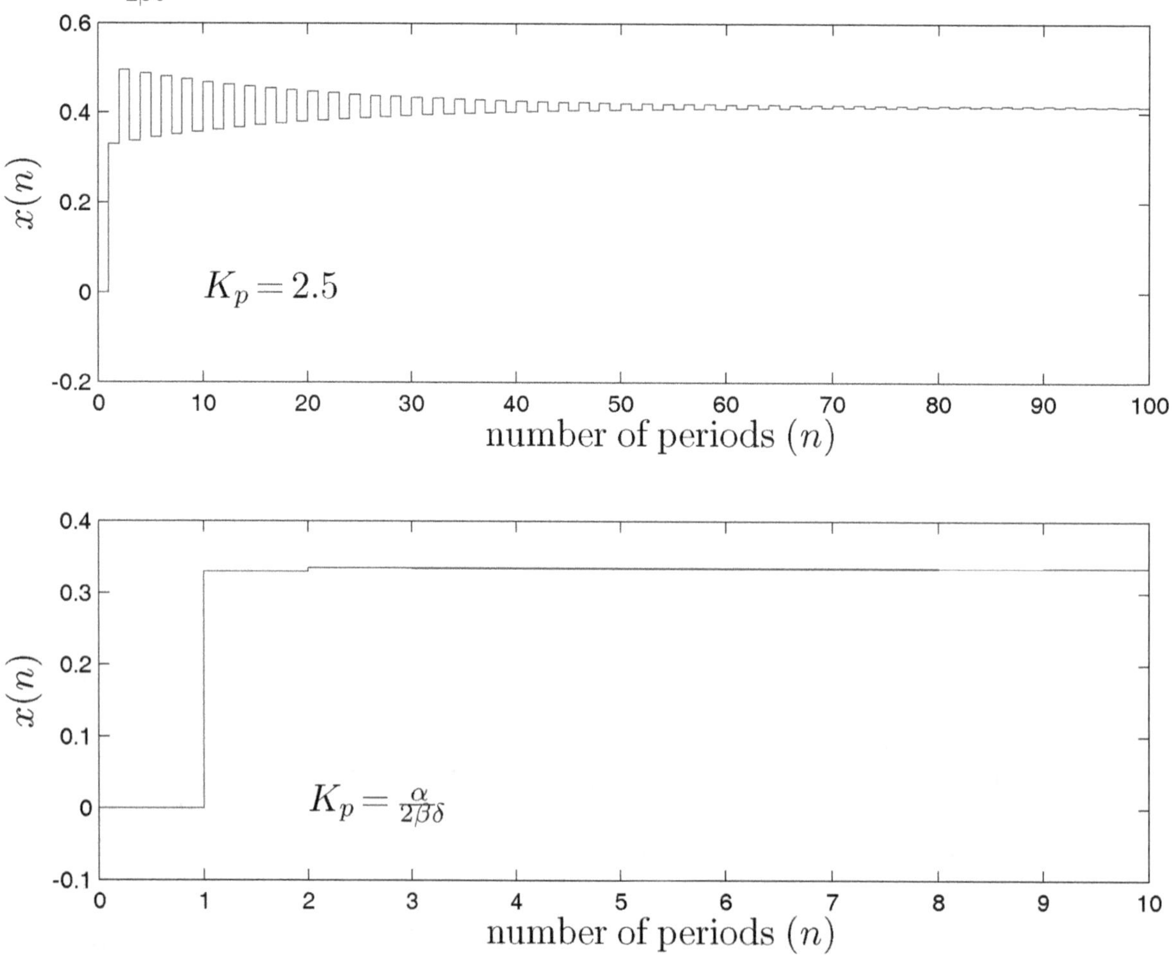

In the sequel, we attempt to bring an answer to the following question. Will the current $x(n)$ escape to infinity if the stability conditions are not met, that is in the case of the proportional controller, condition (7) is not satisfied? The answer is no. Suppose that is large enough to saturate the duty cycle. Hence, if the current is varying in region (see Figure 5) that is $x(n) > I_{ref} + \frac{1}{2K_p}$, then the duty cycle saturates at zero, $d(n) = 0$, and the system has a stable virtual fixed point $x^* = -1$ that is $x^* \notin R_0$ and vice-versa if the current is varying in region R_1, $x(n) < I_{ref} - \frac{1}{2K_p}$, then $d(n)=1$ and the system has a stable virtual fixed point $x^* = 1$ that is $x^* \notin R_1$. This leads to conclude that as K_p is increased the current will oscillate back and forth between different state regions R_0, R_1, or R_2.

Indeed, if K_p is increased further than the upper limit, then the Floquet multiplier crosses the unit circle at $z = -1$ yielding to a flip bifurcation. Figure 5 depicts a bifurcation diagram where we also show the three regions R_0, R_1, and R_2 and the boundaries limiting them.

Clearly, when K_p is larger than $\frac{\alpha+1}{2\beta\delta} = 2.55$, then the current becomes periodic of period $2T$. When calculating the corresponding discrete map we may easily check that the $2T$ behavior is stable as far as $K_p \leq \frac{\alpha + \frac{1}{\alpha}}{2\beta\delta} = 3.3$. Besides, it can be checked that the $2T$ cycle hits the boundary $I_{ref} - \frac{1}{2K_p}$ when

Figure 5. Bifurcation diagram of the controlled H-bridge inverter

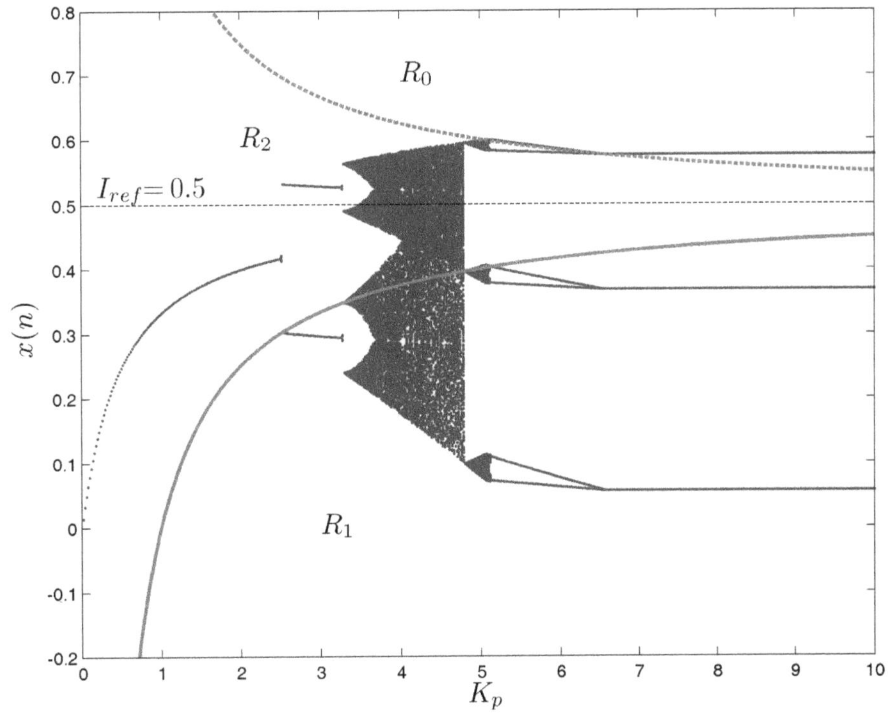

$K_p = \frac{\alpha+1}{2\beta\delta}$. Therefore, the obtained flip bifurcation is non smooth and is also called a boundary collision bifurcation.

3.2 FPIC Based Stabilization

The fixed point induced control technique is based on modifying the control input so that the eigenvalues of the Jacobian matrix are forced to lie inside the unit circle. An important property of this technique is that the fixed point of the original system is not changed (El Aroudi et al., 2009; Angulo et al., 2005).

Theorem 1. *Let us consider a nonlinear map in the following form:*

$$x(n+1) = f(x(n), d(n, K_0)) \tag{8}$$

where $X \in R^n$, $f : R^{n+m} \to R^n$, $d \in R^m$ is the vector of the control input and K_0 is a control parameter. Assume that the nonlinear map has a stable fixed point $(x^\star, d^\star)$ for a limited parameter value $K_0 \leq K_{\max}$. Define the so called fixed point induced controller as follows:

$$d_{FPIC}(n, K_0, \eta) = \frac{d(n, K_0) + \eta d^\star}{\eta + 1}, \tag{9}$$

where $\eta \geq 0$ is a new design parameter. Then the system:

$$x(n+1) = f(x(n), d_{FPIC}(n, K_0, \eta)) \tag{10}$$

has the same fixed point $(x^\star, d^\star)$ which becomes stable for $K_0 \leq K_{\max}(\eta)$ with $K_{\max}(\eta) > K_{\max}(0) = K_{\max}$

Remark 1. *If η=0, the FPIC control is disabled.*
Remark 2. *At steady state we have*

$$d^\star_{FPIC} = \frac{d^\star + \eta d^\star}{\eta + 1} = d^\star \tag{11}$$

which implies that $d^*_{FPIC} = d^*$, meaning that the fixed point of the system is not changed under the application of the FPIC technique.

A proof of the theorem may be obtained through linear analysis (Angulo et al., 2005). Let's consider the special case of the H-bridge inverter. We recall the discrete map:

$$x(n+1) = \alpha x(n) + \beta(2\sinh(\delta d(n)) - \sinh(\delta)) \tag{12}$$

and we consider the FPIC controller having the following expression:

$$d_{FPIC}(n) = \frac{\frac{1}{2} + K_p(I_{ref} - x(n)) + \eta d^\star}{\eta + 1}, \tag{13}$$

Then, it can easily be noted that the Floquet multiplier is given by:

$$z = \alpha - 2\beta\delta \frac{K_p}{\eta + 1} \cosh(\delta d^\star) \tag{14}$$

and thus, the system is stable as far as the design parameters K_p and η are chosen to satisfy:

$$(\eta + 1)\frac{\alpha - 1}{2\beta\delta} < K_p < (\eta + 1)\frac{\alpha + 1}{2\beta\delta} \tag{15}$$

which implies that K_p can be increased to any desired value, provided that η is increased to satisfy the stability condition (15). Consequently, if a maximum steady state error is fixed at $\epsilon_{s,\max}$, then given the steady state error expression:

$$e_s = \frac{1 - \alpha}{2\beta\delta K_p - \alpha + 1} I_{ref} \tag{16}$$

the controller parameters K_p and η should be chosen to satisfy:

$$K_p > \frac{1 - \alpha}{2\beta\delta\epsilon_{s,\max}}(I_{ref} - \epsilon_{s,\max}) \tag{17}$$

$$\eta > \frac{1 - \alpha}{1 + \alpha} \frac{I_{ref} - \epsilon_{s,\max}}{\epsilon_{s,\max}} - 1 \tag{18}$$

Figure 6 shows the behavior of the H-bridge inverter under the FPIC control action with the parameters chosen to obtain a steady state error less than $\epsilon_{s,\max} = 1\%$ while having a dead beat response. That is in addition to satisfying (17) and (18), the Floquet multiplier (14) is chosen to be zero, thus yielding to:

Figure 6. Current behavior of the controlled H-bridge inverter with $\epsilon_{s,\max} = 1\% I_{ref}$ and dead beat response. $K_p = \frac{1-\alpha}{2\beta\delta\epsilon_{s,\max}}(I_{ref} - \epsilon_{s,\max})$, $\eta = \frac{2\beta\delta K_p}{\alpha} - 1$

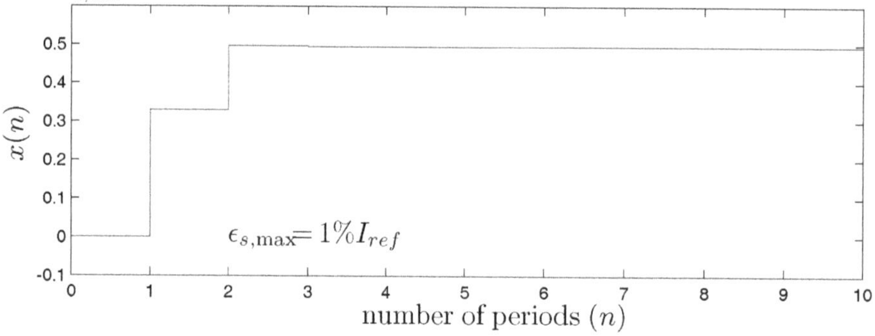

$$K_p = \frac{1-\alpha}{2\beta\delta\epsilon_{s,\max}}(I_{ref} - \epsilon_{s,\max}), \text{ and } \eta = \frac{2\beta\delta K_p}{\alpha} - 1$$

We note that the steady state value is reached in two switching periods although the system is of order one, that should be explained by the nonlinear aspect of the inverter model.

A two dimensional bifurcation diagram is depicted in Figure 7 and shows the steady state behavior of the H-bridge inverter for different pairs (K_p, η), it can be easily deduced that as η is increased, then the period one (stable) behavior shown in brown color is extended for higher values of K_p permitting hence to obtain arbitrary small staedy state errors.

3.3 Time-Delayed Feedback Controller

The TDFC controller has been proposed by Pyragas (1992); Pyragas (1995) to stabilize a predetermined unstable periodic orbit of a chaotic system. TDFC is known for its robustness and simplicity of construction. Besides, to achieve stabilization, only the length τ of the UPO to be stabilized is used. The control signal is proportional to the difference between the current state and the τ-delayed state. The TDFC controller does not modify the fixed point and once the desired UPO is stabilized the TDFC signal vanishes. This method has been successfully applied to the control of Chua's circuit (Fourati et al., 2010), discrete chaotic systems (Konishi and Kokame, 1998) and to the control of power converters (Kaoubaâ et al., 2012; Robert et al., 2006, 2004).

The aim of this section is to stabilize the UPO of length T equal to the switching period. Using the discrete-time notation, this corresponds to stabilizing the fixed point of order one for higher values of K_p thus the TDFC based control expression is given by:

$$d_{TDFC}(n) = d_0 + K_p(I_{ref} - x(n)) + K_d(x(n) - x(n-1)) \tag{19}$$

Figure 7. Two dimensional bifurcation diagram of the FPIC controlled H-bridge inverter

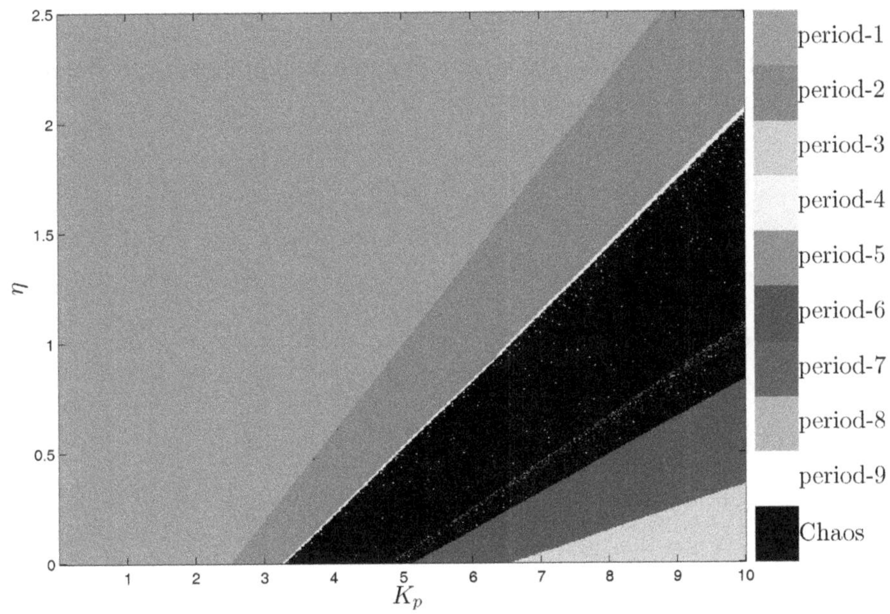

where K_d is the TDFC gain to be adjusted. When TDFC is applied, the discrete system defined by (3) and (19) becomes second order since it involves $x(n)$ and $x(n-1)$. Let us define $x_1(n) = x(n-1)$, and $x_2(n) = x(n)$, using these new state variables, the H-bridge discrete model becomes:

$$x_1(n+1) = x_2(n) \tag{20}$$

$$x_2(n+1) = \alpha x_2(n) + \beta\big(2\sinh(\delta d(n)) - \sinh(\delta)\big) \tag{21}$$

We investigate the local stability by deriving the Jacobian evaluated at the fixed point:

$$J(d^*) = \begin{bmatrix} 0 & 1 \\ J_1(K_d, d^*) & J_2(K_d, K_p, d^*) \end{bmatrix} \tag{22}$$

Since at equilibrium we have $x(n)=x(n-1)$, then the TDFC signal is null. Thus $d^*$is again the fixed point of the controlled system and it follows that:

$$J_1(K_d, d^*) = -2\beta\delta K_d \cosh(\delta d^*)$$
$$J_2(K_d, K_p, d^*) = \alpha + 2\beta\delta(K_d - K_p)\cosh(\delta d^*)$$

The characteristic equation of the linearized system is given by:

$$z^2 - J_2 z - J_1 = 0 \tag{23}$$

According to Jury's stability criterion, the system is stable if the following conditions are satisfied:

$$K_d < \frac{1}{2\beta\delta\cosh(\delta d^*)}, \tag{24}$$

$$K_p > \frac{\alpha - 1}{2\beta\delta\cosh(\delta d^*)} \tag{25}$$

$$K_d > \frac{K_p}{2} - \frac{\alpha + 1}{4\beta\delta\cosh(\delta d^*)}. \tag{26}$$

We should note that $\alpha - 1 < 0$, then it is sufficient to choose a positive proportional gain $K_p > 0$, in order to satisfy condition (25). Using conditions (24), (25) and (26), a stability zone in the parameters space (K_p, K_d) is defined and depicted in Figure 8.

Figure 8. Stability zone in the (K_p, K_d) space.8 Stability zone in the (K_p, K_d) space

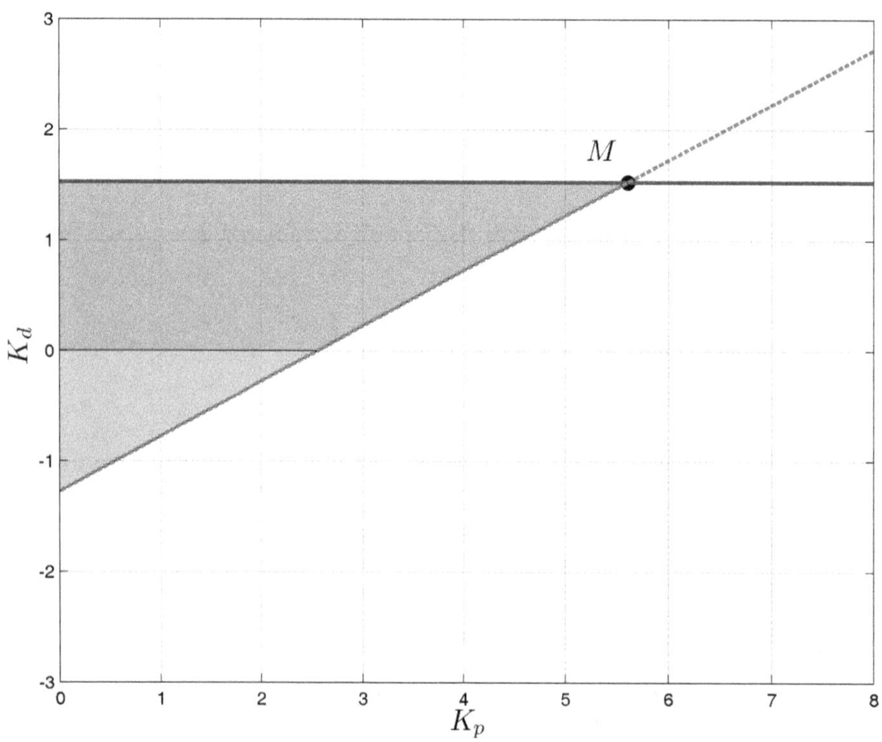

The triangular zone with $K_d < 0$ has a special interest if we suppose that the reference is constant. Indeed, equation (19) can be written in the following form:

$$d_{TDFC}(n) = d_0 + K_p(I_{ref} - x(n)) - K_d\left((I_{ref} - x(n)) - (I_{ref} - x(n-1))\right).$$

It can easily be noted that the control signal is equivalent to a Proportional-Derivative (P.D.) controller when K_d is negative yielding to a positive proportional and derivative gains. It is known from linear control theory that the larger is the proportional gain K_p the less is the steady state error. In view of Figure 8, K_p is maximized when is maximized and the limit of stability is the point M which is the intersection of boundary lines:

$$L_1 : K_d = \frac{1}{2\beta\delta} \quad \text{and} \quad L_2 : K_d = \frac{K_p}{2} - \frac{1+\alpha}{4\beta\delta}$$

obtained from equations (24) and (26) with the approximation $\cosh(\delta d^*) \simeq 1$. The coordinates of M are:

$$K_{d\max} \simeq \frac{1}{2\beta\delta} = 1.5 \tag{27}$$

Figure 9. 2-D bifurcation diagram showing the steady state behavior of the H-bridge inverter under a TDFC controller with different gain pairs (K_p, K_d)

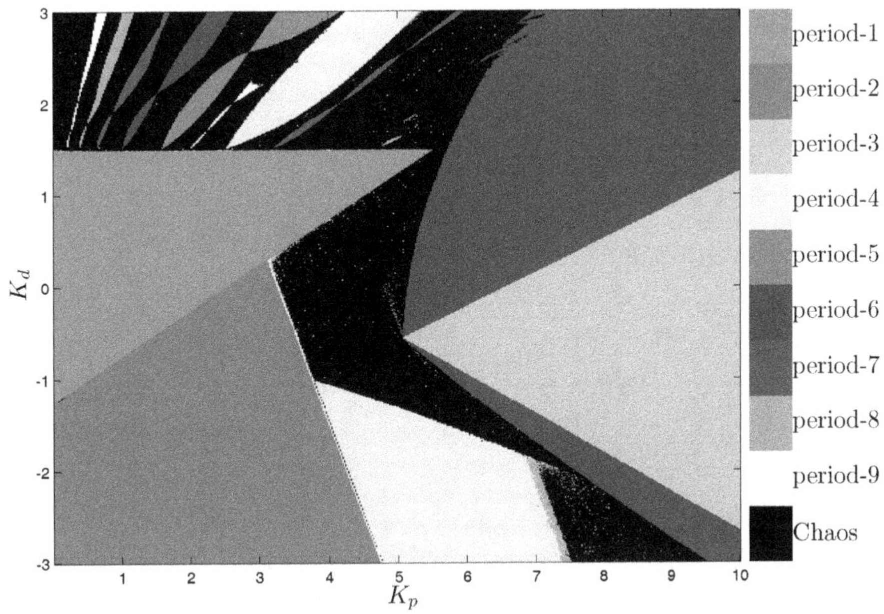

$$K_{p\max} \simeq \frac{3+\alpha}{2\beta\delta} = 5.6 \tag{28}$$

The two-dimensional bifurcation diagram depicted in Figure 9 delineates in brown the zone of the period one mode and it matches the stability zone shown in Figure 8. Zones of higher periods as well as chaotic zones are also shown as parameters K_p and K_d vary. Particularly, when $K_d = 0$ we observe that as K_p increases the output current is of period one then becomes period two and for a short interval of K_p becomes period four before it goes into chaotic mode. As K_p increases further the output becomes period six and finally arrives to period three mode and this matches very well with the one-dimensional bifurcation diagram shown on Figure 5.

We also notice that as K_d increases that is a TDFC controller is acting then the period one mode extends for higher values of K_p as expected. Figure 9 shows that the fixed point of order one that was embedded in a zone of period two mode $(2.5 \lesssim K_p \lesssim 3.3)$ and in a chaotic attractor $(3.3 \lesssim K_p \lesssim 5.0)$ when $K_d = 0$ has been stabilized by the application of the TDFC in conjunction with the proportional controller.

The main concern of the previous design was to obtain stability of the fixed point of order one for larger values of K_p which reduces the steady state error $I_{ref}-x(n)$. In what follows, we show that a drawback of TDFC is to degrade the dynamical response by increasing the settling time. Indeed, when referring to the characteristic equation (23), and by using the approximation of a second order system settling time $T_s = \frac{4}{\xi\omega_0}$ where ξ is the damping ration and ω_0 is the natural frequency and by using the relation

Figure 10. Time-responses to the step with and different values of TDFC gain

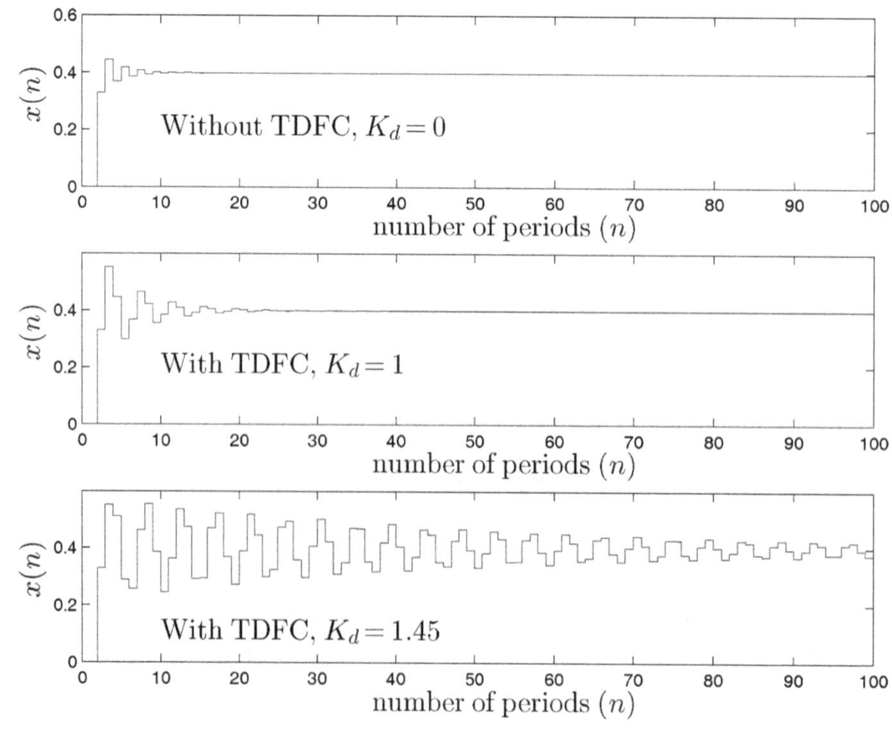

between the continuous time system and the sampled second order system, we may deduce that the settling time is given by $T_s = -\dfrac{8}{\ln 2\beta\delta K_d}$. Now, from the stability condition (24), we have $2\beta\delta K_d < 1$, thus we obviously deduce that T_s increases when K_d increases. Figure 10 shows the output current resulting from a step reference of magnitude $I_{ref} = 0.5$ with $K_p = 2$ and different values of K_d. We note that the transient time is increased when the TDFC is applied and when K_d tends to K_{dmax}, the transient time becomes too large.

To circumvent this problem we present an adaptive law to calculate a value of K_d that leads to a fast response for each value of K_p. Knowing that a second order system has a fast response when its damping ration is equal to $\sqrt{2}/2$, then K_d and K_p should satisfy the following equation:

$$\arccos\left(\frac{\alpha + 2\beta\delta(K_p - K_d)}{2\sqrt{2\beta\delta K_d}}\right) + \ln\left(\sqrt{2\beta\delta K_d}\right) = 0 \tag{29}$$

Figure 11 depicts the effect of adapting the TDFC gain K_d according to the value of the proportional gain K_p. Indeed, for $K_p = 4.5$ the proportional gain in its own leads to a chaotic output. However, if the TDFC is chosen at $K_d = \dfrac{K_{d\max} + K_{d\min}}{2} = 1.25$, that is furthest from limit values, the fixed point of order one is stabilized and the step reference is tracked with a certain steady state error. Nevertheless,

Figure 11. Time-responses to the step with K_p = 4.5 and different values of TDFC gain K_d

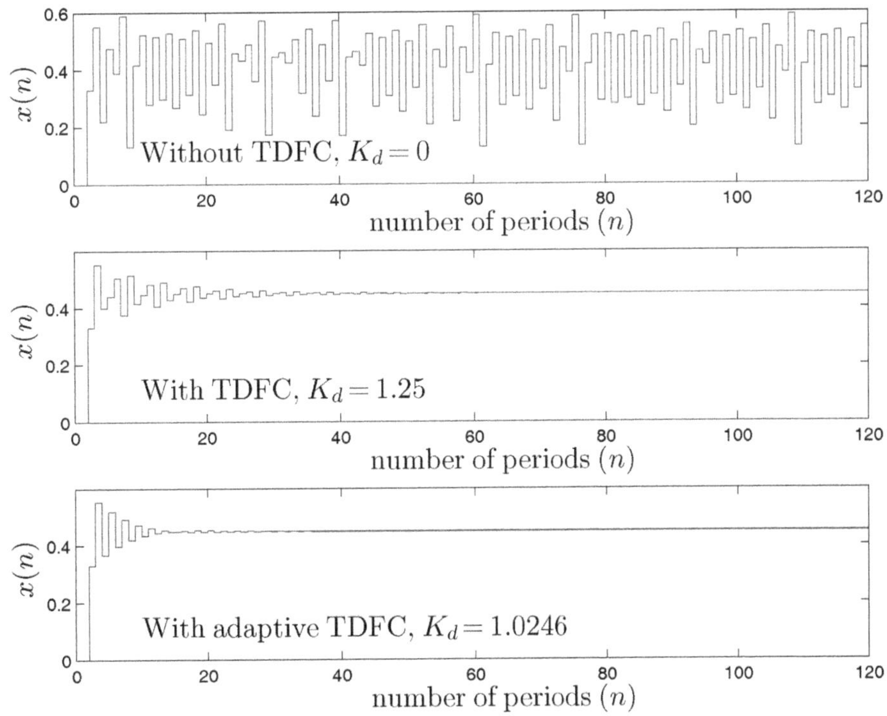

the settling time is significantly long; about 40 switching period. Eventually, the adapted TDFC $K_d =$ 1.0246 was used and yielded to the stability of the fixed point with a considerably shorter settling time of about 15 switching periods.

3.4 Linear Dynamic Controller

To improve the performance of the controlled H-bridge inverter by eliminating the steady state error, we propose a dynamic controller of general form:

$$y(n+1) = Ay(n) + B(I_{ref} - x(n)) \tag{30}$$

$$d(n) = Cy(n) + D(I_{ref} - x(n)) \tag{31}$$

where $y(n)$ is the state of the dynamic controller, and A, B, C and D are parameters to be determined in the sequel.

Combining with the system model (3), the fixed point of the system is the solution of the following equations:

$$x^\star = I_{ref} - \frac{1-A}{B} y^\star \tag{32}$$

$$0 = (\alpha - 1)x^\star + \beta(2\sinh(\delta d^\star) - \sinh(\delta)) \tag{33}$$

Clearly, the fixed point is the reference current if and only if A=1. Moreover, we can easily check through a linear analysis that the Jacobian matrix is given by:

$$J = \begin{pmatrix} \alpha - 2\beta\delta D & 2\beta\delta C \\ -B & A \end{pmatrix}$$

The characteristic equation is thus given by:

$$z^2 - (\alpha - 2\beta\delta D + A)z + A(\alpha - 2\beta\delta D) + 2\beta\delta BC = 0 \tag{34}$$

Using Jury's stability criterion, we may define the stability zone limits such as:

$$| A(\alpha - 2\beta\delta D) + 2\beta\delta BC | < 1 \tag{35}$$

$$1-(\alpha-2\beta\delta D+A)+A(\alpha-2\beta\delta D)+2\beta\delta BC>0 \quad (36)$$

$$1+(\alpha-2\beta\delta D+A)+A(\alpha-2\beta\delta D)+2\beta\delta BC>0 \quad (37)$$

In Figure 12, we delineate the stability zone, obtained from Jury's criterion, in the (D,BC) plane since BC come always as a product. Next, we will fix $C = 1$ and only consider the variation of the parameters B and D. In the PI controller case, we have $A = C = 1$, $B = K_i$ and $D = K_p$. Moreover, since the steady state error is zero for all $K_i > 0$, then there is no need to increase the value of K_p and hence we just need to choose the controller parameter that lead to fastest response which is the dead beat response in this case. Using the characteristic equation (34), and in order to have zero Floquet multipliers, the parameter B and D should satisfy:

$$D=\frac{\alpha+A}{2\beta\delta}, \qquad B=A\frac{2\beta\delta D-\alpha}{2\beta\delta} \quad (38)$$

In Figure 13, we sketch the dead beat time response of the H-bridge inverter under the action of the PI controller. Again here we notice that the dead beat response does not achieve the staedy state value within two switching periods as it should be for a second order system but it only attained the final value within four periods which is explained by the nonlinear aspect of the inverter model.

Figure 12. Stability zone of the H-bridge inverter obtained using Jury's criterion

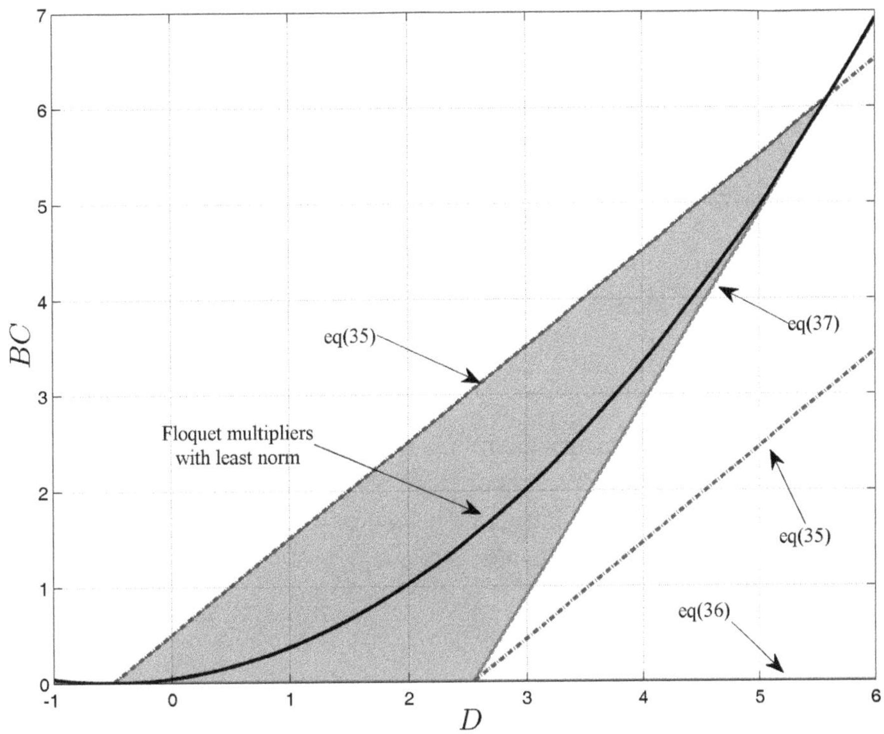

Figure 13. Dead beat time response of the H-bridge inverter under the action of the PI controller

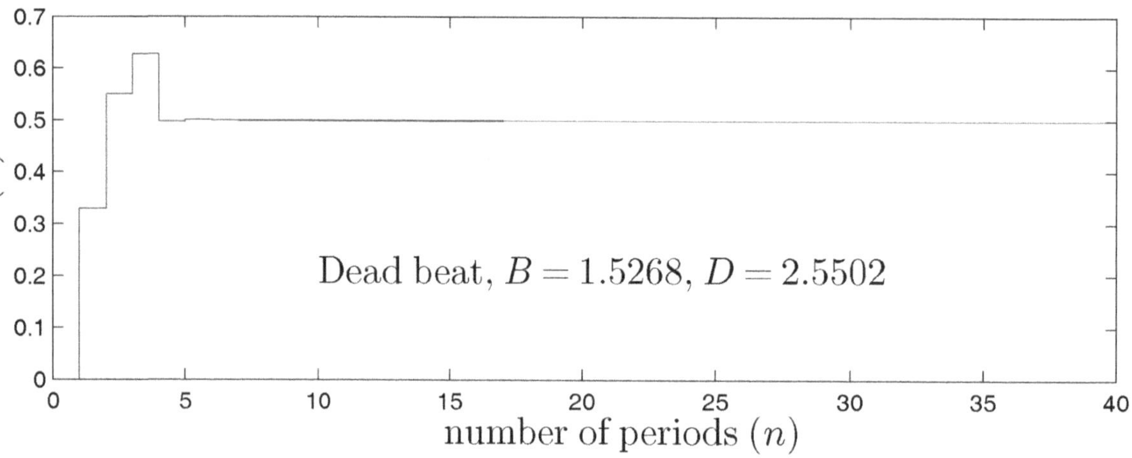

Figure 14. 2-D bifurcation diagram showing the steady state behavior of the H-bridge inverter under a linear dynamic controller with different gain pairs (D,BC)

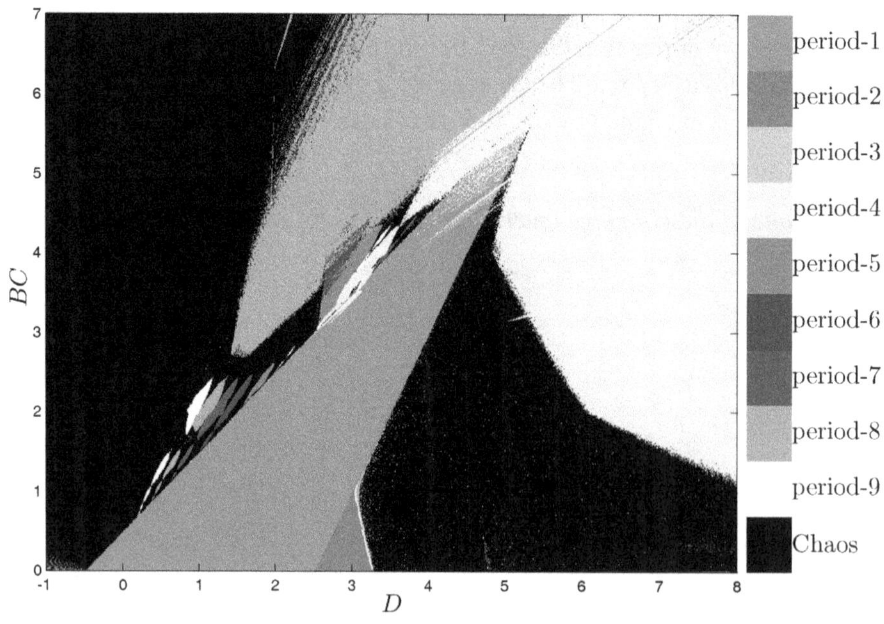

In Figure 14, we present a two dimensional bifurcation diagram where we sketched in brown color the stability zone. We notice that for high values of D, we obtain an interference between the $1T$ periodic motion and the $4T$ periodic motion. This intriguing behavior which was not actually expected through the linear analysis is explored in depth in the next section.

4. BIFURCATION AND COMPLEX BEHAVIOR

Motivated by the fact that a $4T$ periodic behavior has been obtained from control pairs that have been shown to lead to $1T$ periodic behavior, and since the $4T$ periodic behavior is nothing but the extension of the phase locked behavior that has been observed outside the stability region, we present in this section an thorough analysis of the H-bridge inverter behavior when the control parameters are chosen outside the stability region.

Let us first consider the case of the proportional controller. We have already noted that when the proportional gain K_p is increased beyond the stability limit value, then the Floquet multiplier crosses the unit circle at $z=-1$ yielding to a non smooth flip bifurcation that results in a stable $2T$ periodic current wave form. If we further increase the gain K_p, a second flip bifurcation occurs yielding to an unstable four periodic behavior which results in a four cyclical chaotic sets that merge into two cyclical chaotic sets as K_p in increased further. The resulting unstable flip bifurcation is called a generate flip bifurcation Gardini et al. (2011).

4.1 Complex Behavior under Dynamic Controller

In the case where we have used the dynamic controller, the system dimension has been increased by one and this has added several complex behaviors to the H-bridge inverter. For instance, since the Floquet multipliers have a least norm when they are equal, that is when the discriminant of the characteristic equation is zero, that is when the control parameters satisfy:

$$(1-\alpha+2\beta\delta D)^2-8\beta\delta B=0$$

then we may easily deduce that for pairs below the curve depicted on Figure 12, the discreminant of the characteristic equation is positive and the resulting Floquet multipliers are real thus when crossing the stability zone limits a flip bifurcation occurs which results in a $2T$ periodic motion when it enters the red area and results in a generate flip bifurcation when it enters the black area.

On the other hand, for pairs above the optimal pairs curve, the discriminant is negative and the Floquet multipliers are complex conjugate, and they would cross the unit circle while being complex conjugate, hence, a Neimark-Sacker bifurcation occurs. Indeed, when looking at Figure 15, we clearly observe that at the limit of the stability D=0.391 and B=0.8944, as expected, the system exhibits a sustained oscillation that does neither converge nor diverge. When D=0.391 and B=0.9346, expectedly the system is unstable, however the system exhibits a sustained oscillation with different amplitude with respect to the stable limit one and is called quasi-periodic behavior (see Figure 15). This implies that the Neimark-Sacker bifurcation has a supercritical character since a 2D torus was born here. Figure 16 depicts the limit cycle and the 2D torus for the foregoing mentioned cases; there we sketch transient part in black and steady state in red. The limit cycle of length $8T$ is characterized by an eight points periodic cycle on the Poincaré map and the torus is characterized by a closed curve.

It is well known that tori are born when two incommensurate frequencies exist in a single system. In our case, we have the switching frequency of normalized period T=1 and the natural frequency of a second order with complex conjugate poles. It is easy to verify that on the limit of stability we have an oscillating period that depends on D as follows:

Figure 15. Time response of the H-bridge inverter controlled using the dynamic controller. D=0.391 and B=0.8944 (upper curve) B=0.9346(lower curve)

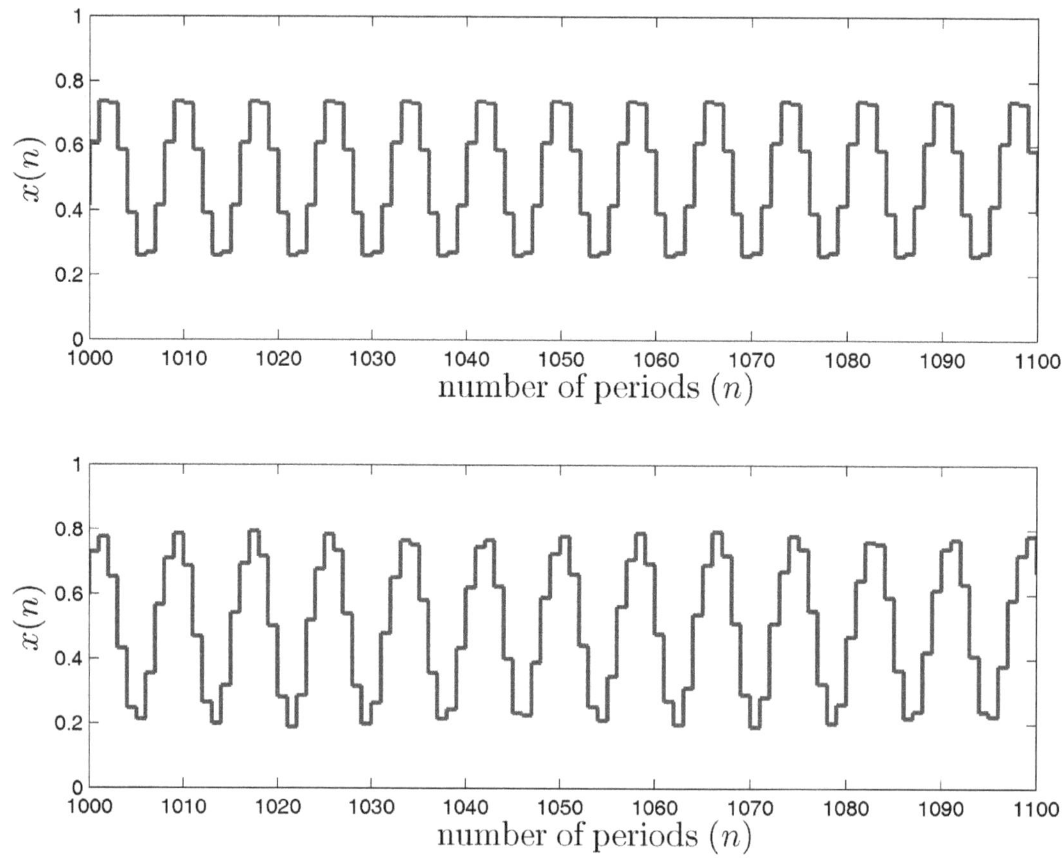

Figure 16. Limit cycle of order 8T with D=0.391 and B=0.8944.(left curve). 2D Torus of octagonal shape with D=0.391 and B=9346.(right curve). Transients are in black and steady state is in red

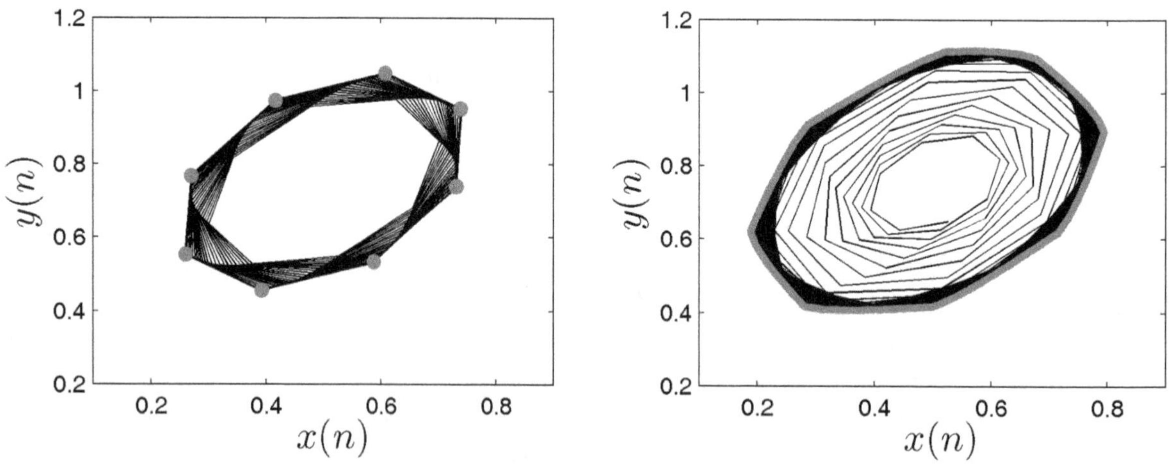

Figure 17. Chaotic behavior and chaotic attractor of the H-bridge inverter.D=2.5 and B=3.36

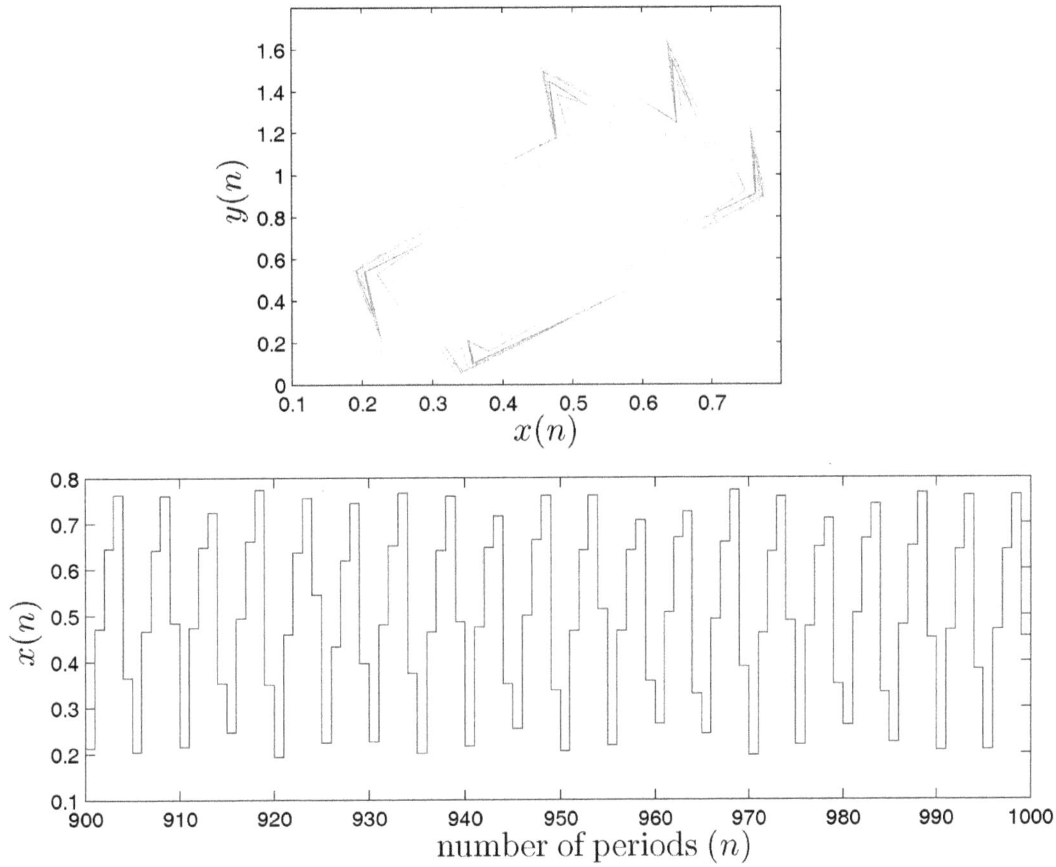

$$T_0 = \frac{2\pi}{\arccos\left(\dfrac{1+\alpha-2\beta\delta D}{2}\right)} T \tag{39}$$

with $D = 2.5$ we have $T_0 = 4.0423T$ from which we notice the incommensuarity of both frequencies. If we increase B a bit further in the unstable region, then the 2D Torus would erupt giving birth to a chaotic motion as shown on Figure 17.

On Figure 14 we also notice that several periodic behavior of higher harmonics exist very close to the stability region of the T periodic behavior. Indeed, these behaviors occurs due to phase locking of higher harmonics that start on the limit of stability when the parameter pair (D,B) yields to a periodic motion such as:

$$B = \frac{1 - A(\alpha - 2\beta\delta D)}{2\beta\delta C}$$

where D takes different values for different periodic behavior.

$$T_0 = 3T \quad \text{if} \quad D = \frac{\alpha + 2}{2\beta\delta}$$

$$T_0 = 4T \quad \text{if} \quad D = \frac{\alpha + 1}{2\beta\delta}$$

$$T_0 = 6T \quad \text{if} \quad D = \frac{\alpha}{2\beta\delta}$$

Although, colored periodic tongues are intermitted by black regions synonym to chaotic behavior, we should remark here that in fact the behavior between phase locked periodic behaviors is rather quasi-periodic and can be confirmed by numerical calculation of the largest Lyapunov exponents (not shown here) which is zero for quasi-periodic motion and positive for chaotic motion.

Figure 18. Coexistence of different periodic behavior. Initial condition (x(0),y(0))=(0.1,0), a 1T periodic behavior is obtained; initial condition (x(0),y(0))=(0,0), a 4T periodic behavior is obtained

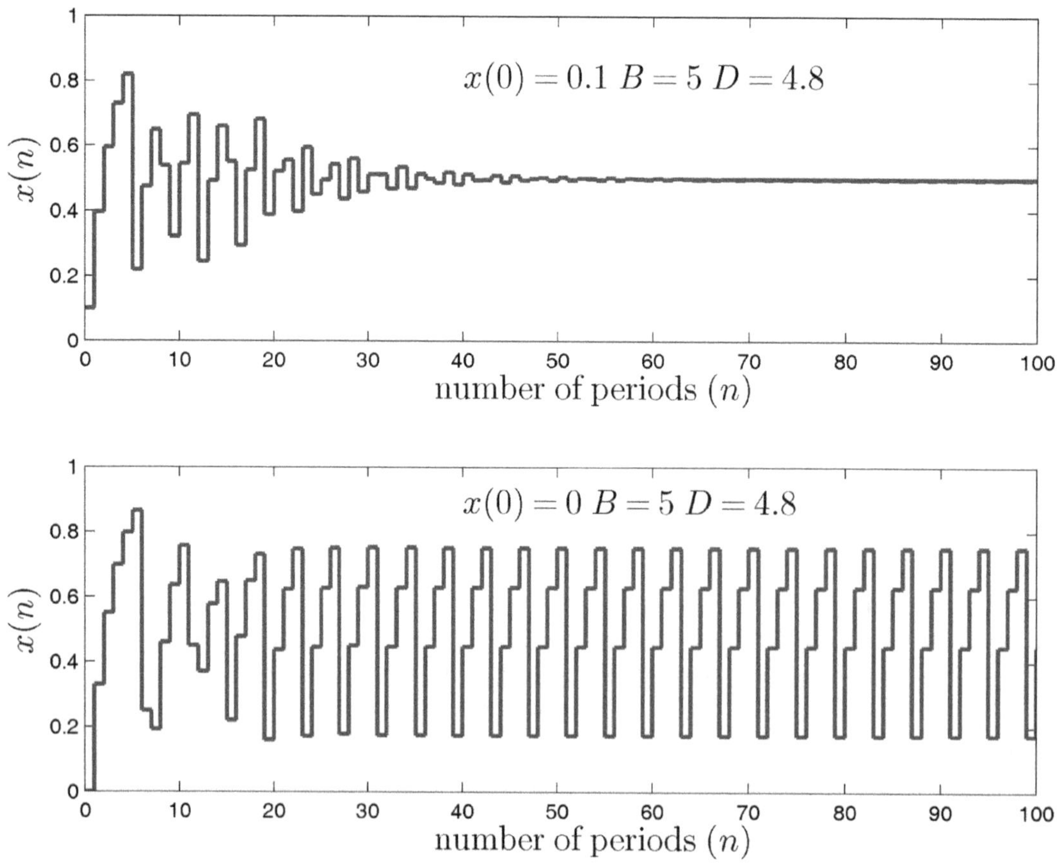

In the case of period 4T, we notice that this behavior invades the previously analyzed to be a region giving a T periodic current. In fact, in that case we have noticed that a coexistence between both periodic behaviors (T and 4T) may exist due to the saturation of the duty cycle. Figure 18 depicts two time responses of the H-bridge inverter controlled using the dynamic controller with parameters fixed at A=1, B=5, C=1 and D=4.8. We notice when the system initial condition is (x(0), y(0))=(0, 0), a 4T periodic behavior is obtained, however when the system initial condition is (x(0), y(0))=(0.1, 0), a 1T periodic behavior is obtained.

4.2 Complex Behavior under TDFC Controller

Similar behavior has been detected under the action of the TDFC controller which also increases the system dimension by one. In fact, when quitting the stability region from above, that is for $K_d > K_{d\max} = \frac{1}{2\beta\delta}$ the Floquet multipliers cross the unit circle while being complex and hence yielding to a Neimark-Sacker Bifurcation. Periodic behavior alternated with quasi periodic behavior can be detected above the stability region due to phase locking. However, the periodic behavior is limited down to the 4T behavior and no phase locking occurs for the 3T behavior. Instead, the 3T behavior leads to a three cyclic chaotic behavior that coexists with the 1T periodic one in a region that has been analyzed to lead to a stable 1T behavior. When compared to Figure 9, we notice In Figure 19, that the stability region has shrunk for the high values of K_p and a chaotic behavior exists instead due to the change of initial conditions.

Figure 19. 2-D bifurcation diagram showing the steady state behavior of the H-bridge inverter under a TDFC controller with a chaotic behavior invading the stability region

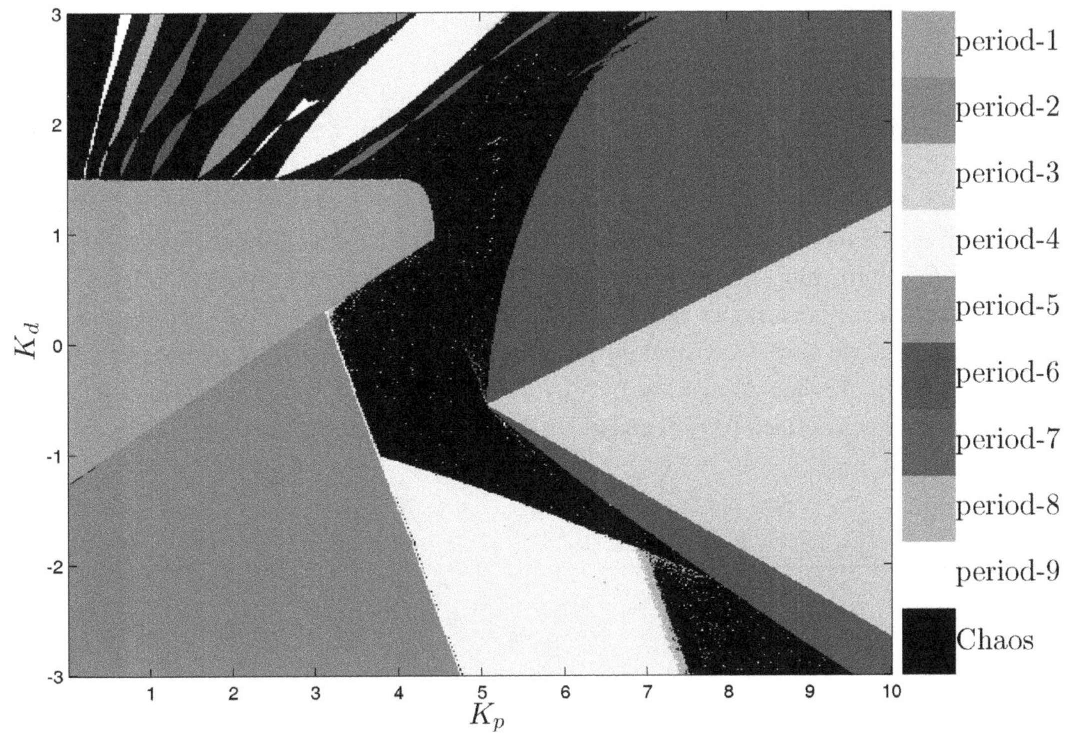

Figure 20. Bifurcation diagram and Lyapunov exponent with TDFC control $K_d = 1.45$, and initial condition $x_2(0) = 0$

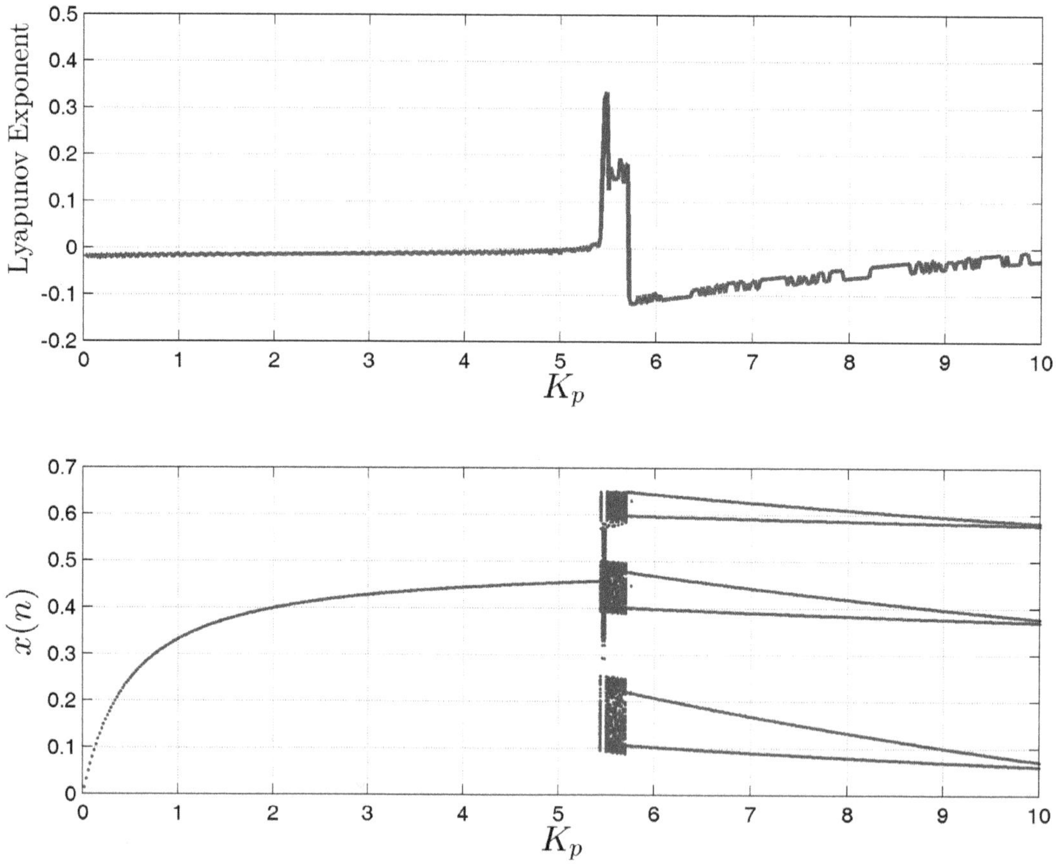

To confirm the coexistence of the chaotic and periodic behavior, we delineate in Figures 20 and 21 the bifurcation diagrams for $K_d = 1.45$ and K_p being the bifurcation parameter. The largest Lyapunov exponent is positive for $5.4 < K_p < 5.7$ when the initial condition is $x_2(0) = 0$ as it can be deduced from Figure 9. However, when the initial condition is $x_2(0) = 0.15$, the largest Lyapunov exponent is positive for $4.25 < K_p < 5.7$ confirming the chaotic behavior in this case.

Finally, to explain the coexistence of the chaotic and periodic behavior, we should not rely on the linear analysis, indeed, we need to consider the switching behavior of the H-bridge inverter among three different models due to the saturation of the duty cycle. In fact, according to the values of K_d and K_p the saturation limits are lines defined by (40) and (41) sketched in blue solid and dashed lines respectively, in Figure 22.

$$\mathcal{L}_0 : I_{ref} - x_2(n) = \frac{-d_0 - (K_p - K_d)(I_{ref} - x_1(n))}{K_d} \tag{40}$$

Figure 21. Bifurcation diagram and Lyapunov exponent with TDFC control $K_d = 1.45$, and initial condition $x_2(0) = 0.15$

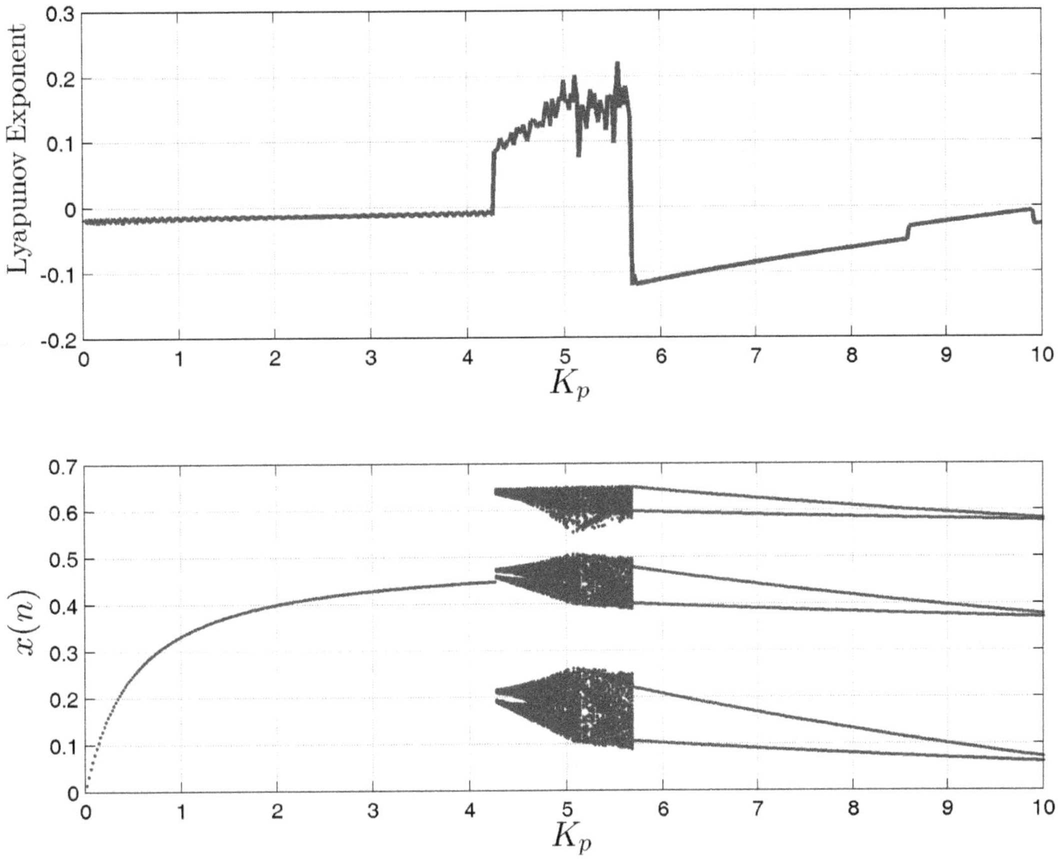

Figure 22. H-bridge behavior with TDFC control; initial condition $x_2(0) = 0.15$ in magenta and $x_2(0) = 0$ in blue. (a) $K_d = 1.4$ and $K_p = 2.55$. (b) $K_d = 1.4$ and $K_p = 4.45$

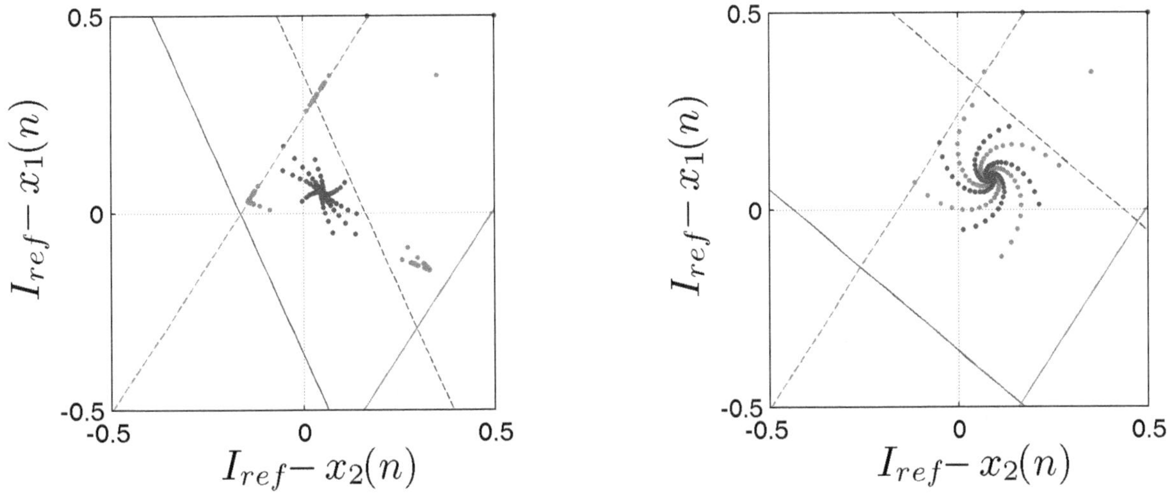

$$\mathcal{L}_1 : I_{ref} - x_2(n) = \frac{1 - d_0 - (K_p - K_d)(I_{ref} - x_1(n))}{K_d} \tag{41}$$

In addition, when the duty cycle saturates, the inverter current varies along a line in the state space towards the virtual equilibrium point occurring in the opposite saturation region. The lines defined by (42) and sketched in red solid and dashed shown on Figure 22 depict the behavior lines of the saturation at $d(n)=0$ and $d(n)=1$ respectively.

$$\Delta : I_{ref} - x_2(n) = \frac{(I_{ref} - x_1(n)) - (1 - \alpha)(I_{ref} - x^*)}{\alpha} \tag{42}$$

From Figure 22, we notice that as K_p gets larger (left side curve), the non saturation region gets narrower so that the one step evolution of the current results in model switching and hence the creation of the three region chaotic behavior for some initial conditions. On the contrary, for lower values of K_p the non saturation region gets larger (right side curve) and the system behavior heads toward the fixed point for any initial condition.

5. DISCUSSION

The main results presented in this chapter can be structured into two folds. The first one is the design of a suitable controller to drive the H-bridge inverter and the second is to explore the side effects of the controller. By suitable controller we mean one that can be simply constructed and efficiently applied to achieve the desired output and by side effects of the controller we mean its effects in presence of parametric variations or uncertainties.

Four different controllers were proposed to drive the H-bridge inverter. Starting by the proportional controller which is the simplest, we have shown that its gain is upper bounded and hence resulting into a lower bounded steady state error of order 25%. Being unsatisfactory, the remedy was either to modify the controller to be able to widen the domain of stability of the proportional gain or to apply a PI controller to annihilate the steady state error. Following the first idea, two modifications were proposed, the first one consists in using the FPIC controller and the second one uses the TDFC controller.

The main difference between the FPIC and the TDFC strategies is that the former is a static controller that does not increase the dimension of the system whereas the latter can be considered as dynamic and requires a memory cell to be constructed. Mathematically, each controller provides the designer with an additional degree of freedom, a new parameter to be designed that renders the parameter space two dimensional. The analysis carried in the third section have proved that the domain of stability in the parameter space was open in the case of FPIC controller and was on the contrary closed for the TDFC controller. Therefore, the obtained steady state error, although decreased, is still lower bounded with the TDFC strategy. However, the FPIC controller can achieve any predetermined steady state error. In addition, a dead beat response can be achieved using the FPIC controller while that was not possible with the adaptive TDFC which can only achieve a relatively fast response by fixing the damping ratio at the critical value $\left(\xi_c = \sqrt{2}/2\right)$.

The PI controller is a particular case of a linear dynamic controller that also provides the designer with an additional degree of freedom. Although it has the characteristic to achieve a zero steady state error, its main drawback is to have a slow response and hence a large settling time. Particularly, in the case of saturation of the control signal, which is in our case the duty cycle ratio that actually saturates at the lower and upper values zero and one respectively. The mathematical analysis carried out in the third section have shown that the fastest response achieved using the PI controller has a settling time of four switching periods.

Eventually, the comparison between the proposed controllers can fairely rule out the proportional and the TDFC controllers. The PI and the FPIC controllers have shown high performance in terms of fast response and precision though the PI has an inportant overshoot.

Concerning the side effects of the controllers, an analysis of the system behavior when the control parameters are pushed out of the stability region has been carried. In fact, many power electrical researchers do not perceive the importance of such analysis. They are usually interested in the stability region and once determined, their final constructed circuit includes auxiliary security circuits to avoid slipping outside the stability region. In our analysis, we aim to investigate every peculiar behavior that may exist around the stability region in order to justify or relinquish the use of the auxiliary security circuits.

In the case of the proportional controller, the increase of the proportional gain was justified by the decrease of the steady state error. However, increasing the gain pushes the Floquet multiplier towards the left hand half of the unit circle which qualitatively changes the behavior of the system response. Indeed, if a dynamic system has a pole placed on the left hand half of the unit circle, then its behavior would oscillate with the highest frequency which is equal to half of the sampling frequency. Therefore, having a pseudo period of $2T$, and when the pole reaches the critical value $z = -1$, a flip bifurcation occurs. When the gain is pushed further outside the unit cicrle, then streching of the system trajectory occurs and hence it either diverges to infinity or if the nonlinearity of the dynamic system is a folding one (which is the case of the saturation function here) then it will prevent the state from diverging and may stimulate the appearance of a chaotic behavior through multiple stretching and folding. The use of the FPIC controller, does not modify the behavior of the dynamic system since the gain of the FPIC reduces the magnitude of the system pole and hence by tuning the controller parameters, the global qualitative behavior does not change.

On the other hand, theTDFC controller and the dynamic linear controller increases by one the system dimension which becomes second order. The main interesting change in this case, is when the system has two complex conjugate poles which means the system has an intrinsic natural frequency. In this particular case, when the control parameters are modified, the Floquet multipliers will hit the unit circle at complex conjugate values leading to a Neimark-Sacker bifurcation. Then due to the coexistence of two frequencies in a single system, namely the switching frequency and the natural frequency, the system behavior may either be periodic if the ratio of both frequencies is rational or it may behave quasi-periodically if the ratio of the frequencies is irrational. Although the constant rational ratio can only form a hyperbolic curve, the periodic behavior of a given order extends to a whole area around the hyperbolic curve due to the phase locking phenomenon. Thus near the stability zone, we observe an intermittence between quasi-periodicity and periodic behaviors of decreasing order as the proportional gain is increased.

Finally, the novel result that we present in this chapter is the coexistence of two periodic behaviors namely the $1T$ and the $4T$ periodic behaviors in the case of the linear dynamic controller and the coexistence of the $1T$ periodic behavior and the chaotic behavior in the case of the TDFC controller application. As a matter of fact, the coexistence behavior cannot be deduced from the linear analysis. Actually, it is necessary to consider the saturation of the duty cycle in order to explain the coexistence phenomenon. The lower and upper saturation of the duty cycle define two boundaries in the state space dividing it into three regions, R_0 in which saturation occurs at zero, R_1 in which saturation occurs at one and a middle region R_2 in which no saturation occurs. It is worth noting that corresponding to each region the H-bridge inverter has a different linear mathematical model. Besides, only region R_2 contains the fixed point of its corresponding model, whereas the fixed point in region R_0 (R_1) corresponds to the model describing the behavior of the system in region R_1 (R_0). If for a chosen control parameters, the non saturation region is larger than the evolution steps of the state variables, then as the state variable enters inside R_2, the system is confined to it and it eventually tends to the stable fixed point of order one. However, if for the chosen controller parameters, region R_2 is narrow enough so that a one step evolution of the state variable can push the system outside of R_2, then depending on the initial condition the system may be attracted towards the fixed point of R_2 if it comes close enough to it, or it will wonder back and forth between different regions thus leading to a periodic motion of higher order or to a chaotic motion if the behavior is closer to the unstable three periodic motion. We shall add that the coexistence of different behaviors was not predictable and has been observed in several other converters such as the multi-cell buck converter.

Eventually, it is worth noting that the given analysis and simulations prove that the behavior of the H-bridge inverter outside the stability region should not be harmful to the driven device as the behavior would affect the quality of the produced signal in such a way that it makes it comparable to a perturbed signal. That should degrade the efficiency from the energy point of view and would not have a significant effect on the global behavior. Investigations and experiments along these lines are being prepared to confirm our postulates.

6. CONCLUSION

In this chapter, we have presented an analysis of the dynamic behavior of the H-bridge single phase inverter under the control of a proportional regulator. Then several controllers have been appended to the proportional controller in order to improve its performance; these range from FPIC, TDFC to dynamic controller. We have shown that the FPIC controller can achieve any desired steady state error with a dead beat performance which outperforms the TDFC controller. However, for the dynamic controller, although a zero steady state error is achieved several new complex behaviors have shown up. In addition to the regular behavior, the Neimark-Sacker bifurcation, quasi-periodicity and coexistence of periodic behavior have been observed. The dynamic analysis in presence of duty cycle saturation was essential to explain the coexistence of different attractors depending on the initial conditions.

Based on the present analysis, we aim to extend the obtained results to three phased inverters which we are convinced that they will exhibit such rich complex behaviors. Next, we aim to explore if additional complex behavior may appear if the driven load is not a simple RL load but an electric motor such as the DC motor and the synchronous motor. Finally, the effect of a chaotically perturbed sine wave applied to the motor shall be investigated.

REFERENCES

Albea, C., Gordillo, F., & de Wit, C. C. (2011). Adaptive control design for a boost inverter. *Control Engineering Practice*, *19*(1), 32–44. doi:10.1016/j.conengprac.2010.08.007

Alvarez-Ramirez, J., & Espinosa-Pérez, G. (2002). Stability of current-mode control for dc-dc power converters. *Systems & Control Letters*, *45*(2), 113–119. doi:10.1016/S0167-6911(01)00169-4

Angulo, F., Fossas, E., Ocampo, C., & Olivar, G. (2005). Stabilization of chaos with fpic, application to zad-strategy buck converters. In *Proceedings of 16th IFAC World Congress*. Prague: IFAC.

Baek, J., & Park, M. (2012). Fuzzy bilinear state feedback control design based on ts fuzzy bilinear model for dc-dc converters. *International Journal of Electrical Power & Energy Systems*, *42*(1), 710–720. doi:10.1016/j.ijepes.2012.04.008

Bensaada, M., & Stambouli, A. B. (2013). A practical design sliding mode controller for dcdc converter based on control parameters optimization using assigned poles associate to genetic algorithm. *International Journal of Electrical Power & Energy Systems*, *53*, 761–773. doi:10.1016/j.ijepes.2013.05.043

Chen, Q., Ren, X., & Oliver, J. A. (2012). Identier-based adaptive neural dynamic surface control for uncertain dcdc buck converter system with input constraint. *Communications in Nonlinear Science and Numerical Simulation*, *17*(4), 1871–1883. doi:10.1016/j.cnsns.2011.08.028

Deane, J., & Hamill, D. (1990). Instability, subharmonics and chaos in power electronic systems. *IEEE Transactions on Power Electronics*, *5*(3), 260–268. doi:10.1109/63.56516

Di Bernardo, M., & Tse, C. (2002). Chaos in power electronics: An overview. In G. Chen & T. Ueta (Eds.), *Chaos in circuits and systems* (Vol. 11, pp. 317–340). New York: World Scientific. doi:10.1142/9789812705303_0016

Di Bernardo, M., & Vasca, F. (2000). On discrete time maps for the analysis of bifurcations and chaos in dc/dc converters. *IEEE Transactions on Circuits and Systems I*, *47*(2), 130–143. doi:10.1109/81.828567

El Aroudi, A., Angulo, F., Olivar, G., Robert, B., & Feki, M. (2009). Stabilizing a two-cell dc-dc buck converter by fixed point induced control. *International Journal of Bifurcation and Chaos in Applied Sciences and Engineering*, *19*(6), 2043–2057. doi:10.1142/S0218127409023895

Feki, M., Robert, B., & Iu, H. (2004). A proportional plus extended time-delayed feedback controller for a pwm inverter. In *Proceedings of IEEE Annual Power Electronics Specialists Conference, PESC Record*, (pp. 3317–3320). Achen, Germany: IEEE. doi:10.1109/PESC.2004.1355061

Fernandez-Vargas, J., & Ledwich, G. (2010). Variable structure control for power systems stabilization. *International Journal of Electrical Power & Energy Systems*, *32*(2), 101–107. doi:10.1016/j.ijepes.2009.06.017

Fourati, A., Feki, M., & Derbel, N. (2010). Stabilizing the unstable periodic orbits of a chaotic system using model independent adaptive time-delayed controller. *Nonlinear Dynamics*, *62*(3), 687–704. doi:10.1007/s11071-010-9754-y

Gardini, L., Fournier-Prunaret, D., & Chargé, P. (2011). Border collision bifurcations in a two-dimensional piecewise smooth map from a simple switching circuit. *Chaos (Woodbury, N.Y.)*, *21*(2), 1–13. doi:10.1063/1.3555834 PMID:21721748

Gnanasaravanan, A., & Rajaram, M. (2012a). Articial neural network for monitoring the asymmetric half bridge dcdc converter. *International Journal of Electrical Power & Energy Systems*, *43*(1), 788–792. doi:10.1016/j.ijepes.2012.05.057

Gnanasaravanan, A., & Rajaram, M. (2012b). Dynamic response analysis and output voltage control of asymmetric half bridge dcdc converter for low voltage applications. *International Journal of Electrical Power & Energy Systems*, *43*(1), 774–778. doi:10.1016/j.ijepes.2012.06.038

Guesmi, K., Essounbouli, N., & Hamzaoui, A. (2008). Systematic design approach of fuzzy PID stabilizer for dcdc converters. *Energy Conversion and Management*, *49*(10), 2880–2889. doi:10.1016/j.enconman.2008.03.012

Hamill, D., & Jefferies, D. (1988). Subharmonics and chaos in a controlled switched-mode power converter. *IEEE Transactions on Circuits and Systems. I, Fundamental Theory and Applications*, *35*(8), 1059–1061. doi:10.1109/31.1858

Hamza, R., Feki, M., Derbel, N., Robert, B. G. M., & Aroudi, E. L. (2011). Controller design and analysis for a two-cell dc-dc converter in the presence of saturation. *International Journal of Bifurcation and Chaos in Applied Sciences and Engineering*, *21*(1), 341–361. doi:10.1142/S0218127411028465

Kaoubaâ, K., Pelaez-Restrepo, J., Feki, M., Robert, B., & El Aroudi, A. (2012). Improved static and dynamic performances of a two-cell dc-dc buck converter using a digital dynamic time-delayed control. *Int. J. Circ. Theor. Appl.*, *40*(4), 395–407. doi:10.1002/cta.735

Kawashima, H., Wardi, Y., & Taylor, D., & M., E. (2012). Optimal switching control of a step-down dc-dc converter. In *Proceedings of American Control Conference*. Montreal, Canada: Academic Press.

Konishi, K., & Kokame, H. (1998). Observer-based delayed-feedback control for discrete-time chaotic systems. *Physics Letters [Part A]*, *248*(5-6), 359–368. doi:10.1016/S0375-9601(98)00673-2

Luo, F. L., & Ye, H. (2004). *Advanced DC/DC Converters. Power electronics and applications series*. CRC Press.

Mazouz, N., & Midoun, A. (2011). Control of a dc/dc converter by fuzzy controller for a solar pumping system. *International Journal of Electrical Power & Energy Systems*, *33*(10), 1623–1630. doi:10.1016/j.ijepes.2011.06.016

Miladi, Y., Feki, M., & Derbel, N. (2014). Using genetic algorithm for chaos control in a piecewise linear hybrid systems with application to power electronics. In *Proceedings of the Annual Seminar on Automation, Industrial Electronics and Instrumentation (SAAEI'14)*. SAAEI.

Nachidi, M., Hajjaji, A. E., & Bosche, J. (2013). An enhanced control approach for dc-dc converters. *International Journal of Electrical Power & Energy Systems*, *45*(1), 404–412. doi:10.1016/j.ijepes.2012.09.003

Olalla, C., Queinnec, I., Leyva, R., & Aroudi, A. E. (2011). Robust optimal control of bilinear dcdc converters. *Control Engineering Practice*, *19*(7), 688–699. doi:10.1016/j.conengprac.2011.03.004

Pinard, M. (2007). *Convertisseurs et Electronique de Puissance Commande, description, mise en oeuvre*. DUNOD.

Pyragas, K. (1992). Continuous control of chaos by self-controlling feedback. *Physics Letters [Part A]*, *170*(6), 421–428. doi:10.1016/0375-9601(92)90745-8

Pyragas, K. (1995). Control of chaos via extended delay feedback. *Physics Letters [Part A]*, *206*(5-6), 323–330. doi:10.1016/0375-9601(95)00654-L

Rashid, M. (2001). *Power Electronics Handbook*. Academic Press.

Robert, B., Feki, M., & Iu, H. (2006). Control of a pwm inverter using proportional plus extended time-delayed feedback. *International Journal of Bifurcation and Chaos in Applied Sciences and Engineering*, *16*(1), 113–128. doi:10.1142/S0218127406014629

Robert, B., Iu, H., & Feki, M. (2003). Widening the stability range of a pwm inverter using a robust chaos control. In *Proceedings of the 10th European Conference on Power Electronics and Applications (EPE2003)*. Toulouse, France: EPE.

Robert, B., Iu, H., & Feki, M. (2004). Adaptive time-delayed feedback for chaos control in a pwm single phase inverter. *Journal of Circuits, Systems, and Computers*, *13*(3), 519–534. doi:10.1142/S0218126604001568

Robert, B., & Robert, C. (2002). Border collision bifurcations in a one-dimensional piecewise smooth map for a pwm current-programmed h-bridge inverter. *International Journal of Control*, *75*(16-17), 1356–1367. doi:10.1080/0020717021000023771

Robert, B. G. M., & El Aroudi, A. (2006). Discrete time model of a multi-cell dc/dc converter: Nonlinear approach. *Mathematics and Computers in Simulation*, *71*(4-6), 310–319. doi:10.1016/j.matcom.2006.02.022

Rodriguez, H., Ortega, R., Escobar, G., & Barabanov, N. (2000). A robustly stable output feedback saturated controller for the boost dc-to-dc converter. *Systems & Control Letters*, *40*(1), 1–8. doi:10.1016/S0167-6911(99)00113-9

Sira-Ramirez, H., Perez-Moreno, R., Ortega, R., & Garcia-Esteban, M. (1997). Passivity-based controllers for the stabilization of dc-to-dc power converters. *Automatica*, *33*(4), 499–513. doi:10.1016/S0005-1098(96)00207-5

Tan, S., Lai, Y., Tse, C., & Martinez-Salamero, L. (2007). Special family of PWM-based sliding-mode voltage controllers for basic dc-dc converters in discontinuous conduction mode. *IET Electric Power Applications*, *1*(1), 64–74. doi:10.1049/iet-epa:20060067

Tan, S.-C., Lai, Y., Cheung, M., & Tse, C. (2005). On the practical design of a sliding mode voltage controlled buck converter. *IEEE Transactions on Power Electronics*, *20*(2), 425–437. doi:10.1109/TPEL.2004.842977

Tse, C., & Di Bernardo, M. (2002). Complex behavior in switching power converters. *Proceedings of the IEEE*, *90*(5), 768–780. doi:10.1109/JPROC.2002.1015006

Wai, R.-J., & Shih, L.-C. (2011). Design of voltage tracking control for dc-dc boost converter via total sliding-mode technique. *IEEE Transactions on Industrial Electronics*, *58*(6), 2502–2511. doi:10.1109/TIE.2010.2066539

KEY TERMS AND DEFINITIONS

Border Collision Bifurcation: A non smooth flip bifurcation that occurs when the system dynamics hit a switching border.

Co-Existence of Behaviors: A special behavior that can happen in dynamic systems with multiple stable attractors. The limit behavior is determined by the initial state.

H-Bridge Inverter: An electronic circuit that enables a voltage to be applied across a load in either direction. The main idea is to use four controlled electronic switches that toggle states pairwisely. Hence, the overall obtained dynamic is piecewise continuous and may exhibit a rich variety of behaviors regardless the type of load it is driving. These circuits are often used in robotics and other applications to allow DC motors to run forwards and backwards.

Fixed Point Induced Controller: A controller introduced by F. Angulo et al that intends to enlarge the stability region of a given behavior by reducing the the modulus of the corresponding Floquet multiplier. The idea is based on the use of the value of the fixed point if it is available.

Flip Bifurcation: A period doubling bifurcation in a discrete dynamical system. It is a bifurcation in which the system switches to a new behavior with twice the period of the original system. That is, there exists two points such that applying the dynamics to each of the points yields the other point. Such bifurcation occurs when the corresponding Floquet multiplier quits the unit circle at z=-1of the complex plain when it varies with respect to a given system parameter denoted as the bifurcation parameter.

Neimark-Sacker Bifurcation: The birth of a closed invariant curve from a fixed point in dynamical systems with discrete time (iterated maps), when the fixed point changes stability via a pair of complex eigenvalues with unit modulus. The bifurcation can be supercritical or subcritical, resulting in a stable or unstable (within an invariant two-dimensional manifold) closed invariant curve, respectively. When it happens in the Poincare map of a limit cycle, the bifurcation generates an invariant two-dimensional torus in the corresponding ODE.

Quasi-Periodicity: An oscillating behavior that could appear when an oscillator is forced by an oscillating time-dependent input. Should the ratio of the intrinsic periodicity and the external period be irrational (incommensurate) then the behavior is called quasi periodic. However, if the ratio is rational then the behavior is rather periodic.

Time Delayed Feedback Control: A controller introduced by K. Pyragas that intends to stabilize the periodic behavior of a chaotic system. The control action is proportional to the error between the present state and the T delayed state where T is the period of the intended behavior.

Section 3
Intelligent Control Applications

Chapter 12
Fuzzy Adaptive Controller for Uncertain Multivariable Nonlinear Systems with Both Sector Nonlinearities and Dead-Zones

Abdesselem Boulkroune
University of Jijel, Algeria

ABSTRACT

This chapter presents two fuzzy adaptive variable structure controllers for a class of uncertain multi-input multi-output nonlinear systems with actuator nonlinearities (i.e. with sector nonlinearities and dead-zones). The design of the first controller concerns systems with symmetric and positive definite control-gain matrix, while the design of the second one is extended to the case of non-symmetric control-gain matrix thanks to an appropriate matrix decomposition, namely the product of a symmetric positive-definite matrix, a diagonal matrix with diagonal entries +1 or -1, and a unity upper triangular matrix. An appropriate adaptive fuzzy-logic system is used to reasonably approximate the uncertain functions. A Lyapunov approach is adopted to derive the parameter adaptation laws and prove the stability of the closed-loop control system. Finally, some simulation results are carried out to show the effectiveness of the proposed controllers.

INTRODUCTION

Problems of process control are more and more complex as the involved systems are multivariable in nature and exhibit uncertain nonlinear behaviours. This explains the fact that only few engineering solutions are available. Thanks to the universal approximation theorem (Wang, 1994), some adaptive fuzzy control systems have been developed for a class of multivariable nonlinear uncertain systems (Ordonez & Passino, 1999; Chang, 2000; Tong et al., 2000; Chekireb et al., 2003; Golea et al., 2003; Li & Tong,

DOI: 10.4018/978-1-4666-7248-2.ch012

2003; Tong & Li, 2003; Labiod et al., 2005; Tong et al., 2005; Essounbouli et al., 2006). The stability of the underlying control systems has been investigated using a Lyapunov approach. The robustness issues with respect to the fuzzy approximation error and external disturbances have been enhanced by appropriately modifying the available adaptive fuzzy controllers. The corner stone of such a modification consists in a robust control term which is conceived using a sliding mode control design (Ordonez & Passino, 1999; Chekireb et al., 2003; Golea et al., 2003; Tong & Li, 2003; Labiod et al., 2005; Tong et al., 2005) or an H^{∞} based robust control design (Chang, 2000; Tong et al., 2000; Li & Tong, 2003; Essounbouli et al., 2006). A key assumption in the available fuzzy adaptive control systems (Ordonez & Passino, 1999; Chang, 2000; Tong et al., 2000; Chekireb et al., 2003; Golea et al., 2003; Li & Tong, 2003; Tong & Li, 2003; Labiod et al., 2005; Tong et al., 2005; Essounbouli et al., 2006) is that the actuator dynamics may be reasonably approximated by a linear model. This is more an exception than a rule in the engineering practice.

The control problem of uncertain multi-input multi-output (MIMO) systems with input nonlinearities has received a remarkable attention because of those ubiquitous actuator nonlinearities, namely saturation, quantization, backlash, dead-zone and so on (Gutierrez & Ro, 1998; Hsu et al., 2004). It is well known that the existence of input nonlinearities may leads to poor performance or even instability of the control system. It is thereby more advisable to take into account the actuator nonlinearities in the control design as well as the stability analysis. Decentralized variable-structure controllers have been proposed in (Hsu, 1999; Shyu et al., 2003; Shyu et al., 2005) for a class of systems with input sector nonlinearities and/or dead-zones. In (Hsu et al., 2004; Niu & Ho, 2006), the authors have designed sliding mode control systems for nonlinear MIMO systems subject to both sector nonlinearities and dead-zones. The underlying results suffer from two fundamental limitations. Firstly, the considered class of systems is relatively reduced. Secondly, the gain reduction tolerances of the nonlinear dead-zones and upper bounds of uncertain nonlinear functions are required to be known.

More recently, adaptive neural or fuzzy control systems have been respectively developed for a special class of multivariable nonlinear systems with unknown dead-zones and gain signs in (Zhang & Ge, 2007; Zhang & Yi, 2007). These contributions suffer from two restrictive modelling assumptions motivated by technical purposes regarding the stability analysis and control design. The first one consists in assuming a lower triangular structure for the system under control while the second one concerns the boundedness of the high-frequency control gains. Moreover, a fuzzy adaptive control for a class of multivariable nonlinear systems with unknown dead-zones has been designed in (Boulkroune et al., 2010a). Note that, in (Zhang & Ge, 2007; Zhang & Yi, 2007; Boulkroune et al., 2010a), simple dead-zones having linear and nonlinear functions outside the dead-band have been considered.

In this chapter, one aims at designing two fuzzy adaptive controllers for two different classes of uncertain nonlinear multivariable systems containing both sector nonlinearities and dead-zones. These classes of systems differ only by the structure of the control-gain matrix, namely a symmetric positive definite matrix or a matrix with known signs of its leading principal minors. Though this work borrows from the available results, it presents a fundamental contribution to the fuzzy adaptive control of uncertain multivariable nonlinear systems from an applicability point of view. Unlike in contributions (Hsu et al., 2004; Niu & Ho, 2006), the class of the considered systems is relatively larger and the gain reduction tolerances of the nonlinear dead-zones and the upper bounds on uncertain nonlinear functions are not assumed to be known. These bounds are indeed estimated using adaptive fuzzy systems. And compared with contributions (Zhang & Ge, 2007; Zhang & Yi, 2007; Boulkroune et al., 2010a), there are three features that are of practical interest. Firstly, the considered class of systems is more larger as the modelling assumptions made in (Zhang & Ge, 2007; Zhang & Yi, 2007) are relatively restrictive, namely *lower triangular control structure with bounded high-frequency control gains.* Such modeling requirements are principally motivated by control design and stability analysis purposes. Secondly, the considered model of the input nonlinearity includes sector nonlinearities and dead-zones and is hence relatively larger than the one considered in (Zhang & Ge, 2007; Zhang & Yi, 2007; Boulkroune et al., 2010a). Thirdly, the control-gain matrix is appropriately decomposed into a product of a symmetric definite-positive matrix, a diagonal matrix with +1 or -1 on the diagonal and a unity upper triangular matrix. It is worth noticing that the diagonal matrix elements are nothing than the ratios of the signs of the leading principal minors of the control-gain matrix.

PROBLEM STATEMENT AND PRELIMINARIES

Consider the following class of nonlinear MIMO systems described by:

$$y_1^{(r_1)} = f_1(x) + \sum_{j=1}^{p} g_{1j}(x)\Phi_j(u_j) + \Lambda_1(x),$$

$$\vdots$$

$$y_p^{(r_p)} = f_p(x) + \sum_{j=1}^{p} g_{pj}(x)\Phi_j(u_j) + \Lambda_p(x), \tag{1}$$

where $x = \left[y_1, \dot{y}_1, \ldots, y_1^{(r_1-1)}, \ldots, y_p, \dot{y}_p, \ldots, y_p^{(r_p-1)}\right]^T \in R^r$ is the overall state vector which is assumed to be available for measurement and $r = r_1 + \ldots + r_p$, $u = \left[u_1, \ldots, u_p\right]^T \in R^p$ is the control input vector, $y = \left[y_1, \ldots, y_p\right]^T \in R^p$ is the output vector, $f_i(x)$ and $g_{ij}(x)$, $i, j = 1, \ldots, p$ are unknown continuous nonlinear functions. $\Lambda_i(x)$, $i = 1, \ldots, p$ are unknown dynamical external disturbances. $\Phi(u) = \left[\Phi_1(u_1), \ldots, \Phi_p(u_p)\right]^T$ is a nonlinear input function vector satisfying some properties which will be given later.

Let us denote

$$y^{(r)} = \left[y_1^{(r_1)} \ldots y_p^{(r_p)}\right]^T, \; F(x) = \left[f_1(x) \ldots f_p(x)\right]^T,$$

$$\Lambda(x) = \left[\Lambda_1(x), \ldots, \Lambda_p(x)\right]^T, \text{and } G(x) = \begin{bmatrix} g_{11}(x) & \cdots & g_{1p}(x) \\ \vdots & \ddots & \vdots \\ g_{p1}(x) & \cdots & g_{pp}(x) \end{bmatrix}.$$

Then, the system (1) can be rewritten in the following compact form:

$$y^{(r)} = F(x) + G(x)\Phi(u) + \Lambda(x) \tag{2}$$

where $F(.) \in R^p$ and $G(.) \in R^{p\times p}$.

The objective of this chapter is to design a stable control system allowing the system output vector y to follows a specified desired trajectory $y_d = \left[y_{d1},\ldots,y_{dp}\right]^T \in R^p$ assuming that the vector $x_d = \left[y_{d1},\dot{y}_{d1},\ldots,y_{d1}^{(r_1-1)},y_{d1}^{(r_1)},\ldots,y_{dp},\dot{y}_{dp},\ldots,y_{dp}^{(r_p-1)},y_{dp}^{(r_p)}\right]^T$ is continuous, bounded and available for measurement. Then $x_d \in \Omega_{x_d} \subset R^{r+p}$ with Ω_{x_d} is a known compact set.

Remark 1: The system (2) represents a relatively large class of MIMO nonlinear systems with input nonlinearities $\Phi(u)$ (i.e. with dead-zone and sector nonlinearities). These input nonlinearities exist widely in the practical control systems, such as hydraulic servo-valves, mechanical connections, piezo-positioners and biomedical systems, but they are commonly neglected in the control design for simplicity purposes. Also, the MIMO system (2) without input nonlinearities has been considered by many literatures (Ordonez & Passino, 1999; Chang, 2000; Tong et al., 2000; Chekireb et al., 2003; Golea et al., 2003; Li & Tong, 2003; Tong & Li, 2003; Labiod et al., 2005; Tong et al., 2005; Essounbouli et al., 2006), and is of higher practical significance.

Let us define the tracking error as

$$\begin{aligned} e_1 &= y_1 - y_{d1} \\ &\vdots \\ e_p &= y_p - y_{dp} \end{aligned} \tag{3}$$

and the filtered tracking error as

$$S = \left[S_1,\ldots,S_p\right]^T \tag{4}$$

with

$$S_i = \left[\frac{d}{dt} + \lambda_i\right]^{r_i - 1} e_i \text{ for } \lambda_i > 0 \ \forall i = 1,\ldots,p. \tag{5}$$

Then, one can rewrite equation (5) as follows

$$S_i = \lambda_i^{r_i-1} e_i + (r_i - 1)\lambda_i^{r_i-2} \dot{e}_i + \cdots + (r_i - 1)\lambda_i e_i^{(r_i-2)} + e_i^{(r_i-1)} \text{ with } i = 1,\ldots,p \tag{6}$$

Notice that if one chooses $\lambda_i > 0$, with $i = 1,\ldots,p$, then the roots of the polynomial $H_i(s) = \lambda_i^{r_i-1} + (r_i - 1)\lambda_i^{r_i-2} s + \cdots + (r_i - 1)\lambda_i s^{r_i-2} + s^{r_i-1}$ related to the characteristic equation of $S_i = 0$ are all in the open left-half plane.

The relation (6) can be rewritten in the following compact form

$$S_i = C_i^T E_i \tag{7}$$

where

$$E_i = [e_i \;\; \dot{e}_i \;\; \ldots e_i^{(r_i-2)} \;\; e_i^{(r_i-1)}]^T \tag{8}$$

and

$$C_i^T = [\lambda_i^{r_i-1} \;\; (r_i - 1)\lambda_i^{r_i-2} \;\; \ldots \;\; (r_i - 1)\lambda_i \;\; 1] \tag{9}$$

Consequently, the vector S takes the following form:

$$S = C^T E \tag{10}$$

with

$$C^T = diag\left[C_1^T C_2^T \;\; \ldots \;\; C_p^T\right]_{(p\times r)}, \text{ and} \tag{11}$$

$$E = \left[E_1^T E_2^T \;\; \ldots \;\; E_p^T\right]_{(r\times 1)}^T \tag{12}$$

The dynamics of S_i are described by:

$$\dot{S}_i = C_{ri}^T E_i + e_i^{(r_i)}, \text{ and } i = 1,\ldots,p \tag{13}$$

where

$$C_{ri}^T = [0 \;\; \lambda_i^{r_i-1} \;\; (r_i - 1)\lambda_i^{r_i-2} \;\; \ldots \;\; 0.5(r_i - 1)(r_i - 2)\lambda_i^2 \;\; (r_i - 1)\lambda_i] \tag{14}$$

The dynamics of S can be rewritten into the following compact form

$$\dot{S} = C_r^T E + e^{(r)} \tag{15}$$

where

$$C_r^T = diag\left[C_{r1}^T C_{r2}^T \; \cdots \; C_{rp}^T\right]_{(p\times r)} \tag{16}$$

$$e^{(r)} = \left[e_1^{(r_1)} \; e_2^{(r_2)} \; \cdots \; e_p^{(r_p)}\right]^T \tag{17}$$

with

$$e^{(r)} = y^{(r)} - y_d^{(r)} \tag{18}$$

where $y^{(r)} = \left[y_1^{(r_1)} \; y_2^{(r_2)} \; \cdots \; y_p^{(r_p)}\right]^T$ is previously defined and

$$y_d^{(r)} = \left[y_{d1}^{(r_1)} \; y_{d2}^{(r_2)} \; \cdots \; y_{dp}^{(r_p)}\right]^T, \tag{19}$$

From (18), one can write (15) as follows

$$\dot{S} = C_r^T E + y^{(r)} - y_d^{(r)} \tag{20}$$

Next, Equation (20) will be used in the development of the proposed controllers and the stability analysis.

Description of the Fuzzy Logic System

The basic configuration of a fuzzy system consists of a fuzzifier, a fuzzy inference engine and a defuzzifier, as presented in Figure 1. The fuzzy inference engine uses the fuzzy IF-THEN rules to perform a mapping from an input vector $\underline{x}^T = [x_1, x_2, ..., x_n] \in R^n$ to an output $\hat{f} \in R$. The ith fuzzy rule is written as

$$R^{(i)} : \text{ if } x_1 \text{ is } A_1^i \text{ and ... and } x_n \text{ is } A_n^i \text{ then } \hat{f} \text{ is } f^i \tag{21}$$

where A_1^i, A_2^i,..., and A_n^i are fuzzy sets and f^i is the fuzzy singleton for the output in the *i*th rule. By using the singleton fuzzifier, the product inference and the center-average defuzzifier, the output of the fuzzy system can be expressed as follows:

Figure 1. The basic configuration of a fuzzy logic system

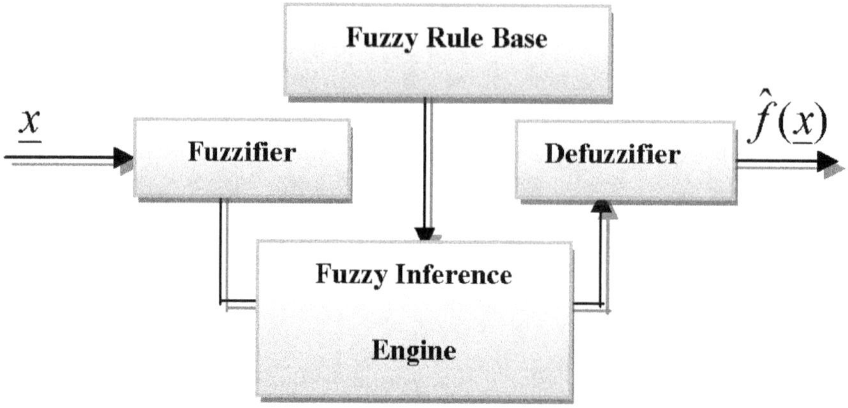

$$\hat{f}(\underline{x}) = \frac{\sum_{i=1}^{m} f^i \left(\prod_{j=1}^{n} \mu_{A_j^i}(x_j) \right)}{\sum_{i=1}^{m} \left(\prod_{j=1}^{n} \mu_{A_j^i}(x_j) \right)} = \theta^T \psi(\underline{x}) \tag{22}$$

where $\mu_{A_j^i}(x_j)$ is the degree of membership of x_j to A_j^i, m is the number of fuzzy rules, $\theta^T = [f^1, f^2, \ldots, f^m]$ is the adjustable parameter vector (composed of consequent parameters) and $\psi^T = [\psi^1 \ \psi^2 \ldots \psi^m]$ with

$$\psi^i(\underline{x}) = \frac{\left(\prod_{j=1}^{n} \mu_{A_j^i}(x_j) \right)}{\sum_{i=1}^{m} \left(\prod_{j=1}^{n} \mu_{A_j^i}(x_j) \right)} \tag{23}$$

being the *fuzzy basis function (FBF)*. Throughout this chapter, it is assumed that the FBFs are selected so that there is always at least one active rule (Wang, 1994; Azar, 2010a; Azar, 2010b; Azar, 2012; Boulkroune & M'saad, 2012a; Boulkroune et al., 2012c; Boulkroune et al., 2012d; Boulkroune et al., 2014a), i.e. $\sum_{i=1}^{m} \left(\prod_{j=1}^{n} \mu_{A_j^i}(x_j) \right) > 0$.

It is worth mentioning that the fuzzy system (22) is generally used in control applications. Following the universal approximation results, the fuzzy system (22) is able to approximate any nonlinear continuous function on a compact operating space to an arbitrary degree of accuracy (Wang, 1994). Of particular importance, it is assumed that the structure of the fuzzy system, namely the pertinent inputs, the number of membership functions for each input and the number of rules, and the membership function parameters, are properly specified beforehand. But, the consequent parameters are then determined by appropriate parameter adaptation algorithms.

Input Nonlinearity

The model of the input nonlinearity (i.e. the sector nonlinearity and dead-zone) under consideration is assumed to be described by

$$\Phi_i(u_i) = \begin{cases} \varphi_{i+}(u_i)(u_i - u_{i+}), & u_i > u_{i+}, \\ 0, & -u_{i-} \leq u_i \leq u_{i+}, \\ \varphi_{i-}(u_i)(u_i + u_{i-}), & u_i < -u_{i-}, \end{cases} \tag{24}$$

where $\varphi_{i+}(u_i) > 0$ and $\varphi_{i-}(u_i) > 0$ are nonlinear functions of u_i, and $u_{i+} > 0$ and $u_{i-} > 0$ are known constants. The involved modeling assumption is:

Assumption 1: The input nonlinearity $\Phi_i(u_i)$ satisfies the following important properties (Boulkroune et al., 2011b; Boulkroune & M'saad, 2011c; Boulkroune & M'saad, 2012b; Boulkroune et al., 2014b):

$$(u_i - u_{i+})\Phi_i(u_i) \geq m_{i+}^*(u_i - u_{i+})^2, \qquad u_i > u_{i+},$$

$$(u_i + u_{i-})\Phi_i(u_i) \geq m_{i-}^*(u_i + u_{i-})^2, \qquad u_i < -u_{i-}, \tag{25}$$

where m_{i+}^* and m_{i-}^* are unknown constants which are called gain reduction tolerances.

It is worth mentioning that the model (24)-(25) allows to consider both dead-zones and sector nonlinearities with reduced prior knowledge. Indeed the gain reduction tolerances m_+^* and m_-^* are unknown, unlike in (Hsu, 1999; Shyu et al., 2003; Hsu et al., 2004; Shyu et al., 2005; Niu & Ho, 2006), and the input nonlinearity is only characterised throughout the property (25) together with the knowledge of the constants u_+ and u_-. By the way, notice that the dead zone considered in the contribution (Zhang & Yi, 2007; Boulkroune et al., 2010) is a particular case of the above general form.

DESIGN OF FUZZY ADAPTIVE CONTROLLERS

Two fuzzy adaptive variable-structure control designs are proposed up to an assumption on the structure of the control-gain matrix. In the first one, the control-gain matrix is assumed to be symmetric and positive definite. Such an assumption is relaxed in the second design where the control-gain matrix may not be symmetric and positive definite provided that the signs of its leading principal minors are known.

The First Fuzzy Adaptive Controller

In the following, one presents a fuzzy adaptive variable-structure controller for the class of system (2) under the following assumption on the control-gain matrix.

Assumption 2: The control-gain matrix is symmetric, positive definite and of class C^1 with the following property: $\partial g_{ij}(x) \big/ \partial y_i^{(r_i-1)} = 0$, $\forall\ i = 1,2,...,p$ and $j = 1,2,...,p$.

Such an assumption is not too restrictive in nature as there are many physical (mechanical and electrical) systems whose dynamics can be described by Equation (2) with symmetric, positive definite and class C^1 control-gain matrix (Boulkroune et al., 2010a; Boulkroune et al., 2010b). The positive-definite property of the control-gain matrix, which is closely related to the system controllability, has been already required in recent fundamental contributions on adaptive (fuzzy or neural) control of MIMO systems. Moreover the required property on the partial derivatives of the control-gain matrix ensures that the time derivative of its inverse, i.e. $dG^{-1}(x) / dt$, depends only on the state vector x (i.e. it ensures that $dG^{-1}(x) / dt$ does not depend on the system input u). It is worth noticing that the following fundamental results could be adapted, up to some appropriate modifications, to the case where the control-gain matrix is symmetric and negative definite.

By substituting (2) into (20), the dynamics of S become

$$\dot{S} = C_r^T E + F(x) + G(x)\Phi(u) - y_d^{(r)} + \Lambda(x) \tag{26}$$

Letting $G_1(x) = G^{-1}(x)$, one has

$$G_1(x)\dot{S} = G_1(x)[C_r^T E - y_d^{(r)} + F(x) + \Lambda(x)] + \Phi(u) \tag{27}$$

For stability analysis and control design simplicity purposes, it is more advisable to rewrite the dynamics of S as follows

$$\frac{1}{2}\dot{G}_1(x)S + G_1(x)\dot{S} = \frac{1}{2}\dot{G}_1(x)S + G_1(x)[C_r^T E - y_d^{(r)} + F(x) + \Lambda(x)] + \Phi(u)$$

$$= \alpha(x, v, S) + \Phi(u) \tag{28}$$

with

$$\alpha(x, v, S) = [\alpha_1(x, v, S),\ ...,\alpha_p(x, v, S)]^T = (1/2)\dot{G}_1(x)S + G_1(x)[v + F(x) + \Lambda(x)]$$

and

$$v = C_r^T E - y_d^{(r)}$$

This form makes it possible to introduce a further assumption that has to be made to get the fundamental result we are concerned by, namely

Assumption 3: There exists an unknown continuous positive function $\bar{\alpha}_i(x)$ such that: $\left|\alpha_i(x,v,S)\right| \le \eta\bar{\alpha}_i(x)$, $\forall x \in \Omega_x \subset R^r$ with $\eta = \min_i\left\{\eta_i\right\}$ where $\eta_i = \min\left\{m_{i+}^*, m_{i-}^*\right\}$.

Notice that such an assumption is not restrictive since $\bar{\alpha}_i(x)$ is assumed to be unknown. Moreover, since v and S are functions of (x, x_d), $x_d \in L_\infty$ and $\alpha_i(x,v,S)$ is a continuous function, such a function always exists.

Remark 2: Note that the presence of the unknown nonlinear function $\alpha(x,v,S)$ and the actuator nonlinearities $\Phi(u)$ makes very difficult the control system design to practically stabilize the dynamics (28). Thereafter, to overcome such problems, we will use

- An adaptive fuzzy system to approximate the unknown nonlinear function $\alpha(x,v,S)$,
- A variable structure control law to treat the effect of the input nonlinearities $\Phi(u)$ (the dead-zone and sector nonlinearities).

The unknown continuous nonlinear function $\bar{\alpha}_i(x)$ can be hence approximated, on a compact set Ω_x, by the fuzzy systems (22) as follows:

$$\hat{\bar{\alpha}}_i(x,\theta) = \theta_i^T\psi_i(x) \tag{29}$$

where $\psi_i(x)$ is the FBF vector, which is fixed a priori by the designer, and θ_i is the adjustable parameter vector of the fuzzy system.

Let us define

$$\theta_i^* = \arg\min_{\theta_i}\left[\sup_{x\in\Omega_x}\left|\bar{\alpha}_i(x) - \hat{\bar{\alpha}}_i(x,\theta_i)\right|\right] \tag{30}$$

as the optimal values of θ_i which are mainly introduced for analysis purposes as their values are not needed when implementing the controller.

Define

$$\tilde{\theta}_i = \theta_i - \theta_i^* \text{ and } \varepsilon_i(x) = \bar{\alpha}_i(x) - \hat{\bar{\alpha}}_i(x,\theta_i^*) \text{ with } \hat{\bar{\alpha}}_i(x,\theta_i^*) = \theta_i^{*T}\psi_i(x) \tag{31}$$

as the parameter estimation error and the fuzzy approximation error, respectively.

As in the open literature (Wang, 1994; Boulkroune et al., 2008, Boulkroune et al., 2009; Boulkroune et al., 2010b; Boulkroune & M'saad, 2011a; Boulkroune et al., 2014c), one assumes that the used fuzzy systems do not violate the universal approximator property on the compact set Ω_x, which is assumed large enough so that the input vector of the fuzzy system remains in Ω_x. It is hence reasonable to assume that the fuzzy approximation error is bounded for all $x \in \Omega_x$, i.e. $\left|\varepsilon_i(x)\right| \le \bar{\varepsilon}_i$, $\forall x \in \Omega_x$, where $\bar{\varepsilon}$ is an unknown constant.

From the above analysis, one has

$$\hat{\bar{\alpha}}_i(x,\theta_i) - \bar{\alpha}_i(x) = \hat{\bar{\alpha}}_i(x,\theta_i) - \hat{\bar{\alpha}}_i(x,\theta_i^*) + \hat{\bar{\alpha}}_i(x,\theta_i^*) - \bar{\alpha}_i(x),$$

$$= \hat{\bar{\alpha}}_i(x,\theta_i) - \hat{\bar{\alpha}}_i(x,\theta_i^*) - \varepsilon_i(x),$$

$$= \tilde{\theta}_i^T \psi_i(x) - \varepsilon_i(x). \tag{32}$$

The following fuzzy adaptive variable-structure controller is proposed to achieve the control objective.

$$u_i = \begin{cases} -\rho_i(t) sign(S_i) - u_{i-}, & S_i > 0, \\ 0, & S_i = 0, \\ -\rho_i(t) sign(S_i) + u_{i+}, & S_i < 0, \end{cases} \tag{33}$$

with

$$\rho_i(t) = k_{0i} + k_{1i}\left|S_i\right| + \theta_i^T \psi_i(x),\ \forall i = 1,\dots,p$$

$$\dot{k}_{0i} = -\gamma_{0i}\sigma_{0i}k_{0i} + \gamma_{0i}\left|S_i\right| \text{for } k_{0i}(0) > 0 \tag{34}$$

$$\dot{\theta}_i = -\gamma_{1i}\sigma_{1i}\theta_i + \gamma_{1i}\left|S_i\right|\psi_i(x) \text{ with } \theta_{ij}(0) > 0 \tag{35}$$

where $\gamma_{0i}, \gamma_{1i}, \sigma_{0i}, \sigma_{1i}, k_{1i} > 0$ are design constants, and k_{0i} and θ_i are the online estimates of the uncertain terms $k_{0i}^* = \bar{\varepsilon}_i$ and θ_i^*, respectively.

It is important to remark that if $k_{0i}(0) > 0$ and $\theta_{ij}(0) > 0$ for $i = 1,\dots,p$ and $j = 1,\dots,m$, it follows from the adaptive laws (34-35) that their respective solutions satisfy $k_{0i}(t) > 0$ and $\theta_{ij}(t) > 0$, for $t > 0$.

Multiplying (28) by S^T and using Assumption 3, one has

$$\frac{1}{2\eta} S^T \dot{G}_1(x) S + \frac{1}{\eta} S^T G_1(x) \dot{S} = \frac{1}{\eta} S^T \alpha(x,v,S) + \frac{1}{\eta} S^T \Phi(u)$$

$$\leq \sum_{i=1}^{p} \left|S_i\right| \bar{\alpha}_i(x) + \frac{1}{\eta} S^T \Phi(u) \tag{36}$$

And using (32) and (36) yields

$$\frac{1}{2\eta}S^T\dot{G}_1(x)S + \frac{1}{\eta}S^T G_1(x)\dot{S} \le \sum_{i=1}^{p}\left|S_i\right|\bar{\alpha}_i(x) + \frac{1}{\eta}S^T\Phi(u)$$

$$\le -\sum_{i=1}^{p}\left|S_i\right|\tilde{k}_{0i} - \sum_{i=1}^{p}\left|S_i\right|\tilde{\theta}_i^T\psi_i(x) + \sum_{i=1}^{p}\left|S_i\right|k_{0i} + \sum_{i=1}^{p}\left|S_i\right|\theta_i^T\psi_i(x) + \frac{1}{\eta}S^T\Phi(u) \tag{37}$$

where $\tilde{\theta}_i = \theta_i - \theta_i^*$ and $\tilde{k}_{0i} = k_{0i} - k_{0i}^* = k_{0i} - \bar{\varepsilon}_i$.

This allows to state the fundamental result concerning the first fuzzy adaptive variable-structure control system.

Theorem 1: Consider the system (2) subject to Assumptions 1-3. Then, the control law defined by equations (33) to (35) guarantees the following properties:

- All the variables involved in the control system are uniformly ultimately bounded (UUB).
- The tracking error of the control system exponentially converges to an adjustable domain.

Proof of Theorem 1: Let us consider the following Lyapunov function candidate:

$$V = \frac{1}{2\eta}S^T G_1(x)S + \frac{1}{2}\sum_{i=1}^{p}\frac{1}{\gamma_{0i}}\tilde{k}_{0i}^2 + \frac{1}{2}\sum_{i=1}^{p}\frac{1}{\gamma_{1i}}\tilde{\theta}_i^T\tilde{\theta}_i\ . \tag{38}$$

Its time derivative is given by

$$\dot{V} = \frac{1}{\eta}S^T G_1(x)\dot{S} + \frac{1}{2\eta}S^T\dot{G}_1(x)S + \sum_{i=1}^{p}\frac{1}{\gamma_{0i}}\tilde{k}_{0i}\dot{k}_{0i} + \sum_{i=1}^{p}\frac{1}{\gamma_{1i}}\tilde{\theta}_i^T\dot{\theta}_i\ . \tag{39}$$

One can easily show from (33) that $u_i < -u_{i-}$ for $S_i > 0$ and $u_i > u_{i+}$ for $S_i < 0$. This makes it possible to conclude from (25) and (33) that for $S_i > 0$ one has

$$(u_i + u_{i-})\Phi_i(u_i) = -\rho_i(t)sign(S_i)\Phi_i(u_i) \ge m_{i-}^*\rho_i^2(t)[sign(S_i)]^2 \ge \eta\rho_i^2(t) \tag{40}$$

and for $S_i < 0$ one has

$$(u_i - u_{i+})\Phi_i(u_i) = -\rho_i(t)sign(S_i)\Phi_i(u_i) \ge m_{i+}^*\rho_i^2(t)[sign(S_i)]^2 \ge \eta\rho_i^2(t) \tag{41}$$

Then, for $S_i < 0$ and $S_i > 0$ one has

$$-\rho_i(t)sign(S_i)\Phi_i(u_i) \ge \eta\rho_i^2(t) \tag{42}$$

And using (42) and the fact that $S_i^2 > 0$ and $S_i sign(S_i) = \left|S_i\right|$ yields

$$-\rho_i(t)S_i^2 sign(S_i)\Phi_i(u_i) \geq \eta\rho_i^2(t)S_i^2$$

$$= \eta\rho_i^2(t)\left|S_i\right|^2 \tag{43}$$

Finally for all S_i one has

$$S_i\,\Phi_i(u_i) \leq -\eta\rho_i\,(t)\left|S_i\right| \tag{44}$$

And using the expressions (34), (35), (44) and (37), (39) becomes

$$\dot{V} \leq \sum_{i=1}^{p}\left|S_i\right|k_{0i} + \sum_{i=1}^{p}\left|S_i\right|\theta_i^T\psi_i(x) + \frac{1}{\eta}\sum_{i=1}^{p}S_i\Phi_i(u_i) - \sum_{i=1}^{p}\sigma_{0i}\tilde{k}_{0i}k_{0i} - \sum_{i=1}^{p}\sigma_{1i}\tilde{\theta}_i^T\theta_i$$

$$\leq \sum_{i=1}^{p}\left|S_i\right|k_{0i} + \sum_{i=1}^{p}\left|S_i\right|\theta_i^T\psi_i(x) + \sum_{i=1}^{p}-\rho_i\,(t)\left|S_i\right| - \sum_{i=1}^{p}\sigma_{0i}\tilde{k}_{0i}k_{0i} - \sum_{i=1}^{p}\sigma_{1i}\tilde{\theta}_i^T\theta_i$$

$$= -\sum_{i=1}^{p}k_{1i}S_i^2 - \sum_{i=1}^{p}\sigma_{0i}\tilde{k}_{0i}k_{0i} - \sum_{i=1}^{p}\sigma_{1i}\tilde{\theta}_i^T\theta_i \tag{45}$$

One can henceforth easily check that

$$-\sigma_{0i}\tilde{k}_{0i}k_{0i} \leq -\frac{\sigma_{0i}}{2}\tilde{k}_{0i}^2 + \frac{\sigma_{0i}}{2}k_{0i}^{*2}$$

$$-\sigma_{1i}\tilde{\theta}_i^T\theta_i \leq -\frac{\sigma_{1i}}{2}\left\|\tilde{\theta}_i\right\|^2 + \frac{\sigma_{1i}}{2}\left\|\theta_i^*\right\|^2$$

Using the previous inequalities, (45) becomes

$$\dot{V} \leq -\sum_{i=1}^{p}k_{1i}S_i^2 - \sum_{i=1}^{p}\frac{\sigma_{0i}}{2}\tilde{k}_{0i}^2 - \sum_{i=1}^{p}\frac{\sigma_{1i}}{2}\left\|\tilde{\theta}_i\right\|^2 + \sum_{i=1}^{p}\frac{\sigma_{0i}}{2}k_{0i}^{*2} + \sum_{i=1}^{p}\frac{\sigma_{1i}}{2}\left\|\theta_i^*\right\|^2 \tag{46}$$

Taking into account the Assumption 2, there exits a positive scalar $\tilde{A}_{g0}$ such that $G(x) \geq \sigma_{g0}I_p$ and henceforth

$$S^TG^{-1}(x)S = S^TG_1(x)S \leq \frac{1}{\sigma_{g0}}\left\|S\right\|^2 \tag{47}$$

And using (46) and (47) yields

$$\dot{V} \le -\mu V + \pi \tag{48}$$

with

$$\pi = \sum_{i=1}^{p} \frac{\sigma_{0i}}{2} k_{0i}^{*2} + \sum_{i=1}^{p} \frac{\sigma_{1i}}{2} \left\|\theta_i^*\right\|^2 \text{ and } \mu = \min\left\{\min_i\left\{2\eta\sigma_{g0}k_{1i}\right\}, \min_i\left\{\gamma_{0i}\sigma_{0i}\right\}, \min_i\left\{\gamma_{1i}\sigma_{1i}\right\}\right\}.$$

Multiplying (48) by $e^{\mu t}$ yields

$$\frac{d}{dt}\left(Ve^{\mu t}\right) \le \pi e^{\mu t} \tag{49}$$

And integrating (49) over [0,*t*], one has

$$0 \le V(t) \le \frac{\pi}{\mu} + \left(V(0) - \frac{\pi}{\mu}\right) e^{-\mu t} \tag{50}$$

Therefore, all variables of the control system, i.e. k_{0i}, θ_i, S_i, E and x, are uniformly ultimately bounded. And hence the input u_i is bounded.

With bearing in mind of (38), one can define $V(0)$ as follows

$$V(0) = \frac{1}{2\eta} S(0)^T G_1(x(0)) S(0) + \frac{1}{2}\sum_{i=1}^{p} \frac{1}{\gamma_{0i}}\left(k_{0i}(0) - k_{0i}^*\right)^2 + \frac{1}{2}\sum_{i=1}^{p} \frac{1}{\gamma_{1i}}\left(\theta_i(0) - \theta_i^*\right)^T \left(\theta_i(0) - \theta_i^*\right) \tag{51}$$

Since $G_1(x)$ is symmetric and positive definite, i.e. *there exists an unknown positive constant* σ_{g1} such that: $G_1(x) \ge \sigma_{g1} I_p$, it follows from (50) and (38) that:

$$\left|S_i\right| \le \left(\frac{2\eta}{\sigma_{g1}}\left(\frac{\pi}{\mu} + \left(V(0) - \frac{\pi}{\mu}\right)e^{-\mu t}\right)\right)^{1/2}$$

Then, the solution of S_i exponentially converges to a bounded adjustable domain defined as follows

$\Omega_{S_i} = \left\{ S_i \;\middle|\; \left|S_i\right| \le \left(\frac{2\eta}{\sigma_{g1}}\frac{\pi}{\mu}\right)^{1/2} \right\}$. This ends the proof of the theorem.

The Second Fuzzy Adaptive Controller

The adaptive fuzzy controller presented above is only applicable for nonlinear systems with a symmetric and positive definite control-gain matrix, e.g. robotic systems and some electrical machines. Such a property is not satisfied for several physical systems, e.g. the visual servoing and the automotive thermal management systems described (Hsu et al., 2007; Boulkroune et al., 2014b). In the following, one presents a fuzzy adaptive controller in the case where the control-gain matrix is no longer symmetric and positive definite using a suitable matrix decomposition that has been already introduced in (Boulkroune et al., 2010b; Boulkroune et al., 2014b).

If the control-gain matrix is with non-zero leading principal minors, it can be always decomposed into the product of a symmetric positive definite matrix, a diagonal matrix whose elements are +1 or -1 and a unity upper triangular matrix as pointed out by the following result.

Lemma 1: Any real matrix $G(x) \in R^{p\times p}$ with non-zero leading principal minors can be decomposed as follows (Boulkroune et al., 2010b; Boulkroune et al., 2014b):

$$G(x) = G_s(x)DT(x) \tag{52}$$

where $G_s(x) \in R^{p\times p}$ is a symmetric positive definite matrix, $D \in R^{p\times p}$ is a diagonal matrix whose elements are +1 or -1 and $T(x) \in R^{p\times p}$ is a unity upper triangular matrix.

Proof of Lemma 1: See (Hsu et al., 2007).

Remark 3: Notice that the decomposition (52) of the matrix $G(x)$ is very useful. The symmetric positive definite matrix $G_s(x)$ will be particularly used in the stability analysis. The unity upper triangular matrix $T(x)$ allows for algebraic loop free sequential determination of the control components. The diagonal elements are nothing than the ratios of the signs of the leading principal minors of the control-gain matrix. This implies the following facts when the control-gain matrix has non-zero leading principal minors.

- If $G(x)$ is positive definite, then $D = I_p$,
- If $G(x)$ is negative definite, then $D = -I_p$, and
- If $G(x)$ is indefinite, the diagonal elements of the matrix D are +1 and -1.

Consider the system (2) subject to Assumption 1 and the following assumption.

Assumption 4: The control-gain matrix is of class C^1, has non-zero leading principal minors of known signs and satisfies the following property: $\partial g_{ij}(x) / \partial y_i^{(r_i-1)} = 0$, $\forall\ i = 1, 2, ..., p$ and $j = 1, 2, ..., p$.

Using the matrix factorization (52) and Equation (20), the dynamics of S get:

$$G_s^{-1}(x)\dot{S} = G_s^{-1}(x)[v + F(x) + \Lambda(x)] + DT(x)\Phi(u) \tag{53}$$

where

$$v = C_r^T E - y_d^{(r)}.$$

Letting $G_{s1}(x) = G_s^{-1}(x)$ and $F_1(x,u) = G_s^{-1}(x)[v + F(x) + \Lambda(x)] + [DT(x) - D]\Phi(u)$, (53) becomes

$$G_{s1}(x)\dot{S} = F_1(x,u) + D\Phi(u) \tag{54}$$

Now, let introduce the following change of the variables for stability analysis and control design simplicity

$$\overline{S} = D^{-1}S \tag{55}$$

or $\overline{S}_i = d_{ii}S_i$ as $D = D^T = D^{-1}$ and $d_{ii} = +1$ or -1.

This allows to rewrite (54) under the form

$$G_{s2}(x)\dot{\overline{S}} = D^{-1}F_1(x,u) + \Phi(u) \tag{56}$$

where $G_{s2}(x) = D^{-1}G_{s1}(x)D.$

Due to the special forms of the matrices D and $G_{s1}(x)$, $G_{s2}(x)$ preserves the important properties of the original matrix $G_{s1}(x)$ (or $G_s(x)$). Indeed, one can easily show that $G_{s2}(x)$ is also symmetric and positive definite. This property is of fundamental interest when investigating the control system stability.

Similarly to the first fuzzy adaptive control system, (56) can be rearranged as follows

$$G_{s2}(x)\dot{\overline{S}} + \frac{1}{2}\dot{G}_{s2}\overline{S} = \alpha(\overline{z}) + \Phi(u) \tag{57}$$

where $\alpha(\overline{z}) = \left[\alpha_1(\overline{z}_1), \alpha_2(\overline{z}_2), \ldots, \alpha_p(\overline{z}_p)\right]^T = D^{-1}F_1(x,u) + \frac{1}{2}\dot{G}_{s2}(x)\overline{S}$, with $\overline{z} = \left[\overline{z}_1^T, \overline{z}_2^T, \ldots, \overline{z}_p^T\right]^T$.

By carefully examining the expression of $F_1(x,u)$ and $\alpha(\overline{z})$, the elements of the vector $\overline{z}$ can be selected as follows:

$$\overline{z}_1 = \left[x^T, S^T, u_2, \ldots, u_p\right]^T$$

$$\overline{z}_2 = \left[x^T, S^T, u_3, \ldots, u_p\right]^T$$

$$\vdots$$

$$\bar{z}_{p-1} = \left[x^T, S^T, u_P\right]^T$$

$$\bar{z}_p = \left[x^T, S^T\right]^T \tag{58}$$

It is clear from the propriety of the matrix $DT(x) - D$ that $\bar{z}_1$ depends on control inputs $u_2,\ldots, u_p$, $\bar{z}_2$ depends on $u_3,\ldots, u_p$ and so on. The nonlinearities $\alpha(\bar{z})$ has an upper triangular control structure, allowing thereby for algebraic loop free sequential determination of the control variables.

Let us define the following compact sets

$$\Omega_{\bar{z}_i} = \left\{ [x^T, S^T, u_{i+1},\ldots, u_p]^T \middle| x \in \Omega_x \subset R^r, x_d \in \Omega_{x_d} \right\} \text{for } i = 1, 2,\ldots, p-1$$

and

$$\Omega_{\bar{z}_p} = \left\{ [x^T, S^T] \middle| x \in \Omega_x \subset R^r, x_d \in \Omega_{x_d} \right\}$$

and introduce the following assumption

Assumption 5: There exists an unknown continuous positive function $\bar{\alpha}_i(\bar{z}_i)$ such that: $\left|\alpha_i(\bar{z}_i)\right| \leq \eta\bar{\alpha}_i(\bar{z}_i)$, $\forall \bar{z}_i \in \Omega_{\bar{z}_i}$ where $\eta = \min_i\left\{\eta_i\right\}$.

The choice of the vectors $\bar{z}_i$, i.e. the input arguments of the unknown functions $\bar{\alpha}_i$, is not unique. Indeed, since S and u are functions of state x and x_d, then it can be easily seen that all vectors $\bar{z}_i$ are functions of x and x_d, e.g. one can choose $\bar{z}_i = [x^T, x_d^{\ T}]^T$ or $\bar{z}_i = [x^T, E^T]^T$ for $i = 1, 2,\ldots, p$. Moreover, since x_d is bounded, one can choose $\bar{z}_i = x$.

The unknown continuous function $\bar{\alpha}_i(\bar{z}_i)$ can be approximated by a fuzzy system as follows

$$\hat{\bar{\alpha}}_i(\bar{z}_i, \theta) = \theta_i^T \psi_i(\bar{z}_i) \tag{59}$$

where $\psi_i(\bar{z}_i)$ is the FBF vector, which is fixed a priori by the designer, and θ_i is the adjustable parameter vector of the fuzzy system. Furthermore, the functions $\bar{\alpha}_i(\bar{z}_i)$ can approximated optimally as follows:

$$\bar{\alpha}_i(\bar{z}_i) = \hat{\bar{\alpha}}_i(\bar{z}_i, \theta_i^*) + \varepsilon_i(\bar{z}_i) = \theta_i^{*T} \psi_i(\bar{z}_i) + \varepsilon_i(\bar{z}_i) \tag{60}$$

and the fuzzy approximation error is assumed to be bounded as usual (Wang, 1994; Boulkroune et al., 2008, Boulkroune et al., 2009; Boulkroune et al., 2010b; Boulkroune & M'saad, 2011a; Boulkroune et al., 2014c), i.e. $\left|\varepsilon_i(\bar{z}_i)\right| \leq \bar{\varepsilon}_i$, $\forall \bar{z}_i \in \Omega_{\bar{z}_i}$, where $\bar{\varepsilon}_i$ is an unknown constant.

From the above analysis, one has

$$\hat{\bar{\alpha}}_i(\bar{z}_i,\theta_i)-\bar{\alpha}_i(\bar{z}_i)=\hat{\bar{\alpha}}_i(\bar{z}_i,\theta_i)-\hat{\bar{\alpha}}_i(\bar{z}_i,\theta_i^*)+\hat{\bar{\alpha}}_i(\bar{z}_i,\theta_i^*)-\bar{\alpha}_i(\bar{z}_i),$$

$$=\hat{\bar{\alpha}}_i(\bar{z}_i,\theta_i)-\hat{\bar{\alpha}}_i(\bar{z}_i,\theta_i^*)-\varepsilon_i(\bar{z}_i),$$

$$=\tilde{\theta}_i^T\psi_i(\bar{z}_i)-\varepsilon_i(\bar{z}_i). \tag{61}$$

The following fuzzy adaptive variable-structure controller is proposed to perform the required tracking control objective:

$$u_i=\begin{cases}-\bar{\rho}_i(t)sign(\bar{S}_i)-u_{i-}, & \bar{S}_i>0,\\ 0, & \bar{S}_i=0,\\ -\bar{\rho}_i(t)sign(\bar{S}_i)+u_{i+}, & \bar{S}_i<0,\end{cases} \tag{62}$$

with

$$\bar{\rho}_i(t)=k_{0i}+k_{1i}\left|\bar{S}_i\right|+\theta_i^T\psi_i(\bar{z}_i)\ \forall i=1,\dots,p$$

$$\dot{k}_{0i}=-\gamma_{0i}\sigma_{0i}k_{0i}+\gamma_{0i}\left|\bar{S}_i\right|,\ k_{0i}(0)>0 \tag{63}$$

$$\dot{\theta}_i=-\gamma_{1i}\sigma_{1i}\theta_i+\gamma_{1i}\left|\bar{S}_i\right|\psi_i(\bar{z}_i),\ \theta_{ij}(0)>0 \tag{64}$$

where $\gamma_{0i}, \gamma_{1i}, \sigma_{0i}, \sigma_{1i}, k_{1i}>0$ are design constants, and k_{0i} and θ_i are the online estimates of the uncertain terms $k_{0i}^*=\bar{\varepsilon}_i$ and θ_i^*, respectively.

Multiplying (57) by $\bar{S}^T$ and using Assumption 5 yields

$$\frac{1}{2\eta}\bar{S}^T\dot{G}_{s2}(x)\bar{S}+\frac{1}{\eta}\bar{S}^TG_{s2}(x)\dot{\bar{S}}=\frac{1}{\eta}\bar{S}^T\alpha(\bar{z})+\frac{1}{\eta}\bar{S}^T\Phi(u)$$

$$\leq\sum_{i=1}^{p}\left|\bar{S}_i\right|\bar{\alpha}_i(\bar{z}_i)+\frac{1}{\eta}\bar{S}^T\Phi(u) \tag{65}$$

And taking into account (61) and (65), one has

$$\frac{1}{2\eta}\bar{S}^T \dot{G}_{s2}(x)\bar{S} + \frac{1}{\eta}\bar{S}^T G_{s2}(x)\dot{\bar{S}} \leq \sum_{i=1}^{p} \left|\bar{S}_i\right| \bar{\alpha}_i(\bar{z}_i) + \frac{1}{\eta}\bar{S}^T \Phi(u)$$

$$\leq -\sum_{i=1}^{p} \left|\bar{S}_i\right| \left(\tilde{k}_{0i} + \tilde{\theta}_i^T \psi_i(\bar{z}_i)\right) + \sum_{i=1}^{p} \left|\bar{S}_i\right| \left(k_{0i} + \theta_i^T \psi_i(\bar{z}_i)\right) + \frac{1}{\eta}\bar{S}^T \Phi(u) \quad (66)$$

where $\tilde{\theta}_i = \theta_i - \theta_i^*$ and $\tilde{k}_{0i} = k_{0i} - k_{0i}^* = k_{0i} - \bar{\varepsilon}_i$.

Theorem 2: Consider the system (2) subject to Assumptions 1, 4 and 5. Then, the control law defined by equations (62) to (64) ensures the following properties:

- All the variables involved in the control system are uniformly ultimately bounded.
- The tracking error of the control system exponentially converges to an adjustable domain.

Proof of Theorem 2: It is carried out using the following Lyapunov function candidate

$$V = \frac{1}{2\eta}\bar{S}^T G_{s2}(x)\bar{S} + \frac{1}{2}\sum_{i=1}^{p} \frac{1}{\gamma_{1i}}\tilde{\theta}_i^T \tilde{\theta}_i + \frac{1}{2}\sum_{i=1}^{p} \frac{1}{\gamma_{0i}}\tilde{k}_i^2 . \quad (67)$$

Since this proof is very similar to that of Theorem 1, it is omitted.

Remark 4: In this chapter, since the function approximation property of fuzzy systems is only guaranteed within a compact set, the stability results proposed are semi-global in the sense that, for any compact set, there exists a fuzzy controller with sufficiently large number of fuzzy rules such that all the closed-loop signals are bounded when the initial states are within this compact set. However, when a large compact set is considered, a large number of fuzzy rules may be required. On the other hand, by increasing the number of rules, the computational complexity will be significantly increased. Therefore, the designer must find a tradeoff between number of fuzzy rules and the computational complexity. This implies that the approximation capability is limited, and a certain fuzzy approximation error is unavoidable. In practical applications, based on the knowledge and complexity of physical system, an appropriate compact set may be considered such that state trajectories do not leave this compact set under the designed controller. In the literature of fuzzy adaptive control, this fuzzy approximation error has been compensated or attenuated respectively by adding to the fuzzy adaptive controller (the main control term) a sliding mode control term (Ordonez & Passino, 1999; Chekireb et al., 2003; Golea et al., 2003; Tong & Li, 2003; Labiod et al., 2005; Tong et al., 2005) or an H^∞ based robust control design (Chang, 2000; Tong et al., 2000; Li & Tong, 2003; Essounbouli et al., 2006). In this chapter, this unavoidable fuzzy approximation error is cancelled by adding the robust control term $k_{0i} sign(.)$ to the fuzzy controller.

Remark 5: According to the definition of μ and π, it can be seen that the size of μ depends on the controller design parameters γ_{1i}, γ_{0i}, σ_{0i}, σ_{1i} and k_{1i} (which must be selected strictly positive) and that of π depends on the controller design parameters σ_{0i} and σ_{1i}. It is clear that if we increase γ_{1i}, γ_{0i}, and k_{1i} and decrease σ_{0i} and σ_{1i}, it will help to reduce the term $\left(2\eta\pi/\sigma_{g1}\mu\right)^{1/2}$. This implies that the filtered tracking errors S_i (or $\bar{S}_i$) and the tracking error E can be made arbitrary small by appropriately choosing those design parameters.

Remark 6:

1. In the case where $u_{i+} = u_{i-} = u_{i0}$, the expressions (33) and (62) are respectively simplified

$$u_i = -\left(\rho_i(t) + u_{i0}\right) sign(S_i) \tag{68}$$

$$u_i = -\left(\bar{\rho}_i(t) + u_{i0}\right) sign(\bar{S}_i) \tag{69}$$

where $\rho_i(t) = k_{0i} + k_{1i}\left|S_i\right| + \theta_i^T \psi_i(x)$ and $\bar{\rho}_i(t) = k_{0i} + k_{1i}\left|\bar{S}_i\right| + \theta_i^T \psi_i(\bar{z}_i)$.

2. Moreover, to eliminate the chattering effect caused by the discontinuous control term in (68) and (69), the function *sign*(.) must be replaced by any equivalent smooth function such as: *tanh*(.), *arctan*(.), *sat*(.).
3. Note that the control law (68) and (69) is mainly composed of the three control terms: a fuzzy adaptive term which is used to cancel the uncertain nonlinearities, a robust control term $k_{0i}Sign(.)$ which is introduced to compensate for the fuzzy approximation errors, and a linear control term $k_{1i}S_i$ (or $k_{1i}\bar{S}_i$) which is used for the stability purposes.

Remark 7: Fuzzy logic systems used in this chapter represents a class of linearly parameterized approximators, and can be replaced by any linearly parameterized networks without any technical difficulty such as RBF, polynomial, splines and wavelet networks. However, only the fuzzy systems, by design, provide a systematic and efficient framework to incorporate the linguistic fuzzy information from human experts. Moreover, even a nonlinearly parameterized fuzzy approximator (note that used fuzzy system (22) becomes nonlinearly parameterized fuzzy systems, if we consider the parameters of the FBFs as unknown parameters) can be always linearized as linearly parameterized approximator, with the higher order terms of Taylor series expansions being taken as part of the modeling error, as used in many literatures (Wang, 1994). The stability and performance properties of the adaptive system using nonlinearly parameterized approximator can be analyzed following the similar procedures therein. It is omitted here for simplicity and conciseness.

Remark 8: We can show without any technical difficulty that the proposed controllers remain applicable for MIMO system (2) free of the input nonlinearities (i.e. in the case where $u_{i+} = u_{i-} = 0$, and $\varphi_{i+}(u_i) = \varphi_{i-}(u_i) = 1$).

DISCUSSION

In this section, we present the important features and drawbacks of the proposed controllers, and some useful recommendations for future research.

The most important characteristics of the proposed controllers can be summarized as follows:

1. The class of MIMO nonlinear systems (2) considered here is relatively large and very complicated. Many reasons make its control a challenge, namely:
   a. In the controller design, the model (2) is assumed to be completely unknown, except it relative degree. The uncertainties in the model are estimated online by using adaptive fuzzy systems.

b. The control gain matrix can be non-symmetric and indefinite (i.e. it can be not positive definite and not negative definite). In order to facilitate the controller design, this matrix is generally assumed to be symmetric and positive-definite, in many literatures (Ordonez & Passino, 1999; Chang, 2000; Tong et al., 2000; Chekireb et al., 2003; Golea et al., 2003; Li & Tong, 2003; Tong & Li, 2003; Labiod et al., 2005; Tong et al., 2005; Essounbouli et al., 2006). In this chapter, to deal with the problem of the symmetry and the sign of the control-gain matrix, a suitable decomposition of the control-gain matrix has been exploited to carry out the controller design and the stability analysis.

c. The system is assumed to be subject to unavoidable actuator nonlinearities (i.e. sector-nonlinearities and dead-zone). It is well known that the existence of dead-zone nonlinearities may lead to notable performance degradations or even instability of the control system. It is thereby more advisable to take into account these actuator nonlinearities in the control design as well as the stability analysis. Herein, the problem of dead-zone and sector nonlinearities has been successfully treated by designing a variable structure controller.

d. The system is assumed to be affected by dynamical unknown external disturbances. They are considered as a uncertain function and approximate by a fuzzy adaptive system.

2. In our proposed controllers, the requirements of an off-line learning phase, the persistent excitation condition and the prior knowledge of an upper bound on the unknown ideal parameters are not required at all.
3. Our proposed fuzzy adaptive controllers can be considered as indirect schemes. They are free of singularity and with a simple structure compared to many indirect adaptive schemes ((Ordonez & Passino, 1999; Chang, 2000; Tong et al., 2000; Golea et al., 2003; Li & Tong, 2003; Tong & Li, 2003; Labiod et al., 2005; Tong et al., 2005; Essounbouli et al., 2006). Moreover, less fuzzy systems (p fuzzy systems) used in our schemes help to make the proposed algorithms computationally more efficient.

The main drawbacks of the proposed controllers may be summarized as follows:

1. In our controller, the state vector is assumed to be completely available for measurement. But this measurement requirement is more an exception than a rule in the engineering practice. That is why observer-based controllers or output-feedback controllers are most used in practice.
2. In the stability analysis, we have shown that the resulting closed-loop control system are only semi-global UUB stable.
3. In the design, the control directions (i.e. the signs of the leading principal minors of the control-gain matrix) are assumed to be known a priori. However, in the general case, this assumption is by no means realistic. When there is no a priori knowledge about the control directions, the design of the fuzzy adaptive controllers for unknown nonlinear systems subject to dead-zone and sector nonlinearities becomes more challenging.
4. The number of the adaptation laws depends on the number of the fuzzy rules. With an increase of the fuzzy rules, the number of parameters to be estimated will increase significantly.

Now, we attempt to give some directions for our future research:

1. The use of more general (nonlinear) parameterization for the fuzzy systems in order to improve the approximation ability.
2. The proposition of an algorithm allowing to optimize the number of the fuzzy rules.
3. The design of an high-gain observer to estimate the state vector.
4. The extension of the proposed controllers to discrete-time systems and others classes of continuous-time systems (as non-affine or strict-feedback systems).
5. The incorporation of the Nussbaum function in the proposed variable structure controllers in order to deal with the unknown sign of the control gain matrix.

SIMULATION RESULTS

Simulation studies are carried out to show the effectiveness of the proposed adaptive fuzzy controllers. Two control problems are considered to this end. The first one concerns a two-link rigid robot manipulator (its model has naturally a symmetric control-gain matrix), while the second one concerns an academic multivariable nonlinear system (its model has a non-symmetric and indefinite control-gain matrix).

Example 1

In the following, one presents simulation results showing the performances of the first controller applied to a two-link rigid robot manipulator which moves in a horizontal plane. The dynamic equations of this MIMO system are given by (Boulkroune et al., 2014b):

$$\begin{pmatrix} \ddot{q}_1 \\ \ddot{q}_2 \end{pmatrix} = \begin{pmatrix} M_{11} & M_{12} \\ M_{21} & M_{22} \end{pmatrix}^{-1} \left\{ \begin{pmatrix} \Phi_1(u_1) \\ \Phi_2(u_2) \end{pmatrix} - \begin{pmatrix} -h\dot{q}_2 & -h(\dot{q}_1 + \dot{q}_2) \\ h\dot{q}_1 & 0 \end{pmatrix} \begin{pmatrix} \dot{q}_1 \\ \dot{q}_2 \end{pmatrix} \right\}, \tag{70}$$

where

$$M_{11} = a_1 + 2a_3 \cos(q_2) + 2a_4 \sin(q_2), \quad M_{22} = a_2, \quad M_{21} = M_{12} = a_2 + a_3 \cos(q_2) + a_4 \sin(q_2),$$
$$h = a_3 \sin(q_2) - a_4 \cos(q_2),$$

with

$$a_1 = I_1 + m_1 l_{c1}^2 + I_e + m_e l_{ce}^2 + m_e l_1^2, \; a_2 = I_e + m_e l_{ce}^2, \; a_3 = m_e l_1 l_{ce} \cos(\delta_e), \; a_4 = m_e l_1 l_{ce} \sin(\delta_e).$$

Note that $\Phi_i(u_i)$, for $i = 1,2$, are the considered input nonlinearities. The robot parameters are: $m_1 = 1,\; m_e = 2,\; l_1 = 1,\; l_{c1} = 0.5,\; l_{ce} = 0.6,\; I_1 = 0.12,\; I_e = 0.25,\; \delta_e = 30°$.

Let $[y_1, y_2] = [q_1, q_2],\; u = [u_1, u_2]^T,\; \Phi(u) = [\Phi_1(u_1), \Phi_2(u_2)]^T,\; x = [x_1, x_2, x_3, x_4]^T = [q_1, \dot{q}_1, q_2, \dot{q}_2]^T$. One can rewrite the robot model with actuator nonlinearities as follows:

$$\ddot{y} = F(x) + G(x)\Phi(u) \tag{71}$$

where

$$F(x) = \begin{pmatrix} f_1(x) \\ f_2(x) \end{pmatrix} = -M^{-1} \begin{pmatrix} -h\dot{q}_2 & -h(\dot{q}_1 + \dot{q}_2) \\ h\dot{q}_1 & 0 \end{pmatrix} \begin{pmatrix} \dot{q}_1 \\ \dot{q}_2 \end{pmatrix},$$

and

$$G(x) = \begin{pmatrix} g_{11}(x) & g_{12}(x) \\ g_{21}(x) & g_{22}(x) \end{pmatrix} = M^{-1} = \begin{pmatrix} M_{11} & M_{12} \\ M_{21} & M_{22} \end{pmatrix}^{-1}.$$

The input nonlinearities are described as in (Shyu et al., 2005; Boulkroune et al., 2014b):

$$\Phi_1(u_1) = \begin{cases} (u_1 - 3)(1.5 - 0.3e^{0.3|\sin(u_1)|}), & u_1 > 3, \\ 0, & |u_1| \le 3, \\ (u_1 + 3)(1.5 - 0.3e^{0.3|\sin(u_1)|}), & u_1 < -3, \end{cases}$$

$$\Phi_2(u_2) = \begin{cases} (u_2 - 3)(1.3 - 0.2e^{0.1|\cos(u_2)|}), & u_2 > 3, \\ 0, & |u_2| \le 3, \\ (u_2 + 3)(1.3 - 0.2e^{0.1|\cos(u_2)|}), & u_2 < -3, \end{cases}$$

The control objective consists in allowing the system outputs q_1 and q_2 to respectively track the sinusoidal desired trajectories $y_{d1} = \sin(t)$ and $y_{d2} = \sin(t)$. Two square waves ($\Lambda_1(t)$ and $\Lambda_2(t)$) having an amplitude ± 1 with a period of $2\pi(s)$ are added to system states as external disturbances. Then, the system (71) becomes

$$\ddot{y} = F(x) + G(x)\Phi(u) + \Lambda(t)$$

Recall that the system nonlinearities *F*(*x*) and *G*(*x*) and the external dynamical disturbances ($\Lambda_1(t)$ and $\Lambda_2(t)$) are assumed to be completely unknown except the symmetry property and the sign of the matrix *G*(*x*), when designing the fuzzy adaptive controller. The fuzzy systems $\theta_1^T \psi_1(x)$ and $\theta_2^T \psi_2(x)$ conceived to estimate the uncertainties have only the state vector $x = [x_1, x_2, x_3, x_4]^T = [q_1, \dot{q}_1, q_2, \dot{q}_2]^T$ as input. The fuzzy membership functions are defined for each variable x_j ,for *j*=1,2,3,4 as follows:

$$\mu_{A_j^1}(x_j) = \frac{1}{1 + \exp\left(3(x_j + 0.25)\right)}, \quad \mu_{A_j^2}(x_j) = \exp\left(-\frac{1}{2}\left(\frac{x_j^2}{0.6}\right)\right) \text{ and}$$

$$\mu_{A_j^3}(x_j) = \frac{1}{1 + \exp\left(-3(x_j - 0.25)\right)}.$$

Thus, the number of the fuzzy rules is $3^4 = 81$ (because our fuzzy rule base is a completed base, all possible combinations of rules are taken into account). The consequent parameters are estimated online via the adaptive laws (35), they are initialized as follows: $\theta_{1i}(0) = 0.005$ and $\theta_{2i}(0) = 0.005$, for $i = 1,...,81$.

The design parameters used in this simulation are specified as follows: $\gamma_{01} = \gamma_{02} = 30$, $\gamma_{11} = \gamma_{12} = 700$, $\sigma_{01} = \sigma_{02} = 0.001$, $\sigma_{11} = \sigma_{12} = 0.001$, $\lambda_1 = \lambda_2 = 2$ and $k_{11} = k_{12} = 2$. The initial conditions are arbitrary chosen as follows: $x(0) = [0.5 \quad 0 \quad 0.5 \quad 0]^T$, and $k_{01}(0) = k_{02}(0) = 0.01$. It is worth noticing that the sign function has been replaced by a smooth function $\tanh(k_{si}S_i)$, with $k_{si} = 20$.

Figure 2 shows proposed fuzzy adaptive controller performs well in spite of uncertainties. Indeed the tracking errors for both links are bounded and converge towards small values as pointed out by Figure 2a and Figure 2b. Figures 2c and 2d show that the control variables as well as the norm of the estimated parameters are bounded.

In summary, in spite of the presence of the input nonlinearities, external dynamical disturbances and uncertain dynamics, the MIMO system (70) remains stable and with a weak-tracking error. Also, we can see clearly that the control signals are smooth and admissible.

Example 2

In the following, one presents simulation results showing the performance of the second fuzzy adaptive control system involving an academic multivariable nonlinear system having a non-symmetric control-gain matrix. The dynamics of this system is given by:

Figure 2. Simulation results of Example 1. (a) Tracking errors of link 1: e_1 (dotted line) and $\dot{e}_1$ (solid line). (b) Tracking errors of link 2: e_2 (dotted line) and $\dot{e}_2$ (solid line). (c) Control input signals: u_1 (dotted line) and u_2 (solid line) (d). Norm of fuzzy parameters: $\|\theta_1\|$ (dotted line), $\|\theta_2\|$ (solid line)

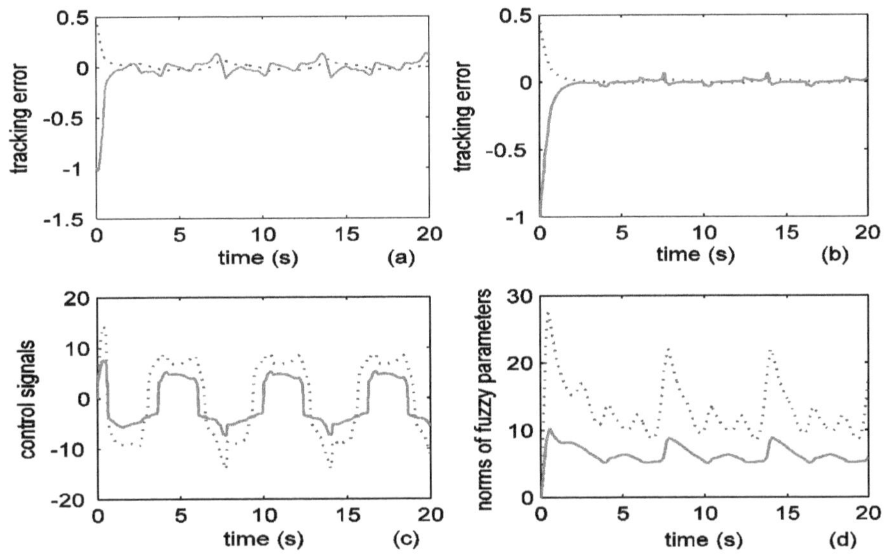

$$\begin{cases} \dot{x}_1 = x_2, \\ \dot{x}_2 = x_3 + x_2^2 + (2+\cos(x_1))\Phi_1(u_1) + (1+(\sin(x_3))^2)\Phi_2(u_2) + \Lambda_1(x), \\ \dot{x}_3 = x_4, \\ \dot{x}_4 = x_4^3 + e^{x_1} + x_2^2 - 0.5\Phi_1(u_1) - (1+(\sin(x_3))^2)\Phi_2(u_2) + \Lambda_2(x), \\ y_1 = x_1, \quad y_2 = x_3. \end{cases} \tag{72}$$

Let $y = [y_1, y_2]^T$, $u = [u_1, u_2]^T$, $\Phi(u) = [\Phi_1(u_1), \Phi_2(u_2)]^T$, and $x = [x_1, x_2, x_3, x_4]^T$. Then, the system (72) can be given in the following form:

$\ddot{y} = F(x) + G(x)\Phi(u) + \Lambda(x)$ (73)where

$$F(x) = \begin{pmatrix} x_{21} - 0.3\sin(x_{11}x_{12}) + x_{12}^2 \\ x_{22}^3 + e^{x_{11}} + x_{12}^2 \end{pmatrix}, \quad G(x) = \begin{pmatrix} (2+\cos(x_{11})) & 1+(\sin(x_{21}))^2 \\ -0.5 & -(1+(\sin(x_{21}))^2) \end{pmatrix}, \text{ and}$$

$$\Lambda(x) = \left[\Lambda_1(x),\ \Lambda_2(x)\right]^T = \left[-0.3\sin(x_1x_2),\ 0.5\sin(x_3)\right]^T,$$

The input nonlinearities $\Phi_i(u_i)$ for $i = 1,2$ are described by:

$$\Phi_i(u_i) = \begin{cases} (u_i - 3)(1 - 0.3\sin(u_i)), & u_i > 3, \\ 0, & -3 \le u_i \le 3, \\ (u_i + 3)(0.8 - 0.3\cos(u_i)), & u_i < -3, \end{cases}$$

It is worth noting that the matrix $G(x)$ is non-symmetric and indefinite. Recall that the system nonlinearities, i.e. $F(x)$, $G(x)$, the dynamical disturbance $\Lambda(t)$ and $\Phi(u)$ are assumed to be unknown when designing the fuzzy adaptive controller, except the signs of the leading principal minors of $G(x)$, the property (25) of $\Phi(u)$ and constants u_{i+} and u_{i-}.

The control objective consists in allowing the system outputs y_1 and y_2 to track the sinusoidal desired trajectories $y_{d1} = \sin(t)$ and $y_{d2} = \sin(t)$, respectively.

The fuzzy system $\theta_1^T\psi_1(\bar{z}_1)$ has the vector $[x^T, u_2]^T$ as input, while the fuzzy systems, $\theta_2^T\psi_2(\bar{z}_2)$, has the state vector *x* as input. The fuzzy membership functions are defined for each variable x_j, for *j*=1,2,3,4 as follows:

$$\mu_{A_j^1}(x_j) = \frac{1}{2}\left(1 + Tanh\left(-\frac{x_j + 0.5}{0.4}\right)\right), \quad \mu_{A_j^2}(x_j) = \exp\left(-\frac{1}{2}\left(\frac{x_j^2}{0.6}\right)\right), \text{ and}$$

$$\mu_{A_j^3}(x_j) = \frac{1}{2}\left(1 + Tanh\left(\frac{x_j - 0.5}{0.4}\right)\right).$$

But for the variable u_2, they are defined as:

$$\mu_{A_j^1}(x_j)=\frac{1}{2}\left(1+Tanh\left(-\frac{x_j+4}{2}\right)\right),\ \mu_{A_j^2}(x_j)=\exp\left(-\frac{1}{2}\left(\frac{x_j^2}{12}\right)\right), \text{ and}$$

$$\mu_{A_j^3}(x_j)=\frac{1}{2}\left(1+Tanh\left(\frac{x_j-4}{2}\right)\right).$$

The design parameters used in this simulation are chosen as follows: $\gamma_{01}=\gamma_{02}=20$, $\gamma_{11}=\gamma_{12}=1500$, $\sigma_{01}=\sigma_{02}=0.01$, $\sigma_{11}=0.001, \sigma_{12}=0.001$, $\lambda_1=\lambda_2=2$ and $k_{11}=k_{12}=0.5$. The initial conditions are selected as: $x(0)=[0.5 \quad 0 \quad 0.5 \quad 0]^T$, $k_{01}(0)=k_{02}(0)=0.01$, $\theta_{1i}(0)=0.005$ and $\theta_{2i}(0)=0.005$. Moreover, to simulate the practical situations, the system states have been corrupted by an additive noise. the sign function has been replaced by an equivalent smooth function, i.e. $\tanh(k_{si}\bar{S}_i)$, with $k_{si}=20$.

Figure 3 shows that the proposed fuzzy adaptive control system performs well from tracking, parameter adaptation and noise measurement sensitivity points of view. In fact, Figure 3a and Figure 3b illustrate the boundedness and convergence of the tracking-errors for both subsystems. The boundedness of the

Figure 3. Simulation results of Example 2. (a) Tracking errors: e_1 (dotted line) and $\dot{e}_1$ (solid line). (b) Tracking errors: e_2 (dotted line) and $\dot{e}_2$ (solid line). (c) Control input signals: u_1 (dotted line) and u_2 (solid line) (d). Norm of fuzzy parameters: $\|\theta_1\|$ (dotted line), $\|\theta_2\|$ (solid line)

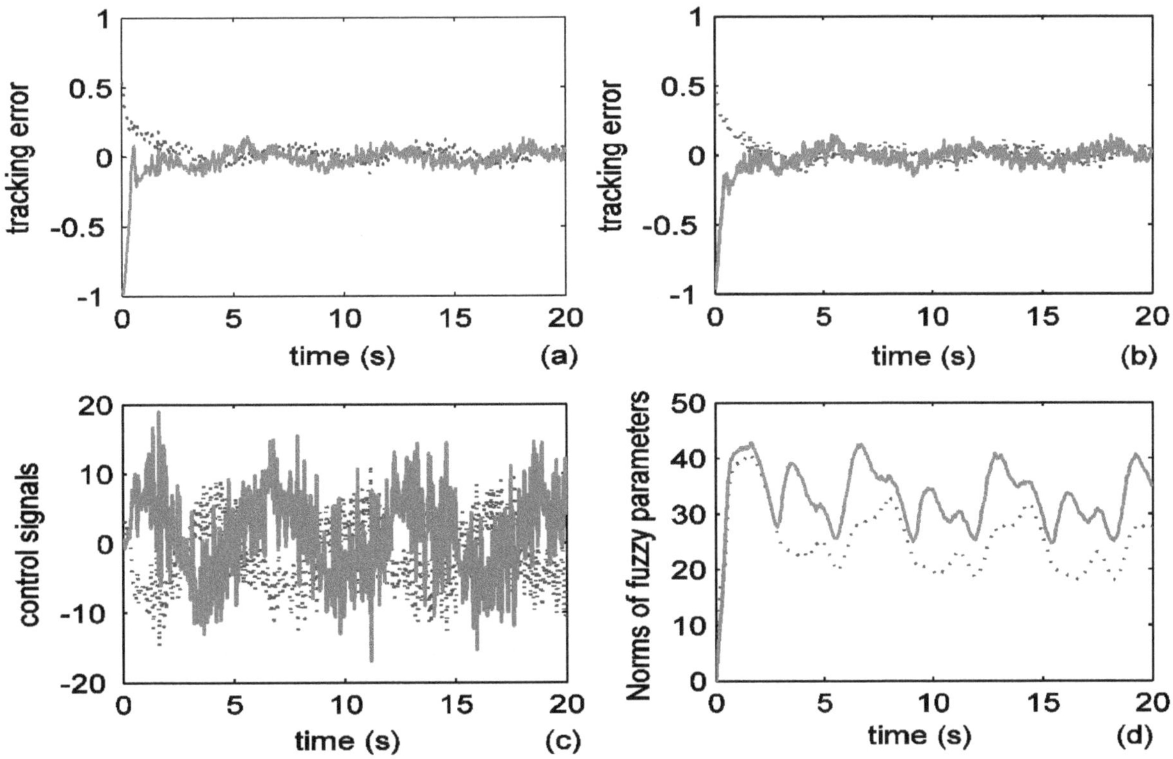

corresponding control signals as well as of the estimated fuzzy parameters is well illustrated in Figure 3c and Figure 3d, respectively. So, clearly, those simulation results verify our theoretical findings.

CONCLUSION

In this chapter, two fuzzy adaptive variable-structure controllers for a class of multivariable unknown nonlinear systems subject to sector nonlinearities and dead-zones have been proposed. The first one has been designed for systems having a symmetric and positive-definite control-gain matrix, while the second one has been particularly designed for systems with a non-symmetric or symmetric control gains matrix whose leading principal minors are non-zero. A suitable decomposition of the control-gain matrix has been fully exploited to carry out the second control design. Of fundamental interest, it has been proven that the proposed control systems are stable and their underlying tracking errors converge exponentially to an adjustable domain. Simulation results have been given to emphasize the effectiveness of the proposed controllers.

REFERENCES

Azar, A. T. (2010a). Adaptive neuro-fuzzy systems. In A. T. Azar (Ed.), *Fuzzy systems*. Vienna, Austria: IN-TECH. doi:10.5772/7220

Azar, A. T. (2010b). *Fuzzy systems*. Vienna, Austria: IN-TECH.

Azar, A. T. (2012). Overview of type-2 fuzzy logic systems. *International Journal of Fuzzy System Applications*, *2*(4), 1–28. doi:10.4018/ijfsa.2012100101

Boulkroune, A., Bounar, N., M'saad, M., & Farza, M. (2014a). Indirect adaptive fuzzy control scheme based based on observer for nonlinear systems: A novel SPR-filter approach. *Neurocomputing*, *135*, 378–387. doi:10.1016/j.neucom.2013.12.011

Boulkroune, A., & M'saad, M. (2011c). A fuzzy adaptive variable-structure control scheme for uncertain chaotic MIMO systems with sector nonlinearities and dead-zones. *Expert Systems with Applications*, *38*(12), 4744–14750. doi:10.1016/j.eswa.2011.05.006

Boulkroune, A., & M'saad, M. (2012a). On the design of observer-based fuzzy adaptive controller for nonlinear systems with unknown control gain sign. *Fuzzy Sets and Systems*, *201*, 71–85. doi:10.1016/j.fss.2011.12.005

Boulkroune, A., & M'saad, M. (2012b). Fuzzy adaptive observer-based projective synchronization for nonlinear systems with input nonlinearity. *Journal of Vibration and Control*, *18*(3), 437–450. doi:10.1177/1077546311411228

Boulkroune, A., M'saad, M., & Chekireb, H. (2010a). Design of a fuzzy adaptive controller for MIMO nonlinear time-delay systems with unknown actuator nonlinearities and unknown control direction. *Information Sciences*, *180*(24), 5041–5059. doi:10.1016/j.ins.2010.08.034

Boulkroune, A., M'saad, M., & Farza, M. (2011b). Adaptive fuzzy controller for multivariable nonlinear state time-varying delay systems subject to input nonlinearities. *Fuzzy Sets and Systems*, *164*(1), 45–65. doi:10.1016/j.fss.2010.09.001

Boulkroune, A., M'saad, M., & Farza, M. (2012c). Adaptive fuzzy tracking control for a class of MIMO nonaffine uncertain systems. *Neurocomputing*, *93*, 48–55. doi:10.1016/j.neucom.2012.04.006

Boulkroune, A., M'saad, M., & Farza, M. (2012d). Fuzzy approximation-based indirect adaptive controller for multi-input multi-output non-affine systems with unknown control direction. *IET Control Theory & Applications*, *17*(17), 2619–2629. doi:10.1049/iet-cta.2012.0565

Boulkroune, A., M'saad, M., & Farza, M. (2014b). State and output feedback fuzzy variable structure controllers for multivariable nonlinear systems subject to input nonlinearities. *International Journal of Advanced Manufacturing Technology*, *71*(1-4), 539–556. doi:10.1007/s00170-013-5453-4

Boulkroune, A., & Msaad, M. (2011a). A practical projective synchronization approach for uncertain chaotic systems with dead-zone input. *Communications in Nonlinear Science and Numerical Simulation*, *16*(11), 4487–4500. doi:10.1016/j.cnsns.2011.02.016

Boulkroune, A., Tadjine, M., M'saad, M., & Farza, M. (2008). How to design a fuzzy adaptive control based on observers for uncertain affine nonlinear systems. *Fuzzy Sets and Systems*, *159*(8), 926–948. doi:10.1016/j.fss.2007.08.015

Boulkroune, A., Tadjine, M., M'saad, M., & Farza, M. (2009). Adaptive fuzzy controller for non-affine systems with zero dynamics. *International Journal of Systems Science*, *40*(4), 367–382. doi:10.1080/00207720802436919

Boulkroune, A., Tadjine, M., M'saad, M., & Farza, M. (2010b). Fuzzy adaptive controller for MIMO nonlinear systems with known and unknown control direction. *Fuzzy Sets and Systems*, *161*(6), 797–820. doi:10.1016/j.fss.2009.04.011

Boulkroune, A., Tadjine, M., M'saad, M., & Farza, M. (2014c). Design of a unified adaptive fuzzy Observer for uncertain nonlinear systems. *Information Sciences*, *265*, 139–153. doi:10.1016/j.ins.2013.12.026

Chang, Y. C. (2000). Robust tracking control for nonlinear MIMO systems via fuzzy approaches. *Automatica*, *36*(10), 1535–1545. doi:10.1016/S0005-1098(00)00083-2

Chekireb, H., Tadjine, M., & Bouchaffra, D. (2003). Direct adaptive fuzzy control of nonlinear system class with applications. *Control and Intelligent Systems*, *31*(2), 1–11.

Essounbouli, N., Hamzaoui, A., & Zaytoon, J. (2006). An improved robust adaptive fuzzy controller for MIMO systems. *Control and Intelligent Systems*, *34*(1), 12–21. doi:10.2316/Journal.201.2006.1.201-1350

Golea, N., Golea, A., & Benmahammed, K. (2003). Stable indirect fuzzy adaptive control. *Fuzzy Sets and Systems*, *137*(3), 353–366. doi:10.1016/S0165-0114(02)00279-8

Gutierrez, H. M., & Ro, P. I. (1998). Sliding-mode control of a nonlinear-input system: Application to a magnetically levitated fast-tool servo. *IEEE Transactions on Industrial Electronics*, *45*(6), 921–927. doi:10.1109/41.735336

Hsu, K. C. (1999). Decentralized sliding mode controller for uncertain time-delayed systems with series nonlinearities. *Journal of Dynamic Systems, Measurement, and Control*, *121*(4), 708–713. doi:10.1115/1.2802539

Hsu, K. C., Wang, W. Y., & Lin, P. Z. (2004). Sliding mode control for uncertain nonlinear systems with multiple inputs containing sector nonlinearities and deadzones. *IEEE Transactions on Systems, Man and Cybernetic. Part-B*, *34*(1), 374–380.

Hsu, L., Costa, R. R., & Lizarralde, F. (2007). Lyapunov/passivity-based adaptive control of relative degree two MIMO systems with an application to visual servoing. *IEEE Transactions on Automatic Control*, *52*(2), 364–371. doi:10.1109/TAC.2006.890381

Labiod, S., Boucherit, M. S., & Guerra, T. M. (2005). Adaptive fuzzy control of a class of MIMO nonlinear systems. *Fuzzy Sets and Systems*, *151*(1), 59–77. doi:10.1016/j.fss.2004.10.009

Li, H. X., & Tong, S. C. (2003). A hybrid adaptive fuzzy control for a class of nonlinear MIMO systems. *IEEE Transactions on Fuzzy Systems*, *11*(1), 24–34. doi:10.1109/TFUZZ.2002.806314

Niu, Y., & Ho, D. W. C. (2006). Design of sliding mode control for nonlinear stochastic systems subject to actuator nonlinearity. *IEE Proceedings. Control Theory and Applications*, *153*(6), 737–744. doi:10.1049/ip-cta:20050194

Ordonez, R., & Passino, K. M. (1999). Stable multi-input multi-output adaptive fuzzy/neural control. *IEEE Transactions on Fuzzy Systems*, *7*(3), 345–353. doi:10.1109/91.771089

Shyu, K. K., Liu, W. J., & Hsu, K. C. (2003). Decentralized variable structure control design for uncertain large scale systems containing a deadzone. *IEE Proceedings. Control Theory and Applications*, *150*(5), 467–475. doi:10.1049/ip-cta:20030758

Shyu, K. K., Liu, W. J., & Hsu, K. C. (2005). Design of large-scale time-delayed systems with dead-zone input via variable structure control. *Automatica*, *41*(7), 1239–1246. doi:10.1016/j.automatica.2005.03.004

Tong, S. C., Chen, B., & Wang, Y. (2005). Fuzzy adaptive output feedback control for MIMO nonlinear systems. *Fuzzy Sets and Systems*, *156*(2), 285–299. doi:10.1016/j.fss.2005.06.011

Tong, S. C., & Li, H. X. (2003). Fuzzy adaptive sliding model control for MIMO nonlinear systems. *IEEE Transactions on Fuzzy Systems*, *11*(3), 354–360. doi:10.1109/TFUZZ.2003.812694

Tong, S. C., Tang, J., & Wang, T. (2000). Fuzzy adaptive control of multivariable nonlinear systems. *Fuzzy Sets and Systems*, *111*(2), 153–167. doi:10.1016/S0165-0114(98)00058-X

Wang, L. X. (1994). *Adaptive fuzzy systems and control: design and stability analysis*. Englewood Cliffs, NJ: Prentice-Hall.

Zhang, T. P., & Ge, S. S. (2007). Adaptive neural control of MIMO nonlinear state time-varying delay systems with unknown dead-zones and gain signs. *Automatica*, *43*(6), 1021–1033. doi:10.1016/j.automatica.2006.12.014

Zhang, T. P., & Yi, Y. (2007). Adaptive fuzzy control for a class of MIMO nonlinear systems with unknown dead-zones. *Acta Automatica Sinica*, *33*(1), 96–99. doi:10.1360/aas-007-0096

KEY TERMS AND DEFINITIONS

Actuator Nonlinearities: They are due to some physical limitations and include friction, dead zone, backlash, hysteresis and saturation.

Adaptive Control: The control parameters in this approach are continuously and automatically adjusted in response to be measured process variables to achieve near-optimum system performance.

Fuzzy Control: A means of expressing an operator's knowledge of controlling a process with a set of fuzzy IF-THEN rules and linguistic variables.

Fuzzy Universal Approximation Theorem: Any sufficiently smooth function can be approximated arbitrarily closely on a compact set using a fuzzy system with appropriate structure.

Lyapunov's (Second) Approach: The method of stability assessment that relies on the use of energy-like functions without resorting to direct solution of the associated evolution equations.

Matrix (SDU) Decomposition: A matrix factorization into a product of a symmetric positive definite matrix, a diagonal matrix whose elements are +1 or -1 and a unity upper triangular matrix.

Multi-Input–Multi-Output System (MIMO): Also known as multivariable system. A system that can transform two or more input signals to two or more output signals.

Variable Structure System: It consists of a set of continuous subsystems together with a suitable switching logic.

Chapter 13
Fault Detection and Isolation for an Uncertain Takagi-Sugeno Fuzzy System using the Interval Approach

Hassene Bedoui
University of Monastir, Tunisia

Atef Kedher
University of Tunis Manar, Tunisia

Kamel Ben Othman
University of Tunis Manar, Tunisia

ABSTRACT

This work deals with the fault detection and localization in the case of uncertain nonlinear systems. The presented method uses the diagnosis based on mathematical models. To model nonlinear systems, the multiple model approach is used. This method uses the Takagi-Sugeno fuzzy systems principle to obtain a nonlinear system named multiple models. This modeling principle has the advantage of obtaining a general model that can describe any class of nonlinear systems. This modeling principle also allows one to obtain the generalization of many results that are already obtained for linear systems to the nonlinear systems. To model the system uncertainties, the interval approach is used because the faults or disturbances are generally unknown, but it is possible to know their upper and lower bounds. The proposed technique is insensitive to measurement uncertainties and highly reliable in case of a fault affecting the outputs system.

1. INTRODUCTION

State estimation plays an important role in a context of system diagnosis. It.is an.analytical source of redundancy used to generate system failure symptoms by making a.comparison between the real behavior

DOI: 10.4018/978-1-4666-7248-2.ch013

signals of the system and the estimated signals. A not desired variation between these signals indicates the possible presence of faults affecting the system operation.

An accurate representation of the system model is the challenge faced by the scientific community. Several methods have been proposed to model the systems behavior in order to ensure proper supervision. Indeed, in many industrial fields, which must give product with good precision or which are dangerous, such as nuclear power plants or aerospace, all faults must be detected to ensure the materials and human security and the product precision (Bedoui et al, 2013). Hardware redundancy is often considered to detect and locate faults solution, or any material is ideal. Each instrument has an uncertainty which is often indicated by the manufacturer. Hardware redundancy consists in using several sensors to measure one output. This technique allows being sure of the output measure. The disadvantage of this technique is the installation blocking and considerable additional cost.

The diagnosis is an important field that allows detecting and locating the origin of faults or failure and permits to locate and track their evolution. It is based on the principle of residuals generation. The principle of residual generation is to compare the evolution of the system state to the state estimated. Computing the state estimation error leads to conclude if the system is faulty or no. If the residual is null, the system has a good behavior and no faults are affecting the system. In the opposite case it is possible to detect a fault affecting the system. These residuals are only functions of the measurement of noise and faults, and are determined, in this work, in the case of sensor faults.

In this work, a method for diagnosis of nonlinear systems with measurement uncertainties (sensors) based on the interval analysis as tools for state estimation is proposed. This method is used to make the fault detection and location.

In general, faults can be detected by developing performing checks to verify the adequacy of the information provided by the model and the sensors. An inconsistency appears when the residual is out of the vicinity of zero (for filter-based methods (Kalman, 1960), observers (Luenberger, 1964) or space parity (Chow & Willsky, 1984)) or when the parameters derived abnormally in the methods of parametric estimations (Isermann, 1984).

The interval analysis is to define an envelope in which the fault is considered non-existent and any measurement deviation inside this envelope is considered normal. Otherwise, the system is declared faulty. This approach has been explored by Adrot (Adrot et al, 2002; 2000) and recently by Bedoui (Bedoui et al, 2014) and applied to a mechatronic system by Letellier (Letellier, 2012).

Recently, Takagi-Sugeno Fuzzy systems are the subject of many researches by virtue of their approximation capabilities (Witczak et al, 2008 ; Khedher & Othman, 2010). They can represent exactly a nonlinear model (Tanaka et al, 1998; Witczak et al., 2008; Khedher & Othman, 2010). They are constructed by a set of linear models blended together with nonlinear functions holding the convex-sum property. In the case of Takagi-Sugeno Fuzzy systems, state estimation is based on the design of nonlinear observers (multiple observers) using the same nonlinear weighting functions as the Takagi-Sugeno models (Akhenak et al, 2009; 2003; Ichalal et al, 2009a; Khedher et al, 2008; 2009).

Approaches using Takagi-Sugeno models, also named multiple models (Murray-Smith & Johansen, 1997), are the object of many works in different contexts including unknown inputs or parameter uncertainties (Akhenak et al, 2007; Ichalal et al, 2009b). Various studies dealing with the presence of unknown inputs acting on the system were published (Akhenak et al., 2007; Edwards & Tan, 2006; Sharma & Aldeen, 2004). Some of them tried to reconstruct the system state in spite of the unknown input existence. This reconstruction is assured via the elimination of unknown inputs (Guan & Saif, 1991; Sharma & Aldeen, 2004). Other works choose to estimate, simultaneously, the unknown inputs and system state (Akhenak et al., 2007; Edwards & Tan, 2006; Ichalal et al., 2009b; Khedher et al., 2008).

Among the techniques that do not require the elimination of the unknown inputs, Wang (S.-H. Wang et al, 1975) has proposed an observer able to entirely reconstruct the state of a linear system in the presence of unknown inputs and in (Lyubchik & Kostenko, 1993), to estimate the state. In the cited work, a model inversion method is used.

In often cases, process can be subjected to disturbances which have as origin the noises due to its environment, uncertainty of measurements, faults of sensors and/or actuators. These disturbances have harmful effects on the normal behavior of the process and their estimation can be used to conceive a control strategy able to minimize their effects (Witczak et al., 2008).

In the case of sensor faults, Edwards (Edwards & Tan, 2006) propose, for linear systems, to use a new state which is a filtrate version of the output, to conceive an augmented system in which the sensor fault appear as an unknown input. This formulation was used by (Zhang et al, 2008), and it is developed in (Khedher et al., 2009) to make sensor faults estimation for systems described by Takagi-Sugeno multiple models. In (Khedher & Othman, 2010) the filtrate version of the output is proposed in the case of Takagi-Sugeno multiple models.

In this work, the proposed method relies on the description of the sensor uncertainties by known compact sets. In the field of diagnosis, robust model-based fault detection of dynamic systems using interval observers has already been addressed. One of the main techniques consists in checking whether the measurements of the output belong to the interval of all possible estimated outputs obtained considering uncertainty on model parameters (Stancu et al, 2005; Ploix et al, 1999; Raıssi et al, 2004). Original applications in the field of flow rate sensor diagnosis and data validation are presented in (Puig et al., 2003; Alhaj-Dibo et al, 2004). Although interval approaches need very little a priori information (only the uncertain parameter bounds), only few published works are dedicated to process diagnosis, like (Planchon & Lunze, 2008; Ragot et al, 2006; Marx et al, 2010) or (Bedoui et al., 2014).

The main contribution in this work is to use the interval analysis to model measurement uncertainties and to propose a method allowing detecting and locating fault in the case of nonlinear uncertain systems described by Takagi-Sugeno multiple models.

This paper is organized as follows. After a brief overview, section 2 recalls the interval analysis in the case of linear systems and its application to an academic example. An elementary background about the Takagi-Sugeno fuzzy models and the extension of the proposed method to this type of models is the subject of section 3. A numerical application of the proposed method is presented in the last section.

2. INTERVAL ANALYSIS

2.1. Interval Analysis Presentation

State estimation is a problem solved if the system is on constant parameters or is subject to disturbances with known statistical characteristics (Othman, 2008). In general case, estimation techniques are based on the knowledge of the mathematical model of the system. In addition, the information collected on a system are from sensors whose reliability can be questioned (Othman et al, 2007), causing disruptive phenomena that undermine the ability of models to accurately reflect the systems operation. A rather attractive method to account for changes in system parameters is to consider bounded perturbations whose boundaries are known in advance.

As every moment disturbances are unknown, it is impossible to determine the actual system state. The solution shall be in the estimation of a domain to which these states belong. This problem is known under the term "set-valued state estimation" (Lin, Zhai, & Antsaklis, 2003; Ploix et al., 1999; Shamma & Tu, 1999; Walter & Piet-Lahanier, 1989) and can be formulated in a rather general way by considering the following model systems (Othman, 2008):

$$S\begin{cases} x(k+1) = f\big(x(k), u(k), v(k)\big) \\ y(k) = h\big(x(k)\big) + w(k) \end{cases} \tag{1}$$

$D_{x,y} = \left\{x \,/ x^-(k) \le x(k) \le x^+(k)\right\}$. Where $x \in {}^n$ is a state of the system, $u \in {}^q$ is the input, $y \in {}^m$ is the measured output; *v(k)* and *w(k)* are bounded disturbances or uncertainties. Knowing the upper and under of limits of the variables *v(k)* and *w(k)*which are respectively ($v^-(k)$ $v^+(k)$) and ($w^-(k+1)$ $w^+(k+1)$), the model S, measures of output and input, it wished to estimate the bounds ($x^-(k)$ $x^+(k)$) of state *x*(*k*) (Benothman et al, 2007). At time *k*, it defined, from the bounds($x^-(k)$ $x^+(k)$), the allowable domain of the state, for example in the form of box:

$$D_{x,y} = \left\{x \,/ x^-(k) \le x(k) \le x^+(k)\right\} \tag{2}$$

At the time *(k+1)*, the new domain $D_{x,k+1}$ is constructed from previous domains $D_{x,k}, \cdots, D_{x,0}$ of the new measure *u(k)*, *y(k+1)*, the bounds ($v^-(k)$ $v^+(k)$) and ($w^-(k+1)$ $w^+(k+1)$).

This definition can be extended to the v-dimension space. An interval vector [x] of $\Re^v$ is a v-dimensional rectangle or box of $\Re^v$ and is the Cartesian product of intervals. The set of all boxes of $\Re^v$ is denoted $\Re^v$ (Benothman, 2007).

The definition of real arithmetic operators and functions is extended to intervals; see some basic examples in Table 1.

Given a function $f : \Re \rightarrow \Re$.

A function $F : \Re \rightarrow \Re$ is an inclusion function of *f* if $\forall x[x],\ f(x)\ F([x])$.

It is clear that this inclusion function is not unique. The natural inclusion function is obtained by substituting all real arguments and elementary functions by their extension to intervals. (Lebbal, 2006; Benothman et al, 2007).

Arithmetic operations are performed on intervals between their bounds.

For nonlinear systems, the analytical determination of the domain $D_{x,k+1}$ is difficult and a numerical method must be resorted for evaluation. The case of linear systems is easier; the domain states can be represented by a polytope (Benothman et al., 2007).

From (1), the system state is estimated and is subsequently corrected by the measures. The initial state *x*(0) of the system (1) is supposed to belong to a set $D_{x,0} \subset \Re^n$. The disturbances bounds *v*(*k*) and *w*(*k*) are assumed to be constant (Benothman et al, 2007).

$$|v(k)| \le \delta_v \ |w(k)| \le \delta_w \tag{3}$$

Table 1. Summary of arithmetic operations between intervals

Operation	Formulation
interval number	$a \in [\, a^- a^+ \,]$, a^-: lower bound, a^+: upper bound $[\, a \,] = [\, a^- \; a^+ \,]$
center, radius	$a_c = \dfrac{\left(a^- + a^+\right)}{2}$, $a_c = \dfrac{\left(a^{\mp} - a^-\right)}{2}$ $a = a_{c+\beta} a_{c\lvert\beta\rvert \leq 1}$
addition, subtraction	$[\, c \,] = [\, a \,] + [\, b \,] = [\, a^- + b^- \; a^+ + b^+ \,]$ $[\, c \,] = [\, a \,] - [\, b \,] = [\, a^- - b^+ \; a^+ - b^- \,]$
multiplication	$[\, c \,] = [\, a \,] \times [\, b \,] = [\, \min(a^- \times b^-, a^- \times b^+, a^+ \times b^-, a^+ \times b^+)$ $\max(a^- \times b^-, a^- \times b^+, a^+ \times b^-, a^+ \times b^+) \,]$
division	$[\, c \,] = [\, a \,] / [\, b \,] = [\, a \,] \times [\, 1/b^+, 1/b^- \,]$ if $0 \notin [\, b \,]$ $= [\, a \,] \times [\, 1/b^+, +\infty \,]$ if $b^- = 0$ and $b^+ > 0$ $= [\, a \,] \times [\, -\infty, 1/b^- \,]$ if $b^- < 0$ and $b^+ = 0$ $= [\, a \,] \times [\, -\infty, +\infty \,]$ if $b^- < 0$ and $b^+ > 0$ $= \varnothing$ if $[\, b \,] = [\, 0 \; 0 \,]$

At time k, the set of the state values $x\,(k)$ compatible with the available information is:

$$\left\{D_{x,0}, \left\{u(i-1), y(i), \left|v(i)\right| \leq \delta_v, \left|w(i)\right| \leq \delta_w\right\}_{i=1}^{k}\right\} \tag{4}$$

At the same time k, it is noted $D_{y,k}$, the set of the system output values knowing the measures $y(k)$ and the bound of the measurement errors:

$$D_{y,k} = \left\{y \,/\, \left|y - y(k)\right| \leq \delta_y\right\} \tag{5}$$

At the instant $(k+1)$, it has a new measure $(u\,(k), y\,(k))$ of the torque input / output. Knowing $D_{x,k}$ and the new input $u\,(k)$, it is possible to calculate the domain $D^{+}_{x,k}$, containing the prediction of the system state thanks to its model (1).As the output $y\,(k+1)$ is dependent on $x(k+1)$. It is also possible to infer a domain $D^{y}_{x,k+1}$, of the state $x\,(k+1)$, using equation S (Benothman et al, 2007). These two state estimations are defined by two domains:

$$\begin{aligned} D^{y}_{x,k+1} &= \left\{x \in \mathbb{R}^n \,/\, h(x) \in D_{y,k+1}\right\} \\ D^{+}_{x,k} &= \left\{f\left(x, u(k), v(k)\right) / \, x \in D_{x,k}, \left|v_k\right| \leq \delta_v\right\} \end{aligned} \tag{6}$$

The intersection of these two domains, if not empty, provides the states domain $D_{x,k+1}$ at the moment $(k+1)$:

2.2. Explanatory Example

A first-order system submitted to two uncertainties, is considered (Benothman et al, 2007):

$x(k+1) = [\ 0.5\ 0.6\ ]\ x(k) + 0.5\ ;\ x(0) \in [\ 0.2\ 0.3\ ]$

$y(k) = x(k) + w(k)\ ;\ |\ w(k)\ | \leq 0.05$

With the measures $y\ (1) = 0.5$, $y\ (2) = 0.82$, it comes:

The initial domain $D_{x,0} = \begin{bmatrix} 0.3 & 0.5 \end{bmatrix} = D_{x,0}^{+}$

- - The outputs and state domains: $D_{y,1} = D^{y}_{x,1} = [\ 0.45\ 0.55\ ]$.

Then, by intersection, the state domain: $D_{x,1} = [\ 0.45\ 0.50\ ]$.
By prediction, the domain of allowable states is $D^{+}_{x,1} = [\ 0.725\ 0.800\ ]$.
With the second measurement, the domain is defined by: $D_{x,2} = [\ 0.77\ 0.80\ ]$.

3. ELEMENTARY BACKGROUND TO THE TAKAGI-SUGENO MODEL

3.1. Preliminary

The multiple model approach idea is to apprehend the global system behavior by local models set. Each local model (also named sub-model) can be a linear time invariant system valid around an operating point. The local models are then aggregated using an interpolation mechanism. The total nonlinear system behavior is the sum of the local models balanced by weighting functions associated to each of them. These weighting functions are used to gradually quantify the membership of the system current operation point at a zone of operation.

In this modeling approach, two main structures of multiple models can be considered according to nature from the coupling between local models (Orjuela et al, 2009). In the first one, the sub-models have the same state-space and consequently the multiple models are composed of homogeneous sub-models (Ichalal et al., 2009a). In the second structure, decoupled multiple structure, the sub-models do not share the same state-space and the multiple model uses heterogeneous sub-models. The first structure, called the Takagi-Sugeno structure, is the most used in multiple model class representation. The main advantage of T-S structure is its simplicity because it originates from the interpolation between linear systems. Thus, analysis and design methods developed for linear systems can be generalized to nonlinear systems (Akhenak et al, 2004).

Takagi–Sugeno fuzzy models are nonlinear systems described by a set of if–then rules; which gives local linear representations of an underlying system (Akhenak et al., 2009; Ichalal et al., 2009b; Kawamoto et al, 1992; Yoneyama, 2010). Such models can approximate a wide class of nonlinear systems (Yoneyama, 2010). They can even describe exactly some nonlinear systems (Azar, 2010a; Witczak et al., 2008; Yoneyama, 2010).

Each nonlinear dynamic system can be simply described by a Takagi-Sugeno fuzzy model (Korbicz, 2004; Takagi & Sugeno, 1985). A Takagi-Sugeno fuzzy model, is the fuzzy fusion of many linear models (Akhenak et al., 2009; 2007; 2004; Ichalal et al., 2009a) and (Orjuela et al., 2009), each of them represents the local system behavior around an operating point. A Takagi-Sugeno model is described by fuzzy IF-THEN rules which represent local linear input/output relations of the non-linear system (Witczak et al., 2008). It has a rule base of *M* rules, each having *P* antecedents (Azar, 2010b).

Recently, Takagi-Sugeno Fuzzy systems, sometimes named multiple models (Murray-Smith & Johansen, 1997), are the subject of many researches by virtue of their approximation capabilities (Witczak et al., 2008; Yoneyama, 2010). They can represent exactly a nonlinear model (Tanaka et al., 1998; Witczak et al., 2008; Yoneyama, 2010). They are constructed by a set of linear models blended together with nonlinear functions holding the convex-sum property (Khedher & Othman, 2010) and (Azar, 2010a).

It becomes clear that the multiple model approach simplifies the study of nonlinear systems. It is able to meet an exact manner the behavior of complex system (Azar, 2012). Moreover Lebbal showed that the nonlinear representation of a system where the multi models are affine functions is a limited representation and does not lead to a reliable diagnostic procedure (Lebbal, 2006)

A Takagi-Sugeno fuzzy model is the fuzzy fusion of many linear models. Each of these linear models represents the local system behavior around the operating point. A Takagi-Sugeno model is described by fuzzy IF-THEN rules which represent local linear *I/O* relations of the non-linear system. It has a rule base of *M* rules, each having *p* antecedents, where i^{th} rule is expressed as:

R^i: IF ξ_1 is F_i^1 and ... and ξ_p is F_i^p

$$\text{THEN}: \begin{cases} \dot{x}_i(t) = ((A_i)x_i(t) + (B_i)u(t)) \\ y(t) = (C_i)x_i(t) \end{cases} \tag{7}$$

In which $i = 1, ...,M$, F_i^j (j = 1, ..., p) are fuzzy sets and $\xi = \xi_1, \xi_2, \ldots, \xi_p$ is a known vector of premise variables which may depend on the state, the input or the output.

3.2. Coupled Structure

A multi model can be written in the coupled form:

$$\begin{cases} \dot{x}_i(t) = \sum_{i=1}^{M} \mu_i(\xi(t))((A_i)x_i(t) + (B_i)u(t)) \\ y(t) = \sum_{i=1}^{M} \mu_i(\xi(t))(C_i)x_i(t) \end{cases} \tag{8}$$

where x_i are the state vectors of the local models, *y* is the system output vectors and *u* is the system input. A_i are the state evolution matrices, they represent its dynamic behavior, B_i are the matrices of control and C_i are the matrices of observation. The matrices A_i, B_i, and C_i are matrices with known coefficients and appropriate dimensions.

$\mu_i\left(\xi\left(t\right)\right)$, $i \in \left\{1,\ldots,M\right\}$ are the activation functions (named also weighting functions) and $\left(t\right)$ is the decision variable vector. The decision variable vector can depend on the state (which can be measurable or not), the system input $u(t)$ or the system output $y(t)$.

The weighting functions satisfy the sum convex property expressed in the following equations:

$$0 \leq \mu_i\left(\xi\left(t\right)\right) \leq 1 \text{ and } \sum_{i=1}^{M} \mu_i\left(\xi\left(t\right)\right) = 1 \tag{9}$$

If the measurement system is considered linear $C_1 = C_2 = \ldots = C_M = C$, the system of equations (8) is rewritten as follows:

$$\dot{x}_i\left(t\right) = \sum_{i=1}^{M} \mu_i\left(\xi\left(t\right)\right)\left(\left(A_i\right)x_i\left(t\right) + \left(B_i\right)u\left(k\right)\right) \tag{10}$$

$$y\left(t\right) = \left(C\right)x\left(t\right) \tag{11}$$

3.3. Decoupled Structure

Similarly, a complex system can be represented by a multiple model with a decoupled structure.

$$\begin{cases} \dot{x}_i\left(t\right) = \left(A_i\right)x_i\left(t\right) + \left(B_i\right)u\left(t\right) \\ y_i\left(t\right) = \left(C_i\right)x_i\left(t\right) \\ y\left(t\right) = \sum_{i=1}^{M} \mu_i\left(\xi\left(t\right)\right) y_i\left(t\right) \end{cases} \tag{12}$$

$i \in \left\{1,\ldots,M\right\}$

where x_i are the state vectors of the local models, y_i are the output vectors of the local models, y is the system output vector and u is the system input. A_i are the state evolution matrices, they represent its dynamic behavior, B_i are the matrices of control and C_i are the matrices of observation. The matrices A_i, B_i, and C_i are matrices with known coefficients and appropriate dimensions.

$\mu_i\left(\xi\left(t\right)\right)$, $i \in \left\{1,\ldots,M\right\}$ are the activation functions (named also weighting functions) and $\xi\left(t\right)$ is the decision variables vector. With this structure, the notion of local state, corresponding to a functioning area, appears much clear.

It is worth noting that local variables $x_i(t)$ do not necessarily have a physical meaning, and measures $y_i(t)$ are artificial outputs, allowing the description of the nonlinear behavior of the real system. The matrices A_i, B_i and C_i, and activation functions $\mu_i\ (\xi(t)\,)$ are calculated in the same way as before (coupled structure). This structure can be seen as the parallel connection of M affine sub-model weighted by their respective weights.

This structure allows decoupling the dynamic parts of sub-models. The sub-models can then have deferent structures, linear or nonlinear and have different states, while maintaining compatibility of dimensions vectors outputs.

The decoupled structure tolerates adjustment complexity of sub-models relating to the characterized operation area, since local conditions are perfectly independent.

The same representation (12) can be written in the discreet case:

$$\begin{cases} x_i(k+1) = (A_i)x_i(k) + (B_i)u(k) \\ y_i(k) = (C_i)x_i(k) \\ y(k) = \sum_{i=1}^{M} \mu_i\ (\xi(k)\,)\ y_i(k) \end{cases} \tag{13}$$

$i \in \{1,\ldots,M\}$

where the activation functions $\xi(t)$, the variables x, y, u, ΔA_i, ΔB_i, and ΔC_i, and the matrices A_i, B_i and C_i, are defined as above.

4. UNCERTAIN TAKAGI-SUGENO MULTIPLE MODEL

In the majority of cases, systems can present some measures or model uncertainties. We take into account the uncertainties that can reach the system and exploit the diagnostic techniques studied in the linear systems literature (interval approach, sliding mode observer ...)

A system may have disturbances on its inputs or its outputs such as the modeling uncertainties.

4.1. Coupled Structure

An uncertain multi model can be written in the coupled form:

$$\begin{cases} \dot{x}_i(t) = \sum_{i=1}^{M} \mu_i(\xi(t))((A_i + \Delta A_i)x_i(t) + (B_i + \Delta B_i)u(t)) \\ y(t) = \sum_{i=1}^{M} \mu_i\ (\xi(t)\,)(C_i + \Delta C_i)x_i(t) \end{cases} \tag{14}$$

If the measurement system is considered linear [C_1] = [C_2] = … = [C_M] = [C], the system of equations (8) is rewritten as follows:

$$\dot{x}_i(t) = \sum_{i=1}^{M} \mu_i(\xi(t))\left((A_i + \Delta A_i)x_i(t) + (B_i + \Delta B_i)u(k)\right) \quad (15)$$

$$y(t) = (C + \Delta C)x(t) \quad (16)$$

where x_i are the state vectors of the local models, y is the system output vectors and u is the system input. A_i are the state evolution matrices, they represent its dynamic behavior, B_i are the matrices of control and C_i are the matrices of observation. The matrices A_i, B_i, and C_i are matrices with known coefficients and appropriate dimensions.

$\mu_i((t))$, $i \in \{1, \ldots, M\}$ are the activation functions (also named weighting functions) and (t) is the decision variables vector

The variables ΔA_i, ΔB_i, and ΔC_i represent the uncertainties that respectively affect the matrices A_i, B_i and C_i. These variables are bounded.

The intervals matrices [A_i], [B_i] and [C_i] characterize the system described by the equations (10) and (11)as:

$$[A_i] = (A_i + \Delta A_i),\ [B_i] = (B_i + \Delta B_i),\ [C_i] = (C_i + \Delta C_i) \quad (17)$$

Several causes can be the origin of the temporal evolution of the system parameters. For example, aging or wear of the system components are sources of change of parameter values of the system (Alhaj-Dibo, 2005).

The same representation (8) can be written in the discrete case:

$$\begin{cases} x_i(k+1) = \sum_{i=1}^{M} \mu_i(\xi(k))\left((A_i + \Delta A_i)x_i(k) + (B_i + \Delta B_i)u(k)\right) \\ y(k) = \sum_{i=1}^{M} \mu_i(\xi(k))(C_i + \Delta C_i)x_i(k) \end{cases} \quad (18)$$

$i \in \{1, \ldots, M\}$

where the activation functions $\mu_i(\xi(t))$ and the decision variables vector $\xi(t)$ are defined as above. A_i are the evolution matrices of state, representative of its dynamic behavior, B_i are the matrices of control and C_i are the matrices of observation. Likewise, the variables ΔA_i, ΔB_i, and ΔC_i represent the bounds uncertainties that affect respectively A_i, B_i and C_i matrices. $x_i(k)$, $y_i(k)$ and $u(k)$ are respectively the state vectors, the output and input vectors of the system.

4.2. Decoupled Structure

Similarly, in the uncertain case, a complex system can be represented by a multiple model with decoupled structure.

$$\begin{cases} \dot{x}_i(t) = (A_i + \Delta A_i)x_i(t) + (B_i + \Delta B_i)u(t) \\ y_i(t) = (C_i + \Delta C_i)x_i(t) \\ y(t) = \sum_{i=1}^{M} \mu_i(\xi(t))\, y_i(t) \end{cases} \tag{19}$$

$$i \in \{1, \ldots, M\}$$

With this structure, the notion of local state, corresponding to a functioning area, appears much clear.

Remember that local variables $x_i(t)$ do not necessarily have a physical meaning, and measures $y_i(t)$ are artificial outputs, allowing the description of the nonlinear behavior of the real system. The matrices A_i, B_i and C_i, and activation functions $\mu_i(\xi(t))$ are calculated in the same way as before (coupled structure).

x_i are the state vectors, y_i are output vectors for the ith sub-model, $u(t)$ is the input of the system and $y(t)$ is the system outputs.

Matrices A_i are the evolution matrices of state, representative of its dynamic behavior, B_i are the matrices of control and C_i are the matrices of observation. $\xi(t)$ is the decision variables vector dependent measurable state variables and possibly of control $u(t)$.

The intervals matrices $[A_i]$, $[B_i]$ and $[C_i]$ characterize the system described by the equations in (12) as:

$$[A_i] = (A_i + \Delta A_i),\ [B_i] = (B_i + \Delta B_i),\ [C_i] = (C_i + \Delta C_i) \tag{20}$$

The variables ΔA_i, ΔB_i, and ΔC_i represents the uncertainties that affect respectively A_i, B_i and C_i matrices. These variables are bounded.

This structure allows to decouple the dynamic parts of sub-models. The sub-models can then have deferent structures, linear or nonlinear and have different states, while maintaining compatibility of dimensions vectors outputs.

The decoupled structure tolerates adjustment complexity of sub-models relating to the characterized operation area, since local conditions are perfectly independent.

The same representation (12) can be written in the discrete case:

$$\begin{cases} x_i(k+1) = (A_i + \Delta A_i)x_i(k) + (B_i + \Delta B_i)u(k) \\ y_i(k) = (C_i + \Delta C_i)x_i(k) \\ y(k) = \sum_{i=1}^{M} \mu_i(\xi(k))\, y_i(k) \end{cases} \tag{21}$$

$i \in \{1, \ldots, M\}$

where the activation functions $\xi(t)$, the variables x, y, u, ΔA_i, ΔB_i, and ΔC_i, and the matrices A_i, B_i and C_i, are defined as above .

5. PROBLEM FORMULATION

5.1. Preliminaries

Most of the techniques used for diagnosis are based on the development of a mathematical model. The problem that may arise is that the model developed from the system redundancy may not reflect the real behavior of the physical system in view of the modeling uncertainties which may affect the system.

Few studies have been developed in this context, however, the works of Adrot (Adrot, 2000), Lebbal (Lebbal, 2006), Alhaj-Dibo (Alhaj-Dibo,2005), and Letellier (Letellier, 2012; Letellier et al., 2011) can be cited. These works are mainly interested in the regular linear systems, which is the least real case.

The modeling approaches of model uncertainties have been developed using stochastic variables (Walter & Piet-Lahanier, 1987) or using the bounded approach (L Jaulin et al., 2001; Walter et al., 1996). The bounded approach assumes that it only knows the extreme limits of uncertainties.

The hypothesis of measurement accuracy is not always valid because the sensors of measures can be not perfectly accurate. This technological imprecision is often mentioned by the manufacturer (Alhaj-Dibo, 2005).

Consider the nonlinear system with uncertain outputs described by a multiple model with decoupled structure based on measurable decision variables $\xi(k) = u(k)$ (Filev, 1991; Orjuela et al., 2008; Uppal et al., 2006)

$$x_i(k+1) = A_i x_i(k) + B_i u(k) \tag{22}$$

$$y_i(k) = C_i(\eta_{Ci}(k)) x_i(k) \tag{23}$$

$$y_m(k) = \sum_{i=1}^{M} \mu_i(u(k)) y_i(k) \tag{24}$$

where $x_i \in \Re^n$, $u \in \Re^q$ and $y_i \in \Re^m$ are respectively the state vectors, the input and output of the ith sub-model. The scalar M represents the number of the local models. The weighting functions μ_i are nonlinear and depend on the decision variable $\xi(k)$ assumed to be measurable. In this work, the decision is variable $\xi(k)=u(k)$. Whence $\mu_i(u(k)) \in [0,1]$ represent the activation functions depending on the input.

The matrices C_i are taken as follows:

$C_i(\eta_{Ci}(k)) \in \Re^{m \times n}$ is the uncertain matrix i^{th} sub-model defined by:

$$C_i\left(\eta_{Ci}\left(k\right)\right) = C_i^0 \pm \Delta_{Ci} \otimes \eta_{Ci} \qquad (25)$$

$$C_i^0 \in \Re^{m\times n},\ \Delta_{Ci} \in \Re^{m\times n}, \eta_{Ci} \in \Re^{m\times n}$$

where $\otimes$ is the operator of multiplication of two matrices element by element. C_i^0 is the nominal matrix of $C_i\left(\eta_{Ci}\left(k\right)\right)$. Δ_{Ci} is the amplitude matrix of the uncertainties of the elements of the matrix $C_i\left(\eta_{Ci}\left(k\right)\right)$.$\eta_{Ci}$ is the matrix consisting of bounded and standard variables of ith sub-model. $i\in\{1, \ldots, M\}$.

The uncertain matrix $C_i\left(_{Ci}\left(k\right)\right)$ can be represented by its extreme limits, where it has the following form:

$$C_i\left(\eta_{Ci}\left(k\right)\right)\left[\mathrm{C}_i\right] = \left[\underline{\mathrm{C}}_i, \bar{\mathrm{C}}_i\right], \left|\eta_{Ci}\left(k\right)\right| \le 1$$

We note that the relations (22), (23) and (24) describe a nonlinear system represented by M uncertain sub-models related by weighting functions $\mu_i(u(k))$ satisfying the properties of convexity $\sum_{i=1}^{M}\mu_i(u(t)) = 1$, and $0 \le \mu_i \le 1$ (Tanaka & Wang, 2004; Wang et al., 1996).

5.2. State Estimation

From equation (23), we can assume that:

$$\hat{x}_i(k) = C_i^{-1}(\eta_{c_i}(k))y_i(k) \qquad (26)$$

The uncertain matrices $C_i\left(\eta_{Ci}\left(k\right)\right)$ are considered square, observable, and bounded $C_i\left(\eta_{Ci}\left(k\right)\right)\left[\mathrm{C}_i\right] = \left[\underline{\mathrm{C}}_i, \bar{\mathrm{C}}_i\right]$, them they can be replaced in (26) by their lower and upper bounds:

$$\left[\hat{x}_i^{(1)}\ (k)\right] = [C_i\ ]^{-1}y_i(k) \qquad (27)$$

In (27) the estimated states $\left[\hat{x}_i^{(1)}\ (k)\right]$ are obtained from a measurement system, as originally assumed valid.

To solve this equation, there are several methods such as Gaussian elimination, the Gauss-Seidel iteration and the Krawczyk iteration (Luc Jaulin, 2001; Neumaier, 1991) which have already been used by Alhaj-Dibo (Alhaj-Dibo, 2005; Alhaj-Dibo et al., 2004) to solve the same problem in the linear systems case.

The optimal solution of $\left[\hat{x}_i^{(1)}(k)\right]$ is the intersection of the three options provided by the three methods mentioned below:

$$\left[\hat{x}_i^{(1)}(k)\right]=\left[\hat{x}_{iG}^{(1)}(k)\right]\cap\left[\hat{x}_{iGS}^{(1)}(k)\right]\cap\left[\hat{x}_{iK}^{(1)}(k)\right] \tag{28}$$

Returning to the equation system (22) and replacing it $[x_i(k)]$ by its estimated intervals calculated in(27), an estimated variable can be calculated at times k and $k+1$

$$\left[\hat{x}_i^{(2)}(k+1)\right]=A_i\left[\hat{x}_i^{(1)}(k)\right]+B_i u(k) \tag{29}$$

In this case, the estimates $[\hat{x}_i^{(2)}(k+1)]$ are obtained from the system model (27) and measures(29).

It is concluded that the state variable estimates at time $k+1$ are calculated twice (from the system at time k and from measurements at time $(k+1)$. So the final local estimates $[\hat{x}_i(k+1)]$ are obtained by performing the intersection between the two estimates

$$\left[\hat{x}_i(k+1)\right]=\left[\hat{x}_i^{(2)}(k+1)\right]\cap\left[\hat{x}_i^{(1)}(k+1)\right] \tag{30}$$

$$i\in\{1,\ldots,M\}$$

In the case of multiple models with uncertain measures, the state estimation is made locally on each sub-model.

5.3. Fault Detection and Localization

At this stage, the overall system is considered faulty if a fault is detected on one of the sub models.

To isolate the fault, the residuals which allow detecting and locating the fault must be generated.

For this reason, the local coherence between the estimates $\left[\hat{x}_i^{(2)}(k+1)\right]$ and $\left[\hat{x}_i^{(1)}(k+1)\right]$ will be is tested at every time for $i\in\{1,\ldots,M\}$.

Suppose that measurements are good at the instant k. The estimate obtained from the system model and measurement model can be judged at the same time $\left[\hat{x}_i^{(2)}(k+1)\right]$. This estimate is coherent with the system model, which allows considering it as a reference. Then it is compared with the estimated obtained from measurements at time $k+1$ $\left[\hat{x}_i^{(1)}(k+1)\right]$.

From the above assumption, it is possible to judge the state of the local estimated, in case of safe functioning and faulty functioning

$$\left[\hat{x}_i\ (k+1)\right] = \left[\hat{x}_i^{(2)}\ (k+1)\right] \cap \left[\hat{x}_i^{(1)}(k+1)\right] \tag{31}$$

for the safe functioning.

$$\left[\hat{x}_i\ (k+1)\right] = \left[\hat{x}_i^{(2)}\ (k+1)\right] \tag{32}$$

for the faulty functioning.

Referring to $\hat{x}_i^{(2)}(k+1)$, the estimated interval calculated from equation (30) is analyzed. Two cases are possible:

- If $\left[\hat{x}_i\ (k+1)\right] = \left[\hat{x}_i^{(2)}\ (k+1)\right] \cap \left[\hat{x}_i^{(1)}(k+1)\right] \neq \varnothing$, it is possible to confirm that the estimated intervals $\left[\hat{x}_i^{(2)}\ (k+1)\right]$ and $\left[\hat{x}_i^{(1)}(k+1)\right]$ are coherent, which measures sub-model, at the same time $k+1$, are safe.
- If $\left[\hat{x}_i(k+1)\right] = \left[\hat{x}_i^{(2)}\ (k+1)\right] \cap \left[\hat{x}_i^{(1)}(k+1)\right] = \varnothing$, the two estimated intervals $\left[\hat{x}_i^{(2)}\ (k+1)\right]$ and $\left[\hat{x}_i^{(1)}(k+1)\right]$ are not coherent where the measures of the overall system at the same time $k+1$, contain one or more faults. The ith sub-model is then used to locate the fault.

In this case, it is considered that the true values of the state variables belong only to the reference $\left[\hat{x}_i^{(2)}\ (k+1)\right]$ only, and the estimated interval cannot be obtained from the system model and measurements at time $k+2$. $\left[\hat{x}_i^{(2)}\ (k+2)\right]$, is a reference to compare with $\left[\hat{x}_i^{(1)}\ (k+2)\right]$. That must be corrected first.

Thus, by referring to the system model (22), the states variables are replaced by their estimated considered as reference to the same time $\left[\hat{x}_i^{(2)}\ (k+1)\right]$ hence:

$$\left[\hat{x}_i^{(2)}\ (k+2)\right] = A_i\left[\hat{x}_i^{(2)}\ (k+1)\right] + B_i u(k+1) \tag{33}$$

To identify measures affected by a fault. The residual intervals for each sub-model $[r_i(k+1)]$ is generated. From equation (23), replaces the state variable at time $(k+1)$ by their estimated considered as reference at the same time $\left[\hat{x}_i^{(2)}\ (k+1)\right]$ and the matrix $C_i\left(\eta_{Ci}\left(k\right)\right)$ by its value interval $[C_i\ ]$ whence:

$$\left[r_i\left(k+1\right)\right] = [C_i\ ]\left[\hat{x}_i^{(2)}\ (k+1)\right] - y_i(k+1) \tag{34}$$

A measure is affected by a fault if the corresponding residual is abnormal i.e. the corresponding residual interval does not contain the value 0.

6. NUMERICAL APPLICATION

We consider a nonlinear system described by Takagi-Sugeno multiple model with three uncertain outputs. The considered system can be described by the multimodel (23) and (24), where the number of sub-models is *M*=2.

A_i and B_i are known constant matrices with appropriate dimensions. $A_i \in \mathfrak{R}^{3\times3}$ and $B_i \in \mathfrak{R}^{3\times1}$ $i\{1,2\}$. The numerical values of the matrices A_i and B_i are given below:

$$A_1 = \begin{bmatrix} -0.264 & -0.270 & 0.218 \\ -0.073 & -0.580 & -0.080 \\ 0.120 & -0.119 & 0.401 \end{bmatrix} A_2 = \begin{bmatrix} -0.100 & 0.290 & 0.216 \\ 0.325 & 0.670 & -0.080 \\ 0.120 & -0.131 & 0.390 \end{bmatrix}$$

$$B_1 = \begin{bmatrix} 0.50 \\ -0.31 \\ 0.60 \end{bmatrix} B_2 = \begin{bmatrix} 1.98 \\ -1.78 \\ -1.32 \end{bmatrix}$$

Ci are uncertain matrices with appropriate dimensions, only their bounds are known. Ci∈$\mathfrak{R}$3×3, i∈ { 1, 2 } . These matrices are given:

$$C_1(\eta_{C_1}(k)) \in [\underline{C}_1 \quad \overline{C}_1] = \begin{bmatrix} [-5.24 \ \ -4.25] & [-1.64 \ \ -1.35] & [1.02 \ \ 1.11] \\ [0.58 \ \ 0.75] & [0.20 \ \ 0.22] & -5 \\ [-2.00 \ \ -1.8] & [-1.16 \ \ -0.98] & [1.07 \ \ 1.28] \end{bmatrix}$$

$$C_2(\eta_{C_2}(k)) \in [\underline{C}_2 \quad \overline{C}_2] = \begin{bmatrix} [-6.75 \ \ -6.2] & [-1.35 \ \ -1.28] & [1 \ \ 1.08] \\ [1.85 \ \ 2.05] & [0.14 \ \ 0.15] & [-5.2 \ \ -4.5] \\ [-2.1 \ \ -2] & [-1.31 \ \ -1.2] & [1.91 \ \ 2.09] \end{bmatrix}$$

The activation functions $\mu_i(u(k))$ should be normalized to ensure the convexity property. They are represented in Figure 1.

A control *U (k)* with uniform distribution is applied to the system. This control is given by figure 2.

Figure 1. Activations functions normalized depending on the input

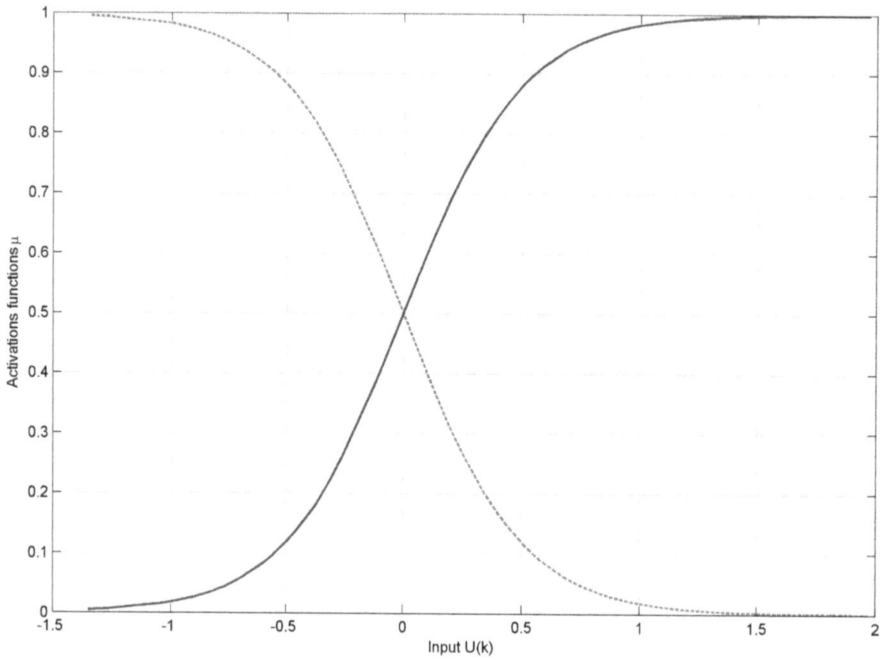

Assume that the three sensors are affected by faults at the instants [10, 20], [40, 50] and [60, 70] respectively. In figure 3, both, the healthy system outputs and the faulty outputs are represented

Referring to relations (8), (10) and (11), it is possible to generate an estimated interval of the state variables. This is considered as limited by the upper bound and the lower bound of the estimated area. From the results, it is possible to readily determine an estimate of the output. The system measures should be included in the estimation of the output domain. This means that if the measurements exceed the upper or lower bound of the estimation domain, the sensor will be declared faulty. Figure 4 shows the estimates of the three outputs as a closed area, and the three measured outputs (dashed line). In the same figure, a fault indicator is illustrated.

It goes from low level to high level every time when the system measurements exceed the estimated values, an alarm signal is triggered.

Note that, at the instants [10 20], [40 50] and [60 70], the measures $y1$, $y2$ and $y3$ exceed the limit of the estimation domain, where a fault is detected in the precise moment of the output concerned.

It can be noticed at the beginning of the simulation system that takes some time to be controlled. Meanwhile, all the estimated and measured outputs are not consistent which causes some false alarms. Similarly, during the simulation, some false alarms are triggered for brief moments (second and third outputs), they shall not be considered. A fault is considered certain that if the fault indicator provides successive signals during a short time interval.

To better locate faults which affect the system, the indicator signals that are also called residuals are generated. These signals will be in the form of a closed envelope which must contain the value zero if the system is not affected by fault. Otherwise, zero will be located outside the envelope. In Figure 5, the envelopes residuals of the three outputs are represented.

Figure 2. Command U (k) applied to the system

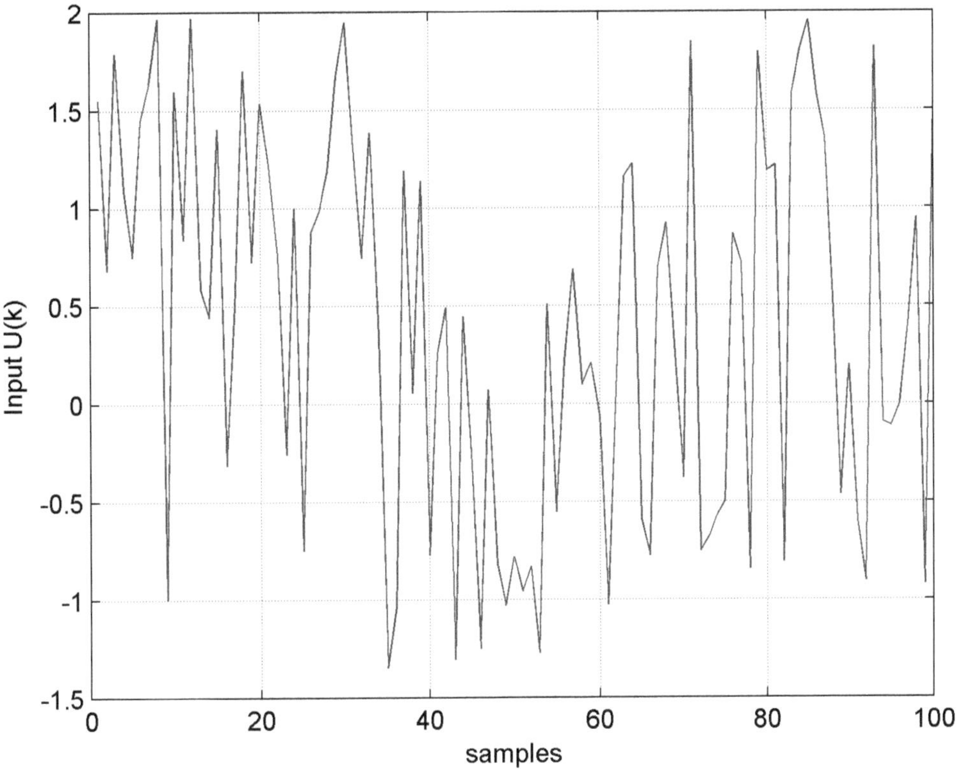

Figure 3. Outputs tainted with their corrections

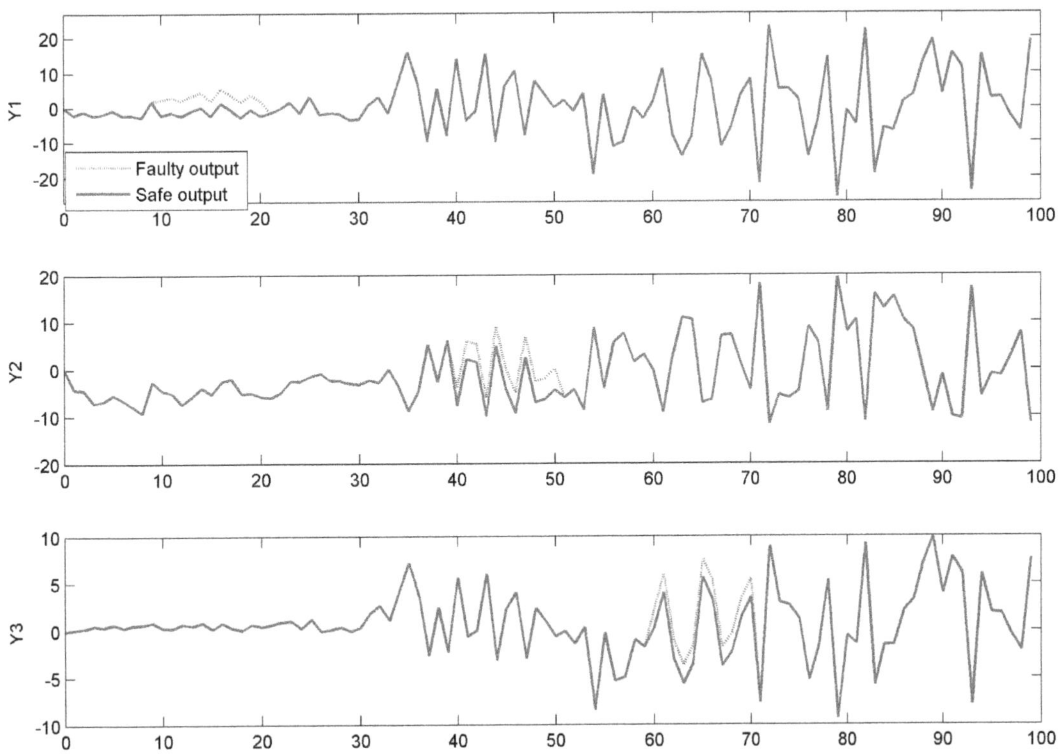

Figure 4. Estimated and measured outputs

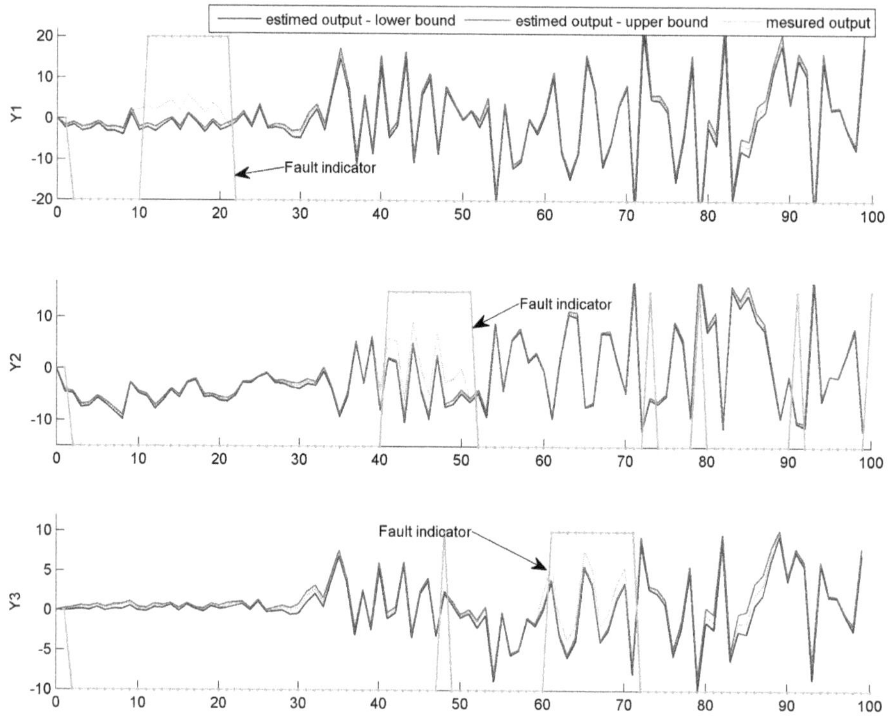

Figure 5. Residuals

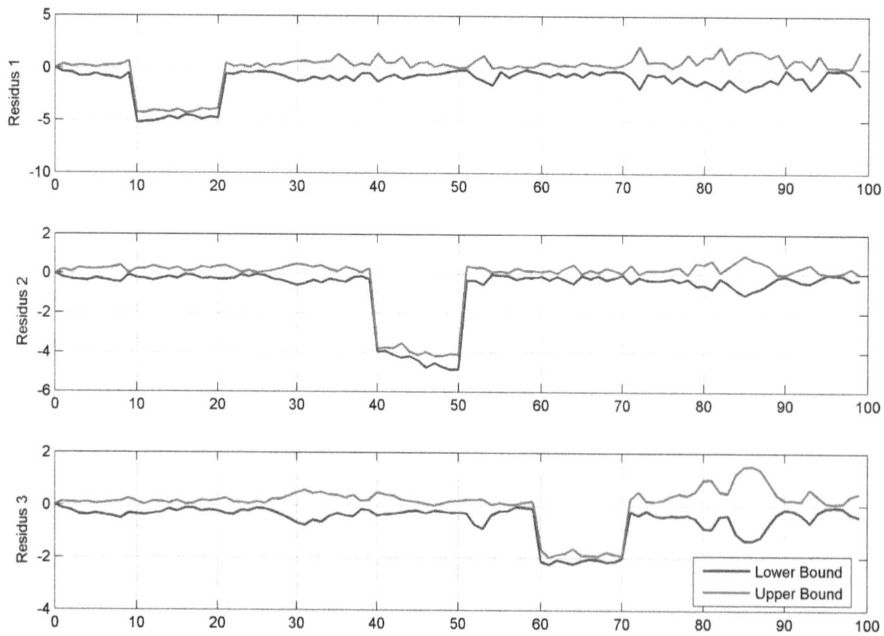

It is noted that each fault impact the residual at the right time. In this way, it's got to perfectly locate faults. Iterating between 10 and 20, it has a fault that affects the first sensor, between 40 and 50, another affects the second output, and between 60 and 70 a third fault taints the last output.

Residuals generation is the most important step in the technique of diagnosis based on the use of mathematical models. This step allows generating fault indicators which allow detecting and locating faults affecting the system. This step must be done with very good precision to avoid the problem of false alarms. The technique presented in this work can be used to generate residuals with a satisfactory precision. Using this method it is possible to detect and locate faults affecting all the sub-models.

7. DISCUSSION

The advantage of the proposed method in this work is to offer the possibility to detect and locate faults affecting the system. Using the multiple model approach, the process can be modeled by several local models aggregated by weighting function to form the global model. It is possible that some local model are affected by faults, the proposed method can detect them.

This approach is very important since it make the generalization of the proposed method to the class of nonlinear system.

In general, works deals with nonlinear systems choose a particular class to apply their theories, it is impossible to use an approach for all the nonlinear systems classes. Using the multiple model approach, this problem is solved because all the nonlinear systems classes can be modeled using this approach.

In most cases sensor are affected by faults. This produces faulty measures and faulty results. The proposed method allows detecting and locating sensor faults which is very important to ensure a good and efficient analysis and diagnosis and make correct system supervision.

By exploiting the numerical application, it appears that the proposed approach is interesting because it could deal with cases of uncertain fuzzy systems.

Based on interval analysis, a state estimation is made for a nonlinear system with uncertain outputs, which was not done in the cited studies such as that proposed by Letellier (Letellier et al., 2011) who processed by the same approach, the case of uncertain linear systems and he ignored the case of non-linear uncertain systems.

The proposed technique has proven its effectiveness in the detection and isolation of faults affecting the output, which has been shown in the previous example where the fault indicator (residuals) remains insensitive to bounded uncertainty and detects a fault as soon as his appearance on one of the sub models.

8. CONCLUSION

No sensor can be considered perfect. To ensure a robust diagnosis, it was tried, through this work, to characterize the measurement uncertainties using the interval analysis. An innovative method used for linear systems, but never with nonlinear systems is used in this work, the proposed technique is applied in the case of a particular class of nonlinear system, this class concerns the nonlinear systems described by Takagi-Sugeno multiple models. This particular class can include all the nonlinear systems as it was shown in the literature. The application of the proposed method to a numerical example shows the effectiveness of this technique to model multiple models with uncertain outputs. Also, this technique allows estimating output and detecting and locating faults in case of their presence. The generated fault indicators are not sensitive to measurement uncertainties.

The proposed technique is effective with a low cost of calculation and can be generalized to the case of uncertainty modeling or command. This work can give many ideas for future works, such as using the same technique for the case of systems affected by model uncertainties, or actuator uncertainties. It is also possible to use the presented technique to estimate faults using some particular observer. State and fault estimation can be used after that to generate fault tolerant control. Fault tolerant control is a relatively new idea in the research literature which allows having a control loop that fulfills its objectives when faults appear.

REFERENCES

Adrot, O. (2000). *Diagnostic à base de modèles incertains utilisant l'analyse par intervalles: l'approche bornante*. (PhD thesis). Institut National Polytechnique de Lorraine, France.

Adrot, O., Janati-Idrissi, H., & Maquin, D. (2002). *Fault detection based on interval analysis*. Paper presented at the 15th IFAC World Congress on Automatic Control. Barcelona, Spain. Doi:10.3182/20020721-6-ES-1901.00762

Adrot, O., Maquin, D., & Ragot, J. (2000). *Diagnosis of an uncertain static system*. Paper presented at the 39th IEEE Conference on Decision and Control. Sydney, Australia. Doi:10.1109/CDC.2000.912366

Akhenak, A., Chadli, M., Maquin, D., & Ragot, J. (2003). *State estimation via multiple observer with unknown inputs: Application to the three tank system*. Paper presented at the 5th IFAC Symposium on Fault Detection, Supervision and Safety for Technical Processes, Safeprocess. Washington, DC.

Akhenak, A., Chadli, M., Maquin, D., & Ragot, J. (2004). *State estimation of uncertain multiple model with unknown inputs*. Paper presented at the 43rd IEEE Conference on Decision and Control. Paradise Island, Bahamas. Doi:10.1109/CDC.2004.1429265

Akhenak, A., Chadli, M., Ragot, J., & Maquin, D. (2007). *Design of sliding mode unknown input observer for uncertain Takagi-Sugeno model*. Paper presented at the Mediterranean Conference on Control & Automation. Athens, Greece. Doi:10.1109/MED.2007.4433753

Akhenak, A., Chadli, M., Ragot, J., & Maquin, D. (2009). *Design of Fuzzy Observers for Takagi-Sugeno Fuzzy Models for Fault Detection and Isolation*. Paper presented at the 7th IFAC Symposium on Fault Detection, Supervision and Safety of Technical Processes SAFEPROCESS. Doi:10.3182/20090630-4-ES-2003.00183

Alhaj-Dibo, M. (2005). *Validation de données et diagnostic des systèmes incertains à l'aide de l'analyse par intervalles*. (PhD thesis). Institut National Polytechnique de Lorraine, France.

Alhaj-Dibo, M., Maquin, D., & Ragot, J. (2004). *Data validation of uncertain dynamic systems*. Paper presented at the 15th International Workshop on Principles of Diagnosis. Carcassonne, France.

Azar, A. (2010a). Fuzzy Systems. In-Tech.

Azar, A. T. (2010b). Adaptive Neuro-Fuzzy Systems. In A. T. Azar (Ed.), *Fuzzy Systems*. Vienna, Austria: IN-TECH.

Azar, A. T. (2012). Overview of Type-2 Fuzzy Logic Systems. *International Journal of Fuzzy System Applications*, *2*(4), 1–28.

Bedoui, H., Bouguila, N., Othman, K. B., & Hassani, M. (2013). *Modeling of the injection and potential decay of a dielectric placed in an environment with high voltage*. Paper presented at the International Conference on Control, Decision and Information Technologies. Hammamet, Tunisia. Doi:10.1109/CoDIT.2013.6689583

Bedoui, H., Khedher, A., & Ben Othman, K. (2014). State estimation of uncertain nonlinear systems using the interval approach: Application to the diagnostic. *Wulfenia Journal*, *21*(4), 369–379.

Benothman, K., Maquin, D., Ragot, J., & Benrejeb, M. (2007). Diagnosis of uncertain linear systems: an interval approach. *International Journal of Sciences and Techniques of Automatic Control & Computer Engineering*, *1*(2), 136-154.

Chow, E., & Willsky, A. (1984). Analytical redundancy and the design of robust failure detection systems. *IEEE Transactions on Automatic Control*, *29*(7), 603–614.

Desai, M., Deckert, J., Deyst, J., Willsky, A., & Chow, E. (1976). Dual redundant sensor FDI techniques applied to the NASA F8C DFBW aircraft. In *Proceedings of Guidance and Control Conference*. American Institute of Aeronautics and Astronautics.

Dibo, M. A., Maquin, D., & Ragot, J. (2004). *Validation de données des systemes dynamiques incertains*. Paper presented at the Conférence Internationale Francophone d'Automatique. Douz, Tunisia.

Edwards, C., & Tan, C. P. (2006). A comparison of sliding mode and unknown input observers for fault reconstruction. *European Journal of Control*, *12*(3), 245-260.

Filev, D. (1991). Fuzzy modeling of complex systems. *International Journal of Approximate Reasoning*, *5*(3), 281–290.

Guan, Y., & Saif, M. (1991). A novel approach to the design of unknown input observers. *IEEE Transactions on Automatic Control*, *36*(5), 632–635.

Ichalal, D., Marx, B., Ragot, J., & Maquin, D. (2009a). *Simultaneous state and unknown inputs estimation with PI and PMI observers for Takagi Sugeno model with unmeasurable premise variables*. Paper presented at the 17th Mediterranean Conference on Control and Automation. Thessaloniki, Greece. Doi:10.1109/MED.2009.5164566

Ichalal, D., Marx, B., Ragot, J., & Maquin, D. (2009b). *State and unknown input estimation for nonlinear systems described by Takagi-Sugeno models with unmeasurable premise variables*. Paper presented at the 17th Mediterranean Conference on Control and Automation. Thessaloniki, Greece. Doi:10.1109/MED.2009.5164542

Isermann, R. (1984). Process fault detection based on modeling and estimation methods—a survey. *Automatica*, *20*(4), 387–404.

Jaulin, L. (2001). *Applied interval analysis: with examples in parameter and state estimation, robust control and robotics*. Springer Verlag.

Jaulin, L., Kieffer, M., Didrit, O., & Walter, E. (2001). Applied interval analysis. London: Springer.

Kalman, R. E. (1960). A new approach to linear filtering and prediction problems. *Journal of Basic Engineering*, *82*(1), 35–45.

Kawamoto, S., Tada, K., Onoe, N., Ishigame, A., & Taniguchi, T. (1992). *Construction of exact fuzzy system for nonlinear system and its stability analysis.* Paper presented at the 8th Fuzzy System Symposium. Hiroshima, Japan.

Khedher, A., Benothman, K., Maquin, D., & Benrejeb, M. (2008). *State and unknown input estimation via a proportional integral observer with unkbown inputs.* Paper presented at the 9th international conference on Sciences and Techniques of Automatic control and computer engineering, STA. Sousse, Tunisia.

Khedher, A., Benothman, K., Maquin, D., & Benrejeb, M. (2009). *State and sensor faults estimation via a proportional integral observer.* Paper presented at the 6th International Multi-Conference on Systems, Signals and Devices. Djerba, Tunisia. Doi:10.1109/SSD.2009.4956795

Khedher, A., & Othman, K. B. (2010). Proportional Integral Observer Design for State and Faults Estimation: Application to the Three Tanks System. *International Review of Automatic Control, 3*(2), 115-124.

Korbicz, J. (2004). *Fault diagnosis: Models, artificial intelligence, applications*. Springer.

Lebbal, M. (2006). *Contribution à la modélisation et au diagnostic des systèmes à commutations*. (PhD thesis). Université du Havre, Le Havre, France.

Letellier, C. (2012). *Diagnostic robuste des systèmes incertains. Application à un système mécatronique pour l'automobile.* PhD thesis, Université de Rouen, France.

Letellier, C., Hoblos, G., & Chafouk, H. (2011). *Robust Fault Detection based on Multimodel and Interval Approach. Application to a Throttle Valve.* Paper presented at the IEEE MED. Corfu, Greece. Doi:10.1109/MED.2011.5983198

Lin, H., Zhai, G., & Antsaklis, P. J. (2003). Set-valued observer design for a class of uncertain linear systems with persistent disturbance and measurement noise. *International Journal of Control*, *76*(16), 1644–1653.

Luenberger, D. G. (1964). Observing the state of a linear system. *IEEE Transactions on Military Electronics*, *8*(2), 74–80.

Lyubchik, L., & Kostenko, Y. T. (1993). *The output control of multivariable systems with unmeasurable arbitrary disturbances-The inverse model approach.* Paper presented at the European Control Conference. Groningen, Netherlands.

Marx, B., Maquin, D., & Ragot, J. (2010). *State estimation and fault detection of uncertain systems based on an interval approach.* Paper presented at the Conference on Control and Fault-Tolerant Systems (SysTol). Nice, France. doi:10.1109/SYSTOL.2010.5676036

Murray-Smith, R., & Johansen, T. (1997). *Multiple Model Approaches to Nonlinear Modelling and Control*. CRC Press.

Neumaier, A. (1991). *Interval methods for systems of equations* (Vol. 37). Cambridge University Press.

Orjuela, R., Marx, B., Ragot, J., & Maquin, D. (2008). State estimation for non-linear systems using a decoupled multiple model. *International Journal of Modelling Identification and Control*, *4*(1), 59–67.

Orjuela, R., Marx, B., Ragot, J., & Maquin, D. (2009). On the simultaneous state and unknown input estimation of complex systems via a multiple model strategy. *IET Control Theory & Applications*, *3*(7), 877–890.

Othman, K. B. (2008). *Diagnostic, disponibilité et Approches Non Conventionnelles pour l'Etude de Systèmes Complexes*. Tunisia: Université de Tunis El Manar.

Othman, K. B., Ragot, J., Manquin, D., & Rejeb, M. B. (2007). *Robust identification in the presence of outliers*. Paper presented at the international conference on Sciences and Techniques of Automatic control and computer engineering IJ-STA. Hammamet,Tunisia.

Planchon, P., & Lunze, J. (2008). Diagnosis of linear systems with structured uncertainties based on guaranteed state observation. *International Journal of Control Automation and Systems*, *6*(3), 306–319.

Ploix, S., Adrot, O., & Ragot, J. (1999). *Parameter uncertainty computation in static linear models*. Paper presented at the 38th IEEE Conference on Decision and Control. Doi:10.1109/CDC.1999.830916

Puig, V., Quevedo, J., Figueras, J., Riera, S., Cembrano, G., Salamero, M., et al. (2003). *Fault detection and isolation of rain gauges and limnimeters of Barcelona's sewer system using interval models*. Paper presented at the 5th Symposium on Fault Detection Supervision and Safety for Technical Processes, Safeprocess. Washington, DC.

Ragot, J., Maquin, D., & Adrot, O. (2006). Parameter uncertainties characterisation for linear models. *The Computer Journal*, *49*(1), 16–20.

Raıssi, T., Ramdani, N., & Candau, Y. (2004). Set membership state and parameter estimation for systems described by nonlinear differential equations. *Automatica*, *40*(10), 1771–1777.

Shamma, J. S., & Tu, K.-Y. (1999). Set-valued observers and optimal disturbance rejection. *IEEE Transactions on Automatic Control*, *44*(2), 253–264.

Sharma, R., & Aldeen, M. (2004). Estimation of unknown disturbances in nonlinear systems. In C. R. Burrows, K. A. Edge, D. N. Johnston, & University of Bath (Eds.), *Centre for Power transmission and Motion Control., Bath Workshop on Power Transmission and Motion Control* (PTMC 2004). University of Bath.

Stancu, A., Puig, V., Cugueró, P., & Quevedo, J. (2005). Benchmarking on Approaches to Interval Observation Applied to Robust Fault Detection. In C. Jermann, A. Neumaier, & D. Sam (Eds.), *Global Optimization and Constraint Satisfaction* (Vol. 3478, pp. 171–191). Springer Berlin Heidelberg.

Takagi, T., & Sugeno, M. (1985, January-February). Fuzzy identification of systems and its applications to modeling and control. *IEEE Transactions on Systems, Man, and Cybernetics*, *SMC-15*(1), 116–132. doi:10.1109/TSMC.1985.6313399

Tanaka, K., Ikeda, T., & Wang, H. O. (1998). Fuzzy regulators and fuzzy observers: Relaxed stability conditions and LMI-based designs. *IEEE Transactions on Fuzzy Systems*, *6*(2), 250–265.

Tanaka, K., & Wang, H. O. (2004). Fuzzy control systems design and analysis: a linear matrix inequality approach. Wiley.com.

Uppal, F. J., Patton, R. J., & Witczak, M. (2006). A neuro-fuzzy multiple-model observer approach to robust fault diagnosis based on the DAMADICS benchmark problem. *Control Engineering Practice*, *14*(6), 699–717.

Walter, E., Norton, J., Piet-Lahanier, H., & Milanese, M. (1996). *Bounding approaches to system identification*. Perseus Publishing.

Walter, E., & Piet-Lahanier, H. (1987). *Exact and recursive description of the feasible parameter set for bounded error models*. Paper presented at the 26th IEEE Conference on Decision and Control. Los Angeles, CA. Doi:10.1109/CDC.1987.272849

Walter, E., & Piet-Lahanier, H. (1989). Exact recursive polyhedral description of the feasible parameter set for bounded-error models. *IEEE Transactions on Automatic Control*, *34*(8), 911–915.

Wang, H. O., Tanaka, K., & Griffin, M. F. (1996). An approach to fuzzy control of nonlinear systems: Stability and design issues. *IEEE Transactions on Fuzzy Systems*, *4*(1), 14–23.

Wang, S.-H., Wang, E., & Dorato, P. (1975). Observing the states of systems with unmeasurable disturbances. *IEEE Transactions on Automatic Control*, *20*(5), 716–717.

Witczak, M., Dziekan, L., Puig, V., & Korbicz, J. (2008). A fault-tolerant control strategy for Takagi-Sugeno fuzzy systems. In *Proceedings of the 17th IFAC World Congress* (Vol. 17, pp. 7387-7392). COEX. Doi:10.3182/20080706-5-KR-1001.01249

Yoneyama, J. (2010). Robust H∞ control of uncertain fuzzy systems under time-varying sampling. *Fuzzy Sets and Systems*, *161*(6), 859–871.

Zhang, K., Jiang, B., & Cocquempot, V. (2008). Adaptive observer-based fast fault estimation. *International Journal of Control Automation and Systems*, *6*(3), 320–326.

KEY TERMS AND DEFINITIONS

Fault Detection: Fault detection determines whether the physical system operates normally and signaling the presence of a fault.

Fault Isolation: Fault isolation aims at the location of a detected fault occurring on the actuators, sensors, instrumentation, control or system controlled by specifying which organ or component is affected by it.

Fault: A fault is any difference between the observed feature on the device and the reference feature.

Interval Analysis: A mathematical tool resulted from the ensemblist approach to perform arithmetic and logic operations between intervals.

Multiple Model Approach: This approach is based on the reduction of the system complexity by the decomposition of its operation space in a finished number of operation zones. Multiple models is a set of submodels aggregated by an interpolation mechanism for characterizing the overall dynamic behavior of a system.

Nonlinear System: Nonlinear system is a system which does not satisfy the superposition principle – meaning that the output of a nonlinear system is not directly proportional to the input.

Sensor Faults: A sensor fault causes a false value of `physical quantities measured.

Uncertain Parameter: Uncertain parameter is a variable whose value is not known.

Chapter 14
Gesture-Driven System for Intelligent Building Control

Marcin Denkowski
Maria Curie-Sklodowska University in Lublin, Poland

Łukasz Sadkowski
SIM Ltd., Poland

ABSTRACT

In recent years, intelligent building systems have become standard in residential buildings as well as commercial buildings. This chapter presents a novel system for controlling such buildings using natural user interfaces. Due to advanced logic, it is capable of identifying people occupying defined areas and recognizing their gestures. The system cooperates with a Domatiq System – Intelligent Building Management System, to provide an overall control of the building. This allows the user to control particular building components in a more natural way. Results shows that for the elderly or disabled people a more natural method of controlling some appliances is more convenient than, for example, by remote controllers. This system can also be used to control and monitor such people, giving them a sense of comfort and safety in their homes while respecting their privacy.

INTRODUCTION

Intelligent buildings are buildings which, by their physical structure and specialized computing devices adapt to the changing needs of its users and organizations that use that building. Such a building should provide specific services to residents, manage the administration, operation and maintenance (Heckman, 2008). Basically, smart homes are equipped with advanced automatic systems of sensors and actuators for various pre-programmed functions and tasks such as lighting, temperature control, multimedia, window and door operations, etc. Smart buildings allow residents to control various aspects of the behavior of a building by using a computer system. Users are able to control various devices such as the audio-visual media and other functions using a computer, touch pad, tablet or smartphone. Unfortunately, the main problem with this solution is that, although they improve convenience compared to a conventional home, users still have to physically interact with these devices and provide some techniques for the user

DOI: 10.4018/978-1-4666-7248-2.ch014

authentication through a password, biometrics or RFID chips. On the other hand, humans communicate and interact with each other using both verbal and non-verbal languages. Verbal language is one of the most natural forms of human-human interaction. It has been used in many virtual systems (Gratch, 2007) for human-computer interaction. However, speech recognition with its low accuracy and subtle rules of syntax and grammar of various languages is a very difficult task. These negative factors limit the practical use of verbal language. However, non-verbal communication such as gestures and body movements, as a secondary means of communication can be employed as another human-computer communication way to enhance the interaction performance. It is therefore logical to use this method of communication in dealing with the intelligent building. But, to our knowledge, few works pay attention to this topic in the virtual reality community (Gobron, 2012).

In this context, several research projects have been developed. The DOMUS (Carner, 2009) project requires a smart control to gather information and impart instructions, one or more home automation devices to be controlled, and an internal communication network that allows the exchange of messages among the control system and the devices. A system described in (Ben Hadj Mohamed, 2013) uses a Microsoft Kinect to detect a few simple gestures but does not provide a system to control the entire building or even a single room. 'Soft Remote Control System' described in (Yang, 2006) utilizes three CCD cameras to control multiple home appliances based on hand gesture. This human-friendly gesture-based interface system is able to recognize 10 different hand gestures but is very sensitive to lightness and the overall scene tint, and so more is simple and very restricted. Another very interesting smart home system is developed by Think Big Factory. Their Openarch (Openarch, 2013) system provides the user with a possibility to control the house's components by the movement of the body, also to hold conferences from home, to know the power consumption at any time and activate any electrical household appliance from work or even share a live video. But it is still a prototype. A wery interesting and promising idea is presented at (Pu, 2013) where the authors transforms Wi-Fi into a gesture-recognition sensor. Their recognition system leverages wireless signals to enable whole-home sensing and recognition of human gestures. Since wireless signals do not require line-of-sight and can traverse through walls, WiSee can enable whole-home gesture recognition using few signal sources. Their results in a 2-bedroom apartment show that WiSee can extract a rich set of gesture information from wireless signals and enable whole-home gesture recognition using only two wireless sources placed in the living room. There are also a simpler systems that only interprets a 3D hand gestures like (Kellogg, 2014; Kristensson, 2012; Takanashi, 2012). They can be used to control a selected devices from a near range or they can be installed in a specific place, for example at wheelchairs, helping disabled people to control things by only a hand.

The proposal presented in this article assumes that the interface between the user and the building can be simplified and the process of communication can become more natural with the introduction of gesture-based control. However, it is a challenging task to apply this new way of interaction and to provide a good user experience and usability along with it.

The objective of this project was to design and implement an automated tracking and user identification system for use in a smart home environment. This project seeks to eliminate conventional inconveniences of controlling smart homes and enable users to issue commands to their homes through simple non-verbal way.

The introduction of a PrimeSense PS1080 chip-based senor (PrimeSense, 2010) made a depth sensor be easily available for every user, enabling higher levels of a human-computer interaction by the use of gestures and body movements. In our system we use Asus Xtion and Microsoft Kinect sensors to provide control over the building components together with the Domatiq System (Domatiq, 2014).

Such a solution enables the possibility to detect the occupancy of the room and to implement occupancy dependent lighting control system or gives the users the possibility to control their audio-visual media just through gestures. If the room is not occupied, the lighting or media can be switched off automatically if someone has neglected to switch it off manually. Such a system can significantly benefits older or disabled people, to whom this imperceptible and natural way of controlling and monitoring provides both a sense of security, privacy respect and autonomy in their living.

SYSTEM OVERVIEW

The whole system consists of two subsystems communicating with each other (see Figure 1). The first is a Domatiq System (Domatiq, 2013). It is a solution for intelligent building automation installation in any building. The Domatiq is a system with a distributed architecture based on dedicated modules that communicate on the common CAN bus in a propriety protocol language. There is the number of modules divided into sensors, actors, controllers and others. All of them and implement definable functions which are necessary to properly control electrical or electronic devices and can act independently because of distributed nature of this system. At the higher level of abstraction, they are represented in the form of virtual operator points. System management is realized by a dedicated configuration application which allows to easily define even sophisticated scenarios for these operator points to adjust the functionality of the system at any time. For example, one can order the house to enter the night mode from a single location: switch off the lights, close a walk gate, close roller shutters, etc. The Domatic System can be controlled by the touch pad, the LCD touch button panel, the PC application, web application or an infrared remote controller.

The second subsystem is a NIU Service responsible for the detection and recognition of gestures, their interpretation and requesting execution of specific commands to the Domatiq System. The main task of the NIU Service is to provide knowledge about the exact location of people in a defined area. The system is able to track users and detect gestures made by them. The NIU Service distinguishes exactly three events: (1) the entry of a user into the field of view of the sensors, (2) the exit of the user from the field of view of the sensor, (3) the detection and recognition of a particular gesture performed by the tracked user. Communication component of the NIU Service is able to map a specific user's actions to the appropriate functions of the Domatiq System through exchanging data frames with Domatiq system using Domatiq panel application. The NIU Service consists of imaging devices (based on PrimeSense 1080 chip) and a control unit on which this service runs. A single instance of the NIU Service is able to monitor the area spanned by the field of view of their sensors. These sensors are placed in the specific and fixed positions in the room. Room configuration, operator points and the NIU sensors are defined in an external software system supplied with the Domatiq System.

Message sent from NIU service to the Domatiq System via the Domatic panel is called the NIU Action. Every Action message consists of:

1. **Type of an Action:** Allows to choose one from four types of an action:
   a. **Point:** The Action executed for a specific operator point on the Domatiq plan, associated with a particular module,
   b. **Scenario:** The Action executed for a specific scenario associated with a global Domatiq macro,

Figure 1. Component diagram of NIU Service and its cooperation with Domatiq System

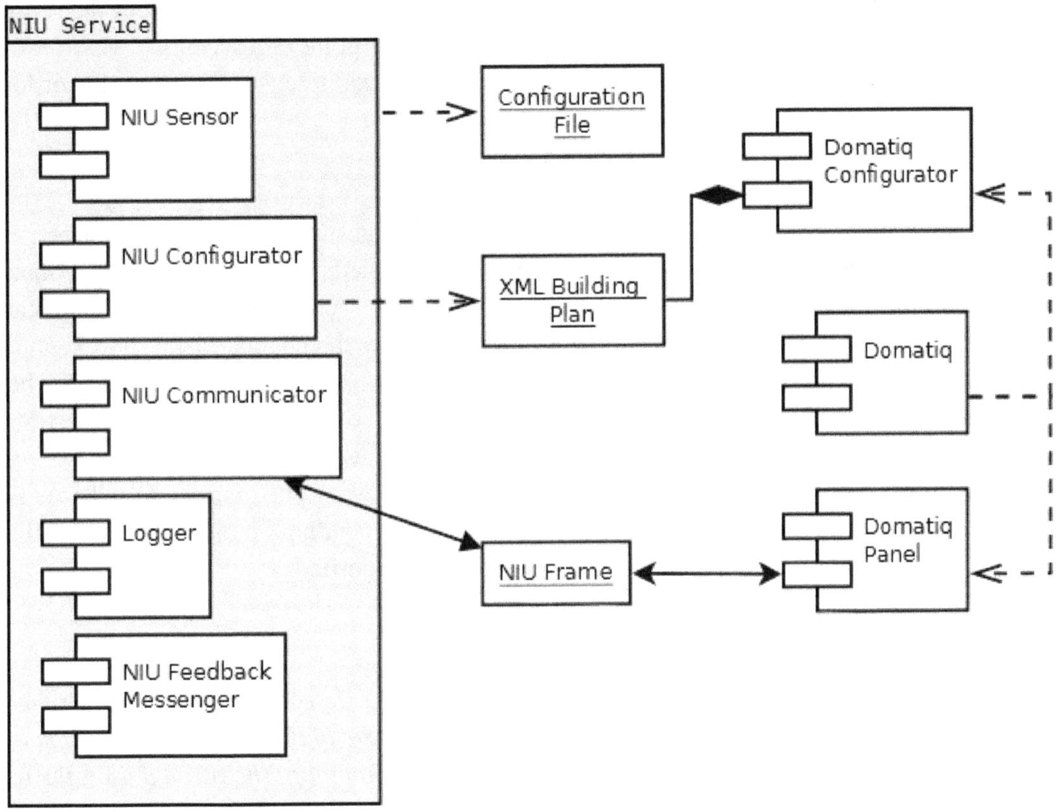

Figure 2. Example coverage of an area of size approximately 800x400cm by three sensors of a PrimeSense type

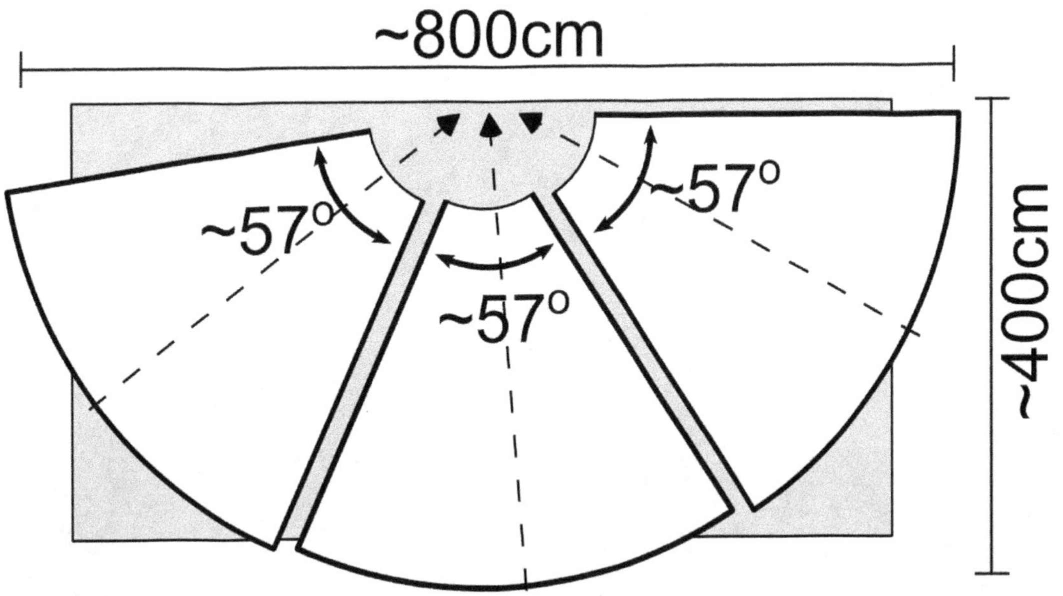

   c. **User Entry:** The Action executed on the entry of any user in the field of view of the sensor,
   d. **User Exit:** The Action executed when the user exits from the field of view of the sensor.
2. **Operation:** Associates the Action with a particular operator point or a scenario.
3. **Function:** Particular function of the Domatiq module, for example: switch on a relay or lower a value on a dimmer, etc.
4. **Gesture:** A defined activity gesture assigned to this Action.

Feedback about the system state, users, currently executing gestures, identification states, etc. is a considerable challenge for the system designer. The simplest methods for this type of feedback are audible or visual messages. Since addition of the speech synthesis would unnecessarily complicate the whole system, therefore the NIU system confirms every action by the characteristic short sound only. There were four such sounds differentiated, (1) when the user logs into a Command Mode, (2) when the user logout from a Command Mode, (3) when the user selects a specific operator point, (4) when the user executes a proper gesture, e.i. an action was executed. If the user installed additional LCD panel, the diagram of the room, including all operator points and users would be displayed on it (see Figure 3). Icons of these users are highlighted in the appropriate color depending on their state, and if the user selects a specific virtual point, an icon of this point is slightly magnified.

Figure 3. Example view of a Domatiq panel application. Locations: left top edge - a single PrimeSense sensor; top right and left bottom vents operator points; in the middle of the top and bottom light operator points. There are also visible three users in different states: (from top) the tracked user, the user in a command mode, and the user who selected an operator point.

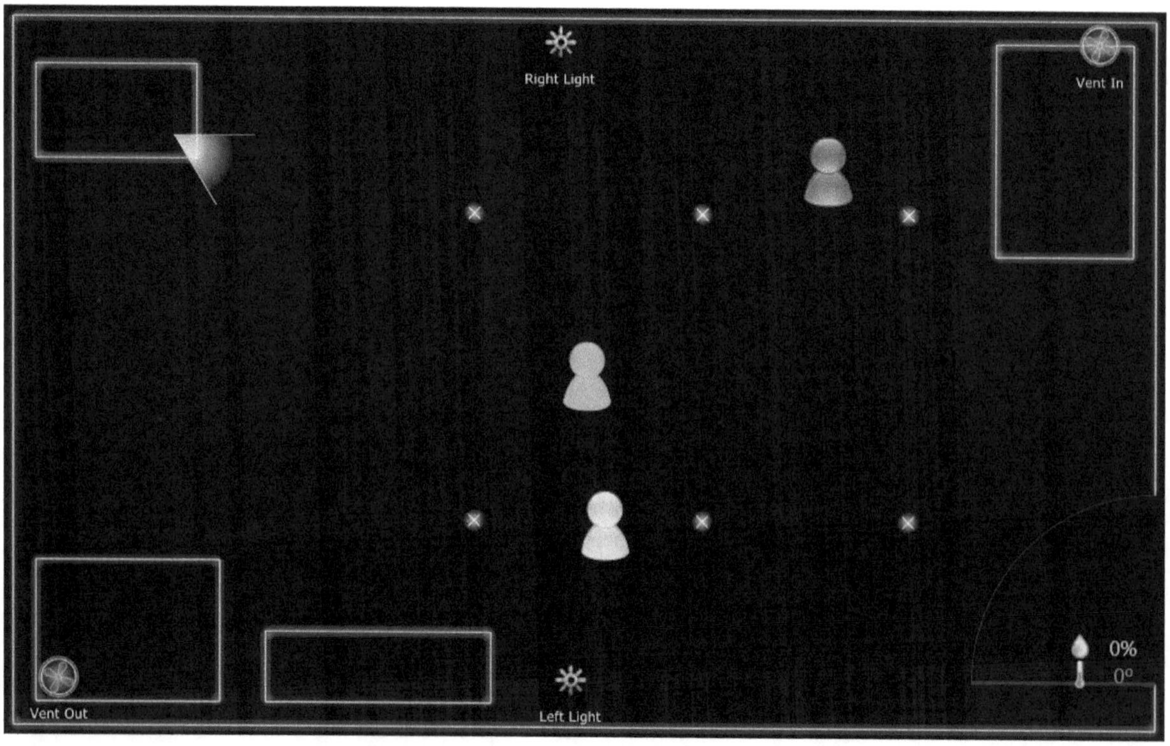

PRIMESENSE SENSOR

In a prototype of NIU system devices based on PrimeSense PS1080 chipset as visual sensors where used for user tracking. At the moment, there are two such devices freely available: Microsoft Kinect and ASUS Xtion. Popularity of these devices exploded against the releases of computer games controlled with gestures for Microsoft XBOX reaching to the awareness of a larger group of people (Bleiweiss, 2010). Many of them started projects not connected to the games industry like for example in medicine (Johnson, 2011) or in robotics (Ma, 2013).

Depth sensor of these devices consists of an infrared laser projector and IR camera which create depth image according to the structured light principle (PrimeSense, 2010). The depth camera, in both cases, runs in a native resolution of 640x480 with 30 frames per second. The range of the field of view is 43° in vertical and 57° in horizontal. The operational range of the depth sensor is approximately from 50 cm to 400 cm. Presented system prototype is capable of using at most three of such devices in a single instance. Due to this limitation there is a possibility to monitor an area as large as a room of a size 800 cm x 400 cm (see Figure 2).

The result of the analysis of projected points (analysis of user tracking) depends directly proportional on the intensity of lighting conditions in the room. Better results will be achieved with darker conditions. However, the system should not have problems to work properly in the standard light (about 200-500 lux). If there are two sensors in one room, they should be arranged in such a way that the IR points projected by one of the sensors are not lighted into the field of view of the other sensor. Moreover, one sensor should not light directly onto the other sensor. It should be also avoided to target the sensor into the strong light source (such as a window). Best results will be obtained for sensors placed at a height of 1.5 m to 2.5 m, facing slightly downward to cover the entire human posture that is in their field of view.

Sensor devices, from the software side, are based on OpenNI library developed by a non-profit organization called OpenNI (Open Natural Interaction) run by PrimeSense. This library is provided along with motion tracking middleware called NiTE. NiTE is able to track the human body giving the 3D positions of 15 specific body joints of each tracked user. A single sensor can track up to six users.

However, OpenNI libraries don't define any functionality to recognize and interpret gestures or to locate and recognize faces. A several gesture recognition systems (Wobbrock, 2007; Kratz, 2011; Martínez del Rincón, 2011; Shotton, 2011; Hasan, 2012) where taken into consideration but non of them did not meet the project assumptions. For such functionality a FUBI framework was selected (Kistler, 2012; Kistler, 2011; FUBI, 2014). FUBI (Full Body Interaction Framework) is a framework for recognizing full body gestures and postures in real time using the data from a depth sensor. Main functionality provided by FUBI can be reduced to: (1) static postures recognition – a specific configuration of several joints (positions or orientations) without any movement, (2) tracking of linear movements of several joints – linear movement of several joints with specific direction and speed, (3) combination of these two gesture categories – combines sets of 1 and 2 in a sequence of states and movements with specific time constraints. We also use an OpenCV library (OpenCV, 2014) as a main tool for all image processing tasks.

GESTURES DEFINITION

In the NIU system a gesture should be understood as a specific movement of a hand (or forearm) or just holding it in a certain position. All gestures can be performed using both left and right hand. To increase the accuracy of the gesture recognition procedure all movements should be smooth.

There is also a problem that the user can be anywhere and in any direction in the room. However, the gesture analyzer requires that the user is in a particular position in space, and is facing forward in the direction of the Z axis. To overcome this problem, after the OpenNI library has designated all the joints of the user, the virtual character is translated into the center of the coordinate system. Then, it is rotated around the X axis to compensate the tilt of the sensor relative to the plane of the floor on which the user is standing. Finally, the character is rotated around a vertical axis at the angle between the axis of the sensor's imaging and the normal vector defining the user's front. With this treatment we obtain the independence from the mutual relation between the sensor and the user.

There are a relatively small number of gestures defined inside the system, which was dictated by practical reasons – the users don't have to learn and remember too many moves. Anyhow, the knowledge of the combination of gestures, together with operators and functions, is quite a large database of actions to be remembered. However, the gestures were carefully selected to imitate the natural movements of the human, such as turning on/off lights or increasing/decreasing the volume of audio devices.

Another important selection criterion were the analysis of the typical, everyday hand movements of users at home environment. On one hand, the gestures used to control could not be too simple, which might result in the frequent accidental execution or on the other hand, too complicated, which would lead to problems in their everyday use. Based on that assumption all gestures which were just a single movement of the hand and a lot of hand gestures requiring holding the hand in a particular place were rejected. To eliminate the randomness of gestures a static gesture called Command Mode which enters the command execution mode has also been introduced.

Complete list of defined gestures is presented in Table 1. All that gestures where defined in respect to the nowadays human interface knowledge (Microsoft, 2013; Bhuiyan, 2009; Bhuiyan, 2011). Thanks to a method for modular definitions of gestures contained in the XML file used by FUBI library this list can be modified according to your needs. Gestures were distinguished into two types: (1) *Command Gestures* (CG) and (2) *Activity Gestures* (AG). Command Gestures always perform the same command, whereas Activity Gestures can be assigned to any system Action or Scenario by users. We have defined two Command Gestures: (1) *Command Mode* which is used to activate and deactivate the Command Mode and (2) *Point Selection* which is used to select a specific operator point virtually located in the room. There are also four Activity Gestures defined: (1) Circle Gesture, (2) Up/Down Gesture, (3) Left/Right Gesture and (4) a gesture of Zorro. Circle, Up/Down and Left/Right gestures occur in two versions: clockwise/counterclockwise, move up/down and move to left/to right respectively. Users can assign these gestures to any Action or Scenario from the configuration menu.

The user must be in the field of view of the sensor while performing a gesture, and should be careful while moving the hand so it will not be obscured by any obstacles including user's body. What is more, too fast execution of the gesture might not be registered by the sensor because of too slow frame rate of the camera. It is also recommended to perform consecutive gestures at intervals of at least 1 second. After the sensors start, a tracked user can activate command mode by performing a Command Mode gesture. Changing to this mode is indicated by a sound signal and a color change of a user icon to green

Table 1. A list of defined gestures

	**Command Mode (CG)** To activate command mode, user should lift a hand higher than the head joint so that the forearm is in the vertical position. Deactivation of the command mode is possible after 3 seconds after enabling command mode with the same gesture. There is no difference whether the gesture is performed by a left hand or a right hand. Proper execution of the gesture will be signaled by a system. Duration of a gesture: minimum 2 seconds.
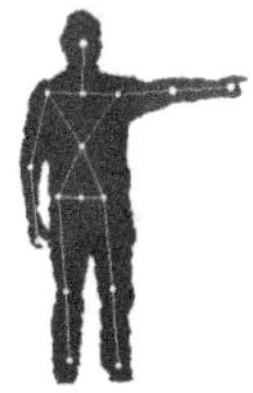	**Point Selection (CG)** In order to select a desired operator point, for which next activity gesture will be performed, user should aim a hand in the direction of that point, in such a way that the whole arm would be straighten on the height of a shoulder. System will select an operator point located nearest to the line connecting the head joint with the hand joint but not farther than an adjustable angle threshold. There is no difference whether the gesture is performed by a left hand or a right hand. Proper execution of the gesture will be signaled by a system. Duration of a gesture: 1 second.
	**Circle (AG)** Circle gesture consists of two clockwise or counterclockwise revolutions of a forearm. The move of the hand joint should start from the position nearest to the body and take two full circles. The radius of a circle should be greater than 15 cm and smaller than 50 cm. There is no difference whether the gesture is performed by a left hand or a right hand but the direction of revolutions determines if the gesture will be Left Circle or Right Circle. Duration of a gesture: minimum 3 seconds, maximum 6 seconds.
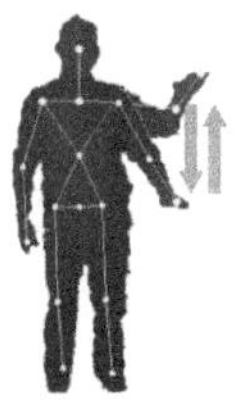	**Up/Down (AG)** The Up gesture consists of two movements of a forearm from the bottom to the top and back. In the initial phase of a gesture, the hand should be placed below the elbow, and in the final above it. The Down gesture is analogous to the Up gesture but it starts from the top position above the elbow. The length of a movement should be about 50 cm. There is no difference whether the gesture is performed by a left hand or a right hand but the direction of movements determines if the gesture will be the Up or Down gesture. Duration of a gesture: minimum 1 second, maximum 3 seconds.
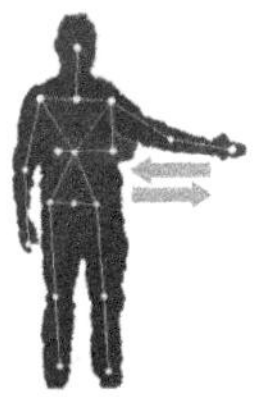	**Left/Right (AG)** The Left gesture consists of two movements of a forearm from the right to the left and back. In the initial phase of the gesture the hand should be placed on the right of the shoulder and in the final left of it. The Right gesture is analogous to the Left gesture but it starts on the left of the shoulder and ends on the right of it. The length of the movement should be about 50 cm. There is no difference whether the gesture is performed by a left hand or a right hand but the direction of movements determines if the gesture will be the Left or the Right gesture. Duration of a gesture: minimum 1 second, maximum 3 seconds.
	**Zorro (AG)** The Zorro gesture consists of three hand movements arranged in a "Z" letter. The gesture starts with a hand placed on the height of a shoulder and goes right horizontally then moves left down to the hip and again goes right horizontally. The length of a movement should be about 50 cm. There is no difference whether the gesture is performed by a left hand or a right hand. Duration of a gesture: minimum 1 second, maximum 3 seconds.

on a Domatiq panel application. Depending on the type of the action, there are two ways to issue a command by gestures and two by pseudo-gestures:

1. In case of an action of *Point* type, after activation of the command mode, the user has to select a specific operator point (by the Point Selection gesture) for which a command action will be

performed. After choosing the correct operator point the Domatiq panel highlights this point on a plan and highlights the user icon in yellow. This selection is also confirmed by a sound signal. Then, the system waits seven seconds for the user to execute the Activity Gesture. If the system recognizes such a gesture, it is assigned to the defined action and the system executes that action. The execution of the action is confirmed by a sound signal and the operator point is deselected. If a user does not execute any Activity Gesture the point will be deselected after seven seconds.

2. In case of an action of *Scenario* type only an Activity Gesture is required. After detection of such a gesture, the system assigns it to the defined scenario and then executes that scenario. The whole procedure is confirmed by a sound signal.
3. In case of a user entry (pseudo-gesture), after detection of a new user, the system checks if there are any scenarios assigned to that user entry event, and if there are any, then a defined scenarios are executed.
4. In case of a user exit (pseudo-gesture), after losing a user from the sensors field of view the system checks if there is any scenario assigned to that user lost event and if there are any, a determined scenarios are executed.

After executing a command a user can log out by performing the Command Mode gesture or just stay in that state ready to issue another commands. Experiments were conducted in a showroom presented on a schema on Figure 3. The whole room was 5m x 3m in size. There were two wardrobes and a desk with a chair. The sensor was placed in the corner of the room, opposite the entrance at a height of approximately 1.9 m. Its field of view covered about 70% of the whole area of the room. Moreover, there was set four relay modules, two of them turn on/off the upper left and upper right lights, and the other two switch on/off the air vents (see Figure 3).

SHOWROOM CONFIGURATION AND TESTS

For the purpose of the tests 12 users, 9 men and 3 women were recruited. All users ranged in age from 25 to 48 with an average of 32. Three of them were classified as advanced users, i.e. those who have already dealt with gesture interaction in a virtual environment, 7 other users had to deal with gestures for the first time, or occasionally.

For each user 20-30 minutes training explaining how the system works, what are the possible command gestures and how the feedback from the system works was conducted. This time users had the opportunity to familiarize with the showroom configuration and test and train all of the gestures. After the training, each user had to perform several tasks:

1. Enter the room, activate the command mode, use a gesture to turn on and turn off the right light;
2. Enter the room, activate the command mode, sit on a chair, use a gesture to turn on the air-in vent;
3. Enter the room, sit on a chair, activate the command mode, use a gesture to turn on the air-out vent;
4. Use a gesture to turn on every module from 6 marked points in the showroom while being in a standing position;
5. Use a gesture to turn off all the lights and vents, and deactivate the command mode;
6. Tasks 1 – 5 with 5 other users in the room.

While performing these tasks all the users' gestures were registered together with the effect of their execution. This allows to objectively assess the accuracy of gestures interaction in virtual environment and their accidental execution. After completing the scenario, the participants filled in a questionnaire in which we asked for a subjective opinion about:

1. The gesture is easy to perform: a) Command Mode, b) Point Selection, c) Circle, d) Left/Right, e) Up/Down, f) Zorro;
2. The gesture is natural: a) Command Mode, b) Point Selection, c) Circle, d) Left/Right, e) Up/Down, f) Zorro;
3. Gestures are easy to learn/remember;
4. Gesture control is reliable,
5. Gestures seem to be tiring;
6. Gestures are often executed accidentally;
7. System feedback.

And two open questions:

1. Which gestures, body poses, and user positions give false results;
2. Suggest any changes to the system.

First seven questions were answered by rating the statements on a ten point Likert scale ranging from 1 (completely disagree) to 10 (completely agree). Questions 8 and 9 were formulated as open questions to make it possible for the testers to provide feedback concerning the problems that may occur while using the system and to make any suggestions for changes or improvements.

RESULTS

Figure 4 shows the accuracy for individual gestures divided between all users and advanced users. The Command Mode gesture and the Point Selection gesture were used in every case by all users, whereas the Activity Gestures were used according to their preference. In every case, advanced users performed gestures more accurately, reaching the reliability from 4% to 25% (with average 14%) higher compared to non-advanced users. The average reliability of gestures performed by all users was 78%, performed by the non-advanced users was 70%, while advanced users reached 84%.

Graph 5 shows the reliability of gestures depending on the standing or sitting pose. For each gesture performed in a sitting position worse results of reliability were achieved (which is from 3% to 12%) in comparison to the standing position, and for the Zorro Gesture even more than 40%. The average reliability for seated gestures was 68% whereas for a standing poses reliability reached 80%.

Figure 6 shows the popularity of individual gestures chosen by the users. The most popular Activity Gesture is a Circle Gesture, which is selected in more than half of the cases. Slightly less popular is the Up/Down Gesture chosen in over twenty to one hundred cases. The Zorro Gesture and the Left/Right Gesture are selected in only about 15% of the cases.

The results of objective tests are confirmed by the subjective feelings of the users gathered in a questionnaire, which was completed after the tests. The subjective feelings of the users concerning the

Figure 4. Accuracy for individual gestures divided for All Users, Non-advanced Users and Advanced Users

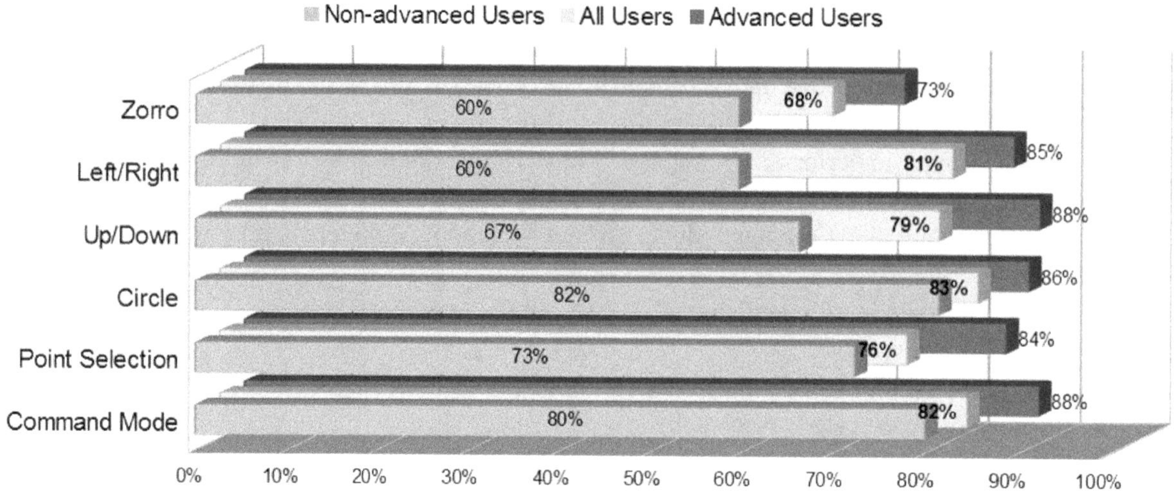

Figure 6. Popularity of Activity Gestures chosen by the users

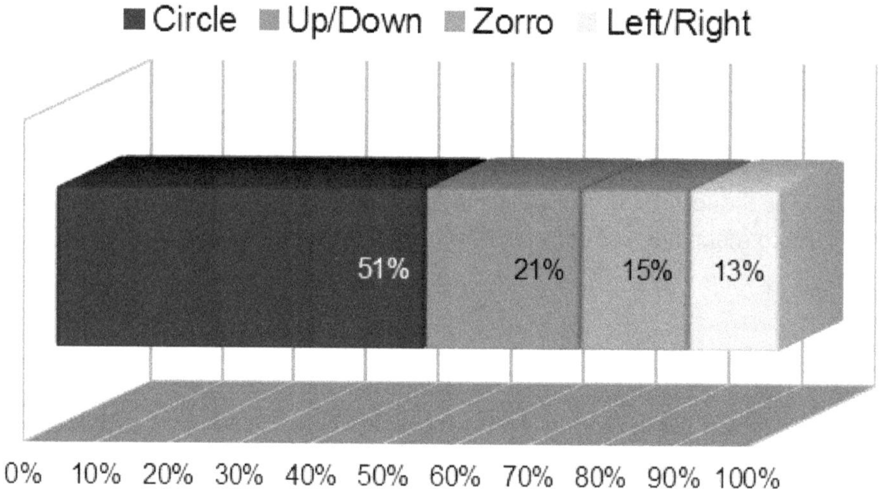

Figure 5. Reliability of gestures depending on the Sitting or the Standing Pose

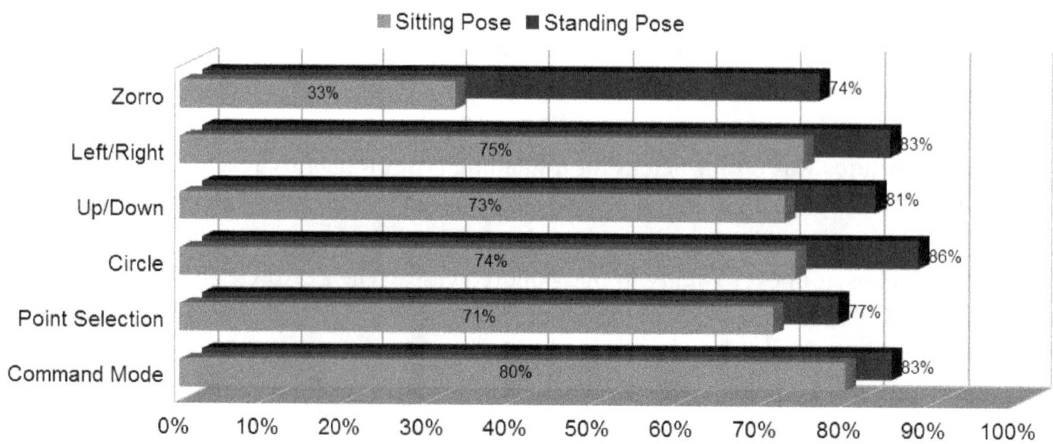

reliability of the gesture control is around 7.0 on a scale out of 10, while the ease and naturalness of the gestures are 7.5 and 7.9 respectively on the same scale. What is more, the respondents reported that they had no major problems with the learning of the gestures – the easiness of gesture learning rated as 8.3. However, the respondents assessed gesture control as tiring at 3.0, whereas the accidental execution of the gestures, according to the testers, is believed to be very rare, which was also assessed at 2.1 in both cases on a scale from 1 to 10. In most cases the respondents found that the system communicates its status sufficiently and gives useful feedback (average 8.11 on a scale from 1 to 10).

In the open question concerning the difficulties with the system usage, the main problem indicated by the users was referred to the recognition of the gestures, which were made in the line of sight of the sensor approximately behind the body. Another observed problem was connected with tracking the gestures in case when the user was located on the edge of the field of view of the sensor. Moreover, several users also signaled difficulties with tracking and recognizing hand gestures in a sitting position. Responders often suggested to increase the field of view of the sensor, and to introduce some new gestures. The most interesting idea was to introduce a gesture that inquiries about the status of the system.

PERFORMANCE ANALYSIS

The demand for computing power of the system is quite large and the system requires at least an Intel I3 class processor. This is due to the depth image analysis, continuous tracking of the users, detecting their gestures and managing the transfer of data between the NIU system and the Domatiq panel application. Furthermore, the NIU system requires properly configured Domatiq system with at least one actor module. A control unit of the NIU system connects and communicates with the Domatiq by a specialized USB or Ethernet module, which translates high level commands at hardware level to a physical data of a propriety protocol of a CAN Bus.

The main application was tested on a desktop computer equipped with an Intel I5 processor and 4GB RAM. For instance, running the application in an idle mode, when there is no user to track, takes about 6% of the overall CPU time. This is due to the NiTE library continuously search for new user's skeletons. What is more, tracking and recognizing gestures use another 2-3% of the CPU time for every user's skeleton. Adding more sensors increases the CPU usage by further 5-6% for each sensor. With three active sensors, NiTE library is able to track up to 18 users, consuming on average 70% of the CPU time. In the case of detection or the loss of a user, there is a temporary but significant increase in the CPU usage reaching up to 50% of free CPU resources.

The PrimeSense sensor itself is demanding in the context of the USB bus bandwidth and saturates it in over 60 percent (USB version 2.0), which enforces the following sensors to be connected to a separate USB controllers. Recent controllers still does not function properly when connected to the USB in version 3.0 and that problems limit the number of used devices to three per a single NIU instance.

Considering memory requirements, the most demanding component is the NIU Sensor responsible for managing every single Kinect or Xtion device, as well as for detecting, tracking and gesture recognition. Each instance of this component takes about 250MB RAM. Each tracked skeleton takes an average of about 25MB. Therefore, with three sensors and the maximum number of tracked users, the system needs almost 1.5GB of free RAM.

DISCUSSION

It must be admit that control our environment by gestures is very tempting prospect, but studies show that this form of interaction with the environment has a number of disadvantages. However, in many cases, it may be important or even necessary replacement for standard controllers. Tests show quite clearly that when a user gets used to and becomes fluent in performing gestures and avoiding of specific situations for which the gesture recognition gives false results, the overall accuracy of the gesture control procedure significantly (almost 15%) improves. Users, which used the NIU System for some time claimed that they quickly used to control certain specific devices, doing it in a natural and automatic way, making less and less faulty gestures. But, from the other hand, many users also reserved that the gesture control is quite tiresome in a longer term and they prefer to use for example a remote controller than executing physical exercises using their whole arms.

A comparison of the control gestures in the standing position and the sitting position require a separate analysis. The gestures are recognized the most effective in a standing position, when the user body is fully seen by the sensor. When the user is only partially visible, the the gesture prediction efficiency largely depends on estimation algorithms used in the NiTE library. This situation also occurs where the user is in a sitting position. The decrease in the quality of gesture recognition in a sitting position is largely due to difficulties in determination of the main body joints by the NiTE library at partial or complete occlusion of the lower parts of the body. A similar situation occurs when the user is on the edge of the field of view of the sensor or on the border between them. In such a case the sensor also analyzes only some joints of the body extrapolating the positions of the rest joints which can lead to faulty gesture recognition.

The whole system is rather complicated because it needs a professional installation of a smart house system (in that case a Domatiq System) by a trained technician. But the NIU System can be connected to any of the smart house systems because there is a dedicated API defined that allows to communicate with external systems. Unfortunately, such a connection needs a device or a module which translates NIU API calls into a native protocol used by such an external system. The cost of the whole system can also be high but it heavily depends to the selected smart house system and the cost of only a NIU System with three sensor devices is at price of a good desktop PC system.

CONCLUSION AND FUTURE WORK

In this chapter, we presented a system based on state-of-the-art intelligent building technology. Our approach covers all aspects of controlling an intelligent building with gestures, ranging from image analysis, recognition and interpretation of gestures to finally controlling devices. The results are very promising. This indicates, that it may be worthwhile designing new and more natural ways of interaction and controlling the intelligent buildings. People's lives can be greatly enhanced by these home-based assistive technologies, especially those who may be physically or cognitively challenged. This includes older persons, who especially should experience the benefits from this technologies (Picking, 2010; Fleming, 2014).

For the further study, we will focus on enhancing the robustness of the system by using user-adaptation technology especially in recognizing gestures in the sitting pose. We also have done some initial work to recognize and identify users based on their face using face recognition algorithms with the database of their faces. Some drawback of our system is also a limited number of sensors used simultaneously. There is some work done in (Ben Hadj Mohamed, 2013; Ben Hadj Mohamed, 2012) to link Kinect to a network via the ethernet or wireless WiFi with a mesh topology. Also the quality of modern depth sensors and the quality of user and gesture recognizing algorithms are still in its infancy but the fashion and demand for this type of equipment is still growing giving the supposition that in the near future, the quality will reach an absolutely sufficient level.

REFERENCES

Ben Hadj Mohamed, A., Val, T., Andrieux, L. & Kachouri1, A. (2013). Assisting people with disabilities through kinect sensors into a smart house. In *Proceedings of ICCMA*, (pp. 1-5). ICCMA. doi:10.1109/ICCMA.2013.6506149

Ben Hadj Mohamed, A., Val, T., Andrieux, L., & Kachouri, A. (2012). Using a Kinect WSN for home monitoring: principle, network and application evaluation. In *Proceedings of ICWCUCA*, (pp. 1-5). ICWCUCA. doi:10.1109/ICWCUCA.2012.6402487

Bhuiyan, M., & Picking, R. (2009). Gesture Control User Interface, What Have We Done and What's Next? In *Proceedings of the Fifth Collaborative Research Symposium on Security, E-Learning, Internet and Networking (SEIN 2009)*, (pp. 59-60). SEIN.

Bhuiyan, M., & Picking, R. (2011). A Gesture Controlled User Interface for Inclusive Design and Evaluative Study of Its Usability. *Journal of Software Engineering and Applications*, *4*(9), 513–521. doi:10.4236/jsea.2011.49059

Bleiweiss, A., Eshar, D., Kutliro, G., Lerner, A., Oshrat, Y., & Yanai, Y. (2010). Enhanced interactive gaming by blending full-body tracking and gesture animation. In *Proceedings of the 3rd ACM SIGGRAPH Conference and Exhibition on Computer Graphics and Interactive Techniques in Asia*, (pp. 34:1—34:2). ACM. doi:10.1145/1899950.1899984

Carner, P. (2009). Project Domus: Designing Effective Smart Home Systems. BSc in Information Systems and Information Technology (DT249) to the School of Computing, Faculty of Science, Dublin Institute of Technology.

Domatiq Main Site. (2014). Retrieved from http://www.domatiq.pl

Fleming, R., & Sum, S. (2014). Empirical studies on the effectiveness of assistive technology in the care of people with dementia: A systematic review. *Journal of Assistive Technologies*, *8*(1), 14–34. doi:10.1108/JAT-09-2012-0021

FUBI. (2014). *Full Body Interaction Framework*. Retrieved from http://www.hcm-lab.de/fubi.html

Gobron, S., Ahn, J., Garcia, D., Silvestre, Q., Thalmann, D., & Boulic, R. (2012). An event-based architecture to manage virtual human non-verbal communication in 3d chatting environment. *Articulated Motion and Deformable Objects. LNCS*, *7378*, 58–68. doi:10.1007/978-3-642-31567-1_6

Gratch, J., Wang, N., Okhmatovskaia, A., Lamothe, F., Morales, M., Werf, R. J. V. D., & Morency, L. P. (2007). Can virtual humans be more engaging than real ones? *Human-Computer Interaction, HCI Intelligent Multimodal Interaction Environments. LNCS, 4552*, 286–297. doi:10.1007/978-3-540-73110-8_30

Hasan, M., & Mishra, P. (2012). Robust Gesture Recognition Using Gaussian Distribution for Features Fitting. *International Journal of Machine Learning and Computing*, *2*(3), 266–273. doi:10.7763/IJMLC.2012.V2.128

Heckman, D. (2008). *A Small World. Smart Houses and the Dream of the Perfect day*. London, UK: Duke University Press. doi:10.1215/9780822388845

Johnson, R., O'Hara, K., Sellen, A., Cousins, C., & Criminisi, A. (2011). Exploring the potential for touchless interaction in image-guided interventional radiology. In *Proceedings of CHI 2011*, (pp. 3323–3332). ACM. doi:10.1145/1978942.1979436

Kellogg, B., Talla, V., & Gollakota, S. (2014). Bringing Gesture Recognition to All Devices. In *Proceedings of NSDI'14*, (pp. 303-316). NSDI.

Kistler, F., Endrass, B., Damian, I., Dang, C., & André, E. (2012). Natural interaction with culturally adaptive virtual characters. Journal on Multimodal User Interfaces, 6(1-2), 39-47. Doi:10.1007/s12193-011-0087-z

Kistler, F., Sollfrank, D., Bee, N., & André, E. (2011). Full Body Gestures enhancing a Game Book for Interactive Story Telling. Lecture Notes in Computer Science, 7069, 207-218. Doi:10.1007/978-3-642-25289-1_23

Kratz, S., & Rohs, M. (2011). Protractor3d: A closed-form solution to rotation-invariant 3d gestures. In Proceedings of IUI'11, (pp. 371-374). ACM.

Kristensson, P., Nicholson, T., & Quigley, A. (2012). Continuous Recognition of One-Handed and Two-Handed Gestures using 3D Full-Body Motion Tracking Sensors. In Proceedings of IUI'12, (pp. 89-92). IUI.

Ma, B., Xu, W., & Wang, S. (2013). A Robot Control System Based on Gesture Recognition Using Kinect. *TELKOMNIKA Indonesian Journal of Electrical Engineering*, *11*(5), 2605–2611. doi:10.11591/telkomnika.v11i5.2493

Martínez del Rincón, J., Makris, D., Orrite Uruñuela, C., & Nebel, J. (2011). Tracking human position and lower body parts using Kalman and particle filters constrained by human biomechanics. *IEEE Transactions on Systems, Man, and Cybernetics: Part B, Cybernetics*, *41*(1), 26–37. doi:10.1109/TSMCB.2010.2044041 PMID:20388598

Microsoft. (2013). *Human interface guidelines, Kinect for Windows v1.8*. Retrieved from http://msdn.microsoft.com/en-us/library/jj663791.aspx

Open Source Computer Vision. (2014). Retrieved from http://opencv.org

Openarch Project. (2013). Retrieved from http://www.openarch.cc

Picking, R., Robinet, A., Grout, V., McGinn, J., Roy, A., Ellis, S., & Oram, D. (2010). A Case Study Using a Methodological Approach to Developing User Interfaces for Elderly and Disabled People. *The Computer Journal*, *53*(6), 842–859. doi:10.1093/comjnl/bxp089

PrimeSense Ltd. (2010). *US Patent Application Publication No. US2010/0118123*. Retrieved from http://www.freepatentsonline.com/20100118123.pdf

Pu, Q., Gupta, S., Gallakota, S., & Patel, S. (2013). Whole-Home Gesture Recognition Using Wireless Signals. In *Proceedings of Mobicom'13*, (pp. 27-38). ACM.

Shotton, J., Fitzgibbon, A., Cook, M., Sharp, T., Finocchio, M., Moore, R., et al. (2011). Real-time human pose recognition in parts from single depth images. In Proceedings of IEEE Computer Vision and Pattern Recognition (CVPR). IEEE Press. doi:10.1109/CVPR.2011.5995316

Takanashi, N., Sato, Y., Ohta, M., Hiroi, N., Okumura, F., & Yamada, A. (2012). Human-Computer Interaction Technology Using Image Projection and Gesture-Based Input. *NEC Technical Journ*al, Special Issue on Smart Device Solutions. *NEC Corporation*, *7*(3), 122–126.

Wobbrock, J. O., Wilson, A. D., & Li, Y. (2007). Gestures without libraries, toolkits or training: a $1 recognizer for user interface prototypes. In Proceedings of UIST '07, (pp. 159-168). UIST.

Yang, S., Do, J., Jang, H., Jung, J., & Bien, Z. (2006). Advanced Soft Remote Control System in Human-friendliness. *SCIS & ISIS*, *2006*(1), 218–222.

KEY TERMS AND DEFINITIONS

Distributed System Architecture: A system in which components located on a common network communicate and coordinate their actions bypassing messages.

Domatiq System: An intelligent building management system designed to monitor and control the mechanical, security, fire and flood safety, lighting and humidity in a building developed by SIM Ltd.

Gesture Recognition: A topic in computer science and language technology with the goal of interpreting human gestures via mathematical algorithms.

Human-Computer Interaction: An interaction between a man and a computer taking place through the user interface.

Intelligent Building: A building controlled by an advanced, smart Building Management System (often referred as Smart Building).

Natural User Interface: A system for human-computer interaction that the user operates through intuitive actions related to natural, everyday human behavior.

Structured Light: The process of projecting a known pattern of light (for example: lines, grids, bars, etc.) on to a scene.

Section 4

Robotics and Control Applications

Chapter 15
Distributed Coordination Architecture for Cooperative Task Planning and Execution of Intelligent Multi-Robot Systems

Gen'ichi Yasuda
Nagasaki Institute of Applied Science, Japan

ABSTRACT

This chapter provides a practical and intuitive way of cooperative task planning and execution for complex robotic systems using multiple robots in automated manufacturing applications. In large-scale complex robotic systems, because individual robots can autonomously execute their tasks, robotic activities are viewed as discrete event-driven asynchronous, concurrent processes. Further, since robotic activities are hierarchically defined, place/transition Petri nets can be properly used as specification tools on different levels of control abstraction. Net models representing inter-robot cooperation with synchronized interaction are presented to achieve distributed autonomous coordinated activities. An implementation of control software on hierarchical and distributed architecture is presented in an example multi-robot cell, where the higher level controller executes an activity-based global net model of task plan representing cooperative behaviors performed by the robots, and the parallel activities of the associated robots are synchronized without the coordinator through the transmission of requests and the reception of status.

INTRODUCTION

Nowadays, advanced manufacturing systems are generally composed of the following control levels: planning, scheduling, coordination, and low-level control. At the planning level, based on definition of tasks and due dates, resource requirements are specified and an actual time period and resources are assigned to each task and subtask. The monitoring of real time events and modification and adaptation of the cell configuration after detection of an equipment failure are supported by a knowledge-based system. The coordination level is naturally decomposed into two sublevels: the flow of parts through

DOI: 10.4018/978-1-4666-7248-2.ch015

the buffers and the coordination of a manufacturing operation. The former can be viewed as producer/consumer processes. The flow control and robot synchronization are performed by a multitasking operating system. The coordination level receives a rather static schedule from the upper level and is provided with dynamic information about particular states of a system from the lower level. It sends commands to low-level controllers and may ask for a rescheduling. Thus, the coordinated system behaves concurrently and includes cooperation and competition relations among its subsystems. This concurrency with links to both upper and lower levels should be modeled formally in the control system design. Currently, control systems for the coordination of specific low-level controllers according to the schedule of operations have been mainly developed in an ad hoc manner to fit just one particular system, which requires many man-years of hard programming again and again to build the control systems because specifications are not sufficiently complete and the system environment changes frequently.

The development of industrial techniques makes sequential operation control for robotic systems more large-scale and complicated one, in which some robots or subsystems operate concurrently and synchronously based on sensing information by external sensors and/or inter-robot communications as cooperative multi-robot systems (Yasuda et al., 1991). Multiple robots have the possibility to perform complex and large scale tasks more efficiently by cooperating in some way than a single robot does. Although there are many tasks that could be constructed for a single complex robot, in many cases there are advantages to using multiple robots, such as robustness and less complexity. However, to perform complex robotic tasks such as cooperative fixing and mating of separate parts by two or more robots, new distributed autonomous control architecture through the integration and cooperation for intelligent multi-robot tasks should be developed (Hörmann, 1991). In such intelligent robotic tasks, synchronization between associated robots is crucial, and cooperation based on ad hoc or temporary master-slave or leader-follower relationships or ideal cooperation through mutual coordination with synchronous communication based on simple reflexive behavioral control on a decentralized control architecture is required (Duffee, 1996). By specifying a conditional sequence of actions and goals or joint plan involving the subject and others, interlocking coordination where each robot adjusts its action selection based on the evolution of the ongoing interaction is achieved. In the distributed planning, the cooperative interaction consists of a series of actions including communication acts such that a robot causes a change in the environment that triggers the precondition of the next action of a sequence.

Besides the algorithm in use, the performance of cooperative tasks depends on the actual implementation. Non-linear systems control theory does not provide tools to analyze and synthesize such complex tasks with respect to output control performance. On the other hand, research on real-time operating systems does not deal with the impact of the organization of the controller on the controlled process, except scheduling, fault tolerance, or liveness. In this context, in order to realize cooperative multi-robot systems, it is necessary to provide effective tools for describing process specifications and control algorithms and developing validation procedures before implementation in a clear and consistent manner. Among the existing modeling formalisms, Petri nets provide a powerful framework to model and analyze robotic manufacturing systems, characterized by their ability to naturally represent the synchronization of processes, the concurrence of activities, the presence of conflicts, causality, resource sharing, mutual exclusion, etc. (Hasegawa, 1996), which are inherent characteristics of multi-robot systems as well as manufacturing systems. Since one can directly convert Petri net definitions into control code for supervisory control of robotic systems after correct Petri net models are constructed, the systematic modeling of robotic systems becomes very important (Caccia, 2005). In this chapter, based on the hierarchical and distributed structure of the manufacturing system, a structural design methodology through a top-down and bottom-up relationship between abstract specification and practical implementation of a behavior-based architecture in the context of multi-robot cooperative task planning and execution is presented to perform the distributed autonomous control of cooperative interactions and reactive behaviors without supervisors so that robots and other machines can synchronize activities and avoid harmful conflicts.

BACKGROUND-RELATED WORKS

From a hardware-oriented point of view, a robot itself is composed of modularized components for different functions, such as vehicle, manipulator, end-effector, hand-eye camera, etc., which are computationally independent from each other and are controlled in parallel; independent subcontroller for each component (Ejiri et al., 1994). The communication between the subcontrollers, through a high-speed bus architecture or point-to-point architecture, makes cooperation between the components possible. By allocating one processor to each component, the components can work in parallel and the execution time is reduced, although unused components (subcontrollers) bring about low scalability.

For a high degree-of-freedom component, the main control loop is decomposed into several control loops; one loop (feedback controller) for each joint of a kinematic chain (Zheng et al., 1993). Because the parallel processing computer architecture reflects the robot hardware architecture, a clear overview of the functionalities of the system and an easy and efficient software development environment are provided (Lepuschitz et al., 2011). With a scalable algorithm for redundant manipulators, an additional joint could easily be controlled by adding another processor. Further each control loop is broken down into simpler subtasks such as sensing, controlling, acting, coordinates transformation, for a high sample rate by pipelining the functions of the control loop with separate processors. Parallelization at the control level provides another advantage of easy software development because the input and output functions are separated from the algorithmic control function (Fatikow et al., 1994).

A control architecture makes a control system from control components and determines the interrelationships between the components and the mechanisms for coordination as a crucial point for establishing the limitations and possibilities of the system. In conventional control structures for automated manufacturing systems, the control components are separated from the manufacturing components and are interconnected by their control interrelationships, and each control component is regarded as a single processing element. For robot control architectures, function and behavior oriented approaches are distinguished, where each function or behavior can be performed by a separate processor (Brooks, 1987; Fleury et al., 1994).

For complex and large scale robotic manufacturing tasks, the conventional, strictly centralized coordination and planning or scheduling is difficult due to huge amount of communication and information to be processed, as well as in an unforeseeable environment. At one extremity, interactions of non-cooperative robots with a decentralized architecture may result in emergent global behavior without central planning and communicating with each other (Stilwell et al., 1993; Lazinica, 2005), where the processing machines broadcast messages for material handling requirements to the listening robots. Although they show adaptability, robustness, extensibility and reactivity, the system is subject to deadlocks and is less efficient due to the limitations to local decision capabilities (Doty et al., 1993). Thus, in such applications, global communication can be advantageous for optimization problems (Sousa et al., 1996). Holonic architectures for multi-robot cells allow the robots to independently plan and execute their own tasks, and the time uncertainty is negotiated on the task with the scheduler (Duffie, 1990; Prabhu et al., 1995;

Van Brussel et al., 1998). Following the concept of non-hierarchical control, based on message sending and receiving with local coordinator processes, global synchronization with distributed semaphores and multi-event communication with vectored semaphores and broadcast messages should be realized. The first scientifically conceptualized non-centralized system was proposed for artificial biomimetic self-reproducing processes in 1971 and applied to develop self-coordinating robots with non-centralized neural networks and grouping robots or agents without centralized control mechanism (Yasuda, 1971). Our system may have been the first to apply the non-centralized system concept seriously to automation systems, and also one of the first mobile robots specifically designed for distributed autonomous group intelligence.

For overall robot control, functional decomposition of control such as global planning, local planning, perception, mission execution and hardware control, was implemented and the coordination and integration of several real-time activities was performed through communication and sensor fusion via parallel blackboards for robot navigation (Liscano et al., 1995). In accordance with the concept of logical sensors generalized with their own computing capabilities, independent autonomous modules are created through holonic evolution with high extensibility and dynamic reconfiguration, so that the layered control architecture can be produced according to the degree of abstraction of processed data and response time, like the evolution of living systems. In a conventional hierarchical structure, the tasks in one layer can be executed simultaneously and synchronously, different layers work in parallel and asynchronously. For shorter response time at the lower levels, connections across the architecture, twin hierarchy or double parallelism can be added. Although hierarchical structures offer high efficiency and can optimize problems, they have to deal with the communication bottlenecks, additional sensors integration, the reaction capacity and the robustness. In this context, novel coordination algorithms should be designed to be suitable for highly computational parallelism by giving simultaneously working robots some independence. Because multiple robot systems essentially have the features of concurrency or parallelism, first of all, discrete event system view should be set up. Further, control software architecture should be matched with the hardware structure of the system. Petri nets are efficient for modeling situations that may involve concurrency and conflict for control and synchronization of operations. They have become quite popular because a model can be graphically represented, directly executed and formally analyzed. However, within the robotics area, Petri nets have been mainly applied to the study of conventional flexible manufacturing systems, where the major concerns are the modeling, performance analysis and scheduling. For the coordination among multiple robots, appropriate representation methods and analysis methods of the control mechanisms have not sufficiently been established.

PETRI NET MODELING OF ROBOTIC ACTIVITIES

The Petri nets are a general purpose graphical-mathematical tool for modeling the dynamic behavior of distributed systems in which multi-condition process concurrency and synchronization play a major role that is the case of multiple robot systems. A Petri net comprises two types of nodes, places and transitions. Objects such as transitions and places, can be associated with different semantics or meaning according to the purpose of the model. The simple Petri net associates events with transitions and conditions with places. Because Petri nets are synthesizable and permit the construction of models applying top-down and bottom-up approaches, the essential features of a system can be adequately represented by the selection of an appropriate detail level. The same Petri net model can be used for the analysis of behavioral properties, as well as for the systematic construction of discrete event controllers. A Petri net can be represented by a bipartite graph where places are denoted by circles and transitions by bars, and places and transitions are linked by directed edges called arcs. The presence of a token in a place indicates that the corresponding condition is true. The marking therefore describes the state of the net in terms of conditions which are true. When all of the input places of a transition become marked, the transition is enabled. When the transition eventually fires, the corresponding event occurs, and the firing removes a token from each of the input places, making the pre-conditions false, and deposits a token in each of the output places, making the post-conditions true (Desel et al., 1998). Due to locality of states and actions, system monitoring can be provided in real time. The behavior of processes executing in parallel can be verified using reachability analysis.

A robotic action is represented by a Petri net model of a network of communicating modules, where closed-loop control algorithms are encapsulated in a reactive logical behavior. Robotic actions are then logically and temporally composed to build more complex actions, such as operations, subtasks, tasks and missions. The practical efficiency of a robotic action composed of closed-loop control algorithms is essentially determined by its implementation. Temporal attributes such as the sampling rate and the latency between a sensor measurement and an actuation applied to the robotic system may lead to temporal inconsistencies and unsafe behavior. Structural deadlocks are due to synchronization structure and may be associated with circularity in inter-module communication. In a net composed of three periodic modules, each module may be blocked while waiting to read. Other structural deadlocks emerge when the order in which a module communicates, reading and writing with synchronous communication, over multiple ports is incorrectly specified. A Petri net is live with respect to an initial marking if there exists for each transition a legal firing sequence leading to a marking in which that transition is enabled. As deadlock arises when no transition can be longer enabled, liveness implies deadlock freedom. The Petri net can be described by algebraic equations, thus formal properties of the model can be analyzed to identify specific characteristics like overflows and deadlocks. A synthesis method to build deadlock free, partially synchronized net models is given as follows. For a marked graph, the token count in a directed circuit is invariant under any firing. Therefore, a marked graph is live if and only if in the initial marking at least one token is placed on each directed circuit. Since the set of arcs in a directed circuit is a support for a minimal p-invariant, which represents an unchanging truth about a set of conditions (places), an initial marking is live when each minimal p-invariant is assigned at least one token. An integrated software package was implemented to build a Petri net model from the GUI, search the p-invariants using linear programming with integers and perform the liveness analysis. The graphical tool can avoid complex specifications with ambiguous textual description or specific mathematical notation for the effective visual communication between designers and customers.

For applications such as discrete parts manufacture and automated assembly by robots, the overall process can be decomposed into a set of distinct operations; an operation means a single stage of the process which is executed independently from the other parts of the system. A robotic task can be described as the set of possible sequences of operations the robot can perform in the system. The task has an initial and a final state. Choices are allowed in the task but infinite iterations are not. Thus the net model of a robotic task is a state machine. Operations can be performed concurrently, but precedence relations among the operations impose a partial ordering on the operation sequence. Each operation requires a set of resources, such as raw materials, fixtures, and other tools. A set of discrete resource states is associated with each resource representing the discrete conditions for making state-dependent sequencing decisions including precedence relations, the availability of resources, and the initiation of error recovery cycles. To model the discrete decision and control structure of a system, Petri nets contain two types of places corresponding to the operations and discrete resource states, which alternate along any directed path in the Petri net. A token present in an operation place indicates the operation is in progress, and a token in a resource state indicates the resource is in the corresponding discrete resource state. Transitions are partitioned with respect to operations into two classes: input transitions and output transitions. Each transition is connected into a single operation place. All arcs from resource places enter input transitions and all arcs from output transitions enter resource places. To represent more than one set of conditions for the execution of an operation, multiple input transitions are mapped into the operation place. As with the input transitions, multiple transitions can be associated with an operation place to indicate the different sets of discrete resource states which can result from the execution of the operation or conditional results, including unsuccessful status due to mechanical failure.

For the use in simulation and real-time control, Petri nets must be extended so that transition firing can be realized deterministically. The extended Petri net adopts the following elements as input and output interfaces which connect the net to its environment: gate arcs and output arcs. A gate arc connects a transition with a status signal source, and depending on the signal, it either permits or inhibits the occurrence of the event. External gate arcs from the machines are connected to the transitions of the net when needed, for example, to synchronize and coordinate operations. An output arc connects a place with an external machine and sends a command signal to the machine. When a token enters a place that represents a subtask, the machine defined by the machine code is informed to execute a subtask with specified parameters. These communication interfaces are represented by transitions which represent the sending and receiving activities with its environment. The hardware controller perceives its external environment outside the system through sensors and updates tokens in the places representing the states of sensors. Sensory data is fed into the places according to the processing result. Then tokens are entered in the places, and the information is transmitted to the target transitions through the gate arcs to trigger the firing of the transitions. On the other hand, the controller executes control policies coded in the tokens in the places through output arcs to actuators. The internal and external gate arcs are extended with more special functions associated with a transition and a place for triggering firing and executing in the external controller, respectively. These functions are implemented by conventional computer code and integrated into the hardware control system.

Sending and receiving activities to/from other robots are also defined as transitions. A sending action is executed by entering a token in the output place of the transition representing sending activity, while a receiving action is executed by deleting a token in the input place of the transition representing receiving activity. For simulation, the two associated places are glued together; they are physically different places but logically only one place forming a fusion set. Therefore, all the places belonging to a fusion set have always the same marking. Figure 1 shows an example of extended Petri net model of robotic action control by transition firing with permissive and inhibitive gate arcs. In the control specification, a primitive task is described to control the robot through the hardware controller using the command/response concept and represented as composed of the two transitions representing the start and end events and one place representing the robotic process. A token inside the place indicates that the robotic process goes on. The robot controller starts an action based on signals from the upper level and/or other robots, and sends the commands through output arcs to the hardware controller, and receives the status signals from the external sensors through permissive gate arcs.

In order to implement directly complex control systems, Petri nets are further modified in the directions of hierarchical net modeling, which is the powerful mechanism allowing the top-down developments of

Figure 1. Extended Petri net model of robotic action control by transition firing with permissive and inhibitive gate arcs

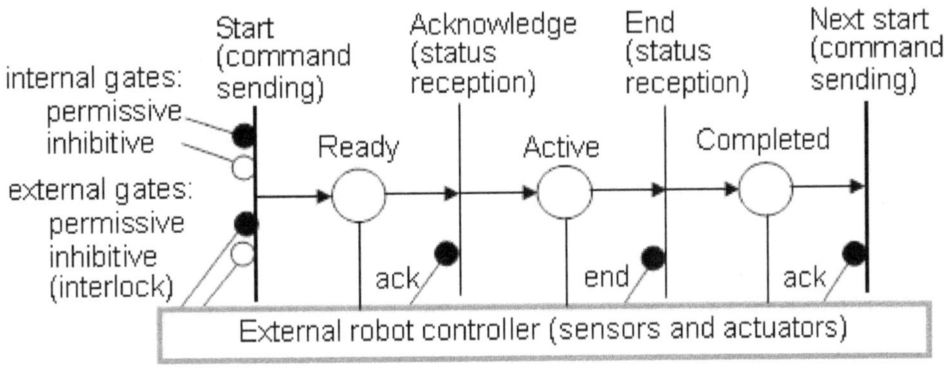

net models, making possible hiding some details in a consistent way and separation into reusable components. The invocation is the only type of hierarchy and corresponds to procedure call in programming. A transition can invoke the execution of another lower-level net by adding a token to an input place of the lower-level net. The transition continues its firing when a token is added to an output place. As other types of hierarchy, the substitution corresponds to macro mechanism that means a subnet is plugged into an upper-level net. The fusion allows transitions or places from various nets to be fused together to permit resource sharing and synchronization across nets.

In case that the activity which a transition represents is a robotic task, the transition represents the manufacturing event: the start of transition firing indicates the start of the manufacturing process and the completion of the firing of the transition indicates the finishing of the manufacturing process. A task can be seen as some connection of more detailed subtasks, which are composed of actions. Subtasks are defined to coordinate the activities of the robot with respect to other robots and the system controller. Because a cooperative task consists of several consecutive phases, coordination can be achieved by synchronizing the transitions of consecutive phases, where all concerned controllers agree on the transitions and are able to end the current phase and start the new phase consistently. In a coordinated movement such as grasping or taking over an object, different movement phases for different control systems for force, contact area, and distance can be distinguished (Osswald et al., 2003). The coordination mechanism must check for internal and external events and synchronize the different robot activities accordingly for the most relevant phases and transitions. Parallel activities begin at the firing of a common transition and end with the firing of another transition. Two transitions are said to be concurrent if they are causally independent, where one transition may fire before or after or in parallel with the other. Pure synchronization between tasks can be naturally represented, which is one of the key solutions to realize the cooperation in multi-robot systems, as shown in Figure 2. Two events are in conflict if both events can occur (are enabled) but not both. They are concurrent if both events can occur in any order without conflicts. If two events are concurrent while each of the transitions is in conflict with another event, such a situation where conflict and concurrency are mixed is called a confusion (Murata, 1989). In a manufacturing environment, utilization of a common resource may be subject to several constraints such as machine availability or operation precedence and the concept of monitors as a programming tool should be extended to organize access to shared functions with multi-condition constraints. Since in the distributed design a task is the entity for resource ownership, multi-event synchronization with Petri nets and global synchronization semaphores are emulated using coordinator processes.

For example, the pick and place operation, that is, transferring an object from a start position to a goal position is a sequence of the following subtasks; searching and detecting the specified object using a vision sensor, picking the object, moving to the goal position, and putting it on the specified place. Each subtask includes planning and execution phases under its environment with several intelligent production machines, such as conveyors, working tables, machine tools, etc. Failure detection and recovery action using external sensors are also required. So, more detailed descriptions are needed for real-time control. For example, when a robot picks a part up from a conveyor, it should synchronize its activity with the conveyor, because the pickup place on the conveyor is considered as a shared resource. Figure 3 shows a Petri net representation of sequential resource assignment for tasks processed in a shared station by two different robots.

A robot, like a manipulator, is only capable of performing one task at a time. So, any activity by a single robot should be implemented as a procedure executed by a single resource. If multiple tasks are requested to the multi-robot system, the robots represent resources shared between the different tasks. Since more than one task may require access to the same resource, but each resource can only serve one task at a time, deadlock between different tasks may occur. The shared resource problem (Zhou et al., 1992) is addressed to synthesize live, bounded, and reversible Petri nets. A net model for mutual exclusion

Figure 2. Petri net model of generic cooperative task where transitions are synchronized by the coordinator

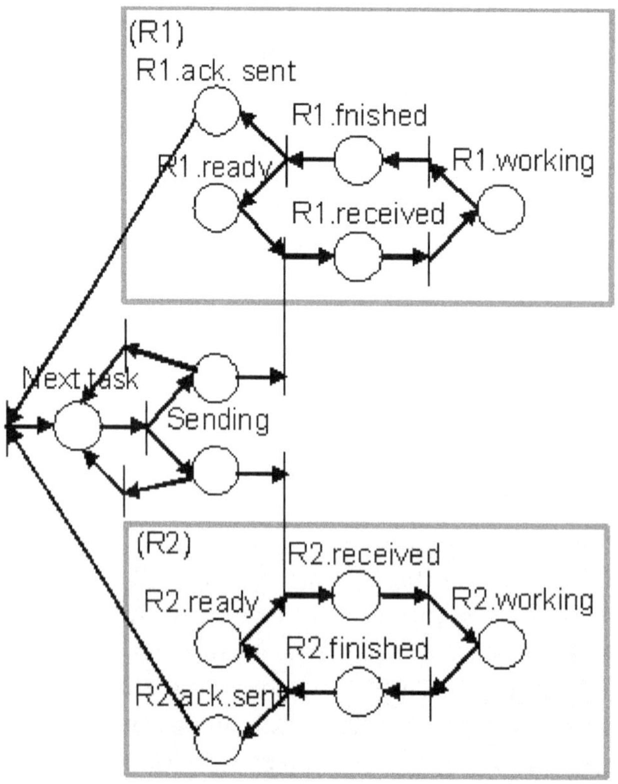

Figure 3. Petri net representation of sequential resource assignment for robotic tasks

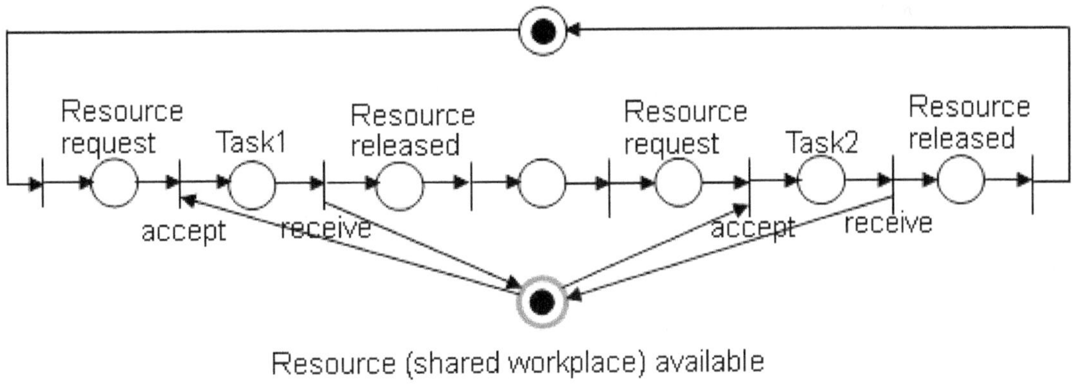

consists of a place marked initially with one token to model a single shared resource and a set of pairs of transitions to model associated tasks as shown in Figure 3. The net models for resource specification are combined with the net models for task specification to design a net based system controller.

CONTROL SPECIFICATION OF ROBOTIC PROCESS PLAN

In the hierarchical and distributed control system composed of one system controller and several robot controllers, the coordination level activities of the system, or subtasks, are defined through a net model in the specified order corresponding to the flow of subtasks processed for a workpiece or part considering the task specification. The conceptual net model, with the places representing the robots or other resources used for the subtasks and their associated arcs, is allocated to the system controller, while the detailed net models of subtasks are allocated to the local robot controllers. Transitions representing the start and end of subtasks are called global transitions, and other transitions are called local transitions. System coordination using the information of the global transitions is performed through communication between the local robot controllers without the intervention by the system controller, so that the distributed autonomous control architecture can be realized based on hierarchical Petri nets.

The example robotic cell has two robots, one station, one machining center, and two conveyors, where one is for carrying in and the other one is for carrying out, as shown in Figure 4. The main execution of the system is indicated as the following task specification. A workpiece is carried in by the conveyor CV1, then loaded to the station STA by the robot R1. Then it is loaded to the table MT of the machining center MC by the robot R2. After processing by the machining center MC, the workpiece is unloaded from the table MT to the station STA by the robot R2, and then unloaded to the conveyor CV2 by the robot R1. Finally the workpiece is carried out by the conveyor CV2. The part flow control algorithms and handshaking processes required for part insertion and removal operations were developed on the CAD computer running UNIX for 3D graphic motion simulation. For real-time response to part and gripper handling the local controllers are driven by interrupts from the machine programs. The flow of parts in the task specification is shown in Figure 5, where each distribution site of a workpiece is indicated by a macro place, while each robotic activity processed on a workpiece is indicated by a macro transition, based on the extended Petri net.

Figure 4. Configuration of example robotic cell

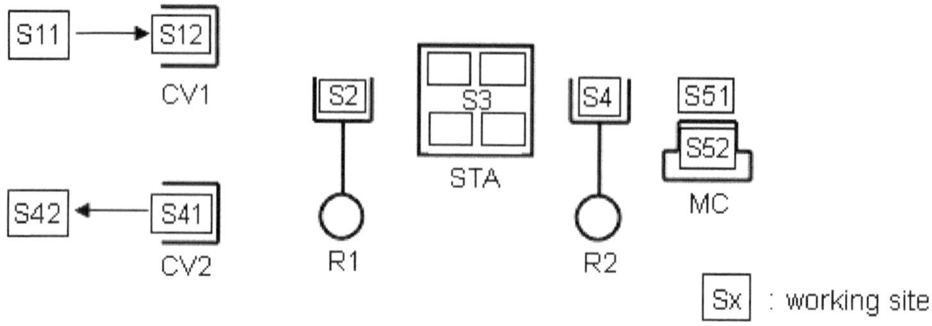

Figure 5. Flow of parts in example task

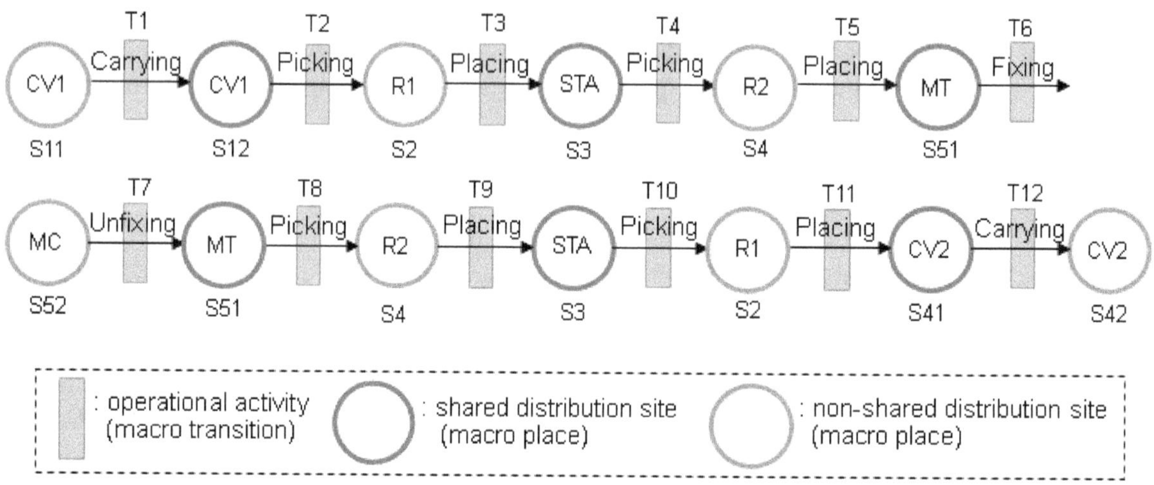

In Figure 5, the distribution sites indicated by a red bold circle are sites shared by preceding and succeeding activities. Thus cooperation between two robots or machines is needed to synchronize these activities and avoid harmful consequence at the end of the preceding activity and the start of the succeeding activity. At the conceptual level the discrete event processes for cooperation are represented by detailed subnets of corresponding activities, as shown in Figures. 6. In the net, a shared transition indicates that the event occurs at the same time in associated robots or machines. Subtask activities such as picking, placing, fixing, unfixing, and transfer are translated into a detailed subnet using primitive actions for the local controllers. After the signal of the status command from the numerical controller that the machining center is ready, the machining task is initiated by a message and then control is transferred to the subtask of the machining center. In order to perform the part loading, the handshaking for part fixing/unfixing and moving is performed through message passing between the machining center and the robot according to the joint evolution of associated net models in the system controller and local controllers. At the completion of machining, control is returned to the system controller, then the unloading subtask is started by message passing, which drives the robot for removal of the part from the machining center, and performed through the handshaking between them. At completion, the subtask returns control to

Figure 6. Detailed net representation of loading and unloading operations: (a) carrying in and picking (site: CV1), (b) placing and picking (site: STA), (c) placing and fixing (site: MT), (d) unfixing and picking (site: MT), (e) placing and picking (site: STA), (f) placing and carrying out (site: CV2)

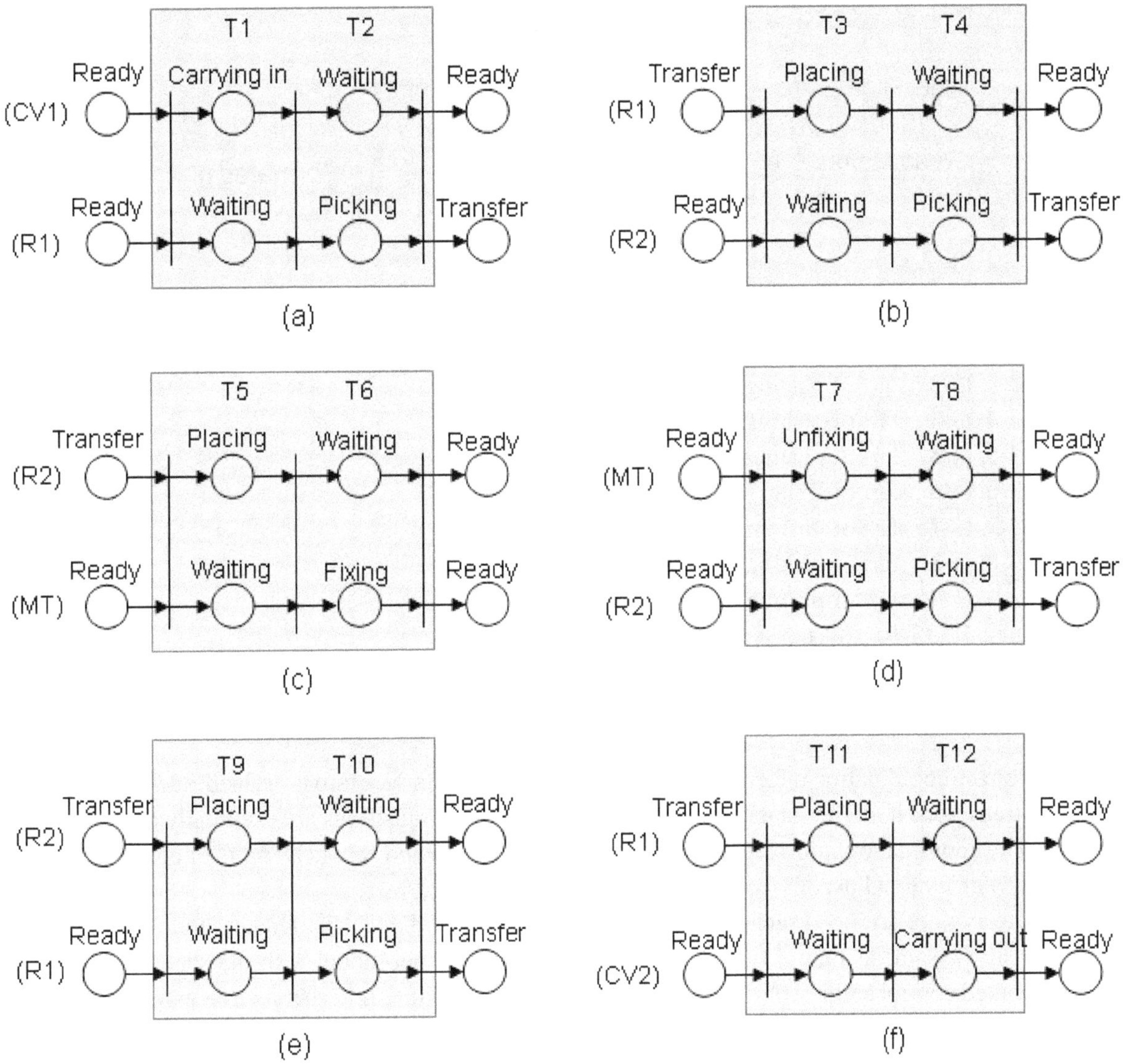

the system controller. For the example system, detailed net models can be automatically generated using the database of net models of general robotic operations.

In the detailed net models assigned to local controllers, using the names of transitions in the controllers, global transitions are defined; for example, G2: T2, t1-2, t2-2 indicates that global transition G2 is composed of transition no.2 of the system controller, transition no.2 of conveyor controller CV1 (no.1), and transition no.2 of robot R1 controller (no.2). The names of the controllers, robot R2, machining center MC, and conveyor CV2 are no. 3, no. 4, and no.5, respectively. Then, the coordination information for the example control system is as follows.

Figure 7. Distributed coordinations for synchronization of concurrent activities using external permissive gate arcs without global synchronization command

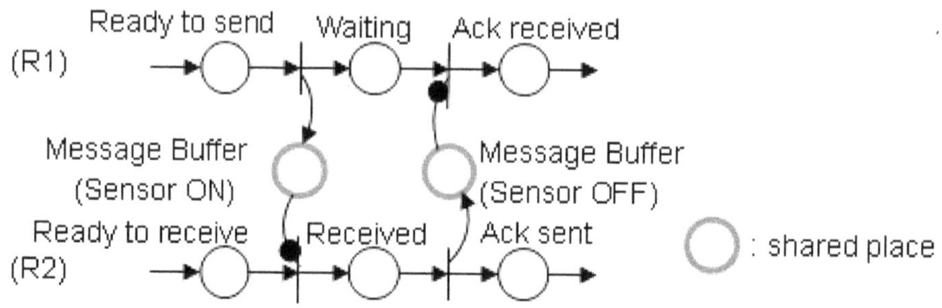

G1: T1, t1-1, t2-1 ; start of carrying in to the cell.
G2: T2, t1-2, t2-2 ; start of picking up from CV1.
G3: T3, t2-3, t3-3 ; start of placing on STA.
G4: T4, t2-4, t3-4 ; start of picking up from STA.
G5: T5, t3-5, t4-5 ; start of placing on MT.
G6: T6, t3-6, t4-6 ; start of fixing on MT.
G7: T7, t3-7, t4-7 ; start of unfixing from MT.
G8: T8, t3-8, t4-8 ; start of picking up from MT.
G9: T9, t2-9, t3-9 ; start of placing on STA.
G10: T10, t2-10, t3-10 ; start of picking up from STA.
G11: T11, t2-11, t5-11 ; start of putting on CV2.
G12: T12, t2-12, t5-12 ; start of carrying out from the cell.

Based on the coordination information, the system coordination program is defined, such that for a set of global transitions if all the transitions are enabled, then these transitions are eventually fired simultaneously. The coordination information is implemented on associated robots by pairs of communication actions through external permissive gate arcs, as shown in Figure 7.

The gate arcs are arranged such that the dynamic behavior of the distributed net model with normal transitions is the same as that of the original centralized, single net model with the shared transitions. The distributed control architecture were implemented with real-time Ethernet interfaces supporting the extended TCP/IP protocol and high-speed dual-port memory running on the real-time OS, as shown in Figure 8. In multi-robot systems, based on the task specification subtasks are assigned to the robots by the system controller that dynamically decides which robot is in the best condition to execute the subtask. The actual coordination between the system controller and robot controllers is performed through communication of events and status based on coordination information of global transitions, as summarized for robotic manufacturing systems (Yasuda, 2010).

IMPLEMENTATION OF PROTOTYPE CONTROL SYSTEM

The software for automatic generation of Petri nets and their execution code is built around a database which contains all the information required to specify the control program. The database contains logical

Figure 8. Distributed control architecture for multi-robot coordination

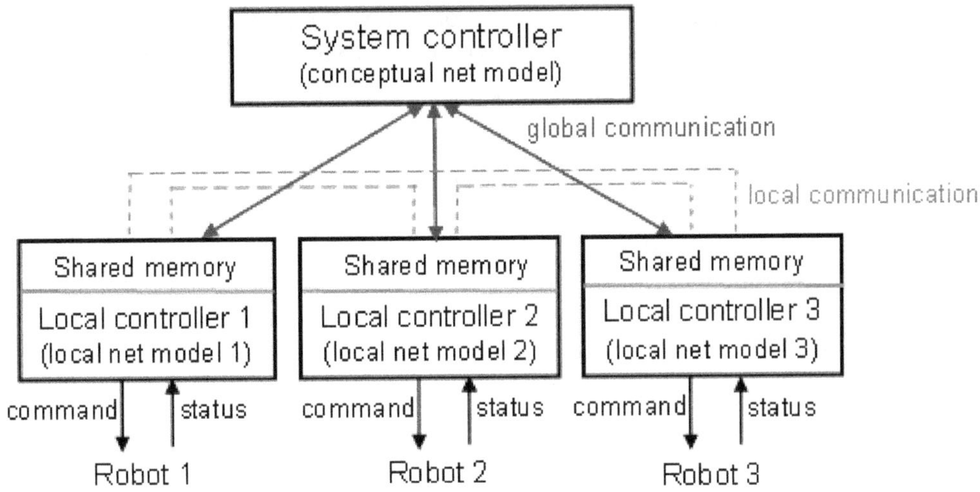

conditions for initiating each operation and the resulting states following the operation with pointers to the related resource states, corresponding to the input and output transitions. The actuators and commands involved in the execution of each operation, and the sensors and sensor states required to determine the resource state are also specified. In the current implementation, binary actuator commands, commands to robot controllers to initiate a program cycle, and binary sensors are supported. Modifying the control logic in the database is simply a matter of changing the relevant data types, which is much easier than changing a conventional computer program. The software that generates control code can eliminate the need for manually encoding the discrete control logic, allowing the system designer to specify and modify the control logic directly via a high-level functional representation of the system. For each operation, the appropriate code is generated to test the specified resource state conditions, conditioned on the states, enable the desired actuator outputs, and update the resultant resource states based on the associated sensor states specified in the database.

Because places and transitions in a Petri net are defined based on local states of the system, distributed implementation of control systems can be naturally realized. For decentralized implementation of the control of the net (Montano et al., 2000; Moreno et al., 2008), the overall detailed net is decomposed into several subnets, each of which is implemented as a separate sequential process, concurrent with the others, respecting the actual concurrency of the system, through behavioral analysis with respect to local access to the input/output modules and machine's local control criteria. Each process integrates both control and operational parts of the subnet and one local coordinator is in charge of executing the corresponding subnet in a centralized form. In a process, transitions and places are classified into private, external, or shared. Every input and output arc of a private place are connected to transitions belonging to the process. A shared place is connected to transitions belonging to different processes. A shared transition models a synchronous communication. A shared place models an asynchronous communication. The net decomposition leads to a set of subnets, a set of shared places and a set of shared transitions, with the appearance of shared places and shared transitions amongst subnets. A shared place is implemented as a protected object with synchronized methods to mark and unmark the place. The communication between distributed subnets is made in asynchronous form using shared places and in a synchronous form using shared transitions to achieve decentralized coordination without the system controller at normal operating situations. The system coordinator is implemented in the system controller to detect deadlocks or abnormal system failure related to the whole net and issue commands for appropriate recovery actions. These communications are also performed using shared places. Each shared place is implemented as a protected object to manage the mark (sending) and unmark (receiving) actions from different coordinators. When a shared place has several input or output transitions distributed in different subnets, the conflict is solved by an arbiter, which is also represented as a detailed net of the shared place, such that the transitions are fired in priority order, considering their operational requirements. In case of synchronous communication, the first task to reach the rendez-vous is blocked until the second one is ready. In case of using asynchronous communication, the sender is running freely and posts messages (or objects) on its output ports. The receiver either takes the message if a new one is available or blocked until the next message is posted. Symmetrically, the receiver runs freely and the sender is blocked to the next request except if a new one was posted since the last receiving. A shared place is implemented in one of the associated controllers. Primitive instructions of synchronous communication are available in ADA code. In order to implement the shared transitions in C++ or Java, a synchronous communication is transformed into two asynchronous communications using shared places. In order to guarantee a bounded time response, real-time communication interface is necessary in the nodes of distributed local coordinators using methods with the appropriate real-time priority. In the first implementation, all the local coordinators in a control computer are implemented as threads on the multithreading kernel, and run simultaneously with the same priority. In the implementation over a real-time communication network such as an industrial Ethernet, where each local coordinator is implemented in a single control computer, threads in charge of communication and conflict resolution among local coordinators were executed at higher priority.

At the task design stage, based on a unified modeling language (UML) diagram, the subtasks sequence net and required resource information for each subtask are integrated to construct compact net models for system control. Resource information includes properties of shared or non-shared, resource release method or condition, and subtask execution priorities as a dispatching rule. Then, the detailed net model for each subtask is developed in a bottom-up way. Although the above design procedure is rather structured and efficient, the construction of a net model for a complex robotic system would be tedious and its graphical representation would lose significance due to the overwhelming number of places and transitions. Moreover, it is difficult to prove a given net model possesses desirable qualitative properties such as liveness and safeness. To construct compact graphs that include information flow, high-level Petri nets, Predicate/Transition nets and their improvement called colored Petri nets, were adopted, which are generally oriented to extensions of ordinary Place/Transition nets to construct more concise graphical models, while they weaken analysis possibilities. The extended Petri net simulator is implemented using a frame-based language, designed as a tool for direct implementation of net models. Each transition is represented with an individual frame. The slots in the frame represent structural and causal relations and attributes. Structural relations link up the frame with mode general one. Causal relations represent input and output arcs. An arc expression, a formal sum of variables with constant coefficients, can be attached to each arc. The expressions of the output arcs are results of attached functions. The guard slot contains predicates representing dependencies among input variables. Each place is also represented with an individual frame. The inverse causal relation defines associated enabled transitions with priorities. The token slot specifies a current marking of the place. Any transition can be associated with an interpretation which specifies the followings with different frames: invocation of lower-level nets, events that must come true before transition firing, commands that are sent to low-level controllers, events that are responses to the above commands, and error handling functions evaluated when an event has not occurred in a given time. Based on the colored Petri nets the firing rule of an enabled transition is modified due to the control interpretation as follows:

1. Remove tokens from input places according to expressions attached to the input arcs and the guard.
2. Verify all pre-events, and if all of them are not verified wait for missing events.
3. Invoke all subnets specified.
4. Send all commands.
5. Verify all post-events and subnets exit, if all of them do not come in time activate the error handling functions.
6. Add tokens to output places according to the output arcs.

The proposed distributed autonomous control framework has been implemented and tested to control different robotic systems. A Petri net simulator for robotic task execution has been developed with a set of tools for designing and debugging tasks. Net models with the extended Petri net are declared using a simple script based language. The Petri net simulator can be implemented on a centralized processor, but emulating a decentralized computation procedure. Petri net models are executed reacting to external events occurring in the environment and to the internal state of the robot. During the real-time control execution, the simulator makes use of a set of functions that can access the state of the robot and return truth values about relevant properties for the execution of the robotic task. Figure 9 shows the Petri net based distributed autonomous control architecture using web service (Komoda, 2006) for cooperative multiple robots (Yasuda, 2011). One robot is the temporary leader and connected to the web service client, which sends a set of commands on the service request specified by the user. The web service server program receives the commands and executes the specified service based on the detailed subnets of activities. Cooperative behavior control between two robots is also established through the message exchange of execution request and notification response, exchanging the roles of the web service client and server.

The controller's server accepts a primitive task execution request, analyzes messages from the client based on the information that the server contains in its task description file, and then launches the monitoring process that monitors primitive task execution to detect the occurrence of events before the primitive task is run. The monitoring process waits until an enabled event is detected and after that it collects the primitive task execution status data. Afterwards the end of primitive task execution is signaled to the client. Primitive task execution is performed by communication between the embedded control driver and the local net model through the software interface and shared memory. For trajectory tracking, the trajectory parameters are input from the net model to the interface, where the trajectory is calculated, filled with the sampled points in the shared memory. The control driver reads these points from the shared memory and uses them as references for the actuators at the sampling rate as a high priority task under a real time OS on a single or multiprocessors target, where multi-tasks programs permit modular programming, software reusability, multi-rate controller design, and optimized computing resources.

Figure 9. Distributed task execution using the web service

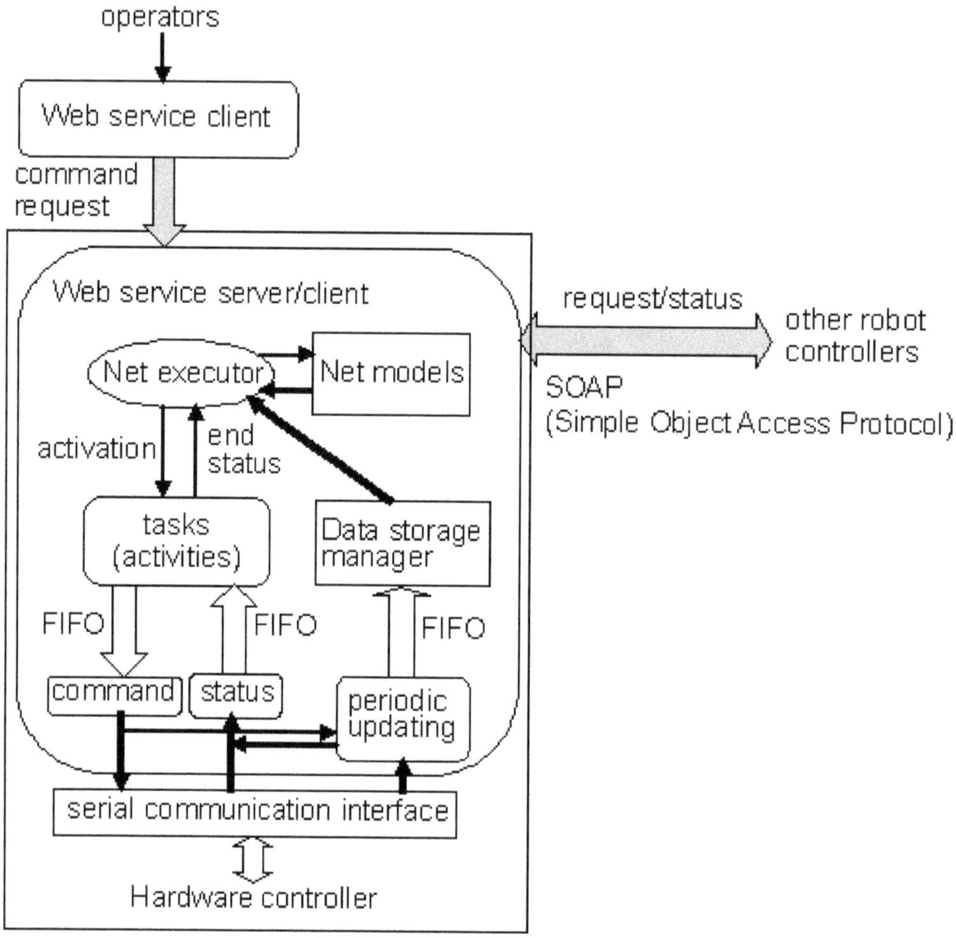

CONCLUSION

A systematic procedure of design and implementation using extended Petri nets to build a distributed autonomous control architecture for multi-robot systems has been proposed. Autonomous cooperation based on synchronous communication has been adopted as a basic framework for the realization of distributed coordination. Cooperation with other robots and requests of shared resource are treated similarly using communication with external gate arcs. By introduction of global transitions based on hierarchical Petri nets modeling, the distributed autonomous coordination mechanism is implemented on an onboard controller. Some fundamental cooperative tasks including synchronized grabbing actions have been implemented and the applicability of the coordination framework for both single-robot and multi-robot tasks has been confirmed. The proposed mechanism enables the programmer to describe coordinated activities easily and to implement such a description in a simple but efficient way on a general PC, microcomputers or PLCs using conventional programming languages.

System coordination through communication via an internet may be deteriorated due to non-real-time characteristics of Ethernet caused by its binary back-off algorithm. The TCP/IP protocol should be improved by extending in a way of limiting the communication traffic or reducing transmission collisions, thus reducing the communication latency (Cho et al., 1998; Santos et al., 2004). Net based controllers are used to coordinate the activities executed by the different concerned subsystems by determining the timing to initiate the particular parts of the cooperative actions. The actual control algorithms are implemented in the actual hardware controller. This separation of control strategies leads to the simpler programming.

Future research directions include the implementation of remote control system on internet environments, for multi-task planning and execution under resource constraints, especially aiming to apply in the hazardous fields such as nuclear power plants. Besides locomotive functions, robust execution of manipulation tasks for various objects should be realized consistently using multiple robots. In the decentralized implementation, deadlocks owing to access to shared resources, should be detected and avoided without any global system coordinator.

REFERENCES

Brooks, R. A. (1987). A hardware retargetable distributed layered architecture for mobile robot control. In *Proceedings of the IEEE International Conference on Robotics and Automation*, (pp. 106-110). IEEE. doi:10.1109/ROBOT.1987.1088016

Caccia, M., Coletta, P., Bruzzone, G., & Veruggio, G. (2005). Execution control of robotic tasks: A Petri Net based approach. *Control Engineering Practice*, *13*(8), 959–971. doi:10.1016/j.conengprac.2004.10.005

Cho, Y.-J., Park, J.-M., Park, J., Oh, S.-R., & Lee, C. W. (1998). A control architecture to achieve manipulation task goals for a humanoid robot. In *Proceedings of the 1998 IEEE International Conference on Robotics and Automation*, (vol. 1, pp. 206-212). IEEE.

Desel, J., & Reisig, W. (1998). Place/transition Petri nets. In W. Reisig & G. Rosenberg (Eds.), Lectures on Petri nets/Basic models (LNCS), (vol. 1491, pp. 122–173). Berlin, Germany: Springer. doi:10.1007/3-540-65306-6_15

Doty, K. L., & Van Aken, R. E. (1993). Swarm robot material handling paradigm for a manufacturing workcell. In *Proceedings of the IEEE International Conference on Robotics and Automation*, (vol. 1, pp. 778-782). IEEE. doi:10.1109/ROBOT.1993.292072

Duffee, N. A., & Prabhu, V. V. (1996). Heterarchical control of highly distributed manufacturing systems. *International Journal of Computer Integrated Manufacturing*, *9*(4), 270–281. doi:10.1080/095119296131562

Duffie, N. (1990). Synthesis of heterarchical manufacturing systems. *Computers in Industry*, *14*(1-3), 167–174. doi:10.1016/0166-3615(90)90118-9

Ejiri, A., Watanabe, I., Okabayashi, K., Hashima, M., Tatewaki, M., Aoki, T., & Maruyama, T. (1994). Satellite berthing experiment with a two-armed space robot. In *Proceedings of the IEEE International Conference on Robotics and Automation*, (pp. 3480-3487). IEEE. doi:10.1109/ROBOT.1994.351035

Fatikow, S., & Magnussen, B. (1994). Intelligent robot hand control system using a tailorable parallel computer concept. In *Proceedings of the IEEE International Conference on Robotics and Automation*, (pp. 216-223). IEEE. doi:10.1109/IROS.1994.407388

Fleury, S., Herrb, M., & Chatila, R. (1994). Design of a modular architecture for autonomous robot. In *Proceedings of the IEEE International Conference on Robotics and Automation*, (pp. 3508-3513). IEEE. doi:10.1109/ROBOT.1994.351031

Hasegawa, K. (1996). Modeling, control and deadlock avoidance of flexible manufacturing systems. In *Proceedings of 11th Brazilian Congress of Automatics*, (pp. 37-51). Academic Press.

Hörmann, A., Meier, W., & Schloen, J. (1991). A control architecture for an advanced fault tolerant robot system. *Robotics and Autonomous Systems*, *7*(2-3), 211–225. doi:10.1016/0921-8890(91)90043-K

Komoda, N. (2006). Service oriented architecture (SOA) in industrial systems. In *Proceedings of the 2006 IEEE International Conference on Industrial Informatics*, (pp. 1-5). IEEE. doi:10.1109/INDIN.2006.275708

Lazinica, A., & Katalinic, B. (2005). Self-organizing multi-robot assembly system. In *Proceedings of the 36th International Symposium on Robotics*, (pp. 42-47). Academic Press.

Lepuschitz, W., Zoitl, A., Vallee, M., & Merdan, M. (2011). Toward self-reconfiguration of manufacturing systems using automation agents. *IEEE Transactions on Systems, Man and Cybernetics. Part C, Applications and Reviews*, *41*(1), 52–69. doi:10.1109/TSMCC.2010.2059012

Liscano, R., Manz, A., Stuck, E. R., Fayek, R. E., & Tigli, J. Y. (1995). Using a blackboard to integrate multiple activities and achieve strategic reasoning for mobile-robot navigation. *IEEE Expert*, *10*(2), 24–36. doi:10.1109/64.395354

Montano, L., García, F. J., & Villarroel, J. L. (2000). Using the time Petri net formalism for specification, validation, and code generation in robot-control applications. *The International Journal of Robotics Research*, *19*(1), 59–76. doi:10.1177/02783640022066743

Moreno, R. P., Tordioli, D., & Salcedo, J. L. V. (2008). Distributed implementation of discrete event control systems based on Petri nets. In *Proceedings of the 2008 IEEE International Symposium on Industrial Symposium on Industrial Electronics*, (pp. 1738-1745). IEEE. doi:10.1109/ISIE.2008.4676963

Murata, T. (1989). Petri Nets: Properties, analysis and applications. *Proceedings of the IEEE*, *77*(4), 541–580. doi:10.1109/5.24143

Osswald, D., Martin, J., Burghard, C., Mikut, R., Woern, H., & Bretthauer, G. (2003). Integrating a robot hand into the control system of a humanoid robot. In *Proceedings of the 2003 International Conference on Humanoid Robots*. Academic Press.

Prabhu, V., & Duffie, N. (1995). Modelling and analysis of non-linear dynamics in autonomous heterarchical manufacturing system control. *CIRP Annals*, *44*(1), 425–428. doi:10.1016/S0007-8506(07)62356-7

Santos, F., Almeida, L., Pedreiras, P., Lopes, L. S., & Facchinnetti, T. (2004). An adaptive TDMA protocol for soft real-time wireless communication among mobile computing agents. In *Proceedings of Workshop on Architectures for Cooperative Embedded Real-time Systems (satellite of RTSS2004). RTSS.*

Sousa, J. B., & Pereira, F. L. (1996). A general control architecture for multiple vehicles. In *Proceedings of the IEEE International Conference on Robotics and Automation*, (pp. 692-697). IEEE. doi:10.1109/ROBOT.1996.503855

Stilwell, D. J., & Bay, J. S. (1993). Toward the development of a material transport system using swarms of ant-like robots. In *Proceedings of the IEEE International Conference on Robotics and Automation*, (pp. 766-771). IEEE. doi:10.1109/ROBOT.1993.292070

Van Brussel, H., Wyns, J., Valckenaers, P., Bongaerts, L., & Peeters, P. (1998). Reference architecture for holonic manufacturing systems: PROSA. *Computers in Industry*, *37*(3), 255–274. doi:10.1016/S0166-3615(98)00102-X

Yasuda, G. (1971). *A Fundamental Study on Reproduction - Application of Graph Theory to Self-Reproducing Processes*. (Master thesis). University of Tokyo, Tokyo, Japan.

Yasuda, G. (2010). Petri net model based hierarchical and distributed control of robotic manufacturing systems. In *Proceedings of 2010 ISFA International Symposium on Flexible Automation*. Tokyo, Japan: ISFA.

Yasuda, G. (2011). Petri net model based design and control of robotic manufacturing cells. In V. Modrak & R. Sudhakara Pandian (Eds.), *Operations Management Research and Cellular Manufacturing Systems: Innovative Methods and Approaches* (pp. 385–400). Hershey, PA: IGI Global. doi:10.4018/978-1-61350-047-7.ch018

Yasuda, G., & Tachibana, K. (1991). A parallel processing control approach to the design of autonomous robotic manufacturing systems, In *Proceedings of the XIth International Conference on Production Research*, (pp.445-449). Academic Press.

Zheng, Z., Lecocq, H., & Jordant, R. (1993). Adaptive manipulator control: A parallel implementation on a network of microprocessors. In *Proceedings of the IEEE International Conference on Robotics and Automation*, (vol. 1, pp. 878-882). IEEE. doi:10.1109/ROBOT.1993.292087

Zhou, M. C., DiCesare, F., & Desrochers, A. A. (1992). A hybrid methodology for synthesis of Petri net models for manufacturing systems. *IEEE Transactions on Robotics and Automation*, *8*(3), 350–361. doi:10.1109/70.143353

ADDITIONAL READING

Floreano, D., & Mattiussi, C. (2008). *Bio-inspired Artificial Intelligence: Theories, Methods, Technologies*. Cambridge, MA: The MIT Press.

Moseman, H., & Whal, F. M. (2001). Automatic decomposition of planned assembly sequences into skill primitives. *IEEE Transactions on Robotics and Automation*, *17*(5), 709–718. doi:10.1109/70.964670

Yasuda, G. (1997). Intelligent Manufacturing and Engineering. In J. Liebowitz (Ed.), *The Handbook of Applied Expert Systems* (pp. 22.1–22.14). Boca Raton, FL: CRC Press.

Yasuda, G. (2012). Modeling and control of production task flows using high-level activity-based Petri nets. In V. Modrák & P. Semančo (Eds.), *Design and Management of Lean Production Systems* (pp. 115–128). Hershey, PA: Business Science Reference, an Imprint of IGI Global.

Yasuda, G., & Tachibana, K. (1986). Development of a microcomputer based CAD/CAM model for industrial robot control. In S. Moriguti, S. Ohtsuki, & T. Furugori (Eds.), *Microcomputers in Secondary Education* (pp. 95–100). Amsterdam: North-Holland.

Zurawski, R., & Zhou, M. C. (1994). Petri nets and industrial applications: A tutorial. *IEEE Transactions on Industrial Electronics*, *41*(6), 567–583. doi:10.1109/41.334574

KEY TERMS AND DEFINITIONS

Centralized System: A system where decisions for the system goal are made in a central mechanism and transmitted to executive components.

Coordination: Decision making and their execution through negotiation processes among the executive components to fulfill a global goal in a distributed system.

Discrete Event Net Modeling: Modeling using a graphical net to capture the operational states and the events or transitions between them and plan a nominal operation sequence and recovery strategies due to uncertainties and tolerances.

Hierarchical Decomposition: Analysis of complex systems by replacing some parts of a model with simple components such that substitution nodes with their surrounding arcs are behaviorally equivalent to the related sub-models. In a discrete event net, either a place or a transition can be a substitution node by associating a sub-model to it.

Hierarchical Modeling: System modeling by describing a set of sub-models which all contribute to a much larger model by interacting with each other in a well-defined way.

Hierarchical Planning: Planning formed and executed by several subsystems organized according to a layered structure for the system goal.

Liveness: Possibility of any event to occur from any reachable state in a discrete event system. In a discrete event net, as deadlock arises when no transition can be longer enabled, liveness implies deadlock freedom.

Manufacturing Planning: Determination of a sequence of operations to be done in order to manufacture a product, such as machining or assembling.

Task Execution: Execution of robotic operations to carry out a manufacturing plan using code generation, involving concurrency, synchronization, resource sharing and real-time requirements.

Task Planning: Translation of a manufacturing plan into the robot operations that may allow the successful execution of the plan, involving the planning of sensory operations, gross-motion planning with collision avoidance, and fine-motion planning with contact motions.

Chapter 16
Robust Unknown Input Observer-Based Fast Adaptive Fault Estimation: Application to Mobile Robot

Olfa Hrizi
Gabes University, Tunisia

Boumedyen Boussaid
Gabes University, Tunisia

Ahmed Zouinkhi
Gabes University, Tunisia

M. Naceur Abdelkrim
Gabes University, Tunisia

ABSTRACT

This chapter studies the problem of fault estimation using a fast adaptive fault diagnosis observer. Note that the advance of observer-based fault diagnosis is outlined and the idea of fault class estimation is introduced and studied. A new form of the estimator bloc considered for this purpose is an Unknown Input Observer (UIO). This observer is designed for an unknown input and fault free system, which is obtained by coordinate transformations of original systems with unknown inputs (disturbance) and faults. Stability of the adaptive estimation is provided by a Lyapunov function ending with solving the Linear Matrix Inequalities (LMI). Due to technological advances in the field of electronic devices, the family of robots is of particular interest. To overcome the drawback of robots' model responses when including a fault, a robust observer is adopted for a Pioneer robot to improve the fault estimation and thereafter to repair its trajectory.

DOI: 10.4018/978-1-4666-7248-2.ch016

INTRODUCTION

In the last several years, considerable attention has been focused on the emerging field of robotics. Therefore, many successful robotic manipulator designs have been introduced thanks to their good terrain adaptability. Many researchers are proposed different ways of control robot design, classified types and ensure performance of robots (Shigeo & Akio, 1990).

A large effort has been devoted by the scientific community especially to the field of mobile robot systems. In particular wheeled robots will be expected to provide many convenient and user friendly transport solutions for both people and objects. The importance of the wheeled mobile robots has long been recognized by the robotics research community, as shown by the numerous robotic competitions and research projects run worldwide in the last decades in the RoboCup federation site at 2008.

The class of unicycle type (mobile) robots, i.e. robots having some forward speed but zero instantaneous lateral motion, is frequently selected for designing and modeling robots. For example many of the robotic competition teams of the last decade selected those robots due to their simplicity and good maneuverability, allowing for example to follow complex trajectories (Pedro et al, Luis et al, 1998). At the same time research was conducted on controllability, feedback linearization and -stabilization (Canudas de Wit, 1997).

However, the right behavior and high performance of robots can be threatened. So, it must be safe against all possible accidents or external and internal failures. In order to avoid this problem, the reliability can be achieved by fault tolerant control (FTC) (Boumedyen et al., 2013) which relies on early detection of faults using fault detection and isolation (FDI) block (Wang & Daley, 1996). Several researches focused on this procedure (FDI) which consists in detecting and isolating faults in a physical system by monitoring its inputs and outputs (Chiang et al., 2001; Boumedyen et al., 2011; Chen & Patton, 2012). A typical system for fault detection and isolation is made of three parts (Ding, 2008): fault detection indicates that there's a mistake in the operating system, i.e., the occurrence of a fault and the time of the fault occurrence (Wu et al., 2000); secondly, fault isolation determines the location and the type of the fault and finally, fault identification determines the size of the fault. Diverse FDI methods have been reported in the literature such as: generating redundancy in the case of physical redundancy between sensors (Frank, 1990) or parity space formulation (Gertler, 1995, 1998; Patton & Chen, 1991; Isermann, 2006). Other approaches based on a Kalman filter have been treated (Doraiswami & Cheded, 2013). After FDI bloc, Fault tolerant control (FTC) systems are needed in order to maintain the performance objectives, or if that turns out to be impossible, to assign achievable objectives so as to avoid catastrophic failures (Jiang & Zhao, 2000; Philippe et al., 2012; Zhang & Jiang, 2000). An introduction given for this term is studied in Jiang (2005), Lunze and Richter (2008). In general, fault tolerance can be achieved in two ways (Youmin & Jin, 2008): Passive FTC deals with a presumed set of system component failures based on the actuator redundancy at the controller design stage. The resulting controller usually has a fixed structure and parameters. However, the main inconvenient of a passive FTCS is that as the number of potential failures and the degree of system redundancy increase. So, controller design could become very complex, and the performance of the resulting controller could become significantly conservative. Moreover, if an unanticipated failure occurs, passive FTC cannot ensure system stability and cannot reach again the nominal performance of the system. Controllers switching underlines the fact that many faulty system representations had to be identified so as to synthesize off-line pre-computed and stabilized controllers. Then, an active FTCS is characterized by an on-line FDI process and a control reconfiguration mechanism (Hsieh, 2002). According to the FDI module, a control reconfiguration mechanism is designed in order to take into account the possibility of fault occurrence (Theilliol et al., 2002). Advanced and sophisticated controllers have been developed with fault accommodation and tolerance capabilities, in order to meet pre-fault reliability and performance requirements as proposed by Gao and Antsaklis (1991) for model matching approaches or by Gao and Antsaklis (1992) to track a trajectory, but also with degraded ones as suggested by Jiang and Zhang (2006).

Since the process used in this work is from the family of mobile robot, it is better to refer to some researches that are interested by both diagnostic term and robotic theme (Duan et al, Steinbauer & Wotawa, 2005; Ji et al, 2003).

On the other hand, if the process's models are precise, the problem of fault detection can be solved by observer bloc or residual generated computed from the inputs and the outputs of the process. Among the approaches of fault diagnosis, some researchers are interested by the adaptive fault diagnosis observer approach (Zhang et al., 2008; Wang et al., 1997). This work focus on the unknown input observer (UIO) knowing that observer design for estimating the state of a system subject to unknown inputs has received considerable attention in the past (Edwards & Pin, 2004; Menighed et al., 2001). However, very little research has been carried out on estimating the unknown inputs. In Park and Stein (1988), unknown inputs are estimated by differentiating the output measurement. In Xiong and Saif (2003), the problem of unknown, constant or slowly varying input estimation using a proportional integral observer (PIO) is discussed.

The contribution in this work is to propose a robust unknown input observer (disturbance decoupling) which is also a fast adaptive fault estimator to enhance the rapidity and certainty of fault estimation. It's in fact a following of a previous work developed in Hrizi et al. (2013).

The present chapter begins with an introduction detailing the special terms used for this work (robotic field, FDI, FTC, UIO). The outline is organized as follows: Section 1 describes the linear systems with actuator fault and the background of the standard adaptive observer for fault estimation. In section 2, a robust UIO based on FAFE algorithm is proposed. Section 3 is reserved to the simulation results of a unicycle robots model followed by some concluding remarks in the last section.

RELATED WORK

The idea of the present work involves different principal terms such as fault estimation which is discussed in some recent searches:

In Park and Lee (2004), a simple and systematic design approach for process fault detection, isolation and reconfigurable control system for linear systems subject to unknown inputs has been proposed. In this work, a numerical method including the design of a reduced order observer and algebric fault estimation is used and verified by an aircraft in the vertical plane. Otherwise, in this chapter, a mobile robot is introduced to verify the novel algorithm of FAFE.

Then, in Zhang et al. (2008), a model based fault estimation scheme was analyzed and a general framework for fast fault estimation based on adaptive observer was developed. However, the application results to a VTOL system are not really efficient to complete the step of fault accommodation strategy based fault-tolerant controller due to the weakness of observer used and thereafter the performances of robot will not be improved. So, a robust adaptive observe is necessary to ameliorate the system's performances.

Another more recent work was interested also by the fault diagnosis written by Wang (2010) where the author studies a fault class isolation as a newer concept in contrast with fault detection and isolation.

Together with this research topic "fault estimation", the present chapter details also the way used to estimate i, e the unknown input observer. It has huge interest for several researches.

In an old work Bhattacharyya (1978), the author tried to present a constructive solution to the problem of designing a Luenberger observer to evaluate a given set of linear functions of the state of a linear system subject to unknown or disturbance inputs. Hence, the idea of Unknown Input Observer. Then, it will be more developed like in Darouach (2000) who presents straightforward derivations of the functional observers design for linear time-invariant multivariable systems. But, this work was limited by necessary conditions for the existence and stability of the observer given.

Furthermore, the work more related to this chapter is developed by Zhang et al. (2008) which study the problem of fault estimation using adaptive fault diagnosis observer. But, that paper developed a proportional and integral estimator. In fact, it explained the standard fault adaptive estimation algorithm.

1. BACKGROUND

This section introduces the preliminary materials and background of the work. It will present two parts: the first one is a reminder of the classic fault estimation. Then, the second contains the development of the structure of a robust unknown input observer based on the novel fast adaptive fault estimation algorithm.

A. Plant Description

Consider the Multiple Input-Multiple Output (MIMO) linear system with actuator fault:

$$\dot{x}(t) = Ax(t) + Bu(t) + F_x f(t) + D_x d(t) \tag{1}$$

$$y(t) = Cx(t)$$

where $x(t) \in \mathbb{R}^n$ is the state vector, $u(t) \in \mathbb{R}^m$ is the input vector and $y(t) \in \mathbb{R}^p$ is the output vector. A, B and C are known parameter matrices of dimensions $n \times n$, $n \times m$ and $p \times n$, respectively.

Many researchers studied the different types of failures, fault and disturbance like in Powell, 1992, also in Basseville and Nikiforov, 1993.

Taxonomy of faults is developed in Avizienis et al. (2004) which classify faults as:

- **Development Faults:** Include all fault classes occurring during development.
- **Physical Faults:** Include all fault classes that affect hardware.
- **Interaction Faults:** Include all external faults, means, faults originate outside the system boundary and propagate errors into the system by interaction or interference.

Then, according to Isermann et al. (2005), the faults can be differentiated by their evolutions, natures and types:

- **Evolution:** The fault can be with different form and behavior (Szigeti, 1991). In fact, it could be an actual fault or appearing, with low or high amplitude, be sudden or happen gradually rather as slow drifts.
- **Nature:** The faults can be classified into multiplicative fault corresponding to the correlation of the output signal, as well as changes in the system dynamics. Or additive fault modeled as additive terms in the system's model affecting state or output.
- **Type:** Three types of fault may affect different part of system such as the actuator, sensor or process fault.

In this work, an additive actuator fault $f\left(t\right) \in \mathbb{R}^{r}$ will be introduced where F_{x} is a known matrix with dimension $n \times r$ and $d(t) \in \mathbb{R}^{q}$ is the unknown input (or disturbance) vector where D_{x} is the known disturbance matrix with dimension $n \times q$. Assuming that the pair $\left(A, C\right)$ is observable.

B. Standard Adaptive Fault Estimation Design

The problem of state estimation is important for reconfigurable control law or the development of a diagnostic strategy. Observability refers to the study of the conditions under which it is possible to uniquely infer the state of a dynamical system from measurements of its output.

Fault estimation is usually performed to accomplish one or more of the following tasks: fault detection (or monitoring), to indicate the fault occurrence; fault isolation, to determine the exact location of the fault and fault identification, to estimate the fault magnitude. The design of an observer-based fault diagnosis system is normally contains three basic steps:

- **Residual Generation:** This is the most necessary element of fault estimation. It gives as result the estimating process output by using a Luenberger observer (Lunberger, 1971) or unknown input observer. The kalman filter is also used for estimation in a stochastic setting. The estimation error (or innovation in the stochastic case) is defined as the residual. Observer based fault detection makes use of the disturbance decoupling principle, in which the residual is computed assuming the decoupling of the effects of faults on different inputs. For the purpose of fault isolation, it is also assumed that the effect of a fault is decoupled from the effects of other faults. A way to achieve this is by using the so-called unknown input observer (UIO). It will be developed in the next section.
- **Residual Evaluation:** The residual is examined in terms of the likelihood of a fault, and a logical decision-making process is then applied aiming at to decide if the fault has occurred and avoid wrong decisions, such as false alarm and fault ignored.
- **Fault Analysis:** This step is to determine the magnitude and the behavior of fault. It is crucial for the control reconfiguration block.

Assuming the following residual generator named Luenberger observer:

$$\dot{\hat{x}}\left(t\right) = A\hat{x}\left(t\right) + Bu\left(t\right) + L\left(y\left(t\right) - \hat{y}\left(t\right)\right)$$

$$\hat{y}(t) = C\hat{x}(t)$$

The residue is $r_y(t) = e_y(t) = y(t)\_\hat{y}(t)$ where L is a observer gain calculated to assure the convergence of estimate to state $x(t)$.

The state estimation error is defined by

$$e_x(t) = \hat{x}(t) - x(t)$$

Then the state error dynamic is described as follow:

$$\dot{e}_x(t) = \dot{\hat{x}}(t) - \dot{x}(t) = (A - LC)e_x(t) - F_x f(t) - D_x d(t)$$

If the initial conditions are null, Laplace transform gives:

$$se_x(s) = (A - LC)e_x(s) - F_x f(s) - D_x d(s)$$

$$e_x(s) = [sI - (A - LC)]^{-1} F_x f(s) - [sI - (A - LC)]^{-1} D_x d(s)$$

Note that this error is sensitive to faults $f(s)$ and could therefore be used to generate a fault indicator vector. However, the state is unknown. So, let's define the output estimation error:

$$\begin{aligned} e_y(s) &= \hat{y}(s) - y(s) = Ce_x(s) = C[sI - (A - LC)]^{-1} F_x f(s) - C\left[sI - (A - LC)\right]^{-1} D_x d(s) \\ &= G_f(s) f(s) + G_d(s) d(s) \end{aligned}$$

It can be seen that the estimation error is sensitive to faults f (s), and generally the unknown input d (s). Consider the residual vector

$$r(s) = Q(s)e_y(s) = Q(s)G_f(s)f(s) + Q(s)G_d(s)d(s)$$

The matrix $Q(s)$ is obtained by checking the two relations:

$$Q(s)G_f(s) \neq 0$$

$$Q(s)G_d(s) = 0$$

The residual vector $r(s)$ obtained is sensitive only to faults. But, in practice, disturbances do not have a negligible effect. Hence, the development of indicators insensitive to perturbations defects seems essential. That's why; special attention has been given to unknown input observer (UIO) approaches (Patton, 1997).

The unknown input observer (UIO) is a generalization of the Luenberger observer. His expression adopted in this work is given with a linear transformation as follow:

$$\dot{z}(t) = Mz(t) + Nu(t) + P_y y(t) + EF_x \hat{f}(t) \tag{2}$$

$$\hat{x}(t) = z(t) - L_y y(t)$$

where M, N, P_y, L_y are matrices that will be designed such that the unknown input will be decoupled from other inputs. $z \in \mathbb{R}^{n \times 1}$ is the state of UIO, obtained by the linear transformation $z = Ex$ and $\hat{x}$ is the estimated state vector. The state estimation error is defined by:

$$e_x(t) = \hat{x}(t) - x(t) \tag{3}$$

$$= z(t) - L_y y(t) - x(t)$$

$$= z(t) - (I + L_y C)x(t)$$

Assuming $E = I + L_y C$, so the state estimation error can be written now as:

$$e_x(t) = z(t) - Ex(t) \tag{4}$$

Then the state error dynamic is described as follow:

$$\dot{e}_x(t) = \dot{z}(t) - E\dot{x}(t) \tag{5}$$

$$= Mz(t) + Nu(t) + P_y y(t) + EF_x \hat{f}(t) - \mathrm{E}\mathrm{A}x(t) - EBu(t) - EF_x f(t) - ED_x(t)d(t)$$

$$\text{Or } z(t) = e_x(t) + Ex(t) \tag{6}$$

So substituting (6) in (5) yields

$$\dot{e}_x(t) = Me_x(t) + MEx(t) + Nu(t) + P_y y(t) + EF_x\hat{f}(t) - \mathrm{EA}x(t) - EBu(t) - EF_x f(t) - ED_x(t)d(t)$$

$$= Me_x(t) + (ME - EA + P_y C)x(t) + (N - EB)u(t) + EF_x(\hat{f}(t) - f(t)) - ED_x(t)d(t) \quad (7)$$

Let's define the following errors according to the state $x(t)$, the output $y(t)$ and the fault $f(t)$, respectively:

$$e_x(t) = \hat{x}(t) - x(t) \quad (8)$$

$$e_y(t) = \hat{y}(t) - y(t) \quad (9)$$

$$e_f(t) = \hat{f}(t) - f(t) \quad (10)$$

If M is Hurwitz matrix and the following relationships are true:

$$ME + P_y C = EA \quad (11)$$

$$N = EB \quad (12)$$

$$ED_x = 0 \quad (13)$$

Then, the Equation (7) becomes

$$\dot{e}_x(t) = Me_x(t) + EF_x e_f(t) \quad (14)$$

Theorem 1

If there exist symmetric positive definite matrices P, $Q \epsilon \mathbb{R}^{n \times n}$, $M \epsilon \mathbb{R}^{n \times n}$, and a matrix $F \epsilon \mathbb{R}^{r \times p}$ which check up the following conditions:

$$P^T M + M^T P = -Q \tag{15}$$

$$F_x^T P = FC \tag{16}$$

Then the adaptive fault estimation algorithm:

$$\dot{\hat{f}}\left(t\right) = -\psi F e_y(t) \tag{17}$$

insure the convergence of the error state and the error fault, i.e., $\lim_{t\to\infty} e_x\left(t\right) = 0$, $\lim_{t\to\infty} e_f\left(t\right) = 0$

Note $\o \in \mathbb{R}^{rxr}$ is the learning rate matrix.

Due to (17), the actuator fault estimation obtained is given by (18)

$$\hat{f}\left(t\right) = -\psi F \int_{t_f}^{t} e_y\left(\tau\right) d\tau \tag{18}$$

In fact, this expression of fault estimator includes only the integral term. However, this fault representation is suitable only for constant faults. Therefore, it should improve this common adaptive fault estimation design to include the time varying fault dynamic.

The proof of Theorem 1 can be referred to Bin et al. (2006).

UNKNOWN INPUT ADAPTIVE OBSERVER BASED FAFE DESIGN

A. Description of FAFE Algorithm

To implement this algorithm with successful result, it must follow this assumption and lemma wich is given also to verify the linear matrix inequality LMI.

- **Assumption:** $rank\left(CD_x\right) = q$,
- **Lemma (Bin et al., 2002):** Given a scalar $\mu \succ 0$ and a symmetric positive definite matrix P which justify the following inequality:

$$2x^T y \leq \frac{1}{\mu} x^T P x + \mu y^T P^{-1} y \tag{19}$$

In the previous theorem, there's no field to consider a time varying fault. So, with $\dot{f}(t) \neq 0$ a novel FAFE algorithm is proposed to ameliorate performances of time varying actuator fault estimation: rapidity, stability and accuracy.

Theorem 2

Under Assumption and conditions (11), (12) and (13) verified, given scalars $\sigma,\ \mu,\ \lambda > 0$, if there exist symmetric positive definite matrices $P \in \mathbb{R}^{n\times n}$, $G \in \mathbb{R}^{r\times r}$ and matrices $M \epsilon\, \mathbb{R}^{n\times n}$, $F \epsilon\, \mathbb{R}^{r\times p}$ such that (16) and the following condition hold,

$$\begin{pmatrix} M^T P + PM & 2PEF_x \\ -2\frac{1}{\sigma}F_x^T PM - 2F_x^T P & -2\frac{1}{\sigma}F_x^T PEF_x + \frac{1}{\sigma\mu}G \end{pmatrix} < 0 \tag{20}$$

then the FAFE algorithm

$$\dot{\hat{f}}(t) = -\psi F(\dot{e}_y(t) + \sigma e_y(t)) \tag{21}$$

can realize $e_x(t)$ and $e_f(t)$ uniformly ultimately bounded.

Proof

Lyapunov techniques exist for testing asymptotic stability of the overall system. So, consider the following Lyapunov function:

$$V(t) = e_x^T(t) Pe_x(t) + \frac{1}{\sigma} e_f^T(t)\psi^{-1} e_f(t) \tag{22}$$

Its derivative with respect to time is

$$\dot{V}(t) = \dot{e}_x^T(t) Pe_x(t) + e_x^T(t) P\dot{e}_x(t) + 2\frac{1}{\sigma} e_f^T(t)\psi^{-1}\dot{e}_f(t) \tag{23}$$

$$= (Me_x(t) + EF_x e_f(t))^T Pe_x(t) + e_x^T(t) P\left(Me_x(t) + EF_x e_f(t)\right) + 2\frac{1}{\sigma} e_f^T(t)\psi^{-1}(\dot{\hat{f}}(t) - \dot{f}(t))$$

$$= e_x^T(t)M^T Pe_x(t) + e_f^T(t)F_x^T E^T Pe_x(t) +$$
$$e_x^T(t)PMe_x(t) + e_x^T(t)PEe_f(t) - 2\frac{1}{\sigma}e_f^T(t)F(\dot{e}_y(t) + \sigma e_y(t)) - 2\frac{1}{\sigma}e_f^T(t)\psi^{-1}\dot{f}(t)$$

With (16), one deduced

$$-2\frac{1}{\sigma}e_f^{\mathrm{T}}(\mathrm{t})FC(\dot{e}_x(t) + \sigma \mathrm{e}_x(\mathrm{t}))\text{is equal to } -2\frac{1}{\sigma}e_f^{\mathrm{T}}(\mathrm{t})F_x^T P(\dot{e}_x(t) + \tilde{A}\mathrm{e}_x(\mathrm{t}))$$

Substituting (14) into (23) yields:

$$\dot{V}(t) = e_x^{\mathrm{T}}(\mathrm{t})M^T Pe_x(t) + e_f^{\mathrm{T}}(\mathrm{t})F_x^T E^T Pe_x(t) + e_x^{\mathrm{T}}(\mathrm{t})PMe_x(t) + e_x^{\mathrm{T}}(\mathrm{t})\mathrm{PE}e_f(t)$$
$$-2\frac{1}{\sigma}e_f^{\mathrm{T}}(\mathrm{t})F_x^T P(\dot{e}_x(t) + \sigma \mathrm{e}_x(\mathrm{t})) - 2\frac{1}{\sigma}e_f^{\mathrm{T}}(\mathrm{t})\psi^{-1}\dot{f}(\mathrm{t})$$

$$\dot{V}(t) = e_x^{\mathrm{T}}(\mathrm{t})(M^T P + PM)e_x(t) + 2e_x^{\mathrm{T}}(\mathrm{t})\mathrm{PE}F_x e_f(t) - 2e_f^{\mathrm{T}}(\mathrm{t})F_x^T P\left(\frac{1}{\sigma}M + I\right)e_x(t) -$$
$$2\frac{1}{\sigma}e_f^{\mathrm{T}}(t)F_x^T PEF_x e_f(t) - 2\frac{1}{\sigma}e_f^{\mathrm{T}}(\mathrm{t})\psi^{-1}\dot{f}(t) \tag{24}$$

From Lemma 1, it can acquired that

$$\frac{1}{\sigma}e_f^{\mathrm{T}}(\mathrm{t})\psi^{-1}\dot{f}(\mathrm{t}) \leq \frac{1}{\sigma\mu}e_f^{\mathrm{T}}(\mathrm{t})\mathrm{G}e_f(t) + \frac{\mu}{\sigma}\psi^{-1}\mathrm{G}^{-1}\psi^{-1}\dot{f}(\mathrm{t})$$

$$\leq \frac{1}{\sigma\mu}e_f^{T}(t)Ge_f(t) + \frac{\mu}{\sigma}f_1^2\lambda_{\max}\psi^{-1}G^{-1}\psi^{-1} \tag{25}$$

Substituting (25) into (24), it is obtained hereafter

$$\dot{V}(t) \leq \xi^T(\mathrm{t})\check{z}\,\xi(t) + \delta \tag{26}$$

where

$$\Xi = \begin{bmatrix} M^T P + PM & 2PEF_x \\ -2\frac{1}{\sigma}F_x^T PM - 2F_x^T P & -2\frac{1}{\sigma}F_x^T PEF_x + \frac{1}{\sigma\mu}G \end{bmatrix} \tag{27}$$

and $\xi(t) = \begin{bmatrix} e_x(t) \\ e_f(t) \end{bmatrix}$, $\delta = \frac{\mu}{\sigma} f_1^2 \lambda_{\max} \psi^{-1} G^{-1} \psi^{-1}$

It follows that $\dot{V}(t) < 0$ which means that $(e_x(t),\ e_f(t))$ converges to a small set according to Lyapunov stability theory. Therefore, estimation errors of the fault and the state are uniformly bounded. Where, the thorem2 are demonstrated. It's easy to show now from (21) the fault estimate's expression

$$\hat{f}(t) = -\psi F(e_y(t) + \sigma \int_{t_f}^{t} e_y(\tau) d\tau) \tag{28}$$

Remark

Solving conditions in Theorem 2 needs the LMI toolbox in MATLAB. So, it is easy to solve (20). But, there are some difficulties in solving (16) and (20) simultaneously to extract P, F and G. So, it must transform (16) into the following optimization problem (Martin & Jay, 1998).

Minimize η subject to (20) and

$$\begin{bmatrix} \eta I & F_x^T P - FC \\ (F_x^T P - FC)^T & \eta I \end{bmatrix} > 0 \tag{29}$$

2. APPLICATION TO MOBILE ROBOT

A. Unicycle Robot Model

A unicycle type robot is in general a robot moving in a 2D world, having some forward speed but zero instantaneous lateral motion. In other words, it is a non-holonomic system. Despite the unicycle name, it describes in general carts or cars having usually two parallel driven wheels, one mounted of each side of their center. This model comprises many known differential drive robots and approximates in many situations even the four wheeled cars.

Modeling unicycle type robots comprises studying their kinematics and dynamics, as usual with most of the physical systems. Kinematics modeling describes the trajectories that the mobile robots follow when subject to commanding speeds. The dynamics modeling complements the kinematics by accounting to the commanding forces and intrinsic frictions, actually defining the commanding speeds.

i. Kinematic Model

To develop this model, an example of mobile robot is given: the Pioneer robot. It is a three wheeled robot shown in Figure 1. The front two wheels are actuated independently thereby enabling differential steering. The rear wheel is a passive castor. The kinematics of the Pioneer is described by a simple non-linear model:

Figure 1. The Pioneer: a mobile robot

$$\begin{cases} \dot{x} = \vartheta cos(\theta) \\ \dot{y} = \vartheta sin(\theta) \\ \dot{\theta} = \omega \end{cases} \text{with} \begin{cases} \vartheta = \frac{R}{2}(\omega_D + \omega_G) \\ \omega = \frac{R}{E_a}(\omega_D - \omega_G) \end{cases}$$

which are the translational and rotational velocities respectively, where (x, y) are the coordinates of robot position, θ is the yaw angle of robot, ω_G and ω_D are respectively the rotational speeds on the left and right wheels, R is the radii of wheel and E_a is the axle length. These kinematic quantities are shown in Figure 2. As known, to linearize a mobile robot model, the process must be controllable. In the rest, the work will be continued by a linearized model around a mass point. Let define the linear model by this expression:

$\dot{x} = Ax + Bu$, $x(t) \epsilon\ \Re^n$ and the controllability matrix: $\Gamma = [B AB \cdots A^{n-1}B]$

The system is completely controllable (CC) if it is possible to go from any initial state to any final state. Therefore, the system is CC if and only if $rank\left(G\right) = n$.

ii. Dynamic Model

A dynamic model for unicycle robots allows including physical limitations of the robot or forces, binaries or friction. According to the 2nd Newton's Law, the translational and rotational dynamics of the unicycle robot can be described by

$$M\dot{\vartheta} = F - B_\vartheta \vartheta$$

$$J\dot{\omega} = T - B_\omega \omega$$

Figure 2. The robot kinematics

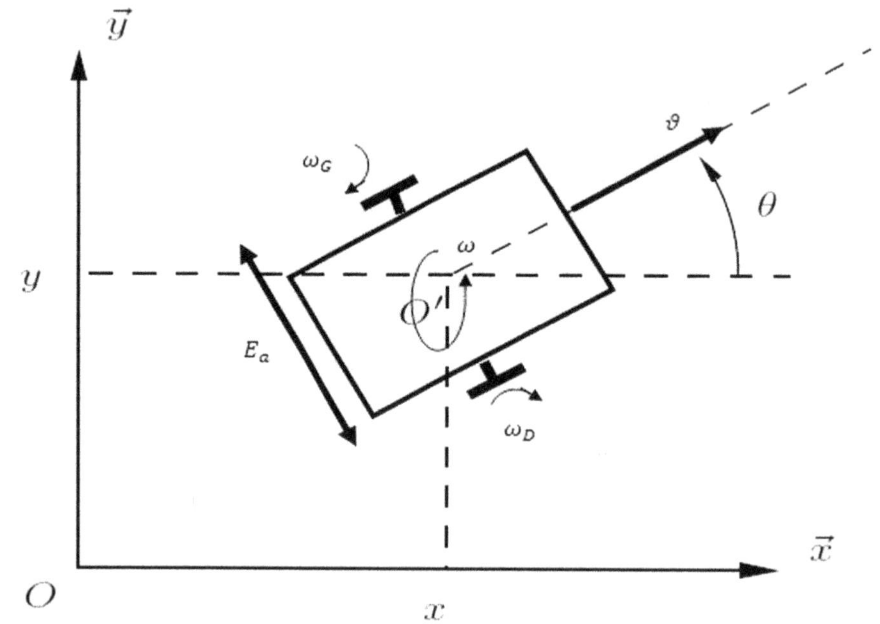

where M is the robot's mass, J is the Inertia moment, F are the forces applied on the system, T is the wheel axis binary, B_ϑ is the translational friction coefficient and B_ω is rotational friction coefficient. Assuming that these values are constant for the typical values of velocities of a moving robot.

Considering the forces acting on the left and right wheels, F_L and F_R, one has that these forces are equivalent to the force F applied in the point r_F, referred to the center of the wheel's axis. More precisely, one has $F = F_R + F_L$ and $T = l(F_R - F_L)$, where l represents half the length of the axis between the wheels.

Similarly, considering the voltages of the left and right wheel motors, one can define also the mean (average) and differential voltages, e_{am} and e_{ad}, and finally obtain the relationship between voltages and forces, $F = K_m e_{am} - K_\vartheta \vartheta$ and $T = K_d e_{ad} - K_\omega \omega$, where K_m, K_ϑ, K_d and K_ω are constants. Concluding, the dynamic model for the unicycle is given as follow:

$$\begin{cases} M\dfrac{\partial \vartheta}{\partial t} = -K_\vartheta \vartheta + K_m e_{am} \\ J\dfrac{\partial \omega}{\partial t} = -K_\omega \omega + K_d e_{ad} \end{cases}$$

iii. Numerical Example

A linearized dynamic model of a unicycle robot in nominal case (Ti-Chung et al. 2001; Magnus, 2002), is given as state space formulation as

$$\dot{x}(t) = Ax(t) + Bu(t) \quad (30)$$
$$y(t) = Cx(t)$$

where $x(t) = \begin{pmatrix} \vartheta \\ \omega \\ \varphi \\ \dot{\varphi} \end{pmatrix}$ is the state vector, $u(t) = \begin{pmatrix} \tau_L \\ \tau_R \end{pmatrix}$ is the input vector and $(A,\ B, C)$ are the system matrices with appropriate dimensions.

$$A = \begin{pmatrix} 0 & 0 & 2.16 & 0 \\ 0 & 0 & 0 & 0 \\ 0 & 0 & 0 & 1 \\ 0 & 0 & 72.49 & 0 \end{pmatrix},\ B^T = \begin{pmatrix} -1.64 & -1.64 & 0.000 & 0 \\ 0.029 & -0.029 & -24.15 & -24.15 \end{pmatrix},\ C = \begin{pmatrix} 1 & 0 & 0 & 0 \\ 0 & 1 & 0 & 0 \\ 0 & 0 & 1 & 0 \\ 0 & 0 & 0 & 1 \end{pmatrix}$$

B. Faulty Case

To develop this work, an actuator fault will occur in the input channel and an unknown disturbance will be taken. So, the robot model is written in faulty case as follow:

$$\dot{x}(t) = Ax(t) + Bu(t) + F_x f(t) + D_x d(t) \quad (31)$$
$$y(t) = C_1 x(t)$$

where $F_x = \begin{pmatrix} 1 & 0 \\ 0.5 & 0 \\ 2 & 0 \\ 1 & 0 \end{pmatrix},\ D_x = \begin{pmatrix} 0.05 & 0 \\ 0.2 & 0.3 \\ 2 & 0 \\ 0 & 1 \end{pmatrix}, C_1 = C$

It is assumed that the pair (A, C_1) is observable and verifying that $rank(C_1 D_x) = q = 2$

Thus, the proposed design is applicable.

C. Numerical Results

Solving the parameters of unknown input observer by satisfying the conditions (11), (12) and (13), L_y, E, N are obtained:

$$L_y = -D_x\left(C_1D_x\right)^+ = \begin{pmatrix} -0.0006 & -0.0023 & -0.0248 & 0.0007 \\ -0.0023 & -0.0909 & -0.0909 & -0.2727 \\ -0.0248 & -0.0909 & -0.9903 & 0.0273 \\ 0.0007 & -0.2727 & 0.0273 & -0.9182 \end{pmatrix},$$

where $(X)^+$ is the pseudo inverse matrix of X.

$$E = eye\left(4\right) + L_yC_1 = \begin{pmatrix} 0.9994 & -0.0023 & -0.0248 & 0.0007 \\ -0.0023 & 0.9091 & -0.0909 & -0.2727 \\ -0.0248 & -0.0909 & 0.0097 & 0.0273 \\ 0.0007 & -0.2727 & 0.0273 & 0.0818 \end{pmatrix},$$

$$N = EB = \begin{pmatrix} -1.6855 & -1.6854 \\ 6.6166 & 6.5638 \\ -0.6195 & -0.6142 \\ -1.9850 & -1.9691 \end{pmatrix},$$

assuming $M = \begin{pmatrix} -1.7 & 0 & 0 & 0 \\ 0 & -2 & 0 & 0 \\ 0 & 0 & -1.1 & 1 \\ 0 & 0 & 0 & -1.8 \end{pmatrix}.$

Then, choosing $\mu = 1, \sigma = 1, \lambda = 1$ and solving the linear matrices inequalities in (20) and (29), these results are obtained subsequently:

$$P = 1.0e + 007 \times \begin{pmatrix} 1.3398 & -0.1015 & -0.6311 & -0.2442 \\ -0.1015 & 1.5613 & -0.1244 & -0.0869 \\ -0.6311 & -0.1244 & 2.8093 & -0.1790 \\ -0.2442 & -0.0869 & -0.1790 & 1.6083 \end{pmatrix},$$

$$F = 1.0e + 007 \times \begin{pmatrix} -0.2173 & 0.3435 & 4.7463 & 0.9625 \\ 0 & 0 & 0 & 0 \end{pmatrix},$$

$$G = 1.0e + 007 \times \begin{pmatrix} -5.3913 & 0 \\ 0 & -4.4126 \end{pmatrix}.$$

With a sampling time $T = -1s$, the system is subjected to some boots:

The initial reference value $r(t) = \begin{pmatrix} 1 \\ 0 \\ 0 \\ 0 \end{pmatrix}$, the initial state $x(t) = \begin{pmatrix} 0 \\ 0 \\ 0 \\ 0 \end{pmatrix}$, the reference input $u(t) = \begin{pmatrix} 1 \\ 1 \end{pmatrix}$.

Compiling the FAFE design, a simulation results will be shown in the following section.

D. Simulation Results and Interpretation

i. Nominal Case

The simulation result illustrated in Figure 3 and Figure 4 shows the responses of nominal system and its control signal.

In Figure 5 and Figure 6, the disturbance vector will be included to the system. This signal is given by Figure 7.

ii. Faulty Model

Considering a partial actuator failure occurring at time $t_f = 5s$, the faulty-model response is shown in Figure 8 and Figure 9.

Figure 3. Output responses of nominal system

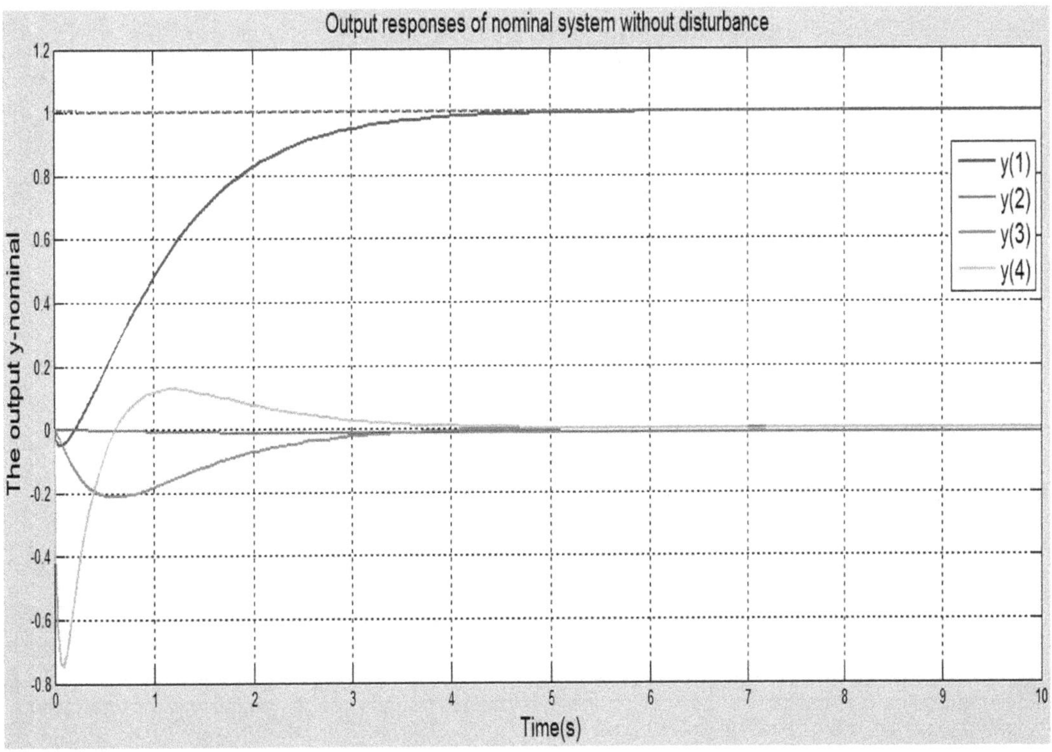

Figure 4. Signal Control in nominal case

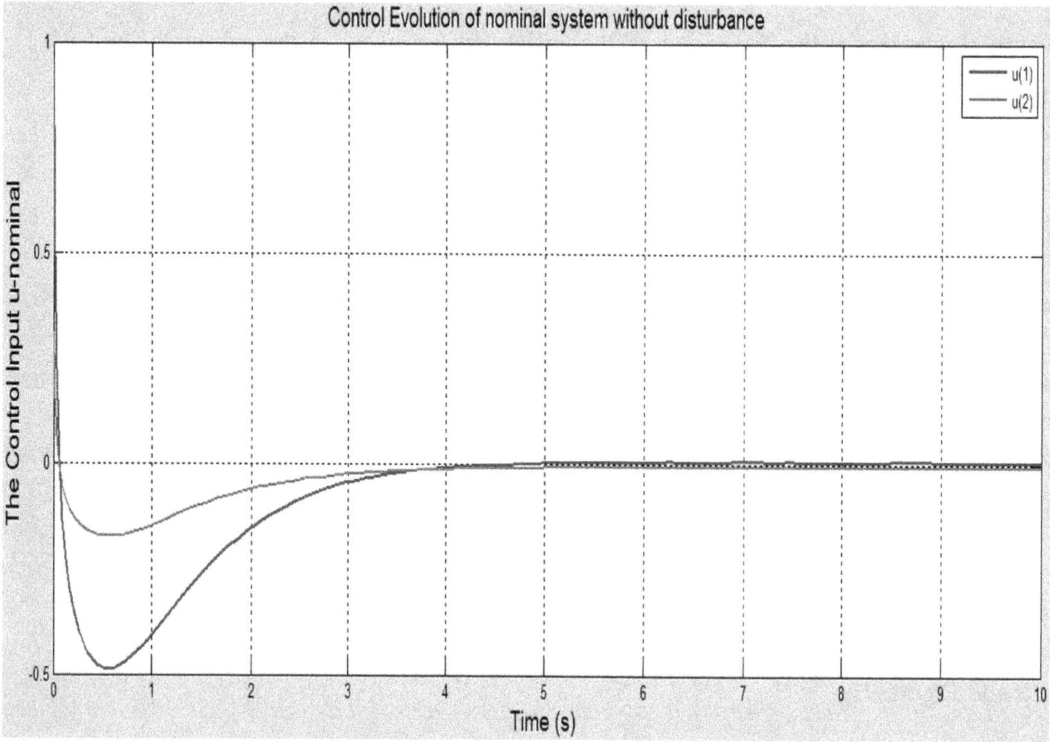

Figure 5. Output responses with unknown input disturbance

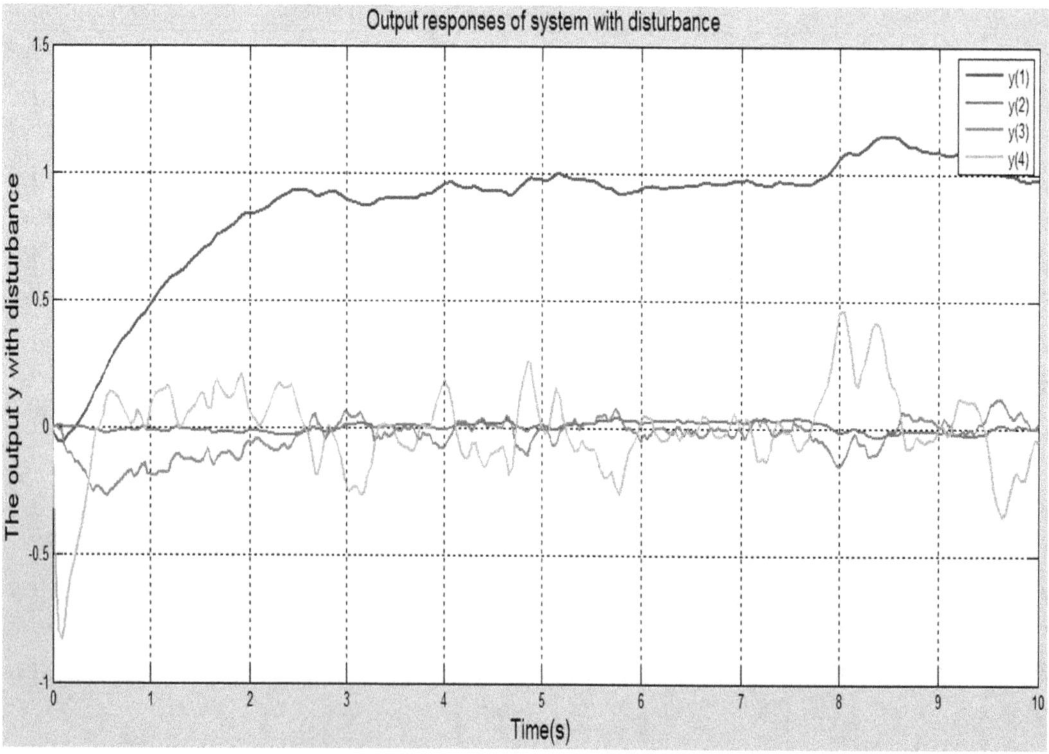

Figure 6. Signal Control with unknown input disturbance

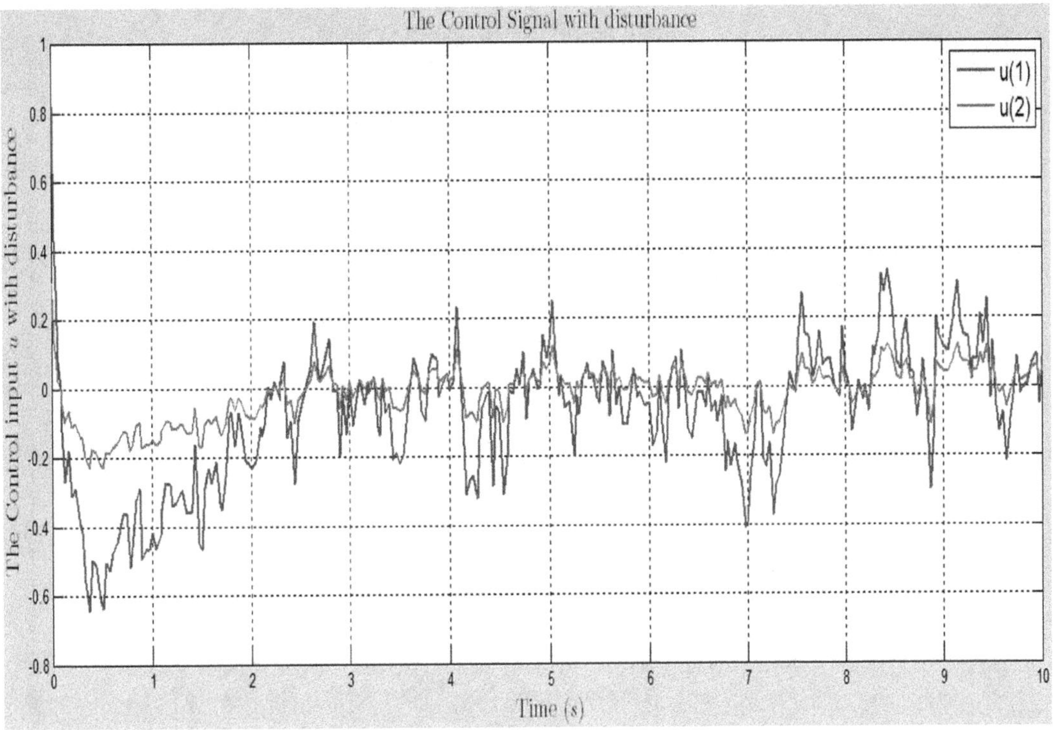

Figure 7. The disturbance evolution

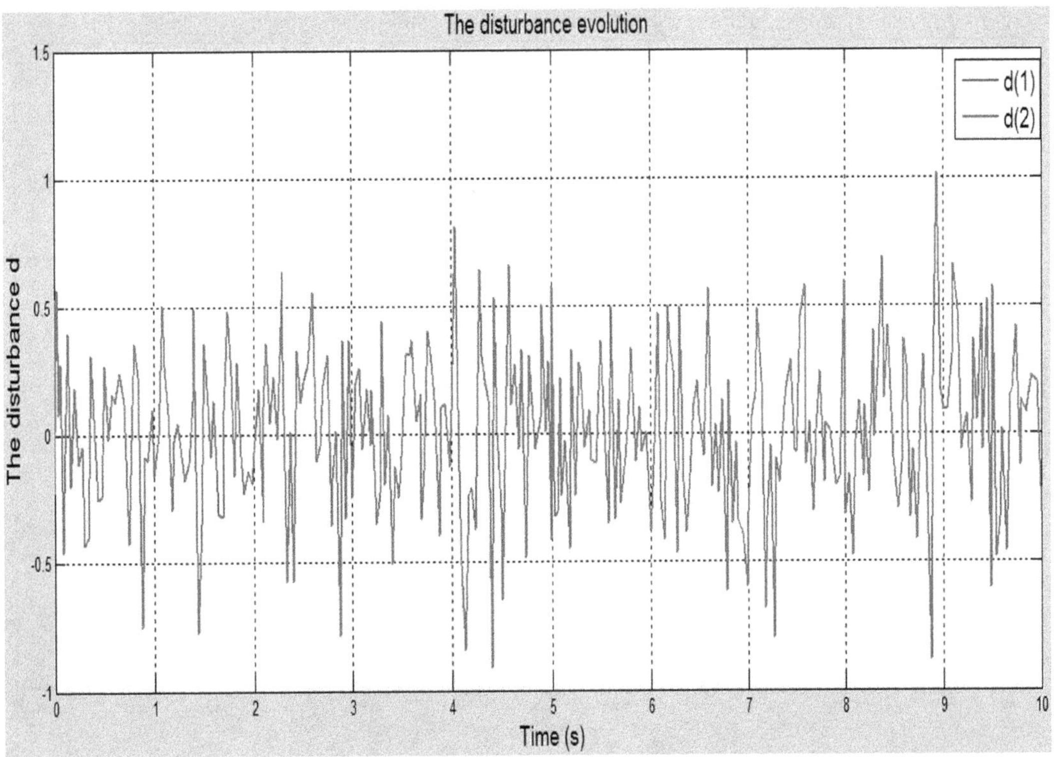

Figure 8. Output responses in faulty case with decoupling the disturbance

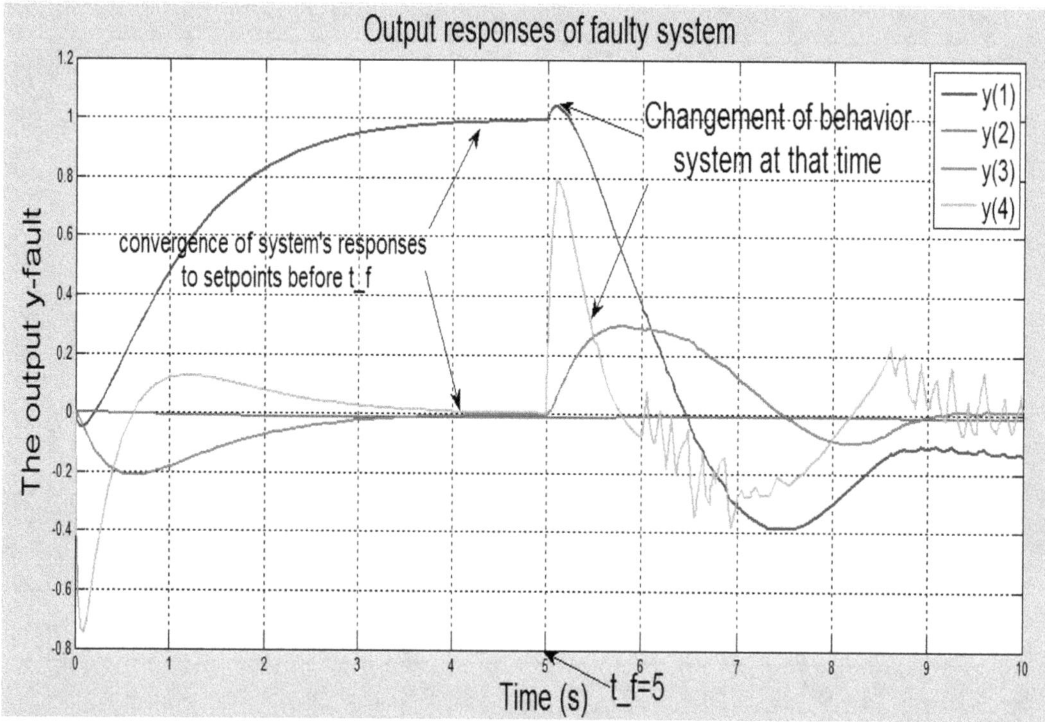

Figure 9. Signal Control in faulty case with decoupling the disturbance

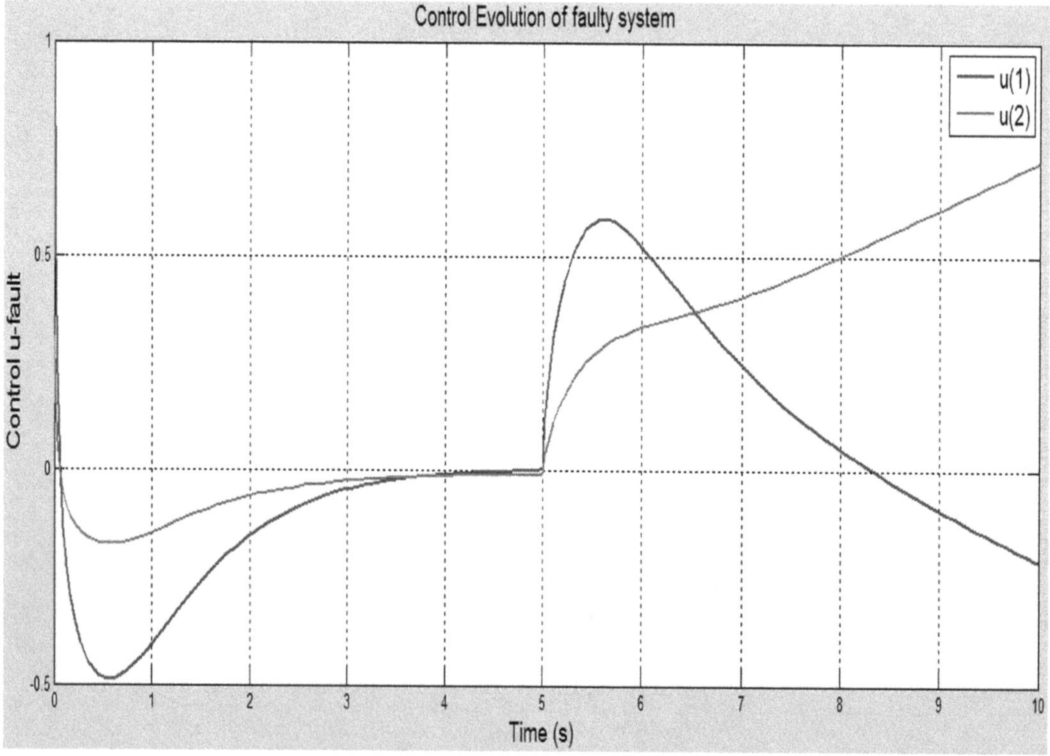

Figure 10. Output responses in faulty case with disturbance input

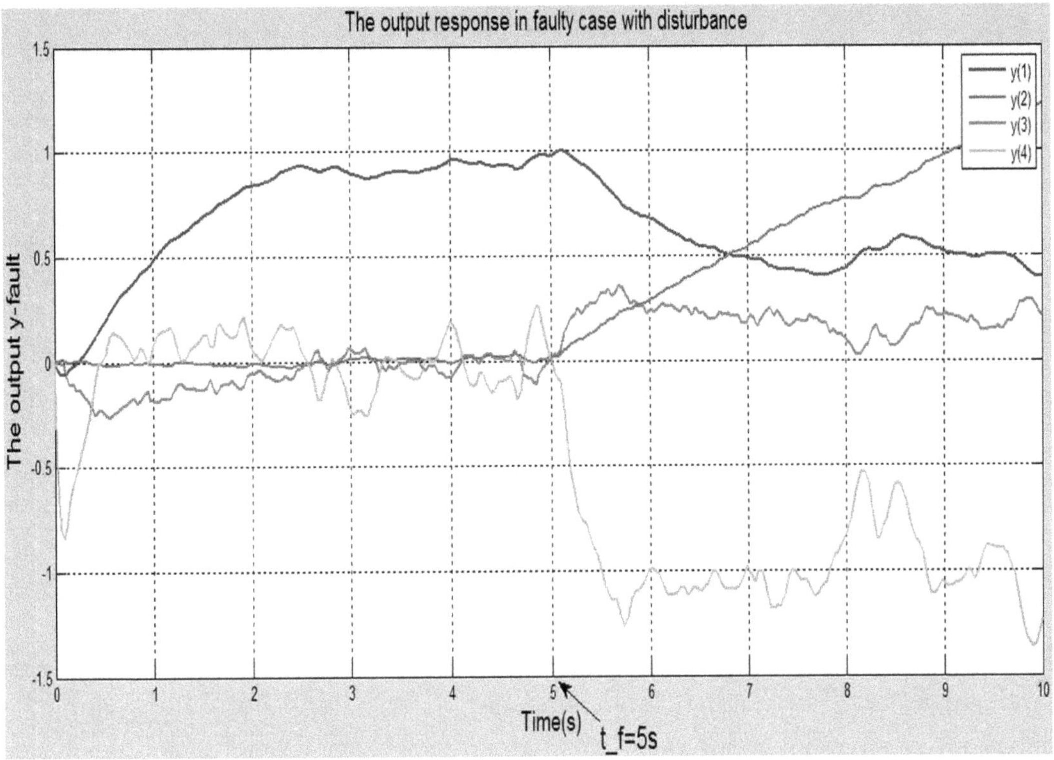

Figure 11. Signal Control in faulty case with disturbance input

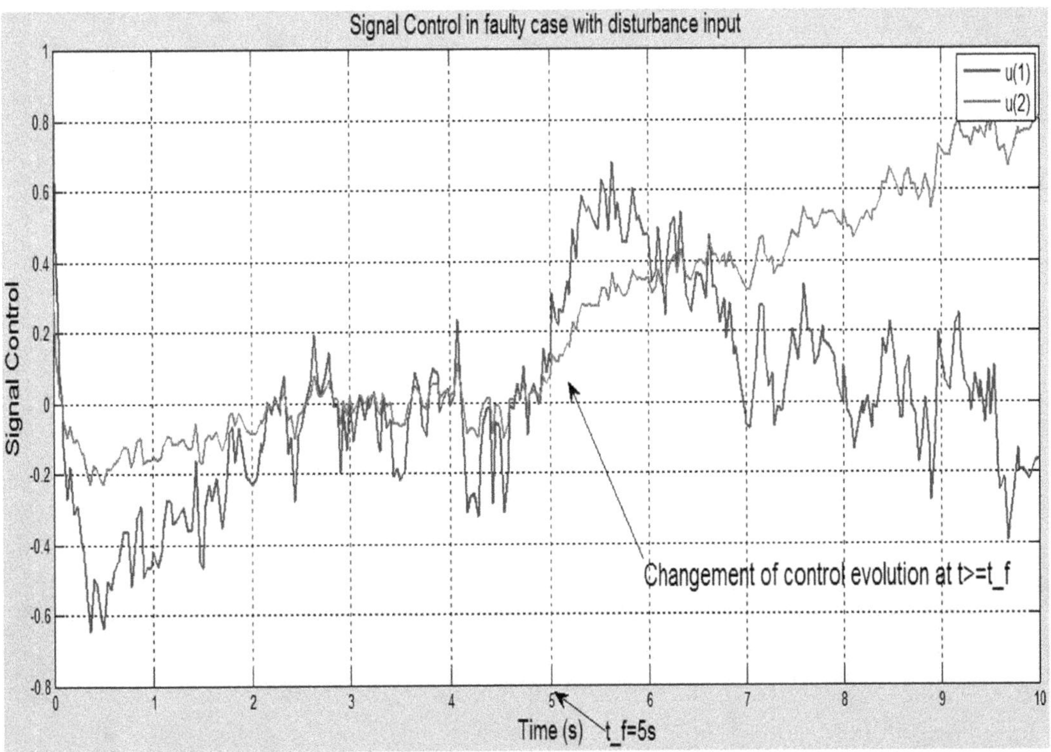

Figure 12. Null fault estimation

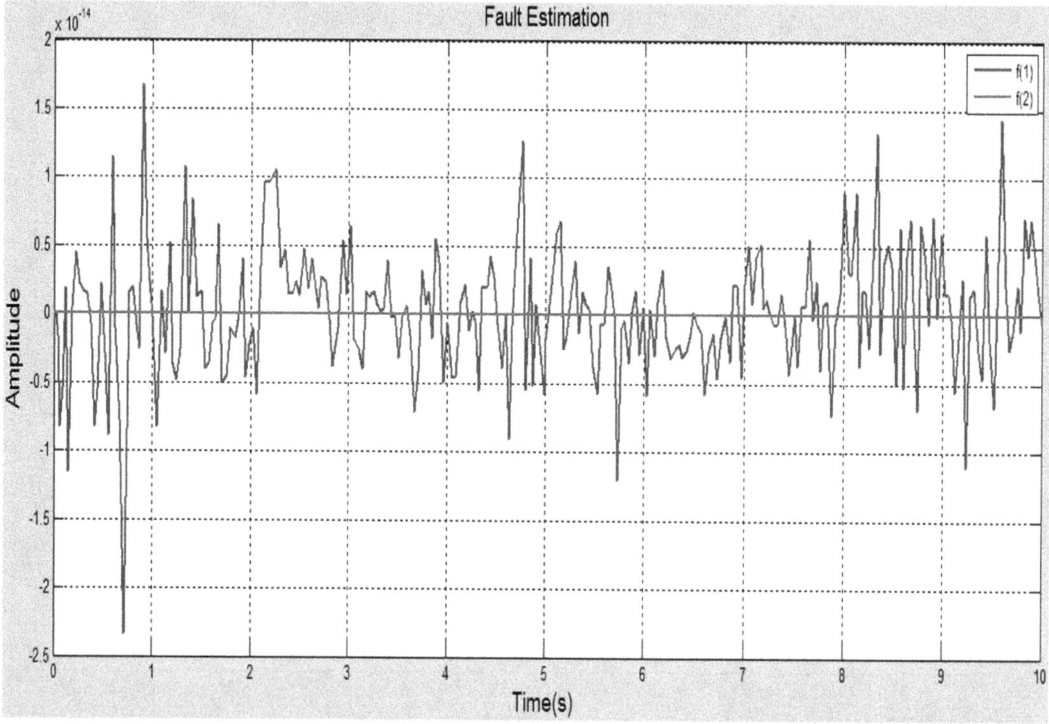

Figure 13. Evolution of fault estimation with FAFE algorithm

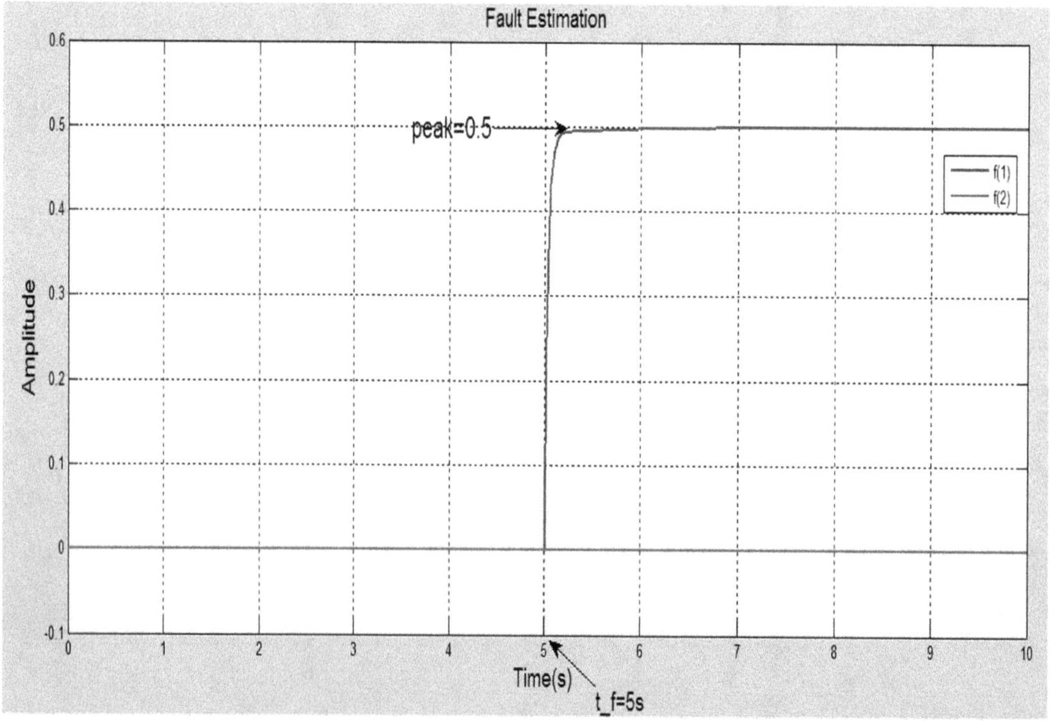

In Figure 10 and Figure 11, in addition to fault, unknown input vector include also the disturbance vector.

iii. Fault Estimation

Hereafter, one will highlight the fault estimation obtained thanks to a robust unknown input observer. Firstly, if there is no fault occurred and under the influence of disturbance, the fault estimation is shown in Figure 12. So, it's the order of 10^{-14} near to zero.

Secondly, when introducing a fault vector $f(t) = \begin{bmatrix} f_1^T(t) \\ f_2^T(t) \end{bmatrix}$, where $f_1(t) = \begin{cases} 0.5 \; if \; t \geq t_f \\ 0 \quad if \; t < t_f \end{cases}$ and $f_2(t) = 0$ (taken in this example $t_f = 5s$), It's easy from Figure 13 to select the peak achieved by the additive actuator fault thanks to the robust estimator used despite of the disturbance vector. Therefore, this estimation is characterized by the strength of fault estimation's response and the clear evolution of fault included. Once the fault estimation phase is determined, the correction step is possible now.

In the case of Pioneer robot, diagnosis module must be provided by the technique based on a fast and robust adaptive observer to drive the robot to desired path despite of the inclusion of an actuator fault.

Discussion about Results Obtained

In the previous section, the simulation results validate the FAFE algorithm developed.

From Figure 3 and Figure 4, it can be noted clearly that the system's response converges to the setpoint in nominal conditions in absence of fault or any unknown input disturbance.

Then Figure 5 and Figure 6 show that there are some oscillations in the trajectory followed and control signal because of input disturbance shown in Figure 7. These oscillations can damage a mobile robot because of their electronic composition which can be affected by the vibration throughout the trajectory. That's why, disturbances must be decoupled.

Otherwise, in the faulty case detailed in Figure 8 and Figure 9, the deviation takes a few instants. So, the system's behavior and the control's evolution were changed at instant of fault included. This behavior can be more damaged when there is disturbance like in Figure 10 and Figure 11. It's clear that the disturbance vector has a great impact on the system's responses. That's why, it is necessary to have a robust observer after to decouple the unknown input shown in Figure 7.

Comparing these results with obtained one in Hrizi et al. (2013), it's clear that disturbance vector can harm the system's responses even in the presence of a Lunberger observer. Therefore, this work exceeds this problem by the robust unknown input observer which not only repairs the trajectory despite of disturbance but also estimates the fault included perfectly. This may engender nice performances when correcting system.

Thus, the advantage of the fault diagnosis method given is that it provides a good estimate of faults, not only for the estimation step but also for the reconfiguration of control law.

FUTURE RESEARCH DIRECTIONS

Further research work includes some aspects. The first one is that fault accommodation strategy-based fault-tolerant controller will be designed to compensate for these faults using the FAFE algorithm, which can guarantee the stability and reliability of control systems.

The aim of this aspect is to propose a method that may be used in the synthesis of fault tolerant control systems (Fan & Song, 2010; Zhang & Jiang, 2001, 2003). The loop correction will be summarized in this figure:

A fault-tolerant control system (FTCS) design technique against stuck actuators is investigated using the previous estimator. The principle of fault-tolerant control is to design a reconfigurable controller using estimated system states, relying on control input adjustments on the redundant actuators to compensate for the effects of stuck actuators (Jiang, 1994). The amount of adjustment is updated based on the transient of fault compensation. The observer provides the information about the estimates of the system states. Another issue in the next work that even multiple faults can also be dealt with. The fault compensation can be carried out swiftly due to the robustness of the observer. It is shown then that FTCS ensures that the system follows the reference model under both normal conditions and with some stuck actuators.

Since most of industrial systems are uncertain and nonlinear, extension of the proposed method to robust fault diagnosis for uncertain nonlinear systems is another interesting issue. To more access for the non linearity principle, it is advisable to see Berdjag et al, 2006.

This work can be also extended to cooperative diagnosis by using a multi interconnected mobile robots (Cao et al., 1997).

As, some elements work in the same regulation loop, it is expected that they are dysfunctional. Therefore, it's better to include the cooperative control that ensures dependability and guarantees system stability when there is a fault and maintain the nominal system performance if possible.

When a group of mobile robot agents attempts to perform a coordinated behavior such as flocking, leader following or a cooperative operation, each agent in the group must identify its neighbor agents and estimate their positions and orientations. It's the same for the case of two or more interconnected robot via wireless network to validate the principle of cooperative diagnosis shown in Figure 15.

Some works validated the diagnosis theme into robotic systems (Stergios et al, 1998).

Figure 14. Fault tolerant control concept

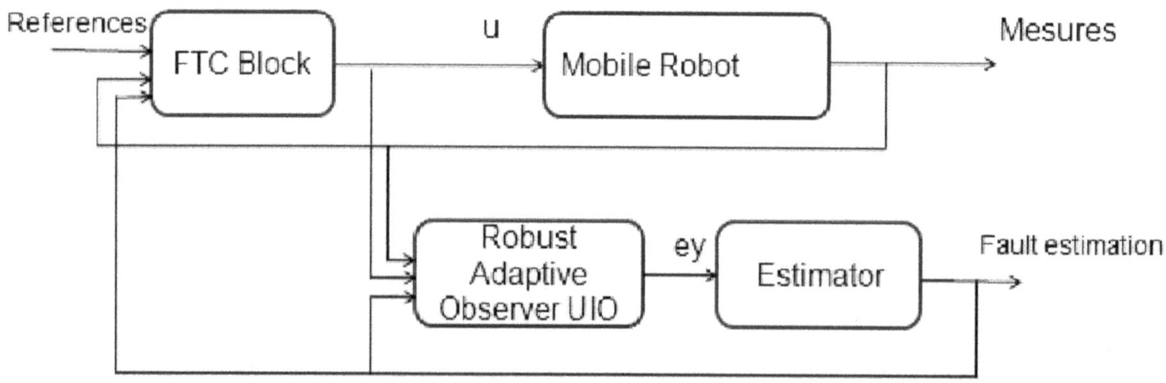

Figure 15. Principle of cooperative diagnosis

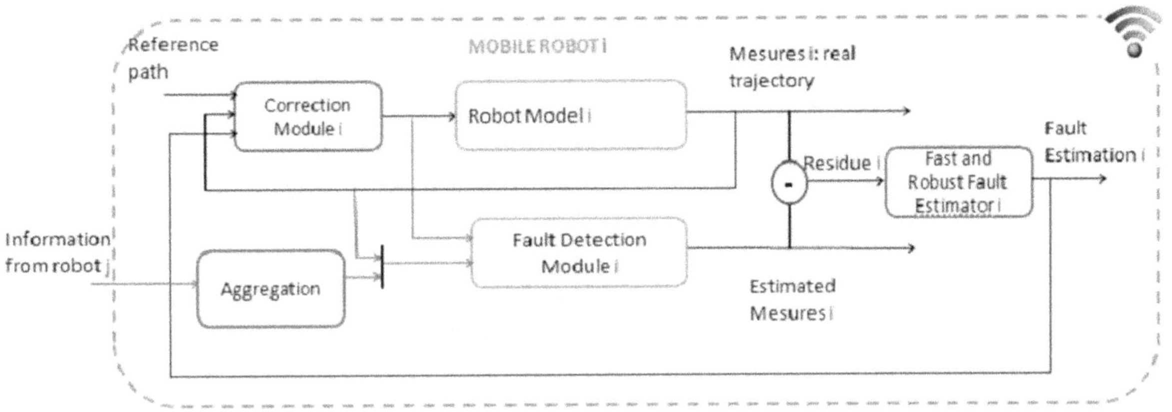

CONCLUSION

In this paper an adaptive observer technique for deterministic system has been developed for estimation of actuator fault and to guarantee the robustness. One would like the diagnostic system to be robust to various noise and uncertainties. One would like its performance to degrade gracefully instead of failing totally and abruptly. Robustness precludes deterministic isolability tests where the thresholds are placed close to zero. In the presence of noise, these thresholds may have to be chosen conservatively. Thus, as noted earlier, robustness needs are to be balanced with those of performance.

It is obvious that the FAFE algorithm can improve performances of fault estimation, including constant and time-varying fault. The application of this scheme to a unicycle robot model shows that actuator fault can be estimated with satisfactory rapidity and accuracy.

REFERENCES

Avizienis, A., Laprie, J. C., Randell, B., & Landwehr, C. (2004). Basic Concepts and Taxonomy of Dependable and Secure Computing. *Dependable and Secure Computing*, *1*(1), 11–33. doi:10.1109/TDSC.2004.2

Basseville, M., & Nikiforov, I. V. (1993). *Detection of Abrupt Changes: Theory and Application*. Englewood Cliffs, NJ: Prentice-Hall.

Berdjag, D., Christophe, C., Vincent, C., & Jiang, B. (2006). Nonlinear model decomposition for robust fault detection and isolation using algebraic tools. *International Journal of Innovative Computing, Information, & Control*, *2*(6), 1337–1354.

Bhattacharyya, S. (1978). Observer design for linear systems with unknown inputs. *IEEE Transactions on Automatic Control*, *23*(3), 483–484. doi:10.1109/TAC.1978.1101758

Bin, J., Jian, L. W., & Yeng, C. S. (2002). An adaptive technique for robust diagnosis of faults with independent effects on system outputs. *International Journal of Control*, *75*(11), 792–802. doi:10.1080/00207170210149934

Bin, J., Marcel, S., & Vincent, C. (2006). Fault accommodation for nonlinear dynamic Systems. *IEEE Transactions on Automatic Control*, *51*(9), 1578–1583. doi:10.1109/TAC.2006.878732

Blanke, M., Kinnaert, M., Lunze, J., & Staroswiecki, M. (2003). *Diagnosis and Fault Tolerant Control*. Berlin: Springer. doi:10.1007/978-3-662-05344-7

Boumedyen, B., Christophe, A., Jiang, J., & Mohamed, N. A. (2013). Fault Tolerant Control approach with actuator saturation avoidance based on reference management. *International Journal of Robust and Nonlinear Control*, *23*(2), 1–17.

Boumedyen, B., Christophe, A., Mohamed, N. A., & Kamel, B. (2011). Performance Evaluation based Fault Tolerant Control with Actuator Saturation Avoidance. *International Journal of Applied Mathematics and Computer Science*, *21*(3), 457–466.

Canudas de Wit, C., Kennouf, H., Samson, C., & Sordalen, O. J. (1997). Nonlinear control design for mobile robots. In Y.-F. Zheng (Ed.), Recent Trends in Mobile Robots (pp. 121–156). Academic Press.

Cao, Y. U., Fukunaga, A. S., & Kahn, A. (1997). Cooperative mobile robotics: Antecedents and directions. *Autonomous Robots*, *4*(1), 7–27. doi:10.1023/A:1008855018923

Chen, J., & Patton, R. J. (2012). *Robust Model-Based Fault Diagnosis for Dynamic Systems*. Springer Verlag.

Chiang, L. H., Richard, D. B., & Russell, E. L. (2001). *Fault Detection and Diagnosis in Industrial Systems*. London. doi:10.1007/978-1-4471-0347-9

Darouach, M. (2000). Existence and design of functional observers for linear systems. *IEEE Transactions on Automatic Control*, *45*(5), 940–943. doi:10.1109/9.855556

Ding, S. X. (2008). *Model-based Fault Diagnosis Techniques*. Springer Verlag.

Doraiswami, R., & Cheded, L. (2013). A unified approach to detection and isolation of parametric faults using a Kalman filter residual-based approach. *Journal of the Franklin Institute*, *350*(5), 938–965. doi:10.1016/j.jfranklin.2013.01.005

Duan, Z., Cai, Z., & Yu, J. (2005). Fault diagnosis and fault tolerant control for wheeled mobile robots under unknown environments. In *Proceedings of the 2005 IEEE International Conference on Robotics and Automation (ICRA),* (pp. 3439–3444). IEEE. DOI: doi:10.1109/ROBOT.2005.1570640

Edwards, C., & Pin, T. C. (2004). A comparison of sliding mode and unknown input observers for fault reconstruction. *European Journal of Control*, *12*(3), 5279–5284.

Fan, L. L., & Song, Y. D. (2010). On Fault-tolerant Control of Dynamic Systems with Actuator Failures and External Disturbances. *Acta Automatica Scinica*, *36*(11), 1620–1625.

Frank, P. M. (1990). Fault diagnosis in dynamic systems using analytical and knowledge based redundancy- a survey and some new results. *Automatica*, *26*(3), 459–474. doi:10.1016/0005-1098(90)90018-D

Frank, P. M. (1990). Analytical and qualitative model-based fault diagnosis - a survey and some new results. *European Journal of Control*, *2*(1), 6–28. doi:10.1016/S0947-3580(96)70024-9

Gao, Z., & Antsaklis, P. J. (1991). Stability of the pseudo-inverse method for reconfigurable control systems. *International Journal of Control*, *53*(3), 717–729. doi:10.1080/00207179108953643

Gao, Z., & Antsaklis, P. J. (1992). Reconfigurable control systems design via perfect model following. *International Journal of Control*, *56*(4), 783–798. doi:10.1080/00207179208934343

Gertler, J. (1995). Diagnosing parametric faults: From parameter estimation to parity relations. In *Proceedings of the 1995 IEEE American Control Conference (ACC)*, (pp. 1615–1620). IEEE. DOI: doi:10.1109/ACC.1995.529780

Gertler, J. (1998). *Fault Detection and Diagnosis in Engineering Systems*. Hershey, PA: CRC Press.

Hrizi, O., Boumedyen, B., Christophe, A., & Mohamed, N. A. (2013). Fast Adaptive Fault Estimation Algorithm: Application to Unicycle Robot. In *Proceedings of the 2nd International Conference on Control and Fault-Tolerant Systems (SYSTOL)*. Nice, France: SYSTOL.

Hsieh, C.-S. (2002). Performance gain margins of the two-stage LQ reliable control. *Automatica*, *38*(11), 1985–1990. doi:10.1016/S0005-1098(02)00114-0

Huang, C. Y., & Stengel, R. (1990). Restructuable control using proportional integral implicit model following. *Journal of Guidance, Control, and Dynamics*, *13*(2), 303–309. doi:10.2514/3.20550

Isermann, R. (2005). Model-based fault detection and diagnosis status and applications. *Annual Reviews in Control*, *29*(1), 71–85. doi:10.1016/j.arcontrol.2004.12.002

Isermann, R. (2006). *Fault-Diagnosis Systems: An Introduction from Fault Detection to Fault Tolerance*. doi:10.1007/3-540-30368-5

Ji, M., Zhang, Z., Biswas, G., & Sarkar, N. (2003). Hybrid fault adaptive control of a wheeled mobile robot. *IEEE Transactions on Mechatronics*, *8*(2), 226–233. doi:10.1109/TMECH.2003.812823

Jiang, J. (1994). Design of reconfigurable control system using eigenstructure assignments. *International Journal of Control*, *59*(2), 395–410. doi:10.1080/00207179408923083

Jiang, J. (2005). Fault tolerant control systems - an introductory overview. *Automatica*, *31*(1), 161–174.

Jiang, J., & Zhang, Y. (2006). Accepting performance degradation in fault-tolerant control system design. *IEEE Transactions on Control Systems Technology*, *14*(2), 284–292. doi:10.1109/TCST.2005.860515

Jiang, J., & Zhao, Q. (2000). Design of reliable control systems possessing actuator redundancies. *Journal of Guidance, Control, and Dynamics*, *23*(4), 709–718. doi:10.2514/2.4587

Luenberger, D. (1971). An introduction to observers. *IEEE Transactions on Automatic Control*, *16*(6), 596–602. doi:10.1109/TAC.1971.1099826

Luis, A., Alexandre, M., & Pedro, F. (1998). Comparing control strategies for line tracking robots. In *Proceedings of the 1998 IEEE Advanced Motion Control (AMC)*, (pp. 542-547). IEEE. DOI: doi:10.1109/AMC.1998.743594

Lunze, J., & Richter, J. (2008). Reconfigurable Fault-Tolerant Control: A Tutorial Introduction. *European Journal of Control*, *14*(5), 359–386. doi:10.3166/ejc.14.359-386

Magnus, E. (2002). Motion description languages for multi-modal control in robotics. In A. Bicchi, H. Christensen, & D. Prattichizzo (Eds.), *Control Problems in Robotics* (pp. 75–90). Las Vegas, NV: Academic Press.

Martin, C., & Jay, T. (1998). State and input estimation for a class of uncertain systems. *Automatica*, *34*(6), 757–764. doi:10.1016/S0005-1098(98)00013-2

Menighed, K., Yamé, J., & Aubrun, C. (2011). Modification of an adaptive observer to improve its performances in actuator fault detection. In *Proceeding of the 18th Mediterranean Conference on Control and Automation (MED)*. Hammamet, Tunisia: MED.

Park, T. G., & Lee, K. S. (2004). Process fault isolation for linear systems with unknown inputs. *IEE Proceedings. Control Theory and Applications*, *151*(6), 720–726. doi:10.1049/ip-cta:20040796

Park, Y., & Stein, J. L. (1988). Closed-loop state and input observer for systems with unknown inputs. *International Journal of Control*, *48*(3), 1121–1136. doi:10.1080/00207178808906239

Patton, R. (1997). Fault tolerant control systems: the 1997 situation. In *Proceeding of IFAC Safe Process* (pp. 1033-1054). IFAC.

Patton, R., & Chen, J. (1991). A review of parity space approaches to fault diagnosis. In *Proceedings of the 1991 IFAC/IMACS Symposium on Fault Detection, Supervision and Safety of Technical Processes (SAFEPROCESS),* (pp. 239–255). IFAC.

Patton, R. J., & Chen, J. (1997). Observer-based fault detection and isolation: Robustness and applications. *Control Engineering Practice*, *5*(5), 671–682. doi:10.1016/S0967-0661(97)00049-X

Pedro, A., Joao, F., Pedro, R., & Pedro Lima, B. (1998). A modular autonomous vehicle. In *Proceedings of the 3rd IFAC Symposium on Intelligent Autonomous Vehicles*. Lisbon, Portugal: IFAC.

Philippe, W., Boumedyen, B., Ahmed, K., Didier, T., & Christophe, A. (2012). Reconfigurable Control Design with integration of a reference governor and reliability indicator. *International Journal of Applied Mathematics and Computer Science*, *22*(1), 139–148.

Powell, D. (1992). Failure Mode Assumptions and Assumption Coverage. In *Proceedings of the 1992 IEEE Symposium on Fault-Tolerant Computing (FTCS)*, (pp. 386-395). IEEE. DOI: doi:10.1109/FTCS.1992.243562

Rouche, N., Habets, P., & Laloy, M. (1977). *Stability Theory by Lyapunov's Direct Method*. Applied Mathematical Sciences. doi:10.1007/978-1-4684-9362-7

Shigeo, H., & Akio, M. (1990). Design and control of a mobile robot with an articulated body. *The International Journal of Robotics*, *9*(2), 99–114. doi:10.1177/027836499000900208

Steinbauer, G., & Wotawa, F. (2005). Detecting and locating faults in the control software of autonomous mobile robots. In *Proceedings of the 16th International Workshop on Principles of Diagnosis*, (pp. 13–18). Monterey, CA: Academic Press.

Stergios, R., Gaurav, S., & George, B. (1998). Fault detection and identification in a mobile robot using multiple model estimation. In *Proceedings of the 1998 IEEE International Conference on Robotics and Automation*, (pp. 2223-2228). IEEE. DOI: doi:10.1109/ROBOT.1998.680654

Szigeti, F. (1991). Kalman's rank condition for infinite-dimensional time-dependent systems. In *Proceedings of the 1991 Differential Equations Conference (equadiff)*, (pp. 927-931). Academic Press.

Theilliol, D., Noura, H., & Ponsart, J. C. (2002). Fault diagnosis and accommodation of a three-tank system based on analytical redundancy. *Instrumentation, Systems, and Automation Transactions*, *41*(3), 365–382. PMID:12160349

Ti-Chung, L., Kai-Tai, S., Ching-Hung, L., & Ching-Cheng, T. (2001). Tracking control of unicycle modeled mobile robots using a saturation feedback controller. *IEEE Transactions on Control Systems Technology*, *9*(2), 305–318. doi:10.1109/87.911382

Wang, H., & Daley, S. (1996). Actuator fault diagnosis: An adaptive observer based technique. In Proceedings of Automatic Control. IEEE Transactions onAutomatic Control, 41(7), 1073-1078.

Wang, H., Huang, Z. J., & Daley, S. (1997). On the use of adaptive updating rules for actuator and sensor fault diagnosis. *Automatica*, *33*(2), 217–225. doi:10.1016/S0005-1098(96)00155-0

Wang, Z. (2010). Unknown Input Observer Based Fault Class Isolation and Estimation. In *Proceedings of the 29th Chinese Control Conference (CCC)*, (pp. 3963-3968). CCC.

Wu, N. E., Zhang, Y., & Zhou, K. (2000a). Detection, estimation, and accommodation of loss of control effectiveness. *International Journal of Adaptive Control and Signal Processing*, *14*(7), 775–795. doi:10.1002/1099-1115(200011)14:7<775::AID-ACS621>3.0.CO;2-4

Xiong, Y., & Saif, M. (2003). Unknown disturbance inputs estimation based on a state functional observer design. *Automatica*, *39*(8), 1389–1398. doi:10.1016/S0005-1098(03)00087-6

Zhang, K., Jiang, B., & Vincent, C. (2008). Adaptive Observer-based Fast Fault Estimation. *International Journal of Control, Automation, and Systems*, *6*(3), 320–326.

Zhang, Y., & Jiang, J. (2001). Integrated Design of Reconfigurable Fault-Tolerant Control Systems. *Journal of Guidance, Control, and Dynamics*, *24*(1), 133–136. doi:10.2514/2.4687

Zhang, Y., & Jiang, J. (2002). An active fault-tolerant control system against partial actuator failures. *IEE Proceedings. Control Theory and Applications*, *149*(1), 95–104. doi:10.1049/ip-cta:20020110

Zhang, Y., & Jiang, J. (2003a). Bibliographical Review on Reconfigurable Fault-Tolerant Control Systems. In *Proceedings of the 5th IFAC Symposium on Fault Detection, Supervision and safety for Technical Processes*, (pp. 265–276). IFAC.

Zhang, Y., & Jiang, J. (2003b). Fault tolerant control system design with explicit consideration of performance degradation. *Aerospace and Electronic Systems*, *39*(3), 838–848. doi:10.1109/TAES.2003.1238740

Zhang, Y., & Jiang, J. (2006). Issues on integration of fault diagnosis and reconfigurable control in active fault-tolerant control systems. In *Proceedings of the 6th IFAC Symposium on Fault Detection, Supervision and Safety of Technical Processes (SafeProcess'06)*, (pp. 1513–1524). IFAC. DOI: doi:10.3182/20060829-4-CN-2909.00240

KEY TERMS AND DEFINITIONS

Actuator Fault: A failure is defined as a deviation of system parameters (Blanke et al., 2003). An actuator fault is a type of failure affecting the system inputs. Due to abnormal operation or material aging, actuator faults may occur in the system. An actuator can be represented by additive and/or multiplicative faults. Actuator failures may drastically change the system behavior, resulting in degradation or even instability.

Cooperative Diagnosis: Based on the communication between a numbers of interconnected systems.

Disturbance: An unknown input included in the system's model. It can be incertitude, noise or a nonlinearity term. It has a considerable affect in process's performances. It is an event that occurs when the delivered service deviates from correct service. A service fails either because it does not comply with the functional specification, or because this specification did not adequately describe the system function.

Fast Adaptive Fault Estimation (FAFE): Proposed to provide much better performances for the process. It is based on adaptive observer (or estimator) composed of a proportional term and an integral one to guarantee both satisfactory dynamical and steady state performances.

Fast Fault Estimation: A good strategy only using the measurable input and output vector. It is proposed to enhance the rapidity of fault estimation and ameliorate the system's performances.

Fault Tolerance: Means to avoid service failures in the presence of faults. It is noteworthy that repair and fault tolerance are related concepts; (Avizienis et al. 2004); the distinction between fault tolerance and maintenance is that maintenance involves the participation of an external agent, e.g., a repairman, test equipment, remote reloading of software. Furthermore, repair is part of fault removal (during the use phase), and fault forecasting usually considers repair situations. In fact, repair can be seen as a fault tolerance activity within a larger system that includes the system being repaired and the people and other systems that perform such repairs.

Linear Matrix Inequalities (LMI): A formulation which gives an effective way to calculate the design parameters.

Mobile Robots: There are a set of processes working in the same cooperative network. They will be providing a better life not only to common people but especially to elderly and impaired.

Rapidity: In this work, it means the rapidity of convergence to the detection threshold.

Robustness: I.e., dependability with respect to external faults, which characterizes a system reaction to a specific class of faults. It is ability of a system to avoid service failures that are more frequent or more severe than is acceptable.

Stability: The ability of properties to converge to existence conditions. It is provided by a parameter-dependent Lyapunov function (Rouche et al, 1977).

Unknown Input Observer (UIO): This observer is subsequently used for a robust fault detection scheme and also as an adaptive detection scheme for a certain class of actuator faults wherein the time instance and characteristics of an incipient actuator fault are detected.

Chapter 17
Navigation Control of a Mobile Robot under Time Constraint using Genetic Algorithms, CSP Techniques, and Fuzzy Logic

Tlijani Hayet
National Engineering School of Tunis, Tunisia

Tlijani Hatem
National Engineering School of Tunis, Tunisia

Knani Jilani
National Engineering School of Tunis, Tunisia

ABSTRACT

This chapter proposes a two-input-and-one-output fuzzy controller for a two-wheeled mobile robot. The robot moves following a hybrid plan based on Genetic Algorithms and Constraint Satisfaction Problem techniques. The path yielded by this hybrid plan is structured by artificial beacons. In each position of these beacons, the current velocity of the controlled robot is by itself an input for the fuzzy controller. The second input is the time delay between the current position and the next sub goal to be reached. The proposed fuzzy controller aims to converge the time delay between positions in the path tracking closely to a time window. It gives the appropriate acceleration so that the time delay becomes closer to the time window.

1. INTRODUCTION

To enclose the capacity of an intelligent and autonomous agent (Hsu & Fu, 2000; Siegwart, Nourbaksh & Scaramuzza, 2011), a mobile robot needs a decision-making ability to react correctly in its environment. It should be in agreement with the state of its environment and with the received commands so that it operates successfully. The question that arises here is: how one can involve this capacity of

DOI: 10.4018/978-1-4666-7248-2.ch017

decision-making into a mobile robot and essentially into its navigation program? To reply to this question one has to think about the architectural design of the mobile robot (Low, Leow & Ang, 2002). The robot architectural design is defined in three levels: the perception, the decision and the action. Then, it comes the level that consists of the decision making which is the part that focuses on how to answer the question. Decision making is essential for a mobile robot to know the best adopted strategy for its navigation. This is of course related to the navigation plan.

Planning a path tracking of a mobile robot problem depends on satisfying a number of criteria (Khatib, Morris, Morris, Rossi, Sperduti & Venable, 2007; Cheng & Chen, 2012). Essentially, it should optimize a cost expressed not only in terms of the distance crossed by the robot in its configuration space but also by the duration or of the energy necessary for the execution of its movements. Sometimes, more than one criterion is considered. In this case, the approaches of that deal with the proposed problem are combined to yield a new hybrid way to solve the problem of the path tracking plan. Several cases of the mobile robot's navigation need to work in scenarios. It means that the plan must be associated with distance and duration. Indeed, temporal constraints allow one to describe in a natural way problems arising in the navigation of the mobile robot (Breivik & Fossen, 2008; Dong, Chen, Chu, & Tarn, 2012).

In literature, a wide variety of problems involve the manipulation of temporal information. This constraint can be viewed as having a bounded delay to execute a plan. It allows vehicles like mobile robot to move under time constraint. Such problems are the Vehicle: Routing Problem with Time Windows (VRPTW). The VRPTW can be described as the problem of designing the least cost routes from one depot to a set of geographically scattered points. The routes must be designed in such a way that each point is visited only once by exactly one vehicle within a given time interval, all routes start and end at the depot, and the total demands of all points on one particular route must not exceed the capacity of the vehicle (Hoong, Melvyn, & Kwong, 2003).

To deal with the problem of temporal constraint in the time window, one has to define the navigation of the mobile robot within the time window by finding the nearest position of the goal to be reached from the current one respecting the imposed time window that may be the only option. In this case Classical techniques of Constraint Satisfaction Problem are the best way to solve this problem.

In general, it seems that classical techniques proved to be insufficient to describe the planning problem in such definite terms as Constraint Satisfaction Problem. A simple example is the expression of preferences among the set of assignments. In recent years, increased efforts have been centered on developing intelligent control systems. These include the development of non-analytical methods of soft computing such as evolutionary computation and fuzzy logic. Those methods have proved to be effective in both fields: designing intelligent control systems and handling real-time uncertainty (Miguel & Shen, 2003).

The navigation under constraints sometimes engenders ambiguous solutions. This urges us to think of a strong and intelligent strategy as the fuzzy logic. This approach of artificial intelligence was used for the command of a small robot, to insure the follow-up of a complex and pre-calculated trajectory. It was also the basic of the strategy developed for the positioning of a vehicle directed from outside (Garcia-Pérez, Garcia-Alegre, Ribeiro, & Guinea, 2003). The fuzzy logic paradigm allows associating a preference to a decision making. So that, the mobile robot can decides according to some preferences while respecting temporal constraint. Herein, the mobile robot can get a human reasoning ability. It improves the results of tracking the mobile robot so that it goes closely to the strait path of the navigation plan (Wong, Wang, Li, & Cheng, 2007)

In this work, a controller based on the fuzzy logic is adopted on the model of the two-wheeled mobile robot described in (Tlijani, M'sirdi, & Knani, 2011). The objective of this work is to respect as close as possible the time window solved in the previous work (Tlijani, Knani, & M'sirdi, 2012).

Indeed, the problem of time window is solved using the techniques of the Constraint Satisfaction Problem. The solution realized is based on the genetic algorithms, so that it yields a minimized path length. Other problems of plantification are treated using genetic algorithms such the study of the work presented in (Zeng, Zhang, & Wei, 2012).

The definition of the time constraint is integrated and resolved in the crossover operator of the generated algorithm as mentioned in (Tlijani, M'sirdi, & Knani, 2012). In this solver the velocity is calculated in purpose to correspond to the time delay between two successive positions. The value of the time delay between two consecutive positions in the path tracking mustn't exceed time window fixed according to the navigation of the mobile robot. Then, it's noticed that the results given by the developed solver can be improved. The idea is to better the time delay and make it closer to the value of the time window. For this reason, it's necessary to change the parameters that impacted the navigation of the mobile robot by acting on the acceleration. The use of the paradigm of the fuzzy logic has enhanced the solution given by the solver of the temporal Constraint. Besides it makes it more natural and efficient.

The presented work will be organized as following. Firstly, it states some related work. Secondly, it describes the problem statement in section 3. Then, it defines in section 4, the new way to use the fuzzy logic approach with Constraint Satisfying Problem. After that, comes the section 5 that deals with the new GA-plan. Then, the presented work gives an example of the executed plan and some results of the simulation of the Fuzzy controller in the section 6. In the end of this work, discussions of the different results yielded in section 6, are exposed in the section 7.

2. RELATED WORK

Intelligent behavior for mobile robot is essential to guarantee its autonomous character. For this reason various paradigms of artificial intelligence are adopted to improve the decision making for a mobile robot. Actually, a big effort is granted to the problem of path planning. Several studies and research have dealt with the use of the intelligent approaches such as fuzzy logic and heuristics methods like evolutionary algorithms to give reliable algorithms that control navigation of mobile robot. IN (Sugiarto, Tung & Rahman, 2008) the authors use the fuzzy logic to solve the problem of an autonomous mobile robot by tracking the path within a maze. This work deals with the real-time navigation of a mobile robot to provide it with the human being ability of reasoning and avoiding collision.

The fuzzy logic and genetic algorithm are also for autonomous mobile robot control in (Narvydas, Simutis & Raudonis, 2007) to design an efficient control algorithm for autonomous mobile robot movement in an unknown and a changing environment with obstacles and walls. Furthermore, the fuzzy logic is adopted for walking robot (Ivanescu, 2001). This work yielded a fuzzy controller that assisted these types of robots to move correctly within intercepted target. The use of fuzzy logic on robot control has been studied in the literature (Sugeno, Murofushi, Mori, Tatematasu & Tanaka, 1989). Its work consists of developing a fuzzy control system for a car model. The car is equipped by ultrasonic sensors to measure its orientation and its distance to the fences. The fuzzy controller aims at guiding the car in parallel to the fence and to turn in the corners. Hence, this car could drive within a fenced-in track. Fuzzy logic has been used to develop a fuzzy control algorithm for mobile robots which were moving next to the obstacle boundaries, avoiding the collisions with them (Viorel & Mircea, 2008).

A fuzzy logic based behavior control for a mobile robot has been studied in a dynamic environment (Bansal, 2011). The main idea is to coordinate the difficulties encountered in navigation by using fuzzy sets and fuzzy rules. Other application of fuzzy logic to robots in the literature appeared with (Song & Tay, 1992; Khatib, 1986; Yan, Ryan & Power, 1994). Their studies dealt with the combination with modeling and planning techniques in order to offer reactive guidance in the navigation of a mobile robot.

Many fields have used Fuzzy system essentially when it was hard to use mathematical model for one system. In those cases, Fuzzy controller has been the best tool for industrial processes. Varied methods were used to design motion control for a mobile robot. GA-method fuzzy controller was proposed in (Ching-Chang, 2007) to determine the velocities of the left-wheeled engine and the right-wheeled engine.

The combination between Genetic Algorithms and Fuzzy Logic has been proposed in different works. The approach of Genetic Algorithms based on fuzzy control design is distinguished in three categories described in (Homaifar & McCormick, 1995). GA-based methods that belong to the first category, considered only the membership functions to resolve an optimal fuzzy system. The second category focused only on the fuzzy rules to determine an optimal fuzzy rule base. Some examples of the use of this category were exhibited in (Chen &Wong, 2002). The third category of GA-based methods, fuzzy rules and fuzzy membership functions were established at once. The application of that category was described in (Homaifar & McCormick, 1995; Nawa & Furuhashi, 1999).

3. PROBLEM STATEMENT

The problem to be solved here is the navigation with time window for a mobile robot using a fuzzy controller. The time window was solved using Constraint Satisfaction Problem and it was considered as a temporal constraint (Tlijani, 2012). The result was a path tracking of a mobile robot that respected time window and has the minimum distance. The optimization was studied in (Tlijani & Knani, 2009) through using the genetic algorithms. However, the solver of the time constraint represented by the time window was realized after combining the genetic algorithm based plan with Constraint Satisfaction Problem techniques. The solver as guided by a meta-heuristic of neighborhood. The objective of the present work is to change the velocity of the mobile robot so that the time between the two positions in the path will be as closer as possible to the time window. For this reason we would like to realize a fuzzy controller to fulfill this goal. The inputs are the time delay between the current position and the following one as well as the velocity of the mobile robot. The output will be the necessary acceleration to correct the time delay in order to pass from one position to the following on the path tracking.

Thus in a previous work, the problem of the time window has been resolved using the combination between genetic algorithms and Constraint Satisfaction Problem techniques (Tlijani, 2012). And in the present work, the hybrid algorithm will be emphasized by the fuzzy logic. This intelligent paradigm improves the decision making of the studied mobile robot. This section will give the different parameters and formulations used for the proposed solution.

Figure 1. Posture definition of a two wheeled mobile robot

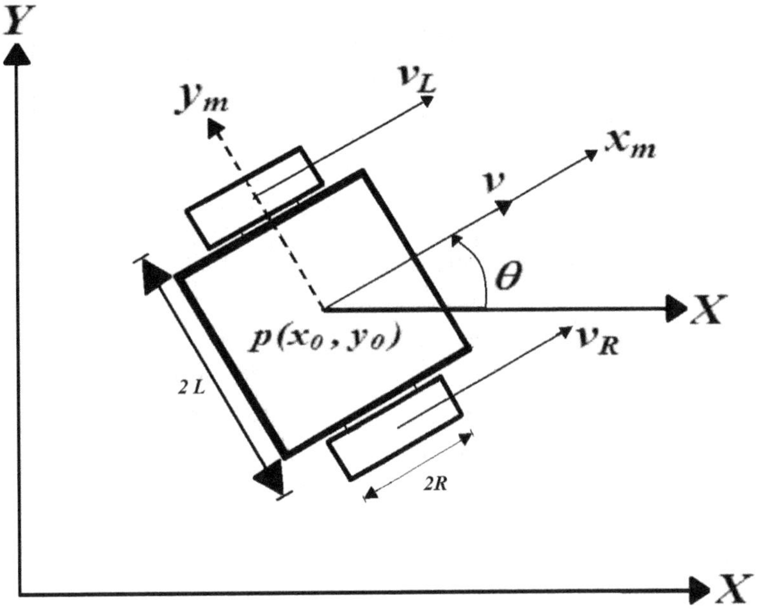

3.1 Localization and Posture of the Mobile Robot

The studied mobile robot is a wheeled mobile robot as described in (Teslic, Škrjanc & Klanèar, 2011). And in (Tlijani, 2011) the model of the used mobile robot has been specified. This model is used for all work of research as the present one. The structure of this mobile robot is described in Figure 1, where X-Y is the global coordinates and xm-ym is the local coordinates which are fixed to the robot with its center p as the origin. As shown in the next figure Figure 1, the body of the mobile robot is of a symmetric shape and the center of the mass is at the geometric center p of the body. R is the radius of the wheel and L is the displacement from the center of robot to the center of the wheel. The set (xo, yo) represents the position of the geometric center p in the world X-Y coordinates, and the angle θ indicates the orientation of the robot. The angle θ is taken counter clockwise from the X-axis to the xm-axis.

The two wheels are fixed and controlled independently by two motors. For this reason, the considered wheeled robot is equipped with motors that are driven by an on board computer. It is assumed that the wheeled mobile robot is made of a rigid frame, with non-deformable wheels that do not slip, and it is moving on a horizontal plane. Besides, we assume that this mobile robot is one particle in its configuration space.

According to this description, and the mobile robot localization using artificial landmarks studied in (Glover, 1986) the posture of the used mobile robot can be defined by the following equation:

$$X_{k+1} = \begin{bmatrix} x_k + v_k \cos\theta \, dt \\ y_k + v_k \sin\theta \, dt \\ v_{k+1} \\ \theta_k + \dot{\theta} dt \\ \dot{\theta}_{k+1} \end{bmatrix} \quad (1)$$

Where:

- $\mathbf{X}_{k+1}$**:** The vector of the posture at the state k+1 generated in function of the previous state;
- $(\mathbf{x}_k,\mathbf{y}_k)$**:** The position of the center of the mobile robot at the state k;
- v_{k+1} **:** The linear velocity of the mobile robot at the state k+1;
- v_k **:** Linear velocity in the position k;
- θ_k **:** The angular position of the center of the mobile robot at the state k;
- $\dot{\theta}_{k+1}$ **:** The angular velocity of the mobile robot at the state k+1.

This vector describes the posture of the mobile robot in its configuration space referring to its previous one. So that we can specify at any time the different coordinates of the studied mobile robot. Indeed, this vector will be used after in the algorithm of the crossover operator for the genetic algorithm, to calculate the position that respects the temporal constraint. Furthermore, it's used to determine the velocity in each beacon across the path tracking for the fuzzy controller.

3.2 Formulation of the Problem

The movement of the two-wheeled mobile robot in its workspace is characterized by a graph. The graph is a set of n knots corresponding to the artificial beacons' positions in the configuration space of the mobile robot. These beacons mark the sites to be observed by the studied mobile robot. Two successive knots k and k+1 represent an arc of the graph that defines the movement of the mobile robot from one objective to the other.

This graph was used to generate a planner of a road that is estimated as a reference trajectory for the studied mobile robot. The cost estimated for this trajectory corresponds to the crossed distance. A heuristics of optimization is used to have a minimal length for this trajectory. It was the genetic algorithms approach that has been carried out in a previous work (Tlijani & Knani, 2009).

In this graph, the state of the mobile robot at a fixed moment is defined by the following data:

- The position of artificial beacon to which the mobile robot is going.
- t_k **:** Time in the position k;
- t_{k+1} **:** Time in the position k+1;
- The posture of the mobile robot (position and speed): described in the previous section of this paper.
- **T:** Time between two successive positions: it determines the time windows.
- **n:** Number of the artificial beacons in the space of work.
- **k:** Current position of the mobile robot; k=1,.., n.

The linear velocity is defined by Equation 2:

$$v_{k+1} = v_k + a\,dt \tag{2}$$

Where:

- v_k : Linear velocity in the position k;
- v_{k+1} : Linear velocity in the position k+1;
- $a\,a$: Acceleration needed so that dt will be closer to the value of the time window;
- dt : Time delay between two positions of the mobile robot (k+1 and k).

And the temporal constraint is defined as following in Equation 3:

$$t_{k+1} - t_k \leq T \tag{3}$$

k =1, ..., n.

The solution presented by the solver of the previous works that was based on the method of neighborhood, allowed to have the closest following position respecting the constraint of Equation 3.

In the present solution one shall be anxious to fit the time delay (dt) to the value of T by acting on the velocity of the mobile robot by the acceleration a.

So that the proposed solution should answer to the cases of the kind:

- If dt close to T then the robot has to keep the same speed;
- If dt far from T then the robot has to accelerate;
- If dt very close to T then the robot has to slow down;
- If dt very far from T then the robot has to accelerate.

The best way to answer these choices is the use of fuzzy logic. In this case a fuzzy controller will be realized. The different beacons that structure the path solution of the solver of time constraint will be the data source to the fuzzy controller. It aims to improve the time delay between two positions of the path tracking. In each position, the current velocity of the controlled mobile robot is an input for the fuzzy controller. The second input is the time delay between the current position and the next sub goal to reach.

4. PROPOSED FUZZY LOGIC APPROACH CONTROL

In robotics, designers like often designing a robot that acts more as a human being than a machine. The use of the traditional logic seems insufficient to include human abilities and adjust them to a robot, essentially decision making. Since this type of logic is based on two values, one (true) and zero (false). Therefore, fuzzy logic is a method that uses the entire interval between zero and one, and can as such be used to imitate human reasoning (Harisha, Ramkanth, Krishna & Sharma, 2008).

The studies on fuzzy logic theory have increased tremendously over the last few decades (Zimmerman, Zadeh & Gaines 1984). The application of fuzzy logic theory to a control system has moved forward to the Fuzzy Logic real world applications (Mamdani & Assilian, 1999). The implementation of a fuzzy controller requires the affectation of the values of the function of membership for the variables of the inputs and outputs since these later are linguistic variables having no quantitative value. For the variables of inputs their values of membership function are often measured and are affected before being used by the interference engine. This section is called the fuzzification. The inference engine is based on a set of rules: If-Then describing linguistic variables. As a result of the interference engine, the variable inputs will have numerical values gathered in a fuzzy set. Later on comes the defuzzification, this third section affects a single value to the output of the fuzzy controller.

The theory of fuzzy sets and fuzzy logic won't be detailed in this paper. However, the reader can refer to some publications in literature that deal with this paradigm as Fuzzy sets (João, 2005; Alkhazaleh.& Salleh, 2012 ; Cornelis, Deschrijver & Kerre, 2004).

In addition, it is interesting to explore some novel ideas defining a fuzzy system and its application. This has been discussed well in the recent literature (Azar, 2010; Azar, 2012). The up-to-date approach such as artificial neural networks, improves the classic modeling of a fuzzy system. Herein, the author proposed to adjust parameters to the fuzzy inference system that were tuned by a learning procedure. In that overview the author explained the use case of the Type-2 Fuzzy Logic Systems and its contribution compared with type-1 Fuzzy Logic System.

The design procedure of the proposed fuzzy logic controller is given in detail in the next section. Indeed, Classical Constraint Satisfaction Problem has been extended to incorporate different types of intelligent techniques frequently found in real problems. One successful example is fuzzy logic. The present application of this paradigm is a new way to the use of fuzzy logic in the resolved constrained problem. Because in this work, the fuzzification is adopted to give results according to the range of values that satisfy the temporal constraint defined by Equation 3.

4.1 Fuzzy Controller Model

A Fuzzy Logic Controller has been considered as a control law implementing an error function in order to secure the desired performances of the system. It contains three main components: the fuzzifier, the inference system and the defuzzifier. Figure 2 exhibits the different blocs that must be in the structure of a fuzzy controller.

The used structure of a fuzzy logic controller begins with a fuzzifier then an inference system. It ends with a defuzzifier. As shown in Figure 3, the fuzzy system to be developed is a two-input-and-one-output fuzzy system. There are two inputs to the fuzzy inference system. One is the velocity of the mobile robot in the position of the current beacon, which refers to a surveyed site, the other one is the time needed to reach the next goal. These two inputs are firstly fuzzified and converted to fuzzy membership values. The output corresponds to the acceleration so that the mobile robot reaches the next surveyed site in a time close to the time window.

The inference system consists of establishing the fuzzy rule base. This later may also be called as the fuzzy decision table, it is the unit mapping two crisp inputs, velocity and time to the fuzzy output space defined on the acceleration. The fuzzy rules represent the knowledge and the abilities of a human operator who makes necessary adjustments to operate the system with needed velocity and to respect the time window.

After that the fuzzy logic controller proceeds to the defuzzification to give the crisp value of the output. The output resultant from the fuzzy logic controller is the acceleration. In reality, it models the

Figure 2. Components of Fuzzy Logic Controller

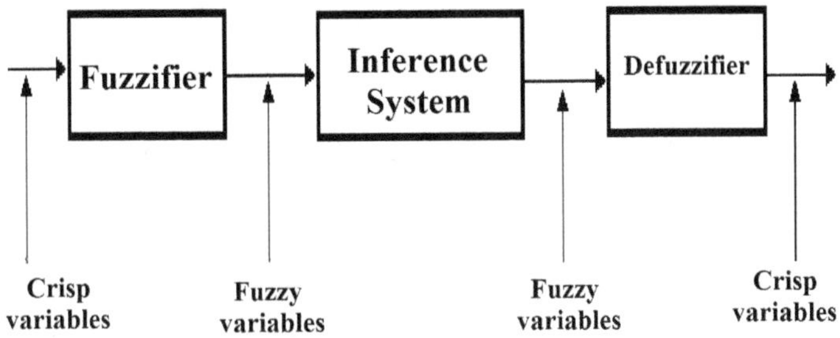

Figure 3. Diagram block of the fuzzy controller system for a two-wheeled mobile robot

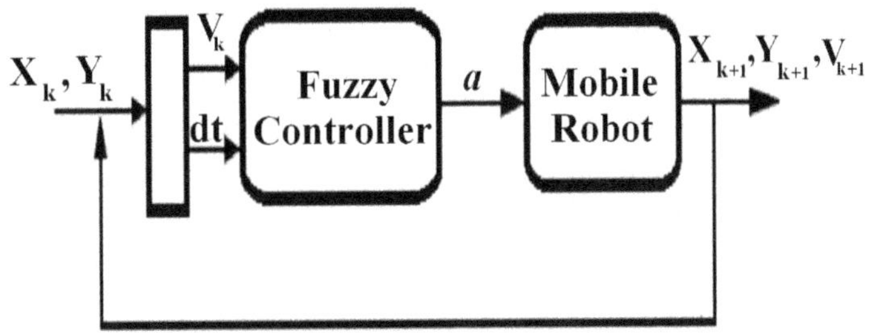

actions that a human operator would decide: whether the velocity of the mobile robot is to be increased or decreased according to the time remained to reach the next beacon.

4.2 Fuzzification

The fuzzification converts the measurements of the crisp variables into fuzzy data. In our case, crisp variables are: the velocity (v_k), the time delay (*dt*) and the acceleration (*a*).

Observing the realized fuzzy controller, the velocity, time delay and acceleration can be described using linguistic terms as following:

- **The Velocity** (v_k): The Big (B), the Very Big (VG), the Very Low (VL), the Low (L);
- **The Time Delay** (dt): The Small (S) and the Very Small (VS) when the value of the time delay is near to the value of the time window T. The Big (B) and the Very Big (VB) when the value of the time delay is far from the value of the time window T;
- **The Acceleration** (a): The Big Negative (BN), the Medium Negative (MN) and the Small Negative (SN) when the mobile robot must decelerate. The Zero (Z) when the mobile robot must keep its velocity constant. The Small Positive (SP), the Medium Positive (PM) and the Big Positive (BP) when it should accelerate.

These linguistic terms, are represented by fuzzy sets partitioned into crisp universes. They are described by Membership functions. A Membership functions is a curve that defines how each point in the input space is mapped to a membership value between 0 to1. Mathematically it is defined as shown in Equation 4 (Sourabh, Sadhna & Vineet, 2013).

$$A = \{(x, \mu_A(x)) \mid x \in X\} \tag{4}$$

Where:

- x is an element in X and $\mu_A(x)$ is the membership function of set A which defines the membership of fuzzy set A in the universe of discourse, X (Lamba, 2008).

The figures Figure 4, Figure 5 and Figure 6 indicate the functions member of triangular membership used to describe fuzzy sets variables of inputs (v_k, *dt*) and output (*a*), respectively.

The membership functions are defined as a Triangular membership functions for all input and output variables of the developed system.

Triangular type membership functions are used for partitioning the crisp universes into fuzzy subsets. Different types of membership functions such as Gaussian, trapezoidal, and bell could have also been used. Each one of these membership functions has its own effects on the fuzzy logic controller output (Altas, 1998). However, triangular membership functions are more convenient for expressing the concept because it is easier to intercept membership degrees from a triangle. Therefore the following function of Equation 5 is used to represent the fuzzy triangular membership functions.

$$\mu(x) = \max\left[\min\left(\frac{x - x_1}{x_2 - x_1}, \frac{x_3 - x}{x_3 - x_2}\right), 0\right] \tag{5}$$

where, x is the crisp values from one of the three universes v, dt, and a. x1 is the left end point on the corresponding crisp universe as the x2 and x3 are the crisp points corresponding peak and right end points, respectively.

Figure 4. Membership function of the velocity v

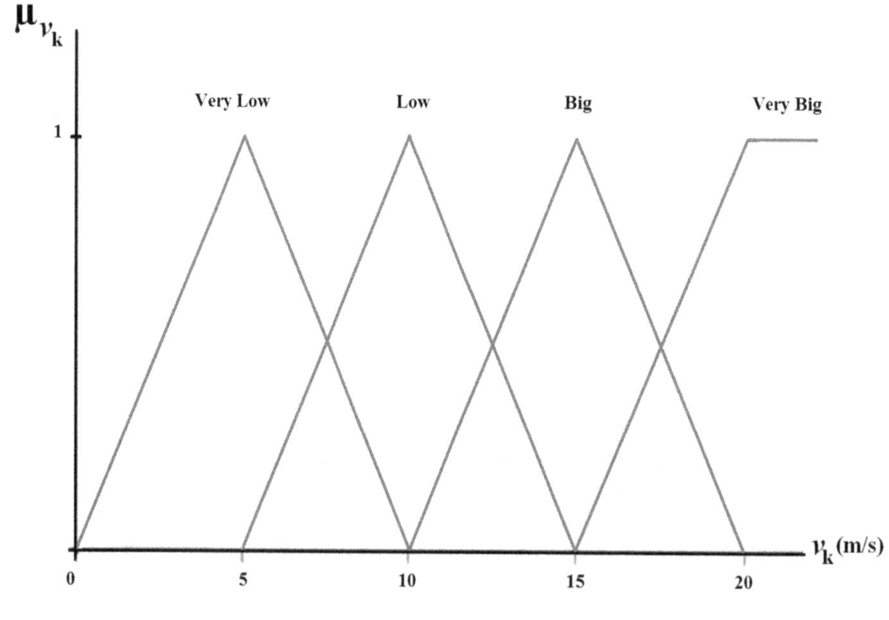

Figure 5. Membership function of the time delay dt

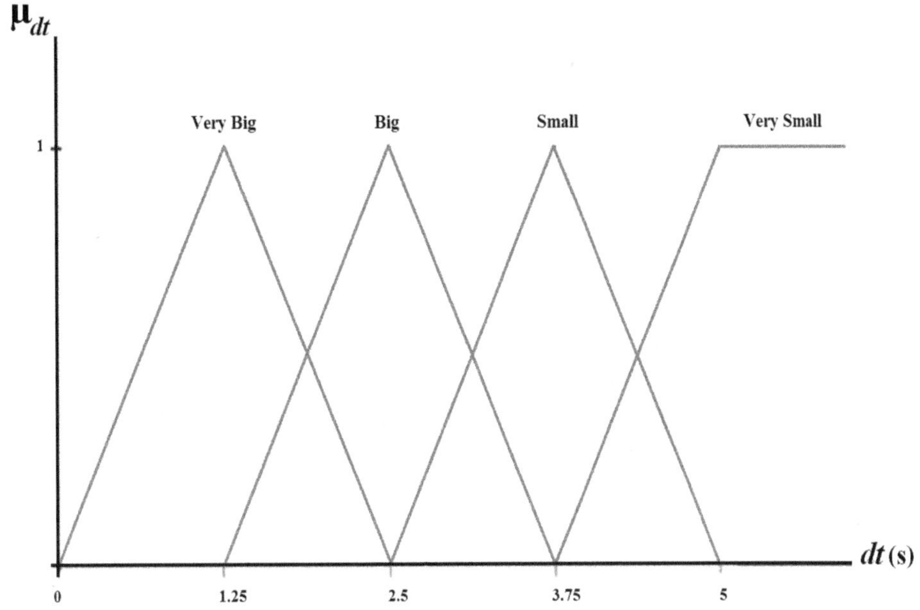

Figure 6. Membership Function of the acceleration a

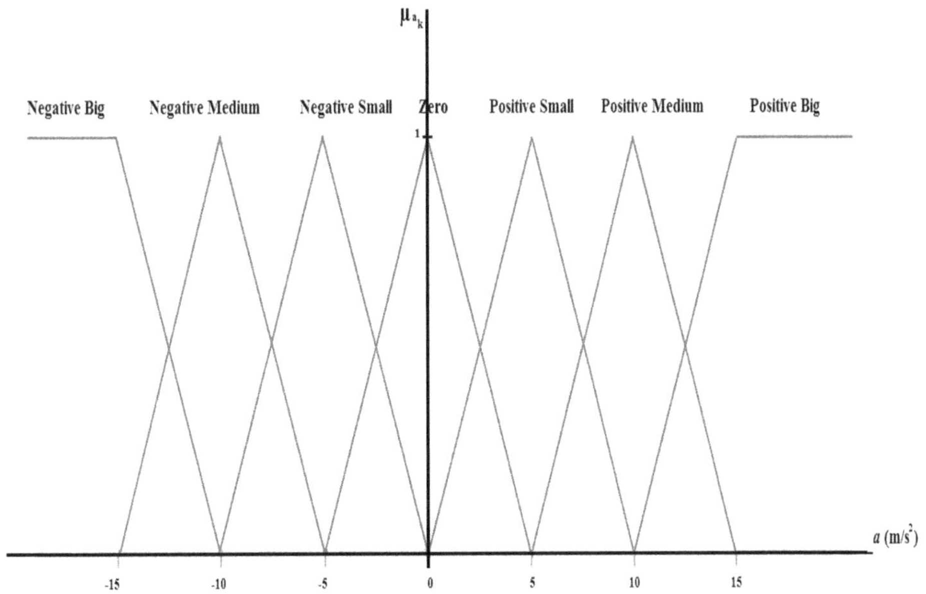

4.3 Inference

Fuzzy inference systems have been applied successfully in fields such as automatic control, data classification, decision analysis, expert system (Lamba, 2008). The inference is the step that maps the inputs

Table 1. Fuzzy Base rule of the fuzzy controller system for the mobile robot

a		v_k			
		VL	**L**	**B**	**VB**
dt	**VB**	PB	PB	PM	PS
	B	PB	PM	PS	Z
	S	PM	PS	NS	NM
	VS	PS	Z	NB	NB

to outputs using fuzzy logic. As a result of this mapping, we get the decision based on the value of the input. The process of the fuzzy inference involves the process of the if–then rules and the decision table.

For the realized fuzzy logic controller, the following rule decision table is defined. The decision table can be formed as shown in Table 1. This table is built manually.

The mode of use of this table is illustrated by the fuzzy rules proposed by the system described as follows:

R1: If (((dt is VB) and ((v_k is VL) or (v_k is L)) or ((dt is B) and (v_k is VL)) Then (a is Positive Big).

This rule indicates that if the time delay is lower than the value of the time window T and the mobile robot moves are slowed down so this robot must accelerate. Consequently, the acceleration of the mobile robot is increasing.

R2: If (((dt is VB) and (v_k is L)) or ((dt is VB) and ($v_k$is B))) Then (a is Positive Medium).

Which means that if the time delay is lower than the time window and the speed of the mobile robot is low or high, the mobile robot has to accelerate with medium value of a.

R3: If (((dt is B)) and (v_k is B)) or ((dt is VB)) and (v_k is VB)) or ((dt is S) and (v_k is L)) or ((dt is very S) and (v_k is VL))) Then (a is Positive Small).

The third rule treats the case when the time delay is both lower than the time window or nearer to its value and the mobile robot moves with low speed. The decision to make is accelerating with small value of a.

R4: If (((dt is B)) and (v_k is VB))) or ((dt is VS) and (v_k is L))) Then (a is Zero).

In the present rule the mobile robot should keep its velocity constant.

R5: If ((dt is S) and ((v_k is B)) Then (a has is Negative Small).

However, in this rule the fuzzy system let the mobile robot decelerate slowly. In fact, the mobile robot moves fast and it's near to next sub-goal.

R6: If ((dt is S) and ((v_k is VB)) Then (a is Negative Medium)).

In this rule, also the mobile robot must decelerate. Since its velocity is high and the time delay to reach the sub-goal is near to the time window T.

R7: If ((dt is very S) and ((v_k is VB) or (v_k is B)) Then (a is Negative Big).

In the 7th rule, the acceleration of the mobile robot must decrease when the time delay to attend the next sub-goal is small and its velocity is big.

These seven rules will form the base rule of the studied fuzzy controller. One can notice the use of the logical operators "AND" and "OR" in these rules. The realization of these logical operations is connected to the method of the used inference. In the present case the method Max-Min as it's mentioned in the section 4.2, should be used. This method of inference realizes the conditions connected by OR are used to the maximum however the conditions connected by AND are used to the minimum.

4.4 Defuzzification

After establishing the base of rules and specifying the membership functions of variables, comes the defuzzification. This stage allows converting the linguistic variables of the fuzzy controller into the physical variables. For this reason the Defuzzification is defined as the process of producing a quantifiable result in fuzzy logic, with given fuzzy sets and corresponding membership degrees. It is typically needed in fuzzy control systems. These will have a number of rules that transform a number of variables into a fuzzy result. For example, in the studied fuzzy controller, the rules designed to decide about how much acceleration to be applied might result in "Big Negative, Zero, Small Positive". The defuzzification is interpreting the membership degrees of the fuzzy sets into a specific decision or a real value.

The simplest but the least useful defuzzification method is to choose the set with the highest membership. The problem with this approach is that it is not accurate as it loses information. The rules that called for decreasing or maintaining velocity might not be there in this case.

A common and useful defuzzification technique is the center of gravity. First, the results of the rules must be added together in some way. The most typical fuzzy set membership function has the graph of a triangle. Now, if this triangle had to be cut in a straight horizontal line somewhere between the top and the bottom, and the top portion had to be removed, the remaining portion would form a trapezoid.

There are many different methods of defuzzification such as: AI (adaptive integration), BADD (basic defuzzification distributions), COA (center of area), COG (center of gravity), ECOA (extended center of area), EQM (extended quality method), FCD (fuzzy clustering defuzzification), and FM (fuzzy mean) …

$$a = \frac{\sum_{i=1}^{7} \mu_a(\alpha_i) * \alpha_i}{\sum_{i=1}^{7} \mu_a(\alpha_i)} \tag{6}$$

5. NEW GENETIC ALGORITHM

Genetic programming (Solc & Honzik, 2002) has been verified in recent times to be a practical approach for mobile robot control and navigation (Lee & Zak, 2002). The robots designers specified the behaviors that were desired by encoding the fitness function for the population to evolve (Lazarus & Hu, 2001). That approach addressed the design of fuzzy logic controllers handling temporal constraint in a Genetic Algorithms-based plan.

5.1 The Basic Algorithm

The first idea is to look for a first eligible optimal solution by applying an evolutionary algorithm (genetic algorithms). The basic algorithm is an application of the genetic algorithms. It was studied; developed and tested at the research presented in (Tlijani & Knani, 2009). The proposed algorithm applies a recombination of two better parents to every generation. If the obtained individual is better than the best parent it will be inserted. A mutation is applied to the worst parent if the resultant individual is worse than both parents. After an iteration of this process, it should yield the optimal path of reference according to the objective function. This function minimizes the distance of the path crossed by the mobile robot in its work space. This process is described in the following pseudo-code:

```
RandomInit-Population ()
For (i = 1 to n) do
Select (individual1)
Select (individual2)
Individual-child=Crossover (indivdual1, individual2)
If (fitness (individual-child) <=fitness (individual2) then
Replace (individual2, individual-child)
Else
Mutate (individual2)
End if
Sort-Population ()
End for
```

5.2 The Genetic Algorithms Operators

Genetic algorithm is particularly easy to implement and promises substantial gains in performance. Its basic idea is to keep several sub-populations that are processed by genetic algorithm. Furthermore, a migration mechanism produces a chromosome exchange between the sub-population. The main characteristics of the proposed GA consist of the following:

- A string of number encoding a feasible path joining all detected artificial beacons in the work space;
- The artificial beacons mark the work space of the mobile robot and refer to the sites to survey;
- A gene in the chromosome is an artificial beacon in the configuration space of the mobile robot;

- A chromosome represents an individual and this later is a path in work space of the mobile robot;
- Each path is an individual that has a certain length;
- The length of the path is a fitness degree of the individual;
- **Selection:** The individual chromosomes are selected for reproduction according to the elitist selection. This type of selection allows choosing the individual that has the best fitness;
- **Crossover:** New populations of chromosomes are generated using the algorithm of recombination based on the KFP operator and the meta-heuristic of neighborhood. The algorithm of this new crossover is founded in (Tlijani et al, 2012) and is detailed in (Tlijani, Knani, & M'sirdi, 2014). The method of neighborhood solves the problem of time windows T imposed in the navigation of the mobile robot to survey the sites referred to by the artificial beacons in its work space;
- **Mutation:** A perturbation is applied to the individuals by reversing the path between two positions.

5.3. The New Algorithm

The new algorithm described in the Figure 7, is a combination between the genetic algorithms; that is the approach of the Constraint Satisfaction Problem and the approach of the fuzzy logic. Through the execution of this new algorithm generates the value of the necessary acceleration in every position of the optimized path. This later is yielded from the hybridization of the genetic algorithms and Constraint Satisfaction Problem techniques conducted on and defined in the research work of (Tlijani et al., 2012) and (Tlijani et al., 2014).

In fact, the evolutionary approach gives the shortest path to be followed crossing the different beacons of the work space. And the Constraint Satisfaction Problem techniques bound the time of the passage way

Figure 7. Organigram: combination between the genetic algorithms, Constraint Satisfaction Problem approach and fuzzy logic

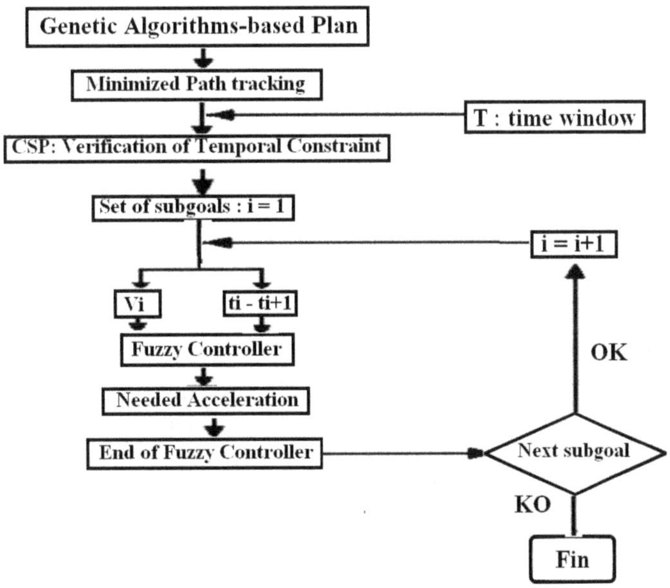

Table 2. Coordinates of artificial beacons' positions

Artificial Beacon Code	0	1	2	3	4
Axe X	7.0	9.0	5.0	3.0	5.0
Axe Y	8.0	1.0	2.0	4.0	6.0

between two successive beacons by a time window. By using the fuzzy logic, the time delayed between every two consecutive beacons is corrected across the path tracking of the tow-wheeled mobile robot. Consequently, the result is a minimal path tracking that respects as much as possible the time window.

6. RESULTS

The simulation of the fuzzy controller of the acceleration is realized by using the programming language C ++ under the development environment of the Microsoft Visual Studio 2010. To run the program, the mobile robot is assumed to a one particle in its configuration space. This later is a horizontal plan equipped by a number of artificial beacons (n = 5). Each one represents an objective to reach. The time window has for value T=5mn and the number of neighborhood is fixed to 3.

The coordinates related to the positions of the artificial beacons are represented in the Table 2.

Figure 8 represents the path tracking solution of the generated algorithm which is encoded by the integer sequence: 04321. The different positions of the artificial beacons across this path which are 0, 4, 3 and 2, will be used for the studied fuzzy controller. In each position, the two inputs will be given to the fuzzy controller. These inputs are: the current velocity of the mobile robot besides the real time delay

Figure 8. Path tracking: result of the hybrid algorithm combining genetic algorithms and Constraint Satisfaction Problem techniques

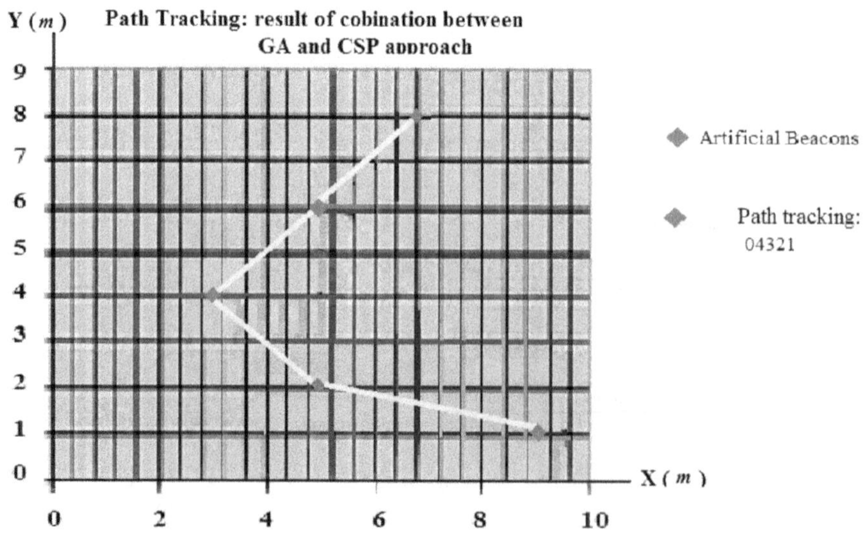

lasting to reach the next site to survey. The role of the fuzzy controller as it's mentioned all through this work is to return back to the needed acceleration. This later will both adjust the velocity of the mobile robot and consequently the real time delay (*dt*) will be corrected to converge to the value of the Time Window.

Since the instance of the genetic algorithm used for the optimization of the path of the mobile robot is programmed using C++, we have used the same programming language to develop the inference system of the realized fuzzy controller. The definition of the linguistic variables inputs and output: *dt*, v_k and *a*, are defined in a class called: RM_En_Flou. In this class defines the various membership functions of the linguistic variables as member functions of this class. The inference is concretized by a class conscripted by RB_Moteur_Inference. In this class declares an array corresponding to the inference table (Table 1). Among the member functions in the class of inference, the function of inference Max-Min (Equation 3) and the function used for the defuzzification were programmed. They are respectively based on the Equation 3 and Equation 4. Running this implemented inference fuzzy system, the fuzzy controller calculates the values of the acceleration. The values of the acceleration correspond to the different points marked by artificial beacons in the resultant trajectory. These values are recapitulated in the following Table 3. The trajectory is yielded by the new algorithm that combines the approach of the genetic algorithms and the techniques of the Constraint Satisfaction Problem.

According to the results existing in Table 3, the studied points in the trajectory are only 0, 4, 3, and 2. It's obvious that the point encoded by 1 of this trajectory is absent. Since this point marks the end of the path tracking it has no successor and the mobile robot will stop its investigation of the surveyed sites at this point.

7. DISCUSSIONS

The generated algorithm was run on a PC (Personal Computer) with dual core Intel. The running time respects full processor utilization. The results of this running were exposed in: Table 3 and represented by the figures Figure 8 and Figure 9.

Figure 8 describes the path that respects the temporal constraint that is having minimal distance after executing the plan based on genetic algorithms. This run applied a population size of 10 and a number of iteration of five runs. The previous work described in (Tlijani, 2014) demonstrated that the path solution yielded is optima which meant that no local minima was given by the proposed algorithm. The genetic programming instance adopted in this work was run under a number of parameters. Ones are related to the generated algorithm based on genetic algorithms such as population size and the value of the time

Table 3. The whole decision for the acceleration value using fuzzy logic

Code Position	0	4	3	2
v_k (m/s)	2	20	13.44	67
dt (mn)	3	2.75	2	4.3
a (m/s²)	6	-1.14	6.9	-9.7

Figure 9. Variation of the speed and the acceleration of the Mobile Robot by using the fuzzy logic controller along the trajectory crossing the surveyed sites and respecting a temporal window

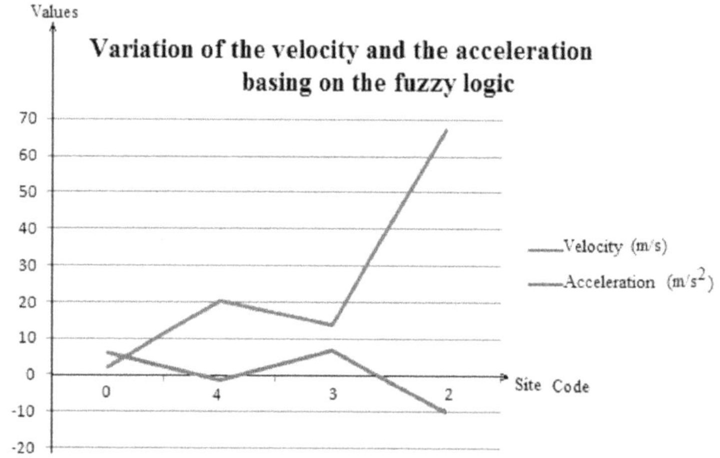

window. The others are related to mobile robot and its work space. These parameters concern in one hand the initial velocity and its initial posture. On the other hand it presents the position of the different artificial beacons (Table 2) that mark the work space and refer to the surveyed sites. In the presented solution mobile robot can start its investigation by a random surveyed site. For instance, in the path of Figure 4, it begins from the site marked by the artificial beacon encoded by 0. The coordinates of this beacon are (7.0, 8.0).

Figure 9 shows the variation of the velocity treated by the fuzzy logic controller and the acceleration that is provided as result. One can notice here that the velocity and the acceleration have opposite variation. This result approves of the realized fuzzy logic controller that executes correctly the inference mapped by the table decision in Table 1 and conducted by the rules defined in section 4.2. All of these inference rules conclude that when the velocity is increasing the acceleration must decrease. Furthermore, it defines the contrast case so when the velocity is decreasing the acceleration must increase.

Figure 9 represents respectively the variation of the velocity and the acceleration along the resultant path tracking (04321). It establishes the curves of the different values obtained for the acceleration then corrected and used for the velocity.

8. CONCLUSION

A mobile robot is a complex system. In fact, it has to satisfy several requirements during its navigation. This requires a reliable and very precise decision-making. A typical example is to execute an optimal plan which respects temporal constraints. In order to reach the various purposes described along this plan, we aim at specifying as much as possible the parameters of the movements of the mobile robot: the acceleration and the velocity. These arbitrations were insured by some control approaches such as the genetic algorithms, the techniques of Constraint Satisfaction Problem and the fuzzy logic. This hybridization shows a new way to treat a constraint defined on a system with fuzzy logic. Actually, this new method approves of the fact that the solver can be guided by human preferences.

The presented work has described the use of the fuzzy logic that affects the parameters related to the movement of the studied mobile robot. It was successfully applied to give the appropriate acceleration so that the time delay becomes closer to the time window.

The future work intends to study the effects of the use of the artificial potential field. This stage aims at involving the ability of correcting the direction in the real navigation of the mobile robot. As it is known, the techniques of the artificial potential field that includes: the attractive function and the repulsive function. This approach is used to avoid the obstacles.

REFERENCES

Alkhazaleh, S., & Salleh, A. R. (2012). Fuzzy soft multiset theory. *Abstract and Applied Analysis, 2012*, 1–20. doi:10.1155/2012/350603

Altas, I. H. (1998). The effects of fuzziness in fuzzy logic controllers. In *Proceedings of the 2nd International Symposium on Intelligent Manufacturing Systems*, (pp. 211-220). Sakarya, Turkey: Academic Press.

Azar, A. T. (2010). *Fuzzy Systems*. Vienna, Austria: In-Tech.

Azar, A. T. (2012). Overview of type-2 fuzzy logic systems. *International Journal of Fuzzy System Applications*, *2*(4), 1–28. doi:10.4018/ijfsa.2012100101

Bansal, P., Maheshwari, S., Kumar, M., & Kumar, U. (2011). Behavior based mobile robot navigation in a dynamic environment: Fuzzy logic approach. *Journal of Active & Passive Electronic Devices*, *6*, 293–303.

Breivik, M., & Fossen, T. I. (2008). Guidance laws for planar motion control. In *Proceedings of the 47th IEEE Conference on Decision and Control*, (pp. 570-577). Cancun, Mexico: IEEE.

Chen, C. C., & Wong, C. C. (2002). Self-generating rule-mapping fuzzy controller design using a genetic algorithm. *IEE Proceedings of Control Theory and Applications*, *149*(2), 143–148. doi:10.1049/ip-cta:20020253

Cheng, P. Y., & Chen, P. J. (2012). Navigation of mobile robot by using D++ algorithm. *Intelligent Service Robotics*, *5*(4), 229–243. doi:10.1007/s11370-012-0120-4

Cooper, M. C. (2003). Reduction operations in fuzzy or value constrained satisfaction. *Fuzzy Sets and Systems*, *134*(3), 311–342. doi:10.1016/S0165-0114(02)00134-3

Cornelis, C., Deschrijver, C., & Kerre, E. E. (2004). Implication in intuitionistic and interval-valued fuzzy set theory: Construction, classification, application. *International Journal of Approximate Reasoning*, *35*(1), 55–95. doi:10.1016/S0888-613X(03)00072-0

Dong, D., Chen, C., Chu, J., & Tarn, T. (2012). Robust quantum-inspired reinforcement learning for robot navigation. *IEEE Transactions on Mechatronics*, *17*(1), 86–97. doi:10.1109/TMECH.2010.2090896

Garcia-Pérez, L., Garcia-Alegre, M. C., Ribeiro, A., & Guinea, D. (2003). Fuzzy control for an approaching maneuver with a car like vehicle in outdoor environments. In *Proceedings of WAF03*. Alicante, Spain: WAF.

Glover, F. (1986). Future paths for integer programming and links to artificial intelligence. *Journal of Computers and Operations Research, 13*(5), 533–549. doi:10.1016/0305-0548(86)90048-1

Harisha, S. K., Ramkanth, K. P., Krishna, M., & Sharma, S. C. (2008). Fuzzy logic reasoning to control mobile robot on pre-defined strip path. *World Academy of Science. Engineering and Technology, 2*(6), 618–622.

Homaifar, A., & McCormick, E. (1995). Simultaneous design of membership functions and rule sets for fuzzy controllers using genetic algorithms. *IEEE Transactions on Fuzzy Systems, 3*(2), 129–139. doi:10.1109/91.388168

Hoong, C. L., Melvyn, S., & Kwong, M. T. (2003). Vehicle routing problem with time windows and a limited number of vehicles. *European Journal of Operational Research, 148*(3), 559–569. doi:10.1016/S0377-2217(02)00363-6

Hsu, F. Y., & Fu, L. C. (2000). Intelligent robot deburring using adaptive fuzzy hybrid position/force control. *IEEE Transactions on Robotics and Automation, 16*(4), 325–335. doi:10.1109/70.864223

Ivanescu, M. (2001). Moving target interception for walking robot by fuzzy controller. In *Proceedings of the Fourth International Conference on Climbing and Walking Robotics*, (pp. 363-376). Karlsruhe, Germany: Academic Press.

Khatib, L., Morris, P., Morris, R., Rossi, F., Sperduti, A., & Venable, K. B. (2007). Solving and learning a tractable class of soft temporal constraints: Theoretical and experimental results. *AI Communication, 20*, 181–209.

Khatib, O. (1986). Real-time obstacle avoidance for manipulators and mobile robots. *The International Journal of Robotics Research, 5*(1), 90–98. doi:10.1177/027836498600500106

Lamba, V. K. (2008). *Neuro Fuzzy Systems*. New Delhi: Laxmi Publications.

Lazarus, C., & Hu, H. (2001). Using genetic programming to evolve robot behaviours. In *Proceedings of the 3rd British Conference on Autonomous Mobile Robotics and Autonomous Systems*. Manchester, UK: Academic Press.

Lee, Y., & Zak, S. H. (May 2002). Genetic fuzzy tracking controllers for autonomous ground vehicles. In *Proceedings of American Control Conference*, (pp. 2144-2149). Anchorage, AK: Academic Press.

Low, K. H., Leow, W. K., & Ang, M. H. (2002). A hybrid mobile robot architecture with integrated planning and control. In *Proceedings of the First International Joint Conference on Autonomous Agents and Multiagent Systems*, (pp. 219-226). New York, NY: Academic Press. doi:10.1145/544741.544797

Mamdani, E. H., & Assilian, S. (1999). An experiment in linguistic synthesis with a fuzzy logic controller. *International Journal of Human-Computer Studies, 51*(2), 135–147. doi:10.1006/ijhc.1973.0303

Miguel, I., & Shen, Q. (2003). Fuzzy rrDFCSP and planning. *Journal of Artificial Intelligence, 148*(1-2), 11–52. doi:10.1016/S0004-3702(03)00020-1

Narvydas, G., Simutis, R., & Raudonis, V. (2007). Autonomous mobile robot control using fuzzy logic and genetic algorithm. In *Proceedings of 4th IEEE Workshop on Intelligent Data Acquisition and Advanced Computing Systems: Technology and Applications*. Dortmund, Germany: IEEE. doi:10.1109/IDAACS.2007.4488460

Nawa, N. E., & Furuhashi, T. (1999). Fuzzy system parameters discovery by bacterial evolutionary algorithm. *IEEE Transactions on Fuzzy Systems*, *7*(5), 608–616. doi:10.1109/91.797983

Siegwart, R., Nourbakhsh, I. R., & Scaramuzza, D. (2011). *Introduction to Autonomous Mobile Robots*. Cambridge, MA: MIT Press.

Solc, F., & Honzik, B. (2002). Modeling and control of a soccer robot. In *Proceedings of the 7th IEEE International Workshop in Advanced Motion Control* (pp. 506-509). Maribor, Slovenia: IEEE. doi:10.1109/AMC.2002.1026972

Song, K. Y., & Tay, J. C. (July 1992). Fuzzy navigation of a mobile robot. In *Proceedings of the 1992 IEEE/RSJ International Conference on Intelligent Robots and Systems*, (pp. 112-132). Raleigh, NC: IEEE.

Sourabh, J., Sadhna, M., & Vineet, R. (2013). An Approach for Image Matching based on ACO fuzzification. *International Journal of Modern Engineering & Management Research*, *1*(2), 28–33.

Sugeno, M., Murofushi, T., Mori, T., Tatematasu, T., & Tanaka, J. (1989). Fuzzy algorithmic control of a model car by oral instructions. *International Journal of Fuzzy Sets and Systems*, *32*(2), 207–219. doi:10.1016/0165-0114(89)90255-8

Sugiatro, I., Tung, L. L. U., & Rahman, M. I. (2008). Implementation of fuzzy logic in FPGA for maze tracking. *Jurnal Teknik Elektro*, *8*(2), 96–102.

Teslic, L., Škrjanc, I., & Klanèar, G. (2011). EKF-based localization of a wheeled mobile robot in structured environments. *Journal of Intelligent & Robotic Systems*, *62*(2), 187–203. doi:10.1007/s10846-010-9441-8

Tlijani, H., & Knani, J. (2009). Optimization of the mobile robot's track using genetic algorithms. In *Proceedings of the 10th International Conference on Sciences and Techniques of Automatic Control & Computer Engineering*. Hammamet, Tunisia: Academic Press.

Tlijani, H., Knani, J., & M'sirdi, K. N. (2012). Optimization of a mobile robot's path tracking with time windows using Hybridization of genetic algorithms and CSP approach. In *Proceedings of International Conference on METAheuristics and Natural Inspired Computing*. Sousse, Tunisia: Academic Press.

Tlijani, H., Knani, J., & M'sirdi, K. N. (2014). Hybrid algorithm based on genetic algorithms and CSP approach to plane an optimized path for a mobile robot moving under time window. *IOSR Journal of Computer Engineering*, *16*(3), 74–83. doi:10.9790/0661-16337483

Tlijani, H., M'sirdi, N. K., & Knani, J. (2011). Optimization of the mobile robot's path tracking: combination of genetic algorithm, artificial potential field and backtracking algorithm. In *Proceedings of the IEEE International Conference on Communications, Computing and Control Applications*, (pp. 97-102). Hammamet, Tunisia: IEEE.

Wong, C. C., Wang, H. Y., Li, S. A., & Cheng, C. T. (2007). Fuzzy controller designed by GA for two-wheeled mobile robots. *International Journal of Fuzzy Systems*, *9*(1), 22–30.

Yan, I., Ryan, M., & Power, I. (1994). *Using Logic Towards Intelligent Systems*. Prentice Hall.

Zeng, C., Zhang, Q., & Wei, X. (2012). GA-based global path planning for mobile robot employing A* algorithm. *Journal of Computers*, *7*(2), 470–474. doi:10.4304/jcp.7.2.470-474

Zimmerman, H. J., Zadeh, L. A., & Gaines, B. R. (1984). Fuzzy sets and decision analysis. *Studies in the Management Sciences*, *20*, 3–8.

KEY TERMS AND DEFINITIONS

Constraint Satisfaction Problems: Mathematical problems defined as a set of objects whose state must satisfy a number of constraints or limitations.

Fuzzy Controller: A control system based on fuzzy logic.

Fuzzy Logic: A form of many-valued logic; it deals with reasoning that is approximate rather than fixed and exact.

Genetic Algorithms: A search heuristic that mimics the process of natural selection. This heuristic is used to generate useful solutions to optimization and search problems.

Mobile Robot: An automatic machine that is capable to move around in its environment and is not fixed to one physical location.

Path Tracking: A reference path followed by a mobile robot.

Time Window: a bound time calculated for a vehicle to go from two sites.

Section 5
Time-Delay and Switched Control Systems

Chapter 18
On Stability Analysis of Switched Linear Time-Delay Systems under Arbitrary Switching

Marwen Kermani
National Engineering School of Monastir, Tunisia

Anis Sakly
National Engineering School of Monastir, Tunisia

ABSTRACT

This chapter focuses on the stability analysis problem for a class of continuous-time switched time-delay systems modelled by delay differential equations under arbitrary switching. Then, a transformation under the arrow form is employed. Indeed, by using a constructed Lyapunov function, the aggregation techniques, the Kotelyanski lemma associated with the M-matrix properties, new delay-dependent sufficient stability conditions are derived. The obtained results provide a solution to one of the basic problems in continuous-time switched time-delay systems. This problem ensures asymptotic stability of the switched time-delay system under arbitrary switching signals. In addition, these stability conditions are extended to be generalized for switched systems with multiple delays. Noted that, these obtained results are explicit, simple to use, and allow us to avoid the problem of searching a common Lyapunov function. Finally, two examples are provided, with numerical simulations, to demonstrate the effectiveness of the proposed method.

1. INTRODUCTION

A switched system is composed of an indexed family of subsystems described by continuous or discrete-time dynamics and a rule orchestrating the switching between them. As a special class of hybrid systems, switched systems have strong engineering background in various areas and are often used as a unified modeling tool for a great number of real-world systems such as power electronics chemical processes,

DOI: 10.4018/978-1-4666-7248-2.ch018

mechanical systems, automotive industry, aircraft and air traffic control and many other fields (Morse, 1997; Liberzon & Morse, 1999; Liberzon, 2003; Sun & Ge; 2005, Sun & Ge, 2011; Shorten et al., 2003).

There has been an increasing research interest in stability analysis and control design for switched systems during the past two decades, Therefore, it is well known that stability of switched systems depends not only on the dynamics of every subsystem but also on the property of the switching signal. Thus, it is commonly recognized that there are mainly located in three basic types of problems considering the stability and stabilisation issues of switched systems (Lin & Antsaklis, 2009; Yang, Xiang & Lee, 2012; Hu et al., 2013):(i) asymptotical stability of the switched system with arbitrary switching, (ii) stability for certain useful classes of switching sequences, and (iii) construction of asymptotically stabilizing switching signals for a switched system. Specifically, stability analysis under arbitrary switching problem (i) which will be focused in this article is fundamental in the analysis and design of switched systems (Liberzon & Morse, 1999; Sun & Ge, 2011; Shorten et al., 2003; Zhang & Yu, 2008; Kim et al., 1999; Shorten et al., 2006; Liberzon & Tempo, 2004, Shorten et al., 2007). This problem deals with the case that all the subsystems are stable. Indeed, they exist many examples where all subsystems are stable but inappropriate switching rules can make the whole system unstable. In addition, if we know that a switched system is stable under arbitrary switching, then we can consider higher control specifications for the system. In this framework, it is well known that the existence of a common Lyapunov function for all subsystems is a sufficient condition for such systems to be asymptotically stable under arbitrary switching (Shorten et al., 2003; Shorten & Narendra, 2003; Narendra & Mason, 2003; Sun et al., 2006; Liberzon & Tempo, 2004). However, this method is usually very difficult to apply even for continuous-time switched linear systems (Liberzon, 2003; Shorten et al., 2007).

Frequently, to avoid the conservatism related to the existence of a common Lyapunov function, some attention has been widely interested for seeking conditions that guarantee stability of the switched systems under restricted switching problems (ii) and (iii). Although many efficient approaches and important results have been proposed for this alternative such as the multiple Lyapunov function approach (Branicky, 1998) and average dwell time method (Ishii, 2002; Hespanha, 1999), stability under arbitrary switching which is considered in this work remains most preferable for practical systems. Indeed, it offers great flexibility and it allows us to achieve other performances for designing a control law along stability maintained.

On the other hand, time-delay phenomenon is very common in various practical systems. The presence of delay may degrade systems performance and lead to instability. Indeed, many engineering systems always involve time delay phenomenon. Therefore, switched time-delay systems have various applications in practical engineering systems for instance, power systems (Meyer et al., 2004), power electronics (Sun & Ge, 2002) and networked control systems (Zhang & Yu, 2008). Hence, considerable discussions on delays and their impact on performance as well as on the analysis and design problems in control systems have attracted the interest of numerous investigators in recent years. It is generally advised to incorporate the delays to avoid design flaws and incorrect analysis conclusions. The available results can be categorized into two broad categories: delay-independent and delay-dependent. In the former category, the solutions of stability and stabilization problems are feasible irrespective of the size of the delay. In the latter category, methods are developed to take information about time delays into consideration in the process of controller design. Generally, delay-dependent methods are regarded as more practical and yield less conservative designs.

The stability analysis problem of switched time-delay systems has attracted a lot of attention from many researchers (Hien et al., 2009; Colaneri et al., 2008; Shao, 2009; Wu et al., 2013; Qi & Sun, 2013). However, the presence of delay makes this problem much more complicated. Indeed, the main approach for stability analysis under arbitrary switching relies on the use of Lyapunov-Krasovskii functional and LMI approach for constructing common Lyapunov function (Hien et al., 2009; Kim et al., 2006). Indeed, getting common Lyapunov function becomes more complicated when switched time-delay systems are considered. This drawback has motivated us to look for another approach to guarantee stability under arbitrary switching.

In this paper, based on the construction of a common Lyapunov function as well the use of the Kotelyanski lemma (Kotelyanski, 1952; Elmadssia et al., 2011; Elmadssia et al., 2013; Jaballi et al., 2013; Kermani & Sakly, 2014; Kermani et al., 2014; Benrejeb et al., 2008; Filali et al., 2012; Benrejeb, & Borne, 1978; Sfaihi, & Benrejeb, 2013; Benrejeb et al., 2006) associated with the $M-$ matrix properties (Araki, 1975; Robert, 1966; Gantmacher, 1966), new delay-dependent sufficient stability conditions for continuous-time switched time-delay systems under arbitrary switching are deduced. In addition, this obtained result is extended to be generalized for switched systems with multiple delays. Noted that, this proposed method can guarantee stability under arbitrary switching and allows us to avoid the search of a common Lyapunov function which is very difficult matter.

Within the frame of studying the stability analysis, this approach was introduced in (Elmadssia et al., 2011; Elmadssia et al., 2013) for continuous-time delay systems and in our previous work (Jaballi et al., 2013; Kermani & Sakly, 2014; Kermani et al., 2014) for discrete-time switched time-delay systems.

This paper is organized as follows. Section 2 formulates the problem and presents some definitions. An overview about the exiting results on stability study of switched systems are presented in Section 3. Some stability conditions deducted by using the overvaluing continuous systems are presented in section 4. The main results of this paper is given in Section 5. Section 6 is devoted to derive new delay-dependent conditions for asymptotic stability of switched systems defined by differential equations. Then, this result is extended for switched systems with multiple delays in section 7. Two numerical examples are provided to illustrate the design results in Section 8. More discussion about the obtained results are given in Section 9. Concluding remarks are given in Section 10.

Notations

The notations in this paper are fairly standard; if not explicitly stated, matrices are assumed to have compatible dimensions. I is an identity matrix with appropriate dimension. Let $\Re^n$ denoted an n dimensional linear vector space over the reals $\|.\|$ stands for the Euclidean norm of vectors. For any $u=\left(u_i\right)_{1\leq i\leq n}$, $v=\left(v_i\right)_{1\leq i\leq n}\in\Re^n$, we define the scalar product of the vector u and v as: $\left\langle u,v\right\rangle=\sum_{i=1}^{n}u_iv_i$. $\lambda\left(M\right)$ denote the set of eigenvalues of matrix $M=\left(m_{i,j}\right)_{1\leq i,j\leq n}$, M^T its transpose and M^{-1} its inverse and we denote $M^*=\left(m^*_{i,j}\right)_{1\leq i,j\leq n}$ with $m^*_{i,j}=m_{i,j}$ if $i=j$ and $m^*_{i,j}=\left|m_{i,j}\right|$ if $i\neq j$ and $\left|M\right|=\left|m_{i,j}\right|$, $\forall i,j$.

2. PROBLEM FORMULATION AND PRELIMINARIES

2.1. Problem Formulation

Consider the following switched time-delay system formed by N subsystems given by:

$$\begin{cases} \dot{x}(t) = A_{\sigma(t)}x(t) + D_{\sigma(t)}x(t-h) \\ x(t) = \phi(t) \quad t \in [-h, 0] \end{cases} \tag{1}$$

where $x(t) \in \Re^n$ is the state vector, $h > 0$ is the time delay, A_i and D_i are constant matrices of appropriate dimensions for $i = 1, \ldots, N$. $\phi(t)$ is the continuous initial function and $\sigma(t)$ is the switching rule defined by $\sigma(t): \Re_+ \to \underline{N} = \{1, \ldots, N\}$, assumed to be available in real time. Therefore, the switched system is composed of N subsystems which are expressed as:

$$\dot{x}(t) = A_i x(t) + D_i x(t-h), \quad i \in \underline{N} \tag{2}$$

Let $v(t) = [v_1(t), \ldots, v_N(t)]^T$ be the indication function which will be used in order to describe the system. Therefore, the components of $v(t)$: $v_i(t): \Re_+ \to M = \{0, 1\}$ are an exogenous functions that depends only on the time and not on the state, they are defined through:

$$v_i(t) = \begin{cases} 1 \text{ if } \sigma(t) = i; \\ 0 \text{ otherwise.} \end{cases} \tag{3}$$

it is obvious that $\sum_{i=1}^{N} v_i(t) = 1$.

2.2. Preliminary

In this section, some definitions are firstly introduced.

Definition 1: The equilibrium point of system (1) is said to be:

1. Uniformly asymptotically stable if for any $\varepsilon > 0$, there is a $\delta(\varepsilon) > 0$ such that $\max_{-h \le t \le 0} \|\phi(t)\| < \delta$ implies $\|x(t, \phi)\| \le \varepsilon$, $t \ge 0$. For arbitrary switching $\sigma(t)$, and there is also a δ' such that $\max_{-h \le t \le 0} \|\phi(t)\| < \delta'$ implies $\|x(t, \phi)\| \to 0$ as $t \to \infty$ for arbitrary switching signal.
2. Locally exponentially stable, if there exist $M > 0$ and $\gamma > 0$ and $\varepsilon > 0$ such that if $\|x(t_0)\| < \gamma$ the solution $x(t, t_0, x(t_0))$ of (1) satisfies:

$$\left\| x\left(t, t_0, x\left(t_0\right)\right) \right\| \leq M e^{-\varepsilon\left(t-t_0\right)},\ t \geq t_0 \tag{4}$$

Globally exponentially stable, if there exist $M > 0$, $\gamma > 0$ and $\varepsilon > 0$ such that the solution $x\left(t, t_0, x\left(t_0\right)\right)$ of (1) satisfies:

$$\left\| x\left(t, t_0, x\left(t_0\right)\right) \right\| \leq M \left\| x\left(t_0\right) \right\| e^{-\varepsilon\left(t-t_0\right)},\ t \geq t_0 \tag{5}$$

3. STABILITY STUDY OF SWITCHED SYSTEMS

In what follows, basic results on stability analysis of switched systems are given. Therefore, these results are classed in two categories: stability under arbitrary switching and stability under constrained switching laws. Indeed, stability analysis under arbitrary switching is mainly based on the common Lyapunov function method. Then, stability under constrained switching is built essentially on the average dwell-time method. Some basic results for switched systems without delay and switched time-delay systems are presented in the next two subsections.

3.1. Stability Analysis of Switched Systems without Delay

In this subsection, let consider the following switched system:

$$\dot{x}\left(t\right) = A_{\sigma(t)} x\left(t\right) \tag{6}$$

Therefore, this considered switched system is composed of N subsystems given as below:

$$\dot{x}\left(t\right) = A_i x\left(t\right),\quad i \in \underline{N} \tag{7}$$

3.1.1. Stability Analysis under Arbitrary Switching

For this problem, it is necessary to require that all the subsystems are asymptotically stable. However, even when this above mentioned condition is satisfied, it is still possible to construct a divergent trajectory from any initial state for such a switched system. Therefore, in general, the above subsystems stability assumption is not sufficient to assure stability for the switched systems under arbitrary switching. In this context, it is well known that, if there exists a common Lyapunov function for all the subsystems, the switched system is stable under arbitrary switching. This provides us with a possible way to solve this problem, and many efforts were focused on the common quadratic Lyapunov functions.

Common Quadratic Lyapunov Functions

The existence of a common quadratic Lyapunov function for all the subsystems (7) assures the quadratic stability of system (6). Indeed, quadratic stability is a special class of exponential stability, which implies asymptotic stability. Therefore, quadratic stability has attracted a lot of research efforts due to its importance in practice it is well known that the conditions for the existence of a common quadratic Lyapunov function can be expressed as linear matrix inequalities. Namely, there exists a positive definite symmetric matrix P, $P \in \Re^{n \times n}$, verified the following inequality (Lin & Antsaklis, 2009):

$$PA_i + A_i^T < 0,\ i \in \underline{N} \tag{8}$$

3.1.2. Stability Analysis under Restricted Switching

Switched systems may be stable under restricted switching signals. Indeed, restricted switching may arise naturally from the physical constraints of the system, for example, in the automobile gear switching. Moreover, there are cases when one may have some information about possible switching of the switched system, partitions of the state space and their induced switching rules. This information may leads restrictions on the switching signals. For example, here must exist certain bound on the time interval between two successive switching, which may be due to the fact that the state trajectories have to spend some finite length of time in traveling from the initial set to certain guard sets, if these two sets are separated.

In the next; the definitions of dwell time and average dwell time are presented.

Definition 2: A positive constant $T_d \in \Re_+$ called the dwell time of the switching signal $\sigma(t)$ if the time interval between any two consecutive switching is no smaller than T_d.

It can be shown that it is always possible to maintain stability when all the subsystems are stable and the switching is slow enough, in order to have that T_d sufficiently large (Lin & Antsaklis, 2009). Actually, it really does not matter if one occasionally have a smaller dwell time between switching, provided this does not occur too frequently. This concept is captured by average dwell-time.

Definition 3 (Liberzon, 2003): For any $T_2 > T_1 \geq 0$, let $N_\sigma(T_2, T_1)$ denote the number of switching $\sigma(t)$ over (T_1, T_2). If $N_\sigma(T_2, T_1) \leq N_0 + \dfrac{(T_2 - T_1)}{T_a}$ holds for $T_a > 0$, $N_0 \geq 0$. Then T_a is called the average dwell time. It is commonly assumed in the literature that $N_0 = 0$.

It was shown in (Lin & Antsaklis, 2009) that if all the subsystems are exponentially stable then the switched system remains exponentially stable provided that the average dwell time is sufficiently large.

Theorem 1 (Lin & Antsaklis, 2009): Assume that all subsystems in a switched linear system are exponentially stable. There exists a scalar $T_a^* > 0$ such that the switched system is exponentially stable if the average dwell time is larger than T_a^*.

3.2. Stability Analysis of Switched Time-Delay System

This section focus on stability analysis problems of continuous-time switched time-delay system. Indeed, for stability under restricted switching problem the abodes mentioned dwell time and average dwell time approaches still play an important role in this context

However, for stability under arbitrary switching, the common Lyapunov function method remainders essential in this context by using other varieties of Lyapunov function that are well suited for switched delay system such as Lyapunov-Krasovskii functions (Hien et al., 2009; Kim et al., 2006).

Hence, to guarantee stability of system (1) under arbitrary switched law, one of the proposed idea consist to use the following Common Lyapunov-Krasovskii functional:

$$V(x,t) = x^T(t)Px(t) + \int_{t-h}^{t} x^T(\theta)Qx(\theta)d\theta \tag{9}$$

where matrices P and Q are symmetric and positive definite. Then, the choice of this Lyapunov-Krasovskii functional implies the following proposition.

Proposition 1 (Oliveira & Peres, 2006 System): (1) is asymptotically stable there exist real symmetric matrices $P > 0$ and $Q > 0$, such that:

$$\begin{bmatrix} A_i^T P + PA_i + Q & PD_i \\ D_i^T P & -Q \end{bmatrix} < 0 \quad \forall i \in \underline{N} \tag{10}$$

is satisfied.

Although, several common Lyapunov-Krasovskii functions were presented in the literature which are solved in them of linear matrix inequalities conditions in order to guarantee stability analysis under arbitrary switching, unfortunately these conditions proved very conservative.

Therefore, to get around this drawback, new delay-dependent asymptotic stability under arbitrary switching are established that are presented in the squeal.

4. STABILITY CONDITIONS USING THE OVERVALUING FOR CONTINUOUS-TIME SYSTEMS

Let us consider the following continuous-time system:

$$\dot{x}(t) = Ax(t) \tag{11}$$

where $x(t)$ is an n vector of the state space E defining the system at the time t, $A = (a_{ij})_{1 \le i,j \le n}$ represents an $n \times n$ matrix. Let $x(t_0) = x_0$ is the initial condition.

Definition 4 (Borne & Benrejeb, 2008): Let $E = \Re^n$ be a vector space and $E_1, E_2, ..., E_k$ subspaces of E which verify: $E = E_1 \cup E_2 \cup ... E_k$. Let $x \in E$ be vector defined on E with a projection in the subspace E_i denoted by x_i, $x_i = p_i x$, where p_i is a projection operator from E into E_i, p_i is a scalar norm $\left(i = 1, ..., k\right)$ defined on the subspace E_i and p denotes a vector norm of dimension k and with i^{th} component, $p_i\left(x\right) = p_i\left(x_i\right)$, $p_i\left(x\right): \Re^n \to \Re_+^k$ where $p_i\left(x_i\right)$ is a scalar norm of x_i.

In the following, some notion and properties relative to the comparison system are presented.

Definition 5 (Borne & Benrejeb, 2008): The matrix T_c is said a pseudo-overvaluing matrix of system (11), with respect to the vector norm p, when the inequality below is met:

$$D^+ p\left(x\right) \leq T_c p\left(x\right) \tag{12}$$

where D^+ denotes the right hand derivative.

$\forall x \in E$ and $t > 0$ is verified for each corresponding component. Consequently, the stability of the comparison system: $\dot{z}\left(t\right) = T_c z\left(t\right)$ with the initial conditions such as $z_0 = p\left(x_0\right)$, implies the same property for the system (11).

When the pseudo-overvaluing matrix T_c of system (11) is defined with respect to regular vector norms, the following properties can be considered:

- The off-diagonal elements of matrix T_c are non-negative;
- If we denote by $\text{Re}\left(\lambda_M\right)$ the real part of the eigenvalue of maximum real part of T_c it comes the inequalities: $\text{Re}\left(\lambda_A\right) = \text{Re}\left(\lambda_M\right)$;
- When all the real parts of the eigenvalues of T_c are negative, this matrix is the opposite of an $M-\text{matrix}$. It allows an inverse whose elements are all non-positive.
- When the inverse of T_c is an irreducible matrix, then T_c admits an eigenvector $\mu\left(x, t\right)$ relative to the eigenvalues λ_M and whose components are strictly positive.

The following lemma and stability criterion play an important role in this paper.

Let us consider the linear process described in state space by $\dot{x}\left(t\right) = Ax\left(t\right)$. For this system, Kotelyanski establishes a particular lemma, this lemma is well adapted for stability study when off diagonal elements of matrix A are non-negative.

- **Kotelyanski Lemma (Kotelyanski, 1952):** The real parts of the eigenvalues of matrix A, with non-negative off-diagonal elements, are less than a real number μ if and only if all those of matrix M, $M = \mu I_n - A$, are positive, with I_n the n identity matrix. When successive principal minors of matrix $\left(-A\right)$ are positive, Kotelyanski lemma permits to conclude on stability property of the system characterized by A.

- **Borne-Gentina Practical Stability Criterion (Borne & Benrejeb, 2008):** Let us consider the linear process described in state space by $\dot{x}\left(t\right) = Ax\left(t\right)$. Therefore, if the non-constant elements of overvaluing matrix T_c are isolated in only one row, the verification of the Kotelyanski condition enables to conclude to the stability of the initial system. The Kotelyanski condition enables to conclude to the stability of the initial system. As an example, if the non-constant elements are isolated in only one row of A, Kotelyanski lemma applied to the overvaluing matrix obtained by the use of the n regular vector norm $p\left(x\right)$ with $x = \left[x_1, x_2, \ldots, x_n\right]^T$, such that: $p\left(x\right) = \left[\left|x_1\right|, \left|x_2\right|, \ldots, \left|x_n\right|\right]^T$, leads to the following stability conditions of initial system:

$$a_{1,1} < 0, \quad \begin{vmatrix} a_{1,1} & \left|a_{1,2}\right| \\ \left|a_{2,1}\right| & a_{2,2} \end{vmatrix} > 0, \ldots, \left(-1\right)^n \begin{vmatrix} a_{1,1} & \left|a_{1,2}\right| & \cdots & \left|a_{1,n}\right| \\ \left|a_{2,1}\right| & a_{2,2} & \cdots & \left|a_{2,n}\right| \\ \vdots & \vdots & \cdots & \vdots \\ \left|a_{n,1}\right| & \left|a_{n,2}\right| & \cdots & a_{n,n} \end{vmatrix} > 0 \tag{13}$$

The Borne-Gentina practical criterion applied to continuous systems generalizes the Kotelyanski lemma for nonlinear systems and defines large classes of systems, either for the initial system or for its comparison system (Borne, Benrejeb, 2008).

Below, the theorem of $M - \mathrm{matrix}$ properties is introduced.

Theorem 2 (Elmadssia et al., 2013): The matrix $A = \left(a_{ij}\right)_{1 \le i,j \le n}$ is said an $M - \mathrm{matrix}$ if the following properties are verified:

- All the eigenvalues of A have a positive real part;
- The trace of A is positive;
- The principal minors of A are positive:

$$\left(A\right)\begin{pmatrix} 1\ 2\ \ldots\ j \\ 1\ 2\ \ldots\ j \end{pmatrix} > 0 \ \forall j \in 1, \ldots, n \tag{14}$$

- For any positive vector $x = \left(x_1, \ldots, x_n\right)^T$ the algebraic equations Ax have a positive solution $w = \left(w_1, \ldots, w_n\right)^T$

Definition 6: $A = \left(a_{ij}\right)_{1 \le i,j \le n}$ is the opposite of an $M - \mathrm{matrix}$ if $\left(-A\right)$ is an $M - \mathrm{matrix}$.

Remark 1: A continuous-time system characterized by a matrix A is stable if A is the opposite of an $M - \mathrm{matrix}$. Indeed, if such condition is verified, then the principal minors of $\left(-A\right)$ are positive and the Kotelyanski lemma permits to conclude on stability property of the system characterized by A.

After this formulation, the main results of this paper are presented in the next section.

5. MAIN RESULTS

In the following, new delay-dependent conditions for global asymptotic stability of system (1) under arbitrary switching signal (3) are established.

Theorem 3: The switched time-delay system (1) is globally asymptotically stable under arbitrary switching rule (3) if matrix T_c is the opposite of an $M-\text{matrix}$; where:

$$T_c = \max_{1\le i\le N}\left(T_{\sigma(t)}\right) \tag{15}$$

and:

$$T_{\sigma(t)} = \left(A_{\sigma(t)} + D_{\sigma(t)}\right)^{*} + h\left(\left|A_{\sigma(t)}D_{\sigma(t)}\right| + \left|D_{\sigma(t)}^{2}\right|\right) \tag{16}$$

$$A_{\sigma(t)} = \begin{bmatrix} \sum_{i=1}^{N} v_i(t)\left(a_i^{11}\right) & \cdots & \cdots & \sum_{i=1}^{N} v_i(t)\left(a_i^{1n}\right) \\ \vdots & \vdots & \vdots & \vdots \\ \vdots & \vdots & \vdots & \vdots \\ \sum_{i=1}^{N} v_i(t)\left(a_i^{n1}\right) & \cdots & \cdots & \sum_{i=1}^{N} v_i(t)\left(a_i^{nn}\right) \end{bmatrix} \tag{17}$$

and:

$$D_{\sigma(t)} = \begin{bmatrix} \sum_{i=1}^{N} v_i(t)\left(d_i^{11}\right) & \cdots & \cdots & \sum_{i=1}^{N} v_i(t)\left(d_i^{1n}\right) \\ \vdots & \vdots & \vdots & \vdots \\ \vdots & \vdots & \vdots & \vdots \\ \sum_{i=1}^{N} v_i(t)\left(d_i^{n1}\right) & \cdots & \cdots & \sum_{i=1}^{N} v_i(t)\left(d_i^{nn}\right) \end{bmatrix} \tag{18}$$

Thus, sufficient conditions for asymptotic stability of system (1) under arbitrary switching are derived. Proof of Theorem 3 is given in the Appendix.

6. APPLICATION TO SWITCHED SYSTEMS DEFINED BY DIFFERENTIAL EQUATIONS

The purpose of this section is to apply this previous result to switched systems modelled by the following functional delay differential equation (Elmadssia et al., 2013):

$$\begin{cases} y^{n}(t)+\sum_{i=1}^{N} v_i(t)\left(\sum_{p=0}^{n-1} a_i^p y^{(p)}(t)\right)+\sum_{i=1}^{N} v_i(t)\left(\sum_{p=0}^{n-1} d_i^p y^{(p)}(t-h)\right)=0\ ,\ h>0 \\ y^{(p)}(t)=\varphi_p(t)\ t\in[-h\ \ 0],\ p=1,\ldots,n-1 \end{cases} \tag{19}$$

The presence of delay renders the stability study of this problem very difficult. However to solve it, the proposed solution consists to use the following change of variable (Elmadssia et al., 2011; Elmadssia et al., 2013; Jaballi et al., 2013; Kermani & Sakly, 2014; Kermani et al., 2014):

$$x_{p+1}(t)=\frac{dy^{(p)}}{dt^{(p)}},\ p=0,\ldots,n-1 \tag{20}$$

By substituting (20) into (19), we obtain:

$$\begin{cases} \dot{x}_p(t)=x_{p+1}(t) \\ \dot{x}_n(t)=-\sum_{i=1}^{N} v_i(t)\left(\sum_{p=0}^{n-1} a_i^p x_{p+1}(t)+\sum_{p=0}^{n-1} d_i^p x_{p+1}(t-h)\right) \end{cases} \tag{21}$$

or under matrix form, we obtain the following state representation:

$$\begin{cases} \dot{x}(t)=\sum_{i=1}^{N} v_i(t)\left(A_i x(t)+D_i x(t-h)\right) \\ x(t)=\phi(t)\quad t\in[-h,0] \end{cases} \tag{22}$$

where $x(t)$ is the state vector, whose these components are $x_p(t)$, $p=1,\ldots,n$, $v_i(t)$ is the switching signal given in (3), and matrices A_i and D_i are given as follows (Jaballi et al.,, 2013):

$$A_i=\begin{bmatrix} 0 & 1 & \cdots & 0 \\ 0 & 0 & \ddots & \vdots \\ \vdots & \vdots & \ddots & 1 \\ -a_i^0 & -a_i^1 & \cdots & -a_i^{n-1} \end{bmatrix} \tag{23}$$

$$D_i = \begin{bmatrix} 0 & 0 & \cdots & 0 \\ 0 & 0 & \ddots & \vdots \\ \vdots & \vdots & \ddots & 0 \\ -d_i^0 & -d_i^1 & \cdots & -d_i^{n-1} \end{bmatrix} \tag{24}$$

where a_i^p is a coefficient of the instantaneous characteristic polynomial $P_i(\lambda)$ of matrix A_i given by:

$$P_i(\lambda) = \lambda^n + \sum_{q=0}^{n-1} a_i^q \lambda^q \tag{25}$$

and d_i^p is a coefficient of the instantaneous characteristic polynomial $Q_i(\lambda)$ of matrix D_i defined such as:

$$Q_i(\lambda) = \sum_{q=0}^{n-1} d_i^q \lambda^q \tag{26}$$

Now, we will introduced the polynomial $\psi_i(\lambda)$, it is given as follows (Elmadssia et al., 2013):

$$\psi_i(\lambda) = Q_i(\lambda)\lambda + d_i^{n-1} P_i(\lambda) = \sum_{q=0}^{n-1} \varsigma_i^q \lambda^q \tag{27}$$

where the parameters ς_i^q, $q = 0, ..., n-1$ are deduced such as:

$$\begin{cases} \varsigma_i^0 = d_i^{n\text{-}1} a_i^0 \\ \varsigma_i^q = d_i^{n\text{-}1} a_i^q - d_i^{q\text{-}1}, \quad q = 1, ..., n-2 \\ \varsigma_i^{n-1} = d_i^{n\text{-}1} a_i^{n-1} \end{cases} \tag{28}$$

Therefore, to simplify the application of the Kotelyanski lemma a change to base of system (19) into the arrow matrix form will be considered.

Leads the new state vector (Elmadssia et al., 2011; Elmadssia et al., 2013; Jaballi et al., 2013; Kermani & Sakly, 2014; Kermani et al., 2014):

$$z(t) = Px(t) \tag{29}$$

where:

$$P=\begin{bmatrix} 1 & 1 & \cdots & 1 & 0 \\ \alpha_1 & \alpha_2 & \cdots & \alpha_{n-1} & 0 \\ (\alpha_1)^2 & (\alpha_2)^2 & \cdots & (\alpha_{n-1})^2 & \vdots \\ \vdots & \vdots & \cdots & \vdots & 0 \\ (\alpha_1)^{n-1} & (\alpha_2)^{n-1} & \cdots & (\alpha_{n-1})^{n-1} & 1 \end{bmatrix} \tag{30}$$

and $\alpha_j\left(j=1,\ldots n-1\right)$, $\alpha_j \neq \alpha_q\ \forall\ j \neq q$, $i=1,\ldots,N$ are free real parameters, that can be chosen arbitrary.

Leads to the following space representation (Elmadssia et al., 2011; Elmadssia et al., 2013; Jaballi et al., 2013; Kermani & Sakly, 2014; Kermani et al., 2014):

$$\begin{cases} \dot{z}(t)=\sum_{i=1}^{N} v_i(t)\left(M_i z(t)+N_i z(t-h)\right) \\ z(t)=P\phi(t) \quad t\in[-h,0] \end{cases} \tag{31}$$

where M_i and N_i are in the arrow form (Elmadssia et al., 2011; Elmadssia et al., 2013; Jaballi et al., 2013; Kermani & Sakly, 2014; Kermani et al., 2014) given as follows:

$$M_i=P^{-1}A_iP=\begin{bmatrix} \alpha_1 & 0 & \cdots & 0 & \beta_1 \\ 0 & \ddots & \ddots & \vdots & \vdots \\ \vdots & \ddots & \ddots & 0 & \vdots \\ 0 & \cdots & 0 & \alpha_{n-1} & \beta_{n-1} \\ \gamma_i^1 & \cdots & \cdots & \gamma_i^{n-1} & \gamma_i^n \end{bmatrix} \tag{32}$$

$$N_i=P^{-1}D_iP=\begin{bmatrix} 0_{n-1,n-1} & \cdots & & 0_{n-1,1} \\ \delta_i^1 \cdots & & \delta_i^{n-1} & \delta_i^n \end{bmatrix} \tag{33}$$

with:

$$\beta_j=\prod_{\substack{q=1\\ q\neq j}}^{n-1}\left(\alpha_j-\alpha_q\right)^{-1},\ \forall\ j=1,\ldots,n-1 \tag{34}$$

Then, the elements of matrices M_i, $i=1,\ldots,N$ are:

$$\begin{cases} \gamma_i^j = -P_i\left(\alpha_j\right) & , \forall\ j = 1,\ldots,n-1 \\ \gamma_i^n = -a_i^{n-1} - \sum_{j=1}^{n-1} \alpha_j \end{cases} \tag{35}$$

and the elements of the matrices N_i, $i = 1,\ldots,N$ are:

$$\begin{cases} \delta_i^j = -Q_i\left(\alpha_j\right), \forall\ j = 1,\ldots,n-1 \\ \delta_i^n = -d_i^{n-1} \end{cases} \tag{36}$$

Taking into account the previous relations, the matrices T_i, $i = 1,\ldots N$ will be define as following:

$$T_i = \begin{bmatrix} \alpha_1 & 0 & \cdots & 0 & \left|\beta_1\right| \\ 0 & \ddots & \ddots & \vdots & \vdots \\ \vdots & \ddots & \ddots & 0 & \vdots \\ 0 & \cdots & 0 & \alpha_{n-1} & \left|\beta_{n-1}\right| \\ t_i^1 & \cdots & \cdots & t_i^{n-1} & t_i^n \end{bmatrix} \tag{37}$$

and matrix $T_{\sigma(t)}$ is given by:

$$T_{\sigma(t)} = \begin{bmatrix} \alpha_1 & 0 & \cdots & 0 & \left|\beta_1\right| \\ 0 & \ddots & \ddots & \vdots & \vdots \\ \vdots & \ddots & \ddots & 0 & \vdots \\ 0 & \cdots & 0 & \alpha_{n-1} & \left|\beta_{n-1}\right| \\ \sum_{i=1}^{N} v_i\left(t\right)t_i^1 & \cdots & \cdots & \sum_{i=1}^{N} v_i\left(t\right)t_i^{n-1} & \sum_{i=1}^{N} v_i\left(t\right)t_i^n \end{bmatrix} \tag{38}$$

with:

$$\begin{cases} t_i^j = \left|\gamma_i^j + \delta_i^j\right| + h\left(\left|\psi_i\left(\alpha_j\right)\right| + \left|\delta_i^j d_i^{n-1}\right|\right), & j = 1,\ldots,n-1 \\ t_i^n = \gamma_i^n + \delta_i^n + h\left(\left|a_i^{n-1}d_i^{n-1} - d_i^{n-2}\right| + \left(d_i^{n-1}\right)^2\right) \end{cases} \tag{39}$$

Now, we can deduce the following theorem.

Theorem 4: The system (22) is globally asymptotically stable under arbitrary switching rule (3) if there exist $\alpha_j < 0$ $\left(j = 1,\ldots,n-1\right)$, $\alpha_j \neq \alpha_q$, $\forall\ j \neq q$, satisfying the following condition:

$$-\bar{t}^{n}+\sum_{j=1}^{n-1}\bar{t}^{j}\left|\beta_{j}\right|\alpha_{j}^{-1}>0 \tag{40}$$

where:

$$\begin{cases}\bar{t}^{n}=\max\limits_{1\le i\le N}\left(t_{i}^{n}\right)\\ \bar{t}^{j}=\max\limits_{1\le i\le N}\left(t_{i}^{j}\right),\ j=1,\dots,n-1\end{cases} \tag{41}$$

Proof: (Elmadssia et al., 2011; Elmadssia et al., 2013). For an arbitrary choice $\alpha_{j}<0\ \left(j=1,\dots,n-1\right)$, $\alpha_{j}\neq\alpha_{q}$, $\forall\ j\neq q$ and according to the Kotelyanski lemma. In this case, it suffices to verify that $T_{c}=\max\limits_{1\le i\le N}\left(T_{v(t)}\right)$ is the opposite of an $M-\text{matrix}$. These conditions require having all the principal minors of T_{c} are negative.

The $n-1$ first conditions are checked because the α_{j} are negative, however the last condition yields to: $\det\left(-T_{c}\right)=\chi\prod_{j=1}^{n-1}\alpha_{j}>0$, where $\chi=\bar{t}^{n}-\sum_{j=1}^{n-1}\bar{t}^{j}\left|\beta_{j}\right|\alpha_{j}^{-1}$.

It comes $\chi<0$, then condition (40) is verified.

Theorem 4 can be simply in the following corollary.

Corollary 1: The system (22) is globally asymptotically stable under arbitrary switching (3) if there exist $\alpha_{j}<0\ \left(j=1,\dots,n-1\right)$, $\alpha_{j}\neq\alpha_{q}$, $\forall\ j\neq q$, satisfying the following condition:

$$\bar{t}^{n}+\sum_{j=1}^{n-1}\alpha_{j}<0 \tag{42}$$

Proof (Elmadssia, Saadaoui, Benrejeb, 2013): For an arbitrary choice $\alpha_{j}<0\ \left(j=1,\dots,n-1\right)$, $\alpha_{j}\neq\alpha_{q}$, $\forall\ j\neq q$ and assuming that. T_{c} is the opposite of an $M-\text{matrix}$. In this case, it is simply to verify that $\text{trace}\left(T_{c}\right)<0$.

In such conditions we can deduce relation (42).

7. STABILITY ANALYSIS OF SWITCHED SYSTEMS WITCH MULTIPLE DELAYS

In this section, a generalization of the previous results will be given to switched systems with multiple delays.

Consider a class of switched systems with multiple delays formed by N subsystems; it is described by (Elmadssia et al., 2013):

$$\begin{cases} \dot{x}(t) = \sum_{i=1}^{N} v_i(t)\left(A_i x(t) + \sum_{l=1}^{m} D_{l,i} x(t - h_l)\right) \\ x(t) = \phi(t) \quad t \in \left[-\max_{1 \le l \le m} h_l, 0\right] \end{cases} \tag{43}$$

where $v_i(t)$ is the switching signal given in (3), $A_i \in \Re^{n \times n}$ and $D_{l,i} \in \Re^{n \times n}\ (l = 1, \ldots m)$ are constant matrices of appropriate dimensions for $i = 1, \ldots, N$.

Generalizing the Lyapunov function introduced in the Appendix, it is easy to obtain the following sufficient stability conditions for system (43).

Theorem 5: The system (43) is globally asymptotically stable under arbitrary switching (3) if matrix $T_{m,c}$ is the opposite of an $M-\text{matrix}$;

where $T_{m,c} = \max_{1 \le i \le N}\left(T_{m,\sigma(t)}\right)$; and:

$$T_{m,\sigma(t)} = \left(A_{\sigma(t)} + \sum_{l=1}^{m} D_{l,\sigma(t)}\right)^{*} + \sum_{l=1}^{m} h_l \left(\left|A_{\sigma(t)} D_{l,\sigma(t)}\right| + \left|D_{l,\sigma(t)}^2\right|\right) \tag{44}$$

where $A_{\sigma(t)}$ is given in (17) and $D_{l,\sigma(t)}$ for each $l \in \{1, \ldots m\}$ are given as below:

$$D_{l,v(t)} = \begin{bmatrix} \sum_{i=1}^{N} v_i(t)\left(d_{l,i}^{11}\right) & \cdots & \cdots & \sum_{i=1}^{N} v_i(t)\left(d_{l,i}^{1n}\right) \\ \vdots & \vdots & \vdots & \vdots \\ \vdots & \vdots & \vdots & \vdots \\ \sum_{i=1}^{N} v_i(t)\left(d_{l,i}^{n1}\right) & \cdots & \cdots & \sum_{i=1}^{N} v_i(t)\left(d_{l,i}^{nn}\right) \end{bmatrix} \tag{45}$$

Proof (Elmadssia et al., 2013): It suffices to choose the Lyapunov function $V(t)$ given as below and follow the same steps as described in the proof of Theorem 3 (Appendix).

$$V(t) = V_1(t) + V_2(t) + V_3(t) + V_4(t) \tag{46}$$

where:

$$V_1(t) = \left\langle \left|x(t)\right|, w \right\rangle \tag{47}$$

$$V_2(t) = \sum_{l=1}^{m} \left\langle \left| A_{\sigma(t)} D_{l,\sigma(t)} \right| \int_h^0 \int_{t+\theta}^t \left| x(s) \right| d\theta ds, w \right\rangle \tag{48}$$

$$V_3(t) = \sum_{l=1}^{m} \left\langle \left| D_{l,\sigma(t)}^2 \right| \int_h^0 \int_{t+\theta}^t \left| x(s-h) \right| d\theta ds, w \right\rangle \tag{49}$$

$$V_4(t) = \sum_{l=1}^{m} h_l \left\langle \left| D_{l,\sigma(t)}^2 \right| \int_{t-h}^t \left| x(s) \right| ds, w \right\rangle \tag{50}$$

Now, an application of Theorem 5 will be performed to switched systems with multiple delays modelled by the following functional differential equation (Elmadssia et al., 2011; Elmadssia et al., 2013):

$$\begin{cases} y^n(t) + \sum_{i=1}^{N} v_i(t) \left(\sum_{p=0}^{n-1} a_i^p y^{(p)}(t) \right) + \sum_{i=1}^{N} v_i(t) \sum_{l=1}^{m} \left(\sum_{p=0}^{n-1} d_{l,i}^p y^{(p)}(t-h_l) \right) = 0, \ h_l > 0, \ l \in \{1,...m\} \\ y^{(p)}(t) = \varphi_p(t), \ p \in \{0,...n-1\}, \ t \in \left[-\max_{1 \le l \le m} (h_l \ ) \ 0 \right] \end{cases} \tag{51}$$

Then, characteristic polynomial $P_i(\lambda)$ of matrix A_i is given in (23) and for each $l = 1,...,m$, we define the new polynomials $Q_{l,i}(\lambda)$ and $\psi_{l,i}(\lambda)$ as follows:

$$Q_{l,i}(\lambda) = \sum_{q=0}^{n-1} d_{l,i}^q \lambda^q \tag{52}$$

and:

$$\psi_{l,i}(\lambda) = Q_{l,i}(\lambda)\lambda + d_{l,i}^{n-1} P_i(\lambda) = \sum_{q=0}^{n-1} \varsigma_{l,i}^q \lambda^q \tag{53}$$

Now, we will applied the same idea performed for system (22) in order to obtain stability conditions for system (51). Then, to achieve this aim, the following change of variables is considered: $x_{p+1}(t) = \dfrac{dy^{(p)}}{dt^{(p)}}$, $p = 0,...,n-1$.

It yields to the following state representation:

$$\begin{cases} \dot{x}(t) = \sum_{i=1}^{N} v_i(t)\left(A_i x(t) + \sum_{l=1}^{m} D_{l,i} x(t - h_l)\right) \\ x(t) = \phi(t) \quad t \in \left[-\max_{1 \le l \le m}(h_l) \ \ 0\right] \end{cases} \tag{54}$$

where $x(t)$ is the state vector, whose these components are $x_p(t)$, $p = 1, \ldots, n$, $v_i(t)$ is the switching signal given in (3), A_i are given in (23) and $D_{l,i}$ for each $l = 1, \ldots, m$ and $i = 1, \ldots, N$ are given as below:

$$D_{l,i} = \begin{bmatrix} 0 & 0 & \cdots & 0 \\ 0 & 0 & \ddots & \vdots \\ \vdots & \vdots & \ddots & 0 \\ -d_{l,i}^{0} & -d_{l,i}^{1} & \cdots & -d_{l,i}^{n-1} \end{bmatrix} \tag{55}$$

By introducing the same variable change given in (29), system (44) will be represented in the arrow form as follows:

$$\begin{cases} \dot{z}(t) = \sum_{i=1}^{N} v_i(t)\left(M_i z(t) + \sum_{l=1}^{m} N_{l,i} z(t - h_l)\right), \quad (l = 1, \ldots, m) \\ z(t) = P\phi(t) \quad t \in \left[-\max_{1 \le l \le m}(h_l) \ \ 0\right] \end{cases} \tag{56}$$

where M_i are introduced in (32) and $N_{l,i}$ $(l = 1, \ldots, m)$ are given as follows:

$$N_{l,i} = P^{-1} D_{l,i} P = \begin{bmatrix} 0_{n-1,n-1} & \cdots & & 0_{n-1,1} \\ \delta_{l,i}^{1} & \cdots & \delta_{l,i}^{n-1} & \delta_{l,i}^{n} \end{bmatrix} \tag{57}$$

with:

$$\begin{cases} \delta_{l,i}^{j} = -Q_{l,i}(\alpha_j), \ \forall\ j = 1, \ldots, n-1, \ \ l = 1, \ldots, m \\ \delta_{l,i}^{n} = -d_{l,i}^{n-1} \end{cases} \tag{58}$$

and finally matrix $T_{m,v(t)}$ is deduced as below:

$$T_{m,v(t)} = \begin{bmatrix} \alpha_1 & 0 & \cdots & 0 & |\beta_1| \\ 0 & \ddots & \ddots & \vdots & \vdots \\ \vdots & \ddots & \ddots & 0 & \vdots \\ 0 & \cdots & 0 & \alpha_{n-1} & |\beta_{n-1}| \\ \sum_{i=1}^{N} v_i(t) t_{l,i}^1 & \cdots & \cdots & \sum_{i=1}^{N} v_i(t) t_{l,i}^{n-1} & \sum_{i=1}^{N} v_i(t) t_{l,i}^{n} \end{bmatrix} \tag{59}$$

with:

$$\begin{cases} t_{l,i}^j = \left|\gamma_i^j + \sum_{l=1}^{m} \delta_{l,i}^j\right| + \sum_{l=1}^{m} h_l \left(|\psi_{l,i}| + |\delta_{l,i}^j d_{l,i}^{n-1}|\right), & j = 1,\ldots,n-1,\ l = 1,\ldots,m \\ t_{l,i}^n = \gamma_i^n + \sum_{l=1}^{m} \delta_{l,i}^n + \sum_{l=1}^{m} h_l \left(\left|a_i^{n-1} d_{l,i}^{n-1} - d_{l,i}^{n-2}\right| + \left(d_{l,i}^{n-1}\right)^2\right), & l = 1,\ldots,m \end{cases} \tag{60}$$

Therefore, $T_{m,c}$ can be deduced as below:

$$T_{m,c} = \begin{bmatrix} \alpha_1 & 0 & \cdots & 0 & |\beta_1| \\ 0 & \ddots & \ddots & \vdots & \vdots \\ \vdots & \ddots & \ddots & 0 & \vdots \\ 0 & \cdots & 0 & \alpha_{n-1} & |\beta_{n-1}| \\ \bar{t}^1 & \cdots & \cdots & \bar{t}^{n-1} & \bar{t}^n \end{bmatrix} \tag{61}$$

where $\bar{t}^j$, $j = 1,\ldots,n$ are given in (31).

Now, by considering Theorem 4, we can deduce the following delay-dependent stability conditions.

Corollary 2: The system (54) is globally asymptotically stable under arbitrary switching (3) if there exist $\alpha_j < 0$ $(j = 1,\ldots,n-1)$, $\alpha_j \neq \alpha_q$, $\forall\ j \neq q$, satisfying the following condition:

$$\bar{t}^n + \sum_{j=1}^{n-1} \alpha_j < 0 \tag{62}$$

Proof: For an arbitrary choice $\alpha_j < 0$ $(j = 1,\ldots,n-1)$, $\alpha_j \neq \alpha_q$, $\forall\ j \neq q$ and assuming that $T_{m,c}$ is the opposite of an $M-\text{matrix}$. It suffices to verify that $\text{trace}(T_{m,c}) < 0$.

8. NUMERICAL EXAMPLES

In this section, two numerical examples are given to illustrate the performance of the proposed approach.

Example 1: Consider the following continuous-time switched time-delay system:

$$\begin{cases} \dot{x}(t) = \sum_{i=1}^{2} v_i(t)\left(A_i x(t) + D_i x(t-h)\right) \\ x(t) = \phi(t) \quad t \in [-h, 0] \end{cases}$$

where: $A_1 = \begin{bmatrix} 0 & 1 \\ -1 & -10 \end{bmatrix}$, $A_2 = \begin{bmatrix} 0 & 1 \\ -0.2 & -10 \end{bmatrix}$, $D_1 = \begin{bmatrix} 0 & 0 \\ -2 & 0 \end{bmatrix}$, $D_2 = \begin{bmatrix} 0 & 0 \\ -1 & 0 \end{bmatrix}$ and $h > 0$ is the time delay.

Now, by using the algebraic properties given in (32), (33), (34), (35) and (36) the matrices M_1, M_2, N_1 and N_2 in the arrow form are the following:

$$M_1 = \begin{bmatrix} \alpha & \beta \\ -(\alpha^2 + 10\alpha + 1) & -10 - \alpha \end{bmatrix}, \quad M_2 = \begin{bmatrix} \alpha & \beta \\ -(\alpha^2 + 10\alpha + 0.2) & -10 - \alpha \end{bmatrix},$$

$$N_1 = \begin{bmatrix} 0 & 0 \\ 2 & 0 \end{bmatrix} \text{ and } N_2 = \begin{bmatrix} 0 & 0 \\ 1 & 0 \end{bmatrix}.$$

According to Corollary 1, for particular choice $\alpha = -1$, $\beta = 1$. Second condition of Corollary 1 is verified as following: $\max(-9 + 2h, -9 + h) - 1 = -10 + 2h < 0$. This implies that: $h < 5s$.

By considering the switched signal depicts by Figure 1, when 1 implies that subsystem (1) is active and 2 that subsystem (2) is active.

Now, according to Figure1, the final time t_f is fixed at $20s$ and the switched time t_1 is fixed at $10s$. The delay is $h = 2s$. The simulation results are shown in Figure 2, Figure 3, and Figure 4 where the initial function $\phi(t) = \begin{bmatrix} 1 & -2 \end{bmatrix}^T$. Figure 2 shows the state responses, the state trajectories are depicted in Figure 3 and Figure 4 present the state's norm which show the stability of the system given in this example.

In the next, the following example give a system with two different delays.

Example 2: Consider the following continuous-time switched system with two delays:

$$\begin{cases} \dot{x}(t) = \sum_{i=1}^{2} v_i(t)\left(A_i x(t) + \sum_{l=1}^{2} D_{l,i} x(t - h_l)\right) \\ x(t) = \phi(t) \quad t \in \left[-\max_{1 \le l \le 2}(h_l), 0\right] \\ y(t) = \sum_{i=1}^{2} v_i(t) C_i x(t) \end{cases}$$

Figure 1. Switching signal between subsystems 1 and 2

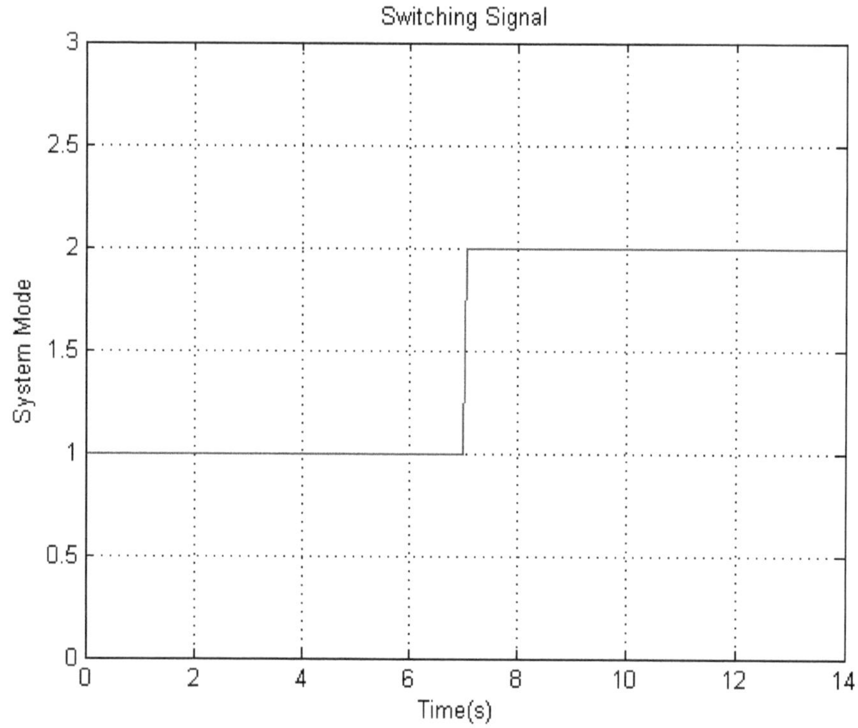

Figure 2. State responses of the system given in the example 1

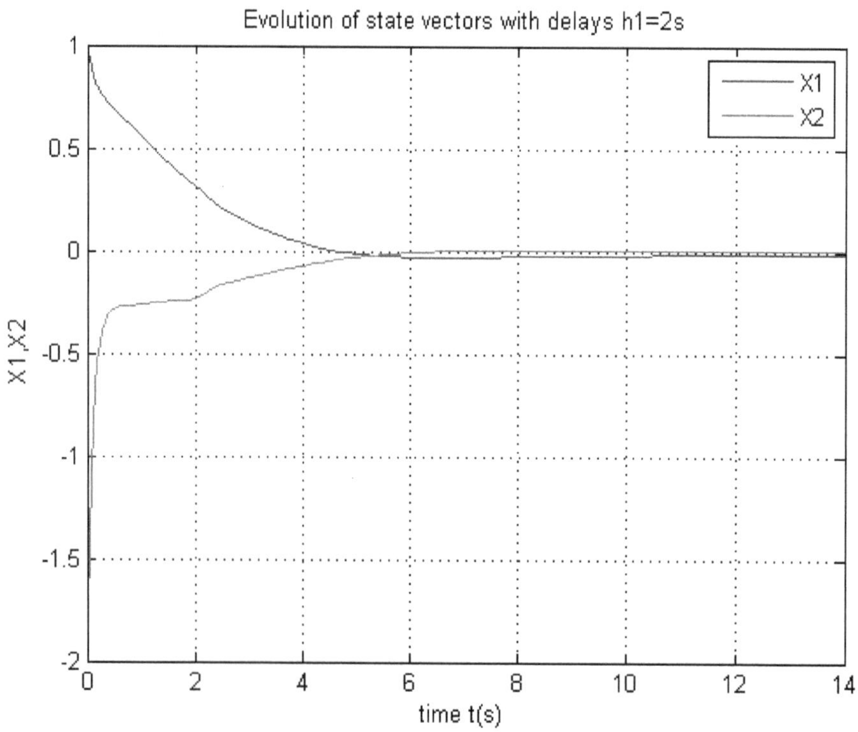

Figure 3. Trajectory response of the system given in the example 1

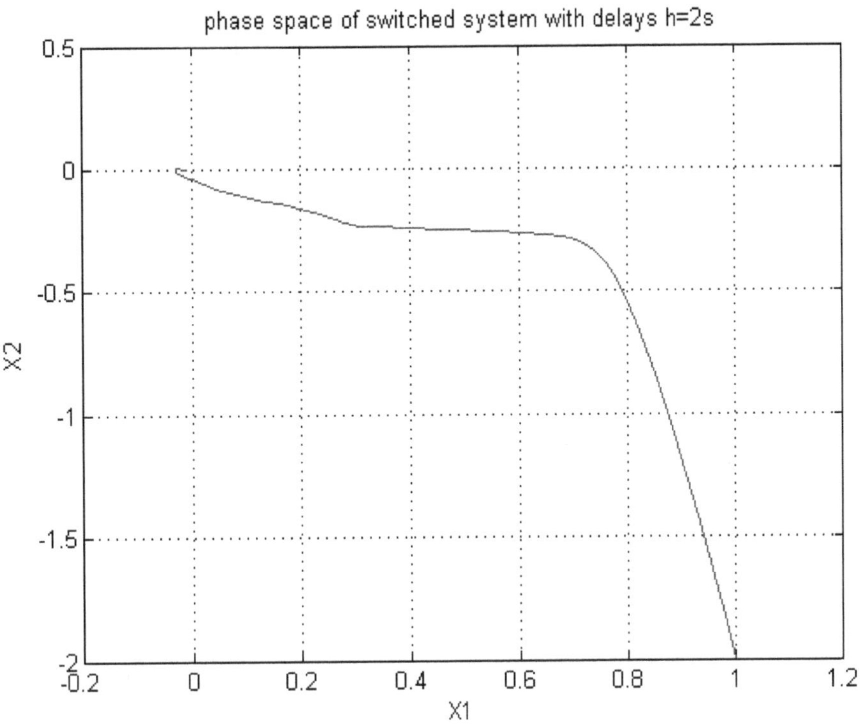

Figure 4. State's norm for example 1 with h=2s

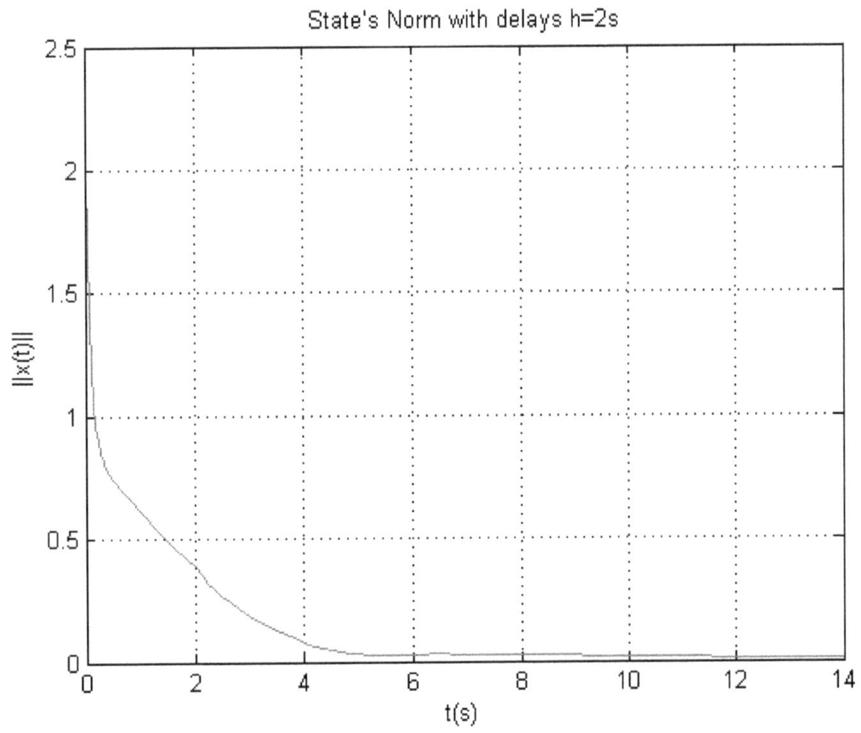

where $y(t)$ is the output, $v_i(t)$ is the switched function. The parameters of this system the following:

$$A_1 = \begin{bmatrix} 0 & 1 \\ -4 & -3 \end{bmatrix},\ A_2 = \begin{bmatrix} 0 & 1 \\ -3 & -3 \end{bmatrix},\ D_{1,1} = \begin{bmatrix} 0 & 0 \\ -1 & 0 \end{bmatrix},\ D_{1,2} = \begin{bmatrix} 0 & 0 \\ -0.5 & 0 \end{bmatrix},$$

$$D_{1,2} = \begin{bmatrix} 0 & 0 \\ -1 & 0 \end{bmatrix},\ D_{2,2} = \begin{bmatrix} 0 & 0 \\ -0.2 & 0 \end{bmatrix},\ C_1^T = \begin{bmatrix} 1 \\ 1 \\ \overline{2} \end{bmatrix},\ C_2^T = \begin{bmatrix} 1 \\ 1 \\ \overline{2} \end{bmatrix},\ h_1 > 0$$

and $h_2 > 0$ are the two delays.

As was done for example 1, due to (37), (38), (39), (57) and (58), the matrices M_1, M_2, $N_{1,1}$, $N_{2,1}$, $N_{1,2}$ and $N_{2,2}$ in arrow form are the following:

$$M_1 = \begin{bmatrix} \alpha & \beta \\ -(\alpha^2 + 3\alpha + 4) & -3-\alpha \end{bmatrix},\ M_2 = \begin{bmatrix} \alpha & \beta \\ -(\alpha^2 + 3\alpha + 3) & -(3+\alpha) \end{bmatrix},$$

Figure 5. Stability domain given for example 2 obtained from corollary 2

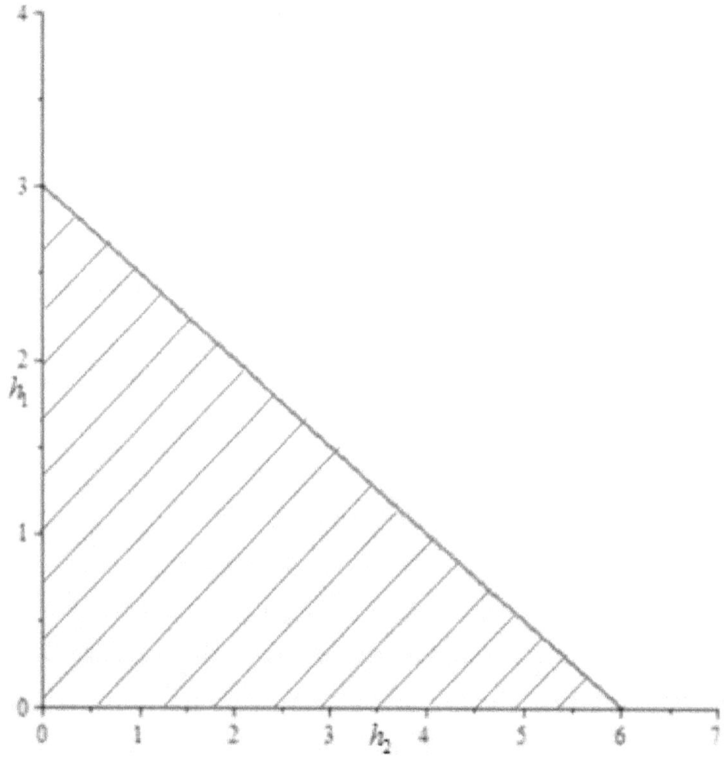

$$N_{1,1} = \begin{bmatrix} 0 & 0 \\ 1 & 0 \end{bmatrix},\ N_{2,1} = \begin{bmatrix} 0 & 0 \\ 0.5 & 0 \end{bmatrix},\ N_{1,2} = \begin{bmatrix} 0 & 0 \\ 1 & 0 \end{bmatrix} \text{and } N_{2,2} = \begin{bmatrix} 0 & 0 \\ 0.2 & 0 \end{bmatrix}.$$

Then, by choosing $\alpha = -1$ and $\beta = 1$. From corollary 2, we deduce the following stability conditions: $h_1 < 3 - 0.5h_2$

Consequently, the stability domain given by the delay h_1 relative to the delay h_2 is given in Figure 5.

Now, taking into consideration the switching sequence given in Figure 6. The simulation result of the system given in this example for two extreme values of delays included in the domain given in Figure5, $h_1 = 2.8s$ and $h_2 = 0.1s$, with the initial function $\phi(t) = \begin{bmatrix} -2 & 1 \end{bmatrix}^T$ are shown in Figures 7, 8, 9 and 10, respectively which correspond to the evolution of states with respect to time, the trajectory response, the norm of system and the system response given in example 2. Figures 11, 12, 13 and 14 gives us respectively the states evolution with respect to time, the trajectory response, the norm of system and the system response of example 2 for des delays values $h_1 = 0.1s$ and $h_2 = 5.3s$.

Figure 6. Switching signal between subsystems 1 and 2

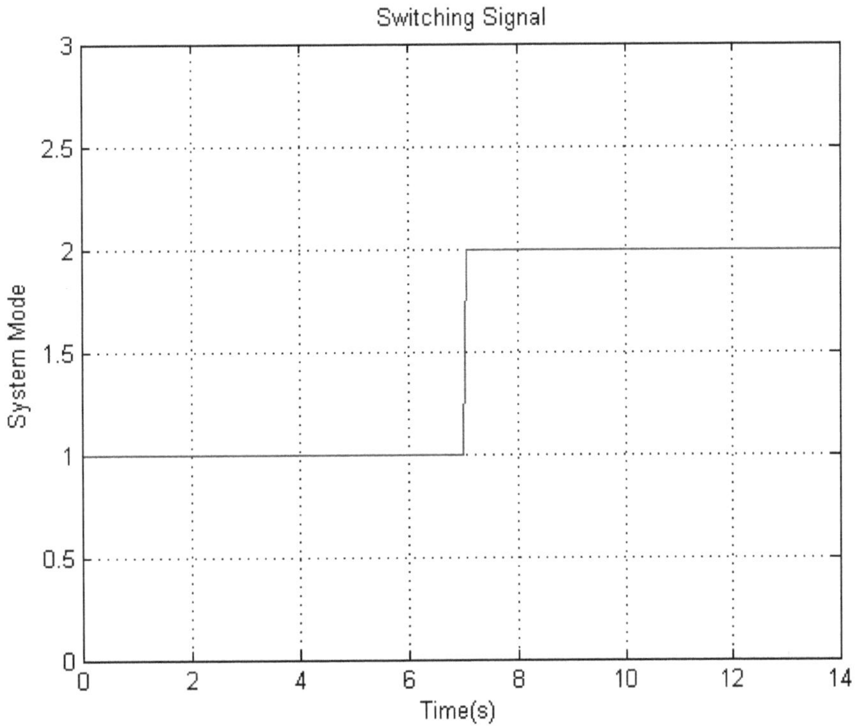

Figure 7. State response of system given in example 2 with h_1=2.8s and h_2=0.1s

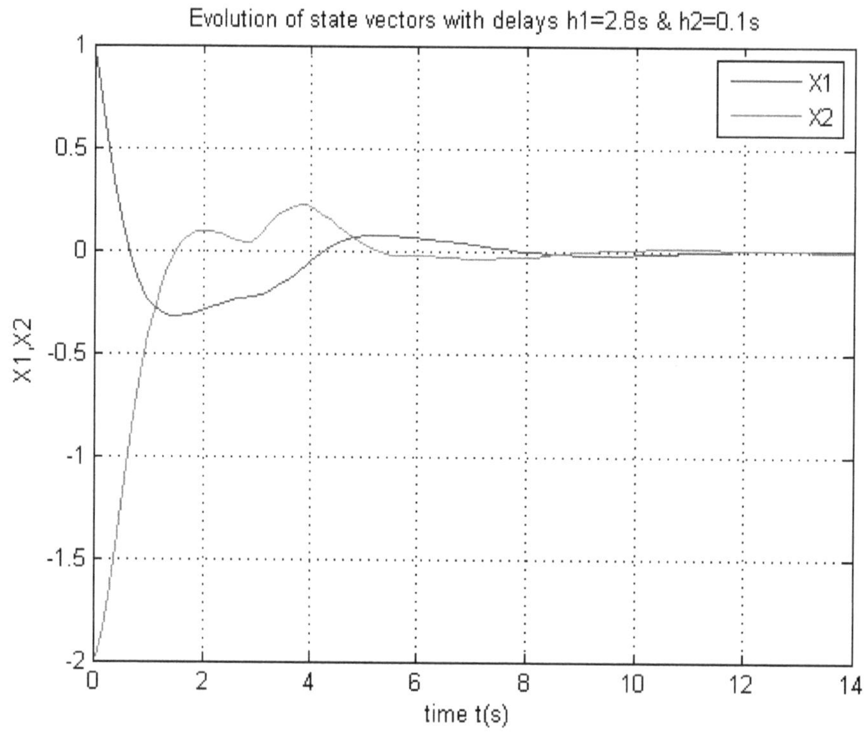

Figure 8. State space for example 2 with h_1= 2.8s and h_2=0.1s

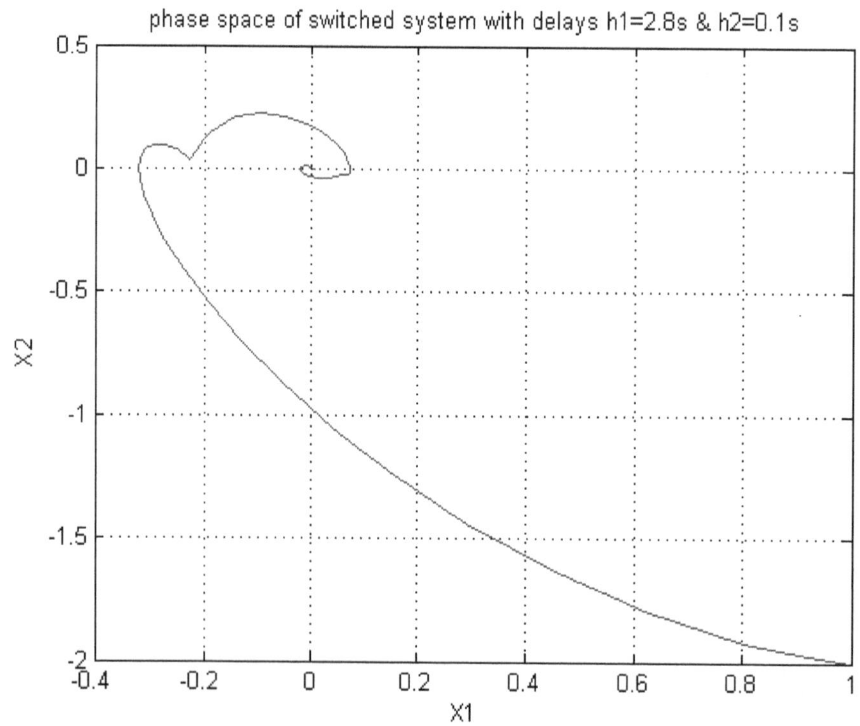

Figure 9. State's norm for example 2 with h_1= 2.8s and h_2=0.1s

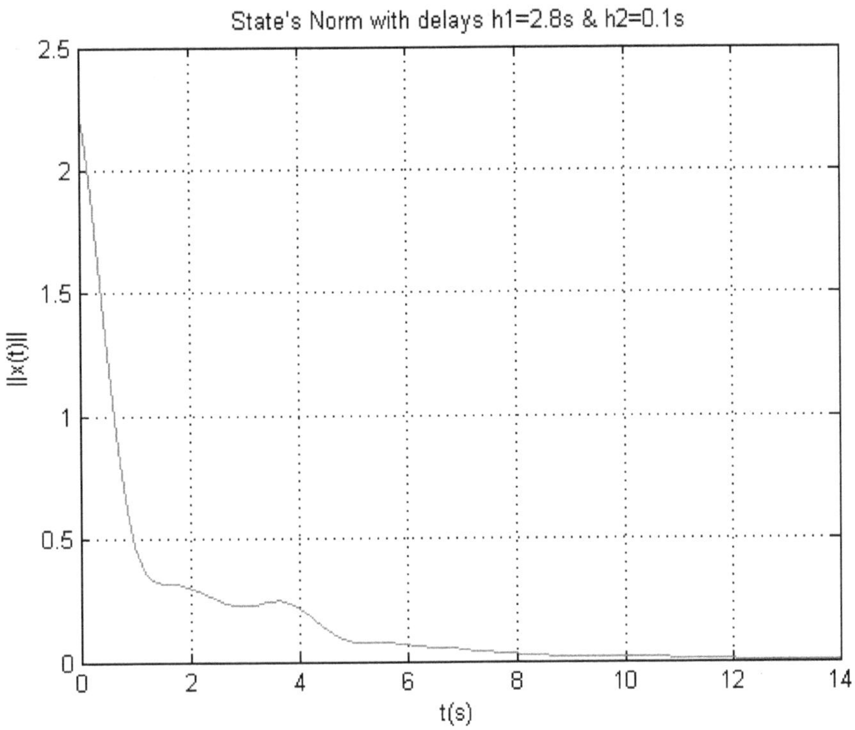

Figure 10. System response for example 2 with h_1= 2.8s and h_2=0.1s

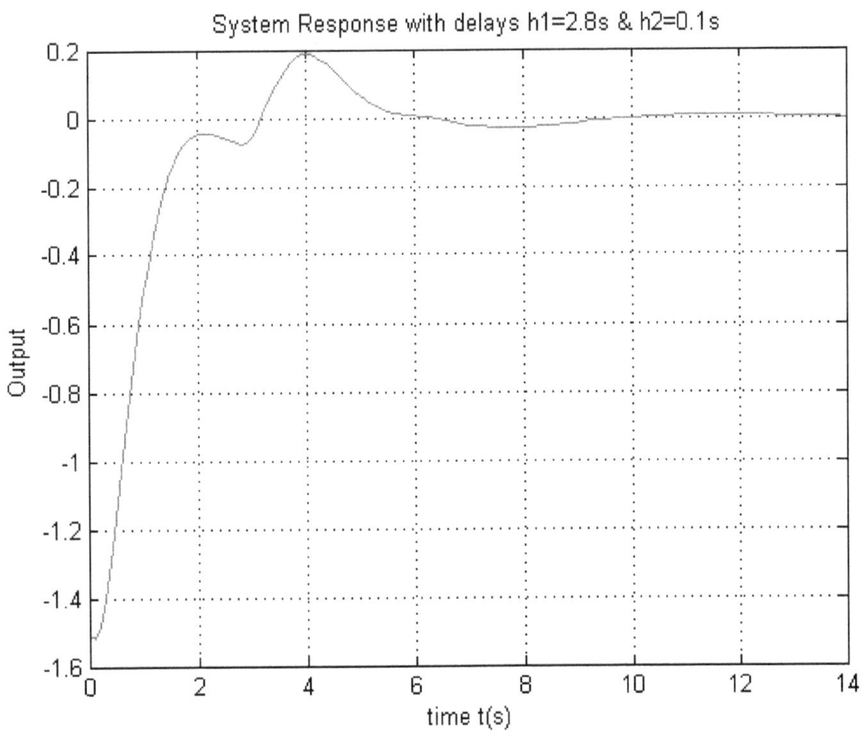

Figure 11. State response of example 2 with h_1=0.1s and h_2=5.3s

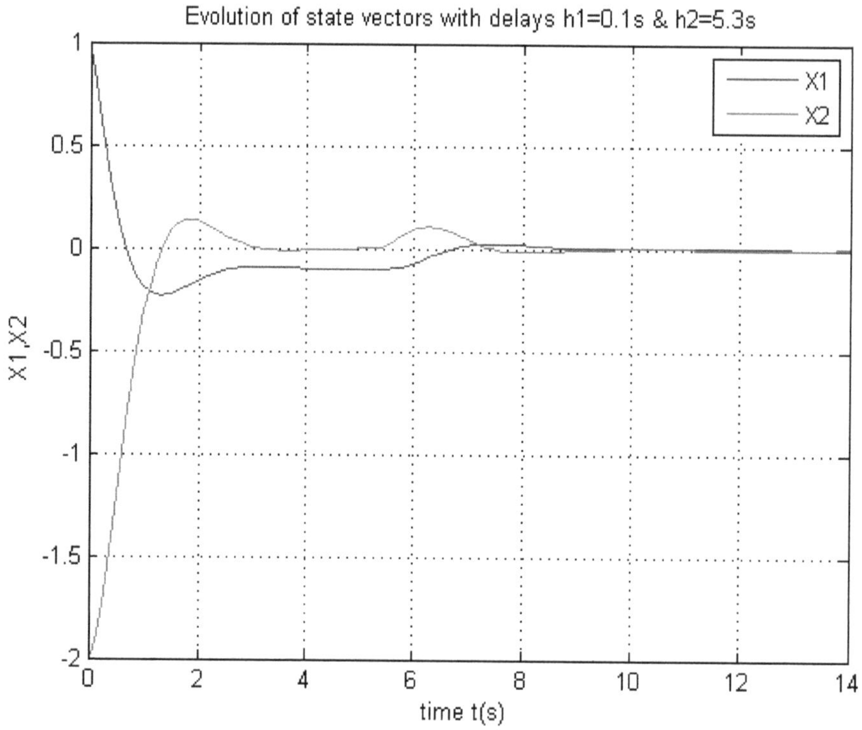

Figure 12. State space for example 2 with h_1=0.1s and h_2=5.3s

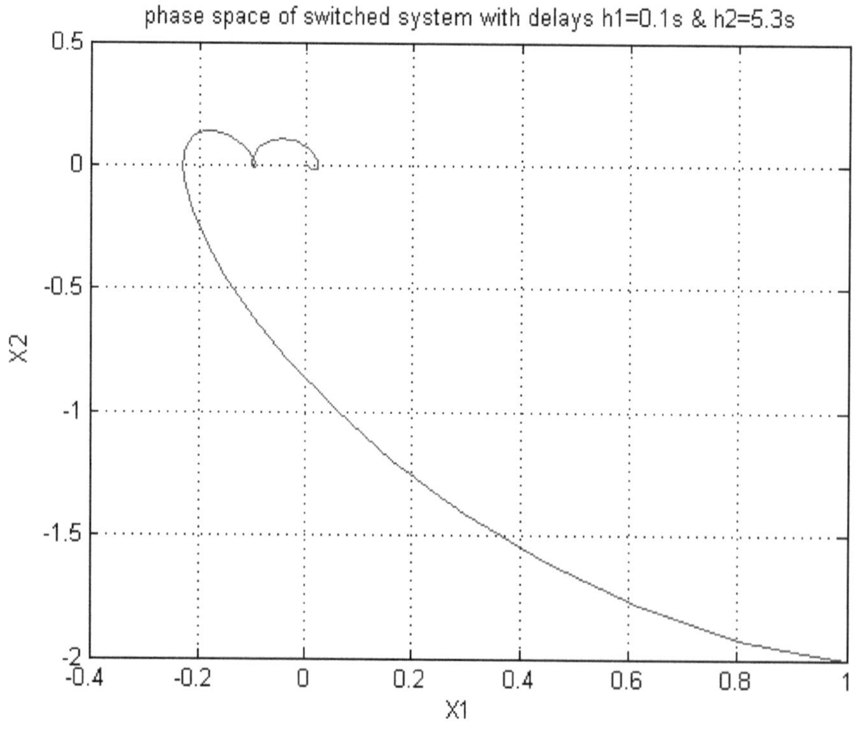

Figure 13. State's norm for example 2 with h_1=0.1s and h_2=5.3s

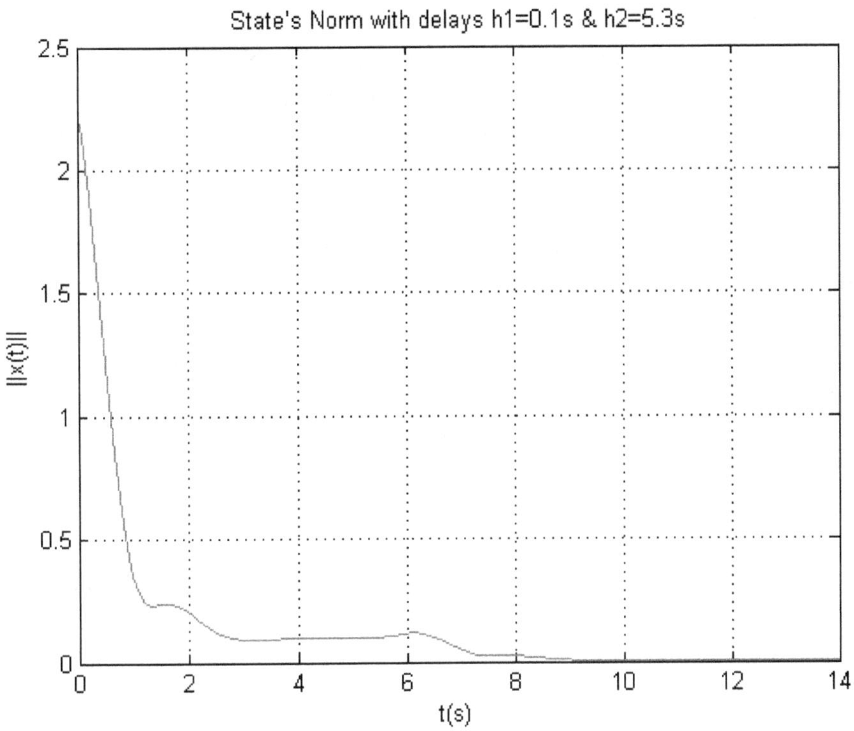

Figure 14. System response for example 2 with h_1=0.1s and h_2=5.3s

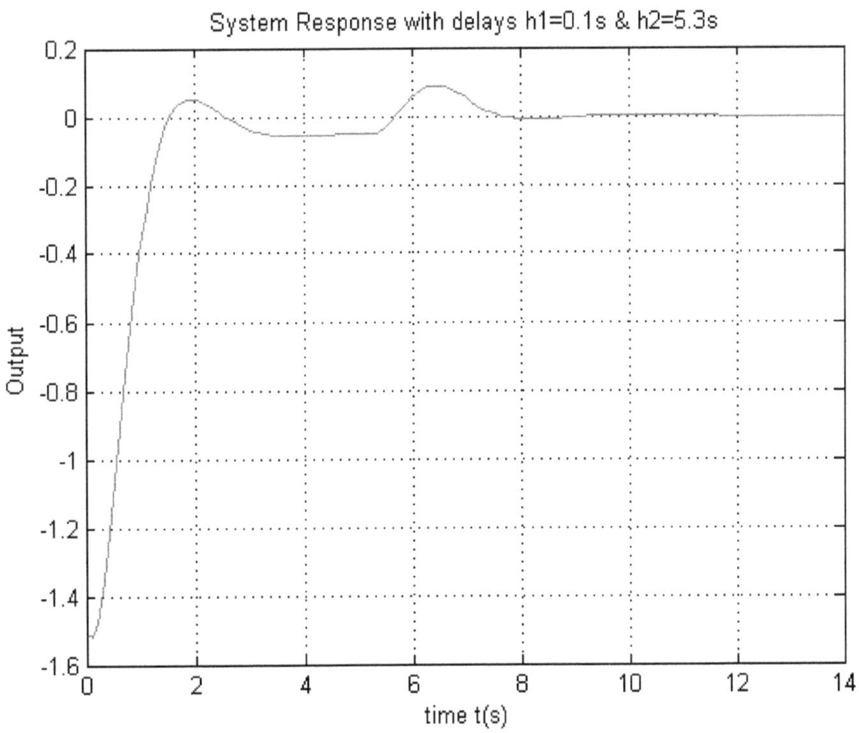

9. DISCUSSION

The stability under arbitrary switching is mainly investigated by employing the common Lyapunov function approach but it should be noted that this may lead to conservatism. The stability under controlled switching is well studied using the average dwell time approach. Although, stability under stability under arbitrary switching is essential topic, the existing results in this context are restricted to the common Lyapunov function.

Basically, current efforts to achieve stability in time delay systems can be divided into two categories, namely, delay-independent criteria and delay-dependent criteria. For the delay dependent criteria, the Common Lyapunov-Krasovskii functional technique is most suitable for application. Thus, to reduce conservatism related to the application of this method, In view of stability under arbitrary switching in this work, new delay-dependent stability criteria for switched time delay systems are established. Therefore, two examples are provided, with numerical simulations, to demonstrate the effectiveness of the proposed method. Indeed, by considering the example 1, it is simple to see that those stability conditions are explicit and simple to manipulate.

On the other hand, into account the complexity of the subsystems, it is very difficult to find a common Lyapunov function for all the subsystems in order to grantee stability under arbitrary switching for this example. On the other side, example 2 was treated to show the delay limit values in order to guarantee the asymptotic stability under arbitrary switching, moreover to see that these stability conditions are sufficient and closely necessary.

10. CONCLUSION AND FUTURE WORK

In this paper, new delay-dependent stability conditions under arbitrary switching for continuous-time switched time-delay systems are established. These stability conditions were derived from the contraction of an appropriate Lyapunov function associated with the application of the Kotelyanski lemma, the $M-\mathrm{matrix}$ properties, the aggregation techniques and the vector norms notion. In addition, these obtained conditions are extended to be generalized for switched systems with multiple delays. Compared with the existing results, the benefit of this method is that; it can avoid the research of a common Lyapunov function which is usually very difficult, or even not possible. Simulation results have been presented to illustrate the effectiveness of the developed method.

On the other side, this proposed approach could be further used as a constructive solution to the problems of state and static output feedback stabilization. As well this method motived us to include stability analysis of switched inputs delays systems.

The limit of this paper is that it has been confined to the boundaries of numerical examples. It would be beneficial to extend the research further so as to include real systems.

REFERENCES

Araki, M. (1975). Application of M-matrices to the stability problems of composite dynamical systems. *Journal of Mathematical Analysis and Applications*, *52*(2), 309–321. doi:10.1016/0022-247X(75)90099-2

Benrejeb, M., & Borne, P. (1978). On an algebraic stability criterion for non-linear processes. Interpretation in the frequency domain. In *Proceedings of the Measurement and Control International Symposium MECO'78* (pp. 678–682). Athens, Greece: MECO.

Benrejeb, M., Sakly, A., Ben Othman, K., & Borne, P. (2008). Choice of conjunctive operator of TSK fuzzy systems and stability domain study. *Mathematics and Computers in Simulation*, *76*(5-6), 410–421. doi:10.1016/j.matcom.2007.04.003

Benrejeb, M., Soudani, D., Sakly, A., & Borne, P. (2006). New Discrete Tanaka Sugeno Kang Fuzzy Systems Characterization and Stability Domain. *International Journal of Computers, Communications & Control*, *1*(1), 9–19.

Borne, P., & Benrejeb, M. (2008). On the Representation and the Stability Study of Large Scale Systems. *International Journal of Computers, Communications & Control*, (2), 55–66.

Branicky, M. S. (1998). Multiple Lyapunov functions and other analysis tools for switched and hybrid systems. *IEEE Transactions on Automatic Control*, *43*(4), 475–482. doi:10.1109/9.664150

Colaneri, P., Geromel, J., & Astolfi, A. (2008). Stabilization of continuous-time switched nonlinear systems. *Systems & Control Letters*, *57*(1), 95–103. doi:10.1016/j.sysconle.2007.07.001

Elmadssia, S., Saadaoui, k. & Benrejeb, M. (2011). New delay-dependent stability conditions for linear systems with delay. In *Proceedings of International Conference on Communications Computing and Control Applications*. Hammamet, Tunisia: Academic Press. doi:10.1109/CCCA.2011.6031533

Elmadssia, S., Saadaoui, K., & Benrejeb, M. (2013). New delay-dependent stability conditions for linear systems with delay Systems. *Science. Control. Engineering*, *1*(1), 37–41.

Filali, R. L., Hammami, S., Benrejeb, M., & Borne, P. (2012). On synchronization, anti-synchronization and hybrid synchronization of 3D discrete generalized H12 non map. *Nonlinear Dynamics and Systems Theory*, *12*(2), 81–96.

Gantmacher, F. R. (1966). *Théorie des matrices*. Ed. Dunod.

Hespanha, J. P., & Morse, A. S. (1999). Stability of switched systems with average dwell-time. In *Proceedings of IEEE Conference on Decision and Control* (pp. 2655–2660). Phoenix, AZ: IEEE. doi:10.1109/CDC.1999.831330

Hien, L. V., Ha, Q. P., & Phat, V. N. (2009). Stability and stabilization of switched linear time-delay systems. *Journal of the Franklin Institute*, *346*(6), 611–625. doi:10.1016/j.jfranklin.2009.03.001

Hu, H., Jiang, B., & Yang, H. (2013). Non-fragile H2 reliable control for switched linear systems with actuator faults. *Signal Processing*, *93*(7), 1804–1812. doi:10.1016/j.sigpro.2013.01.011

Ishii, H., & Francis, B. A. (2002). Stabilizing a linear system by switching control with dwell time. *IEEE Transactions on Automatic Control*, *47*(12), 1962–1973. doi:10.1109/TAC.2002.805689

Jaballi, A., Kermani, M., & Sakly, A. (2013). A New Stability Analysis for Discrete-Time Switched Time-Delay Systems. In *Proceedings of International Conference on Control, Engineering & Information Technology* (pp. 200-204). Sousse, Tunisia: Academic Press.

Kermani, M., Jaballi, A., & Sakly, A. (2014). A. A New Stability Analysis for Discrete-time Switched Time-delay Systems. *International Journal of Control. Energy and Electrical Engineering*, *1*(1), 23–30.

Kermani, M., & Sakly, A. (2014). Stability analysis for a class of switched nonlinear time-delay systems. *Systems. Science. Control Engineering*, *2*(1), 80–89.

Kim, S., Campbell, S. A., & Liu, X. (1999). Stability of a class of linear switching systems with time delay. *IEEE Transactions on Circuits and Systems*, *53*(1), 384–393.

Kotelyanski, D. M. (1952). Some properties of matrices with positive elements. *Matematicheskii Sbornik*, *31*(73), 497–505.

Liberzon, D. (2003). *Switching in Systems and Control*. Boston, MA: Birkhäuser. doi:10.1007/978-1-4612-0017-8

Liberzon, D., & Morse, A. S. (1999). Basic problems in stability and design of switched systems. *IEEE Control Systems Magazine*, *19*(5), 59–70. doi:10.1109/37.793443

Liberzon, D., & Tempo, R. (2004). Common Lyapunov functions and gradient algorithms. *IEEE Transactions on Automatic Control*, *49*(6), 990–994. doi:10.1109/TAC.2004.829632

Lin, H., & Antsaklis, P. (2009). Stability and Stabilizability of Switched Linear Systems: A Survey of Recent Results. *IEEE Transactions on Automatic Control*, *54*(2), 308–322. doi:10.1109/TAC.2008.2012009

Meyer, C., Schroder, S., & Doncker, R. W. D. E. (2004). Solid-state circuit breakers and current limiters for medium-voltage systems having distributed power systems. *IEEE Transactions on Power Electronics*, *19*(5), 1333–1340. doi:10.1109/TPEL.2004.833454

Montagner, V. J., Oliveira, R. C., & Peres, P. L. (2006). State feedback control of switched linear systems: AnLMI approach. *Journal of Computational and Applied Mathematics*, *53*(2), 384–393.

Morse, A. S. (1997). *Control Using Logic-Based Switching. Control and Information Sciences* (pp. 201–207). London: Springer-Verlag. doi:10.1007/BFb0036078

Qi, J., & Sun, Y. (2013). Global exponential stability of certain switched systems with time-varying delays. *Applied Mathematics Letters*, *26*(7), 760–765. doi:10.1016/j.aml.2013.02.010

Robert, F. (1966). Recherche d'une M-matrice parmi les minorantes d'un opérateur liéaire. *Numerische Mathematik*, *9*(2), 188–199.

Sfaihi, B., & Benrejeb, M. (2013). On stability analysis of nonlinear discrete singularly perturbed T-S fuzzy models. *Journal of Dynamical and Control Systems*, *1*(1), 20–31.

Shao, H. (2009). New delay-dependent stability criteria for systems with interval delays. *Automatica*, *45*(3), 744–749. doi:10.1016/j.automatica.2008.09.010

Shorten, R., & Narendra, K. (2003). On Common Quadratic Lyapunov Functions for Pairs of Stable LTI Systems Whose System Matrices are in Companion Form. *IEEE Transactions on Automatic Control*, *48*(4), 618–621. doi:10.1109/TAC.2003.809795

Shorten, R., Narendra, K., & Mason, O. (2003). A Result on Common Quadratic Lyapunov Functions. *IEEE Transactions on Automatic Control*, *48*(1), 110–113. doi:10.1109/TAC.2002.806661

Shorten, R., Narendra, K., & Mason, O. (2003). A result on common quadratic Lyapunov functions. *IEEE Transactions on Automatic Control*, *48*(1), 110–113. doi:10.1109/TAC.2002.806661

Shorten, R., Wirth, F., Mason, O., Wulf, K., & King, C. (2007). Stability criteria for switched and hybrid systems. *SIAM Review*, *49*(4), 545–592. doi:10.1137/05063516X

Sun, Y. G., Wang, L., & Xie, G. (2006). Stability of switched systems with time-varying delays: delay-dependent common Lyapunov functional approach. In *Proceedings of the American Control Conference* (pp.1544–1549). Philadelphia, PA: Academic Press.

Sun, Z., & Ge, S. S. (2002). *Switched Linear Systems Control and Design*. New York: Springer.

Sun, Z., & Ge, S. S. (2005). *Switched linear systems: Control and design. Communications and Control Engineering* (pp. 223–260). London: Springer-Verlag. doi:10.1007/1-84628-131-8_6

Sun, Z., & Ge, S. S. (2011). *Stability Theory of Switched Dynamical Systems*. London, UK: Springer. doi:10.1007/978-0-85729-256-8

Wu, Z. G., Park, J. H., Su, H., & Chu, J. (2013). Delay-dependent passivity for singular Markov jump systems with time-delays. *Communications in Nonlinear Science and Numerical Simulation*, *18*(3), 669–681. doi:10.1016/j.cnsns.2012.08.017

Yang, Y., Xiang, C., & Lee, T. H. (2012). Sufficient and necessary conditions for the stability of second-order switched linear systems under arbitrary switching. *International Journal of Control*, *85*(12), 1977–1995. doi:10.1080/00207179.2012.713987

Zhang, W., & Yu, L. (2008). Modelling and control of networked control systems with both network-induced delay and packet-dropout. *Automatica*, *44*(12), 3206–3210. doi:10.1016/j.automatica.2008.09.001

KEY TERMS AND DEFINITIONS

Arrow Form Matrix: A matrix which has non-constant elements isolated in only one row.

Common Lyapunov Function: The existence of a common Lyapunov function for all the subsystems guarantees the stability of the switched system under arbitrary switching.

Continuous-Time Switched Time-Delay Systems: A switched system is composed of an indexed family of delay subsystems described by continuous-time dynamics and a rule orchestrating the switching between them.

Kotelyanski Lemma: Kotelyanski elaborates a particular lemma, well adapted for stability study when off diagonal elements of matrix A are non-negative.

M-Matrix: A matrix which has all positive principal minors.

Stability under Arbitrary Switching: The stability of switched systems under arbitrary switching signal is guaranteed by the existence of a common Lyapunov function.

APPENDIX

Proof of Theorem 3 (Elmadssia et al., 2011; Elmadssia et al., 2013): Let's consider system (1) under the switched signal (3), let $\left(w_l > 0, \forall l = 1,\ldots,n\right)$. By the Leibniz–Newton formula: $\int_{t-h}^{t} \dot{x}\left(s\right) ds = x\left(s\right) - x\left(s-h\right)$, we get:

$$x\left(s-h\right) = x\left(s\right) - \int_{t-h}^{t} \left[\sum_{i=1}^{N} v_i\left(t\right)\left(A_i x\left(s\right) + D_i x\left(s-h\right)\right)\right] ds \tag{63}$$

Then, system (1) becomes:

$$\dot{x}\left(t\right) = \left[\sum_{i=1}^{N} v_i\left(t\right) A_i + \sum_{i=1}^{N} v_i\left(t\right) D_i\right] x\left(t\right) - \left[\sum_{i=1}^{N} v_i\left(t\right) A_i\right]\left[\sum_{i=1}^{N} v_i\left(t\right) D_i\right] \int_{t-h}^{t} \dot{x}\left(s\right) ds$$

$$-\left[\sum_{i=1}^{N} v_i\left(t\right) D_i\right]^2 \int_{t-h}^{t} \dot{x}\left(s-h\right) ds \tag{64}$$

Combining (63) with (64) leads to:

$$\dot{x}\left(t\right) = \left(A_{v(t)} + D_{v(t)}\right) x\left(t\right) - \left(A_{v(t)} D_{v(t)}\right) \int_{t-h}^{t} \dot{x}\left(s\right) ds - \left(D_{v(t)}\right)^2 \int_{t-h}^{t} \dot{x}\left(s-h\right) ds \tag{65}$$

For system (1), define the following Lyapunov functional as follows:

$$V\left(t\right) = V_1\left(t\right) + V_2\left(t\right) + V_3\left(t\right) + V_4\left(t\right) \tag{66}$$

where:

$$V_1\left(t\right) = \left\langle \left|x\left(t\right)\right|, w \right\rangle \tag{67}$$

$$V_2\left(t\right) = \left\langle \left|A_{\sigma(t)} D_{v(t)}\right| \int_{h}^{0} \int_{t+\theta}^{t} \left|x\left(s\right)\right| d\theta ds, w \right\rangle \tag{68}$$

$$V_3(t) = \left\langle \left|D_{\sigma(t)}^2\right| \int_h^0 \int_{t+\theta}^t \left|x(s-h)\right| d\theta ds, w \right\rangle \tag{69}$$

$$V_4(t) = h\left\langle \left|D_{\sigma(t)}^2\right| \int_{t-h}^t \left|x(s)\right| ds, w \right\rangle \tag{70}$$

it is easy to see that $V(t_0) < \infty$

Now, the right derivative $V(t)$ along the trajectory of system (1) allows to:

$$\frac{d^+V(t)}{dt^+} = \frac{d^+V_1(t)}{dt^+} + \frac{d^+V_2(t)}{dt^+} + \frac{d^+V_3(t)}{dt^+} + \frac{d^+V_4(t)}{dt^+} \tag{71}$$

where:

$$\frac{d^+V_1(t)}{dt^+} = \left\langle \frac{d^+\left|x(t)\right|}{dt^+}, w \right\rangle = \left\langle \operatorname{sgn}\left(Dx(t)\right) \frac{d^+x(t)}{dt^+}, w \right\rangle \tag{72}$$

and:

$$\operatorname{sgn}\left(Dx(t)\right) = \begin{pmatrix} \operatorname{sgn}\left(Dx_1(t)\right) & & \\ & \ddots & \\ & & \operatorname{sgn}\left(Dx_n(t)\right) \end{pmatrix} \tag{73}$$

Then:

$$\frac{d^+V_1(t)}{dt^+} = \Big\langle \operatorname{sgn}\left(Dx(t)\right)\left(A_{\sigma(t)} + D_{\sigma(t)}\right)x(t)$$

$$-\left(A_{v(t)}D_{v(t)}\right) \int_{t-h}^t x(s)\,ds \left[-\left(D_{\sigma(t)}\right)^2 \int_{t-h}^t x(s-h)\,ds\right], w \Big\rangle \tag{74}$$

$$< \left\langle \left(A_{\sigma(t)} + D_{\sigma(t)}\right)^* \left|x(t)\right| + \left(\left|A_{\sigma(t)}D_{\sigma(t)}\right|\right) \int_{t-h}^t x(s)\,ds + \left[\left|D_{\sigma(t)}\right|^2 \int_{t-h}^t x(s-h)\,ds\right], w \right\rangle \tag{75}$$

it follows that:

$$= \left\langle \left(A_{v(t)} + D_{v(t)} \right)^{*} \left| x(t) \right|, w \right\rangle + \left\langle \left(\left| A_{v(t)} D_{v(t)} \right| \right) \int_{t-h}^{t} x(s)\, ds, w \right\rangle + \left\langle \left| D_{v(t)} \right|^{2} \int_{t-h}^{t} x(s-h)\, ds, w \right\rangle \tag{76}$$

$$\frac{d^{+}V_{2}(t)}{dt^{+}} = \left\langle \left(\left| A_{v(t)} D_{v(t)} \right| \right) \left(h \left| x(t) \right| - \int_{t-h}^{t} \left| x(s) \right| ds \right), w \right\rangle \tag{77}$$

$$= h \left\langle \left| A_{v(t)} D_{v(t)} \right| \left| x(t) \right|, w \right\rangle - \left\langle \left| A_{v(t)} D_{v(t)} \right| \int_{t-h}^{t} \left| x(s) \right| ds, w \right\rangle \tag{78}$$

$$\frac{d^{+}V_{3}(t)}{dt^{+}} = \left\langle \left(\left| D_{v(t)}^{\;2} \right| \right) \left(h \left| x(t-h) \right| - \int_{t-h}^{t} \left| x(s-h) \right| ds \right), w \right\rangle \tag{79}$$

$$= \left\langle h \left| D_{v(t)}^{\;2} \right| \left| x(t-h) \right|, w \right\rangle - \left\langle \left| D_{v(t)}^{\;2} \right| \int_{t-h}^{t} \left| x(s-h) \right| ds, w \right\rangle \tag{80}$$

$$\frac{d^{+}V_{4}(t)}{dt^{+}} = \left\langle h \left| D_{v(t)}^{\;2} \right| \left(\left| x(t) \right| - \left| x(t-h) \right| \right), w \right\rangle \tag{81}$$

$$= \left\langle h \left| D_{v(t)}^{\;2} \right| \left| x(t) \right|, w \right\rangle - \left\langle h \left| D_{v(t)}^{\;2} \right| \left| x(t-h) \right|, w \right\rangle \tag{82}$$

From (76), (78), (80) and (72), yield to the following inequality:

$$\frac{d^{+}V(t)}{dt^{+}} < \left\langle \left(A_{v(t)} + D_{v(t)} \right)^{*} \left| x(t) \right|, w \right\rangle + \left\langle h \left| A_{v(t)} D_{v(t)} \right| \left| x(t) \right|, w \right\rangle + \left\langle h \left| D_{v(t)}^{\;2} \right| \left| x(t) \right|, w \right\rangle$$

$$= \left\langle T_{v(t)} \left| x(t) \right|, w \right\rangle \tag{83}$$

and finally, we obtain:

$$\frac{dV(t)}{dt} < \left\langle T_c \left|x(t)\right|, w\right\rangle \tag{84}$$

where T_c is introduced in (4).
Knowing that:

$$\left\langle T_c \left|x(t)\right|, w\right\rangle = \left\langle T_c^T w, \left|x(t)\right|\right\rangle \tag{85}$$

On the other hand, by supposing that T_c is the opposite of an $M-\text{matrix}$, by according to the $M-\text{matrix}$ proprieties, we can find a vector $\rho \in \Re_+^{*n} \left(\rho_l \in \Re_+^* \quad l = 1,\ldots,n\right)$ satisfying the following relation $\left(T_c\right)^T w = -\rho, \forall w \in \Re_+^{*n}$.
So, it easily to follows that:

$$\left\langle \left(T_c\right)\left|x(t)\right|, w\right\rangle = \left\langle \left(T_c\right)^T w, \left|x(k)\right|\right\rangle = \left\langle -\rho, \left|x(t)\right|\right\rangle \tag{86}$$

Substituting (76) into (74) leads to:

$$\frac{dV(t)}{dt} < \left\langle -\rho, \left|x(t)\right|\right\rangle = -\sum_{l=1}^{n} \rho_l \left|x(t)\right| < 0 \tag{87}$$

This completes the proof of Theorem 3.

Chapter 19
Discrete-Time Approximation of Multivariable Continuous-Time Delay Systems

Bemri H'mida
National Engineers School of Tunis BP 37, Tunisia

Mezlini Sahbi
National Engineers School of Tunis BP 37, Tunisia

Soudani Dhaou
National Engineers School of Tunis BP 37, Tunisia

ABSTRACT

Many works are related to the analysis and control of either continuous or else discrete time-delay systems. In general, discrete-time controls have become more and more preferable in engineering because of their easy implementation and simple computation. However, the available discretization approaches for the systems having time delays increase the system dimensions and have a high computational cost. The case studies in this chapter support the efficiency of the two methods. However, the discretization of continuous time-delay systems has not been sufficiently/extensively studied in many works. In this work, the authors present two methods of the effective discretization approach for the continuous-time systems with an input and output delays. Sampled-data time-delay systems with internal and external point delays are described by approximate discrete time-delay systems in the discrete domain. These approximate discrete systems allow the hybrid control of time-delay systems.

1. INTRODUCTION

Systems with delays abound in the world. One reason is that nature is full of transparent delays. Another reason is that time-delay systems are often used to model a large class of systems with delays that frequently appearing in engineering. Typical examples of time-delay systems are communication networks, chemical processes, teleoperation systems, biosystems, and underwater vehicles and so on. The presence

DOI: 10.4018/978-1-4666-7248-2.ch019

of delays makes system analysis and control design much more complicated. The emphasis is on systems with a single input/output delay although the delay-free part of the plant can be multi-input, multi-output (MIMO) when the delays in different channels are the same. The major tools used in this chapter are State-transition method and a Method based on the trapezoidal rule for integration.

With the pace of development in the field of microelectronics, analog controllers yield more places for their digital computers. Indeed, and given the importance of these control systems, it uses methods and numerical models to analyze and/or control industrial processes. To implement such a control structure and ensure the desired objectives, modeling in discrete-time analog systems is required.

Two types of representation are available to model continuous or discrete systems namely the external representation that uses input-output (or transfer function matrix) relations and the internal representation of dynamic system which is based on the concept of state.

Digital control of physical systems usually requires developing discrete models. Several modeling strategies, developed in the literature reflecting a meaningful description of dynamical systems to study, led to mathematical tools generally resulting in linear or nonlinear systems with or without delays whose behavior is close to the real system models. These models, which are described by relations between input and output variables, can be modified by inputs considered secondary (disturbances) that still exist in practice. Knowing the benefits given by the state description of dynamical systems, opt for following such descriptions.

A discrete time delay system is defined as an operator between two discrete-time signals that involve a time called delays. A physical system for which $u(k)$ is the general term of the sequence input and $x(k+1)$ is the general term of the sequence and state of delayed h.

Initial modeling of a system to discrete time delays often leads to writing a recurrent equation between different terms of the input and output sequences . This formulation of the recursive equation is well suited for numerical calculation. This is the form in which these algorithms are control methods. The system is fully defined and recurrent equation can be solved if the initial conditions are specified. These forms can be simplified mathematically using the lag operator and allow formalizing recurrent equations as follows:

$$x(k+1) = Ax(k) + Bx(k-q) + Cu(k) \ ; h = qT \tag{1}$$

where A, B and C are matrices of suitable dimensions, and q is assumed to be constant delay.

Time delay is the property of a physical system by which the response to an applied force (action) is delayed in its effect (Chen & Latchman,1994;Shinskey, 1967). Whenever material, information or energy is physically transmitted from one place to another, there is a delay associated with the transmission. The value of the delay is determined by the distance and the transmission speed. Some delays are short, others are very long. The presence of long delays makes system analysis and control design much more complex. What is worse is that some delays are too long to perceive and the system is misperceived as one without delays.

The presence of time-delays in a practical system may induce instability, oscillation and poor performance and complicates the analysis and design of feedback controllers. Control systems with time-delays exhibit complex behavior because of their infinite dimensionality, even in the case of linear time-invariant systems that have constant time-delays in the input or states have infinite dimensionality when expressed in the continuous-time domain. For these reasons, during the last few decades it was not possible to apply a controller design technique having any time-delays in the variables. Thus it is necessary to develop a control system design scheme that resolves these time-delays.

Generally speaking, there are two kinds of time-delays: input-delay and state-delay. Input-delay is caused by the transmission of a control signal over a long distance, or the delay is sometimes built into a system deliberately for control purposes. State-delay is due to transmission or transport delay among interacting elements in a dynamic system. As for the state-delay system, many works have been done.

However, most methods were developed for continuous-time systems rather than discrete-time systems and they seem to be impractical, not only is the implementation difficult, but also the developed control laws or design approaches involve too much analytical mathematics.The stability of state delay systems has been studied by investigating the properties of a Lyapunov functional (Suh & Lee, 1999). There have been also studies with a neural-network based sliding-mode control scheme by integrating neural-network approximation and the Lyapunov theory into the sliding-mode technique (Niu & Wang, 2003; Stojanović, Debeljković, Mladenović, 2007). The neural-network solves the problem of robust control for state-delayed systems in the presence of mismatched parametric uncertainties, unknown nonlinearity and external disturbance.

Systems with Delays appear in various contexts such as biological, ecological, economic, social, and engineering systems. Typical examples of time-delay systems are communication networks, chemical processes, teleoperation systems, bio-systems, underwater vehicles and so on (Sinha & Lastman, 1981; Sinha, 1972). In many physical, industrial and engineering processes, delays occur due to the finite capabilities of information processing and data transmission among various parts of the system. Delays can arise from inherent physical phenomena such as mass transport flow or recycling. Also, they can be byproducts of computational delays and can be constant or time-varying, known or unknown, deterministic or stochastic, depending on the system under consideration. In all these cases, the effect of time -delay on the evolution of systems leads to poor performance.

Therefore, the subject of time-delay systems has been investigated in the form of functional differential equations over the past three decades (Michiels & Niculescu, 2007; Franklin, Powell, & Workman, 1997; Kailath, 1980).The engineering literature dealing with time-delayed systems is very extensive. Most of the approaches proposed so far deal with linear time-delay control systems and, in particular, with the stability analysis and behavior of such systems with constant and/or uncertain time-delays (Palanisamy & Bhattacharya, 1981; Martens, 1969; Normey-Rico, 2007). However, it should be mentioned that conventional numerical techniques, such as the Euler and Runge-Kutta methods, have been employed in order to obtain a sampled-data representation of the original continuous-time delay-free system (Hung, Liu, & Chan, 1980; Rao & Srinivasan, 1978). All of these approaches require a very small time step in order to be deemed accurate, however this may not be the case in control applications where large sampling periods are inevitably introduced due to physical and technical limitations. Furthermore, the use of too many integration steps may cause unacceptable integration times and even the excessive accumulation of errors.

With the rapid advances in the large-scale integration of semi-conductor devices and the resulting availability of inexpensive computers, there is renewed interest in the discrete-time approximation of continuous-time multivariable systems. Such models have applications in the digital simulation of these systems as well as in the identification of the system through samples of the input- output data. Further applications of these methods are possible in digital adaptive control and computer control of complex processes (Harmuth,1969; Ratchagit & Phat, 2010). Systems including time delay, due to the system dynamics, are widely present in the industry which imposes a lot of constraints that make the control and computer programming of such system difficult (Liu, Lu, & Chen, 2014; Sannuti, 1977).In addition of a system with a time delay is generally difficult; due to the constraints imposed by the time delay. These constraints can cause performance deterioration that leads the instability of the process especially when operating in closed loop (Fang, 2010; Zhong, 2008).Most physical systems, from a macroscopic point of view, are continuous. In modern control systems, information is digitally processed which requires sampling signals.

For this reason the discretization of continuous linear delay systems is needed. Other works are devoted to the study of discrete delay systems (Sinha, 1983; Sinha & Rozsa, 1976; Mahmoud, 2000).

The objective of this chapter is to extend the ideas in the just cited references in order to analyze them. Two different methods are considered to obtain the discrete-time delay approximation:

1. State-transition method.
2. Method based on the trapezoidal rule for integration.

The remainder of the chapter is organized as follow: The next section discusses the discretization of systems with external point delay; Section 3 discusses the discretization of system with internal and external point delays; Section 4 provides a numerical example and a comparison between the two methods and Section 5 consolidates the ideas and presents concluding remarks.

2. DISCRETIZATION OF SYSTEMS WITH EXTERNAL POINT DELAY

State space theory deals with dynamical models describing both the internal dynamics of a given physical process and the interaction of the process with the outside world.

The discrete-time state representations and pseudo-continuous time are equivalent since the bilinear Tustin transform is invertible and preserves algebraic properties. The results established in the continuous-time systems (clean but not strictly) are used in the representations as a pseudo-continuous-time systems with discrete time. A method of synthesis of discrete-time control is therefore to:

Calculate the discrete time model of the system which was controlled (recursive equation and Z transfer or state representation).
Derive the model pseudo- continuous-time (w transfer or state representation).
Calculate the continuous pseudo- command time (transfer w or state representation).
Derive the discrete time control by reversing the bilinear transformation Tustin.

These representations permit the construction of programs for numerical computation:

- By recurrent equations outputs to the inputs.
- Either by internal representation (state model) or transfer functions.

Why Study Models Where the Weather Changes Discretely?

The main reasons:

- The discrete models are a tool to solve the differential equations.
- They may be more appropriate models in some areas: digital electronics, economics, finance, population studies.
- They can be used to generate chaotic dynamics since it is easier to generate complex dynamics in the discrete-time case than in the continuous case.
- They are easier to be simulated by the computer: no algorithm numerical integration is required.

In a digital computer, time cannot flow continuously as it is perceived in the physical world. The time is defined on a discrete set of times which are separated by a regular time interval known by one sampling period. It is therefore necessary to define new mathematical tools adapted to discrete time, represent the sampled signals and systems and adapt tools and methods for automatic analog (continuous time) in the design of digital controllers.

Then the problem may be stated as the determination of a discrete time approximation corresponding to the following state-space equations:

$$\begin{cases} \dot{x}(t) = A_0 x(t) + A_1 x(t-h) + B\,u(t) \\ y(t) = C\,x(t) \end{cases} \tag{2}$$

Where:

- $x(t) \in \mathbb{R}^n$ denotes the state variables; $u(t) \in \mathbb{R}^m$ is control inputs.
- $h > 0$ is a constant denoting delay of state.
- $A_0 x(t)$ **:** Original Term state.
- $A_1 x(t-h)$ **:** Delayed Term state.
- $h = qT$ **:** A multiple delay the sampling period is an integer q.
- T **:** The sampling period assumed chosen suitably.
- $x, u\ \ and\, y$ respectively are the state vector, the vector and the control vector output.
- A_0, A_1, B and C are constant matrices with appropriate dimensions.

In the development of the discrete-time models to assume a suitable sampling interval, denoted by T. A suitable criterion for the choice of T is that wT be less than *0.5*, where w is the magnitude of the Eigen value of farthest from the origin of the s-plane (Harmuth, 1969; Fridman & Shaked, 2005). Among the main concerns of the sampling was one that is not to lose information in the temporal discretization of the continuous signal. For this to be possible, a condition is to be fulfilled. The signal that must be sampled should have a finite spectral width; it is called a spectrum of low-pass type. Since the spectral width of such a signal is defined by the interval [0, F_{max}], where fmax is the highest frequency present in the frequency spectrum of this signal, this condition arises from the phenomenon of aliasing. In reality, all physically realizable analog signals spectral width is "great", although it is necessarily finite energy. This may arise, for example, the presence of additive noise or interference with amplitude spectra which is not negligible at high frequencies. In other words, fmax is very large which necessarily leads to the choice of an even greater F_e. If there is aliasing, it is not possible to return to the original signal spectrum. In this case, the operation changes the sampling characteristics of the input signal. Thus, if must be satisfied the information relative to the signal which is sampled, is not to be lost the condition: (*Fe*> $2F_{max}$) This condition is better known by the name, *Shannon's theorem* (Dorf & Bishop, 2008). A discrete system is a system that involves information that is taken into account only at specific times. Generally these instants are spaced by a constant length called sampling period. In practice the sampling period (*Ts*) depends on the type of process (chemical, thermal, mechanical, etc.), and should be chosen small in relation to the response time (*Tr*) of the closed loop system. Practical value is that the system must be sampled between 4 and 10 *times* the response time. $N_{ech} = T_r / T_s = [4 \sim 10]$.A large sampling period (sub-sampling) results in:

- Depletion of performance because the control is locked an important time, and does not just react when a disturbance occurs.
- If the output is noisy it is convenient to add a filter to reduce the noise before sampling (anti -aliasing filter).

The filter introduces a delay in the control loop. A small sampling period results in:

- Closer to the continuous time, but much more demanding calculation.
- Difficulties inherent to the calculation methods of the order, which can lose strength when $T_s \to 0$ because the control signals can tend to infinity.
- In practice it is recommended to choose the sampling rate in the range of about *6* to *24* times the cut-off frequency of the process.
- Sampling leads to a loss of information in terms of the continuous signal. This loss of information is even greater than the frequency $f = 1/T_s$ is small.

Ideally, it should therefore be sampled at infinite frequency; however, the choice of the sampling period depends on the proceeds and the possibilities offered by digital tools. In any event, sampling must respect the Shannon theorem, which states that the sampling frequency $f = 1/T_s$ must be at least twice the highest frequency contained in the spectrum of the signal you want to sample.

2.1. State-Transition Method

The procedure is a first order system as the state transition matrix generalizing.

A state transition matrix $e^{A(t-t_0)}$ (Schultz & Melsa, 1967) defined by Taylor development series:

$$e^{A(t-t_0)} = I + A(t-t_0) + \frac{A^2}{2!}(t-t_0) + \ldots + \frac{A^k}{k!}(t-t_0)^k + \ldots = \sum_{k=0}^{\infty} \frac{A^k}{k!}(t-t_0)^k \quad (3)$$

Where the series converges for all finite $t > 0$ and $x(0)$ is the initial state. Each term in this series is a matrix of size $n \times n$, and the summation of all terms yields another matrix of size $n \times n$.

It is clear that this equation $x(t) = e^{At}x(0)$ gives the response at any time t to an arbitrary set of initial conditions, thus the computation of e^{At} at any t yields the values of all the state variables $x(t)$ directly. The discrete-time models describe the relationships between sampled variables: $x(\mathrm{kT}_s)$, $u(\mathrm{kT}_s)$ and $y(\mathrm{kT}_s)$, $k = 0,1,2,3,\ldots\ldots$ The value $x(\mathrm{kT}_s)$ is kept constant during the sampling interval $\left[kT_s,(k+1)T_s\right]$.

A discrete-time signal can either represent the sampling of a continuous-time signal, or be an intrinsically discrete signal. If the first order difference equation (autonomous system) is considered:

$$\begin{cases} x(k+1) = Ax(k) \\ x(0) = x_0 \end{cases} \tag{4}$$

A solution of this equation is given by: $x(k) = A^k x(0) = A^k x_0$.

Then:

$$\begin{cases} For \;\; k = 0, x(1) = Ax(0) = Ax_0 \\ \quad\;\; k = 1, x(2) = Ax(1) = A^2x_0 \\ \quad\;\; k = 2, x(3) = Ax(2) = A^3x_0 \\ \quad\;\; . \\ \quad\;\; x(k) = A^k x(0) = A^k x_0 \end{cases} \tag{5}$$

It can be verified that the state transition matrix has the same proprieties:

e^{At} Compared with A^k

- $e^{A(t_0 - t_o)} = I \quad A^{(k=0)} = I$
- $\dfrac{d}{dt} e^{A(t-t_0)} = Ae^{A(t-t_0)} \quad A^{k+1} = AA^k$
- $e^A \times e^B = e^{A+B} \quad A^{k+m} = A^k A^m$

$$\int_0^t e^{A\tau} d\tau = A^{-1}[e^{At} - I];$$

- The eigenvectors of A are vector own e^{A}.
- If λ is an eigenvalue of A then e^{λ} is an eigenvalue of e^{A}.

If A is a diagonal matrix then e^{At} is also diagonal, and each element on the diagonal is an exponential in the corresponding diagonal element of the A matrix, that is e^{At} .This property is easily shown by considering the terms A^{k} in the series definition and noting that any diagonal matrix rose to an integer power is diagonal.

The state representation enables the use of computational techniques available in linear algebra. Synthesis tools have been developed powerfully pole placement, linear quadratic optimal control, linear quadratic Gaussian control, regarding concrete achievements of control systems, it is increasingly using digital computers. It is easier to alter a computer program, an electronic circuit, and the computer can perform other operations while it calculates and applies the command.

The description of the system dynamics in the form of differential equations is retained throughout the analysis and design. In fact, if a subsystem is characterized by a transfer function it is often necessary to convert the transfer functions equations in order to proceed by state-space methods. The state-transition matrix which describes how the state $x(\mathrm{T})$ of the system at some time t evolves into the state $x(\mathrm{T})$ at some other time T .For time-invariant systems, the state-transition matrix is the matrix exponential function, which is easily calculated. For most time-varying systems, however, the state-transition matrix, known to exist, cannot be expressed in terms of simple functions .This is the most well known method. If the input is allowed to vary only at the sampling instants and held constant during each sampling interval, the following equation is obtained:

$$x(k+1) = \psi_1 x(k) + \psi_2 x(k-q) + \psi_3 u(\mathrm{k}) \tag{6}$$

Where $x \in X \subset R^{n}$ is the vector of states and X is an open and connected set, $u \in \mathbb{R}$ is the input variable. It is assumed that x and u are real analytic vector fields on X .

An equidistant grid on the time axis with mesh $T = t_{k+1} - t_{K}$ > 0 is considered.

Where:

$$[t_K, t_{k+1}] = [\mathrm{kT}, (\mathrm{k}+1)\mathrm{T}]$$

The sampling interval and T is the sampling period. It is assumed that the system described in equation (6) is driven by an input that is piecewise constant over the sampling interval. $u(t) = u(kT) = u(k) =$ Constant.

$$\forall t \in \left[kT, (k+1)\mathrm{T}\right], u(t) = u(k) \tag{7}$$

Where:

- $\psi_1 \in \mathbb{R}^{n \times n}$ is the system matrix .
- $\psi_2 \in \mathbb{R}^{n \times n}$ is the system matrix concerning state-delay.
- $\psi_3 \in \mathbb{R}^{n \times p}$ is the input matrix .
- $u(\mathrm{k})$ is an input and q is an integer defines a time-delay.
- $x(k)$ is the value of the state vector x at time $t = t_K = KT$.

This discrete-time description related to equation (6) can be obtained as:

$$\psi_1 = e^{A_0 T} \tag{8}$$

$$\psi_2 = e^{A_1 T} \tag{9}$$

$$\psi_3 = \int_0^T e^{A_0 t} B dt + \int_0^T e^{A_1 t} B dt \tag{10}$$

The matrices ψ_1 , ψ_2 and ψ_3 can be conveniently calculated on a digital computer using the following series:

$$\psi_1 = I + A_0 T + \frac{A_0^2 T^2}{2!} + \frac{A_0^3 T^3}{3!} + \ldots\ldots \tag{11}$$

$$\psi_2 = I + A_1 T + \frac{A_1^2 T^2}{2!} + \frac{A_1^3 T^3}{3!} + \ldots\ldots \tag{12}$$

$$\psi_3 = \left(2I + \frac{A_0 T}{2!} + \frac{A_1 T}{2!} + \frac{A_0^2 T^2}{3!} + \frac{A_1^2 T^2}{3!} + \ldots. \right) TB \tag{13}$$

Figure 1. Approximation by the method of trapeze

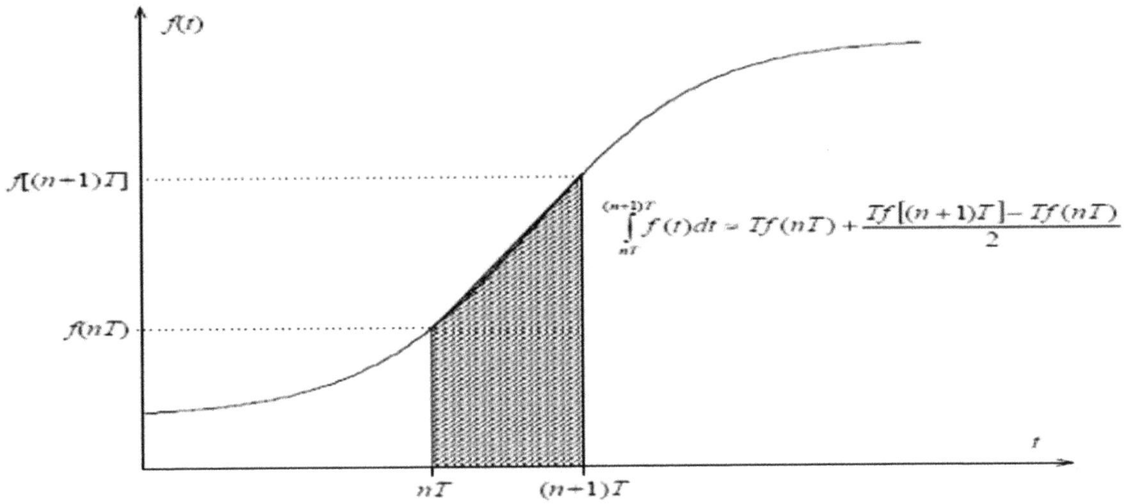

2.2. Method Based on Trapezoidal Rule of Integration

The discretization by the trapezoidal method is to approach the air under the curve $f(t)$ between two successive instants nT and $(n+1)T$ by the air of the trapezoid.

The differential equation:

$$dx(t)\,/\,dt = f(t) \tag{14}$$

Equation (14) can still be written in the form:

$$dx(t) = f(t)\,\mathrm{dt} \tag{15}$$

The problem is to integrate Equation (15) between two successive instants, therefore it is calculated:

$$\int_{x(nT)}^{x[(n+1)T]} dx = \int_{nT}^{(n+1)T} f(t)dt \tag{16}$$

Is therefore:

$$x[(n+1)T] = x(nT) + \int_{nT}^{(n+1)T} f(t)dt \tag{17}$$

Since

$$\int_{nT}^{(n+1)T} f(t)dt = T\frac{f[(n+1)T]+f(nT)}{2} \tag{18}$$

The Equation (17) is defined as follows:

$$x[(n+1)T] = x(nT) + T\frac{f[(n+1)T]+f(nT)}{2} \tag{19}$$

Then:

$$\frac{x[(n+1)T]-x(nT)}{T} = \frac{f[(n+1)T]+f(nT)}{2} \tag{20}$$

The latter relationship is similar to (14) is obtained:

$$\dot{x}(t) = \frac{dx(t)}{dt} = \frac{x[(n+1)T]-x(nT)}{T} = \frac{x(n+1)-x(n)}{T} \tag{21}$$

The determination of a discrete model of a continuous linear or nonlinear process can be envisaged by a transformation on the functional matrix that characterizes the evolution of the autonomous regime of the studied system. It is in this sense that a transformation called homographic (Haykin, 1972), or as transformation trapeze (Shieh et al.1978), this transformation allows to combine the rigorous way to a continuous system and a discrete system. One may use the trapezoidal rule for integrating the state equation over the interval $kT < t < (k+1)T$ to obtain:

- $\dot{x}(t) = \frac{x(k+1)-x(k)}{T}$: Discrete differentiation.
- $x(t) = \frac{x(k+1)+x(k)}{2}$: Average value of the state vector between two sampling times successive.
- $x(t-h) = \frac{x((k+1)T-qT)+x(kT-qT)}{2} = \frac{x(k+1-q)+x(k-q)}{2}$: Average value of the state vector between two sampling times successive.

Replaced $\dot{x}(t)$, $x(t)$ and $x(t-h)$ by their new expressions in the system of equation defined in (2), the new equation obtained is:

$$\frac{x(k+1)-x(k)}{T} = A_0\frac{x(k+1)+x(k)}{2} + A_1\frac{x(k+1-q)+x(k-q)}{2} + Bu(k) \tag{22}$$

Solving equation number (22) for $x(k+1)$ it becomes:

$$x(k+1)=(I-\frac{T}{2}A_0)^{-1}(I+\frac{T}{2}A_0)\;x(k) \\ +(I-\frac{T}{2}A_0)^{-1}\frac{T}{2}A_1\left[x(k+1-q)+x(k-q)\right]+(I-\frac{T}{2}A_0)^{-1}TBu(k) \tag{23}$$

Then

$$x(\mathrm{k}+1)=\varphi_1 x(k)+\varphi_2\left[x(k+1-q)+x(k-q)\right]+\varphi_3\,u(k) \tag{24}$$

Where $x \in X \subset \mathbb{R}^n$ is the vector of states and X is an open and connected set, $u \in \mathbb{R}$ is the input variable. It is assumed that x and u are real analytic vector fields on X.

An equidistant grid on the time axis with mesh $T=t_{k+1}-t_K>0$ is considered.

Where:

$$[t_K, t_{k+1}]=[\mathrm{kT},(\mathrm{k}+1)\mathrm{T}]$$

The sampling interval and T is the sampling period. It is assumed that the system described in Equation (24) is driven by an input that is piecewise constant over the sampling interval. $u(t)=u(kT)=u(k)=$ Constant.

$$\forall t \in \left[kT,(k+1)\mathrm{T}\right], u(t)=u(k) \tag{25}$$

Where:

- $\varphi_1 \in \mathbb{R}^{n\times n}$ is the system matrix .
- $\varphi_2 \in \mathbb{R}^{n\times n}$ is the system matrix concerning state-delay.
- $\varphi_3 \in \mathbb{R}^{n\times p}$ is the input matrix and $\varphi_4 \in \mathbb{R}^{n\times p}$ is the input matrix concerning input -delay.
- $u(\mathrm{k})$ is an input and q is an integer defines a time-delay.
- $x(k)$ is the value of the state vector x at time $t=t_K=KT$.

This discrete-time description related to Equation (24) can be obtained as:

$$\varphi_1=(I-\frac{T}{2}A_0)^{-1}(I+\frac{T}{2}A_0) \tag{26}$$

$$\varphi_2 = (I - \frac{T}{2}A_0)^{-1}\frac{T}{2}A_1 \tag{27}$$

$$\varphi_3 = (I - \frac{T}{2}A_0)^{-1}TB \tag{28}$$

3. DISCRETIZATION OF SYSTEMS WITH INTERNAL AND EXTERNAL POINT DELAYS

Digital computers perform sequences of elementary operations, each of which requires a nonzero finite execution time. Therefore a computer time is not of the same nature as the time the physical system has set. For convenience, this time is considered as a succession of instants, separated by a constant interval, the sampling period Te. The time is then represented by a discrete variable, that is to say defined on a discrete set, isomorphic to the set of relative integers ($k \in Z$). The dynamic aspect of systems is then translated by recurrent equations on numerical sequences. These equations are assumed to be linear here (check the superposition theorem) and stationary (constant coefficients). These assumptions are of the same nature as for continuous time systems.

The problem may be stated as the determination of a discrete time approximation corresponding to the following state-space equations. Time-delay systems with a single state-delay and input delay will be considered:

$$\begin{cases} \dot{x}(t) = A_0x(t) + A_1x(t-h_0) + B_0u(t) + B_1u(t-h_1) \\ y(t) = cx(t) \end{cases} \tag{29}$$

Where:

- $x(t) \in \mathbb{R}^n$: Denotes the state variables; $u(t) \in \mathbb{R}^m$ is a control inputs.
- $h_0 > 0$ and $h_1 > 0$ are a constant denoting delay of state and control respectively.
- $A_0x(t)$: Original Term state.
- $A_1x(t-h_0)$: Delayed Term state.
- $B_0u(t)$: Original Term control.
- $B_1u(t-h_1)$: Delayed Term control.
- $h_0 = q_0T$: A multiple delay of the sampling period and q_0 is an integer.
- $h_1 = q_1T$: A multiple delay of the sampling period and q_1 is an integer.

Note: q_0 and q_1 are not necessarily equal.

$x, u\ and\ y$ respectively are the state vector, the vector and the control vector output.

T : The sampling period is chosen suitably.

A_0, A_1, B_0, B_1 and C are constant matrices with appropriate dimensions.

3.1. State-Transition Method

The concept of a state-transition is considered as important in many theories. The state-transition matrix can be used to obtain the general solution of linear dynamical systems. It is also known as the matrix exponential. In the time-variant case, there are many different functions that may satisfy these requirements, and the solution is dependent on the structure of the system. The state-transition matrix must be determined before analysis on the time-varying solution can be carried on. Assuming that, the input is allowed to vary only at the sampling instants and should be held constant during each sampling interval, the following equation is obtained:

$$x(k+1) = \theta_1 x(k) + \theta_2 x(k - q_0) + \theta_3 u(\mathrm{k}) + \theta_4 u(\mathrm{k} - \mathrm{q}_1) \tag{30}$$

Where $x \in X \subset \mathbb{R}^n$ is the vector of states and X is an open and connected set, $u \in \mathbb{R}$ is the input variable. It is assumed that x and u are real analytic vector fields on X.

An equidistant grid on the time axis with mesh $T = t_{k+1} - t_K > 0$ is considered.

Where:

$$[t_K, t_{k+1}] = [\mathrm{kT}, (\mathrm{k}+1)\mathrm{T}]$$

The sampling interval and T is the sampling period. It is assumed that the system described in Equation (30) is driven by an input that is piecewise constant over the sampling interval.

$u(t) = u(kT) = u(k) =$ Constant.

$$\forall t \in \left[kT, (k+1)\mathrm{T}\right], u(t) = u(k) \tag{31}$$

$$u(t - h_1) = u(k - q_1) \tag{32}$$

Where:

$\theta_1 \in \mathbb{R}^{n \times n}$ is the system matrix, $\theta_2 \in \mathbb{R}^{n \times n}$ is the system matrix concerning state-delay,

$\theta_3 \in \mathbb{R}^{n \times p}$ is the input matrix and $\theta_4 \in \mathbb{R}^{n \times p}$ is the input matrix concerning input –delay.

$u(k)$ and $u(k - q_1)$ are inputs and q_1 is a time-delay.

$x(k)$ is the value of the state vector x at time $t = t_K = KT$.

This discrete-time description related to Equation (30) can be obtained as:

$$\theta_1 = e^{A_0 T} \tag{33}$$

$$\theta_2 = e^{A_1 T} \tag{34}$$

$$\theta_3 = \int_0^T e^{A_0 t} B_0 dt + \int_0^T e^{A_1 t} B_0 dt \tag{35}$$

$$\theta_4 = \int_0^T e^{A_0 t} B_1 dt + \int_0^T e^{A_1 t} B_1 dt \tag{36}$$

The matrices $\theta_1, \theta_2, \theta_3$ and θ_4 can be conveniently calculated on a digital computer using the following series:

$$\theta_1 = I + A_0 T + \frac{A_0^2 T^2}{2!} + \frac{A_0^3 T^3}{3!} + \ldots\ldots \tag{37}$$

$$\theta_2 = I + A_1 T + \frac{A_1^2 T^2}{2!} + \frac{A_1^3 T^3}{3!} + \ldots\ldots \tag{38}$$

$$\theta_3 = \left(2I + \frac{A_0 T}{2!} + \frac{A_1 T}{2!} + \frac{A_0^2 T^2}{3!} + \frac{A_1^2 T^2}{3!} + \ldots. \right) T B_0 \tag{39}$$

$$\theta_4 = \left(2I + \frac{A_0 T}{2!} + \frac{A_1 T}{2!} + \frac{A_0^2 T^2}{3!} + \frac{A_1^2 T^2}{3!} + \ldots. \right) T B_1 \tag{40}$$

3.2. Method Based on Trapezoidal Rule of Integration

The actual control system is increasingly opting for the use of digital computers as they are more efficient and they are becoming less and less expensive.

- Computers have increasingly become rapid, growth rate has doubled during the last 40 years.
- Sizes of equipment and prices have significantly declined.
- Thus the performance of the sampled system becomes very close to that of an equivalent continuous system.
- Ability to handle multiple variables at the same time, treating multivariable systems becomes easier and easier .

  Some application areas of digital commands are:

- Control of machine tools (lathes, milling machines, etc.).
- Complex chemical processes.
- Aerospace.
- Automobile, etc.

This transformation allows to associate rigorous manner to any continuous system, a discrete system. It is possible to deduce a recurrent solution describing state space of the evolution of a vector *x(k)* obtained by discretization *x(t)* at sampling instants. Using the approximations adopted in the homographic transformation matrix discrete model that describes the state space equation, the result obtained is:

- $\dot{x}(t) = \dfrac{x(k+1) - x(k)}{T}$: Discrete differentiation.
- $x(t) = \dfrac{x(k+1) + x(k)}{2}$: Average value of the state vector between two sampling times successive.
- $x(t - h_0) = \dfrac{x((k+1)T - q_0T) + x(kT - q_0T)}{2} = \dfrac{x(k+1-q_0) + x(k - q_0)}{2}$: Average value of the state vector between two sampling times successive.
- $x(t - h_1) = \dfrac{x((k+1)T - q_1T) + x(kT - q_1T)}{2} = \dfrac{x(k+1-q_1) + x(k - q_1)}{2}$: Average value of the state vector between two sampling times successive.

Replacing $\dot{x}(t)$, $x(t)$, $x(t-h_0)$ and $x(t-h_1)$ by their new expressions in the system of equation defined in (29) the result obtained is:

$$\begin{aligned} x(k+1) &= (I - \frac{T}{2}A_0)^{-1}(I + \frac{T}{2}A_0)\ x(k) \\ &+ (I - \frac{T}{2}A_0)^{-1}\frac{T}{2}A_1\left[x(k+1-q_0) + x(k-q_0)\right] \\ &+ (I - \frac{T}{2}A_0)^{-1}TB_0u(k) + (I - \frac{T}{2}A_0)^{-1}TB_1u(k-q_1) \end{aligned} \tag{41}$$

Then:

$$x(k+1) = \sigma_1\, x(k) + \sigma_2\left[x(k+1-q_0) + x(k-q_0)\right] + \sigma_3\, u(k) + \sigma_4\, u(k-q_1) \tag{42}$$

Where:

$x \in X \subset \mathbb{R}^n$ is the vector of states and X is an open and connected set, $u \in \mathbb{R}$ is the input variable. It is assumed that x and u are real analytic vector fields on X.

An equidistant grid on the time axis with mesh $T = t_{k+1} - t_K > 0$ is considered.

Where:

$$[t_K, t_{k+1}] = [kT, (k+1)T]$$

The sampling interval and T is the sampling period. It is assumed that the system described in Equation (42) is driven by an input that is piecewise constant over the sampling interval.

$u(t) = u(kT) = u(k) =$ constant.

$$\forall t \in \left[kT, (k+1)\mathrm{T}\right], u(t) = u(k) \tag{43}$$

$$u(t - h_1) = u(k - q_1) \tag{44}$$

Where:

- $\sigma_1 \in \mathbb{R}^{n \times n}$ is the system matrix .
- $\sigma_2 \in \mathbb{R}^{n \times n}$ is the system matrix concerning state-delay.
- $\sigma_3 \in \mathbb{R}^{n \times p}$ is the input matrix .
- $\sigma_4 \in \mathbb{R}^{n \times p}$ is the input matrix concerning input-delay.
- $u(k)$ and $u(k - q_1)$ are inputs and q_1 is a time-delay.
- $x(k)$ is the value of the state vector x at time $t = t_K = KT$.

This discrete-time description related to Equation (42) can be obtained as:

$$\sigma_1 = (I - \frac{T}{2}A_0)^{-1}(I + \frac{T}{2}A_0) \tag{45}$$

$$\sigma_2 = (I - \frac{T}{2}A_0)^{-1}\frac{T}{2}A_1 \tag{46}$$

$$\sigma_3 = (I - \frac{T}{2}A_0)^{-1}TB_0 \tag{47}$$

$$\sigma_4 = (I - \frac{T}{2} A_0)^{-1} T B_1 \tag{48}$$

4. NUMERICAL EXAMPLES

Modeling is a tool that is increasingly used to configure or analyze systems. The purpose of a model is to inform us about the quantitative and qualitative evaluation as well as the behavior and functioning of a system.

Some advantages that little quote:

- Improved measurement sensitivity.
- Using sensors, transducers and digital computers.
- Decreased sensitivity to noise
- Easy reconfiguration algorithm or control calculation supervision .
- Possibility of more complex algorithms and performing calculations .

The digital computer control with precise additional elements of control systems that modify in part the control loop requires some additional knowledge and is traditionally linked to the discipline of signal processing .

Some of these skills are necessary:

- Loop control (digital sampled, digital) discrete systems.

In this section two different examples of application of the proposed approach are presented. These examples show the validity of the methodology. As a first example, the following system is considered by the state-space equation:

$$\begin{cases} \dot{x}(t) = \begin{pmatrix} 0 & 1 \\ -2 & -3 \end{pmatrix} x(t) + \begin{pmatrix} 0 & 0 \\ -1 & 2 \end{pmatrix} x(t-0.2) + \begin{pmatrix} 0 \\ 1 \end{pmatrix} u(t) \\ y(t) = \begin{pmatrix} 1 & 1 \end{pmatrix} x(t) \end{cases} \tag{49}$$

In this case, an example of a continuous linear system is being treated. A delay is only in the state which was chosen as a sampling period $T = 0.2\ s$ and applying the state-transition method, the following discrete time-delay system is obtained:

$$x(k+1) = \begin{pmatrix} 0.9680 & 0.1493 \\ -0.2987 & 0.5200 \end{pmatrix} x(k) + \begin{pmatrix} 1 & 0 \\ -0.2453 & 1.4907 \end{pmatrix} x(k-q) + \begin{pmatrix} 0.0160 \\ 0.3947 \end{pmatrix} u(\text{k}) \tag{50}$$

The simulation result in the step responses of the continuous system and approximate discrete system is depicted in Figure 2.

The system response does not swing; that allows us to conclude that the system is highly damped. Characteristic desired system performance is specified by a delay time, a rise time and a settling time. Thus, very often the performance characteristics of the controlled system are specified in terms of transient response to a step input, easy to generate. The delay in digital control systems is always a challenge to take into consideration. It introduces a phase shift of the signal as it can induce instability. The evolution of the response in the field below, presents a delay that does not degrade the performance of the system.

Applying the method based on trapezoidal rule of integration to the system considered in the state-space Equation (49), and keeping the same sampling period $T = 0.2\ s$, the following discrete time-delay system is obtained:

$$x(\mathrm{k}+1) = \begin{pmatrix} 0.9697 & 0.1515 \\ -0.3030 & 0.5152 \end{pmatrix} x(k) + \begin{pmatrix} -0.0758 & 0.1515 \\ -0.7576 & 1.5152 \end{pmatrix} \left[x(k+1-q) + x(k-q) \right] + \begin{pmatrix} 0.0152 \\ 0.1515 \end{pmatrix} u(k) \quad (51)$$

The simulation result in the step responses of the continuous system and approximate discrete system is depicted in Figure 3.

The representation of discrete time system, in Figure 1 and Figure 2; shows the continuous time evolution of the system start after a while which is the value of the existing delay in the state equation that defines the system, keeping almost same stability range in both cases.

In this case, the model is insufficient for an accurate representation of the closed-loop response and a more complex approximate discrete system should be considered in order to improve the result. Precision requirements will determine the number of terms to be considered in the exponential series, depending on the application. Anyway, simple discrete systems can be a useful tool, combined with robust control techniques.

The next example shows that one external and one internal delay are considered by state-space equation:

$$\begin{cases} \dot{\mathrm{x}}(t) = \begin{pmatrix} -1 & -1.5 \\ -1 & -2.5 \end{pmatrix} x(t) + \begin{pmatrix} 0.25 & -0.5 \\ 0.5 & 0.5 \end{pmatrix} x(t-0.3) + \begin{pmatrix} 0 \\ 1 \end{pmatrix} u(t) + \begin{pmatrix} 0.5 \\ 0.5 \end{pmatrix} u(t-0.5) \\ y(t) = \begin{pmatrix} 1 & 1 \end{pmatrix} x(t) \end{cases} \quad (52)$$

In this case, the system has a different internal and external delay and, then, combinations of those delays appear in the input of the discrete approximation. Considering $T = 0.1s$ and applying the state-transition method, the following discrete time-delay system is obtained:

Figure 2. Step response of the continuous and discrete time delay system

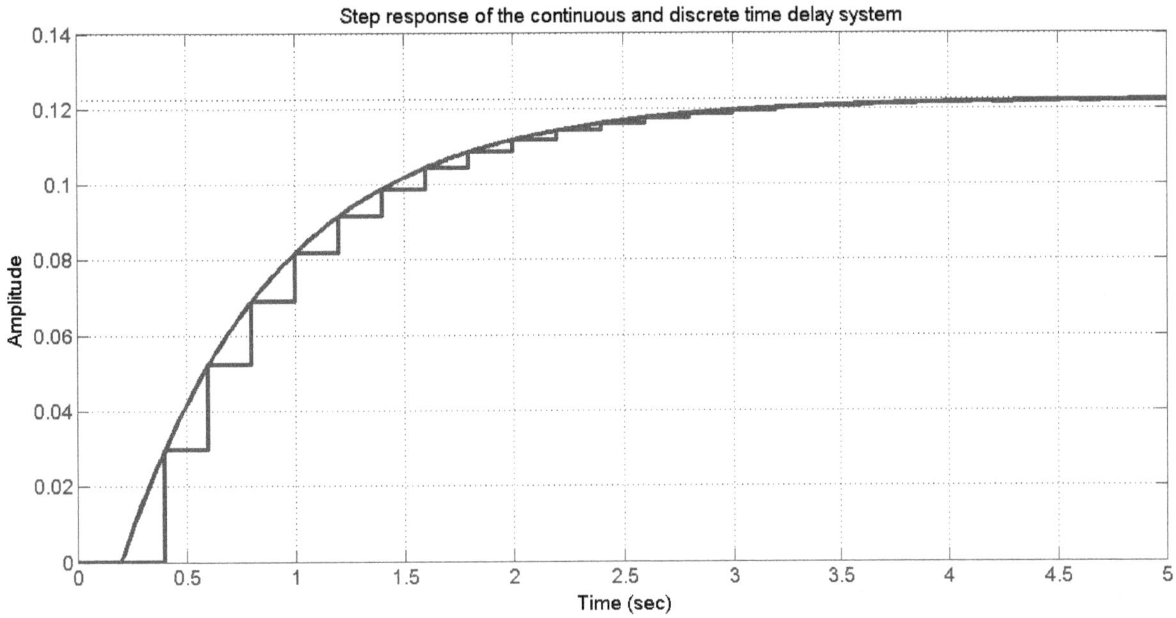

Figure 3. Step response of the continuous and discrete time delay system

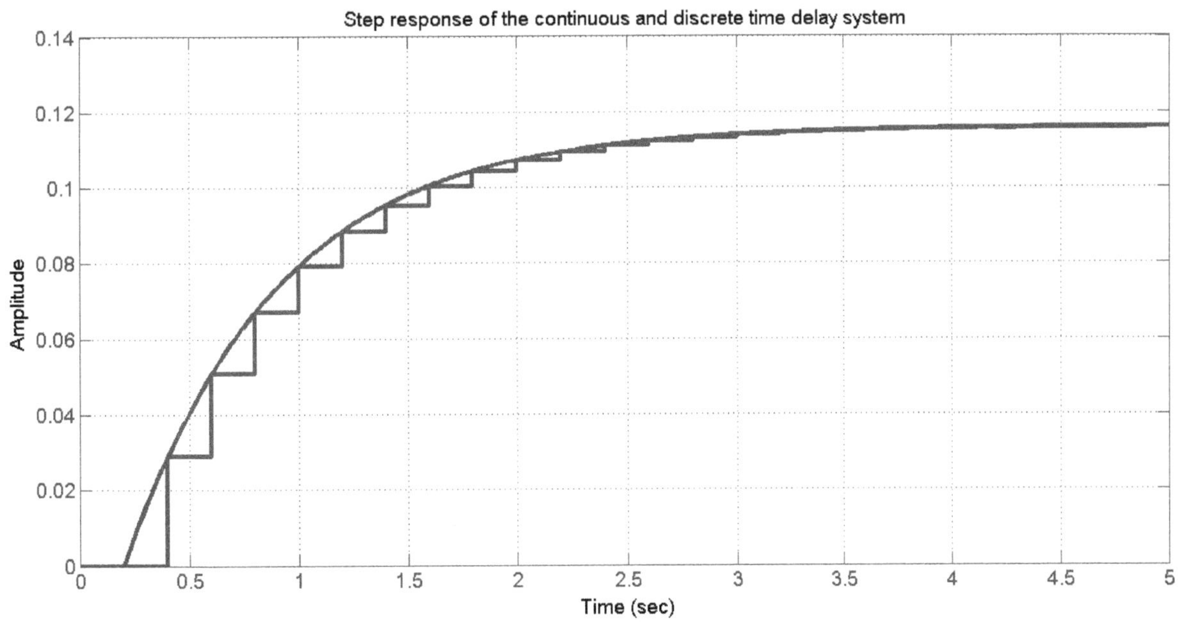

$$x(k+1) = \begin{pmatrix} 0.9112 & -0.1266 \\ -0.0844 & 0.7846 \end{pmatrix} x(k) \tag{53}$$
$$+ \begin{pmatrix} 1.0240 & -0.0519 \\ 0.0519 & -1.0500 \end{pmatrix} x(k-q_0) + \begin{pmatrix} -0.0092 \\ 0.1913 \end{pmatrix} u(k) + \begin{pmatrix} 0.0937 \\ 0.0947 \end{pmatrix} u(k-q_1)$$

The simulation result in the step responses of the continuous system and approximate discrete system is depicted in Figure 4.

In practice, the sampling period Te depends on the type of method (chemical, thermal, mechanical, etc.) and should be chosen small relative to the time of the closed loop system response. Sampling period is too low (sampling) results:

- Closer to the continuous time, but the calculation is much more demanding.
- Difficulties inherent in the importance of the sampling period can be clearly seen.

In the present case the delay is a multiple of the sampling period or of its value which is imposed by the delay. It can see very clearly the importance of the sampling period. A lower sampling period provides monitoring of the response of faithful continuous system with a greater period. The consequence of this is that the system with the lowest sampling period is the one which achieves stability quickly. So, to have the most accurate possible relative to the continuous time result, it will be advantageous not to take a too large sampling period.

Figure 4. Step response of the continuous and discrete time delay system

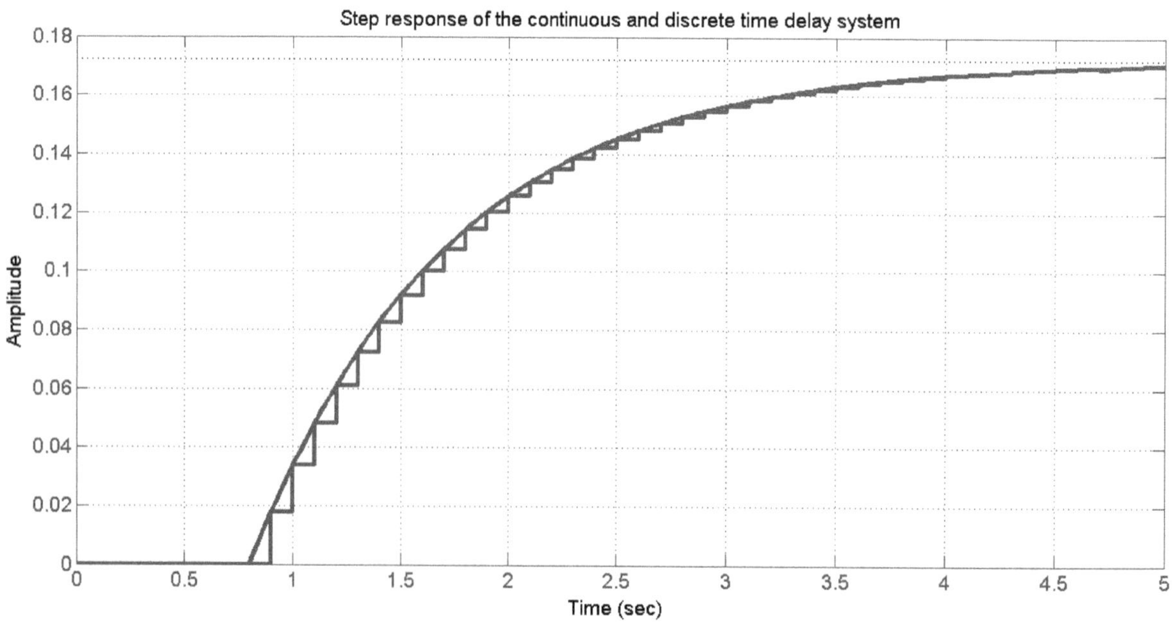

Applying the method based on trapezoidal rule of integration to the system considered in the state-space Equation (52), and keeping the same sampling period *T=0.1s*,the following discrete time-delay system is obtained:

$$\begin{aligned} x(k+1) = &\begin{pmatrix} 0.9108 & -0.1274 \\ -0.0849 & 0.7834 \end{pmatrix} x(k) \\ &+ \begin{pmatrix} 0.2070 & -0.5096 \\ 0.4352 & -0.4671 \end{pmatrix} \left[x(k+1-q_0) + x(k-q_0) \right] \\ &+ \begin{pmatrix} -0.0064 \\ 0.0892 \end{pmatrix} u(k) + \begin{pmatrix} 0.0446 \\ 0.0425 \end{pmatrix} u(k-q_1) \end{aligned} \tag{54}$$

The simulation result in the step responses of the continuous system and approximate discrete system is depicted in Figure 5.

In Figure 4 and Figure 5, the representation of discrete time system follows the continuous time evolution of the system that starts after a certain time which corresponds to the value that is the result of the existing delays in both the state and the control in the equation defined in the system keeping almost the same area of stability, in both cases. But it could not present the same response time and the appearance of a second term; describing the past state of the system during a second sampling instant by the discretization homographic transformation method that requires computation of additive time. The effect of delay on the dynamics of a system depends not only on its value but also on the characteristics of the system. Indeed, the presence of delay can affect the stability of the system, since its presence can

Figure 5. Step response of the continuous and discrete time delay system

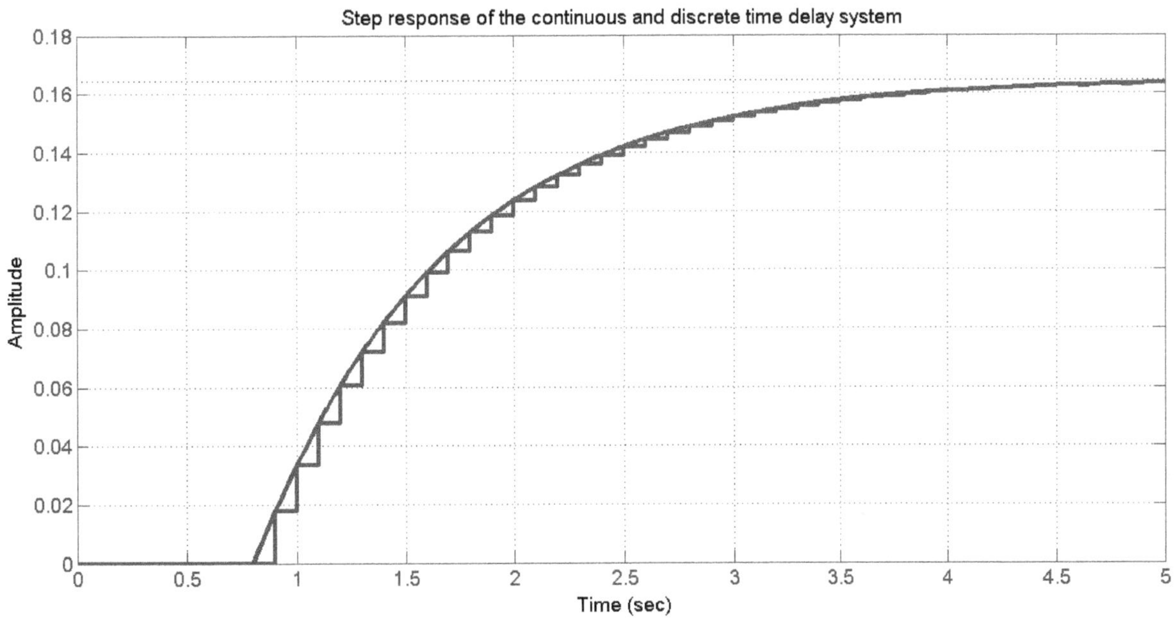

cause complex behaviors namely oscillations, instability and degradation in performance. But sometimes the effect of the delay in some cases can stabilize an initially unstable system. In this case, the delay has a stabilizing effect.

Comparison between Two Methods

Two different methods for obtaining discrete time approximations for continuous-time delay systems have been discussed .The state-transition method has much more complications for calculating than the homographic transformation method. Suggesting that the sampling period should be chosen over the dynamic system, depending on the delay term. The sampling period is chosen as an integer multiple of the delay in addition to the forced choice for dynamic system without delay. It appears that the model using the homographic transformation is best suited for numerical simulation in the case of the continuous system which is defined by a state model and that must be selected in the case of a suitable choice of the sampling period.

It is clear that the value of the delay (*0.2s*) is shown by the simulation by both discretization methods knowing that it did not affect the evolution of the system. Similarly in the second example delays are present in the state (*0.3s*) and in the control (*0.5s*) simulation shows that the amount of delay in the system is the result of two delays (*0.3s +0.5 s*) with a normal evolution of the dynamic system in time without losing performance.

Depending on the value of the delay h, the election of this value as sampling period can be inadequate in the sense of the Shannon (Discretization) Theorem. Then, the sampling period can be chosen $\mathrm{T} = h/j$ for some integer j, where T is a valid value. In general, considering systems given by Equations (49) or (52) and $\mathrm{T} = h/j$, the approximate discrete system is given by the two methods.

5. CONCLUSION AND FUTURE WORK

Process control by computer is spreading through the evolution of microprocessors. The computer is used in a feedback loop regulator to replace the analog. The computer is more flexible employment; setting parameters and structure of the control law is modified by programming rather than by modifying the wiring with an analog controller. But it is mostly possible to develop laws of nonlinear optimal control for better use of technological possibilities real slave process. In the simplest configuration of a single variable servo, the computer provides the functions of the comparator and controller. But the computer is even more efficient in the case where the process of complex knowledge of several representative quantities is necessary. In case of problems on driving, process control can be introduced in the program criteria performance to allow, for example:

- A minimum response time;
- A response minimizing the integral of the squared error.
- Operation minimizing energy consumption;
- Bonding together with a supervisor;
- Fault monitoring...

In this chapter two methods for the discrete time approximation of multi variable continuous-time delay systems have been proposed. If the model is given in the form of state equations, the trapezoidal rule requires only one matrix inversion, and is perhaps the most convenient for digital simulation. It is of interest to compare them for accuracy as well as convenience in computation. It is evident that the accuracy of the approximation will depend, to a large extent, on the choice of the nature of the input applied to the delay continuous linear system.

For example, if the input is a constant or a piece-wise-constant function of time, the state-transition method will give an accurate discrete time model. Therefore the discretization of the continuous delay either by the method of transition state or by the method of homographic transformation matrix linear system shows that the two examples follow the curve form almost continuously, except that the first method has many more complications than the second calculation, suggesting that the sample period should be chosen in addition to the dynamics of the system, depending on the delay term. Usually choosing the sampling period an integer sub multiple of delay in addition to its forced choice for dynamic systems without delay.

Despite the variation of the period allowed in the simulation curves obtained by the two methods are similar domain discretization and the method using homographic transformation is easier to program.

Finally it appears that the model using the homographic transformation is best suited for numerical simulation in the case of the continuous system which is defined by a state model and in the case of a suitable choice of the sampling period.

The research work presented in this chapter is intended for modeling delay systems can be discrete or discredited by different approaches. The research is to be continued and would analyzed the stability and stabilization of these types of systems and study the effect of the delay, the method of discretization and sampling period, and try to do the same work with a variable delay for linear continuous time delay systems and non-linear continuous time delay systems. The method based on Lyapunov - Krasovskii functional technical are to be used. The conditions and solutions to problems related to stability and stabilization are formulated in matrix forms of inequality (LMIs or BMIs).

REFERENCES

Chen, J., & Latchman, H. A. (1994). Asymptotic stability independent of delays: simple necessary and sufficient conditions. In *Proceedings of American Control Conference*, (vol. 12, pp. 1027-1031). Baltimore, MD: IEEE. doi:10.1109/ACC.1994.751903

Dorf, R. C., & Bishop, R. H. (2008). *Modern Control Systems*. Pearson Education.

Fang, M. (2010). Delay-dependent stability analysis for discrete singular systems with time-varying delays. *Acta Automatica Sinica*, *36*, 751–755.

Franklin, G. F., Powell, J. D., & Workman, M. L. (1997). *Digital Control of Dynamic Systems*. Boston: Addison-Wesley.

Fridman, E., & Shaked, U. (2005). Stability and guaranteed cost control of uncertain discrete delay systems. *International Journal of Control*, *78*(4), 235–246. doi:10.1080/00207170500041472

Harmuth, H. F. (1969). Applications of Walsh functions in communications. *IEEE Spectrum*, *6*(11), 82–91. doi:10.1109/MSPEC.1969.5214175

Haykin, S. S. (1972). A unified treatment of recursive digital filtering. *IEEE Transactions on Pattern Analysis and Machine Intelligence*, *3*, 113–116.

Hung, J. C., Liu, C. C., & Chan, P. Y. (1980). An algebraic method for system parameter identification. In *Proceedings of the 14th Asilomar Conference on Circuits, Systems and Computers*, (pp. 277-279). Pacific Grove, CA: IEEE.

Kailath, T. (1980). *Linear systems*. Prentice Hall.

Liu, B., Lu, W., & Chen, T. (2014). New criterion of asymptotic stability for delay systems with time-varying structures and delays. *Neural Networks*, *54*, 103–111. doi:10.1016/j.neunet.2014.03.003 PMID:24686033

Mahmoud, M. S. (2000). *Robust Filtering for Time-Delay Systems*. New York: Marcel Dekker.

Martens, H. R. (1969). A comparative study of digital integration methods. *Simulation*, *12*(2), 87–94. doi:10.1177/003754976901200207

Michiels, W., & Niculescu, S. I. (2007). *Stability and Stabilization of Time-Delay Systems*. Philadelphia, PA: SIAM Press. doi:10.1137/1.9780898718645

Niu, Y., Wang, X., Lam, J., & Ho, D. W. C. (2003). Sliding-mode control for nonlinear state-delayed systems using neural-network approximation. *Proceedings of the IEE Control Theory and Applications*, *150*(3), 233–239. doi:10.1049/ip-cta:20030321

Normey-Rico, J. E. (2007). *Control of Dead-Time Processes*. New York: Springer.

Palanisamy, K. R., & Bhattacharya, D. K. (1981). System identification via block-pulse functions. *International Journal of Systems Science*, *12*(5), 643–647. doi:10.1080/00207728108963772

Rao, G. P., & Srinivasan, T. (1978). Remarks on author's reply to comments on design of piecewise constant gains for optimal control via Walsh functions. *IEEE Transactions on Automatic Control*, *23*(4), 762–763. doi:10.1109/TAC.1978.1101782

Ratchagit, K., & Phat, V. N. (2010). Stability criterion for discrete-time systems. *Journal of Inequalities and Applications*, *10*(6), 444–459.

Sannuti, P. (1977). Analysis and synthesis of dynamic systems via block-pulse functions. *Proceedings of the IEE*, *124*(6), 569–571.

Schultz, D. G., & Melsa, J. L. (1967). *State Functions and Linear Control Systems*. New York: McGraw-Hill.

Shieh, L. S., Yeung, C. K., & Mclnnis, B. C. (1978). Solutions of state-space equations via block-pulse functions. *International Journal of Control*, *28*(3), 383–392. doi:10.1080/00207177808922465

Shinskey, F. G. (1967). *Process Control Systems*. New York: McGraw-Hill.

Sinha, N. K. (1972). Estimation of transfer functions of continuous-time system from sampled data. *Proceedings of the IEE*, *119*(5), 612–614.

Sinha, N. K., & Lastman, G. J. (1981). Transformation algorithm for identification of continuous-time multivariable systems from discrete data. *Electronics Letters*, *17*(5), 779–780. doi:10.1049/el:19810546

Sinha, N. K., & Rozsa, P. (1976). Some canonical forms for linear multivariable systems. *International Journal of Control*, *23*(6), 865–883. doi:10.1080/00207177608922207

Sinha, N. K., & Zhou Qi-Jie, . (1983). Discrete time approximation of multivariable continuous-time systems. *IEE Proceedigs D: Control Theory and Applications*, *130*(3), 103–110. doi:10.1049/ip-d.1983.0019

Stojanović, B. S., Debeljović, D. L., & Mladenović, I. (2007). A Lyapunov-Krasovskii methodology for asymptotic stability of discrete time delay systems. *Serbian Journal of Electrical Engineering*, *4*(2), 109–117. doi:10.2298/SJEE0702109S

Suh, Y. S., & Lee, M. H. (1999).Stability of state delay systems based on a Lyapunov functional. In *Proceedings of the IEEE International Symposium*, (pp. 1093-1098). Bled, Slovenia: IEEE.

Zhong, Q. C. (2008). *Robust Control of Time-Delay Systems*. New York: Springer.

KEY TERMS AND DEFINITIONS

Asymptotic Stability: A time-invariant system is asymptotically stable if all the eigenvalue of the system matrix A have negative real parts. If a system is asymptotically stable, it is also BIBO stable. However the inverse is not true: A system that is BIBO stable might not be asymptotically stable.

Discretization: Discretization is also related to discrete mathematics, and is an important component of granular computing. In this context, discretization may also refer to modification of variable of category granularity, as when multiple discrete variables are aggregated or multiple discrete categories fused.

Exponential Stability: In control theory, a continuous linear time-invariant system is exponentially stable if and only if the system has eigenvalue with strictly negative real parts. A discrete-time input-to-output LTI system is exponentially stable if and only if the poles of its transfer function lie strictly within the unit circle centered on the origin of the complex plane. Exponential stability is a form of asymptotic stability. Systems that are not LTI are exponentially stable if their convergence is bounded by exponential decay.

Linear Matrix Inequalities: There are efficient numerical methods to determine whether an LMI is feasible (e.g., whether there exists a vector y such that $LMI(y) \geq 0$), or to solve a convex optimization problem with LMI constraints. Many optimization problems in control theory, system identification and signal processing can be formulated using LMIs.

Linear Systems: A linear system is a mathematical model of a system based on the use of a linear operator. Linear systems typically exhibit features and properties that are much simpler than the general, nonlinear case. As a mathematical abstraction or idealization, linear systems find important applications in automatic control theory, signal processing, and telecommunications.

Multivariable Control Systems: The control of systems characterized by multiple inputs, which are usually referred to as the controls; or by multiple outputs, which are often the measured variables to be controlled; or by both multiple inputs and outputs. Automobiles, chemical processing plants, aerospace vehicles, biological systems, and the national economy are all examples of multivariable systems which require and receive some form of regulation or control, be it mathematically contrived or not.

Time-Delay Systems: A system can be built with an inherent delay. Delays are units that cause a time-shift in the input signal, but that don't affect the signal characteristics. An ideal delay is a delay system that doesn't affect the signal characteristics at all, and that delays the signal for an exact amount of time. Some delays, like processing delays or transmission delays, are unintentional. Other delays however, such as synchronization delays, are an integral part of a system.

Chapter 20
Robust Iterative Learning Control for Linear Discrete-Time Switched Systems

Ouerfelli Houssem Eddine
University ElManar, Tunisia

Dridi Jamel
University ElManar, Tunisia

Ben Attia Selma
University ElManar, Tunisia

Salhi Salah
University ElManar, Tunisia

ABSTRACT

This chapter aims to study the problem of stability analysis, and robust exponential stabilization for a class of switched linear systems with polytopic uncertainties is reviewed. A sufficient condition based on the average dwell time that guarantees the exponential stability of uncertain switched linear systems is given. First, the iterative learning control is presented to build a formulation ensuring the exponential stability of the given system. The integrated design of this ILC scheme is transformed into a robust control problem of an uncertain 2D Roesser system. The results are obtained through original connection with the notion of stability along the pass for 2D repetitive systems. An overview of the stabilization methods of switched discrete systems found in the literature is outlined. All the given formulations are presented in terms of LMIs. A numerical simulation example is established, showing the effectiveness of the proposed method.

DOI: 10.4018/978-1-4666-7248-2.ch020

1. INTRODUCTION

The stability analysis of switched linear systems is one of the active over the past research due to their wide applications in many areas such (the automotive industry, aircraft and air-traffic control, switched power converters) (Branicky., 1998; Liu et al., 2010). A switched system is a hybrid dynamical system, which consists of a family of continuous-time or discrete-time subsystems and a rule that orchestrates the switching between them (Koutsoukos & Antsklis., 2002; Attia et al., 2012; Attia et al., 2010). For the stability analysis of switched systems there are two issues to be addressed: stability under arbitrary switching (Attia et al., 2009a; Attia et al., 2009b; Attia & Bernussou., 2009c; Xiang & Huang., 2013) and stability under constraint switching, such as dwell time, average dwell time constraints (i.e. minimum time between switching) (Du et al., 2006; Elci et al., 2002; Mitra et al., 2008; Lian et al., 2011 ; Zhang et al., 2009). Although the dwell-time and average dwell time mainly characterize the time-controlled switching signals, the slow switching idea can be generalized to hybrid systems or state-controlled switching signals. In (Mitra et al., 2008) the authors studied the stability analysis problem for a given hybrid automaton (called structured hybrid automaton) via abstracting it into a 'similar' switched system. The similarity is in the sense of preserving the average dwell time property. Actually, the authors developed abstraction schemes to guarantee that the derived switched system has no greater average dwell time than the original hybrid automaton. Under the assumption that all the subsystems are stable, the stability of the abstracted switched systems then implies the original hybrid automaton's stability. However, the stabilization problem of discrete switched systems with classic control law has been studied in (Attia & J. Bernussou., 2009c; Xiang & Huang ., 2013; Shi et al., 2005; Liu et al., 2010; Attia et al., 2012; Fang et al., 2003), by using the common Lyapunov function approach, multiple Lyapunov function approach, piecewise Lyapunov function and switched Lyapunov function (Mehrzi et al., 2007; Attia et al., 2009b; Du et al., 2006). Many results on the issues of stability and control synthesis for the 2D switched systems have been obtained in (Benzaouia et al., 2009; Attia et al., 2012), most of them are based on the dwell time approach. Specifying the minimum dwell time that ensures the exponential stability of the switched systems is an especially attracting problem in this paper. We first introduce necessary notation and review several existing results using the dwell time approach. Given a positive constant τ_a, let $S[\tau_a]$ denote the set of all switching signals with time interval between consecutive switching's being no smaller than τ_a. The constant τ_a is called the "dwell time" (Morse et al., 1996) .When all subsystem matrices A_i are Hurwitz stable, it has been shown in (Morse et al., 1996) that we can choose τ_a sufficiently large so that the switched system is exponentially stable for any switched sequence $\sigma(t) \in S[\tau_a]$. In (Zhat et al., 2001) a dwell time scheme is analyzed for local asymptotic stability of nonlinear switched systems with the activation time being used as a dwell time. The stability issues of such switched systems include several interesting phenomena. For example, even when all the subsystems are exponentially stable, the switched systems may have divergent trajectories for certain switching signals (Decarlo et al., 2000; Liberzon et al., 1999). Another remarkable fact is that one may carefully switch between unstable subsystems to make the switched system exponentially stable.As these examples suggest, the stability of switched systems depends not only on the dynamics of each subsystem but also on the properties of switching signals. On the other hand, from the practical viewpoint, it is important to investigate switched systems which contain uncertain parameters, there are few results concerning the stability of switched linear systems with uncertainties, recently (Attia et al., 2009b; Shi et al., 2005; Du et al., 2006) investigated the exponential stability and stabilizability problems for discrete-time uncertain switched linear systems. Motivated by the results in (Mingxuan et al., 2011; Zhai., 2001), we consider in this paper quadratic stabilizability of switched linear systems that are composed of several uncertain subsystems of polytopic type. It is well known that polytopic uncertainties exist in many real systems, and most uncertain control systems can be approximated by systems with polytopic uncertainties. Although there have been many existing results on switched systems and quadratic stability/ stabilization, to the best of our knowledge, there is no existing result concerning quadratic stabilizability of switched systems with polytopic uncertainties. Firstly, the exponential stability result of general discrete-time switched systems using a Lyapunov function approach is revisited. the most pessimistic one assuming the existence of a common lyaponuv function, even if these condition are easily tractable, they can be used in a very few applications. Based on these, less conservative conditions using multiple lyaponuv function have been proposed a sufficient condition for system is provided using the polyquadratic approach proposed by (Zhang et al., 2010; Mehrzi et al., 2007) for stability analysis and stabilization control. The existence of a common quadratic Lyapunov function (PQLF) for all its subsystems assures the quadratic stability of the switched system. Quadratic stability is a special class of exponential stability, which implies asymptotic stability, and has attracted a lot of research efforts due to its importance in practice. It is known that the conditions for the existence of a PQLF can be expressed as linear matrix inequalities (LMIs) (Gałkowski et al., 2002).The exponential stability criteria for discrete-time polytopic uncertain switched linear systems with average dwell time switching are derived and formulated in terms of a set of linear matrix inequalities, the switching signal design method is proposed and the minimum dwell time stabilization conditions are provided, which guarantee that the uncertain switched linear systems with average dwell time and control are exponentially stablizable. Compared with the different results for synthesis issue by applying the classic control laws, relatively few efforts are made for designing a controller to achieve the exponential stability for uncertain switched repetitive systems. This paper is addressed for the design feedback controllers (ILC) such that the system is stable (Jianming et al., 2008; Dridi et al., 2012; Dridi et al., 2013; Du et al., 2006). Motivated by human learning, the basic idea of (ILC) is to use information from previous executions of the task in order to improve performance from pass-to-pass in the sense that the tracking error is sequentially reduced. Iterative learning control (ILC) is an attractive technique when it comes to dealing with systems that execute the same task repeatedly over a finite time interval. This method of control uses information from previous executions of the task in an attempt to improve performance from repetition to repetition in the sense that the system is robust exponentially stable. The objective of ILC schemes is to use their repetitive process structure (i.e. information propagation from trial-to-trial (or pass-to-pass) and along a trial/pass) to progressively improve the accuracy with which the core operation under consideration is performed, by updating the control input progressively from trial-to-trial. In the literature, there are many iterative learning control laws proposed for different classes of systems: linear or nonlinear systems, time delay systems, cascade systems, stochastic system, etc. These approaches include: P-type learning, D-type learning, PID-type learning, anticipatory learning, robust learning and optimal learning, etc. However, to the best of our knowledge, no one has been studied the iterative learning control for switched systems. This observation motivates the present study. However, to the best of our knowledge, the problems of stability and stabilization for uncertain switched systems via the dwell time approach shows good results in practice, especially for the robust exponential stability problem with (ILC) control for a class of polytopic uncertain linear switched systems via the dwell time approach, which motivates our present study. In this paper, the problem of iterative learning control for a class of linear discrete- time uncertain switched system is studied. In ILC, the 2-D analysis also plays a fundamental role since each ILC creates a 2-D system, with time step and iteration number as independent variables (Elci et al., 2002; Lien et al., 2011; Owens et al ., 2000; Li et al., 2012). This motivates the use of the 2-D system theory, which provides not only mathematical models to describe the entire dynamics of ILC but also useful ways to show the convergence of ILC (Amann et al., 1998; Dridi et al., 2012; Kurek et al., 1993; Lu et al., 1985). Furthermore, due to the finiteness of the time axis, the convergence condition obtained is usually a necessary and sufficient one. In (Li et al., 2005), it has been shown that the 2-D analysis approach to ILC can be extended to general 2D Roesser switched system. The obtained formulation by applying ILC control is transformed into a synthesis problem of a special 2D Roesser switched system (Paszke et al., 2004; Lam et al., 2004; Fang et al., 2003; Attia et al., 2012; Shi et al., 1985; Moore et al., 2008; Moore et al., 2000). In order, to describe and to analyze the performance of the system along the time and along the cycle, new concepts and convergence indices are introduced to describe the dynamical behaviors of the corresponding discrete uncertain switched system in horizontal and vertical propagation directions. Sufficient conditions for stability of the 2D Roesser system are derived (Xiang & Huang., 2013). One is faced with determining the minimum dwell time such that exponential switching is maintained (Zhat et al., 2001). It is well known that the computation of the exact minimum dwell time is a problem which must be overcome. This paper is organized as follows: In section 2, based on the stability and stabilization of 2D Roesser discrete switched systems mathematical settings have been addressed. In section 3, the control law design (ILC) by constructing a sequence of control inputs to a discrete switched system produces 2D repetitive system, and sufficient conditions for the existence of a stabilizing controller (ILC) are derived in terms of a set of matrix inequalities. In section 4, we briefly review some conceptions especially the robust stability and stabilization controller (ILC) for discrete-time switched systems with polytopic uncertainties. In section 5, Numerical examples are presented to illustrate the feasibility and the effectiveness of the proposed design algorithms in this paper. Finally, the conclusion of this paper is given in Section 6.

Notations

Throughout this paper, the superscript "T" denotes the transpose, the notation $\left(X \geq 0, X > 0\right)$ is positive semi-definite (positive definite, respectively). Matrix $\lambda_{\max}\left(P\right)$ and $\lambda_{\min}\left(P\right)$ denote the maximum and minimum eigenvalues of matrix P, respectively. I represents identity matrix with appropriate dimension. The asterisk $*$ in a matrix is used to denote a term induced by symmetry. The set of all nonnegative integers is represented by Z_{+}.

2. EXPONENTIAL STABILITY OF 2D REPETITIVE SWITCHED SYSTEM

Consider the following 2D-discrete linear switched system described by the Roesser model:

$$\begin{bmatrix} x^h\left(i+1,j\right) \\ x^v\left(i,j+1\right) \end{bmatrix} = A_{\sigma(i,j)} \begin{bmatrix} x^h\left(i,j\right) \\ x^v\left(i,j\right) \end{bmatrix} + B_{\sigma(i,j)} u\left(i,j\right) \tag{1}$$

$$x^h(i,j) = \begin{cases} f(j), & 0 < j < r_1 \\ 0, & j > r_1 \end{cases}$$
$$x^h(i,j) = \begin{cases} g(i), & 0 < i < r_2 \\ 0, & i > r_2 \end{cases} \tag{2}$$

where $\{i,j\} \in k := \{0,1...,n\}$, $x^h(i,j) \in IR^{n_1}$ is the horizontal state in IR^{n_1}, $x^v(i,j) \in IR^{n_2}$ is the vertical state in IR^{n_2}, $x(i,j) \in IR^n$ with $n = n_1 + n_2$ is the whole state in IR^n, $u(i,j) \in IR^m$ is the control vector in IR^m, $\sigma(i,j)$ is a switching signal which takes its values in the finite set $N = \{0,..,N\}, i,j \in Z_+$, N is the number of subsystems.

$$A_{\sigma(i,j)} = \begin{bmatrix} A_{11}^{\sigma(i,j)} & A_{12}^{\sigma(i,j)} \\ A_{21}^{\sigma(i,j)} & A_{22}^{\sigma(i,j)} \end{bmatrix}$$

$$A_{11}^{\sigma(i,j)} \in IR^{n_1 \times n_1},\ A_{12}^{\sigma(i,j)} \in IR^{n_1 \times n_2},\ A_{21}^{\sigma(i,j)} \in IR^{n_2 \times n_1},\ A_{22}^{\sigma(i,j)} \in IR^{n_2 \times n_2}$$

$$B = \begin{bmatrix} B_1 \\ B_2 \end{bmatrix},\ B_1 \in IR^{n_1 \times m},\ B_2 \in IR^{n_2 \times m},\ k \in N$$

are known constant matrices with appropriate dimensions.

In the sequel, sufficient condition of exponential stability of 2D discrete switched system described by Roesser structure (1) and (2) is given.

Definition 1 (Xiang & Huang., 2013): System (1) is said to be exponentially stable under $\sigma(i,j)$ if there exists a positive constant c for a given $r \geq 0$, such that:

$$\sum_{i+j=D} \left\| x(i,j) \right\|^2 < \xi e^{-c(D-r)} \sum_{i+j=r} \left\| x(i,j) \right\|^2, D \geq r, \xi \geq 0 \tag{3}$$

holds for all $D \geq r$ and a positive constant ξ.

Remark 1: From the definition 1, it is easy to see that when r is given, $\sum_{i+j=r} \left\| x(i+j) \right\|^2$ will be bounded, and $\sum_{i+j=D} \left\| x(i,j) \right\|^2$ will tend to zero exponentially as D goes to infinity, it means that $\left\| x(i,j) \right\|^2$ will tend to zero.

Definition 2 (Zhai et al., 2000): For any $i + 1 = D \geq r = i_r + j_r$, let $N_\sigma(r,D)$ denote the switching number of $\sigma(i,j)$ on an interval (r,D),if:

$$N_{\sigma}\left(r,D\right)\leq N_{0}+\frac{D-r}{\tau_{a}} \tag{4}$$

with $N_0 \geq 0, \tau_a \geq 0$, τ_a is called the average dwell time, N_0 is the chatter bound.

Now, we present the problem of exponential stability analysis for the 2D discrete linear switched systems described by the Roesser model.

Theorem 1 (Zhai et al., 2000): Consider system (1) and (2) with $u(i,j)=0$,for a given positive constant $\alpha < 1$,if there exist positive definite symmetric matrices X_1^k, X_2^k with appropriate dimensions, $k \in N$ such that.

$$\begin{bmatrix} -\alpha X_1^k & 0 & \left(A_{11}^k X_1^k\right)^T & \left(A_{21}^k X_1^k\right)^T \\ * & -\alpha X_2^k & \left(A_{12}^k X_2^k\right)^T & \left(A_{22}^k X_2^k\right)^T \\ * & * & -\left(X_1^k\right)^T & 0 \\ * & * & * & -\left(X_2^k\right)^T \end{bmatrix} < 0 \tag{5}$$

Then, under the following average dwell time scheme:

$$\tau_a \geq \tau_a^{\ *} = \frac{\ln X}{-\ln \alpha} \tag{6}$$

$$X = \max_{\substack{1\leq k,l\leq N \\ k\neq l}} \frac{\lambda_{\max}\left(X^l\right)}{\lambda_{\min}\left(X^k\right)},\ X^k = diag\left(X_1^k, X_2^k\right) \tag{7}$$

Remark 2: Similar to the 1D switched system case, Definition 2 means that if there exists a positive number τ_a such that the switching signal $\sigma\left(i,j\right)$ has the average dwell time property, the average time interval between consecutive switching is at least τ_a. The average dwell time method is used to restrict the switching number of the switching signal during a time interval such that the stability or other performances of the system can be guaranteed.

In the sequel, we will provide some necessary and sufficient conditions for the exponential stability of switched linear systems under dwell time approach using the iterative learning control. This is a relatively new result, which provides a solution for this long standing problem. It shows that the exponential stability problem for switched linear systems with dwell time approach is equivalent to the robust exponential stability problem for polytopic uncertain linear discrete time switched systems, for which several strong stability conditions exist.

3. EXPONENTIAL STABILIZATION OF LINEAR SWITCHED SYSTEM USING ILC DESIGN

In this section, we propose a discrete iterative learning control algorithm (ILC) ensuring the exponential stability of discrete switched systems described by 2D Roesser models.

Consider the discrete linear switched system described by the following form:

$$x\left(t+1\right)=A_{\sigma(t)}x\left(t\right)+Bu\left(t\right),\ x\left(0\right)=0,\ t\geq 0 \tag{8}$$

where $x\left(t\right)\in IR^{n}$ is the state in IR^{n}, $u\left(t\right)\in IR^{m}$ is the control vector in IR^{m}, $\sigma\left(t\right)$ is a switching signal which takes its values in the finite set $\mathbf{N}=\left\{0,..,N\right\}$, N is the number of subsystems. With $A_{\sigma(t)}$ is constrained to jump among the N vertices of the matrix polytope $\left\{A_1.......A_N\right\}$, B is matrix of appropriate dimension.

In the sequel, if k is defined as the switching number, Thus (8) can be presented in the following 2-D switched form:

$$x_k\left(t+1\right)=A_{\sigma(t)}x_k\left(t\right)+Bu_k\left(t\right),\ x\left(0\right)=0,\ t\geq 0 \tag{9}$$

where, $x_k\left(t\right)$ and $u_k\left(t\right)$ are respectively the state and the control vector of system in switching k. $t=0,....,T$, $k\in IN$.

The stabilizability of the switched discrete time system (8) is investigated by the (ILC) approach, the latter is based on the computing of the difference between control law u at the respectively iteration k and $k+1$ .The system is required to execute the repetitive task over the finite time interval $0<t<T$. In the kth operation/iteration, the control input, $u_k(t)$ may be updated iteratively in a certain way by using the state measurements in the previous operation. The switched system (1) admits a family of solutions that is parameterized both by the initial condition $x_k(0)$ and the switching signal. The proposed ILC algorithm can be viewed as an extension form developed for 2-D switched systems, by considering the average dwell time between switching instants repeated at each iteration. The major contribution of this paper that an iterative learning controller for discrete-time switched systems is investigated when the initial conditions are variable. It is well known that when the initial state value is fixed but not same to the desired initial state value and the ILC algorithm is applied to the system; the converged output trajectory is apart from the desired output trajectory with constant error value. It is obvious that the learning dynamics are more complex when the initial iterative state is different at each iteration (kth switching). Our question is which conditions need to be satisfied when the iterative learning controller is used for the initial conditions with random distribution in each subsystems interval?

According to the principle of ILC, there are two independent dynamic processes: system time t, and learning iteration k during learning process. Every variable can be expressed as a 2-D switched function, such as $x_k(t)$, which represents $x(t)$ in the kth learning iteration (kth switching).

Consider the general ILC law given by:

$$u_{k+1}(t) = u_k(t) + \Delta u_{k+1}(t) \tag{10}$$

where, $\Delta u_{k+1}(t)$ denotes the variation of the control input at the respectively iteration k and $k+1$.

Firstly, an updating law based on the ILC approach described by the following form is used:

$$u_{k+1}(t) - \beta u_k(t) = K_1^p x_k(t) + K_2^p \eta_{k+1}(t) \tag{11}$$

with p is defined as the number of subsystems, if k is defined as the switching number.

Where $\eta_{k+1}(t) = x_{k+1}(t) - \beta x_k(t)$ (12) K_1^p, K_2^p are appropriately dimensioned matrices gains to be designed, and β a positive scalar.

Then clearly (11) and (12) can be written as:

$$\begin{aligned}\eta_{k+1}(t+1) &= x_{k+1}(t+1) - \beta x_k(t+1)\\ &= A_{k+1}x_{k+1}(t) + Bu_{k+1}(t) - \beta\left(A_k x_k(t) + Bu_k(t)\right)\\ &= A_{k+1}\eta_{k+1}(t) + \left(\beta\left(A_{k+1} - A_k\right) + BK_1^p\right)x_k(t) + \left(BK_2^p\eta_{k+1}(t)\right)\\ &= \left(A_{k+1} + BK_2^p\right)\eta_{k+1}(t) + \left(\beta\left(A_{k+1} - A_k\right) + BK_1^p\right)x_k(t)\end{aligned} \tag{13}$$

and

$$x_{k+1}(t) = \eta_{k+1}(t) + \beta x_k(t) \tag{14}$$

The obtained system in the closed-loop is represented by the following form:

$$\begin{cases}\eta_{k+1}(t+1) = \left(A_{k+1} + BK_2^p\right)\eta_{k+1}(t) + \left(\beta\left(A_{k+1} - A_k\right) + BK_1^p\right)x_k(t)\\ x_{k+1}(t) = \eta_{k+1}(t) + \beta x_k(t)\end{cases} \tag{15}$$

Let $A_{11} = \left(A_{k+1} + BK_2^p\right), A_{12} = \left(\beta\left(A_{k+1} - A_k\right) + BK_1^p\right),\ A_{21} = I,\ A_{22} = \beta$

However, the switched system (15) obtained by applying ILC control can be rewritten into 2D discrete linear switched system form described by the Roesser model (1) with:

$$\begin{cases}x^h(i+1,j) = \eta_{k+1}(t+1)\\ x^v(i,j+1) = x_{k+1}(t)\end{cases} \tag{16}$$

Via the Lyapunov characterization of stability along the trial, LMI algorithms are available to guarantee the exponential stability of linear discrete switched systems. The ILC scheme applied to the 2D Roesser switched systems required the feasibility of the following set of LMIs.

Remark 3: The ILC problem is treated from the 2-D switched system viewpoint, which provides a mathematical model to clearly describe both the dynamics of the control system itself and the behavior of the iterative learning process as well. Moreover, benefiting from the 2-D system theory, a necessary and sufficient condition for the convergence of the ILC process is provided.

Theorem 2: Discrete linear switched system (8) under (ILC) control law (11) is exponentially stabilizable, if there exist symmetric positive matrix $X_1^p, X_2^p, R_1^p, R_2^p, p \in Z_+$ and $0 < \alpha < 1$, $0 < \beta < 1$ such that the following LMI is feasible:

$$\begin{bmatrix} -\alpha X_1^p & 0 & \left(X_1^p\right)^T \left(A_{k+1}\right)^T + \left(R_1^p\right)^T (B)^T & \left(X_1^p\right)^T \\ * & -\alpha X_2^p & \left(X_2^p\right)^T \left(\beta\left(A_{k+1}-A_k\right)\right)^T + \left(R_2^p\right)^T (B)^T & \left(\beta X_2^p\right)^T \\ * & * & -\left(X_1^p\right)^T & 0 \\ * & * & * & -\left(X_2^p\right)^T \end{bmatrix} < 0 \tag{17}$$

In this case K_1^p and K_2^p are given by $K_1^p = R_2^p \left(X_2^p\right)^{-1}, K_2^p = R_1^p \left(X_1^p\right)^{-1}$ and the minimum dwell time defined by:

$$\tau_a \geq \tau_a^* = \frac{\ln X}{-\ln \alpha}, \text{ with } X = \max_{\substack{1 \leq k,l \leq N \\ k \neq l}} \frac{\lambda_{\max}\left(X^l\right)}{\lambda_{\min}\left(X^p\right)}, \; X^p = diag\left(X_1^p, X_2^p\right)$$

Proof: Appling theorem1 with,

$$A_{11} = \left(A_{k+1} + BK_2^p\right),\; A_{12} = \left(\beta\left(A_{k+1} - A_k\right) + BK_1^p\right),\; A_{21} = I,\; A_{22} = \beta .$$

and substitute it into (5); we obtained the following LMI condition (17).

The stabilizability problem using the iterative learning control for a class of linear discreet switched system is not study until now. The repetitive processes constitute a distinct class of 2D systems, i.e., systems characterized by information propagation in two independent directions, which are interesting in both theory and applications. They cannot be controlled by a direct extension of the existing techniques from either standard (termed 1D here) or 2D systems theories. Here we give new results on the design of physically based control laws. These results are for a sub-class of discrete linear repetitive processes with switched systems in both independent directions of information propagation. The application of the iterative learning control (ILC) which is a technique for design control systems operating in a repetitive (or pass-to-pass) mode allows to ensure the exponential stability of the closed-loop switched system and the convergence along the time and the pass. Sufficient conditions of exponential stability for the considered discrete switched system are also proposed in (17). It is worth noting that the given conditions in Theorem 2 is obtained by using the average dwell time approach and the parameters α and β play a key role in obtaining the minimum average dwell time τ_a.

4. ROBUST EXPONENTIAL STABILIZATION OF LINEAR UNCERTAIN SWITCHED SYSTEM USING ILC DESIGN

Motivated by the proposed synthesis method. We investigate the exponential robust stabilizability problems of switched linear systems with polytopic uncertain via average dwell time. The considered subsystems using the composite polyquadratic functions. To achieve the control synthesis the switched parameters depend lyaponuv function is introduced .it can be used for both proving the asymptotic stability of switched uncertain system and constructing a switched state feedback when both the dynamic and the input matrix are switched and uncertain.

In this section, the case of two subsystems with polytopic uncertainties is studied the state model is defined by:

$$x_k(t+1) = \hat{A}_k x(t) + Bu_k(t), x(0) = 0, t \geq 0 \tag{18}$$

where, $x_k(t)$ and $u_k(t)$ are respectively the state and the control vector of uncertain system in switching k. $t = 0, \ldots., T, k \in IN$.

By following the preceding section, k is defined as the switching number.

With

$$\{\hat{A}_k : k \in S\}$$

are matrices with appropriate dimension,

$$\hat{A}_k(t) = \sum_{j=1}^{n_k} \lambda_{kj}(t) A_{kj}, \sum_{j=1}^{n_k} \lambda_{kj}(t) = 1, \lambda_{kj}(t) \geq 0,$$

B is matrix of appropriate dimension.

It is clear that the given subsystems are uncertain. Areas of uncertainty have the polytopic form with a different number of peaks corresponding to a more general structure. The equivalent model is given by:

$$x_k(t+1) = \sum_{k=1}^{N} \sum_{j=1}^{n_k} \xi_k \lambda_{kj}(t) A_{kj} x(t) + Bu_k(t) \tag{19}$$

They $\lambda_{kj}(t)(j=1,\dots,n_k)$ are polytopic uncertain parameters for each $k=:\{1,..,n\}$, (k is the switching index), $A_{kj}(j=1,\dots.,n_k)$ are the vertex matrices of the polytopic matrix A_k, and n_k is the number of the vertex matrices. ξ is the indicator arbitrary switching function.

Consider the following notation:

$$\ell=\sum_{k=1}^{N}\xi_k(t)\hat{A}_k=\sum_{k=1}^{N}\sum_{j=1}^{n_k}\xi_k(t)\alpha_{kj}(t)A_{kj} \tag{20}$$

This implies that A is a convex combination of $A_{kj}\in S_1$, S_1 is the set containing all vertices of the system.

Solving the above robust stabilization problem, a new extended formulation of the preceding ILC law (11) is introduced.

$$u_{k+1}(t)-\beta u_k(t)=K_1^{p(j)}x_k(t)+K_2^{p(j)}\eta_{k+1}(t) \tag{21}$$

where p is defined as the number of polytopic subsystem.

Let

$$\eta_{k+1}(t)=x_{k+1}(t)-\beta x_k(t) \tag{22}$$

$K_1^{p(j)},K_2^{p(j)}$ are appropriately dimensioned matrices gains to be designed, and β a positive scalar.

Then clearly (21) and (22) can be written as:

$$\begin{aligned}\eta_{k+1}(t+1)&=x_{k+1}(t+1)-\beta x_k(t+1)\\&=\sum_{j=1}^{n_k}\xi_k\lambda_{kj}(t)\left[A_{k+1j}x_{k+1}(t)-\beta A_{kj}x_k(t)\right]+B\left[u_{k+1}(t)-\beta u_k(t)\right]\\&=\sum_{j=1}^{n_k}\xi_k\lambda_{kj}(t)A_{k+1j}\eta_{k+1}(t)+\left(\beta\sum_{j=1}^{n_k}\xi_k\lambda_{kj}(t)\left(A_{k+1j}-A_{kj}\right)+BK_1^{p(j)}\right)x_k(t)+\left(BK_2^{p(j)}\eta_{k+1}(t)\right)\\&=\left(\sum_{j=1}^{n_k}\xi_k\lambda_{kj}(t)A_{k+1j}+BK_2^{p(j)}\right)\eta_{k+1}(t)+\left(\beta\sum_{j=1}^{n_k}\xi_k\lambda_{kj}(t)\left(A_{k+1j}-A_{kj}\right)+BK_1^{p(j)}\right)x_k(t)\end{aligned} \tag{23}$$

and

$$x_{k+1}(t)=\eta_{k+1}(t)+\beta x_k(t) \tag{24}$$

The obtained system in the closed-loop is represented by the following form:

$$\begin{cases} \eta_{k+1}\left(t+1\right) = \left(A_{k+1j} + BK_2^{p(j)}\right)\eta_{k+1}\left(t\right) + \left(\beta\left(A_{k+1j} - A_{kj}\right) + BK_1^{p(j)}\right)x_k\left(t\right) \\ x_{k+1}\left(t\right) = \eta_{k+1}\left(t\right) + \beta\, x_k\left(t\right) \end{cases} \tag{25}$$

However, the switched system (25) obtained by applying ILC control can be rewritten into 2D discrete linear switched system form described by the Roesser model (1) with:

$$\begin{cases} x^h\left(i+1, j\right) = \eta_{k+1}\left(t+1\right) \\ x^v\left(i, j+1\right) = x_{k+1}\left(t\right) \end{cases} \tag{26}$$

The construction of a synthesis method which characterized the robust control law is based on the poly-quadratic approach. This formulation relies on the use of a switched Lyapunov function depends on uncertain parameters. These functions allow the analysis of the stability and stabilization of linear discrete-time uncertain switched systems. LMI conditions are available to guarantee the exponential stability of linear discrete switched systems. The ILC scheme applied to the 2D Roesser uncertain switched systems required the feasibility of the following set of LMIs.

Theorem 3: Discrete linear uncertain switched system (18) under (ILC) control law (21) is exponentially stabilizable, if there exist symmetric positive matrix $X_1^{p(j)}, X_2^{p(j)}, R_1^{p(j)}, R_2^{p(j)}, \{p, j\} \in Z_+$ and positive parameters $0 < \alpha < 1$, $0 < \beta < 1$ such that the following LMI is feasible:

$$\begin{bmatrix} -\alpha X_1^{p(j)} & 0 & \left(X_1^{p(j)}\right)^T \left(A_{k+1j}\right)^T + \left(R_1^{p(j)}\right)^T (B)^T & \left(X_1^{p(j)}\right)^T \\ * & -\alpha X_2^{p(j)} & \left(X_2^{p(j)}\right)^T \left(\beta\left(A_{k+1j} - A_{kj}\right)\right)^T + \left(R_2^{p(j)}\right)^T (B)^T & \left(\beta X_2^{p(j)}\right)^T \\ * & * & -\left(X_1^{p(j)}\right)^T & 0 \\ * & * & * & -\left(X_2^{p(j)}\right)^T \end{bmatrix} < 0 \tag{27}$$

In this case the gains $K_1^{p(j)}$ and $K_2^{p(j)}$ are computed by $K_1^{p(j)} = R_2^{p(j)}\left(X_2^{p(j)}\right)^{-1}, K_2^{p(j)} = R_1^{p(j)}\left(X_1^{p(j)}\right)^{-1}$ and the minimum dwell time given by:

$$\tau_a \geq \tau_a^* = \frac{\ln X}{-\ln \alpha}, \text{ with } X = \max_{\substack{1 \leq p,l \leq N \\ p \neq l}} \frac{\lambda_{\max}\left(X^l\right)}{\lambda_{\min}\left(X^p\right)},\ X^{p(j)} = diag\left(X_1^{p(j)}, X_2^{p(j)}\right)$$

Proof: The control synthesis problem via a switched state feedback is considered for the following switching uncertain system:

$$x_k(t+1) = \sum_{k=1}^{N}\sum_{j=1}^{n_k} \xi_k \lambda_{kj}(t) A_{kj} x(t) + Bu_k(t)$$

The closed-loop dynamic system with the switched ILC law defined by this structure $u_{k+1}(t) - \beta u_k(t) = K_1^{p(j)} x_k(t) + K_2^{p(j)} \eta_{k+1}(t)$ is described by the given system:

$$\begin{cases} \eta_{k+1}(t+1) = \left(A_{k+1j} + BK_2^{p(j)}\right)\eta_{k+1}(t) + \left(\beta\left(A_{k+1j} - A_{kj}\right) + BK_1^{p(j)}\right)x_k(t) \\ x_{k+1}(t) = \eta_{k+1}(t) + \beta x_k(t) \end{cases}$$

Furthermore, let

$$A_{11} = \left(A_{k+1j} + BK_2^{p(j)}\right), A_{12} = \left(\beta\left(A_{k+1j} - A_{kj}\right) + BK_1^{p(j)}\right), A_{21} = I, A_{22} = \beta$$

In addition, this system can be rewritten into a 2D discrete linear switched system form. Which can be converted into a 2D Roesser model (1) satisfied the following transformation:

$$\begin{cases} x^h(i+1,j) = \eta_{k+1}(t+1) \\ x^v(i,j+1) = x_{k+1}(t) \end{cases}$$

Applying the theorem1, and by introducing $K_1^{p(j)}, K_2^{p(j)}$ into (5) one obtains:

$$\begin{bmatrix} -\alpha X_1^{p(j)} & 0 & \left(X_1^{p(j)}\right)^T (A_{k+1j})^T + \left(K_2^{p(j)} X_1^{p(j)}\right)^T (B)^T & \left(X_1^{p(j)}\right)^T \\ * & -\alpha X_2^{p(j)} & \left(X_2^{p(j)}\right)^T \left(\beta\left(A_{k+1j} - A_{kj}\right)\right)^T + \left(K_1^{p(j)} X_2^{p(j)}\right)^T (B)^T & \left(\beta X_2^{p(j)}\right)^T \\ * & * & -\left(X_1^{p(j)}\right)^T & 0 \\ * & * & * & -\left(X_2^{p(j)}\right)^T \end{bmatrix} < 0 \tag{28}$$

with $R_2^{p(j)} = K_1^{p(j)} X_2^{p(j)}, R_1^{p(j)} = K_2^{p(j)} X_1^{p(j)}$.

This is equivalent to the LMI condition expressed in (27).

Remark 4: The stability of uncertain switched systems has been established by using the approach related to polyquadratic stability. The problem of designing stabilizing gains by applying (ILC) control law is a complex one for switched systems. It is known that (ILC) control design is difficult when the uncertainties are structured ones, which is the case for polytopic uncertainty. In this section, sufficient conditions using polyquadratic stability are investigated to build a robust repetitive control law ensuring the stabilization of uncertain switched systems. The given formulations deal with exponential stability notion under a minimum dwell time. The established results are available under arbitrary switching law. Considering knowledge on switching law, could allow giving less restrictive conditions.

5. NUMERICAL EXAMPLES

To prove the effectiveness of the proposed method, a numerical evaluation is discussed in this section. The studied problem consists in the design of an Iterative learning controller (ILC) stabilizing the discrete switched system. It is assumed that the switched system is operated during a finite time interval repetitively, and then the ILC control scheme can be used to achieve the convergence of the switched system over the whole time interval. A sufficient condition of exponential stability is announced for the considered switched system with ILC control in (11)–(12) minimizing the dwell time needed to overcome the stability analysis problem. Theoretical results are demonstrated by numerical simulations.

5.1. Certain Case

Consider the discrete-time switched system (1) governed by the two instable modes described by the following set of matrices:

$$A_1 = \begin{bmatrix} 5 & 1.5 \\ 1.4 & 2.6 \end{bmatrix}, A_2 = \begin{bmatrix} 8.5 & 5 \\ 3 & 2.5 \end{bmatrix}, \; B = \begin{bmatrix} 9 \\ 3.5 \end{bmatrix}$$

Applying the proposed design ILC scheme to the given system required the solving of the following LMIS (17) in order to obtain the stabilizing gains and the minimum dwell time assuring the exponential stability of the closed loop repetitive switched system.

The obtained results give an idea about the estimation of the minimum dwell time in order to validate the theoretical approach.

Figure 1 represents the evolution of dwell time as a function of the constant β and the convergence rate α.

Choose $\alpha = 0.11, \beta = 0.007$, $\tau_a = 3.3406\, s$, solving the matrix inequalities in Theorem 2 give rise to:

$$K_1^1 = \left[0.004, -0.005\right];\; K_1^2 = \left[-0.0028, -0.0025\right];\; K_2^1 = \left[-0.1393, -1.7132\right];\; K_2^2 = \left[-0.9365, -0.5699\right].$$

Figure 1. Value optimization α, β according to minimum dwell time

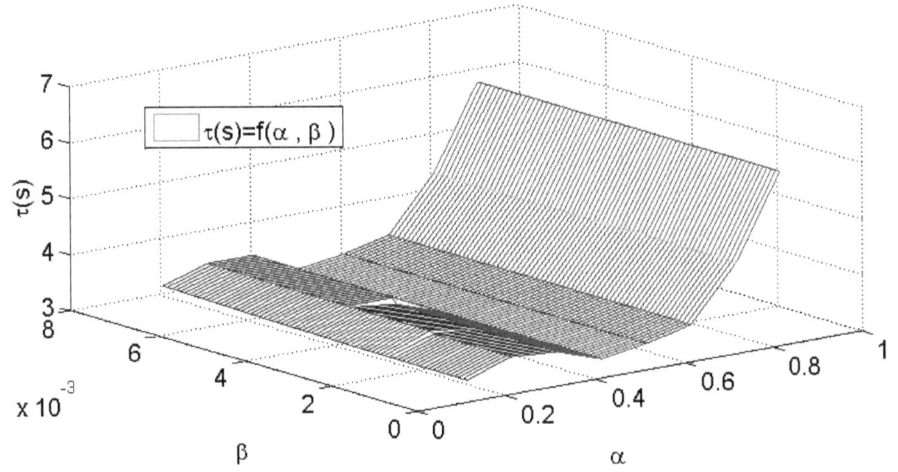

It is worth noting that this condition in Theorem 2 is obtained by using the average dwell time approach, and the parameter α plays a key role in obtaining the infimum of the average dwell time τ_a. From Theorem 2, it is easy to see that a larger α will be favorable to the solvability of inequality (17), which leads to a larger value for the average dwell time τ_a. Considering these, we can first select a larger α to guarantee the feasible solution of inequality (17), and then decrease α to obtain the suitable infimum of the average dwell time τ_a.

Allow the system to switch between these two modes according to the dwell time. For a switching sequence as depicted in Figure 2, an initial condition $x(0) = \begin{bmatrix} 0.5 & 2 \end{bmatrix}^T$, we can see in Figures 2 and 3 that by using our switched (ILC)controller synthesis procedure, that the trajectory of the closed-loop system is stable.

Figure 2. The trajectory of state $x_{k+1}(t)$

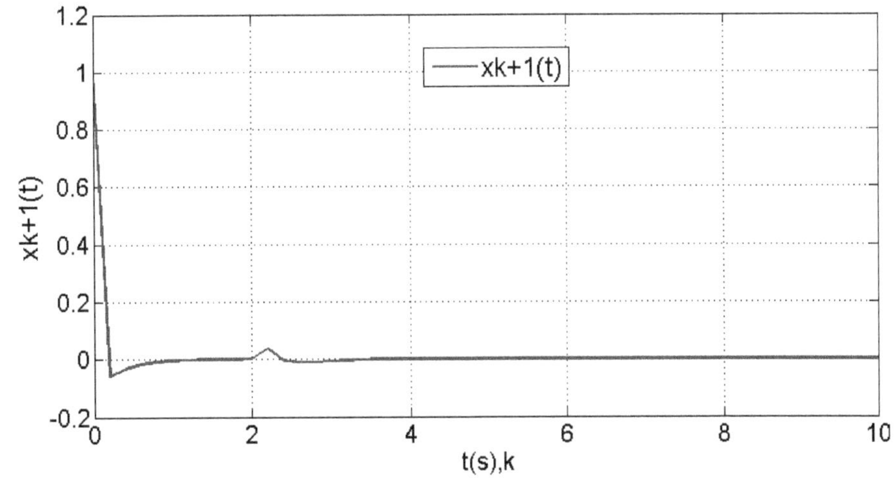

Figure 3. The trajectory of the command $u_{k+1}(t)$

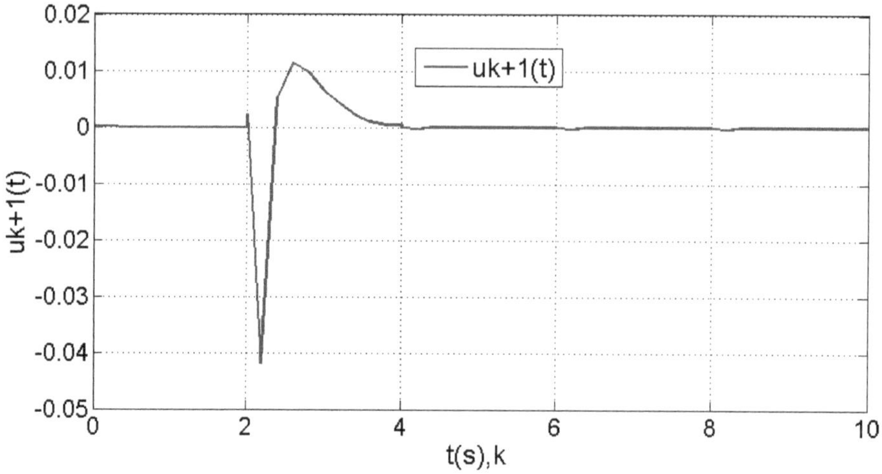

Figure 2 represents the evolution of the command ILC along the time and cycle.

For the linear switched system (8) with the found dwell time, when the ILC algorithm (11) is used, then the tracking state is convergent quickly.

Remark 5: The simulation results shown in Figures 2 and 3, by increasing the iteration number, the switched system is stable along the time and the pass, following selection of K_1^1, K_1^2, K_2^1 and K_2^2. Thus, the resulting ILC process guaranteed the minimum dwell time assuring the exponential stability of the closed loop repetitive switched system. The previous results show the effectiveness of the repetitive control ILC compared to other conventional control. The (ILC) control achieves the exponential stability of the studied closed-loop system.

5.2. Uncertain Case

Described by the state model (18).The polyquadratic stabilization through ILC control law for discrete time switched system is specified to characterize. The ILC laws are two-dimensional (2-D) analysis approach. It is shown that a necessary and sufficient convergence condition can be directly derived using the theory of 2-D switched systems. It is also proved that if the system parameters are subject to polytopic-type uncertainties, this condition can induce a sufficient condition for the robust ILC convergence. A similar example has been proposed for the discrete time switched system with uncertainty polytopic uncertainties to illustrate the effectiveness of the proposed formulation.

An uncertainty is included on the state matrix, especially on $A(2,1)$ and $A(1,2)$.

Consider the uncertain switched discrete time system defined in (18).

Due to the uncertainty on $A_k(2,1)$ and $A_k(1,2)$,with:

$$A_{1j} = \begin{bmatrix} 5 & 1.5 \pm inc\% \\ 1.4 \pm inc\% & 2.6 \end{bmatrix}, \ A_{2j} = \begin{bmatrix} 8.5 & 5 \pm inc\% \\ 3 \pm inc\% & 2.5 \end{bmatrix}, \ B = \begin{bmatrix} 9 \\ 3.5 \end{bmatrix}$$

Figure 4. Value optimization α, β according to minimum dwell time $\left(inc = \pm 10\%\right)$

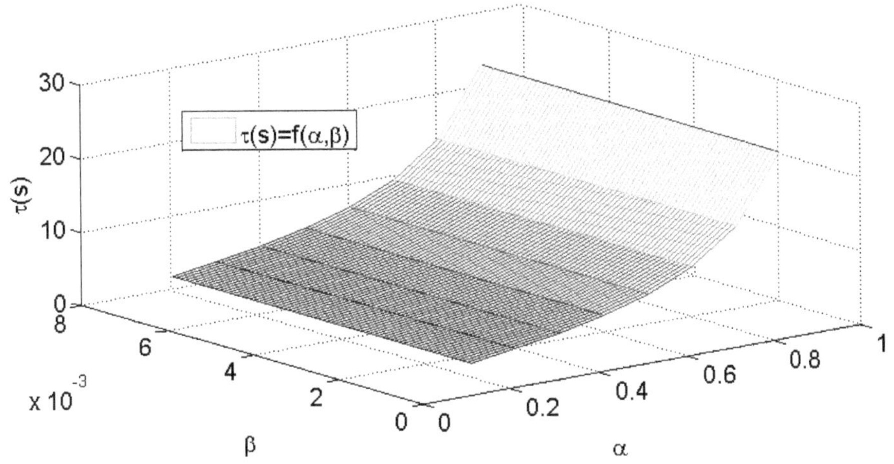

Applying the proposed design ILC scheme to the given system required the solving of the following LMIS (27) in order to obtain the stabilizing gains and the minimum dwell time assuring the exponential stability of the closed loop repetitive uncertain switched system.

The obtained results give an idea about the estimation of the minimum dwell time in order to validate the theoretical approach.

Figure 4 represents the evolution of dwell time as a function of the constant β and the convergence rate α $\left(inc = \pm 10\%\right)$.

Choose $\alpha = 0.11, \beta = 0.0065$, we obtain the value of minimum dwell time $\tau_a = 5.43\,s$.

To stabilize the uncertain switched system using the polyquadratic approach for the value of $\left(inc = \pm 10\%\right)$. The stabilizing gains by solving the LMI conditions in Theorem 3 give rise to:

$$K_1^{1(1)} = \left[-0.035, -1.132\right];\ K_1^{2(1)} = \left[-0.954, -0.611\right];\ K_2^{1(1)} = \left[0.004, -0.002\right];\ K_2^{2(1)} = \left[-0.003, -0.002\right]$$

$$K_1^{1(2)} = \left[0.525, -3.43\right];\ K_1^{2(2)} = \left[-0.954, -0.611\right];\ K_2^{1(2)} = \left[0.001, -0.012\right];\ K_2^{2(2)} = \left[-0.003, -0.002\right]$$

The actuator breakdowns are chosen randomly in time.

Figures 5 and 6 illustrate the trajectories successively of the state $x_{k+1}\left(t\right)$, and the control law $u_{k+1}\left(t\right)$ taken in all the rang of uncertainties.

The convergence state of the trajectory control law is obtained for the best selected value of the relaxed parameter, $\beta = 0.0065$. Note that the parameter α, β play a very important role for the convergence of our algorithm, and then the exponential stability of the system. The convergence to zero of the trajectory of state is very fast in certain and uncertain cases, as it can be seen in Figures 5 and 6. Note that the control law trajectory also converges to zero quickly. It is shown that the iterative learning control law can obtain perfect trajectory performance after some iteration.

Figure 5. The trajectory of state $x_{k+1}(t)\ (inc = \pm 10\%)$

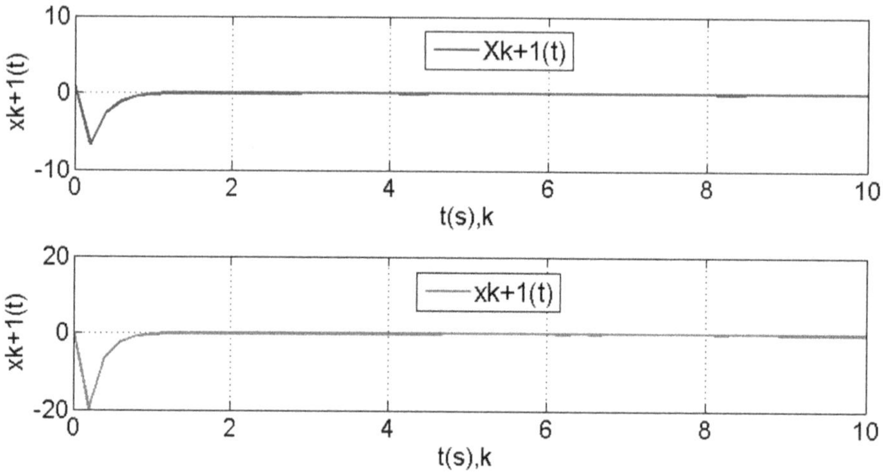

Figure 6. The trajectory of the command $u_{k+1}(t)\ (inc = \pm 10\%)$

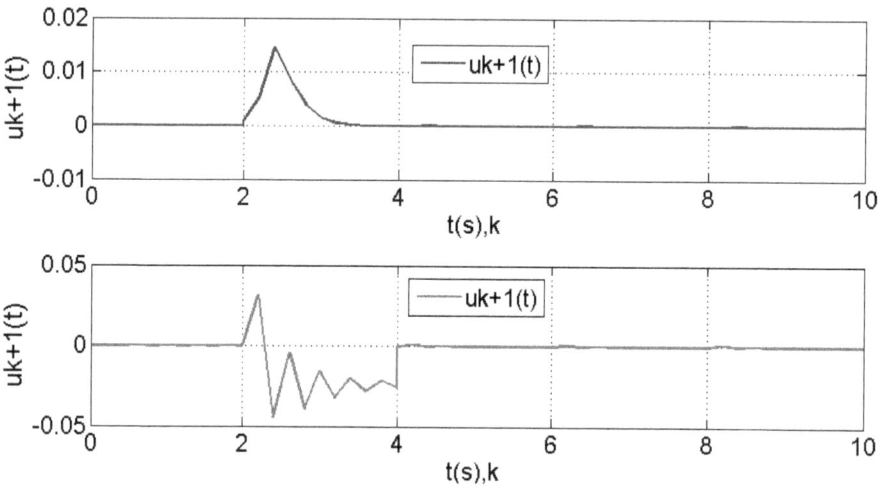

According to the dwell time approach, it's clear that the state systems converge quickly to zero. This is shown by Figure 5. However, by applying the proposed ILC scheme, the robust exponential stabilization of the uncertain switched system is well proved. The ILC law control achieves the stabilization of a discrete switched system if that considered subsystems are unstable which demonstrate the effectiveness and the robustness of the proposed scheme of control.

6. DISCUSSION

For the iterative learning control law given in (21), in the light of the condition $\eta_k(t+1)$, we select the learning gains as $K_1^{p(j)}, K_2^{p(j)}$. Then, the results of simulation are shown in Figure 1, Figure 2, Figure 3, Figure 4, Figure 5 and Figure 6. It can be observed from Figure 1, and Figure4 that by using the proposed iterative learning control law, we can guarantee the minimum dwell time guaranteed the exponential convergence of the output error between the desired state outputs for all $t \in [0,T]$. Figure 2, Figure 3, shows that the convergence of state is perfect after small iterations with minimum dwell time between two modes of systems. To understand the effective of iterative learning control, we also give the input profiles in Figure 3, Figure 6. The ILC control at small iteration is stabilizable along the pass and the time. It is well known that, although the convergence rate α, β of the minimized error from iteration to iteration can be proved to decay monotonically, the super rate α, β may decrease to a huge value before it the state systems converges to the desired level, as shown in Figure 2, Figure 5.

In the literature using the 2-D analysis approach to ILC, the parameter of switched uncertainty system has not been considered explicitly. Usually, it is assumed that there exist parameter uncertainties, and numerical examples are used to illustrate the theoretical results. In contrast to this, section 5 shows that the convergence results of ILC are indeed robust against such explicit system parameter uncertainties.

7. CONCLUSION

In this paper, the problem of synthesis of (ILC) controller for a class of discrete-time linear switched uncertain systems with polytopic uncertainties. A system under time switching laws has been investigated. The control law design (ILC) by constructing a sequence of control inputs to a discrete switched system produces 2D switched repetitive system. This can guarantee the exponential stability of the closed loop switched system and the convergence along the time and the pass. The dwell time approach is used for the stability analysis. Furthermore, a sufficient condition for the robust convergence of ILC can be derived if the switched system parameters are known to lie within an uncertainty polytope. Sufficient conditions for the existence of such controller are formulated in terms of a set of LMI. A numerical example is given to illustrate the effectiveness of the proposed method. The past decade has seen a lot of research activities in the field of switched systems. The results given here are also an important step towards analyzing more general switching in 2D/nD systems theory, such as extension to the more than two switched processes, encompassing the uncertain cases and also to switched differential processes (from pass to pass), which is the subject of on-going research.

This survey is far from a complete review of stability and stabilizability of switched systems. Three issues deserve more attention. The first is on the theoretical comparison of dwell time calculation with the results available in the literature to date. The second is related to the development of numerical algorithms with (ILC) in the case of a switching systems for the solution of the tracking control and applying a PID-type iterative learning control for 2-D repetitive switched systems. The third one is the possible generalization of the stability conditions to cope with nonlinear control design. Although ILC theory has well been developed for linear multivariable systems and uncertain classes of nonlinear systems, most research on ILC mainly focus on the hybrid systems under the same initial iterative conditions. It is not practical to assume that each learning iteration starts at the same point, because it is impossible to repeat the same initial condition in practical engineering application. The initial condition always deviates from the initial condition of last iterative performance. When the initial conditions vary, the learning dynamics becomes much more complex. Hitherto, the effect on the interaction between the learning dynamics and the variable initial conditions remains an open and important issue for ILC research. Viewed the hybrid nature of the involved control (ILC), there is always a very delicate problem of communication between the ILC control and switching systems. These points constitute a real theoretical challenge.

REFERENCES

Amann, N., Owens, D. H., & Rogers, E. (1998). Predictive optimal iterative learning control. *International Journal of Control, 69*(2), 203–226. doi:10.1080/002071798222794

Ben Attia, S., Salhi, S., Bernussou, J., & Ksouri, M. (2009a). Improved LMI formulation for robust dynamic output feedback controller design of discrete-time switched systems via switched Lyapunov function. In *Proceedings of the 3rd International Conference on Signals, Circuits and Systems*. Academic Press.

Ben Attia, S., Salhi, S., Bernussou, J., & Ksouri, M. (2010). Analyse de la stabilité des systèmes à commutation à temps continu en utilisant les systèmes de comparaison. In Proceedings of Sixième IEEE Conférence Internationale Francophone d'Automatique (pp. 15-20). IEEE.

Ben Attia, S., Salhi, S., & Ksouri, M. (2009b). Mode-Independent state feedback design for discrete switched linear. In *Proceedings of European Conference on Modelling and Simultion* (pp. 807-812). Academic Press. doi:10.7148/2009-0807-0812

Ben Attia, S., Salhi, S., & Ksouri, M. (2009c). LMI Formulation for Static Output Feedback Design of Discrete-time Switched Systems. *Journal of Control Science and Engineering, 2009*, 7. doi:10.1155/2009/362920

Ben Attia, S., Salhi, S., & Ksouri, M. (2012). Static Switched Output Feedback Stabilization For Linear Discrete-time Switched Systems. *International Journal of Innovative Computing, Information, & Control, 8*(5A), 3203–3213.

Benzaouia, A., Hmamed, A., Tadeo, F., & Hajjaji, A. E. L. (2011). Stabilization of discrete 2D time switching systems by state feedback control. *International Journal of Systems Science, 42*(3), 479–487. doi:10.1080/00207720903576522

Branick, M. S. (1998). Multiple Lyapunov functions and other analysis tools for switched and hybrid systems. *IEEE Transactions on Automatic Control, 43*(2), 475–482. doi:10.1109/9.664150

Chow, T. W. S., & Fang, Y. (1998). An Iterative Learning Control Method for Continuous-time Systems Based on 2-D System Theory. *IEEE Transactions on Circuits and Systems: Fundamental Theory and Applications, 45*(2), 683–689. doi:10.1109/81.678496

Cui, C., Long, F., & Li, C. (2010). Disturbance attenuation for switched system with continuous-time and discrete-time subsystems: State feedback case. *ICIC Express Letters, 4*(1), 205–212.

Decarlo, R., Branicky, M., Pettersson, S., Lennartson, B., & Antsaklis, P. J. (2000). Perspectives and results on the stability and stabilizability of hybrid systems. *Proceedings of IEEE, 88*(4), 1069–1082. doi:10.1109/5.871309

Dridi, J., Ben Attia, S., Salhi, S., & Ksouri, M. (2012). Repetitive Processes Based Iterative Learning Control Designed by LMIs. *Applications of Mathematics, 2012*, 18.

Dridi, J., Ouerfelli, H., & Salhi, S. (2013).Stability Along the Pass of Differential Linear Repetitive Process Using an LMI. In *Proceedings of IEEE International Conference on Integent Automatition and Robotics*, (pp. 6-10). IEEE.

Du, D., Zhou, S., & Zhang, B. (2006). Generalized H2 output feedback controller design for uncertain discrete-time switched systems via switched Lyapunov functions. *Nonlinear Analysis*, *65*(11), 2135–2146. doi:10.1016/j.na.2005.11.054

Elci, H., Longman, R. W., Phan, M. Q., Juang, J. N., & Ugoletti, R. (2002). Simple Learning Control Made Practical by Zero-phase Filtering: Applications to Robotics. *IEEE Transactions on Circuits and Systems: Fundamental Theory and Applications*, *49*(2), 753–767. doi:10.1109/TCSI.2002.1010031

Fang, Y., Tommy, W., & chow, s. (2003). 2-D analysis for iterative learning controller for discrete-time systems with variable initial conditions. *IEEE Transactions on Circuits and Systems: Fundamental Theory and Application, 50*(5), 722-727.

Gałkowski, K., Rogers, E., Xu, S., Lam, J., & Owens, D. H. (2002). LMIs a fundamental tool in analysis and controller design for discrete linear repetitive processes. *IEEE Transactions on Circuits and Systems. I, Fundamental Theory and Applications*, *49*(6), 768–778. doi:10.1109/TCSI.2002.1010032

Hu, B., & Michel, A. N. (2000). Stability Analysis of Digital Feedback Control Systems with Time-Varying Sampling Periods. *Automati*, *36*(3), 897–905. doi:10.1016/S0005-1098(99)00217-4

Jianming, X., Mingxuan, S., & Li, Y. (2008). LMI-Based Synthesis of Robust Iterative Learning Controller with Current Feedback for Linear Uncertain Systems. *International Journal of Control, Automation, and Systems*, *6*(2), 171–179.

Koutsoukos, D., & Antsklis, J. (2002). Design of stabilizing switching control laws for discrete and continuous-time linear systems using piecewise-linear Lyapunov functions. *International Journal of Control*, *75*(5), 932–945. doi:10.1080/00207170210151076

Kurek, J. E., & Zaremba, M. B. (1993). Iterative Learning Control Synthesis Based on 2-D System Theory. *IEEE Transactions on Automatic Control*, *38*(2), 121–125. doi:10.1109/9.186321

Lam, J., Xu, S. Y., & Zhou, Y. (2004). Robust output feedback stabilization for two-dimensional continuous systems in Roesser form. *Applied Mathematics Letters*, *17*(12), 1331–1341. doi:10.1016/j.aml.2004.04.004

Li, X. T. D., Chow, T. W. S., & Ho, J. K. L. (2005). 2-D System Theory Based Iterative Learning Control for Linear Continuous Systems with Time Delays. *IEEE Transactions on Circuits and Systems. I, Regular Papers*, *52*(7), 1421–1430. doi:10.1109/TCSI.2005.851706

Li, X. T. D., Chow, T. W. S., & Ho, J. K. L. (2008). Iterative Learning Control for a Class of Nonlinear Discrete-time Systems with Multiple Input Delays. *International Journal of Systems Science*, *39*(2), 361–369. doi:10.1080/00207720701778403

Li, X. W., & Gao, H. J. (2012). Robust finite frequency HN filtering for uncertain 2-D Roesser systems. *Automatica*, *48*(6), 1163–1170. doi:10.1016/j.automatica.2012.03.012

Lian, J., Feng, Z., & Shi, P. (2011). Observer design for switched recurrent neural networks: An average dwell time approach. *IEEE Transactions on Neural Networks*, *22*(10), 1547–1556. doi:10.1109/TNN.2011.2162111 PMID:21824843

Liberzon, D., Hespanha, J. P., & Morse, A. S. (1999). Stability of switched linear systems: A Lie-algebraic condition. *Systems & Control Letters*, *37*(3), 117–122. doi:10.1016/S0167-6911(99)00012-2

Lien, C. H., Yu, K. W., Chung, Y. J., Chang, H. C., Chung, L. Y., & Chen, J. D. (2011). Exponential stability and robust H∞ control for uncertain discrete switched systems with interval time-varying delay. *IMA J. Math. Control Inf, 7*(5), 433–444.

Liu, Y., Sun, L. Y., Lu, B. C., & Dai, M. Z. (2010). Feedback control of networked switched fuzzy time-delay systems based on observer. *ICIC Express Letters*, *4*(6), 2369–2376.

Lu, W. S., & Lee, E. (1985). Stability analysis for two-dimensional systems via a Lyapunov approach. *IEEE Transactions on Circuits and Systems*, *32*(1), 61–68. doi:10.1109/TCS.1985.1085596

Mehrzi, E., Bernosso, J., & Mhiri, R. (2007). stability and stabilization for uncertain switched system, a polyquadratic lyaponuv approach. *International Journal on Science and Technique of Automatique Control, 1*(1), 61-74.

Mingxuan, S., Huaying, X., & Hui, W. (2011). Switching-function dynamics designs for uncertain variable structure systems with repetitive control. *International Journal of Control*, *84*(7), 1196–1208. doi :10.1080/00207179.2011.561876

Mitra, S., Liberzon, D., & Lynch, N. (2008). Verifying average dwell time of hybrid systems. *ACM Transactions on Embedded Computing Systems*, *8*(1), 1–37. doi:10.1145/1457246.1457249

Moore, K. L. (2000). A matrix fraction approach to higher-order iterative learning control: 2-D dynamics through repetition-domain filtering. In *Proceedings of the Second International Workshop on Multidimensional (ND) Systems* (pp. 99-104). Academic Press.

Moore, K. L., Hyo-Sung, A., & Chen, Y. Q. (2008). Iteration domain H1-optimal iterative learning controller design. *International Journal of Robust and Nonlinear Control*, *2*(18), 1001–1017. doi:10.1002/rnc.1231

Morse, A. S. (1996). Supervisory Control of Families of Linear Set-Point Controllers-Part 1: Exact Matching. *IEEE Trans. Automat. Control, 41*(10), 1413-1431.

Owens, A., Rogers, D. H., Aman, N., & French, M. (2000). Analysis of Linear Iterative Learning Control Schemes, A 2D Systems/Repetitive Processes Approach. *International Journal of Multidimensional Systems and Signal Processing*, *11*(1-2), 125–177. doi:10.1023/A:1008494815252

Paszke, W., Lam, J., Galkowski, K., Xu, S., & Lin, Z. (2004). Robust stability and stabilization of 2-D discrete state-delayed systems. *Systems & Control Letters*, *51*(3–4), 277–291. doi:10.1016/j.sysconle.2003.09.003

Peleties, P., & De'Carlo, R. A. (1992).Asymptotic stability of m-switched systems using Lyapunov functions. In *Proceedings of the 31th IEEE Conference on Decision and Control* (vol. 18, pp. 3438–3439). Tucson, AZ: IEEE. doi:10.1109/CDC.1992.371213

Roover, D., Bosgra, O. H., & Steinbuch, M. (2000). Internal-model-based design of repetitive and iterative learning controllers for linear multivariable systems. *International Journal of Control*, *73*(10), 914–929. doi:10.1080/002071700405897

Shi, J., Gao, F., & Wu, T. J. (2005). Robust design of integrated feedback and iterative learning control of a batch process based on a 2D Roesser system. *Journal of Process Control*, *15*(2), 907–924. doi:10.1016/j.jprocont.2005.02.005

Xia, Z., Zun-Quan, X., & Yan, G. (2012). Exponential Stabilizability of Switched Systems with Polytopic Uncertainties.Hindawi Publishing Corporation. *Journal of Applied Mathematics*, (853170): 15.

Xiang, Z., & Huang, S. (2013). Stability Analysis and Stabilization of Discrete-Time 2D Switched Systems. *Circuits Syst Signal Process*, *32*(1), 401–414. doi:10.1007/s00034-012-9464-4

Zhai, G. (2001). Quadratic stabilizability of discrete-time switched systems via state and output feedback. In *Proceedings of the 40th IEEE Conference on Decision and Control* (vol. 3, pp. 2165-2166). Orlando, FL: IEEE. doi:10.1109/CDC.2001.980575

Zhang, L. X., & Shi, P. (2009). Stability, l2-gain and asynchronous HN control of discrete-time switched systems with average dwell time. *IEEE Transactions on Automatic Control*, *54*(9), 2192–2199. doi:10.1109/TAC.2009.2026841

Zhat, Y., Bo, H., Yasuda, K., & Michel, A. N. (2001). Stability analysis of switched systems with stable and unstable subsystems: an average dwell time approach. *International Journal of Nonlinear Systems and Science, 32*(8), 1055-1061.

KEY TERMS AND DEFINITIONS

Dwell Time Switching: Logic for orchestrating the switching between controllers in a family of candidate controllers in order to control a process with a highly uncertain model.

Hybrid System: A dynamic system that exhibits both continuous and discrete dynamic behavior a system that can both flow (described by a differential equation) and jump (described by a difference equation or control graph). Often, the term "hybrid dynamic system" is used, to distinguish over hybrid systems such as those that combine neural nets and fuzzy logic, or electrical and mechanical drivelines. A hybrid system has the benefit of encompassing a larger class of systems within its structure, allowing for more flexibility in modelling dynamic phenomena.

Iterative Learning Control (ILC): A method of tracking control for systems that work in a repetitive mode. Examples of systems that operate in a repetitive manner include robot arm manipulators, chemical batch processes and reliability testing rigs. In each of these tasks the system is required to perform the same action over and over again with high precision. This action is represented by the objective of accurately tracking a chosen reference signal $r(t)$ on a finite time interval.

Linear Matrix Inequality and Control Theory: The LMI that arise in system and control theory can be formulated as convex optimization problems, and hence are amenable to computer solution.

Polytopic Uncertainties: Parametric uncertainties. The matrix of the state system is a linear combination of convex and belongs to the field type polytopic.

Robust Control: A branch of control theory that explicitly deals with uncertainty in its approach to controller design. Robust control methods are designed to function properly so long as uncertain parameters or disturbances are within some (typically compact) set. Robust methods aim to achieve robust performance and/or stability in the presence of bounded modeling errors.

Switched system: A hybrid dynamical system defined by a family of subsystems and a switching rule orchestrating the switching between subsystems.

Section 6

Power Systems and Engineering Applications

Chapter 21
Electrical Motor Parameters Estimator Improved by a Computational Algorithm

Flah. Aymen
National School of Engineering of Gabes, Tunisia

Habib Kraiem
National School of Engineering of Gabes, Tunisia

Sbita. Lassaâd
National School of Engineering of Gabes, Tunisia

ABSTRACT

In this chapter, two computational algorithms are proposed and applied on an estimation algorithm, in order to improve the global performance of the estimation phase. The proposed system is studied based on the Model Reference Adaptive System (MRAS). The importance of the estimation phase in a large applications number is basically observed on the applications applied on electrical motors, where a lot number of parameters are measured with real measurement equipments, as Tesla Meter, speed shaft, and others. The idea is based generally on the software applications, where it is possible to guarantee the desired estimation phase using a software algorithm. In this chapter the MRAS technique is proposed as the software algorithm, for replacing the measurement materials for online estimate the overall characteristic PMSM parameters. Our approach aims to ameliorate the MRAS technique with intelligent optimization methods called BFO and PSO.

1. INTRODUCTION

In the last decade, the majority applications related to the industrial, militaries, robot and automobile are basing on the electrical motor, due to many causes related to the environment protection and energy saving and especially to the high efficiency. In the literature many motor models and types are presented and each model is requested for a specific application. Electrical vehicles applications will be widely used

DOI: 10.4018/978-1-4666-7248-2.ch021

in the common years due many advantageous as: simplicity of architecture, robustness, and no need for a periodic maintenance and especially for the rapid depletion of the Earth's petroleum resources. This system seems interesting for the major researches related to the electrical motor drive.

As presented by Zhu and Howe (2007), witch is interested in her study by the electrical vehicles. The electrical vehicle can be operated by at less one electrical motor. The same previous cited authors, was present in them work the importance of the electrical motor choice, face the global efficiency of the vehicle. Many motor models are exposed and compared in his paper. The factors basing on to fix his choice are:

- High reliability and robustness.
- High efficiency over wide speed range.
- High torque density and power density.
- High torque for starting at low speeds and hill climbing.
- Wide speed range.
- Motor size.

In the electrical vehicle application, the most used motors type are the induction motor, switched reluctance motor and the permanent magnet synchronous motor. Zhu and Howe (2007), accomplish a comparative study between these motors type and he proves the efficiency of the Permanent magnet synchronous motors face to the other types. In this context, many researches are established in order to known the control strategy methods, the advantage, disadvantage, difficulty and the problems related to this system. In the control applications, the major new theory, searches the robustness control target, face the external and internal perturbation. Generally, the external perturbation, as load torque, temperature, vibration or dust will affect the system robustness. Those decrease the system performance. In the conventional control methods, the overall system is based on a lot of sensors, for measuring speed, temperature or in some application the vibration. These sensors upgrade the totally cost and increase the architecture complexity. Therefore, many techniques are appeared to reduce the cost of the totally PMSM control loop and to upgrade the control techniques as explain by Lakshmi and all (2013), Kraiem and Messaoudi (2009).

In the electrical vehicle applications, the used motors can be operated at different speed range, and the desired speed can touch the maximum one as demonstrate by Bae and all (2003). However in this running mode, the major problems are appeared in the PMSM parameters variation as discuss by Emadi and all (2008). This is due to the cited previous problems due to the external environment characteristics. Also Flah and Sbita (2012), prove the influence of the high speed mode on the internal motor parameters. These PMSM parameters variations influence are also tested and verified of the identification technique by Chen and all (2012), on the control loop by Lakshmi and all (2013), Stefan and Tomasz. (2011) and Ben Hamed and Sbita (2008) and on the field weakening algorithm by Nalepa and Orlowska-Kowalska (2012) and Tursini and all (2010) etc. So, refers to these reasons, PMSM parameters identification are extremely required. Many methods are proposed aiming to bring up high parameter tracking algorithms. Bolognani in his work was presented the recursive parameters identification method to estimate the PMSM parameters as explain by Bolognani and all (1997).

The adaptive identification model of PMSM parameters based on predictive current control technique was presented by Sozerand all (2010). The adaptive algorithm was used also by Liu and Zhang (2010). Lyapunov and Popov stability criterion was also used to estimate these parameters, as presented by Liu and Zhan (2010), who proves the dominance of the POPOV stability criterion. More discussion about the estimations algorithms and methods are presented in a special section in this chapter.

In this paper the model reference adaptive system (MRAS) technique was proposed as an online PMSM simultaneous parameters observer. In the MRAS technique there are two models, one called the reference model and the other is the adjustable model. Based on the POPOV stability criterion, the error between the estimator and the real models is adjusted to estimate the unknown parameters. These parameters equations depend on PI adapting mechanism; moreover, it is difficult to fix them. This process fails according to the some mismatch errors in these values that can yields to an unstable identification response. In the literature one cannot find a specific way to tune these mechanism parameters as explain by Romero and all (2011). Many approaches based on recurrent neural network, or fuzzy logic was used to replace this PI regulators, and then to eliminate the PI parameters adjusting problems as described by Mir and all (1998) explain. However, these techniques require defining an optimum PI and a large data base for the learning step in the neural network case, or to known the limit margin in the fuzzy case.

For these previous reasons, intelligent optimization techniques are proposed for adjusting these sensitive parameters. Refers to the published works, the most optimization used methods are based on the genetic algorithm, simulated annulling or the particle swarm optimization (PSO) algorithms. These algorithms are widely used in the optimization problems because of its simplicity and reliability. But the precession problems are it inconvenient. Therefore, we propose a novel method based on the bacterial foraging algorithm (BFO). This algorithm principle is to simulate the movement of the bacteria in food research and how they optimize their moving path as presented by Majhi and Panda (2007), Ajith and Amit (2009), Te-Jen and all. (2010). Many applications based on this algorithm for adjusting the parameters of PI controllers, neural network weights or rules number of the fuzzy logic techniques or to use for identification problems, as presented by Sağlam and all (2013), Lokriti and all (2013), Balluq and Atkinson (2004), Ajith. and Amit (2009) and Madasu and Prakash (2009) prove the BFO performance. In short, to find the best optimisation method for the algorithm a comparative analysis between BFO and PSO is achieved.

This chapter is organised as follow: in the first section a general introduction is proposed in order to describe the outlines of the presented work. Section II presents a brief description of the Permanent magnet synchronous motor properties. A simplified mathematical PMSM model is developed in order to use in the control loop and MRAS building. Estimator and identification algorithms in the literatures are presented in section III. Section IV, describe the PMSM parameters identification with an explanation of the model reference adaptive system (MRAS) architecture. Section V, starts with a presentation of the Optimisation techniques in literatures, detailed the PSO and BFO optimization algorithms. This section is ended by a presentation of the overall control loop with an online implementation of the desired estimator algorithm. The objective of the last section is to evaluate the PMSM parameters estimation results. A description of the PMSM control scheme using the MRAS estimator with classical version, the PSO and BFO algorithms are presented. And finally a comparative study with a simulation results are illustrated and discussed.

2. THE PERMANENT MAGNET SYNCHRONOUS MOTOR

A permanent magnet synchronous motor (PMSM) is a motor that uses permanent magnets to produce the air gap magnetic field rather than using electromagnets. These motors have significant advantages, attracting the interest of researchers and industry for use in many applications [Krishnan].

2.1 Permanent Magnet Synchronous Motor Properties

The properties of the permanent magnet material will affect directly the performance of the motor and proper knowledge is required for the selection of the materials and for understanding the permanent magnet motors. The earliest manufactured magnet materials were hardened steel. Magnets made from steel were easily magnetized. However, they could hold very low energy and it was easy to demagnetize. In recent years other magnet materials such as Aluminum Nickel and Cobalt alloys (Alnico), Strontium Ferrite or Barium Ferrite (Ferrite), Samarium Cobalt (First generation rare earth magnet) (SmCo) and Neodymium Iron-Boron (Second generation rare earth magnet) (NdFeB) have been developed and used for making permanent magnets.

The rare earth magnets are categorized into two classes: Samarium Cobalt (SmCo) magnets and Neodymium Iron Boride (NdFeB) magnets. SmCo magnets have higher flux density levels but they are very expensive. With the latest development in rare-earth PM material, it is now possible to develop high efficiency, high power factor, and high power density PM motors for industrial and residential purposes. The neodymium-iron-boron (NdFeB), has 1.4 T of residual flux, 11 kA/m of coercive force and 13.6kA/m intrinsic coercive force, is most used in the modern presented motors. The dropping cost and lower temperature sensitivity have made NdFeB an ideal material for developing high performance PM synchronous motors. A flux density versus magnetizing field for these magnets is illustrated in Figure 1.

2.2 Classification of Permanent Magnet (PM) Motors

A typical PMSM drive system consists of a motor, an inverter, a shaft encoder and microcontroller. A PMSM has a similar structure to a DC excited synchronous motor except that the magnetic field is supplied by the permanent magnet material in a PM machine.

Figure 1. Flux density vs. magnetizing field of permanent magnetic materials

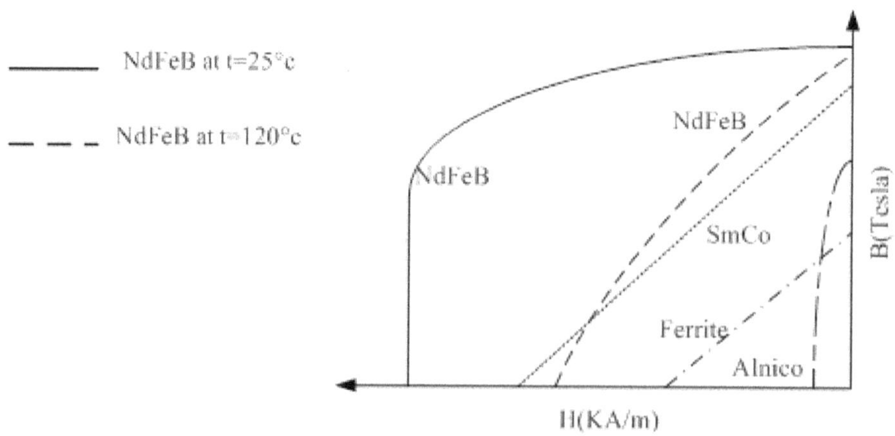

Figure 2. Surface permanent magnet motor

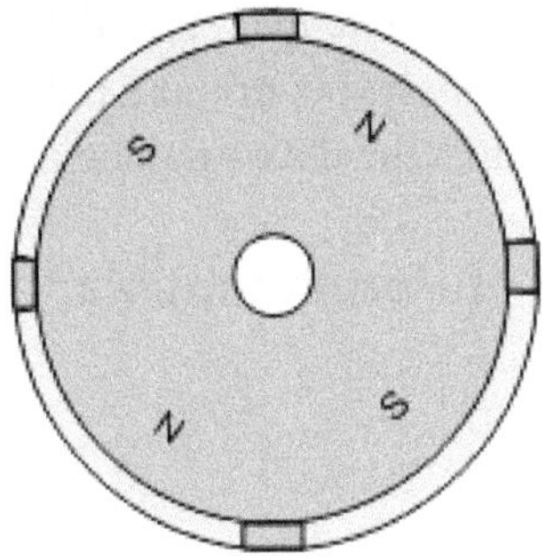

Figure 3. Interior permanent magnet motor

The stator structure of the PMSM is similar to that of an induction machine. Depending on its application, the winding distribution could be near sinusoidal distribution if a smoother torque is desired at low speed. In order to avoid cogging torque, skewed stator slots may be used.

The rotor structure of a PMSM can be built either with surface mounted magnets, interior magnets. Typical rotor structures are shown in Figures 2 and 3.

Among the major types of rotor structures, rotors with surface mounted magnets as shown in Figure 2 are commonly used in the PMSM for their simplicity. The drawback is that the magnets may easly peel-off from the structure if they are originally glued on the rotor surface. Rare earth permanent magnet materials are fragile and scratch easily. There for, the life span of magnets is reduced if they are exposed directly to air. The author has experienced several times the magnets flying off the rotor surfaces during the operation of the PMSM. One effective but costly way to avoid the mechanical weakness of surface mounted motor is to bond the magnet with a cylindrical sleeve made of high strength alloy as shown Figures 2 and 3. Surface mounted require arc shaped magnets which greatly increases the manufacturing cost. For this type of structure permanent magnet motor $L_d = L_q$.

Segmented rectangular strips may be used to reduce the manufacturing cost.

Permanent magnet motors with magnets buried just below the surface have a mechanical advantage over surface mounted motors especially at high speed. Although the possible use of rectangular magnets eases the manufacture of magnets, the manufacture of rotor laminations is more complicated than those for surface mounted magnets

It is possible to build the motor with interior permanent magnet as shown in Figure 3. The overall cost can be significantly reduced by using rectangular shaped magnets. The interior permanent magnet structure is normally suitable for machines with six or more poles. These motors are considered to have saliency with q axis inductance greater than the d axis inductance ($L_q > L_d$).

2.3 Mathematical Expressions of Permanent Magnet (PM) Motors

2.3.1 PMSM Modelling Assumption

In this modelling phase some assumption conditions are taken in consideration in order to facilitate this phase. The first condition is: the saturation is neglected, the second one is related to the induced EMF, where we suppose it sinusoidal and the last one is eddy currents and hysteresis losses are negligible

2.3.2 Three Phases to Two Phases Frame Transformation

The known mathematical model of the three phase PMSM motor is described generally in the three phase frame $\left[x_{abc}\right]$. But the obtained mathematical model is commonly not used in the major control algorithms, due to the complexity of the obtained one. Where three presented equations must be resolved. The idea refers to transform these equations from the three phase frame to the two phases frame, using Clark and Park transformation matrix.

The transformation architecture is given as:

First Step: Clark transformation

$$\begin{bmatrix} x_\alpha \\ x_\beta \\ x_o \end{bmatrix} = \frac{2}{3}\begin{bmatrix} 1 & -\frac{1}{2} & -\frac{1}{2} \\ 0 & \frac{\sqrt{3}}{2} & -\frac{\sqrt{3}}{2} \\ \frac{1}{2} & \frac{1}{2} & \frac{1}{2} \end{bmatrix}\begin{bmatrix} x_a \\ x_b \\ x_c \end{bmatrix}$$

Second Step: Park transformation

$$\begin{bmatrix} x_d \\ x_q \\ x_o \end{bmatrix} = \begin{bmatrix} \cos\theta & \sin\theta & 0 \\ -\sin\theta & \cos\theta & 0 \\ 0 & 0 & 1 \end{bmatrix}\begin{bmatrix} x_\alpha \\ x_\beta \\ x_o \end{bmatrix}$$

2.3.3 PMSM Equations in the Two Phases Frame

The dynamic behaviour of a synchronous machine is described by the following equations written in space vector notation and in the d-q referential.

Stator Voltage Equations:

$$\begin{cases} \vec{v}_s = R\vec{i}_s + \dot{\vec{\lambda}}_s \\ v_d = Ri_d + \dfrac{d\lambda_d}{dt} - \omega_r \lambda_q \\ v_q = Ri_q + \dfrac{d\lambda_q}{dt} + \omega_r \lambda_d \end{cases} \tag{1}$$

Stator Flux Equations:

$$\begin{cases} \vec{\lambda}_s = L_s \vec{i}_s + \vec{\lambda}_m \\ \lambda_d = L_d i_d + \lambda_m \\ \lambda_q = L_q i_q \end{cases} \tag{2}$$

Electromagnetic Torque:

$$\begin{cases} T_e = \dfrac{3}{2} n_p \left(\vec{i}_s \times \vec{\lambda}_s' \right) \\ T_e = \dfrac{3}{2} n_p \left(i_q \lambda_d - i_d \lambda_q \right) \\ T_e = \dfrac{3}{2} n_p \left(\left(L_d - L_q \right) i_d i_q + \lambda_m i_q \right) \end{cases} \tag{3}$$

Mechanical Equation:

$$J \frac{d\omega_r}{dt} + f_v \omega_r = T_e - T_l \tag{4}$$

where: $\vec{v}_s = \left[v_d \; v_q \right]^T$ is the stator voltage vector, $\vec{i}_s = \left[i_d \; i_q \right]^T$ is the stator current vector, $\vec{\lambda}_s = \left[\lambda_d \; \lambda_q \right]^T$ is the stator flux vector, $\vec{\lambda}_m = \left[\lambda_m \; 0 \right]^T$ is the permanent magnet flux linkage vector, R is the stator resistance, L_d and L_q are the d and q axis self stator inductances, n_p the number of pole pairs, T_e and T_l are the electromagnetic and load torques, J is the moment of inertia, ω_r is the rotor speed and f_v is the friction factor.

3. ESTIMATOR AND IDENTIFICATION ALGORITHMS IN THE LITERATURES

In the engineering applications, related to electrical motors, the identification or estimation phase is extremely required to assure the performance of the global control scheme. Generally, the identification phase is executed in the outline function mode before running the system at the control closed loop. The estimation phase can be executed in the online mode, where it is possible to make some measurement parameters without stopping the system. Refers to the literatures, related to this subject, lot of number of the presented estimators are interested by the speed parameters. Where, the researchers find to replace the real speed shaft, by a software one.

As presented by Bilal and all (2004) and Ouhrouche (2002), the Extended Kalman Filter (EKF) algorithm, give a solution for the general estimation problems. The EKF is a stochastic state observer, where nonlinear equations are linearized in every sampling period. An interesting feature of the EKF is related for its ability to estimate simultaneously the states and the parameters of a dynamic process. This is generally useful for both the control and the diagnosis of the process. In [Ouhrouche] the authors used the EKF algorithm to simultaneously estimate variables and parameters of the IM in healthy cases and under different electrical motors faults. Sbita and Ben Hamed (2007) and Kyo and Frede (2006) used the Luenberger Observer for state estimation of electrical motors. The Extended Luenberger Observer (ELO) is a deterministic observer which also linearizes the equations in every sampling period. There is other type of methods for state estimation that is based on the intelligent techniques is used in the recent years by many authors as Faa Jeng and all (2007), Messaoudi and all. (2007). Fuzzy logic and neural networks has been a subject of growing interest in recent years. Neural network and fuzzy logic algorithms are quite heavy for basic microprocessors as used by Saad (2013). In addition, several papers provide sensorless control of electrical motors that are based on the variable structure technique as explain by Lascu and all (2004) and the High Gain Observer (HGO) by Said and all (2010) that is a powerful observer that can estimate simultaneously variables and parameters of a large class of nonlinear systems and doesn't require a high performance processor for real time implementation. Another method based on the Model Reference Adaptive System MRAS technique is used in sensorless electrical motors drives, at a first time, by Schauder (1992). Since this, it has been a topic of many publications by Zhifu and all (2009), and Pradeep and all (2013). As presented in Figure 4, the error between the reference and the adaptive model is used as an online input to the adaptive law in order to make the right recession. Many authors, was used a different versions of the MRAS, basing on the input signal of the adaptive law. Reactive power, active power or flux parameters are used in different literatures as presented respectively by Maiti and all (2008), Kojabadi (2009), Suman and Chakrabatory (2010). All these MRAS versions are implemented in order to online estimate the real speed rotation of the used motor. These estimation algorithms prove the effectiveness for each application with a little advantage and disadvantage for each method. The problem appeared in some of these algorithm is related to the internal motor characteristics, as the physical parameters (resistance inductance and flux). These elements variation or deviation can influence on these algorithms effectiveness in the online running mode, then on the global performance of the system. Starting form this difficulty, the idea is to find the best method can resolve this complexity.

As described in the introduction section, the PMSM is used in this application. The theoretical studies demonstrate that three essential parameters characterise or define the used motor: the stator resistance, the stator inductance and the permanent magnet flux produced by the magnet installed on the rotor. So the idea is to estimate the all three main parameters and use them in the implemented control or speed estimation algorithms to make them adaptive for any possible motor parameters variation. The motor parameters estimator algorithms are also presented in the literatures with different versions, basing on intelligent techniques, as neural networks, fuzzy logic as explain by Faa-Jeng and all (2002) or on the traditional mathematical algorithm as the MRAS system. Basing on the results presented on these applications, these algorithms can be classified in two categories. The MRAS is characterised by the time running rapidity, but without precession. The other algorithms, basing on the intelligent techniques are specified by the precession but with a lot of time calculation.

The running time rapidity is an important factor in each control loop, to make the decision in the exact time. Therefore the MRAS algorithm is taken in consideration in this work. The MRAS problems are generally related to the adaptive law characterise the adaptation phase of the adaptive model. Where, the idea is to guarantee the stability, basing on a stability theory.

As presented in the next section, the adaptive law will be expressed using the PID equations. But the difficulty is related to the P, I and D parameters must be adjusted, without a specific method. Generally, these parameters are adjusted online basing on the trial and error method. However as presented in Mir and all (1998) and Zerdali.and Barut (2013) the precession of the obtained results are not supported. In this situation, the presented solution in this work can solve the MRAS related problem.

4. PMSM PARAMETERS IDENTIFICATION

As described in the introduction section, the motor parameters variation can influence on the performance of the desired application based on the PMSM. Therefore the motor parameters variation identification is extremely required. As a solution for this problem, the model reference adaptive system MRAS is elaborated for online estimate these variations.

4.1 Parallel MRAS Architecture

Firstly, the basic scheme of the parallel MRAS configuration is presented in Figure 4.

The scheme consists of reference and adjustable (adaptive) models and an adaptation mechanism. The block "reference model" represents the actual system having unknown parameter values. The block "adaptive model" has the same structure of the reference one, but with adjustable parameters instead of the unknown ones. The block "adaptive law" estimates the unknown parameter using the error between

Figure 4. Model reference adaptive system

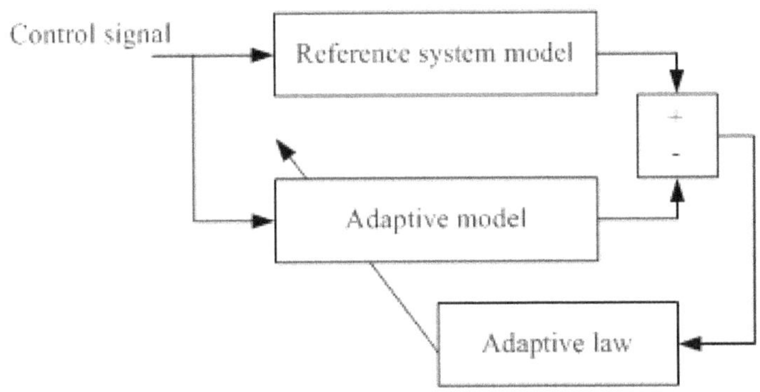

the reference and the adjustable models. It updates the adjustable model with the estimated parameter until the desired performance is achieved. The PMSM motor model was considered to be the reference model. Basing on POPOV theory, the adaptive law is built.

4.2 PMSM-MRAS Implementation

Using the online measurement of current, voltage and rotor speed, the stator resistance, inductance and the rotor flux linkage will be estimated simultaneously. The mathematical analysis of MRAS algorithm can be start by using the d and q stator current components as state variables. The PMSM state system takes the following form:

$$\begin{bmatrix} \dot{i}_d \\ \dot{i}_q \end{bmatrix} = \begin{bmatrix} -\tau & \omega \\ -\omega & -\tau \end{bmatrix} \begin{bmatrix} i_d \\ i_q \end{bmatrix} + \begin{bmatrix} c & 0 \\ 0 & c \end{bmatrix} \begin{bmatrix} v_d \\ v_q \end{bmatrix} + \begin{bmatrix} 0 \\ -e_f \end{bmatrix} \tag{5}$$

with $c = \frac{1}{L_s}$, $I_f = \frac{\lambda_m}{L_s}$, $e_f = \omega I_f$, $\frac{1}{\tau} = \frac{L_s}{R_s}$.

The adjustable parameter state system is given by:

$$\dot{\hat{X}} = \hat{A}\hat{X} + \hat{B}U + \hat{C} + G(\hat{X} - X) \tag{6}$$

with, $\hat{A} = \begin{bmatrix} -\hat{\tau} & \omega \\ -\omega & -\hat{\tau} \end{bmatrix}$; $\hat{B} = \begin{bmatrix} \hat{c} & 0 \\ 0 & \hat{c} \end{bmatrix}$; $\hat{C} = \begin{bmatrix} 0 \\ \omega \hat{I}_f \end{bmatrix}$ and $G = \begin{bmatrix} k_1 & 0 \\ 0 & k_2 \end{bmatrix}$.

G represents the correction gain matrix, where k_1, k_2 are two limited positive real, manually fixed. By subtracting the adjustable parameter system, equation (5), from the state system (4), the error function system can be defined in equation (6):

$$\dot{e} = \Big(\underbrace{A - \hat{A}}_{\Delta A}\Big) e + \Big(\underbrace{B - \hat{B}}_{\Delta B}\Big) U + \Big(\underbrace{C - \hat{C}}_{\Delta C}\Big) + G\, e \tag{7}$$

So, the non linear time variable feedback system which includes a feed forward linear model and a non linear feedback system expression is:

$$\dot{e} = \left(A + G\right) e + w \tag{8}$$

where, the non linear feedback system is:

$$w = -\Delta A \hat{X} - \Delta B\, U - \Delta C \tag{9}$$

The study of the system stability can be resolved upon the POPOV stability theory, where two necessary conditions must be satisfied like demonstrated by Liu and Zhang (2010) and Flah et al. (2011):

✓ The transfer function matrix of the linear forward block must be real and strictly positive. According to [Liu] the gain matrix G presented in equation (7) ensures this condition.

The nonlinear feedback block meets POPOV integral inequality presented as: $\int_0^{t1} w^T e dt \geq -\gamma^2$, where γ^2 is a positive constant at any time t.

By decompressing the second POPOV condition, as described in (10), three new circumstances are obtained as in (11).

$$\int_0^{t1} w^T e dt = \int_0^{t1} e^T w dt = -\int_0^{t1} e^T \left[\left(\Delta A \hat{X} \right) + \left(\Delta B\, U \right) + (\Delta C) \right] \geq -\gamma^2 \tag{10}$$

$$\begin{cases} \int_0^{t1} e^T (\hat{A} - A) \hat{X} dt \geq -\gamma_A^2 \\ \int_0^{t1} e^T (\hat{B} - B) U dt \geq -\gamma_B^2 \\ \int_0^{t1} e^T (\hat{C} - C) dt \geq -\gamma_C^2 \end{cases} \tag{11}$$

As demonstrated by Flah and all (2011), these lasts equations presented in the system (11) are satisfied if the adaptive laws of parameter designed as a PI style are applied as demonstrated in following equations:

$$\begin{cases} \Delta A = \int_0^{t} (e^T \hat{X}.f_{ir}. + e^T \hat{X}.f_{pr}) dt \\ \Delta B = \int_0^{t} (e^T.U.f_{il}. + e^T.U.f_{pl}) dt \\ \Delta C = \int_0^{t} (e^T.f_{ic}. + e^T.f_{pc}) dt \end{cases} \tag{12}$$

So, the adaptive parameters expressions are:

$$\begin{cases} \dfrac{\hat{R}_s}{\hat{L}_s} = -\left(k_{pr} + \dfrac{k_{ir}}{s} \right) \left(\hat{i}_d e_d + \hat{i}_q e_q \right) + \dfrac{\hat{R}_s}{\hat{L}_s}(0) \\ \dfrac{1}{\hat{L}_s} = \left(k_{pl} + \dfrac{k_{il}}{s} \right) \left(v_d e_d + v_q e_q \right) + \dfrac{1}{\hat{L}_s}(0) \\ \dfrac{\hat{\lambda}_m}{\hat{L}_s} = -\left(k_{pf} + \dfrac{k_{if}}{s} \right) \left(\omega e_q \right) + \dfrac{\hat{\lambda}_m}{\hat{L}_s}(0) \end{cases} \tag{13}$$

Three PI adaptation mechanisms are presented in the system (13). Their parameters are denoted by k_{pf}, k_{if}, k_{pl}, k_{il}, k_{pr} and k_{ir}, respectively to permanent magnet flux, stator inductance and resistance identifiers.

The saturation problem is commonly presented in the PI mechanism, therefore an anti-windup PI is used to eliminate this problem.

5. PSO AND BFO OPTIMIZATION ALGORITHMS

5.1 Optimisation Techniques in Literatures

Since the 90s, the numerically optimization problem are explored, due to the evolutionary of computer and calculator. All the problems of optimisation are related generally to the desired objective function must be optimised. The idea is to search the optimal solution, of all possible solutions, that can be implicated to maximize or minimize the objective studied function f.

In the literature different optimisation algorithms are studied and exposed for uses in diverse type of applications. Fouskakis was investigated three different type of stochastic optimisation algorithms in Fouskakis D. and Draper D. (2002): Simulated annealing (SA), genetic algorithm (GA) and Tabu search (TS), were he presents a comparative studying between these three versions. Fouskakis, was demonstrating the importance of each algorithm in a specified theme of application and presenting the advantage and disadvantage in each type of application presented. As example, the Tabu Search algorithm is comported the best in the problems involving the reconfiguration of radial distribution electric power networks. Genetic algorithm is the best in terms of the median energy of the solutions located. Etc…

Iztok and all (2013), was take in his work review a description of the different type of nature and Bio inspired optimisation algorithm as ant colony optimization, particle swarm optimization, cuckoo search, bat algorithm, and firefly algorithm. The authors were classifying these algorithms into two categories, physics algorithms and chemistry algorithms. Another list of all the optimisation algorithms in term of swarm intelligence or Bio inspired algorithms is also presented.

Recently, other optimisation algorithms are exposed for uses. As presented by Gheraibia and all (2013) the author takes in consideration the Penguins search optimization algorithm in the research of the space solution problems. The writer indicates that this version of the meta- heuristic algorithm performs better then other known methods. Other authors believe that the meta-heuristic optimization algorithms are comported the best one in all the optimisation presented problems and search the necessary modification can increase or ameliorate the global performance of this algorithm.

Yan and all (2012) present a novel version of the Bacterial forging Optimisation algorithm, where he base his novel BFO version on the social leaning to enhance the information sharing between bacteria. The author proves the efficiency of the proposed BFOLS face the Canonical BFO, PSO, and GA algorithms.

Basing on these researches and taken in consideration the simplicity and the ruining time of these algorithms, our choice is basing on the meta-heuristic algorithms in order to resolve the optimisation problems presented in our application. The BFO and PSO are elaborated, programmed and tested. A detailed description for each algorithm is given in the next sections for more explanation for the running principle.

5.2 Objective Function Related to the Estimation Problem

The optimisation problem characterised the application presented in this chapter, is based on two essential variables: the error and the overshoot of the studying signals. These parameters are fixed, after discussed the necessary parameters must be minimised and can represent the system performance. Different performance evolutionary criteria are proposed in the literature in order to depict or to define the objective function which can present the problem to be resolved. The integral square error (ISE) is characterised by the longer settling time, Integral of absolute value of error (IAE) allows larger deviation than ISE. Integral Time-weighted of absolute value of error (ITAE) is characterized by weights errors occurring later more heavily.

Basing on the comparison presented by Yogendra and all (2013), the related optimization problem is formulated, used the common several optimization cost functions, called: the Integral of the Squared Error (ISE) and the formulated objective functions will be minimized in this application are given in (14).

$$\begin{cases} \min f_{R_s} = \alpha_1 \int e_{R_s}^2 + \beta_1 \times overshoot_{R_s} \\ \min f_{L_s} = \alpha_2 \int e_{L_s}^2 + \beta_2 \times overshoot_{L_s} \\ \min f_{\lambda_m} = \alpha_3 \int e_{\lambda_m}^2 + \beta_3 \times overshoot_{\lambda_m} \end{cases} \tag{14}$$

Where, e_{Rs}, e_{Ls} and $e_{\lambda m}$ are respectively the errors between the reference and the estimated permanent magnet flux, stator resistance and inductance. The overshoots denoted by $overshoot_{\lambda m}$, $overshoot_{Rs}$ and $overshoot_{Ls}$ are respectively those related to permanent magnet flux, stator resistance and inductance estimation results. α and β are two positive value.

5.3 Particle Swarm Optimization Algorithm: PSO

5.3.1 Generality on PSO

In 1995, Eberhart was defined the social-psychological metaphor, presented in the birds of ships and insects movement to discover her food. This birds technique movement and it communication low are transformed by Kennedy and Russell (1995), into an optimization algorithm, called the particle swarm optimization algorithm. Effectively, this one is characterised by a population, called a swarm and each member of them is called a particle. In the PSO, each particle is placed in the parameter space of the problem and each one evaluates the fitness at its current location. So, each particle determines its movement through the space parameter by combining some aspect of the history of its own fitness values with those of swarm members. This movement technique can be illustrated by the equation (15).

$$\begin{cases} v_i^{k+1} = w_i v_i^k + c_1 . r_1 \times \left(pbest_i - x_i^k \right) + c_2 r_2 \times \left(gbest_i - x_i^k \right) \\ x_i^{k+1} = x_i^k + v_i^{k+1} \end{cases} \tag{15}$$

Where v_i^k, is the current velocity of the particle "i-th" at iteration "k", v_i^{k+1} is the updated velocity of the particle "i-th", x_i^{k+1} and x_i^k the updated and current position, w_i is the inertia weight of the "i-th" particle c_1, c_2, two positive constants, r_1, r_2, two random numbers between 0 and 1, "$pbest_i$ " is the best position on the "i-th" particle and "g_{best}" is the global best position.

5.3.2 PSO Principle

The algorithm principle is to minimize or maximize a desired parameter, called the objective function. So, starting with a randomly positions and velocities, each particle in the swarm flies through the searching space, to define the best position, that equivalent to the best objective function. This corresponding best particle informs the other entire particles by its best position. Then they adjust their own position and velocity based on this best location and the historical behaviour of the particle them self. More explanation is illustrated in the previous work presented by Flah and Sbita (2012).

5.4 Bacteria Foraging Optimization Algorithms: BFO

Since it is proposed by Passino, Bacterial optimization algorithm has drawn the attention in the major of optimization problems. Especially, face the conventional optimisation algorithms, like Genetic Algorithms (GA), Evolutionary Programming (EP) and evolutionary Strategies (ES). This algorithm (BFO) seems is global function mode as searching for nutrients in a manner to maximize or minimize a desired objective goal. BFO is characterised by the communication between the elements by sending signals. A bacterium takes foraging decisions after considering two previous factors, basing on its movement type.

According to the Escherichia coli (E coli), the bacterial foraging optimization support a new method based on foraging behaviour for solving optimization problems. So according to the bacteria behaviour law, all the elements support the best species which have the ability for searching food and eliminate the other. The bacteria foraging strategy can be expressed generally by three essential processes as explain by Madasu and Prakash (2009).

The first one is the chemotaxis step; the second is called the reproduction and the last is the elimination-dispersal step

5.4.1 Chemotaxis Step

This step is extremely important in the explanation of the bacterial movement, where it is possible to have two different ways. This bacterium can swim in one direction without turn, or it can be tumble in many directions. But it is possible that this bacterium can alternate these two movement type. So we can conclude that, the bacterium is said swimming if it moves in a predefined direction and tumbling if it moves in an altogether different direction.

E. Coli was defined the bacterium position parameter by $\theta_i(j, k, l)$, that represents the position of the i-th bacterial in the "j-th" chemotaxis step, the "k-th" reproduction step and the "l-th" elimination step.

The updated bacterial position $\theta_i\left(j+1,k,l\right)$ can be expressed as a sum of the last bacterium location $\theta_i\left(j,k,l\right)$ and the variation of the bacterium position δ_θ, which depends to the movement method, swim or tumble. The bacterium parameter variation δ_θ depends essentially on C(i) factor which can be expressed as $C(i)=C(i)+d_1$ in the swim phase and $C(i)=C(i)-d_2$ in the tumble phase. Where, the parameters "d_1" and "d_2" are two positive elements.

Finally, the corresponding global bacterium position expression can be then expressed as indicated in formula number (16).

$$\theta_i(j+1,k,l)=\theta_i(j,k,l)+C(i)\frac{\Delta s(i)}{\sqrt{\Delta s^T(i)\Delta s(i)}} \tag{16}$$

Δs (i) depicts the vector direction of the "i-th" chemotaxis step. When the bacterial movement is run or swim, Δs(i) takes the same that was available in the last chemotaxis step; otherwise, Δs(i) is a random vector elements in the interval [-1, 1]. This swim action is continued as long as it continues to reduce the cost, but only up to a maximum number of steps, Ns. If not, this action will be stopped.

5.4.2 Reproduction Step

After assured that all the bacteria are executing the entire chemotaxis step, the reproduction phase will be start. Effectively, this action is characterized by, the reproduction of the best bacteria half. Where, the worst bacteria half will be die. However, the other best half split. The obtained copies are placed at the same location as their parent.

5.4.3 Elimination and Dispersal Step

After the reproduction step achieved, the last one which called "elimination-dispersal" will start. Effectively, with a probability "P_{ed}", an elimination and dispersion for each bacterium (this keeps the number of bacteria in the population constant) occur as explain by Ajith and Amit (2009), Madasu and Prakash (2009). The elimination and dispersal events assist chemotaxis progress by placing the bacteria to the nearest required values. Each bacterium according to a fixed probability dispersed from their original position and move to best position within the search space.

5.4.4 BFO Algorithm Description

Figure 6 shows the global identification scheme, when the optimisation algorithm is used for adjusting the MRAS parameters. Generally, the first step is the randomly choosing of the bacterium initial positions, the bacteria number, the chemotaxis step number "j" and the elimination and reproduction steps numbers, "l" and "k" respectively. In the first elimination loop, the algorithm will start and take in consideration the first chemotaxis step in the first reproduction phase. For each bacteria, the corresponding objective function "J" will be evaluated and saved in algorithm memoir. The bacterium will start moving now with the tumbling mode, until the tumbling steps are finished, then the bacterium turn to the swim mode. In each swim action, the objective function will be evaluated and compared to the last know objective function value. If the new obtained objective function is smaller then the last one, this new value is saved and the last one is deleted. This action will continue until the swim steps are finished. After evaluated the objective function for each bacteria, the reproduction step will start, where all the objective function will be classed in increasing whey, from the smaller to the bigger. Then the algorithm will remove the first half of bacterium that corresponds to the bigger value of "J". These actions will continue until the algorithm executes all the entire loops of chemotaxis, reproduction and elimination phase. Figure 5, shows the BFO architecture.

5.5 Optimization Algorithm Implemented on MRAS

The implementation of the desired computational algorithm on the MRAS system can be illustrated as presented in Figure 6. The optimisation algorithm will update online the PI parameters K_i and K_p and the MRAS algorithm will adjust the equation (13).

6. PMSM PARAMETERS ESTIMATION RESULTS

6.1 Description of the PMSM Control Scheme

In this part, the obtained results for the estimation phase, are exposed and discussed, using the MRAS estimator with the PSO and BFO algorithms. As presented in Figure 7, the vector control strategy is used as the motor control drive. Due to the simplicity via torque control method and the precession via the scalar control method. The vector control technique is generally characterised, by three essential PI blocs. The first one is related to the Speed or torque control and the others for the quadrature and direct stator currents. Effectively, after compared the real and the desired speed value, a PI controller is added for generating the transversal stator reference current. This current signal is also compared with the measured one. The obtained error is injected in another PI controller for generating the transversal reference voltage. Generally, the reference direct stator current is fixed to zero. This last will compared to the real one and the obtained error is considered as the input signal for the third PI controller to generate the direct reference stator voltage. The obtained reference voltages in the rotating reference frame are noted by: v_d* and v_q*, where " d and q" present the rotating reference frame. A transformation to the fixed reference frame, " α and β" is applied by a conversion bloc. The obtained voltage signals are injected in the Pulse with Modulation bloc, PWM, for generating the six reference signal to the inverter. This last generate the three reference voltages, "v_a*, v_b* and v_c* " to the motor. The speed information will obtained with a speed sensor. The current signals, in the rotating reference frame, are obtained after measured the three real stator currents, with a transformation bloc, from the fixed frame to the rotating one. Digital simulation test based on Matlab/Simulink software package is carried out.

The characteristic of the used motor in this work are as follow: The motor power is 400W, the rated current is equivalent to 2A and the maximum supported current can reach two time of the rated one. The rated speed is 2500rpm and the inertia coefficient is 85e-6kg.m^2. The motor parameters characteristics are given also by the constructor, the stator resistance is 5.2Ω, the permanent magnet flux is 0.24wb and the stator inductance is 21.5mH. It is important to indicate that the motor is a salient one. For the inverter parameter, the DC bus voltage, U_{DCmax} =660V and the switching frequency is 10 KHz.

Figure 7, presents the overall control system with the online MRAS estimator. The implemented MRAS bloc is as presented in Figure 6.

6.2 Results Discussions

After implemented the MRAS model in parallel with the used motor, the idea is to affect some study analysis on the PMSM parameters variation under the nominal speed. Therefore, the estimation tests are exposed under a variable given speed as illustrated in Figure 8, where the speed target start from 0rpm to the rated one, 2500rpm. PMSM parameters modifications are implemented on the system as follow: The stator resistance is modified at t=0.14 sec, where the stator resistance value is increased to 7.7 Ω. At t= 0.1sec, the stator inductance is modified to 0.0225 H and the magnet flux to 0.239 Wb.

6.2.1 Results in Two PI Cases

As explained previously, the difficulty is to fix the six PI parameters presented in equation (13). In this case the industrial chemical technique is basing on, for solve the cited difficulty. This technique is known by trial and error method. Its principle is to online adjust the PI parameters and observe the system output evolution. Basing on these outputs, the operator must adjust these PI values.

The results, illustrated in Figures 9, 10 and 11, are obtained under two PI parameters examples. These results show the MRAS estimation comportments. Table 1 presents the used PI parameters in the three equations in the system (13).

The obtained results are characterised by the large charting or by the large time convergent, when the PMSM parameters are varied. In Figure 9 and at the first case test, the estimate stator resistance touch the 9.4Ω at t=0.14 sec, however the real stator resistance modification is 7.7Ω. In the second case it is clearly observed, that the time convergence is high than 0.5 sec.

In Figure 10, the second case condition show good results with a minimum chattering at t=0.1 sec, which indicate that the selected PI parameters are acceptable. However in the second case the large chattering in the system output are observable, and the time response can be accepted in this case.

In Figure 11, the permanent magnet flux estimation is illustrated; the obtained results can be accepted for the two test cases. But these presented chattering must be minimised.

These results can be accepted if the used or the controlled system is not characterised by the rapidity and the precession. However, in many control application, as the electrical vehicle, militaries applications (radar), the rapidity and the precession are essential to assure the good performance and the exactly convergence.

6.2.2 BFO Results

In the last part, the obtained MRAS results at two PI cases are discussed. The influence of these PI parameters on the MRAS estimator comportment is clearly. In this part, the idea is to apply the cited previous optimisation algorithms (BFO) to resolve the problem. Before running the algorithm, two essential steps must be adjusted. Firstly it is important to define the optimisation functions then the algorithm must be initialised. In the objective function section, the optimisation equations are defined. In Table 2, the BFO configurations are made.

For the given BFO configuration, it's clear that four chemotaxis steps are presented and six bacteria are chosen. For more understanding the BFO algorithm running it is important to present and discuss, the bacteria positions movement face the measured objective function. In the Figure 12, we visualize the obtained objective function value for each bacterium in each chemotaxis step in the stator resistance

Figure 5. Architecture of the BFO algorithm

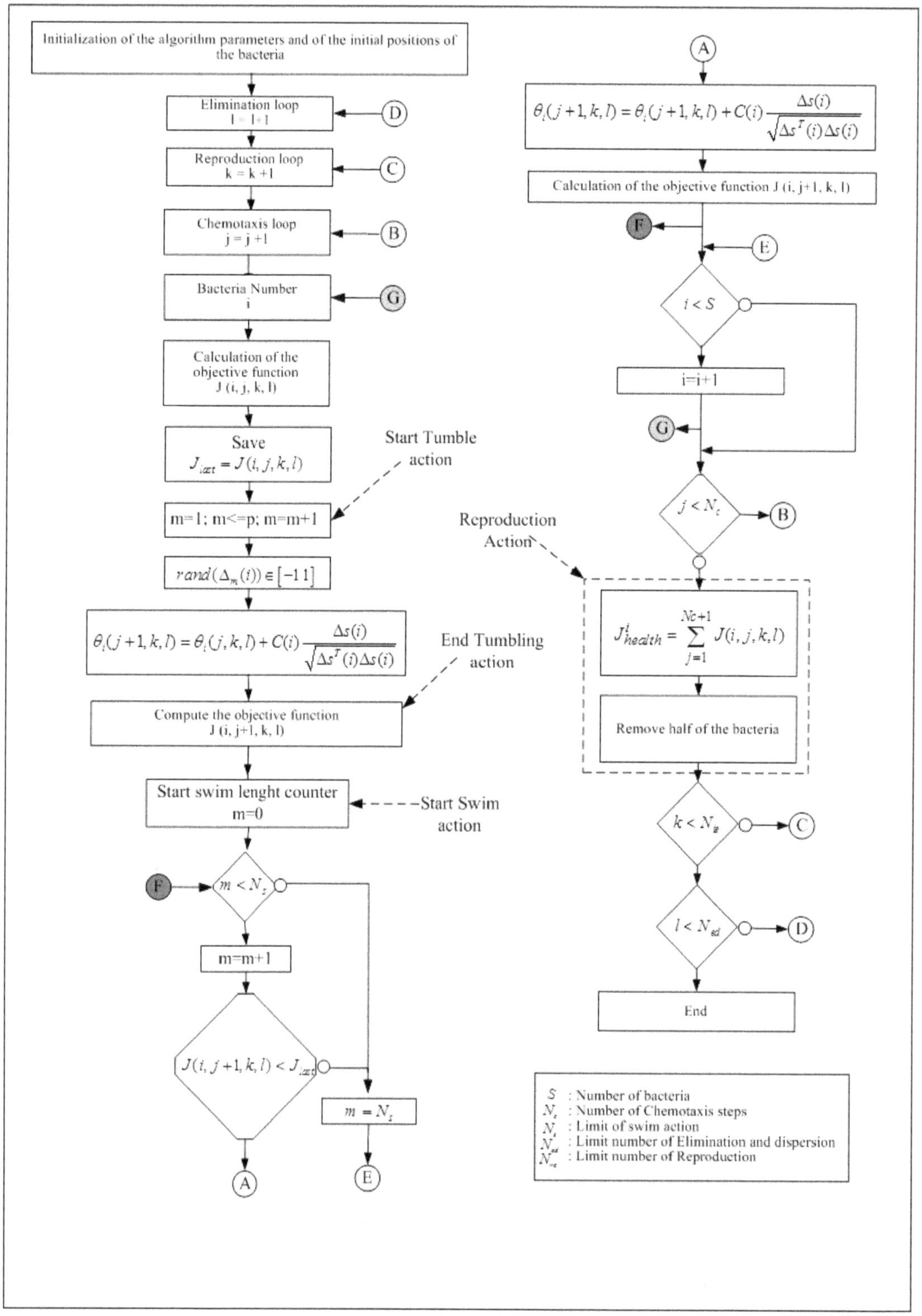

Figure 6. Intelligent MRAS basing on a computational algorithm

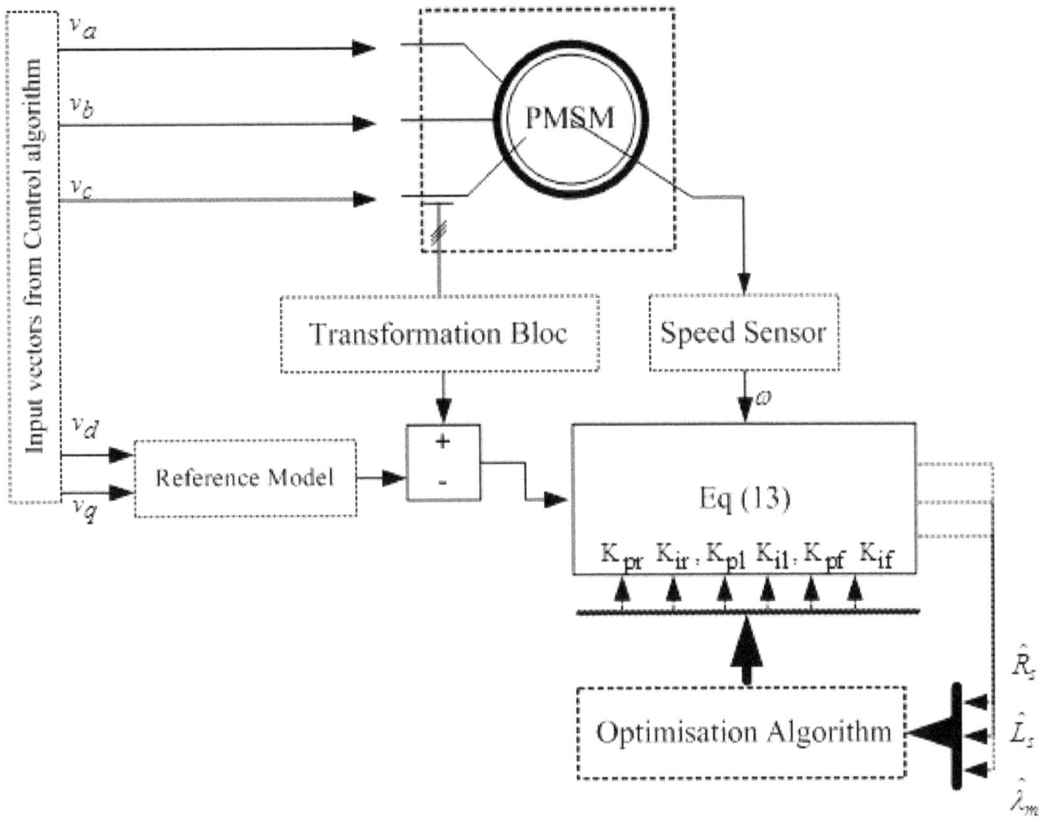

Figure 7. Block diagram of the PMSM control system

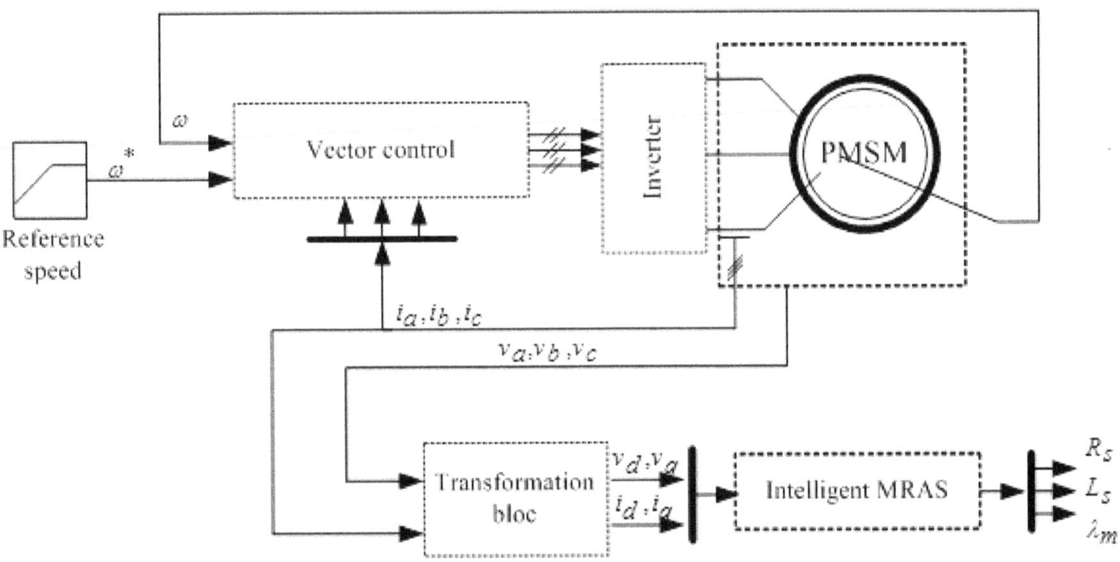

Figure 8. Behaviour of a PMSM speed results

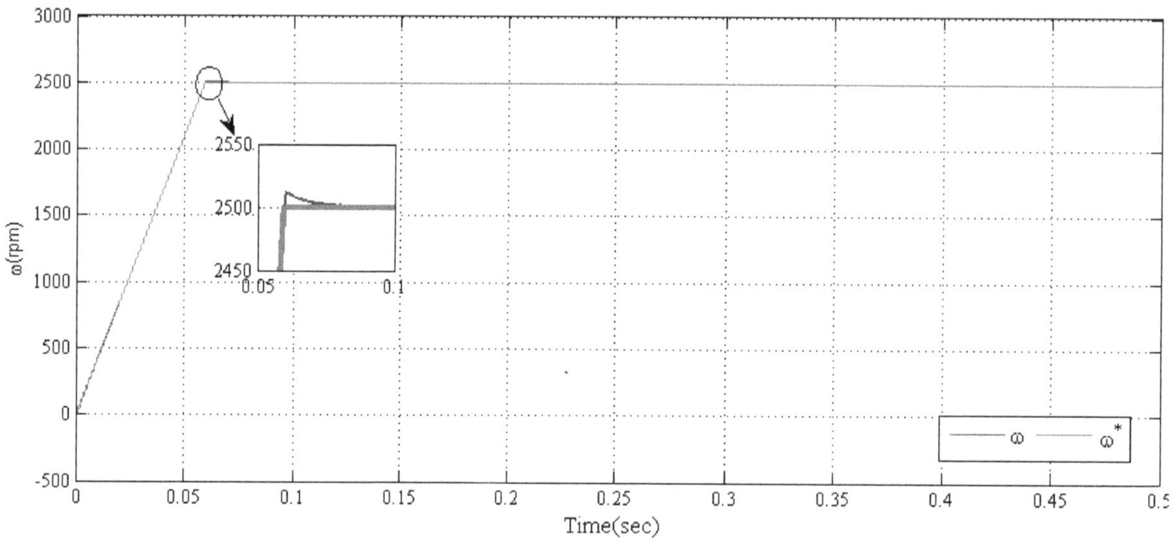

Figure 9. Stator resistance

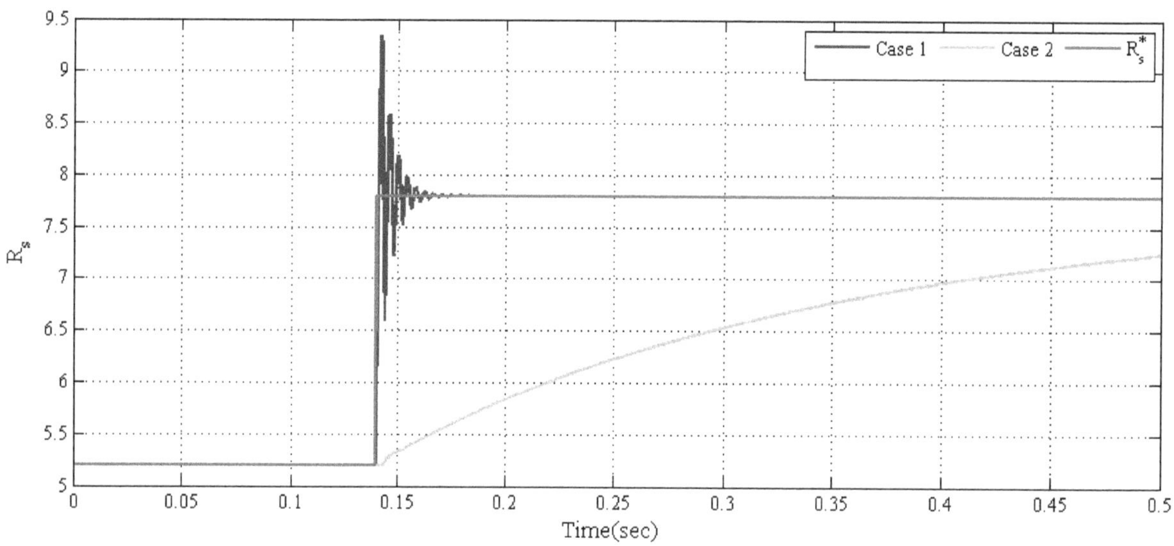

estimation case. The objective function value is minimised during the running algorithm. Each bacterium calculates the corresponding objective function, and the fourth bacterium presents the best comportment in each chemotaxis step. Therefore, the corresponding Bacteria position will take, as the best PI value. Figure 13, shows the corresponding PI parameters variation, that equivalent to the bacterium positions.

Figure 14, shows the objective function evolution in the stator inductance optimisation equation. It's clear, that the objective function value is approximated to zero. This is due to the initial bacteria position close to the optimum one. In the last chemotaxis step, it is observed that the sixth corresponding bacterium gives the best solution. The corresponding PI parameters variation is illustrated in Figure 15.

Figure 10. Stator inductance

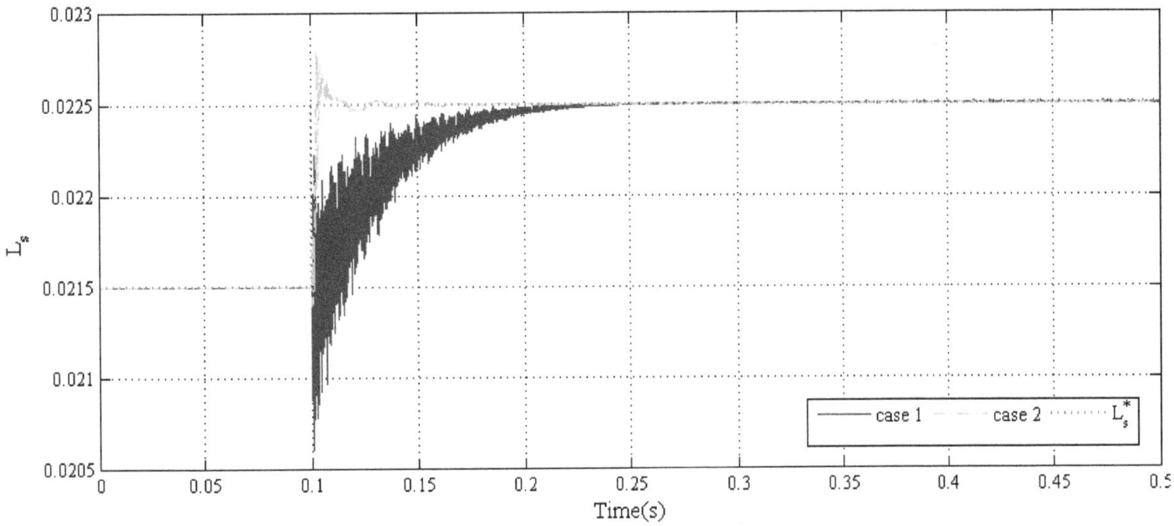

Figure 11. Permanent magnet

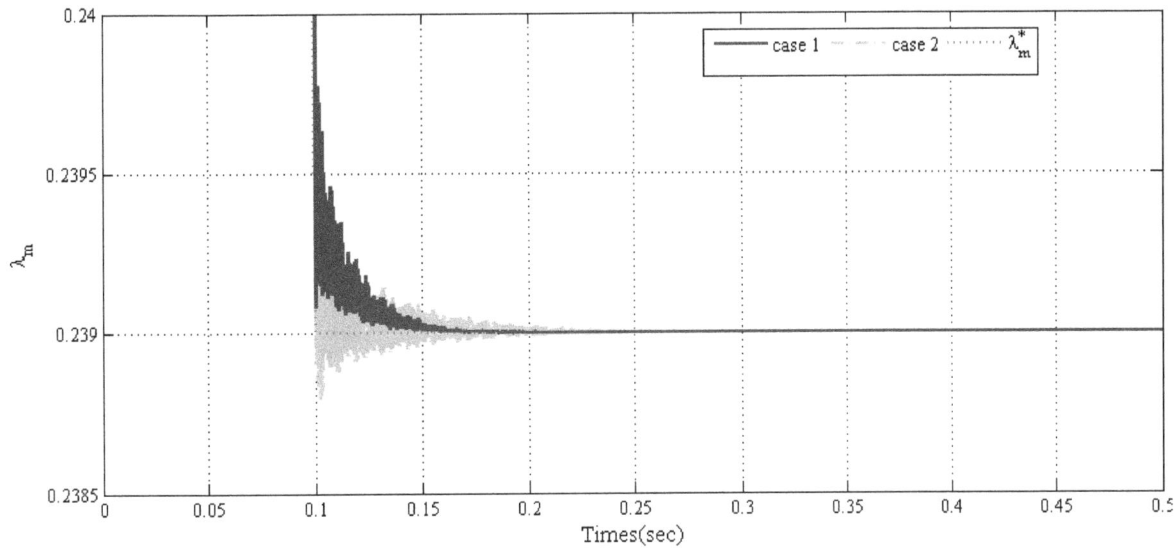

Table 1. PI parameters adjusting by trial and error method

	First case		Second Case	
	K_p	K_i	K_p	K_i
R_s	50000	10	500	2
L_s	1000	0.5	100	0.001
λ_m	10	0.1	15	0.5

Table 2. BFO parameters initialisation

Bacterial Foraging Optimisation	
Bacteria	s = 6
Chemotaxis	Ne = 4
Swim length	Ns = 4
Reproduction	Nre = 2
Elimination-dispersal	Ned = 2
Probability	P_{ed}=0.25
Run length	c= 0.05
d_1	0.01
d_2	0.01
α1, α2, α3	10
β1, β2, β3	10

Figure 12. Objective function corresponding to the stator resistance case

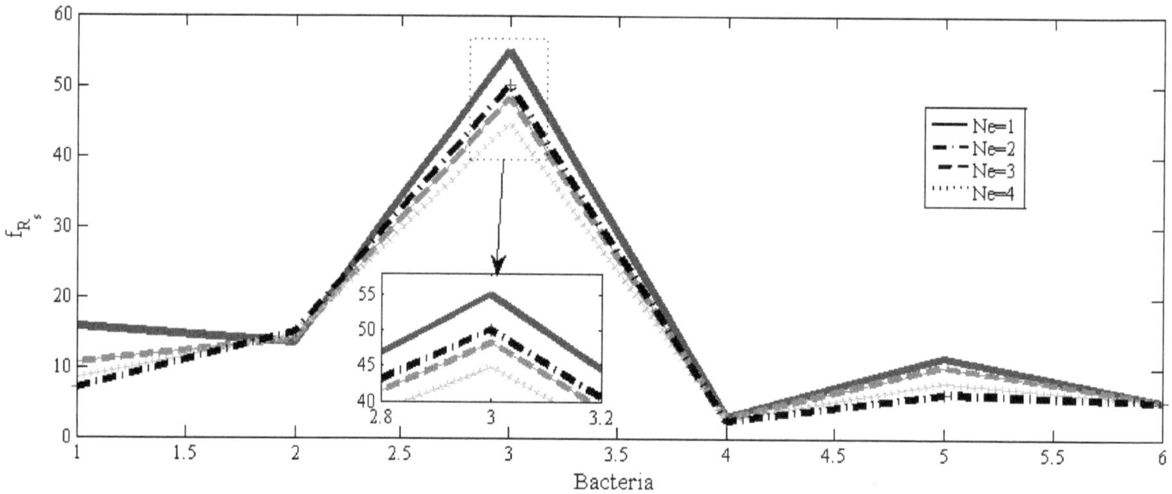

In Figure 16, the objective function evolution, proportional to the chemotaxis step, is illustrated. As previously, the objective function is stable to 0.1004; it is approximately, the minimum find objective function. A less deviation is presented by the bacteria, but without influence. In Figure 17, the corresponding PI parameters are pointed up.

6.2.3 PSO Results

As explained previously, the used algorithm must be adjusted before running. The PSO initialisation is done as explained in Table 3. Three Particles are used for one hundred steps. Each particle will execute 100 steps to define the best objective function. Contrary to the previous case; each bacterium can give her corresponding objective function, in this case the idea is to search the best global objective function in each step.

Figure 13. PI Positions corresponding to the objective function related to the stator resistance case

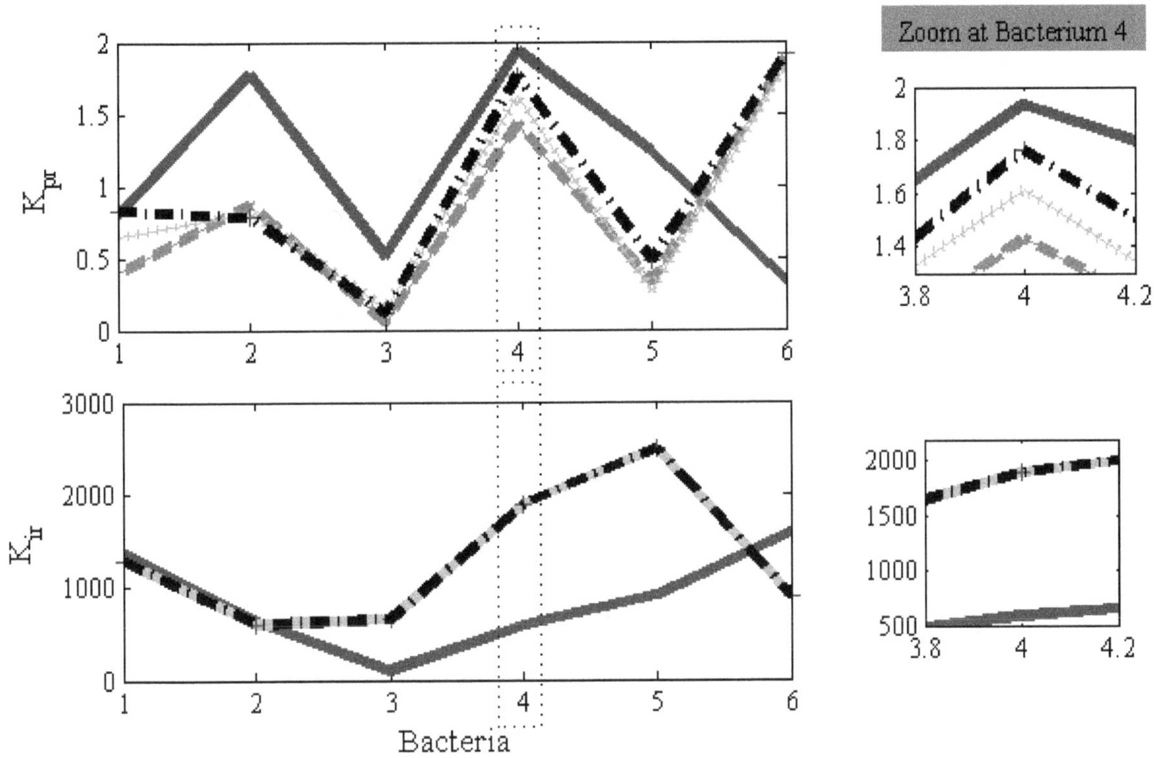

Figure 14. Objective function corresponding to the stator inductance case

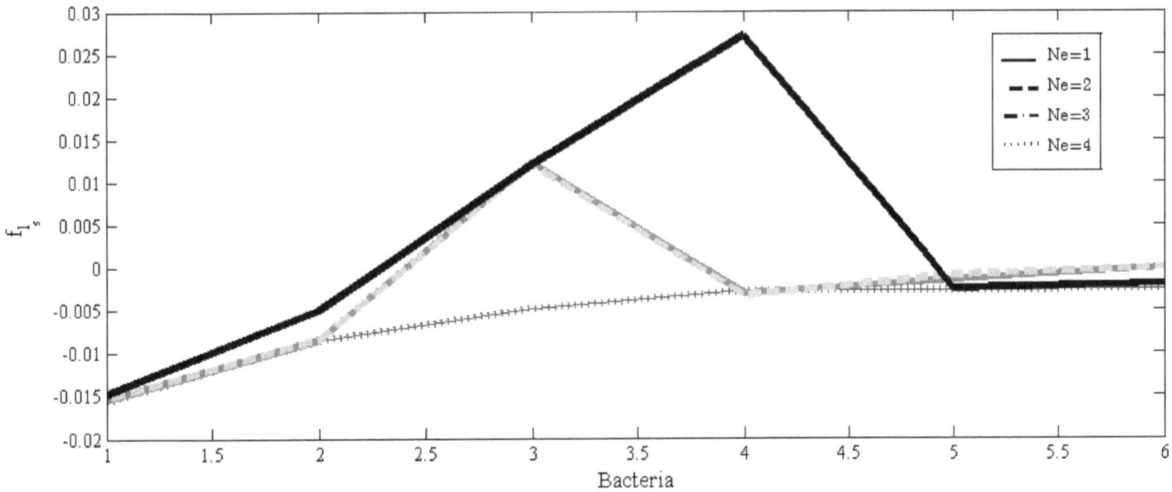

In the programming phase, we have found that the PSO algorithm is simpler than the BFO in the architecture. Where only one loop is required, however in the BFO case three loops characterise our algorithm, which make the algorithm complex. After implemented this method, the results are classified into two parts, the objective function results and the corresponding PI parameters modification.

Figure 15. PI Positions corresponding to the stator inductance case

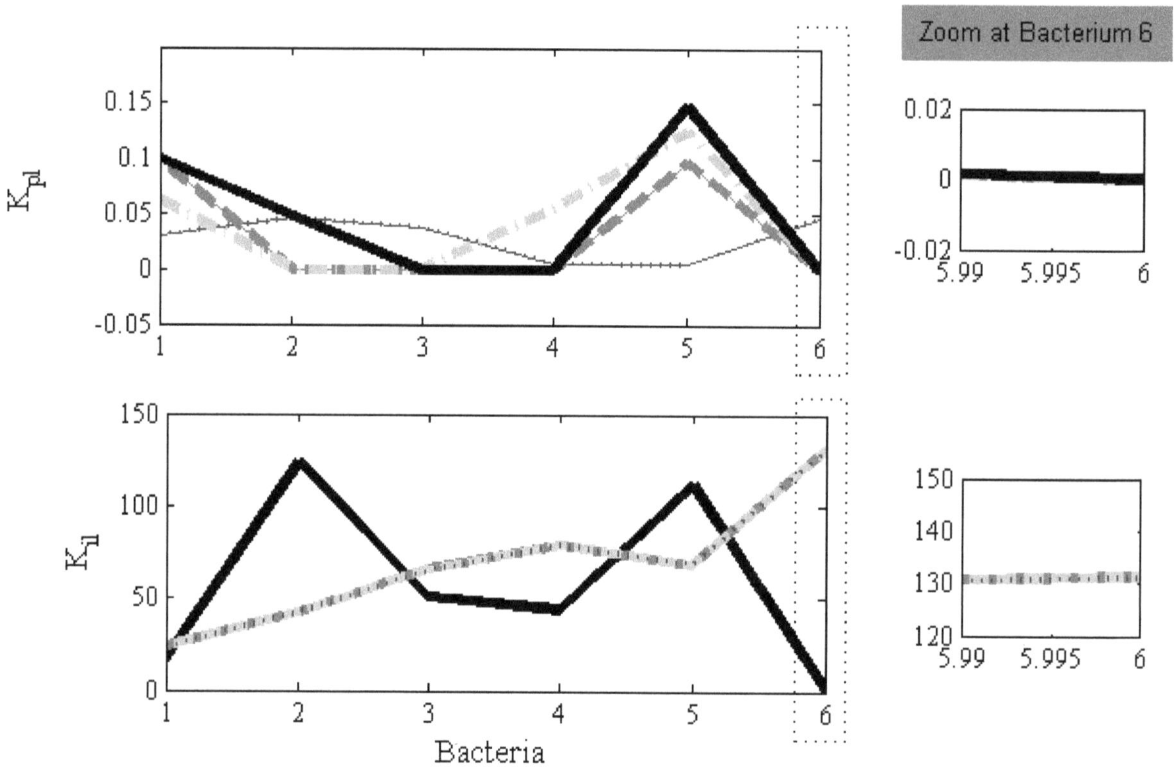

Figure 16. Objective function corresponding to the magnet flux case

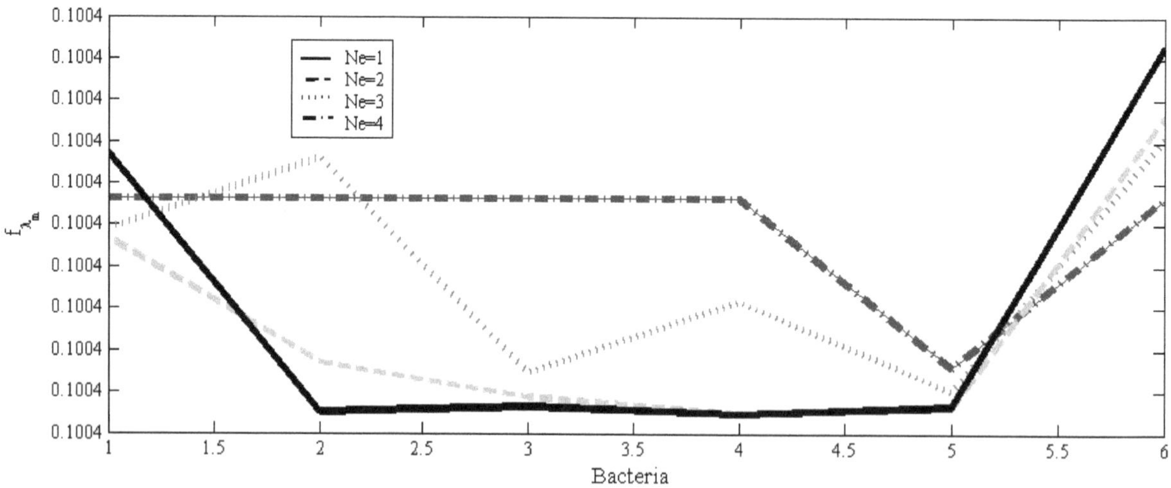

Figure 18, shows respectively the evolution of the objective function value for the stator resistance, stator inductance and magnet flux estimation cases. It is remarkable that the objective function related to the stator resistance and inductance estimation cases is smaller, then the BFO method. However in the magnet flux case, the BFO comportment is look better then the PSO. Mathematically, it is difficult

Figure 17. PI Positions corresponding to the magnet flux case

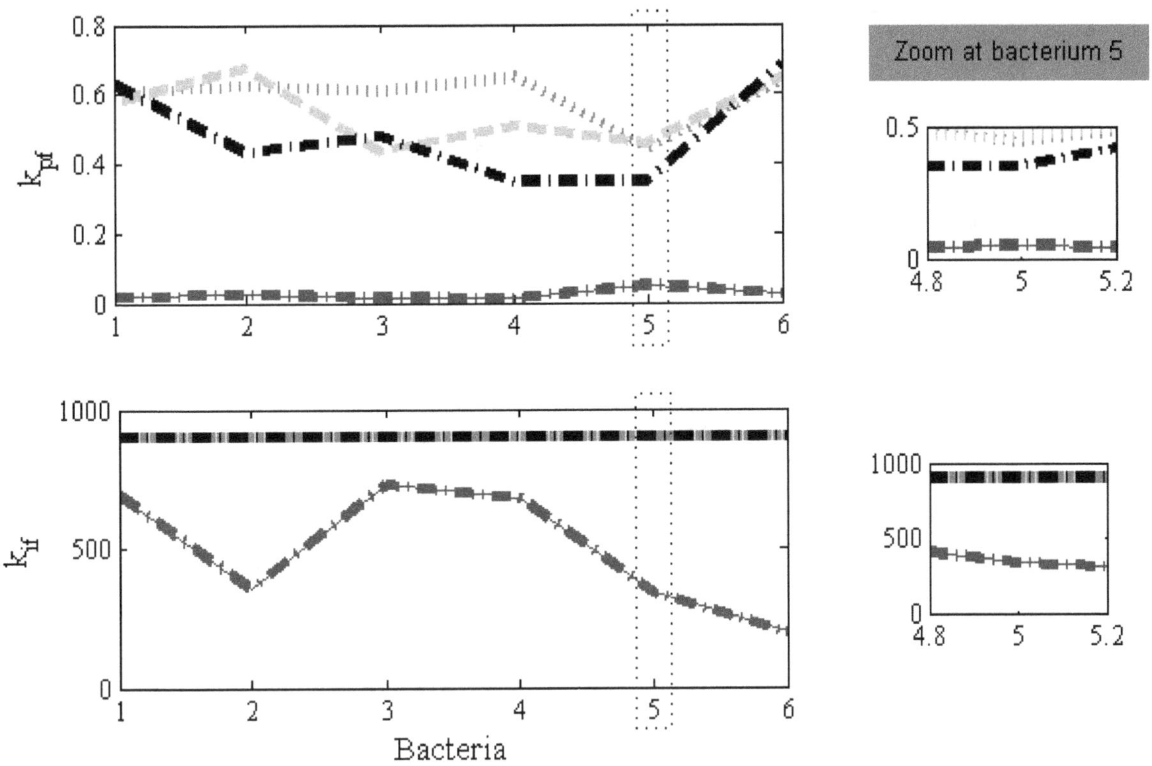

Table 3. PSO parameters initialisation

Particle Swarms Optimisation	
Particles	n=3
Birds steps	100
c1	1.2
c2	0.012
Inertia for each particle	w=0.9
r1	random
r2	random
α1, α2, α3	10
β1, β2, β3	10

to demonstrate these results. But after many tests, the conclusion can be depict off, is that the particles and the steps number, have a direct influence on the algorithm comportment. Where, when the steps and bacterium number are increased, the objective function results will be ameliorated.

The corresponding PI parameters evolutions are illustrated in Figure 19. It is clear that from the fifteen steps the particles found the best position that equivalent to the minimum objective function in her situation. These results demonstrate the rapidity and the stability of this algorithm.

Figure 18. Objective function cost with the PSO algorithm

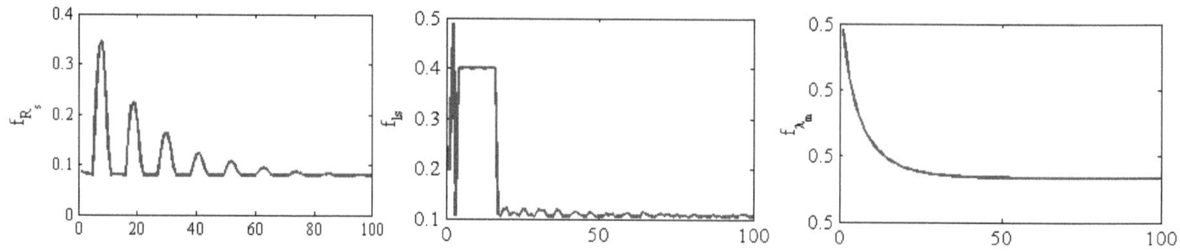

Figure 19. PI parameters evolution corresponding to the Objective function cost with the PSO algorithm

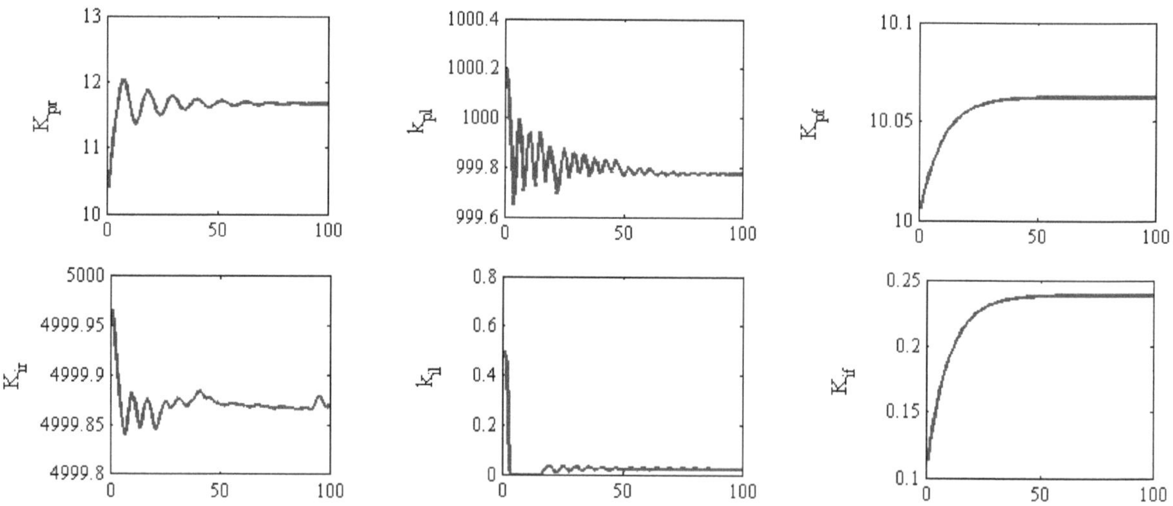

6.2.4 BFO and PSO Methods on MRAS: Comparative Discussion

In this part, the obtained results on the estimation phase after applied the BFO and the PSO algorithms are discussed. The total running time system is 0.5sec. At t= 0.14 sec a stator resistance variation is made in order to touch the 7.3 Ω. At t=0.1sec, the stator inductance modification is applied. The new value comes to 0.0225 H. The magnet flux is also modified at t=0.2sec. Really it is impossible that the magnet flux increase due to its physical characteristics, therefore this value is decreased by 0.01wb.

After applied the MRAS-PSO and MRAS-BFO versions on the global control system under these conditions and under the given speed, the obtained results are presented in Figures 20, 21 and 22. In Figure 20, the obtained results look more acceptable in rise time with the PSO method then with the BFO. In this case, the PSO algorithm seems better then the BFO. In the other results, Figure 21 and Figure 22, the BFO comportment seems better then the PSO.

This discussion is related to the precession of the estimation phase; however other terms are important to find the best method applied in this estimator. Table 4; compare theses two methods in numerical numbers.

Figure 20. Stator resistance estimation with BFO and PSO methods

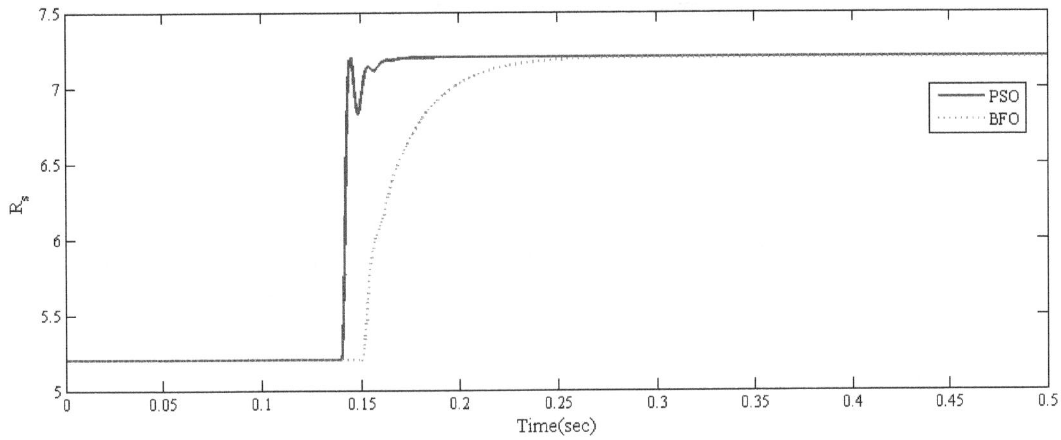

Figure 21. Stator inductance estimation with BFO and PSO methods

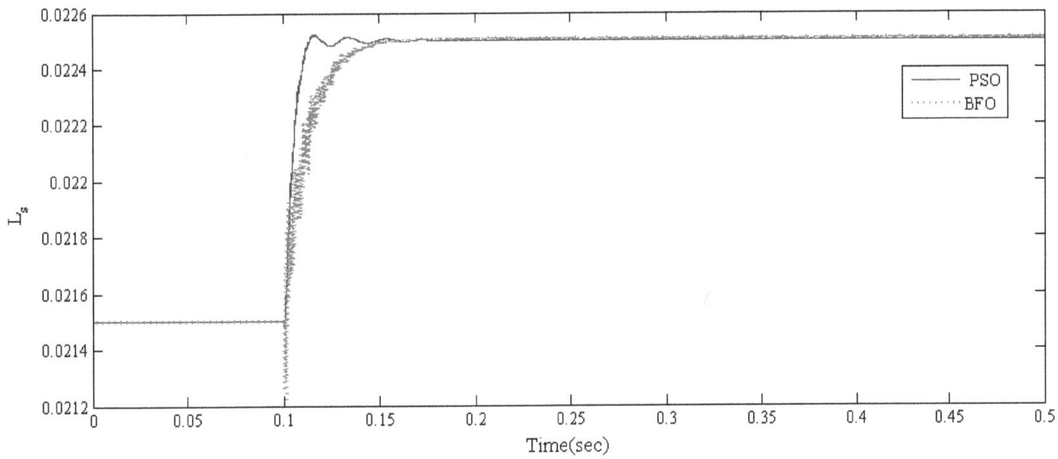

Figure 22. Permanent magnet flux estimation with BFO and PSO methods

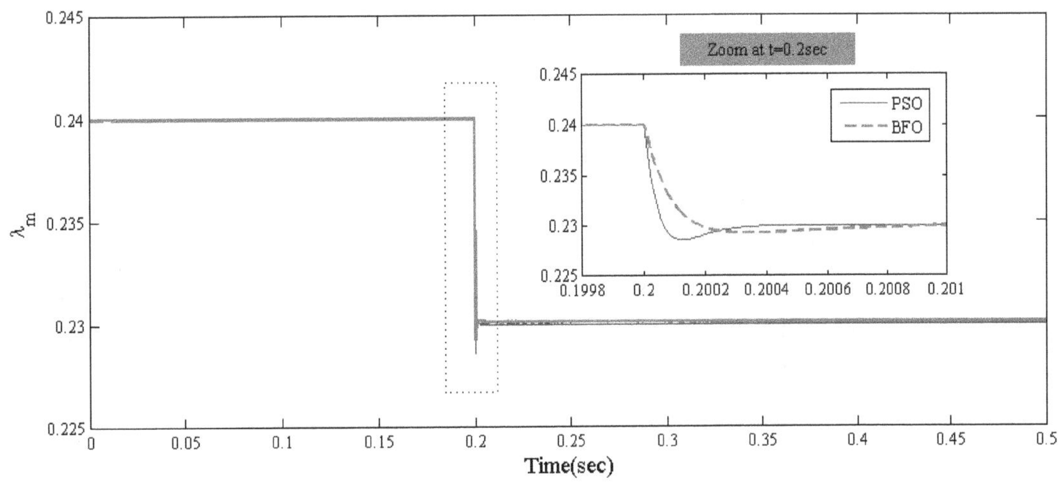

Table 4. BFO and PSO comparison

	PSO	BFO
Rise time	0.01	0.05
overshoot	0.0001	0
Running time	≈1800 sec	≈7200 sec

The efficiency of the BFO face to the PSO algorithm is proved in precession factor, but a running time disadvantage make this method is hardly applied in the real time system. Therefore an outline application is commonly required with the BFO method.

The BFO algorithm proves its efficiency to found an optimum PI values. The BFO rise time problem can be more ameliorated if a less variation on the previous cited BFO configuration, as bacteria number, chemotaxis number, swims length or reproduction loop number is applied.

7. CONCLUSION

In this chapter, an estimation technique is exposed and used on the motor parameters estimation problems. The studied motor is the permanent magnet synchronous motor, where three essential parameters characterise it. The stator resistance, the stator inductance and the permanent magnet flux on the rotor. As presented in this chapter, the used method for the estimation phase is the Model Reference Adaptive System. The problems related to this estimator, PI parameters adjusting, are taken in consideration. Therefore, in this chapter a new version of this estimator basing on an optimisation algorithm is presented, for resolving the cited previous problems. The Bacterial Foraging optimization algorithm and the Particle swarm optimization algorithm are these two techniques. The efficiency of the PSO and BFO methods is validated with the obtained results. The dominance of the BFO method, in the precision, and the superiority of the PSO algorithm in the time running are the general characteristics of these computational algorithms on this application. Generally, this work can be more ameliorated, with other comparative study as the genetic algorithm, neurone techniques, etc. And the exposed MRAS adjusting method can be also implemented on the Model Reference Adaptive Control, MRAC, for resolving the control problem related to this theory.

REFERENCES

Ajith, A., & Amit, K. (2009). On Stability of the Chemotaxis Dynamics in Bacterial-Foraging Optimization Algorithm. *IEEE Transactions on Systems, Man, and Cybernetics. Part A, Systems and Humans*, *39*(3), 670–679. doi:10.1109/TSMCA.2008.2011474

Bae, B., Patel, N., & Schulz, S. (2003), New Field Weakening Technique for high saliency Interior Permanent Magnet Motor, *Industry Applications Conference, 2003. 38th IAS Annual Meeting*. vol.2. 898-905. doi:10.1109/IAS.2003.1257641

Balluq, E., Acarnley, P. P., & Atkinson, D. J. (2004). Adaptive tuning of a PI speed controller for a brushless DC motor: optimum speed control using a neural network. *IEEE International Symposium on Industrial Electronics,* vol.1, pp. 459 - 464.

Ben Hamed M., Sbita L. (2008). Fractional order speed observer for sensorless induction motor drives, *International Review of Electrical Engineering IREE,* vol. 3,(1).

Bilal, A., Umit, O., Aydin, E., & Mehrded, E. (2004, November). A Comparative Study on Non-Linear State Estimators Applied to Sensorless AC Drives: MRAS and Kalman Filter. *30th Annual Conference of IEEE Industrial Electronics Society*, pp. 2148 - 2153 vol. 3, Busan, Korea

Bolognani, S., Zigliotto, M., & Unterkofler, K. (1997). On-line parameter commissioning in sensorless PMSM drives, *Proceedings of the IEEE International Symposium on Industrial Electronics,* Vol.2, pp 480-484, Guimaraes. doi:10.1109/ISIE.1997.648995

Chen, Z., Yang, L., Zhang, Y., & Li, Z. (2012, July). A control method of PMSM current-loop based on on-line parameter identification, *7th IEEE Conference on Industrial Electronics and Applications (ICIEA),* pp.417 – 422, Singapore. doi:10.1109/ICIEA.2012.6360763

Emadi, A., & Rajashekara, K.Young Joo Lee. (2008). Power Electronics and Motor Drives in Electric, Hybrid Electric, and Plug-In Hybrid Electric Vehicles. *IEEE Transactions on Industrial Electronics, 55*(6), 2237–2245. doi:10.1109/TIE.2008.922768

Faa-Jeng, L., Chih-Hong, L., & Po-Hung, S. (2002). Self-constructing fuzzy neural network speed controller for permanent-magnet synchronous motor drive. *IEEE Transactions on Fuzzy Systems, 9*(5), 751–759. doi:10.1109/91.963761

Faa-Jeng, L., Po-Kai, H., & Wen-Der, C. (2007). A genetic algorithm based recurrent fuzzy neural network for linear induction motor servo drive. *Journal of the Chinese Institute of Engineers., 30*(5), 801–817. doi:10.1080/02533839.2007.9671307

Flah A., Ben hamed M. and Sbita L. (2011). Online MRAS-PSO PMSM parameters estimation. *International review on modeling and simulations*, 4 (3), 980-987.

Flah A., Sbita L. (2012). An adaptive high speed PMSM control for electrical vehicle application. *Journal of electrical engineering* 12 (3).

Fouskakis, D., & Draper, D. (2002). Stochastic optimization. *International Statistical Review, 70*(3), 315–349. doi:10.1111/j.1751-5823.2002.tb00174.x

Gheraibia, Y., & Moussaoui, A. (2013). *Penguins Search Optimization Algorithm (PeSOA)* (Vol. 7906, pp. 222–231). Recent Trends in Applied Artificial Intelligence Lecture Notes in Computer Science.

Iztok J., Fister Y., Xin-She F., Iztok B. and Duˇsan F., (2013). A Brief Review of Nature-Inspired Algorithms for Optimization *ELEKTROTEHNI ˇSKI VESTNIK*. 80(3), 116–122.

Kennedy, J., & Russell, E. (1995, Nov/Dec). Particle Swarm Optimization. Proceedings of the *IEEE International Conference on Neural Networks*, vol.4, pp.1942-1948, Perth, WA.

Kojabadi, H. M. (2009). Active power and MRAS based rotor resistance identification of an IM drive. *Simulation Modelling Practice and Theory*, *17*(2), 376–389. doi:10.1016/j.simpat.2008.09.014

Kraiem H. and Messaoudi M. (2009). Ekf- Based Sensorless Direct Torque Control Of Permanent Magnet Synchronous Motor: Comparison Between Two Different Selection Tables, *Transaction in Machine, Power electronics and Drives*, 1(7). 159-165.

Krishnan, R. (2002). *Electric motor drives Modeling, Analysis, and Control*. Upper Saddle River, New Jersey: Prentice Hall.

Kumar, R., & Gupta, R. A. and Ajay Kr., (2007). Identification and Control of PMSM Using Artificial Neural Network. *IEEE International Symposium on Industrial Electronics*, pp. 30 – 35, Vigo. doi:10.1109/ISIE.2007.4374567

Kyo, B. L., & Frede, B. (2006). Reduced-Order Extended Luenberger Observer Based Sensorless Vector Control Driven by Matrix Converter With Nonlinearity Compensation. *IEEE Transactions on Industrial Electronics*, *53*(1), 66–75. doi:10.1109/TIE.2005.862313

Lakshmi, G. S., Kamakshaiah, S., & Das, T. R. (2013), Closed loop PI control of PMSM for hybrid electric vehicle using three level diode clamped inverter for optimal efficiency. *International Conference on Energy Efficient Technologies for Sustainability (ICEETS)*. 754 – 759, Nagercoil. doi:10.1109/ICEETS.2013.6533479

Lascu, C., Boldea, I., & Blaabjerg, F. (2004). Variable-Structure Direct Torque Control - A Class of Fast and Robust Controllers for Induction Machine Drives. *IEEE Transactions on Industrial Electronics*, *51*(2), 354–363.

Liu, K., Zhang, Q., Zhu, Z.-Q., Zhang, J., Shen, A.-W., & Stewart, P. (2010). Comparison of two novel MRAS based strategies for identifying parameters in permanent magnet synchronous motors. *International Journal of Automation and Computing*, *7*(4), 516–524. doi:10.1007/s11633-010-0535-3

Lokriti, A., Salhi, I., Doubabi, S., & Zidani, Y. (2013). Induction motor speed drive improvement using fuzzy IP-self-tuning controller. A real time implementation. *ISA Transactions*, *52*(3), 406–417. doi:10.1016/j.isatra.2012.11.002 PMID:23317661

Madadi, H., & Ghribi, M. (2006, June), MRAS based adaptive speed estimator in PMSM drives, IEEE *International Workshop on Advanced Motion Control*, 569 – 572, Istanbul.

Madasu, H., & Prakash, V. (2009). A Novel Optimal Fuzzy System For Color Image Enhancement Using Bacterial Foraging. *IEEE Transactions on Instrumentation and Measurement*, *58*(8), 2867–2879. doi:10.1109/TIM.2009.2016371

Maiti, S., Chakrabatory, C., Hori, Y., & Ta, M. C. (2008). Model reference adaptive controller-based rotor resistance and speed estimation techniques for vector controlled induction motor drive utilizing reactive power. *IEEE Transactions on Industrial Electronics*, *55*(2), 594–601. doi:10.1109/TIE.2007.911952

Majhi, B., & Panda, G. (2007), Bacterial foraging based identification of nonlinear dynamic system, *IEEE Congress On Evolutionary Computation*, 1636,-1641, Singapore. doi:10.1109/CEC.2007.4424669

Messaoudi, M., Sbita, L., & Abdelkrim, M. N. (2007). A robust nonlinear observer for states and parameters estimation and on-line adaptation of rotor time constant in sensorless induction motor drives. *International. The Journal of Physiological Sciences; JPS*, *2*(8), 217–225.

Mir, S., Elbuluk, M. E., & Zinger, D. S. (1998). PI and Fuzzy Estimators for Tuning the Stator Resistance in Direct Torque Control of Induction Machines. *IEEE Transactions on Power Electronics*, *13*(2), 279–287. doi:10.1109/63.662841

Morimoto S., Takeda Y. and Hirasa T.(1990). Flux Weakening Control Method for Surface Permanent Magnet Synchronous Motors, IPEC. Tokyo'90, 942-949.

Nalepa, R., & Orlowska-Kowalska, T. (2012). Optimum Trajectory Control of the Current Vector of a Nonsalient-Pole PMSM in the Field-Weakening Region. *IEEE Transactions on Industrial Electronics*, *59*(7), 2867–2876. doi:10.1109/TIE.2011.2116755

Ouhrouche, M. A. (2002). Estimation of speed, rotor flux and rotor resistance in cage induction motor using the EKF-algorithm. *International Journal of Power and Energy Systems*, *22*(2), 103–109.

Pradeep M., S. Sirisha and M. Chandramouly, (2013). Design of PMSM based on DTC method with MRAS, M.Pradeep Kumar. *International. Journal of Engineering Research and Application*, 3(5), 01-05.

Quntao An. and Sun L. (2008, September). Online parameters identification for vector controlled PMSM drive using adaptive algorithm, *IEEE Conference Vehicle Power and Propulsion, VPPC '08*, Harbin.

Romero, J. A., Sanchis, R., & Balaguer, P. (2011). PI and PID auto-tuning procedure based on simplified single parameter optimization. *Journal of Process Control*, *21*(6), 840–851. doi:10.1016/j.jprocont.2011.04.003

Saad, M. (2013). Online stator resistance estimation using artificial neural network for direct torque controlled induction motor drive. *IEEE 8th Conference on Industrial Electronics and Applications*, ICIEA 2013. 1486-1491.

Sağlam, G., Tutum, C. C., & Kurtulan, S. (2013). A new PI tuning method for an industrial process: A case study from a micro-cogeneration system. *Journal of Energy Conversion and Management*, *67*(2), 226–239. doi:10.1016/j.enconman.2012.11.009

Said, S. H., & Mimouni, F., MŚahli, F. and Farza, M. (2010). High gain observer based on-line rotor and stator resistances estimation for IMs. *IEEE International Conference on Control Applications (CCA)*, 1927-1932, Yokohama.

Sbita, L., & Ben Hamed, M. (2007). An MRAS–based full order Luenberger observer for sensorless DRFOC of induction motors. *International Journal on Automatic Control and System Engineering*, *7*(1), 11–20.

Schauder, C. (1992). Adaptive Speed Identification for Vector Control of Induction Motors without Rotational Transducers. *IEEE Transactions on Industry Applications*, *28*(5), 1054–1061. doi:10.1109/28.158829

Sozer, Y., Torrey, D. A., & Mese, E. (2010), An adaptive predictive current control technique for permanent magnet synchronous motors, IEEE Energy Conversion Congress and Exposition (ECCE), 2819-2826, Atlanta, GA.

Stefan, B., & Tomasz, P. (2011), Energy Efficient Open Loop Control of Permanent Magnet Synchronous Motors for Fan Applications. *17th International Conference on Electrical Drives and Power Electronics*, Slovakia.

Suman M. and Chakrabatory C. (2010), A new instantaneous reactive power based MRAS for sensorless induction motor drive, *Simulation modelling and practice Theory-Elsevier*,18(2), 1314-1326.

Te-Jen, S., Li-Wei, C., Chia-Jung, Y., & Jui-Chuan, C. (2010), Fuzzy PID controller design using self adaptive bacterial foraging optimization, *Proceedings of SICE Annual Conference* 2010, pp. 2604-2607, Taipei.

Tursini, M., Chiricozzi, E., & Petrella, R. (2010). Feed forward Flux- Weakening Control of Surface-Mounted Permanent-Magnet Synchronous Motors Accounting for Resistive Voltage Drop. *IEEE Transactions on Industrial Electronics*, *57*(1), 440–448. doi:10.1109/TIE.2009.2034281

Yan, X., Zhu, Y., Zhang, H., Chen, H., & Niu, B. (2012). An Adaptive Bacterial Foraging Optimization Algorithm with Lifecycle and Social Learning, *Discrete Dynamics in Nature and Society*, 2012. *Article ID*, *409478*, 1–20.

Yogendra, K., & Rajest, B. (2013). BF-PSO optimized PID Controller design using ISE, IAE, IATE and MSE error criteria. *International Journal of Advanced Research in Computer Engineering and Technology*, *2*(7), 2333–2336.

Zerdali, E., & Barut, M. (2013). Sensorless MRAS based real-time speed-sensorless control of induction motor with optimized fuzzy-PI controller, *IEEE International Symposium on Control for Electrical Drives and Predictive Control of Electrical Drives and Power Electronics*.1-5, Munich. doi:10.1109/SLED-PRECEDE.2013.6684500

Zhifu, W., Qizhi, T., & Chengning, Z. (2009). Speed identification about PMSM with MRAS, *IEEE 6th International Conference on Power Electronics and Motion Control, IPEMC '09.*, pp.1880 – 1884, Wuhan.

Zhu, Z. Q., & Howe, D. (2007). Electrical machines and drives for electrical, *hybrid and fuel cell vehicles. Proceedings of the IEEE*, *95*(4), 746–765. doi:10.1109/JPROC.2006.892482

Zhu, Z. Q., & Zhu, X. (2007). Estimation of winding resistance and PM flux-linkage in brushless AC machines by reduced-order extended Kalman filter, *In proceedings of IEEE International Conference on Networking, sensing and control*, 740-745, London.

KEY TERMS AND DEFINITIONS

Adaptive Control: The control method used by a controller which must adapt to a controlled system with parameters which vary, or are initially uncertain.

Algorithm: A process or set of rules to be followed in calculations or other problem-solving operations, especially by a computer.

BFO: Bacterial foraging optimization algorithm.

Electric Motor: An electric machine that converts electrical energy into mechanical energy.

Model Reference Adaptive Control: Model reference adaptive controllers incorporate a reference model defining desired closed loop performance.

PMSM: Permanent magnet synchronous motor.

PSO: Particle swarm optimization algorithm, which is an evolutionary method used in soft-computing.

APPENDIX: NOMENCLATURE

v_d, v_q: Direct and quadrature stator voltages
λ_d, λ_q: Direct and quadrature stator field
i_d, i_q: Direct and quadrature stator currents
L_d, L_q: Direct and quadrature stator inductances
L_s: Stator inductance
λ_m: Magnet flux
R_s: Stator resistance
T_l: Load torque
np: Poles numbers
J: Rotor inertia coefficient
e: Error between the estimated and the real current
a: Positive real constant

Chapter 22
A New Robust H∞ Control Power

Samir Abdelmalek
Unité de Développement des Equipements Solaires (UDES), Centre de Développement des Energies Renouvelables (CDER), Algeria

Hocine Belmili
Unité de Développement des Equipements Solaires (UDES), Centre de Développement des Energies Renouvelables (CDER), Algeria

ABSTRACT

Attention has been paid by many researchers to address the various challenges of grid connection of DFIG-based Wind Energy Conversion Systems (WECS). This chapter focuses on the design of a robust H_∞ controller for the power flow between the stator of the Doubly-Fed Induction Generator (DFIG) and the grid. The robust H_∞ controller design is formulated as a mixed-sensitivity problem. A mathematical model of the DFIG written in an appropriate d-q reference frame is established to carry out simulations. The proposed power control scheme is elaborated and compared with a conventional Proportional-Integral (PI) controller based on vector control technique. The results show interesting performance of the controlled system in terms of the power reference tracking (the active and reactive power) and robustness against parameter variations compared with the conventional PI controller.

1. INTRODUCTION

Due to depletion of fossil fuels and increase of polluted emissions, renewable energy production is rapidly growing. Wind energy is one of the most promising renewable energy sources due to the fact that it is more efficient, availability and low cost (Shafiullah et al., 2013). It has several advantages, such as: is the most environment-friendly, 100% clean energy resource, energy-efficient, cost-efficient. Hence, wind energy has begun to be used as the panacea for solving the global warming problem. Energy of the wind has been used for thousands of years for water pumping, grinding grain, and other low-power applications. There were several early attempts to build large-scale wind powered systems to generate electrical energy (Gálvez-Carrillo & Kinnaert, 2011).

DOI: 10.4018/978-1-4666-7248-2.ch022

Squirrel cage induction machine is used in several wind energy conversion systems. This machine has proven its efficiency when it is directly connected to the grid due to qualities such as robustness, low cost and simplicity (Pena et al., 1996). However, the wind turbine must be designed to keep the machine's speed constant near the synchronous speed. All this constraint reduces the possibility to increase the electrical energy produced for high wind speeds. A converter can be used between the stator of the machine and the grid but it is crossed by the full power and must be correctly cooled (Schreiber, 2001).

Among the existing aerogenerators, Doubly Fed Induction Generators (DFIGs) are the most widely employed in wind power systems (Babu & Mohanty, 2009; Hansen et al., 2004). DFIG-based wind turbines are the most widely employed in wind power systems, they are used for large-scale wind power generation systems due to their interesting characteristics such as, their ability to operate at variable speed, the capacity to control the active and reactive powers into four quadrants by means of field orientation, besides, they have controllable power factor, improved system efficiency and reduced converter rating, which is typically 30% of the generator rating and, hence, decreases the cost and power loss (Hansen et al., 2007; Rothenhagen et al., 2009; Nian et al., 2011). The power control of DFIG in wind turbine is traditionally based on stator-voltage oriented (Hopfensperger et al., 2000), stator-flux-oriented (Chowdhury & Chellapilla, 2006).

In Leonard (2001), explains the vector control technique used for the independent control of torque and excitation current. The converter design and control technique are well explained in (Mohan et al., 1989). The DFIG with PI controllers and its performance under normal operation conditions has been discussed in a number of publications (Carrasco et al., 2006; Petersson et al., 2005; Andreas & Thiringer, 2004; Xu et al., 2009; Arbi et al., 2009). In Yamamoto & Motoyoshi (1990), a control approach has been proposed and applied using a rotating reference frame fixed on the gap flux of the generator and the control active and reactive powers is carried out independently, the results have been proved by experiment. A detailed design of the DFIG using back-to-back PWM voltage source converters in the rotor circuit and they also validated the system experimentally considering a grid connected system has been studied in Pena et al. (1996). In Roncero-Sanchez et al. (2005), a comparative study in discrete time to achieve a decoupled control of the active and reactive powers and to obtain a deadbeat system with two control methods (a classical PI controller and a predictive-integral controller).

In Poitiers et al. (2009), a control of stator active and reactive powers strategy of DFIG based on wind energy systems has been presented and discussed with different controllers: Proportional Integral (PI), polynomial RST based on pole placement theory and Linear Quadratic Gaussian (LQG) regulators. In Abu-Rub et al. (2004) and Santana et al. (2008), predictive control techniques have been applied in machine drives and inverters. The proportional plus integral (PI) controllers and the stator flux-oriented applied to the DFIG power control has been presented in Tapia et al. (2003). The voltage profile and active power losses in the power system can be optimized with reactive power sources, what motivates studies related to the reactive power control for DFIG (Sguarezi Filho & Ruppert, 2010; Sguarezi Filho et al., 2011). Fuzzy controllers and intelligent optimization algorithms have also been used for control of DFIG (Calderaro et al., 2008; Karimi et al., 2009; Pooja & Bharti, 2013).

Recently, traditional Proportional-Integral-Derivative (PID), Proportional-Integral (PI) and vector control techniques have been widely used in most industrial applications, wind power production control systems can be handled very well using these controllers due to their flexibility and simplicity in practical applications (Tapia et al., 2003; Shaheen et al., 2010; Jin-Sung et al., 2011). However, in case of machine parameters change or variation due especially to the natural machine phenomena (saturation, temperature, and skin effect), disturbance uncertainties on the performance of the control system,

system nonlinearities and the cross coupling on DFIG terms in the whole operating range and tuning of the gains conventional PI or PID control performances may be seriously deteriorated (Hansen et al., 2007; Casella et al., 2013). Performance of the PI controllers is decreased and the applications of such methods are limited.

Replacing conventional PI controllers with more advanced controllers to achieve better performance, stability, improve the robustness of the power control system and to remedy the mentioned problems. Considerable advancement has been made in field of robust H_∞ control synthesis since its inception by (Zames, 1981; Doyle et al., 1989; Glover, 1991). One can find a number of theoretical advantages of the methodology such have high disturbance rejection, high stability. It has been widely used to address different practical and theoretical problems. Mixed weight robust H_∞ controllers provide a closed loop response of the system according to the design specifications such as model uncertainty, disturbance attenuation at higher frequencies, required bandwidth of the closed loop plant etc. Practically, robust H_∞ controllers are of high order which, may lead to large control effort requirement. Moreover, the design may also depend on specific system and can require its specific analysis. When H∞-optimal control approach is applied to a plant, additional frequency dependent weights are incorporated in the plant and are selected to show particular stability and performance specifications relevant to the design objective defined in beginning. Robust H_∞ control approach has successful applications when there are uncertainties of model parameters and disturbances. It has been shown in Gu et al. (2005) and Surinkaew et al. (2014) that these controllers are effective and efficient for the control complex systems.

The DFIG-based WECS control system consists of two parts: the electrical control of the DFIG and the mechanical control of the wind turbine speed. Figure 1 shows a synoptic diagram of the proposed power control of a DFIG based wind turbine system (Tazil et al., 2010). The wind turbine is connected to DFIG through a drive train system, which contains high and low speed shafts, bearings and a gearbox. The DFIG is constructed from a wound rotor induction machine (Ekanayake et al., 2003).The stator of the DFIG is directly connected to the electrical grid, while the rotor is fed from a back-to-back converter via slip rings. The converter rating will be only a fraction of the total power of the system, reducing the cost of the power electronics. The back-to-back converter consists of a rectifier connected to the rotor windings; namely the rotor side converter (RSC) and an inverter connected to the power grid; the so-called grid side converter (GSC) (Leonard, 2001; Boulkroune et al., 2013; Shehata & Gerges, 2013). Control of the DFIG is achieved by control of the variable frequency converter, which includes control of the RSC and control of the GSC. The objective of the RSC is to allow the DFIG wind turbine for decoupled control of active and reactive power. This facilitates high flexibility which enables the turbine to capture maximum energy from wind and at the same time to provide reactive power support to the grid. The objective of the GSC is to keep the DC-link voltage constant regardless of the magnitude and direction of the rotor power.

The contribution of this chapter consists on proposing a new H_∞ based robust control scheme for the DFIG, to control the electrical power flow between the DFIG stator and the grid power by separately controlling the active and the reactive powers. The proposed approach allows the achievement of the required objectives, namely, improved power references tracking and the stability of the control system against the machine's parameters variations. The simulation results, show the effectiveness of the proposed (H_∞) control method, that have better performance in terms of steady state error, decoupling between quantities such as stator active-reactive power, power tracking response is faster than classical Proportional- Integral (PI) controller. This chapter is structured as follows. In section 2, the complete model of the wind turbine generation system is presented. In the first step, the wind profile model is

Figure 1. DFIG-based wind turbine control proposed scheme

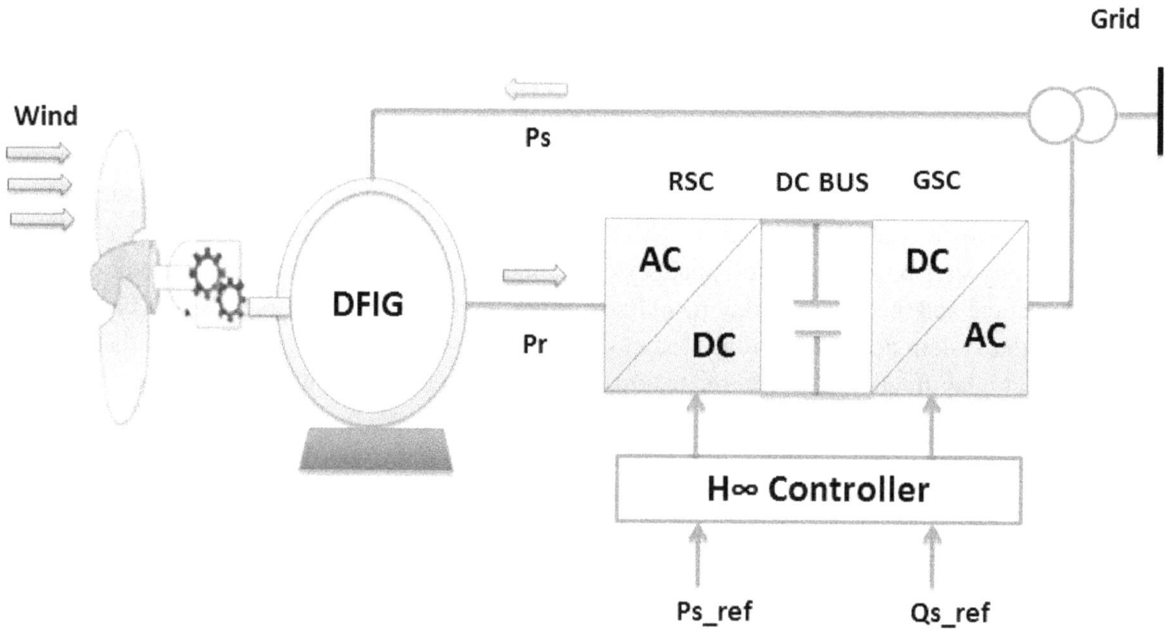

presented, in the second step the aerodynamic system is described then and DFIG is modeled in the final step. In section 3, we have presented a theory of vector control approach. The section 4, deals with the design of the H∞ controller. In the 5 section, some simulation results are presented and compared with those obtained by a conventional PI based vector control scheme. In section 6, concluding remarks and future works are presented.

2. DFIG-BASED WIND TURBINE SYSTEM

The wind turbine is the key device in wind power generation systems, whose basic functionality is to transfer the wind power into the mechanical power on the rotor shaft. The amount of captured wind power is controlled by adjusting blade angle and rotational speed of wind turbine and it is designed to perform different behaviors on different regions of the wind speed.

The wind turbine starts running as soon as wind speed exceeds cut-in speed, usually around 8.5 m/s. While when the wind speed exceeds cut-out speed usually around 19 m/s, the wind turbine stops running to avoid damages. Additionally, when the wind speed is higher than a rated speed at which the generator achieves its rated power, the blade angle of wind turbine is changed to release a part of excess wind energy in order to protect the generator. The wind turbine mostly operates at the region between cut-in speed and rated speed. This section is devoted to the modeling of the wind speed model, modeling mechanical system of a 1.5 MW wind turbine and electrical modeling of the DFIG.

2.1. Wind Speed Model

Wind simulation plays an important role in wind turbine modeling, particularly for power quality analysis of wind farm and their interaction with the grid to which they are connected. The wind models describe the fluctuations in the wind speed, which cause the fluctuations in the power production of the wind turbines. Wind speed varies from one location to another and also fluctuates over the time in a stochastic way (Endusa et al., 2009). One possible approach to generate the wind speed signal on simulations may be to use logs of real measurements of the speed on the real location of the wind turbine generator system. This approach has some evident limitations because it requires a measurement to be done on each place to be simulated. Another choice, proposed in (Slootweg, 2005) is to use a mathematical model which takes some landscape parameters to generate a wind speed sequence for any location. In our simulation case, a wind speed signal has been generated by an autoregressive moving average (ARMA) model described in (Endusa et al., 2009). The wind speed $V_v(t)$ has two constituent parts given by:

$$V_v(t) = V_m(t) + V_t(t) \tag{1}$$

where $V_m(t)$ is the mean wind speed at hub height and $V_t(t)$is the instantaneous turbulent part, whose linear model is composed by a first-order filter excited by Gaussian noise.

$$\dot{V}_t(t) = -\frac{1}{T_v} V_t(t) + \beta_t(t) \tag{2}$$

where T_v is the time constant and β_t is the white noise process with zero mean. The white noise is smoothed by a signal shaping filter, thereby transforming it to colored noise $V_t(t)$, as shown in Figure 2.

The instantaneous turbulence component of wind speed is expressed as (Endusa et al., 2009):

$$V_t(t) = \sigma_t V_t(t) \tag{3}$$

where σ_t is the standard deviation and is the ARMA time series model, which is expressed as (Endusa et al., 2009):

$$V_t(t) = a V_{t-1} - b V_{t-2} + c V_{t-3} + \beta_t - d \beta_{t-1} + e \beta_{t-2} \tag{4}$$

where a, b, and c are the autoregressive parameters and *d* and *e* are moving average parameters whose values being: *a =1.7901, b=0.9087, c=0.0948, d=1.0929 and e = 0.2892*. The wind speed profile of the wind turbine is illustrated in Figure 3. It has a mean value of 8.5 m s^{-1} and max value of 18.5 m s^{-1}.

2.2. Aerodynamics Modeling

Wind turbine electrical generation systems' (WTEGS) power comes from the kinetic energy of the wind, thus it can be expressed as the kinetic power available in the stream of air multiplied by a power

Figure 2. Generation of wind speed by ARMA model in MATLAB/Simulink (Endusa et al., 2009)

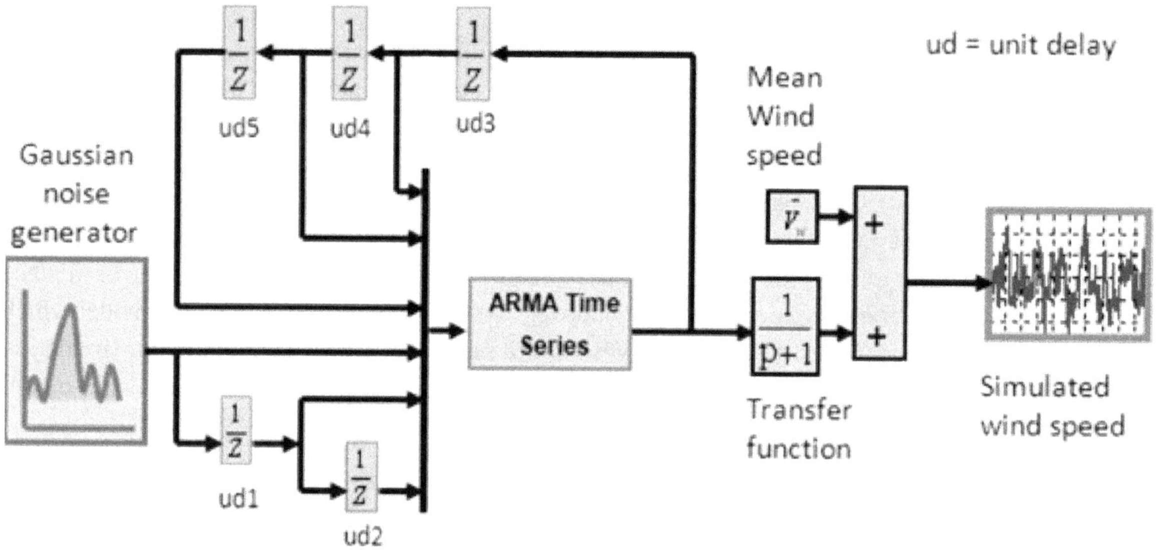

Figure 3. Wind speed profile obtained by using ARMA model

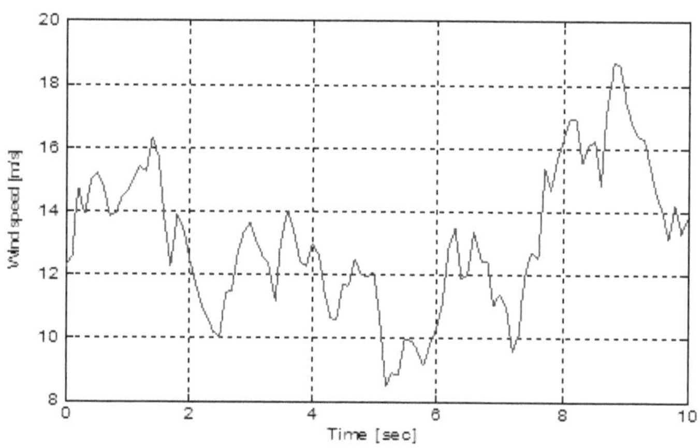

coefficient (C_p) or Betz's factor. C_p mainly depends on the relation between the average speed of the air across the area covered by the wind wheel and its angular speed and geometric characteristics of the turbine (including the instantaneous blade pitch angle configuration). The power extracted by the wind turbine has the following expression (Leonard, 2001; Kamal et al., 2010; Boulkroune et al., 2013):

$$P_m = 0.5 C_p(\lambda, \beta) \rho \pi R^2 V^3 \tag{5}$$

The power coefficient C_p is defined as (Perdana et al., 2004; Chowdhury et al., 2013):

Figure 4. Power coefficient Cp versus λ

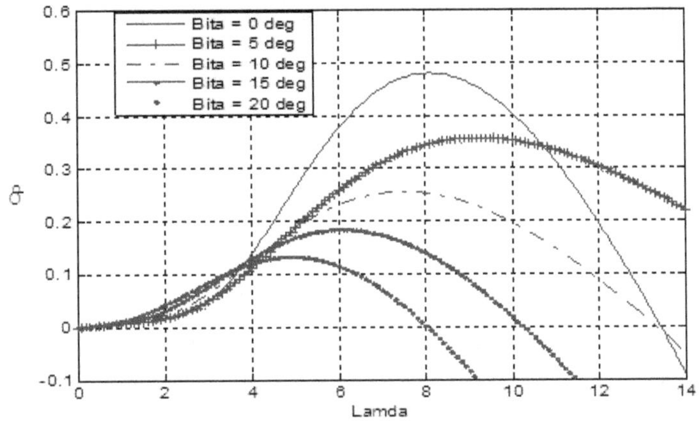

$$C_p = 0.22(\frac{116}{\lambda_i} - 0.4\beta_p - 5)e^{-12.5/\lambda_i} \quad (6)$$

$$\lambda_i = (\frac{1}{\lambda + 0.08\beta_p} - \frac{0.035}{\beta_p^{\ 3} + 1})^{-1} \quad (7)$$

$$\lambda = \Omega_r R / V \quad (8)$$

where C_p is the power coefficient, β is the pitch angle, V is the wind speed, Ω_r is the turbine rotational speed on the low-speed side of the gearbox, R is the rotor-plane radius; ρ is the air density, and λ isthe tip speed ratio (TSR).

The aerodynamic torque on the wind turbine rotor can be obtained using the following expression (Perdana et al., 2004; Kamal et al., 2010; Chowdhury et al., 2013):

$$T_r = 0.5 \cdot C_p(\lambda_{opt}, \beta)\rho \pi R^2 V^3 / \Omega_r \quad (9)$$

A typical relationship between C_p and λ for different values of β is shown in Figure 4.

It is clear from this figure that there is a value of λ for which C_p is maximum ($Cp_{\_max}$ = 0.48for β = 0^0) and that maximize the power for a given wind speed. The peak power for each wind speed occurs at the point where C_p is maximized. To maximize the generated power, it is therefore desirable for the generator to have a power characteristic that will follow the maximum $Cp_{\_max}$ line.

Thus the maximum power captured from the wind is given by:

$$P_{m\_\max} = 0.5 \cdot C_{p\_\max}(\lambda_{opt}, \beta)\rho \pi R^2 V^3 \quad (10)$$

Normally, a variable-speed wind turbine follows the C_{p_max} to capture the maximum power up to the rated speed by varying the rotor speed at Ω_{opt} to keep the TSR at λ_{opt}, T_r is then given by:

$$T_r = K_{opt}\Omega_r^2 \tag{11}$$

where $K_{opt} = 0.5\, C_{p\_max} \rho \pi R^5 / \lambda_{opt}^2$.

2.3. DFIG Model

DFIG is the most commonly used generator in wind power generation systems (Tazil et al., 2010). It is constructed from a wound rotor induction machine; its stator and rotor both have winding structures. The stator windings are directly connected to the three-phase grid, while the rotor windings are connected to the rotor side converter by slip rings and brushes. A voltage is injected into the rotor circuit through slip rings in order to control the rotational speed of the DFIG (Ekanayake et al., 2003). The DFIG operates in two speed regions: super-synchronous and sub-synchronous regions (Aktarujjaman et al., 2008), which are decided by the rotor voltages.

Park's model is the most commonly used model for the DFIG. The mathematical representation of stator voltage and rotor voltage equations as per space vector theory can be described by the equations below (Morren et al., 2008; Boukhezzar & Siguerdidjane, 2009; Kesraoui et al., 2009; Kesraoui et al., 2014):

$$\vec{V}_d = R_s \cdot \vec{I}_s + j \cdot \omega_s \vec{\psi}_s + p \cdot \vec{\psi}_s \tag{12}$$

$$\vec{V}_r = R_r \cdot \vec{I}_r + j \cdot (\omega_s - \omega_r)\vec{\psi}_r + p \cdot \vec{\psi}_r \tag{13}$$

Since the d and q axis are magnetically decoupled, the rotor and stator fluxes are given by:

$$\vec{\psi}_s = L_s \cdot \vec{I}_s + L_m \cdot \vec{I}_r \tag{14}$$

$$\vec{\psi}_r = L_r \cdot \vec{I}_r + L_m \cdot \vec{I}_s \tag{15}$$

Active powers of the stator are defined as:

$$P_s = V_{ds} \cdot I_{ds} + V_{qs} \cdot I_{dr} \tag{16}$$

$$Q_s = V_{qs} \cdot I_{ds} - V_{qs} \cdot I_{dr} \tag{17}$$

Reactive powers of the rotor are defined as:

$$P_r = V_{dr} \cdot I_{dr} + V_{qr} \cdot I_{qr} \tag{18}$$

$$Q_r = V_{dr} \cdot I_{dr} - V_{qr} \cdot I_{dr} \tag{19}$$

The electromagnetic torque C_{em} is defined as:

$$C_{em} = P \cdot (\Phi_{ds} \cdot I_{qs} - \Phi_{qs} \cdot I_{ds}) \tag{20}$$

where R_s, R_r, L_s and L_r are the resistances and leakage inductances of the DFIG stator and rotor windings. Lm is the mutual inductance, V_{ds}, V_{qs}, V_{ds}, V_{qr}, I_{ds}, I_{qs}, I_{dr}, I_{qr}, Φ_{ds}, Φ_{qs}, Φ_{dr} and Φ_{qr} are the d and q components of the space vectors of stator and rotor voltages, currents and fluxes and ω_s and ω_r are the angular frequencies of stator and rotor currents.

3. VECTOR CONTROL STRATEGY

Many contributions have been devoted to power control using classical techniques; vector control strategy is the most used technique to control the DFIG (Yamamoto & Motoyoshi, 1990; Surinkaew et al., 2014). Several assumptions can be found in the literature, for more details see references (Hopfensberger et al., 1999; Slootweg et al., 2001; Forchetti et al., 2002; Cardenas et al., 2004). The rotor-side converter is controlled in a synchronous rotating d–q axis frame, with the d-axis oriented along the stator flux vector position (Pena et al., 1996; Poitiers et al., 2009). In this approach, decoupled control between the stator active and reactive powers is obtained. By choosing a reference frame linked to the stator flux, rotorcurrents will be related directly to the stator active and reactive power. An adapted control of these currents will thus permit to control the power exchanged between the stator and the grid. Ifthe stator flux is linked to the d-axis of the frame we have (Pena et al., 1996; Poitiers et al., 2009):

$$\begin{cases} \psi_{ds} = \Phi_s \\ \psi_{qs} = 0 \end{cases} \tag{21}$$

The torque and consequently the active power only depend on the q-axis rotor current component. If the per phase stator resistance is neglected, which is a realistic approximation for medium power machines used in wind energy conversion, the stator voltage vector is consequently in quadrature advance in comparison with the stator flux vector.

Since the stator is connected to the grid, and the influence of the stator resistance is small, the stator flux can be considered constant. With this consideration, the DFIG model may be written as (Pena et al., 1996; Poitiers et al., 2009):

$$\begin{cases} V_{ds} = 0 \\ V_{qs} = V_s \end{cases} \tag{22}$$

The stator active and reactive powers can be expressed with respect to the rotor currents as:

$$\begin{pmatrix} P_r \\ Q_r \end{pmatrix} = \begin{pmatrix} -V_s \dfrac{L_m}{L_s} & 0 \\ 0 & -\dfrac{V_s L_m}{L_s} \end{pmatrix} \begin{pmatrix} I_{dr} \\ I_{qr} \end{pmatrix} + \begin{pmatrix} 0 \\ \dfrac{V_s^2}{\omega_s L_s} \end{pmatrix} \tag{23}$$

The rotor voltages can be expressed with respect to the rotor currents as:

$$\begin{pmatrix} V_{dr} \\ V_{dr} \end{pmatrix} = \begin{pmatrix} (R_r + \sigma s) & -g.\acute{E}_s\sigma \\ +g.\acute{E}_s\sigma & (R_r + \sigma s) \end{pmatrix} \begin{pmatrix} I_{dr} \\ I_{qr} \end{pmatrix} + \begin{pmatrix} 0 \\ g\dfrac{L_m.V_s}{L_s} \end{pmatrix} \tag{24}$$

With the rotor currents can be expressed by:

$$\begin{pmatrix} I_{dr} \\ I_{qr} \end{pmatrix} = \begin{pmatrix} -V_s \dfrac{L_m}{L_s} & 0 \\ 0 & -\dfrac{V_s L_m}{L_s} \end{pmatrix}^{-1} \left(\begin{pmatrix} P \\ Q \end{pmatrix} - \begin{pmatrix} 0 \\ \dfrac{V_s^2}{\omega_s L_s} \end{pmatrix} \right) \tag{25}$$

In this model, $g = (\omega_s - \omega)/\omega_s$ is defined as the generator's slip, and σ is the dispersion factor, $\sigma = 1 - (L_m * L_m / L_s * L_r)$, V_s is the magnitude of the stator phase voltage [V] and ω_s is the electrical angular velocity of the stator flux [rad/s].

4. ROBUST H∞ CONTROL APPROACH

Various techniques are available in literature for the design of H_∞ controller and H_∞ loop shaping is one of the widely accepted among them as the performance requirements can be embedded in the design stage as performance weights (Zames, 1981; Gu et al., 2005). H_∞ methods are used in control theory to synthesize controllers achieving stabilization with guaranteed performance (Zames, 1981; Doyle et al., 1989; Glover, 1991). To use H_∞ methods, a control designer expresses the control problem as a mathematical optimization problem and then finds the controller that solves this optimization. H_∞ techniques have the advantage over classical control techniques in that they are readily applicable to problems involving multivariate systems with cross-coupling between channels; disadvantages of H_∞ techniques include the level of mathematical understanding needed to apply them successfully and the

Figure 5. Feed back control system

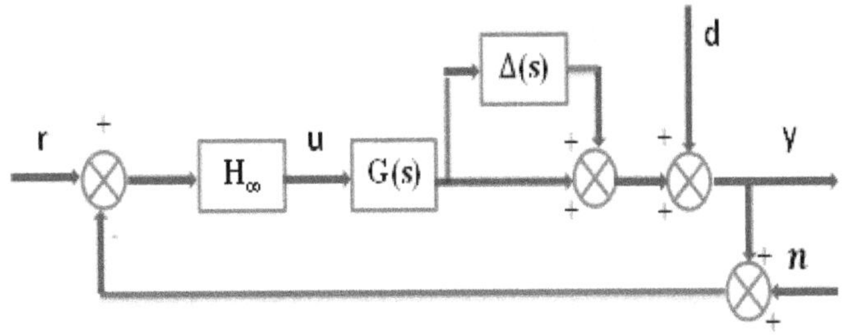

need for a reasonably good model of the system to be controlled. It is important to keep in mind that the resulting controller is only optimal with respect to the prescribed cost function and does not necessarily represent the best controller in terms of the usual performance measures used to evaluate controllers such as settling time, energy expended, etc. Also, non-linear constraints such as saturation are generally not well-handled (William, 1978; Zames, 1981; Allen, 1980).

4.1. H∞ Controller Design

Zames (1981) brought H_∞ norm as a performance requirement. An H_∞ control technique developed by (Doyle et al., 1989; Glover, 1991) not only offers the tradeoff between performance and control effort but also provides the capabilities of accommodating the disturbance and parameter variation. In (Reichert, 1989) was the first to apply H_∞ control and to show how it advantages over classical control in autopilot design. Typical applications for robust control include systems that have high requirements for robustness against parameter variations and high performance (Gu et al., 2005). The controllers synthesized using this algorithm result in robust control systems.

Consider the feedback control system presented in Figure 5. In which, the main transfer functions (closed loop function) of interest when analysing the system are; *G(s)* (the nominal plant) and the synthesized controller *K(s)*. Let's denote (*S*) the sensitivity function, and (*T*) the complementary sensitivity function.

- The sensitivity function (*S*), is often used to describe the influence of the process disturbance d on the output *y*. using the error feedback control scheme, it is also the transfer function from the reference *r* to the control error *e*.

$$S=\left(1+GK\right)^{-1} \tag{26}$$

- The complementary sensitivity function (*T*), is often used to describe the influence of the process disturbance *n* on the output *y*. using the error feedback control scheme, it is also the transfer function from the reference *r* to the control error *e*.

$$T=\left(1+GK\right)^{-1}GK \tag{27}$$

Figure 6. Control problem representation

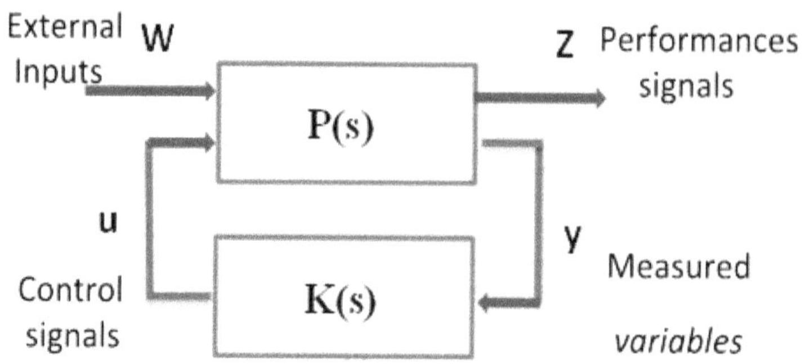

- $\Delta(s)$ is the multiplicative uncertainty, it is assumed that system $\Delta(s)$ is stable with its maximum singular value bounded (Abdelmalek & Sedraoui, 2012; Abdelmalek et al., 2012). As in (25);

$$\sigma_{\max}\left[\Delta(j\omega)\right] < \sigma_{\max}\left[W_3^{-1}\right], \forall\omega \in \left[\omega_{\min} \quad \omega_{\max}\right] \tag{28}$$

4.2. The Optimization Problem

Consider the system given (David, 2011):

$$\begin{cases} \dot{x}(t) = Ax(t) + B_1 W(t) + B_2 u(t) \\ Z(t) = C_1 x(t) + D_{11} W(t) + D_{12} u(t) \\ y(t) = C_2 x(t) + D_{21} W(t) \end{cases} \tag{29}$$

Generally, the vector *W(t)* contains all external inputs to the system (ie. The reference values *r*, the previous disturbances *d* and the measurement disturbances *n*), *x(t)* is the state space vector, *u(t)* is the control input, *Z(t)* is the performance output and *y(t)* is the measured output.

The system defined by (29), is represented by the block diagram in Figure 6.

Where *P* is referred to as the generalized system given by:

$$P = \begin{pmatrix} A & B_1 & B_2 \\ C_1 & D_{11} & D_{12} \\ C_2 & D_{21} & D_{22} \end{pmatrix} \tag{30}$$

The plant transfer function *P(s)* is now divided into four parts, from two inputs *r* and u to the two outputs *Z* and *y*:

$$\begin{pmatrix} Z \\ Y \end{pmatrix} = \begin{pmatrix} P_{11} & P_{12} \\ P_{21} & P_{22} \end{pmatrix} \begin{pmatrix} r \\ u \end{pmatrix} \tag{31}$$

Figure 7. H∞ design model, with weighting functions

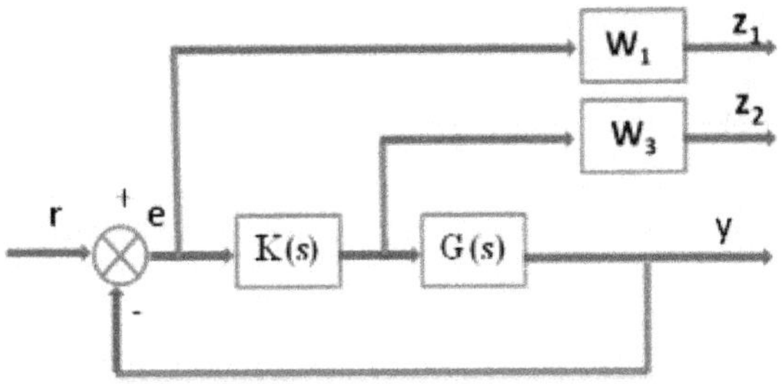

The closed-loop transfer function, from r to Z, then becomes (Gu et al., 2005):

$$Z = \left[P_{11} + P_{12}(I - P_{22}K)^{-1}P_{21}\right]r = F_l(P,K)r \tag{32}$$

This transfer function plays an important role in feedback controller design and is called the Linear Fractional Transformation (LFT) of P and K, by suitably defining r and Z (or P).

$$\text{minimize } \left\|F_l(P,K)\right\|_\infty \tag{33}$$

where, the minimization is over the set of all realisable controllers $K(s)$ that stabilize the closed-loop system. This is known as the H_∞ optimization problem.

The H∞ control is a disturbance attenuation problem, the goal of which is to minimize the gain of the closed-loop system T_{ZW} from disturbance $W(t)$ to output $Z(t)$, so that the impact of the disturbances is minimized in the sense of the H∞norm constraint (McFarlane & Glover,1992; Zhou et al., 1996; Skogestad & Postlethwaite, 1996; Gu et al., 2005). Figure 7 shows the augmented plant model for this design.

Where, the minimization is over all realisable controllers $K(s)$ that stabilize closed-loop system. This is known as the H_∞ optimization problem. Insertion of these matrix elements into $F_l(P,K)$ gives the matrix to minimize (Abdelmalek & Sedraoui, 2012; Abdelmalek et al., 2012):

$$F_l(P,K) = \begin{bmatrix} W_1(jw)(I+GK)^{-1} \\ W_3(jw)(I+GK)^{-1}GK \end{bmatrix} \tag{34}$$

This is known as the *mixed performance and robustness* problem (Grag, 1993). These functions can be recognized as the usual sensitivity, complementary sensitivity. Minimization of ∞- norm of this matrix then will be brought to finding the controller that satisfies the control requirements (Abdelmalek & Sedraoui, 2012; Sedraoui et al., 2012):

$$\left\|F_l(P,K)\right\|_\infty = \left\|\begin{matrix} W_1(jw)\,S(jw) \\ W_3(jw)\,T(jw) \end{matrix}\right\|_\infty \leq \gamma. \tag{35}$$

From this it is seen that *S*, and *T* are shaped as the inverses of their respective weighting functions, scaled by the minimum norm value γ.

4.3. Weighting Functions

To choose the weighting functions, some practical rules are given by (Maciejowski, 1989; Grag, 1993; Maarefors, 1997):

- The weighting function for the sensitivity function, $W_1(s)$ is used to limit the sensitivity function $S(s)$, to define the process disturbance attenuation in low- and mid-frequency range, and to ensure that the performance index of the closed-loop system satisfies the requirements. $W_1(s)$ can also be used to guarantee robustness to uncertainty of the design model and can be used to insert integral action in a control loop, which means low sensitivity by making $W_1(s)$ high at low frequencies. $W_1(s)$ must satisfy:

$$\left\| S(jw)\, W_1(jw) \right\|_\infty \leq 1 \tag{36}$$

- The weighting function for the complementary sensitivity function, $W_3(s)$ can be used to determine the high-frequency roll off. If high-frequency measurements disturbance attenuation should be achieved, $W_3(s)$ should be high at these frequencies. $W_3(s)$ is satisfied by:

$$\left\| T(jw)\, W_3(jw) \right\|_\infty \leq 1 \tag{37}$$

The performance and stability robustness specifications (36) and (37) can be combined into a single infinity norm specification of the form:

$$\left\| \begin{matrix} S(jw)\, W_1(jw) \\ T(jw)\, W_3(jw) \end{matrix} \right\|_\infty \leq 1 \tag{38}$$

It is also important to keep a functional relation in mind. *S* and *T* cannot be made small at the same time; i.e., at frequencies where a high weighting of *T* must be allowed, and vice versa, according to (Grag, 1993; Maarefors, 1997).

$$S(s) + T(s) = 1 \tag{39}$$

It is worth noting that, $S(s)$ and $T(s)$ are just the sensitivity and the complementary sensitivity functions of the closed loop system, respectively. The cost function in (38) can also be interpreted as the design objectives, namely, good tracking or disturbance attenuation and robust stabilization.

5. SIMULATION RESULTS

The proposed robust control method has been evaluated via simulation tests through Matlab/Simulink software. The parameters of a typical 1.5 MW doubly fed induction generator used in the simulation are presented in the Appendix. The performance of the designed robust H_∞ controller are tested and compared with those of a classical PI controller, in terms of power reference tracking (reactive and active powers) responses and the robustness against the machine's parameters variations (rotor resistance and mutual inductance L_m).

Figure 8. Active power reference tracking, with PI and H∞ controllers

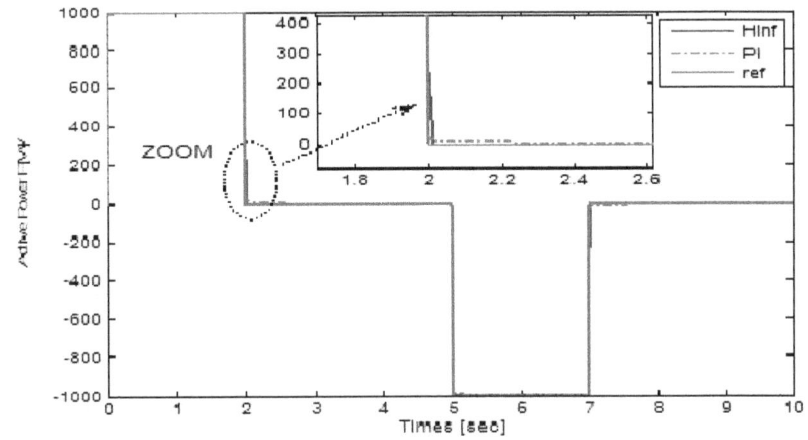

Figure 9. Reactive power reference tracking, with PI and H∞ controllers

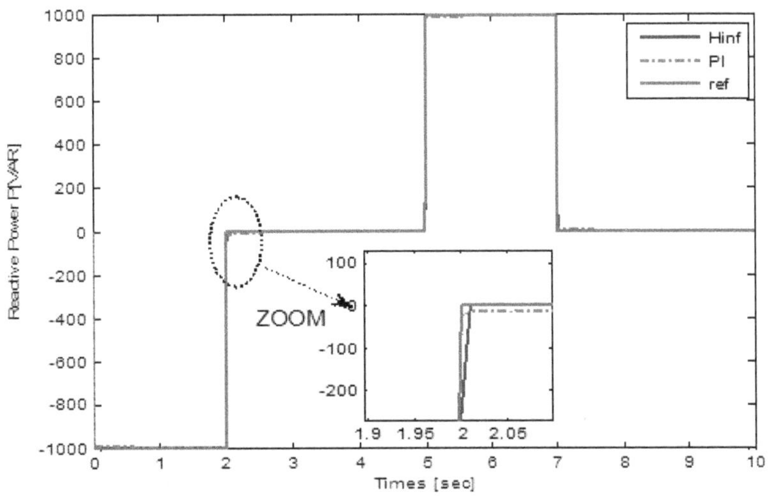

5.1. Reference Tracking

The power control performance is tested with both controllers (robust H∞controller and conventional PI based on vector control methodology). The initial active and reactive power references here are 1kW and -1kVar respectively (At t=0s, both active and reactive power references are changed to 0kW. At t=5s, the active power reference changes to -1kW and the reactive changes to 1kVar). The model is simulated for a simulation time of t=10s and the machine is considered as working under ideal conditions (no parameters variations and no perturbations). Comparison of the simulation results obtained for the active and reactive powers using the proposed approach and the classical PI controller are shown in Figure 8 and Figure 9.

As shown by the simulation result, it is observed that the active and reactive powers generated by the DFIG effectively and rapidly follow the reference values in the case of the proposed approach with no overshoot as opposed to the PI controller. The reference changes made at t=4s and t=6s have been effectively followed by the active and reactive power outputs from the stator. It can be observed in the controller's response to the changing conditions that there is no significantly high overshoot in the response, and the control scheme based on the use of robust H∞ approach have better performance in terms of steady state error, decoupling between quantities such as stator active-reactive power and the speed of tracking power and the speed of tracking power. This clearly means that the robust H∞ controller designed is working properly.

5.2. Robustness against Machines' Parameters Variations

The aim of this test is to analyse the influence of the DFIG parameters variations on the controller's performances. The machine's model parameters have been deliberately modified with excessive variations: the values of the rotor resistance R_r is multiplied by 10 times its nominal value and the value of mutual inductance L_m is decreased by 10% of its nominal value. The effect of parameters variation on the active and reactive powers response for both controllers are shown in Figure 10 and Figure 11.

These results show that parameter variations of the DFIG increase the time-response of the PI controller but not that of the robust H∞ controller, a static tracking error on reactive power appears when the value of the active power is changed but it remains within reasonable values (no more than 5% for the proposed approach). Furthermore, it is shown that the robust H∞ controller has better robustness and is less sensitive to machine parameter variations compared to the PI controller.The results illustrate the effectiveness of the robust H∞ approach than the classical PI controller. Hence, we can conclude that the proposed controller is more powerful than the conventional PI controller.

6. CONCLUSION AND FUTURE WORK

In this chapter, it a new robust H_∞ control scheme which is based on a vector control strategy was proposed and applied to control the flow stator active and reactive power between the DFIG and the grid. The control objective is to improve power reference tracking and robustness against the DFIG parameters variations. The presented analysis shows that the robust H_∞ control strategy is a powerful technique to precisely control the dynamics of active and reactive powers. There are improvements in the dynamic response and stability of the generator system with the robust H_∞ controller compared to the conventional

Figure 10. Robustness against the machine's parameter variations, with PI and H∞ controllers (active power)

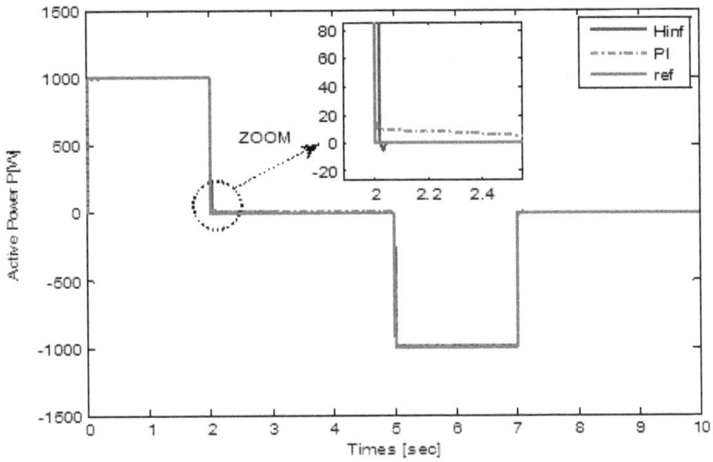

Figure 11. Robustness against the machine's parameter variations, with PI and H∞ controllers (reactive power)

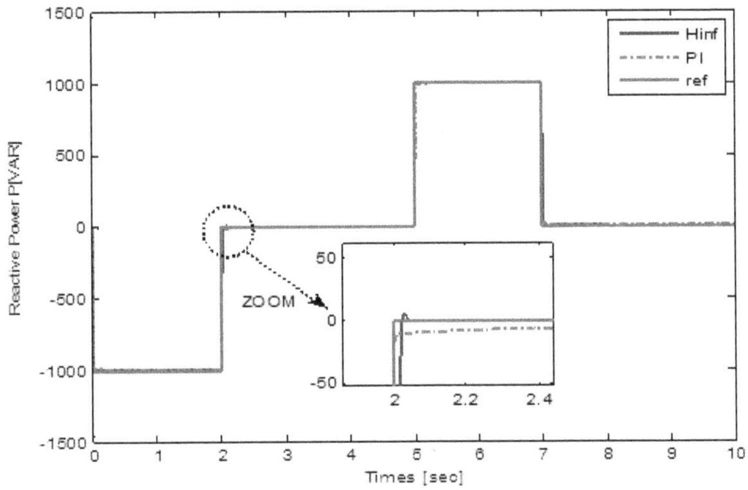

PI Controller. Settling time in active and reactive power response is 0.1sec with the H_∞ controller, whereas in PI it is 0.4 sec. Furthermore, it has been shown that the use of robust H_∞ controller can significantly improve the robustness of the active and reactive powers. As a perspective of this work; we will try to extend the proposed method for nonlinear systems.

REFERENCES

Abdelmalek, S., Chenikher, S., Sedraoui, M., & Soufi, M. (2012, March). Non-integer Order Robust Control of an Electrical Machine. *Proceedings of EVER, 2012*, 1079–1082.

Abdelmalek, S., & Sedraoui, M. (2012, November). Robustness of fractional PID. *Proceedings of ICEECA, 2012*, 159–161.

Andreas, P. S., & Thiringer, L. T. (2004). A DFIG wind-turbine ride through system influence on the energy production. In *Proceedings of NORDIC Wind Power Conference 2004* (pp.1-7). NORDIC.

Arbi, J., Ghorbal, M. B., Slama-Belkhodja, I., & Charaabi, L. (2009). Direct virtual torque control for doubly fed induction generator grid connection. *IEEE Transactions on Industrial Electronics, 56*(10), 4163–4173. doi:10.1109/TIE.2009.2021590

Babu, B. C., & Mohanty, K. B. (2009). Analysis of wind turbine driven double-output induction generator under abnormal condition of the grid. *Proceedings of IPEMC, 2009*, 627–631.

Boukhezzar, B., & Siguerdidjane, H. (2009). Nonlinear control with wind estimation of a DFIG variable speed wind turbine for power capture optimization. *Energy Conversion and Management, 50*(4), 885–892. doi:10.1016/j.enconman.2009.01.011

Boulkroune, B., Gálvez-Carrillo, M., & Kinnaert, M. (2013). Combined Signal and Model-Based Sensor Fault Diagnosis for a Doubly Fed Induction Generator. *IEEE Transactions on Control Systems Technology, 21*(5), 1771–1783. doi:10.1109/TCST.2012.2213088

Calderaro, V., Galdi, V., Piccolo, A., & Siano, P. (2008). A fuzzy controller for maximum energy extraction from variable speed wind power generation systems. *Electric Power Systems Research, 78*(6), 1109–1118. doi:10.1016/j.epsr.2007.09.004

Cardenas, R. J., Pena, R. S., Asher, J., Asher, G. M., & Clare, J. C. (2004). Sensorless control of a doubly-fed induction generator for Stand-alone Operation. In *Proceedings of IEEE 35th Annual Power Electronics Specialists Conference 2004* (pp. 3378-3383). IEEE. doi:10.1109/PESC.2004.1355072

Carrasco, J., Franquelo, L., Bialasiewicz, J., Galvan, E., Guisado, R., & Prats, M. et al. (2006). Power-electronic systems for the grid integration of renewable energy sources: A survey. *IEEE Transactions on Industrial Electronics, 53*(4), 1002–1016. doi:10.1109/TIE.2006.878356

Casella, A. J., Capovilla, C. E., Azcue, J. L., & Ruppert, E. (2013). An Adaptive Neuro-Fuzzy Strategy for a Wireless Coded Power Control in Doubly-Fed Induction Aerogenerators. In T. Abrão (Ed.), Search Algorithms for Engineering Optimization (pp. 175-199). InTech.

Chowdhury, B. H., & Chellapilla, S. (2006). Double fed induction generator control for variable speed wind power generation. *Electric Power Systems Research, 76*(9-10), 786–800. doi:10.1016/j.epsr.2005.10.013

Chowdhury, M. A., Shen, W. X., Hosseinzadeh, N., & Pota, H. R. (2013). A novel aggregated DFIG wind farm model using mechanical torque compensating factor. *Energy Conversion and Management, 67*, 265–274. doi:10.1016/j.enconman.2012.12.001

David, H. T. (2011). *Commande Robuste de Générateurs Électrochimiques Hybrides*. (PhD thesis). Department of Electrical Engineering, Grenoble University, France.

Doyle, J., Glover, K., Khargoeker, P., & Fracis, B. (1989). State space solution to standard H_2 and H_∞ control problem. *IEEE Transactions on Automatic Control, 34*(8), 831–847. doi:10.1109/9.29425

Ekanayake, J., Holdsworth, L., & Jenkins, N. (2003). Control of DFIG wind turbines. *Power Engineering, 17*(1), 28–32. doi:10.1049/pe:20030107

Endusa, B. M., Tomonobu, S., Aki, U., Toshihisa, F., & Chul-Hwan, K. (2009). LQG Design for Megawatt-Class WECS With DFIG Based on Functional Models' Fidelity Prerequisites. *IEEE Transactions on Energy Conversion, 24*(4), 893–904. doi:10.1109/TEC.2009.2025338

Forchetti, D., Garcia, G., & Valla, M. I. (2002). Vector control strategy for a doubly-fed stand-alone induction generator. In *Proceedings of IEEE 28th Annual Conference of the Industrial Electronics Society 2002*, (vol. 2, pp. 991-995). IEEE. doi:10.1109/IECON.2002.1185407

Gálvez-Carrillo, M., & Kinnaert, M. (2011). Sensor fault detection and isolation in doubly-fed induction generators accounting for parameter variations. *Renewable Energy, 36*(5), 1447–1457. doi:10.1016/j.renene.2010.10.021

Glover, K., Limebeer, D., Doyle, J., Kasenally, E. M., & Sofonov, M. (1991). A characterization of all solutions to four block distance problem. *SIAM Journal on Control and Optimization, 29*(2), 283–324. doi:10.1137/0329016

Grag, S. (1993). Robust integrated flight/propulsion control design for a STOVL aircraft using H∞ control design techniques. *Automatica, 29*(1), 129–145. doi:10.1016/0005-1098(93)90177-U

Gu, D. W., Petkov, P. H., & Konstantinov, M.M. (2005). Robust Control Design with MATLAB (3rd ed.). New York: Springer-Verlag.

Hansen, A. D., Gabriele, M., Poul, S., & Torsten, L. (2007). Co-ordinated voltage control of DFIG wind turbines in uninterrupted operation during grid faults. *Wind Energy (Chichester, England), 10*(1), 51–68. doi:10.1002/we.207

Hansen, A. D., Iov, F., Blaabjerg, F., & Hansen, L. H. (2004). Review of contemporary wind turbine concepts and their market penetration. *Journal of Wind Energy, 28*(3), 247–263.

Hopfensberger, B. (1999). Stator flux oriented control of a cascaded doubly-fed induction machine. *Electric Power Applications, 146*(6), 597–605. doi:10.1049/ip-epa:19990590

Hopfensperger, B., Atkinson, D. J., & Lakin, R. A. (2000). Stator-flux-oriented control of a doubly-fed induction machine with and without position encoder. *Electric Power Applications, 147*(4), 241–250. doi:10.1049/ip-epa:20000442

Jin-Sung, K., Jonghyun, J., & Hoon, H. (2011). Design of adaptive PID for pitch control of large wind turbine generator. *Proceedings of EEEIC, 2011*, 1–4.

Kamal, E., Koutb, M., Sobaih, A., & Abozalam, B. (2010). An intelligent maximum power extraction algorithm for hybrid wind-diesel-storage system. *Electrical Power and Energy Systems, 32*(3), 170–177. doi:10.1016/j.ijepes.2009.07.005

Karimi, H., Sheikholeslami, A., Livani, H., & Karimi, D. (2009). Fuzzy Logic control of doubly fed induction generator wind turbine. *World Applied Sciences Journal, 6*(4), 499–508.

Kesraoui, M., Chaib, A., Meziane, A., & Boulezaz, A. (2014). Using a DFIG based wind turbine for grid current harmonics filtering. *Energy Conversion and Management, 78*, 968–975. doi:10.1016/j.enconman.2013.07.090

Kesraoui, M., Toutaoui, S., & Azira, A. (2009). Wind energy conversion system based on a doubly fed induction generator: Study and simulation. In *Proceedings of 2nd EPE Wind Energy Chapter Seminar*. Stockholm, Sweden: EPE.

Leonard, W. (2001). *Control of Electrical Drives* (3rd ed.). New York: Springer-Verlag. doi:10.1007/978-3-642-56649-3

Maarefors, M. (1997). Application of H∞ robust control to the RM12 Jet Engine. *Control Engineering Practice*, *5*(9), 1189–1201. doi:10.1016/S0967-0661(97)84358-4

Maciejowski, J. M. (1989). *Multivaribale Feedback Design*. Wokingham, UK: Addision-Wesley.

McFarlane, D. C., & Glover, K. (1992). A Loop Shaping Design Procedure using Synthesis. *IEEE Transactions on Automatic Control*, *37*(6), 759–769. doi:10.1109/9.256330

Mohan, N., Undeland, T. M., & Robbins, W. P. (1989). *Power Electronics: Converters, Applications and Design*. Oxford, UK: Clarendon Press.

Morren, J., De Haan, S. W. H., Bauer, P., & Pierik, J. T. G. (2003). Comparison of Complete and Reduced Models of a Wind Turbine Using Doubly-Fed Induction Generator. In *Proceedings of the 10th European Conference on Power Electronics Applications*. Toulouse, France: Academic Press.

Nian, H., Song, Y., Zhou, P., & He, Y. (2011). Improved Direct Power Control of a Wind Turbine Driven Doubly Fed Induction Generator During Transient Grid Voltage Unbalance. *IEEE Transactions on Energy Conversion*, *26*(3), 976–986. doi:10.1109/TEC.2011.2158436

Pena, R., Clare, J. C., & Asher, G. M. (1996). Doubly fed induction generator using back-to-back PWM converter and its application to variable-speed wind-energy generation. *Electric Power Applications*, *143*(3), 231–241. doi:10.1049/ip-epa:19960288

Perdana, A., Carlson, O., & Persson, J. (2004). Dynamic response of grid-connected wind turbine with doubly fed induction generator during disturbances. In Proceedings of Nordic Workshop on Power and Industrial Electronics. Academic Press.

Petersson, A., Thiringer, T., Harnefors, L., & Petru, T. (2005). Modeling and experimental verification of grid interaction of a DFIG wind turbine. *IEEE Transactions on Energy Conversion*, *20*(4), 878–886. doi:10.1109/TEC.2005.853750

Poitiers, F., Bouaouiche, T., & Machmoum, M. (2009). Advanced control of a doubly-fed induction generator for wind energy conversion. *Electric Power Systems Research*, *79*(7), 1085–1096. doi:10.1016/j.epsr.2009.01.007

Pooja, D., & Bharti, S. D. (2013). Grid Connected Doubly Fed Induction Generator Wind Energy Conversion System Using Fuzzy Controller. *International Journal of Innovative Technology and Exploring Engineering IJITEE*, *2*(2), 2278–3075.

Reichert, R. T. (1989). Application of H∞ control to missile autopilot design. In *Proceedings of AIAA Guidance Navigation and Control Conference* (pp.1065-1072). AIAA.

Roncero-Sanchez, P. L., Feliu-Batlle, V., & Cerrade, A. G. (2005). Active and reactive power control for a wind generator connected to the grid using a Predective-Integral control and a PI control Schemes. In *Proceedings of the European Conference on Power Electronics and Applications*. Academic Press. doi:10.1109/EPE.2005.219407

Rothenhagen, K., & Fuchs, F. W. (2009). Doubly Fed Induction Generator Model-Based Sensor Fault Detection and Control Loop Reconfiguration. *IEEE Transactions on Industrial Electronics*, *28*(10), 4229–4238. doi:10.1109/TIE.2009.2013683

Santana, E. S. (2008). A predictive algorithm for controlling speed and rotor flux of induction motor. *IEEE Transactions on Industrial Electronics*, *55*(12), 4398–4407. doi:10.1109/TIE.2008.2007376

Schreiber, D. (2001). State of the art of variable speed wind turbines. In *Proceedings of 11th International Symposium on Power Electronics*. Novi Sad, Yugoslavia: Academic Press.

Sedraoui, M., Gherbi, S., & Abdelmalek, S. (2012). A Robust Controller Based on Fractional Structure for MIMO Plant with Multiple Delays. *Control and Intelligent Systems*, *40*(2), 83–94. doi:10.2316/Journal.201.2012.2.201-2216

Sguarezi Filho, A. J., de Oliveira Filho, M. E., & Filho, E. R. (2011). A Predictive Power Control for Wind Energy.Sustainable Energy. *IEEE Transactions on Sustainable Energy*, *2*(1), 97–105.

SguareziFilho, A.J., &Ruppert, E. (2010). A Deadbeat Active and Reactive Power Control for Doubly Fed Induction Generator. *Electric Power Components and Systems*, *38*(14), 592–602.

Shafiullah, G.M., & Amanullah, M.T.O., Shawkat Ali, A.B.M., & Peter, W. (2013). Potential challenges of integrating large-scale wind energy into the power grid: A review. *Renewable & Sustainable Energy Reviews*, *20*, 306–321. doi:10.1016/j.rser.2012.11.057

Shaheen, S. A., Hasanien, H. M., & Badr, M. A. (2010, December). Study on doubly fed induction generator control. *Proceedings of MEPCON*, *2010*, 627–633.

Shehata, E. G., & Gerges, M. (2013). Direct power control of DFIGs based wind energy generation systems under distorted grid voltage conditions. *Electrical Power and Energy Systems*, *53*, 956–966. doi:10.1016/j.ijepes.2013.06.006

Skogestad, S., & Postlethwaite, I. (1996). *Multivariable Feedback Control: Analysis and Design*. New York: John Wiley and Sons.

Slootweg, J. G. (2005). Reduced-order modelling of wind turbines. In *Wind Power in Power Systems*. New York: Wiley. doi:10.1002/0470012684.ch25

Slootweg, J. G., Polinder, H., & Kling, W. L. (2001). Dynamic modelling of a wind turbine with doubly fed induction generator. *IEEE Power Engineering Society Summer Meeting*, *1*, 644-649.

Surinkaew, T., & Ngamroo, I. (2014). Robust power oscillation damper design for DFIG-based wind turbine based on specified structure mixed H_2/H_∞ control. *Renewable Energy*, *66*, 15–24. doi:10.1016/j.renene.2013.11.060

Tapia, A., Tapia, G., Ostolaza, J. X., & Saenz, J. R. (2003). Modeling and control of a wind turbine driven doubly fed induction generator. *IEEE Transactions on Energy Conversion*, *18*(2), 194–204. doi:10.1109/TEC.2003.811727

Tazil, M., Kumar, V., Bansal, R. C., Kong, S., Dong, Z. Y., Freitas, W., & Mathur, H. D. (2010). Three-Phase doubly fed induction generators: An overview. *IET Proceedings of Electric Power Application*, *4*(2), 75-89.

William Helton, J. (1978). Orbit structure of the Mobius transformation semi group action on H-infinity broadband matching. *Advanced in Math. Suppl. Stud*, *3*, 129–197.

Xu, L., Zhi, D., & Williams, B. (2009). Predictive current control of doubly fed induction generator. *IEEE Transactions on Industrial Electronics*, *56*(10), 4143–4153. doi:10.1109/TIE.2009.2017552

Yamamoto, M., & Motoyoshi, O. (1990). Active and reactive Power control for doubly fed wound rotor induction generator. *IEEE Transactions on Power Electronics*, *6*(4), 624–629. doi:10.1109/63.97761

Zames, G. (1981). Feedback and optimal sensitivity model reference transformation, multiplicative semi norms and approximate inverses. *IEEE Transactions on Automatic Control*, *26*(2), 301–302. doi:10.1109/TAC.1981.1102603

Zhou, K., Doyle, J., & Glover, K. (1996). *Robust and Optimal Control participation factor analysis participation factor analysis*. Englewood Cliffs, NJ: Prentice-Hall.

KEY TERMS AND DEFINITIONS

Active Power: The product of the voltage across a branch of an alternating current circuit and the component of the electric current that is in phase with the voltage.

Doubly Fed Induction Generator: Are electric generators that have windings on both stationary and rotating parts, where both windings transfer significant power between shaft and electrical system. Usually the stator winding is directly connected to the three-phase grid and the three-phase rotor winding is fed from the grid through a rotating or static frequency converter.

Grid: (Also referred to as an electricity grid or electric grid) is an interconnected network for delivering electricity from suppliers to consumers.

PI Controller: A PI Controller (proportional-integral controller) is a special case of the PID controller in which the derivative (D) of the error is not used.

Reactive Power: Such as inductors and capacitors dissipate zero power, yet the fact that they drop voltage and draw current gives the deceptive impression that they actually do dissipate power.

Robust Controller: Is such that its properties do not change much if applied to a system slightly different from the mathematical one used for its synthesis. This specification is important: no real physical system truly behaves like the series of differential equations used to represent it mathematically. Typically a simpler mathematical model is chosen in order to simplify calculations; otherwise the true system dynamics can be so complicated that a complete model is impossible.

Robuste H∞: (i.e. "H-infinity") methods are used in control theory to synthesize controllers achieving stabilization with guaranteed performance.

Vector Control: Also called field-oriented control (FOC), is a variable frequency drive (VFD) control method which controls three-phase AC electric motor output by means of two controllable VFD inverter output variables: (Voltage magnitude, Frequency).

APPENDIX

Rated voltage (V_s): 690 V; Frequency (f): 50 Hz; Rated speed (ω_s): 1500 rpm; Number of pole pairs: 2; Stator resistance R_s: 1.2pu; Stator leakage inductance L_s: 0.1554pu; Rotor resistance R_r: 1.8 pu; Rotor leakage inductance L_r: 0.1568pu; Stator and rotor mutual inductance L_m: 0.15 pu.

List of Acronyms

WTEG: Wind turbine electrical generation system
WECS: Wind Energy Conversion System
DFIG: Doubly Fed Induction Generator
ARMA: Autoregressive Moving Average
LFT: Linear Fractional Transformation
TSR: Tip speed ratio
LQG: Linear Quadratic Gaussian
PID: Proportional Integral Derivative
PI: Proportional Integral
RSC: Rotor Side Converter
GSC: Grid Side Converter
Cp: Power Coefficient

Chapter 23
Design of a Controller with Time Response Specifications on STM32 Microcontroller

Maher Ben Hariz
Université de Tunis El Manar, Faculté des Sciences de Tunis, Ecole Nationale d'Ingénieurs de Tunis, LR11ES20, Laboratoire Analyse, Conception et Commande des Systèmes, Tunisia

Faouzi Bouani
Université de Tunis El Manar, Faculté des Sciences de Tunis, Ecole Nationale d'Ingénieurs de Tunis, LR11ES20, Laboratoire Analyse, Conception et Commande des Systèmes, Tunisia

Mekki Ksouri
Université de Tunis El Manar, Faculté des Sciences de Tunis, Ecole Nationale d'Ingénieurs de Tunis, LR11ES20, Laboratoire Analyse, Conception et Commande des Systèmes, Tunisia

ABSTRACT

The basic idea of this chapter is to implement a fixed low order controller on an electronic system using the STM32 microcontroller. The principal aim of this controller is to guarantee some time response specifications as the settling time and the overshoot. The controller parameters are obtained by minimizing a non-convex optimization problem while taking into account the desired closed-loop performances. Accordingly, the resolution of this sort of problem with classical optimization method may lead to a local solution and the achieved control law is not optimal. Hence, with an aim of obtaining an optimal solution that will be able to satisfy the desired specifications, the use a global optimization method is recommended. In this chapter, the Generalized Geometric Programming (GGP) method is exploited. The practical implementation, on a fast electronic system, of the designed control law and a Proportional Integral (PI) controller illustrates the effectiveness of the proposed algorithm.

DOI: 10.4018/978-1-4666-7248-2.ch023

INTRODUCTION

Nowadays the use of microcontroller miniature devices is becoming remarkable. In fact, a microcontroller can be described as a computer on a chip thanks to the features that it contains. Indeed, it has the features of a computer such as central processor, non volatile and volatile memories, input and output ports with special particularities like serial communication, analog to digital conversion and, recently, signal processing. This device is employed in several industrial applications such as medicine, automotive systems and transportation, aerospace, etc. Indeed, the progression of microcontrollers and the features that they combined with their speed, allow them to be more suitable for a wide variety of control applications. There is some works that have used fast devices to implement control algorithms (Ling, Yue & Maciejowski, 2006; Jayaraman & Ravindran, 2008; Ling, Wu & Maciejowski, 2008). One example of the advancement in the embedded technology field is the apparition of STM32 microcontrollers. In fact, they combine several advantages such as low cost, low consumption and good performance. It is in this context that is made in this work which consists on the implementation of a linear system control algorithm on a STM32 microcontroller. There are some works that used the STM32 microcontroller in the control field. Zhen and Yan (2013) proposed a control law of the temperature for the hot runner system. So, they used the STM32 microcontroller to implement their fuzzy PID (Proportional Integral Derivative) control algorithm. They concluded and affirmed that the fuzzy PID controller presents the features of high anti interference and an acceptable adaptability. The STM32 processor was also employed by Lian (2011) as a component of sending frequency control code in a drill hole depth measurement system for coal mines. Wang, Li, Liu, and Ning (2011) proposed the application of STM32 microcontroller in the design of mine DC electrical prospecting instrument. Zhang and Kang (2013) designed an embedded signal acquisition system based on a STM32 microcontroller depending on the mechanical failure appeared with high frequency in the rotating machines. A Radio Frequency (RF) data acquisition system based on STM32 and FPGA was designed by Zhang and Zhao (2011) in order to collect the RF signal in high speed. In the present work, the STM32 microcontroller is used in order to implement a control law that will be used with a real electronic process. Indeed, this control law is obtained by discretizing a continuous time controller. The controller parameters are obtained by resolving a non convex optimization problem. In point of fact, the optimization is found in several real problem solving processes. For that reason, the resolution of optimization problems has gained the attention of eminent researchers in various fields. With an aim of find the global minimum Toksari (2009) proposed an Ant Colony Optimization (ACO) algorithm. The author has tested his proposed algorithm on some standard functions and he has also compared it with other algorithms. The Practical Swarm Optimization (PSO) was used by Zhou, Xiao, Wang, Liang and Hassanien (2013) in the control algorithm with the purpose of allowing robots to navigate towards remote frontier after exploring the region. Abu-Seada, Mansor, Bendary, Emery, and Moustafa Hassan (2013) have also employed the PSO method in order to find an optimal tuning of PID controller parameters for an automatic voltage regulator system of a synchronous generator. With the aim of solving unconstrained optimization problems, Toledo, Oliveira, and França (2014) have used the Genetic Algorithm (GA) with hierarchically structured population. Mamdoohi, Abas, Samsudin, Hisham Ibrahim, Noor Hidayat, and Mahdi (2012) proposed the implementation of GA in an embedded microcontroller based polarization control system. The controller measures the signal intensity. These measures will be used to estimate the genetic value. The GA controls this process. To achieve optimum performance, the best genetic parameters optimize the code such that the fastest execution time can be obtained. Valdez, Melin, and Castillo (2014) presented a hybrid approach for optimization combining

PSO and GAs using fuzzy logic in order to integrate the results. They affirmed that the proposed method, in their work, combines the advantages of PSO and GA to provide an improved Fuzzy PSO (FPSO) and Fuzzy GA (FGA) hybrid method. Fuzzy logic is employed in order to combine the results of the GA and PSO in the best possible way. Fuzzy logic is also exploited with an aim of adjusting parameters in FPSO and FGA. Jiang, Ji, and Shen (2014) proposed a hybrid approach to solve economic emission load dispatch problems considering various practical constraints by employing Hybrid Particle Swarm Optimization and Gravitational Search Algorithm (HPOS-GSA). Their algorithm provided a combination between the PSO and the GSA and adopted co-evolutionary technique to update its particle position in the swarm with the cooperation of PSO and GSA. In order to find the global minimum in the numerical, Servet Kiran, Gunduz, and Kaan Baykan (2012) proposed a hybrid algorithm which is based on PSO and ACO. This algorithm is called hybrid ant particle optimization algorithm. The ACO and the PSO work separately at each iteration and generate their solutions. The best solution is, then, chosen as the global best of the system and its parameters are used to choose the new position of particles and ants at the next iteration. Tabakhi, Moradi, and Akhlaghian (2014) proposed a method based on ACO, called unsupervised feature selection method based on ACO in order to find an optimal solution to the feature selection problem. Geometric branch and bound techniques are widely known solution algorithms for non convex continuous global optimization problems. Several approaches can be found in the literature. Schöbel and Scholz (2014) proposed an extension of geometric branch and bound methods for mixed integer optimization problems. In fact, they presented some general bounding operations. Thus, in order to solve mixed integer optimization problems they provided a general algorithm.

The chapter is organized as follows: Section 2 presents the related work. Then, in section 3, the problem statement is exposed. Subsequently, the design of controllers is developed in section 4. Thereafter, in section 5, the Generalized Geometric Programming (GGP) method is introduced. Then, section 6 describes the hardware and the software development tools employed in this application. To bring out the effectiveness of the proposed controller, a practical implementation on a real system is exposed in section 7. A comparison between the fixed low order controller and the Proportional Integral (PI) controller was been also made in this section. Section 8 discusses and analyses the obtained experimental results. The last section is devoted to conclude this paper and to present the future direction.

RELATED WORK

Pörn, Bjork, and Westerlund (2007) proposed an approach to obtain global solution for non convex Mixed Integer Non Linear Programming (MINLP) problems that include signomial expressions. They showed that it is possible to decompose any non convex optimization problem consisting of an objective function and inequality constraints to a convex MINLP problem by the application of some variable transformation techniques. Tsai, Lin, and Hu (2007) presented a technique for treating both positive and non positive variables with integer powers in GGP problems. Choi and Briker (1996) demonstrated the efficiency of the geometric programming approach in machining economics problems compared with the results of generalized reduced gradient. Tsai (2009) proposed an approach for the treatment of free variables in GGP problems. Manaras and Floudas (1997) proposed a deterministic global optimization algorithm to find the global minimum of generalized geometric (signomial) problems. Kheriji, Bouani, and Ksouri (2011a) used the GGP method to solve a non convex min–max optimization problem to compute a predictive controller for class of constrained linear Multi Input Multi Output (MIMO) systems in presence of parametric uncertainties in the state space model. Bjork, Lindberg, and Westerlund (2003) presented an optimization technique based on the convexification of signomial terms. The proposed approach is based on the solution of a sequence of convexified sub problems.

Kim, Keel, and Bhattacharyya (2003) developed an approach to directly control the transient response of LTI control systems. They presented the methodology of fixing the desired closed loop characteristic equation by the user. Jin and Kim (2008) developed an algorithm for designing a fixed, low order controller with time response specifications for a linear time invariant (LTI), single input single output (SISO) plant. Ben Hariz, Bouani, and Ksouri (2012) presented the synthesis of a robust controller for LTI uncertain systems. To describe the system dynamic behavior, a linear parametric uncertainties model was used. The controller design was formulated as a min max optimization problem which took into account the desired time response specifications and the uncertainties on the model parameters. Then the controller parameters represent the best solution for the worst case of all possible models. In these works, the controller parameters were obtained by resolving a non convex optimization problem by the mean of a local optimization method.

Since the controller parameters are obtained by minimizing a non convex optimization problem, the use of global optimization method is suggested. In this work, the GGP method will be applied in order to resolve this optimization problem. The key idea of this method is to transform a non convex optimization problem to a convex one by means of variable transformations. Ben Hariz, Chagra, and Bouani (2013) presented the design of fixed low order controller for decoupled MIMO systems. The simplified decoupling was used as a decoupling system technique. They used the GGP method to obtain the controller parameters. The synthesized controller must guarantee some time response performances. In their work, the designer can set the desired closed loop performances of each output independently. Kheriji, Bouani, Ksouri, and Ben Ahmed (2011b) proposed an efficient framework for implementation of Generalized Predictive Control (GPC) in the performed STM32 microcontroller. The STM32 Keil starter kit based on a JTAG interface and the STM32 board was used to implement the proposed GPC firmware. Kheriji, Bouani, and Ksouri (2011c) presented an efficient firmware for the implementation of constrained model predictive control in the STM32 microcontroller using interior point method. The main contribution of this work is the implementation of a fixed low order controller with time response specifications on a real electronic system by the use of the STM32 microcontroller.

The proposed control law was implemented using Keil development tool specially engineered for ARM processor based microcontroller devices.

PROBLEM STATEMENT

A SISO system can be represented by a transfer function *G(s)*. The closed loop system is delineated in Figure 1.

$$G(s) = \frac{N(s)}{D(s)} = \frac{n_m s^m + n_{m-1}{}^{m-1} s^{m-1} + \cdots + n_0}{d_l s^l + d_{l-1} s^{l-1} + \cdots + d_0} \quad (1)$$

Figure 1. A feedback control system

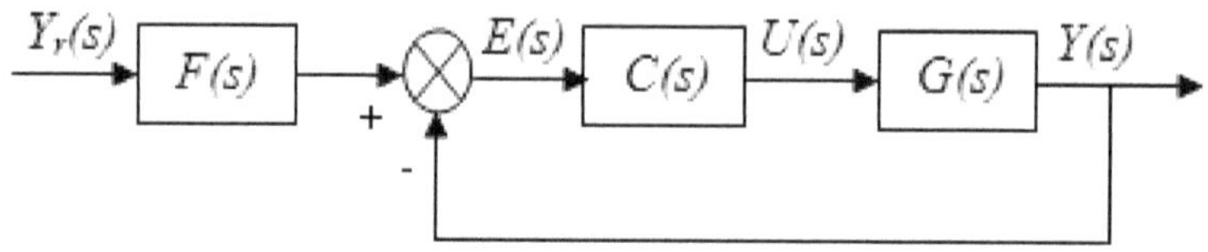

The controller is given by:

$$C(s) = \frac{B(s)}{A(s)} \tag{2}$$

where

$$\begin{aligned} A(s) &= s^t + a_{t-1}s^{t-1} + \cdots + a_1 s + a_0 \\ B(s) &= b_r s^r + b_{r-1}s^{r-1} + \cdots + b_1 s + b_0 \end{aligned} \tag{3}$$

In the case of low order controller, the following relation must be verified: $r \le t < l - 1$.

In order to obtain a closed loop transfer function with a unit static gain, a polynomial *F(s)* will be introduced as follows:

$$F(s) = f_q s^q + f_{q-1}s^{q-1} + \cdots + f_1 s + f_0 \tag{4}$$

By using (1) and (2) the closed loop transfer function will be given by:

$$H(s) = \frac{F(s)B(s)N(s)}{A(s)D(s) + B(s)N(s)} \tag{5}$$

The closed loop characteristic equation is given by the following relation:

$$\begin{aligned} \delta(s) &= A(s)D(s) + B(s)N(s) \\ &= \delta_n s^n + \delta_{n-1}s^{n-1} + \cdots + \delta_1 s + \delta_0 \end{aligned} \tag{6}$$

where $n = l + t$.

The methodology of fixing the desired closed loop characteristic equation which was detailed in (Kim et al., 2003; Jin, & Kim, 2008; Ben Hariz et al., 2012) will be now explained.

Let consider the following polynomial:

$$\delta(s) = \delta_n s^n + \delta_{n-1}s^{n-1} + \cdots + \delta_1 s + \delta_0 = \sum_{j=0}^{n} \delta_j s^j \tag{7}$$

The generalized time constant τ and the characteristic ratios α_j $(j = 1,\dots,n-1)$ were defined by (Kim et al., 2003; Jin & Kim, 2008) as follows:

$$\tau = \frac{\delta_1}{\delta_0} \tag{8}$$

$$\alpha_j = \frac{\delta_j^2}{\delta_{j-1}\delta_{j+1}},\ j = 1,\dots,n-1 \tag{9}$$

Thus, the coefficients of the polynomial $\delta(s)$ can be written as a function of τ and α_j by:

$$\delta_1 = \delta_0\tau,\ \delta_2 = \frac{\delta_0\tau^2}{\alpha_1},\dots,\delta_n = \frac{\delta_0\tau^n}{\alpha_{n-1}\alpha_{n-2}^2\cdots\alpha_1^{n-1}} \tag{10}$$

The characteristic polynomial can be defined as the polynomial possessing the following characteristic ratios:

$$\begin{cases} \alpha_1 > 2 \\ \alpha_k = \dfrac{\sin\left(\dfrac{k\pi}{n}\right) + \sin\left(\dfrac{\pi}{n}\right)}{2\sin\left(\dfrac{k\pi}{n}\right)}\alpha_1 \quad for \quad k = 2,\dots,n-1 \end{cases} \tag{11}$$

Then, the characteristic polynomial can be computed by the fixation of α_1 only for a given τ and δ_0 by employing relations (10) and (11).

The target, now, is to design a controller which will be able to ensure some time response performances such as the overshoot and the settling time. The description of this target is the subject of the next section.

DESIGN OF CONTROLLERS

In this section, the low order controller design method will be presented. It is noted that the proposed controller must guarantee some time closed loop time response specifications.

Let the controller parameters vector be defined by:

$$x = \begin{bmatrix} b_0 & \cdots & b_r & a_0 & \cdots & a_{t-1} \end{bmatrix}^T \tag{12}$$

The coefficients vectors of the closed loop characteristic polynomial δ and the closed loop desired polynomial $\bar{\delta}$ are respectively given by:

$$\delta = \begin{bmatrix} \delta_0 & \delta_1 & \cdots & \delta_{n-1} & \delta_n \end{bmatrix}^T \tag{13}$$

$$\bar{\delta} = \begin{bmatrix} \bar{\delta}_0 & \bar{\delta}_1 & \cdots & \overline{\delta_{n-1}} & \bar{\delta}_n \end{bmatrix}^T \tag{14}$$

Using relation (6), the coefficient vector of the closed loop characteristic polynomial δ can be expressed as a function of x by:

$$\delta = Px + q \tag{15}$$

where

$$P = \begin{bmatrix} n_0 & 0 & \cdots & 0 & d_0 & 0 & \cdots & 0 \\ n_1 & n_0 & \cdots & 0 & d_1 & d_0 & \cdots & 0 \\ \vdots & \vdots & & \vdots & d_2 & d_1 & \cdots & 0 \\ n_m & n_{m-1} & \cdots & n_{m-r} & \vdots & \vdots & & \vdots \\ \vdots & \vdots & & \vdots & d_l & d_{l-1} & \cdots & d_{l-t+1} \\ 0 & 0 & \cdots & n_m & \vdots & \vdots & & \vdots \\ \vdots & \vdots & & \vdots & 0 & 0 & \cdots & d_l \\ 0 & 0 & \cdots & 0 & 0 & 0 & \cdots & 0 \end{bmatrix}$$

$$q = \begin{bmatrix} 0 & \cdots & 0 & d_0 & \cdots & d_{l-t} & \cdots & d_l \end{bmatrix}^T$$

$$P \in R^{(n+1)\times(r+t+1)},\ x \in R^{r+t+1},\ q \in R^{n+1}$$

The controller parameters are determined so that the difference between δ and $\bar{\delta}$ is minimal. This can be obtained by the following weighted cost function:

$$\begin{aligned} f(x) &= \left[\delta - \bar{\delta}\right]^T W \left[\delta - \bar{\delta}\right] \\ &= x^T \left[P^T W P\right] x + 2\left[\left(q - \bar{\delta}\right)^T WP\right] x + \left(q - \bar{\delta}\right)^T W \left(q - \bar{\delta}\right) \end{aligned} \tag{16}$$

where W is a weighting matrix.

It was clarified by Kim et al., (2003, 2004) that the coefficients of lower powers of s in the transfer function are the most related to the step response. Then, the weighting matrix may be selected such that the weights for the low powers of s have greater values than those for higher powers.

The Diophantine equation (6) cannot be resolved for any employed low order controller. Thus, with the aim of getting an approximate solution, the Partial Model Matching (PMM) method will be used. The PMM approach can be considered as a least square estimation problem in the sense of minimizing $f(x)$ compared to x.

So, the controller parameters can be computed by the resolution of the following optimization problem:

$$\min_{x} f(x) \tag{17}$$

such that the function $f(x)$ is given by the following expression:

$$f(x) = x^T \left[P^T W P\right] x + 2\left[\left(q - \bar{\delta}\right)^T WP\right] x + \left(q - \bar{\delta}\right)^T W \left(q - \bar{\delta}\right)$$

The present optimization problem is non convex. That is why its resolution by a local approach may lead us to obtain a local solution and consequently the resulting control law will be not optimal. Therefore, to deal with this problem, the use of global optimization method is suggested. Accordingly, the presentation of the GGP method which is the global optimization method used in this work is the subject of the next section.

GENERALIZED GEOMETRIC PROGRAMMING METHOD

Mathematical Formulation of the GGP Method

GGP problems are appeared commonly in engineering design, management and chemical process industry (Nand, 1995; Choi, & Dennis, 1996; Maranas, & Floudas, 1997; Pörn et al., 2008; Tsai, 2009; Kheriji et al., 2011a, Ben Hariz et al., 2013).The distinctive peculiarity of this method that it is dedicated to solve a class of non convex non linear programming problems with the objective function and the constraints are in polynomial forms. The mathematical formulation of a GGP problem with free variables is expressed as follows (Tsai et al., 2007):

$$\min_{X} Z(X) = \sum_{p=1}^{T_0} c_p z_p \tag{18}$$

where

$$z_p = x_1^{\alpha_{p_1}} x_2^{\alpha_{p_2}} \ldots x_n^{\alpha_{p_n}}, \quad p = 1,\ldots,T_0 \,,$$
$$X = (x_1, x_2, \ldots, x_m, x_{m+1}, \ldots, x_n), \; \underline{x_i} \le x_i \le \overline{x_i}, \tag{19}$$
$$x_i > 0, \text{ for } 1 \le i \le m \text{ and } x_i \le 0, \text{ for } m+1 \le i \le n,$$

$$c_p \in \Re \; \alpha_{pi} \in \Re \text{ for } 1 \le i \le m,$$

α_{pi} is integer for $m+1 \le i \le n$ and $\underline{x_i}$ and $\overline{x_i}$ are, respectively, lower and upper bounds of continuous variables x_i. It should be noted that convexification strategy that will be introduced later is only applied for positive variables x_i because of the logarithmic/exponential transformation. Accordingly, this transformation necessitates substituting x_i by $\exp(y_i)$. Therefore, x_i must be strictly positive. Nevertheless, this is not a handicap because a simple translation of the variables to accomplish specifications for variables originally having negative values can be used (Bjork et al., 2003).

The GGP method has been suggested to find the global optimum based on variable transformations. This transformation technique allows the convexification of the objective function.

To begin with, some definitions are necessitated before presenting the convexification propositions and property.

Definition 1: A "monomial" function is a product of power terms and it can be given by the following expression:

$$f(X) = c\prod_{i=1}^{n} x_i^{p_i} \tag{20}$$

where c is a real constant and p_i can be negative or positive power for $1 \le i \le n$.

Definition 2: A "signomial" function is constituted of a sum with products of power terms, where each product with power terms is multiplied by a real constant (Liberti & Maculan, 2006):

$$f(X) = \sum_{j=1}^{T} c_j \prod_{i=1}^{n} x_i^{p_{i,j}} \tag{21}$$

The constants c_j and powers $p_{i,j}$ for $1 \le i \le n$ and $1 \le j \le T$ can be positive or negative.

Definition 3: The function $f(X)$ is called a "posynomial", when all constants c_j, for $1 \le j \le T$, in a signomial function of equation (21) are positive.

Optimization problems that possess only signomial terms are called GGP problems.

Some convex analysis results before exposing the convexification rules in next subsection will be introduced:

Proposition 1 (Tsai et al., 2007): A twice differential function $f(X) = c\prod_{i=1}^{n} x_i^{p_i}$ is convex in $\Re_+^n$ for $c \geq 0$ if $p_i \leq 0$.

Proposition 2 (Tsai et al., 2007): A twice differential function $f(X) = c\prod_{i=1}^{n} x_i^{p_i}$ is convex in $\Re_+^n$ for $c \leq 0$ if $p_i \geq 0$ and $(1 - \sum_{i=1}^{n} p_i) \geq 0$.

Property (Pörn et al., 2008): The function $c\exp\left(\sum_{i=1}^{n} p_i x_i\right)$ is convex in $\Re_+^n$ if $c \geq 0$ and $p_i \in \Re$.

Convexification Strategy of the GGP Method

The fundamental concept in the convexification strategy is to perform variable transformations to problem (18) that permit to convexify each monomial of the signomial depending on their signs. In this work the objective function of the controller possesses only variables with positive powers. Therefore, in the transformation rules, only this case will be considered.

- **Positively Signed Term (*c>0*):** Let consider the function $f(X) = c\prod_{i=1}^{n} x_i^{p_i}$, where $p_i>0$, new variable y_i are presented according to $x_i=\exp(y_i)$, $i=1,2,\ldots,n$. Then the following equivalence establish can be established:

$$f(X) = c\prod_{i=1}^{n} x_i^{p_i} = c\exp\left(\sum_{i=1}^{n} p_i y_i\right) \tag{22}$$

According to the property previously presented, the signomial in the right hand is convex relatively to y_i. This transformation is called the exponential transformation.

- **Negatively Signed Term (*c<0*):** Let consider the function $f(X) = c\prod_{i=1}^{n} x_i^{p_i}$, where $p_i>0$ and $\left(1 - \sum_{i=1}^{n} p_i\right) < 0$, new variable z_i are presented according to $x_i = z_i^{\frac{1}{\beta}}$, $i=1,2,\ldots,n$, where $\beta = \sum_{i-1}^{n} p_i$.

Then the resulting equality can be established:

$$f(X) = c\prod_{i=1}^{n} x_i^{p_i} = c\prod_{i=1}^{n} z_i^{\frac{p_i}{\beta}} \tag{23}$$

According to Proposition 2, the signomial in the right hand is now convex with respect to z_i, since all exponents are positive and their sum is equal to 1. Another solution to convexify *f(X)* is to simply choose $\beta > \sum_{i=1}^{n} p_i$. This transformation is referred to as power transformation.

STM32 MICROCONTROLLER AND KEIL DEVELOPMENT TOOL

In this section, the hardware as well as the software development tool used in this work will be presented.

STM32F100RB Microcontroller

The choice to use the STM32 is based on the compromise between price (the STM32 discovery costs few dollars), low power consumption and performance. In fact, the STM32F100RB value line family includes the high performance ARM Cortex™-M3 32 bit RISC core working at a 24 MHz frequency, high speed embedded memories (Flash memory of 128 Kbytes size and SRAM of 8 Kbytes size), and an extensive range of enhanced peripherals and I/Os connected to two APB buses. All devices offer two 12bit DACs, one 12 bit ADC, up to six general purpose 16 bit timers plus one PWM timer, as well as standard and advanced communication interfaces: two Serial Peripheral Interface SPIs and I2Cs, three Universal Synchronous Asynchronous Receiver Transmitter USARTs, an USB and a CAN (STMicroelectronics, 2012). In order to load the program into the STM32 device, the ST LINK is used. The STM32F100 architecture is presented in Figure 2.

Keil Development Tool

Keil which is a software development tool includes C/C++ compilers, debuggers, integrated environments, middleware, Real Time Operating System (RTOS), simulation models, and evaluation boards for ARM, Cortex M, Cortex R4, 8051, C166, and 251 processor families. The used version, in this work, is the μVision 4. The μVision 4 screen supplies a menu bar for command entry, windows for source files, and a toolbar where the command buttons dialog boxes, and information displays can be chosen. The μVision 4 can open and view multiple source files simultaneously.

This version has two operating modes: Build Mode and Debug Mode.

- **Build Mode:** It is the standard working mode. It lets us to convert all the application files and to generate executable programs.
- **Debug Mode:** It supplies an effective debugger for testing applications.

In both operating modes, it is possible to use the source editor of μVision 4 in order to modify the source code. The Debug mode adds additional windows and stores an own screen layout.

EXPERIMENTAL RESULTS

In order to show the performances of the presented approach, a practical implementation of the proposed controller on an electronic system is made by using an STM32 microcontroller.

Real System Presentation

The electronic system used in this work is a 3rd order low pass filter as shown in Figure 3. The schematic of this filter is depicted in Figure 4.

Figure 2. STM32F100 architecture

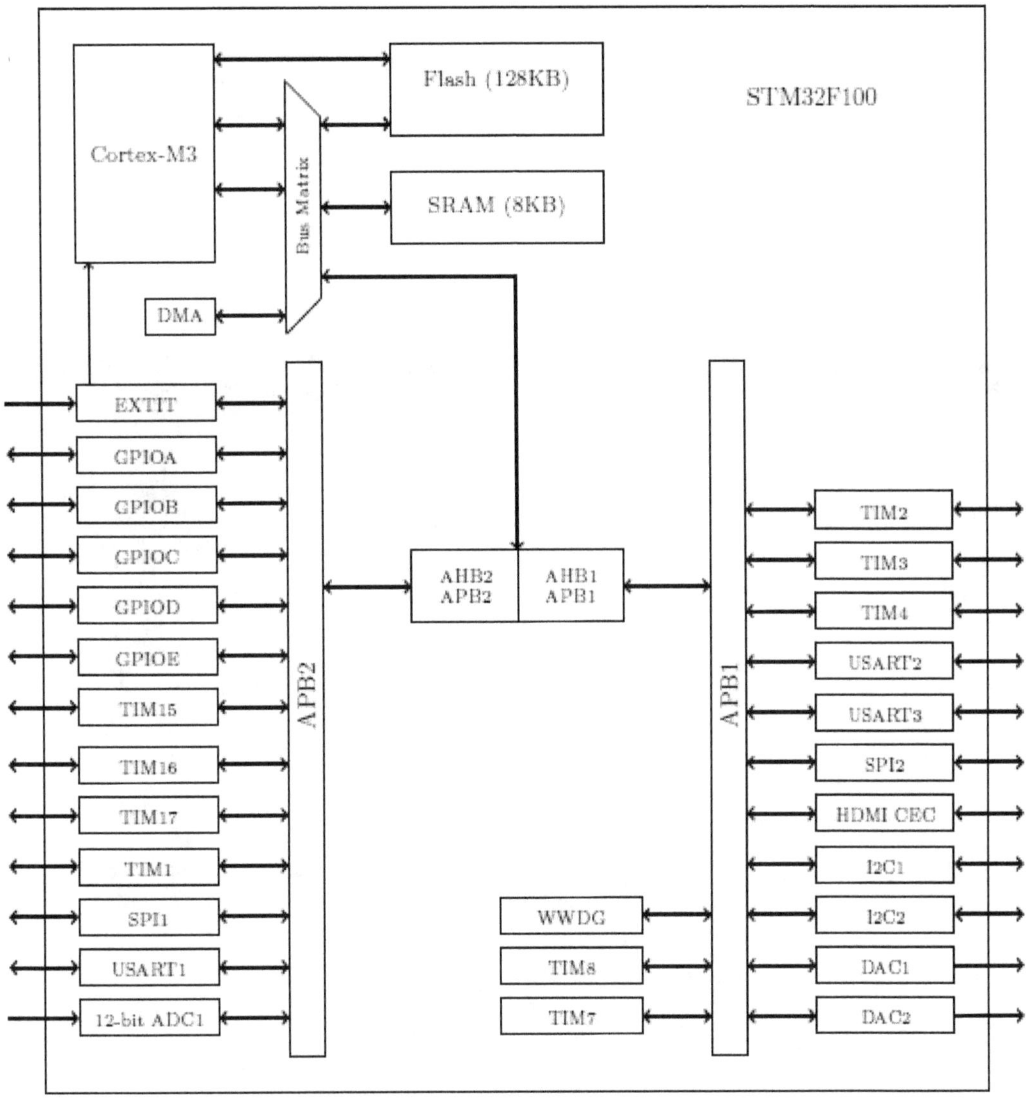

In this work, the chosen values of resistors and capacitors are given by: $R = 47\ K\Omega$ and $C = 50\ \mu F$. So, by adopting these values, the system transfer function will be given by the following expression:

$$G(s) = \frac{Y(s)}{U(s)} = \frac{0.07705}{s^3 + 2.128s^2 + 1.086s + 0.07705} \tag{24}$$

Figure 3. Schematic of the 3rd order low pass filter

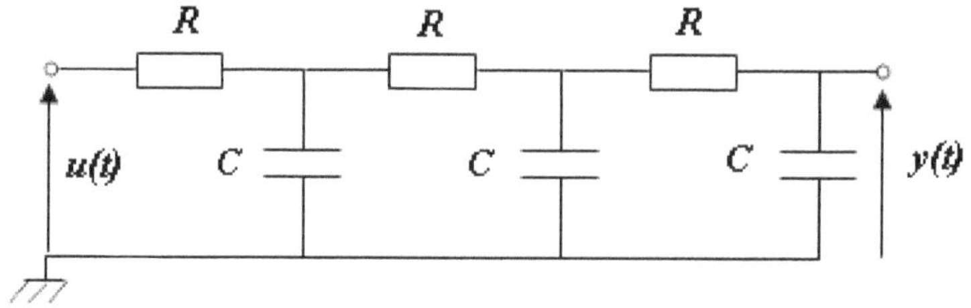

Figure 4. Photo of the real system

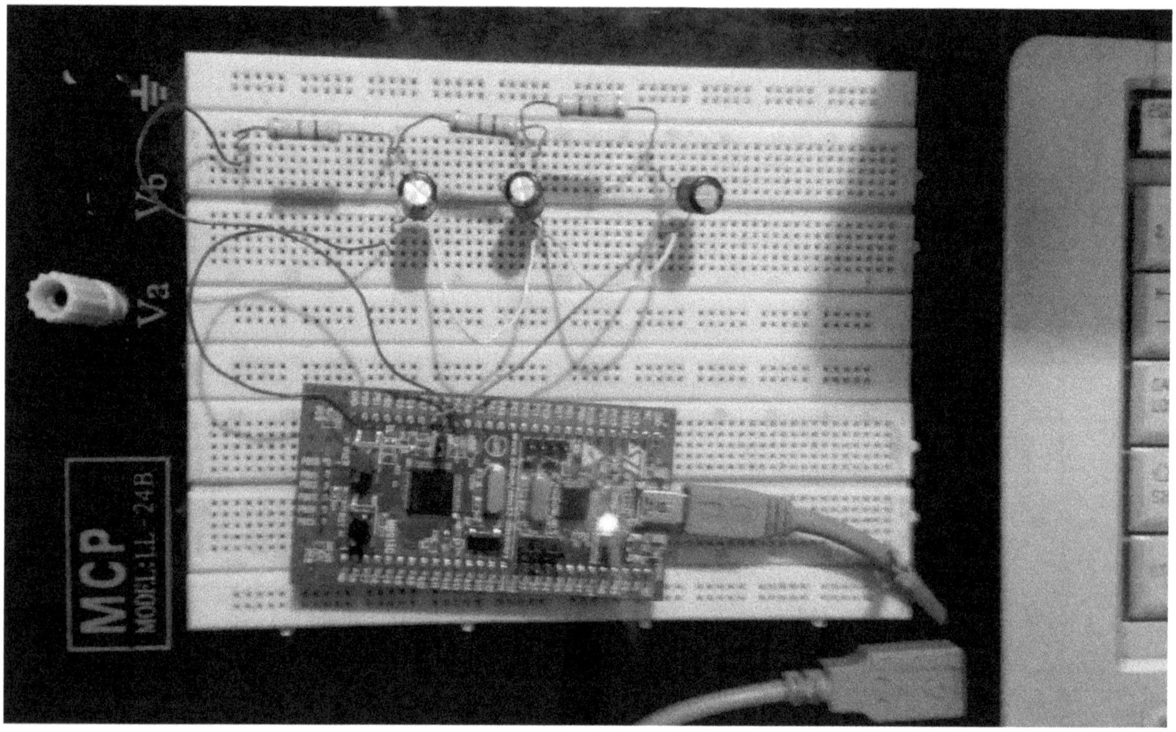

Controllers Design

In order to compare the closed loop performances, the design of two different types of digital controllers will be presented. These controllers are:

- PI controller.
- Fixed low order controller.

By determining the system poles, it is noted that the dominant pole of the studied system is $p_1 = -0.084$. Thus, the time constant τ_1 of this system can be deduced. So, τ_1 is given by the following equation:

$$\tau_1 = -\frac{1}{p_1} = 12s \tag{25}$$

For the implementation of these different types of controllers, the sampling period T_s will be chosen such that it is much lower than τ_1.

Control Law Algorithm

The control law algorithm is explained by the following steps:

Step 1: Algorithm initialization.
Step 2: Initialization of the iteration counter.
Step 3: Acquisition of the system output $y(k)$.
Step 4: Computation of the error between the set point $y_c(k)$ and the system output $y(k)$: $e(k)=y_c(k)-y(k)$.
Step 5: Computation of the control $u(k)$.
Step 6: Sending the control to the system.
Step 7: Waiting the end of the sampling period.
Step 8: Increment the counter value.
Step 9: Go back to step 3 and continue the iteration process and repeat until the stop condition is reach.

In order to compute the error between the set point $y_c(k)$ and the system output $y(k)$ in step 4, the ADC (Analog to Digital Convert) is used to transform the analog system output value to the corresponding digital value using the following relation:

$$y = V_{REF} \frac{\text{analog value}}{4095} \tag{26}$$

In step 6, it is required to use the DAC (Digital to Analog Converter) in order to convert the control signal $u(k)$ and to send it the real system. The DAC output voltage is linear between 0 and V_{REF}. In this work, V_{REF} is equal to 3V.

PI Controller

With an aim of determining the PI controller parameters, the Ziegler-Nichols first method will be used. This method takes into account the step response of the open loop system (Lequesne, 2006) as shown in Figure 5.

The parameters L and T are computed using the open loop step response. The controller parameters Kp and Ki are then determined by using the following expressions: $Kp = 0.9\frac{T}{L}$ and $Ki = \frac{L}{0.3}$.

The PI controller is expressed by:

Figure 5. Response curve for Ziegler-Nichols first method

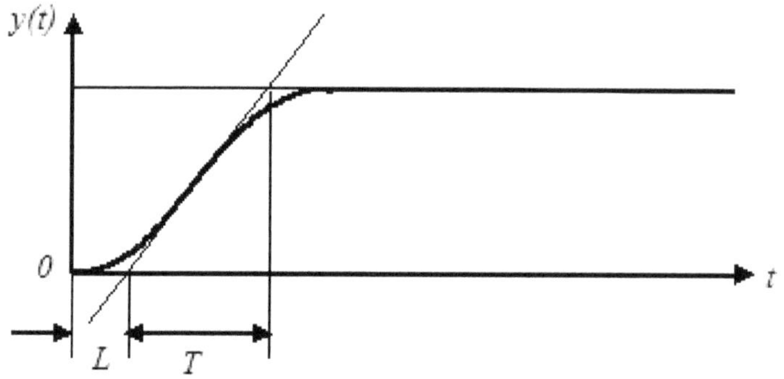

$$C(s) = K_p + K_i \frac{1}{s} \tag{27}$$

By using this method, the PI controller will be given by the following expression:

$$C_1(s) = \frac{8.738s + 1.568}{s} \tag{28}$$

By considering this continuous PI controller and by applying the first order hold (foh) discretization method, the following digital PI controller can be determined:

$$C_1(z) = \frac{8.816z - 8.66}{z - 1} \tag{29}$$

Then, the control law obtained by this controller is given by the following relation:

$$u(k) = u(k-1) + 8.816e(k) - 8.66e(k-1) \tag{30}$$

The evolutions of the set point, the output signal and the control signal, obtained with the PI controller are depicted in Figures 6 and 7.

From Figure 6, it is observed that the system output fluctuates around the set point and the control signal presents many variations. It is also noted that this PI controller allows us to obtain a fast dynamic response. It is remarkable too that the control signal which is plotted in Figure 7 is clipped because of the saturation conditions imposed to protect the microcontroller.

Since the control signal has several variations, it was decided to adjust the PI controller parameters in order to obtain a better signal control than that obtained with the Ziegler-Nichols first method. The obtained controller by adjusting the parameters is given by the following relation:

$$C_2(s) = \frac{0.3s + 0.00017}{s} \tag{31}$$

Figure 6. Set point and output obtained with the first PI controller

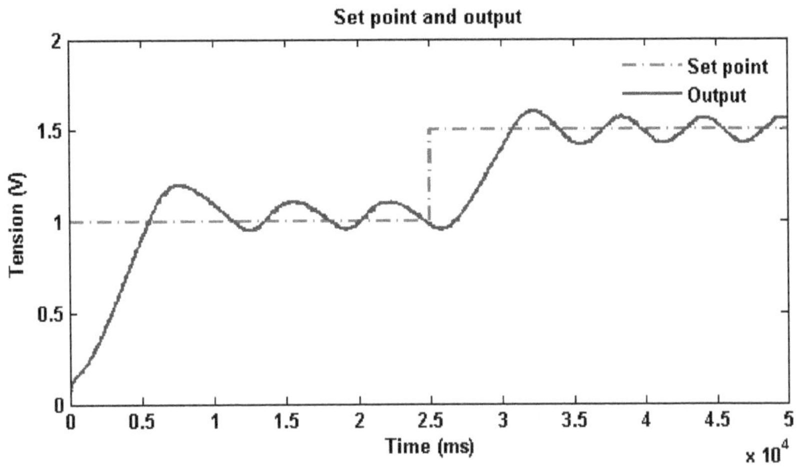

Figure 7. Control signal obtained with the first PI controller

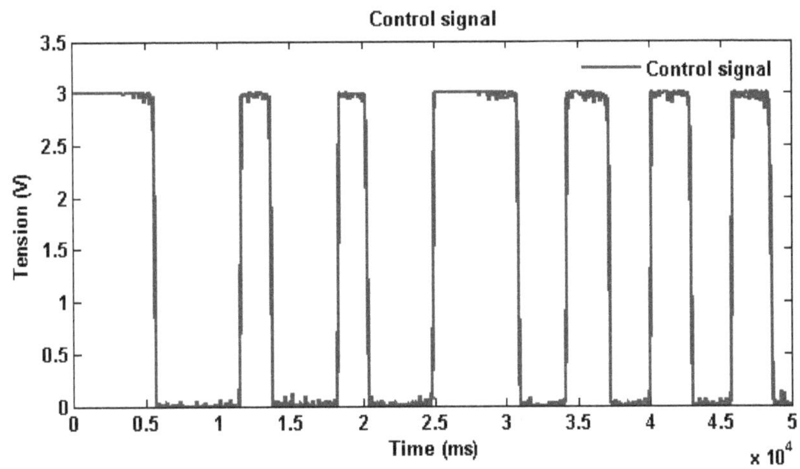

By adopting this continuous PI controller and by applying the Euler approximation method, the following digital PI controller can be obtained:

$$u(k) = u(k-1) + 0.300017e(k) - 0.3e(k-1) \tag{32}$$

The evolutions of the set point, the output signal and the control signal, obtained with this second PI controller are plotted in Figures 8 and 9 respectively.

Figure 8 shows that the system output has a high overshoot. It also illustrates that the system response presents a slow dynamics. Indeed, the required time for the system output so that it has an overshoot less than or equals to 2% is about 100s. It is also observed from Figure 9 that the control signal obtained with this second PI controller is better than that obtained with the first one. In fact, this control signal does not attain the maximum value and it does not present many fluctuations.

Figure 8. Set point and output obtained with the second PI controller

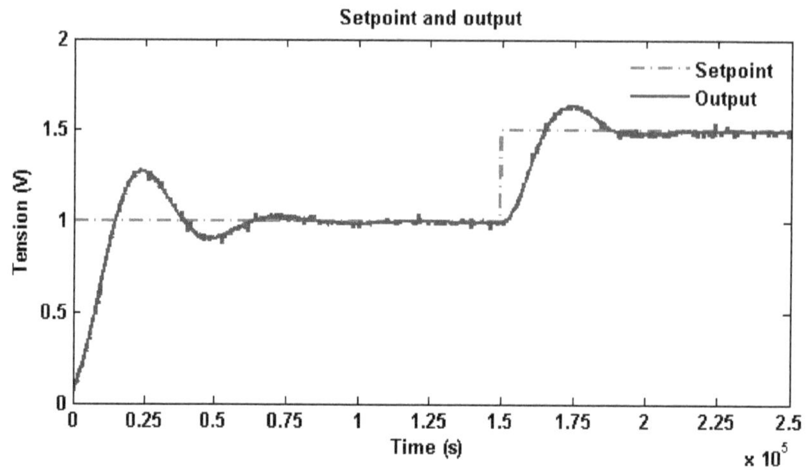

Figure 9. Control signal obtained with the second PI controller

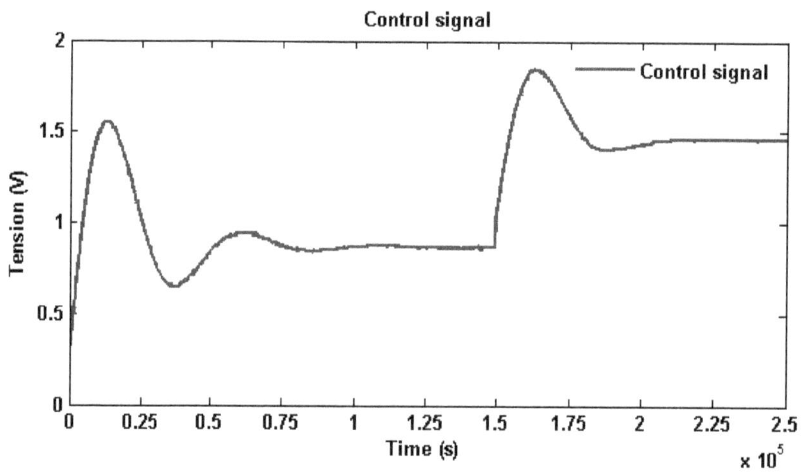

After implementing the two PI controllers presented previously, the design of two fixed low order controllers is the subject of the next subsection.

Fixed Low Order Controller

Let consider the following fixed second order controller:

$$\begin{cases} A(s) = s^2 + a_1 s + a_0 \\ B(s) = b_0 \end{cases} \tag{33}$$

Assume that the objective is to synthesize a controller for the real system with the following specifications:

1. Overshoot $\leq 2\%$.
2. 2% settling time ≤ 16 s.

These performances can be achieved by choosing the parameters $\alpha_1 = 3.3$ and $\tau = 7$ which satisfy the design specifications. The resolution of the optimization problem by using the GGP method leads to the following controller:

$$C_2(s) = \frac{22.18}{s^2 + 6.874s + 18.815} \tag{34}$$

This continuous fixed low order controller is discretized by using the 'foh' method. So, the digital fixed low order controller will be written as follows:

$$C_2(z) = \frac{0.0311z^2 + 0.105z + 0.022}{z^2 - 1.369\ z + 0.503} \tag{35}$$

The recursive equation of the proposed controller is given by the following relation:

$$u(k) = 1.369u(k-1) - 0.503u(k-2) + 0.031e(k) + 0.105e(k-1) + 0.022e(k-2) \tag{36}$$

The evolutions of the output, the set point and the control signal, obtained by applying this fixed second order controller to the real system, are shown in Figures 9 and 10 respectively.

It is remarkable from Figure 10 that, in spite of the variation of the set point, the designed controller meets the specified performances which are a response with an overshoot about 2% and a settling time less than or equals to 16s.

Figure 10. Set point and output signals obtained with the first fixed low order controller

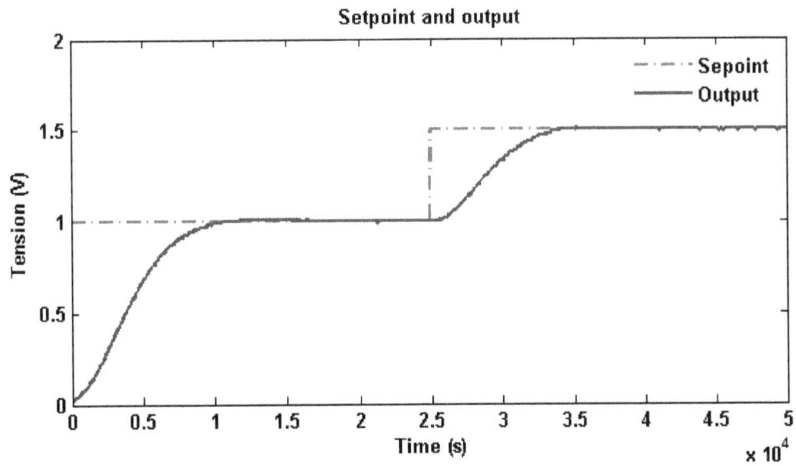

Figure 11. Control signal obtained with the first fixed low order controller

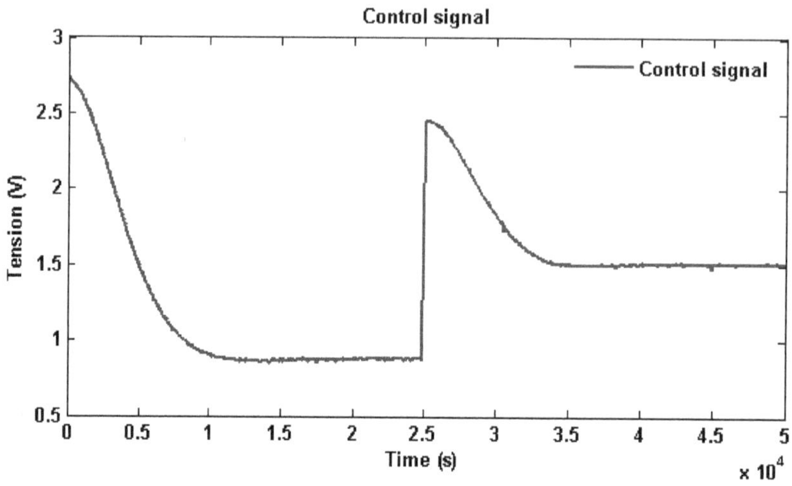

It is observed from Figure 11 that the control signal obtained with this fixed low order controller provides smoother variations compared to the control signals obtained with a PI controller. It is also noted that this control signal does not attain the maximum value that can supported by the microcontroller.

Let the purpose be the design of a controller which will be able to guarantee the following specifications:

1. Overshoot $\leq$ 10%.
2. 2% settling time $\leq$ 15 s.

With an aim of achieving these specifications, the following parameters $\alpha_1 = 2.88$ and $\tau = 4.9$ are chosen. The resolution of the non convex optimization problem by applying the GGP method allows us to obtain the following controller:

$$C_3(s) = \frac{21.9}{s^2 + 0.365s + 6.16} \tag{37}$$

The digital fixed low order controller obtained by using the 'foh' method will be expressed by:

$$C_3(z) = \frac{0.809z^2 + 2.85\ z + 0.736}{z^2 + 0.597\ z + 0.833} \tag{38}$$

Then, the control law obtained by using this controller is given by the following relation:

$$u(k) = -0.597u(k-1) + 0.833u(k-2) + 0.809e(k) + 2.85e(k-1) + 0.736e(k-2) \tag{39}$$

The output signal and the control signal which are plotted, respectively, in Figures 12 and 13, are obtained by using this controller with the real system.

Figure 12. Set point and output signals obtained with the second fixed low order controller

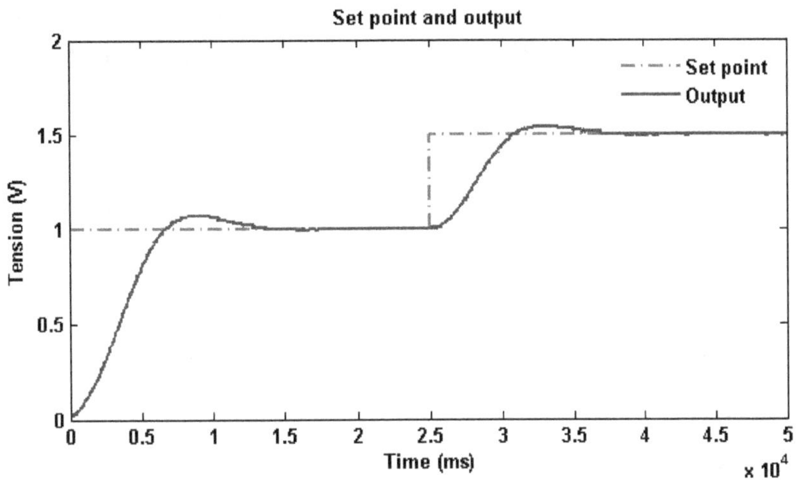

Figure 13. Control signal obtained with the second fixed low order controller

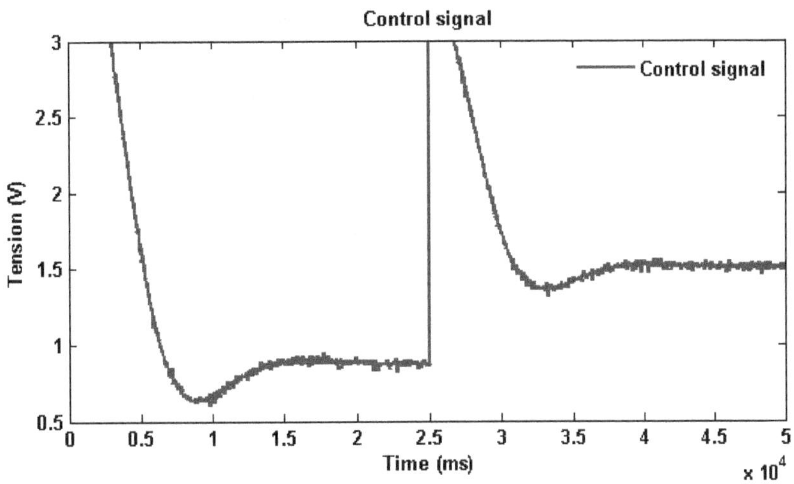

From Figure 12, it is noted that the implemented control law has been able to ensure the desired closed loop performances which are a system output with an overshoot about 10% and a settling time less than or equals to 16s.

DISCUSSION

The practical implementation of a fixed low order controller on a real electronic system by the mean of STM32 microcontroller was presented in this work. In order to find the controller parameters; a non convex optimization problem must be solved. The resolution of such type of problem by the use of a local optimization method may lead us to obtain a local solution. Therefore, the resulted control law is not optimal. So, with the purpose of overcoming this problem and find a global solution, a global optimization approach is used. This method is called the GGP. In fact, the GGP method consists in making some mathematical transformations to the non convex optimization problem. So, by the mean of the GGP the optimization problem, initially non convex, is becoming convex. Therefore, its resolution by a local optimization method allows us to find the controller parameters. Consequently, the achieved control law is optimal. It should be noted that the GGP method relies on a mathematical foundation. The use of the GGP method does not require parameters setting. There are also other types of optimization methods that are metaheuristic optimization methods such as the tabu search or as the GA. It should be noted that the success of genetic algorithms depends on the probabilities of crossover and mutation. In this work, the resolution of the optimization problem is performed offline. Accordingly, only the control law is implemented and computed online. As a result, the implemented control law does not occupy memory space. In fact, you need only a few values of the control signal u(k) and the error e(k) as indicated in relations (34) and (37). In order to shed light the efficiency of the proposed controller; a comparison with the PI controller was made. The real implementation of the proposed control law shows that the fixed low order controller allow us, despite varying the set point, to guarantee the desired closed loop time response specifications which are the overshoot and the settling time for the system output. It is, also, observed that it is possible to specify different time response performances for the same system. Practical results show, also, that the PI controller may not guarantee the required performances. It should be noted that it is relatively difficult, by using the Ziegler Nichols first method, to determine the PI controller parameters values which will allow the achievement of the desired closed loop performances. It was shown that the system response is fast but it fluctuates around the set point. Control signal obtained with the PI controller presents many variations contrary to the one obtained with the low order controller. Since the control signal has several variations, it was decided to adjust the PI controller parameters in order to obtain a better signal control than that obtained with the Ziegler-Nichols first method. Given the fact that the control signal obtained with the first PI controller presents several variations, the idea was to synthesize another PI controller that will allow us to obtain a better control signal. A second PI controller is found by the parameters adjustment. By implementing the designed control law, it was illustrated that the system response obtained with the second PI controller is slow. It is also noted that the control signal is smoother than that obtained by the first PI controller. Indeed, it does not present many fluctuations around the set point and also it does not attain the maximum value that can support the microcontroller.

CONCLUSION AND FUTURE RESEARCH DIRECTIONS

The implementation of a fixed low order controller by the use of the STM32 microcontroller has been the object of this work. This controller is designed so as to ensure certain closed loop performances. The controller parameters were obtained by solving a non convex optimization problem. So, the resolution of such type of problems with local optimization methods may lead to local solution. Therefore, with an aim of obtaining an optimal control law, a global optimization method was used to resolve this optimization problem.

Moreover, the PI controller has been tested with the same microcontroller and compared with the proposed controller. Experimental results show the effectiveness of the fixed low controller.

The considered system in this work presents fixed parameters that do not vary over time. But, on one hand, many real systems are nonlinear and they cannot be described by systems with fixed parameters.

On the other hand, this type of system can be modeled by a transfer function with uncertain parameters. Therefore, the control of such systems will be the subject of the future work.

REFERENCES

Abu-Seada, H. F., Mansor, W. M., Bendary, F. M., Emery, A. A., & Moustafa Hassan, M. A. (2013). Application of Particle Swarm Optimization in Design of PID Controller for AVR System. *International Journal of System Dynamics Applications*, *2*(3), 1–17. doi:10.4018/ijsda.2013070101

Ben Hariz, M., Bouani, F., & Ksouri, M. (2012). Robust controller for uncertain parameters systems. *ISA Transactions*, *51*(5), 632–640. doi:10.1016/j.isatra.2012.04.007 PMID:22749294

Ben Hariz, M., Chagra, W., & Bouani, F. (2013). Controllers design for MIMO systems with time response specifications. In *Proceedings of International Conference on Control, Decision and Information Technologies* (CoDIT). (pp. 573-578). GoDIT. doi:10.1109/CoDIT.2013.6689607

Bjork, K., Lindberg, P., & Westerlund, T. (2003). Some convexifications in global optimization of problems containing signomial terms. *Computers & Chemical Engineering*, *27*(5), 669–679. doi:10.1016/S0098-1354(02)00254-5

Choi, C., & Dennis, L. (1996). Effectiveness of a geometric programming algorithm for optimization of machining economics models. *Computers & Operations Research*, *23*(10), 957–961. doi:10.1016/0305-0548(96)00008-1

Jayaraman, Y., & Ravindran, U. (2008). Fpga implementation of predictive control strategy for power factor correction. *In proceedings of World Academy of Science. Engineering and Technology*, *15*, 199–204.

Jiang, S., Ji, Z., & Shen, Y. (2014). A novel hybrid particle swarm optimization and gravitational search algorithm for solving economic emission load dispatch problems with various practical constraints. *International Journal of Electrical Power & Energy Systems*, *55*(1), 628–644. doi:10.1016/j.ijepes.2013.10.006

Jin, L., & Kim, Y. C. (2008). Fixed, low-order controller design with time response specifications using non-convex optimization. *ISA Transactions*, *47*(4), 429–438. doi:10.1016/j.isatra.2008.05.001 PMID:18606409

Kheriji, A., Bouani, F., & Ksouri, M. (2011a). A GGP approach to solve non convex min-max predictive controller for a class of constrained MIMO systems described by state-space models. *International Journal of Control, Automation, and Systems*, *9*(3), 452–460. doi:10.1007/s12555-011-0304-2

Kheriji, A., Bouani, F., & Ksouri, M. (2011c). A Microcontroller Implementation of Constrained Model Predictive Control. *International Journal of Electrical and Electronics Engineering*, *5*(3), 199–206.

Kheriji, A., Bouani, F., Ksouri, M., & Ben Ahmed, M. (2011b). A Microcontroller Implementation of Model Predictive Control. *World Academy of Science. Engineering and Technology*, *53*(1), 68–74.

Kim, Y., Kim, K., & Manabe, S. (2004). Sensitivity of time response to characteristic ratios. In *Proceeding of the 2004 American Control Conference*, (pp. 2723-2728). Academic Press.

Kim, Y. C., Keel, L. H., & Bhattacharyya, S. P. (2003). Transient response control via characteristic ratio assignment. *IEEE Transactions on Automatic Control*, *48*(12), 2238–2244. doi:10.1109/TAC.2003.820153

Lequesne, D. (2006). Régulation PID Analogique-numérique-floue. Paris: Hermès, Lavoisier.

Lian, J. (2011). The Design of Gas Drainage Holes' Opening Parameters Intelligent Measurement and Control System for Coal Mine. *Procedia Earth and Planetary Science*, *3*(1), 331–337. doi:10.1016/j.proeps.2011.09.102

Liberti, L., & Maculan, N. (2006). *Global Optimization From Theory to Implementation*. Springer.

Ling, K., Wu, B., & Maciejowski, J. (2008). Embedded model predictive control (mpc) using a fpga. In *Proceedings of the 17th International Federation of Automatic Control, Seoul World Congress* (IFAC), (pp. 15250-15255). IFAC.

Ling, K. V., Yue, S. P., & Maciejowski, J. M. (2006). A fpga implementation of model predictive control. In *Proceedings of the 2006 American Control Conference*, (pp. 1930 – 1935). Academic Press. doi:10.1109/ACC.2006.1656502

Mamdoohi, G., Abas, A. F., & Samsudin, K. (2012). Implementation of genetic algorithm in an embedded microcontroller based polarization control system. *Engineering Applications of Artificial Intelligence*, *25*(4), 869–873. doi:10.1016/j.engappai.2012.01.018

Maranas, C., & Floudas, C. (1997). Global optimization in generalized geometric programming. *Computers & Chemical Engineering*, *21*(4), 351–369. doi:10.1016/S0098-1354(96)00282-7

Nand, K. (1995). Geometric programming based robot control design. *Computers & Industrial Engineering*, *29*(1), 631–635.

Pörn, R., Bjork, K., & Westerlund, T. (2008). Global solution of optimization problems with signomial parts. *Discrete Optimization*, *5*(1), 108–120. doi:10.1016/j.disopt.2007.11.005

Schöbel, A., & Scholz, D. (2014). A solution algorithm for non-convex mixed integer optimization problems with only few continuous variables. *European Journal of Operational Research*, *232*(1), 266–275. doi:10.1016/j.ejor.2013.07.003

Servet Kiran, M., Gunduz, M., & Kaan Baykan, O. (2012). A novel hybrid algorithm based on particle swarm and ant colony optimization for finding the global minimum. *Applied Mathematics and Computation*, *219*(1), 1515–1521. doi:10.1016/j.amc.2012.06.078

STMicroelectronics. (2012). *Datasheet stm32f100x4, stm32f100x6, stm32f100x8, stm32f100xb*. Doc ID 16455 Rev 7. Retrieved from http://www.st.com

Tabakhi, S., Moradi, P., & Akhlaghian, F. (2014). An unsupervised feature selection algorithm based on ant colony optimization. *Engineering Applications of Artificial Intelligence*, *32*(1), 112–123. doi:10.1016/j.engappai.2014.03.007

Toksari, M. D. (2009). Minimizing the multimodal functions with Ant Colony Optimization approach. *Expert Systems with Applications*, *36*(3), 6030–6035. doi:10.1016/j.eswa.2008.06.077

Toledo, C. F. M., Oliveira, L., & França, P. M. (2014). Global optimization using a genetic algorithm with hierarchically structured population. *Journal of Computational and Applied Mathematics*, *261*(1), 341–351. doi:10.1016/j.cam.2013.11.008

Tsai, J. (2009). Treating free variables in generalized geometric programming problems. *Computers & Chemical Engineering*, *33*(1), 239–243. doi:10.1016/j.compchemeng.2008.08.011

Tsai, J., Lin, M., & Hu, Y. (2007). On generalized geometric programming problems with non positive variables. *European Journal of Operational Research*, *178*(1), 10–19. doi:10.1016/j.ejor.2005.11.037

Valdez, F., Melin, P., & Castillo, O. (2014). Modular Neural Networks architecture optimization with a new nature inspired method using a fuzzy combination of Particle Swarm Optimization and Genetic Algorithms. *Information Sciences*, *270*(20), 143–153. doi:10.1016/j.ins.2014.02.091

Wang, K., Li, P., Liu, J., & Ning, D. (2011). Application of μc/os - II in the Design of Mine dc Electrical Prospecting Instrument. *Procedia Earth and Planetary Science*, *3*(1), 485–492. doi:10.1016/j.proeps.2011.09.124

Zhang, H., & Kang, W. (2013). Design of the Data Acquisition System Based on STM32. *Procedia Computer Science*, *17*(1), 222–228. doi:10.1016/j.procs.2013.05.030

Zhang, H., & Zhao, J. (2011). The design of RF data acquisition system based on STM32 and FPGA. In *Proceedings of International Conference on Multimedia Technology (ICMT)*, (pp. 832 – 834). ICMT. doi:10.1109/ICMT.2011.6002030

Zhen, Z., & Yan, G. (2013). Design of the fuzzy PID controller for the hot runner temperature control system. In *Proceedings of 32nd Chinese Control Conference (CCC)*, (pp. 3464 – 3469). CCC.

Zhou, Y., Xiao, K., Wang, Y., Liang, A., & Hassanien, A. E. (2013). A PSO-Inspired Multi-Robot Map Exploration Algorithm Using Frontier-Based Strategy. *International Journal of System Dynamics Applications*, *2*(2), 1–13. doi:10.4018/ijsda.2013040101

KEY TERMS AND DEFINITIONS

Generalized Geometric Programming: Global optimization method which consists in transforming an optimization problem originally non convex to a convex one by mean of some mathematical transformations.

Global Optimization: It is a branch of numerical analysis and applied mathematics that deals with the optimization of an objective function with respect to some criteria.

Low Order Controller: A controller with a reduced order.

Non Convex Optimization Problem: It is an optimization problem which may have multiple locally optimal points.

Practical Implementation: It consists in the implementation of an algorithm on a real system.

STM32 Microcontroller: It is 32-bit Flash microcontrollers based on the ARM Cortex™-M3 processor.

Time Response Specifications: They represent specifications for the time response of the closed loop system.

APPENDIX A

Let consider the following function *f(x)* defined by (Liberti, & Maculan, 2006):

$$f(x) = x^4 - 3x^3 - 1.5x^2 + 10x \quad (A.1)$$

where $-2 \le x \le 3$.

Figure 14 shows the function *f(x)* to be optimized over the definition interval of the variable *x*.

Figure 14. Representation of functions f(x) and f(x1)

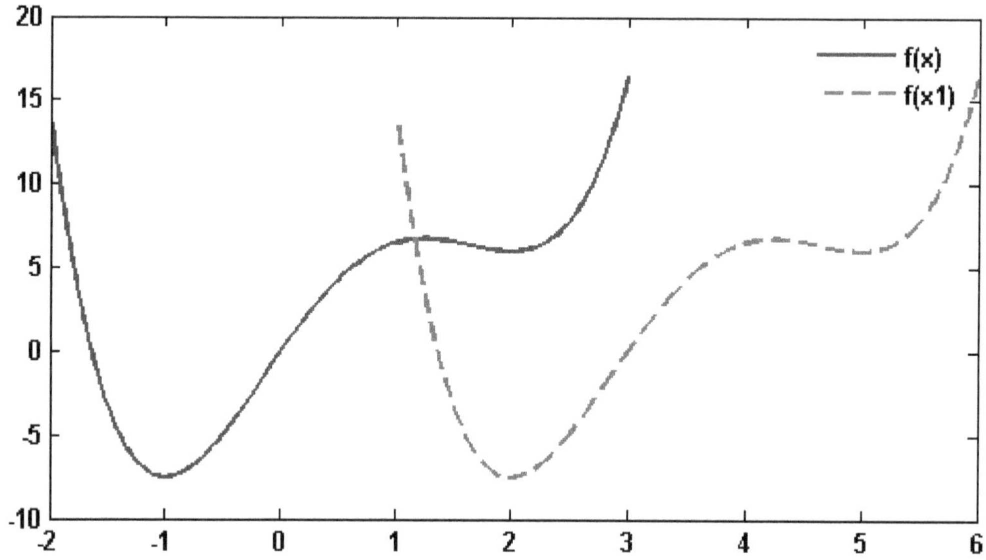

It is noted from Figure 14, that the function *f(x)* has a local minimum for *x=2* and a global minimum for *x=-1*. Afterwards, the methodology of the application of the GGP method that was presented in this work will be applied in order to obtain the global minimum of this function.

First, it is observed that the definition domain of the function *f(x)* has negative values. Then, a transformation of this area is suggested in order to use the proposed optimization approach. So, it is required to make a translation by introducing a new variable $x_1 = x + 3$, and the function *f(x)* will be written as follows:

$$f(x_1) = (x_1 - 3)^4 + 25.5x_1^2 - 62x_1 + 37.5 - 3x_1^3 \quad (A.2)$$

where $1 \le x_1 \le 6$.

Figure 14 also illustrates the translation applied to the function *f(x)*.

Based on the analysis of the function given in the relation (A.2) by the GGP method, it is remarkable that the monomials "$(x_1-3)^4$", "$25.5x_1^{\,2}$" and "$-62x_1$"are convex whereas the monomial "$-3x_1^{\,3}$" is non convex according to proposition 2.

Consequently, if a local optimization method such that "*fmincon*" is applied and an initial point like $x_1=4.5$ is chosen, the found minimum in this case is equal to $x_1=5$, which is a local minimum. By considering another initial point such as $x_1=3$, this optimization algorithm will converge to the global minimum that is $x_1=2$.

So with an aim to convexify this function, some transformation techniques will be applied to the non convex monomial. In the present case, power transformation is necessary, in order to convexify the monomial "$-3x_1^{\,3}$". New variable z_1 is introduced such that: $x_1 = z_1^{\frac{1}{3}}$.

Then a new convex function is obtained and is given by the following expression:

$$f(x_1, z_1) = \left(x_1 - 3\right)^4 + 25.5x_1^2 - 62x_1 + 37.5 - 3z \tag{A.3}$$

In order to find the global minimum of the function $f(x_1,z_1)$, it is sufficient to resolve the following constrained optimization problem:

$$\begin{aligned} &\min\left(x_1 - 3\right)^4 + 25.5x_1^2 - 62x_1 + 37.5 - 3z \\ &\text{subject to} \begin{cases} z = x_1^3 \\ 1 \le x_1 \le 6 \\ 1 \le z \le 216 \end{cases} \end{aligned} \tag{A.4}$$

The resolution of this optimization problem with a local optimization method and whatever the initial point chosen, leads to following solution: $x_1=2$. Finally, to find the global minimum of the original function $f(x)$, it suffices to do the reverse translation on the variable is x_1 which is $x=x_1-3$. This translation allows us to obtain the global solution which is $x=-1$.

APPENDIX B

The objective function to minimize in order to obtain the first fixed low order controller is given by the following expression:

$$f(X) = c_1x_1x_2 + c_2x_2x_3 + c_3x_1^2 + c_4x_2^2 + c_5x_3^2 - c_6x_1 - c_7x_2 - c_8x_3 + c_9 \quad \text{(B.1)}$$

where $c_i>0$ for $i=1,\ldots,9$.

It is noted that the first two terms of second member of (B.1) are non convex. By applying the convexification strategy of the GGP method previously presented, it is observed that:

$c_1x_1x_2$ and $c_1x_2x_3$ have a positively signed term, so the exponential transformation will be applied. New variables y_i are introduced for $i=1,\ldots,4$ such that:

$$x_i = \exp(y_i) \quad \text{(B.2)}$$

A new objective function $f(X)$ can be obtained. This function which is given by equation (B.3) is now convex according to section 5.

$$f(X) = c_1 \exp(y_1)\exp(y_2) + c_2 \exp(y_2)\exp(y_3) + c_3x_1^2 + c_4x_2^2 + c_5x_3^2 - c_6x_1 - c_7x_2 - c_8x_3 + c_9 \quad \text{(B.3)}$$

In order to obtain the global minimum; the following constrained optimization problem that is given by (B.5) must be solved. The solution corresponds to the controller parameters given by equation (28).

$$\begin{aligned} &\min\ f(X) = c_1 \exp(y_1)\exp(y_2) + c_2 \exp(y_2)\exp(y_3) + \\ &c_3x_1^2 + c_4x_2^2 + c_5x_3^2 - c_6x_1 - c_7x_2 - c_8x_3 + c_9 \\ &\text{subject to } y_i = \ln(x_i) \end{aligned} \quad \text{(B.4)}$$

Chapter 24
Analysis of Renewable Energy Power Systems:
Reliability and Flexibility during Unbalanced Network Fault

Rabeh Abbassi
University of Tunis, Tunisia

Salem Saidi
University of Tunis, Tunisia

Manel Hammami
University of Tunis, Tunisia

Souad Chebbi
University of Tunis, Tunisia

ABSTRACT

Due to different concerns, renewable energies stand out as an opportunity for a sustainable future. Wind/Photovoltaic Hybrid Systems are one important type of renewable energy power systems. In order to increase renewable energies integration rate, this chapter proposes a utility interactive grid-connected wind-PV hybrid system with storage batteries. Facing modern electricity industries, the hybrid system considers the dispatch-ability and the quality of its power injection into the grid. The adopted modeling approach has been based on energy exchange considerations taking into account the converters structures. Particular attention has been given to cases where the hybrid system is connected to an unbalanced grid. This chapter develops a new control strategy, which aims to isolate the hybrid system of the adverse impact of the grid fault. The results point out thoroughly the applicability of the proposed control scheme under unbalanced grid conditions.

DOI: 10.4018/978-1-4666-7248-2.ch024

1. INTRODUCTION

Nowadays, the polluting effects associated with the growth of energy consumption in all conventional forms are at the heart of concerns about the future of the planet to deal with the requirements of sustainable development and environmental care (Chao & Chuang-lin, 2013; Draghicescu et al., 2010). In this sense and at the onset of the liberalization of electricity markets, new offerings of electricity generation are being developed. A major advantage was developed by renewable energies which intervene increasingly in the electricity market.

In addition, various renewable sources of electricity generation have been massive investments to increase their penetration rates: the most widespread are wind Energy Conversion Systems (WECS) (Zuher & Mehrdad, 2013) and Photovoltaic Systems (PVS). Whether wind or photovoltaic sources, these electricity generating systems have fluctuating nature. In order to reduce the random and unpredictable nature of a renewable source potential, the multiplication of various kinds of sources is needed as a convincing solution.

Among the combinations of hybrid systems, that of wind-PV is particularly distinguished giving the availability, the ecological balance favor and the ability to combine them with other sources. Indeed, the advantage of a hybrid system compared to pure wind or photovoltaic pure system depends on many fundamental factors: the type of load, the wind speed, the solar radiation, the cost and the availability of energy, the relative cost of wind machine, the photovoltaic panels, the storage system efficiency and other factors.

Assessment methods of stand-alone (Qi et al., 2011; Valenciaga et al., 2005 ; Kellogg et al., 1998; Shi et al., 2008) or grid-connected (Mohammadi et al., 2012; Abbassi 1 & Chebbi, 2012; Kim et al., 2008; Chen et al., 2007) wind-photovoltaic hybrid system and their advantages have grown dramatically in recent years. But given the constraints that limit their use, namely production capacity often uncertain and fluctuating uncorrelated to load changes, optimization of the wind-photovoltaic energy depends heavily on economic models of each system taken separately.

The Technologies of hybrid renewable energy systems (HRES) are becoming among the most promising production technologies driving the demand for distributed generation. Indeed, the combination of renewable energy sources can provide a solution for generating electricity in remote areas and can fully optimize the power generation systems, both from a technical and economic point of view. In addition, wind-solar hybrid systems are distributed generation which could be integrated into the grid.

In addition, the increasing penetration of distributed power generation systems made it necessary to think more to conditions of their connection to balanced or unbalanced network. One appropriate requirement is that distributed power generation systems should provide systems service even under grid imbalances. Indeed, facing the behavior of different sources of renewable energy, we thought to use the power electronics that initiated the revolution of hybrid systems. We intend to operate in exploiting the possibilities of controlling power converters that are interfaces between decentralized production units and the network. This is essential because if these systems are not properly controlled, their connection to the utility grid can produce problems on the consumers side (Abbassi 1 & chebbi, 2012; Xiong et al., 2010; Yazdani, 2009). Therefore, considerations of the production of electric power, safe operation and synchronization of the network must be made before connecting these systems to the utility grid (Saccomando et al., 2001; Meddeb et al., 2010; Suul et al., 2012); (Karimi et al., 2012; Karimi et al., 2004).

The presented work focuses on this application and aims to achieve a beneficial impact on energy production in terms of cost and availability, and to solution the problem of the impact of energy on renewable power systems. The insertion of large-scale means Decentralized production in medium voltage networks, cause new problems that are causing the occurrence of instabilities. The first difficulties to be solved involve an assessment of power to be coupled. Moreover, given the very random and unpredictable nature of renewable sources, it is interesting to use a fairly dynamic storage system (SS) able to act according to the fluctuations of the generated and consumed power allowing him to play as a buffer between the source and the consumer (Abbassi 1 & chebbi, 2012; Mashhour et al., 2010; Kim et al., 2008; Zhou et al., 2008; Abbassi 2 & Chebbi, 2012).

To achieve this purpose, a description of the models of the distributed generation sources present in the hybrid system is required. For both sources (WECS and PVS), it is essential to seek a modeling approach based on considerations of energy exchanges to achieve deduct the appropriate control strategies while giving importance to the power electronic converters: elements that are at the intersection of the power axis and the control axis.

In this context, the contribution envisaged by this work is to collaborate to the optimal design of a grid-connected wind-photovoltaic-batteries production system. The first part presents a bibliographic synthesis of research work related to the models studies and control strategies of grid-connected hybrid systems equipped with their cascaded converters. It should also be mentioned that the storage systems, so far relatively little present in electrical networks, are beginning to make their appearance significantly. This new context has been, also, subject of the developed bibliographic study.

In a last phase of the bibliographic study, we have presented a tools development that dealt with the electrical quantities of systems modeled in unbalanced mode (Karimi et al., 2004, Meddeb et al., 2010; Suul et al., 2012); (Karimi et al., 2012; Vechiu et al., 2013). After having raised the main disadvantages that may cause a voltage imbalance occurring at the point of common coupling, we have identified the origins of these nuisances to process them properly.

In order to apply an adequate control for the hybrid system, its modeling is an indispensable task for its study and the mastery of its operation. The second part is devoted to this task. A first sub-part is interested in the WECS which integrates: a small wind turbine, a synchronous generator, and a three-wire two-level AC-DC rectifier. The choice of models of the system has been carefully done to find the configuration that maximizes the delivered power. A second sub-part is devoted to the modeling and control strategies of PV system. The control laws are developed in order to enable the PVS to track the maximum power point by the control adjustment of the introduced boost chopper. The third sub-section focuses on the modeling and the control system of the batteries storage system, necessary especially in the case of standalone installations.

2. RELATED WORK

2.1 Control of Grid Connected VSI in Balanced Mode

For a long time, the topic of VSCs control has started receiving significant attention. An approach based on the concept of Direct Power Control (DPC) was presented in (Saîdi et al. 2014), but similar methods had already been presented in others works. In (Wang et al., 2012), the Direct Torque Control (DTC) modulation technique has been applied for control of grid connected converter was in analogy to the control of induction machines. Other researchers have been proposed the voltage oriented control (VOC) (Blaabjerg et al., 2006).

Figure 1. Overview of cascaded control loops for conventional VSC control system

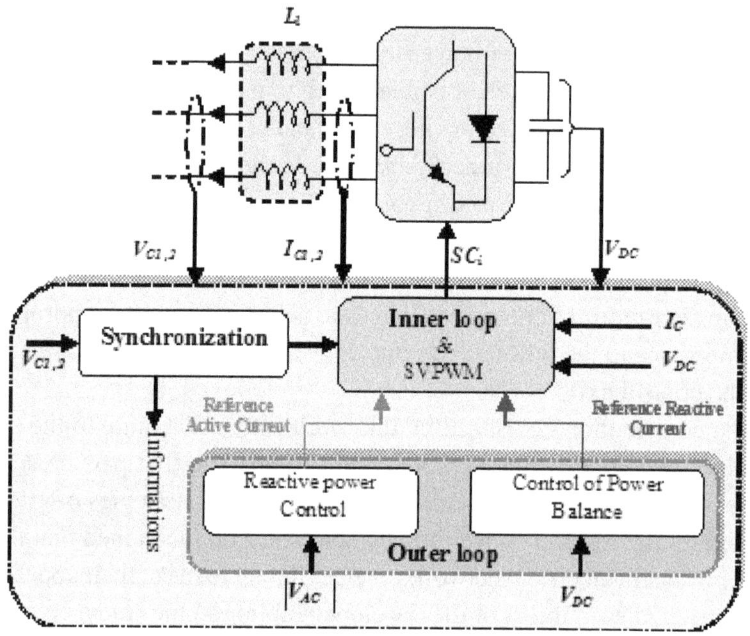

Generally, the principal objective of the control techniques of the inverters, is to allow the obtaining of sinusoidal voltages with adjustable amplitude and fundamental frequency, by eliminating or postponing to higher frequencies the harmonic parasites components resulting from the switching. Their use is equivalent, finally, to the active filtering principle, via the control system, to which will be associated a passive filtering.

For an efficient transfer of power from one system to renewable energy to the electrical network, several types of electronic converters for power, since the converters AC/AC three phase to the DC/DC converter are used. In particular, the of voltage source converters (VSC) are considered, over the past two decades, as the dominant topology for controlled three phase applications.

This development was promoted by the Pulse Width Modulated (PWM) operation, which limits the distortion of the currents and decreases the filtering requirements with the ability to control the power factor and the voltage of the DC-bus (Suul et al., 2012; Meddeb et al., 2010).

Hence, three-phase voltage source converters can be designed and implemented with great flexibility and for a wide range of applications and levels of voltage and power. Nevertheless, the principles of operation and the general structure and functionality of the control system share the general structure based on three main elements (Sul et al., 2012): grid synchronization, inner loop and modulation and outer loop control (Figure 1). These three processes can be described, briefly, as follows:

2.2 Effects of Voltage Imbalance

Knowing that one of the problems which confront the control strategies of hybrid systems is the voltage imbalance of the network to which they are connected, we are interested in what follows to the study of adverse effects that can result from such a fault affecting the utility grid.

The resolution methods of unbalanced three-phase systems (Vechiu et al. 2013; Xia et al., 2012; Meddeb et al., 2010) are based in particular on the theory of Fortescue dealing with complex quantities with predefined real and imaginary part. The main idea of this theory is based on the fact that any unbalanced three-phase voltage (current) system can be decomposed into three balanced sequences: namely the Positive-sequence, the Negative-sequence and the Zero-sequence. Positive-sequence, Negative-sequence and Zero-sequence of three unbalanced voltages vectors $\left[\underline{v}_{ar}\ \underline{v}_{br}\ \underline{v}_{cr}\right]$ are expressed, respectively, according to equations (1), (2) and (3).

$$\begin{bmatrix} \underline{v}_{ar}^{P} \\ \underline{v}_{br}^{P} \\ \underline{v}_{cr}^{P} \end{bmatrix} = \frac{1}{3}.\begin{bmatrix} 1 & a & a^2 \\ a^2 & 1 & a \\ a & a^2 & 1 \end{bmatrix}.\begin{bmatrix} \underline{v}_{ar} \\ \underline{v}_{br} \\ \underline{v}_{cr} \end{bmatrix} \tag{1}$$

$$\begin{bmatrix} \underline{v}_{ar}^{N} \\ \underline{v}_{br}^{N} \\ \underline{v}_{cr}^{N} \end{bmatrix} = \frac{1}{3}.\begin{bmatrix} 1 & a^2 & a \\ a & 1 & a^2 \\ a^2 & a & 1 \end{bmatrix}.\begin{bmatrix} \underline{v}_{ar} \\ \underline{v}_{br} \\ \underline{v}_{cr} \end{bmatrix} \tag{2}$$

$$\begin{bmatrix} \underline{v}_{ar}^{Z} \\ \underline{v}_{br}^{Z} \\ \underline{v}_{cr}^{Z} \end{bmatrix} = \frac{1}{3}.\begin{bmatrix} 1 & 1 & 1 \\ 1 & 1 & 1 \\ 1 & 1 & 1 \end{bmatrix}.\begin{bmatrix} \underline{v}_{ar} \\ \underline{v}_{br} \\ \underline{v}_{cr} \end{bmatrix} \tag{3}$$

Here: $a = e^{j.\frac{2.\pi}{3}}$

In order to apply the theory of Fortescue to instantaneous electrical quantities, we propose to transform the temporal variables in complex quantities in accordance with the principle of phasor writing. This tool, allowing to have the modulus and phase of the electric quantity to be controlled, is a good indicator of its constant or fluctuating state. So, the system tensions rewritten (4):

$$\begin{aligned} \underline{v}_{ar}(t) &= V_a.\left[\cos(\omega_s t + \phi_a) + j.\sin(\omega_s t + \phi_a)\right] \\ \underline{v}_{br}(t) &= V_b.\left[\cos(\omega_s t - \frac{2.\pi}{3} + \phi_b) + j.\sin(\omega_s t - \frac{2.\pi}{3} + \phi_b)\right] \\ \underline{v}_{cr}(t) &= V_c.\left[\cos(\omega_s t + \frac{2.\pi}{3} + \phi_c) + j.\sin(\omega_s t + \frac{2.\pi}{3} + \phi_c)\right] \end{aligned} \tag{4}$$

The direct and quadrature components of the positive, negative and zero sequence of the unbalanced voltages (currents) systems are obtained by application of Park transformation to the real part of these sequences expressed by equations (5), (6) and (7), respectively:

$$\begin{cases} v_{ar}^{P}(t) = \frac{1}{3}.(V_a + V_b + V_c).\cos(\theta_s) \\ v_{br}^{P}(t) = \frac{1}{3}.(V_a + V_b + V_c).\cos(\theta_s - 2.\pi/3) \\ v_{cr}^{P}(t) = \frac{1}{3}.(V_a + V_b + V_c).\cos(\theta_s + 2.\pi/3) \end{cases} \tag{5}$$

$$\begin{cases} v_{ar}^{N}(t) = \frac{1}{3}.(V_a.\cos(\theta_s) + V_b.\cos(\theta_s + 2.\pi/3) + V_c.\cos(\theta_s - 2.\pi/3)) \\ v_{br}^{N}(t) = \frac{1}{3}.(V_a.\cos(\theta_s + 2.\pi/3) + V_b.\cos(\theta_s - 2.\pi/3) + V_c.\cos(\theta_s)) \\ v_{cr}^{N}(t) = \frac{1}{3}.(V_a.\cos(\theta_s - 2.\pi/3) + V_b.\cos(\theta_s) + V_c.\cos(\theta_s + 2.\pi/3)) \end{cases} \tag{6}$$

$$\begin{aligned} v_{ar}^{H}(t) &= v_{br}^{H}(t) \\ &= v_{cr}^{H}(t) \\ &= \frac{1}{3}.(V_a.\cos(\theta_s) + V_b.\cos(\theta_s + 2.\pi/3) +_c .\cos(\theta_s - 2.\pi/3)) \end{aligned} \tag{7}$$

Figure 2 shows a voltage system voltage affected by an imbalance at time t = 0.04 s. The negative and zero sequences are initially zero. At the appearance of the voltage drop, the negative and zero sequences appear. They are all balanced and this is in accordance with the theory of Fortescue. The magnitude of the positive sequence voltage has become equal to 0.7. Va and amplitude of the negative sequence and zero sequence of the sequence are equal to -0.3. Va.

Since it is difficult to control these complex quantities in disturbed mode, we transform these electrical quantities referring to the Park synchronous reference frame rotating at predetermined frequency in accordance with the system of equations (8) (Figure 3).

$$\begin{cases} v_{dr} = \frac{1}{2}.\sqrt{\frac{2}{3}}.\left[\sum_i V_i + \sum_i V_i.\cos\left[2.\theta_s + 2(i-1).\pi/3\right]\right] \\ v_{qr} = -\frac{1}{2}.\sqrt{\frac{2}{3}}.\left[\sum_i V_i.\sin\left[2.\theta_s + 2(i-1).\pi/3\right]\right] \qquad i = a, b, c \\ v_{or} = \frac{1}{\sqrt{2}}.\sqrt{\frac{2}{3}}.\left[\sum_i V_i.\sin\left[\theta_s + 2(i-1).\pi/3\right]\right] \end{cases} \tag{8}$$

Figure 2. Symmetrical sequences of an unbalanced voltages system

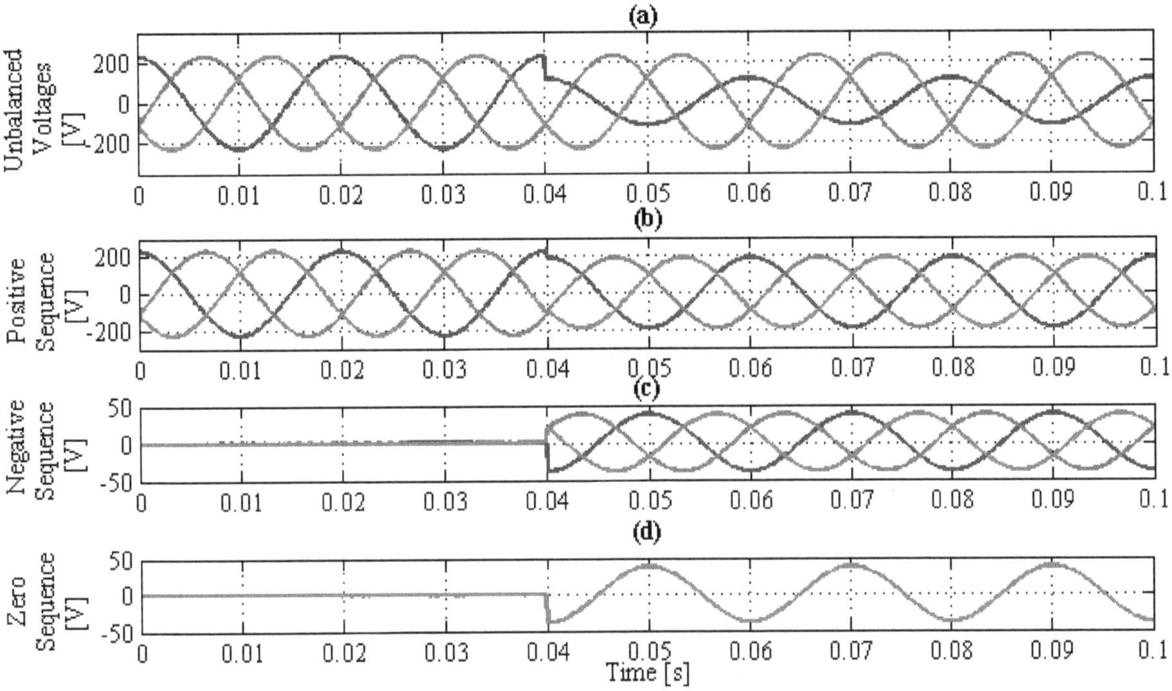

Figure 3. (d, q, o) components of the unbalanced voltages system

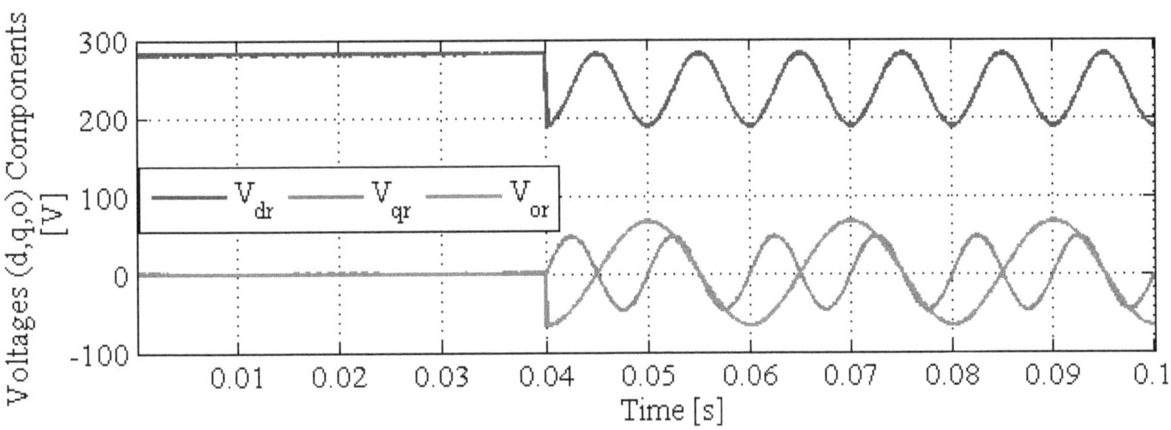

In the stationary frame (α, β), the simulation of the components of the unbalanced system shows the effect of voltage unbalance (Figure 4.).

In order to identify the origins of fluctuations recorded in the d, q components of the unbalanced voltages system (Figure 3.), our analysis was developed referring to two reference frames (Song et al., 1999; Saccomando et al., 2001; Rodriguez et al., 2006; Xu et al., 2007; Meddeb et al., 2010; Xia et al., 2012; Vechiu et al., 2013; Abbassi et al., 2014): the positive reference frame (PRF) and the negative reference frame (NRF) rotating counterclockwise at the speed of $+\omega_s$ and clockwise at $-\omega_s$ respectively.

Figure 4. Concordia components of the unbalanced voltages system

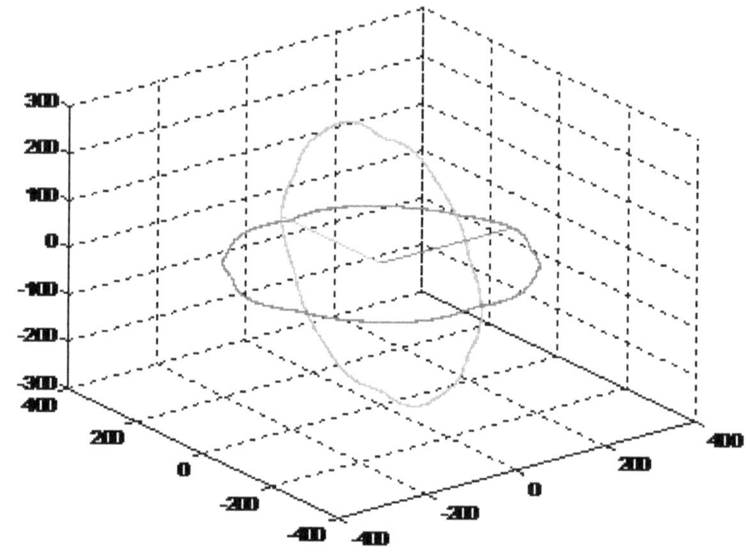

The establishement of the dq components on the PRF and NRF reference frames is ensured, respectively, by the application of two Park transformations given by (Vechiu et al., 2013; Abbassi et al., 2014):

$$\left[P(+\theta_s)\right]_{PRF} = \sqrt{\frac{2}{3}} \cdot \begin{bmatrix} \cos(\theta_s) & \cos(\theta_s - \frac{2\pi}{3}) & \cos(\theta_s + \frac{2\pi}{3}) \\ -\sin(\theta_s) & -\sin(\theta_s - \frac{2\pi}{3}) & -\sin(\theta_s + \frac{2\pi}{3}) \\ \frac{1}{\sqrt{2}} & \frac{1}{\sqrt{2}} & \frac{1}{\sqrt{2}} \end{bmatrix} \tag{9}$$

$$\left[P(-\theta_s)\right]_{NRF} = \sqrt{\frac{2}{3}} \cdot \begin{bmatrix} \cos(\theta_s) & \cos(\theta_s + \frac{2\pi}{3}) & \cos(\theta_s - \frac{2\pi}{3}) \\ \sin(\theta_s) & \sin(\theta_s + \frac{2\pi}{3}) & \sin(\theta_s - \frac{2\pi}{3}) \\ \frac{1}{\sqrt{2}} & \frac{1}{\sqrt{2}} & \frac{1}{\sqrt{2}} \end{bmatrix} \tag{10}$$

In order to facilitate the adopted notations for the various components of the three predefined symmetrical sequences, the subscript represents the direct (d) or quadrature (q) or zero (o) axis; the first superscript is reserved to the positive (P) or negative (N) or zero sequence (Z), while the second superscript (+) denotes the PRF synchronous frame and the superscript (-) represents the NRF frame.

By simulating the temporal representation (Figure 5. (a)), we demonstrate that the (d, q, o) components of the positive sequence are constants in the PRF reference frame. It is the same for the negative sequence components in the NRF reference frame (Figure 5. (e)). Furthermore, v_{dr}^{P-} and v_{qr}^{P-} components

Figure 5. (d q o) components in PRF frame: Positive Sequence (a); Negative Sequence (b); Zero Sequence (c); and in the NRF frame: Positive Sequence (e); Negative Sequence (f); Zero Sequence (g)

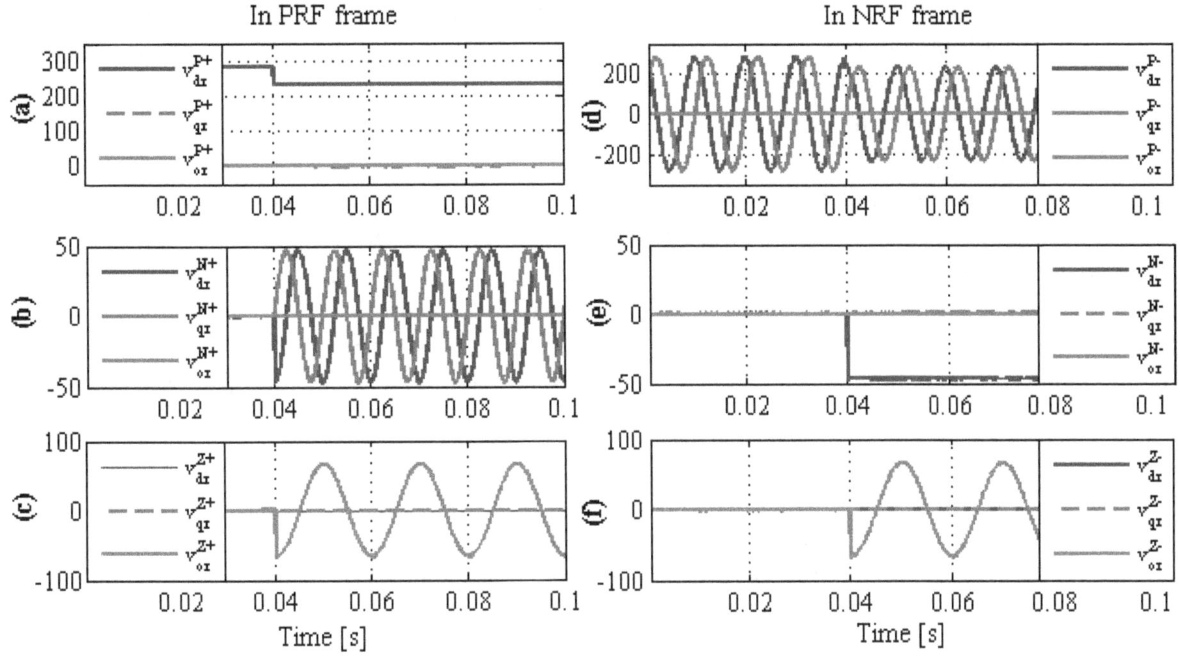

of the one part, and v_{dr}^{N+} and v_{qr}^{N+}, secondly, show fluctuations at $2.\omega_s$ (Figure5. (b), and Figure 5. (d)). In addition, the (o) components of the positive and negative sequences are zero regardless of the reference frame (PRF or NRF) (Figure 5. (a), Figure 5. (b), Figure 5. (d) and Figure 5. (e). It is also shown that the (d, q) components of the zero sequence in both PRF and NRF are zero while the (o) components of the zero sequence (v_{or}^{Z+} and v_{or}^{Z-}), fluctuate at $2.\omega_s$ as shown in Figure 5. (c) and Figure 5. (f).

The developed calculation, allowed us to quantify the fluctuations that occur during the voltage imbalance. The unbalanced (d, q) components are the sum of constant components materializing those direct and quadrature of the positive sequence reported in the PRF frame, to which are superimposed oscillating components at double of grid frequency deriving from the NRF frame (11).

$$\underline{v}_{dqr}^{+} = \underline{v}_{dq}^{P+} + \underline{v}_{dqr}^{N+} = \underline{v}_{dqr}^{P+} + \underline{v}_{dqr}^{N-}.e^{-j.2.\theta_s}$$

$$\underline{v}_{dqr}^{-} = \underline{v}_{dqr}^{N-} + \underline{v}_{dqr}^{P-} = \underline{v}_{dqr}^{N-} + \underline{v}_{dqr}^{P+}.e^{+j.2.\theta_s} \tag{11}$$

When connected to a balanced electrical grid, the hybrid system is performing. However, there is no guarantee that, during grid fault, wind - PV- battery system develops the same performance. It is therefore necessary to establish control algorithms that go beyond both the intermittent renewable energy sources and disturbances that may occur in the utility grid. Having identified the origins of the fluctuations that affect the electrical parameters when a fault occurs in the network connection, a fundamental part of this chapter proposes a new method allowing to achieve this aim.

2.3 Problem Statement

Most often, the grid side converter (GSC) is exposed to transient disturbances and interruptions which can be propagated through the electrical power system and affect the generators. Much of these disturbances may introduce conditions of asymmetric voltage imbalances. In particular, the asymmetrical voltage dips, the most encountered in practice are usually caused by asymmetric grid faults (Xiong et al., 2010; Xia et al., 2012).

The conventional approach was to disconnect decentralized energy generators from the power system to protect the GSC converter when the voltage unbalance occurs and reconnect it when the defects are eliminated. For this reason, this chapter contributes to the requirement to operate decentralized energy generation during such voltage disturbances and grid voltage unbalance.

Under unbalanced conditions, the exchanged active and reactive power flow is characterized by an average value and an oscillating component at double of the grid frequency (Wang et al., 2011; Meddeb et al., 2010). The regulators used in the outer regulation loop must be slower than the inner loop, and only control the average active and reactive power flow. Therefore, an intermediate level between the external and internal control loops is needed to operate during unbalanced conditions, where the active and reactive power flow characteristics must be specified.

In order to increase the integration rate of distributed renewable energy generation, it has become important for the operation of future hybrid systems that the grid side converter be able to limit the influence of faults occurring at the point of common coupling (PCC) (Suul et al., 2012; Abbassi et al., 2014). Thus, control systems should be able to avoid active and reactive double frequency oscillations and to prevent them to propagate through the common DC-bus and even to the generators (PVS and WECS ...). These oscillations harmful to the power system and its control system resulting from the appearance of the negative sequence. In addition, the interaction between the of the positive and negative sequences components leads to an undesired behavior of the electrical power system and even to the activation of the protection circuits to prevent instability. The above brought us to study one of the most important issues in the control of grid-connected converters: How to ensure service continuity of distributed renewable energy generation under Faulty Grid Conditions.

3. THE PROPOSED SYSTEM MODELING

The studied hybrid system made up of seven parts: the Wind Energy Conversion System equipped with its AC/DC wind generator side converter (WGSC), the PV system (PVS), with the associate DC/DC PV side converter (PVSC), the storage system (SS) including the reversible DC/DC storage side converter (SSC), and finally, the DC/AC utility grid side converter (GSC) ensuring the interfacing with the utility grid (Figure 6).

3.1. PV System Modeling and Control

3.1.1. PV System Modeling

Photovoltaic systems are clean production sources. But these systems have a relatively high initial cost. To overcome this drawback, it is necessary to consider further studies that allow PV systems to better

Figure 6. Configuration of the grid-connected wind–PV-battery power system

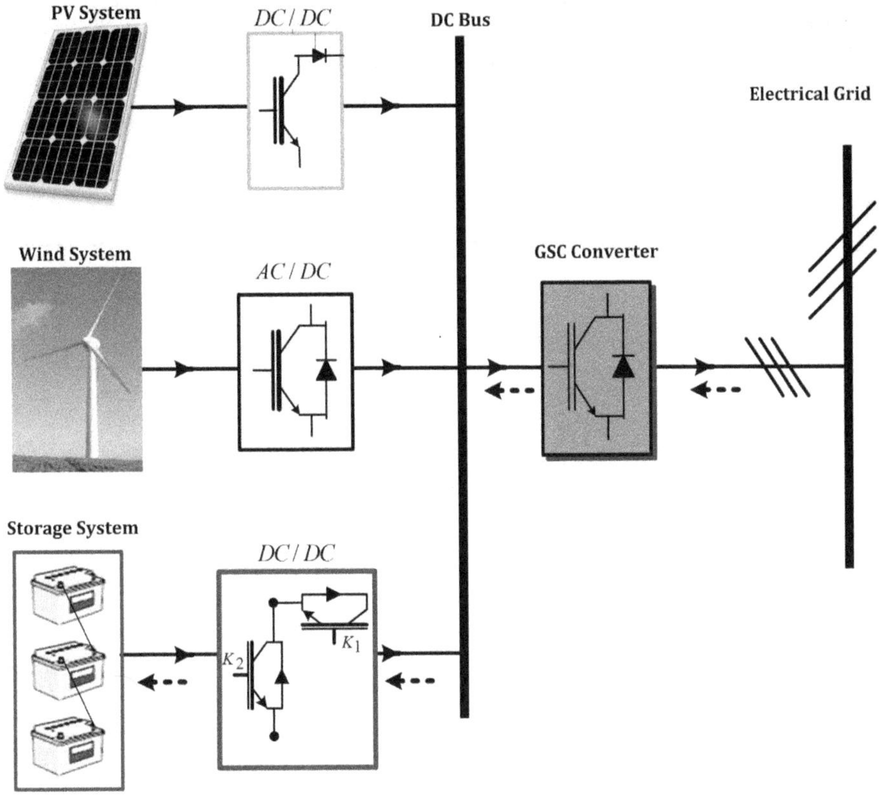

contribute to power the electrical grid. During the last 25 years, many researches have been addressed to modeling and analyzing the behavior of photovoltaic modules. The simple diode model, the most commonly used, offers good performance with a lower difficulty setting equation [ALON'2003], (Esram et al., 2007; Sera et al., 2007; Lee et al., 2008; Onat et al., 2010; Taherbaneh et al., 2010; Mohammadi et al., 2012). For our study, we have used crystalline PV technology that remains currently the dominant technology and represent about 50% of the market despite the relatively high manufacturing costs.

In this work, the PV system is formed by the association of 45 GT PV 216 poly PV panels (216 W for each one), and a boost DC-DC converter (PV Side Converter PVSC). The PVSC converter is controlled by a PWM signal generated through Maximum Power Point Tracking (MPPT) technique (Figure 3). For this type of PV panel, the mathematical model of a typical PV cell is given as:

$$I_C = I_{PH} - I_{Sat}\left[\exp\left(\frac{V_C + I_C.R_s}{n.V_T}\right) - 1\right] - \frac{V_C + I_C.R_s}{R_{sh}} \tag{12}$$

I_{Sat}: Reverse saturation current [A].
n: Dimensionless junction material factor.
I_{PH}: Light-generated current [A].
V_C: Output voltage of the solar cell [V].

Figure 7. Overview of PVS control structure

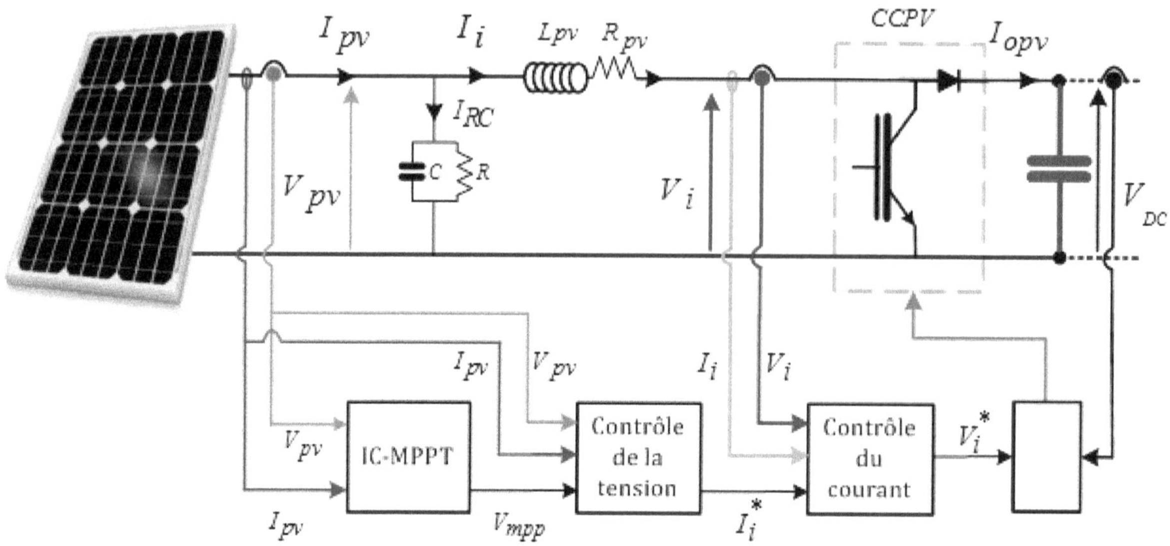

I_C: Output current of the solar cell [*A*].
R_s: Lumped series resistance of the cell [*Ω*].
R_{sh}: Lumped shunt resistance of the cell [*Ω*].

PV cell produces a very low power. In order to meet the voltage and current needs of such an application, N_{cs} cells are associated in series (sum of voltages) constituting branches themselves (N_{cp}) in parallel (total current). For optimum operation of the PV module, it is necessary that ($N_{cs}.N_{cp}$) cells are derived from the same technology, the same batch and subject to the same operating conditions (irradiance, temperature, inclination ...) (Elgendy et al., 2013; Abbassi et al., 2014). The model of the PV module is as follows (Figure 8):

I_{mPV} **:** Current generated by the PV module.
V_{mPV} **:** PV module output voltage.
N_{cp} **:** Number of PV cells connected in parallel.
N_{cs} **:** Number of PV cells connected in series.

The thermal model of the PV cell obeys the relation (13).

$$\frac{dT_C(t)}{dt}=\frac{1}{m.C_p}.\left\{\begin{matrix}\left[\frac{V_C(t)-v_d(t)}{R_s}\right]^2+\left[\frac{V_C(t)}{R_{sh}}\right]^2+ \\ i_d(t).v_d(t)+(1-\rho-\tau-\frac{I_{ph}.v_d(t)}{A.E(t)}).A.E(t)-h_c(t).A.\left[T_C(t)-T_a\right]\end{matrix}\right\} \quad (13)$$

Figure 8. Simplified model of PV module

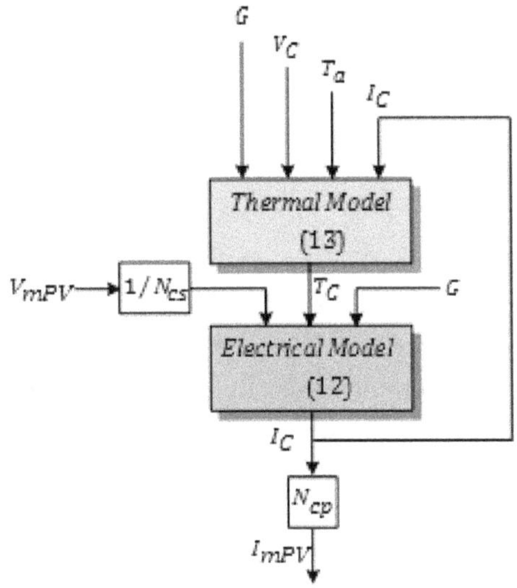

$$I_{PV} = N_p.I_{mPV}$$
$$= N_p.I_{mPH} - N_p.I_{mSat}.\left[\exp\left(\frac{\frac{V_{mPV}}{N_S} + \frac{I_{mPV}.R_s}{N_p}}{n.V_T}\right) - 1\right] - \frac{\frac{N_p.V_{mPV}}{N_S} + I_{mPV}.R_s}{R_{sh}} \tag{14}$$

$$P_{PV} = N_s.V_{mPV}.N_p.I_{mPV} \tag{15}$$

I_{PV} : GPV generated Current.
N_p : Number of PV modules connected in parallel.
V_{PV} : Output voltage of a PV module.
N_s : Number of PV modules connected in series.

The PV systems studies have proven that the current is directly proportional to the irradiation. By contrast, the voltage does not deteriorate in a meaningful way due to the irradiation decrease. The previous studies of PV systems have proven that the current is directly proportional to the irradiation. By contrast, the voltage does not deteriorate in a meaningful way due to the irradiation decrease. This is justified by the fact that the short-circuit current is a linear function of the irradiation while the open-circuit voltage is a logarithmic function. In accordance with this, we established the mathematical expressions that reflect the effects of irradiation and temperature on the electrical parameters of PV module. Herein, we refer to the standard conditions (SC) described by the irradiance G_{SC}= 1000 W/m^2 and the temperature T_{SC} = 25 ° C.

$$I_{mPH} = I_{mPH,CS} \cdot \frac{G_{SC}}{G} \tag{16}$$

$$V_{moc} = V_{moc,SC} + k_v \cdot \left(T - T_{SC}\right) \tag{17}$$

The parameter k_v is the temperature coefficient of the open-circuit voltage and T_{CS} is the temperature under standard.

PV array has non-linear I-V characteristic. Figure 9 shows that there is a point on I-V, P-V characteristic curve of PV, where the PV system produces its maximum output power. This maximum Power Point (MPP) location changes with changing weather conditions, as temperature (Figure 9. (a), Figure 9. (b)), solar irradiation (Figure 9. (c), Figure 9. (d)).

3.1.2. MPPT Control

In order to ensure the efficient operation of the PV generator, the PVSC converter is controlled to track the maximum power point (MPP) and allow to operate the PV system (PVS) to produce permanently the maximum power whatever the weather conditions and the DC-link terminals voltage (Onat et al., 2010; Sera et al., 2013).

This section presents a detailed analysis of the most well-known maximum power point tracking (MPPT) algorithms. The purpose of the analysis is to clarify some common misconceptions in the literature regarding these trackers, therefore helping the selection process for a suitable MPPT. Among all

Figure 9. Irradiance and temperature effects on I-V and P-V characteristics

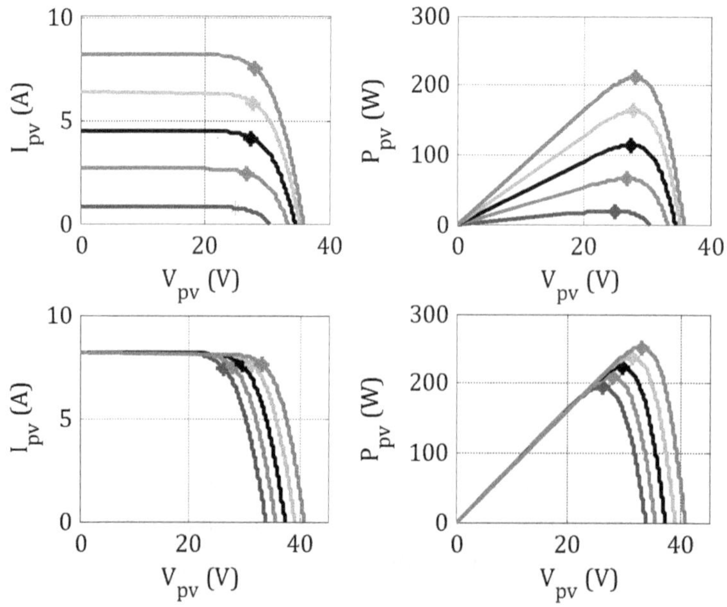

these algorithms, two methods are thoroughly analyzed from many points of view in (Sera et al., 2013). Their analysis reveals that there is no difference between them.

Furthermore, we cite the incremental conductance MPPT used with direct control method. The main difference of this technique to existing MPPT systems includes elimination of the proportional-integral control loop and investigation of the effect of simplifying the control circuit. We have given a particular focus to the IC algorithm because it is efficient, cost-effective and capable of tracking MPPs accurately and rapidly without steady-state oscillation, and also, its dynamic performance is satisfactory. It performs precise control under rapidly changing atmospheric conditions (Elgendy et al. 2013).

In order to adjust the solar operating voltage close to MPP under changing environmental conditions and to continuously gather the maximum PVS output power, despite of the inhomogeneous change in environmental conditions, the IC-MPPT was adopted following the algorithm depicted in Figure 10. Herein, the PPM point is extracted by comparing the instantaneous conductance (I/V) to incremental

Figure 10. IC-MPPT algorithm

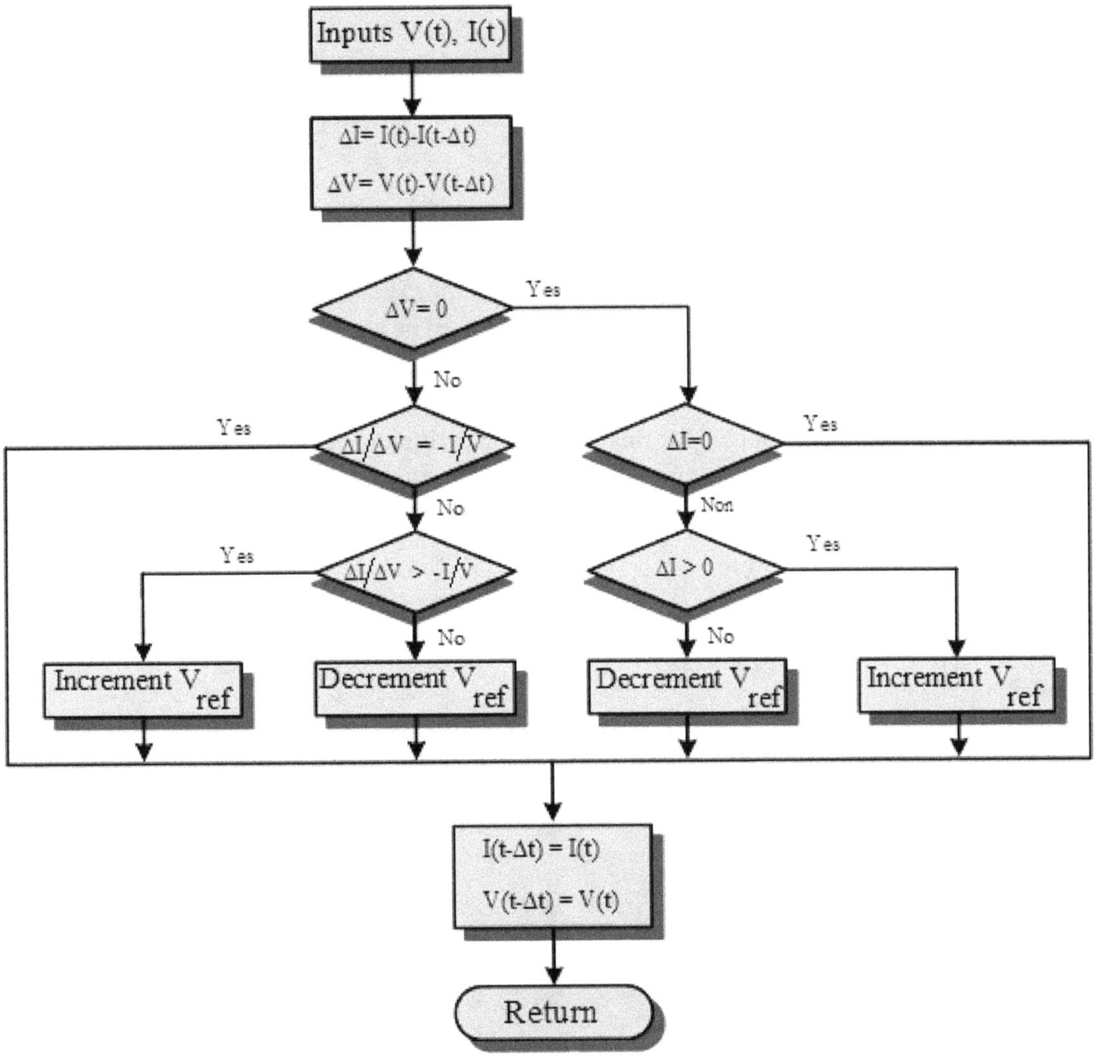

conductance (ΔI/ΔV). The slope of the P-V curve of the PV panel is equal to zero at the MPP, negative on its side and positive on its left side (Kim et al., 2008; Abbassi 1 & Chebbi 2012; Elgendy et al. 2013).

Once the MPP is tracked, the operation of the PV array is maintained. If a change of ΔI will occur, then the technique MPPT should allow the new reached MPP. Knowing that the I_{PV} component is generated by IC-MPPT algorithm, the current I_i at the input PVSC chopper can be controlled as follows.

In this paper, the PV system is the association of 32 PV panels type GT 300 poly with a peak power of 300 W connected to the common DC-link via a MPPT-controlled boost DC-DC converter (PV Side Converter PVSC). Herein, the optimization calculations reported for the PV system were performed with the PVSOL software. In real conditions, the ordinary differential equations governing the operation of the PVSC chopper are as follows:

$$\begin{cases} L_{pv}.\dfrac{dI_i}{dt} = V_{pv} - V_{DC}.(1-\alpha_{pv}) - R_{pv}.I_i \\ C \cdot \dfrac{dV_{pv}}{dt} = I_{pv} - I_i \end{cases} \tag{18}$$

L_{pv} is the inductance which limits the current ripple, R_{pv} is the filter resistance.

3.2. WECS System Modeling and Control

3.2.1. WECS System Modeling

Because of their capability to extract more energy than fixed speed generators and their reduced mechanical stress, variable speed wind generators have long been used as a good choice in wind systems more than fixed speed systems (Nouali et al., 2010; Abo-Khalil et al., 2008; Gharedaghi et al., 2011; She et al., 2011; Hoseynpoor et al., 2011). In this work, we opted for a synchronous generator (SG). The IEEE 4th order state model has been adopted for this machine as written in (19) (Abbassi 1 & Chebbi, 2012; Abbassi 2 & Chebbi., 2012).

$$\begin{cases} \dfrac{dE'_q}{dt} = \dfrac{1}{T'_{do}}.\left[E_{fd} - E'_q + (X_d - X'_d).i_d\right] \\ \dfrac{dE'_d}{dt} = \dfrac{1}{T'_{qo}}.\left[-E'_d + (X'_q - X_q).i_q\right] \\ \dfrac{d\omega}{dt} = \dfrac{1}{M}.\left[P_{t-MPPT} - P_w - D.(\omega - 1)\right] \\ \dfrac{d\delta}{dt} = \omega - 1 \end{cases} \tag{19}$$

where

E'_d, E'_q: d and q axis transient *emf.*
X_d, X_q: d and q axis reactance's.
X'_d, X'_q: d and q axis transient reactance's.
w: Mechanical speed.
M: Moment of inertia.
$P_{t\text{-}MPPT}$: Maximum available power extracted by the turbine.
P_w, Q_w: SG Active and reactive powers.
D: Friction coefficient.
δ: Load angle.
$v_{abc\text{-}w}$: Voltages delivered by the wind generator.
α: Phase angle of the $\bar{\upsilon}_{dq\text{-}w}$ voltage.
$v_{abc\text{-}WSC}$: Voltages at the input of WSC rectifier.
R_w, L_w: Resistance and inductance of the wind side filter.
I_w: Output current delivered by the wind system.

After calculating of the instantaneous (d, q) components of voltages of the generator's terminals and currents as detailed in (Abbassi 1 & Chebbi, 2012), the SG generator active and reactive powers are chosen to be:

$$\begin{cases} P_w = \dfrac{E'_q . V_{dq-w} . \sin(\delta - \alpha)}{X'_d} + V^2_{dq-w} . \dfrac{\sin(2.\delta - \alpha)}{X'_{qd}} \\ Q_w = \dfrac{V^2_{dq-w}}{X'_d} - \dfrac{V_{dq-w} . E'_q . \cos(\delta - \alpha)}{X'_d} + V^2_{dq-w} . \dfrac{1 - \cos(2.\delta - 2.\alpha)}{X'_{qd}} \\ \dfrac{1}{X'_{qd}} = \dfrac{1}{X'_q} - \dfrac{1}{X'_d} \end{cases} \tag{20}$$

3.2.2. MSC Converter Control

As detailed in (Abbassi 1 & Chebbi, 2012), the synchronous generator based WECS control is based on variable speed Maximum Power Point Tracking (MPPT) control. Thus, having set the reference speed for extracting maximum power, by MPPT technique, the SVPWM controlled WSC rectifier is responsible for the speed control of the wind turbine generator (Abbassi 1 & Chebbi, 2012; Chen et al., 2009; Teodorescu et al., 2011; Zghal et al., 2011; Sürgevil et al., 2005) (Figure 11).

We recall here that by controlling the phase and magnitude of the WGSC converter voltages ($v_{abc\text{-}WSC}$), both active and reactive powers are controlled.

Figure 11. Scheme of the local power controller of the WECS system

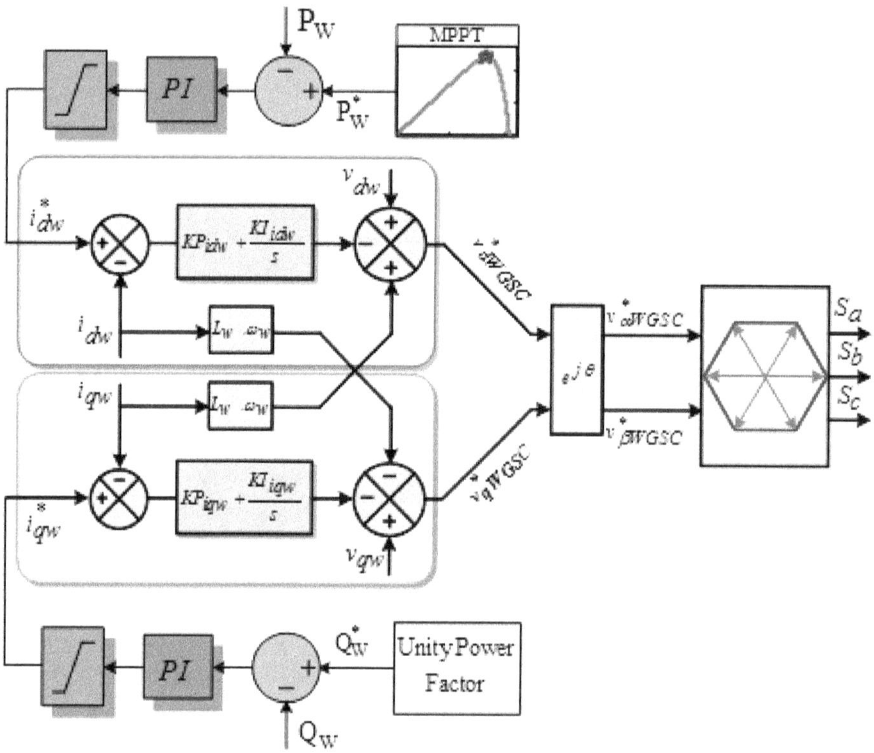

3.3. Storage System Modeling and Control

3.3.1. Battery Storage Modeling

In this section, the goal was to make active the wind/PV hybrid system that despite the availability of its energy sources, its production is fluctuating and intermittent. When the wind/PV hybrid system cannot meet the demand all the time, the storage system (SS) was involved to ensure the availability of energy and thus provide ancillary services to the power system manager.

The batteries storage system stores, in chemical form, the energy received from the hybrid system (charger) and releases it in direct current needs-based power system. In some previous applications (Valenciaga & Puleston, 2005), the storage system was connected directly to the common DC bus without interfacing converter. This configuration requires more batteries in the case of using a converter, and this may therefore reduce system cost. Furthermore, the lifetime of the batteries deteriorates due to the lack of any appropriate control of batteries currents and voltages. In order to solve this problem and to extend the life expectancy of the batteries, we proposed in our case study that the storage system is connected to the common DC bus via a bidirectional buck-boost converter (side storage converter SSC) (Figure 12) (Kim et al., 2008; Abbassi 1 & Chebbi, 2012). The topology of the storage system consists on the association of 12 batteries (200 Ah/12 V). This number of included batteries is required by the sizing study taking into account the terminal voltage of each battery, the level of the common DC-bus voltage and the powers transferred through the same DC-bus.

Figure 12. Block diagram of the storage system

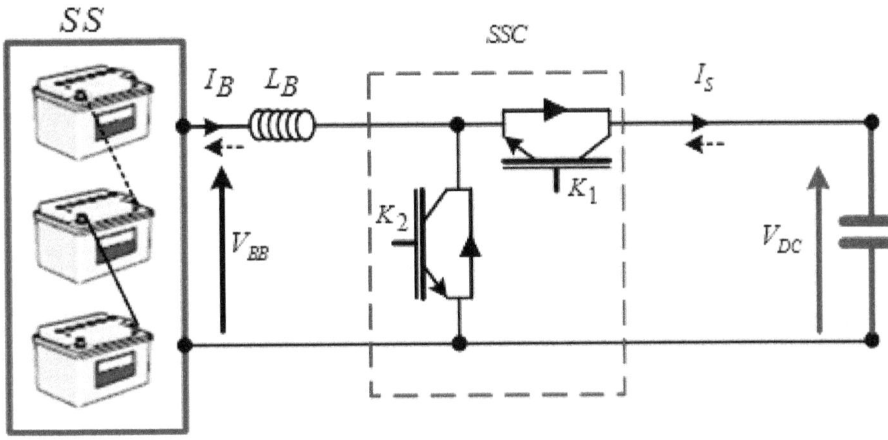

The most important battery parameter to be known and controlled is the State of Charge (*SoC*) or the Depth of Discharge (*DoD*). It has been established as in (González, et al., 2012; Abbassi et al., 2014) by:

$$SoC = 100 \cdot \left(1 - \frac{1}{E_S^n} \cdot \sum P_S \cdot t\right) \tag{21}$$

Based on the status of the balance between the powers generated and absorbed instantaneously, the battery *SoC* helps to make a decision concerning the mode of charge/discharge of the batteries bank.

In fact, depending upon weather conditions, the total generated wind-PV power (P_{wpv}) can be less or greater than that demanded at PCC. Indeed, excepting the balance circumstance where the PCC demand and the total generated power are equal, it is reasonable to materialize the gap between these quantities. This is in order to find a control structure that optimizes the performance of the storage system while satisfying operating constraints. Accordingly, the optimized control has been founded to oversee, in real time, the charge and discharge modes of storage batteries. In these two modes, the physical constraints of storage batteries are taken into account, imperatively (22) (Abbassi et al., 2014).

$$\begin{cases} P_S^{ch-\max}(k+1) = \min(P_S^{\max}, \frac{E_S^n}{\Delta t} \cdot (\frac{100 - SoC(k)}{100})) \\ P_S^{disch-\max}(k+1) = \min(P_S^{\max}, \frac{E_S^n}{\Delta t} \cdot (\frac{100 - SoC(k)}{100})) \end{cases} \tag{22}$$

Herein, $P_S^{\max}$ represents the maximum charge and discharge battery power under normal working conditions; and Δt is the interval between two simulation steps.

For the SSC control, we begin by evaluating the current I_S exchanged by the storage system (DC-link side) in accordance with (23):

$$I_s = \frac{P_S^*}{V_{dc}} = \frac{P_{PCC}^* - P_{wpv}}{V_{dc}} \tag{23}$$

where

P_{PCC}^* : Point of common coupling reference power.
P_S^* : Storage system reference power.
P_{wpv} : Power generated by wind-PV system.
V_{dc} : Common DC-link voltage.

Therefore the current exchanged by number of N_{sB} batteries (I_B) is given by:

$$\begin{cases} I_B = \left(\dfrac{V_{dc}^*}{N_{sB} \cdot V_B + L_B \cdot \dfrac{dI_B}{dt}} \right) \cdot I_S; & \text{in charge} \\ I_B = \left(1 - \dfrac{N_{sB} \cdot V_B + L_B \cdot \dfrac{dI_B}{dt}}{V_{dc}^*} \right) \cdot I_S; & \text{in discharge} \end{cases} \tag{24}$$

3.3.2. SSC Converter Control

The reversible chopper shown in Figure 12 gathers a buck chopper and a boost chopper. Assuming the continuous conduction, the instantaneous model of the Buck -Boost Storage System Converter (SSC) is derived by applying Kirchhoff's law as follows:

$$\begin{cases} C_{dc}.\dfrac{dV_{dc}}{dt} = (1-u).I_B - I_{os} \\ L_B.\dfrac{dI_{Bat}}{dt} = V_{BB} - (1-u).V_{dc} \end{cases} \tag{25}$$

Based on the calculation of the storage system current and voltage, the power received/restored is determined, in the two modes (charge or discharge), through (26).

$$P_S = I_B.V_{BB} = I_B.V_B.N_{sB} \tag{26}$$

where

V_{BB} : Banc Batteries voltage.

N_{sB} : Number of batteries connected in serie.

Here, it is important to note that the control of the storage system is depending on the state of various components of the hybrid system. This was to ensure the power-balancing at the common DC-link and the point of common coupling (PCC).

In order to manage the power flow, a hierarchical control architecture which is composed of different levels has been applied to the hybrid system. The First level has been designated for each electronic power converter integrated in the power system. It involves the generation of control signals by means of the SVPWM modulation technique. For the second level, the control is dedicated to each source and its power conversion system. In this case the control algorithms calculate the reference signals for generating the desired quantities. Finally, the third level control allows the operating optimization by the centralized supervision of the entire hybrid system (Abbassi 1 & Chebbi, 2012). Herein, the implemented algorithms have taken into account the interactions between the various components of the hybrid system, the electrical state of the utility grid and the availability of generators and storage system.

4. PROPOSED CONTROL STRATEGY

The occurrence of such an asymmetrical fault causes the appearance of the negative sequence voltages at the PCC. This gives rise to the ripple of the DC bus voltage; the power injected into the network itself reflects the oscillations at twice the frequency of the power system (Karimi et al., 2004; Saccomando et al., 2001; Meddeb et al., 2010; Suul et al., 2012); (Karimi et al., 2012; Vechiu et al., 2013). Therefore, considerations of power generation, safety and operation of network synchronization must be done before connecting these systems to the network service.

4.1. Synchronization Method

In this paper, it was imperative, especially with the obligation to identify precisely the angle of affected grid voltages, to use a synchronization technique that is able to filter the grid voltage from harmonics, which are by nature present because PWM technique, and extract the positive sequence. In order to ensure GSC connection safe and harmonious with the mission to inject/draw the desired active and reactive powers, the correct detection of the phase angle, the amplitude and the frequency of the grid voltage are crucial information for precise control of the GSC converter. The general structure of phase lock loops (Phase Looked Loop PLL) is formed by a phase detector, a loop filter and voltage-controlled oscillator (Rodriguez et al., 2006; Rodriguez et al., 2007; ; Yazdani et al., 2009; Trintis et al., 2011; Suul et al., 2012; . Among the most popular methods used for detection of positive sequence, our choice of the Double Second Order Generalized Integrator PLL (DSOGI-PLL) has been greatly justified thanks to its high harmonic rejection capacity and to be less oscillatory during transients than others methods (Suul et al., 2012; Karimi et al., 2012; Trintis et al., 2011). The structure of the DSOGI-PLL, shown in Figure 13, rests on a pair of SOGI which are based on adaptive filters in order to generate the in-quadrature voltages. Then, the Positive Negative Sequence Compensation bloc (PNSC) provides the positive sequence components for phase and frequency outputs generation (Suul et al., 2012; Xiong et al., 2010).

Figure 13. DSOGI-PLL algorithm

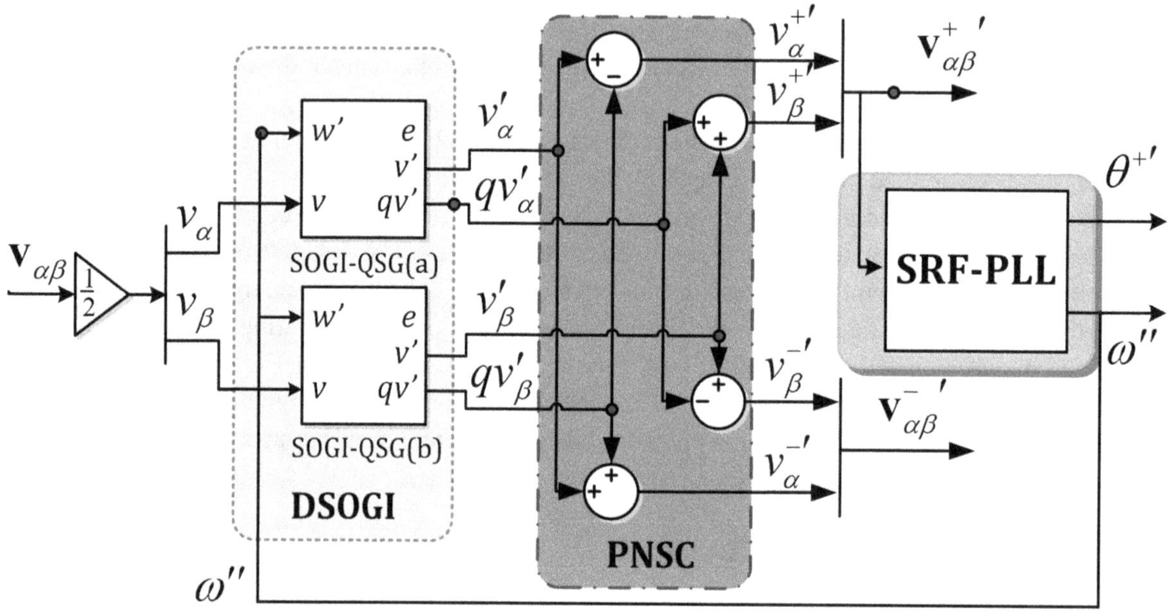

4.2. Proposed SCCS Control Strategy

The main aim of this section is to develop the proposed Symmetrical Components Control Strategy (SCCS) for GSC converter during its connection to an unbalanced electrical grid (Figure 14.). Our option to involve an additional leg, to consider a three-phase four-leg inverter, expands the control capabilities of the grid side voltage source inverter since that the fourth leg provides a path for the neutral current during asymmetrical faults in the connection grid. Based on EN 50160 STD (Antoni & Marta, 2013), a particular attention has been dedicated to one of the major abnormal grid conditions: the voltage unbalance.

Furthermore, the main objective is, firstly, the isolation of generation sources of fault occurrence adverse effects and, secondly, the generation of balanced currents ensuring the service continuity of the power system. The electrical quantities have been assessed after decomposition into symmetrical components (Vechiu et al., 2013; Suul et al., 2012). This was in order to assign a great reliable stability to controlled magnitudes decomposed through the concept of symmetrical components.

The concept of symmetrical components, has been presented by Fortescue, and it becomes well known for the unbalanced power systems, and was then applied for transient analysis in the time domain (Suul et al., 2012). In the literature, a several of works have treated the separation techniques of symmetrical sequences. Depending on the type of the adopted control and the topology of three phase converters, the most published works (Suul et al., 2012; Wang et al., 2011) used only the Positive and Negative Sequences reported in the αβ stationary reference frame. However, other work, (Vechiu et al., 2013; Abbassi et al., 2014) used the three symmetrical sequences in the dq synchronous reference frame.

In this chapter, we have adopted the theory of Fortescue thanks to its simplicity and effectiveness. In addition, we proceeded to establish complex expressions of the processed electrical quantities by recourse to the principle of phasor writing.

Figure 14. Block diagram of the proposed SCCS control strategy

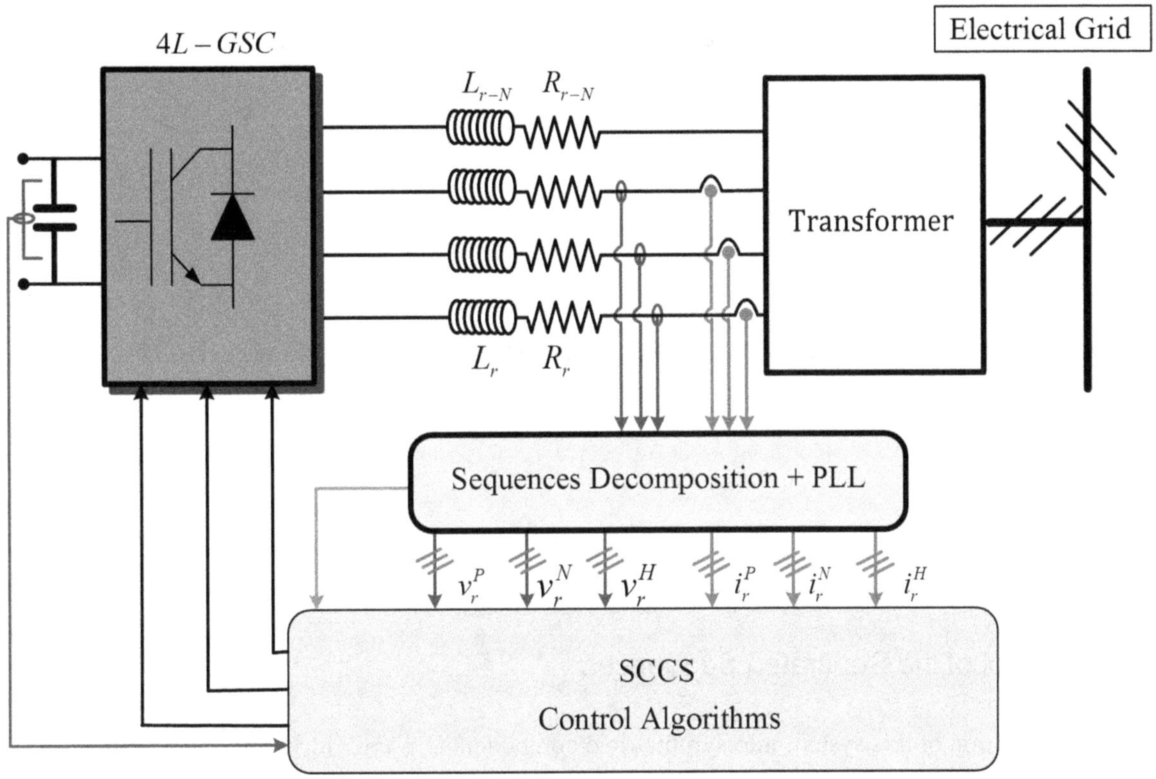

4.2.1. Phasor Writing

Various approaches was presented for the implementation of the required phase shift (Lee et al., 1999; Meddeb et al., 2010; Suul et al., 2012). Explicit delay corresponding to a quarter period of the fundamental frequency was proposed in (Vechiu et al., 2013). The implementation of this phase-shifting corresponding to 90° delay of fundamental frequency sinusoidal components using phasor writing is illustrated in Figure 15.

T: period of the grid electrical quantities.

In accordance with Figure 15, the sum of the measured voltages system and the delayed voltages system gives:

$$\begin{aligned}
\underline{v}_{ar}(t) &= V_a.\left[\cos(\omega_s t + \phi_a) + j.\sin(\omega_s t + \phi_a)\right] \\
\underline{v}_{br}(t) &= V_b.\left[\cos(\omega_s t - \frac{2.\pi}{3} + \phi_b) + j.\sin(\omega_s t - \frac{2.\pi}{3} + \phi_b)\right] \\
\underline{v}_{cr}(t) &= V_c.\left[\cos(\omega_s t + \frac{2.\pi}{3} + \phi_c) + j.\sin(\omega_s t + \frac{2.\pi}{3} + \phi_c)\right]
\end{aligned} \quad (27)$$

Figure 15. Phasor writing of the voltage system at the PCC

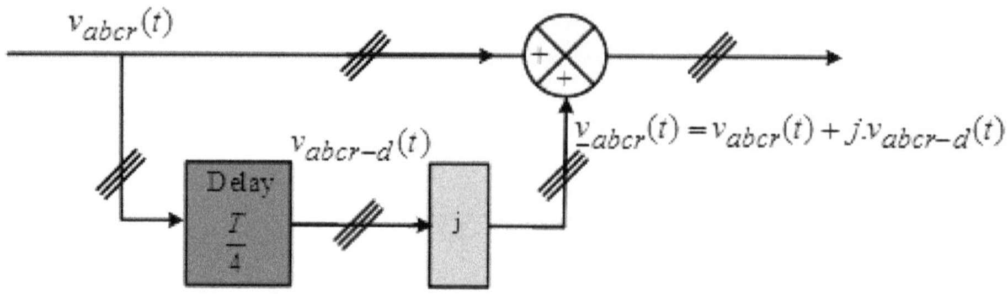

These relationships represents phasor writing that can be rewritten as follows:

$$\begin{bmatrix} \underline{v}_{ar} \\ \underline{v}_{br} \\ \underline{v}_{cr} \end{bmatrix} = \begin{bmatrix} V_a.e^{j(\omega_s t+\phi_a)} \\ V_b.e^{j(\omega_s t-\frac{2.\pi}{3}+\phi_b)} \\ V_c.e^{j(\omega_s t+\frac{2.\pi}{3}+\phi_c)} \end{bmatrix} \tag{28}$$

4.2.2. Control of the Separated Sequences

The decomposition of this system into symmetrical components by applying Fortescue matrixes $\left[F_P\right]$, $\left[F_N\right]$ and $\left[F_z\right]$, allows to obtain the three sequences (positive, negative, and zero), expressed by relations (29), (30) and (31) respectively.

$$\begin{bmatrix} \underline{v}_{ar}^{P} \\ \underline{v}_{br}^{P} \\ \underline{v}_{cr}^{P} \end{bmatrix} = \begin{bmatrix} V_P.e^{j(\omega_s t+\phi_P)} \\ V_P.e^{j(\omega_s t-\frac{2.\pi}{3}+\phi_P)} \\ V_P.e^{j(\omega_s t+\frac{2.\pi}{3}+\phi_P)} \end{bmatrix} \tag{29}$$

$$\begin{bmatrix} \underline{v}_{ar}^{N} \\ \underline{v}_{br}^{N} \\ \underline{v}_{cr}^{N} \end{bmatrix} = \begin{bmatrix} V_N.e^{j(\omega_s t+\phi_N)} \\ V_N.e^{j(\omega_s t+\frac{2.\pi}{3}+\phi_N)} \\ V_N.e^{j(\omega_s t-\frac{2.\pi}{3}+\phi_N)} \end{bmatrix} \tag{30}$$

$$\begin{bmatrix} \underline{v}_{ar}^{Z} \\ \underline{v}_{br}^{Z} \\ \underline{v}_{cr}^{Z} \end{bmatrix} = \begin{bmatrix} V_Z.e^{j(\omega_s t+\phi_Z)} \\ V_Z.e^{j(\omega_s t+\frac{2.\pi}{3}+\phi_Z)} \\ V_Z.e^{j(\omega_s t-\frac{2.\pi}{3}+\phi_Z)} \end{bmatrix} \tag{31}$$

After the application of the Fortescue transformation, the Positive-, Negative- and Zero-sequence component $\underline{v}_{abcr}^{P}$, $\underline{v}_{abcr}^{N}$ and $\underline{v}_{abcr}^{Z}$ are established. We note that, unlike the Positive and Negative sequences, zero component (0) of the Zero sequence is oscillating at the grid frequency, both in PRF (Figure 5. (c)) and NRF frames (Figure 5. (g)). For this reason, further transformation is imperative. This transformation provides, from the voltages vectors of the Zero-sequence, a balanced system (with 120° phase shift and the same amplitude) (Vechiu et al., 2013; Abbassi et al., 2014). Our analysis has been developed referring to two reference frames: the positive reference frame (PRF) rotating counter clockwise at the speed of $+\omega_s$ and the negative reference frame (NRF) rotating clockwise at $-\omega_s$. After that, the Positive-sequence has been dealt according to the PRF frame, while the Negative-sequence and Zero-sequence were processed referring to NRF frame. This is in order to attributes a high reliability to the studied variables stability.

Reported to aforementioned synchronous frames, voltages and currents of each of the three sequences are deprived of fluctuating components and readily to be regulated through the based PI control loops and this improves the accuracy of the adopted algorithms (Figure 16.).

Positive-sequence has been dealt according to the PRF frame, while the Negative-sequence and Zero-sequence were processed referring to NRF frame. The recourse to this type of control attributes a high reliability to the studied variables stability. Indeed, voltages and currents of each of the three sequences are deprived of fluctuating components and readily to be regulated through the based PI control loops.

In order to make an inquiry concerning performances of the adopted strategy and test the adopted models, the presented system was implemented on MATLAB/Simulink.

5. SYSTEM PERFORMANCE

Before presenting the performances of the SCCS control strategy, a first scenario was simulated in order to evaluate the consequences of the unbalance state on the power system. Here, the aim is to inject the entire P_{wpv} power to grid without recourse to the batteries storage system. The classical control strategy (CCS) (Abbassi 1 & Chebbi, 2012) has been applied to a three-phase three-leg voltage source inverter.

Figure 16. Bloc diagram and regulation loops of the SCCS control strategy

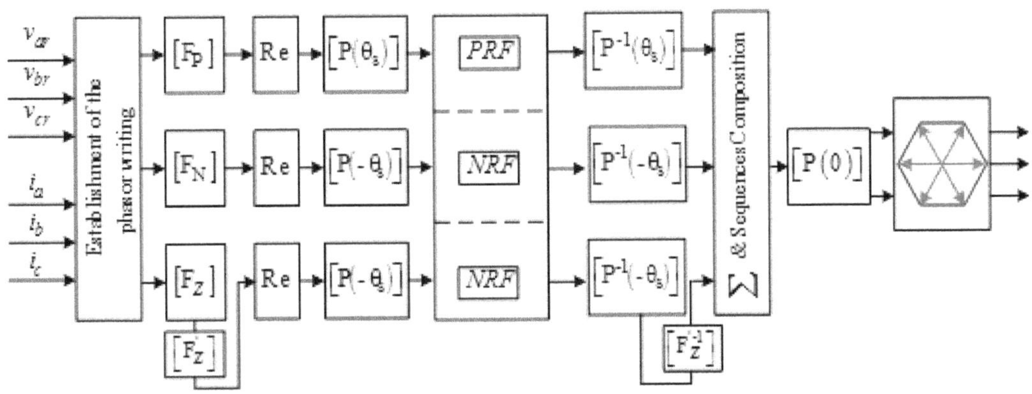

For this case, a single-phase voltage dip of amplitude 90% of the rated voltage and a time duration of 0.6 sec was applied at time 30 sec. DSOGI-PLL performances which are of paramount importance for controlled converter are shown in Figure 17.

Figure 17 shows the behavior of the studied system when the Type B voltage dip occurs (80% of the rated voltage; time duration of 0.6 sec). It highlighted clearly that, as soon the fault occurred, the voltage on the DC-bus capacitor's terminals is subject to oscillations at twice the frequency ($2.\omega_s$) of the electrical grid well as over/under voltage which exceeds 11.5% of the rated DC-bus voltage (Figure 18. (b)). This DC-bus voltage state confirms that similarly, when the default occurs, it turns out that the power P_g injected to the grid is oscillating at twice the frequency ($2.\omega_s$) of the electrical grid around the P_{wpv} power to which is assigned as a reference.

Regarding the currents injected to the grid, they are affected by the fluctuating nature of production sources of generation side. In addition, the simulated disturbance causes an imbalance of these currents evidenced by increased amplitudes and poor quality waveforms (Figure 18. (c)). It is well notable the highest current injected in the phase with the lowest voltage.

The second scenario is focused on the analysis behavior of wind-PV hybrid system with the participation of the storage system, face the previous fault. However, the GSC converter was piloted to inject constant average power ($P^*_{PCC} = P^*_g$ =16.04 kW). This in order to enquire about performances of the

Figure 17. Natural and stationary frame components, voltage vector trajectory and phase angle

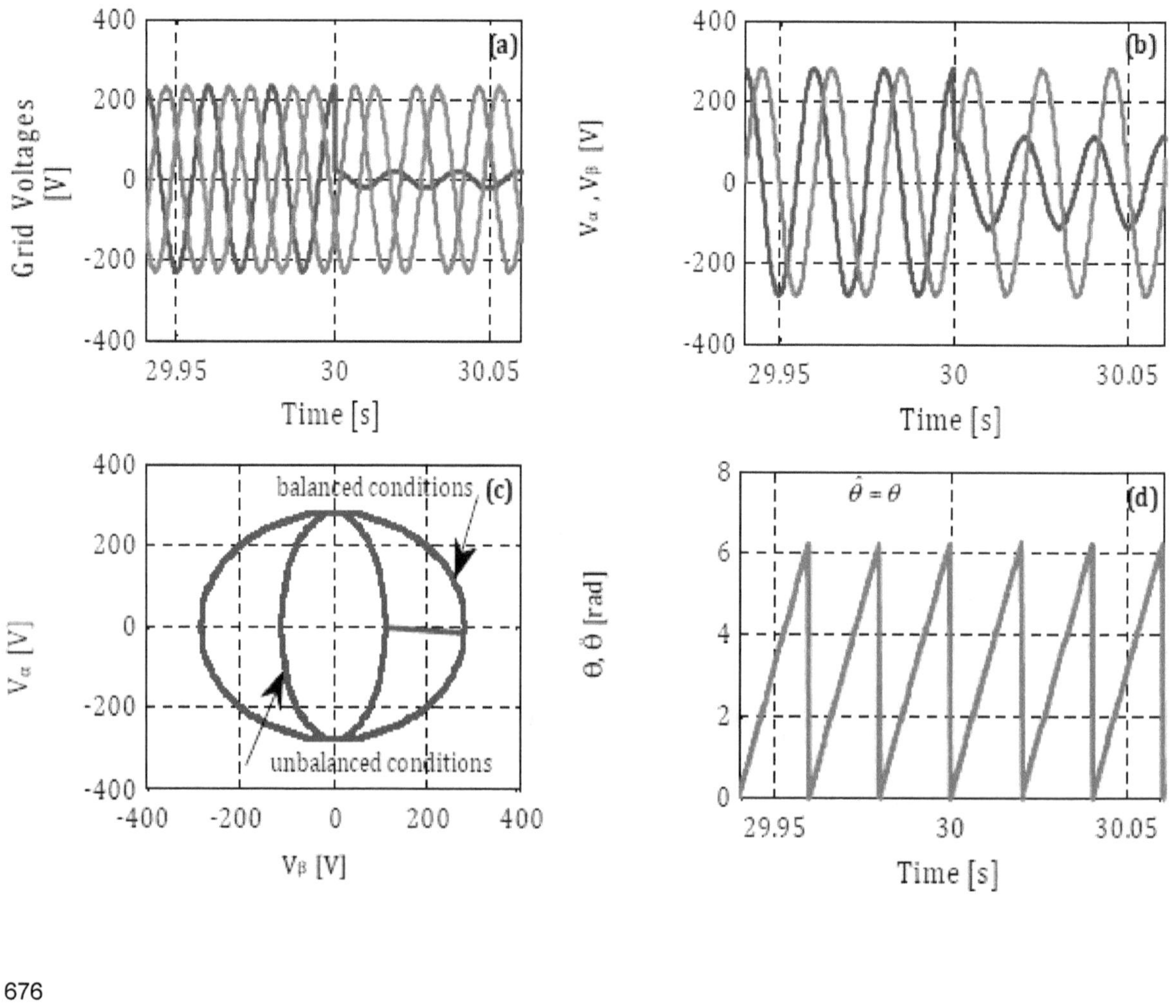

Figure 18. Grid-connected wind-PV system behaviors during a Type B voltage dip (with CCS control strategy)

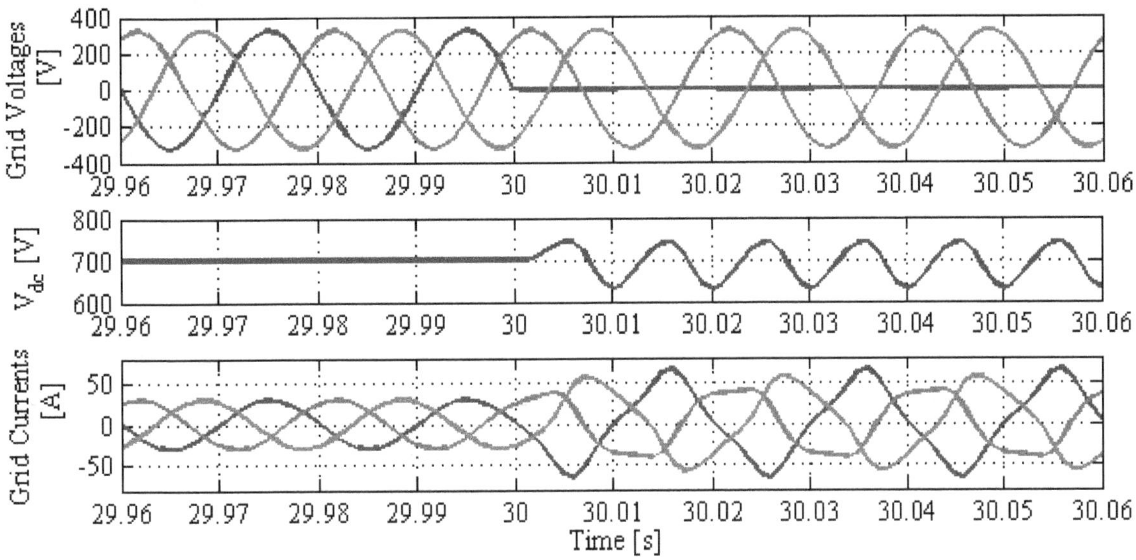

storage system contribution as a buffer between fluctuating generation sources (P_{wpv}) and fixed reference power P_g^*.

As expected, since the batteries banc goal is to achieve both perfect regulation of the DC-link voltage and power balancing at the same bus, a clear proof showing the storage system ability to achieve elimination of V_{dc} voltage oscillations ($2.\omega_s$) is shown in Figure 19. (b) and its zoom.

With adequate proposed control of the SSC reversible chopper, the storage system offsets successfully the oscillations of grid-side power. However, practically and at these frequency oscillations ($2.\omega_s$), it is unreliable to compel the storage system to operate under these extremely unfavorable operating conditions.

At this stage, the adverse consequences of an unbalance voltages fault are proved both on the storage system and the injected power quality, it is well confirmed that a classical CCS control is less efficient under grid disturbances.

The proposed strategy performances will be presented in the third scenario. It will be proved that faced with the same fault conditions, the injected power, the DC-link voltage and the grid currents are perfectly controlled by the proposed SCCS strategy (Figure 20).

Except of short transitional regime obedient to demands requirement, the DC-bus voltage regulation has been improved to some extent compared with the classic control strategy. In comparison with Figure 18 (b) and Figure 19 (b), the waveforms of DC-link voltage show smaller over/under voltage (± 2.28%). As depicted in Figure 20 (b), the DC-bus voltage presents negligible oscillations at the moments of appearance and disappearance of the voltage dip. This is expected under the effect of the amount of negative-sequence grid currents and voltages appearing therein.

Figure 20 (c) presents the waveforms of the injected currents. It is well shown that SCCS control allows generating symmetrical and balanced grid currents and guarantees controlled current values for the three phases. The value of the peak current is mitigated by the reference adequate calculation. So, line over current and filter saturation are avoided during the fault.

Figure 19. Complete system behavior during a Type B voltage dip

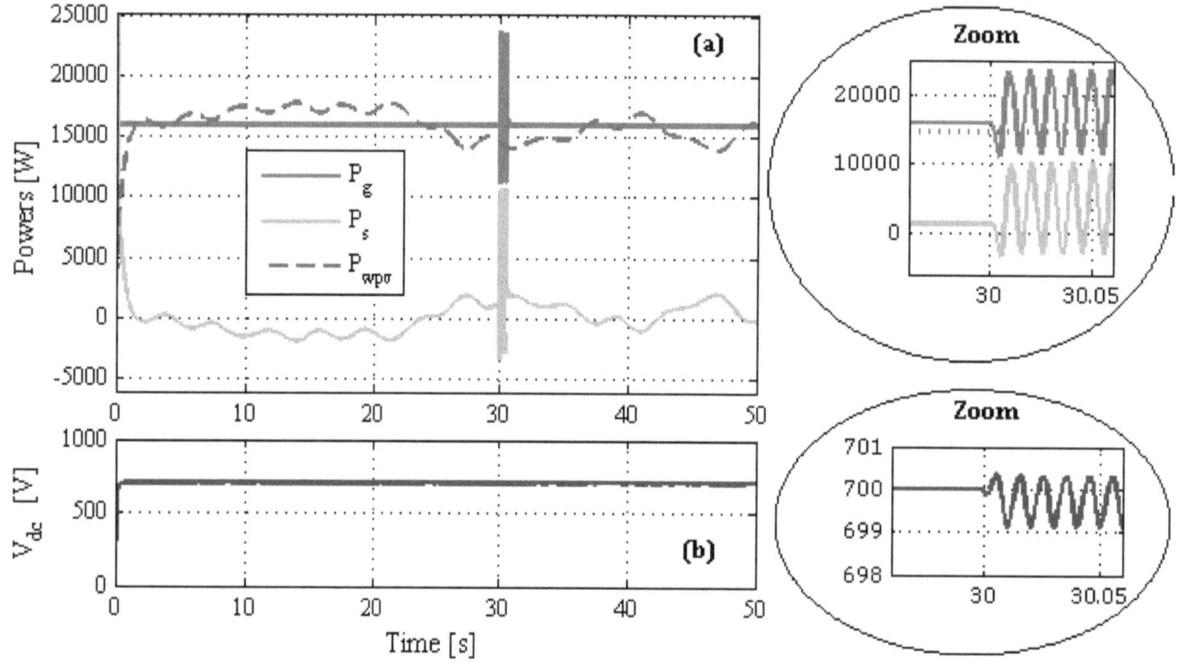

Figure 20. Complete system behavior when controlled by SCCS control during a Type B voltage dip

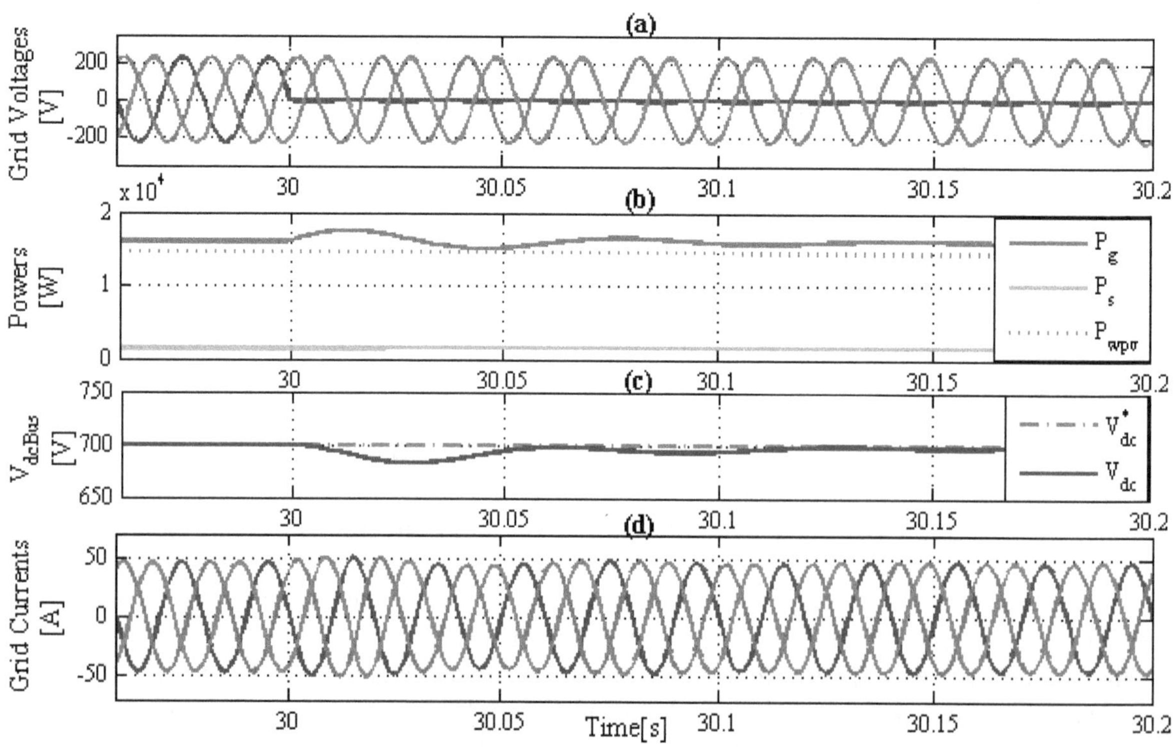

Therefore, even in case of fault appearance, the injected power is well controlled thanks to dynamic response performances of the power system based on the proposed control algorithms. Also for this study case, the storage system remains confident to its double role and its exchanged power suitably follows its optimized reference.

6. DISCUSSION

We present in this section a comparative study of the simulation results of the proposed SCCS control strategy and BPSC control strategy. The BPSC strategy has been proposed for power control of a classical converter to maintain balanced three-phase sinusoidal currents under unbalanced conditions (Wang et al., 2011; Suul et al, 2012). For these simulations, we will focus on the evolution of the output voltages of the 4L-GSC converter, generated current and active and reactive power injected to the utility grid affected by an asymmetric voltage fault.

For this scenario, a voltage dip of 90% of the amplitude of the grid voltages is occurred between t_1=0.2s and t_2= 0.7s. Following the above disturbance, we simulated the temporal evolution of the unbalanced voltages system and the voltages generated at the output of the 4L-GSC before and during fault conditions (Figure 21.). It is well shown that under these conditions, the four leg grid side converter (4L-GSC) controlled by SCCS technique provides a balanced voltage system. The perfect balance of voltage system provided by the 4L-GSC interfacing converter justifies that the behaviour of the control system is stable.

The simulation results depicted in Figure 22 validate the performances of the proposed control strategy. Indeed, imbalance potential is of no consequence on the current provided by the converter 4L-GSC when controlled by the SCCS strategy. These currents are balanced and sinusoidal although minor variations arise damped at their amplitude at the appearance and disappearance of the voltage dip (Figure 22. (b)). Moreover, the currents classical three phase converter when controlled by BPSC strategy are sinusoidal but they undergo a remarkable increase in their amplitudes throughout the phase of the fault (Figure 22. (a)).

Regarding the active power exchanged with the grid, $P_{g\text{-}SCCS}$ supplied by the 4L-GSC converter is of satisfactory quality (Figure IV.22. (c)). We note only the existence of a power jump in $P_{g\text{-}SCCS}$ power (about 8.79% of the reference power) at the appearance and disappearance of the voltages dip. This can be explained by the transition of the control variables to join a new equilibrium state identical to the initial state. However, in the case of the BPSC strategy, the temporal evolution of the active power shows the presence of double frequency oscillations. In this context, there is shown the amplitude of the component can reach to 142.71%. This demonstrates the advantage of the proposed SCCS strategy to provide the grid active power of satisfactory quality.

Moreover, the SCCS strategy allows the control of the reactive power at the PCC. Figure 22 (d) shows the temporal evolution of reactive power $Q_{g\text{-}SCCS}$; this power is adjusted to its zero reference value. However, in this case, the SCCS control strategy eliminates the power jump (0.62%) recorded at the times of appearance and disappearance of the fault (t_1 = t_2 = 0.2s and 0.7s). On the other hand, in the case of the BPSC control, Figure 22. (c) shows that the system contains unstable modes. Reactive power $Q_{g\text{-}SCCS}$ Waveforms prove that the inadequacies of this method (BPSC) is that negative consequences appear as unamortized fluctuations, at double frequency. The fluctuations amplitude reaches 46.76% of the reference active power

Figure 21. Generated voltages system at the output of the 4L-GSC and those at the PCC

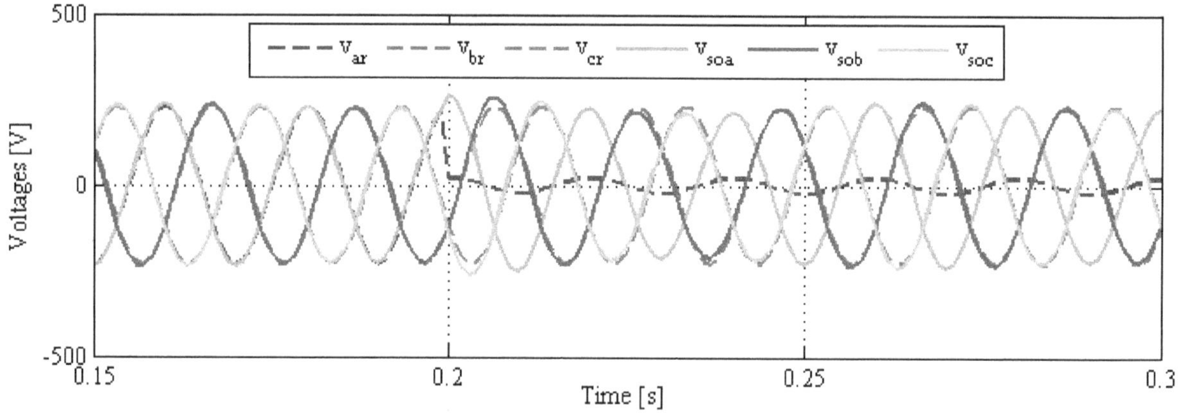

Figure 22. Comparative study of BPSC and SCCS strategies performances

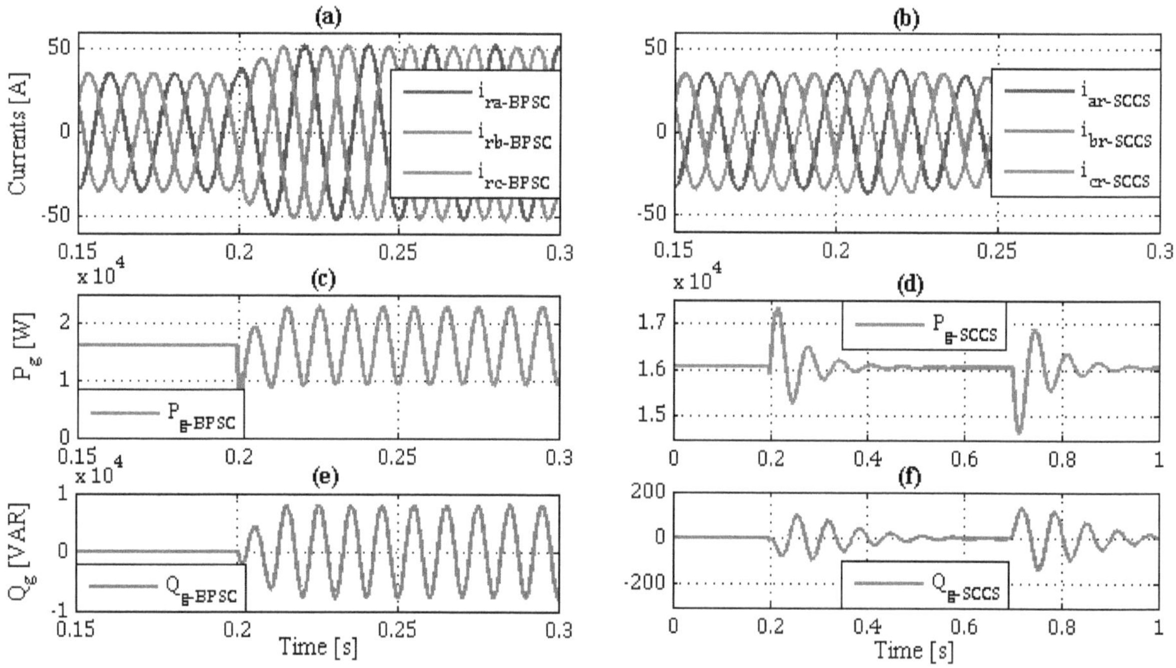

Finally, the simulation of the proposed SCCS control algorithms allowed us to confirm their performances distinctly improved. Indeed, we have evaluated, perfectly, the extents of variations of the components which occur during a defect while being based on the one hand on the behavior of the state variables of the studied system and on the other hand on the power balancing both at the common DC-bus and at the point of common coupling.

7. CONCLUSION AND FUTURE WORK

This chapter is a contribution to the study of an integrated renewable energy conversion system associated with a storage system under unbalanced voltage conditions. This skill is helpful to guarantee a versatile energy system operating under unbalanced conditions. In fact, the proposed control strategy serves to reduce the ripple in the common DC-link voltage and the important electrical quantities double frequency oscillations under a grid voltage unbalance. Both storage system and grid-side-converter have controlled the power flow and the DC-link voltage regulation. The simulation results showed the improvement of the hybrid system operating performance brought about by the proposed control scheme.

Although the application of the SCCS control strategy represents an impressive contribution to planning more profitable hybrid systems connected to an unbalanced grid since it ensures the operating safety and the service continuity of electrical power systems while responding utility grid requirements, the obtained results allow to build immediate and other distant prospects namely:

- The hybrid system was designed for a small scale energy generation. The control system have been developed, tested and validated by simulations in Matlab environment. In order to obtain an industrial and commercial prototype, an implementation in a larger scale energy generation is suggested to consider the problems that can occur with higher voltage and higher current and judge actually algorithms efficiency and potentials in real time.
- In this chapter, the DC-coupled hybrid system architecture was chosen because of their different benefits. In this case, all energy sources must be close to each other. However, sometimes the energy sources are dispersed for geographical reasons. In this sense, it will also be very useful to test the AC-coupled structure.

REFERENCES

Abbassi, R., & Chebbi, S. (2012a). Energy Management Strategy for a Grid–Connected Wind-Solar Hybrid System with Battery Storage: Policy for Optimizing Conventional Energy Generation. International Review of Electrical Engineering, 7(2), 3979-3990.

Abbassi, R., Hammami, M., & Chebbi, S. (2014). A Control Method of Grid-Connected Wind-Solar-Batteries Hybrid System to Mitigate Voltage Disturbances. In *Proceedings of International Conference on Control, Engineering & Information Technology* (CEIT'2014). Sousse, Tunisia: CEIT.

Abbassi, R., & Chebbi, S. (2012b). Optimal Energy Management Strategy for Wind Photovoltaic Hybrid System with Battery Storage. In Proceedings of the 16th IEEE Mediterranean Electrotechnical Conference (MELECON' 12), (pp. 810 – 813). IEEE. DOI: .10.1109/MELCON.2012.6196553

Abo-Khalil, A. G., & Lee, D.-C. (2008). MPPT Control of Wind Generation Systems Based on Estimated Wind Speed Using SVR. *IEEE Transactions on Industrial Electronics*, *55*(3), 1489–1490. doi:10.1109/TIE.2007.907672

Antoni, K., & Marta, B.-P. (2013). *Application Note - Standard EN 50160/Voltage characteristics of electricity supplied by public electricity networks*. ECI Publication No Cu0147. Available from www.leonardo-energy.org/node/145851

Blaabjerg, F., Teodorescu, R., Liserre, M., & Timbus, A. V. (2006). Overview of control and grid synchronization for distributed power generation systems. *IEEE Transactions on Industrial Electronics*, *53*(5), 1398–1409. doi:10.1109/TIE.2006.881997

Chen, Y. M., Liu, Y. C., Hung, S. C., & Cheng, C. S. (2007). Multi-input inverter for grid-connected hybrid PV/wind power system. *IEEE Transactions on Power Electronics*, *22*(3), 1070–1077. doi:10.1109/TPEL.2007.897117

Chen, Z., Guerrero, J. M., & Blaabjerg, F. (2009). A Review of the State of the Art of Power Electronics for Wind Turbines. *IEEE Transactions on Power Electronics*, *24*(8), 1859–1875. doi:10.1109/TPEL.2009.2017082

Draghicescu, M. (2010). Concerns about the non-polluting energy sources in the European Union: Renewable energy sources in Romania. In *Proceedings of the 2nd International Conference on Chemical, Biological and Environmental Engineering* (ICBEE), (pp. 235–237). DOI: doi:10.1109/ICBEE.2010.5652877

Elgendy, M. A., Zahawi, B., & Atkinson, D. J. (2013). Assessment of the Incremental Conductance Maximum Power Point Tracking Algorithm. *IEEE Transactions on Sustainable Energy*, *4*(1), 108–117. doi:10.1109/TSTE.2012.2202698

Esram, T., & Chapman, P. L. (2007). Comparison of Photovoltaic Array Maximum Power Point Tracking Techniques. *IEEE Transactions on Energy Conversion*, *22*(2), 439–449. doi:10.1109/TEC.2006.874230

Gharedaghi, F., Jamali Di, H. M., & Khalili, A. (2011). Maximum Power Point Tracking of Variable speed Wind Generation System Connected to Permanent Magnet Synchronous Generator. *International Review on Modelling and Simulations*, *4*(3), 1044–1049.

González, I., Ramiro, A., Calderón, M., Calderón, A. J., & González, J. F. (2012). Estimation of the state-of-charge of gel lead-acid batteries and application to the control of a stand-alone wind-solar test-bed with hydrogen support. *International Journal of Hydrogen Energy*, *37*(15), 11090–11103. doi:10.1016/j.ijhydene.2012.05.001

Karimi-Gh, M., & Iravani, M. R. (2004). A Method for Synchronization of Power Electronic Converters in Polluted and Variable-Frequency Environments. *IEEE Transactions on Power Systems*, *19*(3), 1263–1270. doi:10.1109/TPWRS.2004.831280

Karimi-Gh, M., Khajehoddin, S. A., Jain, P. K., & Bakhshai, A. (2012). Problems of Startup and Phase Jumps in PLL Systems. *IEEE Transactions on Power Electronics*, *27*(4), 1830–1838. doi:10.1109/TPEL.2011.2169089

Kellogg, W. K., Nehrir, M. H., Venkataramanan, G. G., & Greez, V. (1998). Generation Unit Sizing and Cost Analysis for Stand-Alone Wind, Photovoltaic, and Hybrid Wind-PV Systems. *IEEE Transactions on Energy Conversion*, *13*(1), 70–75. doi:10.1109/60.658206

Kim, S.-K., Jeon, J.-H., Cho, C.-H., Ahn, J.-B., & Won, S.-H. (2008). Dynamic Modeling and Control of a Grid-Connected Hybrid Generation System with Versatile Power Transfer. *IEEE Transactions on Industrial Electronics*, *55*(4), 1677–1688. doi:10.1109/TIE.2007.907662

Lee, J.-P., Min, B.-D., Kim, T.-J., Yoo, D.-W., & Yoo, J.-Y. (2008). A Novel Topology for Photovoltaic DC/DC Full-Bridge Converter with Flat Efficiency under Wide PV Module Voltage and Load Range. *IEEE Transactions on Industrial Electronics*, *55*(7), 2655–2663. doi:10.1109/TIE.2008.924165

Mashhour, E., & Moghaddas-Tafreshi, S. M. (2010). Mathematical modeling of electrochemical storage for incorporation in methods to optimize the operational planning of an interconnected micro grid. *Journal of Zhejiang University-Sciences C*, *11*(9), 737–750. doi:10.1631/jzus.C0910721

Meddeb, A., Abbassi, R., Jemai, K., & Chebbi, S. (2010). Improvement of the Adaptation of a Permanent Magnet Synchronous Generator with an Unbalanced Electrical Supply Network. *International Review on Modelling and Simulations*, *3*(1), 48–56.

Mohammadi, M., Hosseinian, S. H., & Gharehpetian, G. B. (2012). Optimization of hybrid solar energy sources/wind turbine systems integrated to utility grids as microgrid (MG) under pool/bilateral/hybrid electricity market using PSO. *Solar Energy*, *86*(1), 112–125. doi:10.1016/j.solener.2011.09.011

Nouali, S., & Ouali, A. (2010). Sensorless Control strategy for Wind Energy Conversion System based on Doubly Fed Twin Stator Induction Generator. *International Review on Modelling and Simulations*, *3*(6), 1209–1216.

Onat, N. (2010). Recent Developments in Maximum Power Point Tracking Technologies for Photovoltaic Systems. International Journal of Photoenergy, Article ID 245316.

Qi, W., Liu, J., Chen, X., & Christofides, P. D. (2011). Supervisory Predictive Control of Standalone Wind/Solar Energy Generation Systems. *IEEE Transactions on Control Systems Technology*, *19*(1), 199–207. doi:10.1109/TCST.2010.2041930

Rodriguez, P., Luna, A., Ciobotaru, M., Teodorescu, R., & Blaabjerg, F. (2006). Advanced Grid Synchronization System for Power Converters under Unbalanced and Distorted Operating Conditions. In *Proceedings of the 2006 IEEE 32nd Annual Conference on Industrial Electronics* (IECON), (pp. 5173 – 5178). IEEE. DOI: doi:10.1109/IECON.2006.347807

Rodríguez, P., Pou, J., Bergas, J., Candela, J. I., Burgos, R. P., & Boroyevich, D. (2007). Decoupled Double Synchronous Reference Frame PLL for Power Converters Control. *IEEE Transactions on Power Electronics*, *22*(2), 584–592. doi:10.1109/TPEL.2006.890000

Rodriguez, P., Teodorescu, R., Candela, I., Timbus, A. V., & Blaabjerg, F. (2006). New positive-sequence voltage detector for grid synchronization of power converters under faulty grid conditions. In *Proceedings of the 37th IEEE Power Electronics Specialists Conference* (PESC '06). IEEE. DOI: doi:10.1109/PESC.2006.1712059

Saccomando, G., & Svensson, J. (2001) Transient Operation of Grid Connected Voltage Source Converter under Unbalanced Voltage Conditions. In *Proceedings of IEEE Industry Applications Conference, 36th LAS Annual Meeting*, (pp. 2419-2424). IEEE. DOI: doi:10.1109/IAS.2001.955960

Saïdi, S., Abbassi, R., & Chebbi, S. (2014). Virtual Flux Based Direct Power Control of Shunt Active Filter. *Scientica Iranica, Transaction D: Computer Science & Engineering and Electrical Engineering*. In press.

Sera, D., Mathe, L., Kerkes, T. & Spataru, S.V. (2013). On the Perturb-and-Observe and Incremental Conductance MPPT Methods for PV Systems. *IEEE Journal of Photvoltaic, 3*(3), 1070-1078.

Sera, D., Teodorescu, R., Hantschel, J., & Knoll, M. (2007). Optimized Maximum Power Point Tracker for Fast-Changing Environmental Condition. *IEEE Transactions on Industrial Electronics, 55*(7), 2629–2637. doi:10.1109/TIE.2008.924036

She, Y., She, X., & Baran, M. E. (2011). Universal Tracking Control of Wind Conversion System for Purpose of Maximum Power Acquisition under Hierarchical Control Structure. *IEEE Transactions on Energy Conversion, 26*(3), 766–775. doi:10.1109/TEC.2011.2159607

Shi, J.-H., Zhong, Z.-D., Zhu, X.-J., & Caco, G.-Y. (2008). Robust design and optimization for autonomous PV-wind hybrid power systems. *Journal of Zhejiang University Sciences A, 9*(3), 401–409. doi:10.1631/jzus.A071317

Song, H.-S., & Nam, K. (1999). Dual Current Control Scheme for PWM Converter under Unbalanced Input Voltage Condition. *IEEE Transactions on Industrial Electronics, 46*(5), 953–959. doi:10.1109/41.793344

Sürgevil, T., & Akpinar, E. (2005). Modelling of a 5-kW wind energy conversion system with induction generator and comparison with experimental results. *Renewable Energy, 30*(6), 913–929. doi:10.1016/j.renene.2004.09.002

Suul, J. A., Luna, A., Rodriguez, P., & Undeland, T. (2012). Virtual-Flux-Based Voltage-Sensor-Less Power Control for Unbalanced Grid Conditions. *IEEE Transactions on Power Electronics, 27*(9), 4071–4087. doi:10.1109/TPEL.2012.2190301

Taherbaneh, M., Rezaie, A. H., Ghafoorifard, H., Rahimi, K., & Menhaj, M. B. (2010). Maximizing Output Power of a Solar Panel via Combination of Sun Tracking and Maximum Power Point Tracking by Fuzzy Controllers. International Journal of Photoenergy, Article ID 312580.

Teodorescu, R., Liserre, M., & Rodríguez, P. (2011). *Grid Converters for Photovoltaic and Wind Power Systems*. John Wiley & Sons, Ltd. doi:10.1002/9780470667057

Trintis, I. (2011). *Grid Converters for Stationary Battery Energy Storage Systems*. PhD, Faculty of Engineering & Science at Aalborg University.

Valenciaga, F., & Puleston, P. F. (2005). Supervisor Control for a Stand-Alone Hybrid Generation System Using Wind and Photovoltaic Energy. *IEEE Transactions on Energy Conversion, 20*(2), 398–405. doi:10.1109/TEC.2005.845524

Vechiu, I., Ciprian, B., & Gelu, G. (2013). Three-Phase Four-Wire Active Power Conditioners for Weak Grids. *International Journal of Computer and Electrical Engineering, 5*(2), 173–178. doi:10.7763/IJCEE.2013.V5.689

Wang, L.-J., Yang, T., Zhang, D.-M., & Lu, Z.-. (2011). A high performance simulation methodology for multilevel grid-connected inverters. *Journal of Zhejiang University Sciences C, 13*(7), 544–551. doi:10.1631/jzus.C1100315

Wang, X., Xing, Y., Zhipeng, H., & Yan, L. (2012). Research and Simulation of DTC Based on SVPWM of PMSM. *Procedia Engineering, 9*, 1685–1689. doi:10.1016/j.proeng.2012.01.195

Xia, Y., & Mandic, D. P. (2012). Widely Linear Adaptive Frequency Estimation of Unbalanced Three-Phase Power Systems. *IEEE Transactions on Instrumentation and Measurement*, *61*(1), 74–83. doi:10.1109/TIM.2011.2159409

Xiong, F., Yue, W., Ming, L., Ke, W., & Wanjun, L. (2010). A Novel PLL for Grid Synchronization of Power Electronic Converters in Unbalanced and Variable-Frequency Environment. In *Proceedings of the 2nd IEEE International Symposium on Power Electronics for Distributed Generation Systems* (PEDG), (pp. 466-471). IEEE. DOI: doi:10.1109/PEDG.2010.5545832

Xu, L., & Wang, Y. (2007). Dynamic Modelling and Control of DFIG-Based Wind Turbines under Unbalanced Network Conditions. *IEEE Transactions on Power Systems*, *22*(1), 314–323. doi:10.1109/TPWRS.2006.889113

Yazdani, D., Mojiri, M., Bakhshai, A., & Joos, G. A. (2009). Fast and Accurate Synchronization Technique for Extraction of Symmetrical Components. *IEEE Transactions on Power Electronics*, *24*(3), 674–684. doi:10.1109/TPEL.2008.2010321

Zghal, W., Kantchev, G., & Kchaou, H. (2011). Optimization and management of the energy produced by a wind energizing system. *Renewable & Sustainable Energy Reviews*, *15*(2), 1080–1088. doi:10.1016/j.rser.2010.11.021

Zhou, W., Yang, H., & Fang, Z. (2008). Battery Behavior Prediction and Battery Working States Analysis of a Hybrid Solar–Wind Power Generation System. *Renewable Energy*, *33*(6), 1413–1423. doi:10.1016/j.renene.2007.08.004

Zuher, A., & Mehrdad, K. (2013). An analytical literature review of stand-alone wind energy conversion systems from generator viewpoint. *Renewable & Sustainable Energy Reviews*, *28*, 597–615. doi:10.1016/j.rser.2013.08.027

KEY TERMS AND DEFINITIONS

Battery Storage: It is a technical solution and an economic opportunity for the guarantee of the supply/demand balance of electricity required for increasing the energy systems reliability.

Grid Synchronization: The purpose of synchronization is to monitor, access, enable, and automatically take the control action to prevent the abnormalities of voltage and frequency by the minimization of difference in voltage, frequency and phase angle between the corresponding phases of the generator output and grid supply.

Grid Voltage Unbalance: It is regarded as a power quality problem of significant concern at the electricity distribution level. It is caused by unequal system impedances and unequal distribution of single -phase loads.

MPPT: In photovoltaic (PV) systems, maximum power point trackers (MPPT) is power conditioning electronics that constantly adjust the voltage in order to maximize power output.

Renewable Energies: Derived from resources that are naturally inexhaustible, such as wind energy, heat (geothermal, solar), moving water (hydro, tidal, and wave power), and biomass.

SVPWM Converters: Power electronics converter controlled by space vector pulse width modulation technique.

Symmetrical Components Separation: A mathematical theory which breaks down an unbalanced system into three balanced systems. These three are called positive sequence, negative and zero sequence systems. For a perfectly balanced system both negative and zero sequence systems would be absent.

Chapter 25
Modelling and Realisation of a Three-Level PWM Inverter Using a DSP Controller to Feed an Asynchronous Machine

Denoun Hakim
Mouloud Mammeri University, Algeria

Benamrouche Nacereddine
Mouloud Mammeri University, Algeria

Benyahia Nabil
Mouloud Mammeri University, Algeria

Salah Haddad
Mouloud Mammeri University, Algeria

Zaouia Mustapha
Mouloud Mammeri University, Algeria

Sadek Ait Mamar
Mouloud Mammeri University, Algeria

ABSTRACT

Multilevel inverters have seen an increasing popularity in the last few years for medium- and high-voltage applications. The most popular has been the three-level neutral clamped inverter. Multilevel inverters synthesize output voltage from more than two voltage levels. Thus, the output signal spectrum is significantly improved in comparison with the classical two-level converters. This chapter discusses modelling and control of a Neutral Point Clamped (NPC) inverter which operates with the PWM switching pattern using a DSP. The mathematical model of the NPC inverter is carried out using conversion and connection functions for an easier understanding of the system operation. Simulation results using MATLAB program are reported, and it is shown that the performances obtained for driving an asynchronous motor using this inverter are very promising. Finally, analysis of the theoretical and the experimental results is carried out in order to validate the effectiveness of the proposed control solution.

1. INTRODUCTION

The three level inverters often referred to as Neutral Point Clamped (NPC) inverters offer several advantages over the more common two level inverters. As compared to two level inverters, three level

DOI: 10.4018/978-1-4666-7248-2.ch025

Figure 1. NPC inverter

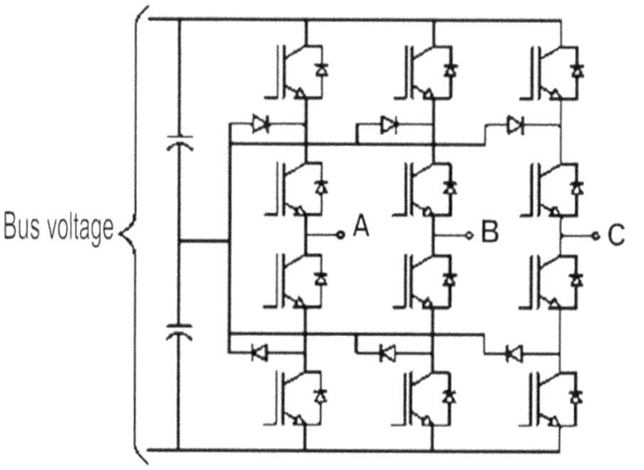

inverters have smaller output voltage steps that mitigate motor issues due to long power cables between the inverter and the motor. These issues include surge voltages and rate of voltage rise at the motor terminals and motor shaft bearing currents. In addition, the cleaner output waveform provides an effective switching frequency twice that of the actual switching frequency. Should an output filter be required, the components will be smaller and less costly than for an equivalent rated two level inverter. Most often the NPC inverter is used for higher voltage inverters. Since the IGBTs are only subjected to half of the bus voltage, lower voltage IGBT modules can be used.

Figure 1 shows the circuit configuration of the NPC inverter. Each leg has four IGBTs connected in series. The applied voltage on the IGBT is one-half that of the conventional two level inverters. The bus voltage is split in two by the connection of equal series connected bus capacitors. Each leg is completed by the addition of two clamp diodes.

This topology has been used traditionally for medium voltage industrial applications. In addition to the capability of handling higher voltages, the NPC inverter has several favorable features including lower line-to-line and common mode voltage steps and lower output current ripples for the same switching frequency as that used in a two level inverter.

An attempt is made to present the most common multilevel (three-level) inverters and briefly describe their advantages and disadvantages based on a literature survey. The literature survey is intended to compare the inverters based on their performance aspects and the challenges and restrictions they could impose for application in a given area.

The fundamental purpose of considering multilevel inverters instead of the popular two-level inverters is due to the high value of DC-link voltage that needs to be employed in future developments. With the increase in the DC-link voltage, it is aimed that the performance of the inverter in use does not degrade and hence, multilevel inverter options have been considered.

The increase in power-handling capabilities of the power electronic switches has made the use of multilevel inverters feasible for high-power applications. For high voltage and high power systems, instead of using switches with high voltage ratings, it is beneficial to connect the switches having low-voltage ratings in series (Tolbert et al., 1997). This allows the latter to be switched faster than the switches having higher voltage ratings, thus resulting in switching harmonics of higher frequencies which can be filtered easily.

Due to the dynamic voltage sharing problem when the switches are connected in series, multilevel power inverters have come into existence and they are being used due to several advantages that they offer over the conventional two-level converters. However, in principle, with increased DC-link voltage, there are different ways in which the semiconductor devices can be connected, that is

- Using switching devices with higher voltage rating. However, this involves high *dV/dt* stress and renders slower switching due to high voltage level.
- Series connection of devices, however, dynamic voltage sharing among the devices represents a big challenge for this option.
- Adopting the multilevel solution which is more preferable due to the limited voltage ratings for the semiconductor devices.

There are a number of multilevel inverters introduced since the year 1975 but the three basic and most well-known topologies are the Cascaded H-Bridge Multilevel Converter (CHB), Neutral Point-Clamped Multilevel Converter (NPC) and the Flying Capacitor Multilevel Converter (FCC). These three basic topologies have been widely accepted for industrial applications (Rodriguez et al., 2002).

In order to control the switches of multilevel inverters, a number of new and modified modulation techniques have been developed (Colak et al., 2011). These schemes include the Sinusoidal Pulse Width Modulation (SPWM), Space Vector Modulation (SVM), Selective Harmonic Elimination (SHE), etc. In addition, a number of other modified control schemes based on these basic schemes have also been proposed (Urmila et al., 2010).

Multilevel converters are based on the neutral point clamped inverter topology proposed by Nabae (Nabae, 1981). They are recently applied in many industrial applications such as power supplies, static VAR compensatory, drive systems, etc.

One of the most significant advantages of multilevel configuration is the harmonic reduction in the output waveform without increasing frequency, thus three level inverters can be operated at lower switching frequencies ($f_{sw} < 500$ Hz) leading to smaller commutation losses and hence higher efficiencies. Moreover they have great interests in overcoming the series connections problems (Bows et al., 2000). In fact, the three-level inverter is able to generate voltages without output transformer and its harmonic components are fewer than those of conventional two-level inverters at the same switching frequency. In addition, when the blocking voltage of each switching device is half of the dc-link voltage, it is easy to produce high voltage and power inverter systems (Meynard et al., 2002; Zhou et al., 2007).

In recent years huge advances have been made in digital control strategies leading to improving the performances of multi-level inverters and thus spreading their use (Meynard et al., 1997; Wanfeng et al., 2006).

In this contribution, we start by giving the operating principle of a thre level PWM inverter before proceeding to its modeling and simulation. The control is carried out using a TMS320LF2407 DSP controller. Emphasis is made on elaborating the PWM control strategy using this DSP programmed using software developed by Texas Instruments "Code composer".(Girar et al., 2006; Tzung et al., 2009).

Furthermore, its experimental implementation is performed and the built system is tested. The experimental results are used to validate those obtained theoretically. In addition, the performances of this inverter are analyzed with respect to its output voltage contents. Finally, the NPC inverter is used to feed an asynchronous motor. The performances of this drive system are evaluated.

2. BASIC MULTILEVEL INVERTER TOPOLOGIES

A power electronic circuit that could operate in either inverter or rectifier mode is called a converter. Since the present project concentrates only on the inverter mode of operation of the converter, henceforth, only the term 'inverter' will be used.

In the case of a multilevel inverter, the DC-link constitutes more than one capacitor bank as opposed to that in a conventional two-level inverter which consists of only one capacitor bank. The multiple DC sources formed with the series capacitors are aggregated with the commutation of the power semiconductor switches thus giving a high voltage at the output, and at the same time, each semiconductor switch has to withstand only the reduced level of voltage that appears across each capacitor. The connected DC voltage source determines the rating of the power semiconductor switches. It is difficult to connect a single power semiconductor switch directly to high voltage DC link, as very high voltage rating of the switch has to be chosen by giving allowance to the voltage overshoot due to the stray inductances present in the semiconductor switch and the module (read as a phase leg of the inverter).

The number of levels in a multilevel inverter, in effect, can be defined as the number of levels of phase voltage with respect to the negative terminal of the inverter.

If '*m*' denotes the number of steps (or, levels) of phase voltage with respect to the negative terminal of the inverter, then the number of steps in the voltage between the two phases of the load (or the line voltage at the load side) is denoted by '*k*' which is given by

$$k= 2m+1 \tag{1.1}$$

The number of levels in the phase voltage of a three-phase Y-connected load is denoted by '*l*' which is given by

$$l= 2k - 1 \tag{1.2}$$

The above relationships show that the number of steps of voltage seen by the load is increased in multilevel inverter when compared to the two-level inverter, thus providing a smoother output voltage (nearly sinusoidal) with reduced harmonic distortion.

In general, the advantages of utilizing multilevel inverters over the shortcomings of a conventional two-level inverter can be summarized as follows (Rodreguez et al., 2007):

- The output current of a multilevel inverter has lower distortion when compared to a two-level inverter.
- The multilevel inverter makes better utilization of the DC-bus voltage when compared to the two-level inverter.
- The smaller common-mode voltage of multilevel inverter reduces the stress in the bearings of a motor that is connected to a multilevel motor drive.
- Due to higher number of output voltage levels, the *dV/dt* stress gets reduced which in-turn reduces the electromagnetic compatibility (EMC) issues.
- Multilevel inverters can be operated at fundamental as well as low or high switching frequency PWM; high switching frequency renders higher switching losses thus reducing the efficiency of the inverter.

There are a few disadvantages of multilevel inverters which can be summarized as follows:

- As the number of the required output level increases, the number of the power semiconductor switches required per phase also increases.
- Though the voltage rating of the switches required is lower than that of a two-level inverter, each switch demands a related gate drive circuit and this in turn makes the overall system more complex and expensive.

Since 1975, many research works have been dedicated to multilevel inverter topologies and their modulation schemes. The most promising topologies have been the cascaded H-Bridge inverter with separate DC sources, Neutral Point-Clamped (or Diode-Clamped) inverter and the Flying Capacitors (or Capacitor-Clamped) inverter.

The Cascaded H-Bridge multilevel inverter was first introduced in 1975. The idea was to connect the separate DC-sourced full-bridge cells in series so as to synthesize a staircase AC output voltage waveform. Later in 1980s, the Diode-Clamped multilevel inverter was introduced with diodes blocking the sources. It got its alternate name as Neutral-Point Clamped inverter because the mid-voltage level in three-level inverter is defined as the "neutral" point. The Capacitor-Clamped multilevel inverter was introduced in the 1990s. The latter possesses a similar structure as that of the Diode-Clamped multilevel inverter; except for replacing the clamping diodes by the clamping capacitors. The duty of the clamping device is to clamp the voltage level of the switch to that of the DC-link capacitors. Hence, each level in the output staircase waveform represents the voltage of each DC-link capacitor. Therefore, the aggregate of the capacitor voltages gives the output levels (or steps) in the staircase AC waveform. For applications involving high voltage and high power, it is favorable to utilize Diode-Clamped or Capacitor-Clamped multilevel inverters to replace the full-bridge cell in a cascaded inverter so as to reduce the number of separate DC sources. Moreover, in applications with limitations in space allowance, and those demanding reduced weight, it is favorable to choose the Diode-Clamped multilevel inverter as the system would be bulky if capacitors are used for clamping purposes.

2.1 Cascaded H-Bridge Multilevel Inverter

In a Cascaded H-Bridge inverter, separate H-bridges are connected in series in each phase depending on the number of levels that are desired at the output. The three-phase structure of a Cascaded H-bridge inverter is shown in Figure 2. Separate DC sources are connected to each single-phase full-bridge inverter. The AC output of each level is then connected in series such that the overall output voltage waveform of the multilevel inverter is the sum of the individual inverter output.

From the knowledge of single-phase full-bridge inverters, each inverter level with V_{dc} the DC voltage for each full-bridge can generate three different voltage outputs; $+V_{dc}$, 0 and $-V_{dc}$. In the case of the inverter represented in Figure 2, all the levels from $-2V_{dc}$ to $+2V_{dc}$ are present constituting 5 levels for the phase. This is made possible by connecting the DC sources (or capacitors) sequentially to the AC (output) side via the four power switches present in each cell.

For an m-level cascaded inverter, the number of DC voltage sources '*s*' is related to the number of levels by

$$m = 2s+1 \tag{1.3}$$

Figure 2. Three-phase Cascaded H-Bridge multilevel inverter

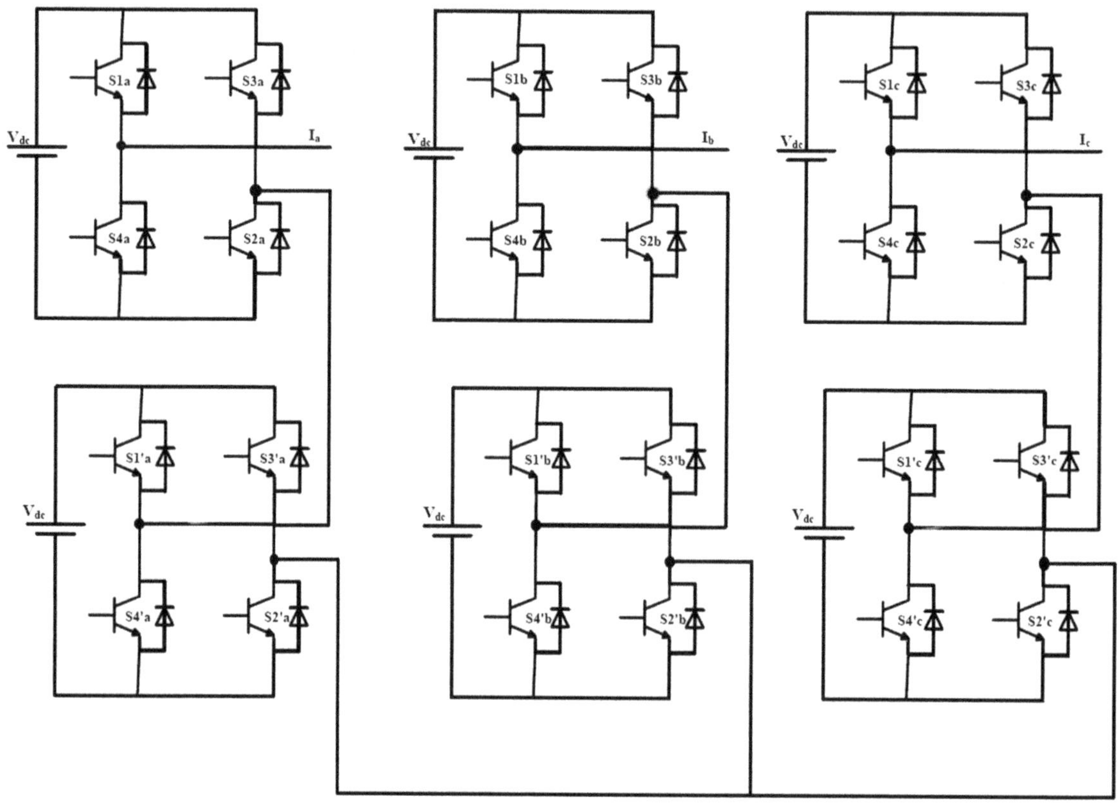

Cascaded inverters have been proposed for use as the main traction drive in electric vehicles where the inverter could serve as a rectifier/charger for the batteries of the vehicle when it is connected to an AC supply. For a vehicle that uses regenerative braking, this inverter can act as a rectifier.

2.1.1 Advantages

- The possible output voltage level is more than twice the number of DC sources.
- Modularity of the series H-bridges make the layout and packaging simple.

2.1.2 Disadvantages

- The main restriction of a cascaded H-bridge inverter is that separate DC sources are required for each of the H-bridges thus limiting its application.
- Batteries or ultra-capacitors make the inverter very bulky with higher levels.
- The problem of charge equalization for the separate bridges is an important issue to be taken care of.

2.2 Flying Capacitor Multilevel Inverter

The Flying Capacitor multilevel inverter came into existence in 1992. The three phases of the three-level Flying Capacitor inverter is shown in Figure 3. The three phases of the inverter share a common DC bus. Similar to the three-level Diode-Clamped multilevel inverter, the Capacitor-Clamped multilevel inverter has two series-connected capacitors, C1 and C2 dividing the DC bus voltage into three levels; namely $+V_{dc}/2$, 0, and $-V_{dc}/2$.

These voltage levels appear at the output of each phase of the inverter by appropriate switching of the power semiconductor devices. In place of the "clamping diodes" present in the Diode-Clamped multilevel inverter, the Capacitor-Clamped multilevel inverter consists of "clamping capacitors". Each clamping capacitor clamps the device voltage to one DC-link capacitor voltage level (Ramani et al.,2009).

In order to obtain three levels across 'a' and 'n', there are three switch combinations as follows:

- Turn on upper switches, S1 and S2, in order to obtain $V_{an}=+V_{dc}/2$.
- Turn on switches, (S1,S1') or (S2, S2'), in order to obtain an=0.
- Turn on lower switches, S1' and S2', in order to obtain $V_{an}=-V_{dc}/2$.

During the zero output voltage, the clamping capacitor C3 charged when the switch pair (S1, S1') is turned on; and C3 is discharged when the switch pair (S2, S2') is turned on.

2.2.1 Advantages

- Availability of switching redundancy for capacitor voltage balance.
- Increased number of capacitors allows the inverter to ride through short duration outages.

Figure 3. Three-phase Capacitor-Clamped multilevel inverter

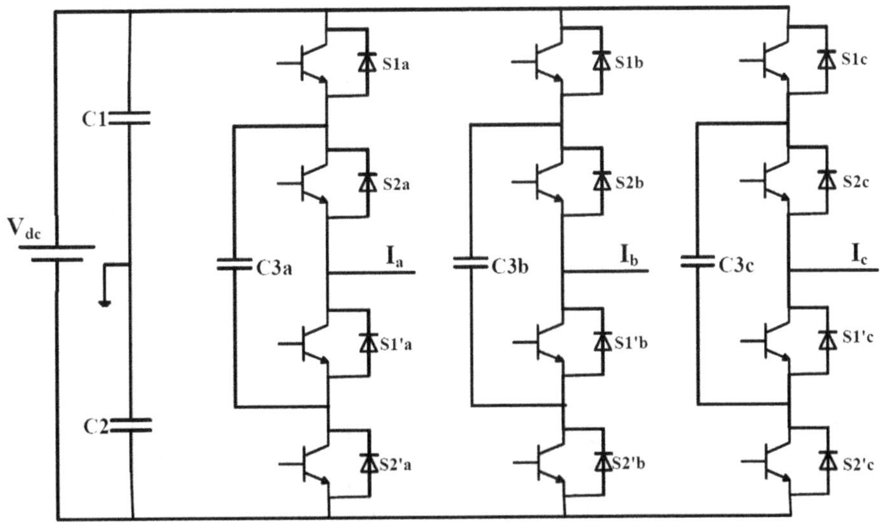

2.2.2 Disadvantages

- Increased number of capacitors makes the system very bulky with increased losses.
- Capacitor voltage levels have to be maintained at all times; the capacitors require a separate pre-charge circuit.
- Cost increases manifold due to the increased number of capacitors in the Capacitor Clamped topology when compared to the clamping diodes in the Diode-Clamped topology.

2.3 Neutral Point-Clamped Multilevel Inverter

Since its proposal in 1981, this inverter has found its use in a number of industrial applications. The three-phase structure of a three-level Neutral Point-Clamped inverter is shown in Figure 4 (Nabae, 1981). The three phases of the inverter share a common DC bus. The three-level Neutral Point-Clamped inverter consists of two series-connected capacitors, C1 and C2. The DC-link capacitors divide the DC bus voltage into three levels; namely $+V_{dc}/2$, 0 and $-V_{dc}/2$. These voltage levels appear at the output of each phase of the inverter by appropriate switching of the power semiconductor devices. The middle point of the two capacitors is denoted as 'n' which is the neutral point. There are two complementary switch pairs (S1, S1') and (S2, S2') and two clamping diodes (D1, D1') per phase present in this inverter. The outer two switches are the main switching devices (S1, S2') that operate for pulse width modulation while the inner two switches are the auxiliary switching devices (S2, S1') that clamp the output terminal potential to the neutral point potential along with the help of the two clamping diodes.

Figure 4. Three-phase, three-level Neutral Point-Clamped inverter

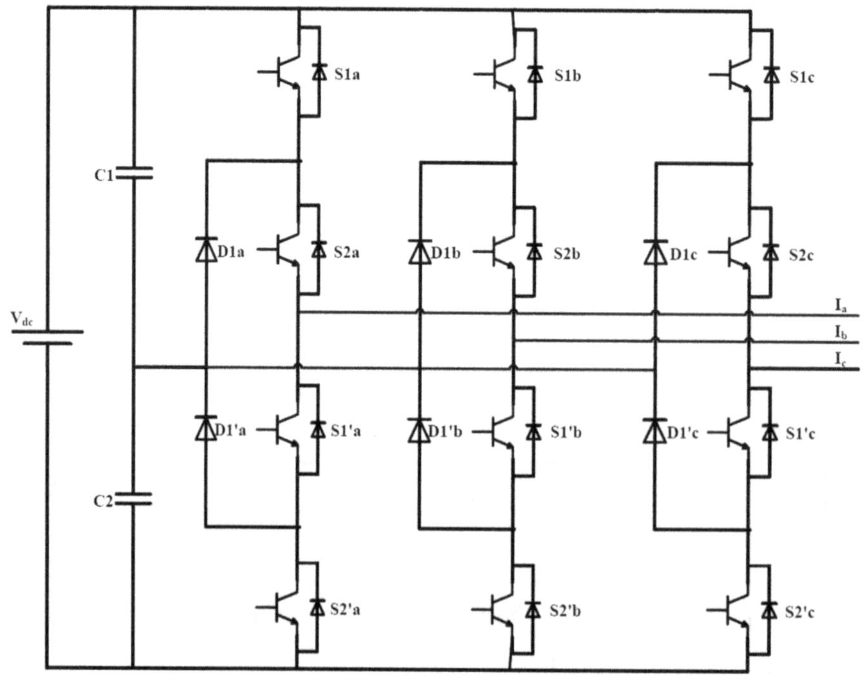

When both the upper switches S1 and S2 turn on, the voltage across 'a' (the first phase) and '0' (the negative inverter terminal), also called the pole voltage, is V_{dc}. The lower clamping diode, D1' balances out the voltage sharing between the two lower switches, S_1' and S_2'. While the switch S_1'blocks the voltage across C1, the switch S_2' blocks the voltage across C2. The voltage between 'a' and '0' is the DC voltage whereas the voltage between 'a' and 'n' is the AC voltage. It is because the voltage appearing with respect to the negative inverter terminal ('0') is the voltage across each capacitor and the voltage appearing with respect to the neutral point of the inverter ('n') is the aggregate of the capacitor voltages; giving an AC waveform.

In order to obtain three levels across 'a' and 'n', there are three switch combinations as follows:

- Turn on upper switches, S1 and S2, in order to obtain $V_{an} = + V_{dc} / 2$,
- Turn on middle switches, S2 and S1', in order to obtain $V_{an} = 0$,
- Turn on lower switches, S1' and S2', in order to obtain $V_{an} = - V_{dc} / 2$.

Each power device is required to block a voltage level of (V_{dc}/m-1) and if we assume that each blocking diode also has the same voltage rating as the active device, the number of diodes required for each phase will be {(m-1)x(m-2)}. There is a quadratic dependency for the number of diodes with the number of levels of the multilevel inverter.

2.3.1 Advantages

- The three phases share a common DC-bus minimizing the capacitance requirements.
- The DC-link capacitors can be pre-charged, as a group.
- High efficiency for fundamental frequency switching.

2.3.2 Disadvantages

- Increased number of clamping diodes.
- Neutral point control for balanced voltages across the DC-link capacitors should be achieved for all conditions of operation.

From the abovementioned review it is clear that the NPC is the most popular inverter and presents many advantages. Therefore, it is thoroughly investigated in this work. In fact, we start by explaining its operating principle before dealing with its modeling and simulation. An effort is made to improve its control by the introduction a DSP.

3. OPERATING PRINCIPLE OF NPC INVERTER

The NPC inverter can produce three voltage levels on the output: the DC bus positive voltage, the zero voltage and the DC bus negative voltage. The two level inverters can only connect the output to either the plus bus or the negative bus. (Refer to Figure 5 for the following example.) For a one phase operation, when IGBTs $B_{11}$and $B_{12}$are turned on, the output is connected to Vp; when $B_{12}$and $B_{13}$are on, the output is connected to V0; and when B_{13} and $B_{14}$are on, the output is connected to Vn.

Figure 5. Single leg

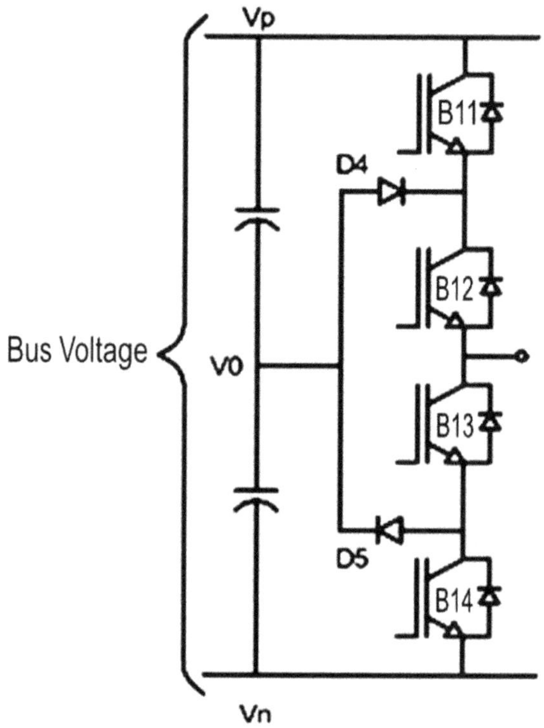

Table 1. Switching states

IGBT	Vout =Vp	Vout = V0	Vout = Vn
B11	On	Off	Off
B12	On	On	Off
B13	Off	On	On
B14	Off	Off	On

Switching states for the four IGBTs are listed in Table 1. Clamp diodes D4 and D5 provide the connection to the neutral point. From the switching states, it can be deduced that IGBTs B_{12} and B_{13} are on for most of the cycle, resulting in greater conduction loss than B_{11} and B_{14} but far less switching loss. In addition, the free wheel diodes for B_{12} and B_{13} are for most cases, soft switched as the IGBT parallel to the diode is on, thus holding the recovery voltage across the diode to that of the IGBT Vce.

The DC bus capacitors are connected in series and establish V0, the mid-point voltage. Due to available capacitor voltage rating, series connected capacitors are generally required in inverters rated for 480V and 600V services. In NPC inverters, maintaining the voltage balance between the capacitors is important to the proper operation of the NPC topology.

Figure 6 is a graph of the leg output voltage and Figure 7 is a graph of the phase-to-phase output voltage.

Figure 6. Leg output voltage

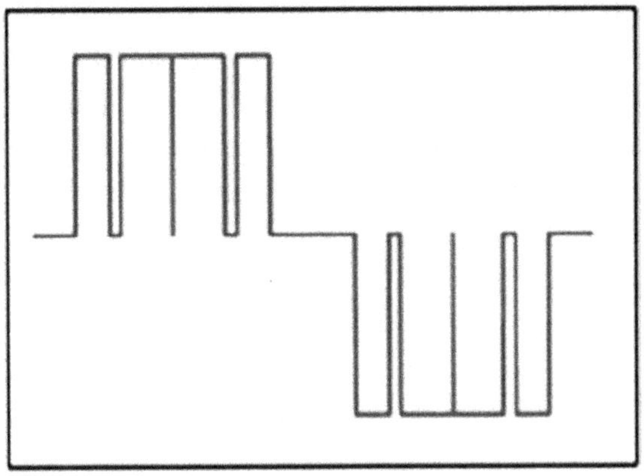

Figure 7. Phase-to-phase output voltage

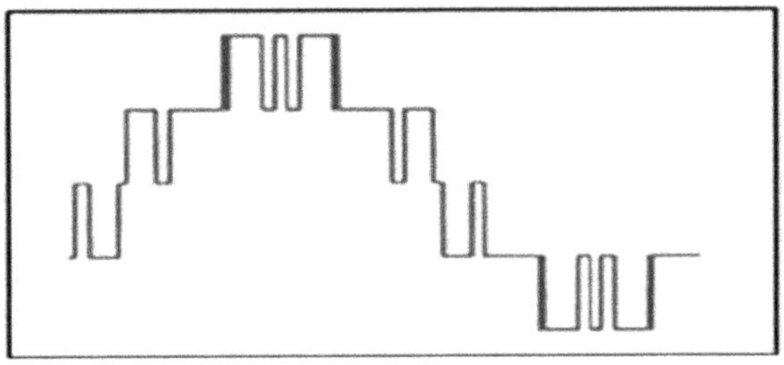

Careful observation shows that the effective switching frequency of the phase-to-phase voltage in Figure 7 is twice that of the phase voltage shown in Figure 6. A two level inverter is required to use two times the switching frequency of an NPC inverter in order to achieve the same ripple in the output current. This simple fact coupled with the intermediate voltage steps of the NPC inverter offers two advantages over the two level inverters. First, there are far less switching losses in the NPC inverter and second, if an output filter is required, the filter components will be smaller in both value and size than the filter components for a two level inverter.

Multilevel converters can be operated at high voltages without the need for series connection of the switching devices. There are different ways in which the devices can be connected so that the dynamic voltage sharing problem posed by the series connection of the devices is no more confronted. The capability of the inverter to operate at high power level with lower harmonic distortion and lower voltage stress across the switches makes it a strong candidate for drives applications.

The features that make a multilevel converter an attractive candidate are (Khomfoi et al., 2007):

- **Staircase Waveform Quality:** Multilevel inverters can generate output voltage with low distortion and reduced *dV/dt* stresses, resembling a near sinusoidal waveform with increase in the number of levels.
- **Switching Frequency:** These inverters can be operated at both fundamental and high switching frequency pulse width modulation (PWM). Lower switching frequency would yield lower switching losses and thus improving the efficiency.

- **Common-Mode Voltage:** In an inverter-driven A.C. machine, there exists a common-mode voltage as the VSI does not constitute an ideal balanced source. The parasitic capacitances in an A.C. motor become much relevant when this motor is driven by a PWM VSI. High *dV /dt of* the common mode voltage applied across the stator and the ground of the motor (in a 3-φ, 4 wire system) causes pulsed currents (the common-mode currents) to flow through these capacitances thus producing the common-mode voltage. In multilevel inverter-fed motor drive, due to smaller *dV/dt* when compared to the two-level inverter-fed drive, there is smaller common-mode voltage at the motor bearing terminals.

4 MODELING OF A THREE LEVEL INVERTER

4.1 General Structure

The NPC multilevel inverter uses capacitors in series to divide the DC bus voltage into different voltage levels. To produce m-levels in the phase voltage, an m-level NPC inverter needs m-1 capacitors on the de bus. A three- phase three-level NPC inverter is shown in Figure 1. The dc bus consists of two capacitors C (Meynard, 2006). For a dc bus voltage Uc, the voltage across each capacitor is Uc/2 and each device voltage stress is limited to one capacitor voltage level Uc/2 through clamping diodes (Yo-Han Lee et al., 1996; Rioual et al., 1996).

Figure 8. Power circuit of NPC

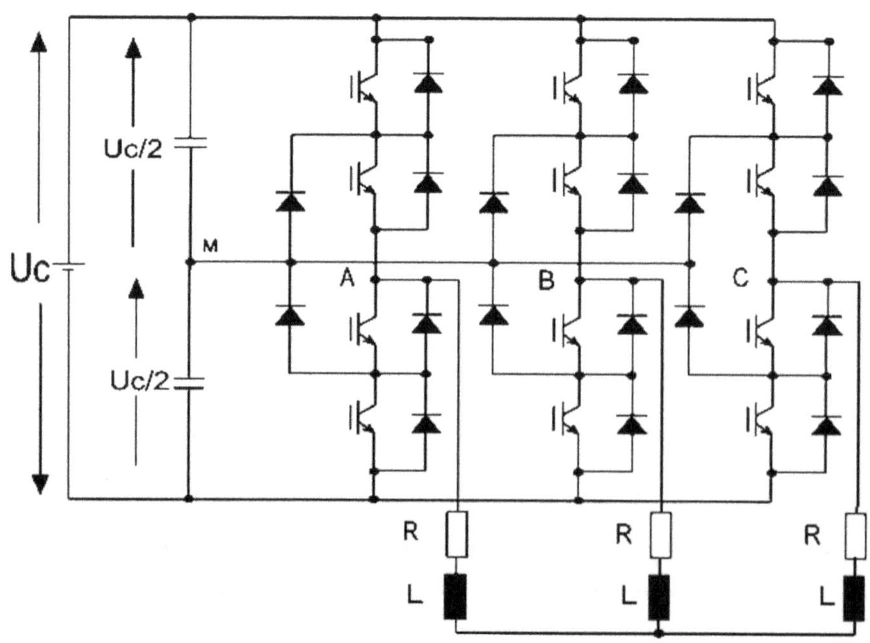

Figure 9. Different configurations of a leg

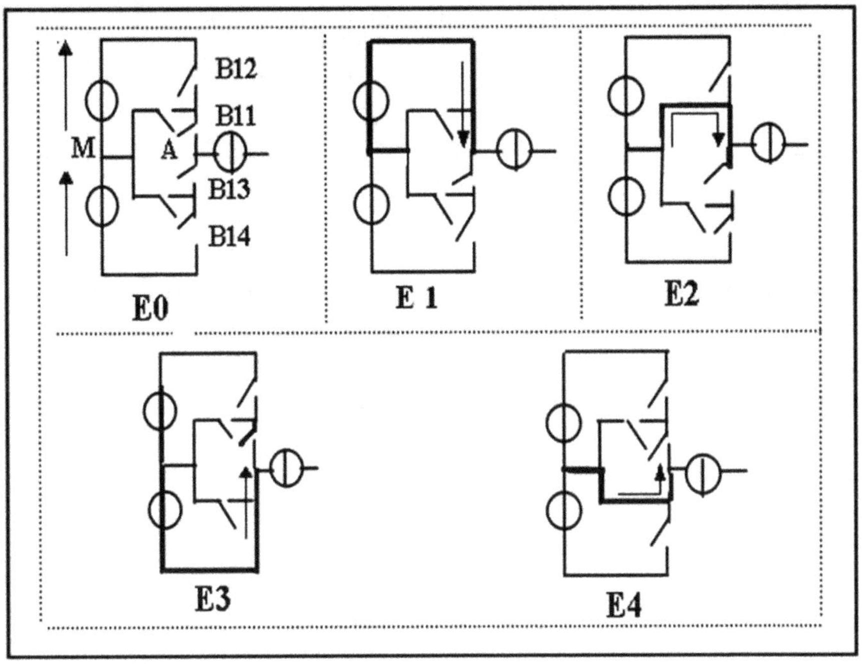

4.2 Different Configurations of the Inverter

To describe the different sequences of the inverter function, let us consider the possible states of the first leg switches, Figure 9.

The three-level inverter has the advantages that the blocking voltage of each switching device is one half of dc-link voltage whereas full dc-link voltage for two- level inverter.

4.3 Complementary Commands

We define the complementary control of the inverter leg as follows (Berkouk, 1996):

$$\begin{cases} B_{K1} = \overline{B}_{K4} \\ B_{K2} = \overline{B}_{K3} \end{cases} \tag{1.4}$$

It is demonstrated that the command given by relationship (1.4) is the one which gives the three levels U_c, O, $-U_c$, in a optimum way.

Table 2 shows the states of the switches and the corresponding output voltage of the inverter, where V_k (k = a, b, c) is the potential of the leg.

Table 2. Switches states and the corresponding voltages

B_{k1}	B_{k2}	B_{k3}	B_{k4}	V_k
0	0	1	1	-Uc2
0	1	0	1	Unknown
1	0	1	0	0
1	1	0	0	Uc1

4.4. Model Control of the Three-Level Inverter: Connection Functions

These functions define the state of the switches. They are equal to 1 if the switch is ON and 0 if the switch is OFF. They are defined by the following relationship.

$$F_{k1} = 1 - F_{k4}$$

$$F_{k2} = 1 - F_{k3}\ (k = 1, 2, 3) \tag{1.5}$$

Therefore, the branch voltages V_{AM}, V_{BM}, V_{CM} are expressed as follow:

$$V_{AM} = F_{11}.F_{12} \cdot U_{C1} - F_{13}.\ F_{14}.\ U_{c2}$$

$$V_{BM} = F_{21}.F_{22} \cdot U_{C1} - F_{23}.\ F_{24}.\ U_{c2}$$

$$V_{CM} = F_{31}.F_{32} \cdot U_{C1} - F_{33}.\ F_{34}.\ U_{c2}$$

However, the output phase voltages of the inverter can be deduced from equation (1.5) as follows:

$$V_A = (2V_{AM} - V_{BM} - V_{CM}) / 3$$

$$V_B = (2V_{BM} - V_{CM} - V_{AM}) / 3 \tag{1.6}$$

$$V_C = (2V_{CM} - V_{AM} - V_{BM}) / 3$$

And these equations can be rewritten using the voltage branches which give:

$$\begin{bmatrix} V_A \\ V_B \\ V_C \end{bmatrix} = \frac{1}{3}\begin{bmatrix} 2 & -1 & -1 \\ -1 & 2 & -1 \\ -1 & -1 & 2 \end{bmatrix}\left\{\begin{bmatrix} F_{11}F_{12} \\ F_{21}F_{22} \\ F_{31}F_{32} \end{bmatrix} U_{CI} - \begin{bmatrix} F_{13}F_{14} \\ F_{23}F_{24} \\ F_{33}F_{34} \end{bmatrix} U_{C2}\right. \tag{1.7}$$

5. SIMULATION RESULTS WITH PWM STRATEGY

The parameters of the PWM control technique are defined as the amplitude modulation index m_a and the frequency ratio m_f.

$$m_a = A_m / A_c \tag{1.8}$$

$$m_f = f_c / f_m \tag{1.9}$$

where A_m is the peak amplitude of the control signal, while A_c is the peak amplitude of a triangular signal (carrier); m_f is the ratio between the carrier and the control frequency.

The PWM three level inverter algorithm of one carrier is given by:

$$\begin{aligned} &\left(\left|V_{ref.1}\right| \le U_p\right) \Rightarrow B_{11} = 1, B_{12} = 0 \\ &\left(\left|V_{ref.1}\right| > U_p\right)\ et\ V_{ref.1} \langle 0 \Rightarrow B_{11} = B_{12} = 0 \\ &\left(\left|V_{ref.1}\right| > U_p\right)\ et\ V_{ref.1} \rangle 0 \Rightarrow B_{11} = B_{12} = 1 \end{aligned} \tag{1.10}$$

where, V_{ref} is called the modulating signal and U_p is the carrier signal.where,

$$A = 1 \text{ when } V_{ref} \rangle 0 \text{ if not } A = 0 \quad B = 1 \text{ when} \left|V_{ref}\right| > U_p \quad \text{if not } B = 0$$

The logical expressions of B_{11} and B_{12} are:

$$\begin{aligned} B_{11} &= \bar{B}_{14} = \bar{B} + A \\ B_{12} &= \bar{B}_{13} = AB \end{aligned} \tag{1.11}$$

Table 3. The truth table of the algorithm (Berkouk, 1996)

A	B	B11	B12
0	0	1	0
0	1	0	0
1	1	1	1
1	0	1	0

The synoptic diagram of the proposed simulation of the strategy is shown in Figure 10 (Gaad et al., 2000, Denoun, 2001).

To illustrate the performance of the three-level PWM inverter, the system is investigated through computer simulation using a passive load by taking the following parameters:

R= 22 Ω, L=340mH, m_f=24, m_a=0.8, Uc/2= 30V, f =50Hz.

These parameters are chosen bearing in mind the components and the hardware available in our laboratory while conducting the experimental investigation.

After running the simulation program under *Matlab simulink* environment the following results are obtained. They are presented in Figures 11 through 17.

Figure 11 shows the simulation sequences of the upper-arm while Figure 12 shows the line voltage V_{ab}, it has five voltage levels (Uc, Uc/ 2,0, -Uc/ 2, -Uc).

Figure 13 shows the output voltage of the first phase of the inverter. Its Fourier decomposition analysis presented in Figure 14 shows that the harmonics are grouped into multiple frequencies centered around the switching frequency families. Harmonic number 23 and 25 are the most important harmonics present in the voltage waveform. Figure 15 shows the voltage V_{AM} with the three levels of output voltage of an arm: Uc/2, 0, -Uc/2. Figure 16 shows the evolution of the amplitude of the fundamental output voltage

Figure 10. Synoptic of the proposed PWM strategy

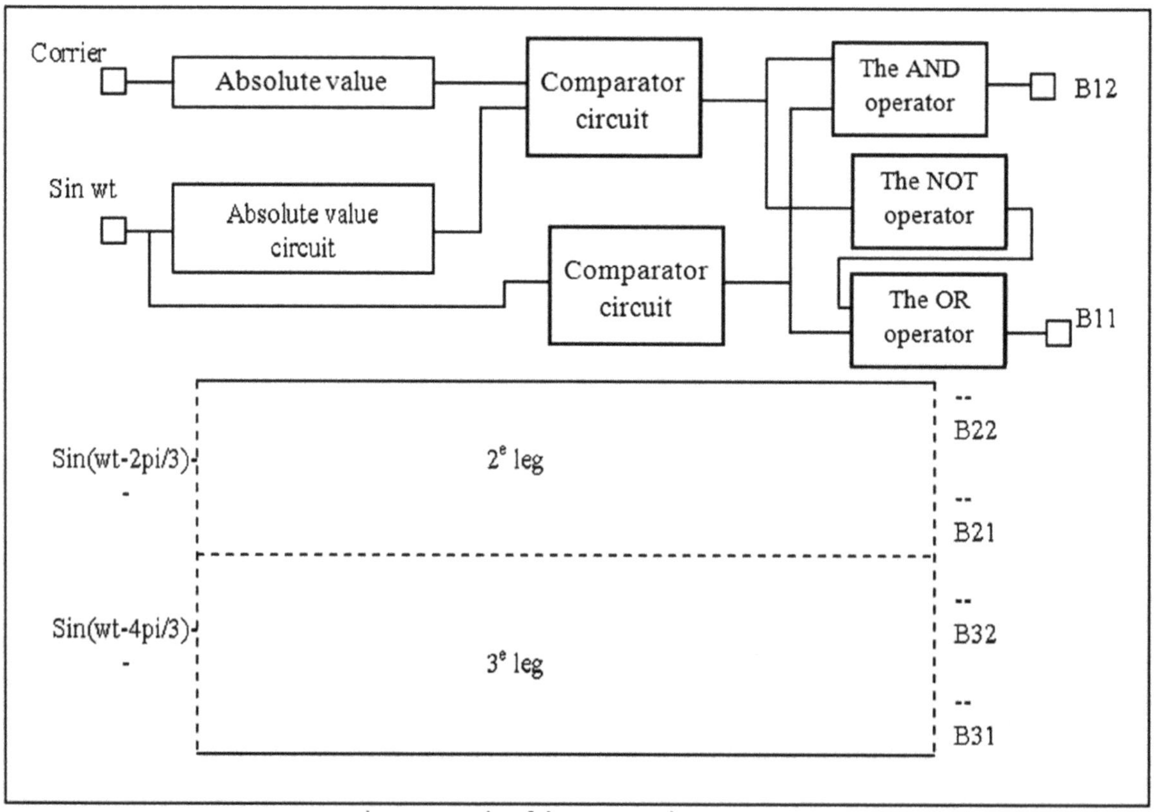

Figure 11. Control pulses for m_f=24, m_a=0.8

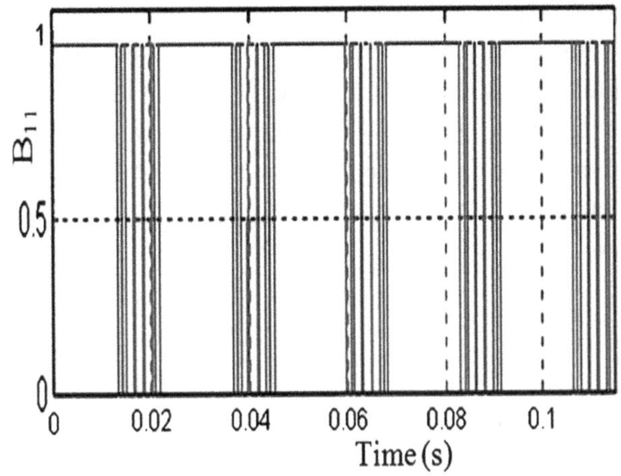

Figure 12. Voltage V_{AB}

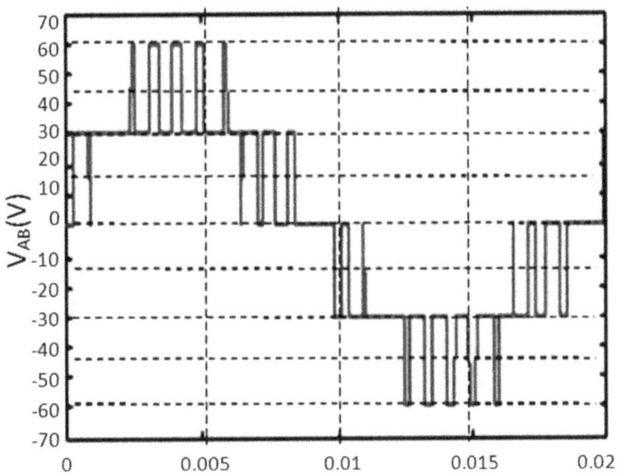

Figure 13. Output voltage of phase V_A

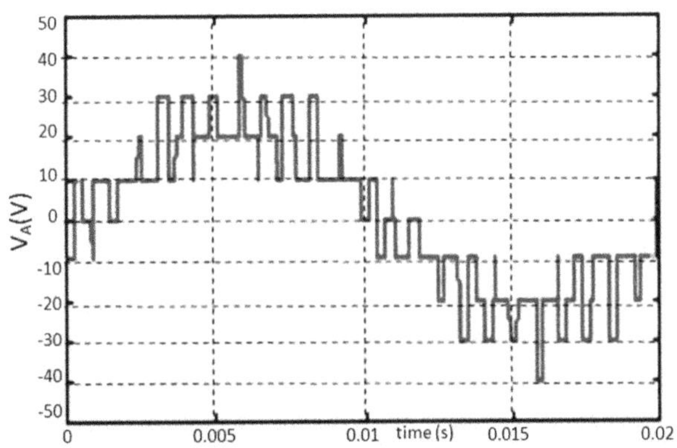

V_A versus the modulation index m_a. Finally, Figure 17 shows the THD versus m_a where it is shown that the harmonics decrease as m_a increases.

These results are much better than those which could be obtained with a conventional two–level inverter as the harmonics content is pushed to their higher order and thus easily filtered. It should be that the comparison of the reference sinusoidal signal with the triangular waveform is performed in the PWM generator of the DSP to generate the control signals for the switching devices along with the inverted signals with the required dead band.

Figure 14. V_A harmonic spectrum and THD

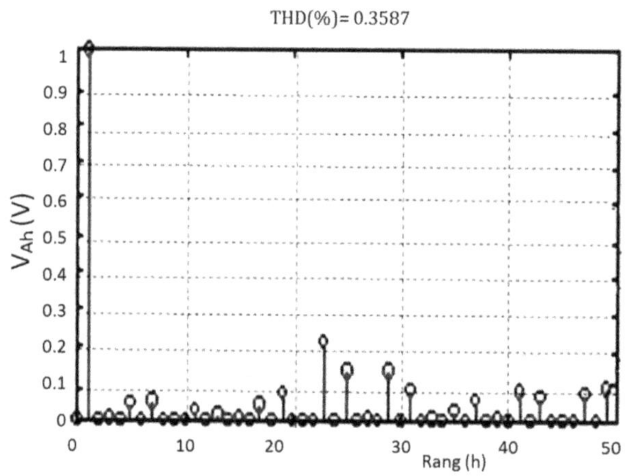

Figure 15. Voltage V_{AM}

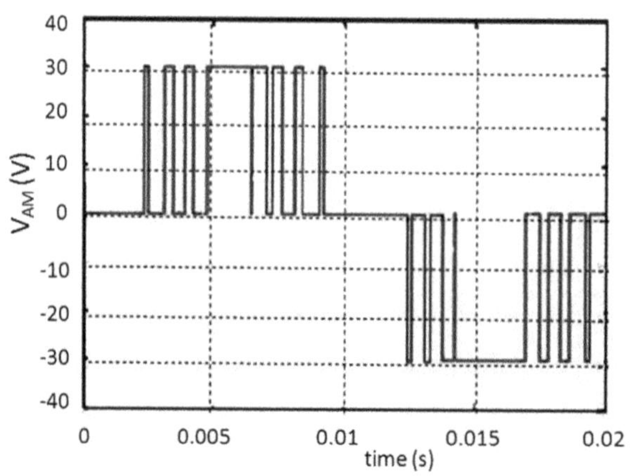

Figure 16. Evolution of the Fundamental voltage V_A

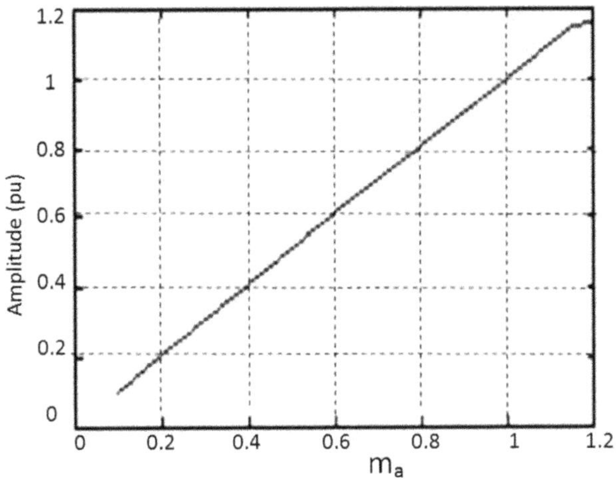

Figure 17. THD of voltage V_A

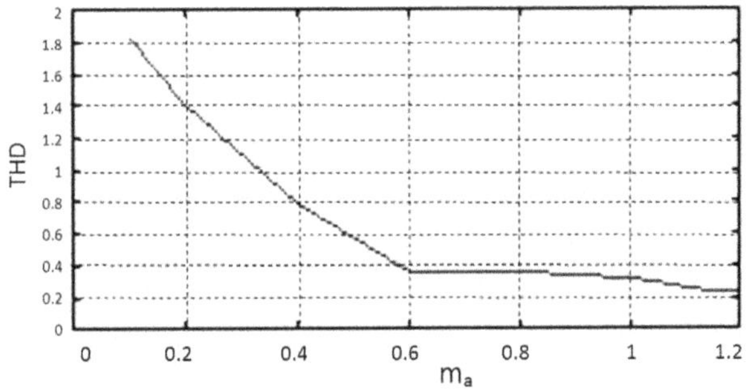

6. ASYNCHRONOUS MACHINE FED BY NPC INVERTER: SIMULATION RESULTS

6.1 Electrical Equations

The equations of an induction machine are defined by the state equation as follow:

$$\frac{dX}{dt} = AX + BU$$

$$\text{where}\,X = \begin{bmatrix} i_{sd} \\ i_{sq} \\ i_{rd} \\ i_{rq} \end{bmatrix};\quad U = \begin{bmatrix} V_{sd} \\ V_{sq} \end{bmatrix} \tag{1.12}$$

$$A = \begin{bmatrix} \frac{-R_s}{\sigma L_s} & \frac{1-\sigma}{\sigma}\omega_r & \frac{M}{\sigma L_s T} & \frac{M}{\sigma L_s}\omega_r \\ \frac{1-\sigma}{\sigma}\omega_r & \frac{-R_s}{\sigma L_s} & -\frac{M}{\sigma L_s}\omega_r & \frac{M}{\sigma L_s T_r} \\ \frac{MR_s}{\sigma L_s L_r} & -\frac{M}{\sigma L_s}\omega_r & -\frac{1}{\sigma T_r} & -\frac{\omega_r}{\sigma} \\ \frac{M}{\sigma L_r}\omega_r & \frac{MR_s}{\sigma L_s L_r} & \frac{\omega_r}{\sigma} & -\frac{1}{\sigma T_r} \end{bmatrix} \tag{1.13}$$

$$B = \begin{bmatrix} \frac{1}{\sigma L_s} & 0 \\ 0 & -\frac{1}{\sigma L_s} \\ -\frac{M}{\sigma L_s L_r} & 0 \\ 0 & -\frac{M}{\sigma L_s L_r} \end{bmatrix} \tag{1.14}$$

6.2. Mechanical Equations

$$Ce = 3/2\, P_p M\, (i_{rd} i_{sq} - i_{sq} i_{rd})$$

$$= 3/2\, P_p \frac{M}{L_r}(\phi_{rd} i_{sq} - i_{sd}\phi_{rq}) \tag{1.15}$$

$$J = \frac{d\omega_m}{dt} + f\omega_m = C_e - C_r \tag{1.16}$$

6.3. Decoupling Control by Compensation

Figure 18 shows the block diagram of a decoupling control which is governed by equations (1.17). Figure 19 is the block diagram of the indirect field oriented induction motor drive system. The power supply is characterised by a PWM voltage source inverter. The motor is controlled according to the principal of the orientation of the rotor flux which is aligned along the d-axis.

Figure 18. Decoupling control

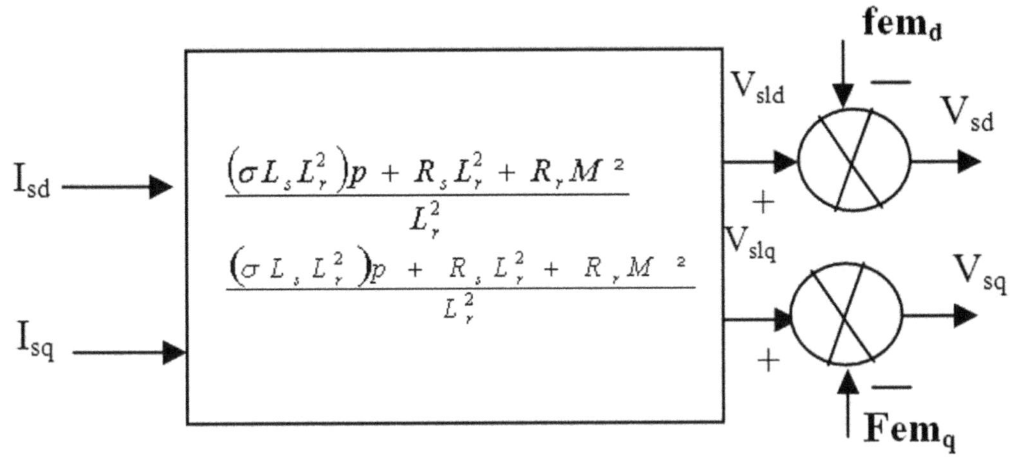

Figure 19. Block diagram of the indirect field oriented induction motor drive system fed by three – level PWM inverter with a PI controller

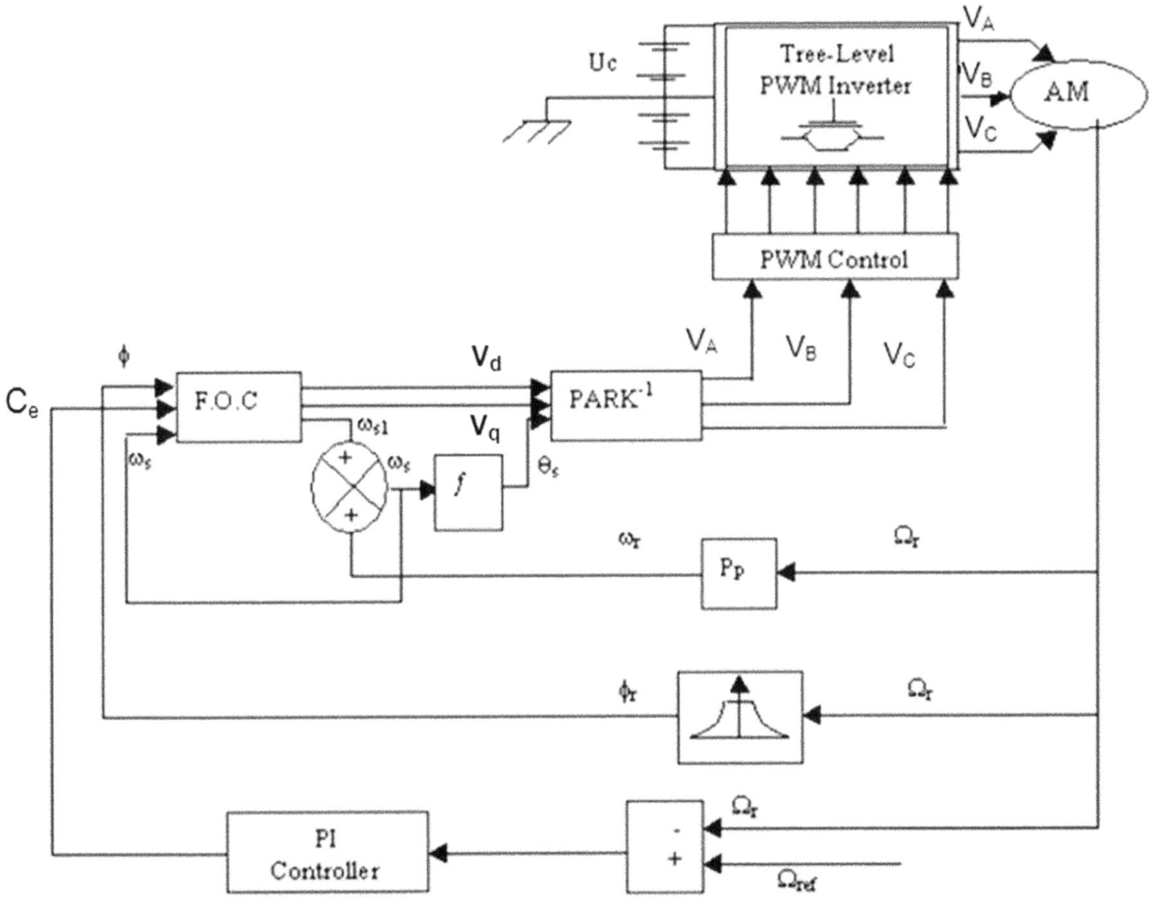

$$\begin{pmatrix} \phi_{dr} & \phi_r \\ \phi_{qr} & 0 \end{pmatrix}$$
$$V_{sd} = V_{sld} - fem_d$$
$$V_{sq} = V_{slq} - fem_q$$

With:

$$V_{sld} = \frac{\left(\sigma L_s L_r^2\right) p + R_s L_r^2 + R_r M^2}{L_r^2} i_{sd}$$

$$V_{slq} = \frac{\left(\sigma L_s L_r^2\right) p + R_s L_r^2 + R_r M^2}{L_r^2} i_{sq} \tag{1.17}$$

$$Fem_d = \omega_s \sigma L_s i_{sq} + \frac{MR_r}{L_r^2} \phi_r$$

$$Fem_q = -\omega_s \phi L_s i_{sd} - \frac{M}{L_r} \omega_s \phi_r + \frac{M^2}{L^2 T_r} i_{sq}$$

$$i_{sd} = \frac{\varphi_r (T_r p + 1)}{M}$$

$$i_{sq} = \frac{2}{3} \frac{C_e L_r}{PpM\varphi_r}$$

$$\omega_{sl} = \frac{M}{\varphi_r T_r} i_{sq} \text{ and } \omega_s = \omega_r + \omega_{sl}$$

To illustrate the performance of the three-level PWM inverter, the systems was investigated through computer simulation under no load. Figure 20 shows the phase voltage for m_f=21 and m_a =0.9 with a sinusoidal triangle PWM and contains harmonics of greater order which can be easily filtered and the results are much better than those which could be obtained with a conventional two level inverter. Figure 21 shows the speed variation at no load.

Figure 20. Phase voltage (VA)

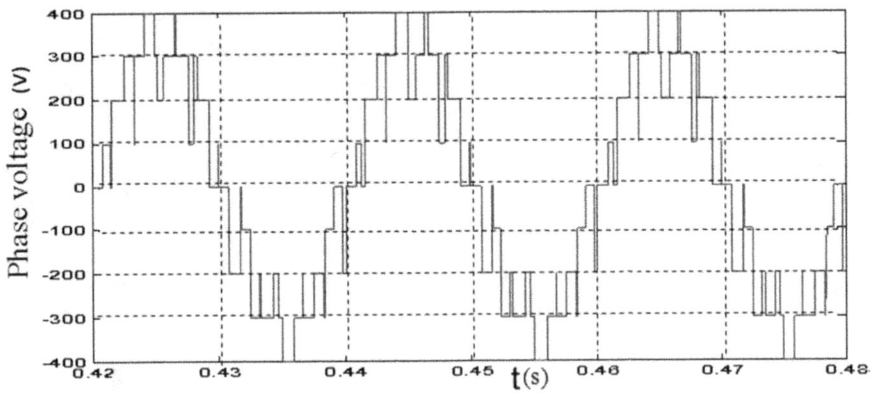

Figure 21. Speed at no load

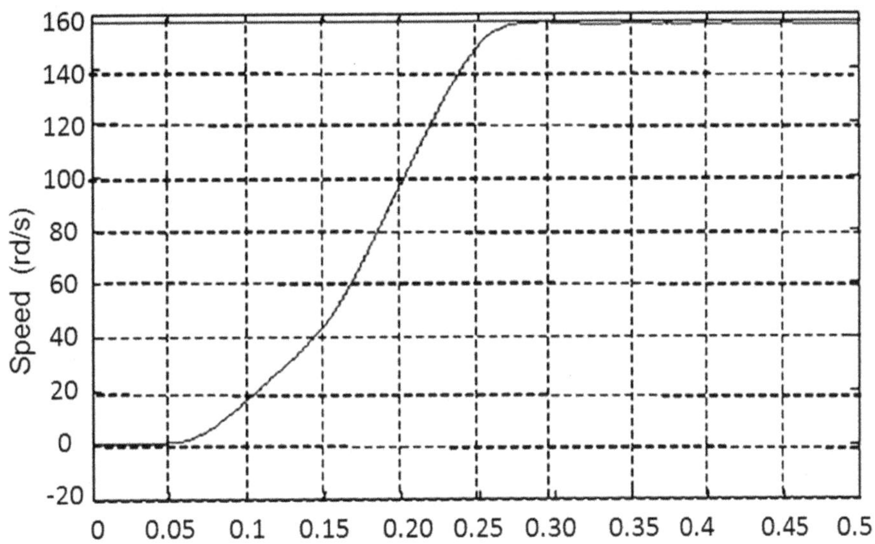

7. EXPERIMENTAL INVESTIGATION OF NPC INVERTER

The schematic diagram of the implemented converter circuit is given by Figure 22. It consists of two parts: the power circuit, and the control circuit. The power circuit is formed by two arms, formed by twelve I.G.B.Ts and eighteen diodes connected in anti-parallel. The generated control signals are obtained using several control cards such as conditioning, monitoring, D/A conversion cards etc.

The shaded part is the control circuit containing the DSP controller TMS320LF2407that generates the PWM signals and also provides soft start functions. The DSP programming software is developed by Texas Instruments (Code Composer). The control algorithm is written in assembly language in order to optimize the whole tasks cooperation and to ensure communication between the DSP, the converter, and the passive load (Denoun et al., 2011; Panoiu et al., 2007; Tatizawa et al., 2009; Ghadbane et al., 2013).

Figure 22. Left: DSP TMS320LF2407 implementation diagram. Right: Experimental bench.

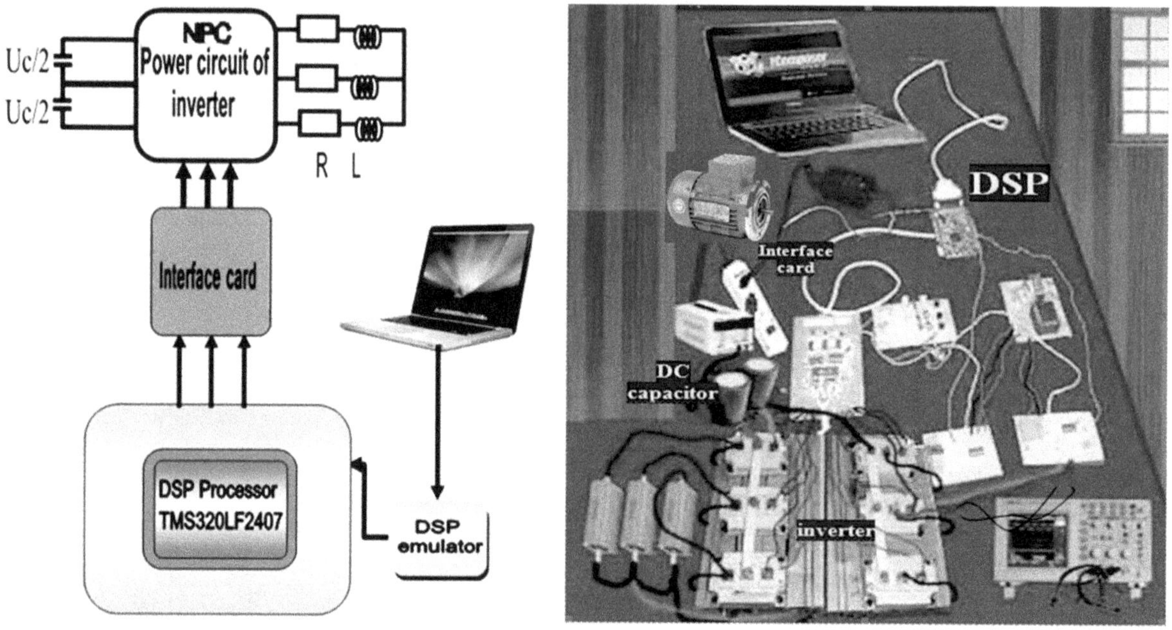

Figure 22 also shows a photograph of the experimental bench including the inverter and all the peripheries.

It should be noted that the parameters used in the experimental investigation are similar to those used earlier in the simulation in order to validate the developed models. Furthermore, for the same triangular-sinusoidal strategy, a frequency ratio equals to 0.8 and an index of modulation equals to 24 is taken.

After the construction of the various electronic circuits and the development of the software, the three-level converter was tested in the laboratory. It was used as an inverter supplying a passive load formed by a resistance and an inductance. The oscillographic results obtained are given in Figures 23, 24, 25, and 26.

Figure 23 clearly shows the V_{AM} voltage variation. The wave form of the voltage is similar to that obtained from the simulation except the small shift observed around the zero level of the voltage. This is may be due to the imbalance of the midpoint.

Figures 24, 25, and 26 show respectively the voltage of the first phase V_A the line voltage and the control pulses obtained experimentally. All the parameters are almost similar to those obtained in the simulation.

8. SIMULATION AND EXPERIMENTAL RESULTS OF AN ASYNCHRONOUS MACHINE FEED PWM

The block diagram of the inverter feeding an asynchronous machine is presented in Figure 27.

In order to verify the obtained simulation results, a reduced power model of the PWM inverter has been realised in our research laboratory. The load is a three-phase 1KW asynchronous machine with the rated parameters given as follows:

Figure 24. Output voltage of phaseVA(15V/Div)

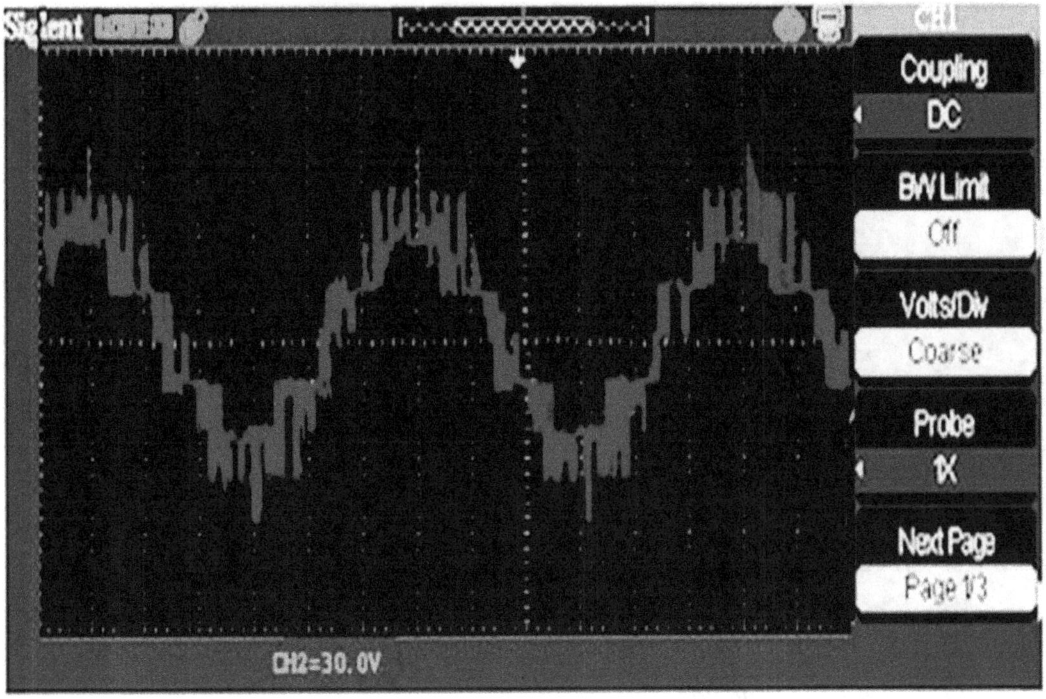

Figure 25. Voltage V_{AB} (30V/Div)

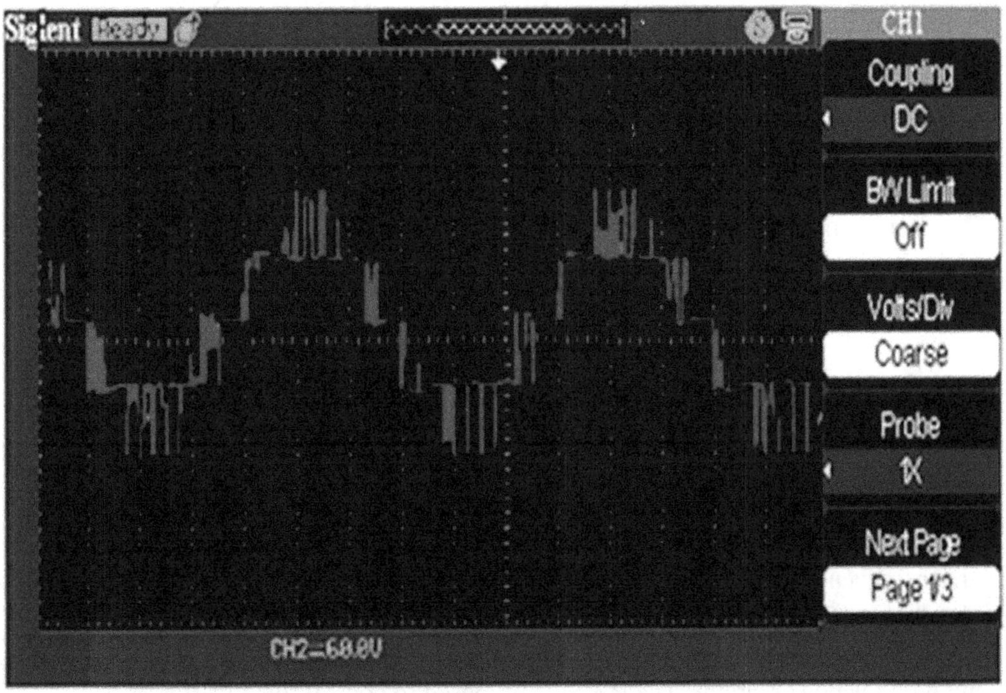

Figure 26. Control pulses (1v/Div)

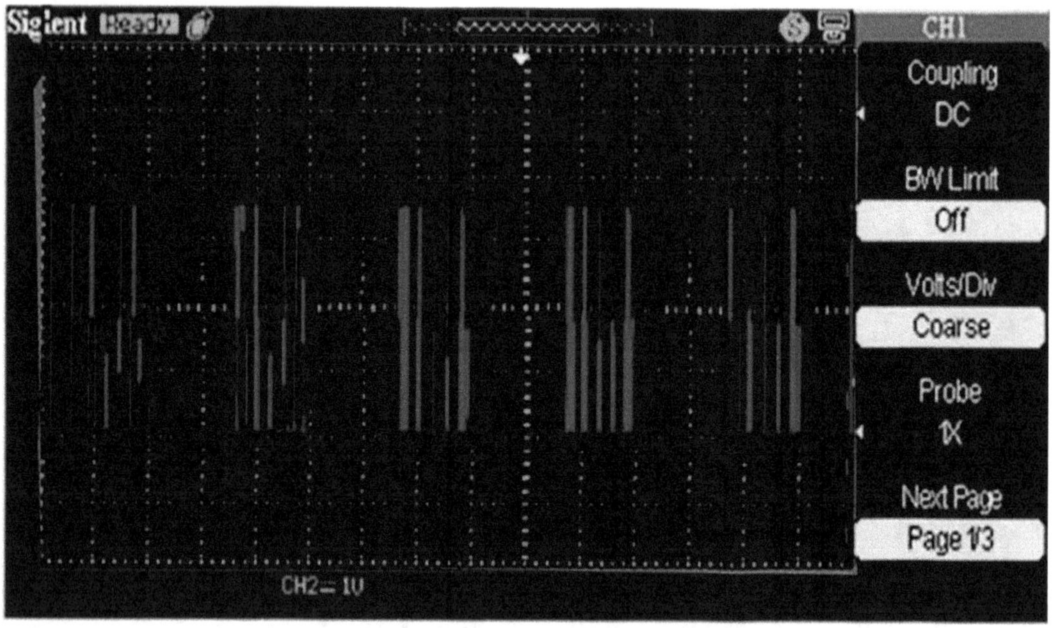

Figure 27. Block diagram of induction motor fed a PWM inverter

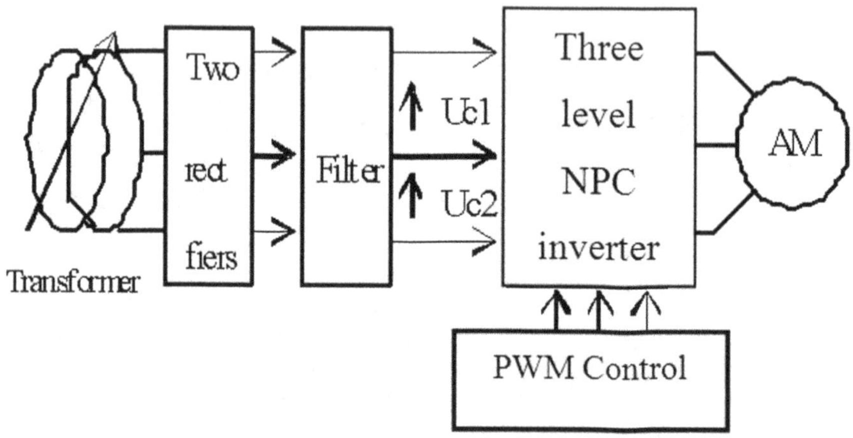

ΔY220/380 V 4, 7A /2,7 A, 2830 Rpm, 50 Hz.

It should be noted that the test was carried out by running the induction motor at no load where the Inverter carrier frequency is taken as 830 Hz, the output phase inverter current $I_a = 2A$, and the output inverter voltage $U_{ab} = 50V$.

The algorithms are implemented more easily by numerical circuits. The proposed control circuits are shown along with the power circuit. The PWM is proposed to control this inverter...

Figure 28. Simulated Load current

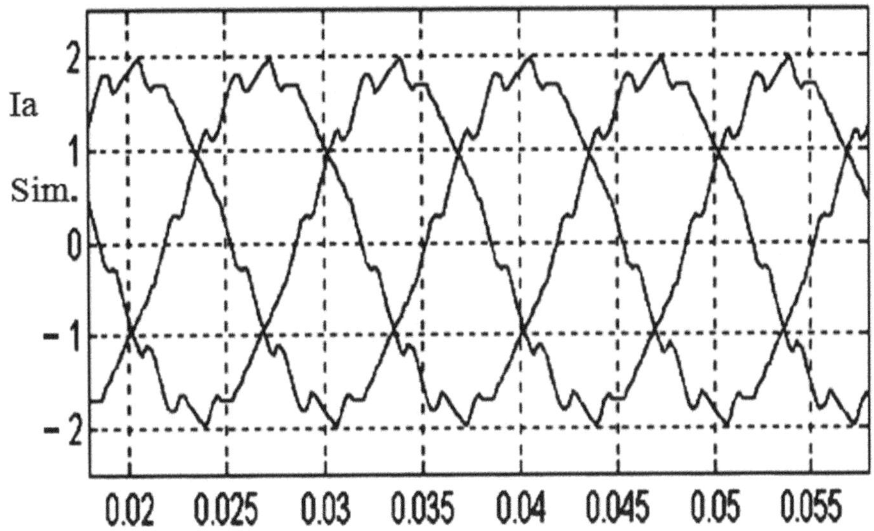

Figure 29. Measured Load current

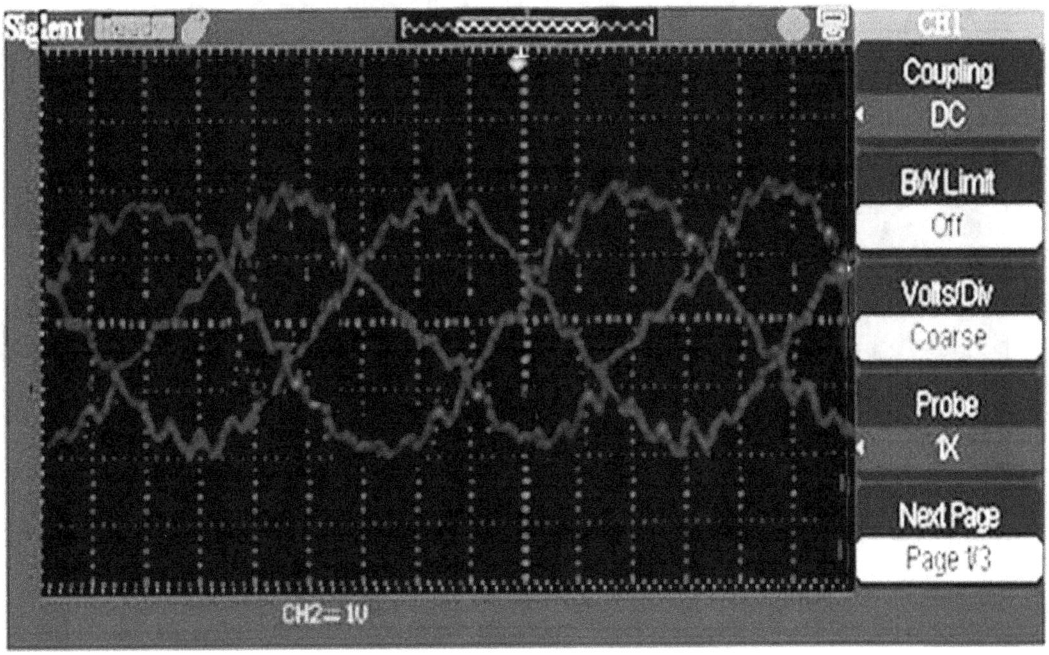

The graphs presented in Figure 28 and Figure 29 are showing respectively the simulation and experimental line current and the phase voltage. It is easily seen that the line current generated by the three-level inverter has low harmonic components even though the switching frequency of the active devices is very low.

Figure 30. Simulated Load voltage

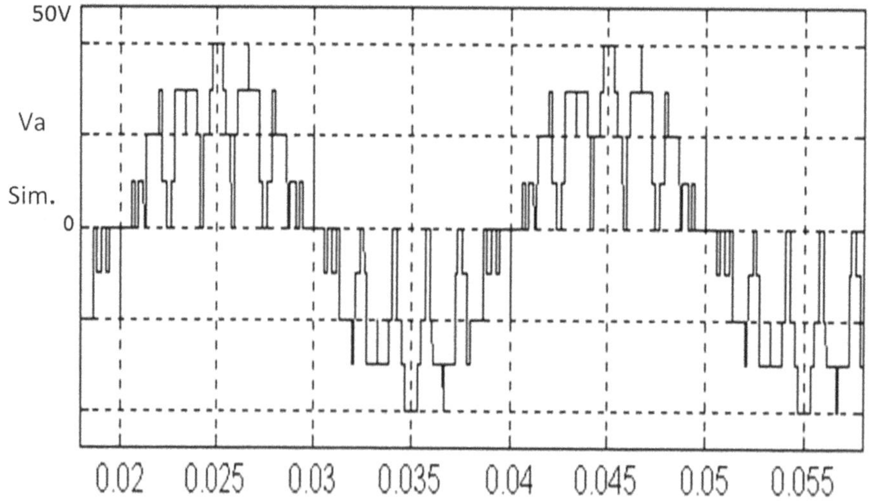

Figure 31. Measured Load voltages

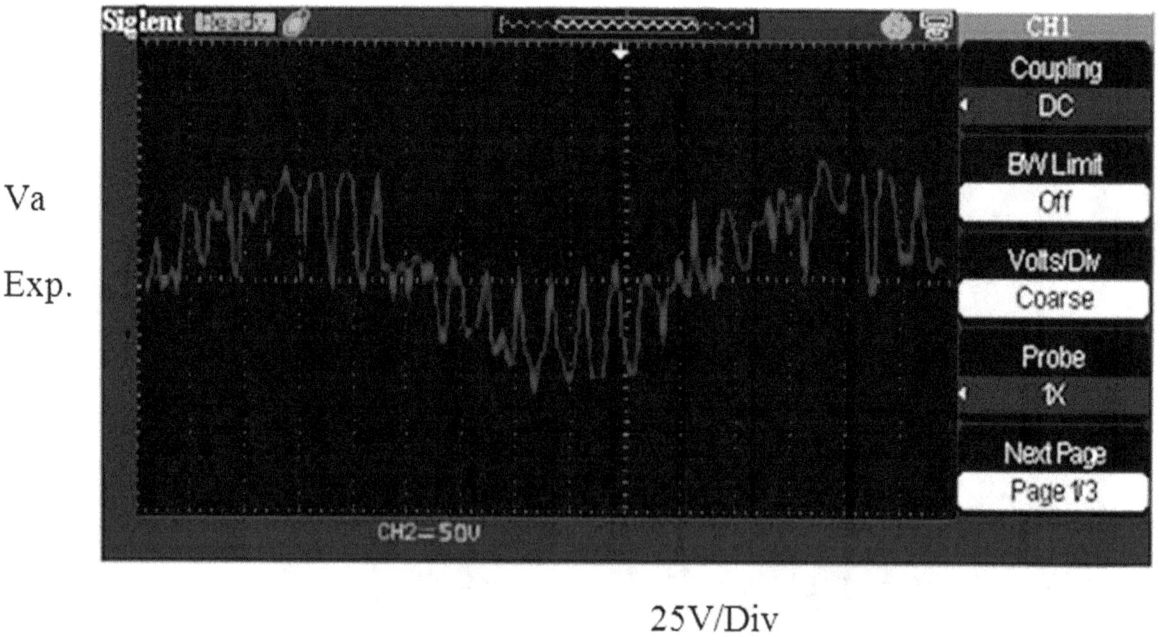

9. CONCLUSION

This contribution starts by giving an overview of the different works available in the literature which aimed to developed multilevel inverters. In fact, the three main topologies were reviewed: these are the cascaded H bridge inverter, the flying capacitor cascaded multilevel inverter and the NPC multilevel inverter. It is shown that the most popular and efficient of these inverters is the NPC inverter which is thoroughly investigated in this work. In fact this contribution presents the analysis and design of a digitally controlled NPC inverter using a DSP. DSP TMS320lf2407 is programmed to produce gate pulses by comparing a triangular carrier wave in high frequency and a sinusoidal reference wave in 50 Hz. Simulation and experimental results show the feasibility of such a system. The low commutation frequency of three-level inverters permits a realisation of an optimal control by relatively simple tools . With a high number of semiconductors devices, current quality is improved and weight reduced by avoiding heavy current filters. The control voltage was controlled by microcontroller in a sophisticated manner. The obtained results are satisfactory because the output harmonic components are fewer than those obtained in conventional two-level inverters at the same switching frequency. *Experimental results using the DSP confirm those obtained in numerical simulation.*

The work is then extended by using this inverter to drive an asynchronous motor. The performances of the drive system are investigated using simulation results which are confronted by testing the whole system.

The work achieved in this contribution opens ways to many other investigations. In fact, work is being undertaken to use this system with five levels inverter to feed an asynchronous machine and the results will be published in the near future.

REFERENCES

Berkouk, E. M. (1996). *Contribution à la conduite des machines asynchrones monophasées et triphasées alimentées par des convertisseurs directs et indirects: Application aux gradateurs et onduleurs multi-niveaux*. (Doctoral thesis). CNAM, Paris, France.

Bouchafaa, F., Berkouk, E. M., & Boucherit, M. S. (2003, September). Analysis and simulation for nine-Level voltage source inverters application to the speed control of the PMSM. *Elmectromotion, 10*(3), 246–251.

Bows, R., Grewal, S., & Holliday, D. (2000). High frequency PWM technique for two and three single-Phases inverters. *IEE Proceedings on Electric Power Applications, 147*(3), 181–191. doi:10.1049/ip-epa:20000330

Colak, I., Kabalci, E., & Bayindir, R. (2011). Review of multilevel voltage source inverter topologies and control schemes. *Energy Conversion and Management, 52*(2), 1114–1128. doi:10.1016/j.enconman.2010.09.006

Denoun, H. (2001). *Contribution à l'étude des convertisseurs AC/DC à MLI à facteur de puissance unitaire, application à la conduite de la machine asynchrone*. (Magistère thesis). EMP, Alger, Algéria.

Denoun, H., Benamrouche, N., Haddad, S., Meziani, S., & AitMamar, S. (2011). A DSP (TMS320lf2407) Based Implementation of PWM for Single-phase AC-DC Bipolar Converter with a Unity Power Factor. *WSEAS International Journal of Circuits. Systems and Signal Processing, 5*(4), 354–361.

Denoun, H., Benamrouche, N., Haddad, S., Meziani, S., & AitMamar, S. (2011). A DSP (TMS320lf2407) Based Implementation of PWM for Single-phase AC-DC Bipolar Converter with a Unity Power Factor. In *Proceeding of the 10th International Conference on signal Processing*. Cambridge, UK: Academic Press.

Gaad, M., Berkouk, E. M., & Aliouane, K. (2000).Contribution à la réalisation de la commande d'un bras d'onduleur à trois niveaux. In *Proceedings of 2[ème]conference internationale sur l'électrotechnique*. Academic Press.

Ghadbane, I., & Benchouia, M. T. (2013). Feed back linearised control based three Phase shunt active power filter. *WSEAS Trans. on Systems, 7*(1), 18-25.

Girard, A., & Pappas, G. J. (2006). Verification using simulation, Hybrid systems Computation and control. *LNCS*, *3927*, 272–286.

Khomfoi, S., & Tolbert, L. M. (2007). *Multilevel Power Converter. Power Electronics Handbook* (2nd ed., pp. 45–482). Elsevier.

Lee, Y-H., & Hyun, D-S. (1996). A Novel PWM Scheme for three-Level Source Inverter With GTO Thyristors. *IEEE Transactions on Industrial Electronics*, *32*(2), 260–268.

Meynard, T., Fadel, M., & Aouda. (1997). Modeling of multi-level converter. IEEE, 914-918.

Meynard, Th. (2006). Flying capacitor multi-cell converters with stored energy. *IEEE Transactions on Industry Applications*, *IA-17*(5), 518–523.

Meynard, T., Foch, H., Forest, F., Turpin, C., Richardeau, L., Delmas, G., … Lefeuvre, E. (2002). Multi-converters: Derived topologies. *IEEE Transactions on Industrial Electronics, 49,* 978-987.

Nabae, A., Takahashi, I., & Akagi, H. (1981). A new neutral-point-clamped PWM inverter. *IEEE Transactions on Industry Applications, 1A-17*(5), 518-523.

Panoiu, C., Baciu, Panoiu, M., & Cuntan, C. (2007). Simulation results on the currents harmonics mitigation on the railway station line feed using a data acquisition system. *WSEAS Trans. on Electronics, 4*(10), 228-236.

Rioual, P., Pouliquen, H., & Louis, J. P. (1996). Regulation of a PWM rectifier in the Unbalanced Network state using Generalized model. *IEEE Transactions on Power Electronics*, *11*(3), 495–502. doi:10.1109/63.491644

Rodriguez, J., Bernet, S., Wu, B., Pontt, J. O., & Kouro, S. (2007). Multilevel voltage source converter topologies for industrial medium-voltage drives. *IEEE Transactions on Industrial Electronics*, *54*(6), 2930–2942. doi:10.1109/TIE.2007.907044

Rodriguez, J., Lai, J. S., & Peng, F. Z. (2002). Multilevel inverters: A survey of topologies, controls and applications. *IEEE Transactions of Industrial Electroics*, *49*(4), 724–738. doi:10.1109/TIE.2002.801052

Tatizawa, H., Netsuo, E.S. (2009).Calibration of high voltage transducers for power quality measurements. *WSEAS Trans. on Systems, 8*(3), 400-409.

Tolbert, L. M., & Peng, F. Z. (1997). Multilevel inverters for large automotive drives. In *Proceedings of All Electric Combat Vehicle Second International Conference*, (pp. 209-214). Academic Press.

Tzung, Y. L., & Ying-Yu, T. (2009). Digital Control of Boost PFC AC/DC Converters With Low THD and Fast Dynamic Response. In *Proceedings of IEEE IPEMC Conference. Rec*, (pp. 1672-1677). IEEE. DOI:10.1109/ *IPEMC.2009.5157660.*

Urmila, B., & Subbarayudu, D. (2010). Multilevel inverters: A comparative study of pulse width modulation techniques. *International Journal of Scientific and Engineering Research*, *1*(3), 1–5.

Wanfeng, Z., Guang, F., Yan-Fei, L., & Bin, W. (2006). New digital control method for power factor correction. *IEEE Transactions on Industrial Electrics, 53*(3), 987-990.

Zhou, G., Wu, B., & Xu, D. (2007). Direct power control of a multilevel inverter based active power filter. *Electric Power Systems Research*, *77*(3), 284–294. doi:10.1016/j.epsr.2006.03.005

KEY TERMS AND DEFINITIONS

Asynchronous Machine: An AC electric motor in which the electric current in the rotor needed to produce torque is induced by electromagnetic induction from the magnetic field of the stator winding.

DSP (Digital Signal Processor): A specialized microprocessor with an architecture optimized for the operational needs of digital signal processing.

IGBT (Insulate- Gate Bipolar Transistor): Combines the simple gate-drive characteristics of MOSFETs with the high-current and low-saturation-voltage capability of bipolar transistors. The IGBT combines an isolated gate FET for the control input, and a bipolar power transistor as a switch, in a single device.

Multi-Level Inverter: A source of high power: used in industrial applications and use either sine or modified sine waves. Instead of using one converter to convert an AC current into a DC current, a multilevel inverter uses a series of semiconductor power converters (usually two to three) thus generating higher voltage.

NPC (Neutral Point Clamping): To convert DC voltage having neutral point into AC voltage

PWM (Pulse-Width Modulation): A modulation technique that conforms the width of the pulse,

THD (Total Harmonic Distortion): A measurement of the harmonic distortion present and is defined as the ratio of the sum of the powers of all harmonic components to the power of the fundamental frequency.

APPENDIX

Table 4.

Machine Parameters	
Rated current	6.4A/3.7A+
Rated voltage	220/380V
Rated Power	1.5Kw
RatedSpeed	1420 rev/mn
R_s	4.85Ω
R_r	3.805Ω
L_s, L_r	0.274Ω
M	0258Ω
J	0.031
F	0.00114
P_p	2
PI Controller Parameters	
P	0.35
I	1.1

Compilation of References

Abbassi, R., & Chebbi, S. (2012a). Energy Management Strategy for a Grid–Connected Wind-Solar Hybrid System with Battery Storage: Policy for Optimizing Conventional Energy Generation. International Review of Electrical Engineering, 7(2), 3979-3990.

Abbassi, R., & Chebbi, S. (2012b). Optimal Energy Management Strategy for Wind Photovoltaic Hybrid System with Battery Storage. In Proceedings of the 16th IEEE Mediterranean Electrotechnical Conference (MELECON' 12), (pp. 810 – 813). IEEE. DOI: .10.1109/MELCON.2012.6196553

Abbassi, R., Hammami, M., & Chebbi, S. (2014). A Control Method of Grid-Connected Wind-Solar-Batteries Hybrid System to Mitigate Voltage Disturbances. In *Proceedings ofInternational Conference on Control, Engineering & Information Technology* (CEIT'2014). Sousse, Tunisia: CEIT.

Abdelmalek, S., Chenikher, S., Sedraoui, M., & Soufi, M. (2012, March). Non-integer Order Robust Control of an Electrical Machine. *Proceedings of EVER, 2012*, 1079–1082.

Abdelmalek, S., & Sedraoui, M. (2012, November). Robustness of fractional PID. *Proceedings of ICEECA, 2012*, 159–161.

Abo-Khalil, A. G., & Lee, D.-C. (2008). MPPT Control of Wind Generation Systems Based on Estimated Wind Speed Using SVR. *IEEE Transactions on Industrial Electronics*, *55*(3), 1489–1490. doi:10.1109/TIE.2007.907672

Aboulnasr, T., & Mayyas, K. (1997). A robust variable step-size LMS-type algorithm analysis and simulations. *IEEE Transactions on Signal Processing*, *45*(3), 631–639. doi:10.1109/78.558478

Abu-Seada, H. F., Mansor, W. M., Bendary, F. M., Emery, A. A., & Moustafa Hassan, M. A. (2013). Application of Particle Swarm Optimization in Design of PID Controller for AVR System. *International Journal of System Dynamics Applications*, *2*(3), 1–17. doi:10.4018/ijsda.2013070101

Adams, R. J., & Bar-on, J. R. (1998). Loop Shaping to Improve the Multivariable Phase Margin. In *Proceedings of the 37th IEEE Conference on Decision & Control. Decision and Control.* Tampa, FL: IEEE. doi:10.1109/CDC.1998.757872

Adrot, O. (2000). *Diagnostic à base de modèles incertains utilisant l'analyse par intervalles: l'approche bornante.* (PhD thesis). Institut National Polytechnique de Lorraine, France.

Adrot, O., Janati-Idrissi, H., & Maquin, D. (2002). *Fault detection based on interval analysis.* Paper presented at the 15th IFAC World Congress on Automatic Control. Barcelona, Spain. Doi:10.3182/20020721-6-ES-1901.00762

Adrot, O., Maquin, D., & Ragot, J. (2000). *Diagnosis of an uncertain static system.* Paper presented at the 39th IEEE Conference on Decision and Control. Sydney, Australia. Doi:10.1109/CDC.2000.912366

Agnew, G. B. (1986). *Random sources from cryptographic systems. In Proceedings of Advances in Cryptology-CRYPTO'85* (pp. 77–81). New York: Springer-Verlag.

Ahmad, N. J., & Khorrami, F. (1999). Adaptive Control of Systems with Backlash Hysteresis at the Input. In *Proceedings of the American Control Conference. American Control Conference*. San Diego, CA: IEEE. doi:10.1109/ACC.1999.782315

Ajith, A., & Amit, K. (2009). On Stability of the Chemotaxis Dynamics in Bacterial-Foraging Optimization Algorithm. *IEEE Transactions on Systems, Man, and Cybernetics. Part A, Systems and Humans*, *39*(3), 670–679. doi:10.1109/TSMCA.2008.2011474

Akhenak, A., Chadli, M., Maquin, D., & Ragot, J. (2003). *State estimation via multiple observer with unknown inputs: Application to the three tank system.* Paper presented at the 5th IFAC Symposium on Fault Detection, Supervision and Safety for Technical Processes, Safeprocess. Washington, DC.

Akhenak, A., Chadli, M., Maquin, D., & Ragot, J. (2004). *State estimation of uncertain multiple model with unknown inputs.* Paper presented at the 43rd IEEE Conference on Decision and Control. Paradise Island, Bahamas. Doi:10.1109/CDC.2004.1429265

Akhenak, A., Chadli, M., Ragot, J., & Maquin, D. (2007). *Design of sliding mode unknown input observer for uncertain Takagi-Sugeno model.* Paper presented at the Mediterranean Conference on Control & Automation. Athens, Greece. Doi:10.1109/MED.2007.4433753

Akhenak, A., Chadli, M., Ragot, J., & Maquin, D. (2009). *Design of Fuzzy Observers for Takagi-Sugeno Fuzzy Models for Fault Detection and Isolation.* Paper presented at the 7th IFAC Symposium on Fault Detection, Supervision and Safety of Technical Processes SAFEPROCESS. Doi:10.3182/20090630-4-ES-2003.00183

Albea, C., Gordillo, F., & de Wit, C. C. (2011). Adaptive control design for a boost inverter. *Control Engineering Practice*, *19*(1), 32–44. doi:10.1016/j.conengprac.2010.08.007

Albu, F., & Nishikawa, K. (September 2013). The Kernel Proportionate NLMS Algorithm. In *Proceedings of EUSIPCO 2013* (no. TU-P6.1). EUSIPCO.

Albu, F. (2011, August). A Proportionate Affine Projection Algorithm using Fast Recursive Filtering and Dichotomous Coordinate Descent Iterations. *Proceedings of SPAMEC*, *2011*, 93–96.

Albu, F. (2012, April). Simplified proportionate affine projection algorithms. In *Proceedings of IWSSIP*, (pp. 382-385). IWSSIP.

Albu, F., Bouchard, M., & Zakharov, Y. (2007). Pseudo Affine Projection Algorithms for Multichannel Active Noise Control. *Audio. IEEE Transaction on Speech and Language Processing*, *15*(3), 1044–1052. doi:10.1109/TASL.2006.881677

Albu, F., Nishikawa, K., Rotaru, M., & Coltuc, D. (September2013). An efficient implementation of the kernel affine projection algorithm. In *Proceedings of IEEE ISPA* (pp. 342-346). IEEE. doi:10.1109/ISPA.2013.6703765

Albu, F., Paleologu, C., & Benesty, J. (2011). A Variable Step Size Evolutionary Affine Projection algorithm. In *Proceedings of ICASSP 2011* (pp. 429-432). ICASSP.

Albu, F., Paleologu, C., Benesty, J., & Ciochina, S. (2010, August). A low complexity proportionate affine projection algorithm for echo cancellation. In *Proceedings of EUSIPCO*, (pp. 6–10). EUSIPCO.

Al-Duwaish, H., & Nazmul Karim, M. (1997). A new method for the identification of Hammerstein model. *Automatica*, *33*(10), 1871–1875.

Alessandria, A., Bagliettob, M., & Battistellic, G. (2008). Moving-horizon state estimation for nonlinear discrete-time systems: New stability results and approximation schemes. *Automatica*, *44*(7), 1753–1765.

Alhaj-Dibo, M. (2005). *Validation de données et diagnostic des systèmes incertains à l'aide de l'analyse par intervalles.* (PhD thesis). Institut National Polytechnique de Lorraine, France.

Alhaj-Dibo, M., Maquin, D., & Ragot, J. (2004). *Data validation of uncertain dynamic systems.* Paper presented at the 15th International Workshop on Principles of Diagnosis. Carcassonne, France.

Al-Hiddabi, S. (2009). Quadrotor control using feedback linearization with dynamic extension. In *Proceedings ofthe Conference on Mechatronics and its Applications.* ISMA. Doi:10.1109/ISMA.2009.5164788

Alkhazaleh, S., & Salleh, A. R. (2012). Fuzzy soft multiset theory. *Abstract and Applied Analysis*, *2012*, 1–20. doi:10.1155/2012/350603

Allgöwer, F. (1997). Approximate input-output linearization of non-minimum phase nonlinear systems. In *Proceedings of the 1997 European Control Conference (ECC)*, (pp. 604–610). ECC.

Altas, I. H. (1998). The effects of fuzziness in fuzzy logic controllers. In *Proceedings of the 2nd International Symposium on Intelligent Manufacturing Systems*, (pp. 211-220). Sakarya, Turkey: Academic Press.

Alvarez, G., & Li, S. (2006). Some basic cryptographic requirements for chaos based cryptosystems. *International Journal of Bifurcation and Chaos in Applied Sciences and Engineering*, *16*(08), 2129–2151. doi:10.1142/S0218127406015970

Alvarez-Ramirez, J., & Espinosa-Pérez, G. (2002). Stability of current-mode control for dc-dc power converters. *Systems & Control Letters*, *45*(2), 113–119. doi:10.1016/S0167-6911(01)00169-4

Amann, N., Owens, D. H., & Rogers, E. (1998). Predictive optimal iterative learning control. *International Journal of Control*, *69*(2), 203–226. doi:10.1080/002071798222794

Andreas, P. S., & Thiringer, L. T. (2004). A DFIG wind-turbine ride through system influence on the energy production. In *Proceedings of NORDIC Wind Power Conference 2004* (pp.1-7). NORDIC.

Andrieu, V., & Praly, L. (2008). Global asymptotic stabilization for nonminimum phase non linear systems admitting a strict normal form. *IEEE Transactions on Automatic Control*, *53*(5), 1120–1132. doi:10.1109/TAC.2008.923657

Andrievskii, B. R., & Fradkov, A. L. (2003). Control of Chaos: Methods and Applications I: Methods. *Automation and Remote Control*, *64*(5), 673–713. doi:10.1023/A:1023684619933

Andrievskii, B. R., & Fradkov, A. L. (2004). Control of Chaos: Methods and Applications. II. Applications. *Automation and Remote Control*, *65*(4), 505–533. doi:10.1023/B:AURC.0000023528.59389.09

Angulo, F., Fossas, E., Ocampo, C., & Olivar, G. (2005). Stabilization of chaos with fpic, application to zad-strategy buck converters. In *Proceedings of16th IFAC World Congress*. Prague: IFAC.

Anishchenko, V. S., Neiman, A. B., & Chua, L. O. (1993). *A paradigm for chaos*. Singapore: World Scientific.

Antoni, K., & Marta, B.-P. (2013). *Application Note - Standard EN 50160/Voltage characteristics of electricity supplied by public electricity networks*. ECI Publication No Cu0147. Available from www.leonardo-energy.org/node/145851

Arablouei, R., Dogançay, K., & Perreau, S. (October 2010). Proportionate affine projection algorithm with selective projections for sparse system identification. In *Proceedings of APSIPA ASC 2010* (pp. 362). APSIPA.

Araki, M. (1975). Application of M-matrices to the stability problems of composite dynamical systems. *Journal of Mathematical Analysis and Applications*, *52*(2), 309–321. doi:10.1016/0022-247X(75)90099-2

Arbi, J., Ghorbal, M. B., Slama-Belkhodja, I., & Charaabi, L. (2009). Direct virtual torque control for doubly fed induction generator grid connection. *IEEE Transactions on Industrial Electronics*, *56*(10), 4163–4173. doi:10.1109/TIE.2009.2021590

Asano, F., & Luo, Z. W. (2009). On efficiency and optimality of asymmetric dynamic bipedal gait. In *Proceedings ofIEEE International Conference on Robotics and Automation*, (pp. 1972-1977). Kobe, Japan: IEEE. doi:10.1109/ROBOT.2009.5152322

Avizienis, A., Laprie, J. C., Randell, B., & Landwehr, C. (2004). Basic Concepts and Taxonomy of Dependable and Secure Computing. *Dependable and Secure Computing*, *1*(1), 11–33. doi:10.1109/TDSC.2004.2

Awrejcewicz, J., & Lamarque, C. H. (2003). *Bifurcation and Chaos in Nonsmooth Mechanical Systems*. Singapore: World Scientific.

Azar, A. T. (2010). *Fuzzy Systems*. Vienna, Austria: In-Tech.

Azar, A. T. (2012). Overview of type-2 fuzzy logic systems. *International Journal of Fuzzy System Applications*, *2*(4), 1–28. doi:10.4018/ijfsa.2012100101

Azarfar, A., Shandiz, T. H., & Shafiee, M. (2013). Adaptive feedback control for linear singular systems. *Turkish Journal of Electrical Engineering & Computer Sciences*, *22*(2014), 132-142.

Babu, B. C., & Mohanty, K. B. (2009). Analysis of wind turbine driven double-output induction generator under abnormal condition of the grid. *Proceedings of IPEMC, 2009*, 627–631.

Bae, B., Patel, N., & Schulz, S. (2003), New Field Weakening Technique for high saliency Interior Permanent Magnet Motor, *Industry Applications Conference, 2003. 38th IAS Annual Meeting*. vol.2. 898-905. doi:10.1109/IAS.2003.1257641

Baek, J., & Park, M. (2012). Fuzzy bilinear state feedback control design based on ts fuzzy bilinear model for dc-dc converters. *International Journal of Electrical Power & Energy Systems*, *42*(1), 710–720. doi:10.1016/j.ijepes.2012.04.008

Balachandran, B., Kalmar-Nagy, T., & Gilsinn, D. E. (2009). *Delay Differential Equations: Recent Advanced and New Directions*. New York: Springer.

Balas, M. J. (1979). Direct velocity feedback control of large space structures. *Journal of Guidance and Control*, *2*(3), 252–253. doi:10.2514/3.55869

Balluq, E., Acarnley, P. P., & Atkinson, D. J. (2004). Adaptive tuning of a PI speed controller for a brushless DC motor: optimum speed control using a neural network.*IEEE International Symposium on Industrial Electronics,*vol.1, pp. 459 - 464.

Banaszuk, A., & Hauser, J. (1993). Least squares approximate feedback linearization: A variational approach. In *Proceedings of the 1993 IEEE Control and Decision Conference (CDC)*, (pp. 2760–2765). IEEE. Doi:10.1109/CDC.1993.325698

Banaszuk, A., & Hauser, J. (1996). Approximate feedback linearization: A homotopy operator approach. *SIAM Journal on Control and Optimization*, *34*(5), 1533–1554. doi:10.1137/S0363012994261306

Banerjee, S., Mitra, M., & Rondoni, L. (2011). *Applications of chaos and nonlinear dynamics in engineering – Part 1*. Berlin: Springer-Verlag. doi:10.1007/978-3-642-21922-1

Banerjee, S., & Verghese, G. C. (2001). *Nonlinear Phenomena in Power Electronics*. New York: IEEE Press. doi:10.1109/9780470545393

Banerjee, T., Biswas, D., & Sarkar, B. C. (2012). Design and analysis of a first order time-delayed chaotic system. *Nonlinear Dynamics*, *70*(1), 721–734. doi:10.1007/s11071-012-0490-3

Bansal, P., Maheshwari, S., Kumar, M., & Kumar, U. (2011). Behavior based mobile robot navigation in a dynamic environment: Fuzzy logic approach. *Journal of Active & Passive Electronic Devices*, *6*, 293–303.

Bara, L. G. (2011). Robust analysis and control of parameter-dependent uncertain descriptor systems. *Elsevier Systems & Control Letters*, *60*(6), 356–364. doi:10.1016/j.sysconle.2011.03.001

Barbot, J. P., Monaco, S., & Normand-Cyrot, D. (1997). Quadratic forms and approximate feedback linearization in discrete time. *International Journal of Control*, *67*(4), 567–586. doi:10.1080/002071797224081

Barbot, J. P., Monaco, S., & Normand-Cyrot, D. (1999). Discrete time approximated linearization of SISO systems under output feedback. *IEEE Transactions on Automatic Control*, *44*(9), 1729–1733. doi:10.1109/9.788541

Bardis, N. G., Markovskyi, A. P., Doukas, N., & Karadimas, N. V. (2009). True random number generation based on environmental noise measurements for military applications. In *Proc. of 8th WSEAS Int. Conf. on Signal Processing, Robotics and Automation*, (pp. 68-73). Cambridge, UK: WSEAS.

Barreiro, A., & Baños, A. (2006). Input output stability of systems with backlash. *Automatica*, *42*(6), 1017–1024. doi:10.1016/j.automatica.2006.02.017

Barreto, E., & Grebogi, C. (1995). Multiparameter control of chaos. *Physical Review E: Statistical Physics, Plasmas, Fluids, and Related Interdisciplinary Topics*, *52*(4), 3553–3557. doi:10.1103/PhysRevE.52.3553 PMID:9963831

Basseville, M., & Nikiforov, I. V. (1993). *Detection of Abrupt Changes: Theory and Application*. Englewood Cliffs, NJ: Prentice-Hall.

Baumann, W. T., & Rugh, W. J. (1986). Feedback control of nonlinear systems by extended linearization. *IEEE Transactions on Automatic Control*, *31*(1), 40–46. doi:10.1109/TAC.1986.1104100

Bedoui, H., Bouguila, N., Othman, K. B., & Hassani, M. (2013). *Modeling of the injection and potential decay of a dielectric placed in an environment with high voltage*. Paper presented at the International Conference on Control, Decision and Information Technologies. Hammamet, Tunisia. Doi:10.1109/CoDIT.2013.6689583

Bedoui, H., Khedher, A., & Ben Othman, K. (2014). State estimation of uncertain nonlinear systems using the interval approach: Application to the diagnostic. *Wulfenia Journal*, *21*(4), 369–379.

Begley, C. J., & Virgin, L. N. (2001). On the OGY Control of an Impact-Friction Oscillator. *Journal of Vibration and Control*, *7*(6), 923–931. doi:10.1177/107754630100700609

Belforte, G., & Gay, P. (2001). Discrete time Hammerstein model identification with unknown but bounded errors. *IEE Proceedings. Control Theory and Applications*, *148*(6), 523–529.

Ben Attia, S., Salhi, S., & Ksouri, M. (2009b). Mode-Independent state feedback design for discrete switched linear. In *Proceedings ofEuropean Conference on Modelling and Simultion* (pp. 807-812). Academic Press. doi:10.7148/2009-0807-0812

Ben Attia, S., Salhi, S., Bernussou, J., & Ksouri, M. (2009a). Improved LMI formulation for robust dynamic output feedback controller design of discrete-time switched systems via switched Lyapunov function. In *Proceedings of the 3rd International Conference on Signals, Circuits and Systems*. Academic Press.

Ben Attia, S., Salhi, S., Bernussou, J., & Ksouri, M. (2010). Analyse de la stabilité des systèmes à commutation à temps continu en utilisant les systèmes de comparaison. In Proceedings of Sixième IEEE Conférence Internationale Francophone d'Automatique (pp. 15-20). IEEE.

Ben Attia, S., Salhi, S., & Ksouri, M. (2009c). LMI Formulation for Static Output Feedback Design of Discrete-time Switched Systems. *Journal of Control Science and Engineering*, *2009*, 7. doi:10.1155/2009/362920

Ben Attia, S., Salhi, S., & Ksouri, M. (2012). Static Switched Output Feedback Stabilization For Linear Discrete-time Switched Systems. *International Journal ofInnovative Computing, Information, & Control*, *8*(5A), 3203–3213.

Ben Hadj Mohamed, A., Val, T., Andrieux, L. & Kachouri1, A. (2013). Assisting people with disabilities through kinect sensors into a smart house. In *Proceedings of ICCMA*, (pp. 1-5). ICCMA. doi:10.1109/ICCMA.2013.6506149

Ben Hadj Mohamed, A., Val, T., Andrieux, L., & Kachouri, A. (2012). Using a Kinect WSN for home monitoring: principle, network and application evaluation. In *Proceedings of ICWCUCA*, (pp. 1-5). ICWCUCA. doi:10.1109/ICWCUCA.2012.6402487

Ben Hamed M., Sbita L. (2008). Fractional order speed observer for sensorless induction motor drives, *International Review of Electrical Engineering IREE,* vol. 3,(1).

Ben Hariz, M., Chagra, W., & Bouani, F. (2013). Controllers design for MIMO systems with time response specifications. In *Proceedings ofInternational Conference on Control, Decision and Information Technologies* (CoDIT). (pp. 573-578). GoDIT. doi:10.1109/CoDIT.2013.6689607

Ben Hariz, M., Bouani, F., & Ksouri, M. (2012). Robust controller for uncertain parameters systems. *ISA Transactions*, *51*(5), 632–640. doi:10.1016/j.isatra.2012.04.007 PMID:22749294

Bendixson, I. (1901). Sur les courbes definies par des equations differentielles. *Acta Mathematica*, *24*(1), 1–88. doi:10.1007/BF02403068

Benesty, J., & Gay, S. L. (2002, May). An improved PNLMS algorithm. *Proceedings ofICASSP*, *2*, 1881–1884.

Benothman, K., Maquin, D., Ragot, J., & Benrejeb, M. (2007). Diagnosis of uncertain linear systems: an interval approach. *International Journal of Sciences and Techniques of Automatic Control & Computer Engineering*, *1*(2), 136-154.

Benrejeb, M., & Borne, P. (1978). On an algebraic stability criterion for non-linear processes. Interpretation in the frequency domain. In *Proceedings of the Measurement and Control International Symposium MECO'78* (pp. 678–682). Athens, Greece: MECO.

Benrejeb, M., Sakly, A., Ben Othman, K., & Borne, P. (2008). Choice of conjunctive operator of TSK fuzzy systems and stability domain study. *Mathematics and Computers in Simulation*, *76*(5-6), 410–421. doi:10.1016/j.matcom.2007.04.003

Benrejeb, M., Soudani, D., Sakly, A., & Borne, P. (2006). New Discrete Tanaka Sugeno Kang Fuzzy Systems Characterization and Stability Domain. *International Journal of Computers, Communications & Control*, *1*(1), 9–19.

Bensaada, M., & Stambouli, A. B. (2013). A practical design sliding mode controller for dcdc converter based on control parameters optimization using assigned poles associate to genetic algorithm. *International Journal of Electrical Power & Energy Systems*, *53*, 761–773. doi:10.1016/j.ijepes.2013.05.043

Bentsman, J., Orlov, Y., & Aguilar, L. T. (2013). Nonsmooth H-infinity Output Regulation with Application to a Coal-Fired Boiler/Turbine Unit with Actuator Deadzone. In *Proceedings of2013 American Control Conference (ACC)*. Washington, DC: IEEE. doi:10.1109/ACC.2013.6580434

Benzaouia, A., Hmamed, A., Tadeo, F., & Hajjaji, A. E. L. (2011). Stabilization of discrete 2D time switching systems by state feedback control. *International Journal of Systems Science*, *42*(3), 479–487. doi:10.1080/00207720903576522

Berdjag, D., Christophe, C., Vincent, C., & Jiang, B. (2006). Nonlinear model decomposition for robust fault detection and isolation using algebraic tools. *International Journal of Innovative Computing, Information, & Control*, *2*(6), 1337–1354.

Bergé, P., Pomeau, Y., & Vidal, C. (1984). *Order within Chaos*. New York: Wiley.

Berkouk, E. M. (1996). *Contribution à la conduite des machines asynchrones monophasées et triphasées alimentées par des convertisseurs directs et indirects: Application aux gradateurs et onduleurs multi-niveaux*. (Doctoral thesis). CNAM, Paris, France.

Bernardini, R., & Cortelazzo, G. (2001). Tools for designing chaotic systems for secure random number generation. *IEEE Transactions on Circuits and Systems*, *48*(5), 552–564. doi:10.1109/81.922458

Bernstein, G. M., & Lieberman, M. A. (1990). Secure random number generation using chaotic circuits. *IEEE Transactions on Circuits and Systems*, *37*(9), 1157–1164. doi:10.1109/31.57604

Bernstein, S. D., & Haddad, M. W. (1989). LQG control with H_∞. *IEEE Transactions on Automatic Control*, *34*(3), 293–305. doi:10.1109/9.16419

Bhattacharyya, S. (1978). Observer design for linear systems with unknown inputs. *IEEE Transactions on Automatic Control*, *23*(3), 483–484. doi:10.1109/TAC.1978.1101758

Bhikkaji, B., & Moheimani, S. O. (2008). Integral resonant control of a piezoelectric tube actuator for fast nanoscale positioning. *IEEE/ASME Transactions on Mechatronics*, *13*(5), 530–537. doi:10.1109/TMECH.2008.2001186

Bhuiyan, M., & Picking, R. (2009). Gesture Control User Interface, What Have We Done and What's Next? In *Proceedings of the Fifth Collaborative Research Symposium on Security, E-Learning, Internet and Networking (SEIN 2009)*, (pp. 59-60). SEIN.

Bhuiyan, M., & Picking, R. (2011). A Gesture Controlled User Interface for Inclusive Design and Evaluative Study of Its Usability. *Journal of Software Engineering and Applications*, *4*(9), 513–521. doi:10.4236/jsea.2011.49059

Biagiola, S. I., & Figueroa, J. L. (2009). Wiener and Hammerstein uncertain models identification. *Mathematics and Computers in Simulation*, *79*(11), 3296–3313.

Bilal, A., Umit, O., Aydin, E., & Mehrded, E. (2004, November). A Comparative Study on Non-Linear State Estimators Applied to Sensorless AC Drives: MRAS and Kalman Filter.*30th Annual Conference of IEEE Industrial Electronics Society*, pp. 2148 - 2153vol. 3, Busan, Korea

Billings, S. A. (1980). Identification of nonlinear systems. *IEE Proceedings Pt. D*, *127*(6), 272-285.

Billings, S. A. (2013). *Nonlinear System Identification NARMAX Methods in the Time, Frequency, and Spatio –Temporal Domains*. Wiley.

Bin, J., Jian, L. W., & Yeng, C. S. (2002). An adaptive technique for robust diagnosis of faults with independent effects on system outputs. *International Journal of Control*, *75*(11), 792–802. doi:10.1080/00207170210149934

Bin, J., Marcel, S., & Vincent, C. (2006). Fault accommodation for nonlinear dynamic Systems. *IEEE Transactions on Automatic Control*, *51*(9), 1578–1583. doi:10.1109/TAC.2006.878732

Binnig, G., & Rohrer, H. (1987). Scanning tunneling microscopy from birth to adolescence. *Reviews of Modern Physics*, *59*(3), 615–625. doi:10.1103/RevModPhys.59.615

Binnig, G., & Smith, D. P. E. (1986). Single-tube three-dimensional scanner for scanning tunneling microscopy. *The Review of Scientific Instruments*, *57*(8), 1688–1689. doi:10.1063/1.1139196

Bjork, K., Lindberg, P., & Westerlund, T. (2003). Some convexifications in global optimization of problems containing signomial terms. *Computers & Chemical Engineering*, *27*(5), 669–679. doi:10.1016/S0098-1354(02)00254-5

Blaabjerg, F., Teodorescu, R., Liserre, M., & Timbus, A. V. (2006). Overview of control and grid synchronization for distributed power generation systems. *IEEE Transactions on Industrial Electronics*, *53*(5), 1398–1409. doi:10.1109/TIE.2006.881997

Blanke, M., Kinnaert, M., Lunze, J., & Staroswiecki, M. (2003). *Diagnosis and Fault Tolerant Control*. Berlin: Springer. doi:10.1007/978-3-662-05344-7

Bleiweiss, A., Eshar, D., Kutliro, G., Lerner, A., Oshrat, Y., & Yanai, Y. (2010). Enhanced interactive gaming by blending full-body tracking and gesture animation. In *Proceedings of the 3rd ACM SIGGRAPH Conference and Exhibition on Computer Graphics and Interactive Techniques in Asia*, (pp. 34:1—34:2). ACM. doi:10.1145/1899950.1899984

Blum, L., Blum, M., & Shub, M. (1986). A simple unpredictable pseudo-random number generator. *SIAM Journal on Computing*, *15*(2), 364–383. doi:10.1137/0215025

Boccaletti, S., Grebogi, C., Lai, Y.-C., Mancini, H., & Maza, D. (2000). The control of chaos: Theory and applications. *Physics Reports*, *329*(3), 103–197. doi:10.1016/S0370-1573(99)00096-4

Boccaletti, S., Kurths, J., Osipov, G., Valladares, D. L., & Zhou, C. S. (2002). The synchronization of chaotic systems. *Physical Review A*, *64*, 821–824. PMID:12443117

Bocci, V., Chiodi, G., Iacoangeli, F., Nobrege, R., Pinci, D., & Rinaldi, W. (2005). Time-multiplexing of signal using highly integrated digital delay: an FPGA implementation. In *Proc. IEEE Nuclear Science Symposium Conference Record*, (pp. 398-402). IEEE. doi:10.1109/NSSMIC.2005.1596279

Bolognani, S., Zigliotto, M., & Unterkofler, K. (1997). On-line parameter commissioning in sensorless PMSM drives, *Proceedings of the IEEE International Symposium on Industrial Electronics,* Vol.2, pp 480-484, Guimaraes. doi:10.1109/ISIE.1997.648995

Bonivento, C., Zanasi, R., & Sandri, M. (1998). Discrete variable structure integral controllers. *Automatica*, *34*(3), 355–36. doi:10.1016/S0005-1098(97)00197-0

Borne, P., & Benrejeb, M. (2008). On the Representation and the Stability Study of Large Scale Systems. *International Journal of Computers, Communications & Control*, (2), 55–66.

Bortoff, S. A., & Spong, M. W. (1992). Pseudolinearization of the acrobot using spline functions. In *Proceedings ofIEEE Control and Decision Conference (CDC)*, (pp. 593–598). IEEE. Doi:10.1109/CDC.1992.371658

Bortoff, S. A. (1997). Approximate state-feedback linearization using spline functions. *Automatica*, *33*(8), 1449–1458. doi:10.1016/S0005-1098(97)00070-8

Bouchafaa, F., Berkouk, E. M., & Boucherit, M. S. (2003, September). Analysis and simulation for nine-Level voltage source inverters application to the speed control of the PMSM. *Elmectromotion*, *10*(3), 246–251.

Boukas, K. E. (2005). Stabilization of stochastic nonlinear hybrid systems. *International Journal of Innovative Computing, Information, & Control*, *1*(1), 131–141.

Boukas, K. E., Shi, P., & Nguang, S. (2003). Robust Hinfini control for linear Markovian jump systems with unknown non linearrities. *Journal of Mathematical Analysis and Applications*, *282*(1), 241–255. doi:10.1016/S0022-247X(03)00144-6

Boukhezzar, B., & Siguerdidjane, H. (2009). Nonlinear control with wind estimation of a DFIG variable speed wind turbine for power capture optimization. *Energy Conversion and Management*, *50*(4), 885–892. doi:10.1016/j.enconman.2009.01.011

Boukhnifer, M., & Ferreira, A. (2005). H Infinity Loop Shaping for Stabilization and Robustness of a Tele-Micromanipulation System. In *Proceedings ofIEEE/RSJ International Conference on Intelligent Robots and Systems* (IROS 2005). IEEE.

Boulkroune, A., Bounar, N., M'saad, M., & Farza, M. (2014a). Indirect adaptive fuzzy control scheme based based on observer for nonlinear systems: A novel SPR-filter approach. *Neurocomputing*, *135*, 378–387. doi:10.1016/j.neucom.2013.12.011

Boulkroune, A., & M'saad, M. (2011c). A fuzzy adaptive variable-structure control scheme for uncertain chaotic MIMO systems with sector nonlinearities and dead-zones. *Expert Systems with Applications*, *38*(12), 4744–14750. doi:10.1016/j.eswa.2011.05.006

Boulkroune, A., M'saad, M., & Farza, M. (2011b). Adaptive fuzzy controller for multivariable nonlinear state time-varying delay systems subject to input nonlinearities. *Fuzzy Sets and Systems*, *164*(1), 45–65. doi:10.1016/j.fss.2010.09.001

Boulkroune, A., & Msaad, M. (2011a). A practical projective synchronization approach for uncertain chaotic systems with dead-zone input. *Communications in Nonlinear Science and Numerical Simulation*, *16*(11), 4487–4500. doi:10.1016/j.cnsns.2011.02.016

Boulkroune, A., & M'saad, M. (2012a). On the design of observer-based fuzzy adaptive controller for nonlinear systems with unknown control gain sign. *Fuzzy Sets and Systems*, *201*, 71–85. doi:10.1016/j.fss.2011.12.005

Boulkroune, A., & M'saad, M. (2012b). Fuzzy adaptive observer-based projective synchronization for nonlinear systems with input nonlinearity. *Journal of Vibration and Control*, *18*(3), 437–450. doi:10.1177/1077546311411228

Boulkroune, A., M'saad, M., & Chekireb, H. (2010a). Design of a fuzzy adaptive controller for MIMO nonlinear time-delay systems with unknown actuator nonlinearities and unknown control direction. *Information Sciences*, *180*(24), 5041–5059. doi:10.1016/j.ins.2010.08.034

Boulkroune, A., M'saad, M., & Farza, M. (2012c). Adaptive fuzzy tracking control for a class of MIMO nonaffine uncertain systems. *Neurocomputing*, *93*, 48–55. doi:10.1016/j.neucom.2012.04.006

Boulkroune, A., M'saad, M., & Farza, M. (2012d). Fuzzy approximation-based indirect adaptive controller for multi-input multi-output non-affine systems with unknown control direction. *IET Control Theory & Applications*, *17*(17), 2619–2629. doi:10.1049/iet-cta.2012.0565

Boulkroune, A., M'saad, M., & Farza, M. (2014b). State and output feedback fuzzy variable structure controllers for multivariable nonlinear systems subject to input nonlinearities. *International Journal of Advanced Manufacturing Technology*, *71*(1-4), 539–556. doi:10.1007/s00170-013-5453-4

Boulkroune, A., Tadjine, M., M'saad, M., & Farza, M. (2009). Adaptive fuzzy controller for non-affine systems with zero dynamics. *International Journal of Systems Science*, *40*(4), 367–382. doi:10.1080/00207720802436919

Boulkroune, A., Tadjine, M., M'saad, M., & Farza, M. (2010b). Fuzzy adaptive controller for MIMO nonlinear systems with known and unknown control direction. *Fuzzy Sets and Systems*, *161*(6), 797–820. doi:10.1016/j.fss.2009.04.011

Boulkroune, A., Tadjine, M., M'saad, M., & Farza, M. (2014c). Design of a unified adaptive fuzzy Observer for uncertain nonlinear systems. *Information Sciences*, *265*, 139–153. doi:10.1016/j.ins.2013.12.026

Boulkroune, A., Tadjine, M., M'saad, M., & Farza, M. (2008). How to design a fuzzy adaptive control based on observers for uncertain affine nonlinear systems. *Fuzzy Sets and Systems*, *159*(8), 926–948. doi:10.1016/j.fss.2007.08.015

Boulkroune, B., Gálvez-Carrillo, M., & Kinnaert, M. (2013). Combined Signal and Model-Based Sensor Fault Diagnosis for a Doubly Fed Induction Generator. *IEEE Transactions on Control Systems Technology*, *21*(5), 1771–1783. doi:10.1109/TCST.2012.2213088

Boumedyen, B., Christophe, A., Jiang, J., & Mohamed, N. A. (2013). Fault Tolerant Control approach with actuator saturation avoidance based on reference management. *International Journal of Robust and Nonlinear Control*, *23*(2), 1–17.

Boumedyen, B., Christophe, A., Mohamed, N. A., & Kamel, B. (2011). Performance Evaluation based Fault Tolerant Control with Actuator Saturation Avoidance. *International Journal of Applied Mathematics and Computer Science*, *21*(3), 457–466.

Bows, R., Grewal, S., & Holliday, D. (2000). High frequency PWM technique for two and three single- Phases inverters. *IEE Proceedings on Electric Power Applications*, *147*(3), 181–191. doi:10.1049/ip-epa:20000330

Branicky, M. S. (1998). Multiple Lyapunov functions and other analysis tools for switched and hybrid systems. *IEEE Transactions on Automatic Control*, *43*(4), 475–482. doi:10.1109/9.664150

Breivik, M., & Fossen, T. I. (2008). Guidance laws for planar motion control. In *Proceedings of the 47th IEEE Conference on Decision and Control*, (pp. 570-577). Cancun, Mexico: IEEE.

Brogan, W. L. (n.d.). *Modern Control Theory*. Prentice Hall.

Brooks, R. A. (1987). A hardware retargetable distributed layered architecture for mobile robot control. In *Proceedings of the IEEE International Conference on Robotics and Automation*, (pp. 106-110). IEEE. doi:10.1109/ROBOT.1987.1088016

Budd, C. J. (1996). Non-smooth dynamical systems and the grazing bifurcation. (C. J. Aston, Éd.) Cambridge, UK: Cambridge University Press.

Buscarino, A., Fortuna, L., Frasca, M., Sciuto, G., & Xibilia, M. G. (2011) Harmonic balance method for time-delay chaotic systems design. In *Proc. World Congress of the International Federation of Automatic Control (IFAC 2011)*, (pp. 5112-5117). Milan, Italy: IFAC.

Buscarino, A., Fortuna, L., Frasca, M., & Sciuto, G. (2011). Design of time-delay chaotic electronic circuits. *IEEE Transactions on Circuits and Systems I, Regular Papers*, *58*(8), 1888–1896. doi:10.1109/TCSI.2011.2107190

Caccia, M., Coletta, P., Bruzzone, G., & Veruggio, G. (2005). Execution control of robotic tasks: A Petri Net based approach. *Control Engineering Practice*, *13*(8), 959–971. doi:10.1016/j.conengprac.2004.10.005

Calderaro, V., Galdi, V., Piccolo, A., & Siano, P. (2008). A fuzzy controller for maximum energy extraction from variable speed wind power generation systems. *Electric Power Systems Research*, *78*(6), 1109–1118. doi:10.1016/j.epsr.2007.09.004

Canudas de Wit, C., Kennouf, H., Samson, C., & Sordalen, O. J. (1997). Nonlinear control design for mobile robots. In Y.-F. Zheng (Ed.), Recent Trends in Mobile Robots (pp. 121–156). Academic Press.

Cao, Y. U., Fukunaga, A. S., & Kahn, A. (1997). Cooperative mobile robotics: Antecedents and directions. *Autonomous Robots*, *4*(1), 7–27. doi:10.1023/A:1008855018923

Cardenas, R. J., Pena, R. S., Asher, J., Asher, G. M., & Clare, J. C. (2004). Sensorless control of a doubly-fed induction generator for Stand-alone Operation. In *Proceedings of IEEE 35th Annual Power Electronics Specialists Conference 2004* (pp. 3378-3383). IEEE. doi:10.1109/PESC.2004.1355072

Carner, P. (2009). Project Domus: Designing Effective Smart Home Systems. BSc in Information Systems and Information Technology (DT249) to the School of Computing, Faculty of Science, Dublin Institute of Technology.

Carrasco, J., Franquelo, L., Bialasiewicz, J., Galvan, E., Guisado, R., & Prats, M. et al. (2006). Power-electronic systems for the grid integration of renewable energy sources: A survey. *IEEE Transactions on Industrial Electronics*, *53*(4), 1002–1016. doi:10.1109/TIE.2006.878356

Casella, A. J., Capovilla, C. E., Azcue, J. L., & Ruppert, E. (2013). An Adaptive Neuro-Fuzzy Strategy for a Wireless Coded Power Control in Doubly-Fed Induction Aerogenerators. In T. Abrão (Ed.), Search Algorithms for Engineering Optimization (pp. 175-199). InTech.

Chaabane, M., Tadeo, F., Mehdi, D., & Souissi, M. (2011). Robust Admissibilization of Descriptor Systems by Static Output-Feedback: An LMI Approach. Hindawi Publishing Corporation, 2012(2012).

Champetier, C., Mouyon, P., & Magni, J. F. (1985). Pseudolinearization of nonlinear systems by dynamic precompensation. In *Proceedings of the 1985 IEEE Control and Decision Conference (CDC)*, (pp. 1371–1372). IEEE. Doi:10.1109/CDC.1985.268735

Champetier, C., Mouyon, P., & Reboulet, C. (1984). Pseudolinearization of multi-input nonlinear systems. In *Proceedings of the 1984 IEEE Control and Decision Conference (CDC)*, (pp. 96–97). IEEE. Doi:10.1109/CDC.1984.272279

Chang, H.-H., & Liu, S.-I. (2005). A wide-range and fast-locking all digital cycle-controlled delay-locked loop. *IEEE Journal of Solid-State Circuits*, *40*(3), 661–670. doi:10.1109/JSSC.2005.843596

Chang, Y. C. (2000). Robust tracking control for nonlinear MIMO systems via fuzzy approaches. *Automatica*, *36*(10), 1535–1545. doi:10.1016/S0005-1098(00)00083-2

Charfeddine, M., Jouili, K., & Benhadj Braiek, N. (2013) The Output Tracking Control of a Nonlinear Non-Minimum Phase System. In *Proceedings of the 2013 International Conference on Control, Engineering & Information Technology (CEIT)*, (pp. 83-88). CEIT.

Charfeddine, M., Jouili, K., & Benhadj Braiek, N. (2013). Output tracking control of inverted cart-pendulum using input-output linearization and predictive control. In *Proceedings of the 2013 IEEE International conference on Sciences and Techniques of Automatic Control & Computer Engineering*, (pp. 84–88). IEEE. Doi:10.1109/STA.2013.6783110

Charfeddine, M., Jouili, K., & Benhadj Braiek, N. (2014). Cascade control for nonlinear non-minimum phase system based on backstepping approach and input output linearization. *International Journal of Control. Energy and Electrical Engineering*, *1*, 43–49.

Charfeddine, M., Jouili, K., Jerbi, H., & Benhadj Braiek, N. (2010). Linearizing Control with a Robust Relative Degree Based on Lyapunov Function: Case of the Ball and Beam system. *International Review on Modelling and Simulations*, *3*(2), 219–226.

Charfeddine, M., Jouili, K., Jerbi, H., & Benhadj Braiek, N. (2011). Output Tracking Control Design for Non-Minimum Phase Systems: Application to the Ball and Beam Model. *International Review of Automatic Control*, *4*(1), 47–55.

Chekireb, H., Tadjine, M., & Bouchaffra, D. (2003). Direct adaptive fuzzy control of nonlinear system class with applications. *Control and Intelligent Systems*, *31*(2), 1–11.

Chen, C. C., & Wong, C. C. (2002). Self-generating rule-mapping fuzzy controller design using a genetic algorithm. *IEE Proceedings of Control Theory and Applications*, *149*(2), 143–148. doi:10.1049/ip-cta:20020253

Chen, D., & Paden, B. (1996). Stable inversion of nonlinear nonminimum-phase systems. *International Journal of Control*, *64*(1), 81–97. doi:10.1080/00207179608921618

Cheng, P. Y., & Chen, P. J. (2012). Navigation of mobile robot by using D++ algorithm. *Intelligent Service Robotics*, *5*(4), 229–243. doi:10.1007/s11370-012-0120-4

Chen, J., & Latchman, H. A. (1994). Asymptotic stability independent of delays: simple necessary and sufficient conditions. In *Proceedings of American Control Conference*, (vol. 12, pp. 1027-1031). Baltimore, MD: IEEE. doi:10.1109/ACC.1994.751903

Chen, J., & Patton, R. J. (2012). *Robust Model-Based Fault Diagnosis for Dynamic Systems*. Springer Verlag.

Chen, Q., Ren, X., & Oliver, J. A. (2012). Identier-based adaptive neural dynamic surface control for uncertain dcdc buck converter system with input constraint. *Communications in Nonlinear Science and Numerical Simulation*, *17*(4), 1871–1883. doi:10.1016/j.cnsns.2011.08.028

Chen, S., & Billings, S. A. (1992). Neural networks for nonlinear dynamic system modelling and identification. *International Journal of Control*, *56*(2), 319–346.

Chen, Y. M., Liu, Y. C., Hung, S. C., & Cheng, C. S. (2007). Multi-input inverter for grid-connected hybrid PV/wind power system. *IEEE Transactions on Power Electronics*, *22*(3), 1070–1077. doi:10.1109/TPEL.2007.897117

Chen, Z., Guerrero, J. M., & Blaabjerg, F. (2009). A Review of the State of the Art of Power Electronics for Wind Turbines. *IEEE Transactions on Power Electronics*, *24*(8), 1859–1875. doi:10.1109/TPEL.2009.2017082

Chen, Z.-M., Djidjeli, K., & Price, W. G. (2006). Computing Lyapunov exponents based on the solution expression of the variational system. *Applied Mathematics and Computation*, *174*(2), 982–996. doi:10.1016/j.amc.2005.05.016

Chen, Z., Yang, L., Zhang, Y., & Li, Z. (2012, July). A control method of PMSM current-loop based on on-line parameter identification, *7th IEEE Conference on Industrial Electronics and Applications (ICIEA)*, pp.417 – 422, Singapore. doi:10.1109/ICIEA.2012.6360763

Chiang, L. H., Richard, D. B., & Russell, E. L. (2001). *Fault Detection and Diagnosis in Industrial Systems*. London. doi:10.1007/978-1-4471-0347-9

Cho, Y.-J., Park, J.-M., Park, J., Oh, S.-R., & Lee, C. W. (1998). A control architecture to achieve manipulation task goals for a humanoid robot. In *Proceedings of the 1998 IEEE International Conference on Robotics and Automation*, (vol. 1, pp. 206-212). IEEE.

Choi, C., & Dennis, L. (1996). Effectiveness of a geometric programming algorithm for optimization of machining economics models. *Computers & Operations Research*, *23*(10), 957–961. doi:10.1016/0305-0548(96)00008-1

Chowdhury, B. H., & Chellapilla, S. (2006). Double fed induction generator control for variable speed wind power generation. *Electric Power Systems Research*, *76*(9-10), 786–800. doi:10.1016/j.epsr.2005.10.013

Chowdhury, M. A., Shen, W. X., Hosseinzadeh, N., & Pota, H. R. (2013). A novel aggregated DFIG wind farm model using mechanical torque compensating factor. *Energy Conversion and Management*, *67*, 265–274. doi:10.1016/j.enconman.2012.12.001

Chow, E., & Willsky, A. (1984). Analytical redundancy and the design of robust failure detection systems. *IEEE Transactions on Automatic Control*, *29*(7), 603–614.

Chow, T. W. S., & Fang, Y. (1998). An Iterative Learning Control Method for Continuous-time Systems Based on 2-D System Theory. *IEEE Transactions on Circuits and Systems: Fundamental Theory and Applications*, *45*(2), 683–689. doi:10.1109/81.678496

Chuang, N., Petersen, I. R., & Pota, H. R. (2011). Robust H-infinity control in fast atomic force microscopy. In *Proc. of the American Control Conference* (pp. 2258–2265). Academic Press.

Chyung, D. H. (1992). Output Feedback Controller for Systems Containing a Backlash. In *Proceedings of the 31st Conference on Decision and Control.*Tucson, AZ: IEEE. doi:10.1109/CDC.1992.371218

Cimen, T. (2010). Systematic and effective design of nonlinear feedback controllers via the state-dependent Riccati equation (SDRE) method. *Annual Reviews in Control*, *34*(1), 32–51. doi:10.1016/j.arcontrol.2010.03.001

Colak, I., Kabalci, E., & Bayindir, R. (2011). Review of multilevel voltage source inverter topologies and control schemes. *Energy Conversion and Management*, *52*(2), 1114–1128. doi:10.1016/j.enconman.2010.09.006

Colaneri, P., Geromel, J., & Astolfi, A. (2008). Stabilization of continuous-time switched nonlinear systems. *Systems & Control Letters*, *57*(1), 95–103. doi:10.1016/j.sysconle.2007.07.001

Collins, S., Ruina, A., Tedrake, R., & Wisse, M. (2005). Efficient bipedal robots based on passive-dynamic walkers. *Science Magazine*, *307*, 1082–1085. PMID:15718465

Cooper, M. C. (2003). Reduction operations in fuzzy or value constrained satisfaction. *Fuzzy Sets and Systems*, *134*(3), 311–342. doi:10.1016/S0165-0114(02)00134-3

Cornelis, C., Deschrijver, C., & Kerre, E. E. (2004). Implication in intuitionistic and interval-valued fuzzy set theory: Construction, classification, application. *International Journal of Approximate Reasoning*, *35*(1), 55–95. doi:10.1016/S0888-613X(03)00072-0

Croft, D., Shedd, G., & Devasia, S. (2000). Creep, hysteresis, and vibration compensation for piezoactuators: atomic force microscopy application. In *Proc. of the American Control Conference* (pp. 2123–2128). Academic Press. doi:10.1109/ACC.2000.879576

Croft, D., & Devasia, S. (1999). Vibration compensation for high speed scanning tunneling microscopy. *The Review of Scientific Instruments*, *70*(12), 4600–4605. doi:10.1063/1.1150119

Cui, C., Long, F., & Li, C. (2010). Disturbance attenuation for switched system with continuous-time and discrete-time subsystems: State feedback case. *ICIC Express Letters*, *4*(1), 205–212.

Cuomo, K. M., & Oppenheim, A. V. (1993). Circuit implementation of synchronized chaos with applications to communications. *Physical Review Letters*, *71*(1), 65–68. doi:10.1103/PhysRevLett.71.65 PMID:10054374

Dai, L. (1989). *Singular Control Systems. Lecture Notes in Control and Information Sciences*. Germany: Springer.

Darouach, M. (2000). Existence and design of functional observers for linear systems. *IEEE Transactions on Automatic Control, 45*(5), 940–943. doi:10.1109/9.855556

Das, S. K., Pota, H. R., & Petersen, I. R. (2012). Resonant controller for fast atomic force microscopy. In *Proc. of the Conference on Decision and Control* (pp. 2471–2476). Academic Press. doi:10.1109/CDC.2012.6426563

Das, S. K., Pota, H. R., & Petersen, I. R. (2013d). (in press). Resonant Controller Design for a Piezoelectric Tube Scanner: A Mixed Negative-Imaginary and Small-Gain Approach. *IEEE Transactions on Control Systems Technology*.

David, H. T. (2011). *Commande Robuste de Générateurs Électrochimiques Hybrides*. (PhD thesis). Department of Electrical Engineering, Grenoble University, France.

Davies, B. (2004). *Exploring Chaos: Theory and experiment*. Oxford, UK: Westview Press – Perseus Books Group.

Davis, D., Ihaka, R., & Fenstermacher, P. (1994). *Cryptographic randomness from air turbulence in disk drives. In Advances in Cryptology* (pp. 114–120). Berlin: Springer-Verlag.

Davison, D. E., & Bortoff, S. A. (1994). Enlarge your region of attraction using high-gain feedback. In *Proceedings of the 1994 IEEE Control and Decision Conference*, (pp. 634–639). IEEE. Doi:10.1109/CDC.1994.410974

de Korte, R. J., Schouten, J. C., & van den Bleek, C. M. (1995). Experimental control of a chaotic pendulum with unknown dynamics using delay coordinates. *Physical Review E: Statistical Physics, Plasmas, Fluids, and Related Interdisciplinary Topics, 52*(4), 3358–3365. doi:10.1103/PhysRevE.52.3358 PMID:9963811

de Paula, A. S., & Savi, M. A. (2009). A multiparameter chaos control method based on OGY approach. *Chaos, Solitons, and Fractals, 40*(3), 1376–1390. doi:10.1016/j.chaos.2007.09.056

de Souza, S. L., & Caldas, I. L. (2004a). Calculation of Lyapunov exponents in systems with impacts. *Chaos, Solitons, and Fractals, 19*(3), 569–579. doi:10.1016/S0960-0779(03)00130-9

de Souza, S. L., & Caldas, I. L. (2004b). Controlling chaotic orbits in mechanical systems with impacts. *Chaos, Solitons, and Fractals, 19*(1), 171–178. doi:10.1016/S0960-0779(03)00129-2

de Souza, S. L., Caldas, I. L., & Viana, R. L. (2007). Damping control law for a chaotic impact oscillator. *Chaos, Solitons, and Fractals, 32*(2), 745–750. doi:10.1016/j.chaos.2005.11.046

de Weger, J. G. (2005). *The Grazing Bifurcation and Chaos Control*. Eindhoven, The Netherlands: Tehnische Universiteit Eindhoven.

Deane, J., & Hamill, D. (1990). Instability, subharmonics and chaos in power electronic systems. *IEEE Transactions on Power Electronics, 5*(3), 260–268. doi:10.1109/63.56516

Decarlo, R., Branicky, M., Pettersson, S., Lennartson, B., & Antsaklis, P. J. (2000). Perspectives and results on the stability and stabilizability of hybrid systems.*Proceedings of IEEE, 88*(4), 1069–1082. doi:10.1109/5.871309

Deng, H., & Doroslovački, M. (2006). Proportionate adaptive algorithms for network echo cancellation. *IEEE Transactions on Signal Processing, 54*(5), 1794–1803. doi:10.1109/TSP.2006.872533

Denoun, H. (2001). *Contribution à l'étude des convertisseurs AC/DC à MLI à facteur de puissance unitaire, application à la conduite de la machine asynchrone*. (Magistère thesis). EMP, Alger, Algéria.

Denoun, H., Benamrouche, N., Haddad, S., Meziani, S., & AitMamar, S. (2011). A DSP (TMS320lf2407) Based Implementation of PWM for Single-phase AC-DC Bipolar Converter with a Unity Power Factor.*WSEAS International Journal of Circuits. Systems and Signal Processing, 5*(4), 354–361.

Desai, M., Deckert, J., Deyst, J., Willsky, A., & Chow, E. (1976). Dual redundant sensor FDI techniques applied to the NASA F8C DFBW aircraft. In *Proceedings ofGuidance and Control Conference*. American Institute of Aeronautics and Astronautics.

Desel, J., & Reisig, W. (1998). Place/transition Petri nets. In W. Reisig & G. Rosenberg (Eds.), Lectures on Petri nets/Basic models (LNCS), (vol. 1491, pp. 122–173). Berlin, Germany: Springer. doi:10.1007/3-540-65306-6_15

Desoer, C. A., & Vidyasagar, M. (1975). *Feedback Systems: Input-Output Properties.* New York: Academic Press.

Deutscher, J. (2005). Input-output linearization of nonlinear systems using multivariable Legendre polynomials. *Automatica, 41*(2), 299–304. doi:10.1016/j.automatica.2004.11.001

Di Bernardo, M., & Tse, C. (2002). Chaos in power electronics: An overview. In G. Chen & T. Ueta (Eds.), *Chaos in circuits and systems* (Vol. 11, pp. 317–340). New York: World Scientific. doi:10.1142/9789812705303_0016

Di Bernardo, M., & Vasca, F. (2000). On discrete time maps for the analysis of bifurcations and chaos in dc/dc converters. *IEEE Transactions on Circuits and Systems I, 47*(2), 130–143. doi:10.1109/81.828567

Dibo, M. A., Maquin, D., & Ragot, J. (2004). *Validation de données des systemes dynamiques incertains.* Paper presented at the Conférence Internationale Francophone d'Automatique. Douz, Tunisia.

Ding, S. X. (2008). *Model-based Fault Diagnosis Techniques.* Springer Verlag.

Domatiq Main Site. (2014). Retrieved from http://www.domatiq.pl

Dong, D., Chen, C., Chu, J., & Tarn, T. (2012). Robust quantum-inspired reinforcement learning for robot navigation. *IEEE Transactions on Mechatronics, 17*(1), 86–97. doi:10.1109/TMECH.2010.2090896

Doraiswami, R., & Cheded, L. (2013). A unified approach to detection and isolation of parametric faults using a Kalman filter residual-based approach. *Journal of the Franklin Institute, 350*(5), 938–965. doi:10.1016/j.jfranklin.2013.01.005

Dorf, R. C., & Bishop, R. H. (2008). *Modern Control Systems.* Pearson Education.

Doty, K. L., & Van Aken, R. E. (1993). Swarm robot material handling paradigm for a manufacturing workcell. In *Proceedings of the IEEE International Conference on Robotics and Automation,* (vol. 1, pp. 778-782). IEEE. doi:10.1109/ROBOT.1993.292072

Doyle, F. J. III, Allgower, F., & Morari, M. (1995). A normal form approach to approximate input-output linearization for maximum phase nonlinear systems. *IEEE Transactions on Automatic Control, 41*(2), 305–309. doi:10.1109/9.481553

Doyle, J., Glover, K., Khargonekar, P. P., & Francis, B. (1989). State space solutions to standards H_2 H_∞. *IEEE Transactions on Automatic Control, 34*(8), 831–847. doi:10.1109/9.29425

Doyle, J., Zhou, K., Glover, K., & Bodenheimer, B. (1994). Mixed H_2 / H_∞. *IEEE Transactions on Automatic Control, 39*(8), 1575–1587. doi:10.1109/9.310031

Dragan, V. (2005). The linear quadratic optimization problem for a class of singularly perturbed stochastic systems. *International Journal of Innovative Computing, Information, & Control, 1*(1), 53–63.

Dragan, V. Sh. P., & Boukas, K. E. (1999). Control of singularly perturbed systems with Markovian jump parameters: An H infini approach. *Automatica, 35*(8), 1369–1378. doi:10.1016/S0005-1098(99)00047-3

Draghicescu, M. (2010). Concerns about the non-polluting energy sources in the European Union: Renewable energy sources in Romania. In *Proceedings of the 2nd International Conference on Chemical, Biological and Environmental Engineering* (ICBEE), (pp. 235–237). Doi:10.1109/ICBEE.2010.5652877

Dressler, U., & Nitsche, G. (1992). Controlling chaos using time delay coordinates. *Physical Review Letters, 68*(1), 1–4. doi:10.1103/PhysRevLett.68.1 PMID:10045097

Dridi, J., Ouerfelli, H., & Salhi, S. (2013).Stability Along the Pass of Differential Linear Repetitive Process Using an LMI. In *Proceedings ofIEEE International Conference on Integent Automatiition and Robotics,* (pp. 6-10). IEEE.

Dridi, J., Ben Attia, S., Salhi, S., & Ksouri, M. (2012). Repetitive Processes Based Iterative Learning Control Designed by LMIs. *Applications ofMathematics, 2012,* 18.

Driver, R. D. (1977). *Ordinary and Delay Differential Equations.* New York: Springer Verlag. doi:10.1007/978-1-4684-9467-9

Duan, R. G. (2010). Analysis and Design of linear descriptor systems. New York, NY: Springer. doi:10.1007/978-1-4419-6397-0

Duan, Z., Cai, Z., & Yu, J. (2005). Fault diagnosis and fault tolerant control for wheeled mobile robots under unknown environments. In *Proceedings of the 2005 IEEE International Conference on Robotics and Automation (ICRA),* (pp. 3439–3444). IEEE. Doi:10.1109/ROBOT.2005.1570640

Duan, S., & Wang, L. (2009). A novel delayed chaotic neural model and its circuitry implementation. *Computers and Mathematics with Application, 57*(11-12), 1736–1742. doi:10.1016/j.camwa.2008.10.017

Dube, R. R. (2008). *Hardware-based computer security techniques to defeat hackers: From biometrics to quantum cryptography.* Wiley.

Du, D., Zhou, S., & Zhang, B. (2006). Generalized H2 output feedback controller design for uncertain discrete-time switched systems via switched Lyapunov functions. *Nonlinear Analysis, 65*(11), 2135–2146. doi:10.1016/j.na.2005.11.054

Duffee, N. A., & Prabhu, V. V. (1996). Heterarchical control of highly distributed manufacturing systems. *International Journal of Computer Integrated Manufacturing, 9*(4), 270–281. doi:10.1080/095119296131562

Duffie, N. (1990). Synthesis of heterarchical manufacturing systems. *Computers in Industry, 14*(1-3), 167–174. doi:10.1016/0166-3615(90)90118-9

Duttweiler, D. L. (2000). Proportionate normalized least-mean-squares adaptation in echo cancellers. *IEEE Transactions on Speech and Audio Processing, 8*(5), 508–518. doi:10.1109/89.861368

Edwards, C., & Tan, C. P. (2006). A comparison of sliding mode and unknown input observers for fault reconstruction. *European Journal of Control, 12*(3), 245-260.

Edwards, C., & Pin, T. C. (2004). A comparison of sliding mode and unknown input observers for fault reconstruction. *European Journal of Control, 12*(3), 5279–5284.

Eielsen, A. A., & Fleming, A. J. (2010), Passive shunt damping of a piezoelectric stack nanopositioner. In *Proceedings ofAmerican Control Conference.* Academic Press. doi:10.1109/ACC.2010.5530938

Ejiri, A., Watanabe, I., Okabayashi, K., Hashima, M., Tatewaki, M., Aoki, T., & Maruyama, T. (1994). Satellite berthing experiment with a two-armed space robot. In *Proceedings of the IEEE International Conference on Robotics and Automation,* (pp. 3480-3487). IEEE. doi:10.1109/ROBOT.1994.351035

Ekanayake, J., Holdsworth, L., & Jenkins, N. (2003). Control of DFIG wind turbines. *Power Engineering, 17*(1), 28–32. doi:10.1049/pe:20030107

El Aroudi, A., Angulo, F., Olivar, G., Robert, B., & Feki, M. (2009). Stabilizing a two-cell dc-dc buck converter by fixed point induced control. *International Journal of Bifurcation and Chaos in Applied Sciences and Engineering, 19*(6), 2043–2057. doi:10.1142/S0218127409023895

Elci, H., Longman, R. W., Phan, M. Q., Juang, J. N., & Ugoletti, R. (2002). Simple Learning Control Made Practical by Zero-phase Filtering: Applications to Robotics. *IEEE Transactions on Circuits and Systems: Fundamental Theory and Applications, 49*(2), 753–767. doi:10.1109/TCSI.2002.1010031

Elgendy, M. A., Zahawi, B., & Atkinson, D. J. (2013). Assessment of the Incremental Conductance Maximum Power Point Tracking Algorithm. *IEEE Transactions on Sustainable Energy, 4*(1), 108–117. doi:10.1109/TSTE.2012.2202698

Elmadssia, S., Saadaoui, k. & Benrejeb, M. (2011). New delay-dependent stability conditions for linear systems with delay. In *Proceedings ofInternational Conference on Communications Computing and Control Applications.* Hammamet, Tunisia: Academic Press. doi:10.1109/CCCA.2011.6031533

Elmadssia, S., Saadaoui, K., & Benrejeb, M. (2013). New delay-dependent stability conditions for linear systems with delay Systems. *Science. Control. Engineering, 1*(1), 37–41.

Emadi, A., & Rajashekara, K.Young Joo Lee. (2008). Power Electronics and Motor Drives in Electric, Hybrid Electric, and Plug-In Hybrid Electric Vehicles. *IEEE Transactions on Industrial Electronics*, *55*(6), 2237–2245. doi:10.1109/TIE.2008.922768

Endusa, B. M., Tomonobu, S., Aki, U., Toshihisa, F., & Chul-Hwan, K. (2009). LQG Design for Megawatt-Class WECS With DFIG Based on Functional Models' Fidelity Prerequisites. *IEEE Transactions on Energy Conversion*, *24*(4), 893–904. doi:10.1109/TEC.2009.2025338

Engel, Y., Mannor, S., & Meir, R. (2004). The Kernel Recursive Least-Squares Algorithm. *IEEE Transactions on Signal Processing*, *52*(8), 2275–2285. doi:10.1109/TSP.2004.830985

Ergün, S. (2007). Truly random number generators based on a non-autonomous chaotic oscillator. *AEÜ. International Journal of Electronics and Communications*, *61*(4), 235–242. doi:10.1016/j.aeue.2006.05.006

Ergün, S., & Özouz, S. (2010). Truly random number generators based on non-autonomous continuous-time chaos. *International Journal of Circuit Theory and Applications*, *38*(1), 1–24. doi:10.1002/cta.520

Esram, T., & Chapman, P. L. (2007). Comparison of Photovoltaic Array Maximum Power Point Tracking Techniques. *IEEE Transactions on Energy Conversion*, *22*(2), 439–449. doi:10.1109/TEC.2006.874230

Essounbouli, N., Hamzaoui, A., & Zaytoon, J. (2006). An improved robust adaptive fuzzy controller for MIMO systems. *Control and Intelligent Systems*, *34*(1), 12–21. doi:10.2316/Journal.201.2006.1.201-1350

Faa-Jeng, L., Chih-Hong, L., & Po-Hung, S. (2002). Self-constructing fuzzy neural network speed controller for permanent-magnet synchronous motor drive. *IEEE Transactions on Fuzzy Systems*, *9*(5), 751–759. doi:10.1109/91.963761

Faa-Jeng, L., Po-Kai, H., & Wen-Der, C. (2007). A genetic algorithm based recurrent fuzzy neural network for linear induction motor servo drive. *Journal of the Chinese Institute of Engineers.*, *30*(5), 801–817. doi:10.1080/02533839.2007.9671307

Fairfield, R. C., Mortenson, R. L., & Coulthart, K. B. (1987). *An LSI random number generator (RNG). In Advances in Cryptology* (pp. 203–230). Berlin: Springer-Verlag.

Fang, Y., Tommy, W., & chow, s. (2003). 2-D analysis for iterative learning controller for discrete-time systems with variable initial conditions. *IEEE Transactions on Circuits and Systems: Fundamental Theory and Application, 50*(5), 722-727.

Fang, H. C., & Chang, R. F. (1993). Analysis of stability robustness for generalized state space systems with structured perturbations. *Systems & Control Letters*, *21*(2), 109–114. doi:10.1016/0167-6911(93)90113-K

Fang, H. C., Lee, L., & Chang, F.-R. (1994). Robust control analysis and design for discrete time singular systems. *Automatica*, *30*(11), 1741–1750. doi:10.1016/0005-1098(94)90076-0

Fang, M. (2010). Delay-dependent stability analysis for discrete singular systems with time-varying delays. *Acta Automatica Sinica*, *36*, 751–755.

Fan, L. L., & Song, Y. D. (2010). On Fault-tolerant Control of Dynamic Systems with Actuator Failures and External Disturbances. *Acta Automatica Scinica*, *36*(11), 1620–1625.

Faria, A. F., Assunçao, E., Teixeira, M., Cardim, R., & da Silva, N. A. P. (2009). Robust state-derivative pole placement LMI-based designs for linear systems. *International Journal of Control*, *82*(1), 1–12. doi:10.1080/00207170801942188

Farmer, J. D. (1982). Chaotic attractor of an infinite-dimensional dynamical system. *Physica D Nonlinear Phenomena*, *4*(3), 366–393. doi:10.1016/0167-2789(82)90042-2

Fatikow, S., & Magnussen, B. (1994). Intelligent robot hand control system using a tailorable parallel computer concept. In *Proceedings of the IEEE International Conference on Robotics and Automation*, (pp. 216-223). IEEE. doi:10.1109/IROS.1994.407388

Feki, M., Robert, B., & Iu, H. (2004). A proportional plus extended time-delayed feedback controller for a pwm inverter. In *Proceedings ofIEEE Annual Power Electronics Specialists Conference, PESC Record*, (pp. 3317–3320). Achen, Germany: IEEE. doi:10.1109/PESC.2004.1355061

Feng, Y., Yagoubi, M., & Chevrel, P. (2010). Extended output Control for Continuous Descriptor Systems. In *Proceedings of 49th IEEE Conference on Decision and Control* (pp. 6016-6021). Atlanta, GA: IEEE. doi:10.1109/CDC.2010.5717866

Feng, Y., Yagoubi, M., & Chevrel, P. (2011). State Feedback H_2 optimal Controllers under regulation constraints for Descriptor Systems. *International Journal of Innovative Computing, Information, & Control*, *7*(10), 5761–5770.

Feng, Y., Yagoubi, M., & Chevrel, P. (2012). Extended H_2 Controller Synthesis for Continuous Descriptor Systems. Automatic Control. *IEEE Transactions*, *57*(6), 1559–1564.

Ferguson, N., & Schneier, B. (2003). *Practical cryptography*. Wiley.

Fernandez-Vargas, J., & Ledwich, G. (2010). Variable structure control for power systems stabilization. *International Journal of Electrical Power & Energy Systems*, *32*(2), 101–107. doi:10.1016/j.ijepes.2009.06.017

Filali, R. L., Hammami, S., Benrejeb, M., & Borne, P. (2012). On synchronization, anti-synchronization and hybrid synchronization of 3D discrete generalized H12 non map. *Nonlinear Dynamics and Systems Theory*, *12*(2), 81–96.

Filev, D. (1991). Fuzzy modeling of complex systems. *International Journal of Approximate Reasoning*, *5*(3), 281–290.

Flah A., Ben hamed M. and Sbita L. (2011). Online MRAS-PSO PMSM parameters estimation. *International review on modeling and simulations*, 4 (3), 980-987.

Flah A., Sbita L. (2012). An adaptive high speed PMSM control for electrical vehicle application. *Journal of electrical engineering* 12 (3).

Fleming, A. J. (2010). Nanopositioning system with force feedback for highperformance tracking and vibration control. *IEEE/ASME Transactions on Mechatronics*, *15*(3), 433–447. doi:10.1109/TMECH.2009.2028422

Fleming, A. J., Aphale, S. S., & Moheimani, S. O. R. (2010). A new method for robust damping and tracking control of scanning probe microscope positioning stages. *IEEE Transactions on Nanotechnology*, *9*(4), 438–448. doi:10.1109/TNANO.2009.2032418

Fleming, R., & Sum, S. (2014). Empirical studies on the effectiveness of assistive technology in the care of people with dementia: A systematic review. *Journal of Assistive Technologies*, *8*(1), 14–34. doi:10.1108/JAT-09-2012-0021

Fleury, S., Herrb, M., & Chatila, R. (1994). Design of a modular architecture for autonomous robot. In *Proceedings of the IEEE International Conference on Robotics and Automation*, (pp. 3508-3513). IEEE. doi:10.1109/ROBOT.1994.351031

Fliess, M., Lévine, J., Martin, P., & Rouchon, P. (1992). On differentially flat nonlinear systems. In Proceedings of the 1992 IFAC symposium on Nonlinear Control Systems Design, (pp. 408–412). IFAC.

Fliess, M., Lévine, J., Martin, P., & Rouchon, P. (1995). Flatness and defect of non-linear systems: Introductory theory and examples. *International Journal of Control*, *61*(6), 1327–1361. doi:10.1080/00207179508921959

Forchetti, D., Garcia, G., & Valla, M. I. (2002). Vector control strategy for a doubly-fed stand-alone induction generator. In *Proceedings of IEEE 28th Annual Conference of the Industrial Electronics Society 2002*, (vol. 2, pp. 991-995). IEEE. doi:10.1109/IECON.2002.1185407

Fortuna, L., Frasca, M., & Xibilia, M. G. (2009). *Chua's Circuit Implementations: Yesterday, Today and Tomorrow*. Singapore: World Scientific.

Fourati, A., Feki, M., & Derbel, N. (2010). Stabilizing the unstable periodic orbits of a chaotic system using model independent adaptive time-delayed controller. *Nonlinear Dynamics*, *62*(3), 687–704. doi:10.1007/s11071-010-9754-y

Fouskakis, D., & Draper, D. (2002). Stochastic optimization. *International Statistical Review*, *70*(3), 315–349. doi:10.1111/j.1751-5823.2002.tb00174.x

Fradkov, A. L., Evans, R. J., & Andrievsky, B. R. (2006). Control of chaos: Methods and applications in mechanics. *Philosophical Transactions of the Royal Society A*, *364*(1846), 2279-2307.

Fradkov, A. L., & Evans, R. J. (2005). Control of chaos: Methods and applications in engineering. *Annual Reviews in Control*, *29*(1), 33–56. doi:10.1016/j.arcontrol.2005.01.001

Fradkov, A. L., & Pogromsky, A. Y. (1996). *Introduction to Control of Oscillations and Chaos*. Singapore: World Scientific.

Franklin, G. F., Powell, J. D., & Workman, M. L. (1997). *Digital Control of Dynamic Systems*. Boston: Addison-Wesley.

Frank, P. M. (1990). Analytical and qualitative model-based fault diagnosis - a survey and some new results. *European Journal of Control*, *2*(1), 6–28. doi:10.1016/S0947-3580(96)70024-9

Frank, P. M. (1990). Fault diagnosis in dynamic systems using analytical and knowledge based redundancy- a survey and some new results. *Automatica*, *26*(3), 459–474. doi:10.1016/0005-1098(90)90018-D

Fraser, A. M. (1989). Information and entropy in strange attractors. *IEEE Transactions on Information Theory*, *35*(2), 245–262. doi:10.1109/18.32121

Fridman, E., & Shaked, U. (2005). Stability and guaranteed cost control of uncertain discrete delay systems. *International Journal of Control*, *78*(4), 235–246. doi:10.1080/00207170500041472

FUBI. (2014). *Full Body Interaction Framework*. Retrieved from http://www.hcm-lab.de/fubi.html

Fu, S. M., Chen, Z. Y., & Zhou, Y. A. (2004). Chaos based random number generators. *Computer Research and Development*, *41*, 749–754.

Gaad, M., Berkouk, E. M., & Aliouane, K. (2000). Contribution à la réalisation de la commande d'un bras d'onduleur à trois niveaux. In *Proceedings of 2ème conference internationale sur l'électrotechnique*. Academic Press.

Gałkowski, K., Rogers, E., Xu, S., Lam, J., & Owens, D. H. (2002). LMIs a fundamental tool in analysis and controller design for discrete linear repetitive processes. *IEEE Transactions on Circuits and Systems. I, Fundamental Theory and Applications*, *49*(6), 768–778. doi:10.1109/TCSI.2002.1010032

Gálvez-Carrillo, M., & Kinnaert, M. (2011). Sensor fault detection and isolation in doubly-fed induction generators accounting for parameter variations. *Renewable Energy*, *36*(5), 1447–1457. doi:10.1016/j.renene.2010.10.021

Gantmacher, F. R. (1966). *Théorie des matrices*. Ed. Dunod.

Gao, Z. W., & So, A. T. (2003). A general doubly coprime factorization for descriptor systems. *Systems & Control Letters*, *49*(3), 213–224. doi:10.1016/S0167-6911(02)00325-0

Gao, Z., & Antsaklis, P. J. (1991). Stability of the pseudo-inverse method for reconfigurable control systems. *International Journal of Control*, *53*(3), 717–729. doi:10.1080/00207179108953643

Gao, Z., & Antsaklis, P. J. (1992). Reconfigurable control systems design via perfect model following. *International Journal of Control*, *56*(4), 783–798. doi:10.1080/00207179208934343

Garcia, M., Chatterjee, A., & Ruina, A. (2000). Efficiency, speed, and scaling of two-dimensional passive-dynamic walking. *Dynamics and Stability of Systems*, *15*(2), 75–99. doi:10.1080/713603737

Garcia, M., Chatterjee, A., Ruina, A., & Coleman, M. (1998). The simplest walking model: Stability, complexity, and scaling. *Journal of Biomechanical Engineering*, *120*(2), 281–288. doi:10.1115/1.2798313 PMID:10412391

Garcia-Pérez, L., Garcia-Alegre, M. C., Ribeiro, A., & Guinea, D. (2003). Fuzzy control for an approaching maneuver with a car like vehicle in outdoor environments. In *Proceedings of WAF03*. Alicante, Spain: WAF.

Gardini, L., Fournier-Prunaret, D., & Chargé, P. (2011). Border collision bifurcations in a two-dimensional piecewise smooth map from a simple switching circuit. *Chaos (Woodbury, N.Y.)*, *21*(2), 1–13. doi:10.1063/1.3555834 PMID:21721748

Gentle, J. E. (2003). *Random Number generation and Monte Carlo method.* New York: Springer.

Geromel, C. J., Peres, D. L. P., & Souza, S. R. (1992). Mixed control for continuous time linear systems. In *Proceedings of 31th IEEE Conference Decision and control* (pp. 3717-3722). IEEE.

Gerosa, A., Bernardini, R., & Pietri, S. (2001). A fully integrated chaotic system for the generation of truly random numbers. *IEEE Transactions on Circuits and Systems I, 49*(7), 993–1000. doi:10.1109/TCSI.2002.800833

Gertler, J. (1995). Diagnosing parametric faults: From parameter estimation to parity relations. In *Proceedings of the 1995 IEEE American Control Conference (ACC)*, (pp. 1615–1620). IEEE. Doi:10.1109/ACC.1995.529780

Gertler, J. (1998). *Fault Detection and Diagnosis in Engineering Systems*. Hershey, PA: CRC Press.

Ghadbane, I., & Benchouia, M. T. (2013). Feed back linearised control based three Phase shunt active power filter. *WSEAS Trans. on Systems, 7*(1), 18-25.

Gharedaghi, F., Jamali Di, H. M., & Khalili, A. (2011). Maximum Power Point Tracking of Variable speed Wind Generation System Connected to Permanent Magnet Synchronous Generator. *International Review on Modelling and Simulations, 4*(3), 1044–1049.

Gheraibia, Y., & Moussaoui, A. (2013). *Penguins Search Optimization Algorithm (PeSOA)* (Vol. 7906, pp. 222–231). Recent Trends in Applied Artificial Intelligence Lecture Notes in Computer Science.

Girard, A., & Pappas, G. J. (2006). Verification using simulation, Hybrid systems Computation and control. *LNCS, 3927*, 272–286.

Giri, F., & Bai, E.-W. (Eds.). (2010). *Block-oriented Nonlinear System Identification*. Lecture Notes in Control and Information Sciences.

Glaria Lopez, T. A., & Sbarbaro, D. (2011). Observer design for nonlinear processes with Wiener structure. In *Proceedings of the IEEE Conference on Decision and Control and European Control Conference* (pp. 12-15). Orlando, FL: IEEE.

Glover, F. (1986). Future paths for integer programming and links to artificial intelligence. *Journal of Computers and Operations Research, 13*(5), 533–549. doi:10.1016/0305-0548(86)90048-1

Glover, K., Limebeer, D., Doyle, J., Kasenally, E. M., & Sofonov, M. (1991). A characterization of all solutions to four block distance problem. *SIAM Journal on Control and Optimization, 29*(2), 283–324. doi:10.1137/0329016

Glück, T., Eder, A., & Kugi, A. (2013). Swing-up control of a triple pendulum on a cart with experimental validation. *Automatica, 49*(3), 801–808. doi:10.1016/j.automatica.2012.12.006

Gnanasaravanan, A., & Rajaram, M. (2012a). Articial neural network for monitoring the asymmetric half bridge dcdc converter. *International Journal of Electrical Power & Energy Systems, 43*(1), 788–792. doi:10.1016/j.ijepes.2012.05.057

Gnanasaravanan, A., & Rajaram, M. (2012b). Dynamic response analysis and output voltage control of asymmetric half bridge dcdc converter for low voltage applications. *International Journal of Electrical Power & Energy Systems, 43*(1), 774–778. doi:10.1016/j.ijepes.2012.06.038

Gobron, S., Ahn, J., Garcia, D., Silvestre, Q., Thalmann, D., & Boulic, R. (2012). An event-based architecture to manage virtual human non-verbal communication in 3d chatting environment. *Articulated Motion and Deformable Objects. LNCS, 7378*, 58–68. doi:10.1007/978-3-642-31567-1_6

Goebel, R. (2004). Stabilizing linear systems with saturation through optimal control. In *Proceedings ofIEEE Conference on Decision and Control. Decision and Control*. IEEE.

Golea, N., Golea, A., & Benmahammed, K. (2003). Stable indirect fuzzy adaptive control. *Fuzzy Sets and Systems, 137*(3), 353–366. doi:10.1016/S0165-0114(02)00279-8

Golub, G. H., & Van Loan, C. F. (1996). *Matrix Computation*. Baltimore, MD: Johns Hopkins University Press.

González, I., Ramiro, A., Calderón, M., Calderón, A. J., & González, J. F. (2012). Estimation of the state-of-charge of gel lead-acid batteries and application to the control of a stand-alone wind-solar test-bed with hydrogen support. *International Journal of Hydrogen Energy*, *37*(15), 11090–11103. doi:10.1016/j.ijhydene.2012.05.001

Goswami, A., Thuilot, B., & Espiau, B. (1996). *Compass-like biped robot. Part I: Stability and bifurcation of passive gaits*. France: INRIA.

Goswami, A., Thuilot, B., & Espiau, B. (1998). A study of the passive gait of a compass-like biped robot: Symmetry and chaos. *The International Journal of Robotics Research*, *17*(12), 1282–1301. doi:10.1177/027836499801701202

Grag, S. (1993). Robust integrated flight/propulsion control design for a STOVL aircraft using H∞ control design techniques. *Automatica*, *29*(1), 129–145. doi:10.1016/0005-1098(93)90177-U

Grassberger, P., & Procaccia, I. (1983). Characterization of strange attractors. *Physical Review Letters*, *50*(5), 346–349. doi:10.1103/PhysRevLett.50.346

Gratch, J., Wang, N., Okhmatovskaia, A., Lamothe, F., Morales, M., Werf, R. J. V. D., & Morency, L. P. (2007). Can virtual humans be more engaging than real ones? *Human-Computer Interaction, HCI Intelligent Multimodal Interaction Environments. LNCS*, *4552*, 286–297. doi:10.1007/978-3-540-73110-8_30

Grebogi, C., & Lai, Y. C. (1997). Controlling chaotic dynamical systems. *Systems & Control Letters*, *31*(5), 307–312. doi:10.1016/S0167-6911(97)00046-7

Grebogi, C., & Yorke, J. (1997). *The impact of chaos on science and society*. United Nations University Press.

Gritli, H., Belghith, S., & Khraief, N. (2012a). Intermittency and interior crisis as route to chaos in dynamic walking of two biped robots. International Journal of Bifurcation and Chaos, 22(3).

Gritli, H., Belghith, S., & Khraief, N. (2012b). Cyclic-fold bifurcation and boundary crisis in dynamic walking of biped robots. International Journal of Bifurcation and Chaos, 22(10).

Gritli, H., Belghith, S., & Khraief, N. (2014a). Control of Chaos in an Impact Mechanical Oscillator. *Proceedings-Copyright IPCO*, 116-122.

Gritli, H., Khraeif, N., & Belghith, S. (2011a). A Period-Three Passive Gait Tracking Control for Bipedal Walking of a Compass-Gait Biped Robot. In *Proceedings of the IEEE-2011 International Conference on Communications, Computing and Control Applications*, (pp. 01-06). Hammamet, Tunisia: IEEE.

Gritli, H., Khraeif, N., & Belghith, S. (2011b). Cyclic-fold bifurcation in passive bipedal walking of a compass-gait biped robot with leg length discrepancy. In *Proceedings ofIEEE International Conference on Mechatronics*, (pp. 851-856). Istanbul, Turkey: IEEE. doi:10.1109/ICMECH.2011.5971233

Gritli, H., Khraeif, N., & Belghith, S. (2011c). Falling of a Passive Compass-Gait Biped Robot Caused by a Boundary Crisis. In *Proceedings of the 4th Chaotic Modeling and Simulation International Conference* (pp. 155-162). Crete, Greece: Academic Press.

Gritli, H., Khraief, N., & Belghith, S. (2014). Further Analysis of the Period-Three Route to Chaos in Passive Dynamic Walking of a Compass-Gait Biped Robot. *Proceedings-Copyright IPCO*, 123-130.

Gritli, H., Khraief, N., & Belghith, S. (2014b). Further Analysis of the Period-Three Route to Chaos in Passive Dynamic Walking of a Compass-Gait Biped Robot. *Proceedings-Copyright IPCO*, 123-130.

Gritli, H., Khraeif, N., & Belghith, S. (2012c). Period-three route to chaos induced by a cyclic-fold bifurcation in passive dynamic walking of a compass-gait biped robot. *Communications in Nonlinear Science and Numerical Simulation*, *17*(11), 4356–4372. doi:10.1016/j.cnsns.2012.02.034

Gritli, H., Khraeif, N., & Belghith, S. (2013). Chaos control in passive walking dynamics of a compass-gait model. *Communications in Nonlinear Science and Numerical Simulation*, *18*(8), 2048–2065. doi:10.1016/j.cnsns.2012.12.009

Gu, D. W., Petkov, P. H., & Konstantinov, M.M. (2005). Robust Control Design with MATLAB (3rd ed.). New York: Springer-Verlag.

Guan, Y., & Saif, M. (1991). A novel approach to the design of unknown input observers. *IEEE Transactions on Automatic Control*, *36*(5), 632–635.

Guardabassi, G. O., & Savaresi, S. M. (2000). Virtual reference direct design method: An off-line approach to data-based control system design. *IEEE Transactions on Automatic Control*, *45*(5), 954–959. doi:10.1109/9.855559

Guardabassi, G. O., & Savaresi, S. M. (2001). Approximate linearization via feedback - An overview. *Automatica*, *37*(1), 1–15. doi:10.1016/S0005-1098(00)00117-5

Guemghar K., Srinivasan, B. & Bonvin,D. (2005). Analysis of cascade structure with predictive control and feedback linearisation. In *Proceedings of the 2005 Control Theory and Applications* (pp. 317-324). IEEE. DOI:.10.1049/ip-cta:20041309

Guesmi, K., Essounbouli, N., & Hamzaoui, A. (2008). Systematic design approach of fuzzy PID stabilizer for dcdc converters. *Energy Conversion and Management*, *49*(10), 2880–2889. doi:10.1016/j.enconman.2008.03.012

Guide, M. (1985). Concept for a high-performance random number generator based on physical random phenomena. *Frequenz*, *39*, 187–190.

Guo, J., Yao, B., Chen, Q., & Wu, X. (2009). High Performance Adaptive Robust Control for Nonlinear System with Unknown Input Backlash. In *Proceedings of Joint 48th IEEE Conference on Decision and Control and 28th Chinese Control Conference*. Shanghai, China: IEEE.

Gustafson, H., Dawson, H. E., Nielsen, L., & Caelli, W. (1994). A computer package for measuring the strength of encryption algorithms. *Journal of Computer Security*, *13*(8), 687–697. doi:10.1016/0167-4048(94)90051-5

Gutiérrez, E., & Arrowsmith, D. K. (2003). Control of a double impacting mechanical oscillator using displacement feedback. *International Journal of Bifurcation and Chaos in Applied Sciences and Engineering*, *14*(9), 3095–3113. doi:10.1142/S021812740401120X

Gutierrez, H. M., & Ro, P. I. (1998). Sliding-mode control of a nonlinear-input system: Application to a magnetically levitated fast-tool servo. *IEEE Transactions on Industrial Electronics*, *45*(6), 921–927. doi:10.1109/41.735336

Haber, R., & Keviczky, L. (1976). Identification of nonlinear dynamics systems – A survey paper. In *Proceedings of the* 4th*International Federation of Automatic Control (IFAC) Symposium on Identification and System Parameter Estimations* (pp. 62-112). Tbilisi.

Haber, R., & Keviczky, L. (1999). *Nonlinear System Identification: Input-output Modelling Approach*. Boston: Kluwer.

Haber, R., & Unbehauen, H. (1990). Structure identification of nonlinear dynamic system – a survey on input/output approaches. *Automatica*, *26*(4), 651–677.

Haddad, W. M., & Chellaboina, V. (n.d.). *Nonlinear Dynamical Systems and Control - A Lyapunov Based Approach*. Princeton University Press.

Hamill, D., & Jefferies, D. (1988). Subharmonics and chaos in a controlled switched-mode power converter. *IEEE Transactions on Circuits and Systems. I, Fundamental Theory and Applications*, *35*(8), 1059–1061. doi:10.1109/31.1858

Hamza, R., Feki, M., Derbel, N., Robert, B. G. M., & Aroudi, E. L. (2011). Controller design and analysis for a two-cell dc-dc converter in the presence of saturation. *International Journal of Bifurcation and Chaos in Applied Sciences and Engineering*, *21*(1), 341–361. doi:10.1142/S0218127411028465

Hansen, A. D., Gabriele, M., Poul, S., & Torsten, L. (2007). Co-ordinated voltage control of DFIG wind turbines in uninterrupted operation during grid faults. *Wind Energy (Chichester, England)*, *10*(1), 51–68. doi:10.1002/we.207

Hansen, A. D., Iov, F., Blaabjerg, F., & Hansen, L. H. (2004). Review of contemporary wind turbine concepts and their market penetration. *Journal of Wind Energy*, *28*(3), 247–263.

Harisha, S. K., Ramkanth, K. P., Krishna, M., & Sharma, S. C. (2008). Fuzzy logic reasoning to control mobile robot on pre-defined strip path. *World Academy of Science. Engineering and Technology*, *2*(6), 618–622.

Harmuth, H. F. (1969). Applications of Walsh functions in communications. *IEEE Spectrum*, *6*(11), 82–91. doi:10.1109/MSPEC.1969.5214175

Hasan, M., & Mishra, P. (2012). Robust Gesture Recognition Using Gaussian Distribution for Features Fitting. *International Journal of Machine Learning and Computing*, *2*(3), 266–273. doi:10.7763/IJMLC.2012.V2.128

Hasegawa, K. (1996). Modeling, control and deadlock avoidance of flexible manufacturing systems. In *Proceedings of 11th Brazilian Congress of Automatics*, (pp. 37-51). Academic Press.

Hasselblatt, B., & Katok, A. (2003). *A first course in dynamics: With a panorama of recent developments*. Cambridge, UK: Cambridge University Press. doi:10.1017/CBO9780511998188

Hauser, J., Sastry, S., & Meyer, G. (1989). Nonlinear controller design for flight control systems. In Proceedings of the 1989 IFAC symposium on Nonlinear Control Systems Design, (pp. 136–141). IFAC.

Hauser, J. (1991). Nonlinear control via uniform system approximation. *Systems & Control Letters*, *17*(2), 145–154. doi:10.1016/0167-6911(91)90041-C

Hauser, J., Sastry, S., & Kokotovic, P. (1992). Nonlinear control via approximates input-output linearization: The ball and beam example. *IEEE Transactions on Automatic Control*, *37*(3), 392–398. doi:10.1109/9.119645

Haweel T.I., (2004). A simple variable step size LMS adaptive algorithm. *International Journal of Circuit Theory and Applications*, (32), 523-536.

Haykin, S. S. (1972). A unified treatment of recursive digital filtering. *IEEE Transactions on Pattern Analysis and Machine Intelligence*, *3*, 113–116.

Heckman, D. (2008). *A Small World. Smart Houses and the Dream of the Perfect day*. London, UK: Duke University Press. doi:10.1215/9780822388845

Hespanha, J. P., & Morse, A. S. (1999). Stability of switched systems with average dwell-time. In *Proceedings ofIEEE Conferenceon Decision and Control* (pp. 2655–2660). Phoenix, AZ: IEEE. doi:10.1109/CDC.1999.831330

Hien, L. V., Ha, Q. P., & Phat, V. N. (2009). Stability and stabilization of switched linear time-delay systems. *Journal of the Franklin Institute*, *346*(6), 611–625. doi:10.1016/j.jfranklin.2009.03.001

Hilborn, R. C. (2000). *Chaos and nonlinear dynamics: An introduction for scientists and engineers*. Oxford, UK: Oxford University Press. doi:10.1093/acprof:oso/9780198507239.001.0001

Hippe, P. (1994). Strictly Doubly Coprime Factorizations and all Stabilizing Compensators Related to Reduced-order Observer. *Automatica*, *30*(12), 1955–1959. doi:10.1016/0005-1098(94)90056-6

Hiskens, I. A., & Pai, M. A. (2000). Trajectory sensitivity analysis of hybrid systems. *IEEE Transactions on Circuits and Systems-Part I. Fundamental Theory and Applications*, *47*(2), 204–220.

Hoang, T.-M., Nguyen, T. D., Nguyen, V. D., Chedjou, J. C., & Kyamakya, K. (2009) Design and simulation of circuit for synchronization of multidelay feeback systems. In *Proc. 25th International Symposium on Theoretical Engineering (ISTET 2009)*, (pp. 1-4). Lubeck, Germany: ISTET.

Holland, J. H. (1998). *Emergence: From chaos to order*. Oxford, UK: Oxford University Press.

Holman, W. T., Connelly, J. S., & Downlatabadi, A. B. (1997). An integrated analog-digital random noise source. *IEEE Transactions on Circuits and Systems I*, *44*(6), 521–528. doi:10.1109/81.586025

Homaifar, A., & McCormick, E. (1995). Simultaneous design of membership functions and rule sets for fuzzy controllers using genetic algorithms. *IEEE Transactions on Fuzzy Systems*, *3*(2), 129–139. doi:10.1109/91.388168

Hong, X., & Harris, C. J. (2001). Nerofuzzy design and model construction of nonlinear dynamical processes from data. *IEE Proceedings Control Theory and Applications*, *148*(6), 530–538.

Hoong, C. L., Melvyn, S., & Kwong, M. T. (2003). Vehicle routing problem with time windows and a limited number of vehicles. *European Journal of Operational Research*, *148*(3), 559–569. doi:10.1016/S0377-2217(02)00363-6

Hopfensberger, B. (1999). Stator flux oriented control of a cascaded doubly-fed induction machine. *Electric Power Applications*, *146*(6), 597–605. doi:10.1049/ip-epa:19990590

Hopfensperger, B., Atkinson, D. J., & Lakin, R. A. (2000). Stator-flux-oriented control of a doubly-fed induction machine with and without position encoder. *Electric Power Applications*, *147*(4), 241–250. doi:10.1049/ip-epa:20000442

Horbelt, W., Timmer, J., & Voss, H. U. (2002). Parameter estimation in nolinear delayed feedback systems from noisy data. *Physics Letters [Part A]*, *299*(5-6), 513–521. doi:10.1016/S0375-9601(02)00748-X

Hörmann, A., Meier, W., & Schloen, J. (1991). A control architecture for an advanced fault tolerant robot system. *Robotics and Autonomous Systems*, *7*(2-3), 211–225. doi:10.1016/0921-8890(91)90043-K

Hoshuyama, O., Goubran, R. A., & Sugiyama, A. (2004, May). A generalized proportionate variable step-size algorithm for fast changing acoustic environments. In *Proceedings of ICASSP*, (pp. 161–164). ICASSP.

Howell, G. W., & Baillieul, J. (1998). Simple controllable walking mechanisms which exhibit bifurcations.*IEEE Conference on Decision and Control*, (pp. 3027–3032). Tampa, FL: IEEE. doi:10.1109/CDC.1998.757956

Hrizi, O., Boumedyen, B., Christophe, A., & Mohamed, N. A. (2013). Fast Adaptive Fault Estimation Algorithm: Application to Unicycle Robot. In *Proceedings of the 2nd International Conference on Control and Fault-Tolerant Systems (SYSTOL)*. Nice, France: SYSTOL.

Hsieh, C.-S. (2002). Performance gain margins of the two-stage LQ reliable control. *Automatica*, *38*(11), 1985–1990. doi:10.1016/S0005-1098(02)00114-0

Hsu, F. Y., & Fu, L. C. (2000). Intelligent robot deburring using adaptive fuzzy hybrid position/force control. *IEEE Transactions on Robotics and Automation*, *16*(4), 325–335. doi:10.1109/70.864223

Hsu, K. C. (1999). Decentralized sliding mode controller for uncertain time-delayed systems with series nonlinearities. *Journal of Dynamic Systems, Measurement, and Control*, *121*(4), 708–713. doi:10.1115/1.2802539

Hsu, K. C., Wang, W. Y., & Lin, P. Z. (2004). Sliding mode control for uncertain nonlinear systems with multiple inputs containing sector nonlinearities and deadzones. *IEEE Transactions on Systems, Man and Cybernetic. Part-B*, *34*(1), 374–380.

Hsu, L., Costa, R. R., & Lizarralde, F. (2007). Lyapunov/passivity-based adaptive control of relative degree two MIMO systems with an application to visual servoing. *IEEE Transactions on Automatic Control*, *52*(2), 364–371. doi:10.1109/TAC.2006.890381

Huang, C. Y., & Stengel, R. (1990). Restructuable control using proportional integral implicit model following. *Journal of Guidance, Control, and Dynamics*, *13*(2), 303–309. doi:10.2514/3.20550

Huaping, L., Wang, S., & Gang, H. (2004). Pseudo-random number generator based on coupled map lattices. *International Journal of Modern Physics B*, *18*(17n19), 2409–2414. doi:10.1142/S0217979204025440

Hu, B., & Michel, A. N. (2000). Stability Analysis of Digital Feedback Control Systems with Time-Varying Sampling Periods. *Automati*, *36*(3), 897–905. doi:10.1016/S0005-1098(99)00217-4

Hübinger, B., Doerner, R., Martienssen, W., Herdering, M., Pitka, R., & Dressler, U. (1994). Controlling chaos experimentally in systems exhibiting large effective Lyapunov exponents. *Physical Review E: Statistical Physics, Plasmas, Fluids, and Related Interdisciplinary Topics*, *50*(2), 932–948. doi:10.1103/PhysRevE.50.932 PMID:9962053

Hu, H. Y., & Wang, Z. H. (2002). *Dynamics of Controlled Mechanical Systems with Delayed Feedback*. Berlin: Springer-Verlag. doi:10.1007/978-3-662-05030-9

Hu, H., Jiang, B., & Yang, H. (2013). Non-fragile H2 reliable control for switched linear systems with actuator faults. *Signal Processing*, *93*(7), 1804–1812. doi:10.1016/j.sigpro.2013.01.011

Hung, J. C., Liu, C. C., & Chan, P. Y. (1980). An algebraic method for system parameter identification. In *Proceedings of the 14th Asilomar Conference on Circuits, Systems and Computers*, (pp. 277-279). Pacific Grove, CA: IEEE.

Hung, J. Y., Gao, W., & Hung, J. C. (1993). Variable structure control: A survey. *IEEE Transactions on Industrial Electronics*, *40*(1), 2–21. doi:10.1109/41.184817

Hunt, L. R., Su, R., & Mayer, G. (1983). Global transformations of nonlinear systems. *IEEE Transactions on Automatic Control*, *28*(1), 24–31. doi:10.1109/TAC.1983.1103137

Hu, Y., Liao, X., Wong, K., & Zhou, Q. (2009). A true random number generator based on mouse movement and chaotic cryptography. *Chaos, Solitons, and Fractals*, *40*(5), 2286–2293. doi:10.1016/j.chaos.2007.10.022

Ichalal, D., Marx, B., Ragot, J., & Maquin, D. (2009a). *Simultaneous state and unknown inputs estimation with PI and PMI observers for Takagi Sugeno model with unmeasurable premise variables.* Paper presented at the 17th Mediterranean Conference on Control and Automation. Thessaloniki, Greece. Doi:10.1109/MED.2009.5164566

Ichalal, D., Marx, B., Ragot, J., & Maquin, D. (2009b). *State and unknown input estimation for nonlinear systems described by Takagi-Sugeno models with unmeasurable premise variables.* Paper presented at the 17th Mediterranean Conference on Control and Automation. Thessaloniki, Greece. Doi:10.1109/MED.2009.5164542

Ikeda, M., Lee, T., & Uezato, E. (2000). A Strict LMI Condition for Control of Descriptor Systems. In *Proceedings of 38th IEEE Conference on Decision and Control* (pp. 4092-4097). Sydney: IEEE.

Ikeda, K., Daido, H., & Akimoto, O. (1980). Optical turbulence: Chaotic behavior of transmitted light from a ring cavity. *Physical Review Letters*, *45*(9), 709–712. doi:10.1103/PhysRevLett.45.709

Ikeda, K., & Matsumoto, K. (1987). High-dimensional chaotic behavior in systems with time-delayed feedback. *Physica D. Nonlinear Phenomena*, *29*(1-2), 223–235. doi:10.1016/0167-2789(87)90058-3

Indri, M., & Tornambe, A. (1997). Application of a PD controller on two mating gears with elasticity and backlash. In *Proceedings of the 36th Conference on Dectston & Control. Decision and Control*. San Diego, CA: IEEE. doi:10.1109/CDC.1997.657892

Iqbal, S., Bhatti, A. I., Akhtar, M., & Ullah, S. (2007). Design and Robustness Evaluation of an H Infinity Loop Shaping Controller for a 2DOF Stabilized Platform. In *Proceedings of the European Control Conference*.Kos, Greece: Academic Press.

Iqbal, S., Zang, X. Z., Zhu, Y. H., & Zhao, J. (2014). Bifurcations and chaos in passive dynamic walking: A review. *Robotics and Autonomous Systems*, *62*(6), 889–909. doi:10.1016/j.robot.2014.01.006

Isermann, R. (1984). Process fault detection based on modeling and estimation methods—a survey. *Automatica*, *20*(4), 387–404.

Isermann, R. (2005). Model-based fault detection and diagnosis status and applications. *Annual Reviews in Control*, *29*(1), 71–85. doi:10.1016/j.arcontrol.2004.12.002

Isermann, R. (2006). *Fault-Diagnosis Systems: An Introduction from Fault Detection to Fault Tolerance.* doi:10.1007/3-540-30368-5

Ishihara, Y. J., Terra, M. H., & Sales, R. M. (2003). The full information and state feedback H_2. *Automatica*, *39*(3), 391–402. doi:10.1016/S0005-1098(02)00243-1

Ishii, H., & Francis, B. A. (2002). Stabilizing a linear system by switching control with dwell time. *IEEE Transactions on Automatic Control*, *47*(12), 1962–1973. doi:10.1109/TAC.2002.805689

Isidori, A. (1990). Nonlinear control systems. London: Spring-Verlag.

Isidori, A., & Byrnes, C. I. (1990). Output regulation of nonlinear systems. *IEEE Transactions on Automatic Control*, *35*(2), 131–140. doi:10.1109/9.45168

Ivanescu, M. (2001). Moving target interception for walking robot by fuzzy controller. In *Proceedings of the Fourth International Conference on Climbing and Walking Robotics*, (pp. 363-376). Karlsruhe, Germany: Academic Press.

Iztok J., Fister Y., Xin-She F., Iztok B. and Duˇsan F., (2013). A Brief Review of Nature-Inspired Algorithms for Optimization *ELEKTROTEHNI ˇSKI VESTNIK*. 80(3), 116–122.

Jaballi, A., Kermani, M., & Sakly, A. (2013). A New Stability Analysis for Discrete-Time Switched Time-Delay Systems. In *Proceedings of International Conference on Control, Engineering & Information Technology* (pp. 200-204). Sousse, Tunisia: Academic Press.

Janczak, A. (2005). *Identification of Nonlinear Systems Using Neural Networks and Polynomial Models: A Block-Oriented Approach*. Lecture Notes in Control and Information Sciences.

Jang, J. O., Lee, P. G., Chung, H. T., & Jeon, G. J. (2003). Output Backlash Compensation of Systems Using Fuzzy Logic. In *Proceedings ofAmerican Control Conference*. Academic Press. doi:10.1109/ACC.2003.1243449

Jaulin, L., Kieffer, M., Didrit, O., & Walter, E. (2001). Applied interval analysis. London: Springer.

Jaulin, L. (2001). *Applied interval analysis: with examples in parameter and state estimation, robust control and robotics*. Springer Verlag.

Jayaraman, Y., & Ravindran, U. (2008). Fpga implementation of predictive control strategy for power factor correction. *In proceedings of World Academy of Science. Engineering and Technology, 15*, 199–204.

Jiang, J. (1994). Design of reconfigurable control system using eigenstructure assignments. *International Journal of Control, 59*(2), 395–410. doi:10.1080/00207179408923083

Jiang, J. (2005). Fault tolerant control systems - an introductory overview. *Automatica, 31*(1), 161–174.

Jiang, J., & Zhang, Y. (2006). Accepting performance degradation in fault-tolerant control system design. *IEEE Transactions on Control Systems Technology, 14*(2), 284–292. doi:10.1109/TCST.2005.860515

Jiang, J., & Zhao, Q. (2000). Design of reliable control systems possessing actuator redundancies. *Journal of Guidance, Control, and Dynamics, 23*(4), 709–718. doi:10.2514/2.4587

Jiang, S., Ji, Z., & Shen, Y. (2014). A novel hybrid particle swarm optimization and gravitational search algorithm for solving economic emission load dispatch problems with various practical constraints. *International Journal of Electrical Power & Energy Systems, 55*(1), 628–644. doi:10.1016/j.ijepes.2013.10.006

Jianming, X., Mingxuan, S., & Li, Y. (2008). LMI-Based Synthesis of Robust Iterative Learning Controller with Current Feedback for Linear Uncertain Systems. *International Journal of Control, Automation, and Systems, 6*(2), 171–179.

Ji, M., Zhang, Z., Biswas, G., & Sarkar, N. (2003). Hybrid fault adaptive control of a wheeled mobile robot. *IEEE Transactions on Mechatronics, 8*(2), 226–233. doi:10.1109/TMECH.2003.812823

Jing, Z., & Wen, C. (2007). Adaptive Inverse Control of a Magnetic Suspension System with Input Backlash. In *Proceedings of16th IEEE International Conference on Control Applications. Control Applications*. IEEE.

Jin, L., & Kim, Y. C. (2008). Fixed, low-order controller design with time response specifications using nonconvex optimization. *ISA Transactions, 47*(4), 429–438. doi:10.1016/j.isatra.2008.05.001 PMID:18606409

Jin, L., Lu, Q. S., & Twizell, E. H. (2006). A method for calculating the spectrum of Lyapunov exponents by local maps in non-smooth impact-vibrating systems. *Journal of Sound and Vibration, 298*(4-5), 1019–1033. doi:10.1016/j.jsv.2006.06.035

Jin-Sung, K., Jonghyun, J., & Hoon, H. (2011). Design of adaptive PID for pitch control of large wind turbine generator. *Proceedings of EEEIC, 2011*, 1–4.

Johnson, R., O'Hara, K., Sellen, A., Cousins, C., & Criminisi, A. (2011). Exploring the potential for touchless interaction in image-guided interventional radiology. In *Proceedings of CHI 2011*, (pp. 3323–3332). ACM. doi:10.1145/1978942.1979436

Jouili, K., Jerbi, H., & Benhadj Braiek, N. (2010). An advanced fuzzy logic gain scheduling trajectory control for nonlinear systems. *Journal of Process Control, 20*(4), 426–440. doi:10.1016/j.jprocont.2010.01.001

Kai, T., & Shintani, T. (2011). A Gait Generation Method for the Compass-type Biped Robot on Slopes via Discrete Mechanics. In *Proceedings of50th IEEE Conference on Decision and Control and European Control Conference (CDC-ECC)*, (pp. 675-681). Orlando, FL: IEEE. doi:10.1109/CDC.2011.6161060

Kailath, T. (1980). *Linear systems*. Prentice Hall.

Kalagnanam, J. R. (1994). Controlling Chaos: The Example of an Impact Oscillator. *Journal of Dynamic Systems, Measurement, and Control, 116*(3), 557–564. doi:10.1115/1.2899253

Kalman, R. E. (1960). A new approach to linear filtering and prediction problems. *Journal of Basic Engineering*, *82*(1), 35–45.

Kalman, R. E., & Bucy, R. S. (1961). New results in linear filtering and prediction theory. *Journal of Basic Engineering*, *83*(1), 95–108.

Kamal, E., Koutb, M., Sobaih, A., & Abozalam, B. (2010). An intelligent maximum power extraction algorithm for hybrid wind-diesel-storage system. *Electrical Power and Energy Systems*, *32*(3), 170–177. doi:10.1016/j.ijepes.2009.07.005

Kamoun, S. (2003). *Contribution to identification and adaptive control of complex systems* (in French). (Ph.D. dissertation). National School of Engineering of Sfax, University of Sfax, Sfax, Tunisia.

Kamoun, S. (2007). Design of parametric and state estimation algorithms for stochastic systems. *International Journal of Sciences and Techniques of Automatic Control & Computer Engineering*, *1*(2), 120–135.

Kamoun, S. (2008). Parametric estimation of non-linear stochastic systems described by input-output mathematical models. *International Review of Automatic Control*, *1*(4), 422–434.

Kang, W. (1994). Approximate linearization of nonlinear control systems. *Systems & Control Letters*, *23*(5), 43–52. doi:10.1016/0167-6911(94)90080-9

Kanter, J. R., Soroush, M., & Seider, W. D. (2001). Continuous time, nonlinear feedback control of stable processes. *Industrial & Engineering Chemistry Research*, *40*(9), 2069–2078. doi:10.1021/ie000406k

Kanter, J. R., Soroush, M., & Seider, W. D. (2002). Nonlinear feedback control of multivariable nonminimum-phase processes. *Journal of Process Control*, *12*(6), 667–686. doi:10.1016/S0959-1524(01)00059-2

Kaoubaâ, K., Pelaez-Restrepo, J., Feki, M., Robert, B., & El Aroudi, A. (2012). Improved static and dynamic performances of a two-cell dc-dc buck converter using a digital dynamic time-delayed control. *Int. J. Circ. Theor. Appl.*, *40*(4), 395–407. doi:10.1002/cta.735

Kaplan, J. L., & Yorke, J. A. (1978). Chaotic behavior of multidimensional difference equations, in functional differential equations and approximation of fixed points. *Lecture Notes in Mathematics*, *730*, 228–237. doi:10.1007/BFb0064320

Karimi-Gh, M., & Iravani, M. R. (2004). A Method for Synchronization of Power Electronic Converters in Polluted and Variable-Frequency Environments. *IEEE Transactions on Power Systems*, *19*(3), 1263–1270. doi:10.1109/TPWRS.2004.831280

Karimi-Gh, M., Khajehoddin, S. A., Jain, P. K., & Bakhshai, A. (2012). Problems of Startup and Phase Jumps in PLL Systems. *IEEE Transactions on Power Electronics*, *27*(4), 1830–1838. doi:10.1109/TPEL.2011.2169089

Karimi, H., Sheikholeslami, A., Livani, H., & Karimi, D. (2009). Fuzzy Logic control of doubly fed induction generator wind turbine. *World Applied Sciences Journal*, *6*(4), 499–508.

Kashiwagi, H., Harada, H., & Toyozawa, Y. (2004). Identification of nonlinear mechatronic servo system having backlash. *SICE Annual Conference in Sapporo*. In *Proceedings of SICE 2004 Annual Conference*. IEEE.

Kawamoto, S., Tada, K., Onoe, N., Ishigame, A., & Taniguchi, T. (1992). *Construction of exact fuzzy system for nonlinear system and its stability analysis*. Paper presented at the 8th Fuzzy System Symposium. Hiroshima, Japan.

Kawashima, H., Wardi, Y., & Taylor, D., & M., E. (2012). Optimal switching control of a step-down dc-dc converter. In *Proceedings ofAmerican Control Conference*. Montreal, Canada: Academic Press.

Kaygisiz, B. H., Erkmen, I., & Erkmen, A. M. (2006). Intelligent analysis of chaos roughness in regularity of walk for a two legged robot. *Chaos, Solitons, and Fractals*, *29*(1), 148–161. doi:10.1016/j.chaos.2005.08.047

Kellogg, B., Talla, V., & Gollakota, S. (2014). Bringing Gesture Recognition to All Devices. In *Proceedings of NSDI'14*, (pp. 303-316). NSDI.

Kellogg, W. K., Nehrir, M. H., Venkataramanan, G. G., & Greez, V. (1998). Generation Unit Sizing and Cost Analysis for Stand-Alone Wind, Photovoltaic, and Hybrid Wind-PV Systems. *IEEE Transactions on Energy Conversion*, *13*(1), 70–75. doi:10.1109/60.658206

Kennedy, J., & Russell, E. (1995, Nov/Dec). Particle Swarm Optimization. Proceedings of the *IEEE International Conference on Neural Networks*, vol.4, pp.1942-1948, Perth, WA.

Kennedy, M. P., Rovatti, R., & Setti, G. (2000). *Chaotic electronics in Telecommunications*. CRC Press.

Kermani, M., Jaballi, A., & Sakly, A. (2014). A. A New Stability Analysis for Discrete-time Switched Time-delay Systems. *International Journal of Control. Energy and Electrical Engineering*, *1*(1), 23–30.

Kermani, M., & Sakly, A. (2014). Stability analysis for a class of switched nonlinear time-delay systems. *Systems. Science. Control Engineering*, *2*(1), 80–89.

Kesraoui, M., Toutaoui, S., & Azira, A. (2009). Wind energy conversion system based on a doubly fed induction generator: Study and simulation. In *Proceedings of 2nd EPE Wind Energy Chapter Seminar*. Stockholm, Sweden: EPE.

Kesraoui, M., Chaib, A., Meziane, A., & Boulezaz, A. (2014). Using a DFIG based wind turbine for grid current harmonics filtering. *Energy Conversion and Management*, *78*, 968–975. doi:10.1016/j.enconman.2013.07.090

Khalil, H. K. (1996). *Nonlinear Systems*. Prentice Hall.

Khatib, L., Morris, P., Morris, R., Rossi, F., Sperduti, A., & Venable, K. B. (2007). Solving and learning a tractable class of soft temporal constraints: Theoretical and experimental results. *AI Communication*, *20*, 181–209.

Khatib, O. (1986). Real-time obstacle avoidance for manipulators and mobile robots. *The International Journal of Robotics Research*, *5*(1), 90–98. doi:10.1177/027836498600500106

Khedher, A., & Othman, K. B. (2010). Proportional Integral Observer Design for State and Faults Estimation: Application to the Three Tanks System. *International Review of Automatic Control, 3*(2), 115-124.

Khedher, A., Benothman, K., Maquin, D., & Benrejeb, M. (2008). *State and unknown input estimation via a proportional integral observer with unkbown inputs.* Paper presented at the 9th international conference on Sciences and Techniques of Automatic control and computer engineering, STA. Sousse, Tunisia.

Khedher, A., Benothman, K., Maquin, D., & Benrejeb, M. (2009). *State and sensor faults estimation via a proportional integral observer.* Paper presented at the 6th International Multi-Conference on Systems, Signals and Devices. Djerba, Tunisia. Doi:10.1109/SSD.2009.4956795

Kheriji, A., Bouani, F., & Ksouri, M. (2011a). A GGP approach to solve non convex min-max predictive controller for a class of constrained MIMO systems described by state-space models. *International Journal of Control, Automation, and Systems*, *9*(3), 452–460. doi:10.1007/s12555-011-0304-2

Kheriji, A., Bouani, F., & Ksouri, M. (2011c). A Microcontroller Implementation of Constrained Model Predictive Control. *International Journal of Electrical and Electronics Engineering*, *5*(3), 199–206.

Kheriji, A., Bouani, F., Ksouri, M., & Ben Ahmed, M. (2011b). A Microcontroller Implementation of Model Predictive Control. *World Academy of Science. Engineering and Technology*, *53*(1), 68–74.

Khomfoi, S., & Tolbert, L. M. (2007). *Multilevel Power Converter. Power Electronics Handbook* (2nd ed., pp. 45–482). Elsevier.

Kim, Y., Kim, K., & Manabe, S. (2004). Sensitivity of time response to characteristic ratios. In *Proceeding of the 2004 American Control Conference*, (pp. 2723-2728). Academic Press.

Kim, S., Campbell, S. A., & Liu, X. (1999). Stability of a class of linear switching systems with time delay. *IEEE Transactions on Circuits and Systems*, *53*(1), 384–393.

Kim, S.-E., Kong, S.-J., & Song, W.-J. (2009). An affine projection algorithm with evolving order. *IEEE Signal Processing Letters*, *16*(11), 937–940. doi:10.1109/LSP.2009.2027638

Kim, S.-K., Jeon, J.-H., Cho, C.-H., Ahn, J.-B., & Won, S.-H. (2008). Dynamic Modeling and Control of a Grid-Connected Hybrid Generation System with Versatile Power Transfer. *IEEE Transactions on Industrial Electronics*, *55*(4), 1677–1688. doi:10.1109/TIE.2007.907662

Kim, Y. C., Keel, L. H., & Bhattacharyya, S. P. (2003). Transient response control via characteristic ratio assignment. *IEEE Transactions on Automatic Control*, *48*(12), 2238–2244. doi:10.1109/TAC.2003.820153

Kistler, F., Endrass, B., Damian, I., Dang, C., & André, E. (2012). Natural interaction with culturally adaptive virtual characters. Journal on Multimodal User Interfaces, 6(1-2), 39-47. Doi:10.1007/s12193-011-0087-z

Kistler, F., Sollfrank, D., Bee, N., & André, E. (2011). Full Body Gestures enhancing a Game Book for Interactive Story Telling. Lecture Notes in Computer Science, 7069, 207-218. Doi:10.1007/978-3-642-25289-1_23

Knuth, D. (1998). *The art of computer programming: Semiemperical algorithms.* Addison Wesley.

Kocarev, L. (2001). Chaos-based cryptography: A brief overview. *IEEE Circuits and Systems Magazine*, *1*(3), 6–21. doi:10.1109/7384.963463

Kocarev, L., & Jakimoski, G. (2003). Pseudorandom bits generated by chaotic maps. *IEEE Transactions on Circuits and Systems. I, Fundamental Theory and Applications*, *50*(1), 123–126. doi:10.1109/TCSI.2002.804550

Kocarev, L., & Lian, S. (2011). *Chaos-Based Cryptography. Theory, Algorithms and Applications.* Berlin: Springer-Verlag.

Koc, C. K. (2009). *Cryptographic Engineering*. New York: Springer. doi:10.1007/978-0-387-71817-0

Kodera, N., Yamashita, H., & Ando, T. (2005). Active damping of the scanner for high-speed atomic force microscopy. *Rev. of Sci. Instrum.*, *76* (5).

Kohda, T., & Tsuneda, A. (1997). Statistics of chaotic binary sequences. *IEEE Transactions on Information Theory*, *43*(1), 104–112. doi:10.1109/18.567654

Kojabadi, H. M. (2009). Active power and MRAS based rotor resistance identification of an IM drive. *Simulation Modelling Practice and Theory*, *17*(2), 376–389. doi:10.1016/j.simpat.2008.09.014

Kolesov, V. V., Belyaev, R. V., & Voronov, G. M. (2001). Digital random-number generator based on the chaotic signal algorithm. *Journal of Communications Technology and Electronics*, *46*, 1258–1263.

Komoda, N. (2006). Service oriented architecture (SOA) in industrial systems. In *Proceedings of the 2006 IEEE International Conference on Industrial Informatics*, (pp. 1-5). IEEE. doi:10.1109/INDIN.2006.275708

Konishi, K., & Kokame, H. (1998). Observer-based delayed-feedback control for discrete-time chaotic systems. *Physics Letters [Part A]*, *248*(5-6), 359–368. doi:10.1016/S0375-9601(98)00673-2

Korbicz, J. (2004). *Fault diagnosis: Models, artificial intelligence, applications*. Springer.

Kortmann, M., & Unbehauen, H. (1988). Structure detection in the identification of nonlinear systems. *RAIRO–APII*, *22*(1), 5–25.

Kotelyanski, D. M. (1952). Some properties of matrices with positive elements. *Matematicheskii Sbornik*, *31*(73), 497–505.

Koutsoukos, D., & Antsklis, J. (2002). Design of stabilizing switching control laws for discrete and continuous-time linear systems using piecewise-linear Lyapunov functions. *International Journal of Control*, *75*(5), 932–945. doi:10.1080/00207170210151076

Kraiem H. and Messaoudi M. (2009). Ekf- Based Sensorless Direct Torque Control Of Permanent Magnet Synchronous Motor: Comparison Between Two Different Selection Tables, *Transaction in Machine, Power electronics and Drives*, 1(7). 159-165.

Kratz, S., & Rohs, M. (2011). Protractor3d: A closed-form solution to rotation-invariant 3d gestures. In Proceedings of IUI'11, (pp. 371-374). ACM.

Kravaris, C., & Daoutidis, P. (1992). Output feedback controller realizations for open-loop stable nonlinear processes. In *Proceedings of the 1992 IEEE American control conference*, (pp. 2596–2601). IEEE.

Kravaris, C., Daoutidis, P., & Wright, R. A. (1994). Output feedback control of nonminimum-phase nonlinear processes. *Chemical Engineering Science*, *49*(13), 2107–2122. doi:10.1016/0009-2509(94)E0009-F

Krener, A. J., Karahan, S., & Hubbard, M. (1988). Approximate normal forms of nonlinear systems. In *Proceedings of the 1988 IEEE Control and Decision Conference*, (pp. 1223–1229). IEEE. Doi:10.1109/CDC.1988.194516

Krener, A. J., Karahan, S., Hubbard, M., & Frezza, F. (1987).Higher order linear approximations to nonlinear control systems. In *Proceedings of the 1987 IEEE Control and Decision Conference*, (pp. 519–523). IEEE. Doi:10.1109/CDC.1987.272895

Krener, A. J. (1973). On the equivalence of control systems and the linearization of nonlinear systems. *SIAM Journal on Control*, *11*(4), 670–676. doi:10.1137/0311051

Krener, A. J. (1984). Approximate linearization by state feedback and coordinate change. *Systems & Control Letters*, *5*(3), 181–185. doi:10.1016/S0167-6911(84)80100-0

Krishnan, R. (2002). *Electric motor drives Modeling, Analysis, and Control*. Upper Saddle River, New Jersey: Prentice Hall.

Kristensson, P., Nicholson, T., & Quigley, A. (2012). Continuous Recognition of One-Handed and Two-Handed Gestures using 3D Full-Body Motion Tracking Sensors. In Proceedings of IUI'12, (pp. 89-92). IUI.

Kruif, B., & Vries, T. (2003). Pruning error minimization in least squares vector support machines. *IEEE Transactions on* Neural Networks, *14*(3), 696–702.

Kuang, Y. (1993). *Delay Differential Equations: With Applications in Population Dynamics*. New York: Academic Press.

Kumar, R., & Gupta, R. A. and Ajay Kr., (2007). Identification and Control of PMSM Using Artificial Neural Network.*IEEE International Symposium on Industrial Electronics*, pp. 30 – 35, Vigo. doi:10.1109/ISIE.2007.4374567

Kuo, A. D. (1995). An optimal control model for analyzing human postural balance. *IEEE Transactions on Bio-Medical Engineering*, *42*(1), 87–101. doi:10.1109/10.362914 PMID:7851935

Kurek, J. E., & Zaremba, M. B. (1993). Iterative Learning Control Synthesis Based on 2-D System Theory. *IEEE Transactions on Automatic Control*, *38*(2), 121–125. doi:10.1109/9.186321

Kwong, R. H., & Johnston, E. W. (1992). A variable step size LMS algorithm. *IEEE Transactions on Signal Processing*, *40*(7), 1633–1642. doi:10.1109/78.143435

Kwon, O. M., Park, J. H., & Lee, S. M. (2011). Secure communication based on chaotic synchronization via interval time varying delay feedback control. *Nonlinear Dynamics*, *63*(1-2), 239–252. doi:10.1007/s11071-010-9800-9

Kyo, B. L., & Frede, B. (2006). Reduced-Order Extended Luenberger Observer Based Sensorless Vector Control Driven by Matrix Converter With Nonlinearity Compensation. *IEEE Transactions on Industrial Electronics*, *53*(1), 66–75. doi:10.1109/TIE.2005.862313

Labiod, S., Boucherit, M. S., & Guerra, T. M. (2005). Adaptive fuzzy control of a class of MIMO nonlinear systems. *Fuzzy Sets and Systems*, *151*(1), 59–77. doi:10.1016/j.fss.2004.10.009

Lakshmanan, M., & Murali, K. (1996). *Chaos in Nonlinear Oscillators: Controlling and Synchronization* (Vol. 13). Singapore: World Scientific.

Lakshmanan, M., & Senthikumar, D. V. (2011). *Dynamics of Nonlinear Time-Delay Systems*. Berlin: Springer. doi:10.1007/978-3-642-14938-2

Lakshmi, G. S., Kamakshaiah, S., & Das, T. R. (2013), Closed loop PI control of PMSM for hybrid electric vehicle using three level diode clamped inverter for optimal efficiency. *International Conference on Energy Efficient Technologies for Sustainability (ICEETS)*. 754 – 759, Nagercoil. doi:10.1109/ICEETS.2013.6533479

Lamba, V. K. (2008). *Neuro Fuzzy Systems*. New Delhi: Laxmi Publications.

Lam, J., Xu, S. Y., & Zhou, Y. (2004). Robust output feedback stabilization for two-dimensional continuous systems in Roesser form. *Applied Mathematics Letters*, *17*(12), 1331–1341. doi:10.1016/j.aml.2004.04.004

Lanzon, A., & Petersen, I. R. (2007). A modified positive-real type stability condition. In *Proceedings ofEuropean Control Conference*. Kos, Greece: Academic Press.

Lanzon, A., & Petersen, I. R. (2008). Stability robustness of a feedback interconnection of systems with negative imaginary frequency response. *IEEE Transactions on Automatic Control*, *53*(4), 1042–1046. doi:10.1109/TAC.2008.919567

Lascu, C., Boldea, I., & Blaabjerg, F. (2004). Variable-Structure Direct Torque Control - A Class of Fast and Robust Controllers for Induction Machine Drives. *IEEE Transactions on Industrial Electronics*, *51*(2), 354–363.

Lawrence, D. A. (1995). A general approach to input-output pseudo-linearization for nonlinear systems. In *Proceedings of the 1995 IEEE Control and Decision Conference*, (pp. 613–618). IEEE. Doi:10.1109/CDC.1995.478980

Lawrence, D. A. (1999). Approximate model matching for nonlinear systems. In *Proceedings of the 1999 IEEE Control and Decision Conference*, (pp. 720–725). IEEE. Doi:10.1109/CDC.1999.832873

Lazarus, C., & Hu, H. (2001). Using genetic programming to evolve robot behaviours. In *Proceedings of the 3rd British Conference on Autonomous Mobile Robotics and Autonomous Systems*. Manchester, UK: Academic Press.

Lazinica, A., & Katalinic, B. (2005). Self-organizing multi-robot assembly system. In *Proceedings of the 36th International Symposium on Robotics*, (pp. 42-47). Academic Press.

Leang, K. K., & Devasia, S. (2007). Feedback-linearized inverse feedforward for creep, hysteresis, and vibration compensation in afm piezoactuators. *IEEE Transactions on Control Systems Technology*, *15*(5), 927–935. doi:10.1109/TCST.2007.902956

Lebbal, M. (2006). *Contribution à la modélisation et au diagnostic des systèmes à commutations*. (PhD thesis). Université du Havre, Le Havre, France.

Lee, Y-H., & Hyun, D-S. (1996). A Novel PWM Scheme for three-Level Source Inverter With GTO Thyristors. *IEEE Transactions on Industrial Electronics*, *32*(2), 260–268.

Lee, C.-H., & Teng, C.-C. (2000). Identification and control of dynamic systems using recurrent fuzzy neural networks. *IEEE Transactions on Fuzzy Systems*, *8*(4), 349–366.

Lee, H. G., & Marcus, S. J. (1986). Approximate and local linearizability of nonlinear discrete-time systems. *International Journal of Control*, *44*(4), 1103–1124. doi:10.1080/00207178608933653

Lee, J.-P., Min, B.-D., Kim, T.-J., Yoo, D.-W., & Yoo, J.-Y. (2008). A Novel Topology for Photovoltaic DC/DC Full-Bridge Converter with Flat Efficiency under Wide PV Module Voltage and Load Range. *IEEE Transactions on Industrial Electronics*, *55*(7), 2655–2663. doi:10.1109/TIE.2008.924165

Lee, Y., & Zak, S. H. (May2002). Genetic fuzzy tracking controllers for autonomous ground vehicles. In *Proceedings of American Control Conference*, (pp. 2144-2149). Anchorage, AK: Academic Press.

Le, L. B., Konishi, K., & Hara, N. (2012). Design and experimental verification of multiple delay feedback control for time-delay nonlinear oscillators. *Nonlinear Dynamics*, *67*(2), 1407–1418. doi:10.1007/s11071-011-0077-4

Lenci, S., & Rega, G. (2004). Regular nonlinear dynamics and bifurcations of an impacting system under general periodic excitation. *Nonlinear Dynamics*, *34*(3/4), 249–268. doi:10.1023/B:NODY.0000013507.01603.a3

Leonard, W. (2001). *Control of Electrical Drives* (3rd ed.). New York: Springer-Verlag. doi:10.1007/978-3-642-56649-3

Leontaritis, I. J., & Billings, S. A. (1985a). Input-output parametric models for non-linear systems. Part 1: Deterministic non-linear systems. *International Journal of Control*, *41*(2), 303–328.

Leontaritis, I. J., & Billings, S. A. (1985b). Input-output parametric models for non-linear systems. Part 2: Stochastic non-linear systems. *International Journal of Control*, *41*(2), 329–344.

Lepuschitz, W., Zoitl, A., Vallee, M., & Merdan, M. (2011). Toward self-reconfiguration of manufacturing systems using automation agents. *IEEE Transactions on Systems, Man and Cybernetics. Part C, Applications and Reviews*, *41*(1), 52–69. doi:10.1109/TSMCC.2010.2059012

Lequesne, D. (2006). Régulation PID Analogique-numérique-floue. Paris: Hermès, Lavoisier.

Letellier, C. (2012). *Diagnostic robuste des systèmes incertains. Application à un système mécatronique pour l'automobile*. PhD thesis, Université de Rouen, France.

Letellier, C., Hoblos, G., & Chafouk, H. (2011). *Robust Fault Detection based on Multimodel and Interval Approach. Application to a Throttle Valve.* Paper presented at the IEEE MED. Corfu, Greece. Doi:10.1109/MED.2011.5983198

Levin, A. U., & Narendra, K. S. (1996). Control of nonlinear dynamical systems using neural networks. Part ii: Observability, identification, and control. *IEEE Transactions on Neural Networks*, *7*(1), 30–42. doi:10.1109/72.478390 PMID:18255556

Lewis, L. F. (1986). Asurvey of linear singular systems. *Circuits System Signal Processing*, *5*(1), 3–36. doi:10.1007/BF01600184

Lian, J. (2011). The Design of Gas Drainage Holes' Opening Parameters Intelligent Measurement and Control System for Coal Mine. *Procedia Earth and Planetary Science*, *3*(1), 331–337. doi:10.1016/j.proeps.2011.09.102

Lian, J., Feng, Z., & Shi, P. (2011). Observer design for switched recurrent neural networks: An average dwell time approach. *IEEE Transactions on Neural Networks*, *22*(10), 1547–1556. doi:10.1109/TNN.2011.2162111 PMID:21824843

Liberti, L., & Maculan, N. (2006). *Global Optimization From Theory to Implementation.* Springer.

Liberzon, D. (2003). *Switching in Systems and Control.* Boston, MA: Birkhäuser. doi:10.1007/978-1-4612-0017-8

Liberzon, D., Hespanha, J. P., & Morse, A. S. (1999). Stability of switched linear systems: A Lie-algebraic condition. *Systems & Control Letters*, *37*(3), 117–122. doi:10.1016/S0167-6911(99)00012-2

Liberzon, D., & Morse, A. S. (1999). Basic problems in stability and design of switched systems. *IEEE Control Systems Magazine*, *19*(5), 59–70. doi:10.1109/37.793443

Liberzon, D., & Tempo, R. (2004). Common Lyapunov functions and gradient algorithms. *IEEE Transactions on Automatic Control*, *49*(6), 990–994. doi:10.1109/TAC.2004.829632

Lien, C. H., Yu, K. W., Chung, Y. J., Chang, H. C., Chung, L. Y., & Chen, J. D. (2011). Exponential stability and robust H∞ control for uncertain discrete switched systems with interval time-varying delay. *IMA J. Math. Control Inf*, *7*(5), 433–444.

Li, H. X., & Tong, S. C. (2003). A hybrid adaptive fuzzy control for a class of nonlinear MIMO systems. *IEEE Transactions on Fuzzy Systems*, *11*(1), 24–34. doi:10.1109/TFUZZ.2002.806314

Li, H.-X. (1999). Identification of Hammerstein models using genetic algorithms. *IEE Proceedings Control Theory and Applications*, *146*(6), 499–504.

Li, K., Soh, Y. C., & Li, Z. G. (2003). Chaotic cryptosystem with high sensitivity to parameter mismatch. *IEEE Transactions on Circuits and Systems. I, Fundamental Theory and Applications*, *50*(4), 579–583. doi:10.1109/TCSI.2003.809808

Lin, C., Wang, D. J., & Soh, B. C. (1997). Robustness of uncertain descriptor systems. *Systems & Control Letters*, *31*(3), 129–138. doi:10.1016/S0167-6911(97)00038-8

Lin, C., Wang, L. J., Yang, H. G., & Lam, J. (2000). Robust stabilization via state feedback for descriptor systems with uncertainties in the derivative matrix. *International Journal of Control*, *73*(5), 407–415. doi:10.1080/002071700219588

Lin, F., & Brandt, R. (1998). An Optimal Control Approach to Robust Control of Robot Manipulators. *IEEE Transactions on Robotics and Automation*, *14*(1), 69–77. doi:10.1109/70.660845

Ling, K. V., Yue, S. P., & Maciejowski, J. M. (2006). A fpga implementation of model predictive control. In *Proceedings of the 2006 American Control Conference*, (pp. 1930 – 1935). Academic Press. doi:10.1109/ACC.2006.1656502

Ling, K., Wu, B., & Maciejowski, J. (2008). Embedded model predictive control (mpc) using a fpga. In *Proceedings of the 17th International Federation of Automatic Control,Seoul World Congress* (IFAC), (pp. 15250-15255). IFAC.

Lin, H., & Antsaklis, P. (2009). Stability and Stabilizability of Switched Linear Systems: A Survey of Recent Results. *IEEE Transactions on Automatic Control*, *54*(2), 308–322. doi:10.1109/TAC.2008.2012009

Lin, H., Zhai, G., & Antsaklis, P. J. (2003). Set-valued observer design for a class of uncertain linear systems with persistent disturbance and measurement noise. *International Journal of Control*, *76*(16), 1644–1653.

Li, Q., Tang, S., & Yang, X. S. (2013). New bifurcations in the simplest passive walking model. *Chaos: An Interdisciplinary Journal of Nonlinear Science*, *23*(4), 043110. doi:10.1063/1.4824975 PMID:24387549

Li, Q., & Yang, X. S. (2012). New walking dynamics in the simplest passive bipedal walking model. *Applied Mathematical Modelling*, *36*(11), 5262–5271. doi:10.1016/j.apm.2011.12.049

Li, S., Mou, X., & Cai, Y. (2001). Pseudo-random bit generator based on coupled chaotic systems and its application in stream-ciphers cryptography. *In Progress in Cryptology – INDOCRYPT 2001. Lecture Notes in Computer Science*, *2247*, 316–329.

Liscano, R., Manz, A., Stuck, E. R., Fayek, R. E., & Tigli, J. Y. (1995). Using a blackboard to integrate multiple activities and achieve strategic reasoning for mobile-robot navigation. *IEEE Expert*, *10*(2), 24–36. doi:10.1109/64.395354

Liu, L., Fukumoto, M., Saiki, S., & Zhang, S. (2009). A variable step-size proportionate Affine Projection Algorithm for identification of sparse impulse response. EURASIP Journal on Advances in Signal Processing, Article ID 150914.

Liu, B., Lu, W., & Chen, T. (2014). New criterion of asymptotic stability for delay systems with time-varying structures and delays. *Neural Networks*, *54*, 103–111. doi:10.1016/j.neunet.2014.03.003 PMID:24686033

Liu, K., Zhang, Q., Zhu, Z.-Q., Zhang, J., Shen, A.-W., & Stewart, P. (2010). Comparison of two novel MRAS based strategies for identifying parameters in permanent magnet synchronous motors. *International Journal of Automation and Computing*, *7*(4), 516–524. doi:10.1007/s11633-010-0535-3

Liu, S. J., Jiang, Z. P., & Zhang, J. F. (2008). Globally output-feedback stabilization for a class of stochastic non-minmum-phase nonlinear systems. *Automatica*, *44*(8), 1944–1957. doi:10.1016/j.automatica.2007.11.011

Liu, W., Pokharel, P. P., & Principe, J. C. (2008). The Kernel Least-Mean-Square Algorithm. *IEEE Transactions on Signal Processing*, *56*(2), 543–554. doi:10.1109/TSP.2007.907881

Liu, W., Principe, J. C., & Haykin, S. (2010). *Kernel Adaptive Filtering*. Hoboken, NJ: Wiley. doi:10.1002/9780470608593

Liu, Y., Sun, L. Y., Lu, B. C., & Dai, M. Z. (2010). Feedback control of networked switched fuzzy time-delay systems based on observer. *ICIC Express Letters*, *4*(6), 2369–2376.

Li, X. T. D., Chow, T. W. S., & Ho, J. K. L. (2005). 2-D System Theory Based Iterative Learning Control for Linear Continuous Systems with Time Delays. *IEEE Transactions on Circuits and Systems. I, Regular Papers*, *52*(7), 1421–1430. doi:10.1109/TCSI.2005.851706

Li, X. T. D., Chow, T. W. S., & Ho, J. K. L. (2008). Iterative Learning Control for a Class of Nonlinear Discrete-time Systems with Multiple Input Delays. *International Journal of Systems Science*, *39*(2), 361–369. doi:10.1080/00207720701778403

Li, X. W., & Gao, H. J. (2012). Robust finite frequency HN filtering for uncertain 2-D Roesser systems. *Automatica*, *48*(6), 1163–1170. doi:10.1016/j.automatica.2012.03.012

Li, Y., Sundararajan, N., & Saratchandran, P. (2000). Analysis of minimal radial basis function network algorithm for real-time identification of nonlinear dynamic systems. *IEE Proceedings Control Theory and Applications*, *147*(4), 476–484.

Li, Y., & Xu, Q. (2009). Design and analysis of a totally decoupled flexurebased xy parallel micromanipulator. *IEEE Transactions on Robotics*, *25*(3), 645–657. doi:10.1109/TRO.2009.2014130

Ljung, L. (2000). *System Identification Toolbox for use with MATLAB*. Natick, MA: The Math Works, Inc.

Ljung, L. (2002). Prediction error estimation methods. *Circuits. Systems and Signal Processing*, *21*(3), 11–21. doi:10.1007/BF01211648

Lokriti, A., Salhi, I., Doubabi, S., & Zidani, Y. (2013). Induction motor speed drive improvement using fuzzy IP-self-tuning controller. A real time implementation. *ISA Transactions*, *52*(3), 406–417. doi:10.1016/j.isatra.2012.11.002 PMID:23317661

Lopez, R., & Yuz, J. (2012). Recursive parameter and state estimation for a mining industry process. In *Proceedings of the 20th Mediterranean Conference on Control & Automation* (pp. 30-35). Barcelona, Spain: Academic Press.

Lorenz, E. N. (1963). Deterministic non-periodic flow. *Journal of the Atmospheric Sciences*, *20*(2), 130–141. doi:10.1175/1520-0469(1963)020<0130:DNF>2.0.CO;2

Low, K. H., Leow, W. K., & Ang, M. H. (2002). A hybrid mobile robot architecture with integrated planning and control. In *Proceedings of the First International Joint Conference on Autonomous Agents and Multiagent Systems*, (pp. 219-226). New York, NY: Academic Press. doi:10.1145/544741.544797

Luenberger, D. (1971). An introduction to observers. *IEEE Transactions on Automatic Control*, *16*(6), 596–602. doi:10.1109/TAC.1971.1099826

Luenberger, D. (1977). Dynamic equations in descriptor form. *IEEE Transactions on Automatic Control*, *22*(3), 312–321. doi:10.1109/TAC.1977.1101502

Luenberger, D. G. (1964). Observing the state of a linear system. *IEEE Transactions on Military Electronics*, *8*(2), 74–80.

Lu, H., & He, Z. (1996). Chaotic behavior in first–order autonomous continuous-time systems with delay. *IEEE Transactions on Circuits and Systems. I, Fundamental Theory and Applications*, *43*(8), 700–702. doi:10.1109/81.526689

Luis, A., Alexandre, M., & Pedro, F. (1998). Comparing control strategies for line tracking robots. In *Proceedings of the 1998 IEEE Advanced Motion Control (AMC)*, (pp. 542-547). IEEE. Doi:10.1109/AMC.1998.743594

Lu, J., Yang, G., Oh, H., & Luo, A. C. (2005). Computing Lyapunov exponents of continuous dynamical systems: Method of Lyapunov vectors. *Chaos, Solitons, and Fractals*, *23*(5), 1879–1892. doi:10.1016/j.chaos.2004.07.034

Lunze, J., & Richter, J. (2008). Reconfigurable Fault-Tolerant Control: A Tutorial Introduction. *European Journal of Control*, *14*(5), 359–386. doi:10.3166/ejc.14.359-386

Luo, F. L., & Ye, H. (2004). *Advanced DC/DC Converters. Power electronics and applications series*. CRC Press.

Lu, W. S., & Lee, E. (1985). Stability analysis for two-dimensional systems via a Lyapunov approach. *IEEE Transactions on Circuits and Systems*, *32*(1), 61–68. doi:10.1109/TCS.1985.1085596

Lyubchik, L., & Kostenko, Y. T. (1993). *The output control of multivariable systems with unmeasurable arbitrary disturbances-The inverse model approach.* Paper presented at the European Control Conference. Groningen, Netherlands.

Maarefors, M. (1997). Application of H∞ robust control to the RM12 Jet Engine. *Control Engineering Practice*, *5*(9), 1189–1201. doi:10.1016/S0967-0661(97)84358-4

Ma, B., Xu, W., & Wang, S. (2013). A Robot Control System Based on Gesture Recognition Using Kinect. *TELKOMNIKA Indonesian Journal of Electrical Engineering*, *11*(5), 2605–2611. doi:10.11591/telkomnika.v11i5.2493

Maciejowski, J. M. (1989). *Multivaribale Feedback Design*. Wokingham, UK: Addision-Wesley.

Mackey, M. C., & Glass, L. (1977). Oscillation and chaos in physiological control system. *Science*, *197*(4300), 287–289. doi:10.1126/science.267326 PMID:267326

Madadi, H., & Ghribi, M. (2006, June), MRAS based adaptive speed estimator in PMSM drives, IEEE *International Workshop on Advanced Motion Control*, 569 – 572, Istanbul.

Madasu, H., & Prakash, V. (2009). A Novel Optimal Fuzzy System For Color Image Enhancement Using Bacterial Foraging. *IEEE Transactions on Instrumentation and Measurement*, *58*(8), 2867–2879. doi:10.1109/TIM.2009.2016371

Magnus, E. (2002). Motion description languages for multi-modal control in robotics. In A. Bicchi, H. Christensen, & D. Prattichizzo (Eds.), *Control Problems in Robotics* (pp. 75–90). Las Vegas, NV: Academic Press.

Ma, H.-J., & Yang, G.-H. (2008). Adaptive regulation of uncertain nonlinear systems with dead-zone. In *Proceedings of the 47th IEEE Conference on Decision and Control.*Cancun, Mexico: IEEE.

Mahmood, I. A., & Moheimani, S. O. R. (2009). Making a commercial atomic force microscope more accurate and faster using positive position feedback control. *Rev. of Sci. Instrum., 80*(6).

Mahmoud, M. S. (2000). *Robust Filtering for Time-Delay Systems.* New York: Marcel Dekker.

Mahmoud, M., & Shi, P. (2003). Robust stability, stabilization and H_∞. *International Journal Robust Nonlinear Control, 13*(8), 755–784. doi:10.1002/rnc.744

Maiti, S., Chakrabatory, C., Hori, Y., & Ta, M. C. (2008). Model reference adaptive controller-based rotor resistance and speed estimation techniques for vector controlled induction motor drive utilizing reactive power. *IEEE Transactions on Industrial Electronics, 55*(2), 594–601. doi:10.1109/TIE.2007.911952

Majhi, B., & Panda, G. (2007), Bacterial foraging based identification of nonlinear dynamic system, *IEEE Congress On Evolutionary Computation*, 1636,-1641, Singapore. doi:10.1109/CEC.2007.4424669

Mamdani, E. H., & Assilian, S. (1999). An experiment in linguistic synthesis with a fuzzy logic controller. *International Journal of Human-Computer Studies, 51*(2), 135–147. doi:10.1006/ijhc.1973.0303

Mamdoohi, G., Abas, A. F., & Samsudin, K. (2012). Implementation of genetic algorithm in an embedded microcontroller based polarization control system. *Engineering Applications of Artificial Intelligence, 25*(4), 869–873. doi:10.1016/j.engappai.2012.01.018

Maranas, C., & Floudas, C. (1997). Global optimization in generalized geometric programming. *Computers & Chemical Engineering, 21*(4), 351–369. doi:10.1016/S0098-1354(96)00282-7

Marino, R., & Tomei, P. (2005). A class of globally output feedback stabilizable nonlinear non minimum phase systems. *IEEE Transactions on Automatic Control, 50*(12), 451–460. doi:10.1109/TAC.2005.858652

Marsaglia, G. (1995). *DIEHARD Statistical Tests.* Retrieved from http://stst.fsu.edu/pub/diehard

Martens, H. R. (1969). A comparative study of digital integration methods. *Simulation, 12*(2), 87–94. doi:10.1177/003754976901200207

Martin, C., & Jay, T. (1998). State and input estimation for a class of uncertain systems. *Automatica, 34*(6), 757–764. doi:10.1016/S0005-1098(98)00013-2

Martínez del Rincón, J., Makris, D., Orrite Uruñuela, C., & Nebel, J. (2011). Tracking human position and lower body parts using Kalman and particle filters constrained by human biomechanics. *IEEE Transactions on Systems, Man, and Cybernetics: Part B, Cybernetics, 41*(1), 26–37. doi:10.1109/TSMCB.2010.2044041 PMID:20388598

Martins-Filho, L. S., & Macau, E. E. N. (2007). Patrol mobile robots and chaotic trajectories. Mathematical Problems in Engineering, Article ID 61541.

Marx, B., Maquin, D., & Ragot, J. (2010). *State estimation and fault detection of uncertain systems based on an interval approach.* Paper presented at the Conference on Control and Fault-Tolerant Systems (SysTol). Nice, France. Doi:10.1109/SYSTOL.2010.5676036

Mashhour, E., & Moghaddas-Tafreshi, S. M. (2010). Mathematical modeling of electrochemical storage for incorporation in methods to optimize the operational planning of an interconnected micro grid. *Journal of Zhejiang University-Sciences C, 11*(9), 737–750. doi:10.1631/jzus.C0910721

Masubuchi, I., Kamitane, Y., Ohara, A., & Suda, N. (1997). control for descriptor systems: A matrix inequalities approach. *Automatica, 33*(4), 669–673. doi:10.1016/S0005-1098(96)00193-8

Matsumoto, M., & Nishimura, T. (1998). Mersenne twister: A 623-dimensionally equidistributed uniform pseudo-random number generator. *ACM Transactions on Modeling and Computer Simulation, 8*(1), 3–30. doi:10.1145/272991.272995

Mazouz, N., & Midoun, A. (2011). Control of a dc/dc converter by fuzzy controller for a solar pumping system. *International Journal of Electrical Power & Energy Systems, 33*(10), 1623–1630. doi:10.1016/j.ijepes.2011.06.016

McFarlane, D., & Glover, K. (1992). A Loop Shaping Design Procedure Using H Infinity Synthesis. *IEEE Transactions on Automatic Control*, *37*(6), 759–769. doi:10.1109/9.256330

McGeer, T. (1990). Passive dynamic walking. *The International Journal of Robotics Research*, *9*(2), 62–68. doi:10.1177/027836499000900206

Meddeb, A., Abbassi, R., Jemai, K., & Chebbi, S. (2010). Improvement of the Adaptation of a Permanent Magnet Synchronous Generator with an Unbalanced Electrical Supply Network. *International Review on Modelling and Simulations*, *3*(1), 48–56.

Medio, A., Pireddu, M., & Zanolin, F. (2009). Chaotic dynamics for maps in one and two dimensions: A geometrical method and applications to economics. *International Journal of Bifurcation and Chaos in Applied Sciences and Engineering*, *19*(10), 3283–3309. doi:10.1142/S0218127409024761

Mehrzi, E., Bernosso, J., & Mhiri, R. (2007). stability and stabilization for uncertain switched system, a polyquadratic lyaponuv approach. *International Journal on Science and Technique of Automatique Control*, *1*(1), 61-74.

Menezes, A. J., Van Oorschot, P. C., & Vanstone, S. A. (1996). *Handbook of applied cryptography*. CRC Press. doi:10.1201/9781439821916

Menighed, K., Yamé, J., & Aubrun, C. (2011). Modification of an adaptive observer to improve its performances in actuator fault detection. In *Proceeding of the 18th Mediterranean Conference on Control and Automation (MED)*. Hammamet, Tunisia: MED.

Messaoudi, M., Sbita, L., & Abdelkrim, M. N. (2007). A robust nonlinear observer for states and parameters estimation and on-line adaptation of rotor time constant in sensorless induction motor drives. *International. The Journal of Physiological Sciences; JPS*, *2*(8), 217–225.

Meyer, C., Schroder, S., & Doncker, R. W. D. E. (2004). Solid-state circuit breakers and current limiters for medium-voltage systems having distributed power systems. *IEEE Transactions on Power Electronics*, *19*(5), 1333–1340. doi:10.1109/TPEL.2004.833454

Meyer, H. E., & Bennewitz, R. (2004). *Scanning Probe Microscopy*. Heidelberg, Germany: Springer. doi:10.1007/978-3-662-09801-1

Meynard, T., Fadel, M., & Aouda. (1997). Modeling of multi-level converter. IEEE, 914-918.

Meynard, T., Foch, H., Forest, F., Turpin, C., Richardeau, L., Delmas, G., … Lefeuvre, E. (2002). Multi-converters: Derived topologies. *IEEE Transactions on Industrial Electronics, 49,* 978-987.

Meynard, Th. (2006). Flying capacitor multi-cell converters with stored energy. *IEEE Transactions on Industry Applications*, *IA-17*(5), 518–523.

Michiels, W., & Niculescu, S. I. (2007). *Stability and Stabilization of Time-Delay Systems*. Philadelphia, PA: SIAM Press. doi:10.1137/1.9780898718645

Microsoft. (2013). *Human interface guidelines, Kinect for Windows v1.8*. Retrieved from http://msdn.microsoft.com/en-us/library/jj663791.aspx

Mielczarski, W., & Zajaczkowsk, A. M. (1994). Nonlinear Field V oltage Control of a Synchronous Generator using Feedback Linearization. *Automatica*, *30*(10), 1625–1630. doi:10.1016/0005-1098(94)90102-3

Miguel, I., & Shen, Q. (2003). Fuzzy rrDFCSP and planning. *Journal of Artificial Intelligence*, *148*(1-2), 11–52. doi:10.1016/S0004-3702(03)00020-1

Miladi, Y., Feki, M., & Derbel, N. (2014). Using genetic algorithm for chaos control in a piecewise linear hybrid systems with application to power electronics. In *Proceedings of the Annual Seminar on Automation, Industrial Electronics and Instrumentation (SAAEI'14)*. SAAEI.

Mingxuan, S., Huaying, X., & Hui, W. (2011). Switching-function dynamics designs for uncertain variable structure systems with repetitive control. *International Journal of Control*, *84*(7), 1196–1208. doi:10.1080/00207179.2011.561876

Mir, S., Elbuluk, M. E., & Zinger, D. S. (1998). PI and Fuzzy Estimators for Tuning the Stator Resistance in Direct Torque Control of Induction Machines. *IEEE Transactions on Power Electronics*, *13*(2), 279–287. doi:10.1109/63.662841

Mishra, D., & Ucar, A. (2006). Pspice realization of the Ucar prototype delay model. *International Journal of Nonlinear Sciences and Numerical Simulation*, *7*, 209–214.

Mitra, S., Liberzon, D., & Lynch, N. (2008). Verifying average dwell time of hybrid systems. *ACM Transactions on Embedded Computing Systems*, *8*(1), 1–37. doi:10.1145/1457246.1457249

Mohammadi, M., Hosseinian, S. H., & Gharehpetian, G. B. (2012). Optimization of hybrid solar energy sources/wind turbine systems integrated to utility grids as microgrid (MG) under pool/bilateral/hybrid electricity market using PSO. *Solar Energy*, *86*(1), 112–125. doi:10.1016/j.solener.2011.09.011

Mohan, N., Undeland, T. M., & Robbins, W. P. (1989). *Power Electronics: Converters, Applications and Design*. Oxford, UK: Clarendon Press.

Montagner, V. J., Oliveira, R. C., & Peres, P. L. (2006). State feedback control of switched linear systems: An-LMI approach. *Journal of Computational and Applied Mathematics*, *53*(2), 384–393.

Montano, L., García, F. J., & Villarroel, J. L. (2000). Using the time Petri net formalism for specification, validation, and code generation in robot-control applications. *The International Journal of Robotics Research*, *19*(1), 59–76. doi:10.1177/02783640022066743

Moore, K. L. (2000). A matrix fraction approach to higher-order iterative learning control: 2-D dynamics through repetition-domain filtering. In *Proceedings of the Second International Workshop on Multidimensional (ND) Systems* (pp. 99-104). Academic Press.

Moore, K. L., Hyo-Sung, A., & Chen, Y. Q. (2008). Iteration domain H1-optimal iterative learning controller design. *International Journal of Robust and Nonlinear Control*, *2*(18), 1001–1017. doi:10.1002/rnc.1231

Moreno, R. P., Tordioli, D., & Salcedo, J. L. V. (2008). Distributed implementation of discrete event control systems based on Petri nets. In *Proceedings of the 2008 IEEE International Symposium on Industrial Symposium on Industrial Electronics*, (pp. 1738-1745). IEEE. doi:10.1109/ISIE.2008.4676963

Morimoto S., Takeda Y. and Hirasa T.(1990). Flux Weakening Control Method for Surface Permanent Magnet Synchronous Motors, IPEC. Tokyo'90, 942-949.

Morren, J., De Haan, S. W. H., Bauer, P., & Pierik, J. T. G. (2003). Comparison of Complete and Reduced Models of a Wind Turbine Using Doubly-Fed Induction Generator. In *Proceedings of the 10th European Conference on Power Electronics Applications*. Toulouse, France: Academic Press.

Morse, A. S. (1996). Supervisory Control of Families of Linear Set-Point Controllers-Part 1: Exact Matching. *IEEE Trans. Automat. Control, 41*(10), 1413-1431.

Morse, A. S. (1997). *Control Using Logic-Based Switching. Control and Information Sciences* (pp. 201–207). London: Springer-Verlag. doi:10.1007/BFb0036078

Müller, K.-R., Adali, T., Fukumizu, K., Principe, J. C., & Theodoridis, S. (2013, July). Special Issue on Advances in Kernel-Based Learning for Signal Processing. *IEEE Signal Processing Magazine*, *30*(4), 14–15. doi:10.1109/MSP.2013.2253031

Muller, P. C. (1995). Calculation of Lyapunov exponents for dynamics systems with discontinuities. *Chaos, Solitons, and Fractals*, *5*(9), 1671–1681. doi:10.1016/0960-0779(94)00170-U

Murata, T. (1989). Petri Nets: Properties, analysis and applications. *Proceedings of the IEEE*, *77*(4), 541–580. doi:10.1109/5.24143

Murray, J. D. (1990). *Mathematical Biology*. New York: Springer.

Murray-Smith, R., & Johansen, T. (1997). *Multiple Model Approaches to Nonlinear Modelling and Control*. CRC Press.

Nabae, A., Takahashi, I., & Akagi, H. (1981). A new neutral-point-clamped PWM inverter. *IEEE Transactions on Industry Applications, 1A-17*(5), 518-523.

Nachidi, M., Hajjaji, A. E., & Bosche, J. (2013). An enhanced control approach for dc-dc converters. *International Journal of Electrical Power & Energy Systems*, *45*(1), 404–412. doi:10.1016/j.ijepes.2012.09.003

Nalepa, R., & Orlowska-Kowalska, T. (2012). Optimum Trajectory Control of the Current Vector of a Nonsalient-Pole PMSM in the Field-Weakening Region. *IEEE Transactions on Industrial Electronics*, *59*(7), 2867–2876. doi:10.1109/TIE.2011.2116755

Namajunas, A., Pyragas, K., & Tamasevicius, A. (1995). An electronic analog of the Mackey-Glass system. *Physics Letters [Part A]*, *201*(1), 42–46. doi:10.1016/0375-9601(95)00208-K

Nam, K., Arapostathis, A., & Lee, S. (1993). Some numerical aspects of approximate linearization of single-input nonlinear systems. *International Journal of Control*, *57*(2), 463–472. doi:10.1080/00207179308934399

Nam, K., Lee, S., & Won, S. (1994). A local stabilizing control scheme using an approximate feedback linearization. *IEEE Transactions on Automatic Control*, *39*(11), 2311–2314. doi:10.1109/9.333783

Nand, K. (1995). Geometric programming based robot control design. *Computers & Industrial Engineering*, *29*(1), 631–635.

Narvydas, G., Simutis, R., & Raudonis, V. (2007). Autonomous mobile robot control using fuzzy logic and genetic algorithm. In *Proceedings of4th IEEE Workshop on Intelligent Data Acquisition and Advanced Computing Systems: Technology and Applications*. Dortmund, Germany: IEEE. doi:10.1109/IDAACS.2007.4488460

Nawa, N. E., & Furuhashi, T. (1999). Fuzzy system parameters discovery by bacterial evolutionary algorithm. *IEEE Transactions on Fuzzy Systems*, *7*(5), 608–616. doi:10.1109/91.797983

Nelles, O. (2001). *Nonlinear System Identification From Classical Approaches to Neural Networks and Fuzzy Models*. Springer.

Neumaier, A. (1991). *Interval methods for systems of equations* (Vol. 37). Cambridge University Press.

Neumann, J. V. (1951). Various techniques used in connection with random digits. In G. E. Forsythe (Ed.), *Applied Mathematical Series-Notes* (Vol. 12, pp. 36–38). National Bureau of Standards.

Nian, H., Song, Y., Zhou, P., & He, Y. (2011). Improved Direct Power Control of a Wind Turbine Driven Doubly Fed Induction Generator During Transient Grid Voltage Unbalance. *IEEE Transactions on Energy Conversion*, *26*(3), 976–986. doi:10.1109/TEC.2011.2158436

Niemiec, M., & Kravaris, C. (2003). Nonlinear model-state feedback control for nonminimum-phase processes. *Automatica*, *39*(7), 1294–1302. doi:10.1016/S0005-1098(03)00103-1

Nijmeijer, H. & Van Der Schaft, A. J. (1990). *Nonlinear Dynamical Control Systems*. Springer-Verlag.

Nishikawa, K., & Makizaki, K. (October 2011). Fixed Order Implementation Method of Kernel Adaptive Filters with Lower Computational Complexity. In *Proceedings of APSIPA ASC 2011*. APSIPA.

Nishikawa, K., Ogawa, Y., & Albu, F. (2013, October). Fixed Order Implementation of Kernel RLS-DCD Adaptive Filters. In *Proceedings of APSIPA ASC*, (pp. 1–6). APSIPA.

NIST. (2001). *Security Requirements for Cryptographic Modules* (FIPS PUB 140-2). Retrieved from http://csrc.nist.gov/publications/fips/fips140-2/ fips1402.pdf

Niu, Y., & Ho, D. W. C. (2006). Design of sliding mode control for nonlinear stochastic systems subject to actuator nonlinearity. *IEE Proceedings. Control Theory and Applications*, *153*(6), 737–744. doi:10.1049/ip-cta:20050194

Niu, Y., Wang, X., Lam, J., & Ho, D. W. C. (2003). Sliding-mode control for nonlinear state-delayed systems using neural-network approximation. *Proceedings of the IEE Control Theory and Applications*, *150*(3), 233–239. doi:10.1049/ip-cta:20030321

Nobale, B., & Daniel, J. W. (1988). *Applied Linear Algebra*. (N. P. Hall, Ed.). Englewood Cliffs, NJ: Prentice Hall.

Nordin, M., & Gutman, P.-O. (2002). Controlling mechanical systems with backlash—a survey. *Automatica*, *38*(10), 1633–1649. doi:10.1016/S0005-1098(02)00047-X

Nørgaard, M., Poulsen, N. K., & Ravn, O. (2000). New developments in state estimation for nonlinear systems. *Automatica*, *36*(11), 1627–1638.

Normey-Rico, J. E. (2007). *Control of Dead-Time Processes*. New York: Springer.

Norquay, S., & Palazoglu, A. (1998). Model predictive control based on Wiener models. *Chemical Engineering Science*, *53*(1), 75–84.

Norris, J. A., Marsh, A. P., Granata, K. P., & Ross, S. D. (2008). Revisiting the stability of 2D passive biped walking: Local behavior. *Physica D Nonlinear Phenomena*, *237*(23), 3038–3045. doi:10.1016/j.physd.2008.07.008

Nouali, S., & Ouali, A. (2010). Sensorless Control strategy for Wind Energy Conversion System based on Doubly Fed Twin Stator Induction Generator. *International Review on Modelling and Simulations*, *3*(6), 1209–1216.

Ogunfunmi, T. (2007). *Adaptive Nonlinear System Identification: The Volterra and Wiener Model Appraches*. Springer.

Oishi, S., & Inoue, H. (1982). Pseudo-random number generators and chaos. *Transactions of the Institute of Electronics and Communication Engineers of Japan E*, *65*, 534–541.

Olalla, C., Queinnec, I., Leyva, R., & Aroudi, A. E. (2011). Robust optimal control of bilinear dcdc converters. *Control Engineering Practice*, *19*(7), 688–699. doi:10.1016/j.conengprac.2011.03.004

Onat, N. (2010). Recent Developments in Maximum Power Point Tracking Technologies for Photovoltaic Systems. International Journal of Photoenergy, Article ID 245316.

Open Source Computer Vision. (2014). Retrieved from http://opencv.org

Openarch Project. (2013). Retrieved from http://www.openarch.cc

Ordonez, R., & Passino, K. M. (1999). Stable multi-input multi-output adaptive fuzzy/neural control. *IEEE Transactions on Fuzzy Systems*, *7*(3), 345–353. doi:10.1109/91.771089

Orjuela, R., Marx, B., Ragot, J., & Maquin, D. (2008). State estimation for non-linear systems using a decoupled multiple model. *International Journal of Modelling Identification and Control*, *4*(1), 59–67.

Orjuela, R., Marx, B., Ragot, J., & Maquin, D. (2009). On the simultaneous state and unknown input estimation of complex systems via a multiple model strategy. *IET Control Theory & Applications*, *3*(7), 877–890.

Oseledec, V. I. (1968). A multiplicative ergodic theorem: Lyapunov characteristic numbers for dynamical systems. *Transactions of the Moscow Mathematical Society*, *19*, 197–231.

Osswald, D., Martin, J., Burghard, C., Mikut, R., Woern, H., & Bretthauer, G. (2003). Integrating a robot hand into the control system of a humanoid robot. In *Proceedings of the 2003 International Conference on Humanoid Robots*. Academic Press.

Osuka, K., & Kirihara, K. (2000). Motion Analysis and Experiments of Passive Walking Robot Quartet II. In *Proceedings ofIEEE International Conference on Robotics andAutomation*, (pp. 3052-3056). San Francisco, CA: IEEE. doi:10.1109/ROBOT.2000.846491

Othman, K. B., Ragot, J., Manquin, D., & Rejeb, M. B. (2007). *Robust identification in the presence of outliers*. Paper presented at the international conference on Sciences and Techniques of Automatic control and computer engineering IJ-STA. Hammamet,Tunisia.

Othman, K. B. (2008). *Diagnostic, disponibilité et Approches Non Conventionnelles pour l'Etude de Systèmes Complexes*. Tunisia: Université de Tunis El Manar.

Ott, E., Grebogi, C., & Yorke, J. A. (1990). Controlling chaos. *Physical Review Letters*, *170*, 421–428. PMID:10041332

Ouhrouche, M. A. (2002). Estimation of speed, rotor flux and rotor resistance in cage induction motor using the EKF-algorithm. *International Journal of Power and Energy Systems*, *22*(2), 103–109.

Owens, A., Rogers, D. H., Aman, N., & French, M. (2000). Analysis of Linear Iterative Learning Control Schemes, A 2D Systems/Repetitive Processes Approach. *International Journal ofMultidimensional Systems and Signal Processing*, *11*(1-2), 125–177. doi:10.1023/A:1008494815252

Ozeki, K., & Umeda, T. (1984). An adaptive filtering algorithm using an orthogonal projection to an affine subspace and its properties. *Electronic Communications Japan*, *67-A*(5), 19–27.

Palanisamy, K. R., & Bhattacharya, D. K. (1981). System identification via block-pulse functions. *International Journal of Systems Science*, *12*(5), 643–647. doi:10.1080/00207728108963772

Paleologu, C., Benesty, J., & Albu, F. (2012). Regularization of the improved proportionate affine projection algorithms. In *Proceedings of ICASSP*, (pp. 169–172). ICASSP.

Panjapornpon, C., Soroush, M., & Seider, W. D. (2003). Model based control of unstable nonminimum-phase nonlinear processes. In *Proceedings of the 2003 IEEE Conference on Decision and Control (CDC)*, (pp. 6151–6156). IEEE. Doi:10.1109/CDC.2003.1272256

Pannocchia, G., Rawlings, J. B., Mayne, D. Q., & Marquardt, W. (2010). On Computing Solutions to the Continuous Time Constrained Linear Quadratic Regulator. *IEEE Transactions on Automatic Control, 55*(9), 2192–2198. doi:10.1109/TAC.2010.2053478

Panoiu, C., Baciu, Panoiu, M., & Cuntan, C. (2007). Simulation results on the currents harmonics mitigation on the railway station line feed using a data acquisition system. *WSEAS Trans. on Electronics, 4*(10), 228-236.

Parayandeh, A., & Prodic, A. (2008). Programmable analog-to-digital converter for low-power DC-DC SMPS. *IEEE Transactions on Power Electronics, 23*(1), 500–505. doi:10.1109/TPEL.2007.913932

Pareschi, F., Rovatti, R., & Setti, G. (2006). Simple and effective post-processing stage for random stream generated by a chaos-based RNG. In *Proceedings of 2006 International Symposium in Nonlinear Theory and its Applications*, (pp. 383-386). Academic Press.

Parker, T. S., & Chua, L. O. (1989). *Practical numerical algorithms for chaotic systems*. New York: Springer-Verlag. doi:10.1007/978-1-4612-3486-9

Park, S., & Miller, K. (1988). Random numbers generators: Good ones are hard to find. *Communications of the ACM, 31*(10), 1192–1201. doi:10.1145/63039.63042

Park, T. G., & Lee, K. S. (2004). Process fault isolation for linear systems with unknown inputs. *IEE Proceedings. Control Theory and Applications, 151*(6), 720–726. doi:10.1049/ip-cta:20040796

Park, Y., & Stein, J. L. (1988). Closed-loop state and input observer for systems with unknown inputs. *International Journal of Control, 48*(3), 1121–1136. doi:10.1080/00207178808906239

Paszke, W., Lam, J., Galkowski, K., Xu, S., & Lin, Z. (2004). Robust stability and stabilization of 2-D discrete state-delayed systems. *Systems & Control Letters, 51*(3–4), 277–291. doi:10.1016/j.sysconle.2003.09.003

Patra, S., & Lanzon, A. (2011). Stability analysis of interconnected systems with "mixed" negative-imaginary and small-gain properties. *IEEE Transactions on Automatic Control, 56*(6), 1395–1400. doi:10.1109/TAC.2011.2122670

Patton, R., & Chen, J. (1991). A review of parity space approaches to fault diagnosis. In *Proceedings of the 1991 IFAC/IMACS Symposium on Fault Detection, Supervision and Safety of Technical Processes (SAFEPROCESS)*, (pp. 239–255). IFAC.

Patton, R. (1997). Fault tolerant control systems: the 1997 situation. In *Proceeding of IFAC Safe Process* (pp. 1033-1054). IFAC.

Patton, R. J., & Chen, J. (1997). Observer-based fault detection and isolation: Robustness and applications. *Control Engineering Practice, 5*(5), 671–682. doi:10.1016/S0967-0661(97)00049-X

Pecora, L. M., & Caroll, T. L. (1990). Synchronization in chaotic signals. *Physical Review Letters, 64*(8), 821–824. doi:10.1103/PhysRevLett.64.821 PMID:10042089

Pedro, A., Joao, F., Pedro, R., & Pedro Lima, B. (1998). A modular autonomous vehicle. In *Proceedings of the 3rd IFAC Symposium on Intelligent Autonomous Vehicles*. Lisbon, Portugal: IFAC.

Pei, L., Wang, Q., & Shi, H. (2011). Bifurcation dynamics of the modified physiological model of artificial pancreas with insulin secretion delay. *Nonlinear Dynamics, 63*(3), 417–427. doi:10.1007/s11071-010-9812-5

Peleties, P., & De'Carlo, R. A. (1992).Asymptotic stability of m-switched systems using Lyapunov functions. In *Proceedings of the 31th IEEE Conference on Decision and Control* (vol. 18, pp. 3438–3439). Tucson, AZ: IEEE. doi:10.1109/CDC.1992.371213

Pena, R., Clare, J. C., & Asher, G. M. (1996). Doubly fed induction generator using back-to-back PWM converter and its application to variable-speed wind-energy generation. *Electric Power Applications, 143*(3), 231–241. doi:10.1049/ip-epa:19960288

Perdana, A., Carlson, O., & Persson, J. (2004). Dynamic response of grid-connected wind turbine with doubly fed induction generator during disturbances. In Proceedings of Nordic Workshop on Power and Industrial Electronics. Academic Press.

Petersen, I. R. (2011). Negative imaginary systems theory in the robust control of highly resonant flexible structures. In *Proceedings ofAustralian Control Conference*. Melbourne, Australia: Academic Press.

Petersen, I. R. (2008). Robust adaptive H infinity control using integral quadratic constraints. *Automatica*, *44*(10), 2661–2668. doi:10.1016/j.automatica.2008.03.007

Petersen, I. R., & Pota, H. R. (2003). Minimax LQG optimal control of a flexible beam. *Control Engineering Practice*, *11*(11), 1273–1287. doi:10.1016/S0967-0661(02)00240-X

Petersen, I., & Lanzon, A. (2010). Feedback control of negative-imaginary systems. *IEEE Control Systems Magazine*, *30*(5), 54–72. doi:10.1109/MCS.2010.937676

Petersson, A., Thiringer, T., Harnefors, L., & Petru, T. (2005). Modeling and experimental verification of grid interaction of a DFIG wind turbine. *IEEE Transactions on Energy Conversion*, *20*(4), 878–886. doi:10.1109/TEC.2005.853750

Pfeiffer, F., & Glocker, C. (1996). *Multibody Dynamics with Unilateral Contacts*. New York: Wiley-VCH Verlag. doi:10.1002/9783527618385

Pham, V.-T., Fortuna, L., & Frasca, M. (2012). Implementation of chaotic circuits with a digital time-delay block. *Nonlinear Dynamics*, *68*(1), 345–355. doi:10.1007/s11071-011-9982-9

Philippe, W., Boumedyen, B., Ahmed, K., Didier, T., & Christophe, A. (2012). Reconfigurable Control Design with integration of a reference governor and reliability indicator. *International Journal of Applied Mathematics and Computer Science*, *22*(1), 139–148.

Picking, R., Robinet, A., Grout, V., McGinn, J., Roy, A., Ellis, S., & Oram, D. (2010). A Case Study Using a Methodological Approach to Developing User Interfaces for Elderly and Disabled People. *The Computer Journal*, *53*(6), 842–859. doi:10.1093/comjnl/bxp089

Pinard, M. (2007). *Convertisseurs et Electronique de Puissance Commande, description, mise en oeuvre*. DUNOD.

Planchon, P., & Lunze, J. (2008). Diagnosis of linear systems with structured uncertainties based on guaranteed state observation. *International Journal of Control Automation and Systems*, *6*(3), 306–319.

Ploix, S., Adrot, O., & Ragot, J. (1999). *Parameter uncertainty computation in static linear models*. Paper presented at the 38th IEEE Conference on Decision and Control. Doi:10.1109/CDC.1999.830916

Poitiers, F., Bouaouiche, T., & Machmoum, M. (2009). Advanced control of a doubly-fed induction generator for wind energy conversion. *Electric Power Systems Research*, *79*(7), 1085–1096. doi:10.1016/j.epsr.2009.01.007

Pooja, D., & Bharti, S. D. (2013). Grid Connected Doubly Fed Induction Generator Wind Energy Conversion System Using Fuzzy Controller. *International Journal of Innovative Technology and Exploring Engineering IJITEE*, *2*(2), 2278–3075.

Pörn, R., Bjork, K., & Westerlund, T. (2008). Global solution of optimization problems with signomial parts. *Discrete Optimization*, *5*(1), 108–120. doi:10.1016/j.disopt.2007.11.005

Pota, H., Moheimani, S. R., & Smith, M. (1999). Resonant controllers for flexible structures. In *Proc. Conf. of Decision and Control*, (vol. 1, pp. 631–636). Academic Press.

Pota, H., Moheimani, S. R., & Smith, M. (2002). Resonant controller for smart structures. *Smart Materials and Structures*, *11*(1), 1–8. doi:10.1088/0964-1726/11/1/301

Powell, D. (1992). Failure Mode Assumptions and Assumption Coverage. In *Proceedings of the 1992 IEEE Symposium on Fault-Tolerant Computing (FTCS)*, (pp. 386-395). IEEE. Doi:10.1109/FTCS.1992.243562

Prabhu, V., & Duffie, N. (1995). Modelling and analysis of non-linear dynamics in autonomous heterarchical manufacturing system control. *CIRP Annals*, *44*(1), 425–428. doi:10.1016/S0007-8506(07)62356-7

Pradeep M., S. Sirisha and M. Chandramouly, (2013). Design of PMSM based on DTC method with MRAS, M.Pradeep Kumar. *International. Journal of Engineering Research and Application*, 3(5), 01-05.

PrimeSense Ltd. (2010). *US Patent Application Publication No. US2010/0118123*. Retrieved from http://www.freepatentsonline.com/20100118123.pdf

Prokhorov, M. D., & Ponomarenko, V. I. (2008). Encryption and decryption of information in chaotic communication systems governed by delay-differential equations. *Chaos, Solitons, and Fractals*, *35*(5), 871–877. doi:10.1016/j.chaos.2006.05.081

Pu, Q., Gupta, S., Gallakota, S., & Patel, S. (2013). Whole-Home Gesture Recognition Using Wireless Signals. In *Proceedings of Mobicom'13*, (pp. 27-38). ACM.

Puig, V., Quevedo, J., Figueras, J., Riera, S., Cembrano, G., Salamero, M., et al. (2003). *Fault detection and isolation of rain gauges and limnimeters of Barcelona's sewer system using interval models.* Paper presented at the 5th Symposium on Fault Detection Supervision and Safety for Technical Processes, Safeprocess. Washington, DC.

Pun, D., Lua, S. L., Law, S. S., & Cao, D. Q. (1998). Forced vibration of a multidegree impact oscillator. *Journal of Sound and Vibration*, *213*(3), 447–466. doi:10.1006/jsvi.1997.1494

Pyragas, K. (1992). Continuous control of chaos by self-controlling feedback. *Physics Letters. [Part A]*, *170*(6), 421–428. doi:10.1016/0375-9601(92)90745-8

Pyragas, K. (1995). Control of chaos via extended delay feedback. *Physics Letters [Part A]*, *206*(5-6), 323–330. doi:10.1016/0375-9601(95)00654-L

Pyragas, K. (2006). Delayed feedback control of chaos. *Philosophical Transactions of the Royal Society A*, *364*(1846), 2309–2334. doi:10.1098/rsta.2006.1827 PMID:16893790

Qi, J., & Sun, Y. (2013). Global exponential stability of certain switched systems with time-varying delays. *Applied Mathematics Letters*, *26*(7), 760–765. doi:10.1016/j.aml.2013.02.010

Qi, W., Liu, J., Chen, X., & Christofides, P. D. (2011). Supervisory Predictive Control of Standalone Wind/Solar Energy Generation Systems. *IEEE Transactions on Control Systems Technology*, *19*(1), 199–207. doi:10.1109/TCST.2010.2041930

Quntao An. and Sun L. (2008, September). Online parameters identification for vector controlled PMSM drive using adaptive algorithm, *IEEE Conference Vehicle Power and Propulsion, VPPC '08*, Harbin.

Ragot, J., Maquin, D., & Adrot, O. (2006). Parameter uncertainties characterisation for linear models. *The Computer Journal*, *49*(1), 16–20.

Raıssi, T., Ramdani, N., & Candau, Y. (2004). Set membership state and parameter estimation for systems described by nonlinear differential equations. *Automatica*, *40*(10), 1771–1777.

Ramasubramanian, K., & Sriram, M. S. (2000). A Comparative study of computation of Lyapunov spectral with different algorithms. *Physica D Nonlinear Phenomena*, *139*(1-2), 72–86. doi:10.1016/S0167-2789(99)00234-1

Rao, C. V., Rawlings, J. B., & Mayne, D. Q. (2003). Constrained state estimation for nonlinear discrete-time systems: Stability and moving horizon approximations. *IEEE Transactions on Automatic Control*, *48*(2), 246–258.

Rao, G. P., & Srinivasan, T. (1978). Remarks on author's reply to comments on design of piecewise constant gains for optimal control via Walsh functions. *IEEE Transactions on Automatic Control*, *23*(4), 762–763. doi:10.1109/TAC.1978.1101782

Rashid, M. (2001). *Power Electronics Handbook*. Academic Press.

Ratchagit, K., & Phat, V. N. (2010). Stability criterion for discrete-time systems. *Journal of Inequalities and Applications*, *10*(6), 444–459.

Ratnam, M., Bhikkaji, B., Fleming, A., & Moheimani, S. (2005). PPF control of a piezoelectric tube scanner. In *Proc. of the 44th IEEE Conference on Decision and Control and European Control Conference* (pp. 1168 – 1173). IEEE. doi:10.1109/CDC.2005.1582316

Reboulet, C., & Champetier, C. (1984). A new method for linearizing nonlinear systems: The pseudolinearization. *International Journal of Control*, *40*(4), 631–638. doi:10.1080/00207178408933297

Reichert, R. T. (1989). Application of H∞ control to missile autopilot design. In *Proceedings of AIAA Guidance Navigation and Control Conference* (pp.1065-1072). AIAA.

Ren, J., & Zhang, Q. (2010). Robust H_∞. *International Journal of Control, 83*(1), 89–96. doi:10.1080/00207170903100222

Renquan, L., Wei, D., & Lin, X. (2008). Robust Control for Descriptor System Based on New Bounded Real Lemma. In *Proceedings of 27th Chinese Control Conference* (pp. 636-646). Yunna, China: Academic Press.

Richard, C., Bermudez, J. C. M., & Honeine, P. (2009). Online Prediction of Time Series Data with Kernels. *IEEE Transactions on Signal Processing, 57*(3), 1058–1067. doi:10.1109/TSP.2008.2009895

Riedinger, P., & Iung, C. (1999). Optimal Control for Hybrid Systems: An Hysteresis Example. ystems, Man, and Cybernetics, 1999. In *Proceedings of IEEE SMC '99 Conference*. IEEE.

Rioual, P., Pouliquen, H., & Louis, J. P. (1996). Regulation of a PWM rectifier in the Unbalanced Network state using Generalized model. *IEEE Transactions on Power Electronics, 11*(3), 495–502. doi:10.1109/63.491644

Ritz, T., Chweinsberg, A. S., & Dressler, U. (1997). Chaos control with adjustable control times. *Chaos, Solitons, and Fractals, 8*(9), 1559–1576. doi:10.1016/S0960-0779(96)00155-5

Robert, B. G. M., & El Aroudi, A. (2006). Discrete time model of a multi-cell dc/dc converter: Nonlinear approach. *Mathematics and Computers in Simulation, 71*(4-6), 310–319. doi:10.1016/j.matcom.2006.02.022

Robert, B., Feki, M., & Iu, H. (2006). Control of a pwm inverter using proportional plus extended time-delayed feedback. *International Journal of Bifurcation and Chaos in Applied Sciences and Engineering, 16*(1), 113–128. doi:10.1142/S0218127406014629

Robert, B., Iu, H., & Feki, M. (2003). Widening the stability range of a pwm inverter using a robust chaos control. In *Proceedings of the 10th European Conference on Power Electronics and Applications (EPE2003)*. Toulouse, France: EPE.

Robert, B., Iu, H., & Feki, M. (2004). Adaptive time-delayed feedback for chaos control in a pwm single phase inverter. *Journal of Circuits, Systems, and Computers, 13*(3), 519–534. doi:10.1142/S0218126604001568

Robert, B., & Robert, C. (2002). Border collision bifurcations in a one-dimensional piecewise smooth map for a pwm current-programmed h-bridge inverter. *International Journal of Control, 75*(16-17), 1356–1367. doi:10.1080/0020717021000023771

Robert, F. (1966). Recherche d'une M-matrice parmi les minorantes d'un opérateur liéaire. *Numerische Mathematik, 9*(2), 188–199.

Rodriguez, P., Luna, A., Ciobotaru, M., Teodorescu, R., & Blaabjerg, F. (2006). Advanced Grid Synchronization System for Power Converters under Unbalanced and Distorted Operating Conditions. In *Proceedings of the 2006 IEEE 32nd Annual Conference on Industrial Electronics* (IECON), (pp. 5173 – 5178). IEEE. Doi:10.1109/IECON.2006.347807

Rodriguez, P., Teodorescu, R., Candela, I., Timbus, A. V., & Blaabjerg, F. (2006). New positive-sequence voltage detector for grid synchronization of power converters under faulty grid conditions. In *Proceedings of the 37th IEEE Power Electronics Specialists Conference* (PESC '06). IEEE. Doi:10.1109/PESC.2006.1712059

Rodriguez, H., Ortega, R., Escobar, G., & Barabanov, N. (2000). A robustly stable output feedback saturated controller for the boost dc-to-dc converter. *Systems & Control Letters, 40*(1), 1–8. doi:10.1016/S0167-6911(99)00113-9

Rodriguez, J., Bernet, S., Wu, B., Pontt, J. O., & Kouro, S. (2007). Multilevel voltage source converter topologies for industrial medium-voltage drives. *IEEE Transactions on Industrial Electronics, 54*(6), 2930–2942. doi:10.1109/TIE.2007.907044

Rodriguez, J., Lai, J. S., & Peng, F. Z. (2002). Multilevel inverters: A survey of topologies, controls and applications. *IEEE Transactions of Industrial Electroics, 49*(4), 724–738. doi:10.1109/TIE.2002.801052

Rodríguez, P., Pou, J., Bergas, J., Candela, J. I., Burgos, R. P., & Boroyevich, D. (2007). Decoupled Double Synchronous Reference Frame PLL for Power Converters Control. *IEEE Transactions on Power Electronics, 22*(2), 584–592. doi:10.1109/TPEL.2006.890000

Romero, J. A., Sanchis, R., & Balaguer, P. (2011). PI and PID auto-tuning procedure based on simplified single parameter optimization. *Journal of Process Control*, *21*(6), 840–851. doi:10.1016/j.jprocont.2011.04.003

Roncero-Sanchez, P. L., Feliu-Batlle, V., & Cerrade, A. G. (2005). Active and reactive power control for a wind generator connected to the grid using a Predective-Integral control and a PI control Schemes. In *Proceedings of the European Conference on Power Electronics and Applications*. Academic Press. doi:10.1109/EPE.2005.219407

Roover, D., Bosgra, O. H., & Steinbuch, M. (2000). Internal-model-based design of repetitive and iterative learning controllers for linear multivariable systems. *International Journal of Control*, *73*(10), 914–929. doi:10.1080/002071700405897

Rotea, A. M., & Khargonekar, P. P. (1991). H2-optimal Control with an H∞ constraint The State Feedback Case. *Automatica*, *27*(2), 307–316. doi:10.1016/0005-1098(91)90079-H

Rotella, F., & Zambettakis, I. (2008). *Commande des systèmes par platitude*. Available at: http://pierre.rouchon.chez-alice.fr/MVA/FlatSystems.pdf

Rothenhagen, K., & Fuchs, F. W. (2009). Doubly Fed Induction Generator Model-Based Sensor Fault Detection and Control Loop Reconfiguration. *IEEE Transactions on Industrial Electronics*, *28*(10), 4229–4238. doi:10.1109/TIE.2009.2013683

Rouche, N., Habets, P., & Laloy, M. (1977). *Stability Theory by Lyapunov's Direct Method*. Applied Mathematical Sciences. doi:10.1007/978-1-4684-9362-7

Rugar, D., Mamin, H. J., Guethner, P., Lambert, S. E., Stern, J. E., McFadyen, I., & Yogi, T. (1990). Magnetic force microscopy: General principles and application to longitudinal recording media. *Journal of Applied Physics*, *68*(3), 1169–1183. doi:10.1063/1.346713

Rugh, W. J. (1984). Design of nonlinear compensators for nonlinear systems by an extended linearization technique. In *Proceedings of the 1984 IEEE Control and Decision Conference (CDC)*, (pp. 69–73). IEEE. Doi:10.1109/CDC.1984.272275

Rugh, W. J. (1986). An extended linearization approach to nonlinear system inversion. *IEEE Transactions on Automatic Control*, *31*(8), 725–733. doi:10.1109/TAC.1986.1104385

Rzepka, D. (September2012). Fixed-budget kernel least mean squares. In *Proceedings of IEEE 17th Conference on Emerging Technologies & Factory Automation* (pp. 1-4). IEEE.

Saad, M. (2013). Online stator resistance estimation using artificial neural network for direct torque controlled induction motor drive. *IEEE 8th Conference on Industrial Electronics and Applications*, ICIEA 2013. 1486-1491.

Saccomando, G., & Svensson, J. (2001) Transient Operation of Grid Connected Voltage Source Converter under Unbalanced Voltage Conditions. In *Proceedings of IEEE Industry Applications Conference,36th LAS Annual Meeting*, (pp. 2419-2424). IEEE. Doi:10.1109/IAS.2001.955960

Safa, A. T., Saadat, M. G., & Naraghi, M. (2007). Passive dynamic of the simplest walking model: Replacing ramps with stairs. *Mechanism and Machine Theory*, *42*(10), 1314–1325. doi:10.1016/j.mechmachtheory.2006.11.001

Safonov, L. A., Tomer, E., Strygin, V. V., Ashkenazy, Y., & Havlin, S. (2002). Multifractal chaotic attractors in a system of delay-differential equations modeling road traffic. *Chaos (Woodbury, N.Y.)*, *12*(4), 1006–1014. doi:10.1063/1.1507903 PMID:12779624

Sağlam, G., Tutum, C. C., & Kurtulan, S. (2013). A new PI tuning method for an industrial process: A case study from a micro-cogeneration system. *Journal of Energy Conversion and Management*, *67*(2), 226–239. doi:10.1016/j.enconman.2012.11.009

Said, S. H., & Mimouni, F., MŚahli, F. and Farza, M. (2010). High gain observer based on-line rotor and stator resistances estimation for IMs. *IEEE International Conference on Control Applications (CCA),* 1927-1932, Yokohama.

Saïdi, S., Abbassi, R., & Chebbi, S. (2014). Virtual Flux Based Direct Power Control of Shunt Active Filter. *Scientica Iranica, Transaction D: Computer Science & Engineering and Electrical Engineering*. In press.

Salapaka, S., Sebastian, A., Cleveland, J. P., & Salapaka, M. V. (2002). High bandwidth nano-positioner: A robust control approach. *The Review of Scientific Instruments*, *73*(9), 3232–3241. doi:10.1063/1.1499533

Salarieh, H., & Alasty, A. (2009). Chaos control in an economic model via minimum entropy strategy. *Chaos, Solitons, and Fractals*, *40*(2), 839–847. doi:10.1016/j.chaos.2007.08.045

Sannuti, P. (1977). Analysis and synthesis of dynamic systems via block-pulse functions. *Proceedings of the IEE*, *124*(6), 569–571.

Santana, E. S. (2008). A predictive algorithm for controlling speed and rotor flux of induction motor. *IEEE Transactions on Industrial Electronics*, *55*(12), 4398–4407. doi:10.1109/TIE.2008.2007376

Santos, F., Almeida, L., Pedreiras, P., Lopes, L. S., & Facchinnetti, T. (2004). An adaptive TDMA protocol for soft real-time wireless communication among mobile computing agents. In *Proceedings of Workshop on Architectures for Cooperative Embedded Real-time Systems (satellite of RTSS2004). RTSS.*

Sari, B., Bachelier, O., & Mehdi, D. (2006). Robust state feedback admissibilization of discrete linear polytopic descriptor systems: a strict linear matrix inequality approach. *IET Control Theory and Applications, 6*(8), 1097-1108.

Savaresi, S. M., Nijmeijer, H., & Guardabassi, G. O. (2000). On the design of approximate nonlinear parametric controllers. *International Journal of Robust and Nonlinear Control*, *10*(3), 137–155. doi:10.1002/(SICI)1099-1239(200003)10:3<137::AID-RNC466>3.0.CO;2-I

Sayed, A. H. (2003). *Fundamentals of Adaptive Filtering*. John Wiley & Sons.

Sbarbaro, D., & Johansen, T. A. (1997). Multiple local Laguerre models for modelling nonliner dynamic systems of the Wiener class. *IEE Proceedings. Control Theory and Applications*, *144*(5), 375–380.

Sbita, L., & Ben Hamed, M. (2007). An MRAS–based full order Luenberger observer for sensorless DRFOC of induction motors. *International Journal on Automatic Control and System Engineering*, *7*(1), 11–20.

Schauder, C. (1992). Adaptive Speed Identification for Vector Control of Induction Motors without Rotational Transducers. *IEEE Transactions on Industry Applications*, *28*(5), 1054–1061. doi:10.1109/28.158829

Scherer, C., & Weiland, S. (2005). Linear Matrix Inequalities in Control, *Lecture notes course. CiteUlike*, *3853526*, 1–232.

Schetzen, M. (1980). *The Volterra and Wiener Theories of Nonlinear Systems*. Wiley.

Schitter, G. (2007). Advanced mechanical design and control methods for atomic force microscopy in real-time. In *Proc. of the American Control Conference* (pp. 3503–3508). Academic Press. doi:10.1109/ACC.2007.4282253

Schitter, G., Stemmer, A., & Allgower, F. (2003). Robust 2 DoF-control of a piezoelectric tube scanner for high speed atomic force microscopy. In *Proc. of the American Control Conference* (*vol. 5*, pp. 3720–3725). Academic Press. doi:10.1109/ACC.2003.1240413

Schitter, G., Menold, P., Knapp, H. F., Allgwer, F., & Stemmer, A. (2001). High performance feedback for fast scanning atomic force microscopes. *The Review of Scientific Instruments*, *8*, 72.

Schöbel, A., & Scholz, D. (2014). A solution algorithm for non-convex mixed integer optimization problems with only few continuous variables. *European Journal of Operational Research*, *232*(1), 266–275. doi:10.1016/j.ejor.2013.07.003

Schöll, E., & Schuster, H. G. (2007). Handbook of Chaos Control (2nd ed.). Weinheim, Germany: Wiley-VCH.

Schreiber, D. (2001). State of the art of variable speed wind turbines. In *Proceedings of11th International Symposium on Power Electronics*. Novi Sad, Yugoslavia: Academic Press.

Schultz, D. G., & Melsa, J. L. (1967). *State Functions and Linear Control Systems*. New York: McGraw-Hill.

Sebastian, A., Salapaka, M. V., & Cleveland, J. P. (2003). Robust control approach to atomic force microscopy. In *Proc. of the 42nd IEEE Conf. Decision and Control* (vol. 4, pp. 3443–3444). IEEE. doi:10.1109/CDC.2003.1271677

Sebastian, A., & Salapaka, S. M. (2005). Design methodologies for robust nano-positioning. *IEEE Transactions on Control Systems Technology*, *13*(6), 868–876. doi:10.1109/TCST.2005.854336

Sedra, A. S., & Smith, K. C. (2003). *Microelectronic Circuits*. London: Oxford University Press.

Sedraoui, M., Gherbi, S., & Abdelmalek, S. (2012). A Robust Controller Based on Fractional Structure for MIMO Plant with Multiple Delays. *Control and Intelligent Systems*, *40*(2), 83–94. doi:10.2316/Journal.201.2012.2.201-2216

Seferiadis, G., Pouchet, M., & Gough, M. P. (2006). FPGA implementation of a delay-line readout system for a particle detector. *Measurement*, *40*(1), 90–99. doi:10.1016/j.measurement.2005.07.003

Sekhavat, P., Sepehri, N., & Wu, Q. (2004). Calculation of Lyapunov exponents using nonstandard finite difference discretization scheme: A case study. *Journal of Difference Equations and Applications*, *10*(4), 369–378. doi:10.1080/10236190310001625244

Sera, D., Mathe, L., Kerkes, T. & Spataru, S.V. (2013). On the Perturb-and-Observe and Incremental Conductance MPPT Methods for PV Systems. *IEEE Journal of Photvoltaic, 3*(3), 1070-1078.

Sera, D., Teodorescu, R., Hantschel, J., & Knoll, M. (2007). Optimized Maximum Power Point Tracker for Fast-Changing Environmental Condition. *IEEE Transactions on Industrial Electronics*, *55*(7), 2629–2637. doi:10.1109/TIE.2008.924036

Seraji, H. (1989). Decentralized Adaptive Control of Manipulators: Theory, Simulation, and Experimentation. *IEEE Transactions on Robotics and Automation*, *5*(2), 183–201. doi:10.1109/70.88039

Servet Kiran, M., Gunduz, M., & Kaan Baykan, O. (2012). A novel hybrid algorithm based on particle swarm and ant colony optimization for finding the global minimum. *Applied Mathematics and Computation*, *219*(1), 1515–1521. doi:10.1016/j.amc.2012.06.078

Seydezadeh, S. M., & Mirzakuchaki, S. (2012). A fast color image encryption algorithm based on coupled two-dimensional piecewise chaotic map. *Signal Processing*, *92*(5), 1202–1215. doi:10.1016/j.sigpro.2011.11.004

Sfaihi, B., & Benrejeb, M. (2013). On stability analysis of nonlinear discrete singularly perturbed T-S fuzzy models. *Journal of Dynamical and Control Systems*, *1*(1), 20–31.

Sguarezi Filho, A. J., de Oliveira Filho, M. E., & Filho, E. R. (2011). A Predictive Power Control for Wind Energy. Sustainable Energy. *IEEE Transactions on Sustainable Energy*, *2*(1), 97–105.

SguareziFilho, A.J., &Ruppert, E. (2010). A Deadbeat Active and Reactive Power Control for Doubly Fed Induction Generator. *Electric Power Components and Systems*, *38*(14), 592–602.

Shafiullah, G.M., & Amanullah, M.T.O., Shawkat Ali, A.B.M., & Peter, W. (2013). Potential challenges of integrating large-scale wind energy into the power grid: A review. *Renewable & Sustainable Energy Reviews*, *20*, 306–321. doi:10.1016/j.rser.2012.11.057

Shaheen, S. A., Hasanien, H. M., & Badr, M. A. (2010, December). Study on doubly fed induction generator control. *Proceedings of MEPCON*, *2010*, 627–633.

Shamma, J. S., & Tu, K.-Y. (1999). Set-valued observers and optimal disturbance rejection. *IEEE Transactions on Automatic Control*, *44*(2), 253–264.

Shannon, C. (1949). Communication theory of secrecy systems. *The Bell System Technical Journal*, *28*(4), 656–725. doi:10.1002/j.1538-7305.1949.tb00928.x

Shao, H. (2009). New delay-dependent stability criteria for systems with interval delays. *Automatica*, *45*(3), 744–749. doi:10.1016/j.automatica.2008.09.010

Sharma, R., & Aldeen, M. (2004). Estimation of unknown disturbances in nonlinear systems. In C. R. Burrows, K. A. Edge, D. N. Johnston, & University of Bath (Eds.), *Centre for Power transmission and Motion Control., Bath Workshop on Power Transmission and Motion Control* (PTMC 2004). University of Bath.

Shehata, E. G., & Gerges, M. (2013). Direct power control of DFIGs based wind energy generation systems under distorted grid voltage conditions. *Electrical Power and Energy Systems*, *53*, 956–966. doi:10.1016/j.ijepes.2013.06.006

She, Y., She, X., & Baran, M. E. (2011). Universal Tracking Control of Wind Conversion System for Purpose of Maximum Power Acquisition under Hierarchical Control Structure. *IEEE Transactions on Energy Conversion*, *26*(3), 766–775. doi:10.1109/TEC.2011.2159607

Shieh, L. S., Yeung, C. K., & Mclnnis, B. C. (1978). Solutions of state-space equations via block-pulse functions. *International Journal of Control*, *28*(3), 383–392. doi:10.1080/00207177808922465

Shigeo, H., & Akio, M. (1990). Design and control of a mobile robot with an articulated body. *The International Journal of Robotics*, *9*(2), 99–114. doi:10.1177/027836499000900208

Shi, J., Gao, F., & Wu, T. J. (2005). Robust design of integrated feedback and iterative learning control of a batch process based on a 2D Roesser system. *Journal of Process Control*, *15*(2), 907–924. doi:10.1016/j.jprocont.2005.02.005

Shi, J.-H., Zhong, Z.-D., Zhu, X.-J., & Caco, G.-Y. (2008). Robust design and optimization for autonomous PV-wind hybrid power systems. *Journal of Zhejiang University Sciences A*, *9*(3), 401–409. doi:10.1631/jzus.A071317

Shin, H. C., Sayed, A. H., & Song, W. J. (2004). Variable Step-Size NLMS and Affine Projection Algorithms. *IEEE Signal Processing Letters*, *11*(2), 132–135. doi:10.1109/LSP.2003.821722

Shinskey, F. G. (1967). *Process Control Systems*. New York: McGraw-Hill.

Shi, P., Boukas, K. E., & Agarwal, K. R. (1999). Control of Markovian jump discrete time systems with norm bounded uncertainty and unknown delay. *IEEE Transactions on Automatic Control*, *44*(11), 2139–2155. doi:10.1109/9.802932

Shorten, R., & Narendra, K. (2003). On Common Quadratic Lyapunov Functions for Pairs of Stable LTI Systems Whose System Matrices are in Companion Form. *IEEE Transactions on Automatic Control*, *48*(4), 618–621. doi:10.1109/TAC.2003.809795

Shorten, R., Narendra, K., & Mason, O. (2003). A Result on Common Quadratic Lyapunov Functions. *IEEE Transactions on Automatic Control*, *48*(1), 110–113. doi:10.1109/TAC.2002.806661

Shorten, R., Wirth, F., Mason, O., Wulf, K., & King, C. (2007). Stability criteria for switched and hybrid systems. *SIAM Review*, *49*(4), 545–592. doi:10.1137/05063516X

Shotton, J., Fitzgibbon, A., Cook, M., Sharp, T., Finocchio, M., Moore, R., et al. (2011). Real-time human pose recognition in parts from single depth images. In Proceedings of IEEE Computer Vision and Pattern Recognition (CVPR). IEEE Press. doi:10.1109/CVPR.2011.5995316

Shu, T. (1995). *Uniform random numbers: Theory and practice*. Kluwer Academic Publishers.

Shyu, K. K., Liu, W. J., & Hsu, K. C. (2003). Decentralized variable structure control design for uncertain large scale systems containing a deadzone. *IEE Proceedings. Control Theory and Applications*, *150*(5), 467–475. doi:10.1049/ip-cta:20030758

Shyu, K. K., Liu, W. J., & Hsu, K. C. (2005). Design of large-scale time-delayed systems with dead-zone input via variable structure control. *Automatica*, *41*(7), 1239–1246. doi:10.1016/j.automatica.2005.03.004

Siegwart, R., Nourbakhsh, I. R., & Scaramuzza, D. (2011). *Introduction to Autonomous Mobile Robots*. Cambridge, MA: MIT Press.

Silva, E. I., & Erraz, D. A. (2006). An LQR Based MIMO PID Controller Synthesis Method for Unconstrained Lagrangian Mechanical Systems. In *Proceedings of the 45th IEEE Conference on Decision & Control. Decision and Control*. San Diego, CA: IEEE. doi:10.1109/CDC.2006.377348

Sinha, N. K. (1972). Estimation of transfer functions of continuous-time system from sampled data. *Proceedings of the IEE*, *119*(5), 612–614.

Sinha, N. K., & Lastman, G. J. (1981). Transformation algorithm for identification of continuous-time multivariable systems from discrete data. *Electronics Letters*, *17*(5), 779–780. doi:10.1049/el:19810546

Sinha, N. K., & Rozsa, P. (1976). Some canonical forms for linear multivariable systems. *International Journal of Control*, *23*(6), 865–883. doi:10.1080/00207177608922207

Sinha, N. K., & Zhou Qi-Jie, . (1983). Discrete time approximation of multivariable continuous-time systems. *IEE Proceedigs D: Control Theory and Applications*, *130*(3), 103–110. doi:10.1049/ip-d.1983.0019

Sira-Ramirez, H., Perez-Moreno, R., Ortega, R., & Garcia-Esteban, M. (1997). Passivity-based controllers for the stabilization of dc-to-dc power converters. *Automatica*, *33*(4), 499–513. doi:10.1016/S0005-1098(96)00207-5

Skogestad, S., & Postlethwaite, I. (1996). *Multivariable Feedback Control: Analysis and Design*. New York: John Wiley and Sons.

Slootweg, J. G., Polinder, H., & Kling, W. L. (2001). Dynamic modelling of a wind turbine with doubly fed induction generator. *IEEE Power Engineering Society Summer Meeting*, *1*, 644-649.

Slootweg, J. G. (2005). Reduced-order modelling of wind turbines. In *Wind Power in Power Systems*. New York: Wiley. doi:10.1002/0470012684.ch25

Solc, F., & Honzik, B. (2002). Modeling and control of a soccer robot. In *Proceedings of the 7th IEEE International Workshop in Advanced Motion Control* (pp. 506-509). Maribor, Slovenia: IEEE. doi:10.1109/AMC.2002.1026972

Song, K. Y., & Tay, J. C. (July 1992). Fuzzy navigation of a mobile robot. In *Proceedings of the 1992 IEEE/RSJ International Conference on Intelligent Robots and Systems*, (pp. 112-132). Raleigh, NC: IEEE.

Song, H.-S., & Nam, K. (1999). Dual Current Control Scheme for PWM Converter under Unbalanced Input Voltage Condition. *IEEE Transactions on Industrial Electronics*, *46*(5), 953–959. doi:10.1109/41.793344

Soroush, M., & Kravaris, C. (1996). A continuous-time formulation of nonlinear model predictive control. *International Journal of Control*, *63*(1), 121–146. doi:10.1080/00207179608921835

Sourabh, J., Sadhna, M., & Vineet, R. (2013). An Approach for Image Matching based on ACO fuzzification. *International Journal of Modern Engineering & Management Research*, *1*(2), 28–33.

Sousa, J. B., & Pereira, F. L. (1996). A general control architecture for multiple vehicles. In *Proceedings of the IEEE International Conference on Robotics and Automation*, (pp. 692-697). IEEE. doi:10.1109/ROBOT.1996.503855

Sozer, Y., Torrey, D. A., & Mese, E. (2010), An adaptive predictive current control technique for permanent magnet synchronous motors, IEEE Energy Conversion Congress and Exposition (ECCE), 2819-2826, Atlanta, GA.

Spong, M. W., & Hutchinson, S. (2006). *Robot Modeling and Control*. John Wiley and Sons.

Sprott, J. (2003). *Chaos and Time-Series Analysis*. Oxford, UK: Oxford University Press.

Sprott, J. C. (2007). A simple chaotic delay differential equation. *Physics Letters [Part A]*, *366*(4-5), 397–402. doi:10.1016/j.physleta.2007.01.083

Sprott, J. C. (2010). *Elegant Chaos: Algebraically Simple Chaotic Flows*. Singapore: World Scientific.

Stancu, A., Puig, V., Cugueró, P., & Quevedo, J. (2005). Benchmarking on Approaches to Interval Observation Applied to Robust Fault Detection. In C. Jermann, A. Neumaier, & D. Sam (Eds.), *Global Optimization and Constraint Satisfaction* (Vol. 3478, pp. 171–191). Springer Berlin Heidelberg.

Stefan, B., & Tomasz, P. (2011), Energy Efficient Open Loop Control of Permanent Magnet Synchronous Motors for Fan Applications.*17th International Conference on Electrical Drives and Power Electronics*, Slovakia.

Steinbauer, G., & Wotawa, F. (2005). Detecting and locating faults in the control software of autonomous mobile robots. In *Proceedings of the 16th International Workshop on Principles of Diagnosis*, (pp. 13–18). Monterey, CA: Academic Press.

Stergios, R., Gaurav, S., & George, B. (1998). Fault detection and identification in a mobile robot using multiple model estimation. In *Proceedings of the 1998 IEEE International Conference on Robotics and Automation*, (pp. 2223-2228). IEEE. Doi:10.1109/ROBOT.1998.680654

Stilwell, D. J., & Bay, J. S. (1993). Toward the development of a material transport system using swarms of ant-like robots. In *Proceedings of the IEEE International Conference on Robotics and Automation*, (pp. 766-771). IEEE. doi:10.1109/ROBOT.1993.292070

STMicroelectronics. (2012). *Datasheet stm32f100x4, stm32f100x6, stm32f100x8, stm32f100xb*. Doc ID 16455 Rev 7. Retrieved from http://www.st.com

Stojanović, B. S., Debeljović, D. L., & Mladenović, I. (2007). A Lyapunov-Krasovskii methodology for asymptotic stability of discrete time delay systems. *Serbian Journal of Electrical Engineering*, *4*(2), 109–117. doi:10.2298/SJEE0702109S

Stojanovski, T., & Kocarev, L. (2001a). Chaos-based random number generators - Part I: Analysis. *IEEE Transactions on Circuits and Systems. I, Fundamental Theory and Applications*, *48*(3), 281–288. doi:10.1109/81.915385

Stojanovski, T., Pihl, J., & Kocarev, L. (2001b). Chaos-based random number generators - Part II: Practical realizations. *IEEE Transactions on Circuits and Systems. I, Fundamental Theory and Applications*, *48*(3), 382–385. doi:10.1109/81.915396

Strogatz, S. H. (1994). *Nonlinear Dynamics and Chaos: with Applications to Physics, Biology, Chemistry, and Engineering*, Massachusetts: Perseus Books, Sun, C., Lin, Y., & Han, M. (2006). Stability and Hopf bifurcation for an epidemic disease model with delay. *Chaos, Solitons, and Fractals*, *30*, 204–216.

Sugeno, M., Murofushi, T., Mori, T., Tatematasu, T., & Tanaka, J. (1989). Fuzzy algorithmic control of a model car by oral instructions. *International Journal of Fuzzy Sets and Systems*, *32*(2), 207–219. doi:10.1016/0165-0114(89)90255-8

Sugiatro, I., Tung, L. L. U., & Rahman, M. I. (2008). Implementation of fuzzy logic in FPGA for maze tracking. *Jurnal Teknik Elektro*, *8*(2), 96–102.

Suh, Y. S., & Lee, M. H. (1999).Stability of state delay systems based on a Lyapunov functional. In *Proceedings of the IEEE International Symposium*, (pp. 1093-1098). Bled, Slovenia: IEEE.

Suman M. and Chakrabatory C. (2010), A new instantaneous reactive power based MRAS for sensorless induction motor drive, *Simulation modelling and practice Theory-Elsevier*,18(2), 1314-1326.

Sundarapandian, V., & Karthikeyan, R. (2012). Hybrid synchronization of hyperchaotic Lorenz and hyperchaotic Chen systems via active control. *Journal of Engineering and Applied Sciences (Asian Research Publishing Network)*, *7*(3), 254–264. doi:10.3923/jeasci.2012.254.264

Sun, Y. G., Wang, L., & Xie, G. (2006). Stability of switched systems with time-varying delays: delay-dependent common Lyapunov functional approach. In *Proceedings of the American Control Conference* (pp.1544–1549). Philadelphia, PA: Academic Press.

Sun, Y., & Wua, C. Q. (2012). Stability analysis via the concept of Lyapunov exponents: A case study in optimal controlled biped standing. *International Journal of Control*, *85*(12), 1952–1966. doi:10.1080/00207179.2012.713985

Sun, Z., & Ge, S. S. (2002). *Switched Linear Systems Control and Design*. New York: Springer.

Sun, Z., & Ge, S. S. (2011). *Stability Theory of Switched Dynamical Systems*. London, UK: Springer. doi:10.1007/978-0-85729-256-8

Su, R. (1982). On the linear equivalents of nonlinear systems. *Systems & Control Letters*, *2*(1), 48–52. doi:10.1016/S0167-6911(82)80042-X

Sürgevil, T., & Akpinar, E. (2005). Modelling of a 5-kW wind energy conversion system with induction generator and comparison with experimental results. *Renewable Energy*, *30*(6), 913–929. doi:10.1016/j.renene.2004.09.002

Surinkaew, T., & Ngamroo, I. (2014). Robust power oscillation damper design for DFIG-based wind turbine based on specified structure mixed H_2/H_∞ control. *Renewable Energy*, *66*, 15–24. doi:10.1016/j.renene.2013.11.060

Suul, J. A., Luna, A., Rodriguez, P., & Undeland, T. (2012). Virtual-Flux-Based Voltage-Sensor-Less Power Control for Unbalanced Grid Conditions. *IEEE Transactions on Power Electronics*, *27*(9), 4071–4087. doi:10.1109/TPEL.2012.2190301

Suykens, J., Brabanter, J. D., Lukas, L., & Vandewalle, J. (2002). Weighted least squares support vector machines: Robustness and sparse approximation. *Neurocomputing*, *48*(14), 85–105. doi:10.1016/S0925-2312(01)00644-0

Szigeti, F. (1991). Kalman's rank condition for infinite-dimensional time-dependent systems. In *Proceedings of the 1991 Differential Equations Conference (equadiff)*, (pp. 927-931). Academic Press.

Tabakhi, S., Moradi, P., & Akhlaghian, F. (2014). An unsupervised feature selection algorithm based on ant colony optimization. *Engineering Applications of Artificial Intelligence*, *32*(1), 112–123. doi:10.1016/j.engappai.2014.03.007

Taherbaneh, M., Rezaie, A. H., Ghafoorifard, H., Rahimi, K., & Menhaj, M. B. (2010). Maximizing Output Power of a Solar Panel via Combination of Sun Tracking and Maximum Power Point Tracking by Fuzzy Controllers. International Journal of Photoenergy, Article ID 312580.

Takaba, K., & Katayama, T. (1997). Robust performance of uncertain descriptor systems. In *Proceeding of European Control Conference*. Brussels, Belgium: Academic Press.

Takaba, K., & Katayama, T. (1998). Output Feedback Control for Descriptor Systems. *Automatica*, *34*(7), 841–850. doi:10.1016/S0005-1098(98)00025-9

Takagi, T., & Sugeno, M. (1985, January-February). Fuzzy identification of systems and its applications to modeling and control. *IEEE Transactions on Systems, Man, and Cybernetics*, *SMC-15*(1), 116–132. doi:10.1109/TSMC.1985.6313399

Takanashi, N., Sato, Y., Ohta, M., Hiroi, N., Okumura, F., & Yamada, A. (2012). Human-Computer Interaction Technology Using Image Projection and Gesture-Based Input. *NEC Technical Journ*al, Special Issue on Smart Device Solutions. *NEC Corporation*, *7*(3), 122–126.

Tamasevicius, A., Mykolaitis, G., & Bumeliene, S. (2006). Delayed feedback chaotic oscillator with improved spectral characteristic. *Electronics Letters*, *42*(13), 736–737. doi:10.1049/el:20061245

Tanaka, K., & Wang, H. O. (2004). Fuzzy control systems design and analysis: a linear matrix inequality approach. Wiley.com.

Tanaka, K., Ikeda, T., & Wang, H. O. (1998). Fuzzy regulators and fuzzy observers: Relaxed stability conditions and LMI-based designs. *IEEE Transactions on Fuzzy Systems*, *6*(2), 250–265.

Tang, K. W., & Tang, W. (2006). A low cost chaos-based random number generator realized in 8-bit precision environment. In *Proceedings of 2006 International Symposium in Nonlinear Theory and its Applications*, (pp. 395-398). Academic Press.

Tang, K. S., Zhong, G. Q., Chen, G., & Man, K. F. (2001). Generation of n-scroll attractors via sine function. *IEEE Transactions on Circuits and Systems I*, *48*(11), 1369–1372. doi:10.1109/81.964432

Tang, Y., Giao, L., & Guan, X. (2010). Identification of Wiener model using Step signals and particle swarm optimization. *Expert Systems with Applications*, *37*(4), 3398–3404.

Tang, Y., Wang, Z. D., & Fang, J. A. (2010). Image encryption using chaotic coupled map lattices with time-varying delays. *Communications in Nonlinear Science and Numerical Simulation*, *15*(9), 2456–2468. doi:10.1016/j.cnsns.2009.09.023

Tan, S.-C., Lai, Y., Cheung, M., & Tse, C. (2005). On the practical design of a sliding mode voltage controlled buck converter. *IEEE Transactions on Power Electronics*, *20*(2), 425–437. doi:10.1109/TPEL.2004.842977

Tan, S., Lai, Y., Tse, C., & Martinez-Salamero, L. (2007). Special family of PWM-based sliding-mode voltage controllers for basic dc-dc converters in discontinuous conduction mode. *IET Electric Power Applications*, *1*(1), 64–74. doi:10.1049/iet-epa:20060067

Tao, G., & Kokotovic, P. V. (1993). Adaptive Control of Systems with Backlash. *Automatica*, *29*(2), 323–335. doi:10.1016/0005-1098(93)90126-E

Tao, G., & Kokotovic, P. V. (1993). Continuous-time Adaptive Control of Systems with Unknown Backlash. In *Proceedings of the American Control Conference. American Control Conference.* San Francisco, CA: IEEE.

Tao, G., & Kokotovik, P. (1995). Adaptive Control of Systems with Unknown Output Backlash. *IEEE Transactions on Automatic Control*, *40*(2), 326–330. doi:10.1109/9.341803

Tao, G., & Ma, X. (1999). Backlash Compensation for Multivariable Nonlinear Systems with Actuator Dynamics. In *Proceedings of the 38th Conference on Decision & Control. Decision and Control*. Phoenix, AZ: IEEE.

Tao, G., Ma, X., & Ling, Y. (2001). Optimal and nonlinear decoupling control of systems with sandwiched backlash. *Automatica, 37*(2), 165–176. doi:10.1016/S0005-1098(00)00153-9

Tapia, A., Tapia, G., Ostolaza, J. X., & Saenz, J. R. (2003). Modeling and control of a wind turbine driven doubly fed induction generator. *IEEE Transactions on Energy Conversion, 18*(2), 194–204. doi:10.1109/TEC.2003.811727

Tatizawa, H., Netsuo, E.S. (2009).Calibration of high voltage transducers for power quality measurements. *WSEAS Trans. on Systems, 8*(3), 400-409.

Tazil, M., Kumar, V., Bansal, R. C., Kong, S., Dong, Z. Y., Freitas, W., & Mathur, H. D. (2010). Three-Phase doubly fed induction generators: An overview. *IET Proceedings of Electric Power Application, 4*(2), 75-89.

Te-Jen, S., Li-Wei, C., Chia-Jung, Y., & Jui-Chuan, C. (2010), Fuzzy PID controller design using self adaptive bacterial foraging optimization, *Proceedings of SICE Annual Conference*2010, pp. 2604-2607, Taipei.

Teodorescu, R., Liserre, M., & Rodríguez, P. (2011). *Grid Converters for Photovoltaic and Wind Power Systems*. John Wiley & Sons, Ltd. doi:10.1002/9780470667057

Teslic, L., Škrjanc, I., & Klanèar, G. (2011). EKF-based localization of a wheeled mobile robot in structured environments. *Journal of Intelligent & Robotic Systems, 62*(2), 187–203. doi:10.1007/s10846-010-9441-8

Theilliol, D., Noura, H., & Ponsart, J. C. (2002). Fault diagnosis and accommodation of a three-tank system based on analytical redundancy. *Instrumentation, Systems, and Automation Transactions, 41*(3), 365–382. PMID:12160349

Thotaa, P., Zhaob, X., & Dankowicz, H. (2006). Co-dimension-Two Grazing Bifurcations in Single-Degree-of-Freedom Impact Oscillators. *Journal of Computational and Nonlinear Dynamics, 1*(4), 328–335. doi:10.1115/1.2338658

Thota, P., & Dankowicz, H. (2006a). Continuous and discontinuous grazing bifurcations. *Physica D. Nonlinear Phenomena, 214*(2), 187–197. doi:10.1016/j.physd.2006.01.006

Thota, P., & Dankowicz, H. (2006b). Analysis of grazing bifurcations of quasiperiodic system attractors. *Physica D. Nonlinear Phenomena, 220*(2), 163–174. doi:10.1016/j.physd.2006.07.005

Tian, H., & Hou, Y. (2009). Robust Stabilization for a class of Uncertain Descriptor Systems via both Proportional and Derivative State Feedback. In *Proceedings of the Second International Conference on Intelligent Computation Technology and Automation* (pp. 874-877). Academic Press. doi:10.1109/ICICTA.2009.218

Ti-Chung, L., Kai-Tai, S., Ching-Hung, L., & Ching-Cheng, T. (2001). Tracking control of unicycle modeled mobile robots using a saturation feedback controller. *IEEE Transactions on Control Systems Technology, 9*(2), 305–318. doi:10.1109/87.911382

Tierno, J. E., Kim, K. Y., Lacy, S. L., & Bernstein, D. S. (2000). Describing Function Analysis of an Anti-Backlash Controller. In *Proceedings of the American Control Conference*. Academic Press. doi:10.1109/ACC.2000.877005

Tlijani, H., M'sirdi, N. K., & Knani, J. (2011). Optimization of the mobile robot's path tracking: combination of genetic algorithm, artificial potential field and backtracking algorithm. In *Proceedings of the IEEE International Conference on Communications, Computing and Control Applications*, (pp. 97-102). Hammamet, Tunisia: IEEE.

Tlijani, H., & Knani, J. (2009). Optimization of the mobile robot's track using genetic algorithms. In *Proceedings of the 10th International Conference on Sciences and Techniques of Automatic Control & Computer Engineering*. Hammamet, Tunisia: Academic Press.

Tlijani, H., Knani, J., & M'sirdi, K. N. (2012). Optimization of a mobile robot's path tracking with time windows using Hybridization of genetic algorithms and CSP approach. In *Proceedings of International Conference on METAheuristics and Natural Inspired Computing*. Sousse, Tunisia: Academic Press.

Tlijani, H., Knani, J., & M'sirdi, K. N. (2014). Hybrid algorithm based on genetic algorithms and CSP approach to plane an optimized path for a mobile robot moving under time window. *IOSR Journal of Computer Engineering*, *16*(3), 74–83. doi:10.9790/0661-16337483

Toksari, M. D. (2009). Minimizing the multimodal functions with Ant Colony Optimization approach. *Expert Systems with Applications*, *36*(3), 6030–6035. doi:10.1016/j.eswa.2008.06.077

Tolbert, L. M., & Peng, F. Z. (1997). Multilevel inverters for large automotive drives. In *Proceedings of All Electric Combat Vehicle Second International Conference*, (pp. 209-214). Academic Press.

Toledo, C. F. M., Oliveira, L., & França, P. M. (2014). Global optimization using a genetic algorithm with hierarchically structured population. *Journal of Computational and Applied Mathematics*, *261*(1), 341–351. doi:10.1016/j.cam.2013.11.008

Tong, S. C., Chen, B., & Wang, Y. (2005). Fuzzy adaptive output feedback control for MIMO nonlinear systems. *Fuzzy Sets and Systems*, *156*(2), 285–299. doi:10.1016/j.fss.2005.06.011

Tong, S. C., & Li, H. X. (2003). Fuzzy adaptive sliding model control for MIMO nonlinear systems. *IEEE Transactions on Fuzzy Systems*, *11*(3), 354–360. doi:10.1109/TFUZZ.2003.812694

Tong, S. C., Tang, J., & Wang, T. (2000). Fuzzy adaptive control of multivariable nonlinear systems. *Fuzzy Sets and Systems*, *111*(2), 153–167. doi:10.1016/S0165-0114(98)00058-X

Tong, X., & Cui, M. (2009). Image encryption scheme based on 3D Baker with dynamical compound chaotic sequence cipher generator. *Signal Processing*, *89*(4), 480–491. doi:10.1016/j.sigpro.2008.09.011

Trcek, D. (2006). *Managing information systems security and privacy*. Berlin: Springer-Verlag.

Trintis, I. (2011). *Grid Converters for Stationary Battery Energy Storage Systems*. PhD, Faculty of Engineering & Science at Aalborg University.

Tsai, J. (2009). Treating free variables in generalized geometric programming problems. *Computers & Chemical Engineering*, *33*(1), 239–243. doi:10.1016/j.compchemeng.2008.08.011

Tsai, J., Lin, M., & Hu, Y. (2007). On generalized geometric programming problems with non positive variables. *European Journal of Operational Research*, *178*(1), 10–19. doi:10.1016/j.ejor.2005.11.037

Tse, C., & Di Bernardo, M. (2002). Complex behavior in switching power converters. *Proceedings of the IEEE*, *90*(5), 768–780. doi:10.1109/JPROC.2002.1015006

Tsuneda, A., Eguchi, K., & Inoue, T. (1999). Design of chaotic binary sequences with good statistical properties based on piecewise linear into maps. In *Proceedings of the 7th International Conference on Microelectronics for Neural, Fuzzy and Bio-Inspired Systems*, (pp. 261-266). Academic Press. doi:10.1109/MN.1999.758873

Tursini, M., Chiricozzi, E., & Petrella, R. (2010). Feed forward Flux- Weakening Control of Surface- Mounted Permanent-Magnet Synchronous Motors Accounting for Resistive Voltage Drop. *IEEE Transactions on Industrial Electronics*, *57*(1), 440–448. doi:10.1109/TIE.2009.2034281

Tzung, Y. L., & Ying-Yu, T. (2009). Digital Control of Boost PFC AC/DC Converters With Low THD and Fast Dynamic Response. In *Proceedings of IEEE IPEMC Conference. Rec*, (pp. 1672-1677). IEEE. DOI:10.1109/*IPEMC.2009.5157660.*

Ucar, A. (2002). A prototype model for chaos studies. *International Journal of Engineering Science*, *40*(3), 251–258. doi:10.1016/S0020-7225(01)00060-X

Ucar, A. (2003). On the chaotic behaviour of a prototype delayed dynamical system. *Chaos, Solitons, and Fractals*, *16*(2), 187–194. doi:10.1016/S0960-0779(02)00160-1

Ueda, Y., & Akamatsu, N. (1981). Chaotically transitional phenomena in the forced negative-resistance oscillator. *IEEE Transactions on Circuits and Systems*, *CAS-28*(3), 217–224. doi:10.1109/TCS.1981.1084975

Uppal, F. J., Patton, R. J., & Witczak, M. (2006). A neuro-fuzzy multiple-model observer approach to robust fault diagnosis based on the DAMADICS benchmark problem. *Control Engineering Practice*, *14*(6), 699–717.

Urmila, B., & Subbarayudu, D. (2010). Multilevel inverters: A comparative study of pulse width modulation techniques. *International Journal of Scientific and Engineering Research*, *1*(3), 1–5.

Valdez, F., Melin, P., & Castillo, O. (2014). Modular Neural Networks architecture optimization with a new nature inspired method using a fuzzy combination of Particle Swarm Optimization and Genetic Algorithms. *Information Sciences*, *270*(20), 143–153. doi:10.1016/j.ins.2014.02.091

Valenciaga, F., & Puleston, P. F. (2005). Supervisor Control for a Stand-Alone Hybrid Generation System Using Wind and Photovoltaic Energy. *IEEE Transactions on Energy Conversion*, *20*(2), 398–405. doi:10.1109/TEC.2005.845524

Van Brussel, H., Wyns, J., Valckenaers, P., Bongaerts, L., & Peeters, P. (1998). Reference architecture for holonic manufacturing systems: PROSA. *Computers in Industry*, *37*(3), 255–274. doi:10.1016/S0166-3615(98)00102-X

Van Vaerenbergh, S., Santamaria, I., Liu, W., & Principe, J. (2010, March). Fixed-Budget Kernel Recursive Least-Squares. In *Proceedings of ICASSP*, (pp. 1882–1885). ICASSP.

Vapnik, V. N. (1995). *The Nature of Statistical Learning Theory*. New York: Springer-Verlag. doi:10.1007/978-1-4757-2440-0

Vechiu, I., Ciprian, B., & Gelu, G. (2013). Three-Phase Four-Wire Active Power Conditioners for Weak Grids. *International Journal of Computer and Electrical Engineering*, *5*(2), 173–178. doi:10.7763/IJCEE.2013.V5.689

Volos, C. K., Kyprianidis, I. M., & Stouboulos, I. N. (2011). Various synchronization phenomena in bidirectionally coupled double scroll circuits. *Communications in Nonlinear Science and Numerical Simulation*, *16*(8), 3356–3366. doi:10.1016/j.cnsns.2010.11.015

Volos, C. K., Kyprianidis, I. M., & Stouboulos, I. N. (2012). A chaotic path planning generator for autonomous mobile robots. *Robotics and Autonomous Systems*, *60*(4), 651–656. doi:10.1016/j.robot.2012.01.001

Volos, Ch. K. (2013b). Chaotic random bit generator realized with a microcontroller. *Journal of Computations & Modelling*, *3*(4), 115–136.

Volos, Ch. K., Kyprianidis, I. M., & Stouboulos, I. N. (2006). Experimental demonstration of a chaotic cryptographic scheme. *WSEAS Transactions on Circuits and Systems*, *5*, 1654–1661.

Volos, Ch. K., Kyprianidis, I. M., & Stouboulos, I. N. (2010). Fingerprint images encryption process based on a chaotic true random bits generator. *International Journal of Multimedia Intelligence and Security*, *1*(4), 320–335. doi:10.1504/IJMIS.2010.039234

Volos, Ch. K., Kyprianidis, I. M., & Stouboulos, I. N. (2013a). Image encryption process based on chaotic synchronization phenomena. *Signal Processing*, *93*(5), 1328–1340. doi:10.1016/j.sigpro.2012.11.008

Volos, Ch. K., Laftsis, I., & Gkogka, G. D. (2009). Synchronization of two chaotic Chua – type circuits with the inverse system approach. In *Proc. 1st Panhellenic Conference on Electronics and Telecommunications*. Patras, Greece: Academic Press.

Volterra, V. (1959). Theory of Functional of Integral and Integro-Differential Equations. New York: Dover.

Von Neumann, J. (1951). Various techniques used in connection with random digits. Applied Mathematics Series, 12, 36-38.

Vörös, J. (2002). Modeling and parameter identification of systems with multisegment piecewise-linear characteristics. *IEEE Transactions on Automatic Control*, *47*(1), 184–188.

Voss, H. U. (2002). Real-time anticipation of chaotic states of an electronic circuit. *International Journal of Bifurcation and Chaos in Applied Sciences and Engineering*, *12*(07), 1619–1625. doi:10.1142/S0218127402005340

Wadikhaye, S., Yong, Y., & Moheimani, S. (2012). Design of a compact serial-kinematic scanner for high-speed atomic force microscopy: An analytical approach. *Micro Nano Letters, IET*, *7*(4), 309–313. doi:10.1049/mnl.2011.0477

Wagg, D. J. (2005). Periodic sticking motion in a two-degree-of-freedom impact oscillator. *International Journal of Non-linear Mechanics*, *40*(8), 1076–1087. doi:10.1016/j.ijnonlinmec.2005.03.002

Wai, R.-J., & Shih, L.-C. (2011). Design of voltage tracking control for dc-dc boost converter via total sliding-mode technique. *IEEE Transactions on Industrial Electronics*, *58*(6), 2502–2511. doi:10.1109/TIE.2010.2066539

Walter, E., & Piet-Lahanier, H. (1987). *Exact and recursive description of the feasible parameter set for bounded error models*. Paper presented at the 26th IEEE Conference on Decision and Control. Los Angeles, CA. Doi:10.1109/CDC.1987.272849

Walter, E., Norton, J., Piet-Lahanier, H., & Milanese, M. (1996). *Bounding approaches to system identification*. Perseus Publishing.

Walter, E., & Piet-Lahanier, H. (1989). Exact recursive polyhedral description of the feasible parameter set for bounded-error models. *IEEE Transactions on Automatic Control*, *34*(8), 911–915.

Wanfeng, Z., Guang, F., Yan-Fei, L., & Bin, W. (2006). New digital control method for power factor correction. *IEEE Transactions on Industrial Electrics*, *53*(3), 987-990.

Wang, H., & Daley, S. (1996). Actuator fault diagnosis: An adaptive observer based technique. In Proceedings of Automatic Control. IEEE Transactions onAutomatic Control, 41(7), 1073-1078.

Wang, Z. (2010). Unknown Input Observer Based Fault Class Isolation and Estimation. In *Proceedings of the 29th Chinese Control Conference (CCC)*, (pp. 3963-3968). CCC.

Wang, D., Chu, Y., & Ding, F. (2010). Auxiliary model based recursive generalized least squares parameter estimation for Hammerstein OEAR systems. *Mathematical and Computer Modelling*, *52*(1-2), 309–317.

Wang, H. O., Tanaka, K., & Griffin, M. F. (1996). An approach to fuzzy control of nonlinear systems: Stability and design issues. *IEEE Transactions on Fuzzy Systems*, *4*(1), 14–23.

Wang, H., Huang, Z. J., & Daley, S. (1997). On the use of adaptive updating rules for actuator and sensor fault diagnosis. *Automatica*, *33*(2), 217–225. doi:10.1016/S0005-1098(96)00155-0

Wang, J., & Rugh, W. J. (1987). Feedback linearization families for nonlinear systems. *IEEE Transactions on Automatic Control*, *32*(10), 935–940. doi:10.1109/TAC.1987.1104470

Wang, J., & Rugh, W. J. (1987). Parameterized linear systems and linearization families for nonlinear systems. *IEEE Transactions on Circuits and Systems*, *34*(6), 650–657. doi:10.1109/TCS.1987.1086191

Wang, J., & Rugh, W. J. (1989). On the pseudo-linearization problem for nonlinear systems. *Systems & Control Letters*, *12*(2), 161–167. doi:10.1016/0167-6911(89)90009-1

Wang, K., Li, P., Liu, J., & Ning, D. (2011). Application of μc/os - II in the Design of Mine dc Electrical Prospecting Instrument. *Procedia Earth and Planetary Science*, *3*(1), 485–492. doi:10.1016/j.proeps.2011.09.124

Wang, L. X. (1994). *Adaptive fuzzy systems and control: design and stability analysis*. Englewood Cliffs, NJ: Prentice-Hall.

Wang, L.-J., Yang, T., Zhang, D.-M., & Lu, Z.-. (2011). A high performance simulation methodology for multilevel grid-connected inverters. *Journal of Zhejiang University Sciences C*, *13*(7), 544–551. doi:10.1631/jzus.C1100315

Wang, S., Feng, J., & Tse, C. K. (2013). Kernal affine projection sign algorithms for combating impulse interference. *IEEE Transactions on Circuits and Wystems II, Express Briefs*, *60*(11), 811–815. doi:10.1109/TCSII.2013.2281712

Wang, S.-H., Wang, E., & Dorato, P. (1975). Observing the states of systems with unmeasurable disturbances. *IEEE Transactions on Automatic Control*, *20*(5), 716–717.

Wang, X. M., Yu, H. G., Zhong, S. M., & Agarwal, R. P. (2010). Analysis of mathematics and dynamics in a food web system with impulsive perturbations and distributed time delay. *Applied Mathematical Modelling*, *34*(12), 3850–3863. doi:10.1016/j.apm.2010.03.024

Wang, X., Teng, L., & Qin, X. (2012). A novel colour image encryption algorithm based on chaos. *Signal Processing*, *92*(4), 1101–1108. doi:10.1016/j.sigpro.2011.10.023

Wang, X., Xing, Y., Zhipeng, H., & Yan, L. (2012). Research and Simulation of DTC Based on SVPWM of PMSM. *Procedia Engineering*, *9*, 1685–1689. doi:10.1016/j.proeng.2012.01.195

Wei, J., Liao, X., Wong, K., & Xiang, T. (2006). A new chaotic cryptosystem. *Chaos, Solitons, and Fractals*, *30*(5), 1143–1152. doi:10.1016/j.chaos.2005.09.005

Wiener, N. (1949). *Extrapolation, Interpolation and Smoothing of Stationary Time Series, with Engineering Applications*. Wiley.

Wiercigroch, M., & de Kraker, B. (2000). *Applied nonlinear dynamics and chaos of mechanical systems with discontinuities*. Singapore: World Scientific.

Wiggins, S. (2003). *Introduction to Applied Nonlinear Dynamical Systems and Chaos (Texts in Applied Mathematics)* (2nd ed.). Heidelberg, Germany: Springer.

Wigren, T. (1993). Recursive prediction error identification using the nonlinear Wiener model. *Automatica*, *29*(4), 1011–1025.

William Helton, J. (1978). Orbit structure of the Mobius transformation semi group action on H-infinity broadband matching. *Advanced in Math. Suppl. Stud*, *3*, 129–197.

Wisse, M., Schwab, A. L., & Van Der Helm, F. C. (2004). Passive dynamic walking model with upper body. *Robotica*, *22*(6), 681–688. doi:10.1017/S0263574704000475

Witczak, M., Dziekan, L., Puig, V., & Korbicz, J. (2008). A fault-tolerant control strategy for Takagi-Sugeno fuzzy systems. In *Proceedings of the 17th IFAC World Congress* (Vol. 17, pp. 7387-7392). COEX. Doi:10.3182/20080706-5-KR-1001.01249

Witvoet, G. (2005). *Control of chaotic dynamical systems using OGY*. Eindhoven, The Netherlands: Technische Universiteit Eindhoven.

Wobbrock, J. O., Wilson, A. D., & Li, Y. (2007). Gestures without libraries, toolkits or training: a $1 recognizer for user interface prototypes. In Proceedings of UIST '07, (pp. 159-168). UIST.

Woft, A., Swift, J., Swinney, H., & Vastano, J. (1985). Determining Lyapunov exponents from a time series. *Physica D. Nonlinear Phenomena*, *16*(3), 285–317. doi:10.1016/0167-2789(85)90011-9

Wong, C. C., Wang, H. Y., Li, S. A., & Cheng, C. T. (2007). Fuzzy controller designed by GA for two-wheeled mobile robots. *International Journal of Fuzzy Systems*, *9*(1), 22–30.

Wu, N. E., Zhang, Y., & Zhou, K. (2000a). Detection, estimation, and accommodation of loss of control effectiveness. *International Journal of Adaptive Control and Signal Processing*, *14*(7), 775–795. doi:10.1002/1099-1115(200011)14:7<775::AID-ACS621>3.0.CO;2-4

Wu, Z. G., Park, J. H., Su, H., & Chu, J. (2013). Delay-dependent passivity for singular Markov jump systems with time-delays. *Communications in Nonlinear Science and Numerical Simulation*, *18*(3), 669–681. doi:10.1016/j.cnsns.2012.08.017

Xia, Y., Shi, P., Liu, G., & Rees, D. (2005). Robust Mixed H_2 / H_∞ Sate feedback Control for Continuous-Time Descriptor Systems with Parameter Uncertainties. *Circuit Systems Signal*, *24*(4), 431–443.

Xiang, Z., & Huang, S. (2013). Stability Analysis and Stabilization of Discrete-Time 2D Switched Systems. *Circuits Syst Signal Process*, *32*(1), 401–414. doi:10.1007/s00034-012-9464-4

Xia, Y., Fu, M., & Shi, P. (2009). Analysis and synthesis of dynamical systems with time-delays. In *Lecture Notes in Control and Information Sciences*. New York: Springer.

Xia, Y., & Jia, Y. (2003a). Robust sliding mode for uncertain time delay systems: An lmi approach. *IEEE Transactions on Automatic Control*, *84*(6), 1086–1092.

Xia, Y., & Jia, Y. (2003b). Robust control of state delayed systems with polytopic type uncertainties via parameter dependent Lyapunov functionals. *Systems & Control Letters*, *50*(3), 183–193. doi:10.1016/S0167-6911(03)00153-1

Xia, Y., & Mandic, D. P. (2012). Widely Linear Adaptive Frequency Estimation of Unbalanced Three-Phase Power Systems. *IEEE Transactions on Instrumentation and Measurement*, *61*(1), 74–83. doi:10.1109/TIM.2011.2159409

Xia, Z., Zun-Quan, X., & Yan, G. (2012). Exponential Stabilizability of Switched Systems with Polytopic Uncertainties.Hindawi Publishing Corporation. *Journal of Applied Mathematics*, (853170): 15.

Xiong, F., Yue, W., Ming, L., Ke, W., & Wanjun, L. (2010). A Novel PLL for Grid Synchronization of Power Electronic Converters in Unbalanced and Variable-Frequency Environment. In *Proceedings of the 2nd IEEE International Symposium on Power Electronics for Distributed Generation Systems* (PEDG), (pp. 466-471). IEEE. Doi:10.1109/PEDG.2010.5545832

Xiong, K., Wei, C. L., & Liu, L. (2012). Robust Kalman filtering for discrete-time nonlinear systems with parameter uncertainties. *Aerospace Science and Technology*, *18*(1), 15–24.

Xiong, Y., & Saif, M. (2003). Unknown disturbance inputs estimation based on a state functional observer design. *Automatica*, *39*(8), 1389–1398. doi:10.1016/S0005-1098(03)00087-6

Xu, L., & Wang, Y. (2007). Dynamic Modelling and Control of DFIG-Based Wind Turbines under Unbalanced Network Conditions. *IEEE Transactions on Power Systems*, *22*(1), 314–323. doi:10.1109/TPWRS.2006.889113

Xu, L., Zhi, D., & Williams, B. (2009). Predictive current control of doubly fed induction generator. *IEEE Transactions on Industrial Electronics*, *56*(10), 4143–4153. doi:10.1109/TIE.2009.2017552

Xu, S., Dooren, V. P., Stefan, R., & Lam, J. (2002). Robust stability and stabilization for singular systems with state delay and parameter uncertainty. *IEEE Transactions on Automatic Control*, *47*(7), 1112–1128.

Xu, S., & Lam, J. (2003). Robust stability for uncertain discrete singular systems with state delay. *Asian Journal of Control*, *5*(3), 399–405. doi:10.1111/j.1934-6093.2003.tb00132.x

Xu, S., & Lam, J. (2004). Robust stability and stabilization for discrete singular systems: An equivalent characterization. *IEEE Transactions on Automatic Control*, *49*(4), 568–574. doi:10.1109/TAC.2003.822854

Xu, S., & Lam, J. (2006). *Robust Control and Filtering of Singular Systems*. Berlin: Springer.

Xu, S., Lam, J., & Yang, C. (2002). Robust H_∞ control for discrete singular systems with state delay and parameter uncertainty. *Dynamic Continuous Discrete Impulsive Systems B*, *9*, 539–554.

Xu, S., & Yang, C. (1999). Stabilization of discrete time singular systems: A matrix inequalities approach. *Automatica*, *35*(9), 1613–1617. doi:10.1016/S0005-1098(99)00061-8

Xu, S., Yang, C., Niu, Y., & Lam, J. (2001). Robust stabilization for uncertain discrete singular systems. *Automatica*, *37*(5), 769–774. doi:10.1016/S0005-1098(01)00013-9

Yagoubi, M., Chevrel, P., & Claveau, F. (2007). An Extended state-feedback for Descriptor Systems. In *Proceedings of the 3rd IFAC Symposium on System, Structure and Control* (pp.386-391). Iguassu Falls, Brazil: IFAC.

Yakoub, Z., Charfeddine, M., Jouili, K., & Benhadj Braiek, N. (2013). A combination of backstepping and the feedback linearization for the controller of inverted pendulum. In *Proceedings of the 2013 IEEE International Multi-Conference on Systems, Signals and Devices (SSD)*. IEEE. DOI: 10.1109/SSD.2013.6564123

Yalcin, M. E., Suykens, J. A. K., & Vandewalle, J. (2004). True random bit generation from a double-scroll attractor. *IEEE Transactions on Circuits and Systems I*, *51*(7), 1395–1404. doi:10.1109/TCSI.2004.830683

Yamamoto, M., & Motoyoshi, O. (1990). Active and reactive Power control for doubly fed wound rotor induction generator. *IEEE Transactions on Power Electronics*, *6*(4), 624–629. doi:10.1109/63.97761

Yang, C., & Wu, Q. (2006). On stabilization of bipedal robots during disturbed standing using the concept of Lyapunov exponents. *Robotica*, *24*(5), 621–624. doi:10.1017/S0263574706002700

Yang, J., & Sobelman, G. E. (2010). Efficient μ. *Electronics Letters*, *47*(2), 73–74. doi:10.1049/el.2010.7937 PMID:21079852

Yang, S., Do, J., Jang, H., Jung, J., & Bien, Z. (2006). Advanced Soft Remote Control System in Human-friendliness. *SCIS & ISIS*, *2006*(1), 218–222.

Yang, T. (2004). A survey of chaotic secure communication systems. *International Journal of Computational Cognition*, *2*, 81–130.

Yang, Y., Xiang, C., & Lee, T. H. (2012). Sufficient and necessary conditions for the stability of second-order switched linear systems under arbitrary switching. *International Journal of Control*, *85*(12), 1977–1995. doi:10.1080/00207179.2012.713987

Yan, I., Ryan, M., & Power, I. (1994). *Using Logic Towards Intelligent Systems*. Prentice Hall.

Yan, X., Zhu, Y., Zhang, H., Chen, H., & Niu, B. (2012). An Adaptive Bacterial Foraging Optimization Algorithm with Lifecycle and Social Learning, *Discrete Dynamics in Nature and Society*, 2012. *Article ID*, *409478*, 1–20.

Yan, Z., Zhang, G., & Wang, J. (2012). Infinite horizon H_2 / H_∞. *Control Theory Application*, *10*(2), 159–165. doi:10.1007/s11768-012-0038-6

Yasuda, G. (1971). *A Fundamental Study on Reproduction - Application of Graph Theory to Self-Reproducing Processes*. (Master thesis). University of Tokyo, Tokyo, Japan.

Yasuda, G., & Tachibana, K. (1991). A parallel processing control approach to the design of autonomous robotic manufacturing systems, In *Proceedings of the XIth International Conference on Production Research*, (pp.445-449). Academic Press.

Yasuda, G. (2010). Petri net model based hierarchical and distributed control of robotic manufacturing systems. In *Proceedings of 2010 ISFA International Symposium on Flexible Automation*. Tokyo, Japan: ISFA.

Yasuda, G. (2011). Petri net model based design and control of robotic manufacturing cells. In V. Modrak & R. Sudhakara Pandian (Eds.), *Operations Management Research and Cellular Manufacturing Systems: Innovative Methods and Approaches* (pp. 385–400). Hershey, PA: IGI Global. doi:10.4018/978-1-61350-047-7.ch018

Yazdani, D., Mojiri, M., Bakhshai, A., & Joos, G. A. (2009). Fast and Accurate Synchronization Technique for Extraction of Symmetrical Components. *IEEE Transactions on Power Electronics*, *24*(3), 674–684. doi:10.1109/TPEL.2008.2010321

Yingwei, L., Sundararajan, N., & Saratchandran, P. (1997). Identification of time-varying nonlinear systems using minimal radial basis function neural networks. *IEE Proceedings Control Theory and Applications*, *144*(2), 202–208.

Yin, W., & Mehr, A. S. (2010). A variable regularization method for affine projection algorithm. *IEEE Transactions on Circuits and Systems II*, *57*(6), 476–480. doi:10.1109/TCSII.2010.2048355

Yogendra, K., & Rajest, B. (2013). BF-PSO optimized PID Controller design using ISE, IAE, IATE and MSE error criteria. *International Journal of Advanced Research in Computer Engineering and Technology*, *2*(7), 2333–2336.

Yoneyama, J. (2010). Robust H∞ control of uncertain fuzzy systems under time-varying sampling. *Fuzzy Sets and Systems*, *161*(6), 859–871.

Yongzhen, P., Shuping, L., & Changguo, L. (2011). Effect of delay on a predator-prey model with parasitic infection. *Nonlinear Dynamics*, *63*(3), 311–321. doi:10.1007/s11071-010-9805-4

Zakharov, Y., & Albu, F. (2005). Coordinate descent iterations in fast affine projection algorithm. *IEEE Signal Processing Letters*, *12*(5), 353–356. doi:10.1109/LSP.2005.843765

Zakharov, Y., & Tozer, T. (2004). Multiplication-free iterative algorithm for LS problem. *Electronics Letters*, *40*(9), 567–569. doi:10.1049/el:20040353

Zakharov, Y., White, G., & Liu, J. (2008). Low complexity RLS algorithms using dichotomous coordinate descent iterations. *IEEE Transactions on Signal Processing*, *56*(7), 3150–3161. doi:10.1109/TSP.2008.917874

Zames, G. (1981). Feedback and optimal sensitivity model reference transformation, multiplicative semi norms and approximate inverses. *IEEE Transactions on Automatic Control*, *26*(2), 301–302. doi:10.1109/TAC.1981.1102603

Zeng, C., Zhang, Q., & Wei, X. (2012). GA-based global path planning for mobile robot employing A* algorithm. *Journal of Computers*, *7*(2), 470–474. doi:10.4304/jcp.7.2.470-474

Zerdali, E., & Barut, M. (2013). Sensorless MRAS based real-time speed-sensorless control of induction motor with optimized fuzzy-PI controller, *IEEE International Symposium on Control for Electrical Drives and Predictive Control of Electrical Drives and Power Electronics*.1-5, Munich. doi:10.1109/SLED-PRECEDE.2013.6684500

Zghal, W., Kantchev, G., & Kchaou, H. (2011). Optimization and management of the energy produced by a wind energizing system. *Renewable & Sustainable Energy Reviews*, *15*(2), 1080–1088. doi:10.1016/j.rser.2010.11.021

Zhai, G. (2001). Quadratic stabilizability of discrete-time switched systems via state and output feedback. In *Proceedings of the 40th IEEE Conference on Decision and Control*(vol. 3, pp. 2165-2166). Orlando, FL: IEEE. doi:10.1109/CDC.2001.980575

Zhang, H., & Zhao, J. (2011). The design of RF data acquisition system based on STM32 and FPGA. In *Proceedings ofInternational Conference on Multimedia Technology (ICMT)*, (pp. 832 – 834). ICMT. doi:10.1109/ICMT.2011.6002030

Zhang, P., Tian, Y., Liu, Z., Yang, S., & Tian, R. (2008). Further research and comparison of gaits for compass-like biped and kneed passive dynamic walker. In *Proceedings ofInternational Conference on Intelligent Robotics and Applications*, (pp. 1216-1226). Wuhan, China: Academic Press. doi:10.1007/978-3-540-88513-9_129

Zhang, Y., & Jiang, J. (2003a).Bibliographical Review on Reconfigurable Fault-Tolerant Control Systems. In *Proceedings of the 5th IFAC Symposium on Fault Detection, Supervision and safety for Technical Processes*, (pp. 265–276). IFAC.

Zhang, Y., & Jiang, J. (2006). Issues on integration of fault diagnosis and reconfigurable control in active fault-tolerant control systems. In *Proceedings of the 6th IFAC Symposium on Fault Detection, Supervision and Safety of Technical Processes (SafeProcess'06)*, (pp. 1513–1524). IFAC. Doi:10.3182/20060829-4-CN-2909.00240

Zhang, H., & Kang, W. (2013). Design of the Data Acquisition System Based on STM32. *Procedia Computer Science*, *17*(1), 222–228. doi:10.1016/j.procs.2013.05.030

Zhang, K., Jiang, B., & Cocquempot, V. (2008). Adaptive observer-based fast fault estimation. *International Journal of Control Automation and Systems*, *6*(3), 320–326.

Zhang, K., Jiang, B., & Vincent, C. (2008). Adaptive Observer-based Fast Fault Estimation. *International Journal of Control, Automation, and Systems*, *6*(3), 320–326.

Zhang, L. X., & Shi, P. (2009). Stability, l2-gain and asynchronous HN control of discrete-time switched systems with average dwell time. *IEEE Transactions on Automatic Control*, *54*(9), 2192–2199. doi:10.1109/TAC.2009.2026841

Zhang, L., Liao, X., & Wang, X. (2005). An image encryption approach based on chaotic maps. *Chaos, Solitons, and Fractals*, *24*(3), 759–765. doi:10.1016/j.chaos.2004.09.035

Zhang, T. P., & Ge, S. S. (2007). Adaptive neural control of MIMO nonlinear state time-varying delay systems with unknown dead-zones and gain signs. *Automatica*, *43*(6), 1021–1033. doi:10.1016/j.automatica.2006.12.014

Zhang, T. P., & Yi, Y. (2007). Adaptive fuzzy control for a class of MIMO nonlinear systems with unknown dead-zones. *Acta Automatica Sinica*, *33*(1), 96–99. doi:10.1360/aas-007-0096

Zhang, W., & Yu, L. (2008). Modelling and control of networked control systems with both network-induced delay and packet-dropout. *Automatica*, *44*(12), 3206–3210. doi:10.1016/j.automatica.2008.09.001

Zhang, Y., & Jiang, J. (2001). Integrated Design of Reconfigurable Fault-Tolerant Control Systems. *Journal of Guidance, Control, and Dynamics*, *24*(1), 133–136. doi:10.2514/2.4687

Zhang, Y., & Jiang, J. (2002). An active fault-tolerant control system against partial actuator failures. *IEE Proceedings. Control Theory and Applications*, *149*(1), 95–104. doi:10.1049/ip-cta:20020110

Zhang, Y., & Jiang, J. (2003b). Fault tolerant control system design with explicit consideration of performance degradation. *Aerospace and Electronic Systems*, *39*(3), 838–848. doi:10.1109/TAES.2003.1238740

Zhang, Z., & Chen, G. (2007). Chaotic liquid shaker: Design, implementation and application. *International Journal of Bifurcation and Chaos in Applied Sciences and Engineering*, *17*(12), 4443–4451. doi:10.1142/S0218127407020105

Zhan, M., Wang, X., Gong, X., Wei, G. W., & Lai, C. H. (2003). Complete synchronization and generalized synchronization of one-way coupled time-delay systems. *Physical Review E: Statistical, Nonlinear, and Soft Matter Physics*, *68*(3), 036208. doi:10.1103/PhysRevE.68.036208 PMID:14524869

Zhao, J., & Kanellakopoulos, I. (2002a). Active identification for discrete-time nonlinear control – Part 1: Output-feedback systems. *IEEE Transactions on Automatic Control*, *47*(2), 210–224.

Zhao, J., & Kanellakopoulos, I. (2002b). Active identification for discrete-time nonlinear control – Part 2: Strict-feedback systems. *IEEE Transactions on Automatic Control*, *47*(2), 225–240.

Zhat, Y., Bo, H., Yasuda, K., & Michel, A. N. (2001). Stability analysis of switched systems with stable and unstable subsystems: an average dwell time approach. *International Journal of Nonlinear Systems and Science*, *32*(8), 1055-1061.

Zhen, Z., & Yan, G. (2013). Design of the fuzzy PID controller for the hot runner temperature control system. In *Proceedings of32nd Chinese Control Conference (CCC)*, (pp. 3464 – 3469). CCC.

Zheng, Z., Lecocq, H., & Jordant, R. (1993). Adaptive manipulator control: A parallel implementation on a network of microprocessors. In *Proceedings of the IEEE International Conference on Robotics and Automation*, (vol. 1, pp. 878-882). IEEE. doi:10.1109/ROBOT.1993.292087

Zhifu, W., Qizhi, T., & Chengning, Z. (2009). Speed identification about PMSM with MRAS, *IEEE 6th International Conference on Power Electronics and Motion Control, IPEMC '09.*, pp.1880 – 1884, Wuhan.

Zhong, G. Q. (1994). Implementation of Chua's circuit with a cubic nonlinearity. *IEEE Transactions on Circuits and Systems*, *41*(12), 934–941. doi:10.1109/81.340866

Zhong, Q. C. (2008). *Robust Control of Time-Delay Systems*. New York: Springer.

Zhou, G., Wu, B., & Xu, D. (2007). Direct power control of a multilevel inverter based active power filter. *Electric Power Systems Research*, *77*(3), 284–294. doi:10.1016/j.epsr.2006.03.005

Zhou, J., Er, M. J., & Wen, C. (2005). Adaptive Control of Nonlinear Systems with Uncertain Dead-zone Nonlinearity. In *Proceedings of the 44th IEEE Conference on Decision and Control, and the European Control Conference. IEEE.* doi:10.1109/CDC.2005.1582254

Zhou, K., Doyle, J. C., & Glover, K. (1996). *Robust and Optimal Control.* (N. P. Hall, Ed.). Upper Saddle River, NJ: Prentice Hall.

Zhou, K., Doyle, J., & Glover, K. (1996). *Robust and Optimal Control participation factor analysis participation factor analysis*. Englewood Cliffs, NJ: Prentice-Hall.

Zhou, K., Glover, K., Bodenheimer, B., & Doyle, J. (1994). Mixed H_2 / H_∞ performance objective I: Robust performance analysis. *IEEE Transactions on Automatic Control*, *39*(8), 1564–1574.

Zhou, M. C., DiCesare, F., & Desrochers, A. A. (1992). A hybrid methodology for synthesis of Petri net models for manufacturing systems. *IEEE Transactions on Robotics and Automation*, *8*(3), 350–361. doi:10.1109/70.143353

Zhou, S.-L., Han, P., Wang, D.-F., & Liu, Y.-Y. (2004). A Kind of Multivariable PID Design Method for Chaos System using H infinity loop shaping design procedure. In *Proceedings of the Third International Conference on Machine Learning and Cybernetics.*IEEE.

Zhou, W., Yang, H., & Fang, Z. (2008). Battery Behavior Prediction and Battery Working States Analysis of a Hybrid Solar–Wind Power Generation System. *Renewable Energy*, *33*(6), 1413–1423. doi:10.1016/j.renene.2007.08.004

Zhou, Y., Xiao, K., Wang, Y., Liang, A., & Hassanien, A. E. (2013). A PSO-Inspired Multi-Robot Map Exploration Algorithm Using Frontier-Based Strategy. *International Journal of System Dynamics Applications*, *2*(2), 1–13. doi:10.4018/ijsda.2013040101

Zhu, Z. Q., & Zhu, X. (2007). Estimation of winding resistance and PM flux-linkage in brushless AC machines by reduced-order extended Kalman filter, *In proceedings of IEEE International Conference on Networking, sensing and control*, 740-745, London.

Zhu, Z. Q., & Howe, D. (2007). Electrical machines and drives for electrical, *hybrid and fuel cell vehicles. Proceedings of the IEEE*, *95*(4), 746–765. doi:10.1109/JPROC.2006.892482

Zimmerman, H. J., Zadeh, L. A., & Gaines, B. R. (1984). Fuzzy sets and decision analysis. *Studies in the Management Sciences*, *20*, 3–8.

About the Contributors

Ahmad Taher Azar has received the M.Sc. degree (2006) in System Dynamics and Ph.D degree (2009) in Adaptive Neuro-Fuzzy Systems from Faculty of Engineering, Cairo University (Egypt). He is currently assistant professor, Faculty of computers and information, Benha University, Egypt. Dr. Azar is the Editor in Chief of three journals published by IGI Global, USA entitled *International Journal of System Dynamics Applications* (IJSDA), *International Journal of Rough Sets and Data Analysis* (IJRSDA) and *International Journal of Service Science, Management, Engineering, and Technology* (IJSSMET). Also, he is the Editor in Chief of *International Journal of Intelligent Engineering Informatics* (IJIEI), Inderscience Publishers, Olney, UK. Dr. Azar is associate editor of *IEEE Trans. Neural Networks and Learning Systems*. Dr. Ahmad Azar has worked in the areas of Control Systems, Dynamical Systems, Chaos Control and Synchronization, Computational Intelligence, Machine learning and has authored/ coauthored over 130 research publications in peer-reviewed reputed journals, book chapters and conference proceedings. He is an editor of many Books in the field of Fuzzy logic systems, modeling techniques, control systems, computational intelligence, Chaos modeling and Machine learning. Dr. Ahmad Azar is closely associated with several international journals as a reviewer. He serves as international programme committee member in many international and peer-reviewed conferences. Dr Ahmad Azar is currently the chair of IEEE Computational Intelligence Society (CIS) Egypt Chapter and Vice President Of System Dynamics Africa Chapter. He is an Academic Member of IEEE Systems, Man, and Cybernetics Society Technical Committee on Computational Collective Intelligence and also a member in KES Focus Group on Agent and Multi-agent Systems.

Sundarapandian Vaidyanathan has received the Doctor of Science (DSc) degree (1996) in the Department of Electrical and Systems Engineering from the School of Engineering and Applied Science, Washington University, St. Louis, Missouri, USA. He is currently the Professor and Dean, Research and Development Centre, Vel Tech University, Avadi, Chennai, India. Dr. Sundarapandian is the Associate Editor of two journals published by IGI Global, USA, titled *International Journal of System Dynamics Applications* (IJSDA) and *International Journal of Rough Sets and Data Analysis* (IJRSDA). He is also in the editorial boards of many international journals on Control Systems, Computer Science, Information Technology, Chaos Theory, and Automation. He has published over 110 Scopus-indexed researched publications in peer-reviewed reputed journals, book chapters, and conference proceedings. He has published three graduate-level textbooks on Computational Science, which are adopted in many universities. He serves as international programme committee member in many international and peer-reviewed conferences. Dr. Sundarapandian has guided many PhD students in Mathematics, Computer Science, and Electrical Engineering. His papers on observer design for nonlinear control systems have received many citations. He is an active researcher in the areas of Chaos and Control and constructed many novel chaotic and hyperchaotic systems in the chaos literature. He has given many plenary lectures in international conferences on Control Engineering, Computer Science, and information Technology. Dr. Sundarapandian's research interests include Control Systems, Dynamical Systems, Stability Theory, Chaos Theory, Computational Science, Mathematical Modelling, Optimal Control, Robotics, Operations Research, and Intelligent Control.

* * *

Rabeh Abbassi received the MSc degree from the University of Tunis, High Engineering School of Tunis (ENSIT), Tunisia, in 2008. Since 2007, he has been with LaTICE laboratory, where he is currently preparing his thesis within the LaTICE laboratory. His research interests include renewable energies, wind turbines control, photovoltaic systems, storage systems, integration of distributed generation, control of power converters, and power conditioning. Rabeh is an Assistant with the Department of Electrical Engineering of ISSATK.

Mohamed Naceur Abdelkrim obtained a Diploma in Technical Sciences in 1980, his Master's Degree in Control in 1981 from the ENSET School of Tunis (Tunisia), his PhD in Control in 1985, and the Doctorate in Sciences Degree (Electrical Engineering) in 2003 from the ENIT School of Tunis. Since 2003, he is Professor at the Electrical Engineering Department (Control) of the National Engineering School of Gabes (Tunisia), and he is manager of the Modeling, Analysis, and Control Systems (MACS) laboratory.

Samir Abdelmalek was born in 1986 in Tebessa, Algeria. He received the Engineer's degree in 2009 from Electrical and Electronics Engineering Faculty of Education of Tebessa University, Algeria. He received Magister's degree in 2012 from Tebessa University, Algeria. Now, he continues his PhD at University of Sciences and Technology Houari Boumediene (USTHB), Algeria, in the field of Electro Energetic Systems Engineering. Between 2010 and 2011, he worked as an Electrical Engineer at private company of Electrical Installation. Following, he has also worked as a Lecturer and Vocational professor at Mohamed Bougerra University (INH) between 2012 and 2013. Currently, he is an Associate Researcher at the Unité de Développement des Equipements Solaires (UDES/CDER), Bou-Ismail, Algérie. His research interest includes artificial intelligence, computer science, advanced robust control systems, hybrid renewable energy systems, diagnosis, and fault tolerant control systems.

Felix Albu obtained the BEng degree in Applied Electronics and the PhD degree in Telecommunications from Politehnica University of Bucharest, Romania, in 1993 and 1999, respectively. He worked as a teaching assistant at Politehnica University of Bucharest (1993-1999), as a postdoctoral researcher in University College Dublin, Ireland (1999-2002), and Aristotle University of Thessaloniki, Greece (2004-2005). In addition, he was a DSP engineer at Lake Communications, Dublin, Ireland (2002-2003) and imaging researcher at Fotonation Romania (2006-2011). Currently, he is an associate professor at Valahia University of Targoviste, Romania. He is a senior member of IEEE. His research interests are the following: liniar and nonlinear adaptive filtering, speech processing, image processing, neural networks, logarithmic number systems, and machine learning.

Flah Aymen was born in Tunisia in Gabès. He is a professor assistant in the National School of Engineering of Gabès. He received his MS degree in 2009 from national school of engineering of Gabès in Automatic and Intelligent techniques specialty, and the PhD at National School of Engineering of Gabès in Tunisia in the field of Electrical Engineering. Between 2009 and 2012, he has worked as a Research Assistant in Wind, Photovoltaic, and Geothermal Lab in the Motor Control Field. From 2012, he has changed his studies to electrical vehicles. Many researchers are working now under his supervision on many fields based on electrical motor. He has extensive experience in motor control applications. In addition, he has elaborated a number of applications using intelligent techniques such as fuzzy logic, neural network, optimization techniques as PSO and BFO. There are many publications in internationals journals and conferences are elaborated by Dr. Flah Aymen from 2009 until now.

Hassene Bedoui was born in Tunisia in 1979. He obtained the Master's degree in Industrial Informatics and Automation (IIA) from the "Ecole Superieur des Sciences et Techniques de Tunis – ESSTT" in 2003 and obtained the Master's degree in Automatic and Industrial Maintenance from the "Ecole Nationale d'Ingenieurs de Monastir" in 2007. He is currently preparing the PhD degree in Automatic and Diagnosis within the framework of "Laboratoire de Recherchce en Automatique, Traitement de Signal et Image LARATSI" from "Ecole Nationale d'Ingenieurs de Monastir – ENIM." His research is related to state and faults estimation for uncertain Takagi-Sugeno fuzzy systems.

Safya Medimagh Belghith has received his engineering diploma in Electricity (1981) and his DEA (1982) and the PhD degree (1985) in "Automatic and Signal Processing" from the High School of Electricity, Lab Signals & Systems, University Paris XI Orsay. In 1997, she received her Status Doctorate in "Physical Sciences" from the Faculty of Sciences of Tunis FST, with collaboration of the Laboratory Signals & Systems, High School of Electricity University Paris XI Orsay. She is presently a Professor at the National School of Engineering Tunis, Tunisia, and a senior researcher at ENIT. Her research interests include dynamical systems, chaos control and synchronization, with applications in secure communication and cryptography, biped robots.

Hocine Belmili was born in 1978 in Algeria. In 1996, he obtain his baccalaureate degree in the exact sciences. In 2001, he has a degree of engineer from the University of Constantine, Algeria. In 2004, he completed his Master's (Magister) degree from the University of Farhat Abbas Setif, Algeria. In 2012, he obtained a Doctorate of Sciences, from the National Polytechnic School in Algiers. Since 2005, he operates as a senior of research in the center for development of renewable energies. His areas of specialization are Renewable Energy Technologies, Power Electronics, and Systems Control.

Faouzi Bouani is a professor at the National Engineering School of Tunis (ENIT). He received his BSc and MSc degrees from the "Ecole Normale Supérieure de l'Enseignement Technique de Tunis" in 1990 and 1992, respectively. He received his PhD thesis degree in 1997 and the "Habilitation Universitaire" in 2007, in Electrical Engineering both from ENIT. He is a member of the Analysis, Conception, and Control Systems laboratory at ENIT and an IEEE member. His research interests include predictive control, robust control, predictive control, and supervision of control neuronal networks algorithms based on fuzzy logic and genetic algorithms.

Abdesselem Boulkroune received his Engineering degree from Setif University in 1995, his Master grade from the Military Polytechnic School (EMP) of Algiers in 2002, his PhD degree in Automatic control from the National Polytechnic School (ENP) of Algiers in 2009. In 2003, he joined the Automatic Control Department at Jijel University, Algeria, where he is currently an associate professor. His research interests are in nonlinear control, high-gain observer, sliding mode observer, output-feedback control, adaptive control, actuators nonlinearities, nonlinear systems with time-delays, chaotic systems, synchronization, drive of doubly fed induction motors. Abdesselem Boulkroune is a member of the LAJ laboratory of the Jijel University.

Boumedyen Boussaid is associate professor at National School of Engineers of Gabes, Tunisia, in Electrical Engineering and Automatic Control Department. He received his PhD degree in 2011 in Automatic Control and Computer Sciences from both University of Nancy and University of Gabes. He received his Engineering Degree in 1997 in Electrical Engineering from National School of Engineers of Tunis. He is currently a member of the Research Centre in Automation of Nancy (CRAN) and the research unit on Modeling, Analysis, and Control of Systems (MACS). His research interests focus on fault tolerant control and fault detection in wireless sensor networks with application to wind firms and robots networks.

Naceur Benhadj Braiek was born in Mahdia, Tunisia, in 1963. He received the Master's degree in Electrical Engineering from the National School of Engineering of Tunis, Tunisia, in 1987, the Doctorate in Automatic Control from the University of Sciences and Techniques of Lille, France, in 1990, and the "Doctorat d'Etat" in Electrical Engineering from the National School of Engineering of Tunis, Tunisia, in 1995. Currently, he is Professor of Automatic Control at the University of Tunis. His domain of research concerns the nonlinear control systems including Modeling, Analysis, and Control.

Monia Charfeddine was born in Mednine, Tunisia, in 1984. She received the Engineer Degree in Electrical Engineering in 2009, and the Master's degree in Automatic Control in 2010, both from the National School of Engineering of Gabes, Tunisia. He is currently completing work on his PhD in Automatic Control at the National School of Engineering of Gabes, Tunisia. She is also Assistant of Electrical Engineering at the University of Gabes, Tunisia. His research includes systems control, nonlinear system, tracking control problem non-minimum phase system, and systems stability.

Souad Chebbi obtained her Doctorate in 1983 in Electrical Engineering. She is the head of the LaTICE Laboratory at High Engineering School of Tunis (ENSIT), University of Tunisia. Her research is in the areas of electrical networks and electrical systems.

Sajal K. Das was born in Rajshahi, Bangladesh. He received his BSc degree in Electrical & Electronic Engineering (EEE) from the Rajshahi University of Engineering & Technology (RUET), Bangladesh, in April 2010. He is currently pursuing the PhD degree in the School of Engineering and Information Technology of the University of New South Wales, Canberra, Australia. He has published numerous journals and conference papers in the field of control theory and application. His research includes control of nanopositioning system for precision positioning, magnetic recording devices, and hard disc drive.

Marcin Denkowski received MS degree in 2003 from Maria Curie-Sklodowska University, Poland, in Physics in the field of Image Segmentation. In 2011, he received PhD at Silesian University of Technology at Gliwice, Poland, in the field of Computer Science in 3D Object Reconstruction. Currently, he is a Lecturer in Maria Curie-Sklodowska University, Poland. He also leads research and development projects at SIM Ltd. in Lublin, Poland. His research interest includes computer graphics and vision, GPU proccessing, human-computer interaction, and gesture recognition.

Soudani Dhaou received the scientific degree in Mathematics and Physics within the Normal School of Tunis (ENT) in 1980. He received the Master's degree in Technical Sciences within the Normal School of Tunis (ENT) in 1982. He began to work as a teacher of second degree within the secondary school Zahra in 1982. He received the certificate of thorough research in automatic within the National School of engineers of Tunis (ENIT) in 1992. He received the Doctoral degree on the determination of explicit solutions has problems in systems analysis substantially disrupted within the national school of engineers of Tunis (in ENIT) in 1997. He received the rank of assistant master at the national school of engineers of Tunis (in ENIT) in 1997. He received the rank of master of conference in 2004. He received the degree professor in 2010. He is a member of the group of researchers in the Laboratory LARA.

Moez Feki received his BSc and an MSc in Electrical and Electronics Engineering from Bilkent University, Turkey, in 1995 and 1997. Next, he obtained a PhD degree in Applied Mathematics from Universit de Metz France in 2001. Until 2004, he was a post-doc student in the EEA department of Universit de Reims Champagne-Ardenne, France. Next, he moved on an Assistant Professor position to the Engineering School of Gabes, Tunisia. He is currently a Professor at the Ecole Superieure des Science et Technologie de Hammam Sousse, Tunisia, and carrying his research activities in the Research group CEMLab of the Engineering School of Sfax. His research interests include non-linear control theory, with emphasis on control and synchronization of chaotic systems.

Hassène Gritli received the engineer's degree and the MSc degree in Automatic and Industrial Computing from the National Institute of Applied Sciences and Technology – INSAT, Tunisia, in 2006 and in 2007, respectively. He received the PhD degree in Electrical Engineering from the National School of Engineers of Tunis – ENIT, Tunisia, in 2013. He is presently a senior researcher at ENIT. From 2008-2012, he was an assistant at the Higher Institute of Applied Sciences and Technology of Mateur, Tunisia. He is currently assistant professor at the Higher Institute of Technological Studies of Kélibia, Tunisia. His research interests include chaos theory, hybrid dynamical systems, biped robots, impact mechanical/electrical systems, and nonlinear control.

Nahla Khraief Haddad was born in Tunis (Tunisia) in 1972. She graduated in Electrical Engineering in 1995 from National School of Engineering Sfax. She obtained a Master's degree in Automatic and Industrial Automation in 1999 from High School of Science and Technology Tunis and a PhD in Robotics from University of Versailles Saint Quentin in Yvelines, France, in 2004. Her research interests include the analysis and the control of the nonlinear dynamic systems. She works particularly on the trajectory generation and the control of humanoid robots, the analysis and the control of the chaos, and bifurcations of the mechanical systems.

Denoun Hakim received his BSc degree in Electrical Engineering in 1997 from the University of Tizi Ouzou, Algeria, and the DEA (Academic year) degree from Paris 6, France, and the Magister degree from Polytechnic School, Algiers, Algeria, in 1998 and 2001, respectively. He received his PhD degree in Electrical Engineering from The University of Mouloud Mammeri, Tizi-Ouzou. He is a researcher in laboratory (Laboratoire des Technologies Avancées du Génie Electrique LATAGE) and a senior lecturer at the same university. He has published articles and participates in many international scientific conferences. His research interests include electrical machines and drives, power electronics, and control systems.

Manel Hammami received the MSc degree in 2012 from ENSIT. Currently, she is preparing a thesis on renewable energies within the LaTICE laboratory.

Maher Ben Hariz was born in Djerba, Tunisia, in 1986. He received his BSc degree in Electronic and Informatics in 2008 and his MSc in 2010, both from the Faculty of Sciences of Tunis (FST). He is currently preparing his PhD thesis in the field of automatic control. He is a member of the Analysis, Conception, and Control Systems laboratory at the National Engineering School of Tunis and an IEEE member. His research interests include robust control of linear and nonlinear systems and implementation of control algorithms.

Tlijani Hatem had an engineering diploma from the National Engineering School of Sfax (ENIS) in 1999. In 2004, he received his Master's degree from the National Engineering School of Tunis (ENIT) of El-Manar University in Tunisia. He had a Doctorate in Electrical Engineering in 2011 from the same engineering school (ENIT). His research interest includes robotics, multi-actuator control, incremental actuators, optimization, artificial intelligence, fuzzy logic, renewable energy, and green energy. Currently, he is teaching in the High Institute of applied Science and Technology of Gafsa (ISSATG), Tunisia.

Tlijani Hayet was a graduate in Computer Science in 2000 from the High Institute of Management (ISG) of Tunis University, Tunisia. Then she was a postgraduate in Computer Science in 2002 from the National Engineering School of Computer Science of Tunis (ENSI), Tunisia. In 2008, she received her Master's degree from the National Engineering School of Tunis (ENIT) of El-Manar University in Tunisia. Now, she is a PhD student in the same engineering school (ENIT). Her research interest includes robotics, robot's path plan, multi-robot, optimization, evolutionary algorithms, meta-heuristics, artificial intelligence, fuzzy logic, and computer science. Currently, she teaches computing in the High Institute of Technology of Gafsa (ISETG), Tunisia.

Bemri H'mida received the scientific degree in Mathematics, Physics, and Computer Science within the Faculty of Sciences of Tunis (FST) in 2003.He received the degree of MA within the Faculty of Sciences of Tunis (FST) in 2005. He received the certificate of Aptitude Teaching of Second Degree (CAPES) in 2008. He has been working as a teacher in 2 Mars 1934 secondary school in Siliana since 2008. He received the degree of Master of Research in Automatic and Signals Processing (ATS) within the National Engineering School of Tunis (ENIT) in 2013. His research interests include modeling and control systems sampled to delays.

Ouerfelli Houssem Eddine was born in 1988. He received the License degree from National School of Engineers of Tunis in 2010. He received MS degree from National School of Engineers of Tunis in 2012. Now, he continues DS/PhD at School Engineering of Tunis in the field of Electric Engineering. Currently, he is an assistant in ISI. He has worked as a Research Assistant in ISI. His research interest includes iterative learning control and hybrid system, the Lyaponuv theory, stability, learning control, H-infinity control, robust control, and tracking control.

Olfa Hrizi is a PhD student at the National Engineering School of Gabes (Tunisia) and a member of Modeling, Analysis, and Control Systems (MACS) research unit. She received the National Engineering Diploma from the National Engineering School of Gabes (ENIG) in Tunisia in 2011. Then, she obtained the Master Degrees from the same institution in 2012 in Automatic and Intelligent Techniques. She received the bachelor degree in 2006 in Experimental Sciences. Currently, her research framework focuses on Cooperative Diagnosis in wireless sensor networks: application to the networks of robots.

Dridi Jamal received his BSc and an MSc in Electrical and Electronics Engineering from FST University of El Manar, Tunisia, in 2010 and 2011, respectively. He is currently a PhD student at the Engineering School of Tunis (ENIT), University of El Manar, Tunisia. His research interests include linear control theory, with emphasis on repetitive learning control and the control of switched linear systems.

Knani Jilani is a full Professor at the National Engineering School of Tunis (ENIT) of El-Manar University in Tunisia. His research interest includes robotics, robot's motion, and control.

Khalil Jouili was born in Mednine, Tunisia, in 1981. He received the Master's degree in Automatic Control in 2007, and the PhD degree in Automatic Control from the National School of Engineering of Gabes, Tunisia, in 2010. Currently, he is Assistant Professor in Automatic Control at the Unversity of Gabes, Tunisia. His current research includes systems control, nonlinear systems, and systems stability.

Samira Kamoun received the PhD and Habilitation Universitaire degrees in Automatic Control and Industrial Computer from the National School of Engineering of Sfax, Sfax University, Tunisia, in 2003 and 2009, respectively. Since 1990, she has been with the Department of Electrical Engineering of the National School of Engineering of Sfax, where she is currently a Professor of Automatic Control. Her research interests include the identification and the adaptive control of complex systems (large-scale systems, nonlinear systems, time-varying systems, stochastic systems), with applications to the automatic control of engineering systems. She has published several articles and presented various communications in international conferences. Professor Kamoun is currently a member of the Sciences and Techniques of Automatic control and computer engineering Laboratory (Lab-STA) of Sfax and a member of the scientific committee of several international conferences (international conference on Sciences and Techniques of Automatic control and computer engineering [STA], etc.).

Marwen Kermani was born in Tunisia in 1983. He received the Electrical Engineering diploma in 2008 from the National Engineering school of Monastir. He is currently a PhD student at National Engineering School of Monastir. His research interests include Analysis and Synthesis Problems of Nonlinear Systems.

Atef Khedher was born in Tunisia in 1980. He obtained the engineer degree in Electro-Mechanical Engineering from the "Ecole Nationale d'Ingenieurs de Sfax" in 2003 and obtained the Master's degree in Automatic and Industrial Maintenance from the "Ecole Nationale d'Ingenieurs de Monastir" in 2005. He obtained PhD degree in Automatic and Computer Science from the "Université de Tunis-Tunisia" in 2011. His research is related to state and faults estimation and the faults tolerant control for Takagi-Sugeno fuzzy systems. He is alson interested to the state and fault estimation for descrete events systems using the Petri Net. At present, he is an assosiate professor in the "Institut Supérieur des Sciences Appliquées et de Technologies de Gafsa."

Habib Kraiem obtained the Doctorate thesis on October 2010 from National Engineering School of Gabes. His PhD thesis titled "Direct Torque Control with Fixed and Variable Frequency Applied to Induction and Synchronous Motors." He works as an Assistant Professor at the Higher Institute of Industrial Systems Gabes Tunisia, recruited from the academic year 2012-2013. His fields of interest include power electronics, machine drives, automatic control, modeling, and sensorless control.

Mekki Ksouri received his BSc degree in Physics from the FST in 1971 and his Diplôme d'Ingénieur from the Ecole Supérieure d'Électricité, Paris, in 1973. He received his DSc and PhD degrees from the university of Pierre and Marie Curie, France, respectively, in 1975 and 1977. He had received the Outstanding Contribution Award for Leadership in the Organization of the CIFA 2004 Symposium, Chevalier dans l'Ordre des Palmes Acadmiques France 2002, Outstanding Contribution Award IEEE SMC 1998. He is a senior member IEEE. His activities focus on automatic control, robust control, and optimization in planning and scheduling, including implementation of fuzzy logic and genetic algorithms. He has also published six books.

Moufida Ksouri is a full professor/PhD supervisor and department head of Automatic Control in National School Engineering of Tunis. His research interests include robust control, h-infinity control, h-infinity control, time-varying systems, and switched systems.

Ioannis M. Kyprianidis received the MS degree in Electronics and PhD degree in Electromagnetism from the Physics Department of the Aristotle University of Thessaloniki, Greece, in 1978 and 1988, respectively. He joined the Physics Department of the Aristotle University of Thessaloniki, Greece, in 1977 as a research assistant, where he was involved in projects concerning the study of electric and magnetic properties of materials. Since 1992, he has been involved in projects concerning the study of nonlinear circuits and systems in the same department, where has been elected as Associate Professor.

Sadek Ait Mamar received his BSc degree in Electrical Engineering in 2000 from the University of Tizi Ouzou, Algeria, and the DEUA (university applicable studies diploma), degree in Electronics from Mouloud Mammeri University of Tizi-Ouzou, Algeria (2000), and studies as a student in PhD at Mouloud Mammeri University, worked in Electrical Engineering laboratory from The University of Mouloud Mammeri, Tizi-Ouzou. He is a student and technician in laboratory (Laboratoire des Technologies Avancées du Génie Electrique – LATAGE). He has participated in many published articles and participates in many international scientific conferences. His research interests include electrical machines and drives, power electronics, and control systems.

Yosra Miladi received her BSc and an MSc in Electrical and Electronics Engineering from the Engineering School of Sfax (ENIS), University of Sfax, Tunisia, in 2011 and 2012, respectively. She is currently pursuing her PhD degree and is a teaching assistant in ENIS. Her research interests include intelligent control methods such as Genetic Algorithms, ACO, PSO, and fuzzy methods with application to the control of switched linear and nonlinear systems with an emphasis on those exhibiting chaotic behavior. Yosra Miladi is a member of the Research group CEMLab of the Engineering School of Sfax.

Zaouia Mustapha received his Engineer degree (equivalent to a BSc) in Electrical Engineering from the University of Tizi Ouzou, Algeria; he received Magister degree from Polytechnic School, Algiers, Algeria, in 2001 and the PhD degree in the Mouloud Mammeri University of Tizi-Ouzou, Algeria. He is Assistant Professor. He is a researcher in laboratory (Laboratoire des Technologies Avancées du Génie Electrique – LATAGE). He has published more than 10 articles and participates in more than 15 scientific conferences. His research interests are the electromagnetic-mechanical modelling and numerical simulation of electromagnetic machines, particularly the linear machines and permanent-magnet machines applied to the electrical and hybrid vehicles.

Benyahia Nabil received his BSc degree in Electrical Engineering in 2002 from the University of Béjaia, Algeria; he received the Magister degree from the Bejaia. He is a researcher in laboratory (Laboratoire des Technologies Avancées du Génie Electrique – LATAGE). Presently, he prepares the PhD degree in the Bejaia University, Algeria, and he is Assistant-Professor in the Electrical Engineering Department, University of Tizi-Ouzou, Algeria. He published many research articles in various referred international journals, international conferences, and participates in many international scientific conferences. His research interests the fuel cell modelling and numerical simulation of renewable energy, particularly the photovoltaic energy, wind energy, and hydrogen energy.

Benamrouche Nacereddine received his BSc degree in Electrical Engineering in 1985 from the University of Tizi Ouzou, Algeria. He then received his PhD in Electrical Engineering from The University of Sheffield, UK in 1990. He worked as a teaching assistant at the University of Leeds in 1990/1991 and as a Head of Department in Najran Technical College of Technology, Saudi Arabia, from 2000-2004. He is currently a professor at the Electrical Engineering Department, University of Tizi-Ouzou, Algeria, where he is director of a research laboratory (Laboratoire des Technologies Avancées du Génie Electrique – LATAGE). He occupied the chair of Vice Chancellor to postgraduate studies and research at the same university from 2007 to 2010. He also chaired the scientific board for the electrical engineering department for many years. His research interests include electrical machines and drives, power electronics, control systems, and renewable energy particularly photovoltaic.

Kiyoshi Nishikawa received the BEng, MEng, and DEng degrees in Electrical Engineering from Tokyo Metropolitan University, Tokyo, Japan, in 1990, 1992, and 1996, respectively. He joined Nippon Steel Ltd., Tokyo, Japan, in 1992, and in 1993, he joined Tokyo Metropolitan University, where he is now an Associate Professor. His current research interests lie in the area of signal processing for adaptive systems and communication systems. Dr. Nishikawa is a senior member of IEICE and a member of IEEE. He served as an associate editor (2006-2007) and as a secretary (2008-2009) for the *Transactions of the IEICE on Fundamentals of Electronics, Communications, and Computer Sciences*. He has been the secretary of IEEE Signal Processing Society Japan Chapter from 2005 to 2006, and the secretary of IEICE Signal Processing TC. He is currently a member of the TCs on Signal Processing of the IEICE.

Kamel Ben Othman was born in Tunisia in 1958. He obtained the engineer degree in Mechanical and Energetic Engineering from the Université de Valencienne, France, in 1981 and obtained PhD degree in Automatic and Signal Processing from the Université de Valencienne, France, in 1984 and the HDR from the Ecole Nationale d'Ingenieurs de Tunis in 2008. He is currently professor at Ecole Nationale d'Ingénieurs de Sousse. His research is related to reliability, fuzzy systems, and diagnosis of complex systems.

Ian R. Petersen was born in Victoria, Australia. He received a PhD in Electrical Engineering in 1984 from the University of Rochester. From 1983 to 1985, he was a Postdoctoral Fellow at the Australian National University. In 1985, he joined the University of New South Wales at the Australian Defence Force Academy where he is currently Scientia Professor and an Australian Research Council Laureate Fellow in the School of Engineering and Information Technology. He has served as an Associate Editor for the *IEEE Transactions on Automatic Control, Systems, and Control Letters, Automatica*, and *SIAM Journal on Control and Optimization*. Currently, he is an Editor for *Automatica*. He was made a Fellow of the IEEE in 1999 and a Fellow of the Australian Academy of Science in 2011. His main research interests are in robust control theory, quantum control theory, and stochastic control theory.

Viet-Thanh Pham graduated in Electronics and Telecommunications in 2005 at Hanoi University of Technology, Vietnam. He received the PhD degree in Electronics, Automation, and Control of Complex Systems Engineering in 2013 from the University of Catania, Italy. Currently, he is Lecture at the School of Electronics and Telecommunications, Hanoi University of Science and Technology, Vietnam, where he has been involved in projects concerning the study of nonlinear circuits and systems. His scientific interest includes applications of nonlinear systems, analysis and design of analog circuits, and FPGA-based digital circuits.

Hemanshu R. Pota received the BE degree from SVRCET, Surat, India, in 1979, the ME degree from the IISc, Bangalore, India, in 1981, and the PhD degree from the University of Newcastle, NSW, Australia, in 1985, all in Electrical Engineering. He is currently an Associate Professor at the University of New South Wales, Australian Defence Force Academy, Canberra, Australia. He has held visiting appointments at the University of Delaware; Iowa State University; Kansas State University; Old Dominion University; the University of California, San Diego; and the Centre for AI and Robotics, Bangalore. He has a continuing interest in the area of power system dynamics and control, and modelling and control of mechanical systems such as flexible structures, acoustical systems, and UAVs.

Łukasz Sadkowski received his BSc degree in Computer Science with Physics in 2010 at Maria Curie-Skłodowska University in Lublin, Poland. In 2012, he graduated from the same university, completing the course in Computer Science with the specialization in Computer Graphics and receiving his MSc degree. He currently works in the Department of Information Technologies at Maria Curie-Skłodowska University in Lublin. His fields of interest include computer science, robotics, and automation.

Mezlini Sahbi received the scientific degree in mathematics and physics preparatory cycle of the National School of Engineers of Gabes (ENIG) in 2000. He received the engineering degree in Electrical Engineering in the National School of Engineers of Monastir (ENIM) in 2003. He received the degree of Master of Research in Automatic and Signal Processing (ATS) within the National Engineering School of Tunis (ENIT) in 2005. His research interests include the command by internal model of discrete system.

Salem Saidi received the diploma in 2007 in Sciences and Techniques from the High Engineering School of Tunis (ENSIT). His first level university studies have achieved by a Master's degree in 2009 from ENSIT. Currently, he is preparing a thesis on the electrical networks active power filtering of the LaTICE laboratory. He is interested in the power electronics and machine modeling, control induction motor drives, electrical network, and active filter.

Anis Sakly is currently a Professor at the Electrical Department, National School of Engineering of Monastir (ENIM). He received the Electrical Engineering diploma in 1994 from ENIM, then the PhD degree in Electrical Engineering in 2005 from National Engineering School of Tunis (ENIT). His research interests are in analysis, synthesis, and implementation of intelligent control systems. Particularly, his current research interests include soft computing-based approaches applied in optimal control, signal and image processing, and renewable energy systems optimization. He served on the technical program committees for several international conferences.

Haddad Salah received his Engineer degree (equivalent to a BSc) in Electrical Engineering from the University of Tizi Ouzou, Algeria, in 1985. He moved then to France where he received his DEA and PhD degrees in Electrical Engineering in 1986 and 1990, respectively, from the University of Nancy. He served as head of Department of Electrical Engineering at Tizi Ouzou University, Algeria, for many years and is currently a professor and a research group leader in a research laboratory (LATAGE). He has conducted many research projects sponsored by the Ministry of Higher Education, Algeria. His research interests include electrical machines and drives, power electronics, control systems, power systems, and wind energy generation.

Houda Salhi received her BS degree in Electrical Engineering and MS degree in Automatic Control and Industrial Computing from the National School of Engineering of Sfax, University (Tunisia) in 2007 and 2009, respectively. Now, she is a PhD candidate of the Laboratory of Sciences and Techniques of Automatic Control and Computers Engineering (Lab-STA) in the National School of Engineering of Sfax (Tunisia). Her main research interests include the identification and the adaptive control of non-linear systems.

Salah Salhi was born in 1972. He received his habilitation degree in Automatic Control from the National School Engineering of Tunis in 2012 and his PhD degree in Automatic Control from the National School Engineering of Tunis in 2005. Currently, he is an associate professor in the Institute Superior in Computer Science. His research interests include robust control, h-infinity control, h-infinity control, time-varying systems, and switched systems.

Lassaâd Sbita obtained the Doctorate thesis on July 1997 in Electrical Engineering from ESSTT of Tunis, Tunisia. He works as a Professor at the Electrical and-Automatic Department of the National

Engineering School of Gabès, University of Gabès, Tunsia. He received the "Habilitation de Diriger des Recherches" (HDR) from the National Engineering School of Sfax, Tunsia, in March 2008. He is the director of the research unit of photovoltaic, wind, and geothermal systems. His fields of interest include renewable energies, power electronics, machine drives, automatic control, modeling, observation, and identification.

Ben Attia Selma was born in 1982. She received her Master's degree from the National School Engineering of Tunis in 2007. She received her engineering degree in electrical branch from the National School Engineering of Tunis in 2005. Currently, she is an assistant in ISET'COM. Her current research interests include switched systems and robust control.

Fernando E. Serrano received a BS in Industrial and Systems Engineering from Universidad Tecnologica Centroamericana, Tegucigalpa, Honduras, in 2003, and a MSc in Electrical Engineering from Florida International University, Miami, FL, USA, in 2007. He is currently a reviewer of several automatic control journals, conferences, and book chapters. He is working in different projects related to the automatic control field, specifically, in topics that deal with robotics and mechatronics for industrial and different kinds of applications. His interests are linear control systems, nonlinear control, sliding mode control, robust control, robot dynamics and kinematic, observer design, and intelligent control.

Ioannis N. Stouboulos received the MS degree in Electronic & Telecommunications in 1982 and PhD degree in Non-Linear Electric Circuits from the Physics Department of the Aristotle University of Thessaloniki, Greece, in 1998. He joined the Physics Department of the Aristotle University of Thessaloniki, Greece, in 1980, as a research assistant, where he was involved in projects concerning the analysis and synthesis of acoustical signals. Since 1995, he has been involved in projects concerning the study of nonlinear circuits and systems in the same department, where he was elected assistant professor in 2008.

Sundarapandian Vaidyanathan received MS degree in 1992 and DSc degree in Electrical and Systems Engineering in 1996 from Washington University in St. Louis, Missouri, USA. In his Doctoral dissertation on observers for nonlinear control systems, he successfully resolved some unsolved problems on local and global observer design for nonlinear control systems. From 2009, he has worked as Professor and Dean of the Research and Development Centre at Vel Tech University, Chennai, India. Formerly, he was a Professor and Acting Director of the Indian Institute of Information Technology and Management-Kerala (IIITM-K), Thiruvananthapuram, India. His research interests include chaos theory, linear and nonlinear control systems, computational science, soft computing, scientific modelling, and simulation. He has published over 100 research papers in international conferences and over 300 research papers in international journals. He has delivered plenary lectures in many international conferences on control theory and computer science. He has also conducted many workshops on computational science and scientific modelling using MATLAB and SCILAB.

Christos Volos received the BS degree in 1999 from Physics Department of Aristotle University of Thessaloniki. He received MSc degree in Electronics and Telecommunications and the PhD degree in Non-Linear Electronic Circuits from the Physics Department of the Aristotle University of Thessaloniki, Greece, in 2002 and 2008, respectively. From 2004 to 2010, he was a Scientific Associate at the Department of Electronics, of Technological Educational Institute of Thessaloniki. He joined the Physics Department of the Aristotle University of Thessaloniki, Greece, in 2014 as a Lecturer. His research interests include the design of analog and mixed-signal electronic circuits, chaotic electronics and their applications (secure communications, cryptography, and robotics). He has authored or co-authored more than 50 journal and conference papers. He has participated, as a researcher, in several (Greek-) national and international (EU) funded projects.

Gen'ichi Yasuda joined Information Science and Control Engineering Course in Department of Mechanical Engineering, Faculty of Engineering, at Nagasaki Institute of Applied Science, Japan, as a Research Assistant, in 1983. He was a Professor in Department of Mechanical Engineering during 1996-2005, a Professor in Department of Human and Computer Intelligence, Faculty of Informatics during 2005-2010, and also a Professor in Division of Instrument and Control, Department of Electronics and Information Technology, Graduate School of Engineering during 1996-2010. He is now a Guest Professor at Nagasaki Institute of Applied Science. He is a recipient of CACS International Automatic Control Conference Best Paper Award in 2009. He also served as session chairman in many international conferences. He originated the frontier researches on non-centralized systems and grouping robots as a Doctoral dissertation at Tokyo Institute of Technology in 1971. His current research projects include multi-microcomputer-based development of automated noodle-making machines.

Rim Zaghdoud was born in Tunis, Tunisia. She received, respectively, from the Gabes National School of Engineering (ENIG), the Engineering diploma in Electrical Engineering in 2006, the Master's degree in Automatic and Technical Intelligent in 2008. Currently, he is preparing the PhD Degree in Electrical Engineering. His research interests include robust control and h-infinity control.

Ahmed Zouinkhi is associate professor at the National Engineering School of Gabes (Tunisia) and a member of Modeling, Analysis, and Control Systems (MACS) laboratory. He received the National Engineering degree from the National Engineering School of Monastir (ENIM), Tunisia, in 1997 in Industrial Computing. He received his PhD degree in 2011 in Automatic Control from the National Engineering School of Gabes (Tunisia) and a PhD degree in Computer Engineering from the Nancy University (France). His research activities are focused on Distributed Systems, Smart Objects Theory and Applications, Ambient Intelligence Systems and Architectures, RFID, and Wireless Sensors Network Concepts and Applications in Manufacturing and Supply Chain.

Index

K

L

M

N

O

P

U

V

W

www.ingramcontent.com/pod-product-compliance
Lightning Source LLC
LaVergne TN
LVHW080616300926
849162LV00005B/1184